To City University

Best wishes

Jim Longfoot

B.C.E.M.C.

December '94

Combustion Engineering and Gas Utilisation

Combustion Engineering and Gas Utilisation

BRITISH GAS SCHOOL OF FUEL MANAGEMENT

Editor: J R Cornforth, BSc, CEng, MIMechE, FInstE. FIGasE.

British Gas

E & FN SPON
An Imprint of Chapman & Hall
London · Glasgow · New York · Tokyo · Melbourne · Madras

Published by British Gas plc
Rivermill House, 152 Grosvenor Road, London SW1V 3JL

E & FN Spon, an imprint of Chapman & Hall,
2-6 Boundary Row, London SE1 8HN

Chapman & Hall, 2-6 Boundary Row, London SE1 8HN, UK

Blackie Academic & Professional, Wester Cleddens Road, Bishopbriggs, Glasgow G64 2NZ, UK

Chapman & Hall, 29 West 35th Street, New York NY10001 – 2291, USA

Chapman & Hall Japan, Thomson Publishing Japan, Hirakawacho Nemoto Building, 6F, 1-7-11 Hirakawa-cho, Chiyoda-ku, Tokyo 102, Japan

Chapman & Hall Australia, Thomas Nelson Australia, 102 Dodds Street, South Melbourne, Victoria 3205, Australia

Chapman & Hall India, R. Seshadri, 32 Second Main Road, CIT East, Madras 600 035, India

---

First edition 1970

Third edition 1992

Typeset and printed in Great Britain by The College of Fuel Technology, London

ISBN 0 419 17670 5

A catalogue record for this book is available from the British Library

Library of Congress Cataloging-in-Publication data available.

Foreword

This is the third edition of 'Combustion Engineering and Gas Utilisation' and I am pleased to have the opportunity to contribute a foreword to a publication that has clearly established itself as an essential work for study and reference. That it is produced by the British Gas School of Fuel Management is not only a tribute to the expertise and commitment of the School's staff but should re-assure those concerned with gas utilisation that the School continues its world-leading work in energy efficiency. This has become so much more important as we recognise the fundamental role of energy efficiency in meeting environmental challenges.

This book offers a comprehensive source of reference on all aspects of gas utilisation in the industrial, commercial and administrative market sectors. It covers process, space-heating and hot-water equipment in detail, Codes of Practice and Standards and the important issue of fuel efficiency.

In commending it to a wide readership, I am conscious of the success of the two previous editions, which attracted a world-wide usage among gas undertakings and academic institutions. I am certain that it will continue as a valued work for reference and education in the gas industry.

Mr R Evans

Chairman and Chief Executive, British Gas plc

Contents

Preface

Since the second edition of this handbook was published in 1983, a whole range of new efficient gas fired space, water and process heating equipment has appeared on the industrial and commercial market. Also British Gas has been privatised and many new Standards, Codes of Practice and Regulations have been introduced. This publication attempts to bring engineers up to date with equipment currently available and sound combustion engineering principles and practice.

In addition, the importance of correct installation, commissioning, servicing and maintenance of gas utilisation equipment for safety and efficiency cannot be stressed too strongly. It is important that only competent and trained staff are employed in such activities. All operators should be aware of the safe light up, shut down and emergency isolation procedures.

The installation of pipework should conform to the Gas Safety (Installation and Use) Regulations 1984 and also to British Gas Publication IM/16, Guidance Notes on the Installation of Pipework in Customer's Premises. Before gas fired equipment is brought into production for the first time it should be properly commissioned, as in the Guide to Commissioning Procedures for Gas Fired Plant document which is being prepared by the Institution of Gas Engineers.

Gas fired plant requires regular attention throughout its life in order to ensure safe and efficient operation. Planned maintenance is therefore essential in order to extend the useful life of components and prevent where possible breakdowns during production periods.

Additionally, wear and tear on mechanically moving parts means that over a period of time pressure settings, air gas ratios and flue gas analyses may tend to drift from their original setting resulting in loss of efficiency and perhaps a reduction in the safety margin employed. For this reason then, it will usually be more economic as well as good practice to regularly inspect and maintain gas fired equipment. Appliances must not be used if they are unsafe in any respect.

The frequency of servicing will be determined by the type of equipment employed, period of use, the environment in which it is installed and the degree of reliability expected. In practice whilst many tests can be carried out annually, there will be some checking which should be done quarterly, weekly or even daily, in order to maintain optimum efficiency and safety. All new equipment should be supplied with a service manual giving an extensive guide to servicing requirements but older equipment should be serviced using sound engineering practice and judgement.

J. R. CORNFORTH

Contributors

The initial script for a large proportion of this textbook was prepared by FEC Consultants Ltd. However, during its preparation there have been many contributors and much advice given by friends and colleagues from within the gas industry and also by equipment suppliers and manufacturers. We hope they will all accept our sincere expression of thanks. In addition, most of the original artwork has had to be redrawn and the whole work standardised and unified. We would like to thank the College of Fuel Technology for their help and guidance here.

The main contributors and the chapters involved are listed below.

Original material for Waste Heat Recovery, Chapter 7.
D K Austin, CEng, LRSC, MInstE

Gas and Air Supply Controls, Chapter 9.
J R Cornforth, BSc, CEng, MIMechE, FInstE, FIGasE

Advice and material for Chapters 9-11.
P Fleming, BSc, CEng, MInstE
G A Butler, BSc

Advice on Space Heating Equipment, Chapter 6.
J D Ford, BSc, CEng, MInstE

Initial scripts for Chapters 1-8 and 12-14.
FEC Consultants Ltd:
J D Mason, BEM, CEng, MInstE, MIGasE,
R Langlands, BSc

Section on Baking Ovens, Chapter 6.
R Proffitt, CEng, FInstE, FIGasE, MIM, FICeram, MBIM

Process Control, Chapter 11.
C E Thompson, BTech, BA, CEng, MInstE

Ignition and Combustion Safeguards, Chapter 10.
D Wilson, CEng, MIGasE, FInstE, FSGT

Editor: J R Cornforth, BSc, CEng, MIMechE, FInstE, FIGasE

Index compiled by A Hopkins

Combustion 1

CHAPTER 1

COMBUSTION

1.1 INTRODUCTION

This section deals with the essential properties of natural gas and liquefied petroleum gases when used in heating processes. In view of the widespread use of dual-fuel burners, fuel oil combustion properties are also covered. The basic combustion reactions are discussed but the detailed mechanism of combustion is not treated in any depth.

The main properties of natural gas relating to flame structure and development are described in detail with particular emphasis on burner design.

To determine combustion performance, the heat losses in flue gases are calculated and presented in a series of graphs. For condensing flue systems a method of calculation is described for assessing combustion performance based on hygrometric charts.

Techniques for the determination of flue gas loss on actual plant are also highlighted.

1.2 GASEOUS FUELS

The commodity which is distributed throughout Britain as a gaseous fuel is natural gas. The term natural gas is generally used to describe the flammable gas which is derived from the earth's crust, usually from deep boreholes and frequently in association with crude petroleum. The main combustible constituent is methane but there are others, notably ethane, propane, butane, pentane and some higher hydrocarbons, in decreasing order of proportion. There may also be other, non-combustible, gases such as nitrogen and carbon dioxide.

The exact composition varies according to the source of the fuel and a detailed analysis of natural gas types is shown in Table 1.1.

Methane forms the major constituent of natural gas and it is from methane that most of the heat of combustion is obtained.

Propane and butane are readily liquefiable gases associated with petroleum distillation which are handled as liquids but find application as gaseous fuels. They are generally refered to as liquefied petroleum gases and are marketed as fuels in their own right. Although described as 'propane' and 'butane', they are in fact mixtures, with the titular constituent comprising about 90%. The combustion properties of methane, ethane, propane and butane are broadly similar although ethane generally occurs only as a constituent of commercially distributed natural gas. Composition analysis of liquefied petroleum gases is shown in Table 1.2.

Table 1.1 Composition of Natural Gas Fuels

		Mean North Sea	Bacton	Easington	Theddlethorpe	St Fergus	Morecambe*
Composition % Volume							
Nitrogen	N_2	2.72	1.78	1.56	2.53	0.47	7.97
Helium	He	—	0.05	—	—	—	—
Carbon Dioxide	CO_2	0.15	0.13	0.54	0.52	0.32	0.55
Methane	CH_4	92.21	93.63	93.53	91.43	95.28	85.26
Ethane	C_2H_6	3.6	3.25	3.36	4.10	3.71	4.45
Propane	C_3H_8	0.9	0.69	0.70	0.99	0.18	1.02
Butanes	C_4H_{10}	0.25	0.27	0.24	0.33	0.03	0.52
Pentanes	C_5H_{12}	0.07	0.09	0.07	0.10	0.01	} 0.23
Higher Hydrocarbons		0.1	0.11	—	—	—	
Composition % Mass							
Nitrogen	N_2	4.38	2.90	2.55	4.05	0.79	12.10
Helium	He	—	0.003	—	—	—	—
Carbon Dioxide	CO_2	0.38	0.33	1.39	1.31	0.84	1.31
Methane	CH_4	84.89	87.37	87.28	83.62	91.10	74.08
Ethane	C_2H_6	6.21	5.69	5.88	7.03	6.65	7.25
Propane	C_3H_8	2.28	1.59	1.80	2.49	0.47	2.44
Butanes	C_4H_{10}	0.83	0.91	0.81	1.09	0.11	1.64
Pentanes	C_5H_{12}	0.29	0.38	0.29	0.41	0.04	} 1.18
Higher Hydrocarbons		0.74	0.827	—	—	—	
Ultimate Composition % Mass							
Carbon		72.16	73.23	72.93	71.98	74.38	66.07
Hydrogen		23.18	23.62	23.51	23.02	24.21	20.88
Nitrogen		4.38	2.91	2.55	4.05	0.79	12.10
Oxygen		0.28	0.24	1.01	0.95	0.62	0.95

* After processing at Barrow

Table 1.2 Composition of Liquefied Petroleum Gases

		Typical Commercial Propane	Typical Commercial Butane
Composition % volume			
Methane	CH_4	—	0.1
Ethane	C_2H_6	1.5	0.5
Propane	C_3H_8	91.0	7.2
Butane	C_4H_{10}	2.5	87.1
Propylene	C_3H_6	5.0	4.2
Composition % mass			
Methane	CH_4	—	0.03
Ethane	C_2H_6	1.02	0.27
Propane	C_3H_8	90.91	5.70
Butane	C_4H_{10}	3.30	90.83
Propylene	C_3H_6	4.77	3.17
Ultimate Composition % mass			
Carbon		82.02	82.79
Hydrogen		17.98	17.21

The main properties of gaseous fuels related to combustion are given in Table 1.3. Calorific Value (specific enthalpy) is the most significant property of a fuel, and for gases is defined as the quantity of heat released by the complete combustion of a standard volume in air and the subsequent cooling of the products of combustion back to standard reference conditions. British Gas standard conditions are 288.7 K and 1 013.7405 mbar absolute and specific enthalpy units are MJ/m^3. (Consult Chapter 12 for further information on gas standard conditions).

Table 1.3 Properties of Gaseous Fuels and Combustion Products

		Mean North Sea	Commercial L.P.G.	
			Propane	Butane
Gas relative density (air = 1)		0.602	1.523	1.941
Gross c.v.	— MJ/m^3	38.63	93.87	117.75
	— MJ/kg	52.41	50.22	49.41
Nett c.v.	— MJ/m^3	34.88	86.43	108.69
	— MJ/kg	47.32	46.24	45.61
Liquid relative density at 15 °C (water = 1)		—	0.510	0.575
Vol. gas/vol. liquid (at 0 °C)		—	274	233
Wobbe number	— MJ/m^3	49.79	76.06	84.52
Stoichiometric air requirements:				
- vol/vol fuel (gaseous)		9.76	23.76	29.92
- mass/mass fuel		16.5	15.6	15.3
- vol/mass fuel	— m^3/kg	13.24	12.73	12.50
Stoichiometric combustion products:				
- CO_2	— % vol	9.7	11.7	12.0
- H_2O	— % vol	18.6	15.4	14.9
- N_2	— % vol	71.7	72.9	73.1
- Dewpoint	— °C	59	55	54
- CO_2 in dry products	— % vol	11.9	13.8	14.1
Theoretical flame temperature °C		1 930	2 000	2 000
Limits of flammability (% vol./vol. in gas/air mixture)	%	5-15	2-10	1.8-9

Notes:
1. Reference conditions 288 K and 1013 mbar absolute unless otherwise stated.
2. Where fuel is liquid at reference conditions, quoted gaseous properties are for ideal-gas volumes
3. Calorific value (saturated) = 0.9826 x calorific value (dry).

Cooling the products to 288 K (15 °C) implies the inclusion of the latent heat of condensation of the water vapour from combustion of the hydrogen content and the calorific value is thus termed the gross value. This latent heat is frequently not available in practice as most heating processes reject flue gases at temperatures above the water dewpoint; the nett calorific value, which exludes the water vapour latent heat, is a more realistic expression of the heat potentially available. For natural gas, the nett calorific value is about 10% lower than the gross; for any fuel, the difference between gross and nett depends on the fuel's carbon : hydrogen ratio. In the UK, the gross calorific value is generally used, but in most other countries the nett value is quoted and care must always be exercised when comparing or collating information from various sources.

The relative density of gases is generally quoted as the ratio of the mass of a unit volume of the gas to the mass of the same volume of air. The relative density of natural gas is a little over half that of air so that natural gas if freely released in air rises and thus more readily disperses. This is important since leaks in reasonably well ventilated spaces clear quickly and gas released from vents and pressure reliefs will disperse with minimum hazard.

The liquefied petroleum gases have densities greater than air and if released into the atmosphere sink to the ground and accumulate in ducts etc. This is a serious disadvantage to their use and represents a potentially hazardous situation. Great care must be taken to avoid leaks in L.P.G. systems and freely discharging relief valves are not acceptable except under very carefully chosen conditions.

1.3 COMBUSTION OF NATURAL GAS

Combustion is a chemical reaction involving the combination of fuel and oxygen to produce heat and combustion products. The most convenient source of oxygen is that which forms almost 21% of atmospheric air. Hence the combustion of natural gas is in most cases carried out in this way although some specialised applications demand the use of pure oxygen.

The combustion reaction can be represented by way of a chemical equation for each constituent present in the fuel. This enables the theoretical air to gas ratio to be determined together with the volumes of combustion products produced. For example the reaction for methane gas is:

$$\text{Methane + oxygen} \rightarrow \text{Carbon dioxide + water}$$

$$CH_4 + 2O_2 \rightarrow CO_2 + 2H_2O$$

However since two volumes of oxygen are taken from 9.52 volumes of air the complete reaction using ambient air is:

$$CH_4 + 2O_2 + 7.52N_2 \rightarrow CO_2 + 2H_2O + 7.52N_2$$

It should be noted that this equation is an over-simplification and represents the terminal condition only. Combustion with oxygen of a fuel containing carbon and hydrogen is initiated by a chain reaction with the intermediate transient existence of many unstable compounds. 'Nitrogen' in air includes very small amounts of other inert gases (e.g. carbon dioxide) but this may be ignored for all except the most exacting calculations. Similarly, the small quantity of water vapour in air is only taken into account when preparing very detailed heat or material balances.

1.3.1 Stoichiometric Combustion Conditions

Stoichiometric combustion conditions are those where the relative fuel and air quantities are the theoretical minimum needed to give complete combustion, and can be readily calculated knowing the analysis of the gaseous fuel by volume or indeed any fuel by ultimate analysis.

For mean Bacton natural gas the calculation procedure is tabulated in Table 1.4.

Column 1 shows the volumetric composition of the gas.

Column 2 contains the required information from the combustion equation concerning stoichiometric oxygen demand. Thus for methane:

$$CH_4 + 2O_2 \rightarrow CO_2 + 2H_2O$$

The equations for the other natural gas constituents are:

Ethane	$C_2H_6 + 3.5O_2 \rightarrow 2CO_2 + 3H_2O$
Propane	$C_3H_8 + 5O_2 \rightarrow 3CO_2 + 4H_2O$
Butane	$C_4H_{10} + 6.5O_2 \rightarrow 4CO_2 + 5H_2O$
Pentane	$C_5H_{12} + 8O_2 \rightarrow 5CO_2 + 6H_2O$
Higher hydrocarbons (assumed C_9H_{20})	$C_9H_{20} + 14O_2 \rightarrow 9CO_2 + 10H_2O$

Table 1.4 Calculation of Stoichiometric Combustion Products from One Volume of Mean North Sea Gas

Constituent		O_2/air	Combustion Products		
			CO_2	H_2O	N_2
1		2	3	4	5
CH_4	0.9221	x 2 = 1.844	x 1 = 0.922	x 2 = 1.844	
C_2H_6	0.0360	x 3.5 = 0.126	x 2 = 0.072	x 3 = 0.108	
C_3H_8	0.0090	x 5 = 0.045	x 3 = 0.027	x 4 = 0.036	
C_4H_{10}	0.0025	x 6.5 = 0.016	x 4 = 0.010	x 5 = 0.013	
C_5H_{12}	0.0007	x 8 = 0.006	x 5 = 0.003	x 6 = 0.004	
C_9H_{20}	0.0010	x 14 = 0.014	x 9 = 0.009	x 10 = 0.010	
CO_2	0.0015	—	0.002	—	
N_2	0.0272	—	—	—	0.027
		2.051			
		N_2 7.712			7.712
	1.000	Air 9.763	1.045	2.015	7.739

Below the total oxygen requirement in Column 2 is the associated amount of nitrogen, on the basis that ambient air is comprised of 3.76 volumes of N_2 to 1 volume of O_2.

Column 3 gives the amount of CO_2 produced on stoichiometric combustion for each of the constituent gases, with columns 4 and 5 giving the same information for water and nitrogen.

Hence for stoichiometric combustion, in vols./vol. fuel:

Theoretical air requirement,	A_o	= 9.763
Wet combustion products,	V_{ow}	= 10.799
Dry combustion products,	V_{od}	= 8.784

The maximum concentration of CO_2 in the dry combustion products is:

$$CO_{2st} = (1.045/8.784) \times 100$$

$$= 11.9\% \text{ vol.}$$

1.3.2 Combustion with Excess Air

Combustion in practice is hardly ever carried out in stoichiometric conditions. Some industrial burners may operate at air/gas ratios which are extremely close to the theoretically correct value but the majority of burners require a measure of air in excess of the stoichiometric quantity to ensure that combustion is carried to completion.

The main reason for the requirement of excess air is failure of the gas and air streams to mix completely in the distance over which combustion is required to take place. The occurrence of combustion is dependent upon the collision of molecules of fuel with molecules of oxygen. If there are deficiencies in the mixing of the two fluids then an excess of oxygen must be provided to increase the incidence of molecular collisions.

A convenient method of relating the proportion of air actually present in a combustion system to the theoretically required amount is as a percentage above requirements or as a factor.

The air factor is expressed as the ratio of the air actually used (vol./vol.fuel) to the stoichiometric air demand (vol./vol.fuel):

$$AF = \text{(air actually used)/(stoichiometric air demand)} \qquad 1.1$$

The percentage of excess air can be expressed as 100 x (AF–1).

1.3.3 Determination of Air Factor

If the flow of fuel and air to a burner system could always be reliably metered the determination of air factor could be readily calculated. This is however seldom practicable particularly as regards the supply of air and thus the air factor must be determined from other measurable parameters. The presence of excess air dilutes the stoichiometric flue gases and thus reduces the % CO_2 and introduces a % O_2. Thus excess air can be determined knowing either the carbon dioxide or oxygen content of the flue products using one of the following equations:

$$AF - 1 = (V_o/A_o)[(CO_{2\,st} - CO_{2\,act})/CO_{2\,act}] \qquad 1.2$$

or

$$AF - 1 = (V_o/A_o)[O_2/(21 - O_2)] \qquad 1.3$$

where
V_o = stoichiometric dry combustion products vol./vol. fuel
A_o = stoichiometric air vol./vol. fuel
$CO_{2\,st}$ = stoichiometric CO_2 in dry combustion products % vol.
$CO_{2\,act}$ = measured CO_2 in dry combustion products % vol.
O_2 = measured O_2 in dry combustion products % vol.

The ratio V_o/A_o varies from 0.89 to 0.98 for various fuels with a value of 0.898 for natural gas. For routine purposes as an approximation it can be taken as unity.

For sub-stoichiometric conditions unburned fuel will be present in the form of products of incomplete combustion, with carbon monoxide as the major constituent. Clearly the quantity of carbon in the flue gases from the combustion of a given fuel is the same whether or not the theoretical air volume is used. Thus to allow for any carbon monoxide (CO) present the excess air equation is modified:

$$AF - 1 = (V_o/A_o)[(CO_{2\,st})/(CO_{2\,act} + CO) - 1] \qquad 1.4$$

where CO = measured CO in dry combustion products % vol.

The variation in combustion products with air factor for natural gas is shown in Figure 1.1.

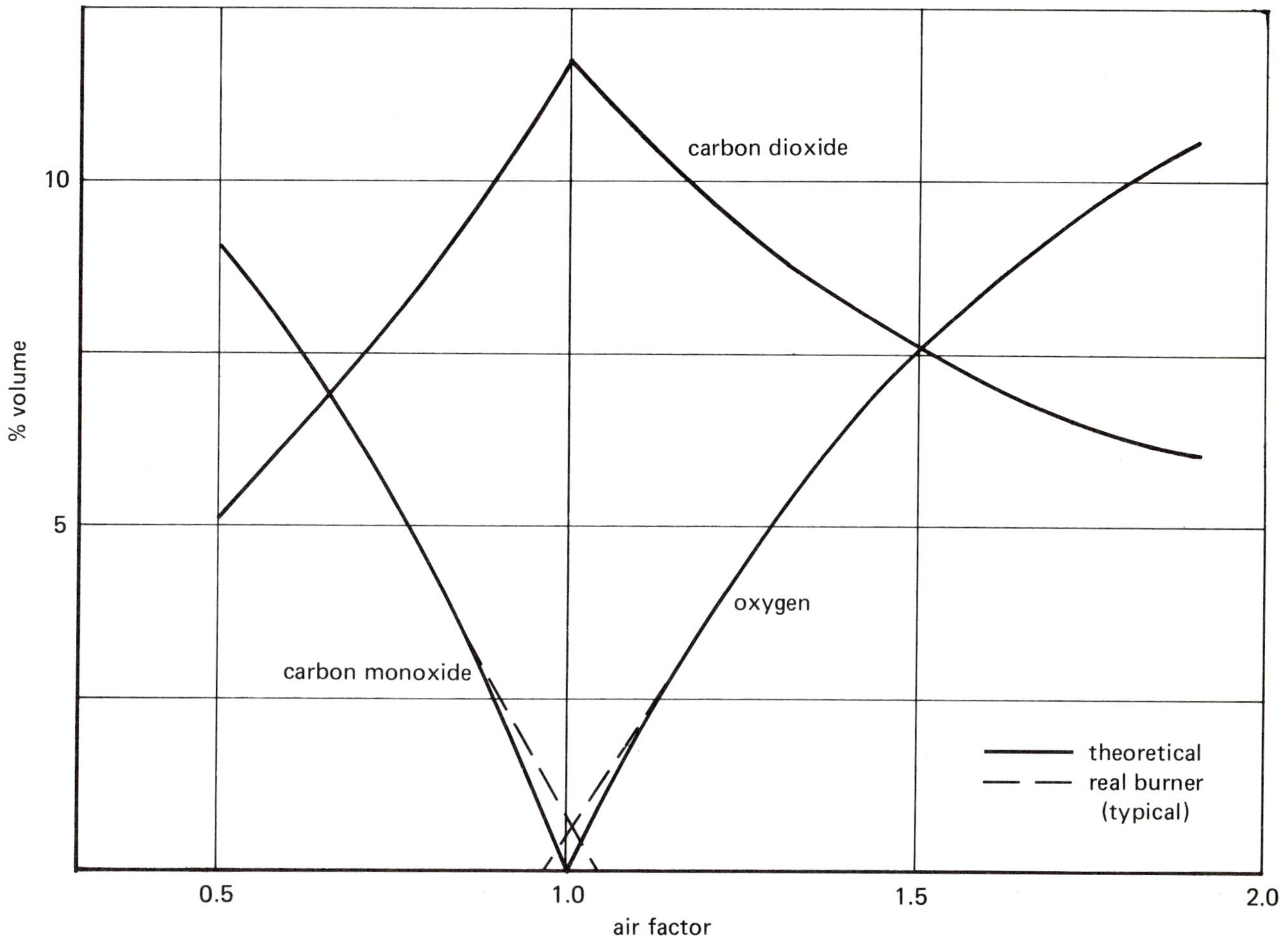

Fig. 1.1 Equilibrium combustion products at 1 200 K from natural gas

1.3.4 Excess Air and CO Production

The ideal graph of combustion products in Figure 1.1 shows that carbon monoxide is completely eliminated at the stoichiometric air/gas ratio and that no free oxygen exists at air factors less than unity. In practice because of inadequate mixing, too small a combustion chamber, and other factors, this idealised situation will rarely be achieved. A typical real burner performance is also shown in Figure 1.1.

This emphasises the need to monitor CO as well as O_2 (or CO_2) when adjusting burners. If oxygen concentration alone is monitored then it is possible to reach a burner setting that apparently has a small amount of free oxygen present but which in fact produces a high level of carbon monoxide.

1.3.5 Wet and Dry Combustion Products Analysis

Products of combustion from a fossil fuel will always include water vapour. The amount of water vapour is dependent on the initial hydrogen content of the fuel. The proportion of water vapour in the flue gases will vary from about 19% with natural gas to 7% with bituminous coal and less than

12% with oil. It is therefore of considerable importance to define whether an analysis is made on a wet or dry basis. The relationship between wet and dry analysis for natural gas is shown in Figure 1.2.

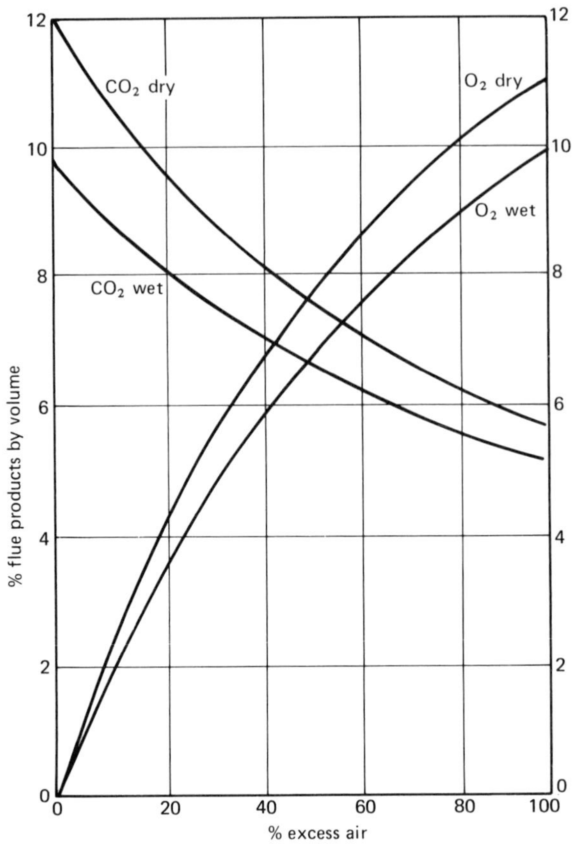

Fig. 1.2 Relationship between wet and dry flue gas analysis

Most instruments used for the analysis of flue gases measure on the dry basis because a sample is extracted and cooled; water vapour is removed either in the instrument or in an intermediate drying column and knock-out pot. An important exception is the zirconia oxygen analyser where the measuring device is inserted directly in the flue: this type of instrumentation measures on the wet basis.

1.4 FLAME PROPERTIES AND STRUCTURE

An understanding of the flame properties of natural gas is important as a basis for burner design and in order to assess the suitability of burner types for particular applications.

1.4.1 Limits of Flammability

These are the lower and upper percentages of fuel in fuel-air mixtures outside which self-sustaining flames will not occur. Clearly any burner system must be capable of giving a reliable air-to-fuel ratio within these limits if a stable flame is to be produced. This is especially important where the establishment of pilot flames is concerned.

For common gases the accepted limits are given in Table 1.5 although they may vary slightly with ambient temperature and pressure conditions. The limits are conventionally expressed as %vol/vol fuel in air, Z_o and Z_u for lower and upper respectively.

Table 1.5 Measured Upper and Lower Flammability Limits in Air

Combustible Gas	Z_o	Z_u
Carbon Monoxide	12.5	74.0
Hydrogen	4.0	75.0
Methane	5.0	15.0
Ethane	3.0	12.5
Propane	2.1	10.1
Butane	1.86	8.41
Ethylene	2.75	28.6
Propylene	2.0	11.1
Butylene	1.98	9.65
Acetylene	2.5	81.0
Hydrogen Sulphide	4.3	45.5
Formaldehyde	7.0	73.0

These limits may be expressed in terms of air factor. For natural gas they are:

Lower limit Z_o : AF = 1.8

Upper limit Z_u : AF = 0.6

It is important to note that these limits apply to the combustion zone; in practice, combustion outside these limits will be found in the furnace or flue due to such factors as hot surroundings or the use of preheated air.

The limits of flammability of gases have important safety connotations. Leaks, either into the general atmosphere or into a combustion chamber, are more easily rendered harmless if the gas in question has narrow flammability limits. A leak of natural gas is more easily diluted, compared with acetylene or hydrogen, to below its lower level of flammability. If the leak is into a closed combustion chamber then the concentration will quickly build up until it has passed through the upper flammability limit.

Flammability limits can be calculated for mixed industrial fuel gases using Le Chatelier's principle:

$$Z_{u,o} = 100/[\Sigma_{i=1}^{i=n} (c_i/Z_{iu,o})] \qquad 1.5$$

where $Z_{u,o}$ = upper and lower flammability limits of the mixture % vol/vol

c_i = concentration of the individual pure components % vol

$Z_{iu,o}$ = upper and lower flammability limits of the ith component %vol/vol

n = no. of components in mixture.

1.4.2 Minimum Ignition Energy

Combustion is initiated by ignition, the purpose of which is to raise the system represented by the fuel and air mixture to a level at which the combustion reaction becomes self-propagating. The concept of ignition temperature postulates a minimum temperature below which ignition cannot occur. Depending on external conditions, development of enough heat to compensate for heat dissipated will permit such a temperature to be reached.

The amount of energy required to initiate the combustion reaction can be measured and is expressed as the minimum ignition energy. For natural gas the value is 0.30 mJ and applies to an air-to-gas ratio of 1.5:1 or approximately 17% excess air. The amount of energy required increases for other ratios as shown in Figure 1.3 and approaches infinity at the limits of flammability. Similarly an increase in proportion of inerts increases the minimum ignition energy. Minimum ignition energy is also related to the laminar burning velocity:

$$Q_{min} \propto c_p (a/v)^3 (t_f - t_o) \qquad 1.6$$

where

Q_{min}	=	minimum ignition energy
c_p	=	specific heat of mixture
a	=	thermal diffusivity of mixture = $K/(\rho c_p)$
k	=	conductivity of air-gas mixture
ρ	=	mixture density
v	=	burning velocity of mixture
t_f	=	flame temperature
t_o	=	initial air-gas mixture temperature

The minimum ignition energies at stoichiometric conditions for propane and butane are 0.30 mJ and 0.38 mJ respectively, compared with 0.34 mJ for natural gas.

Where heat is used to initiate combustion, the minimum ignition temperature is appropriate, which for natural gas is approximately 650 °C.

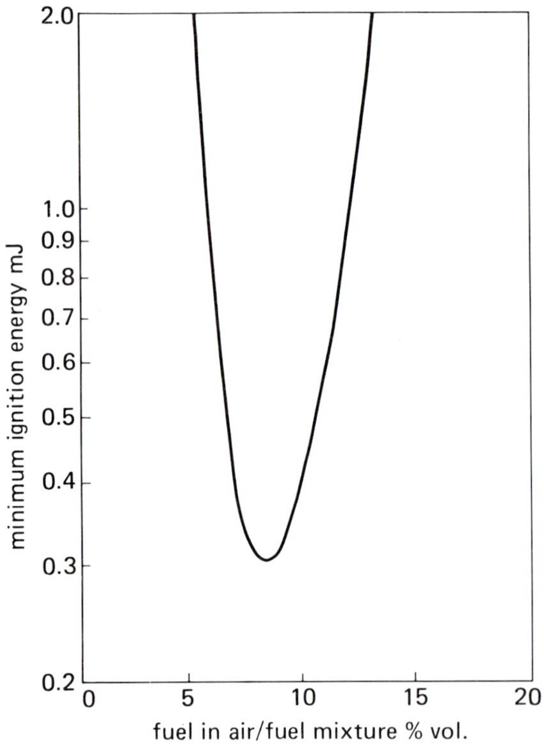

Fig. 1.3 Minimum ignition energy for the methane/air mixture

1.4.3 Burning Velocity

The burning velocity of a gaseous fuel is the rate at which the flame front travels through a completely self-burning mixture of gas and air. The rates obtained from experimental work are influenced by the experimental conditions and any quoted values must be taken only as a guide, since discrepancies may be apparent in values taken from different sources for the same gas.

Burning velocity is important in the design of premix burners since it is the parameter which determines how fast the mixture of gas and air may be pushed out of the burner head before the flame lifts away from the burner and goes out. A gas flame represents a balance between the velocity at which the gas and air mixture is ejected from the burner and the speed with which the flame front travels back towards the mixer. Low burning velocities give low maximum flame port loadings corresponding to a low amount of heat generated in unit time per unit area of burner port. High burning velocities mean relatively high minimum rates of firing to avoid light-back.

For natural gas the maximum value of burning velocity is 0.43 m/s. This maximum value occurs only at mixtures close to stoichiometric while as the level of excess air or air deficiency increases the burning velocity decreases as shown in Figure 1.4. The differences in burning velocity between gaseous fuels is generally attributable to the differences in hydrogen content. The maximum value of burning velocity is 0.47 m/s for propane and 0.45 m/s for butane, while that for hydrogen is 3.5 m/s.

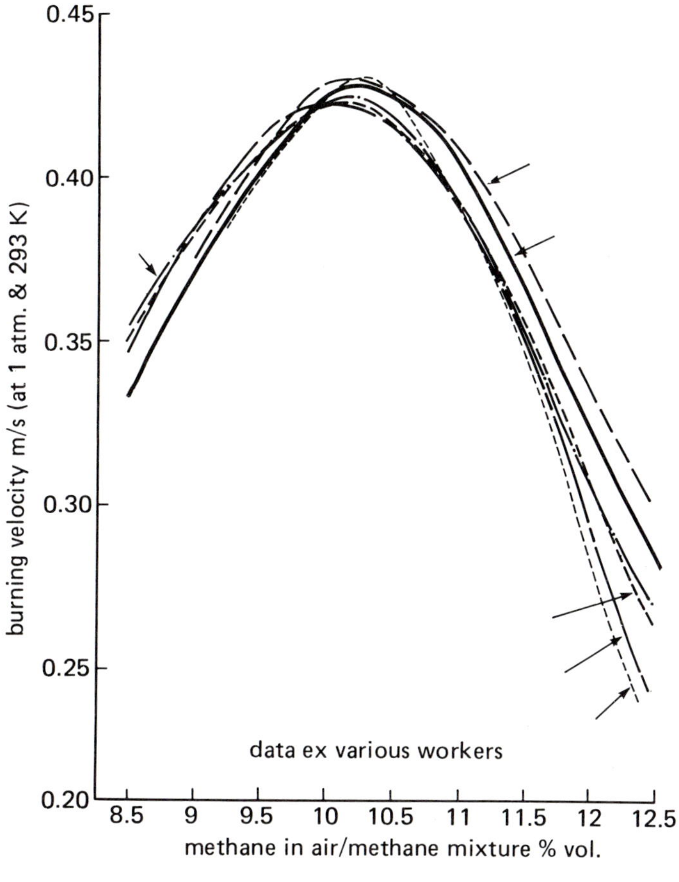

Fig. 1.4 Burning velocity for the methane-air mixture

1.4.4 Wobbe Number

The rate at which heat may be generated in a burner depends upon two factors:

- The volume flow rate of fuel gas through the burner port, q_v.
- The specific enthalpy of the fuel gas, h.

Thus the heat generated, Q, is proportional to the product of the enthalpy value and the flow rate.

Now for a given pressure of fuel gas at the burner the flow rate is inversely proportional to the square root of the relative density of the gas, d. Thus the heat released can be expressed as:

$$Q \propto h/d^{0.5} \qquad 1.7$$

where $h/d^{0.5}$ is the Wobbe Number = $\dfrac{\text{Calorific value of gas}}{\sqrt{\text{Specific gravity}}}$

For the same pressure at the burner, gases of equal Wobbe Number will generate heat at equal rates per unit of burner port area. The burner port area is the area of the orifice through which fuel gas or fuel gas and air mixture flows from a burner to the atmosphere and upon which the flame forms. In a multiport burner it is the sum of the areas of all the ports.

In addition to Wobbe Number other factors influence the rates at which gas may be burned on a burner of given port area, the most important of which is the burning velocity.

For simple atmospheric burners without any form of flame retention device the limiting burning rate before lift-off occurs is 550 to 660 W per m^2 of burner port for natural gas.

1.4.5 Principles of Flame Development

Once having ignited a flow of mixture, a flame is initiated. The development of the flame itself follows very much the same pattern of chemical events as that relative to ignition, and its progress is almost entirely proportional to mixing rates, since the reaction time between fuel particles and oxygen is relatively so small as to be negligible. The precise structure of the visible flame will depend on two factors:

- The aerodynamic situation relative to introduction of air and gas to the point of mixing, and to the intensity with which mixing occurs.
- The thermodynamic and aerodynamic situation of the flame relative to its surroundings.

Industrial application of flames depends for success almost entirely on the ability to produce stable flames. Stability here is the property of location of the concentration of flame reactions in a constant and predictable position. Normally, this condition is best satisfied when the flame appears to be securely seated on the port from which gas, or mixture, is issuing. Departure from this condition can occur with a number of results. The flame can wander away from the issue port, and may do so to an extent at which extinction finally occurs. With premix burners, the flame can transfer itself from the mixture issue location to a point further back where gas alone is issuing from an orifice: this fault is prevalent in ill-designed or badly operated burners using high burning-velocity fuels. Even incipient instability will contribute to difficulties of reliable ignition and monitoring of the flame.

Flame structure

Development of flame structure is the result of mixing between combustible gas (or vapour) and the air (or other oxygen source) which supports combustion. A flame can exist, once a flammable mixture is ignited, providing the balance of energy distribution of the process producing the flame is favourable. In general terms this will entail that a supply of energy is continuously available at the point where it is desired to initiate the flame. This can be provided by several means, including for example:

- Radiation from the flame envelope
- Recirculation of the hot gases in and around the flame envelope.

In the simplest event, a boundary can be imagined between a body of combustible gas and the air which is supporting its combustion.

When combustion is occurring, the boundary layer constitutes a flame front, usually visible. In a flowing system, such as for example in a stream of neat gas containing no added air and issuing into air, the rate of mixing may be relatively slow, since the mixing process is one comprising the diffusion of particles from one stream into the other, with a very slow transfer rate indeed. The boundary between the two gas streams can be imagined as almost intact over relatively long distances: that is, bulk penetration of one stream into the other, as opposed to particle diffusion, is relatively slight. When the relative velocity of one stream compared with the other is increased, the boundary becomes more and more disturbed, and bulk penetration of one stream into the other becomes an important factor in assisting the eventual particle collision which must occur before combustion can take place, largely by increasing the area of any boundary between the component gas streams. These conditions would describe a laminar diffusion flame system, the term laminar being used in its strict sense as a measure of the Reynolds number of the gas flow condition, related to a geometrical dimension of the gas port.

1.4.6 Flame Heat Transmission Characteristics

The emissions of radiation from flames arises from molecules and particles within the flame gases. A yellow flame, due to the presence of a high level of suspended particles, is called a luminous flame, while a blue flame, arising from the presence of the reacting gases only, is described as a non-luminous flame.

Non-luminous flames

Non-luminous flames are produced, especially from gaseous combustibles, in conditions where combustion air is supplied in such a way as to prevent subsequent pyrolysis, or cracking, of hydrocarbons in the combustible gas.

In gas burners, this condition is met when gas and air are mixed prior to discharge from the burner port on which the flame is to be established. The design characteristics of this type of burner are discussed more fully in Chapter 2. The flame which such systems produce is characterised by the familiar double-cone structure, indicating a primary reaction of mixture with secondary air, and a secondary stage in which the products of the first reaction react with further secondary air. The result is a blue flame, whose size characteristics (overall length, height of inner cone, volume) depend on the port diameter on which the flame is seated, and the rate at which mixture is supplied to the port.

When air and gas are supplied under pressure to a burner system in which they issue from their respective discharge nozzles in close proximity (in the so-called nozzle-mix configuration) there is rapid mixing between gas and air streams. Non-luminous blue flames are thus produced so long as air mixture occurs before unmixed gas can be raised to a temperature high enough for cracking to occur. Flame length is consequently reduced, and combustion intensity increased. Nozzle mix burners can be used for heating furnaces and similar installations, in temperature ranges where heat transmission is required to be mainly convective with a minor contribution from radiation from the flame. The furnace lining is directly involved as a recipient of convected heat, which is partly re-radiated to the furnace charge and is itself working at a temperature low enough to tolerate high coefficients of heat transfer. Compactness of furnace, and hence of flame is highly desirable in such situations.

Luminous flames

The principle mode of heat transmission from this type of flame is by radiation from suspended carbon particles, carried in the stream of flame gases. The carbon particles stem from pyrolytic decomposition of carbon compounds in the fuel, e.g. methane, during the period at which fuel gas temperature is high enough for this to occur, but prior to access of combustion air. Control over the pattern of mixing in such a system is therefore a very important factor, and thus the burner configuration is decisive to flame performance.

Diffusion (or neat gas burner) flames and pre-mixed slow burning flames produce larger particles whilst pre-mixed fast burning flames produce smaller and fewer particles. A lesser but still important amount of heat is transmitted from such flames by radiation from tri-atomic gases, principally CO_2 and H_2O. The amount of heat transmitted by this route will depend on both the thickness of the gas stream and its temperature.

The principles of gas and multi-fuel burners are discussed at greater length in Chapters 2 and 3.

1.5 COMBUSTION PERFORMANCE

One of the most important aspects of the application of combustion theory is the determination of the efficiency of the heating process. If stoichiometric combustion could be achieved reliably and the flue gases reduced to room temperature before being exhausted then the maximum use of the heat released is made. In most practical cases, air factors other than 1.0 are encountered and flue gases must be exhausted at above the process temperature. This induces two sources of inefficiency.

Firstly, if the air factor measured is less than 1.0 unburned fuel will be present in the form of products of incomplete combustion with carbon monoxide being the major product. In this event some of the calorific value of the fuel is discharged into the flue without ever being released in the form of heat. On the other hand where excess air is present, heat is being absorbed to increase the temperature of additional air as it passes from burner to flue.

Secondly, flue products leaving the working chamber of equipment at above ambient temperatures take out useful heat which is lost unless recovered in a recuperator, waste heat boiler or other means of heat recovery.

In all applications it is necessary to accept that the full gross calorific value of the fuel cannot be usefully employed and some heat will be lost both from the flue and the heated surfaces of the plant. However, there are obvious advantages in reducing the heat loss to a minimum. Combustion performance is an expression used to determine the amount of heat liberated in the process as a proportion of the calorific value, and is usually derived from measuring the flue gas heat loss: combustion performance % = (100 − % flue gas loss). (It is often termed 'combustion efficiency', which is technically incorrect, and is not the same as thermal efficiency since no account is taken of structural or other losses.)

1.5.1 Heat Losses in Flue Products

The heat lost in the products of combustion from fossil fuel fired plant is a combination of several parameters:

- The total enthalpy of the various component gases at the flue gas temperature.
- The enthalpy due to water vapour of the flue gas temperature.

– The latent heat of vapourisation of the water vapour.
– The loss of available heat due to unburnt combustible components.

For complete combustion, the percentage gross stack losses can be evaluated using the formula:

$$L = 100\,[HV + (h_g - h_n)]\,/h_g \qquad 1.8$$

where H = total sensible heat content of the combustion products, including the sensible heat of water vapour (kJ/kg).

V = quantity of combustion gases produced from a given quantity of fuel (kg/kg).

h_g, h_n = gross and net calorific values of the fuel respectively (kJ/kg).

Thus the amount of heat lost can be calculated from the constituent flue gases and graphs can be drawn showing flue losses plotted against excess air for different flue temperatures. Figure 1.5 shows two such graphs for natural gas.

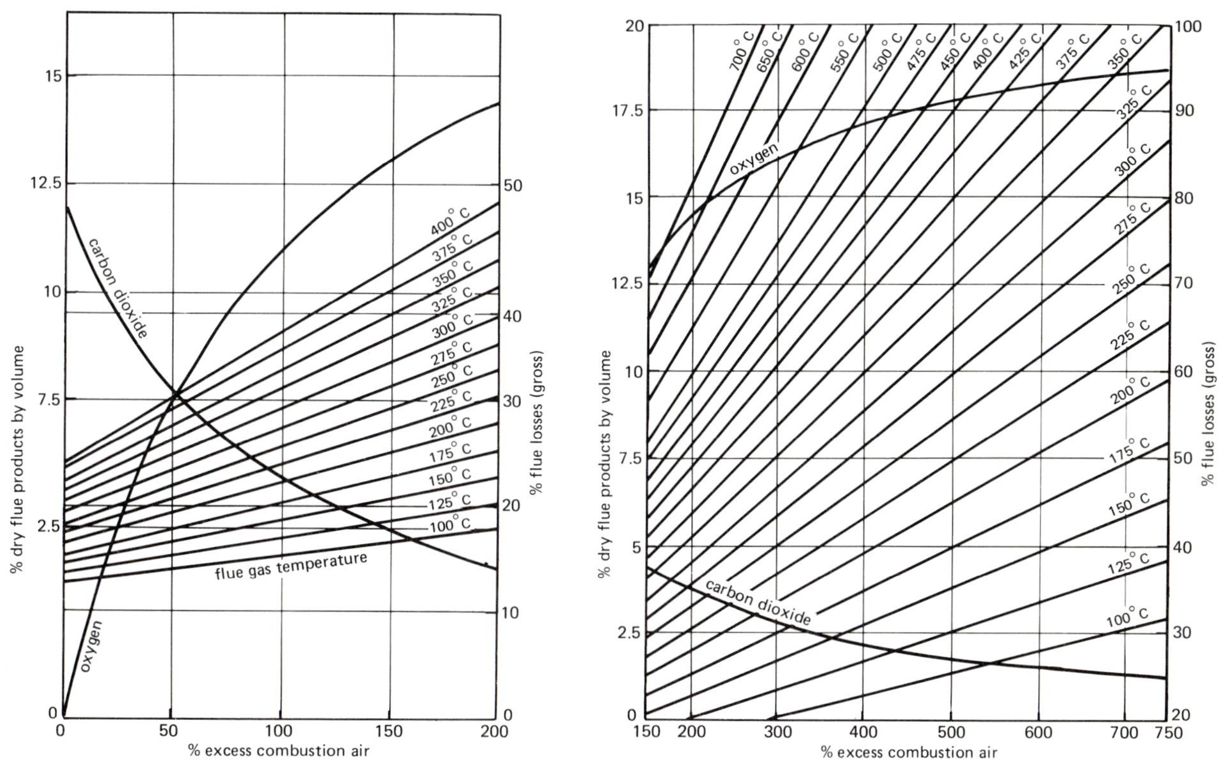

Fig. 1.5 Natural gas flue loss graphs

1.5.2 Flue Losses Due to Products of Incomplete Combustion

Small quantities of unburnt constituents nearly always occur in the flue products from gas and oil fired heating plant. Smoke is a good indication of the incompleteness of combustion on oil fired plant, whereas on gas fired plant it is indicated by the presence of carbon monoxide and any other unburnt constituents in the flue gases.

If it is suspected that significant quantities of unburnt constituents may be present in the flue gases, it is necessary to determine their concentrations in order to obtain a reliable assessment of the flue losses. At low excess air levels, unburnt constituent concentrations of less than 100 ppm may be ignored as their effect on flue losses and excess air/oxygen/carbon dioxide relationship is small compared with normal errors in measuring flue gas concentrations. For unburnt constituent concentrations in excess of this value or for very high dilution with excess air they must be considered.

In order to determine the true excess air level it is necessary to correct the oxygen and/or carbon dioxide concentrations to allow for burn out of the flammable gases. The 'corrected' oxygen (O_2) and carbon dioxide (CO_2) concentrations are determined as follows:

$$\text{Corrected } O_2\ (\%) = \text{Measured } O_2\ (\%) - \tfrac{1}{2}\text{ measured } CO(\%) - \tfrac{1}{2}\text{ measured } H_2\ (\%) \qquad 1.9$$

$$\text{Corrected } CO_2\ (\%) = \text{Measured } CO_2\ (\%) + \text{Measured } CO(\%) \qquad 1.10$$

These approximate relationships apply to the combustion of natural gas in air for unburnt constituent concentrations of 1% (10 000 ppm) or less. The corrected O_2 or CO_2 level can be used with the relevant graph from Figure 1.5 to determine the excess air level for the plant under consideration.

It is also necessary to make allowance for the calorific value of unburnt constituents leaving the plant. This additional loss L_1 can be determined from the following relationship:

$$L_1 = [(\text{Volumes dry products per unit vol. fuel gas}) \times \Sigma(V_i h_i)]/h_{fuel} \qquad 1.11$$

After rearranging and substitution of various constants this expression becomes

$$L_1 = 2.68(11.97\ CO + 12.1\ H_2)/(CO_2 + CO) \qquad 1.12$$

where L_1 = additional loss % gross
CO, H_2, CO_2 = fractional concentration i.e. dry % vol. ÷ 100

Figure 1.6 shows the flue losses represented by various levels of CO up to 10 000 ppm for excess air levels represented by oxygen concentrations up to 7%. The calorific values of carbon monoxide and hydrogen can be considered to be equal. Therefore, where the hydrogen and carbon monoxide concentrations are known, the curves on Figure 1.6 should be used after adding the dry hydrogen and carbon monoxide concentrations to produce a 'total unburnt' figure which is used to estimate the additional gross loss. Where the hydrogen concentration is not known the assumption can be made that in an appliance flue the ratio of carbon monoxide to hydrogen concentrations is approximately 4:1. The additional gross flue loss can then be determined. The error introduced by assuming the CO/H_2 ratio in the above manner is small.

Products of combustion must always be analysed for two components and preferably for three. The variation in combustion products with air factor in Figure 1.1 shows that for a given CO_2 concentration combustion can be occurring with either excess air or with high levels of CO. Ideally flue gases should be tested for carbon dioxide, carbon monoxide and oxygen. The presence of oxygen in the flue gases would seem to eliminate the chance of carbon monoxide being found but this is not necessarily true. There are circumstances in which carbon monoxide and oxygen can exist together. If the reactant gases, which constitute a flame, impinge directly upon a very cold surface such as the water cooled surface of a boiler before mixing and hence combustion is complete, the gases may be cooled to below ignition temperature and so combustion will effectively be extinguished. Under such circumstances, partly burned fuel gas and unreacted oxygen will pass out of the system and both carbon monoxide and carbon dioxide will be in the products of combustion,

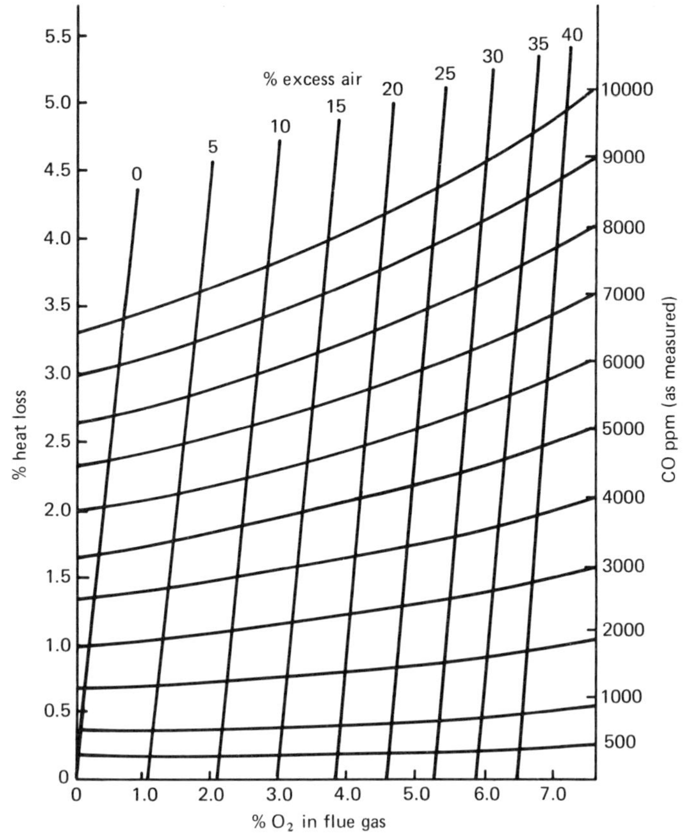

Fig. 1.6 Heat loss due to incomplete combustion

even though excess air may have been provided at the burner. No amount of excess air will resolve this condition.

1.5.3 Flue Losses with Condensing Recuperators

In condensing recuperators as used for heat recovery, or in direct contact water heaters, a spray of water passes through the flue products. If the water is at a temperature below the dewpoint of the flue products and there is sufficient flow rate, then the flue gases are cooled to around the dewpoint provided that there is a sufficient degree of contact between water and flue gas in order to obtain the necessary mass and heat transfer. Dewpoint for natural gas combustion products will be about 50 °C and will vary with excess air, as shown in Figure 1.7. Sensible heat in the flue gases and some latent heat from water formed by combustion are recovered. As some water in the combustion products has been condensed, heat has been recovered and therefore normal flue gas heat loss charts based on gross enthalpy (Figure 1.5) are inapplicable.

In order to calculate flue losses for a condensing system the procedure outlined below should be followed.

Condensing mechanism

The theory of a condensing spray recuperator can be explained with the use of a standard hygrometric chart. Most charts are produced only for temperatures close to ambient and a high temperature version is necessary, as in Figure 1.8. The hygrometric chart is for water in air, not water in flue products, but the error involved is small. Note that enthalpies are quoted above base 0 °C, and must be corrected to base 15 °C or base ambient air temperature, whichever is the reference standard.

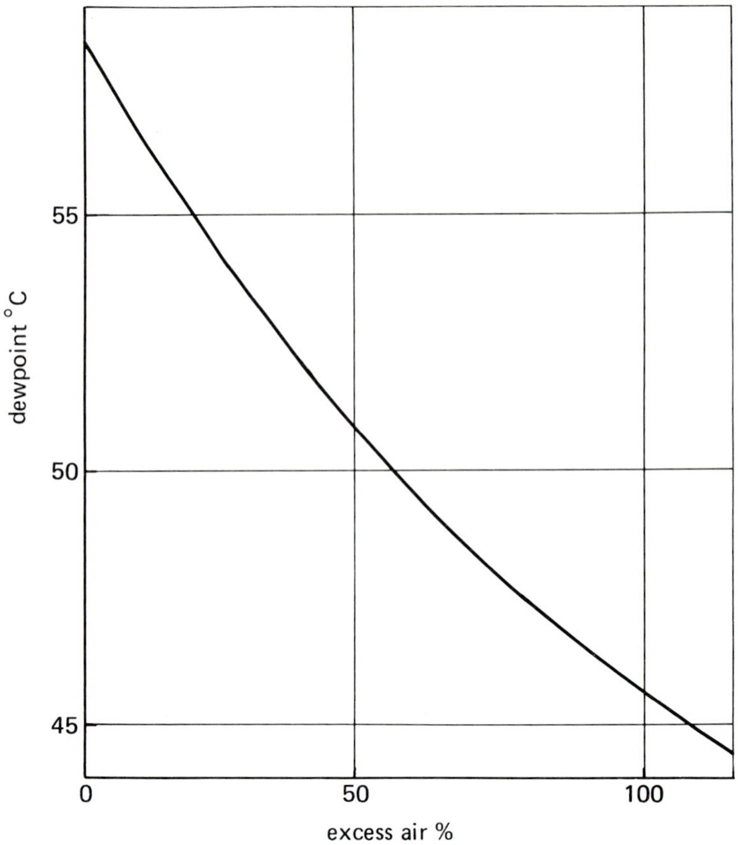

Fig. 1.7 Dew point vs excess air for natural gas/air systems

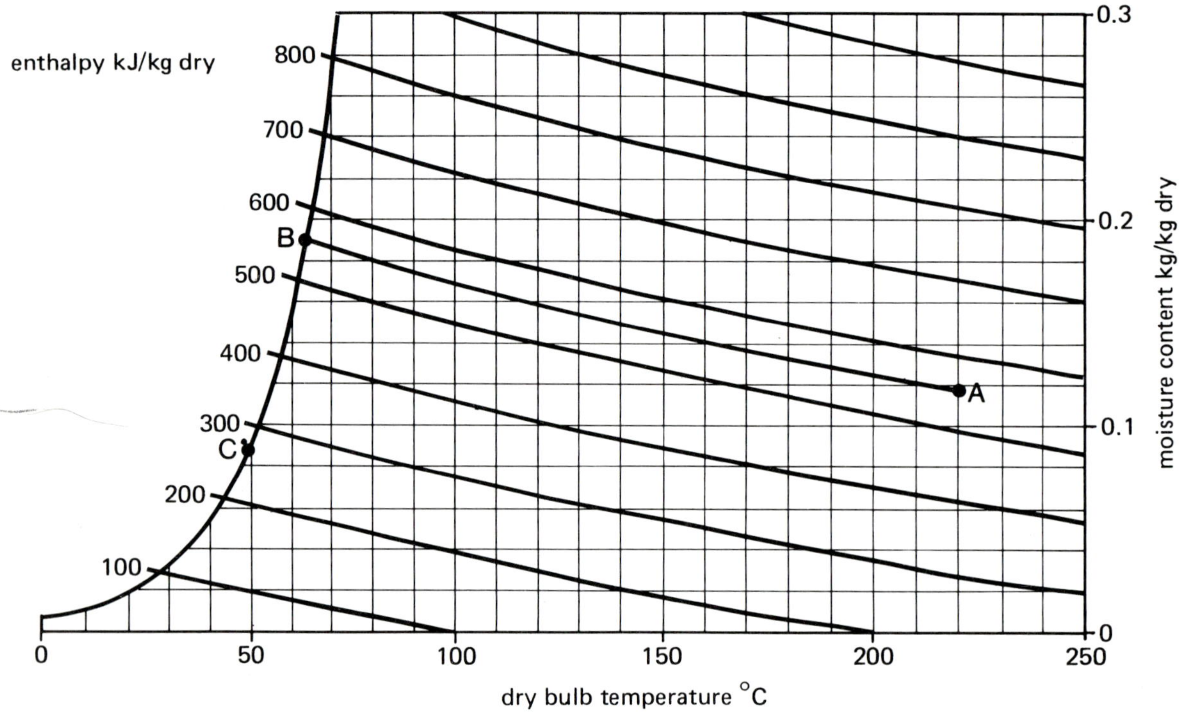

Fig. 1.8 Hygrometric chart for flue gases

The flue gases leaving the heat exchanger can be represented by point A. As water is added this water evaporates and the moisture content increases together with the temperature decreasing. There is no heat loss from the system and therefore the heat taken out of the flue gas appears as heat in water vapour carried in the flue gas. The line A-B is followed.

At point B the flue gases become saturated. As more cold water is added condensation commences and the flue gases are cooled to their final temperature, point C. The flue gases are still saturated at the exit from the spray recuperator.

It should be noted that if insufficient cold water is added the line A-B is followed but saturation is not reached and there is no heat recovery, only a loss of spray water. If the spray water is at too high a temperature no condensation will take place.

Example 1.1 A boiler operates at a gross thermal efficiency of 80%, consuming 340 m^3/h of natural gas. Excess air level is 20% and flue gas stack temperature is 220 °C. If a spray recuperator is fitted, with water entering at 45 °C and leaving at 55 °C, and flue gas temperature is thereby reduced to 50 °C, what gas saving will be made and what will boiler efficiency be raised to? Assume fuel properties as for mean North Sea Gas (Tables 1.1 and 1.3) and combustion air 15 °C and 50% R.H.

		kg/kg fuel	
Combustion air	- stoich	16.5	
	- 20% excess	3.3	
	- total	19.8	
Fuel		1.0	
Total flue gases		20.8	(by virtue of conservation of mass)
Combustion $H_2O = 9 \times H_2$ % mass		2.1	
Dry flue gases by difference		18.7	

From standard psychrometric charts, combustion air has 0.0054 kg H_2O/kg dry air, therefore total H_2O from air is $0.0054 \times 19.8 = 0.1$ kg/kg fuel. Total H_2O in flue gases = 2.1 + 0.1 = 2.2, i.e. 0.118 kg/kg dry gases.

Flue gases entering recuperator (point A, Figure 1.7):

220 °C, 0.118 kg H_2O/kg, i.e. 550 kJ/kg dry above 0 °C.
But combustion air is reference base and has enthalpy of 28.5 kJ/kg dry.
Gross flue gas heat loss = $(550 - 28.5) \times 18.7 \times 100/(52.41 \times 10^3) = 18.6\%$ (the same result as obtained from Figure 1.5).

Flue gases leaving recuperator (point C, Figure 1.7):

50 °C saturated, i.e. 0.085 kg H_2O/kg and enthalpy 268 kJ/kg dry.
Gross flue gas loss $= (268 - 28.5) \times 18.7 \times 100/(52.4 \times 10^3)$
$= 8.5\%$, i.e. 10.1% points reduction.

Savings:

Savings and the new boiler efficiency can be calculated by equating before and after heat balances, but this is laborious. The gain in efficiency equals the drop in flue gas loss (ignoring what should be minor losses within the heat recovery system):

New efficiency = 80 + (18.6 - 8.5)
= 90.1% gross

Savings = (New efficiency - old efficiency) x 100/New efficiency
= (90.1 - 80) x 100/90.1
= 11.2%

Original natural gas consumption was 340 m^3/h.

Gas saving = 0.112 x 340
= 38 m^3/h

1.5.4 Flue Gas Sampling

It has been demonstrated that the heat content of the gases can be derived with a good degree of accuracy from charts showing flue losses plotted against excess air for different flue temperatures. These graphical methods can only be as accurate as the data that is used with them. This means that both temperature and excess air must be measured as accurately as possible.

The sampling position used must give a representative sample. This will normally mean carrying out a traverse of the flue to determine the extent of stratification. It is necessary to avoid dead-spaces to ensure that the point of sampling is being continuously swept by the gas to be analysed. Points in the flue or duct system where reasonable turbulence exists are to be preferred for sampling because the possibility of stratification exisiting is thereby reduced. Reduction in flue cross-sectional area, operation of fans and bends all assist in breaking up stratification patterns. Sampling on bends themselves should be avoided because of the probable existence of dead-space. It should be noted that in ducts of square or rectangular cross-section, the flow of fluid takes up a roughly circular profile so that the corners of these cross-sections will not be subject to much flow and errors of dead-space could arise if sampling is carelessly performed.

The flue gas sample should be rapidly cooled, usually by means of a cooling coil, after withdrawal from the flue. This quenching prevents further reaction between the constituents which would give erroneous results.

Accurate temperature measurement is important. The use of a simple sheathed thermocouple does not give accurate readings of gas temperature. Radiation from the thermocouple to the surrounding flue walls, which are cooler than the flue gases, causes an erroneously low temperature to be indicated. With low temperature plant, such as boilers, this is not too important but a suction pyrometer should be used for measurement at higher temperatures. Determination of the percentage of excess air present is usually done by measuring CO_2 or O_2, preferably together with CO. For maximum accuracy all three should be measured. The excess air based on each parameter should be the same, within experimental error: if it is not; an error in reading must be assumed and both CO_2 and O_2 measurements repeated until acceptable correlation is achieved.

Further details of instrumentation and techniques for flue gas sampling will be found in Chapter 14.

1.6 FURNACE AND PROTECTIVE ATMOSPHERES

1.6.1 Introduction

An important aspect of the heat treatment of materials is the effect of the furnace atmosphere on the stock being heated. In most cases the need is to minimise or eliminate an undesirable effect of the furnace gases, such as oxidation or decarburisation, but there are processes such as the carburising of steels in which the interaction between the gases and the stock is the purpose of the process.

In directly fired processes only a limited control of furnace atmosphere is possible and although this is adequate for some low-temperature heat-treatment processes, for example stress-relieving of low-carbon steels, an increasing number of heat-treatment processes are carried out in indirectly fired furnaces. In these the stock is separated from the heating gases by a muffle or radiant tubes and the working chamber is fed with a separately generated prepared atmosphere.

Protective atmospheres find their widest application in metallurgical heat-treatment processes but are also used for protective and fire-prevention purposes in, for example, the paint, pharmaceuticals and rubber industries and for purging. Coverage in this section will be limited to their application in the metallurgical field.

1.6.2 Effects of Gases on Metals

The terms 'oxidizing', 'neutral' and 'reducing' are often applied to furnace atmospheres. A common misconception is that if a furnace atmosphere contains no free oxygen it is necessarily neutral in its effects. However carbon dioxide and water vapour are both oxidizing agents and furthermore a furnace atmosphere which may be neutral to one material may be highly oxidizing to another. Additionally, the temperature of the process exerts an important influence on the neutrality or otherwise of the furnace atmosphere. Consequently the terms oxidizing, neutral and reducing when applied to furnace gases produced from combustion with excess air, stoichiometric air and substoiciometric air respectively, should not be taken as implying that these effects will occur.

The consideration of metal : gas reactions occurring in the furnace chamber is essentially one of the chemical equilibria existing at the temperature and pressure of the process.

The main reaction to be considered is oxidation and this may be caused by CO_2 and H_2O as well as free oxygen, i.e.

(1) $R + CO_2 \rightleftharpoons RO + CO$

(2) $R + H_2O \rightleftharpoons RO + H_2$

(3) $R + O \rightleftharpoons RO$

The equilibrium constant K for reaction (1) is:

$$K_1 = p_{ro}\, p_{co} / p_r\, p_{co_2} \qquad 1.13$$

where p denotes the partial pressure.

The vapour pressures of the metal and its oxides are negligibly small and Equation 1.13 may be rewritten:

$$K_1 = p_{co} / p_{co_2} \qquad 1.14$$

Consider the reaction between iron and carbon dioxide:

(4) $Fe + CO_2 \rightleftharpoons FeO + CO$

Figure 1.9 shows the equilibrium ratios for this oxidation-reduction reaction.

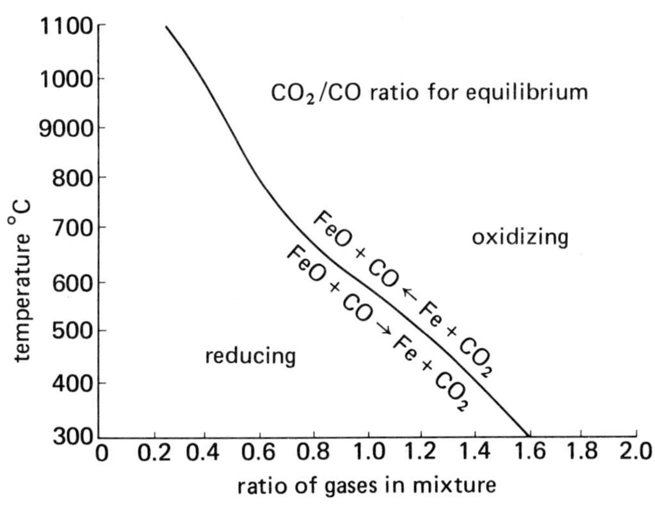

Fig. 1.9 Equilibrium diagram for the oxidation of ferrite

Furnace gases and most prepared atmospheres consist of multi-component gas mixtures and so in practice it is necessary to consider not only the reactions which may take place between the gases and the solids but between the gases themselves. In addition, the metals being treated, in particular nickel and chromium and to a lesser extent iron, may act as catalysts on the reactions between the constituents of the furnace atmosphere. The inter-gas reactions are of course temperature-dependent and it is necessary to consider the atmosphere conditions at all stages of a heat-treatment process, including cooling.

Industrial protective atmospheres are derived from the combustion or reforming of fuel gases and contain some or all of the following constituents: nitrogen, carbon dioxide, carbon monoxide, hydrogen, water vapour and hydrocarbons. The atmospheres of direct gas-fired furnaces will contain these constituents and in addition sulphur dioxide will be present in the case of oil or coal firing. However, an increasing number of metallurgical processes utilise inert gas atmospheres maintained from e.g. liquid nitrogen delivered by industrial gas suppliers.

Industrially the most significant reactions are those between these gases and the constituents of steel, i.e. ferrite (α-Fe) and cementite (Fe_3C) and these are considered below.

The following gas-metal and inter-gas reactions may be involved in the heat processing of ferrous metals:

Oxidation-reduction

(4) $Fe + CO_2 \rightleftharpoons FeO + CO$

(5) $Fe + H_2O \rightleftharpoons FeO + H_2$

Decarburizing-carburizing

(6) $Fe_3C + CO_2 \rightleftharpoons 3Fe + 2CO$

(7) $Fe_3C + H_2O \rightleftharpoons 3Fe + CO + H_2$

(8) $3Fe + CH_4 \rightleftharpoons Fe_3C + 2H_2$

Producer-gas reaction

(9) $C + CO_2 \rightleftharpoons 2CO$

Water-gas reaction

(10) $CO + H_2O \rightleftharpoons CO_2 + H_2$

Thermal decomposition

(11) $CH_4 \rightleftharpoons C + 2H_2$

Others

(12) $H_2O + C \rightleftharpoons CO + H_2$

(13) $H_2O + CH_4 \rightleftharpoons CO + 3H_2$

(14) $CO_2 + CH_4 \rightleftharpoons 2CO + 2H_2$

(15) $C + O_2 \rightleftharpoons CO_2$

The gas : metal reactions are oxidation-reduction and decarburization-carburization. The combined effects of the four gases carbon dioxide, carbon monoxide, water vapour and hydrogen on α-ferrite can be illustrated by constructing equilibrium diagrams for each reaction, based on Figure 1.9. This is shown in Figure 1.10.

Reactions (4) to (7) involving reactions with CO_2 and H_2O cause oxidation and decarburisation in steel. In general it is necessary to keep water vapour and carbon dioxide levels to a minimum in heat treatment processes. Whether the process is one of oxidation or decarburisation CO and H_2 are reaction products and thus an increase in their concentration will inhibit those reactions. It is

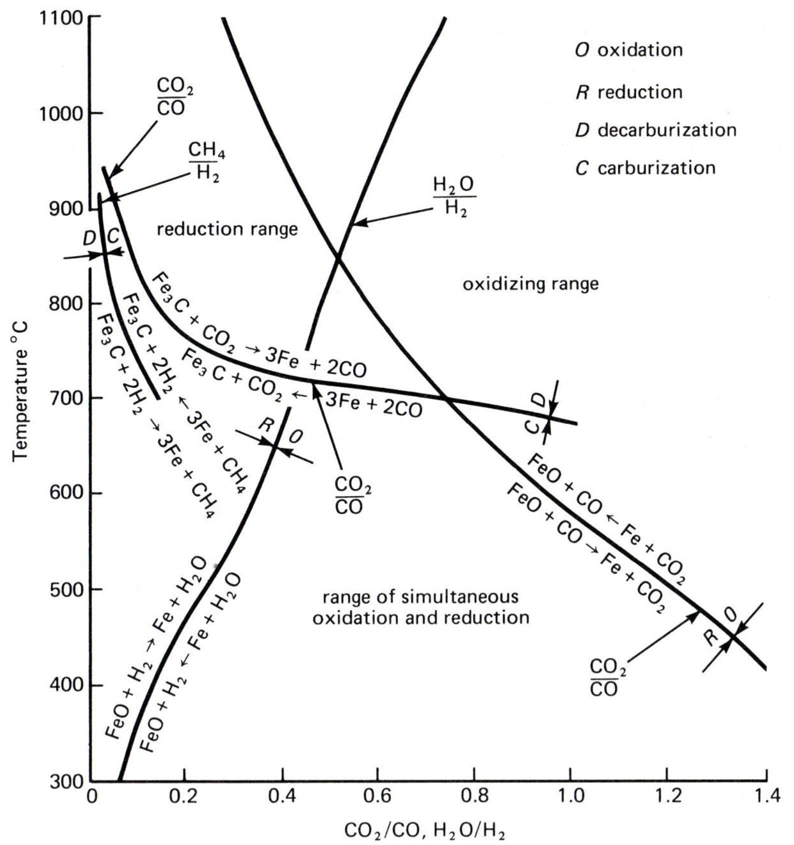

Fig. 1.10 Combined equilibrium diagram showing the reactions between CO_2, CO, H_2O and H_2 gases with ferrite

possible, as Figure 1.10 indicates, to produce an atmosphere which is non-scaling but decarburising, but non-decarburising atmospheres must be non-scaling at all practical temperatures. For high carbon steels and high temperatures the CO/CO_2 ratio will need to be high, primarily to prevent decarburisation, as can be seen from Figure 1.10.

1.6.3 Practical Controlled Atmospheres

Industrial controlled atmospheres are generated from two main sources, single relatively pure gases in cylinders and fuel gases. As an alternative to the use of a prepared atmosphere a vacuum is sometimes employed, particularly for copper annealing. Of the gases drawn from cylinders pure hydrogen is used where extreme reducing conditions are required and argon for processes in which nitrogen is not sufficiently inert.

Atmospheres derived from fuel gases may be termed endothermic or exothermic. The terms refer to the production processes, the former being produced by catalytic reforming of the gas with a small amount of air (the mixture being well beyond the rich limit of flammability) in an externally heated retort and the latter by combustion of the fuel with a controlled amount of air ranging from above-stoichiometric down to just sufficient to maintain combustion. These atmospheres have a wide range of composition which can be further extended and controlled, if required, by selective purification after generation. The processes are flexible and controllable and the gases obtained range from active atmospheres containing free oxygen and highly reducing and carburising atmospheres to inert atmospheres comprising almost pure nitrogen. The typical air/gas ratio requirements for the type of atmosphere produced by a range of common feedstock gases are shown in Table 1.6.

Table 1.6 Typical Air/Gas Ratios Required for Controlled Atmosphere Production

	Approximate Air/Gas Ratio Range vol./vol.		
	Endothermic	Exothermic	
		Rich	Lean
Natural gas	2.0 – 3.0	6.0 – 7.5	8.5 – 10.0
Propane	6.0 – 8.0	12.0 – 16.0	20.5 – 24.0

Production and properties of endothermic atmospheres

Endothermic atmospheres are produced by the catalysed reforming of fuel gases with a small amount of air. Air and gas from a machine-premix system are fed into an externally heated nickel-chrome steel or refractory retort containing nickel-impregnated refractory as a catalyst. The generator temperature is normally about 1 000 °C. The retort is generally heated by tangential-firing air-blast burners or in some instances by natural-draught burners.

Normal air-gas ratios on various fuel gases are: natural gas 2.4:1, propane 7.5:1, butane 9.8:1. The aim is to produce exact conversion to H_2, CO and N_2, which can be written approximately for methane as:

$$(16) \qquad CH_4 + \tfrac{1}{2}O_2 + 2N_2 \rightarrow CO + 2H_2 + 2N_2$$

A typical composition using natural gas (reactor temperature = 1050 °C) is: 0.2% CO_2, 19.8% CO, 39.4% H_2, 0.2% CH_4, balance N_2, dew point 2 °C.

Carbon is deposited on the catalyst and this needs to be burned off periodically by reducing the retort temperature below 700 °C and passing air through the retort. The interval between burning-off runs depends largely on the dew point at which the plant has been running, e.g. for -1 °C to -5 °C the interval is about three weeks. After burning off, the nickel catalyst is partially oxidized and needs to be reduced to metallic nickel by 'conditioning', i.e. passing a mixture with the correct air-gas ratio at the rated throughput for a few hours.

In addition to CO, H_2 and N_2, endothermic gases contain traces of H_2O, CO_2 and hydrocarbons. The quantities of CO_2 and H_2O are very sensitive to changes in air-gas ratio and since even a small increase of these decarburizing agents may be significant, accurate proportioning is required and fuel gases of constant chemical composition are advantageous. In this connection natural gas is a particularly suitable feedstock. Commercial-grade L.P.G. containing varying amounts of unsaturated hydrocarbons is less satisfactory in this respect. Since the dew point of endothermic gas is an accurate and easily measured index of the concentration of decarburizing agents, control is usually based on this property. Alternatively the oxygen potential of the atmosphere may be measured using a solid electrolytic cell.

For gas-carburizing purposes it is usually necessary to enrich endothermic gas by the addition of a few per cent of propane or natural gas to increase its carbon potential. This is defined as the percentage carbon content of the steel that would be in equilibrium with the atmosphere at the temperature of the heat-treatment process. It is discussed in further detail in Chapter 6.

Production and properties of exothermic atmospheres

Exothermic atmosphere generators consist essentially of a burner, combustion chamber and atmosphere cooler. Pre-mix burners are used almost exclusively and, although air-blast burners have been used, the most common arrangement is a machine-premix system which combines relatively accurate gas-air ratio control with a high mixture pressure to overcome the appreciable system resistance.

Exothermic atmospheres may be classified as rich or lean, the limit of the former being dictated by the upper flammability limit. Lean exothermic atmospheres are produced by combustion around stoichiometric proportions, generally slightly below, to produce a gas consisting of CO_2, N_2 and H_2O with a small excess of combustibles, but for a few specialised processes combustion with a very slight excess of air is used. The air-gas ratios on various fuels are typically:

Natural gas	:	lean 8.5 – 10.0,	rich 6.0 – 7.5
Propane	:	lean 20.0 – 24.0,	rich 12.0 – 16.0

The intermediate atmosphere ranges are rarely used.

The method adopted for cooling the combustion products in the generator has a significant effect on the composition of the atmosphere generated and is of particular importance for lean exothermic atmospheres close to stoichiometric composition. In addition to containing N_2, CO_2 and H_2O, equilibrium considerations dictate that at flame temperature stoichiometric combustion products also contain small but significant amounts of CO, H_2 and O_2 together with a variety of atomic species. If the combustion products are cooled slowly, equilibrium composition is maintained throughout the cooling process and the cooled atmosphere consists of N_2, CO_2 and H_2O only. If on the other hand the combustion products are rapidly cooled, for example by a direct water quench, the prepared atmosphere retains the same proportions of the trace constituents as at flame temperature.

For rich exothermic atmospheres, the product analysis varies but for natural gas is typically 15% H_2, 10% CO, 5% CO_2 and 1% CH_4 on a dry basis.

Rich exothermic gas may be used 'raw' or alternatively 'stripped', i.e. after CO_2 and water vapour have been removed. CO_2 and water vapour may be removed by three alternative methods: compression over water, the use of molecular sieves or by the use of organic solvents, e.g. ethanolamines. Stripped rich exothermic gas is extremely reducing and may be used as an alternative to endothermic gases. Near-pure nitrogen may be produced by CO_2 and water removal from lean exothermic atmospheres.

1.7 OXYGEN IN COMBUSTION PROCESSES

The advent of tonnage oxygen plants and large bulk storage has greatly increased the use of oxygen for combustion in recent years.

Oxygen enrichment of combustion air to increase flame temperatures by reducing the proportion of nitrogen used in the combustion process is of particular value to operators of high temperature furnaces, both for improving efficiency but, in particular, as a means of increasing furnace productivity.

As the proportion of oxygen used in combustion is increased, the characteristics of the flame change. Increased levels of oxygen lead to an increase in flame temperature, an increase in burning velocity and a decrease in ignition temperature. Since there is a decrease in the proportion of nitrogen accompanying the combustion process there will also be a reduction in the waste gas volume.

The use of a properly designed burner is essential as rapid deterioration results if parts of a burner become overheated. In general this is not significant for levels of oxygen enrichment approaching 10% in the combustion air. Most oxy/fuel burners are water-cooled and in this respect oil firing has distinct advantages as the burner size and hence the water cooling requirement are considerably less than for oxy/gas burners of equivalent heating capacity.

1.7.1 Flame Temperature

As the proportion of oxygen in the combustion air is increased the flame temperatures are progressively increased as shown in Figure 1.11.

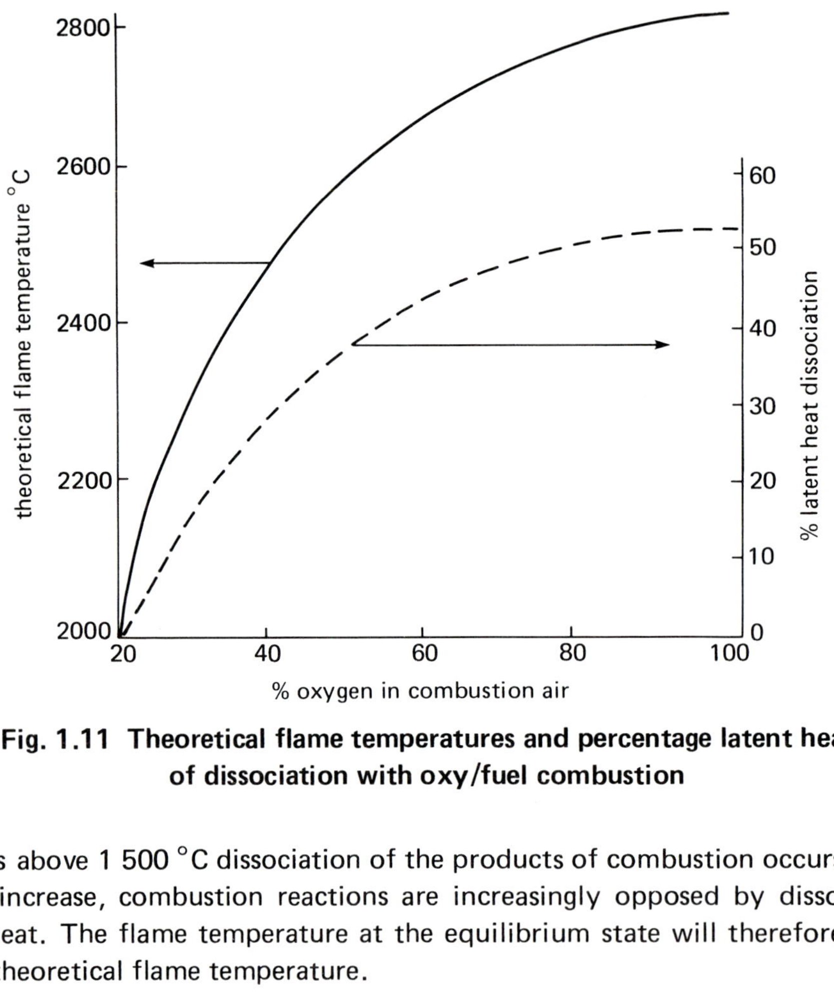

Fig. 1.11 Theoretical flame temperatures and percentage latent heat of dissociation with oxy/fuel combustion

At temperatures above 1 500 °C dissociation of the products of combustion occurs. Thus as oxygen addition levels increase, combustion reactions are increasingly opposed by dissociation reactions which absorb heat. The flame temperature at the equilibrium state will therefore be considerably lower than the theoretical flame temperature.

By adopting the correct technique of adding oxygen to a flame, the heat lost by dissociation can be regained. When the dissociated species contact the relatively lower temperature walls of the furnace and the charge, the radicals recombine and thereby release the heat of dissociation. This percentage latent heat of dissociation against oxygen content of the combustion air is shown in Figure 1.11.

The higher temperatures resulting from oxygen addition will give rise to higher heat transfer rates by radiation, conduction and convection. Clearly the effect on heat transfer rate will depend on factors such as flame type, the degree of oxygen addition and the furnace type. Similarly, accelerated combustion may also markedly increase flame intensity and non-luminous emissivity, thereby leading to a change in the mode of heat transfer within a furnace.

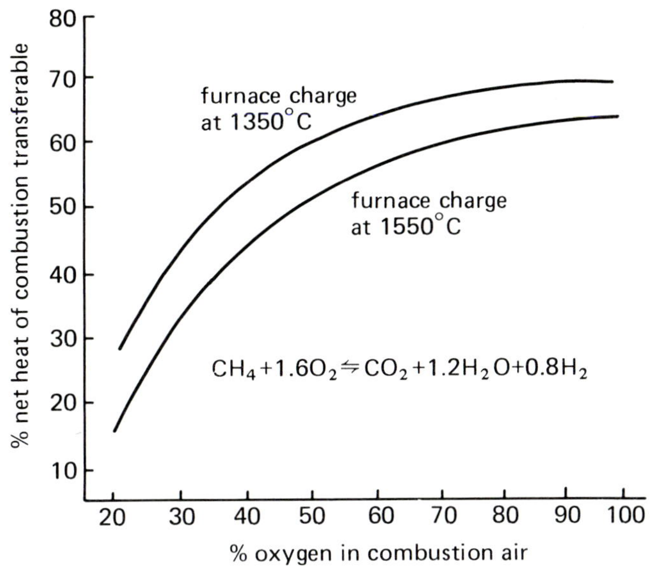

Fig. 1.12 The proportion of net heat of combustion of methane transferable with increasing levels of oxygen

The reduced waste gas volume resulting from oxy/fuel firing may be significant where some heat transfer is effected by convection. Thus the heat loss in a furnace will be lowered due to the reduction in waste gas volume and thus the potential for heat transfer will be increased. The effect of oxygen addition on the potential heat transfer from the combustion of methane is shown in Fig. 1.12.

1.7.2 Burning Velocity

The addition of oxygen gives rise to a higher burning velocity. A comparison of burning velocity of various fuels with air and oxygen is shown in Table 1.7.

Table 1.7 Maximum Burning Velocities in Air and Oxygen (m/s)

Fuel	Air (21% O_2)	100% O_2
Hydrogen	2.8	11.8
Methane	0.4	4.0
Propane	0.5	3.8
Butane	0.4	3.6
Carbon monoxide	1.6	11.3

Oxygen addition results in a reduction in flame dimensions and thereby an increase in heat release per unit volume. Burner design and the selection of the currect technique of oxygen addition is thus important in order to avoid light back. The techniques can be broadly categorised into enrichment, under-flame enrichment and oxy/fuel combustion and they are discussed in greater detail in Chapter 2.

1.7.3 Ignition Temperature

Since oxygen enrichment improves the mixing of fuel and oxidant the temperature at which combustion becomes self-sustaining will tend to reduce. A comparison of ignition temperatures in air and oxygen for various gaseous fuels is shown in Table 1.8.

Table 1.8 Ignition Temperatures (°C) in Air and Oxygen

Fuel	Air (21% O_2)	100% O_2
Hydrogen	572	560
Methane	632	556
Propane	493	468
Butane	408	283
Carbon monoxide	609	588

1.8 COMBUSTION OF FUEL OILS

Crude petroleum as first recovered from the earth's crust is a liquid, from light brown to black in colour and varying very considerably in other properties, according to its origin. It occurs in sedimentary rock strata and is tapped by drilling, often to a depth of 3000 m or more. In the initial stages it may emerge from the well under hydrostatic pressure or under gas pressure. The gas exerting this pressure, which also occurs in large accumulations contiguous to crude oil, is natural gas and natural gas should therefore be regarded as part of the same series of compounds as the constituents of crude petroleum. Petroleum is a mixture of many compounds and is refined into a wide range of saleable products including fuel oils.

1.8.1 Fuel Oils

Fuel oils include a variety of products, from distillates such as gas oil to residual oils and blends of distillates, residuals and bitumens. They have a wide range of physical properties but by careful manipulation of the blending processes the products may be made to conform in properties to specification with comparatively close tolerances. A range of fuel oils may therefore be marketed with each specific product falling into a closely defined category. Table 1.9 lists the salient physical properties of fuel oils sold in the U.K.

1.8.2 Fuel Oil Combustion

With gaseous fuels good mixing of combustion air and gas is possible and combustion can take place to completion with very little excess air.

Oil requires atomisation to achieve the intimacy of mixing required for good combustion and this is therefore a stringent design necessity in an oil burner. It entails the supply of oil droplets at a specific particle size range into a region in which they can be vapourised on ignition, and in which they will continue to vaporise once the flame is established. The vapours resulting from this distillation can then behave as gases and can mix with air and produce flames. At the same time, since solid carbon is a residual from thermal cracking, the combustion process requires adequate air and sufficient time for combustion of these carbon particles to be completed.

The air-oil system is mainly operated with oil droplets and air meeting at their respective discharge ports Pre-mix, in the sense in which it was used to describe an air-gas system, is possible only when

Table 1.9 Physical Properties of Fuel Oils

Class	D	E	F	G
Grade	Gas Oil	Light Fuel Oil	Medium Fuel Oil	Heavy Fuel Oil
BS 2869 : 1983 Specification				
Kinematic viscosity:				
cST at 40 °C – min	1.5			
– max	5.5			
cSt at 100 °C – max		8.2	20.0	40.0
Flashpoint PMCC °C min	56	66	66	66
Water % vol max	0.05	0.50	0.75	1.00
Sediment % mass max	0.01	0.15	0.25	0.25
Ash % mass max	0.01	0.15	0.15	0.20
Sulphur % mass max	0.5	3.5	3.5	4.0
Typical as Supplied				
Water % vol	nil	0.2	0.4	0.5
Ash % mass	nil	0.1	0.1	0.15
Sulphur % mass	0.3	2.2	2.6	3.0
Density at 15°C kg/l	0.840	0.955	0.976	1.000
Mean specific heat capacity between 0 °C & 100°C kJ/kg K	2.05	1.93	1.89	1.89
Min storage temp °C	—	10	25	40
Min handling temp °C	—	10	30	50
Vol correction factor per K	0.00083	0.00072	0.00070	0.00068

a stream of oil particles can be vapourised before emerging from a port. Oil vapourising burners are restricted to using oils which have a limited distillation temperature range, with a minimum residue content especially kerosene.

In the majority of cases air and oil meet at the discharge from the oil atomiser, whether this be operated on oil pressure alone, or with a second atomising fluid (such as steam or compressed air or another compressed gas), or by mechanical means.

The combustion of fuel oil is considered in detail in Chapter 3. The flame produced, as in the case of gas combustion, depends on the mixing properties of the system. This factor predominantly will determine the pattern of heat release, flame dimension, and combustion performance.

Combustion calculations for fuel oils

Calculations of the air requirement and the quantity and composition of flue gases follows the chemical equation requirements of the elements of which the fuel oil is composed. The main properties of fuel oils related to combustion are noted in Table 1.10.

Table 1.10 Typical Thermal and Combustion Properties of Fuel Oils

		D	E	F	G
		Gas Oil	Light Fuel Oil	Medium Fuel Oil	Heavy Fuel Oil
Gross calorific value	- MJ/kg	45.60	43.02	42.46	41.83
	- MJ/l	38.30	41.08	41.44	41.83
Nett calorific value	- MJ/kg	42.80	40.60	40.13	39.57
	- MJ/l	35.95	38.77	39.17	39.37
Ultimate Analysis (dry basis):					
Carbon	% mass	86.1	85.6	85.6	85.4
Hydrogen	% mass	13.6	12.0	11.5	11.2
Sulphur	% mass	0.3	2.2	2.6	3.0
Oxygen, Nitrogen, Ash	% mas	—	0.2	0.3	0.4
Stoichiometric air requirements:					
- mass/mass fuel		14.595	14.064	13.908	13.797
- vol/mass fuel	m^3/kg	11.94	11.51	11.38	11.29
Stoichiometric combustion products:					
- CO_2	% vol	13.29	13.79	13.96	14.08
- H_2O	% vol	12.59	11.59	11.28	11.10
- N_2	% vol	74.10	74.49	74.60	74.64
- SO_2	% vol	0.02	0.13	0.16	0.18
- Water dewpoint	°C	50	50	50	50
- Acid dewpoint	°C	127	143	154	160
- CO_2 in dry products	% vol	15.21	15.60	15.74	15.84
- SO_2 in dry products	% vol	0.02	0.15	0.18	0.20
Theoretical flame temp.	°C	2022	2028	2028	2028

The combustion gas analysis is calculated from the oxygen requirement of each constituent on a mass basis. The volumetric composition of the combustion gas is then derived via density from its weight composition. The volumetric analysis appropriate to each of the main combustible constituents of fuel oil is summarised in Table 1.11.

Table 1.11 Combustion Gas Volumetric Composition

Fuel Element	Reaction Equation	Unit	Demand m^3		Flue Gas Volume m^3			
			O_2	Air	CO_2	H_2O	N_2	Total
C	$C + O_2 = CO_2$	1 kg	1.87	8.91	1.87	-	7.04	8.91
H_2	$H_2 + ½O_2 = H_2O$	1 kg	5.60	26.70	–	11.20	21.10	32.30
S	$S + O_2 = SO_2$	1 kg	0.70	3.33 (SO_2)	0.70	-	2.63	3.33

Thus for any given fuel oil where an ultimate analysis is known curves for CO_2 and the associated O_2 content can be plotted relative to air factor.

Heat losses in flue products

In a directly analogous manner to the treatment of gaseous fuels in Section 1.5.1. the amount of heat lost can be calculated from the constituent flue gases and graphs drawn showing flue losses plotted against excess air for different flue temperatures. Graphs enabling ready assessment of flue gas loss for temperatures up to 400 °C and for gas oil, light fuel oil, medium fuel oil and heavy fuel oil are shown in Figure 1.13.

Fig. 1.13 Fuel oil flue loss graphs.

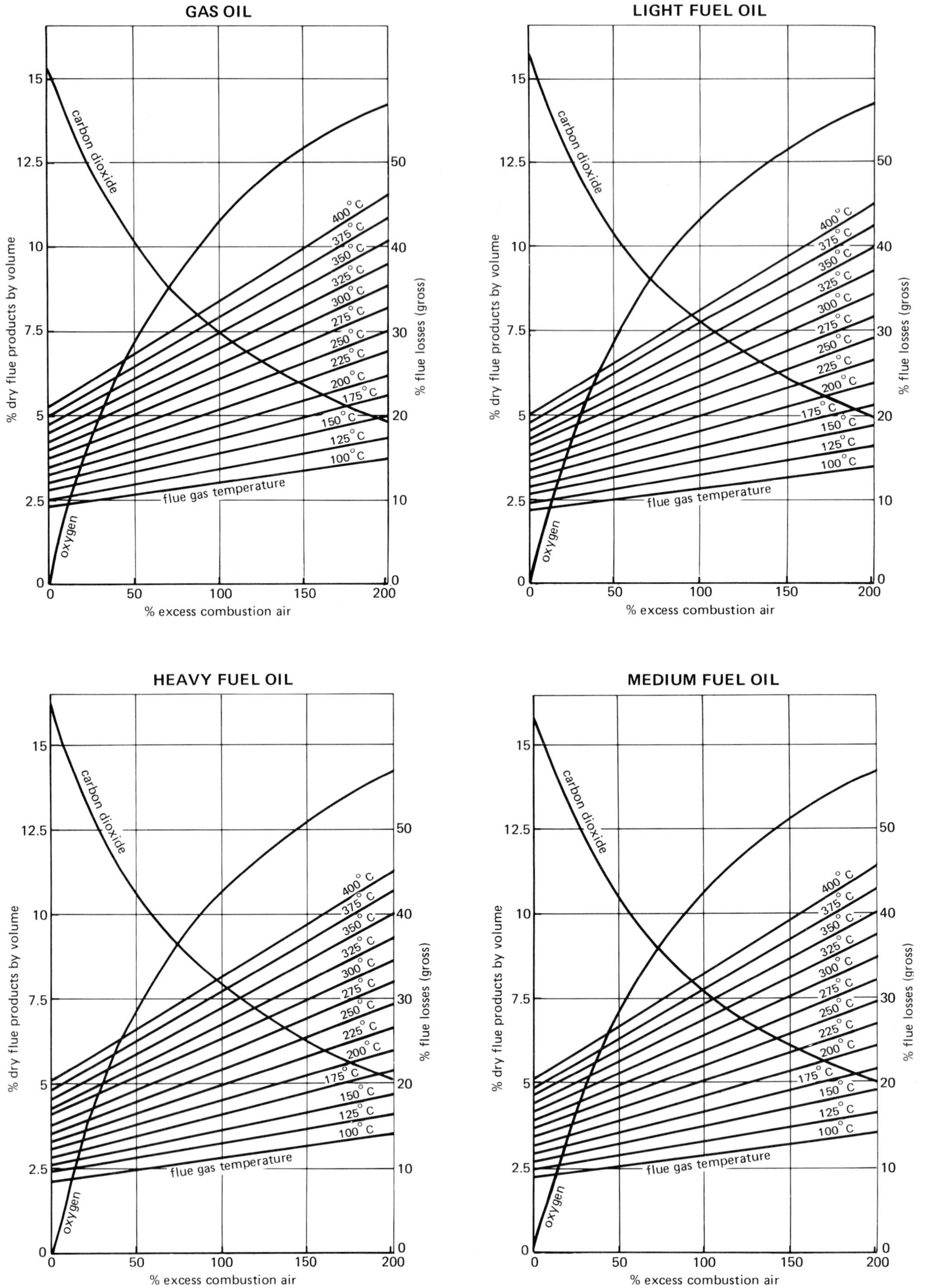

REFERENCES

BRITISH STANDARDS

BS 1756 : 1971 Part 1 Methods for the Sampling and Analysis of Flue Gases
BS 2869 : 1983 Specification for Fuel Oils for Oil Engines and Burners for Non-Marine Use

OTHER PUBLICATIONS

Combustion Flames and Explosions of Gases. Lewis B and von Elbe G. 2nd Edition, Academic Press, 1961

Technical Data on Fuel. Rose J.W. and Cooper J.R. (edited by). 7th Edition, World Energy Conference, 1977

Industrial Gas Utilisation : Engineering Principles and Practice. Pritchard R, Guy J.J., Connor N.E., Bowker, 1977.

Combustion Calculations: Theory, Worked Examples and Problems. Goodger E.M. MacMillan, 1977

North American Combustion Handbook. Reed R.J. (ed), 2nd Edition, North American Mfg. Co. 1978

Flames, Their Structure, Radiation and Temperature. Gaydon A.G. and Wolfherd H.G. 4th Edition, Chapman and Hall, 1979

Engineering Thermodynamics, Work and Heat Transfer. Rogers G.F.C. and Mayhew Y.R., Longman, 1980

Oxygen Injection Cuts Fuel Consumption. Dyer A.O. Factory Equipment News, September 1980

The Use of Oxygen for Energy Saving in High Temperature Combustion Processes. Booker P.I. Presented at the International Flame Research Foundation, October 1980

Fuel and Energy. Harker J.H. and Backhurst J.R. Academic Press, 1981

Combustion. Glassman I. 2nd Edition, Academic Press 1987

Gas Burners and Burner Systems 2

CHAPTER 2

GAS BURNERS AND BURNER SYSTEMS

2.1 INTRODUCTION

The great majority of gaseous fuels are easy to burn and control using relatively simple techniques. The wide range of uses for domestic, commercial and industrial purposes has led to the development of innumerable varieties of gas burners. The smallest burners, as used to provide pilot flames, have ratings as low as 30 watts (W) whilst the larger burners, as for example used in some water-tube boilers, have ratings up to 60 megawatts (MW). This therefore represents a capacity range of two million to one.

Since the mid-1970s all the piped gas supplies in mainland UK have been distributing natural gas with the use of substitute natural gas (SNG) anticipated in the 21st Century. For this reason most of the information in this section relates to equipment suitable for natural gas, but it may be equally applicable to liquefied petroleum gases. Some burners designed for town gas are also satisfactory for use with natural gas but these are not included where they are suitable solely for use with town gas, unless they illustrate an important principle of burner design.

As an introduction to the subject of burner design and selection a general description of the different ways of burning gas is presented together with the reference to gas properties relevant to combustion. The detailed information on burners selected for different applications forms the main part of this Chapter and emphasis is placed on burner principles and the technology used in the differing systems with current examples of the range of burners available and their application.

2.2 GAS BURNERS AND BURNER SYSTEMS

Burners for industrial and commercial applications may be broadly classified under five main headings, governed principally by the means adopted for mixing gas and air.

Diffusion-flame or post-aerated burners

In these burners the gas issuing from the burner or pipe is neat and no premixing of air and gas takes place. The necessary air for combustion arrives at the reaction zone from the surrounding atmosphere by diffusion or entrainment after the burner nozzle. Laminar-flow and turbulent-flow types of burner may be distinguished, and the burners are also known as neat gas, non-aerated or luminous burners.

Atmospheric burners

An obvious improvement on the above system is where the gas entrains a proportion of the air for combustion before entering the burner. This is achieved by means of an atmospheric injector, where the gas under pressure entrains atmospheric air. A shorter more intense flame is produced by this method due to the fact that most of the air for combustion is already present on entering the burner. If the gas is supplied at the normal supply pressure in the range 10 to 20 mbar the primary aeration is normally less than stoichiometric and secondary air is obtained from the atmophere into which the burner is firing. Such burners are normally termed low-pressure aerated burners. The Bunsen burner is the most widely known example. If the gas pressure is higher, stoichiometric proportions are readily achieved and the burner becomes a high pressure gas burner.

The alternative terminology for this type is natural draught or gas blast burner.

Air-blast burners

In these burners all or part of the air required for combustion is supplied at pressures above atmospheric by means of a fan or compressor. This air at higher pressure is used to entrain gas at normal supply pressure, or governed to atmospheric pressure, via an injector in order to obtain stoichiometric mixtures.

Both the natural draught and air blast burners are pre-mixed systems with the air and gas mixed in varying degrees prior to the burner. The same results can also be achieved by a pre-mixing machine, but the initial cost and operating costs are considerably higher.

Nozzle-mix burners

If there is no prior mixing of the gas and air until the burner nozzle, the system is known as nozzle-mixing. The gas and air are proportioned separately by linked valves or other techniques and fed independently to the burner nozzle. The air and gas enter separate manifolds prior to the burner nozzle where their flow patterns are controlled. Mixing by this method is very positive so that intense rates of combustion are developed for high temperature work.

Other burner systems

The preceding four groups comprise of burner systems encountered in industrial and commercial gas usage. Systems such as pulsating combustors and catalytic combustors do not fit conveniently into this classification and these are considered individually together with specialised applications such as radiant tubes, recuperative burners and immersion tube burners.

2.3 INTERCHANGEABILITY OF GASES

Before considering in detail the range of burner types and systems available it is important to understand that there are limits within which satisfactory interchangeability of distributed gases may take place.

Although gas burning equipment is capable of tolerating some variation in the properties of the gas supplied the use of fuel gas mixtures, without regard to their combustion properties, is unlikely to be satisfactory.

Fuel gases are grouped in families, according to their Wobbe numbers, as an aid to the classification and testing of gas burning equipment. There are three families covering the ranges of:

Family	Wobbe Number MJ/m^3
1	24.4 to 28.8
2	48.2 to 53.2
3	72.6 to 87.8

Town gases manufactured from coal or oil usually fall into the first family. Natural gas and substitute natural gases fall into the second family while liquefied petroleum gases are classified under the third family.

2.3.1 Manufactured Town Gases

The main properties of a gas which affect its combustion characteristics are calorific value, relative density and burning velocity. Calorific value and relative density are combined in the Wobbe number (Chapter 1) which determines the heat input to an appliance. The burning velocity is usually expressed in terms of Weaver's flame-speed factor but the accuracy of this is limited, particularly when considering multi-component mixtures.

Using Wobbe numbers and flame speed factors it is possible to construct an interchangeability prediction diagram for a range of gas groups. The interchangeability prediction diagram for appliances adjusted to burn gas of group G4 (Wobbe number 26.6 – 28.8 MJ/m^3) is shown in Figure 2.1. Similar diagrams may be plotted for each of the other gas groups.

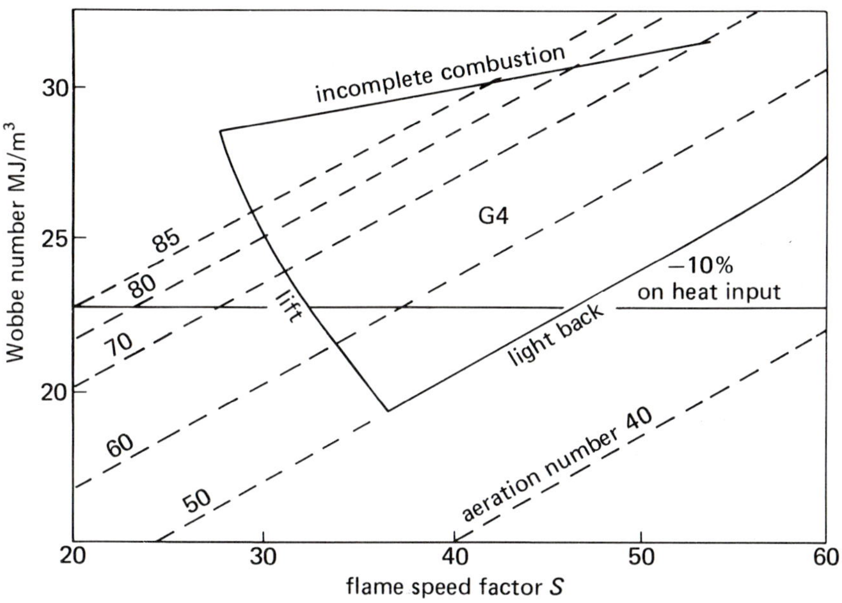

Fig. 2.1 Limits of satisfactory performance for G4-adjusted appliances

The Wobbe number and flame-speed factor of any gas can be calculated from its chemical composition. If the characteristics are then plotted on the diagram and the point lies within the boundaries of satisfactory service then the gas can be regarded as interchangeable with the adjustment gas. The limits within the diagram are based on experimentally determined measurements.

2.3.2 Natural Gas

Compared with first-family gases, natural gas has a higher Wobbe number and is relatively slow burning. The burning velocity thus affects the design of aerated burners and introduces the problem of flame lift on post-aerated burners; however lightback is generally not a problem. The narrower flammability range of natural gas makes it more susceptible to variation and it is considerably more difficult to ignite.

First-family and second-family gases are not interchangeable and gas burners and equipment must be converted.

The natural gas interchangeability prediction diagram is based on a series of test gases classified as NGA, NGB etc. corresponding to limiting compositions as regards incomplete combustion, lightback and other effects. The suitability of the test gases is under continuous review.

The prediction diagram shown in Figure 2.2 is based on the reference gas, NGA, being pure methane saturated with water vapour. The boundaries of the usable area are fixed by the characteristics of the limit test gas as follows:

- Heat input : normal limits, mean Wobbe number 50.7 MJ/m^3 ±5%; emergency limits +8% to −10%.
- Lightback : although lightback presents no problem with natural gas, possible substitute gases may have high hydrogen contents. The limit must pass through the point representing NGC which has an aeration number of 250; the lightback limit therefore is a line representing gases of aeration number 250.
- Lift : NGD(2) has a reversion pressure (measured on an aerated test burner) of 5 mbar and a line corresponding to this reversion pressure represents the lift limit.
- Incomplete combustion : this limit line must pass through NGB.
- Sooting : a tendency to soot is not apparently related to the flame-speed factor and cannot therefore be described by the diagram.

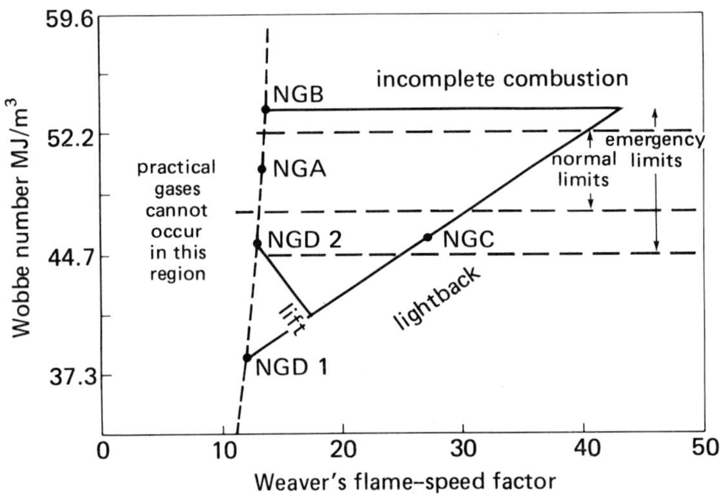

Fig. 2.2 Limits of satisfactory performance for UK natural-gas appliances

The natural-gas prediction diagram, like its town-gas counterpart, is useful but not complete nor sufficiently precise for gases falling close to the boundaries. Major criticisms are the diagram's inability to predict sooting and its reliance on Weaver's flame speed factor.

As a result of research work by British Gas, flame lift has been quantified as 'Lift Number' and expressions developed enabling lift number to be derived from gas composition with considerable reliability.

2.4 DIFFUSION FLAME BURNERS

In these burners neat gas is supplied to the burner head and mixes with the surrounding atmospheric air. In laminar diffusion, or post-aerated flames, the air and gas mix by a process of essentially molecular diffusion while eddy diffusion is the dominant process in turbulent post-aerated flames.

Thus the combustion depends more on the rate of mixing than on the much faster combustion reactions.

The diffusion flame is characterised by its luminous appearance. This is due to the unburnt gas issuing from the burner being cracked by the heat developed by the burning outer flame. Minute carbon particles are formed which on reaching the outer zone of the flame react with oxygen and burn with yellow luminosity. Radiation from the flame will also arise from triatomic molecules such as CO_2 and H_2O contained in the combustion gas and from transient free radicals which exist as intermediate products in combustion reactions.

Historically, in the era of manufactured gas, post-aerated burners dominated the design of most domestic, commercial and low temperature industrial heating plant. However, due to the reduced stability of natural gas diffusion flames and the necessity for flame stabilisation design, there are now few post-aerated burners for natural gas.

2.4.1 Laminar Post-aerated Burners for Natural Gas

There are few commercially available burners of this type designed for natural gas. Post-aerated burners designed for first-family gases are not satisfactory when burning natural gas giving rise to yellow tipping and soot formation. The much decreased flame stability is due principally to the lower burning velocity and narrower range of flammability.

Numerous attempts have been made to stabilise natural gas burners with varying degrees of success. The low burning velocity of natural gas implies that the forward speed of the gas will render difficult the formation of a zone in which mixture occurs with the surrounding air. If at the same time, the range of mixture concentrations which are flammable is also reduced, then stabilisation of self propagating combustion will evidently call for correction of these deficiencies. When this is done, the flame produced is stablilised at the baffle, or bluff body, and combustion proceeds smoothly. This principle whereby a natural gas free flame is stablished by recirculation of hot gases is shown schematically in Figure 2.3.

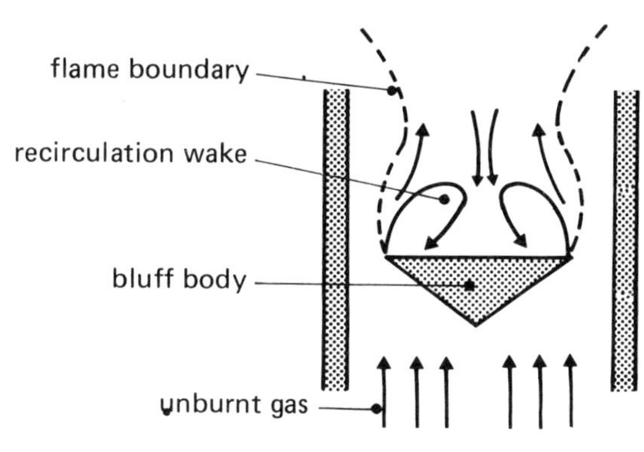

Fig. 2.3 Flame stabilisation of bluff body

This form of wake-eddy stabilisation exists in a variety of designs and the output of such burners can be relatively high. The multi-spud burner shown in Figure 2.4 utilises this principle and its output can be over 10 MW utilising high pressure gas.

Bar burners were often used with town gas for general purpose heating. With natural gas these generally must be pre-aerated to prevent soot formation. However neat gas bar burners are available in which air is entrained as the gas leaves the jet. The flame is stabilised by sudden enlargement and

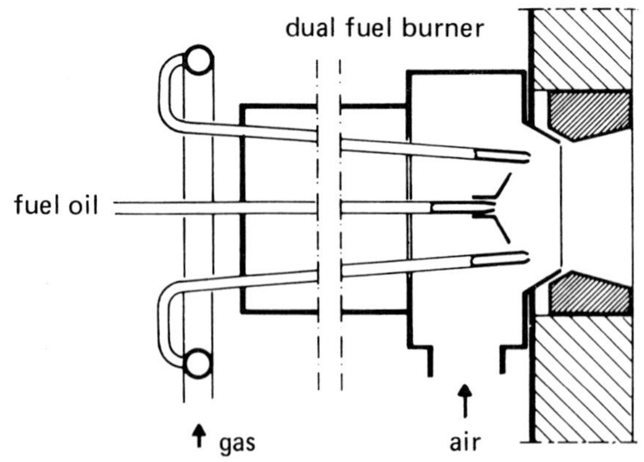

Fig. 2.4 Flame stabilised multispud high pressure gas/fuel oil burner

sits in an angle-formed trough situated above the bar. This has been used successfully in a limited number of applications, for example in small pottery kilns and for firing small cast iron sectional boilers.

Matrix burners

The matrix burner was first developed by British Petroleum and consists of a symmetrical array of short hexagonal-ended holes as shown in Figure 2.5.

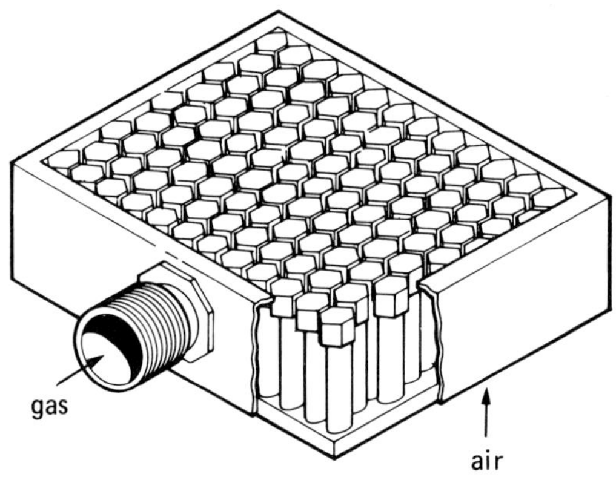

Fig. 2.5 Matrix burner design

Gas fed into the free space between the tubes emerges through the very small inter-connected slots framed between hexagonal tube-ends on the burner top, the bottom ends being sealed. The burner has virtually zero gas pressure at the outlets, the ratings to the burner being controlled by an orifice upstream of the burner head. In the absence of induced air flow the burner gives a large irregular flame, but when air is drawn or blown through the burner a very short compact mat of blue flame burns just above the burner. As with other post-aerated burners, matrix burners suffer from sooting during the initial operation period prior to the burner mass attaining equilibrium temperature.

2.4.2 Turbulent Post-aerated Burners

Suction burners which are used mainly for low-temperature oven heating are generally post-aerated. The air for combustion is induced either by natural draught or by an extraction fan in the exhaust ducting giving rise to a suction pressure of 5 to 10 mbar.

The Firecone burner developed by British Gas and shown in Figure 2.6 is used in conjunction with suction chambers, although it can be supplied in packaged burner form as a pressure fed unit with its own forced draught fan.

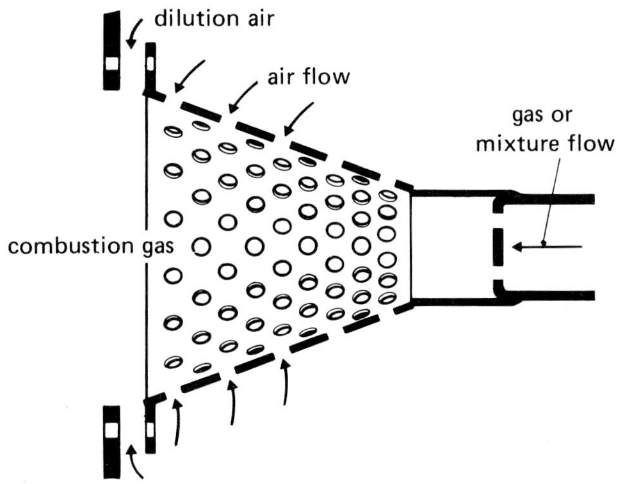

Fig. 2.6 Firecone burner

It may operate on neat gas or a partially aerated mixture with primary air being drawn through the holes in the cone which increase in size as the open end is approached. Diluent air is introduced via a shutter on the burner mounting. Nominal capacities from 20 kW up to 600 kW are available for use with low pressure natural gas.

Post-aerated burners in the form of crosses or wedges, have been specifically developed for air heating applications including make-up air heating and drying ovens. In general the burner sections are manifolded as needed to achieve the desired firing capacity. Typical capacities of up to 475 kW/m can be obtained and the burner turndown range can be up to 25:1.

2.5 ATMOSPHERIC AERATED BURNERS

In an atmospheric, or natural draught, burner primary air is entrained by the gas stream as it issues from a jet into the throat of a mixing tube. The mixing tube then either constitutes, or is connected with, one or a series of ports at which the partial mixture burns, finding air to complete the combustion from the ambient air in which it is situated. Figure 2.7 shows the essential features of an atmospheric burner and also indicates the terminology used.

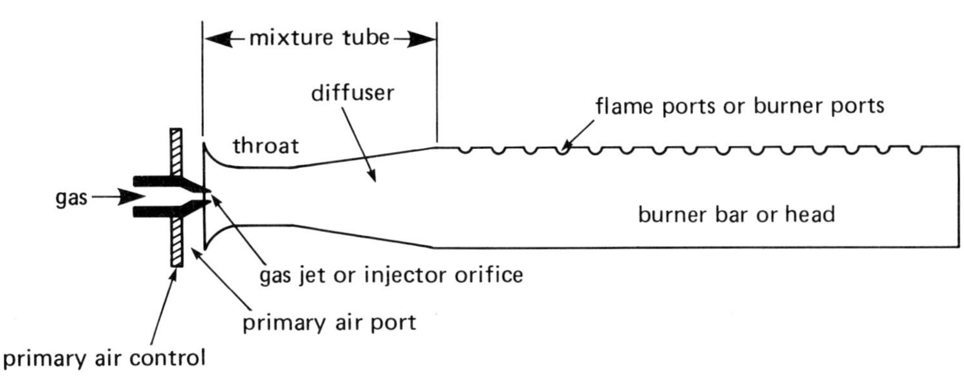

Fig. 2.7 Atmospheric aerated burner

Two categories of atmospheric burners can be distinguished: those in which the gas is used at normal supply pressures, ranging from 6.25 to 20 mbar, and those in which the gas is supplied from a compressor or high pressure supply. Low-pressure atmospheric burners are used in many low-temperature industrial applications and in most domestic appliances. High-pressure natural draught systems are restricted to industrial applications and are not widely adopted in Britain.

The design and experimental investigation of atmospheric burners dates back to 1855 with the development of the Bunsen burner which marked the beginning of a new phase in gas utilisation. The analysis is concerned with the energy exchange which occurs when an inducing gas expands into a throat taking air with it. The ratio of jet diameter to throat diameter will be decisive to the proportions of gas and air in the mixture. Because of the lower burning velocity of natural gas mixtures measures must be adopted to stabilise the flames at the ports when natural gas is the fuel. The principles of burner design, including flame stabilisation techniques, are firstly discussed followed by a survey of the range of commercially available atmospheric burner equipment.

2.5.1 Burner Design Principles

There are a large number of parameters that influence burner design and performance. These may be grouped as follows:

Basic specification:

- Wobbe number, flame speed and stoichiometric air requirement of the gas.
- Gas supply pressure.
- Heat requirement.
- Effective length of the port section of the bar.

Parameters that must be selected by the designer:

- The level of primary aeration.
- The magnitude of port loading (heat input per unit cross section of burner ports).
- Pattern of injector (with or without a diffuser outlet).
- Type and size of burner ports.
- Mode of flame stabilisation.

From the above, the essential burner dimensions and constructional features become:

- Gas jet size.
- Burner port area and number.
- Burner port spacing and number of rows of ports.
- Size of injector throat and burner cross section.

The gas jet or injector orifice

The discharge of gas through the injector orifice is determined from the Bernoulli equation. (Chapter 4). The size of the gas jet is related to the potential heat input as follows:-

$$Q = 1.255\, A_o\, C_{do}\, W\, p_o^{0.5} \qquad 2.1$$

where	Q	=	burner heat release	watts
	A_o	=	injector orifice area	mm^2
	C_{do}	=	orifice discharge coefficient	
	W	=	Wobbe number	MJ/m^3
	p_o	=	static gas pressure at orifice	Pa

In practice, for an orifice length to diameter ratio close to unity and an approach angle between 30° and 60°, C_{do} lies between 0.85 and 0.95 for turbulent flow. For values of R_e less than 2 000, C_{do} reduces markedly and may be as low as 0.6 at low turndown flow rates. Discharge coefficients for a wide range of orifice geometries are available in the literature.

The relationship between jet diameter and heat input for natural gas at a supply pressure of 15 mbar, assuming C_{do} = 0.9, is shown in Figure 2.8.

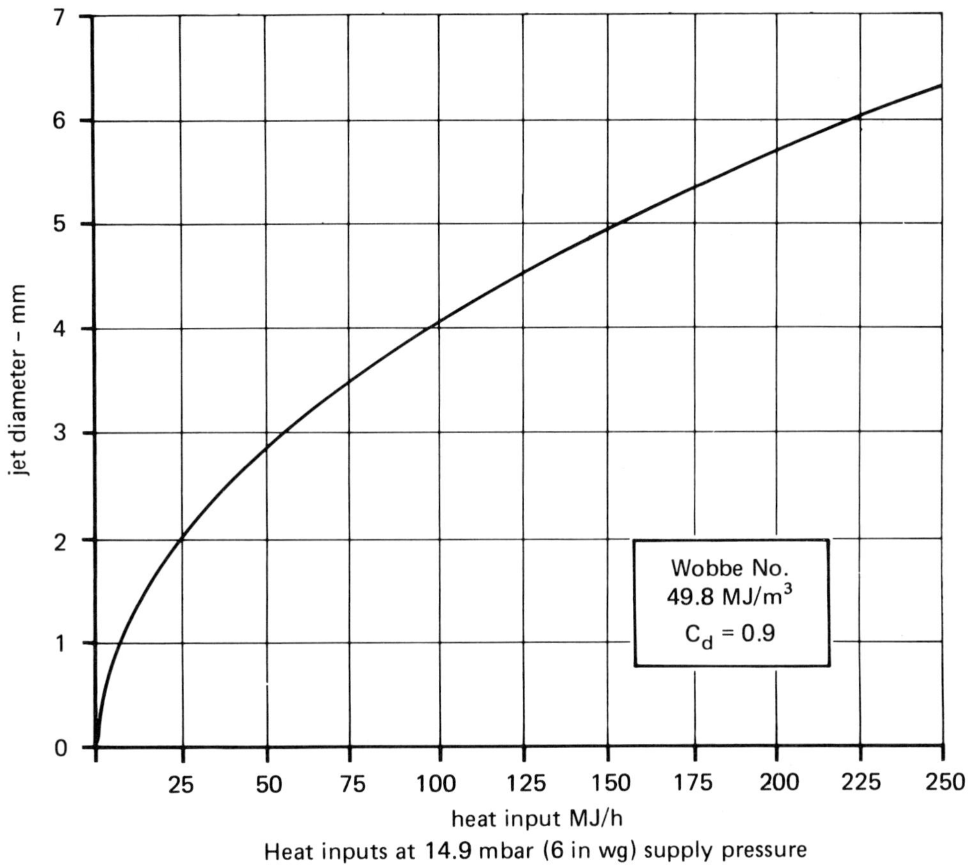

Fig. 2.8 Relationship between jet diameter and heat input for natural gas

Primary aeration

The percentage primary air required and the means by which this aeration is achieved are the principal influences on the design of atmospheric burners. Equating the mixture pressure from the injector to the pressure drop across the burner ports yields a dimensionless relation from which the ratio of the burner port area to the jet orifice area may be predicted from the primary air/gas ratio.

$$A_p/A_o = (r + 1)(r + d)[F(N + C_f)]^{0.5}/dC_{dp} \qquad 2.2$$

where A_p = burner port area mm²
A_o = injector orifice area mm²
r = ratio of entrained to jet fluid i.e. primary air/gas ratio
d = relative density of gas (air = 1)
F = correction factor representing the pressure recovery along the length of the burner bar. Typically, F = 0.9 for an injector with diffuser and 0.7 for a burner with a cylindrical injector.
N = 1 for an injector with diffuser and 2 for an injector without a diffuser
C_f = total friction loss coefficient
C_{dp} = discharge coefficient of the burner ports.

The burner port loadings attained by burners designed to the above procedure can be calculated by combining equations 2.1 and 2.2 to give:

$$Q/A_p = 1.255\ WC_{do}\ (A_o/A_p)p_o^{0.5} \qquad 2.3$$

The results of this calculation for burner-port loading Q/A_p are shown in Figure 2.9.

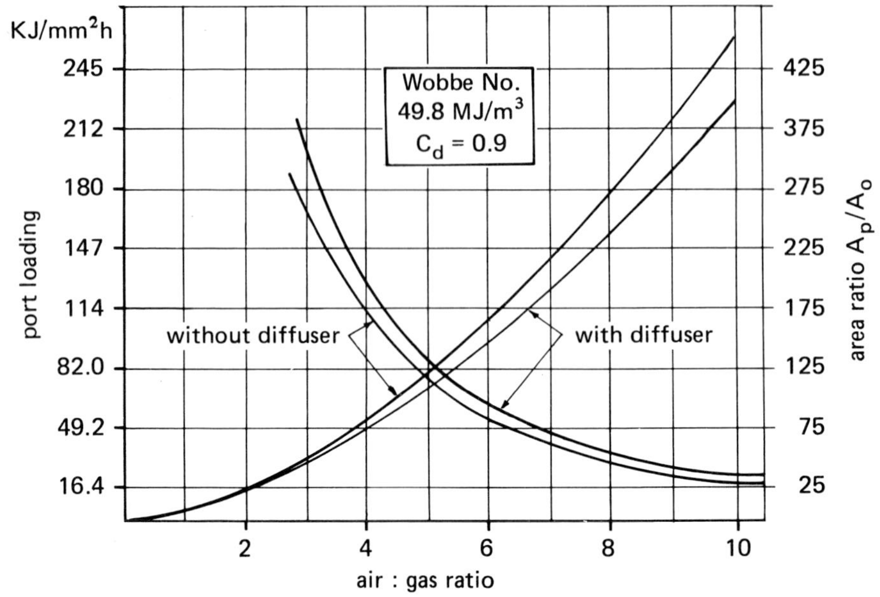

Fig. 2.9 Relationship between the burner port/injector orifice area and between port loading air/gas ratio for natural gas of relative density 0.56

The major factor determining the degree of primary aeration, r, is the total area of the burner ports for a particular duty, r increasing with values of A_p/A_o.

As well as the effects on air entrainment, changing the burner-port loading will also have an effect on flame stability. Thus reducing the burner-port loading, which is equivalent to reducing the mixture velocity at the flame ports, will increase the resistance of the burner to blow-off, and decrease its resistance to lightback particularly if the reduction in burner port loading is achieved by providing larger burner ports.

Burner port design

The stability of the flame on a burner port is the result of the establishment of a predictably positioned zone in the system within which self-propagating combustion can take place. For ignition to commence, a mixture within the flammable range of composition must be raised at least locally to a minimum temperature. For ignition to continue and be self-propagating, this zone must be maintained and supplied with sufficient energy, predominantly as heat. Thus, the stabilisation of a flame is primarily a question of adapting the dynamics of the flow system so that this zone is always present, and capable of being heated to the required temperature. At the same time, the forwards flow of mixture and the reverse motion of the flame front must be balanced for a stationary resultant front to exist. Evidently, at a certain flow rate, the flame front velocity will overcome the mixture velocity, and the flame will enter the mixture tube, i.e. will 'light back'. Conversely, a rate of flow will exist at which the forwards velocity of the mixture overcomes the reverse velocity of the front, and the flame will be 'lifted off'.

The range of working conditions over which this balance can be expected to exist will evidently depend on the burning velocity of the mixture. Thus burners utilising natural gas with its low flame speed would tend to give rise to 'lift off'. The options are to reduce port loading, entailing port enlargement which may affect both the degree of primary aeration and the cone height of the flames, or to introduce flame stabilisation means downstream of the port.

On a drilled burner bar without flame retention, lift-off occurs with natural gas with about 50% primary aeration for port loadings over 9 MW/m^2. For lower primary aeration, port loadings may be extended toward 22 MW/m^2 but care must be taken to avoid yellow tipping and soot formation. For a drilled bar burner without flame retention, the burner port loading should not exceed 9 MW/m^2.

To prevent lift from burners with high burner port loading, two widely used techniques are either to promote the recirculation of hot flame-gas at the base of the flame or alternatively to provide low-velocity auxiliary flames close to the base of the main flame. Flame retention by hot gas recirculation can be achieved using a continuous strip welded at the base of the main flame as shown in Figure 2.10.

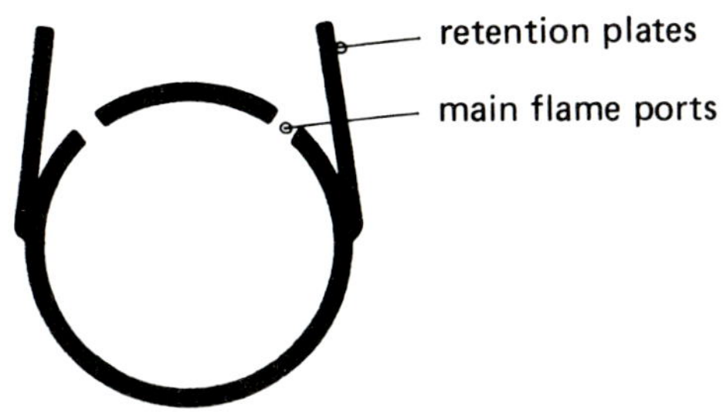

Fig. 2.10 Drilled port burner with flame retention plates

For open ended burners, resistance to blow-off may be improved through sudden enlargement of the mixture flow cross section, which generates a recirculation eddy within which partially or completely combusted gas circulates towards the root of the flame, thus representing a source of heat for ignition: this principle is illustrated in Figure 2.11.

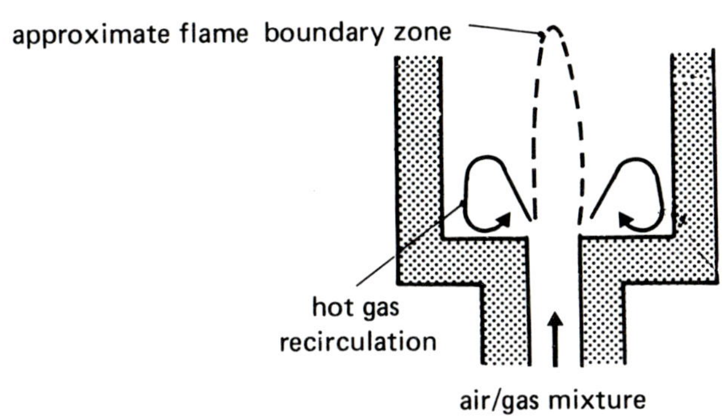

Fig. 2.11 Stabilisation by sudden enlargement

Because of practical drawbacks such as dirt accumulation and buckling of the strip, auxiliary flame retention is the commonest solution to the problem of lift. An example of this method employing pilot flames produced by retention tips is shown in Figure 2.12.

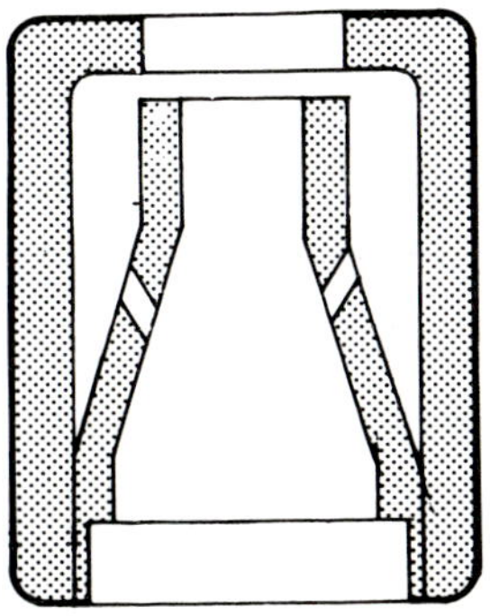

Fig. 2.12 Proprietary non-blow-off tip

Another well known form of auxiliary flame retention is the ribbon bar burner shown in Figure 2.13. The inset ribbon tends to form a combination of large and small burner ports, a small port being formed at each narrowing of the strip. These burners are very versatile and can be used for high and low pressure natural draught injectors as well as air blast and premix systems giving stoichiometric aeration.

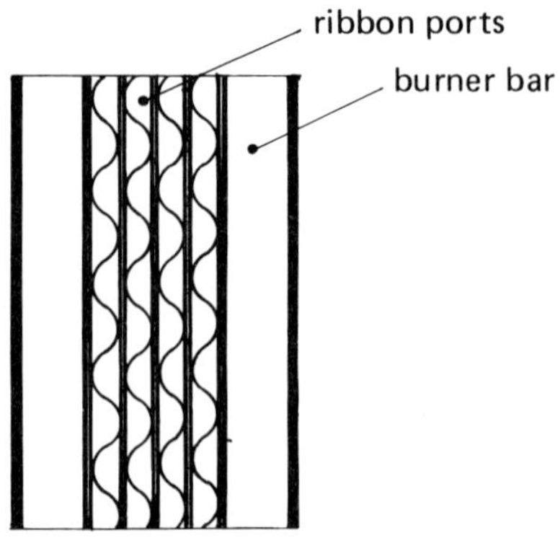

Fig. 2.13 Ribbon bar burner

Design of mixture tube and throat

In designing the mixture tube and throat it is generally desirable to use the minimum diameters consistent with obtaining the desired primary aeration and port loading.

For a cylindrical injector without a diffuser the ratio of the throat area to burner port area is given by:

$$A_t/A_p = C_{dp}[(2 + C_f) + K_{av}]^{0.5} \qquad 2.4$$

where A_t = area of mixture throat (mm^2)

K_{av} = average pressure rise along the bar burner, typically 0.90 to 0.66

For an injector with a diffuser the corresponding expression is:

$$A_t/A_p = C_{dp}[(1 + C_f) + (K_{av}/4)]^{0.5} \qquad 2.5$$

When there is a diffuser fitted, the burner bar diameter should lie between 1.25 and 1.55 times that of the mixture throat to ensure that the dynamic pressure in the injector is converted to static pressure.

When a cylindrical injector without a diffuser is used the alternatives are either to fit a separate parallel throat inside the burner bar, or to choose the burner bar diameter to be equal to the mixture throat diameter. The latter results in a very simple burner construction since the bar itself serves as a mixing throat.

It is generally desirable to provide aeration control on industrial burners and this may be provided by an air shutter, by screwing up the injector orifice close to the mixture tube entry or by means of an adjustable disc in the burner throat.

Application of design procedure

An example of industrial bar burner design is now given to illustrate the application of the foregoing design information.

Example 2.1 Determine the principal dimensions of a drilled tubular natural gas bar burner of high port loading incorporating a diffuser and using auxiliary flame retention for a heat release of 15 kW, a flame run length of 280 mm and a primary aeration of 60 per cent stoichiometric.

Orifice sizing:

Using Equation 2.1 the heat input Q is given by

$$Q = 1.255\, A_o\, C_{do}\, W p_o^{0.5}$$

With
$$W = 49.8 \text{ MJ/m}^3$$
$$p_o = 1\,500 \text{ Pa}$$
$$C_{do} = 0.9$$

Therefore $A_o = 6.89 \text{ mm}^2$

and thus the orifice diameter D_o is 3.0 mm.

Alternatively this value may be read directly from Figure 2.8.

Burner-port area:

Primary aeration at 60% stoichiometric is equivalent to an air gas ratio of 5.8/1. The burner port/orifice jet area ratio can be determined from Equation 2.2:

$$A_p/A_o = (r + 1)(r + d)\ [F(1 + C_f)]^{0.5}\, d\, C_{dp}$$

Given that relative density d = 0.56 for natural gas and assuming C_{dp} = 0.65, F = 0.9 and C_f = 0.3, then

$$A_p/A_o = 129$$

The same result may be obtained from Fig. 2.9. Since A_o = 6.89 mm² then A_p = 889 mm².

Number of burner ports:

Assuming that 20% of the total flow is supplied to the auxiliary port, the area of the main port totals 0.8 x 889 = 711 mm² and the area of the metering orifice for the auxiliary port 711/8.04 = 88 are required.

If 1.6 mm diameter metering orifices are used then the same number is required.

Burner port spacing

The main burner ports may be spaced about 1 diameter apart in two rows giving a flame run length of about 282 mm, as stipulated.

The auxiliary ports would be placed at the same spacing in two rows adjacent to the main ports as in Figure 2.14.

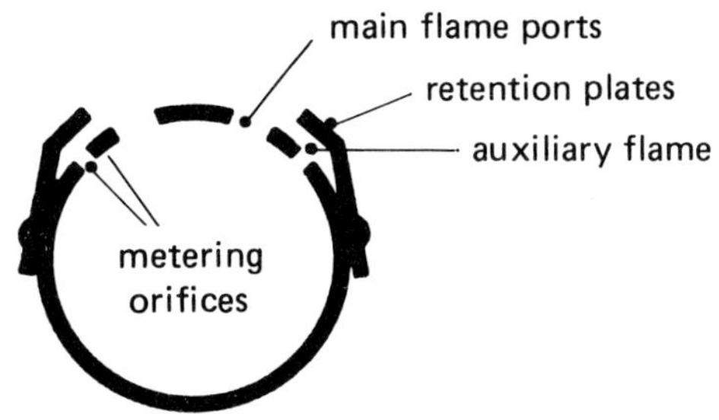

Fig. 2.14 Bar burner design with two rows each of main flame and auxiliary flame ports

Throat sizing:

From Equation 2.5:

$$A_t/A_p = C_{dp}[(1 + C_f) + (K_{av}/4)]^{0.5}$$

Given that K_{av} = 0.90 for an assumed flame run/bar diameter ratio of 10 then:

$$A_t/A_p = 0.65[(1 + 0.3) + (0.25 \times 0.9)]^{0.5}$$
$$= 0.80$$

Thus A_t = 711 mm^2 corresponding to a throat diameter of 30.1 mm.

Burner bar diameter:

The burner bar diameter is normally 1.25 to 1.55 times the throat diameter and thus a nominal pipe size of 38 mm would be chosen.

Burner port loading:

The burner port loading that would be obtained with this burner is (15 kW/889 mm^2) x ($10^3/10^6$) = 16.9 MW/m^2.

2.5.2 Low Pressure Aerated Burners

The use of the design principles detailed in Section 2.5.1 will allow satisfactory burners to be designed for a wide variety of applications. In practice burners are not always designed in this way.

Low pressure atmospheric induction burners, which form a large proportion of the equipment in use range from a simple single-ported burner, such as the Bunsen burner, to quite large assemblies of drilled or ribbon ported bar or ring burners and usually operate in surroundings in which the flame is enveloped in air at ambient conditions. Thus, they are characterised as having numerous flames which are usually relatively small compared with the enclosure in which they are placed. In this context the temperature of the process in which such burners are taking part is relatively low, say 200-400 °C, and heat release patterns from the flames themselves are relatively unimportant, their function being mainly to heat up the gaseous content of an oven, for example, so that some level of convectional motion is created, and heat transfer can take place from the oven gases to the charge by convection. Other versions, and they are numerous, include the heating of open tanks containing liquids (if the more efficient immersion tube method is not suitable) such as solution tanks in electro-plating shops, where the burner is merely positioned in the open air beneath the tank.

There is a wide range of proprietary injectors, burner heads, bars and jets and complete burner assemblies available for the gas engineer to draw on in plant design; the fundamental design features are shown in Figure 2.15.

2.5.3 High Pressure Atmospheric Burners

If gas is available at pressure it may be used in an atmospheric burner with a number of potential advantages over gas at normal supply pressures. The principal advantage is that stoichiometric primary aeration is then easily obtained using smaller burner cross-sections thereby producing high-temperature flames free from the need for secondary air. Thus higher burner port loadings are possible and control valves and pipework sizes may be reduced.

However, many of the same benefits are provided by air-blast systems which by using single stage air fans avoid the complications of high pressure gas systems.

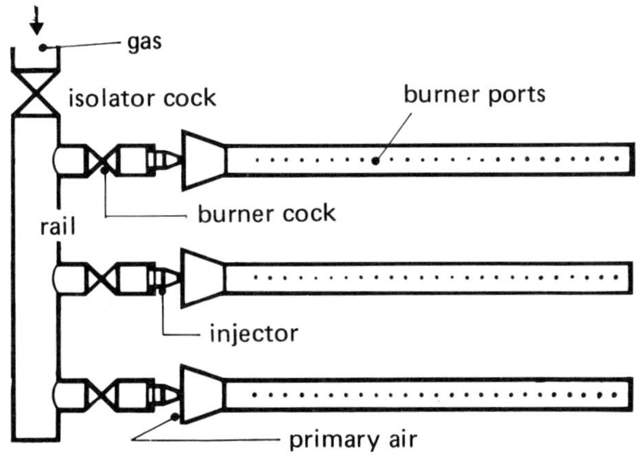

Fig. 2.15 Basic natural draught burner system

Only modest mixture pressures are attainable even with high pressure natural gas. Calculations indicate that a gas pressure of 200 mbar will raise the flame port heat release to about 45 MW/m^2 but the maximum mixture pressure is about 2.5 mbar, which is about one tenth of that available from an air-blast system using fan air at 70 mbar.

Furthermore a large increase in gas pressure is required to produce only a modest increase in burner port heat release. For example a hundred-fold pressure increase from 200 mbar to 20 bar gives only a fivefold increase in heat release. Note that this burner port heat release is still only approximately 50% greater than that obtained from a normal air-blast burner system.

The design principles of high pressure atmospheric burners are essentially the same as those for low-pressure systems. Flame retention provision is an important aspect due to the relatively high burner port loadings.

Jet pumps based on the Coanda effect are finding increasing applications as air movers in combustion systems. The aerodynamic noise produced by a Coanda pump is lower than that of a conventional design and they have been applied to reducing noise levels in high pressure atmospheric burners. The Coanda pump operates by injecting an annular jet of driving fluid radially inwards close to a curved surface. The driving fluid attaches to the surface and is rotated through 90° to form an annular forward flowing jet in the throat of the device, as shown in Figure 2.16.

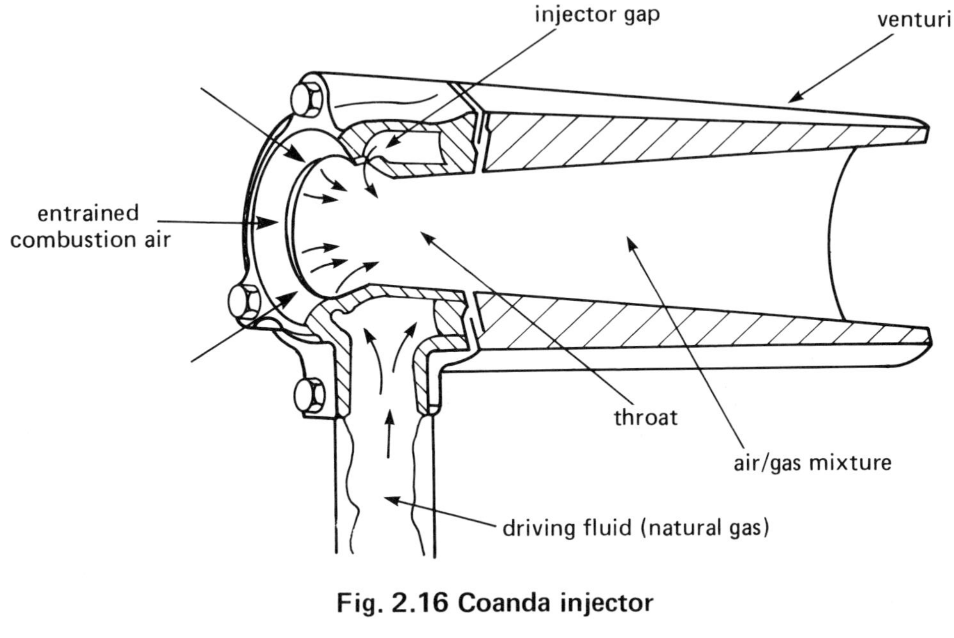

Fig. 2.16 Coanda injector

2.6 AIR BLAST BURNERS

In premix air blast burners air from a centrifugal fan operating at about 70 mbar is allowed to expand through a nozzle into a venturi mixing tube. This is illustrated schematically in Figure 2.17 which also defines the terminology employed in describing air blast burners.

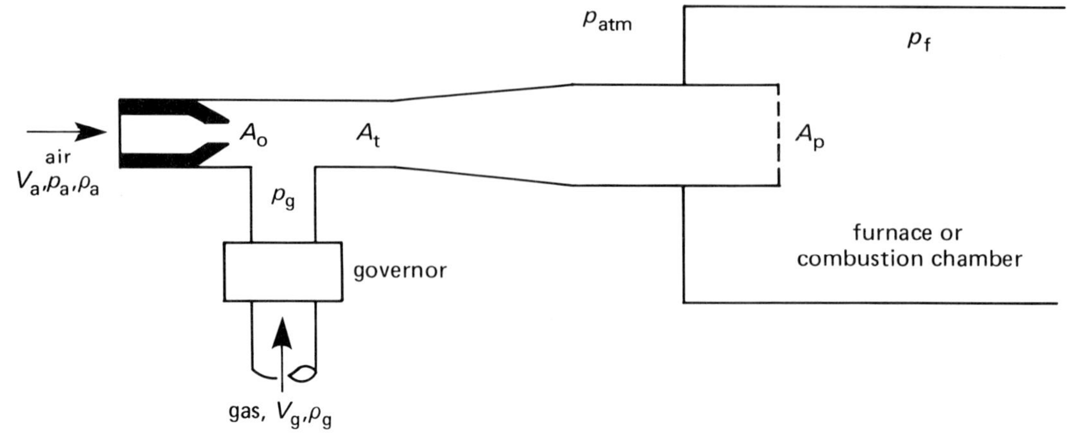

Fig. 2.17 Air blast injector nomenclature

The expanding jet of air entrains gas available at zero pressure which is admitted to the mixing tube near the exit of the air nozzle. This principle gives a fully premixed air/gas mixture giving all the air required for combustion. A very short flame results with rapid rates of heat release and thus these burners are used for high temperature work. Forced draught premix burners have been adopted widely in industrial gas usage for a number of reasons:

- Stoichiometric proportions are easily attained.
- Control of air/gas ratio is relatively simple and precise.
- The system is inherently self-proportioning if the gas is entrained from a supply at atmospheric or combustion chamber pressure.
- Throughput can be varied by a single control valve on the air supply.
- High mixture pressures are obtained using modest air supply pressures thereby resulting in high turndown ratios.
- High burner-port loadings can be achieved.
- Combustion product momentum gives rise to positive hot-gas distribution in the combustion chamber.
- Easy interchangeability of fuel gases.

2.6.1 Burner Design Principles

Air blast pressure burners are conveniently considered as two separate components, the injector assembly and the burner head.

Air-blast injectors

Air at a pressure normally in the range 25 to 75 mbar is fed to the orifice of an injector inducing gas at normal supply pressure. Alternatively if a zero governor is employed the gas is introduced at atmospheric pressure via a side limb. As in natural draught burner systems both simple cylindrical injectors and those incorporating diffusers are employed.

The size of the air orifice can be determined from equation 2.6 by substituting d = 1 and $V_a = rV_g$, where V_a is the desired air flow rate for an air/gas ratio of r:

$$V_a = rV_g = 1.255\ C_{dao}\ A_{ao}\ (p_{ao}/d)^{0.5} \qquad 2.6$$

where V_a = air flow rate m^3/s
r = air/gas ratio
V_g = gas flow rate m^3/s
C_{dao} = orifice discharge coefficient, normally in the range 0.85 to 0.95
A_{ao} = air injector orifice area mm^2
p_{ao} = static air pressure at orifice Pa
d = relative density (air = 1)

The nomenclature is also defined in Figure 2.17.

An analogous equation to that for natural draught burners can be derived expressing the relationship between primary air entrainment, air rate and burner dimensions:

$$[(p_g-p_f)A_t^2/\rho_g V_a^2] - 0.5(r+1)(r+d)[C_f+(1/C_{dp}^2)(A_t/A_p)^2+1] + d(A_t/A_o) = 0 \qquad 2.7$$

where p_g = static gas pressure Pa
p_f = combustion chamber pressure Pa
A_t = throat area m^2
ρ_g = density of fuel gas kg/m^3
V_a = volume rate of flow of gas m^3/s
r = air/gas ratio
d = relative density of air = 1
C_f = friction loss coefficient
C_{dp} = discharge coefficient of burner ports
A_p = burner port area m^2

If the gas pressure and furnace pressure are identical then there is an inherent self-proportioning action and a constant air/gas ratio is given for all throughputs. Throughput is controlled by a single control valve on the air supply.

Where furnace pressure p_f is atmospheric and if air/gas ratio control is required, p_g is reduced to atmospheric via a zero-pressure governor.

Unlike post-aerated and low pressure atmospheric burners, air blast burners are frequently required to fire into combustion chambers in which the pressure differs significantly from atmospheric. Thus to maintain self-proportioning,the reference pressure for the zero governor is not atmospheric but combustion chamber pressure i.e. $p_g = p_f > p_{atm}$. In this case changes in p_f are easily compensated for by back loading the governor diaphragms with the combustion chamber pressure. Alternatively, mixture pressure back-loading may be used which provides complete compensation for changes in back pressure downstream of the mixture tube. However, since the zero-governor pressure must be slightly higher than the back-loading pressure, gas pressure boosting is normally required.

Commercially available designs for an injector with diffuser give a mixture pressure of approximately 40% of the air pressure. Design procedures then usually require a throat cross sectional area of twice the air-nozzle area. The diffuser taper is generally 8 to 15° included angle and the throat length 4 to 6 throat diameters. Optimised parallel tube injector design dimensions for air-blast tunnel burners are discussed below in Section 2.7.

Air-blast burner heads

Air blast burners produce high mixture velocities and therefore means need to be provided to prevent blow-off.

The most common method uses a refractory tunnel. In such tunnel burners, as shown schematically in Figure 2.18, stabilisation results partly from the shielding and heating effect of the tunnel, which becomes incandescent, and partly from the sudden enlargement which exists between the tip of the mixture tube and the back of the tunnel.

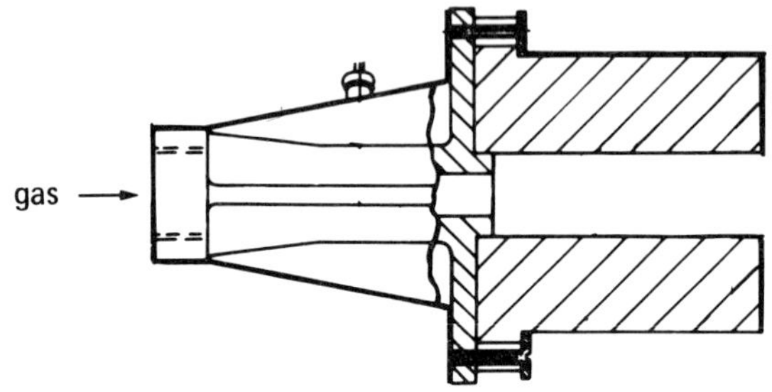

Fig. 2.18 Air-blast burner with refractory tunnel

A further major advantage of tunnel burners which has led to their wide-spread use is that the tunnel dimensions can be designed such that the burner emits a high velocity stream of burned gas at near flame temperature with little or no flame at the tunnel exit. Burners of this type are generally termed high velocity tunnel burners. The terms jet burner and high intensity combustor are also used.

Their major uses are in providing very high convective heat transfer rates for local heating processes, in promoting positive hot gas circulation in furnace chambers and in producing high levels of jet-driven recirculation thereby increasing convective heat transfer and minimising temperature differentials. The specialised use of these burner types is discussed later in this Chapter in Section 2.13.

A second method of stabilisation is the use of a flame retention cup incorporating either auxiliary flames or a sudden enlargement. These methods have already been discussed and illustrated in Section 2.5.

2.6.2 Air Blast Burners

Typical air blast burners are shown schematically in Figures 2.18 and 2.19. The burners are sealed into the furnace wall and used in conjunction with a suitably flared tunnel.

The burner in Figure 2.19 is of heavy cast iron construction and incorporates radiating surfaces outside the furnace. The burner design can, in fact, be used for both air blast and high pressure gas. If changing from town gas to natural gas the burners may require longer quarls to minimise flame blow-off. When air-blast burners are used without refractory quarls, non-blow-off burner tips as illustrated in Figure 2.12 are commonly used.

The main disadvantage of this type of burner is its poor turn-down limitation of about 4:1, which explains why nozzle mix burners are now generally used instead for high temperature applications.

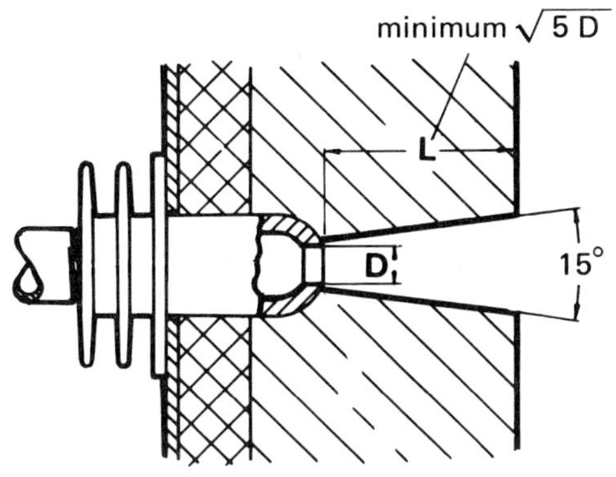

Fig. 2.19 Mont Selas Type 66 air blast burner

2.7 NOZZLE MIX BURNERS

This burner category, known also as tunnel mixing burners or package burners, uses an air supply at pressure which is discharged through an annulus or annuli at the burner head or face. Gas ports, which again may be of many forms, admit gas into the air stream and combustion takes place either totally or partially in a burner tunnel or at the burner face.

In some designs part of the combustion air is admitted into the gas stream prior to the main mixing zone so that the resulting flame is of the air and gas mixture, the remaining air completing combustion. Many designs of nozzle mixing burners are suitable for all types of gases, but some are suitable only for individual gas families. These burners are the most commonly used types for industrial processes and boiler firing up to and including power station boilers.

In nozzle mixing burners air/gas mixing and combustion takes place simultaneously at the burner nozzle, usually in a refractory tunnel or quarl. The tunnel acts as a means of flame retention in that heat stored in the refractory walls of the quarl serves to stabilise the flame, as also does the recirculation of flame gases within it, by forming a heat transport system, keeping the root of the flame at temperatures encouraging ignition. This enables satisfactory operation at quite low throughputs, both in stoichiometric conditions and with a large excess of air. Light-back is unlikely to occur, since combustion and mixing are concurrent.

The type of flame produced results from relatively rapid mixing, and is thus characterised by maximum rates of heat release at relatively short distances from the point of mixing. The flames are thus more intense than diffusion flames, and can also be accommodated in smaller combustion chambers. Because of the more intensive mixing and also because the flame is retained in a tunnel or quarl, it is possible to use air/gas ratios very near to the stoichiometric. This both reduces thermal loss due to excess air in the combustion gases, and also minimises the risk of contact between incompletely burnt combustion gas and the charge. Conversely, because combustion is more intense, temperatures in the immediate vicinity of the burner port may be too high for stock to be placed immediately in front of it. However, the burner is very versatile and in most cases can be used gas rich to give a longer more gentle flame for less intense local heating. The degree of turndown of this type of burner is also greater than the equivalent air blast system.

The utilisation of nozzle-mixing burners has increased considerably in recent years due to a number of advantages possessed over premix air-blast systems, including:

- Lower air pressure may be used.
- Preheated air may be employed.

- Flame stability with respect to lift is greatly improved and burners are capable of operating at high excess air levels.
- Gas interchangeability is facilitated, the only necessary alteration being a change of air/gas ratio.

An extremely wide variety of proprietary nozzle mixing burners are now available having been designed for particular heating requirements. An increasingly popular variant integrates the burner and control system into a packaged unit; such package burners are discussed separately in Section 2.7.2.

Nozzle mixing burners also constitute the gas burner part of dual-fuel burners which are treated in Chapter 3. An advantage of nozzle-mixing burners of particular importance is that they can operate satisfactorily with preheated air. Nozzle-mixing burners incorporating integral recuperators are considered in detail in Section 2.9.

2.7.1 Design Principles and Burner Types

There are a number of methods for designing nozzle mixing burners and for a detailed treatment the reader is referred to standard textbooks such as Industrial Gas Utilisation. Thus for example, design procedures for nozzle mixing burners can be developed by summing the pressure losses through the burner system as follows:

$$(p_g-p_2)/(p_a-p_1) = (r+d)(r+0.7)(T_2/T_1)(A_1/A_2)^2(C_{f_{12}}+2)/r^2 - 2(A_1/A_2) + d(A_1/A_g C_{dg})^2/r^2 \qquad 2.8$$

where
p_g = inlet gas pressure Pa
p_2 = furnace pressure Pa
p_a = air inlet pressure Pa
p_1 = nozzle gas pressure Pa
r = air/gas ratio
d = gas relative density (air=1)
T_2 = combustion product temperature K
T_1 = ambient air temperature K
A_1 = effective air flow area at nozzle mm^2
A_2 = tunnel exit area mm^2
A_g = area of gas flow at nozzle mm^2
C_{f12} = friction loss coefficient along the tunnel
C_{dg} = discharge coefficient of burner port

The nomenclature is defined in Figure 2.20 which also illustrates schematically the features of a nozzle mixing burner.

Equation 2.8 indicates that perfect self proportioning occurs when (p_g-p_2) is zero and all the terms on the right hand side of the equation remain constant at all flow rates. In practice, this is not so: C_g, T_2 and C_{f12} may all vary with flow so that to obtain reasonably constant proportioning over the entire turndown range the ratio-setting valve opening (A_g) is set at high flow rates and the zero-governor outlet pressure (p_g) at low flow rate.

For use with preheated air, the air-nozzle size must be increased for a given flow rate. Conversely the tunnel size may be reduced since the airstream momentum is increased.

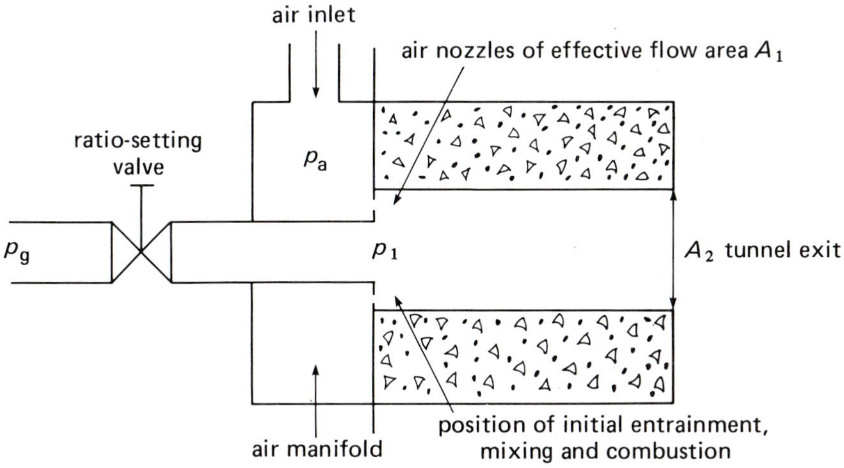

Fig. 2.20 Diagram of nozzle mixing burner showing dimensions and pressures used in the derivation of the design equations

Air/gas ratio control of nozzle-mixing burners

In the case of nozzle mixing burners which are capable of entraining gas from a zero-governed gas supply the pressure-air technique allowing single valve throughput control is applicable. This is identical in principle with that used for premix air-blast systems. Compensation for furnace-chamber pressure fluctuations may be incorporated by back-loading both the zero and appliance governors from the furnace chamber.

For burners which possess no inherent self proportioning action one of the methods normally classified as pressure-air/pressure-gas techniques would be employed. In these the relationship between flow rate, the area of an orifice and the differential pressure across it leads to two basic methods of throughput and air/gas ratio control. First, the linked-valve method, in which constant differential pressures are maintained across linked valves with suitably matched flow characteristics. Secondly, the fixed metering orifice method, in which a constant relation is maintained between the differential pressures across fixed metering orifices in the air and gas lines.

Nozzle-mixing burners

A number of examples are considered from the large range of nozzle mixing burners available in order to illustrate the salient design features.

The design can be divided broadly under two main headings, depending on how mixing of fuel and air is achieved:

- Swirling of the gas or air stream. This is the most common method, particularly in larger burners, and the design consists essentially of mixing the gas issuing from the nozzles by a swirling air stream. The burner flame shape and furnace thermal pattern can be altered appreciably by swirl control.
- Axial. This comprises two concentric tubes, the inner usually carrying the gas flow and the outer the air. Mixing is enhanced by breaking up the streams through orifices.

Combustion tunnels are available in a range of materials dependent on the temperature requirements. For temperatures up to 850 °C alloy flame tubes are generally available. For high temperature applications burners with refractory or vacuum formed fibre quarls and holders would normally be used. Silicon carbide tunnels are ideal for use in high temperature kilns and furnaces, particularly those constructed from ceramic fibre; application temperatures can be up to 1 300 °C. Convergent tunnels can be used to provide medium and high velocity combustion products to promote rapid stirring within furnaces and kilns and thereby increasing the rate of heat transfer. Tunnel configurations (Stordy) are illustrated in Figure 2.21.

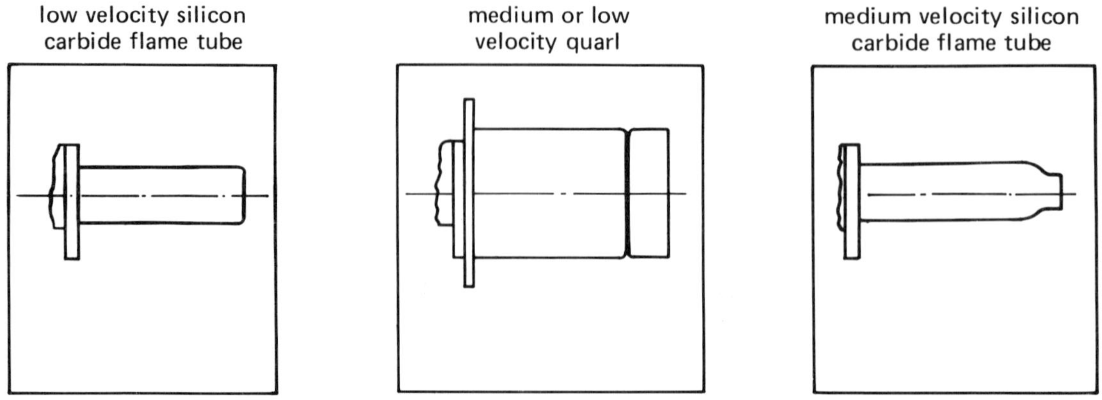

Fig. 2.21 Silicon carbide tunnels

The burner shown in Figure 2.22 utilises opposed gyrations in the two streams, setting up high relative velocities. In spite of this, overall flow from the burner is largely axial, since the gyration of the air stream is relatively low speed. Combustion is stabilised within a divergent refractory quarl due to the fact that material and heat transfer in the swirl system encourage ignition at the entry to the quarl.

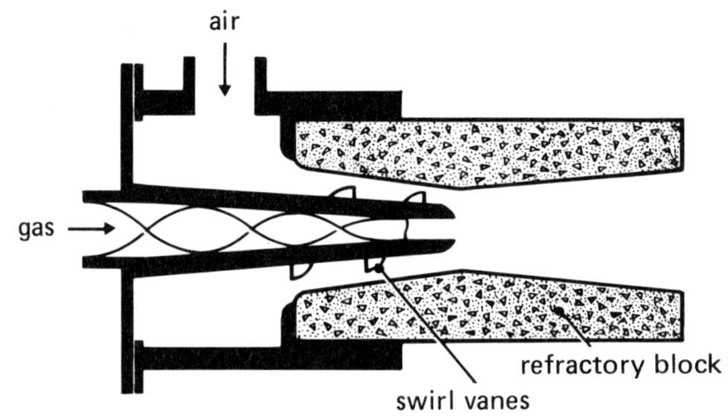

Fig. 2.22 Tunnel mixing burner with swirling flow discharge and stabilisation

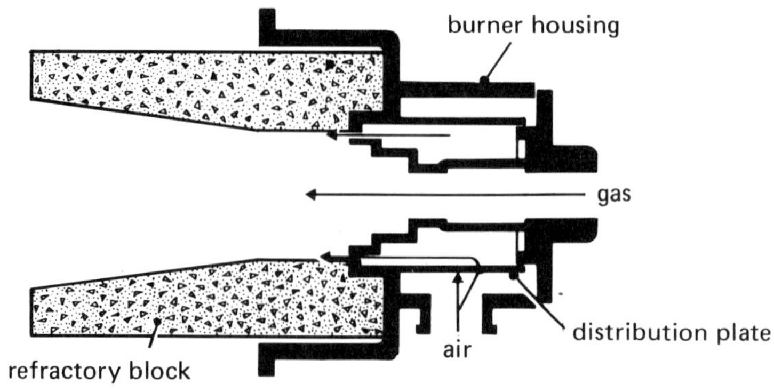

Fig. 2.23 Tunnel mixing burner with parallel flow discharge and stabilisation by hot refractory surface

The Stordy-Hauck NMG burner illustrated in Figure 2.23 provides low velocity gas entry, whilst air enters in three ways:

- Tangentially in a plane at 90° to the flow axis, through a series of holes in the gas pipe.
- Axially, through a ring of small holes situated in a stepped collar surrounding the gas pipe.
- Axially, through an outer annulus formed by the stepped collar of the gas pipe and the burner quarl. This last is the main air flow.

The Urquhart type NM burner illustrated in Figure 2.24 is somewhat simpler in construction.

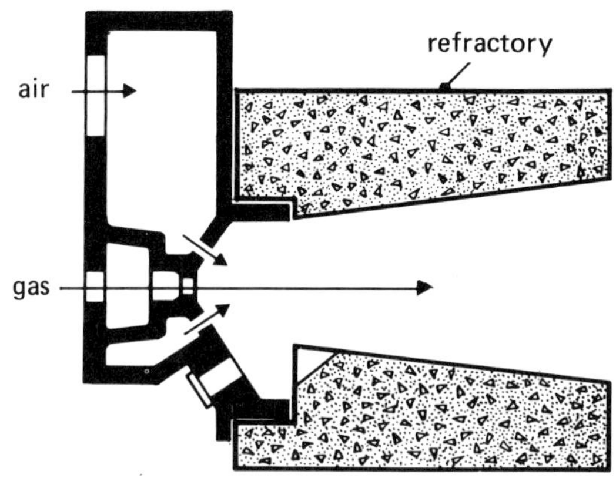

Fig. 2.24 Tunnel mixing burner with opposing gas and air swirls to assist mixing

Air is supplied through a series of air jets inclined to the burner axis, gas entering through a central orifice. Mixing surface area is increased by using a series of air orifices rather than an annulus. The air jets so generated constitute a means for forming stabilising eddies around them.

The burner type illustrated in Figure 2.25 is a British Gas MRS tunnel burner design, which incorporates relative areas of air nozzle and tunnel exit based on calculations which result in substantially stoichiometric injection of the gas from zero pressure over a range of throughputs. The design also incorporates a large area for mixing, since a series of air jets is used surrounding the central gas pipe, all discharging axially into a refractory tunnel, or quarl. Here again, the air jets provide a recirculation stabilising system. Multi-tube versions of this burner are used for large loadings, in excess of 300 kW.

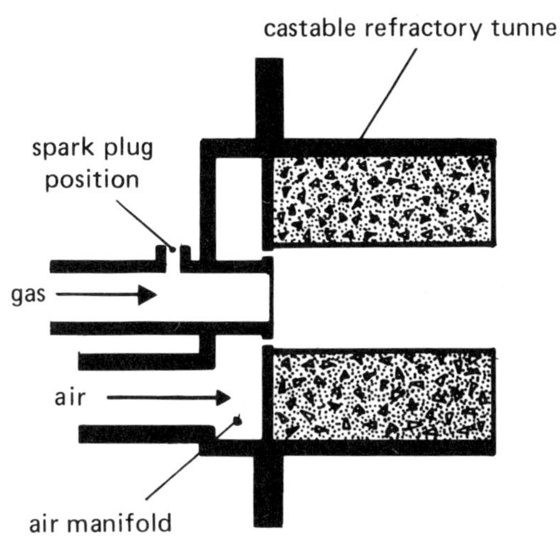

Fig. 2.25 British Gas self proportioning tunnel mixing burner

2.7.2 Automatic Package Burners

A significant development in the firing of boilers and other low-temperature plant has been the evolution of a wide range of forced-draught automatic packaged burners in which the burner, fan, throughput and air/gas ratio controls, ignition and safety systems are integrated into a single unit.

Package burners are almost exclusively of the nozzle-mixing type using metallic burner heads and incorporating a wide variety of stabilising and mixing arrangements. The burner is designed with a view to providing safe and reliable ignition, flame detection, cross lighting, flame stability, acceptable combustion quality and freedom from excess noise. Thus a number of components are integrated into the burner head assembly, including the ignition source, pilot burner and flame detector and their placing and security of fixing are crucial for reliable operation.

Generally, the air and fuel mixing is achieved either by axial flow from concentric tubes or by swirling vanes. There are many variations, but typical systems are illustrated in Figure 2.26.

Packaged burner developments

Until fairly recently smaller packaged burners in the range 150-600 kW (½ M to 2 M Btu/h) were invariably on/off or high low control. Many of these ranges of burners are available now with modulating control fitted as standard, with good air/fuel ratio control over the turndown range. Some manufacturers achieve this by adjustable cam profile valves to control the air and gas rates whilst others use multifunctional control valves in association with air pressure loading systems. Weishaupt have recently introduced a modulating burner in the range of 10-27 kW (30,000–100,000 Btu/h). Multifunctional valves make the overall size of the burner very compact. Figure 2.26B shows the modulating air shutter on a Weishaupt package burner.

Low noise burners have been developed where quietness is essential such as in hospitals, convalescent and old peoples homes. This is achieved by a novel arrangement of mixture tubes and gas and air passage ways to achieve intimate premixing of part of the combustion air and gas. The remaining air is introduced through a perforated plate surrounding the mixture tube. This requires only low air and gas pressures, keeping fan noise to a minimum.

Low NO_x packaged burners have also been developed where a proportion of the flue gas is recirculated back either into the combustion air fan or into a spacing piece between the burner tube and the inlet to the boiler. In the former system oxides of nitrogen are reduced from 62 to 32 ppm.

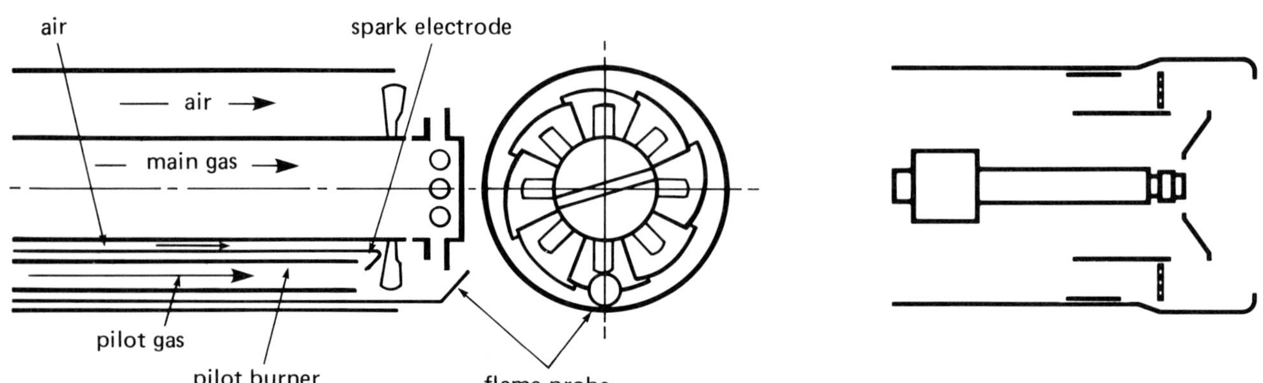

a) On-off burner with air swirl

b) Modulating air shutter - Weishaupt

Fig. 2.26 Typical proprietary package burnerhead assemblies

New and updated designs are continuously appearing on the market.

2.8 RADIANT BURNERS

In radiant burners a high proportion of the energy supplied in the fuel is converted into radiant energy. The temperature and material of construction of the burner radiant surface govern the heat radiated from it. In general, refractory materials from the radiating surfaces of radiant burners and the intensity of the emissions differ appreciably at different wavelengths. In this section the performance of radiant burners is discussed together with a detailed assessment of the wide range of radiant burners now in use.

There is a wide variety of radiant burner types in use. A convenient means of indicating the performance of radiant burners is to use the radiant efficiency, defined as the ratio of the radiation output to the heat input.

The performance of a range of burner types is detailed in Table 2.1.

Table 2.1 Radiation Efficiencies and Surface Temperatures for Various Types of Burner Operating in the Open Air.

Type	Surface Temperature °C	Radiation efficiency based on net CV %		Maximum temperature burner can withstand °C
		Typical	Ideal	
Porous medium	950	54	58	1 030
Schwank	850	45	63	900
Large multiple tunnel burner	900	33	60	1 330
Radiant cup	1 200	10-20	46	1 500-1 700
Radiant tube (recuperative)	1 000	50	56	1 000

The ideal efficiency is calculated on the basis of black body radiation and the absence of other heat losses such as from the sides of the burner.

Radiant burners are generally used in furnaces and for drying/curing of continuous web materials such as paper, textiles or coatings. The main features of each classification in the table together with other types are discussed below.

Porous-medium burners

In this type air/gas mixture either from an atmospheric injector or from an air blast mixing machine flows through a porous tile or refractory medium including ceramic fibre and burns at and below the surface, producing a uniformly radiating area as shown in Figure 2.27 (e.g. Stordy-Marsden radiant plaque). This has a quick heat up and fast cooling time.

Three modes of combustion may be distinguished: free flame combustion in which small multiple flames burn above the panel surface which remains cool, surface combustion which occurs beneath the panel surface and which is the desired mode, and unstable interstitial combustion in which the flame propagates back through the medium until lightback finally occurs. The surface temperature attained is normally 850-950 °C although using different materials higher temperatures are possible.

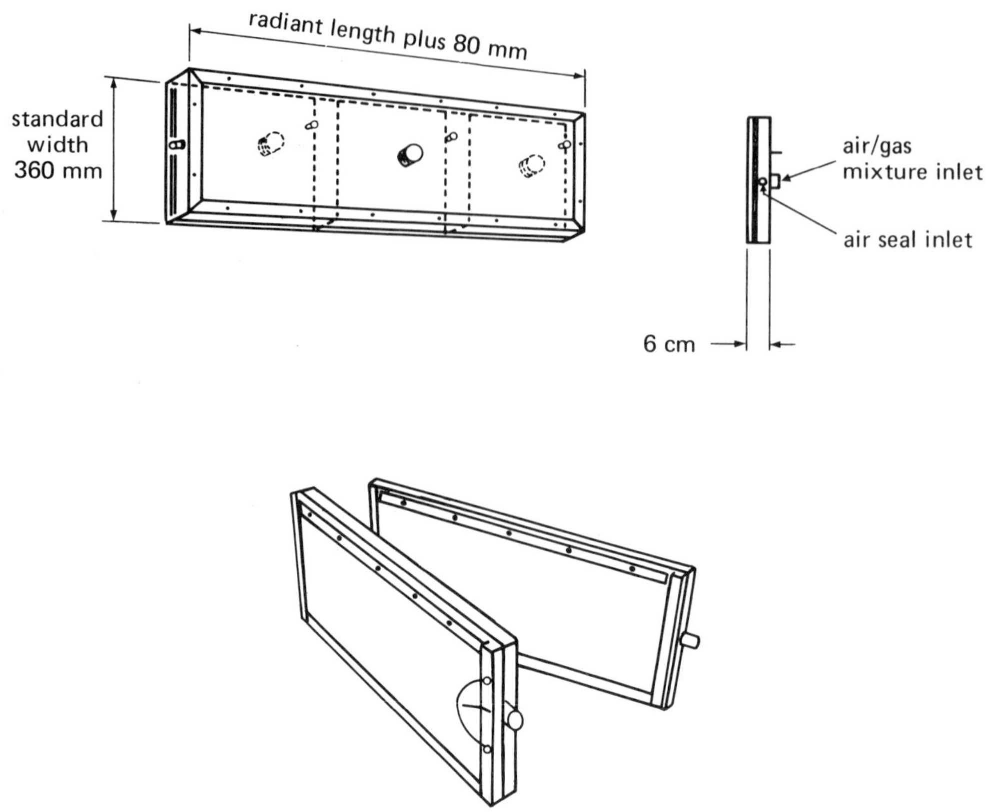

Fig. 2.27 Stordy-Marsden ceramic fibre radiant burner

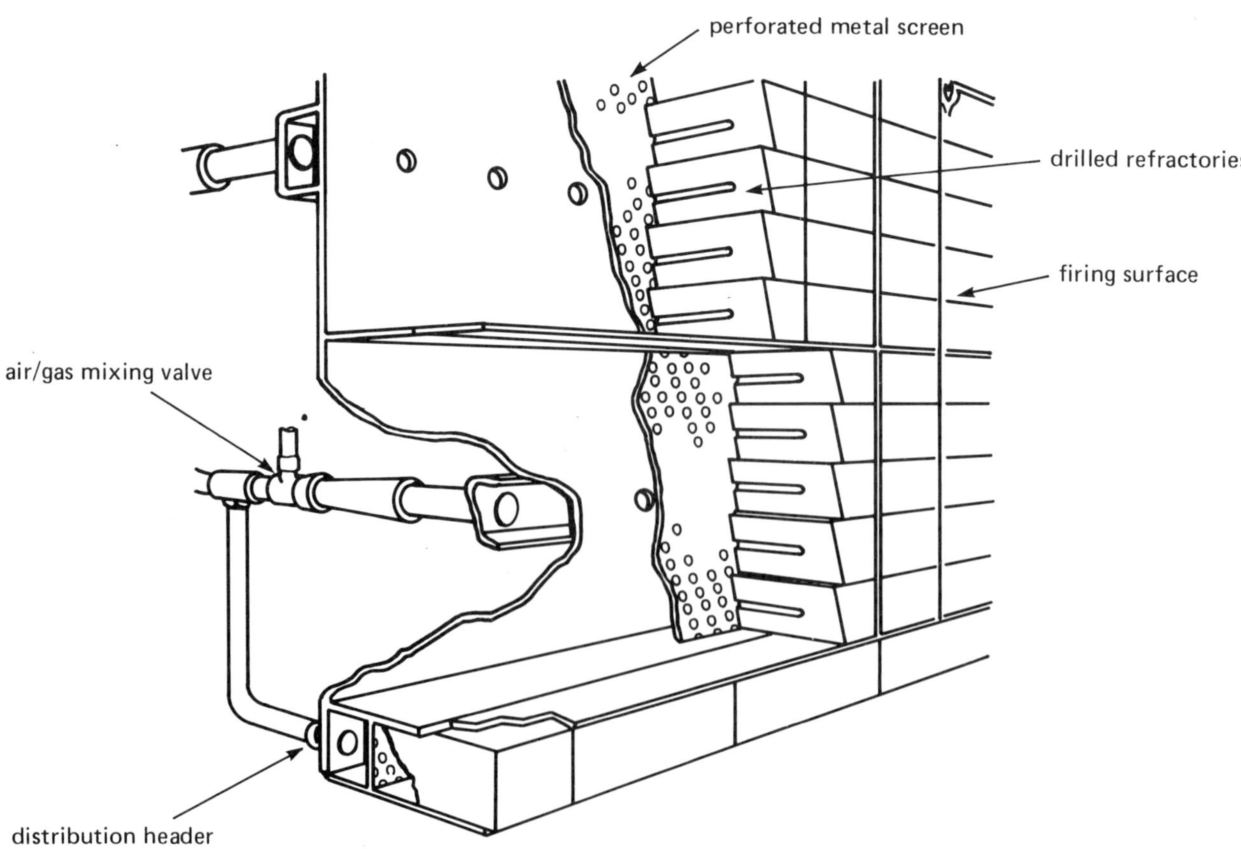

Fig. 2.28 Maywick luminous-wall system

An extension of this idea is the luminous wall design in which a large proportion of the furnace structure is constructed of porous burner bricks. Heat input rates of 250 kW/m^2 are typical although a maximum of 475 kW/m^2 is attainable. The surface temperature is normally limited to 1 100 °C. An example of this system is the Maywick luminous wall shown in Fig. 2.28.

Combustion in channels

In this group combustion occurs in a series of channels or perforations through the burner block. A widely used example is the Schwank burner shown in Figure 2.29 in which the air/gas mixture is supplied from an atmospheric injector and passes through perforations of about 1.4 mm diameter giving a surface temperature of about 850 °C. The radiating area is built up from individual ceramic plates.

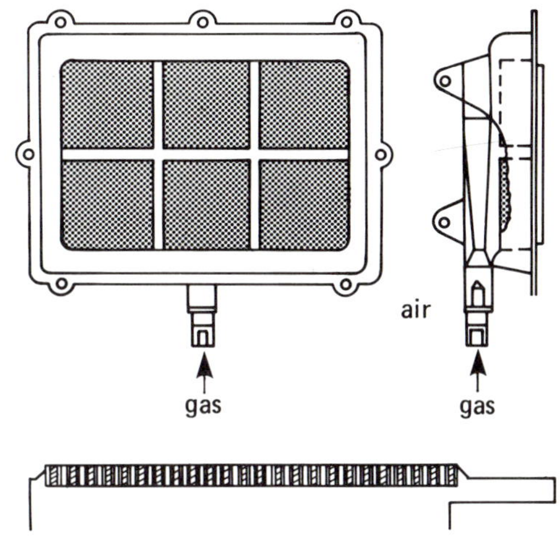

Fig. 2.29 Schwank radiant burner panel

Other multiple-tunnel designs incorporate much larger channels up to 20 mm diameter. The flame is much more apparent in these designs but the burner still transfers much of its heat by radiation. Burners of this type may be operated up to 1 350 °C with a maximum heat input of 1.6 MW/m^2.

Radiant-cup and flat-flame burners

This group differs from the types previously described in that gas/air mixture does not flow through pores or tunnels in the radiating block. The radiating surface is a cup or cone-shaped refractory quarl fed centrally with gas/air mixture. Flat-flame burners of this type are described in Section 2.13 of this Chapter.

Radiant tubes

High temperature radiant tubes can be employed in metal treatment and melting processes where it is required to separate the combustion products from the stock. They are discussed separately in Section 2.10.

Wire-mesh burners

Wire meshes instead of porous blocks have been used in burners for low temperature applications. The burner shown in Figure 2.30 is used in overhead radiant space heaters and consists of an atmospheric ring burner firing onto a ceramic-fibre truncated cone, operating at temperatures up to 1 200 °C. An outer cylindrical shield of stainless steel mesh which operates at 850 °C forms the radiant surface.

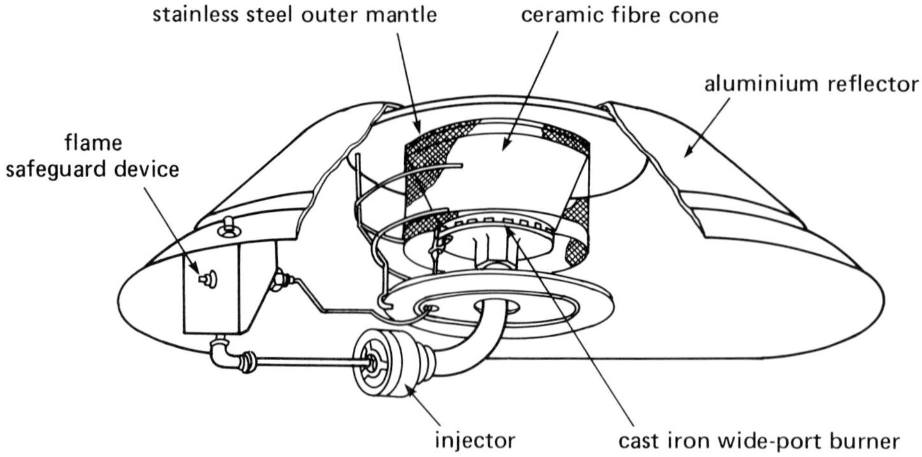

Fig. 2.30 Wire-gauze radiant burner for space heating

Low temperature gas fired tubular radiant heaters are increasingly used for space heating particularly where ceiling heights are in excess of 4 m. A metal tube of either mild steel or stainless steel is used as the radiant source so that the operating temperature is limited by the property of the chosen metal. With heaters working at temperatures in the range 200-650 °C, premix burners are used, generally of the induced draught type, and provide hot combustion products within the tube as shown in Figure 2.31.

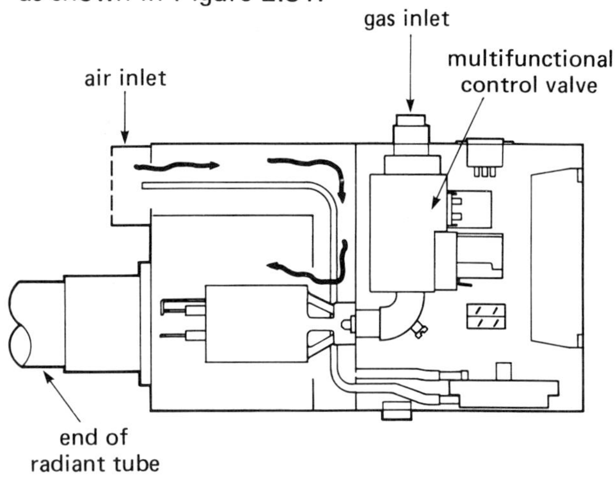

Fig. 2.31a Radiant tube heater burner (Ambi Rad Ltd)

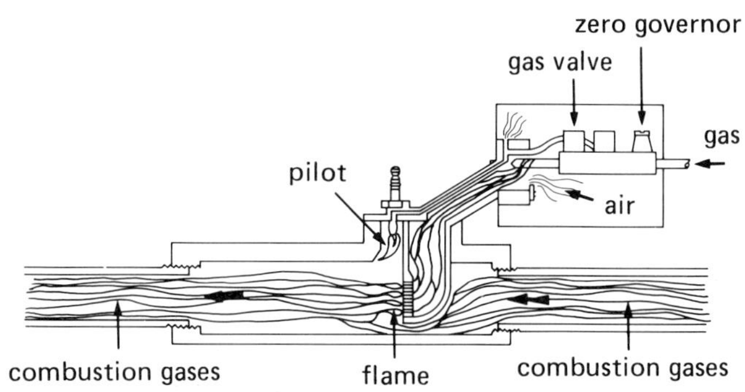

Fig. 2.31b In line radiant tube burner (Nor-Ray Vac Ltd)

Installations can be flued or unflued dependent on the building ventilation level. The tube radiant output is approximately 65% of the gas input.

Catalytic burners

It is possible to burn gases without visible flame at the surface of a catalyst at very low temperatures (about 450 °C for natural gas), but at the same heat release as normal high-temperature combustion. Flueless catalytic space heaters using LPG are widely used for domestic and camping purposes. However there is only limited industrial equipment currently available in the United Kingdom.

In principle the system consists of a porous pad of catalyst-impregnated fibres through which neat gas passes. Oxygen from the atmosphere diffuses through the pad in the reverse direction and reaction occurs at the catalyst surface. Heat is transferred by radiation from the surface and the products of combustion are forced by the gas flow into the atmosphere.

2.9 RECUPERATIVE NOZZLE-MIXING BURNERS

In most industrial heating processes a large proportion of the heat supplied is wasted in the hot flue gases. Waste-heat recovery methods can be employed as described in Chapter 7 but their use has been mainly restricted to large-scale process plant because of the cost and size of the equipment required. However, smaller furnaces and kilns are often the least efficient. This indicates the need for a simple compact low cost system such as the recuperative burners developed by British Gas and which are described in this Section.

2.9.1 General design

An accepted method of heat recovery on high temperature furnaces is the use of the recuperative burner. This combines the functions of a burner, flue and recuperator. The burner incorporates an integral heat recovery unit and is available as a replacement for the conventional burners on industrial high temperature plant. In comparison with separate recuperators, hot gas valves, ducting insulation, and high temperature fans are not required.

In these burners hot flue gases pass countercurrent to the flow of air in a concentric-tube heat exchanger. Recuperative burners have predominantly been natural gas fired but have been developed for oil firing. In operation, preheated air and gas are mixed at the nozzle of an air blast burner. The flame is stabilised in a ceramic tunnel or quarl. Hot combustion products issue from the quarl at a velocity of up to 80 m/s. Combustion is essentially complete before the combustion products pass into the furnace chamber, the high velocities promote good circulation, rapid heat transfer and generally improve temperature uniformity.

The recuperative burner is illustrated in Figure 2.32. The British Gas design of recuperative burner is manufactured by different licencees who manufacture their own range of burners with their own design within the overall concept of the BG development. In all designs an air driven eductor assists the extraction of the exhaust gases through the burner. The eductor is mounted on the burner outlet and comprises a venturi by which a jet of air draws the combustion products out to atmosphere. Control of the eductor air flow will maintain the furnace at the desired pressure.

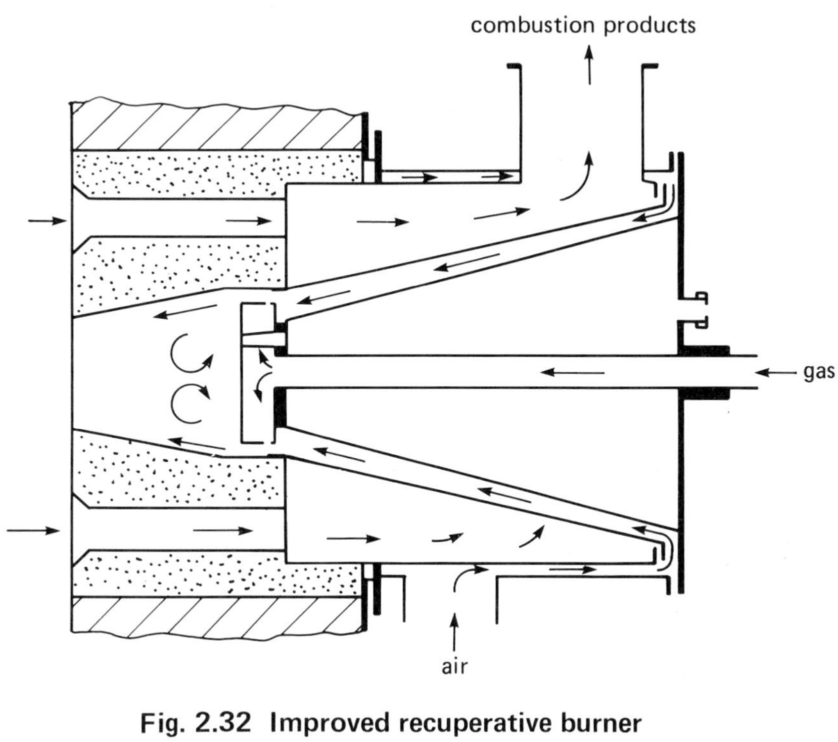

Fig. 2.32 Improved recuperative burner

The hot flue gases, at up to 1 400 °C, pass countercurrent to the flow of air in a compact and leak-proof heat exchanger made of heat resisting nickel chrome alloy. Heat transfer is so efficient that the incoming air temperature can be raised to about 600 °C, depending on process temperature. The design allows the heat exchanger to expand freely on heating up, thus avoiding undue stress.

The recuperative burner is bolted by means of a flange to the furnace casing in a position to suit the heating requirement of the process. The exit velocity of the combustion products often improves heating efficiency and the recirculation flow path generated by carefully siting the recuperative burners can often reduce the number of burners from that originally required. This simplifies pipe-work and allows flame safety equipment to be fitted economically. That portion of the recuperator which is outside the furnace is designed so that the hot flue gases are surrounded by a jacket cooled by the incoming combustion air. By this means the external parts of the recuperator are maintained below 150 °C.

The unit occupies only one tenth of the volume of a conventional flue stack recuperator. It can be fitted as readily as a conventional burner and entirely eliminates the need for costly flue ducts to be built into the furnace structure.

The burner and flow control system can operate from a standard combustion air fan delivering air at a pressure of 70 mbar and a normal gas supply pressure of 20 mbar.

2.9.2 Rating and Performance

The size range of commercially available recuperative burners is 60 kW to 900 kW. They are suitable for process temperatures of up to 1 400 °C and combustion air preheats are up to 600 °C depending on process temperature and flow. Fuel savings are usually at least 30%, although higher savings are common due to other benefits such as improved air/gas ratio control and better recirculation of hot gases within the furnace chamber. If the recuperator carries only 50% of the flue gases the air pre-heat temperature is only reduced by about 10%. This means that fuel savings are little affected by leakage from the furnace structure. A variety of burner nozzles is available. High velocity designs are used for furnace systems requiring a high degree of jet driven recirculation, such as kilns and heat treatment furnaces. Low velocities are suitable for drop forge furnaces where open door working is practiced, and when high velocities can lead to cold air entrainment through the furnace door.

2.9.3 Operating Parameters

In processes operating at 1 300 °C the refractory quarl has to contain flame temperatures of up to 2 200 °C and can itself attain surface temperatures of about 1 800 °C. The metallic nozzle in which the combustion air and gas are mixed also has to withstand heat, both from the flame and the quarl surface. Nozzle cooling is by the preheated air, which may itself be 600 °C, and from the cooler flow of natural gas. It is important to ensure in the design that gas is not preheated in the recuperator as gas cracking readily occurs above 700 °C with the production of soot.

The metallic recuperator is designed to give high convective heat transfer rates from the flue gases, which enter at up to 1 600 °C. The metal chrome alloy used has a maximum service temperature of about 1 000 °C and a melting point of about 1 400 °C. This metal is also cooled by the combustion air which is thus heated up to 600 °C. Sizing of the annular passageways is critical if preheat is to be achieved but excessive metal temperatures are to be avoided. Computer models have been used to optimise the recuperator length, diameter, pressure drop and cost for the full range of sizes.

2.9.4 Prediction of Recuperative Burner Performance

Mathematical models have been developed to predict the overall performance of the heat exchanger and the temperature to which its components would be subjected. The model divides the recuperator into zones. In each zone there are six unknown temperatures (combustion products, combustion air, inner and outer surfaces of the air annulus, inner and outer surfaces of the flue gas annulus) giving six heat balance equations to solve. The model is used to progress from the 'cold' end to the 'hot' end initially using estimates and by means of an iterative procedure the equilibrium temperatures are established. By repeating this process for a range of dimensions and operating conditions, design charts of the type shown in Figure 2.33 can be generated.

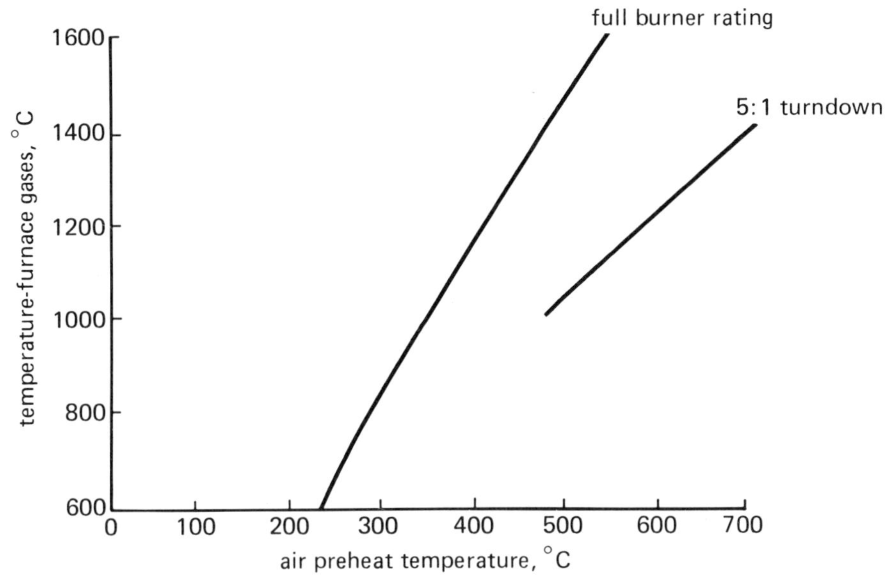

Fig. 2.33 Theoretical performance of recuperator

Recuperative burners have been used in a wide range of applications including rapid billet-heating furnaces, steel-reheating furnaces, aluminium-melting crucible furnaces and intermittent ceramic kilns showing fuel savings of 30 to 50%. Reference to typical examples is made in the appropriate sections of Chapter 7.

In addition, recuperative burners are applied to radiant tubes and these are considered in Section 2.10 of this chapter.

2.9.5 Separate Recuperators

The recuperative burner combines the burner and recuperator within one package with the advantages outlined previously. There are, however, possible disadvantages in the positioning of the burner and flue together. This creates a flow pattern for the combustion products which is not always advantageous for certain processes, e.g. when it is desired to fire at high level but exhaust at low level.

A development of the recuperative burner has been the split burner/recuperator system. This is illustrated in Figure 2.34 which shows a recuperative burner and the recuperator as used separately. This arrangement offers flexibility of application and enables the benefits of heat recovery to be obtained on furnaces where space limitations have precluded the use of recuperative burners.

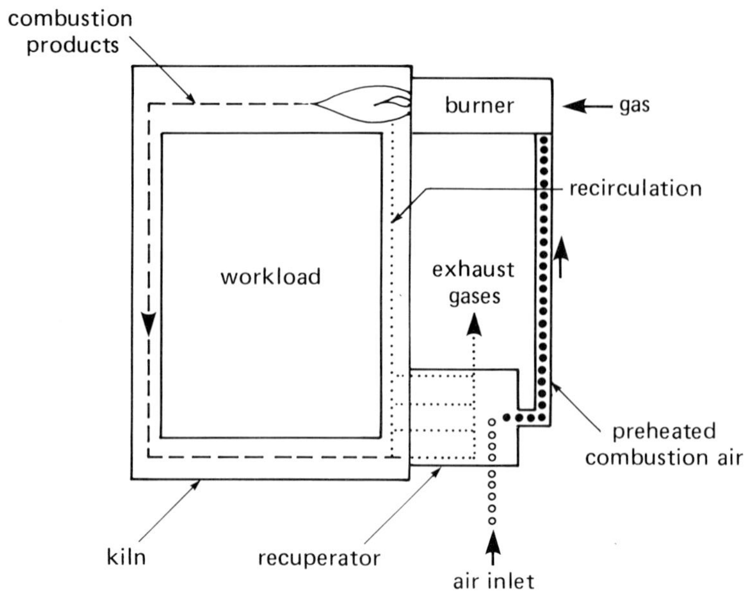

Fig. 2.34 Split burner/recuperator system

2.10 RADIANT TUBE BURNERS

Indirect firing of the heat processing equipment is essential for many heat treatment processes where the work must not come into contact with the products of combustion or the flame. There are a number of ways in which separation can be achieved, such as using a muffle to enclose the work, but in many cases this is not practical or desirable, especially when the process must be carried out in a vacuum or special atmosphere. The gas fired radiant tube offers many advantages as a source of indirect heating and has become firmly established in the metallurgical heat treatment field. For processes requiring special atmospheres or scrupulously clean air conditions in the furnace it provides very even heat distribution at high density. The absence of combustion products in the chamber means that furnace wall linings last much longer and down time for rebricking is greatly reduced.

The tubes are usually made of heat resisting metal and indeed, the temperature limitation of this type of furnace is conditioned by the tube material. It must be emphasised, however, that without going to special materials such as ceramics, metallic tubes offer a wide range of temperatures, from say 300 °C to 1 100 °C, encompassing a large range of industrial processes. It is necessary, however, in many circumstances, to use a tube material very near the critical temperature at which oxidation and creep become serious factors. To ensure a reasonable tube life this means that hot spots must be eliminated or reduced to the minimum and for this reason considerable research and development have gone in to obtaining more even temperature distribution along the length of the tube.

There are many different types of radiant tubes for industrial furnaces but they can generally be sub-divided into two classes:

- Non-recirculating types
- Recirculating radiant tubes

Non-recirculating radiant tubes

The straight through types as illustrated in Figure 2.35 principally comprise straight, curved parabolic, 'U' shaped and 'W' shaped tubes, each of which can be fitted with external recuperators if required. The 'U' and 'W' types provide a greater length of path with consequent increase in heat transfer area and efficiency. In general temperature uniformity is poor.

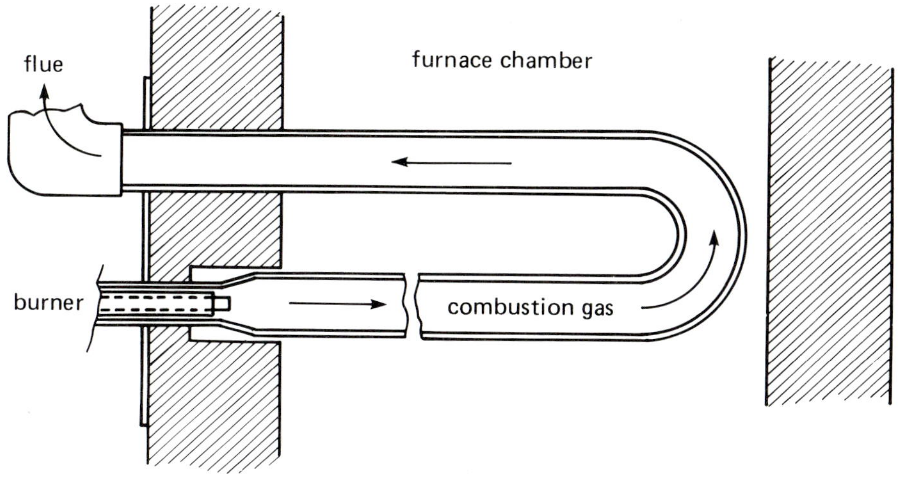

Fig. 2.35 U-shaped non-recirculating radiant tube

A second group of non-recirculating radiant tubes may be described as double-pass single ended types in which the burner and flue are combined at one end and an internal tube transmits the hot gases to the closed end of the outer tube where they turn through 180° and travel back through the annulus. The tube shown in Figure 2.36 is an example of this type and uses a short flame high-velocity burner. Due to the counterflow arrangement the outer tube temperature uniformity is improved compared with the single pass design and tolerances from 10 to 70 °C can be achieved.

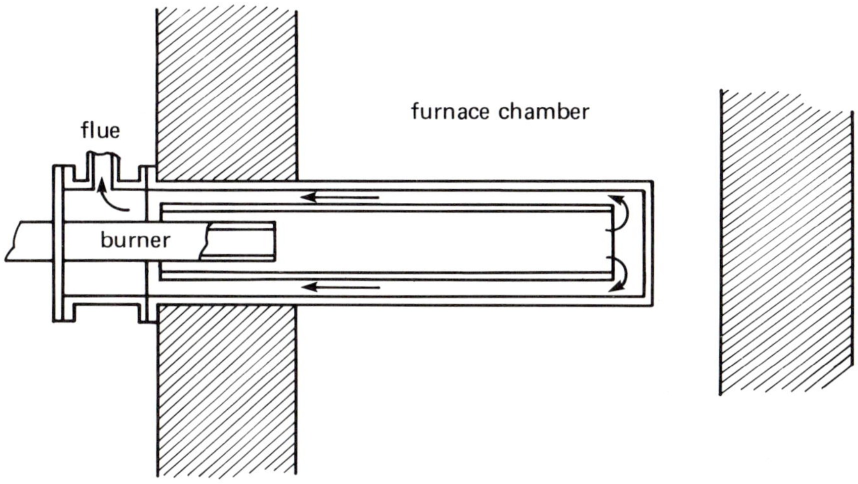

Fig. 2.36 Double-pass single-ended radiant tube

Recirculating radiant tubes

In the recirculating types, combustion of the gas is effectively completed within the burner tunnel. A high velocity jet of combustion products then enters the centrally mounted recirculation tube, entraining the cooler gases returning along the radiant tube as it does so. This recirculation and rapid mixing of the gases within the tube produces fairly even gas temperature distribution throughout the tube. Any lack of uniformity here is offset by the hottest gases being adjacent to the coolest, with the recirculation tube intervening. The radiant tube then receives heat by convection and non-luminous radiation from the gases and also by radiation from the recirculation tube which has also been heated by the recirculating gases. Since nearly uniform conditions exist within the tube and any deviation tends to be compensated in the manner described above, the final radiant tube temperature is very uniform with a tolerance of ±5 °C at surface temperatures up to 1230 °C. A recirculating radiant tube designed by British Gas is illustrated schematically in Figure 2.37.

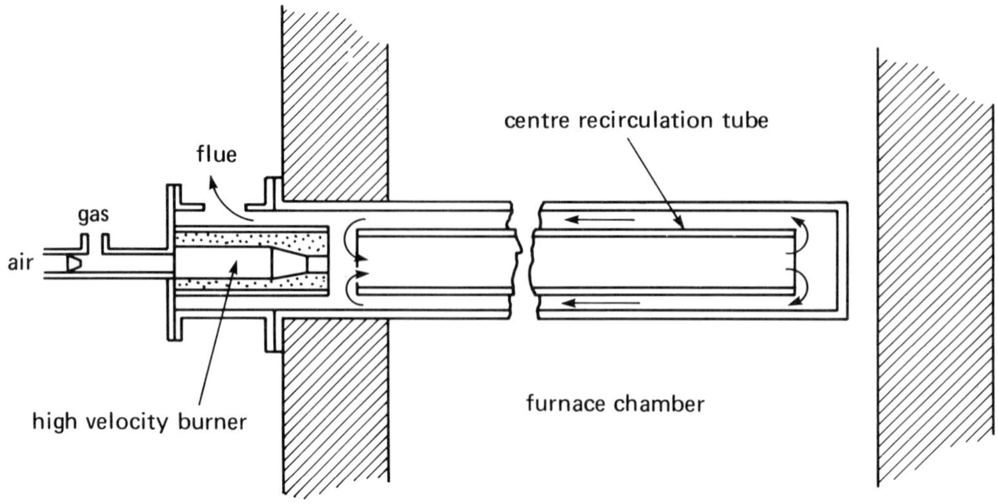

Fig. 2.37 British Gas recirculating radiant tube

A feature, common to all single-ended tubes, is that these units can be inserted into the furnace from one side; alternate side firing is thereby eliminated, and only one mounting aperture is required. Constant-ratio control equipment is necessary to maintain stoichiometric conditions because of the changes in combustion air density during warm-up and high/low firing.

Whilst radiant tubes have found their biggest application in metallurgical heat treatment where special atmospheres are vitally important, they are also used in processes where treatment is carried out in a vacuum, such as vacuum annealing. In the non-ferrous field, radiant tubes are employed for aluminium soaking pits and atmosphere annealing furnaces.

Metallic tubes are generally made of nickel-chrome alloy and are used in metallurgical heat treatment, vitreous enamelling and aluminium soaking pits. They have a maximum tube temperature of 1050 °C giving a process temperature range 400-970 °C. Tube efficiency is around 50% with a heat flux from the tubes of 31.6 kW/m^2.

Ceramic radiant tubes as shown in Figure 2.38 are manufactured from silicon carbide bonded with silicon nitride, and are used in the sintering of ferrous alloys, bright annealing of stainless steel,

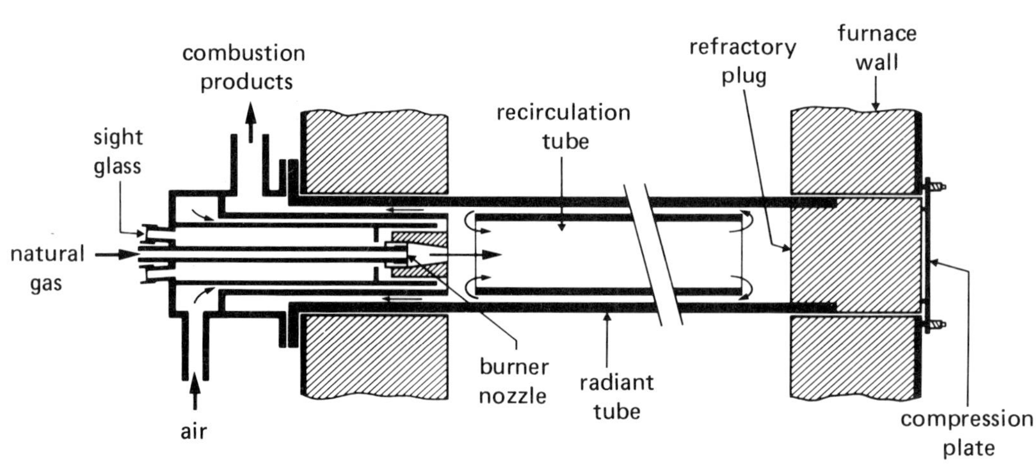

Fig. 2.38 Ceramic radiant tube burner

reheating of special steels, and ceramic wares. They have a maximum tube operating temperature of 1 350 °C giving a working process temperature in the range 980 to 1 250 °C. Tube efficiency is around 50-55% with a heat flux as in the metallic tube. Tube life is around 4000 working hours or one year.

Recuperative types

The logical development of the recirculating radiant tube is to combine its virtues with those of an integral recuperative burner thereby increasing the efficiency to the order of 65 to 70%.

Waste heat recovery can be achieved by either an external heat exchanger, or a recuperator integral with the burner. The integral system shown in Figure 2.39 avoids the losses associated with external pipework leading to central recuperation, but there is the problem of air density variation as the burner warms up. Often this involves air/gas ratio control equipment to compensate for this effect by backloading linked valves, or incorporating volumetric governor and relay combinations.

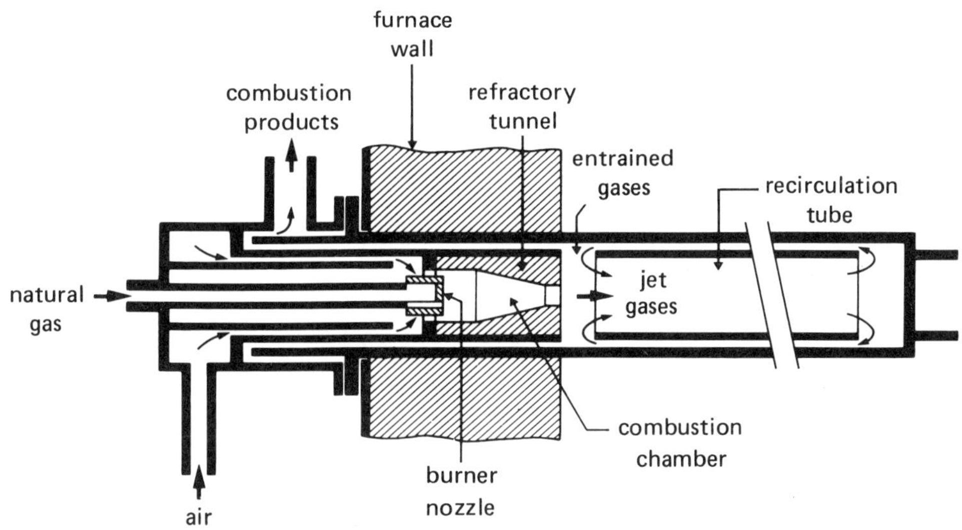

Fig. 2.39 Recuperative single-ended metallic recirculating radiant tube burner

The ceramic recuperative radiant tube consists of a plain ceramic tube held between refractory lined metallic tubular extensions. An axial compressive force is applied, via a spring loaded plate, to reduce tensile stresses in the ceramic material and to ensure an effective refractory to ceramic seal. Suitable applications for ceramic tubed units include bright annealing of steel, steel reheating prior to forging, heat treatment of high speed and tool steels and the firing of ceramic ware.

Recuperative ceramic tube heaters have been developed for galvanising and aluminium holding furnaces. The principles of operation of these immersion tube units, similar to those of the ceramic radiant tube, are shown in Figure 2.40.

The tube shaped like a test tube is made of nitride bonded silicon carbide and is fitted with a small recuperative burner firing vertically downwards. Tubes are available in 150 mm and 200 mm sizes with input ratings between 30 and 65 kW. Significantly the ceramic immersion tube operates at lower temperatures than the ceramic radiant tube, typically between 450 and 750 °C.

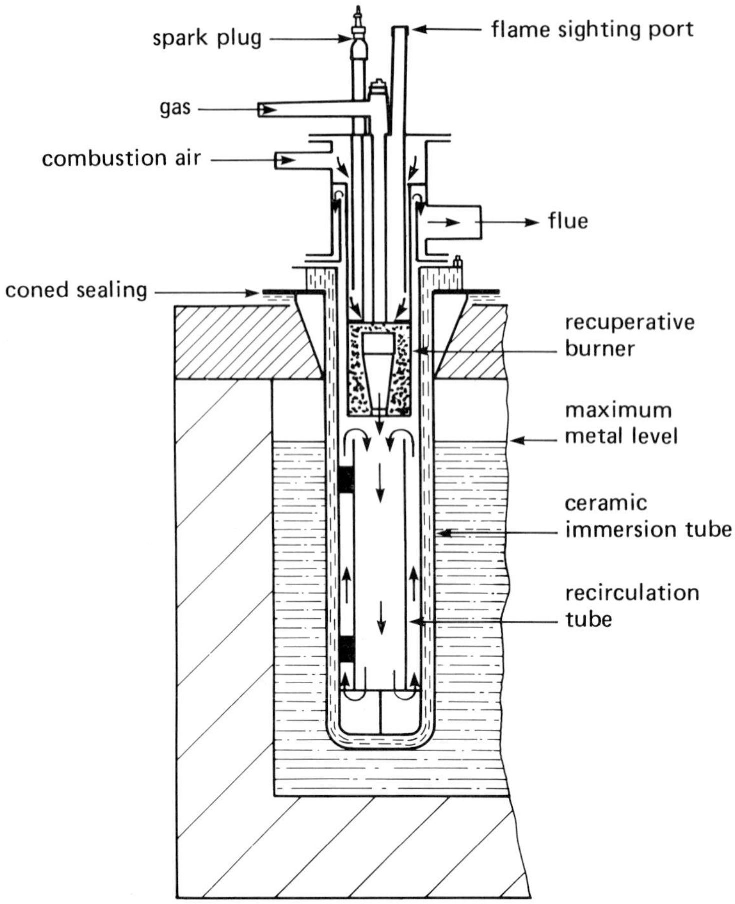

Fig. 2.40 Immersion tube in molten aluminium bale-out furnace

2.11 HIGH TEMPERATURE REGENERATIVE BURNERS

In section 2.10 the use of ceramic materials for very high temperature systems was discussed. These offer higher levels of performance than metallic recuperators allowing higher air preheat temperatures to be obtained. However, high temperature ceramics can be difficult to fabricate into a suitable recuperator construction because of production difficulties in producing the complex shapes necessary for a compact, effective unit. In addition they are often susceptible to failure under low tensile stresses and can be difficult to seal between the two gas-streams.

An alternative approach to increased performance is the regenerator. Regenerators have traditionally been used in large continuously operating high temperature processes such as glass melting furnaces and open hearth furnaces. They have largely been incorporated in plant during its original construction and their size does not easily lend them to the retrofit market. Static regenerators are discussed in Chapter 7.

British Gas have developed a novel compact regenerative burner, jointly developed with a manufacturer. This development has resulted from technological advances in controls, ceramics, burner design and computer aided regenerator design.

The regenerative system shown in Figure 2.41 comprises two burners, each of which has a regenerator chamber containing a packed bed of ceramic shapes of large surface area to provide the heat store. The burners fire alternatively for a few minutes each with the flue gases from the one firing leaving the furnace through the burner tunnel of the other. The hot gases flow through the packed bed heating it almost to the furnace exit gas temperature. On reversal, cold combustion air passes through the packed bed and preheats before entering its associated burner. The cycle is continuous with each burner firing at either full rate or turned down to a minimum firing rate when its combustion chamber is acting as the flue path.

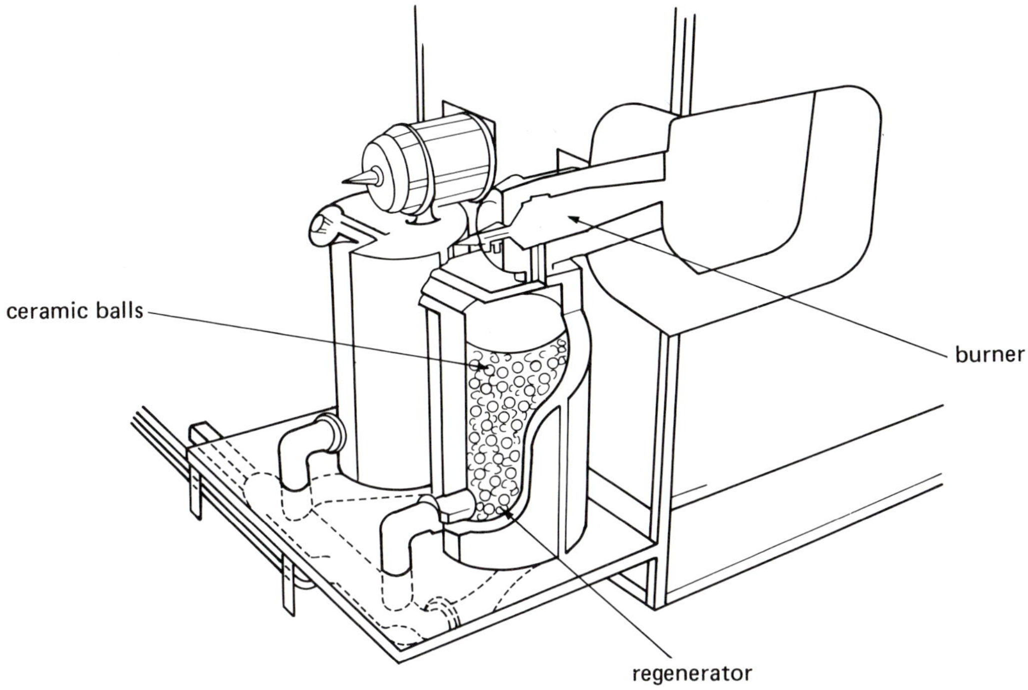

Fig. 2.41 Regenerative burner system

The extensive surface area, short reversal periods and high convective heat transfer coefficients of the ceramic shapes provide an effective, compact design. Laboratory trials have shown that as much as 90% of the available heat in the flue gases can be recovered, in favourable agreement with theoretical predictions. This results in an average outlet gas temperature from the regenerator of about 450 °C when subjected to furnace gases at 1 400 °C. At the same time the peak air preheat temperature approaches the temperature of the inlet exhaust gases. The resistance of the regenerator necessitates an exhaust fan to overcome pressure losses but this can be achieved by the use of a conventional 'cold' air fan due to the relatively low outlet temperatures. Gas and combustion air pressure requirements are similar to those for a recuperative burner, around 5 and 37.5 mbar respectively. Relative performance characteristics are shown in Figure 2.42.

There are many successful installations in traditional 'clean' applications. The system has also been applied in, for example, a glass melting furnace operating at temperatures up to 1 400 °C and using 295 kW of natural gas. The installation has proved reliable and effective and has demonstrated a further advantage of this regenerator system. It offers the facility for operation with 'dirty' atmospheres by ready removal of the ceramic heat exchange shapes for renewal or cleaning in contrast to conventional metallic recuperators and recuperative burners which are susceptible to corrosion and fouling.

As a compact and highly effective system the British Gas regenerator has the potential to virtually eliminate flue gases as a major source of heat loss from high temperature plant. Its successful development provides a wide scope in the retrofit market and its superior performance makes it a more attractive proposition than the recuperative burner.

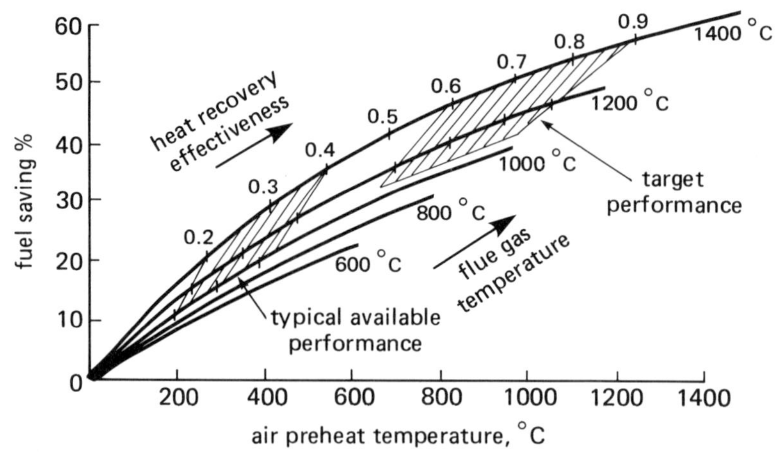

Fig. 2.42 Fuel saving against air preheat temperature

2.12 IMMERSION TUBE BURNERS FOR LIQUID HEATING

Liquid heating is one of a number of industrial processes where steam, generated by a central boiler house, is used as the source of heat. Many engineers have examined such steam systems and concluded that the overall efficiency is low, sometimes being less than 50%. It is therefore clear that considerable fuel savings are possible through the application of efficient direct gas fired systems as described in this section. Gas fired natural draught burners have higher efficiencies, of approximately 70%, but are comparatively bulky.

2.12.1 High Efficiency Immersion Tube Burners

British Gas has developed a system to heat vats and tanks by directly firing hot combustion products down small bore immersion tubes at high velocity and hence high efficiency. In this system, which employs a special design of nozzle mixing burner, heat release rates comparable with conventional steam tubes can be obtained and the overall efficiency is typically 80% with tube length to diameter ratios of approximately 140. The system is shown schematically in Figure 2.43.

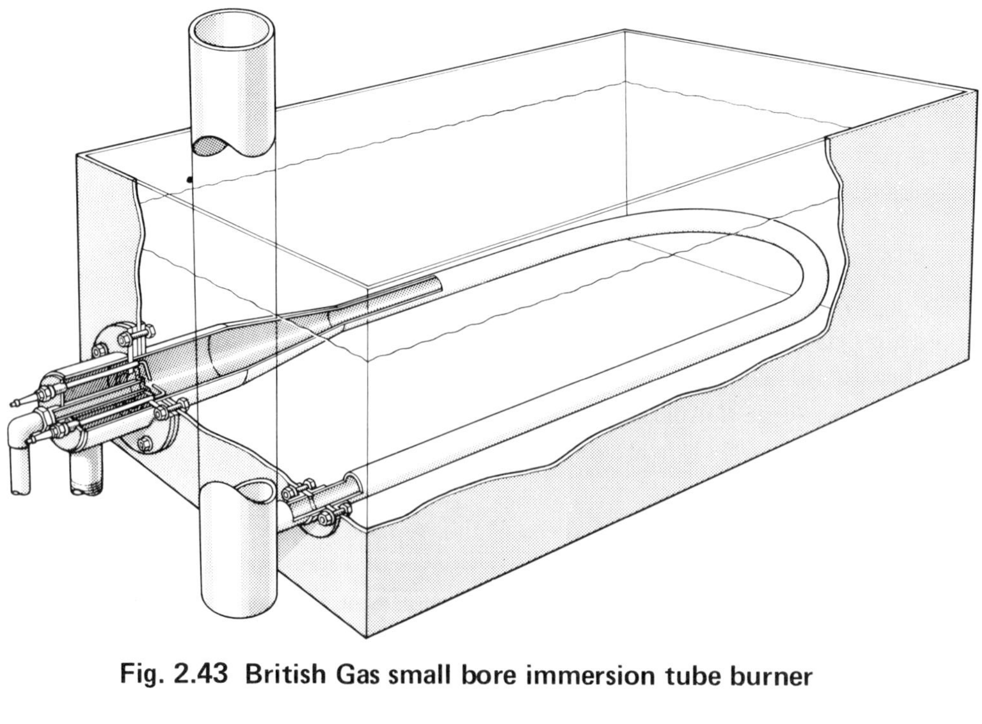

Fig. 2.43 British Gas small bore immersion tube burner

In the burner development, particular attention has been paid to avoiding the pressure fluctuations to which closed, high-intensity combustion systems are particularly prone. Flame stabilisation is achieved by a combination of reduced port loading and hot gas recirculation. Gas flowing through radially drilled ports is progressively supplied with primary air axially by means of a disc arrangement to avoid high air velocities sweeping the flame away. In addition, hot-gas recirculation takes place around the air jets issuing from the holes in the disc. The remainder of the air for combustion is supplied from radial ports in the U-shaped disc. The position of the air diffuser in relation to the entrance to the combustion chamber is crucial in preventing pulsations and incomplete combustion.

For a particular immersion tube bore, there is a greatly increased gas through-put with this forced-draught design compared with natural draught systems. For example a 50 mm diameter tube operating at normal gas supply pressure 20 mbar has a rating of 110 kW. Heat transfer coefficients are typically 100 W/m^2 K, an order of magnitude higher than natural-draught tubes.

A range of units is available in tube sizes ranging from 25 to 150 mm diameter, with heat release rates from 30 kW up to 600 kW. The burners are supplied with fully automatic control systems and have a turndown ratio of about 4:1 by setting the linked valves. Provision is made for thermostatic control and clock control if required. The existing range of burners is readily applicable in the general engineering sector, particularly to aqueous solutions.

2.12.2 Multi-Tubular Heat Exchanger Designs

Certain liquid heating processes require fast heat-up times or high heat inputs into relatively small tanks. To satisfy these particular market applications a multiple tube system has been developed.

One such system, illustrated in Figure 2.44, uses the British Gas immersion tube burner in a compact multi-tubular heat exchanger.

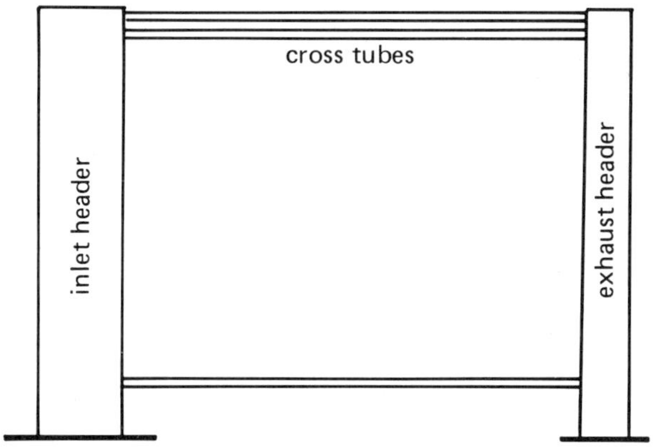

Fig. 2.44 Multi-tube heat exchanger

In an alternative configuration shown in Figure 2.45 burner/heat exchanger units are available which can be directly retrofitted to existing calorifiers.

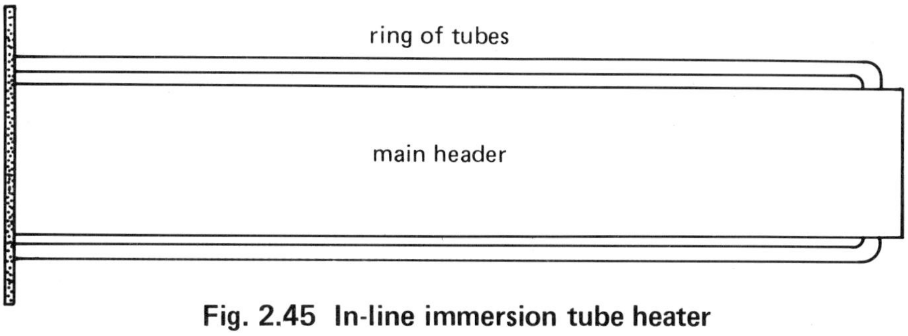

Fig. 2.45 In-line immersion tube heater

Generally overall efficiency at maximum firing rate is over 82% rising to over 85% on low fire setting. Heat inputs are available in the range 47 to 580 kW.

2.12.3 High Temperature Direct Contact Water Heater

It has long been recognised that heating water by direct contact between the combustion products from a burner and the water to be heated is extremely efficient. The method has been used for many years, for example in submerged combustion heating systems discussed in Section 2.13.4.

With direct contact water heating, as the required temperature of the water to be heated rises, more of the available heat is used for evaporation rather than to raise temperature. This means that there is a practical temperature limit of about 65 to 70 °C for the water being heated. However, by using a combination of direct contact heating in a counter-flow recuperator column and indirect heating using an immersion tube in a tank at the base of the column, both high temperatures and high efficiencies can be achieved. A system using a short immersion tube fired by a standard package burner is shown in Figure 2.46a. This is described in more detail in Chapter 6, pp 234-5.

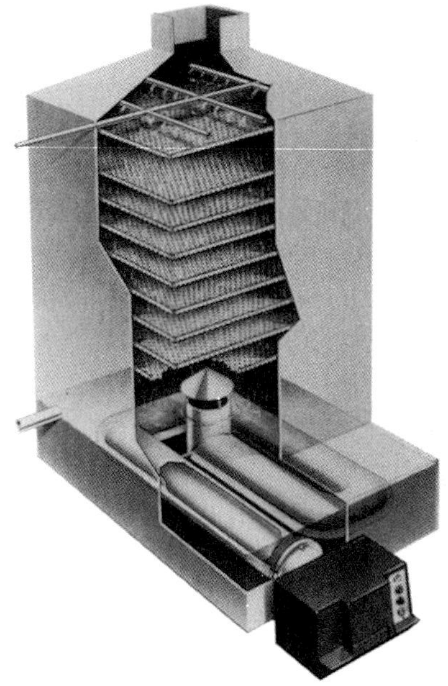

Fig. 2.46a High temperature direct contact water heater based on an immersion tube system

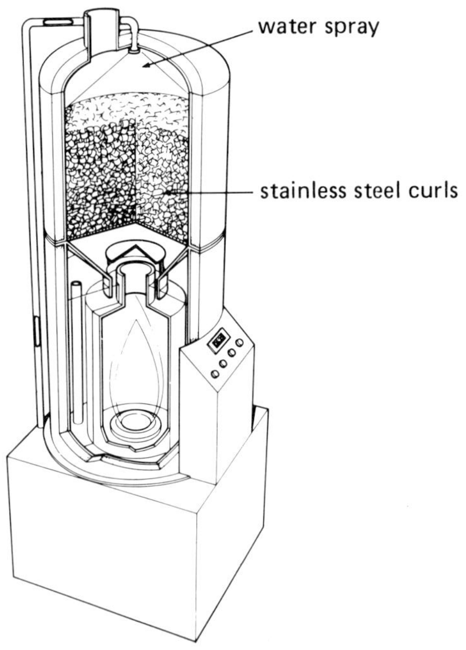

Fig. 2.46b High temperature direct contact water heater with stainless steel curved shapes

About 50% of the heat is transferred by this immersion tube to the water in the tank. The combustion products then pass upwards through stainless steel baffle trays through a spray of water from jets above the top tray. When the combustion products finally leave the unit they are virtually at room temperature, so an efficiency of around 98% is possible with cold water at 15 °C. The latest design shown in Figure 2.46b, uses stainless steel curved shapes to form a packing with high contact area instead of the perforated trays. This burner fires from the base vertically upwards.

Outlet water temperatures of up to 95 °C are available if required. Average gross efficiencies of over 95% have been demonstrated with efficiencies of up to 98% achievable. The system is discussed in further detail in Chapter 6.

2.13 SPECIALISED TYPES OF GAS BURNER

2.13.1 High Velocity Burners

In heating processes which rely on forced convective heat transfer as the main heating medium, the main obstacle to efficiency is the thin film of gas on the surface to be heated. Increasing the velocity of heating gases reduces or removes this insulating film, so greatly improving heat transfer efficiency. A high mass flow rate of hot gases is therefore an essential requirement for efficient heating, but if this is to be achieved, the combustion products must be forced to envelop the work completely. The result is high speed heating with extremely low temperature differentials throughout. Since high temperature gradients are never present, then hot spots cannot be formed and overall uniformity of temperature within the system is markedly enhanced.

Conventional methods of creating high mass flow necessitate the use of special recirculating fans within the working chamber of the furnace, the cost of which at medium to high temperatures would be prohibitive. The alternative is to supply this energy externally by a suitably designed gas burner system. The burner thus becomes the source of energy to move the hot gases at turbulent velocity over the heated surface, as shown in Figure 2.47.

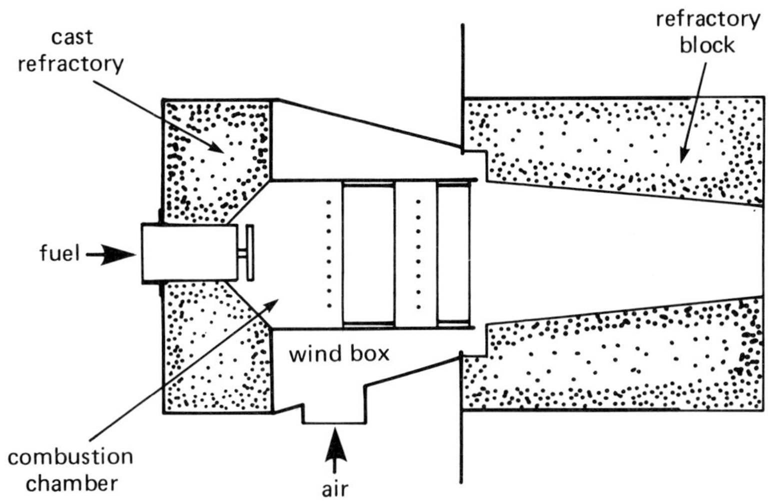

Fig. 2.47 High velocity burner

Complete combustion takes place in a refractory tunnel, air and gas being mixed at the base of the tunnel. This design, because of the special nozzle mixing feature, permits the use of high excess air, thus obviating the need for a cold air jacket. All the air passes through a recuperative system before reaching the combustion chamber. The inherent efficiency of this design of burner is exceptionally good; 65% is a normal efficiency at working chamber temperatures of the order of 1 000 °C.

This type of burner has been used successfully in the stress relieving of pressure vessels. The hot gases from the burner are blown into the interior of the vessel which is externally insulated, thus obviating the need for a stress relieving furnace large enough to receive the vessels. Such burners have been employed in preheating glass furnaces and other large structures.

2.13.2 Flat Flame Burners

This type of burner is used on high temperature furnace applications where impingement of the flame on the stock to be heated must be avoided. It is essentially a nozzle mix burner which can be fitted with the usual pilot burner, spark ignition and UV flame detection.

It is important that on installation the outlet of the refractory quarl is made flush with the inner furnace wall or roof, so that the flame can flow smoothly from the quarl to the adjacent surfaces. These burners generally have a wide range of turndown without flashback and operate with up to 20% excess fuel and 200% excess air for particular applications. Typically, flat flame burners will operate at a gas pressure of 10 mbar or more and outputs range from 60 to 730 kW with the use of gas pressure boosted systems required for the higher release rates.

The mode of operation is air passing over a side orifice plate which produces a spinning action which is further increased by fixed air swirl vanes, as indicated in Figure 2.48. Gas is introduced into the spinning air by axial and radial gas inlet ports. After ignition, the mixture continues to burn as it moves forward and follows the contour of the combustion quarl. The result is a flame which is flat against the furnace wall or roof with no forward velocity of combustion products. There is a low pressure area at the discharge of the burner, thus large volumes of hot gases are drawn to the centre of the burner creating recirculation.

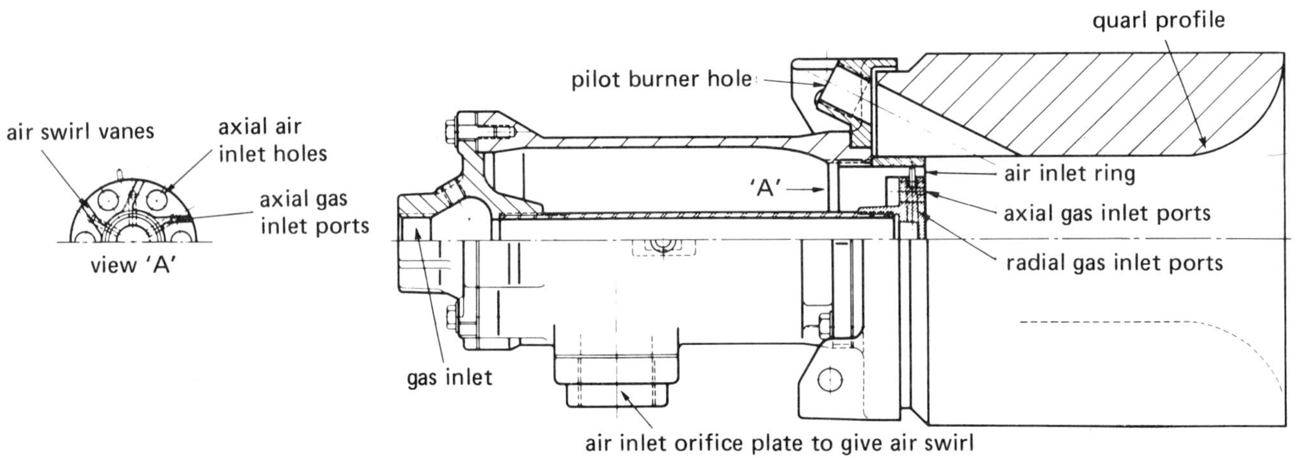

Fig. 2.48 Flat flame burner

Apart from its use in metal annealing, heat treatment furnaces and forge reheat furnaces, it has also been used successfully, for example, in zinc galvanising baths.

2.13.3 Low NO_X Burners

The emission of atmospheric pollutants from combustion equipment is assuming increasing importance. European Community legislation has been drafted covering particulates and SO_2, in addition to NO_X. Thus it is important to minimise the amount of combustion generated NO_X particularly since directly heated air used for drying foodstuffs can react to produce nitrosamines which can constitute a health hazard.

To achieve emission levels of 0.2 ppm combustion added NO_X/% CO_2 involves the production of low temperature flames. Generally this can be achieved by burning the fuel gas at very high excess air levels at a limiting temperature set by the lower limit of flammability of the fuel gas with air.

Thus low NO_X burners operate on a premix air/gas system with the addition of secondary air in the burner manifold as illustrated in Figure 2.49. Flame stabilisers are used with recirculation of combustion products at the root of the flame. Gas is drawn into the air gas mixer tube through a series of holes in the tube after the throat sections.

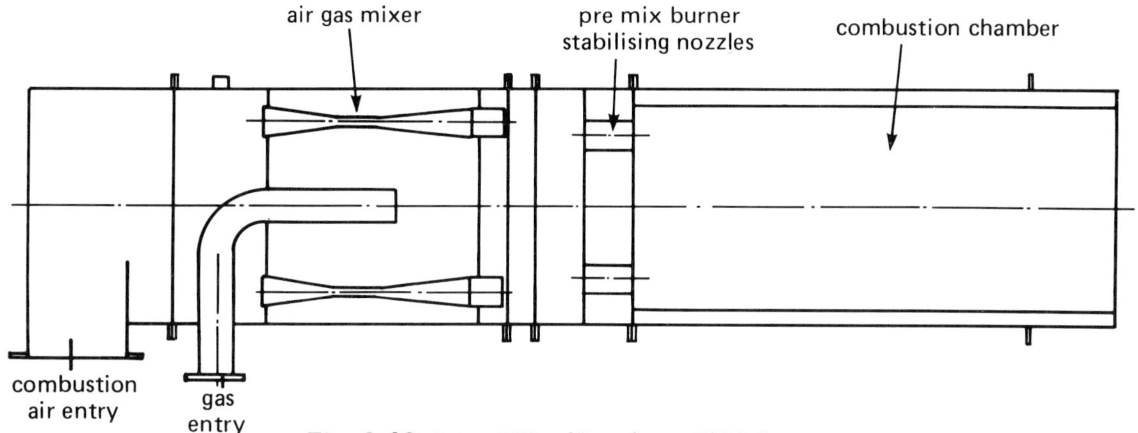

Fig. 2.49 Low NO_X Urquhart CXA burner

2.13.4 Submerged Combustion

In this system the flame itself impinges on the fluid to be heated and the products of combustion themselves percolate through the fluid to the atmosphere. Such a burner employing a nozzle mixing system is shown in Figure 2.50.

Most of the heat liberated by the burner leaves the burner head as sensible heat in the combustion products which takes the form of large numbers of very small bubbles. These present a very large total heat-transfer surface so that the combustion products cool quickly to the temperature of the bath liquid.

For water heating duties there is a practical fluid temperature limit of 65-70 °C as already stated. Submerged combustion has therefore also found particular application in the concentration and crystallisation of dilute solutions, where evaporation is specifically required.

2.13.5 Pulsating Combustors

In pulsating-combustor systems, combustion and thus heat release occur periodically and the pressure waves produced give rise to high sound levels at the frequency of the combustion oscillations. Although oscillatory combustion sometimes occurs in conventional burner systems, particularly those using high combustion intensities, it normally constitutes a serious nuisance.

In pulsating combustors, however, the system is specifically designed to produce oscillations, the energy in the pressure waves being used to induce the combustion air and to eject the combustion products.

Interest in practical applications for pulsating combustion dates back to the 1930s and in more recent years development has been directed towards uses including immersion-tube water heating, portable heaters, ice-melters, vehicle heaters, grain driers and space heating. A wide variety of fuels has been employed including LPG, natural gas and pulverised coal.

The simplest form of pulsating combuustor, the Schmidt Pulse Burner, is shown in Figure 2.51(a) and consists of a resonance tube several metres long and gas and air inlet arrangements.

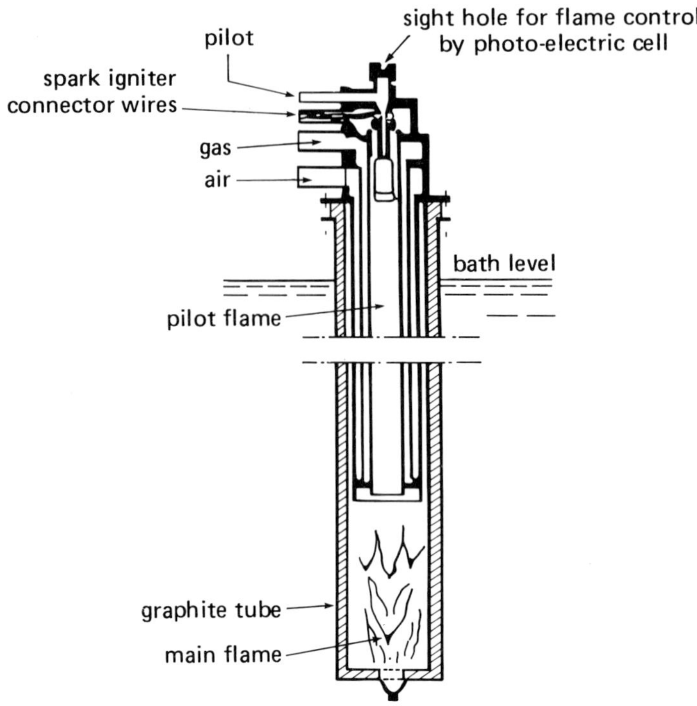

Fig. 2.50 Submerged combustion burner

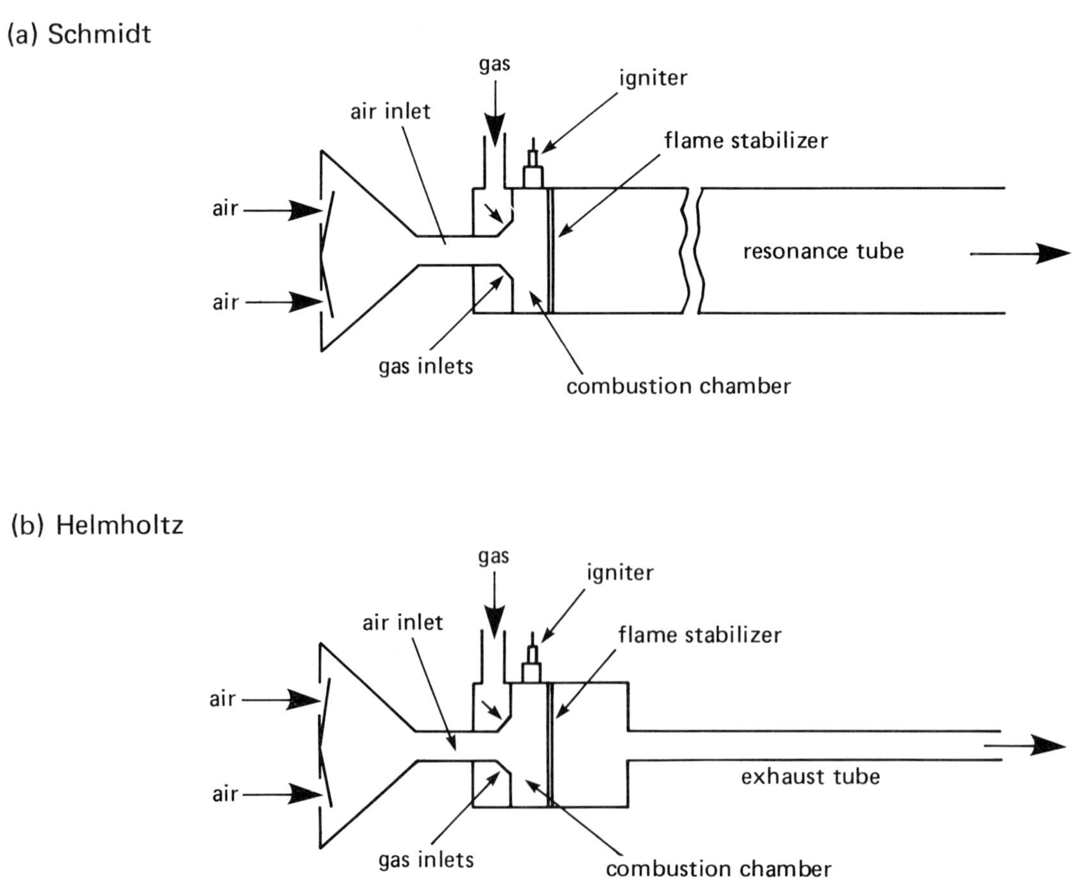

Fig. 2.51 Pulse burners

The inlet end of the resonance tube forms the combustion chamber into which air is drawn from the atmosphere through a valve which may be a flap valve as shown or an aerodynamic valve without moving parts. The flame stabiliser, a small rod fixed diametrically in the chamber, prevents irregular combustion.

The combustion sequence is as follows:

- Gas and air flow into the combustion chamber and are ignited by spark.
- Combustion occurs and the increase in pressure closes the inlet valve and drives the burnt gases out through the resonance tube.
- Inertia creates a low pressure in the combustion chamber which opens the inlet valve and more air and gas enter.
- Re-ignition occurs from the combustion wave and process continues. The ignition spark is not required after this stage.

The frequency of combustion depends upon the length of the tube and is usually 50-200 Hz (cycles/s).

Figure 2.51(b) shows an alternative arrangement, normally designated the Helmholtz burner, in which the exhaust tube is very much narrower than the combustion chamber.

In pulsating combustors a much larger volume of combustion products may be driven through a given size of exhaust tube than in the natural-draught case. Consequently significantly enhanced convective heat transfer is achieved. In addition the presence of the pulsations themselves lead to further increases in convection. The increase in heat transfer is related to the magnitude of the oscillating velocity of the gas stream and hence to the pressure amplitude of the acoustic wave. Increases of over 100% in heat transfer coefficient can be achieved compared with the corresponding steady-flow case.

One of the drawbacks of the system is the relatively high noise level from pulsating combustion. However, exhaust absorption silencers provide sound level attenuation of 15 dB or more. An alternative approach is with two pulsating combustors arranged to run with a phase difference of 180° thereby providing cancellation of pressure fluctuations by interference: sound level reductions of up to 30 dB (decibels) have been achieved by this method.

A schematic diagram of a commercially available pulse combustor is shown in Figure 2.52.

2.14 OXYGEN ENRICHMENT TECHNIQUES FOR GAS BURNERS

Flame characteristics can be modified or enhanced by many techniques, such as air-gas mixing, combustion chamber design, preheating of the combustion air and/or fuel and so on. However, although air may be the simplest choice as the means of supplying the oxygen necessary for combustion, it is not necessarily ideal for every application. It is possible to 'tailor' flame characteristics with advantage by adjusting the relative proportions of the two main gases which constitute the combustion air.

Usually this is achieved by increasing the proportion of oxygen above the normal 21% but it should be borne in mind that an increase in the percentage of nitrogen present can lead to improved operations under certain circumstances. In theory, the level of oxygen in the combustion 'air' can be taken as high as 100%, although for reasons of both safety and commercial viability, the most practical level in conventional burners is about 25%.

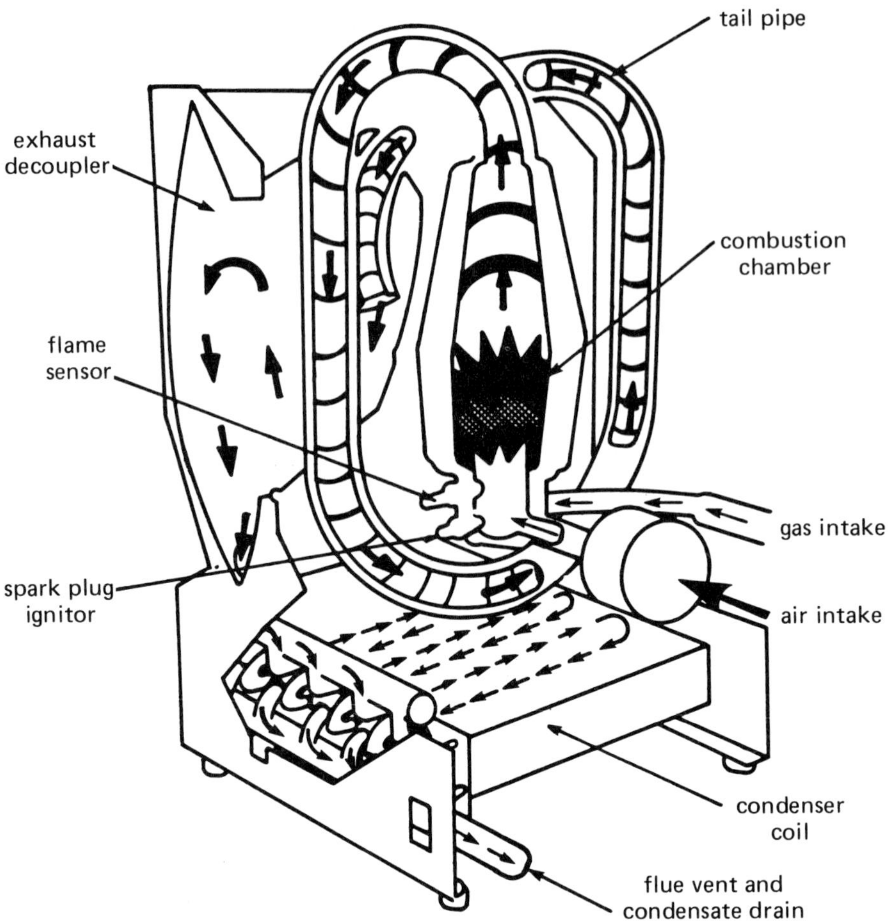

Fig. 2.52 Lennox pulse combustor design

There are, however, purpose-designed burners which operate with a fuel and pure or near pure oxygen. Whatever the level to which the oxygen in the combustion air is raised the following effects will always be noted to an extent dependent upon the degree of oxygen addition:

- Increased flame temperature
- Increased burning velocity
- Reduced ignition temperature
- Less mass of hot products wasted up the flue.

In a conventional air fuel system the reaction between molecules of fuel and oxygen is retarded by the presence of nitrogen. This interferes in two ways. Firstly, it conveys heat away from the reaction and secondly it reduces the opportunity for oxygen and fuel molecules to collide. Thus, a reduction in the proportion of nitrogen present increases the kinetics of the combustion reaction and leads to the effects listed above.

The increase in flame temperature will increase heat transfer to the stock mainly by radiation but also by conduction and convection. This will result in faster heating up rates with the consequent reduction in fuel. Increased burning velocity will result in a reduction in flame dimensions and a greater heat release per unit volume. Because there will be proportionally less nitrogen which takes no part in combustion itself in the combustion products, there will be less heat wasted up the flue, and hence the efficiency will be higher.

There are three basic techniques of oxygen addition. These are general enrichment, underflame enrichment and oxy-fuel burners. In the first system oxygen is added to the combustion air upstream of the burner as shown in Figure 2.53. Levels of addition can vary but would generally fall in the range 2-15% giving total oxygen levels in the combustion air of around 23-35%. The technique is non-directional and is suitable for such applications as batch refractory furnaces, reheating furnaces and soaking pits.

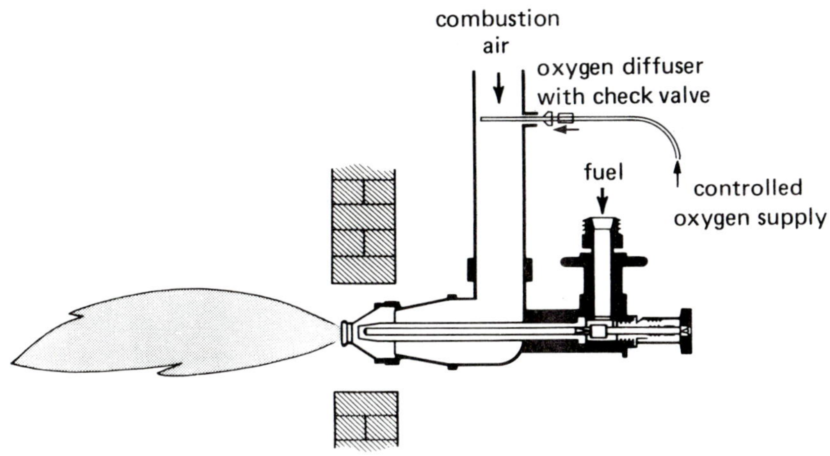

Fig. 2.53 General oxygen enhancement of combustion chamber

In underflame enrichment, the oxygen is injected through a separate lance beneath the flame as illustrated in Figure 2.54. Only the underside of the flame is affected and therefore the technique is directional. One major advantage of adding oxygen in this manner is that the furnace refractories remain unaffected by increased flame temperature, only the furnace charge being affected by the increased heat transfer rates. This technique is suitable for use with rotary kilns, glass melting tanks, reverberatory furnaces and so on.

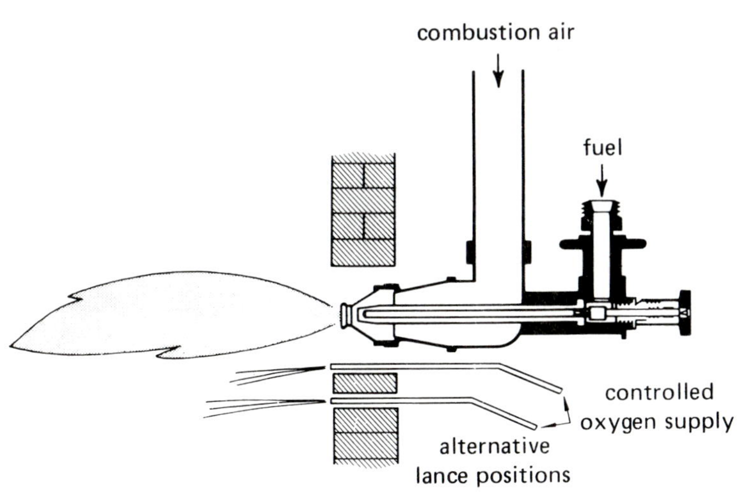

Fig. 2.54 Underflame oxygen enhancement

Oxy-fuel burners are used in melting applications where the large temperature difference between the charge and the flame results in high heat transfer rates. Specially designed water cooled burners are used. The burner is usually operated as the auxiliary heating source and in batch operations is turned off when the charge is 60% melted.

Although premixing the fuel gas and oxygen provides the most effective form of mixing this technique is mainly restricted to individual burner ratings below 30 kW used for welding, cutting or glass working. Since the burning velocity is some four times as great as gas-air flames, lightback is the limiting consideration. Small scale oxy-gas burners are described in Section 2.16. Larger oxy-gas burners are designated face-mixing or post-mixing. Face-mixing burners comprise concentric tubes carrying gas and oxygen as shown in Figure 2.55 and produce long small diameter flames. Other designs of face-mixing burners incorporate a conical divergence at the end of the oxygen tube to encourage reverse flow gas into the root of the oxygen tube and provide satisfactory stabilisation on natural gas.

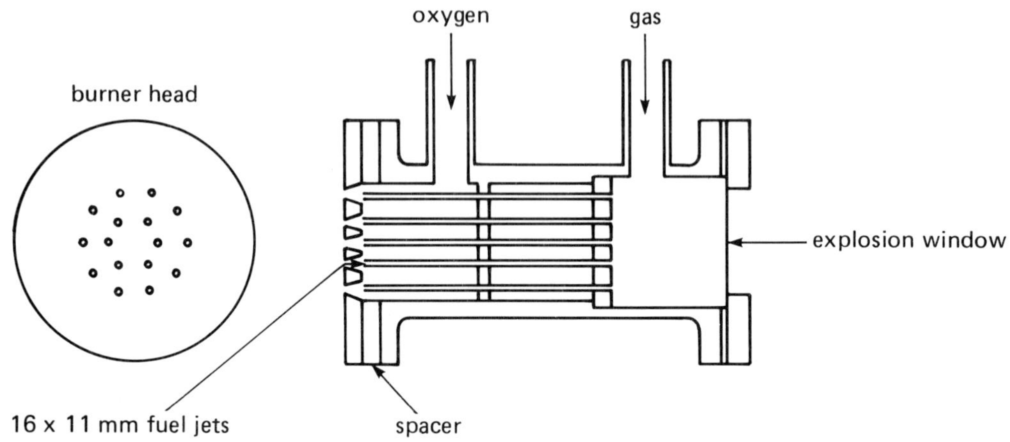

Fig. 2.55 Face-mixing oxygen/gas burner

In post-mixing burners, angled fuel and oxygen jets are arranged to impinge downstream of the burner heat; see Figure 2.56.

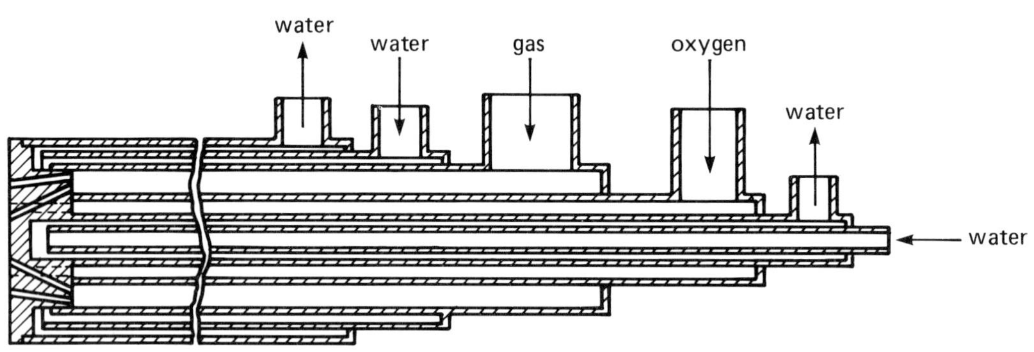

Fig. 2.56 Post-mixing oxygen/gas burner

In summary the advantages of oxygen addition are:

- Increased production resulting from increase in heat transfer efficiency and the opportunity to burn more fuel.
- Reduction in the specific fuel consumption due to less nitrogen in the products of combustion carrying away with it sensible heat.
- Improved product quality.
- Reduced refractory wear due to reduction in flame dimensions resulting in less flame impingement on refractories.
- Increased flexibility.

2.15 PILOT BURNERS

A pilot flame is a subsidiary flame used for the ignition of a main flame. A separate pilot burner may be employed or it may be integral with the main burner but provided with a separate gas supply.

Pilot flames are required where it is desirable to provide a flame of limited thermal input to ignite a main burner either because of the hazards of lighting a main flame directly or for reasons of process intermittency. Pilot type flames are also used as a method of flame stabilisation, when a low velocity flame at the periphery of the main flame is used to produce a zone of continuous re-ignition.

Pilots must be designed and positioned such that the pilot flame will provide immediate and smooth ignition of the main flame under any operative thermal input of both pilot and main flames and under any changes in gas characteristics, pressure and draught as may occur. To avoid gas starvation on pilot burners when main burners are turned on, the gas supply should be taken from upstream of the main governor and be separately governed with ample size piping to reduce pressure losses and develop full capacity.

Although manual gas lighting torches can be used for igniting burners on large furnaces, ovens and other heating plant, spark ignited pilot flames are the most widely used method. In spark ignition, pilot ignition rods utilising 5-10 kV electrical pulses are used for auto-ignition. Burner control methods including spark ignition and flame detection methods are discussed fully in Chapters 9 and 10.

Pilot Burner Design

Pilot burner design is based on the same principles as those for gas burners. The designs range from pre-mix pilots utilising atmospheric natural draught to, more commonly, air blast techniques.

A widespread method used particularly on package burners, is where a pilot flame issues from the same gas ports as used for the main flame, with an ignition electrode for dependable ignition from a single transformer. Figure 2.57 illustrates this method as applied in the NuWay Package Burner.

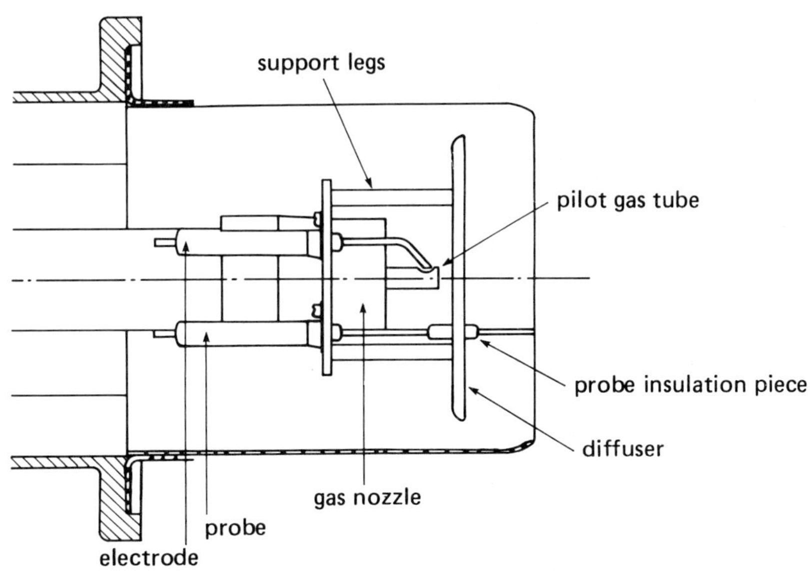

Fig. 2.57 Pilot burner design on a package burner.

Automatic spark ignition pilots can be either intermittent or interrupted. In the former the pilot gas is ignited prior to the burner main flame and remains lit throughout the required burner operating period. Interrupted pilots are extinguished when the main flame has been established and proved.

2.16 WORKING FLAME BURNERS

Working flame burners are by definition flame impingement burners where the flame is used as a tool. In manual operations this could well be a gas heating torch for brazing or soldering. Burners can also be used as instantaneous heating devices for rapid heating on automatic machines, where provision of a furnace or heated enclosure would be impractical. Rapid heating burners can also be switched on or off within the machine cycle time thus giving substantial savings against continuously heated enclosures.

Working flame burners are generally of high intensity, high velocity design with integral low velocity pilot flames to ensure stability of the main flame.

2.16.1 Burner Design

Air-gas and oxygen burners can generally be sub-divided into four basic types, described in increasing combustion intensity as: surface mixing, nozzle mixing, partial pre-mixing and full pre-mixing designs.

Surface mix burners ensure gas and oxygen do not mix until outside the confines of the burner producing safe operation without risk of flashback. They are used where high-low or on-off burner operation is required for automatic machines requiring high temperature flames, with uniform flame characteristics and low noise levels. Multiport surface mix burners can be constructed as in the Nordsea unit shown in Figure 2.58, with square gas tubes fitted through round oxygen ports.

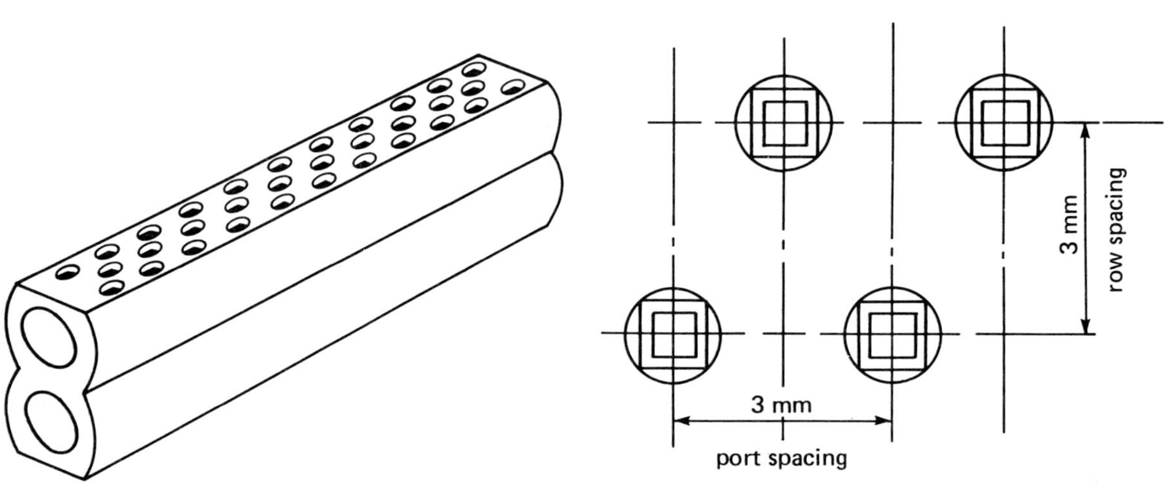

Fig. 2.58 Multiport surface mixing burner

Nozzle mix burners give improved mixing and slightly higher intensities than surface mix burners due to the gas and oxygen flows impinging just within the burner casing; the resulting turbulence generally produces poor flame definition with high noise levels.

Partial pre-mix burners are constructed to enable a proportion of oxygen to be fully mixed with the fuel gas prior to exit from the burner ports. The balance of oxygen flow is introduced into the flame from the surface mixing ports as shown in Figure 2.59. The integral venturi mixing device allows controlled operation from full pre-mix to full surface mix. Intermediate operation in practice gives optimum performance. The resulting flame combines the benefits of both surface mix and pre-mix design. Higher intensity combustion results, giving high flame temperature and low noise levels with a high degree of control over flame characteristics.

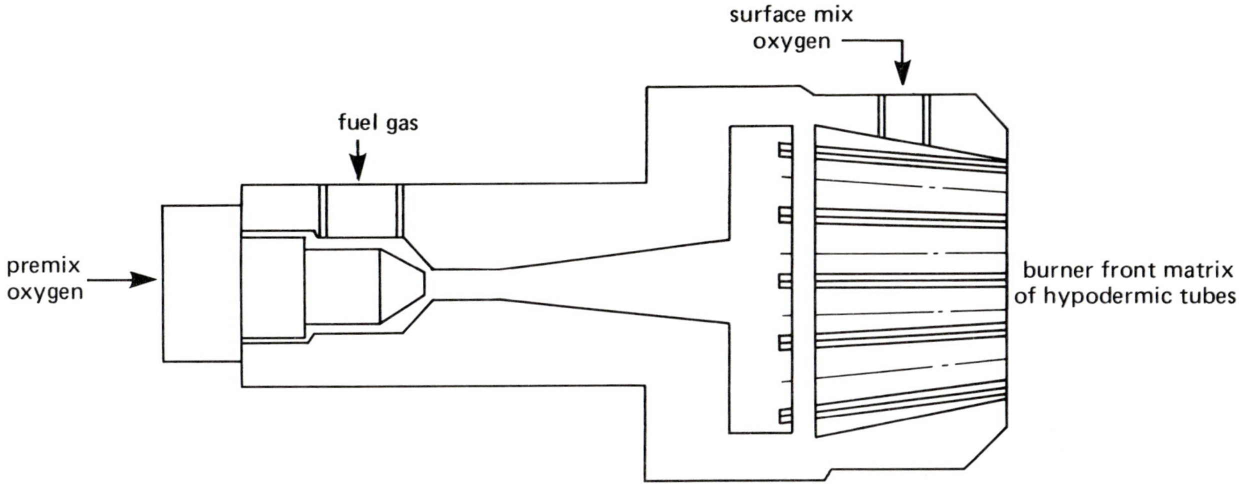

Fig. 2.59 Combined pre-mix and surface mix gas/oxygen burner

Full pre-mix burners are generally used in applications where peak flame temperatures are required and burners are usually of fixed output. Noise levels can be extremely high and short flame lengths tend to restrict the use to very localised heating. Mixing is usually achieved by the use of venturi mixing units (Figure 2.60) where the flow of air or oxygen through a converging jet and diverging throat provides a suction to the mixing gas flow and efficient mixing in the diverging throat prior to entering the burner. Jet and throat sizing is of great importance in maximising mixture pressure and uniformity of mix, particularly for smaller burners. Combined mixer/valve units are available with selection charts for accurate sizing.

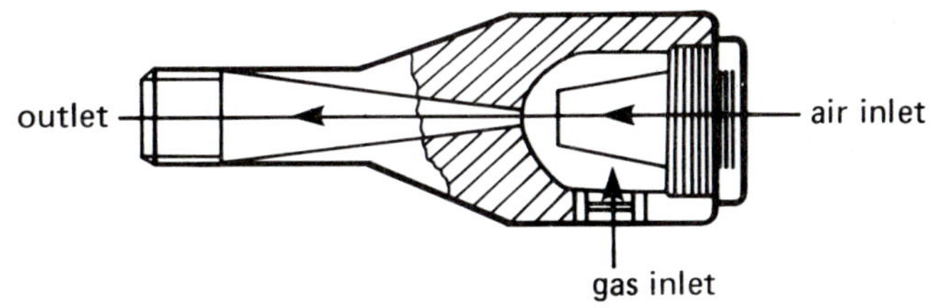

Fig. 2.60 Venturi mixing unit

Air can be introduced into the oxygen flow to any of the above burners to reduce intensity of combustion. As the percentage of air increases, combustion intensity and flame temperature will reduce.

2.16.2 Control of Gas Supplies

Control of air, gas and oxygen flow to the burner often requires the facilities of fine flow control and accurate mixing of the gases to produce consistent flame characteristics, which are essential given the short heating times usually associated with working flame burners. Fine control needle valves which are suitable for oxygen use, with a range of orifice sizes and tapered needle stems, provide the means for controlling flow to the burner or mixing device.

In addition to flow control devices, there is a requirement to prevent reverse flow of air or oxygen, usually supplied at higher pressure than the fuel gas. Reverse flow is obviously a hazardous situation, particularly with oxygen, and can occur if the burner head becomes blocked on say an automatic machine. Approved non-return valves should therefore be installed at each mixing point and certainly at each machine gas supply point for multi-burner situations. (IM/1, 3rd edition, gives further recommendations.)

Flash back can occur in pre-mix burners, usually when flow velocity through the burner port is reduced to below the mixture burning velocity. This often happens when burners are turned on or off and especially if oxygen is being used to enrich the combustion air. The use of flash back arrestors or fire checks can reduce or eliminate the effect of flash backs although high oxygen concentrations can present difficulty, due to the effect of detonation waves rather than flame fronts making flame arrestors inoperative. As a general rule, the mixing device should be as close as possible to the burner to reduce the length of mixture pipe feeding the burner tip.

REFERENCES

BRITISH GAS PLC PUBLICATIONS

IM/1 3rd Edition 1989 : Non-return Valves for Oxy-gas Glass Working Burners.

IM/12 2nd Edition 1989 : Code of Practice for the Use of Gas in High Temperature Plant.

IM/18 1st Edition 1982 with Amendments 1986 : Code of Practice for the Use of Gas in Low Temperature Plant

BRITISH STANDARDS

BS 5885 : 1988 Part 1 Specification for Automatic Gas Burners of Input Rating 60 kW and above.

OTHER PUBLICATIONS

Proceedings of the First International Symposium on Pulsatiung Combustion. Brown D.J. (Ed). University of Sheffield, 1971.

Industrial Gas Heating. Priestley J.J. Ernest Benn, London. 1973.

The Efficient use of Energy. Dryden I.G.C. (Ed). IPC Science and Technology Press. 1975.

Industrial Gas Utilisation: Engineering Principles and Practice. Pritchard R., Guy J.J., Conner N.E. Bowker 1977.

North American Combustion Handbook. R.J. Reed (Ed). North American Mfg. Co. Cleveland. 1978.

Multifuel Burners 3

CHAPTER 3

MULTI-FUEL BURNERS

3.1 INTRODUCTION

A substantial proportion of the gas supplied to industry in Britain is sold on an interruptible basis. This enables the gas industry to match seasonal variation in supply and demand. At times of very high demand such consumers may be requested to cease or reduce gas consumption and operate their plant on an alternative standby fuel. Since the supply profile approximates to the demand profile, interruptions can occur at any time of the year although they are most likely to occur during the winter months. Additionally, interruptible gas allows the gas industry to optimise the operation of its transmission and distribution system. The consumer benefits from the advantageous terms offered compared with 'firm' gas prices.

Provision for interruptible operation may be made by the installation of separate burners for the two fuels, but it is much more frequently made by the use of multi-fuel burners which are designed to burn either gas and oil or, less commonly, gas and coal. The widest area of application is in industrial boiler firing, where dual-fuel burners are frequently used.

Multi-fuel burner design and performance are discussed in this Chapter but with particular consideration given to dual-fuel burners, since these systems are by far the most common. Fuel properties and combustion are considered to the minimum degree needed for continuity, and readers should refer to Chapter 1 for more detailed treatment.

3.2 MULTI-FUEL BURNER SYSTEMS

Combustion systems which can burn more than one fuel either separately or together may be classified as follows:

- Gas/oil burners which can burn more than one fuel simultaneously.
- Combination burners which are capable of burning gas and/or oil (or gas and/or coal) simultaneously or separately.
- Dual-fuel burners which are designed to burn gas or oil (or gas or coal) separately but not together.

Combination burners and gas/oil burners

The performance of these types of burner on individual fuels is poorer than either the dual-fuel burner or the purpose built gas, coal or oil burner.

In particular, difficulties arise in controlling the overall air/fuel ratio and combustion quality because variations both in individual fuel flow rates and fuel quality may occur simultaneously. Although some advantage in luminosity may accrue from combined firing due to the oil or coal particles forming nuclei, the disadvantages of loss of control of the flow and air/fuel ratio, as well as poor mixing and combustion quality and reduced cleanliness of flue products, make this system less suited to boiler firing than the dual-fuel burner.

It should be noted that the difficulties of controlling and maintaining a satisfactory air/fuel ratio when operating with more than one fuel simultaneously can be largely overcome by the introduction of modern oxygen monitoring equipment based on zirconia cells and linked to control systems operating on the air damper mechanism.

Dual-fuel burners

The advantage of the dual-fuel system is that optimum conditions can be obtained for operation with the main fuel, while any deviations in performance from this optimum condition experienced with the substitute fuel could be tolerated over short periods. In general gas is used for at least 95% of the time, the oil being used only during short periods of peak winter demand.

The simplest burner is that in which alternative gas and oil injectors may be fitted. However, the inconvenience and delay which results from changing these components may make this system unsuitable for application to interruptible loads.

With most fixed component dual-fuel burners rapid interchange of fuels can invariably be achieved, with the degree of sophistication varying from manual to fully automatic depending on the cost of the system.

Dual-fuel burners may be conveniently divided into two classes:

- Packaged burners supplied as factory-assembled units incorporating the fan and controls.
- Register burners using separate fans with the burners and controls being assembled on site.

The former group find their widest application in shell boiler firing whilst the latter are generally fitted to water tube boilers.

3.3 DESIGN PRINCIPLES

The design and performance of multi-fuel burners are considerably affected by the properties of the fuels that are fired. The effect of physical state differences between fuels is to increase the complexity of construction of both burner and control equipment from gaseous through liquid to solid fuels while the influence of C/H ratio is evident in combustion, flame and heat transfer characteristics.

The design of multi-fuel burners is based on the integration of individual components derived from familiar gas, oil and coal burners. Consideration here of this aspect is restricted to the design of the burner head of dual-fuel burners.

In general, dual-fuel burners have been developed around existing oil burners. Since the oil atomiser is normally sited on the centreline of the burner assembly, the gas must be introduced in the space between the oil atomiser and the outside of the burner. Similar considerations apply in the much less common case of gas/coal burners in which the pulverised fuel duct lies on the burner axis.

A dual-fuel burner may be regarded as consisting of three parts: the combined burner components; the oil burner components (or coal burner components); and the gas burner components.

The fundamental requirements of automatic control systems suitable for gas firing are the integration of techniques for controlling throughput, air/gas ratio, ignition and flame detection; and with provision for a programmed sequence which includes essential safety interlocks. These topics are discussed in detail in Chapters 9 and 11.

Control schemes for oil and coal are again based on standard practice. Provision must also be made to handle the particular fuel being fired.

Interchange from one fuel to another can be achieved in a variety of ways and with varying degrees of sophistication and this aspect is discussed in Section 3.3.7.

3.3.1 Fuel Properties

Typical values of properties related to burner design are given in Table 3.1 for several fuels.

Table 3.1 Selected Fuel Properties

	Units	Pulverised Coal	Fuel Oil Class				LPG		North Sea Natural Gas
			D	E	F	G	Propane	Butane	
Gross specific enthalpy	MJ/kg	36.6	45.6	43.5	43.1	42.9	49.4	50.2	-
	MJ/m^3	-	-	-	-	-	93.9	117.8	38.6
Density	kg/m^3	-	830	930	950	960	1.87	2.38	0.72
Viscosity	cSt	-	1.4	8.2	20.0	40.0	-	-	-
Theoretical air required for combustion	m^3/MJ	0.323	0.245	0.248	0.249	0.249	0.253	0.254	0.257
Theoretical wet combustion products:									
- Volume	m^3/MJ	0.238	0.261	0.263	0.264	0.264	0.274	0.274	0.279
- Water dewpoint	°C	38	50	50	50	50	55	54	59
- Acid dewpoint	°C	n/a	121	140	141	143	-	-	-
C/H ratio by mass		20.4	6.5	7.3	7.4	7.5	4.8	4.8	3.0
Flame temperature	°C	1958	2022	2028	2028	2028	2000	2000	1930
Sulphur content	% mass	0.7-1.0	0.5	3.5	2.9	3.3	-	-	-
	ppm	-	-	-	-	-	130	170	3-13

The following comments explain the effects these will have on burner design and appliance performance. Fuel properties are discussed in greater detail in Chapter 1.

Physical state

The difference between the solid, liquid and gaseous states implicit in Table 1 has a profound effect on the design and performance of mixing and combustion systems. Gaseous fuels easily intermix with combustion air, while liquid fuels must be atomised to produce fine droplets before mixing commences. Solid fuels need to be pulverised to fire burner systems. The effect of physical state is to increase the complexity of the burner and ancillary plant, from gaseous through to solid fuels.

Density and specific enthalpy

For a specified heat input the density and specific enthalpy influence both the size of the supply pipework and the storage space required for the fuel. Fuels of high density and high specific enthalpy per unit mass can be supplied through small pipelines and compact burner heads. In the case of pulverised coal this may be offset to some extent since it is necessary to transport the fuel as a suspension in air. In order to achieve the same rates of heat release with gaseous fuels higher supply pressures and/or larger size pipework and ports are necessary.

Storage of the higher density, high specific enthalpy fuels is achieved in a more compact manner, but it must be remembered that natural gas is a piped supply requiring no storage facilities on site.

Viscosity

The higher viscosities of liquid fuels, which increase progressively with grade, necessitate pumping facilities for all liquid fuels, and heating, steam tracing and lagging when the heaviest grades are being used. The pressure in the piped gas supply is normally adequate for transporting the gas through the supply lines.

Air requirements

The air required to achieve stoichiometric combustion is virtually constant for all the fuels when calculated on gross heat input. In practice, increasing proportions of excess air are used with liquid and solid fuels.

Combustion product mass flows

For stoichiometric combustion, total mass flows of products of combustion for all fuels, calculated on an equal heat input basis, differ only marginally. In practice these differences are accentuated by the need to supply excess air up to 5% for gas, 5-20% for liquid fuels and up to 30% for solid fuels.

Flame temperature

Although liquid fuels give the highest adiabatic flame temperature, their greater air requirement results in flame temperatures for gas and oil being very similar. Flame temperatures with coal are lower than with oil or gas.

Dew point and sulphur content

Although water dew points are higher for gaseous fuels because of their low C/H ratio, it is the acid dew point which governs the acceptable flue gas exit temperature. Since natural gas contains a negligible proportion of sulphur there is no acid dew point, consequently lower fuel gas temperatures and stack losses may be possible.

The cleanliness of natural gas flue products particularly in respect of sulphur compounds, implies that considerable reduction in chimney heights could well be achieved, providing both economic and aesthetic benefits, for gas only operation. For dual fuel operation the chimney height would have to be that for oil firing.

Carbon/Hydrogen ratio

The influence of C/H ratio is evident in combustion, flame and heat transfer characteristics. Increasing C/H ratios are associated with changes in the mechanism of combustion. For gaseous fuels intimate mixing of fuel and air is possibly followed by rapid combustion. For fuel oil, particularly the heavier grades, and pulverised coal, combustion is achieved by burning discrete particles in large quantities of air. As a result the amount of excess air required to ensure satisfactory combustion increases gradually from gases through liquids to solids. The flames produced with gaseous fuels are generally short and non-luminous although they radiate in the infra-red region as a result of the greater amount of water vapour in the combustion products. The flames produced from the heavier oils and coal which have higher C/H ratios are longer and more luminous. In these flames radiation appears both in the visible spectrum and from the infra-red region. Coal and oil flames possess high emissivity of the order of 0.65 to 0.80, while gases are essentially non-luminous with emissivities of some 0.3 to 0.35. Consequently, there is less heat transferred by radiation from gaseous flames in the early stages of combustion than from coal and oil flames. It follows that oil and coal flames lose temperature more rapidly than gas flames.

The lower C/H ratio of gaseous fuels does, however, lead to higher non-luminous gas radiation from the water vapour present in the combustion products. The non-luminous emissivity increases as the size of the combustion chamber increases and as the temperature falls. This effect is likely to maintain radiant heat transfer further along the flame. The overall effect of the reduction in luminous radiation from gas flames is to alter the mechanism and pattern of heat transfer, but not necessarily to produce a fall in heat transfer commensurate with the fall in emissivity.

3.3.2 Combined Burner Components

The components common to each fuel system include the air-supply duct which in the case of packaged burners is attached to the burner body, vanes or plates to control the air direction, an air manifold or 'windbox' and a refractory tunnel or quarl. In addition a pilot burner or burners must be provided and carefully located so as to ignite both oil and gas reliably. The pilot burner will be integrated into the control programme of both fuel systems when automatic control is used. Flame detectors also need to be particularly carefully located so that the pilot and main gas and oil (or coal) flames are detected. The fact that two fuel-supply systems are incorporated into one burner housing leads to some degree of mutual interference. The gas system is much more tolerant with regard to mixing and combustion than the alternative system and the effect of introducing a small oil atomizer on the centreline of the gas-supply system presents few problems. Conversely the introduction of a relatively much larger gas supply into an oil burner may interfere with the mixing pattern and the solution has generally been to dispose the gas inlets out towards the periphery of the burner.

3.3.3 Oil Burner Components

These may be classified according to the method by which the oil is atomized, the atomizer serving the function of disintegrating the liquid stream and dispersing the resulting droplets in a controlled way into the airstream. A wide variety of atomizing devices exist and a comprehensive treatment is inappropriate in a book of this kind. Coverage will be restricted to a brief description of the types commonly used in dual-fuel burners.

The large dual-fired packaged burners have in general developed around existing rotary-cup atomizers; some medium-pressure air atomizers have been used and pressure-jet atomizers are used in the small sizes. Register burners, which are normally of larger capacity than packaged burners use steam-blast atomization, medium-pressure air or pressure-jet systems.

Since the liquid fuel droplets must either be vaporised if they are light oils, or individually gasified if they are heavy oils, the rates of mixing with air are less rapid than for gaseous fuels and a greater excess of air is required to ensure complete combustion. The characteristics of the flames produced are largely dependent on both the properties of the fuel and the rate at which mixing can be achieved. By the correct selection of fuel quality and mixing system long flames with progressive heat release along the length or short intense flames can be produced.

Pressure-jet atomizer

Figure 3.1 shows a typical design. The oil enters through angled ports into a swirl chamber and the discharge annulus as a conical sheet which breaks up into droplets. Oil pressures are typically 3.5 to 14 bar but may be as high as 75 bar. The oil viscosity for optimum operation is around 15 cSt. Simple pressure-jet systems suffer from a poor turndown range but this is improved in systems incorporating spill-return in which a constant quantity of oil is fed to the swirl chamber regardless of throughput, maintaining rotation at a high level and thereby holding the film thickness at the final orifice at a minimum level. On turndown the surplus oil is returned to the pump suction. Although the droplet size is small on turndown the spray angle increases and this may cause problems. The major problem, however, is the return of hot oil (on heavy-oil units) to the pump suction which presents maintenance problems.

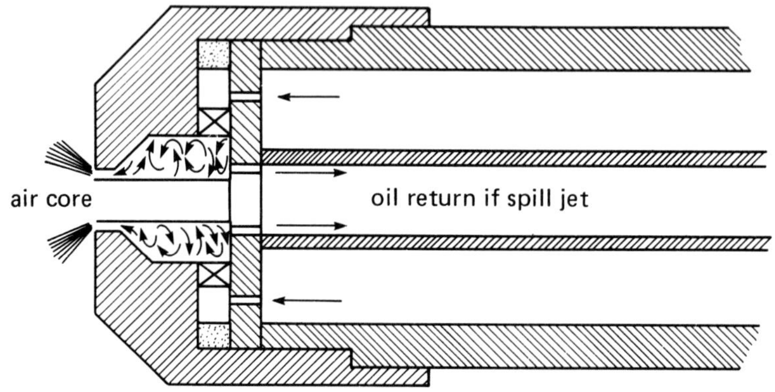

Fig. 3.1 Pressure-jet oil atomizing system

Steam blast atomizers

Some of the earliest atomizers used were steam-blast types in which steam was directed onto the oil jet which it broke up by impact. These rapidly went out of favour with the introduction of the pressure-jet system since the steam consumption was often an appreciable proportion of the boiler output. The design has been greatly improved in recent years and very low steam consumptions (0.2 to 0.5% output) are now possible.

Figure 3.2 shows a modern steam-blast atomizer in which the oil is sheared by the steam blowing across an oil film. The turndown ratio of steam atomizers can be very high (6:1 to 15:1) and the installation is usually cheap since large atomizing air fans or compressors are not required. However the absence of steam on lighting-up complicates matters. This may be overcome by the provision of an air compressor, by the use of a small auxiliary kerosene-fired steam boiler or by the use of a small pressure-jet auxiliary burner.

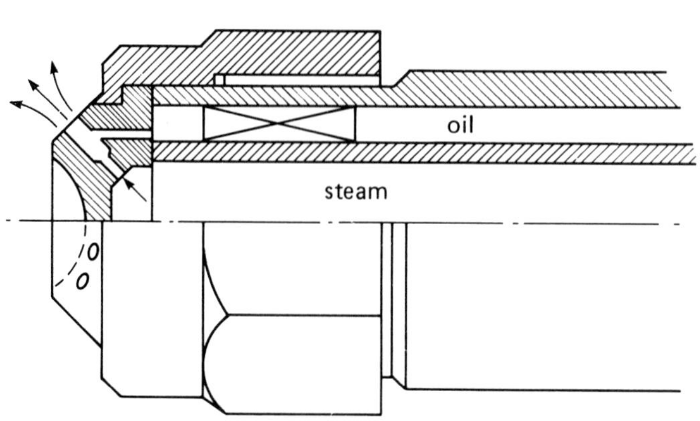

Fig. 3.2 Steam-blast oil atomizing system

Medium and low pressure air atomizers

This is perhaps the most versatile design of atomizer. A wide range of spray angles is possible with a good turndown ratio (5:1 to 10:1). The main disadvantage is the cost of the compressed air, most proprietary designs operating at 0.7 to 1 bar and comprising 1 to 2% of the stoichiometric air requirement. This is satisfactory for relatively small burners but may become expensive in capital and running costs for large units. Higher pressures are in general uneconomical. For shell boilers medium-pressure air burners are technically attractive, particularly for the smaller sizes, since they are capable of achieving high heat-release rates with minimum stack-solids emission and the turndown ratio is good. Figure 3.3 shows a typical design.

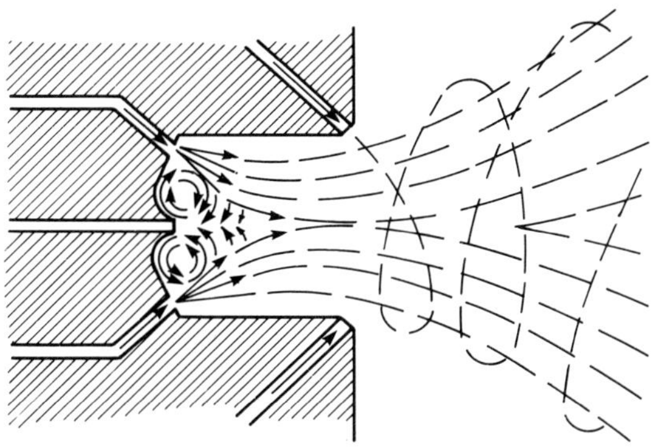

Fig. 3.3 Medium-pressure air atomizer

Low pressure air atomizers use air at around 60 mbar from a single-stage fan for atomization, the atomizing air constituting 15 to 25% of the stoichiometric air requirements. This type is not widely used on boiler plant.

Rotary-cup atomizers

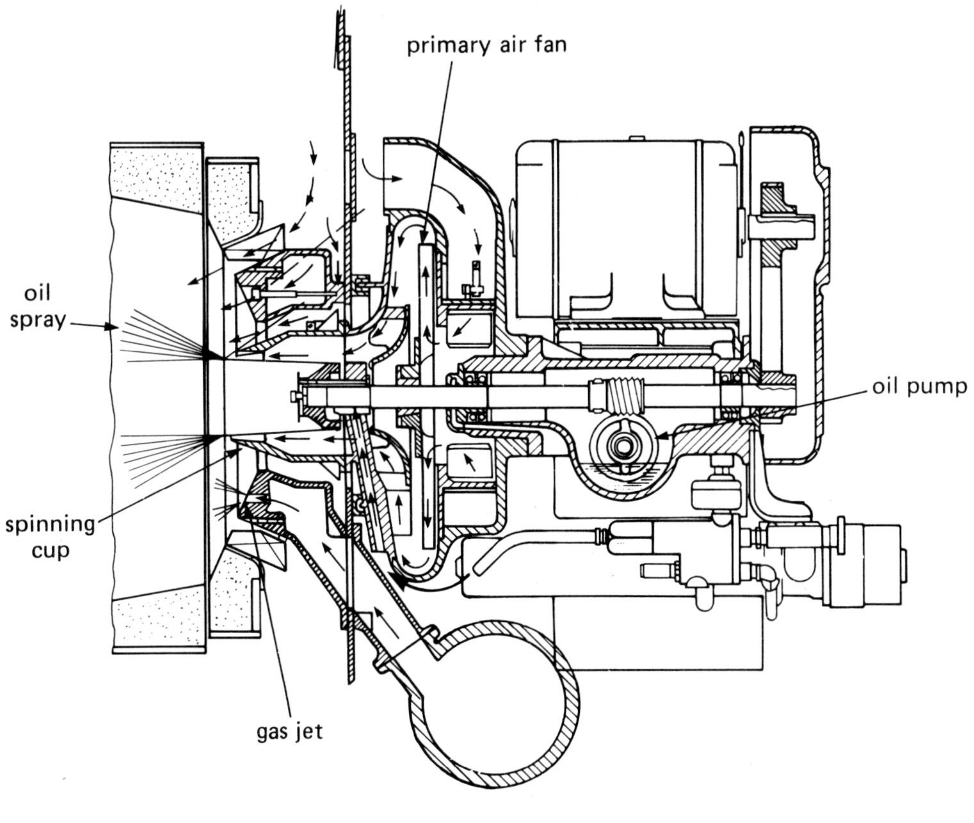

Fig. 3.4 Section through a rotary cup gas/oil burner (Hamworthy)

In these designs the oil is spread onto the surface of a cup rotating at 3 500 to 6 000 rev/min via a hollow shaft. This ensures an even oil distribution and also initiates atomization. Air at low pressure (40 to 50 mbar) from an integral fan atomizes the sheet. Approximately 10 to 20% of the stoichiometric air requirement is provided as atomizing air. The cup may be rotated either by a small air turbine or directly motor driven. Rotary-cup burners have advantages of low cost, relative insensitivity to oil viscosity and good combustion performance. The main disadvantage is that the cup has to be cleaned regularly which makes uninterrupted running difficult. Figure 3.4 shows a typical design.

Stabilisation of Oil Flames

A variety of techniques is employed, all directed towards obtaining a region of low velocity and some degree of reverse flow. The simplest arrangement is a bluff-body baffle mounted just upstream of the atomizer head. If the baffle is sufficiently large, adequate recirculation is created to provide flame stability. Flames tend to be long and narrow, however, and fuel is drawn back onto the face of the baffle possibly leading to overheating and carbon deposition. This may be overcome by perforating the baffle but if carried to excess the reversal zone may be destroyed.

Axial swirlers are sometimes employed in which the air passes through a baffle incorporating a swirl causing a reversal zone centralised on the axis of the burner. The diverging air flow from the swirler also diverges the air flowing around the swirler and increases the diameter of the recirculation zone to greater than the baffle diameter causing the flame to become wider. Radial swirlers incorporate no baffle as such but direct the air tangentially towards the burner axis some distance upstream of the oil nozzle. The final exit path of the air depends on the angle and number of swirl vanes and the air velocity. This provides a method of controlling the flame shape but has the disadvantage on large sizes that the reverse-flow region tends to be wide, extending upstream towards the swirl vanes.

In the case of rotary-cup burners the oil leaves in the form of a hollow cone and reversal may take place within the cone to provide stabilisation.

3.3.4 Gas Burner Components

Gas burner systems used in dual-fuel burners are basically of the nozzle-mixing type in which mixing and combustion take place simultaneously, usually in a refractory quarl; in some types there is partial pre-mixing, gas entering a proportion of the combustion air just prior to the flame base. Unlike normal nozzle-mixing gas burners in which a number of options are open with regard to the introduction of gas and air, in dual-fuel applications there are considerable limitations. As mentioned above, the oil atomizer invariably lies on the centreline and is therefore normally in the centre of the stabilising zone. The gas can be introduced in the centre using an interchangeable gun or an annular supply but this is not possible, for example, with a rotary-cup burner and in other designs central gas feed becomes more difficult as the burner capacity increases. Thus, as previously noted, the gas is commonly introduced by means of ring manifolds or via 'spuds' in the space between the atomizer and the outer periphery of the register. In this position the air flow will probably have a poor stabilising effect and it becomes necessary to introduce local stabilisation means or to use low efflux velocities and ensure good mixing by careful positioning of the gas jets.

A wide variety of individual burner designs spanning a range of throughputs from 30 kW to 60 MW is available. Mixing of gas and air is achieved by axial co-current streams or jets, by swirling the gas or air, or by a combination of these methods.

In the rotary cup burner illustrated in Figure 3.4, gas enters the secondary air cone via a ring of nozzles and is thus partially pre-mixed before emerging from the quarl and encountering the remaining (tertiary) air; the primary air is for supplementary oil atomization only.

The burner in Figure 3.5 illustrates axial mixing and is essentially a tunnel-mixing gas burner with the addition of a low-pressure air atomizer in the centre. Air flows through the outer tube and gas through the inner one. Additional mixing is desirable to achieve flame stabilisation and this is produced by a series of small orifices upstream of the main air/gas interface. It is a combination burner in that it can burn gas and oil separately or simultaneously. It is restricted in use to Class E light distillate fuel oils.

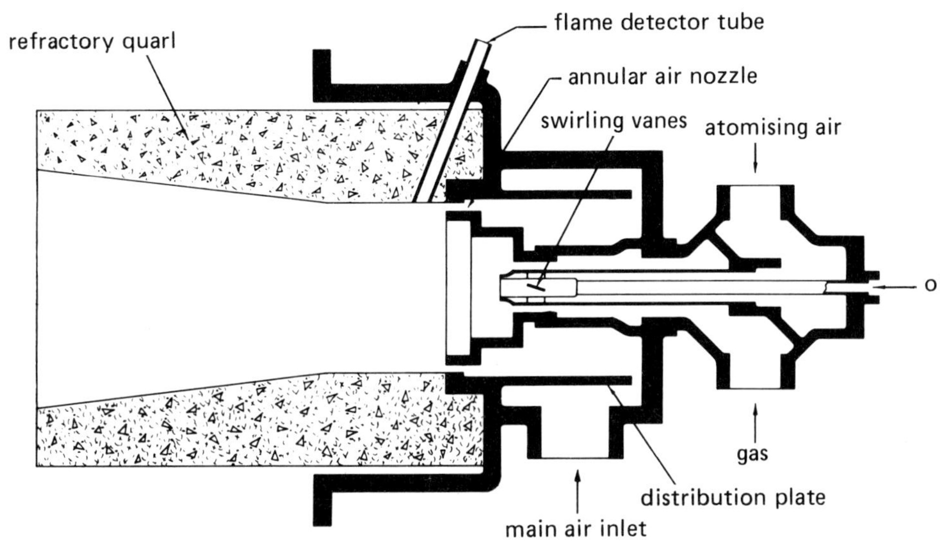

Fig. 3.5 Dual-fuel burner employing an axial annular air nozzle

In the burner illustrated in Figure 3.6 the gas stream rather than the air stream is broken into a series of separate jets to provide the increased surface area for mixing. A series of widely spaced nozzles or spuds with individual baffles surround an oil atomizer with axial swirl.

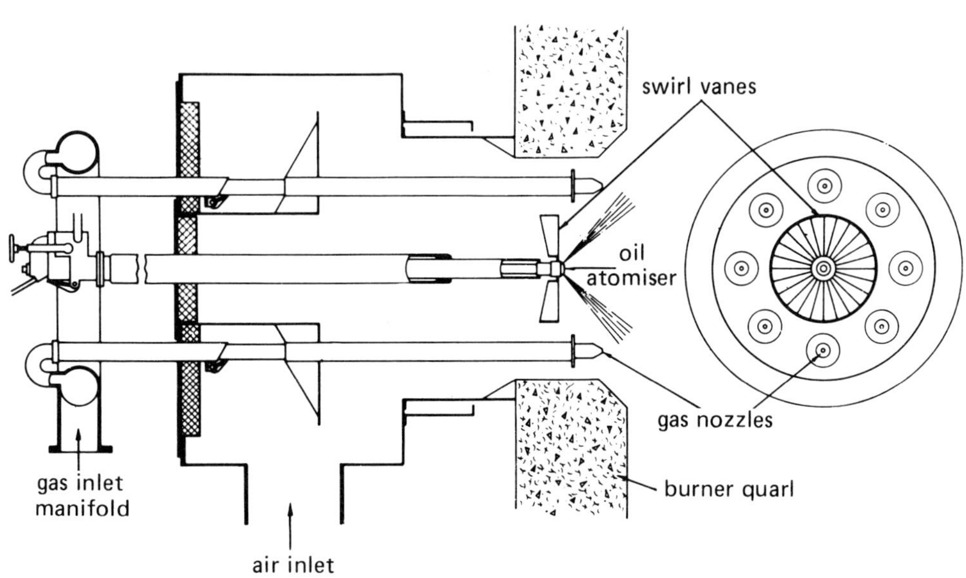

Fig. 3.6 Dual-fuel burner utilising multiple gas nozzles and axial swirl

The intensity of combustion achieved using axial mixing is controlled by the interface area between the two streams and the relative velocities of the gas and the air. Long lambent flames are produced when minimum surface area between streams are used and when relative velocities are low. Increases in interfacial area and relative velocities tend to result in shorter, more intense flames. In general flames tend to be shorter with gas than with other fuels.

Gas flames have several attractive features for boiler firing since they result in rapid and complete combustion with low excess air, eliminating the loss in stack solids, and sooting up in the boiler tubes. The maintenance cost on the boiler and installation as a whole is less than with oil. However the non-luminous flame reduces heat transfer by radiation in the fire tube, but convective transfer in the second and third pass tubes improves with the higher backend temperature to make the overall heat transfer with gas and oil comparable.

Flame stability using axial jets for mixing is achieved either by employing very low velocities through the burner or by producing recirculation eddies at the base of individual gas or air jets. Stability may to some extent be enhanced as the refractory quarl becomes incandescent.

The second important method by which mixing is achieved is to swirl either the air or the gas stream. In practice it is usual to swirl the air stream, and to situate the gas nozzles so that gas is drawn into the recirculating air stream. Properly applied, swirl can not only enhance mixing of the air and gas, it can also control the shape of the flame, depending on the strength of the swirl. At the same time, it creates a large scale recirculation zone which provides a high degree of flame stabilisation.

The burner shown in Figure 3.7 will handle all oil grades and natural gas or equivalents. Gas is admitted via radial jets through the quarl into the central cone where air swirled round the oil burner nozzle promotes rapid mixing. The remaining air requirement is provided by an outer ring of angled jets through the quarl. The burner has a turn-down range of up to 10:1 and can accept preheated combustion air up to 300 °C in the standard version. It may also be used as a combination burner.

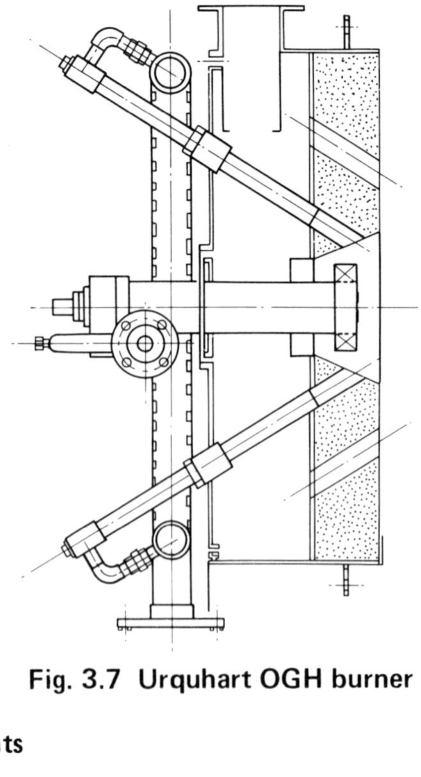

Fig. 3.7 Urquhart OGH burner

3.3.5 Coal Burner Components

Coal for dual-fuel operation is invariably introduced as pulverised coal (pulverised fuel, PF) which is transported in a primary stream of high velocity air as shown in Figure 3.8. Although the need to transport the coal in the primary air stream leads to some degree of premixing, it is usual to supply

20 to 30% excess air to achieve satisfactory combustion. The flames tend to be both long and luminous and carry over of solid matter, either ash or unburnt fuel, is significant.

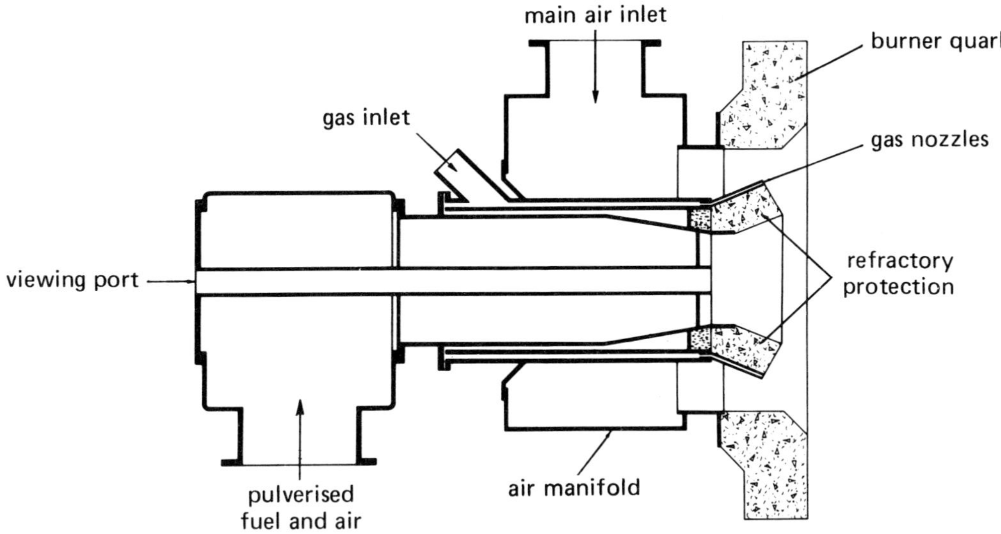

Fig. 3.8 Gas/coal dual-fuel burner

3.3.6 Air Registers

The purpose of the air register is to provide a metering orifice, a method of flame stabilisation and, coupled with the spray angle of the atomizer, a method of controlling mixing and flame shape. Although air registers differ in detail there are two basic forms.

Radial swirl type

The type shown in Figure 3.9 is the classical form, used in the USA and until recently almost exclusively used on the Continent. This form is also used for PF firing. The air is swirled by radial swirl vanes, passes through a convergent throat and diverges via a quarl. There is a stabilising disc on the register axis around the atomiser, which may be bladed but is usually solid except for a few holes or slots. It is claimed that by adjusting the position of the swirl blades, a variation of air swirl can be obtained and thus a wide adjustment in flame shape. High stability can be achieved at low turndown by closing the vanes and increasing the swirl. The burner position is normally adjustable.

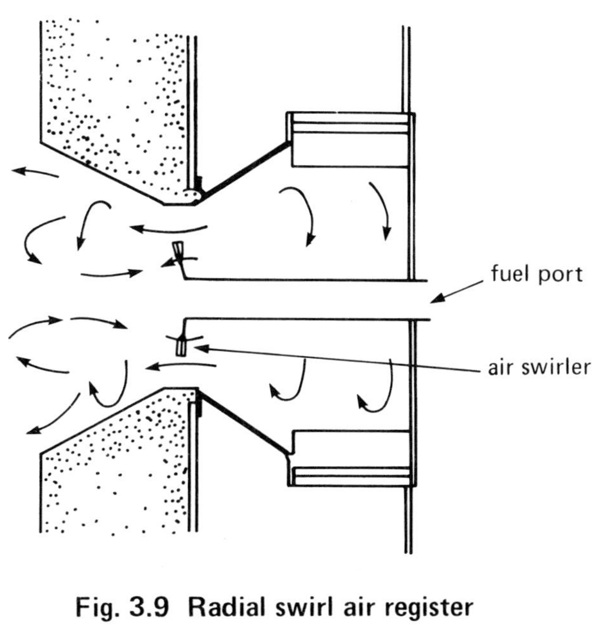

Fig. 3.9 Radial swirl air register

The very features which make this register so flexible are those which, left in the hands of the operators, can produce unwanted effects. In multiple burner installations, for example, if register swirl vanes are put into different positions, the overall coefficient of discharge (C_d) of the registers will vary and hence uneven air flow results. With a high degree of swirl, the reverse flow generated on the register axis causes carbon to form on the stabiliser and also causes overheating of the register. If the degree of swirl is varied and the atomizer spray angle is not matched, it is not possible to achieve constant flow patterns over a wide turndown; however by correct selection it can be done.

This type of register does have the advantage that flame shape, particularly length, is adjustable to some degree and it requires only a shallow windbox, a dubious advantage if looking for very low excess air, although in France registers of this type are achieving flue gas O_2 levels of 0.1% to 0.4% on some large power stations.

Axial swirl type

The register shown in Figure 3.10 is a more recent development and this type with its variants are currently in favour in the UK and the Continent, particularly in Germany.

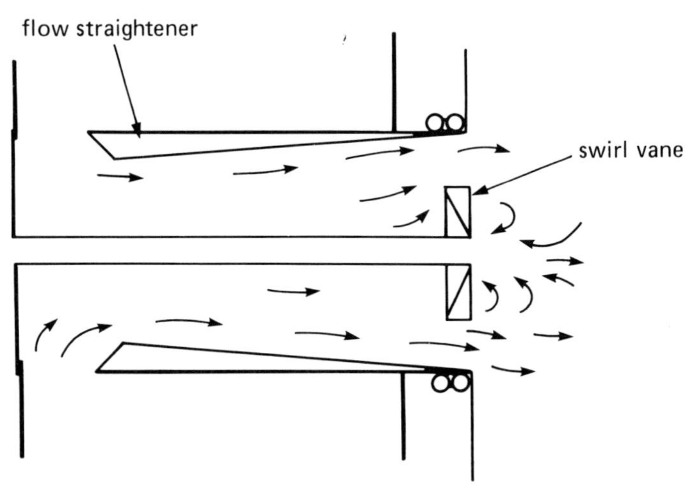

Fig. 3.10 Axial swirl air register

The main feature is that stability and mixing are largely controlled by the axially mounted bladed stabiliser. No quarl is necessary. The C_d is relatively high, enabling easier tube bending. All positions are fixed for any particular design and accurate calibration is possible. The stabiliser throat/diameter ratio can be varied, and with correct selection of spray angle provides some degree of flame diameter variation, as does the final leaving angle of the air register exit. The flow pattern from this register is constant over a wide turndown range. The length/diameter ratio of the register is quite important and this has to be such that flow is fully developed by the time the air passes from the entry to the stabiliser. The air flow in the register is substantially without swirl and in fact most of the problems occur on this type of register if the air from the windbox enters the register with swirling flow, thus starving the centre of the stabiliser. This is particularly likely to happen in individual windboxes where entry conditions are not good and air is almost bound to feed up one side of the casing.

3.3.7 Fuel Changeover

The interchange from oil to gas and vice versa can be achieved in several ways with varying degrees of sophistication. Two basic methods are employed. One is to shut the burner off completely and then to restart on the alternative fuel, while the other is to reduce the flow of one fuel gradually while the flow of the alternative fuel is increased so that there is a short period of overlap of fuels during interchange. Where continuity of heating must be guaranteed, the latter method employing overlapping of the two fuel supplies is favoured, provided adequate precautions are taken to ensure smooth cross ignition and to prevent the burner from running fuel rich.

The rapidity with which changeover can be attained is dependent on the grade of the fuel oil and the readiness of the ancillary plant. With light fuel oils or LPG which could be used as alternative fuels to natural gas, changeover can be made rapidly. With heavier fuel oils rapid interchange can only be achieved if the oil has previously been heated and circulated so that easily pumped low viscosity oil is readily available throughout the system.

The control of interchange may be achieved by entirely manual supervision (i.e. by the operator physically shutting off the fuel to one system, purging and carrying out the ignition sequence on the other fuel), or by changing a switch which then initiates the shut down of one fuel and the automatic ignition of the alternative fuel, or by a fully automatic interchange, resulting from a signal to change fuels.

When the changeover is manual, and in installations in which the oil pump, flange mounted onto the blower driving motor, runs continuously during gas operation, an effective means of protection against the injection of oil into burning gas for this working phase must be used (due to the risk of combustion with air deficiency) and also during the pre-ventilation period.

The signal to change over depends on the type of interruptible load considered being either emergency relief for peak demand or a predetermined programme of demand. For a predetermined programmed demand, a manually operated or timeswitch system can be used, while for emergency or peak relief a remote signalling impulse, as may be derived from a telephone or computer linked network, or selector relay, can be employed. Another method which has been used for peak relief is to connect a pressure switch to the gas supply so that a reduction in pressure below a predetermined level can initiate changeover. An outside air temperature sensing element can also be used.

3.4 BURNER CHARACTERISTICS AND PERFORMANCE

3.4.1 Combustion performance

Satisfactory combustion is attained when the flue products are clean and non toxic, and contain the minimum excess air so that unnecessary sensible heat losses are avoided. Furthermore, the combustion system should not make an excessive noise. The quality of combustion is controlled by the proportioning of fuel and air, mixing and combustion in the burner head, and the available space in which combustion can be completed.

In addition to field experience, tests have been carried out on a variety of burner designs covering a range of ratings from 3.5 MW to 29 MW with regard to flame establishment, stability and off-ratio operation, flame shape and size, combustion quality and minimum excess air requirements.

In general, flame establishment is satisfactory, although in the case of register burners using widely spaced gas spuds the location of the flame was critical.

Good air and gas distribution are extremely important. Considerable stratification may occur in shell boilers at the end of the firetube, where combustion should be complete, if the gas and air distribution at the burner is poor. Thus normal oxygen levels can be present in one quadrant at the end of a circular fire tube, with CO in other quadrants.

Figure 3.11 shows the relationship between the excess air supplied and the CO level in the combustion products on gas firing for a range of burners.

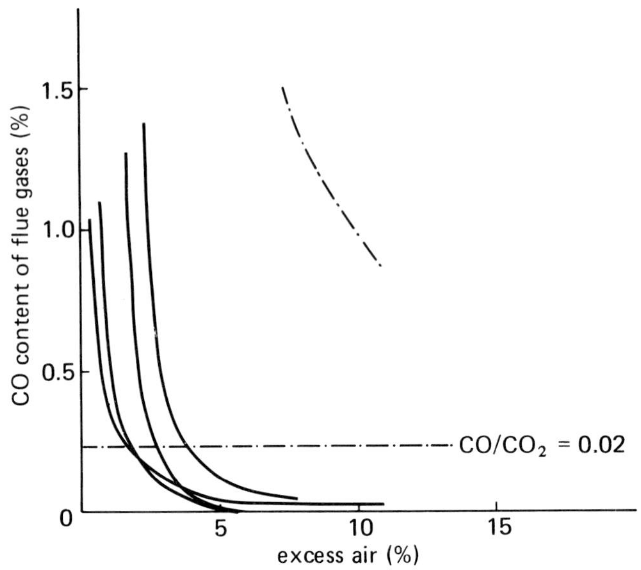

Fig. 3.11 Combustion quality and minimum excess air requirement for gas firing

There is a characteristic sharp increase in CO as the excess air is reduced below a certain value and this takes place over a small change in excess air (3 to 5%).

Similar considerations apply in the case of oil firing in which the performance is normally indicated in terms of stack solids and combustible gases as shown in Figure 3.12. For large water-tube boilers it is possible to obtain less than 0.2% solids emission at 1.5 to 2.5% excess air but for most burners used in industrial water-tube boilers a more realistic range would be 2.5 to 10% excess air.

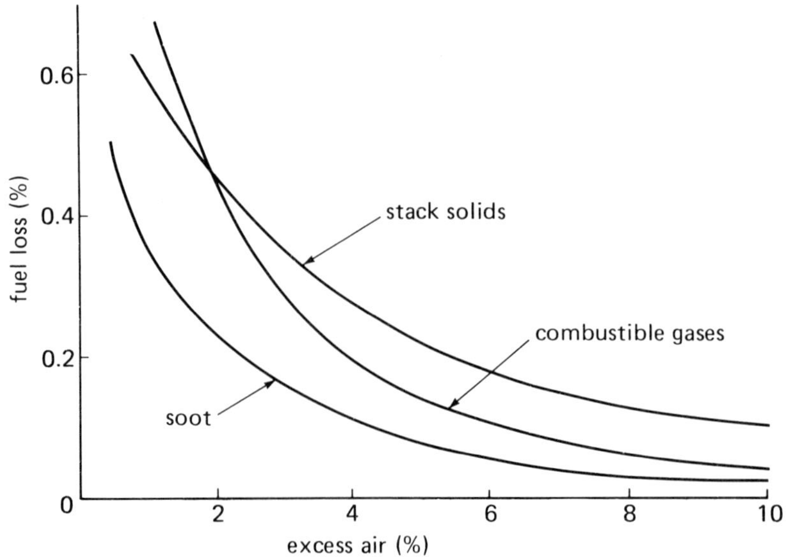

Fig. 3.12 Combustion quality and minimum excess air requirement on oil firing of water-tube boilers using register burners

Clearly, in addition to the burner head design the accuracy of the air-gas ratio control system exerts an important influence on the selection of suitable excess air levels.

Operating excess air levels for a variety of burner head/air-gas ratio control combinations for gas firing range from 3.5 to 8.5% for register burners with closed-loop control to 10 to 18% for packaged burners with linked valves. On oil firing with good designs of register burners it is possible to operate at the same excess air levels as gas but in most packaged systems higher excess levels are required (15 to 30% at maximum load). The difference in excess air requirement between oil and gas on this type of burner presents difficulties in dual-fuel operation. The excess air level may be set for gas and the burner downrated on oil firing, the excess air may be set for oil in which case the performance would be degraded for most of the time, or, as a more satisfactory long term solution, the ratio-control linkages may be designed to set the appropriate excess air level for each fuel automatically.

A higher excess air level is required with oil because combustion is more difficult to achieve with liquid fuels and because variations in oil temperature can lead to a smoky stack. Even more excess air is required for coal flames.

Mixing and combustion are affected by both the quality of the fuel and the design of the burner head. Rapid combustion can readily be achieved with air/gas flames with simple designs of burner while increasing complexity is required for either oil or coal.

Both oil and gas flames can be controlled to give either short intense flames or long lambent flames. Coal flames are generally larger than those for oil or gas. Even the smallest fire tubes with combustion intensities up to 2 MW/m^3 are adequate to ensure complete combustion provided a satisfactory burner has been selected.

When any fuel burns, there is always the possibility that noise may be generated. In dual-fuel systems three types of noise could occur: resonant, combustion and fan noise. The most disturbing is resonant noise which is most prevalent with the high intensity combustion systems and originates from the interaction of burner and boiler. Although alterations in combustion geometry can eliminanate this noise it is usually more practical in existing boilers to reduce the intensity of combustion by reducing the rate of mixing. Combustion noise, although much less of a problem, occasionally causes a nuisance in particularly quiet environments. Reductions in combustion intensity again reduce noise levels. It should be noted that the requirements for complete and rapid combustion and the suppression of noise are in conflict, since the former is favoured by high intensity mixing and combustion while the latter is more suited to slower mixing. Experience over recent years has enabled burner manufacturers to develop designs with which it is possible to achieve satisfactory combustion quality and avoid excessive noise. Air fan noise, which frequently exceeds combustion noise, can be minimised by correct choice and siting of the fan or by conventional silencing techniques.

3.4.2 Burner Turn-down Ratio

The range of input rates within which a burner will operate is specified by the burner turn-down ratio. This is the ratio of the maximum to minimum heat input rates with which the burner will operate satisfactorily. For any burner with fixed air orifices, the turn-down ratio is also the square root of the ratio of maximum to minimum pressure drops across the orifice or burner head. For example, if the maximum supply pressure is 30 mbar and the minimum is 0.75 mbar, then the turn-down ratio is $(30/0.75)^{0.5}$ = 6.3:1. Naturally, limitations on the fuel supply pressure can limit the fuel flow before the maximum air capacity of the burner is reached.

The maximum input rate is limited by a phenomenon known as flame lift-off (which results from mixture velocity exceeding the flame velocity) and by the cost of equipment for developing higher pressures. The minimum input rate is limited by the phenomenon known as flash-back (which results from the flame velocity exceeding the mixture velocity) and by the minimum flow at which the ratio control equipment will function. The former limitation applies to premixing, but not nozzle mixing burners. For low pressure air atomising oil burners, the minimum firing rate is equivalent to the atomising air supply rate. This limits the turndown range of these burners. For example, if the atomising air flow for an oil burner is 35 m^3/h and the required turndown ratio is 5 to 1, then the maximum firing rate must be 5 x 35 = 175 m^3/h of air (combustion and atomising air combined).

Stability is another important characteristic of burners. A stable burner is one which will maintain ignition when cold and at the pressures and ratios ordinarily used. No burner is considered stable merely because it is equipped with a pilot.

3.4.3 Flame Shape and Combustion

For a given burner, changes in the mixture pressure or the amount of primary air will affect the flame shape. For most burner types, an increase in mixture pressure will broaden the flame, and an increase in the per cent primary air will shorten the flame, where the fuel input rate remains the same.

Burner design has much more effect upon flame length and shape than either of the above operating variables. Good mixing, produced by a high degree of turbulence and high velocities, produces a short bushy flame, whereas poor mixing (delayed mixing) and low velocities result in long slender flames. In oil burners, high atomising air pressure may tend to throw the oil flame away from the burner nozzle before it can be heated to its ignition temperature, and thus lengthen the flame. Turbulence and good mixing may be promoted by the use of vanes in the atomising and combustion air streams to impart a swirl to the air.

Boilers

Efficient combustion in any type of steam generator is ensured only in so far as the furnace dimensions are determined so that the flame is formed completely before the products sweep the heating surfaces. In a shell boiler, the volume of the furnace is conditioned by the restrictive diameter of the furnace tube, the flame being compressed into a tube determined by the fact that it is designed within a shell and not by the requirements of flame propagation or combustion conditions. The time for combustion is very short.

A furnace should therefore be suitable for the size and shape of the flame produced and vice versa but an oil jet may enter the furnace under the influence of centrifugal or swirling action, so that the droplets tend to fly outwards toward the cold wall of a flame tube. It is thus necessary to shape the flame by the momentum of the combustion air, bend the natural trajectory of the oil droplets and compress the formation into a flame tube the diameter of which was determined by manufacturing techniques and not by the requirement of flame development. There is not much opportunity for carrying this out within the diameter of the furnace, and the viscosities of air and flame at 1 000 °C and their influence upon adequate turbulence do not make it any easier.

These problems are accentuated in the case when a shell boiler, for example, is highly rated but they are eliminated when it is externally fired or when the boiler design lends itself to fuel oil firing because of the complete adaptability of the furnace, as in the case of the corner-tube boiler for relatively small capacities, and all types of water-tube boiler for large installations.

When burning liquid fuel, the main requirements are: that the fuel shall burn without the flame touching the walls of the furnace or flame tube; that in the case of shell boilers, the fuel shall burn within the length of the flame tube, and the temperature of the gases leaving the flame tube shall not exceed 1 200 °C in order to avoid possible damage to the back tube plate; that in burning, the fuel shall give up the maximum heat possible to heating surface; and that the gases do not remain in a flame tube at a temperature below which they would be more usefully employed in smoke tubes.

3.4.4 Air-fuel Mixing and Air Requirements

The air used through a burner is known as primary air and that used to complete combustion is termed secondary air, tertiary air etc. depending on the number of air stages. Considerable skill is called for in the design of air control details and in their operation, for the quantity of air admitted in the early stages of combustion may be locally excessive and so tend to quench combustion. Air admission may also be too late in that temperature conditions are less favourable for rapid combustion. The correct quantity of air admitted too early or too late could result in restricted combustion and smoke formation. These considerations are the more important in the case of the pressure jet burner because of the absence of primary air.

The desirability of a uniform mixture strength is paramount and burner design is aimed at ensuring that sufficient air is available to the centre core of the flame. Any region of flame which suffers from oxygen deficiency, or where chilling occurs, produces oil globules which are substantially more difficult to burn.

The amount of air theoretically required for the combustion of gas and oil varies with the chemical composition of the fuel but is, on average, between 13 and 16 kg/kg of fuel, that is approximately 11 to 13 m^3/kg fuel. In small or shell boiler practice the efficiency of the plant is considered to be average if the excess air level is below 25%.

However, this is only a guide to excess air levels and they can vary according to the quality of the burner and the throughput control systems, the fuel being burnt, the firing rate and the boiler itself.

A most important part of the installation as far as furnace efficiency is concerned is the method by which air is admitted to the furnace. An air register must impart a swirling motion to the air so that it takes a whirling or spiral path, entraining the particles of oil which follow a similar path and thus creating a good mixing pattern.

Most air registers have throats of fixed dimensions and if the air velocity at high load is adequate and turbulent, then at 30% load it often is not. Air registers consisting of a series of overlapping vanes impart spin to the air passing through. A diffuser is mounted at the furnace end of the burner assembly to protect the oil spray from the high velocity air in order to stabilise the flame and prevent it from being blown away from the burner tip.

The entry opening into the furnace is usually of refractory material although in large boiler installations, water-cooled throats are being installed to reduce refractory maintenance.

The important consideration in all oil firing equipment is that the most frequent cause of smoke is cooling off before combustion is complete, and that it is necessary to prevent the contact of flame with surfaces capable of cooling it before the carbon has been so far consumed as to prevent the formation of smoke. It is unnecessary that, before the cooling surface is reached, complete combustion should have taken place but that the carbon should have been burned to carbon monoxide, which is incapable in this context of smoke formation in its subsequent conversion to carbon dioxide.

Regarding the admission of excess air, in the first place it is doubtful whether it is possible to maintain some types of boiler without air leaks. This applies particularly to large water tube boilers in which attempts to achieve combustion with low excess air are defeated if the boiler casing allows air to leak in, especially in the combustion zone. The corners of a square or rectangular casing are often particularly difficult to seal so that very many small gaps caused by differential expansion appear.

With oil firing, if the flame is incandescent, white and transparent, considerable excess air is present. As the quantity of excess air is reduced, the colour of the flame at the rear of the furnace becomes pale yellow, then yellowish-orange and orange-red. In general, with an efficient installation working with a minimum quantity of excess air, the ends of the flame should be yellowish-orange or golden in colour.

At very high rates of combustion, when the flame completely fills the furnace, the very high furnace temperatures preclude the presence of these temperature colours. An incandescent and dazzling whiteness still indicates excess air, but a reduction in this excess is only indicated in the furnace by softening in the intensity of the white flame.

The presence of smoke does not necessarily mean insufficient air. With an inefficient installation poor atomisation, poor mixture of oil and air, air entering the furnace at the wrong place and cooling and unconsumed gases, unconsumed oil striking cooling surfaces, the presence of water in the oil, may all cause smoke, even when the air supplied is far in excess of that theoretically required.

A non-luminous flame gives little radiant heat while the incandescent particles in a luminous flame give a very considerable amount. It is probably best in order to facilitate the transmission of a portion of the heat as radiant heat that the flame should be luminous and that combustion should be completed by additional air.

However, the question of excess air levels is probably more difficult with natural gas firing since there is no tell-tale smoke, until considerably sub-stoichiometric combustion conditions are reached.

The dangers of operating natural gas fired plant at low excess air levels are extensively discussed in Chapter 10.

Generally then, the excess air levels should be set as low as possible commensurate with safe and reliable operation, in order to:

- Minimise the amount of heat carried out in the waste gases (chimney or stack losses)
- Maintain high flame temperatures when required.
- Minimise, with oil firing, the formation of sulphur trioxide.

However, the excess air levels must not be set so low as to cause any of the following:

- Soot deposition within the boiler and/or chimney emissions, particularly with fuel oil firing.

- Loss of efficiency due to unburnt gases passing to the boiler chimney.
- The possibility, particularly with shell boilers, of structural damage to the boiler itself, due to combustion continuing to take place in the region of the tube plate.

3.5 TRIPLE-FUEL FIRING

Triple-fuel firing is being increasingly catered for in the shell boiler field, with commercially available units capable of burning gas, oil or coal. These are equipped with dual-fuel oil/gas burners, and a fixed or (less commonly) moving grate for coal. As the name burner is conventionally applied only to combustors handling fluid fuels, such triple-fuel boilers cannot be said to be fitted with triple-fuel burners. Boilers of this type are also widely used for burning the more tractable solid wastes, particularly wood offcuts and sawdust generated during furniture manufacture.

Pulverised coal does behave as a fluid, but will rarely be encountered in industry because of the high plant and running costs entailed in producing it. Its use in the UK is predominantly in watertube boilers for large scale power generation, almost invariably as the sole fuel. Some similar boilers in the USA are fitted with true triple-fuel burners which can operate on natural gas, fuel oil or PF.

Less rare in the UK, but still uncommon, are triple-fuel fluidised bed combustors. They have the advantage of requiring little or no fuel preparation before firing and are therefore particularly suited for combustible wastes which are not amenable to conventional firing techniques.

Multi-fuel firing is often encountered where the aim is to establish self-sustaining combustion of the main fuel (after which the ignition burner shuts down) rather than to provide fuel interchangeability. The arrangement typically comprises gas and/or oil ignition burners fitted to, for example, pulverised coal burners, hospital waste incinerators, fluidised bed combustors and wood-waste boilers.

Triple-fuel firing does not necessarily incorporate solid fuel. In integrated steelworks, for example, internally produced as well as externally supplied fuels may be used, with furnace burners capable of firing oil, natural gas and blast furnace or coke oven gas; the relative consumptions of the fuels are adjusted in response to availability and cost to secure the best overall economic advantage.

REFERENCES

BRITISH GAS PLC PUBLICATIONS

IM/18 1st Edition 1982 with Amendment 1986 : Code of Practice for Use of Gas in Low Temperature Plant

BRITISH STANDARDS

BS 5885 : 1988 Part 1 Specification for Automatic Gas Burners of Input Rating 60 kW and above.
BS 5978 : 1983 Safety and Performance of Gas Fired Hot Water Boilers (60 kW to 2 MW input).
BS 6644 : 1986 Installation of Gas Fired Hot Water Boilers of Rated Inputs between 60 MW and 2 MW.

OTHER PUBLICATIONS

Dual Fuel Burners for Boilerplant, I.H.V.E. Symposium 1969. Hoggarth, M.L.

Journal, Institute of Fuel (now Institute of Energy), 1971. Hedley A.B., Naruzzarman A.S.M. and Martin G.F.

Industrial Gas Heating. Priestley J.J. Ernest Benn. London 1973.

The Efficient Use of Energy. Dryden, I.C.G. (Ed). IPC Science and Technology Press 1975.

Industrial Gas Utilisation: Engineering Principles and Practice. Pritchard R, Guy J.J. and Connor N.E. Bowker 1977

Flow of Fluids 4

CHAPTER 4

FLOW OF FLUIDS

4.1 INTRODUCTION

Fluid flow is of major importance to engineers. Virtually all premises, ranging from domestic dwellings to large industrial sites, have one or more fluid flow systems, some self-contained and others like gas and water supplied by local networks, in turn served by regional and national grids: all grist to the working engineer's mill.

The theory of fluid flow is formidable, covered by numerous textbooks (a selection of which is listed in the References) and continually added to as a result of R & D work. This kind of detailed specialised knowledge is of routine interest only to the fundamental designer, engaged for example in the design of a new heat exchanger or in modelling the behaviour of gases in a new type of furnace.

At the practical level, the engineer is required to find answers to questions such as:

- What size pipe is needed to deliver gas from the meter to the appliance at the desired pressure?
- Is the steam distribution system capable of supplying a proposed additional load at its far end?
- Why is the extract ventilation system not performing to specification?

Traditionally, a whole range of empirical aids, rules of thumb, graphs and nomograms, flow calculators and the like is available and routinely used without recourse to basic theory, thus saving a busy engineer's time. Whilst adequate in most cases, this simplification can be overdone and so result in under-performance, over-design, energy wastage, unnecessary cost etc. This has been partly redressed in recent years by the development of programmes for sophisticated calculators and desk-top computers, with the fundamental theory built into the software and offering the desired speed of use.

This chapter, therefore, is for that majority of working engineers and technicians for whom fluid flow is only a part (albeit an important part) of their activities. It seeks to re-familiarise them with the basic theory and provide a more informed background to the use of empirical aids, which will quite properly continue.

4.2 TERMS AND DEFINITIONS

There are three states of matter: solid, liquid and gaseous. Liquids and gases have no permanent shape and if not enclosed they continuously deform; they are therefore termed fluids. When enclosed, fluids adopt the shape of the container, gases expanding to fill the total available volume and liquids maintaining their own volume and presenting a free surface.

'Permanent' gases and 'permanent' liquids are rather imprecise terms based on general experience and are better defined as fluids sufficiently remote from change of state by liquefaction or evaporation as to possess reasonably consistent properties and conform to the basic laws of physics. The properties of fluids under conditions near the change of state have been determined experimentally and reference must then be made to published data for all but the most approximate calculations. Gases near the liquefaction stage are termed vapours.

S.I. units, because they comprise a coherent system based on the minimum number of base units from which all others are derived, have considerably simplified the understanding of technologies such as fluid flow which was previously bedevilled by the ambiguities and contradictions inherent in the foot pound second (fps) system. Some basic terms have, however, become common parlance with consequent retained ambiguity; as their precise use is essential, they are defined below.

Mass and weight are often confused. Mass is expressed in the basic unit kg. Weight is a force that is exerted by gravity on a mass in free fall and is expressed in Newtons. Mass and force are related by the general formula:

$$F = ma \qquad 4.1$$

where F = force, N
m = mass, kg
a = acceleration, m/s^2

The acceleration due to gravity (g), to the degree of accuracy adequate for practical purposes, is equal to 9.81 m/s^2. Formula 4.1 for weight thus becomes $W = 9.81\ m$, i.e. a mass of 1 kg has a weight of 9.81 N.

Density is mass per unit volume in kg/m^3. (The term 'specific weight' should not be used unless it is intended precisely, in which case its value is 9.81 times density). Relative density (formerly called specific gravity) is the ratio of a fluid's density to that of a nominated standard; water for liquids and air for gases are the two most used standards. For example, the relative density of natural gas is about 0.72 (air = 1) and of mercury is 13.6 (water = 1). As fluid volume varies according to temperature and pressure, it is evident that densities must be quoted for given temperature and pressure conditions.

4.3 THE EFFECTS OF TEMPERATURE AND PRESSURE ON FLUID VOLUME

The basic effects are simply stated: volume increases as temperature rises, reduces as pressure rises, and vice versa. When temperature and pressure together vary, the effects are combined. In many cases, physical laws apply and problems can be solved by equations. In others, recourse must be made to tables of properties derived from experimental data. Liquids and gases are affected to differing extent and this difference, as will be seen later, is important for fluid flow.

4.3.1 Liquids

Liquids behave similarly to solids, in that the volumetric changes caused by varying temperature and pressure are usually relatively small. When temperature and pressure differentials are low they can often be ignored altogether and the use of standard densities will suffice; this obviously depends on the properties of the fluid and the degree of accuracy required.

Temperature is the most important variable and its effect soon becomes significant as the range widens. Pressure is rarely significant (i.e. liquids are virtually incompressible) and its effect on volume may be disregarded in the majority of cases. This is well illustrated in the case of water (Figure 4.1), which expands about 4% when raised from 15 °C to 95 °C, whereas to achieve a corresponding volumetric change by compression requires enormous pressures. Water is in most circumstances close to vapourisation or solidification and its properties are not consistent. Exhaustive data on water and steam are contained in the UK Steam Tables, but the simplified tables widely available are usually adequate for day-to-day use.

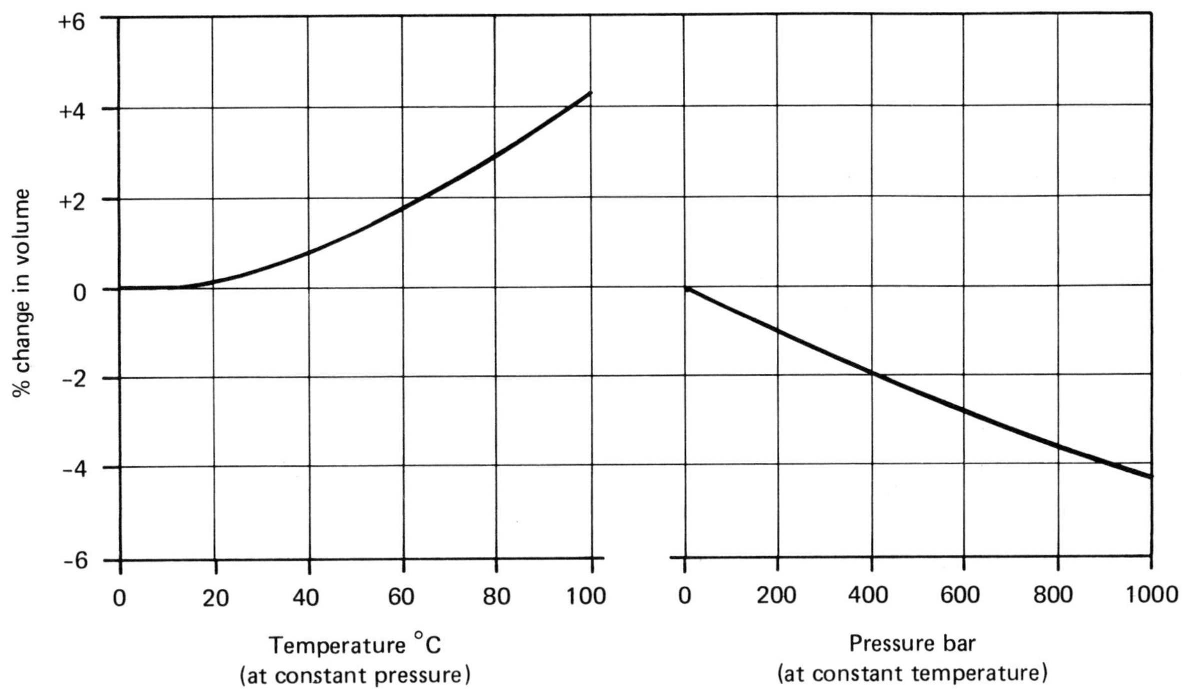

Fig. 4.1 Variation in water volume with temperature and pressure

Example 4.1. A meter in the feedwater line to a steam boiler recorded a flow (q_v), of 7 500 litres per hour at 130 °C and a gauge pressure of 8 bar. What was the mass flow (q_m of water into the boiler?

$$q_m = q_v \times \text{density } (\rho)$$
$$= 7500 \times \rho$$

From the steam tables, which give specific volume (v) rather than density, v for water at 130 °C over a wide pressure range is 1.07 litres per kg.

$$\rho = 1/v$$
$$= 0.935 \text{ kg/l}$$
$$q_m = 7500 \times 0.935$$
$$= \mathbf{7013\ kg/h}$$

With more consistent liquids, calculations may be simplified by using a volume correction factor, i.e. the volume change per degree of temperature variation (positive for temperature rise, negative for temperature drop). Factors may be found by consulting manufacturers' or suppliers' literature, or reference works such as the Chemical Engineers' Handbook.

Example 4.2 The boiler in Example 4.1 is fired by heavy fuel oil which is metered at 40 °C and preheated to 120 °C at the burner. If the meter records a flow (q_v) of 490 l/h, what is the volume flow rate through the burner?

The oil supplier's literature quotes a volume correction factor (α_v) of 0.00068/K. The volume increase from 40 °C to 120 °C is:

$$\Delta V = q_{v15} \times \alpha_v \times \Delta T$$
$$= 490 \times 0.00068 \times (120 - 40)$$
$$= 26.6 \text{ or } 27$$

Volume flow rate at 120 °C is:

$$q_{v120} = q_{v15} + \Delta V$$
$$= 490 + 27$$
$$= \mathbf{517\ l/h.}$$

4.3.2 Gases

The effects of temperature and pressure variation are much greater for gases than for liquids and unless the variation is small, it must be taken into account in calculations. Fortunately, the effects in most cases follow mathematical relationships which greatly simplify procedures. The exceptions are gases near to the liquefaction point and recourse must then be made to published empirical data; steam is the most obvious and important gas in this category.

The basic relationship between volume, temperature and pressure is the Equation of State for gases, derived from gas laws determined experimentally by Boyle and Charles and their development by Avogadro and others as atomic theory took shape. It applies to 'ideal' gases, and while real gases all deviate to some extent, it may be used in calculations to give results to an accuracy adequate for most engineering purposes.

Gas laws and the equations of state

Boyle's Law states that providing the temperature remains constant, the volume (V) of a given mass of gas varies inversely as the absolute pressure (p):

$$V \propto 1/p$$
$$\text{or } Vp = \text{constant}$$

Thus for pressure conditions 1 and 2:

$$V_1 p_1 = V_2 p_2 \qquad 4.2$$

Example 4.3 A compressor takes 7 000 l/min of air at ambient temperature and pressure, raises it to a gauge pressure of 7 bar and aftercools it back to ambient before delivery into the distribution pipework. If the barometric pressure is 1 020 mbar, and assuming no reduction in air moisture content, what is the actual volume of compressed air delivered?

$$V_1 = 7\,000,\ p_1 = 1.02 \text{ bar},\ p_2 = 7 + 1.02 = 8.02 \text{ bar}$$

Using equation 4.2:

$$7\,000 \times 1.02 = V_2 \times 8.02$$
$$V_2 = \mathbf{890\ l/min}$$

Charles' Law states that providing the temperature remains constant, the volume of a given mass of gas varies directly as its absolute temperature (T):

$$V \propto T$$
$$\text{or } V/T = \text{constant}$$

Thus for temperature conditions 1 and 2:

$$V_1/T_1 = V_2/T_2 \qquad 4.3$$

Example 4.4 Gas is metered into a factory at 9 °C. The gas pipework between meter and user passes through areas of high ambient temperature which cause the gas temperature to rise to 28 °C delivered. If 2 950 m^3 is recorded by the meter over a one day period, what would a meter at point of use record, assuming negligible pressure drop?

$$V_1 = 2950, \quad T_1 = 9 + 273 = 282 \text{ K}, \quad T_2 = 28 + 273 = 301 \text{ K}$$

Using equation 4.3:

$$2\,950/282 = V_2/301$$
$$V_2 = \mathbf{3\,150\ m^3} \text{ (to nearest 10 } m^3\text{)}$$

Charles' and Boyle's Laws can be used as a quick mental check to decide whether or not to apply correction in a calculation. At typical atmospheric conditions, for example, it requires a temperature variation of ±3 K or a pressure variation of ±10 mbar to cause a volume change of approximately 1%; few practical engineering problems require this degree of accuracy.

When temperature and pressure vary simultaneously, the two Laws may be used in succession to achieve the desired result, but this is cumbersome and there is a single equation combining volume, temperature and pressure. Its development depends on a fundamental property of ideal gases. This property is based on Avogadro's Hypothesis, which states that under the same pressure and temperature conditions, equal volumes of all gases contain the same number of molecules. Developing and re-stating this in standard units, the relative molecular mass (formerly called 'molecular weight') in grammes of any gas will occupy a volume of 0.02241 m^3 at STP i.e. 273.15 K and 101 325 Pa (1 013.25 mbar). This standard volume is termed the molar volume (V_m) and may be used to obtain an approximation of gas density in the absence of reference data, for example, the relative molecular mass of methane CH_4 is 16, so 16 g will occupy 0.02241 m^3 therefore its ideal density at STP is 16/0.02241 = 714 g/m^3 or 0.714 kg/m^3 which is accurate to better than 1%.

The combined relationship derived from Charles, Boyle and Avogadro can be expressed:

$$pV_m = RT \qquad 4.4$$

where p and T represent STP in Pa and K respectively, and V_m is the molar volume. R is the same for all gases and is termed the Universal Gas Constant, with a numerical value of 8.3144 usually expressed in the dimensions of J/(mol K). Equation 4.4 is termed the Equation of State for gases and in less specific form is shown as:

$$pV = R_G T$$

where R_G = R/M_r, i.e. the Universal Gas Constant divided by the relative molecular mass in consistent units. R_G is the Gas Constant for a particular gas and has a fixed numerical value irrespective of temperature and pressure. The gas constant for air, for example is 287 J/(kg K).

The equation of state is most familiar in the general form:

$$pV/T = \text{constant}$$
$$\text{i.e. } p_1 V_1/T_1 = p_2 V_2/T_2 \qquad 4.5$$

If any five of the variables are known, Equation 4.5 can be used to calculate the remaining unknown. Perhaps its widest applications are for correcting a gas volume to standard conditions or for calculating the density of a gas (defined under standard conditions) at the actual conditions encountered.

Example 4.5 A gas has a quoted density of 1.213 kg/m^3 at STP. What will its density be at 50 °C and an absolute pressure of 2.5 bar?

$V_1 = 1\ m^3$, $p_1 = 1.013$ bar, $T_1 = 273$ K, $p_2 = 2.5$ bar, $T_2 = 50 + 273 = 323$ K

Using equation 4.5:

$$(1.013 \times 1)/273 = (2.5 \times V_2)/323$$
$$V_2 = 0.479\ m^3$$

Mass is unchanged, so V_2 has a mass of 1.213 kg
Density at 50 °C and 2.5 bar is thus:

$$= 1.213/0.479$$
$$= \mathbf{2.532\ kg/m^3}$$

Deviations from the gas laws

At higher pressures real gases progressively deviate from the gas laws. The pressure at which the deviation becomes significant, and the extent of the deviation (which is also temperature related) are specific to the gas under consideration. This deviation is termed compressibility and compressibility factors (z) for the common gases may be found in reference books in tabular form or as families of constant temperature curves plotted against absolute pressure. Figure 4.2 shows 'z' plotted for air. Inspection of such data will indicate whether compressibility needs to be included in any particular calculation.

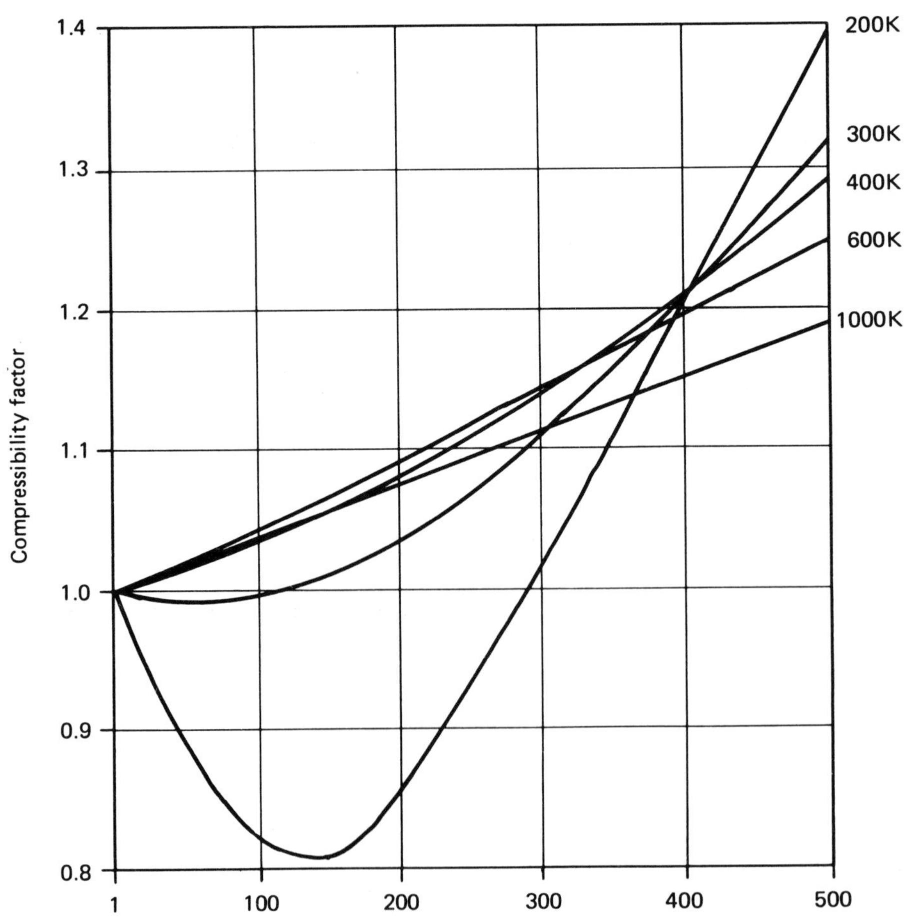

Fig. 4.2 Compressibility Factor for Air

Taking compressibility into account, the equation of state now becomes:

$$pV_m = zRT$$

which gives the complete form of the general equation:

$$p_1 V_1/(T_1 z_1) = p_2 V_2/(T_2 z_2) \quad 4.6$$

When routinely correcting gas volumes to reference temperature and pressure conditions it is convenient to use the compressibility (or deviation) correction factor F_z, which is Z_r/Z_m where Z_r and Z_m are the compressibility factors at reference and metered conditions respectively. For natural gas, for example, the compressibility of which over the atmospheric temperature range becomes significant at gauge pressures over 2 bar, F_z may be rapidly determined from values tabulated in the British Gas publication 'Standard Factors for Temperature and Pressure Correction'.

Example 4.6 If 500 m^3 of dry air at 0 °C is compressed to an absolute pressure of 300 bar and maintained at a temperature of 400 K, what volume will then be occupied?

$p_1 = 0 + 1.01 = 1.01$ (to sufficient accuracy), $V_1 = 500$, $T_1 = 0 + 273 = 273$,

$z_1 = 1$, $p_2 = 300$, $T_2 = 400$, $z_2 = 1.14$ (from Figure 4.2)

Using Equation 4.6:

$$1.01 \times 500/(273 \times 1) = 300 \times V_2/(400 \times 1.14)$$

$$V_2 = \mathbf{2.81\ m^3}$$

Heat, expansion and compression

Extensive experimental work has demonstrated that when ideal gases are expanded or compressed they obey the following equation:

$$pV^n = \text{constant}$$

$$\text{i.e. } p_1 V_1^n = p_2 V_2^n \quad 4.7$$

$$\text{or } p_1/p_2 = (V_2/V_1)^n$$

If combined with the equation of state, it can be shown in the form:

$$p_2/p_1 = (T_2/T_1)^{n/(n-1)}$$

The exponential n is called the polytropic index.

Figure 4.3 shows a cylinder containing a volume of gas, with a piston and a heat source.

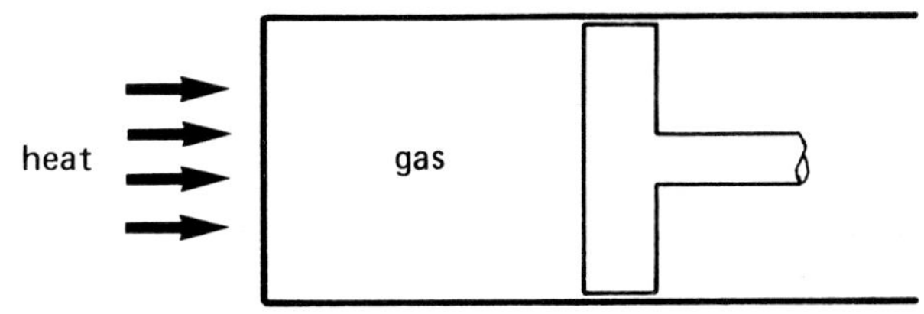

Fig. 4.3

First consider a quantity of heat added to the gas with the piston fixed, i.e. volume is constant and temperature and pressure increase. As the piston has not moved, the gas has done no work and the heat added has all been used to increase the internal energy of the gas:

$$Q = c_v \times m \times \Delta T \qquad 4.8$$

where Q = heat, J
c_v = specific heat capacity at constant volume, J/(kg K)
m = mass of gas, kg
ΔT = temperature rise, K

Next consider the case where heat is added to produce the same temperature rise and the piston allowed to move such that gas pressure remains constant:

$$Q = c_p \times m \times \Delta T \qquad 4.9$$

But this amount of heat is obviously greater than in Equation 4.8 because work has been done in addition to the increase in gas internal energy. c_p is therefore greater than c_v and is termed the specific heat capacity at constant pressure. Effectively, only gases have two specific heats and care must always be exercised to use the correct one in calculations.

When expansion or compression take place at constant temperature (i.e. isothermal) the polytropic index is unity and Equation 4.7 is then identical to Equation 4.2, Boyle's Law.

When expansion or contraction take place with no heat transfer to or from the gas (i.e. isentropic) the polytropic index is the ratio of the two specific heat capacities c_p/c_v, designated γ. Specific heat capacities vary with temperature and pressure and consequently so does γ. Another important relationship is that the difference between c_p and c_v equals the gas constant.

Example 4.7. The specific heat capacity of air is 1005 J/(kg K) at constant pressure and 718 J/(kg K) at constant volume. What is the gas constant for air and what is the isentropic exponent?

Gas constant $R_G = c_p - c_v$
= 1 005 - 718
= **287 J/(kg K)**

Isentropic exponent $\gamma = c_p/c_v$
= 1 005/718
= **1.4** (dimensionless)

4.4 HYDROSTATICS

The science of fluids at rest is termed hydrostatics and its consideration is a necessary preliminary to the study of hydrodynamics, fluids in motion, which is the main purpose of this Chapter.

4.4.1 Pressure

Pressure (p) is defined as force per unit area and is expressed in the standard units of Pascals (Pa) or N/m^2. These units are small and it is often more convenient in practice to use mbar (Pa x 10^2) or bar (Pa x 10^5). Fluid pressure has two important characteristics:

- At any point within the fluid, the pressure is the same in all directions.
- The pressure on the inside walls of the fluid container always acts at right angles to the wall surface.

Absolute and atmospheric pressure

The pressure within a complete vacuum is zero and this is the starting point of the absolute or total pressure scale. This is demonstrable from the definition of pressure as force per unit area:

$$p = F/A$$

from Equation 4.1, $F = ma$

$$\text{so} \quad p = ma/A \qquad 4.10$$

If mass (m) is zero as in a total vacuum, there must obviously be zero pressure.

Consider Figure 4.4, which shows diagrammatically a cross-section through the Earth's atmosphere.

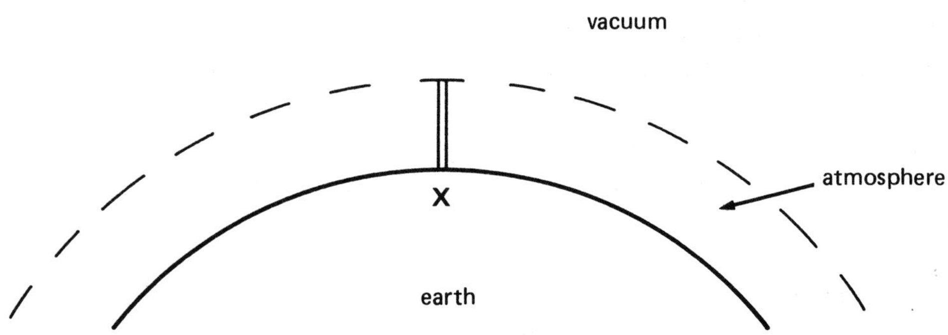

Fig. 4.4

At point X at sea level on the Earth's surface, the total mass of a column of 1 m^2 cross-section extending right through the atmosphere is about 10 200 kg. Using Equation 4.10:

$$m = 10\,200, \quad a = g = 9.81 \text{ m/s}^2, \quad A = 1$$

$$p = (10\,200 \times 9.81)/1$$
$$= \mathbf{100\,000 \text{ Pa or } 1\,000 \text{ mbar}}$$

(Illustrative calculation only; no account taken of the centrifugal force of the Earth's rotation). This pressure is atmospheric or barometric or ambient pressure (p_{amb}) and is the most widely known example of pressure quoted in absolute terms. It varies according to meteorological conditions and is measured by barometer, a pressure gauge which takes a differential measurement against vacuum i.e. zero absolute pressure. It also varies by location, because the Earth is not a perfect sphere and g is not therefore uniformly 9.81 at sea level.

It will be evident from the above that atmospheric pressure decreases with height above the Earth's surface as the height and hence mass of the atmosphere above the reference point becomes less, the effect being enhanced by reducing air density and hence the relationship is not linear. Figure 4.5 shows the percentage reduction of atmospheric pressure at heights of up to 5 km above sea level.

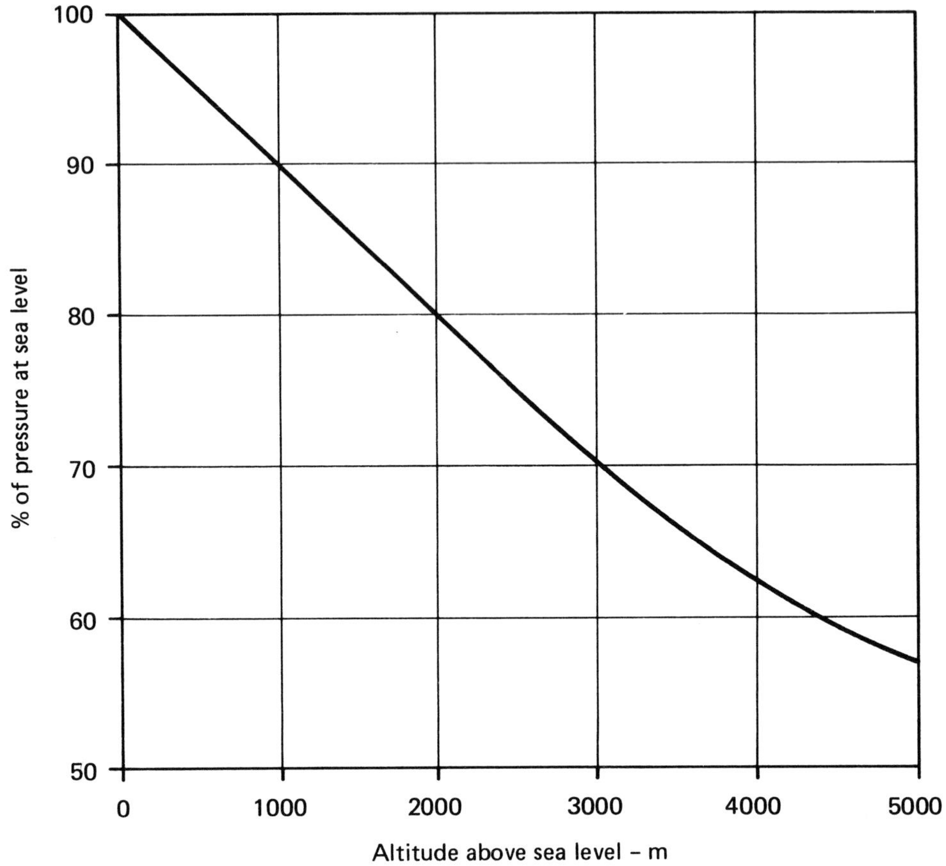

Fig. 4.5 Variation of Barometric Pressure with Altitude

As will be seen from the graph, the relationship does not become significantly non-linear until altitude exceeds 3 000 m and for most practical purposes a useful rule of thumb is to deduct 1% or approximately 10 mbar for each 100 m of height.

Fluids, like all terrestrial objects, are at all times subject to atmospheric pressure and if they are also subjected to further pressure increase or reduction (p_e), their absolute or total pressure (p) will be greater or less than atmospheric. As has already been demonstrated, the use of absolute pressure is essential in many calculations and it is derived by adding the pressure increase or reduction (negative sign) to atmospheric pressure.

Some processes (e.g. distillation) require a controlled absolute pressure and p_e must therefore be regulated to compensate for the variations in atmospheric pressure. Absolute pressure is by definition always positive.

Gauge pressure

Figure 4.6 shows a fluid flowing through a uniform pipe to which is connected a pressure gauge P. Fluid pressure is above atmospheric pressure.

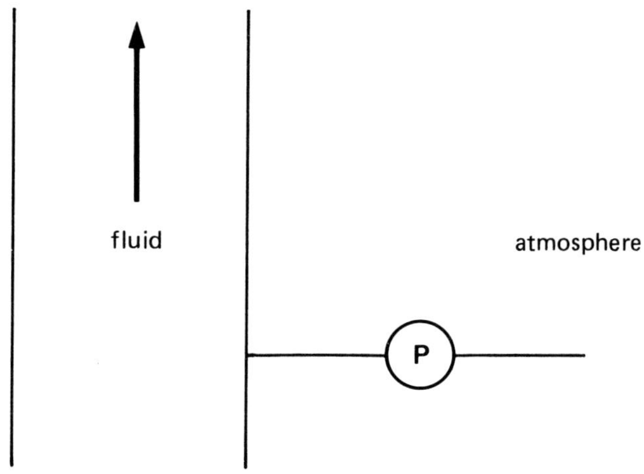

Fig. 4.6

If the outer leg of the pressure gauge is connected to a perfect vacuum, it will record absolute pressure p, i.e. $p_{amb} + p_e$.

If the outer leg of the pressure gauge is left open to atmosphere, the indicated pressure will be added pressure p_e only, termed gauge or static pressure, and is dependent on atmospheric pressure variations. This is the method used in most practical situations. Gauge pressure is positive or negative for pressure added or removed respectively, although negative pressure is sometimes termed suction or vacuum pressure without the minus sign.

4.4.2 Fluid Columns

Consider Figure 4.7, which shows an open-topped vessel with an internal cross-sectional area of 1 m^2, filled with water to a height of 5 m. A pressure gauge with its outer leg open to atmosphere is fitted at each 1 m of depth.

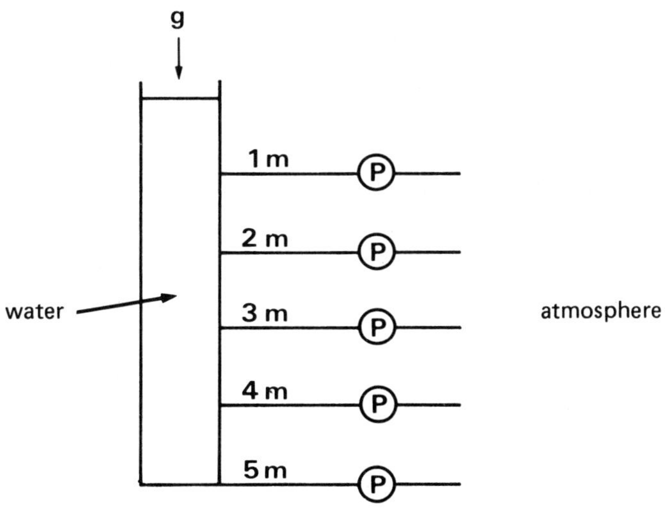

Fig. 4.7

Each 1 m of depth thus has a volume of 1 m^3, which given the density of water as 1 00 kg/m^3 has a mass of 1 000 kg. The force acting on the water is g = 9.81 m/s^2, so from Equation 4.10 the gauge pressure within the water column is:

$$p_e = m\,g/A$$

At 1 m depth, $p_e = 1 \times 1000 \times 9.81/1$

$$= 9810 \text{ Pa}$$

and the pressure at the other depths is similarly determined and plotted against height of water column in Figure 4.8.

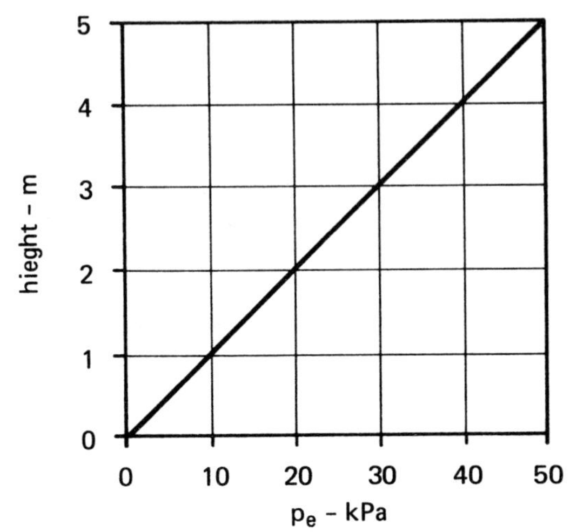

Fig. 4.8

The relationship is clearly linear and illustrates the important point that pressure may also be expressed in units of height. This in turn suggests an alternative form of equation 4.10. The mass of any fluid column is density times volume and as volume is cross-sectional area (A) times height (h) equation 4.10 may be re-written:

$$p = \rho\,g\,A\,h/A$$

i.e. Pressure in Pa $= \rho\,g\,h$ 4.11

In other words, the pressure exerted by a fluid column is independent of its cross-sectional area. The principle is widely applied in the measurement of pressure using liquid-filled manometers. Figure 4.9 shows a simple U-tube manometer connected to a vessel containing gas under pressure.

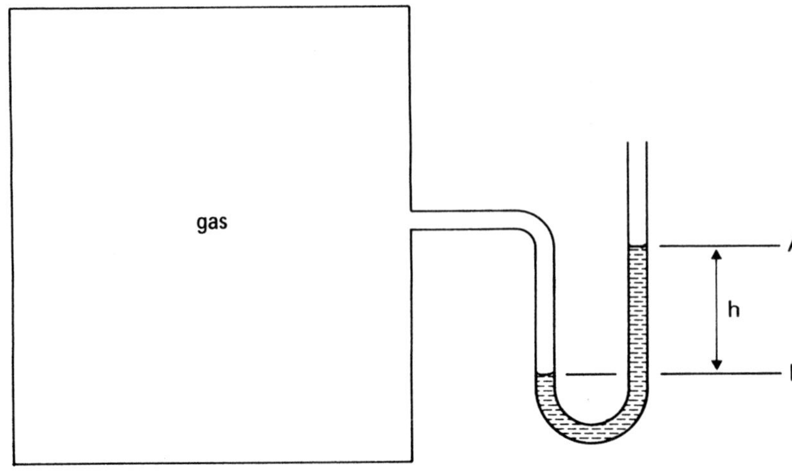

Fig. 4.9

The liquid surface at level A is at atmospheric pressure, and the differential height h between levels A and B is therefore the liquid column equivalent of the gauge pressure in the vessel. If the liquid density is ρ, vessel pressure is ρgh. Strictly speaking, the pressure of the gas column in the left hand leg should also be taken into account, but in the majority of manometer applications the gas density is so much lower than that of the liquid that it is ignored.

Example 4.8 If in Fig. 4.9 the gas is air, the manometer fluid is water of density 1000 kg/m^3 and height h is 0.05 m, what is the gauge pressure?

Using equation 4.11:

$$p_e = 1000 \times 9.81 \times 0.05$$
$$= \mathbf{491\ Pa}$$

Thus a pressure of 491 Pa may be expressed as 0.05 m head of water, or 50 mm H_2O. It will be evident that pressures expressed as height must always specify the column fluid. Had the manometer fluid in Example 4.9 been mercury, for instance, which has a relative density to water of 13.6, height h would have been 0.05/13.6, i.e. 0.0037 m head of mercury or 3.7 mm Hg. Dense fluids like mercury are used for higher pressures to keep manometer dimensions within bounds.

It will also be appreciated that for accurate work the manometer reading must be corrected for any variation in density caused by the effect of temperature on the measuring fluid.

4.4.3 Buoyancy

Figure 4.10 shows a rising main carrying gas at a gauge pressure of 25 mbar at level A. The atmospheric pressure at that level is 1 bar. The absolute pressure of the atmosphere is thus:

$$p = p_{amb} + p_e$$
$$= 100\,000 + 0$$
$$= 100\,000\ \text{Pa}$$

Similarly, the absolute pressure of the gas is:

$$p = 100\,000 + 2500$$
$$= 102\,500\ \text{Pa}$$

It is assumed that the pipe is of uniform bore and has no frictional resistance.

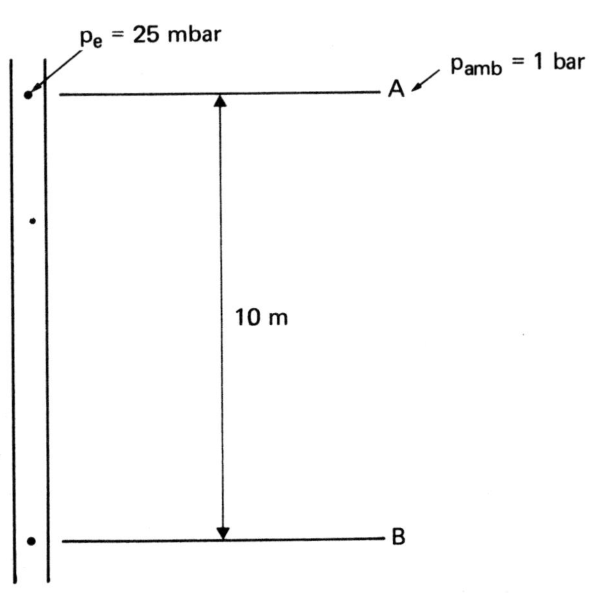

Fig. 4.10

Given air and gas densities (assumed constant) of 1.225 and 0.72 kg/m^3 respectively, pressures 10 m below at level B can be determined using equation 4.11.

Absolute pressure of air = p at A + Δp of the 10 m column
= 100 000 + ρ gh
= 100 000 + (1.225 x 9.81 x 10)
= 100 120 Pa

Similarly, absolute pressure of gas = 102 500 + (0.72 x 9.81 x 10)
= 102 570 Pa.

The gauge pressure of the gas at level B must therefore be:

p_e = p - p_{amb} at B
= 102 570 - 100 120
= 2 450 Pa

which is 50 Pa less than that at level A.

In other words, the gauge pressure of the gas has increased by 50 Pa for a *height* change of 10 m, although its absolute pressure has decreased. This is the basis for the handy rule of thumb that natural gas gauge pressure rises by 0.5 mbar for each 10 m of height. As g and h are constants in both the gas and air equations, the gauge pressure change in a fluid may be more conveniently calculated from the general expression:

$$\Delta p_e = (\rho_{air} - \rho)gh \qquad 4.12$$

where Δp_e is the fluid gauge pressure change, ρ_{air} and ρ are the air and fluid densities respectively and h is positive for height increase and negative for height decrease.

It is evident from Equation 4.12 that the effect is reversed for fluids having densities greater than air, because the expression in brackets becomes negative, and gauge pressure then *decreases* with height.

Example 4.9. If the pipe in Figure 4.10 contains water of density 1000 kg/m^3, what will the gauge pressure be at level B?

ρ_{air} = 1.225, ρ = 1000, g = 9.81, h = -10.

Using Equation 4.12:

Δp_e = (1.225 - 1 000) x 9.81 x (-10)
= (-998.775) x (-98.1)
= + 97 980 Pa

p_e at B = p_e at A + Δp_e
= 2 500 + 97 980
= **100 480 Pa**

(Where the difference in fluid densities is very high as in Example 4.9, air density is ignored in practice for all but the most precise calculations. Ignoring it in this instance affects the answer by only 0.1%.)

This effect is termed 'buoyancy' ('negative buoyancy' for fluids heavier than air). It is of great practical importance for fluid systems, because the increase in gauge pressure can be used as 'free' motive power to promote flow. The start of a lighter-than-air distribution system, for example, should be at the lowest point possible to take advantage of the gauge pressure gain with height and thus supply all users with the minimum expenditure of mechanical power; a principle followed in the days of town gas when gas works were sited wherever possible in the lowest area of the district. Conversely, the start of a heavier-than-air system should be at the highest point, the commonest application being the town water head tank positioned at the top of a building so as to serve all users by gravity flow. Two other widespread applications merit more detailed examination.

The standard density of discharged combustion products is close to that of air, but in most cases they are at higher temperature with an actual density considerably less. Their consequent buoyancy is used via a stack to create a negative gauge pressure at flue level and obviate or minimise the use of a fan. This is illustrated in Figure 4.11, which shows a furnace flue and stack.

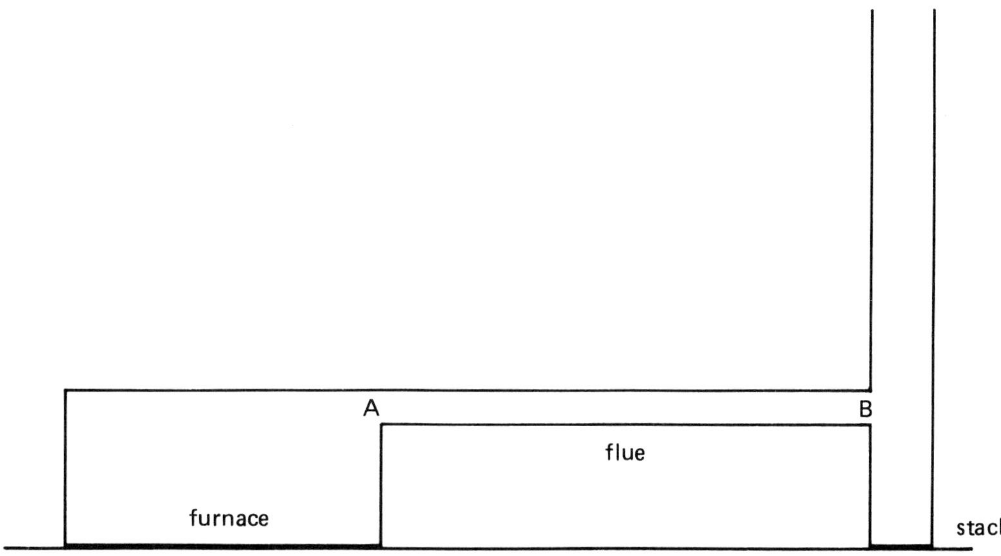

Fig. 4.11 Diagrammatic Stack Effect

Assume for example a low-temperature, balanced-pressure furnace, with a flue system having a pressure loss of 1 mbar discharging flue gases at 120 °C into the stack. Given that the density of the flue gases at 120 °C is 0.86 kg/m³ and of atmospheric air is 1.225 kg/m³, it is possible to calculate the minimum chimney height required. As the furnace atmosphere is at balanced pressure i.e. equal to atmospheric pressure, gauge pressure p_e at point A is zero. Gauge pressure at point B after a pressure loss of 1 mbar is therefore:

$$
\begin{aligned}
& p_e \text{ at A} - 100 \\
&= 0 - 100 \\
&= -100 \text{ Pa}
\end{aligned}
$$

As the gases equalise with atmospheric pressure at the top of the stack and consequently have zero gauge pressure, the stack has to generate a minimum gauge pressure increase of 100 Pa to offset the pressure drop in the flue.

$$\Delta p_e = 100, \quad \rho_{air} = 1.225, \quad \rho_{gas} = 0.86, \quad g = 9.81$$

Using equation 4.12:

$$100 = (1.225 - 0.86) \times 9.81 \times h$$
$$h = 28 \text{ m}$$

Note that this is the *effective height* from point B, not the height above ground level. As will be seen later, the solution is only correct if the cross-sectional areas of the flue and stack are equal and there are no temperature and pressure losses from point B onwards.

Another common application is in low temperature hot water heating systems. Figure 4.12 is a diagrammatic layout of a typical system, simplified to one radiator for ease of illustration, having a flow temperature of 80 °C and a return temperature of 60 °C.

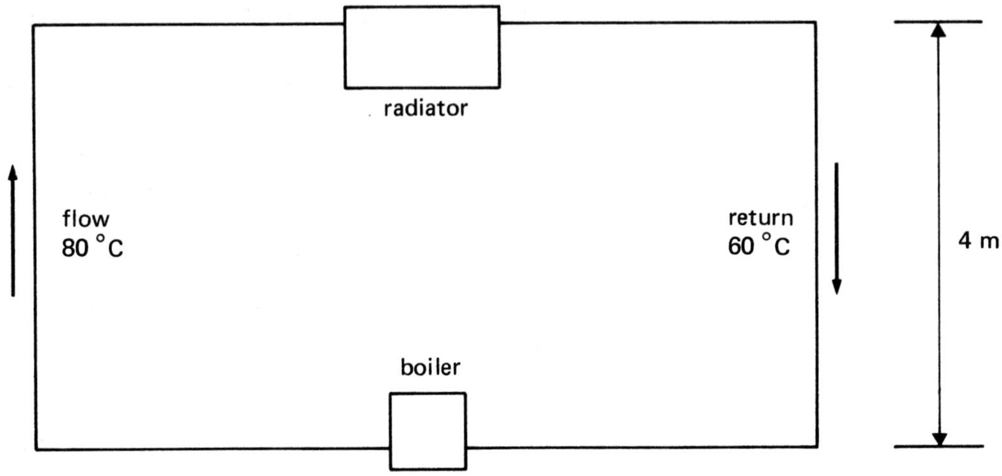

Fig. 4.12 Gravity l.t.h.w. System

In this case, the comparison is made between the flow and return vertical legs assuming the system is at rest and atmospheric pressure is ignored as it has an equal effect on both. Equation 4.12 may then be re-written in the form:

$$\Delta p_e = (\rho_1 - \rho_2)\,gh$$

Where ρ_1 and ρ_2 are the densities of the water in the return and flow legs respectively. If ρ_1 = 983.2 kg/m^3 and ρ_2 = 971.6 kg/m^3, then:

$$\Delta p_e = (983.2 - 971.6) \times 9.81 \times 4$$
$$= 455 \text{ Pa}$$

Gauge pressure at the base of the return leg is thus 455 Pa higher than at the base of the flow leg and this pressure difference (called circulation pressure) will induce water flow if total pressure losses are lower, so activating the system. This constitutes a gravity heating system and while there are many still in use, current practice is to use a circulating pump because of the higher frictional resistance of modern compact boilers and small bore pipework; nevertheless, the circulating pressure makes a valuable contribution to the pressure energy requirement and is taken into account in system design. Being height related, the effect is obviously more marked for radiators and other emitters on upper floors than on lower floors, and local flow rates vary accordingly, which is why upstairs radiators are hotter until 'balanced'.

Convection

The effect just described, where change in temperature by the addition of heat induces buoyancy within the same fluid and thereby creates flow, is termed convection. A gravity system like the one shown, where heat is added at one point, removed at another and fluid flows in an effectively closed loop, is called a thermosyphon. Convection is of fundamental importance in heat transfer and is treated in more detail in Chapter 5.

4.5 MECHANICS OF FLUID FLOW

4.5.1 Energy Balance

If there are no energy losses or gains throughout a flow system, the Law of Conservation of Energy requires that the total energy E of the fluid must be the same at all points in the system. Summations of all the energy components can therefore be made at each desired point, and equated to analyse fluid behaviour. This ideal situation is a useful starting point to derive the equations.

Unit mass of fluid in this ideal system has three energy components: potential, pressure and kinetic energy. It is convenient to consider each with reference to an arbitrary base or datum level.

Potential energy is that available by virtue of the fluid's position in the Earth's gravitational field and is equal to gh where h is the height above datum.

Pressure energy is that required to introduce the fluid into the system without volume change, equal to pV/m (where m = mass). As m/V is density, pressure energy per unit mass is therefore p/ρ.

Kinetic energy (velocity energy) is the energy of fluid motion, and is $u^2/2$ where u = linear velocity in metres per second.

Each of these can change within a system and the following equation is derived by analysis of these changes:

$$p/\rho + u^2/2 + gh = \text{constant} \qquad 4.13$$

This is Bernoulli's equation in energy terms and is a re-statement of the Law oт Conservation of Energy.

The derivation may be illustrated by considering Figure 4.13, where a volume V of liquid flows from point 1 to point 2 through a section of pipe which rises and also increases in cross-sectional area. u, p and h are velocity, pressure and height above datum respectively, subscripted 1 and 2 as appropriate.

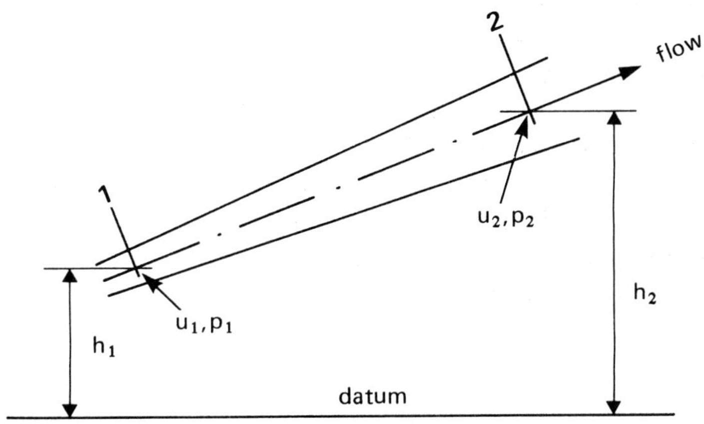

Fig. 4.13

The work done on a liquid to cause it to move is force multiplied by distance moved by the force:

= (pressure)(cross sectional area)(distance)

But area multiplied by distance equals volume displaced, so the energy imparted is p.V.

In flowing from point 1 to point 2, the gain in liquid energy is therefore:

$$p_1V - p_2V = (p_1 - p_2)V.$$

This must equal gain in kinetic energy plus gain in potential energy, with V multiplied by density to convert to unit mass.

Gain in kinetic energy $= (\rho V u_2^2/2) - (\rho V u_1^2/2)$

$= \rho V(u_2^2 - u_1^2)/2$

Gain in potential energy $= \rho g V h_2 - \rho g V h_1$

$= \rho g V(h_2 - h_1)$

Thus $(p_1 - p_2)V = [\rho V(u_2^2 - u_1^2)/2] + \rho g V(h_2 - h_1)$

Removing V and regrouping the terms for points 1 and 2:

$$p_1/\rho + u_1^2/2 + gh_1 = p_2/\rho + u_2^2/2 + gh_2$$

and thus Equation 4.13:

$$p/\rho + u^2/2 + gh = \text{constant}$$

It is used in several derived forms to suit the application. If multiplied throughout by ρ, it becomes:

$$p + \rho u^2/2 + \rho gh = \text{constant} \qquad 4.14$$

which is in pressure terms. If divided throughout by g, Equation 4.13 becomes:

$$p/(\rho g) + u^2/2g + h = \text{constant} \qquad 4.15$$

which is in terms of height, i.e. metres head, and the constant is called total head. This is its most familiar form, but it will be recalled that head must always be specified in terms of the fluid concerned. As S.I. units are increasingly adopted, with pressure expressed exclusively in Pa and derivatives, the use of Equation 4.14 is preferable. Figure 4.14 shows a diagrammatic application of Equation 4.15 to the Figure 4.13 system.

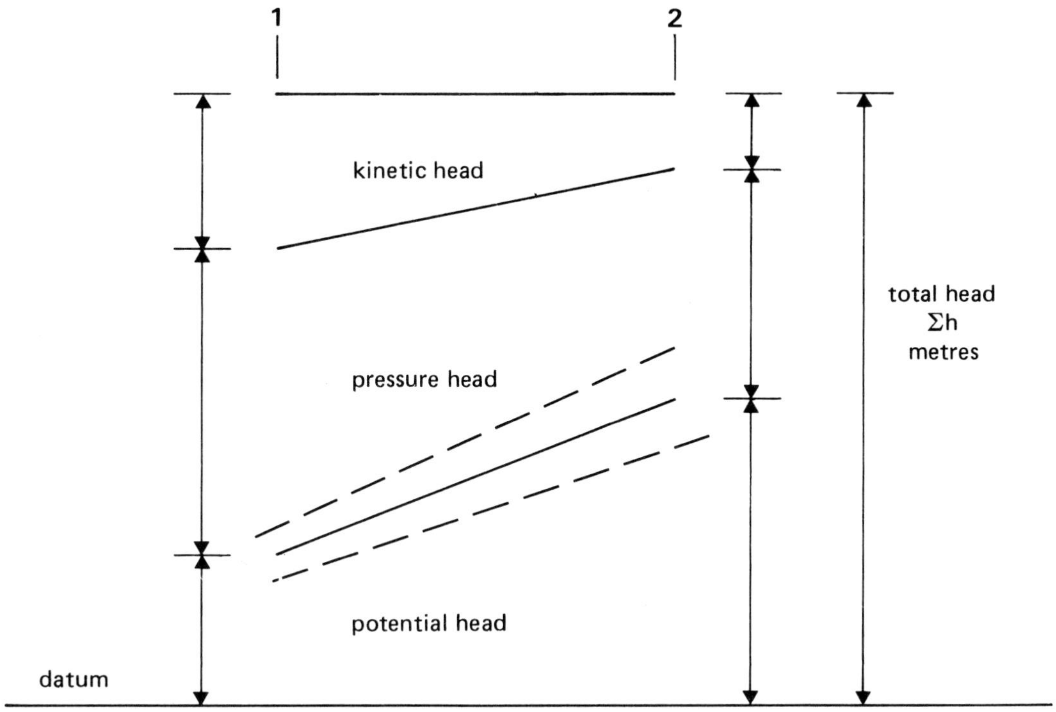

Fig. 4.14 Head Diagram for Fig. 4.13

Figure 4.14 demonstrates how the individual contributions of the energy components to the total vary with changing height and cross-sectional area. In practice, flow regimes are often simpler; a horizontal pipe, for example, is at constant height above datum and potential energy may be omitted from the equation. If pipe bore is unchanged, velocity is constant and kinetic energy may be omitted. Equations 4.13 to 4.15 need further development, however, to reflect conditions in practice.

4.5.2 Practical Energy Balance

As a fluid flows through a system, some of its energy is degraded into heat by friction and is lost from the total. If Δp_f represents the pressure loss by friction between the reference points, Equation 4.14 may be re-written:

$$p_1 + \rho u_1^2/2 + \rho g h_1 = p_2 + \rho u_2^2/2 + \rho g h_2 + \Delta p_f \qquad 4.16$$

If a pump is interposed between the reference points, to compensate for the friction or for insufficient potential pressure or for both, the total energy will be raised accordingly. If Δp_p represents the pressure increase across the pump, Equation 4.16 becomes:

$$p_1 + \rho u_1^2/2 + \rho g h_1 + \Delta p_p = p_2 + \rho u_2^2/2 + \rho g h_2 + \Delta p_f \qquad 4.17$$

This equation is of considerable practical use and, re-arranged to suit the particular problem to be solved, may be applied between any two points in a liquid flow system such as the one shown in Figure 4.15, which illustrates the familiar case of a pipe run negotiating obstacles.

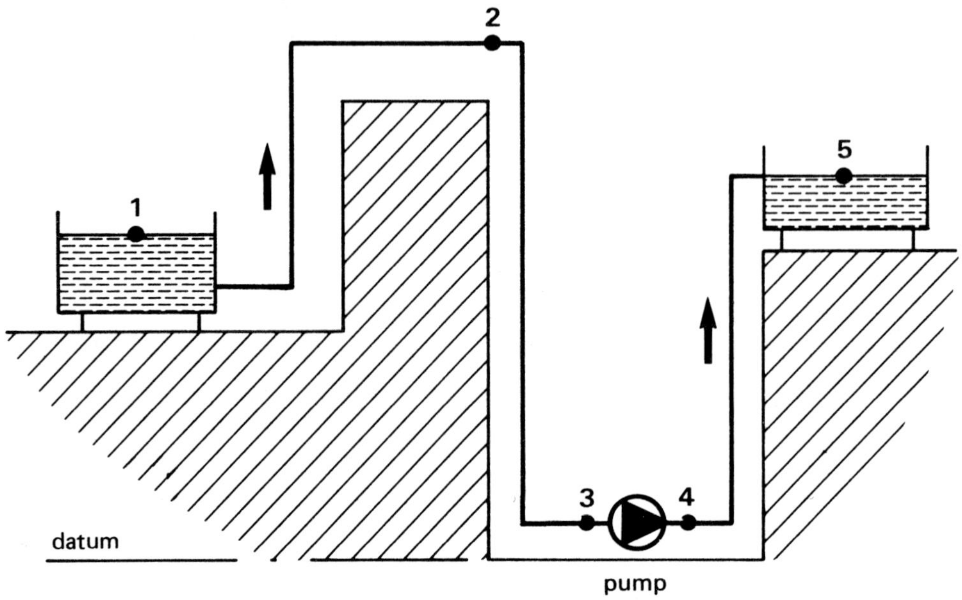

Fig. 4.15 Diagrammatic Pipe Run

The re-arrangement of Equation 4.17 best suited to this is:

$$(p_2-p_1) + \rho(u_2^2-u_1^2)/2 + \rho g(h_2-h_1) + \Delta p_f = \Delta p_p \qquad 4.18$$

Prior inspection will indicate which expressions may be ignored. If the section between reference points does not include the pump (e.g. 1 to 3), Δp_p is zero. Similarly, if the section is horizontal h_2-h_1 is zero and the potential change may be ignored. It is important to assess pressures at the points of lowest value, i.e. at highest h. If p_2 in Figure 4.15, for example, is found to be negative this is clearly impossible as it is an absolute pressure, and the system will not operate. Inoperability, or operational difficulties, may also occur if such pressures are very low; this happens when the pressure approaches the vapour pressure of the liquid and it starts to evaporate, causing partial and eventual total vapour locking. The problem becomes more acute with rising temperature; water at 20 °C, for example, will not boil until the absolute pressure falls to 24 mbar but at 80 °C it boils at 470 mbar. This also applies to pump suctions, where inadequate potential head and/or high preceding friction loss may induce the same condition which in this case is called cavitation. If analysis of a system identifies such problems, it must be re-designed, which usually means increasing the potential head upstream which in turn entails re-positioning plant. If p_2 in Figure 4.15 were found to be negative, for example, tank 1 would need to be resited at a greater height above datum.

A pump suction pressure such as p_3 must thus always be higher than the vapour pressure, i.e. p_3-p_v must always be positive; this differential pressure is called the nett positive suction pressure. It is obviously also important to minimise friction pressure losses caused by for example partly-opened valves in suction lines. A common example is the boiler feedwater pump, drawing hot water from the hotwell, where cavitation is often experienced because of insufficient height of water level above pump inlet, often compounded by high friction loss caused by poorly designed pipework between tank and pump. Another example is the pumping of a hot liquid from a sump, as shown in Figure 4.16.

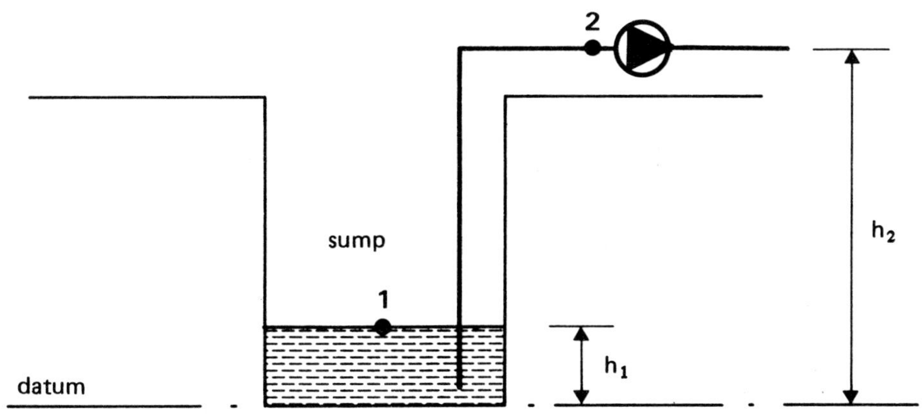

Fig.4.16 Pump and Sump Arrangement

As the effective depth $(h_2 - h_1)$ of the sump increases, the point will eventually be reached where it is impossible to achieve a nett positive suction head and the pump will not function. The solution is to re-site the pump part-way down the sump, or use a sump pump (i.e. the impeller is immersed in the liquid and driven by an extended shaft).

Equation 4.16 and its derivatives are generally applicable where density variation due to pressure change may be ignored. This covers liquids in most practical cases, and gases where the fall in pressure is less than 10% of the initial pressure.

4.5.3 Application to Flow Measurement

As Velocity multiplied by Cross-sectional Area = Volumetric Rate of Flow $(u.A = q_v)$, Bernoulli-based equations can be used to determine flow rates. For an exhaustive treatment of the subject, the interested reader should consult BS 1042, and refer to Chapter 14 for flow measuring equipment. Two examples in simplified form will be considered here, both based on measuring a differential pressure but differing in the way the differential is generated.

Pitot-tube

Figure 4.17 is a diagrammatic representation of pressure measurement by a pitot tube, which in standard form comprises two concentric tubes as shown. In practice, the assembly is of minimum size in relation to the tube or duct bore so as to cause negligible disturbance to flow.

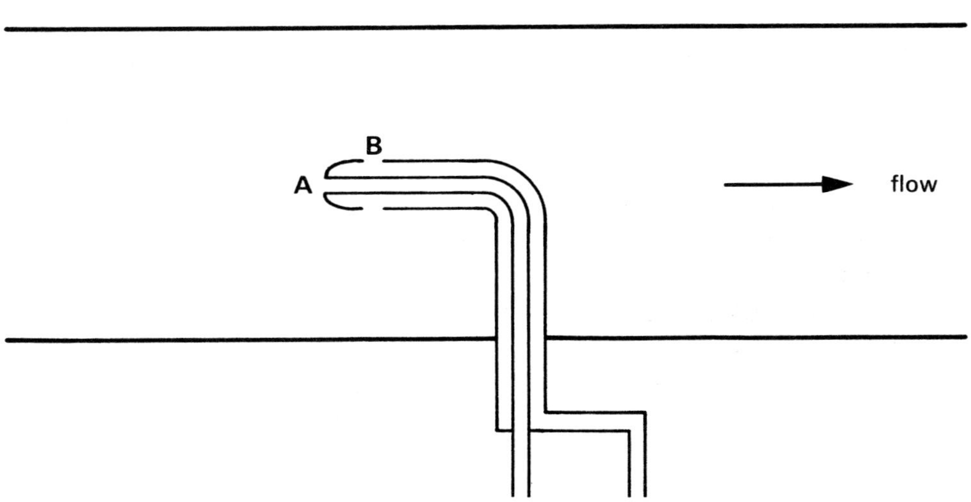

Fig. 4.17 Pitot Tube

Consider now an insignificantly small stream of the fluid coaxial with nozzle A. Upstream of the nozzle may be considered point 1, and from equation 4.14 its total pressure energy will be:

$$p_1 + \rho u_1^2/2 + \rho g h_1$$

When the stream strikes the nozzle, however, it is brought to rest and at zero velocity there must be zero kinetic energy. If the height datum is the nozzle centre line, potential energy is zero throughout and conservation of energy dictates that all the kinetic energy has been converted to pressure energy. Thus:

$$p_1 + \rho u_1^2/2 = p_2 \qquad 4.19$$

Pressure p_2 will therefore be transmitted through the inner tube of the pitot and will register on a gauge connected to the left hand leg. The static pressure only (p_1) will be transmitted through holes B to a gauge connected to the outer annulus via the right hand leg. In practice, a differential gauge is connected across both legs and thus records ($p_2 - p_1$), or Δp. Re-arranging Equation 4.19:

$$(p_2 - p_1) = \Delta p = \rho u_1^2/2$$

$$u_1 = \sqrt{(2\Delta p)/\rho} \qquad 4.20$$

Equation 4.20 is the basic pitot tube formula. Velocity times cross-sectional area will give volumetric flow rate and volume times density will give mass flow rate. As will be shown later in this Chapter, fluid velocities are not constant over a tube or duct cross-section, and Δp must be measured at a number of points and $\sqrt{\Delta p}$ averaged before converting to velocity and flow units. Furthermore, the pitot must be positioned with great care to ensure that the nozzle is precisely aligned with the direction of flow. Whilst of limited application for permanent instrumentation, the pitot tube is widely used for rapid, approximate flow measurement of gases.

Orifice plate

Figure 4.18 shows a horizontal pipe containing a plate normal to the axis in which a sharp edged concentric circular hole has been bored. The fluid stream is contracted and hence accelerated by the orifice, reaching maximum contraction at point 3 called the vena contracta, thereafter diverging and becoming turbulent. As with the pitot tube, the height datum is the tube centre line and potential energy may be ignored. Unlike the pitot, however, the whole cross-section of the stream is considered. As flow rate must be constant throughout, fluid velocity and hence kinetic energy will vary with the changes in cross-sectional area; these variations will have corresponding effects on pressure and a differential pressure will be created which can be measured and used to determine flow rate.

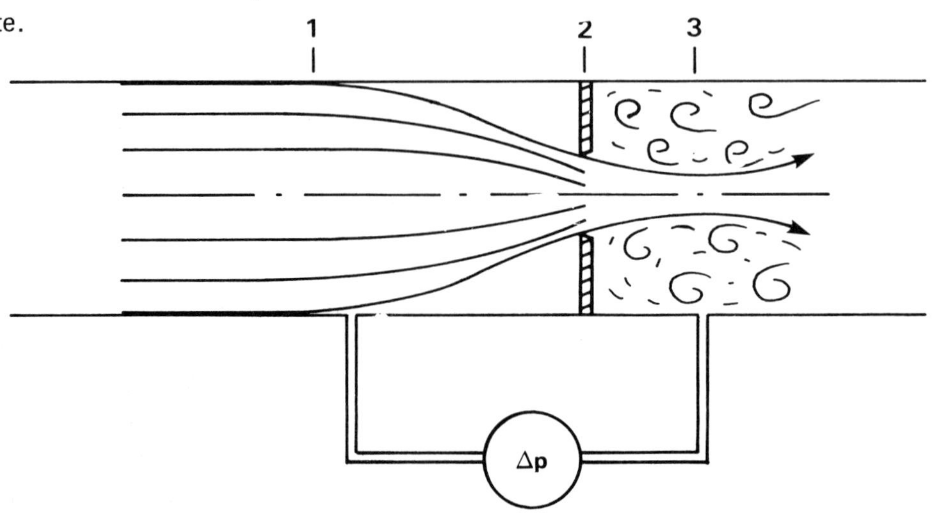

Fig. 4.18 Orifice Plate

Adapting Equation 4.14 and ignoring potential energy:

$$p_1 + \rho u_1^2/2 = p_3 + \rho u_3^2/2 \qquad 4.21$$

$(p_1 - p_3)$ is the differential pressure Δp and Equation 4.21 may be arranged:

$$\Delta p/\rho = u_3^2/2 \times (1 - u_1^2/u_3^2) \qquad 4.22$$

Volumetric flow rate q_v, given no expansion or contraction, must be constant throughout, therefore for fluid cross-sectional area A:

$$q_v = u_1 A_1 = u_2 A_2 = u_3 A_3 \qquad 4.23$$

Combining equations 4.22 and 4.23 and re-arranging the terms, the theoretical rate of flow is:

$$q_v = u_3 A_3 = A_3\sqrt{(2\Delta p)/[\rho(1 - A_3^2/A_1^2)]} \qquad 4.24$$

Equation 4.24 has serious practical limitations. It assumes frictionless flow, which is not the case, and the location and size of the vena contracta vary with the geometry of the system. It is moreover obviously more convenient to limit dimensions to those which are easily ascertainable. The basic orifice plate equation is thus:

$$q_v = c_d A_2\sqrt{(2\Delta p)/[\rho(1-d_2^4/d_1^4)]} \qquad 4.25$$

where c_d is the (dimensionless) discharge coefficient which accounts for friction and layout geometry, and d_2/d_1 is the ratio of the orifice and pipe bores.

Values of c_d are listed in BS 1042, which also gives the derivations of the basic equation to cover vertical pipes, location of pressure tappings, non-concentric and non-circular orifices, and additional coefficients which may need to be included.

As the fluid stream contracts to and through the orifice, its static pressure clearly falls as it accelerates and increases its kinetic energy. When fully expanded again to fill the pipe bore, the initial pressure p_1 is not fully restored because of the loss due to friction. This 'permanent pressure loss' is often expressed as a percentage of the differential pressure created by the orifice plate.

4.6 FRICTION

4.6.1 Viscosity

Friction for fluids differs from friction for solids in that the former has an effect not only at the contact surfaces, but also within the fluid itself. This internal effect occurs because of the fluid characteristic of deformation when in motion, and is termed viscosity. The concept may be illustrated by reference to Figure 4.19, which shows an enlarged cross-section of the inner wall of a tube containing a fluid moving at moderate velocity.

Layer A of the fluid is the boundary layer, retarded by the friction of the tube wall roughness, and is in fact stationary. This is true of all fluids in motion, although in practice the boundary layer may be as little as one molecule thick. As the main body of the fluid is by definition moving in the direction of flow, it follows that it deforms next to the boundary layer and energy is required to overcome the internal shear stress. This shearing occurs in layers throughout (B, C, D etc), each layer moving at a higher velocity as the pipe centre line is neared, the velocity gradient then declining

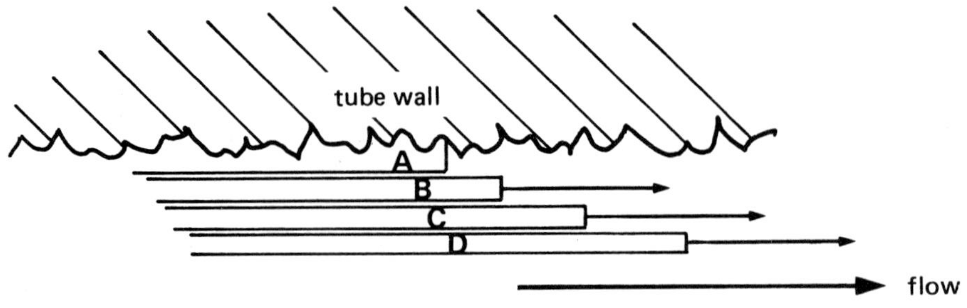

Fig. 4.19 Flow at Tube Wall

as the opposite wall is approached. Viscosity is the measure of a fluid's resistance to shear and is a matter of common experience (e.g. water - low viscosity, treacle - high viscosity).

The flow regime in Figure 4.19 also serves to illustrate the definition of viscosity. If the shear stress is proportional to the velocity gradient, it conforms to Newton's law of viscosity and the constant of proportionality is known as the coefficient of viscosity (μ). Fluids obeying this law, which include those most commonly encountered, are called Newtonian. Those which do not are non-Newtonian; space does not permit further consideration in this chapter and the appropriate textbook should be consulted for more information.

The quantitative definition of viscosity (μ) is: a fluid where a shear stress of 1 Pa is sufficient to cause a velocity gradient of 1 m/s is said to have a viscosity of 1 Pascal-second (1 Pa s). Viscosity is normally expressed in this form, particularly for gases, and is also called dynamic viscosity (μ) to distinguish it from kinematic viscosity (ν).

Kinematic viscosity is dynamic viscosity divided by density, i.e. $\nu = \mu/\rho$, and is thus expressed in m^2/s. It is usually applied to liquids only.

Dynamic and kinematic viscosities are still commonly quoted in the centimetre, gram, second (c.g.s) units of poises and stokes respectively, or the more convenient smaller units of centipoises (cP) and centistokes (cSt). Their relationships to S.I. units are:

		c.g.s.		S.I.
dynamic viscosity	μ	1 P	=	0.1 Pa s
kinematic viscosity	ν	1 St	=	10^{-4} m^2/s

Viscosity varies with temperature and must always be quoted to a temperature reference. Liquid viscosities fall as temperature increases and are little affected by pressure. Gas viscosities rise as temperature increases and are also affected by pressure but to a lesser extent. Some examples of the viscosity/temperature relationship are shown in Figure 4.20.

The existence of a velocity gradient obviously has an important effect on all the equations involving fluid velocity quoted so far in this Chapter. Wherever the velocity (u) of the total fluid stream in a pipe or duct has been stated hitherto it must be understood as being the *mean* velocity over the cross-section.

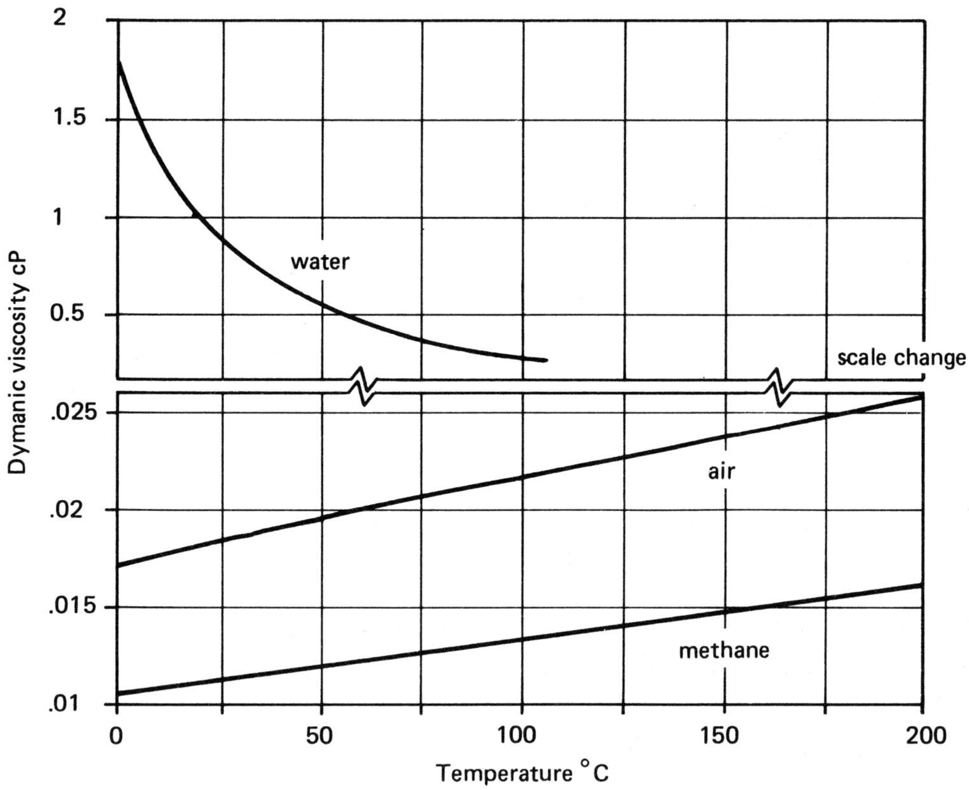

Fig. 4.20 Viscosity v. Temperature at Atmospheric Pressure

4.6.2 Flow Regimes and Reynolds Number

In Reynolds' classic experiments, he injected a thin stream of dye into water flowing under steady conditions in a glass tube, and observed the flow patterns displayed at different velocities. The results are shown diagrammatically in Figure 4.21.

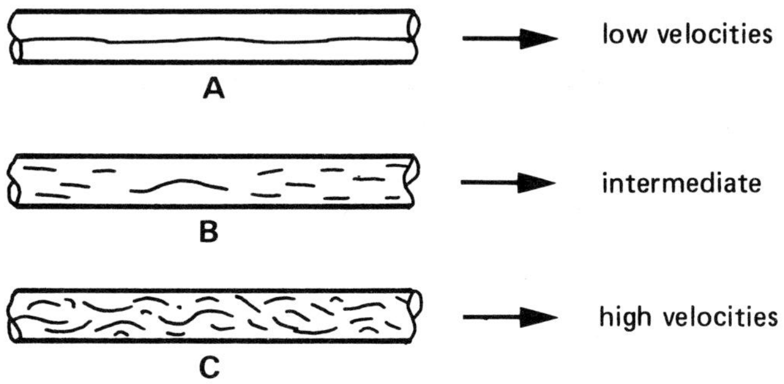

Fig. 4.21 Fluid Flow Pattern

He observed that in case A the dye followed a line parallel to the direction of flow; the fluid stream is in fact a bundle of such lines or streamlines, each at constant velocity and with no transfer from one to the other, i.e. no flow perpendicular to the tube axis. This is called laminar or streamlined or viscous flow and was used for Figure 4.19 to define viscosity.

In case C there are no streamlines and flow within the fluid stream is in random directions with consequent continual fluid mixing. This is turbulent flow and is the type mostly encountered in practical engineering.

Case B is called the transition stage and occurs over a relatively narrow range of flow velocity between the end of laminar flow and the start of turbulent flow.

The experiments have since been repeated for a wide variety of liquids and tube diameters. Their significance becomes fundamental when the results are expressed in terms of the Reynolds number (Re) which is defined for pipe-flow as follows:

$$Re = \rho ud/\mu$$

$$\text{or} \quad Re = ud/\nu \qquad 4.26$$

where ρ = density, u = velocity, d = diameter and μ and ν = dynamic and kinetic viscosities respectively. Inspection of the dimensions of the units in these expressions will show that Re is a dimensionless number.

A first practical use of Reynolds number is to define the flow regime. If Re is below about 2 000, flow is laminar. If Re is above about 4 000, flow is generally turbulent and above Re = 10 000 is usually completely turbulent. For the critical zone Re = 2 000 to 4 000, flow conditions are indeterminate and calculations should not be attempted, or should be referred to a specialist. Due caution should be exercised for calculations involving the transition zone where Re = 4 000 to 10 000. These boundaries are approximate and there are cases (e.g. liquid flow in helical pipe coils) where different ones prevail, but apply to most practical experience.

Velocity gradient

The two basic flow regimes, laminar and turbulent, give rise to different velocity gradients across the pipe cross-section.

In streamline flow, the stationary boundary layer 'shields' the moving liquid stream from the frictional drag of the pipe wall roughness and friction forces are essentially internal to the stream only, i.e. viscosity effects. This gives a velocity distribution which is parabolic.

In turbulent flow, pipe roughness is significant and the velocity gradient is steeper near the wall, but thereafter becomes flatter, increasingly so at higher values of Re.

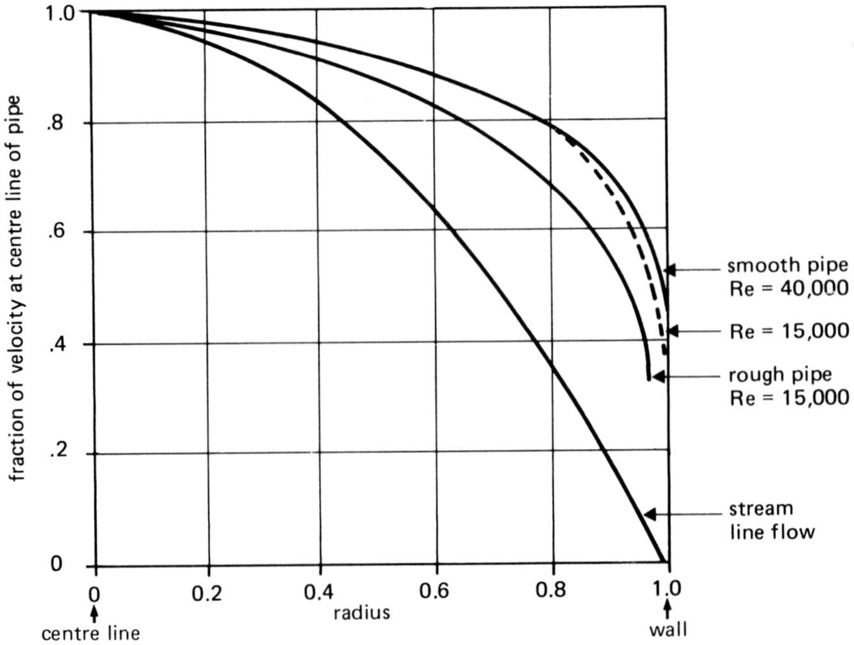

Fig. 4.22 Velocity Distribution in Pipes

Figure 4.22 shows some typical velocity gradients, plotted as the ratio of actual velocity to centre line velocity. This further illustrates the importance of determining mean velocity, especially when measuring flow by point methods such as the pitot tube, or when sampling isokinetically e.g. for grit and dust burden in gases. It is convenient to express mean velocity as a fraction of centre-line velocity, the latter being easiest to measure in practice, and Figure 4.23 shows this ratio plotted over a range of Reynolds number for circular pipes.

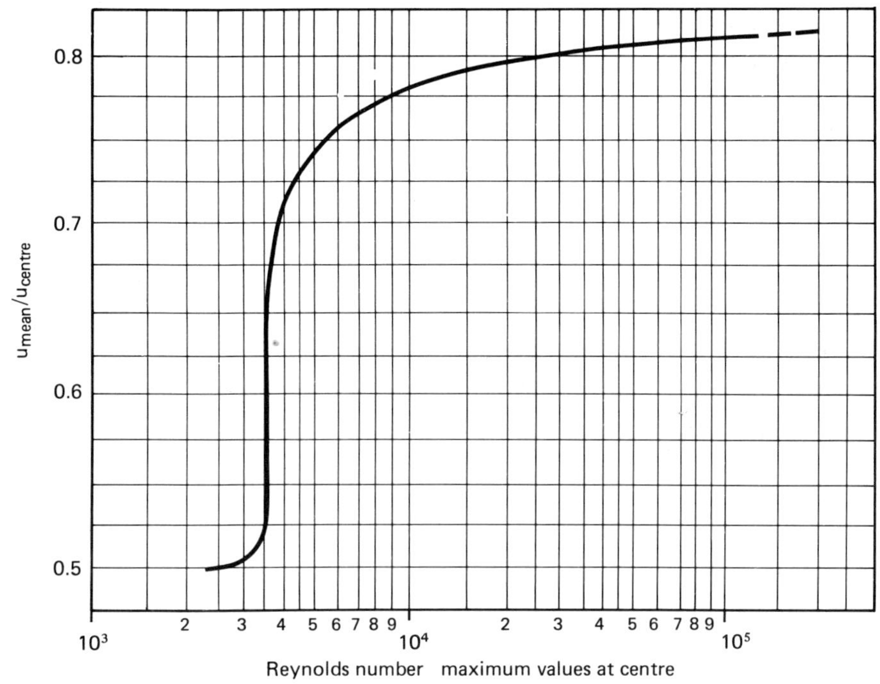

Fig. 4.23 Ratio of Mean to Maximum Velocity in Pipes

The graph illustrates a rule of thumb which is adequate for many practical purposes where high accuracy is not required: for laminar flow, mean velocity is 0.5 centre-line velocity and for turbulent flow it is 0.8 centre-line velocity.

4.6.3 Circular and Non-circular Sections

Hitherto, flow in pipes and ducts of circular cross-section has been considered. The general case for all conduits i.e. including non-circular sections is achieved via the concept of hydraulic mean diameter or depth, which (to oversimplify) is a means of expressing any section shape as the 'diameter' of a circular pipe having equivalent flow properties. Hydraulic mean diameter (d_h) is defined as four times the cross-sectional area (A) divided by the internal (i.e. 'wetted') perimeter (Λ):

$$d_h = 4A/\Lambda \qquad 4.27$$

Thus for the commonest cross-sections:

circular, diameter d,	$d_h = (4\pi d^2/4)/\pi d$	$= d$
square, side a,	$d_h = 4a^2/4a$	$= a$
rectangular, sides a & b	$d_h = 4ab/2(a+b)$	$= 2ab/(a+b)$

Equation 4.26 for Reynolds number, $Re = \rho ud/\mu$, can now be expressed in the general form for any cross-section:

$$Re = \rho u_m d_h/\mu \qquad 4.28$$

As the starting point in a calculation is often the mass flowrate (q_m in kg/s) or the volumetric flowrate (q_v in m^3/s), it is convenient to have Equation 4.28 in these terms also. Substituting $4A/\Lambda$ for d_h:

$$Re = (\rho u_m 4A/\Lambda)/\mu$$

$$\text{or} \quad Re = 4\rho u_m A/\mu\Lambda$$

As $q_v = u_m$, and $q_m = \rho u_m A$:

$$Re = 4\rho q_v/\mu\Lambda \qquad 4.29$$

$$\text{and } Re = 4 q_m/\mu\Lambda \qquad 4.30$$

4.6.4 Wall Roughness

There are two obvious factors governing the pressure loss due to wall roughness. Firstly, its degree: the rougher the pipe or duct wall, the greater its frictional resistance. Secondly, the conduit size: as hydraulic mean diameter increases, the effect of roughness decreases because the ratio of surface area to cross-sectional area declines. Both must therefore be taken into account, using the concepts of absolute and relative roughness.

The analysis of fluid flow adjacent to a roughened surface is extremely complex. Extensive experimental work has demonstrated, however, that for practical purposes the frictional drag may be considered proportional to the average height of the projections causing the roughness and this is the basis of absolute roughness (b). Table 4.1 lists values of b for the commoner pipe and ducting materials encountered.

Table 4.1 Values of Absolute Roughness

Material	b mm
Non-ferrous drawn piping	0.0015
Plastic piping	0.003
Asbestos-cement piping	0.013
Black steel piping (new)	0.046
Black steel piping (rusted)	2.5
Asphalted cast iron piping	0.12
Cast iron piping	0.2
Galvanised steel piping and ducting	0.15
Aluminium ducting	0.05
Cement or plaster ducting	0.25
Brick or concrete ducting (fair-faced)	1.3
Rough brickwork ducting	5.0

Relative roughness (ϵ) is absolute roughness divided by hydraulic mean diameter in mm, b/d_h, and is dimensionless. Its derivation is the starting point for most friction loss calculations. It will be evident that there is considerable scope for error and informed judgement must be exercised when deciding which value of b to apply. In design calculations, for example, where pumps or fans must be sized adequately, the ultimate and not new condition of piping or ducting must be taken into account if there is a likelihood of rusting or other corrosion or deposition occurring under operating conditions.

For pipes, it is important that the actual internal bore, which is not necessarily the same as nominal bore, be used to establish relative roughness. It is often convenient to use ready-calculated tables of relative roughness for standard piping and these may be found in reference works such as the CIBSE Guide. Table 4.2 gives relative roughness for a selection of standard piping over a range of nominal pipe size.

Table 4.2 Relative Roughness Values for Pipes

	Relative Roughness $\epsilon \times 10^3$				
Nominal Size	Copper Table X	UPVC Class E	Cast Iron Class C	Mild Steel	
mm	BS 2871 : Pt 1	BS 3505	BS 1211	Medium Black BS 1387	BS 3600
6	0.31				
10	0.17	0.22		3.7	
25		0.11		1.7	
28	0.057				
50		0.058		0.87	
54	0.029				
100		0.031	1.7	0.44	
108	0.014				
150		0.021	1.3	0.30	
159	0.0097				
300		0.011	0.61		0.15
600					0.078

4.6.5 Coefficient of Friction

The theoretical and empirical aspects of friction developed so far in this section may now be combined to derive the 'coefficient of friction', variously called friction factor and drag coefficient, which may be applied to the calculation of friction loss in straight runs of pipe and ducting and is therefore of direct practical value. The coefficient is defined as follows:

$$c_f = \tau/(\rho u_m^2/2)$$

where c_f is the coefficient of friction, τ is the shear stress at the pipe or duct wall, and $\rho u_m^2/2$ will be recognised as kinetic energy in pressure terms using mean fluid velocity.

If the coefficient of friction is plotted against Reynolds number and a range of relative roughness, the result is the log/log graph shown in Figure 4.2.4, known as the Moody diagram.

A great deal of information is revealed by the graph. The laminar (Re up to 2 000) and turbulent (Re 4 000+) zones are clearly shown. It will be observed that for turbulent flow at higher values of Re, c_f levels out to a constant value for a given relative roughness. This latter effect has given rise to some greatly simplified friction loss equations much used by earlier generations of engineers, but liable to considerable error if applied outside their range as inevitably happens sooner or later, and they should be avoided.

Laminar flow

For laminar flow, c_f is independent of roughness and pipes and annuli are each represented by a

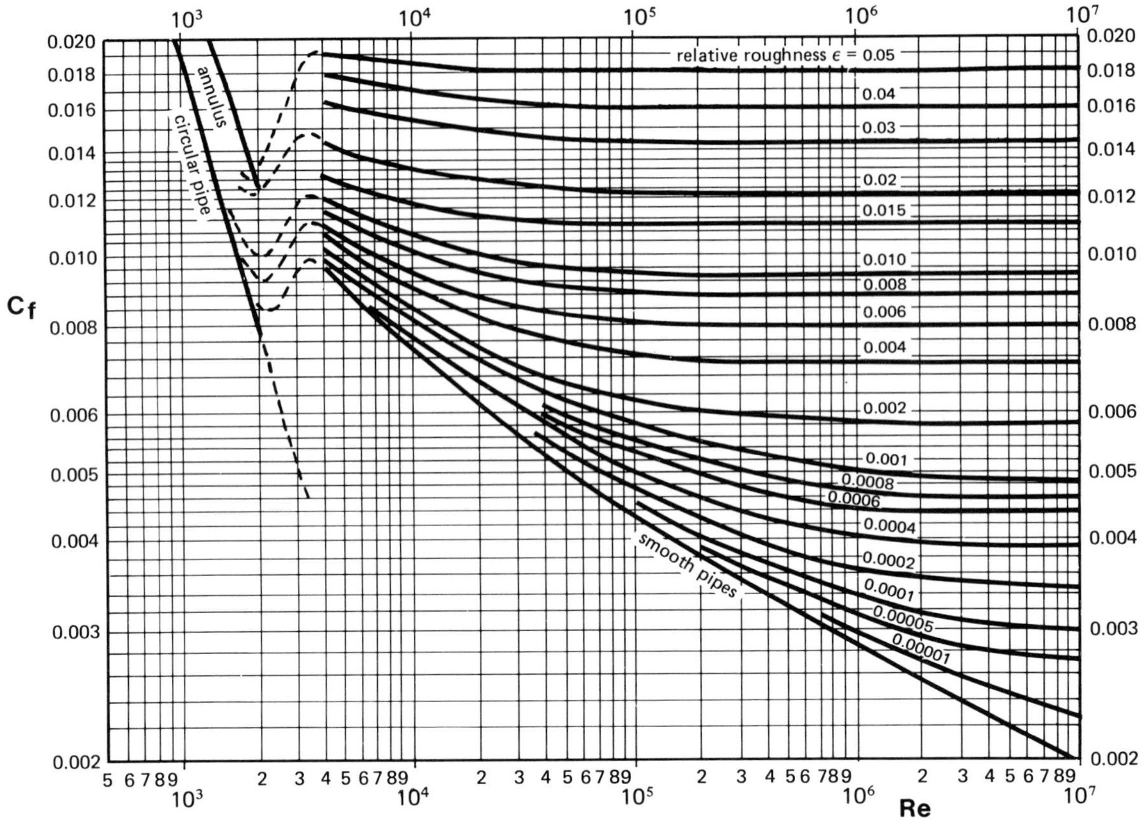

Fig. 4.24 Variation of Friction Factor with Reynolds Number and Relative Roughness

single line, i.e. friction loss is proportional only to Re and thus ultimately viscosity as previously stated. To determine c_f for circular pipes and annuli below the Re range of the diagram, use the equations:

$$c_f = 16/\mathrm{Re} \text{ for circular pipes} \tag{4.31}$$

$$\text{and } c_f = 24/(\beta \mathrm{Re}) \text{ for annuli} \tag{4.32}$$

where β depends on the ratio of the pipe diameters as follows:

diameter ratio	0.1	0.2	0.3	0.4	0.5	0.6 to 1.0
β	1.07	1.04	1.02	1.01	1.01	1.00

For rectangular ducts having sides a and b, use:

$$c_f = 16/(\Psi \mathrm{Re}) \tag{4.33}$$

where Ψ depends on the 'aspect' or side ratio a/b as shown in Figure 4.25.

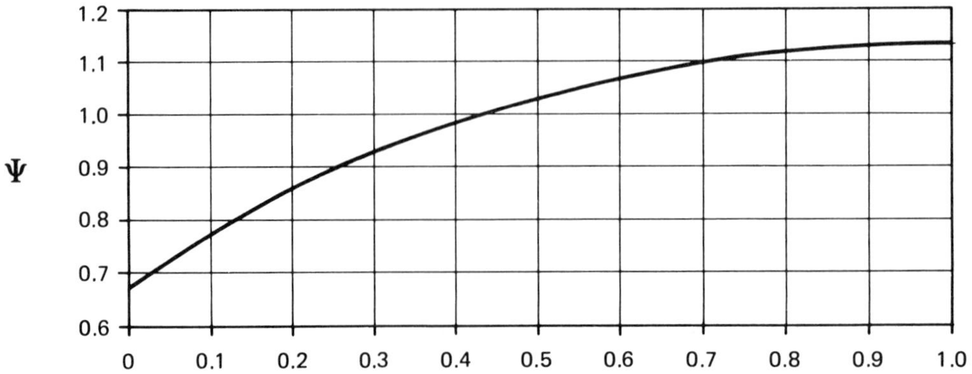

Fig. 4.25 Values of Ψ for Laminar Flow in Rectangular Ducts

Transitional flow

In the critical or transition zone of Re 2 000 to 4 000, there is great doubt as to the appropriate value of c_f. Calculations in this area should therefore be considered approximate only, or left to the specialist.

Turbulent flow

Most engineering problems concern flow in the turbulent region. Equations for c_f have been derived, but entail complexities beyond the scope of this chapter and the semi-graphical method using the Moody diagram, is the one routinely employed by the non-specialist. Summarising the procedure:

- First establish the basic data, i.e. fluid flow rate, density and viscosity, conduit internal dimensions, and calculate the mean fluid velocity.
- Determine the hydraulic mean diameter d_h using Equation 4.27 or its derivatives.
- Calculate Reynolds number Re using Equation 4.28, 4.29 or 4.30.
- Select a *realistic* value of absolute roughness b from data sources such as Table 4.1 and divide by d_h in mm to obtain relative roughness ϵ (or for standard pipes obtain direct from sources such as those from which Table 4.2 was derived).
- For these values of Re and ϵ, determine c_f from the Moody diagram Figure 4.24.

A couple of examples to illustrate the method now follow.

Example 4.10 Water at 10 °C and 2 bar gauge pressure is flowing through 50 mm nominal bore medium grade BS 1387 steel piping at the rate of 3 litres per second. Assuming that the pipe has rusted internally, calculate the coefficient of friction for straight pipe runs.

From appropriate data sources:

dynamic viscosity μ = 0.0013 Pa s, internal dia d = 0.053 m, density ρ = 1000 kg/m^3.
Volumetric flowrate q_v = 3×10^{-3}
= 0.003 m^3/s

Using Equation 4.29 for Reynolds number:

$$\begin{aligned} Re &= 4\,\rho q_v/\mu\Lambda \\ &= (4 \times 1000 \times 0.003)/(0.0013 \times \pi \times 0.053) \\ &= 5.5 \times 10^4 \end{aligned}$$

and flow is therefore turbulent.

For circular sections hydraulic mean diameter d_h = d = 0.053 m

From Table 4.1, absolute roughness b for rusted steel pipe is 2.5 mm, relative roughness ϵ is thus b/(d_h in mm)

$$\begin{aligned} &= 2.5/53 \\ &= 0.047 \end{aligned}$$

For Re = 5.5×10^4 and ϵ = 0.047, the Moody diagram Figure 4.24 gives the coefficient of friction:

c_f = **0.0174**

Example 4.11. Air for space heating is conveyed at 60 °C through a straight run of 1 m x 1.5 m rectangular galvanised steel ducting at a velocity of 9 metres per second. Calculate the coefficient of friction.

From appropriate data sources, for air at 60 °C: dynamic viscosity, $\mu = 0.02 \times 10^{-3}$ Pa s, density = 1.056 kg/m^3.

For rectangular sections, hydraulic mean diameter d_h

$$\begin{aligned} d_h &= 2\,ab/(a+b) \\ &= (2 \times 1 \times 1.5)/(1 + 1.5) \\ &= 1.2 \text{ m} \end{aligned}$$

Using Equation 4.28 for Reynolds number:

$$\begin{aligned} Re &= \rho u_m d_h/\mu \\ &= (1.056 \times 9 \times 1.2)/(0.02 \times 10^{-3}) \\ &= 5.7 \times 10^5 \end{aligned}$$

and flow is therefore turbulent.

From Table 4.1, absolute roughness b for galvanised steel is 0.15 mm, relative roughness ϵ is thus b/(d_h in mm).

$$\begin{aligned} &= 0.15/1200 \\ &= 0.000125 \end{aligned}$$

For Re = 5.7×10^5 and $\epsilon = 0.000125$, the Moody diagram Figure 4.24 gives the coefficient of friction:

$$c_f = \mathbf{0.00036.}$$

IMPORTANT NOTE

The definition and magnitude of the coefficient of friction used here are those found in the reference works likely to be consulted by the reader, notably Technical Data on Fuel and the CIBSE Guide. There is otherwise no general agreement. Unfortunately, the names coefficient of friction, friction factor, drag coefficient and the symbols c_f or f or c_d are found elsewhere with numerical values half, twice or four times those used here. In these cases, of course, the equations or graphs embodying c_f are consistent and contain multipliers or dividers so that the final answer is the same. The danger arises when, as often happens, several works are consulted to build up the information needed to tackle a particular problem: great care must then be exercised to ensure that the coefficient, equations and other data obtained are all consistent. A useful quick check is to examine the Moody diagram in each reference work and compare the values of c_f for given roughness and Reynolds numbers. (The diagram in Industrial Gas Engineering, for example, is plotted for 0.5 c_f.) The general trend does seem to be towards the definition adopted in this chapter and it is to be hoped that standardisation will in due course be widespread.

4.7 CALCULATION OF PRESSURE LOSSES

The equations and methods detailed in this Section are for turbulent flow only, and are for fluids which may be termed 'incompressible', i.e. where density changes are small enough to be ignored. This covers most liquids under conditions normally encountered, and gases where the pressure loss is not greater than 10% of the initial pressure. Outside of these limits, fluids are considered compressible and density change must be taken into account. There are some simplified expressions available for this, but the further complications such as deviation from the ideal gas laws and heat transfer to or from the fluid introduce complexities beyond the scope of this Chapter and are usually undertaken from first principles by the specialist only. Suitable software packages are, however, increasingly available to enable the general engineer to tackle compressible flow problems.

4.7.1 Straight Runs

For straight pipes and ducts of constant cross-section running full, the pressure loss due to friction is:

$$\Delta p_f = (4\, c_f L/d_h) \times (\rho u_m^2/2) \qquad 4.34$$

where

Δp_f	=	pressure loss	Pa
p_f	=	coefficient of friction	
L	=	length of conduit	m
d_h	=	hydraulic mean diameter	m
ρ	=	fluid density	kg/m^3
u_m	=	fluid mean velocity	m/s

This is D'Arcy's equation, derived by substitution in previous equations and in this form may only be used for 'incompressible' fluids.

Example 4.12. If the water in Example 4.10 flows through 30 m of horizontal straight pipe, what will its final pressure be?

$c_f = 0.0174$; $L = 30$ m; $d_h = 0.053$ m; $\rho = 1000$ kg/m^3; $q_v = 0.003$ m^3/s.

$$u_m = q_v/\text{Area} = 0.003/(\pi \times 0.053^2/4) = 1.36 \text{ m/s}$$

Using Equation 4.34:

$$\text{pressure loss } \Delta p_f = (4 \times 0.0174 \times 30/0.053) \times (1000 \times 1.36^2/2)$$
$$= 36\,434 \text{ Pa } (0.364 \text{ bar})$$

Initial pressure was 2 bar, therefore final pressure is:

$$2 - 0.364$$
$$= \mathbf{1.636 \text{ bar.}}$$

It should be noted that Δp_f is strictly the loss in total pressure. In the last example, it was debited against static pressure only because velocity pressure (u and ρ unchanged) and potential pressure (h unchanged) were constant throughout.

4.7.2 Fittings

Re-stating Equation 4.34:

$$\Delta p_f = (4\, c_f L/d_h) \times (\rho u_m^2/2)$$

It will be observed that the second component is velocity pressure. If the first component be expressed as ζ, pressure loss may be re-defined as a factor multiplied by velocity pressure.

Thus: $$\Delta p_f = \zeta \times (\rho u_m^2/2) \qquad 4.35$$

where ζ = Velocity pressure loss factor

u_m = Velocity m/s at a specified cross-section of the fitting (usually downstream)

Values of ζ have been determined empirically for all the pipe and duct fittings encountered in practice; the CIBSE Guide C4, for example, has extensive lists.

Some examples are shown in Tables 4.3 and 4.4

Table 4.3 Velocity Pressure Loss Factors for Pipe Fittings

Fittings		Factor ζ
Maleable 90 ° elbow	- 15 mm	0.80
	- 38 mm	0.70
	⩾ 75 mm	0.60
Welded 90 ° elbow	- 15 mm	0.40
	- 38 mm	0.40
	⩾ 75 mm	0.30
Reducer, A_2/A_1	- 0.1	0.55
	- 0.3	0.45
	- 0.6	0.25
Enlarger, A_1/A_2	- 0.1	0.80
	- 0.3	0.50
	- 0.6	0.15
Full bore globe valve	- 25 mm	7.00
	- 50 mm	5.00
Gate valve	- 25 mm	0.30
	- 50 mm	0.20

A_1, A_2 = initial and final cross-sectional areas respectively. Factor relates to velocity in smaller bore.

Table 4.4 Velocity Pressure Loss Factors for Duct Fittings

Fitting		Factor ζ
Abrupt contraction, A_2/A_1	- 0.2	0.34
	- 0.5	0.23
	- 0.7	0.09
Abrupt enlargement, A_1/A_2	- 0.3	0.49
	- 0.5	0.25
	- 0.7	0.09
Gradual reduction, taper	- 30°	0.02
	- 45°	0.04
	- 60°	0.07
Fully-open damper		0.2 to 0.5

A_1, A_2 = initial and final cross-sectional areas respectively. Factor relates to velocity in smaller bore.

To calculate the total frictional pressure loss in a piping and ducting system, the procedure is as follows:

- Derive ζ for the straight pipe or duct, using overall length inclusive of fittings.
- Obtain ζ for each fitting.
- Add all the values of ζ.
- Use $\Sigma\zeta$ in Equation 4.35 to calculate the total frictional pressure loss.

Note that if the system comprises pipe or duct of differing sizes, the method must be applied individually to each length of a given size, and the individual totals then added together to give the loss for the whole system. The next example illustrates the method.

Example 4.13. Water at ambient temperature enters a pipework system at 5 bar absolute pressure. The pipe is BS 1387 medium black, rusted, and the system comprises: 15 m of 100 mm nominal bore containing two welded elbows followed by 10 m of 50 mm nominal bore containing one elbow and a globe valve. If the flow rate is 4 litres per second, what is the pressure loss due to friction?

It is convenient to set out the solution in tabular form.

		100 mm run	50 mm run	
Dynamic viscosity	Pa s	0.0013	0.0013	
Hydraulic diameter (= internal diameter)	m	0.1051	0.053	
Density	kg/m^3	1 000	1 000	
Flow rate	m^3/s	0.004	0.004	
Reynold no. ex Equation 4.29		3.7×10^4	7.4×10^4	
(Flow is turbulent in both cases)				
Absolute roughness ex Table 4.1	mm	2.5	2.5	
Relative roughness		0.024	0.047	
c_f for straight pipe ex Figure 4.24		0.0132	0.0174	
Velocity = q_v/area	m/s	0.46	1.81	
Length	m	15	10	
Velocity pressure loss factors:				
Straight pipe = $4\, c_f L/d_h$		7.536	13.132	
Elbow		0.3	0.4	
Elbow		0.3	-	
Reducer (A_2/A_1 = 0.25)		-	0.45	
Globe valve		-	5.0	
$\Sigma\zeta$		8.136	18.982	
$\Delta p_f = \Sigma\zeta \times (\rho u_m^2/2)$	Pa	861	31 093	093

$\Sigma\Delta p_f$ = 861 + 31 093

= **31 954 Pa**

In the last example, the velocity pressure has changed because of the different pipe bore, therefore friction loss cannot simply be debited against static pressure. Bernoulli must be used to evaluate the effects of the pressure changes.

Example 4.14. Demonstrate the effect on static pressure of the frictional pressure loss calculated in Example 4.13, assuming that the beginning and end of the pipework are at the same height.

Using Equation 4.16:

$$p_1 + \rho u_1^2/2 + \rho g h_1 = p_2 + \rho u_2^2/2 + \rho g h_2 + \Delta p_f$$

Eliminating ρgh which is unchanged, and re-arranging:

$$\begin{aligned} p_2 &= p_1 + \rho(u_1^2 - u_2^2)/2 - \Delta p_f \\ &= (5 \times 10^5) + [1000 \times (0.46^2 - 1.81^2)/2] - 31\,954 \\ &= (500\,000) - (1\,532) - (31\,954) \\ &= 466\,514 \text{ Pa} \end{aligned}$$

i.e. the final static pressure is 4.665 bar absolute; most of the 0.335 bar loss is due to friction, but a significant proportion is due to conversion to velocity pressure.

In practice, of course, there is often a difference in height between the ends of a system, resulting in a further total pressure change. (Note that intermediate height changes are of no account in this context.)

Example 4.15. Demonstrate the effect on static pressure of the frictional pressure loss calculated in example 4.13. The pipework ends 5.5 m lower than it started.

Using Equation 4.16:

$$p_1 + \rho u_1^2/2 + \rho g h_1 = p_2 + \rho u_2^2/2 + \rho g h_2 + \Delta p_f$$

Using h_2 as datum and re-arranging:

$$\begin{aligned} p_2 &= p_1 + [\rho(u_1^2 - u_2^2)/2] + \rho g(h_1 - h_2) - \Delta p_f \\ &= (5 \times 10^5) + [1000 \times (0.46^2 - 1.81^2)/2] + [1000 \times 9.81 \times (5.5 - 0)] - 31\,954 \\ &= 500\,000 - 1\,532 + 53\,955 - 31\,954 \\ &= 520\,469 \text{ Pa} \end{aligned}$$

i.e. the final static pressure is 5.205 bar absolute. The losses due to friction and conversion to velocity pressure have been more than offset by the conversion of potential pressure to static pressure.

4.7.3 Calculation Aids

In practical experience, it is rarely necessary to use all the methods and procedures described so far. For the common fluids, water, low pressure and compressed air, natural gas and steam, there are more rapid methods available embodied in computer programs, calculators and tabulated systems.

Programs are available for desk-top computers and are continually being updated and extended.

Calculators range from simple slide rules, each usually catering for a particular fluid, to more sophisticated versions such as the Metriflow produced by British Gas plc which may be used for all fluids.

Tabulated Systems are perhaps the best known and most widely used, notably those in the CIBSE Guide C4. The coverage is extensive, comprising all the common fluids and a range of ducting and standard piping. Other advantages are that non-turbulent flow regimes and compressible fluids, particularly steam, are included.

REFERENCES

BRITISH GAS PUBLICATIONS

Metriflow Calculator, 1977
Standard Factors for Temperature and Pressure Correction, 3rd Edition, 1984

BRITISH STANDARDS

BS 1042 : 1984 Measurement of Fluid Flow in Closed Conduits

OTHER PUBLICATIONS

CIBSE Guide. Chartered Institution of Building Services Engineers:
- C1 & C2. Properties of Humid Air, Water & Steam
- C4. Flow of Fluids in Pipes and Ducts.

UK Steam Tables in S.I. Units. UK Committee on the Properties of Steam. Edward Arnold. 1970

Technical Data on Fuel. Rose J.W. & Cooper J.R. (editors). 7th Edition, World Energy Conference. 1977.

Industrial Gas Utilisation: Engineering Principles & Practice, Pritchard R., Guy J.J. and Connor N.E., Bowker, 1977.

Woods Practical Guide to Fan Engineering. Woods of Colchester. 3rd Edition. 1978.

Chemical Engineers' Handbook. Perry R.H., Green D.W. & Maloney J.O. (editors). 6th Edition, McGraw-Hill, 1984.

Heat Transfer 5

CHAPTER 5

HEAT TRANSFER

5.1 INTRODUCTION

When a fuel and air are burnt they produce products of combustion at high temperature. The heat within these combustion products must be transferred to the substance to be heated in some sort of furnace, oven, boiler or other heating plant. An understanding of the rates of heat transfer and the laws governing this process is therefore fundamental in the design of plant and its efficient operation.

The subject of heat transfer is dealt with in a large number of text books. The material which follows does not attempt to replace any of these text books nor is it a complete or rigorous treatment. Instead it concentrates on those areas of heat transfer which have the most direct bearing on correct design and optimum performance of fired equipment.

Heat is energy in transition under the motive force of a temperature difference, and the study of heat transfer deals with the rate at which such energy is transferred. In thermodynamic analyses ideal processes can be conceived during which heat is transferred by virtue of an infinitesimally small temperature difference, without regard to time as a limiting factor. In actual processes some of the available temperature drop, which might be used for the production of work, must be sacrificed to ensure that the required quantity of heat is transferred in a reasonable time across a surface of practicable size. Two broad types of problem are frequently of interest to engineers: the first, met in the design of boilers, heat exchangers and other plant, is concerned with the promotion of the required rate of heat transfer with the minimum possible surface area and temperature difference. The other is concerned with the prevention of heat transfer, i.e. with thermal insulation.

Before considering the various mechanisms by which heat is transferred it is necessary to understand the different ways in which substances are made up from the constituent atoms or molecules. The phases of pure substances (solids, liquids and gases) are associated very closely with the internal energy of the substance.

For the same substance in its various phases, the various thermal properties - specific heat capacity, conductivity etc - will have values of different orders of magnitude. The specific heat capacity of a solid is generally very low, for a liquid it is comparatively high, while a gas usually has an intermediate value. In addition, any change of phase will involve a large amount of energy, often termed latent heat (fusion, vaporisation or sublimation) as shown in Figure 5.1 for the heating of liquid water to steam. Therefore, whenever a body is absorbing or losing heat, special consideration must be given as to whether the change is one of sensible heat, latent heat, or a combination of the two.

5.1.1 Mechanisms of Heat Transfer

There are three modes of heat transfer: conduction, convection and radiation.

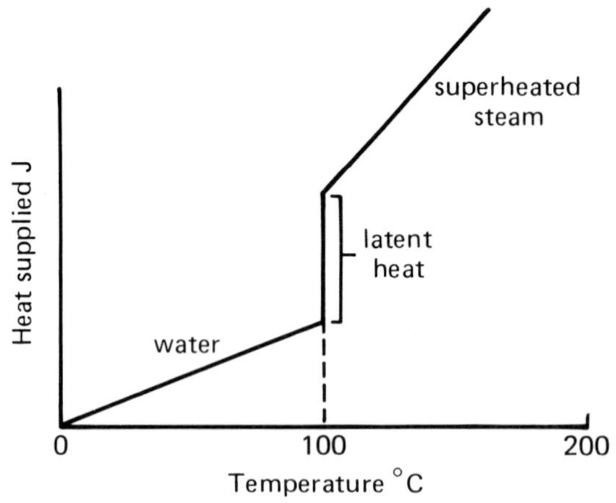

Fig. 5.1 Sensible and latent heating of water

Conduction

Conduction is the transfer of heat from one body to another under the influence of temperature gradients without the appreciable displacement of particles i.e. the transfer of energy is from one molecule to an adjacent molecule. It is minimal in gases, but is the predominant mechanism in opaque solids.

Convection

Convection involves the transfer of heat by the mixing of one parcel of fluid with another. The motion of the fluid may be entirely the result of density changes, as in natural convection, or may be produced by mechanical means, as in forced convection. Therefore discussion of convective heat transfer requires a knowledge of the thermal properties and the fluid flow characteristics of the substances involved and is thus far more difficult to analyse than conduction.

Radiation

Radiation does not depend upon the existence of an intervening medium and involves the heat being transferred in electromagnetic wave form analogous to light. When radiation issues from a source to a receiver, part of the energy is absorbed by the receiver and part is reflected by it. If two bodies, one hotter than the other, are placed in an enclosure there will be a continuous interchange of energy between them. The hotter body radiates more energy than it absorbs, while the cooler one absorbs more than it radiates. Eventually the two reach a steady state, at which the rates of radiation and absorption become equal, but not zero. Radiation therefore involves consideration of optical effects as well as heat transfer.

Often all three forms of heat transfer are involved simultaneously. It is then usual to calculate the rate of heat transfer by each mode separately, adding the separate effects to provide an estimate of the total rate of heat transfer. In this way a complex problem can be resolved and the relative importance of the various modes of heat transfer can be assessed.

Each mechanism together with practical examples is considered in this Chapter.

5.2 CONDUCTION

5.2.1 General Description

Conduction is the simplest form of heat transfer to understand and measure. It is the passing of heat through a material from one particle to the next. Conduction is, therefore, the characteristic mode of heat transfer through a solid, and from one solid object to another if in physical contact. It is present in heat transfer through liquids and gases but the effect is usually small. The rate of conductive heat transfer through a body depends upon its thickness, its cross-sectional area perpendicular to the direction of flow of heat, the temperature gradient and the property of the body called thermal conductivity. This property is controlled both by the nature of the basic material of construction and the type of construction. A loose structure has a very different value of thermal conductivity to that of compact board of the same material. Conduction is primarily of interest where heat flow through a solid material is concerned, such as through furnace walls and insulation.

The application of conduction theory in this Chapter is limited to one dimensional steady state conduction. The concept is not difficult to grasp, and once understood, many practical problems can be solved with reasonable accuracy by making appropriate assumptions. One dimensional means that temperature varies in one direction only. This assumption could be made for example in the case of a furnace wall which has a large area compared with its thickness, and where the temperature of its surface does not vary with position of the surface. It would not be the case at the corner of a structure where heat transfer could be taking place along, as well as through the material thickness. Steady state means that the structure has reached equilibrium, the resulting temperature profile through the material being constant with time. The wall of a building probably never reaches this condition. At the end of a day, for example, the wall may have received heat both from inside the building and from the sun on the outside and the result could be that the wall is giving up heat from both faces inside and outside the building. Nevertheless, the basic equation of conduction is probably the most accurate of all heat transfer formulae. The accuracy depends on the data which is used in the equation.

The rate of conductive heat transfer through a body depends on its thickness, its cross-sectional area perpendicular to the flow of heat, its temperature gradient and the thermal conductivity of the material. This is Fourier's Law and it can be expressed by the equation:

$$\Phi = -kA\,(dt/dd) \qquad 5.1$$

where Φ	=	quantity of heat transferred	W
A	=	area through which it is passing	m^2
dt/dd	=	temperature gradient dt across an infinitesimally small plane layer of thickness dd	K/m
k	=	thermal conductivity of material	W/mK

Materials have at specific temperatures, specific values of thermal conductivity k. Insulating materials have low values of k and metals such as steel have high values because the free electrons in the metallic structure can carry energy easily through the material. Table 5.1 gives values of k for a wide variety of materials.

Under steady state conditions the Fourier equation can be readily solved for single and composite wall structures in addition to other one dimensional structures such as pipes.

Table 5.1 Thermal Conductivity k of Common Materials

Metal	Conductivity W/mK		Non-metal	Conductivity W/mK	
Aluminium	0 °C	202	Asbestos	0 °C	0.151
	200 °C	228		200 °C	0.208
Duralumin	0 °C	159		400 °C	0.223
	200 °C	194	Cardboard	20 °C	0.020
Lead	0 °C	175	Concrete	20 °C	0.80
	200 °C	204	Cork Slab	30 °C	0.04
Iron	0 °C	73	Rubber	20 °C	0.16
	200 °C	62	Wood (Oak)	20 °C	0.18
	400 °C	48	Fuel Ash	80 °C	0.10
	800 °C	36	Boiler Lagging	0.063 -	0.190
Steel	0 °C	43	Calcium-Silicate	300 °C	0.068
(1% carbon)	200 °C	42		400 °C	0.075
	400 °C	36		500 °C	0.082
	800 °C	29		600 °C	0.090
Steel	0 °C	16		700 °C	0.100
(Chrome-nickel)	200 °C	17	Glass	20 °C	1.05
	400 °C	19	Glass Fibre	20 °C	0.003
	800 °C	26	Polystyrene	20 °C	0.035
Steel	0 °C	38	Polyurethane	20 °C	0.023
(manganese)	200 °C	36	Plastics	0.17 -	0.50
	400 °C	34	Firebrick	500 °C	1.00
Steel	0 °C	62		1 000 °C	1.30
(tungsten)	200 °C	54	Silica Brick	500 °C	1.30
	400 °C	45		1 000 °C	1.40
Copper	0 °C	386	Vermiculite	500 °C	0.26
	200 °C	374		1 000 °C	0.29
	400 °C	363	Refractory	400 °C	0.45
Brass 70/30	100 °C	128	concrete	800 °C	0.49
	200 °C	144	Refractory	500 °C	0.29
Constantan	100 °C	22	brick	1 000 °C	0.34
	200 °C	26			

Steady-state heat transfer through plane walls

It follows immediately from equation 5.1 that for steady state heat flow the temperature gradient will be constant if k is constant, i.e. the temperature profile will be linear. Equation 5.1 can then be integrated between the limits d_1 and d_2 to enable the heat flow to be expressed in terms of the surface temperatures t_1 and t_2. This situation is shown in Figure 5.2.

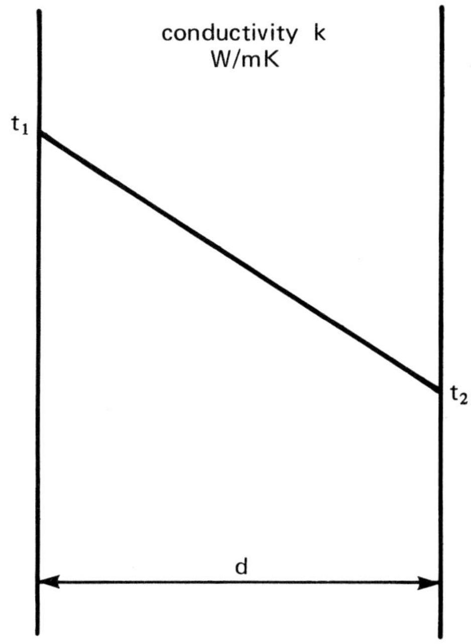

Fig. 5.2 Steady heat conduction through a wall

Single material: for the heat flow through a wall of a single material when the hot and cold face temperatures are known, the basic equation is:

$$\Phi = (k/d)\,A\,(t_1 - t_2) \qquad 5.2$$

where t_1	=	temperature of hot face	°C
t_2	=	temperature of cold face	°C
d	=	thickness of material	m

Composite material: for a wall made up of a number of different materials, the rate of heat transfer is calculated from the formula:

$$\Phi = A(t_1 - t_2)/(d_1/k_1 + d_2/k_2 + d_3/k_3 + ...) \qquad 5.3$$

where t_1 and t_2 are the inside and outside wall face temperatures respectively. The thicknesses and thermal conductivities of the layers of different materials are given in d_1, d_2, d_3... and k_1, k_2, k_3... respectively. Note that the temperatures used in this equation apply to the walls and not to the temperatures of the inside and outside atmospheres which will be different.

Steady-state heat transfer through the walls of a pipe

Single material: The rate of heat transfer through the wall of a pipe with known internal and external face temperatures of t_1 and t_2 is given by:

$$\Phi = 2\pi l k(t_1 - t_2)/\ln(r_2/r_1) \qquad 5.4$$

where l is the length of the pipe, r_1 and r_2 are the inner and outer radii of the pipe and ln is the logarithm to base e.

Composite material: A more general formula applies:

$$\Phi = 2\pi l\,(t_1 - t_2)/[(\ln(r_2/r_1))/k_1 + (\ln(r_3/r_2))/k_2 + (\ln(r_4/r_3))/k_3 + ...] \qquad 5.5$$

5.2.2 Heat Transfer Where Conduction is Controlling Factor

Usually more than one mode of heat transfer is occurring in any set of circumstances. As an example, consider the case of the heat loss through a wall to the surroundings. Heat is transferred to the inner surface of the wall usually by convection and radiation and heat is then transferred through the wall by conduction. On the outer surface heat is transferred by convection and radiation. In practical situations usually only one temperature is known and this makes the calculation tedious, although modern computer techniques enable solutions to be readily obtained. However, simplified procedures are available for a number of common cases including:

- Furnace insulation
- Heat loss from lagged pipe
- Heat loss from buildings
- Scale on boiler tubes

These situations are now described in detail, together with worked examples.

Furnace insulation

Most furnace walls are subject to unsteady state conditions i.e. the temperatures in the walls vary with time as the furnace heating cycle progresses. The temperature profile across the wall is initially non linear, but as the wall absorbs heat the profile approaches linear until, at a steady state, and assuming constant conductivity k, it is fully linear as shown in Figure 5.3.

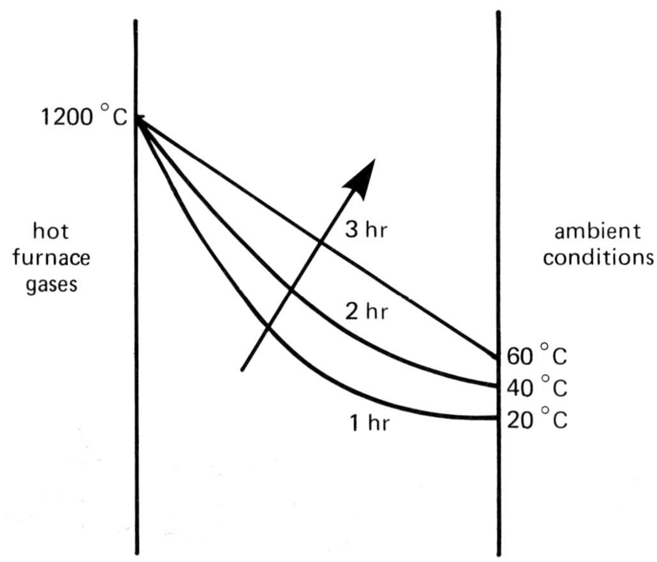

Fig. 5.3 Attainment of steady state conditions in a furnace wall

A significant fraction of the wall heat loss during a furnace heating cycle may be accounted for by considering heat stored in refractory walls. This may, however, be drastically modified by the correct use of insulation.

Example 5.1 Consider a furnace operating at 1 200 °C with an external surface temperature varying from 20 °C up to 60 °C. The insulation may be placed on the outside, a fairly common mode, or on the inside using materials such as alumina fibres.

Case A: 300 mm firebrick with 150 mm external insulation (Fig. 5.4). The firebrick has a relatively high thermal conductivity, and so a steady state linear profile will be reached very early in the cycle. This means that the inner firebrick soon becomes saturated with heat at a high temperature. This of course demands a large amount of heat which is not heating up the load in the furnace.

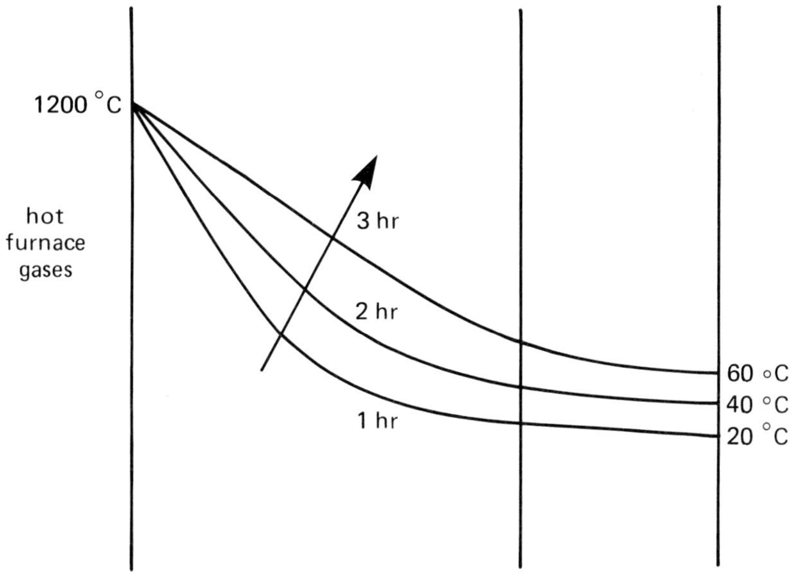

Fig. 5.4 Schematic effect of external insulation for a furnace wall

Case B: 300 mm firebrick with 150 mm internal insulation (Figure 5.5). Because the insulation has a very low conductivity, it would take a considerable time to achieve a linear profile. As a result it will not be achieved in the time available in many types of heating cycle. This means that the majority of the heat stored in the furnace is taken up in the region close to the inner wall, and so the wall has a low overall thermal storage capacity.

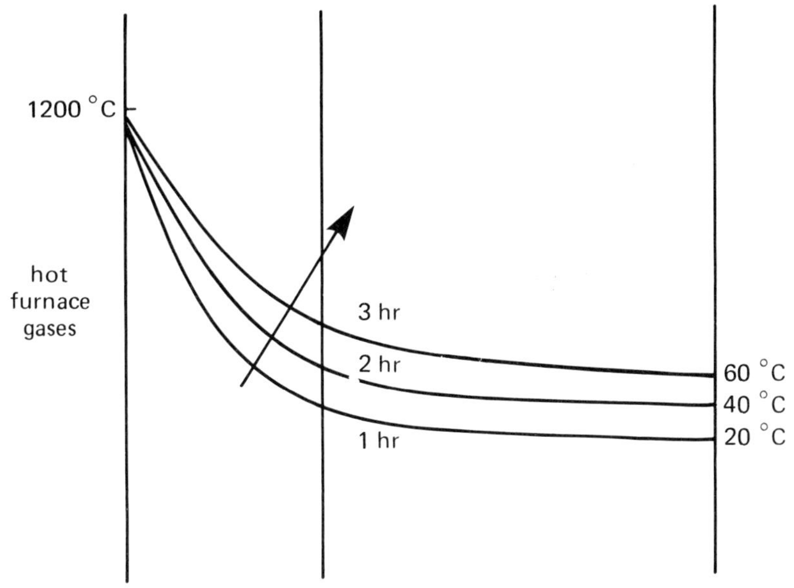

Fig. 5.5 Schematic effect of internal insulation for a furnace wall

From these two cases it can easily be seen that a considerable reduction in heat losses by wall storage can be achieved by using low conductivity insulation material as the inner furnace lining. Consideration should also be given to the density and thermal capacity of the refractory.

Prediction of heat loss and interface temperatures

If temperatures are to be predicted for say, design and material selection, then assumptions must be made and the use of a chart as in Figure 5.6 is recommended. The following assumptions were made when preparing the chart:

- The surroundings have a temperature of 15.6 °C and an emissivity (see Section 5.4 of this Chapter) of 1.0.
- The emissivity of the cold face is 0.9.
- Convection losses occur under turbulent conditions (see Section 5.3 of this Chapter).

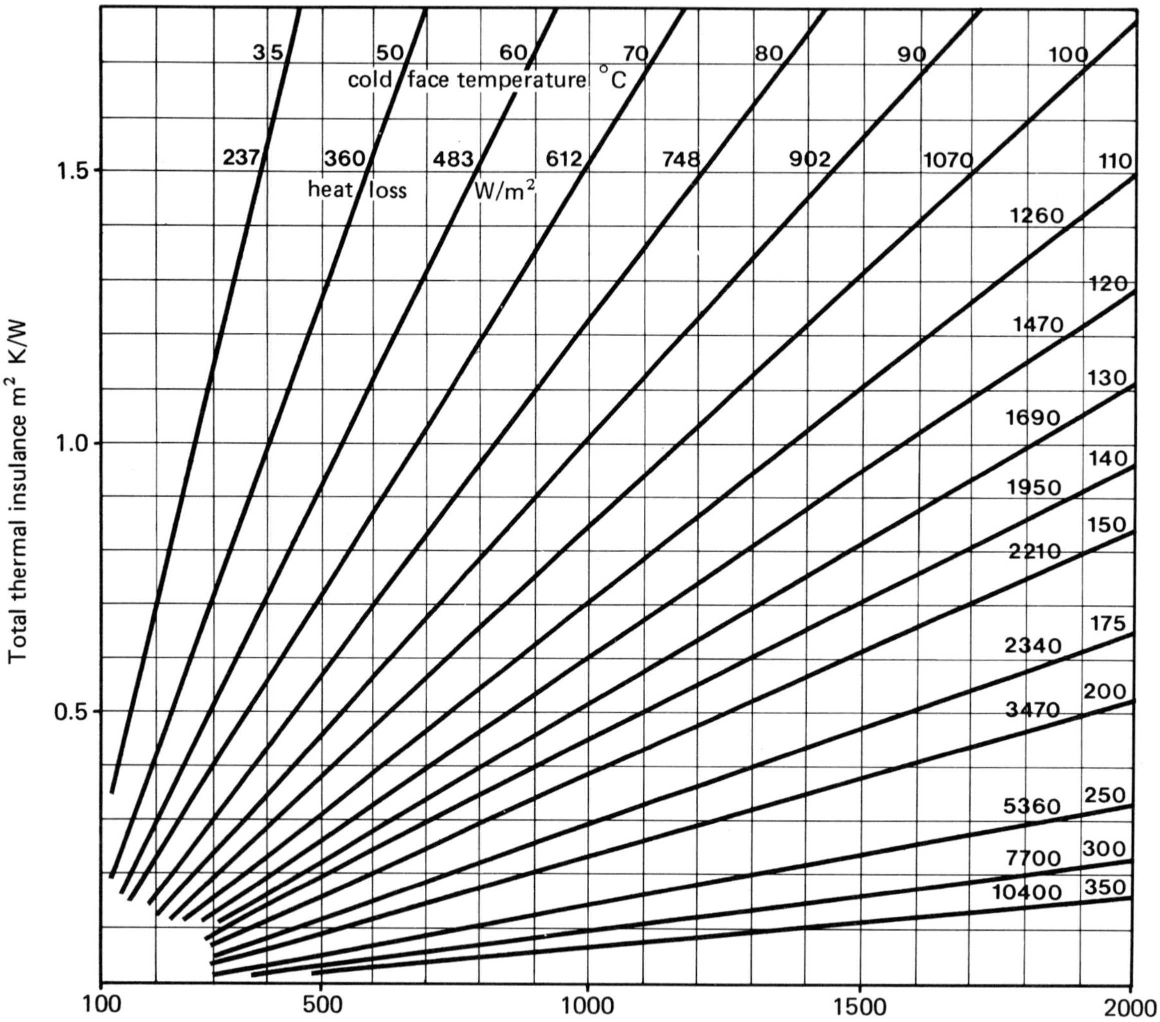

Fig. 5.6 Heat loss through furnace walls

To enable use of the chart it is necessary to introduce a new concept, namely thermal insulance. Thermal insulance (often called thermal resistance) M is defined as:

$$M = d/k \qquad 5.6$$

where	d	=	thickness	m
	k	=	thermal conductivity	W/mK
	M	=	thermal insulance	m^2 K/W

The total thermal insulance of a composite wall is simply the sum of the individual insulances. Flow of heat through a wall is governed by this insulance which, as can be seen, depends on the thickness and the thermal conductivity of the materials and is analogous to resistance in an electrical circuit.

To determine the heat loss through a furnace wall, and the interface temperatures, given the hot face temperature and the materials and dimensions of the wall use the chart in Figure 5.6 as follows:

- Assume a reasonable average temperature for each component and obtain suitable values of thermal conductivity from e.g. Table 5.1.
- Using formula 5.6, derive component insulances and hence total insulance.
- From the chart read the cold face temperature and heat loss corresponding to the given hot face temperature and the total thermal insulance.
- The total heat loss is equal to the heat loss obtained from the chart multiplied by the area of the wall in m^2 and would typically be divided by 1 000 and expressed in kW.

An interface temperature is then derived from the chart as follows:

- Derive the thermal insulance between the cold face and the interface.
- From the chart read the hot face temperature corresponding to the cold face temperature already obtained and the thermal insulance calculated above. This is the required interface temperature.

Heat loss from pipes

Heat loss from hot pipes to the atmosphere can be estimated quite easily if the outside surface temperature of the pipe, whether lagged or unlagged, is known. Heat loss is given by:

$$\Phi = hA(t_s - t_a) \qquad 5.7$$

where	Φ	=	heat loss	W
	h	=	surface heat transfer coefficient obtainable from e.g. Table 5.2	$W/m^2 K$
	A	=	outside surface area i.e. π Dl	m^2
	t_s	=	surface temperature of pipe or lagging	°C
	t_a	=	ambient temperature	°C
	D	=	outside diameter of pipe or lagging	m
	l	=	length of pipe	m

Note that the actual value of h depends on both radiation and convection of heat to the room and accurate calculations are very complex, hence the use of data such as Table 5.2. Certain assumptions about room conditions and the type of surface finish were made when drawing up the table.

Table 5.2 assumes that the surface temperature is known and frequently this can only be assessed by direct measurement. For unlagged pipes conveying steam or hot water, the outisde temperature of the pipe is often assumed to be the steam or water temperature because the heat transfer coefficient on the fluid side of the pipe is very high. In reality the pipe surface temperature must be lower than the temperature of the fluid flowing inside it otherwise no heat transfer would take place, and practical values range from 5K to 25K lower.

Table 5.2 Surface Heat Transfer Coefficient for Lagged and Unlagged Horizontal Pipes ($W/m^2 K$)

Nominal pipe size mm	Temperature difference (°C) between surface and ambient									
	25	50	75	100	125	150	200	300	400	500
15	11.9	13.7	15.2	16.8	18.5	20.2	23.7	32.1	42.5	55.2
25	11.4	13.1	14.6	16.2	17.8	19.4	23.0	31.3	41.7	54.3
50	10.9	12.5	13.9	15.4	17.0	18.6	22.1	30.4	40.6	53.3
100	10.3	11.9	13.3	14.8	16.3	17.8	21.2	29.5	39.6	52.3
200	9.9	11.4	12.6	14.1	15.6	17.1	20.5	28.6	38.8	51.4
300	9.6	11.1	12.3	13.7	15.2	16.7	20.2	28.2	38.3	50.9
600	9.2	10.6	11.8	13.2	14.7	16.1	19.5	27.5	37.6	50.1

In the case of heat loss from a lagged pipe the pipe material is usually of a highly conducting material and its resistance to heat flow can be neglected. Heat therefore can be considered to flow through only the lagging at a mean area (note: the inside and outside areas will be different). Therefore:

$$\Phi = 2\pi k_m l(t_1 - t_2)/[\ln(r_2/r_1)] \qquad 5.8$$

where Φ = heat loss W
k_m = mean thermal conductivity of lagging W/mK
l = length being considered m
t_1 = inside temperature of lagging (assumed to be fluid temperature) °C
t_2 = outside temperature of lagging °C
r_1 = inside radius of lagging m
r_2 = outside radius of lagging (from pipe centre) m
ln = log base e

For design purpose t_2 is often taken as the surrounding air temperature but this cannot be so. In fact t_2 has ro be obtained from a trial and error balance such that

$$\Phi = 2\pi k_m l(t_1 - t_2)/[\ln(r_2/r_1)] = h_2 A (t_2 - t_3) \qquad 5.9$$

where h_1 = heat transfer coefficient from outside surface to its surroundings $W/m^2 K$
A = outside area of pipe m^2
t_3 = surrounding air temperature °C

Thus h_1 and t_2 can be evaluated by balancing the two sides of Equation 5.9 and by the use of Table 5.2 on a trial and error basis. Alternatively, and much more easily, heat losses for both lagged and unlagged pipework can be obtained directly from published tables: see for example CIBSE Guide C3 1986.

When the thickness of lagging is increased its surface area is increased and this can have the effect of increasing heat loss when a critical thickness is exceeded. Nowadays with materials of good insulation properties in nominal sizes it is not usually a problem but it is worth bearing in mind.

The value of the outer radius of insulation for which heat transfer through the wall will be a maximum is given by:

$$r = k/h \qquad 5.10$$

where	r	=	critical radius	m
	k	=	thermal conductivity of lagging	W/mK
	h	=	surface heat transfer coefficient obtainable from table 5.2 and making assumptions for surface temperature	$W/m^2 K$

If the external radius is less than k/h, then increasing the insulation up to a radius of k/h will increase the heat loss from the pipe. The situation is only likely to arise if the value of k is relatively high, that is if the insulating properties are poor.

Heat loss from buildings

The reduction of heat loss from buildings is of great economic importance. The CIBSE Guide A3 1986 gives overall heat transfer coefficients ('U values') for various types of wall, window and roof materials.

Table 5.3 U- Values for Common Constructional Materials

Construction	U value $W/m^2 K$
260 mm cavity wall, 105 mm inner and outer leaves, plus 16 mm light plaster on inner face	1.3
220 mm solid wall, with 16 mm light plaster	1.9
335 mm solid wall, with 16 mm light plaster	1.5
Pitched roof, tiles on battens with roofing felt, roof space, foil backed plasterboard ceiling	1.5
As above, plus 50 mm glass fibre loft insulation	0.5
Window, single glazing, 30% area due to wood frame	4.3
As above, double glazing	2.5
Single glass, no frames	5.6
Double glass, no frames – with 25 mm air gap	2.9
– with 6 mm air gap	3.4
105 mm brickwork with 25 mm mineral fibre, or slate and blocks with plaster internally	0.53

U values are based on theory and empirical results and simplify what would be complex mathematical problems. Examples of their use are given below.

Example 5.2 Use U values to calculate the heat transfer rate through a building structure. The wall area is 110 m^2, window total area 14 m^2, upstairs ceiling area 36 m^2, environmental temperature difference 21 K (20 °C internal, -1 °C external).

335 mm solid brick wall, U = 1.5 $W/m^2 K$; pitched roof with tiles U = 1.5 $W/m^2 K$; single glazed windows, U = 4.3 $W/m^2 K$.

For heat flow through walls, windows and roof:

Heat loss = (temperature difference)(area)(U value)
Heat loss = $21[(110 \times 1.5) + (36 \times 1.5) + (14 \times 4.3)]$
= 5 860 W
= **5.86 kW**

Example 5.3 Consider the same house structure as above with 30 mm foam wall board lining k = 0.026 W/mK; pitched roof as before plus 50 mm glass fibre insulation U = 0.5; double glazed windows U = 2.5 W/m^2 K.

For the heat flow through the walls we use the concept of thermal insulance. Thermal insulance of the insulated wall is the original insulance plus the insulation insulance which is:

$1/1.5 + 0.03/0.026 = 1.82$
New U value $= 1/1.82 = 0.55$

$\Phi = 21[(110 \times 0.55) + (36 \times 0.5) + (14 \times 2.5)] = 2\,380\text{ W} = \mathbf{2.38\text{ kW}}$

A saving of 3.48 kW is achieved.

Scale on the water side of boiler tubes

This is another problem that can be considered using conduction theory. If the tubes are initially clean and free from scale (Figure 5.7 A), then from equation 5.3 the heat transfer rate is:

$$\Phi = (k_w/d)A(t_g - t_w) \qquad 5.11$$

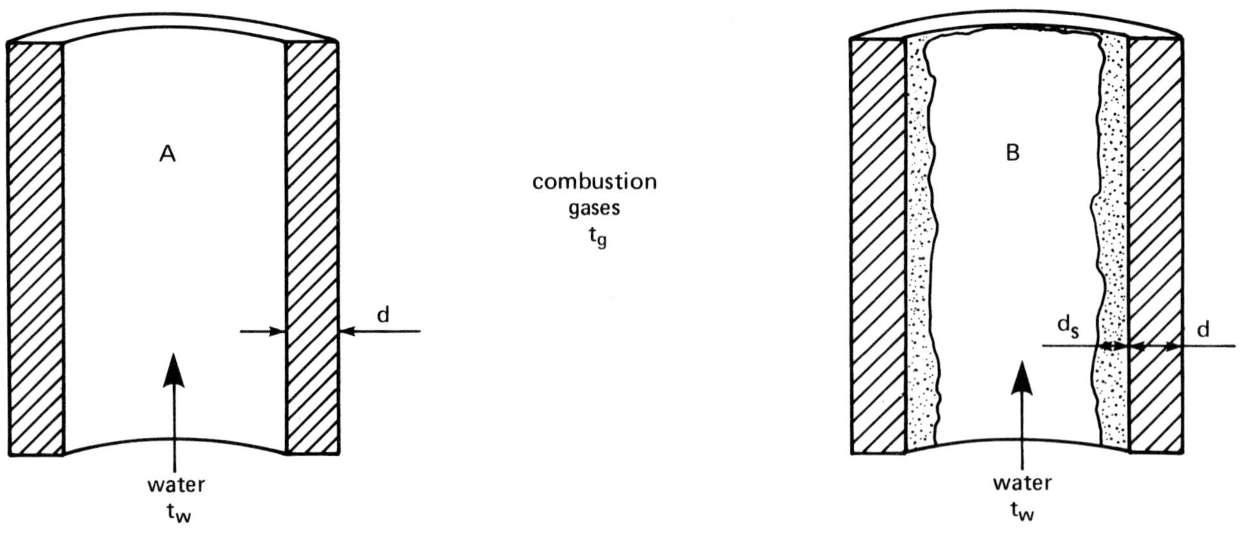

Fig. 5.7 Boiler tube scale formation

After operation over a period of time, scale may develop and build up as shown schematically in Figure 5.7B. The heat transfer then is:

$$\Phi = (k_w/d)\, A\, (t_g - t_i)$$
$$= (k_s/d_s)\, A\, (t_i - t_w) \qquad 5.12$$

where

k_w	=	thermal conductivity of clean tube	W/mK
k_s	=	thermal conductivity of scale	W/mK
d	=	thickness of tube	m
d_s	=	thickness of scale	m
A	=	area of heat transfer	m^2
t_g	=	combustion gas temperature	°C
t_w	=	water temperature	°C
t_i	=	the scale/tube interface temperature	°C

Rearrangement and elimination of t_i from these equations gives:

$$t_g - t_w = (\Phi/A)/[(d/k_w) + (d_s/k_s)] \qquad 5.13$$

$$\Phi = A(t_g - t_w)/[(d/k_w + (d_s/k_s)] \qquad 5.14$$

Comparison of equations 5.14 and 5.11 shows an increase in the denominator and hence a reduction in the heat transfer with scale formation.

Example 5.4 Typical values of gas and water side temperatures and scale characteristics are:

d = 5 mm of stainless steel, k_w = 15.9 W/mK
d_s = 0.6 mm of scale, k_s = 2.2 W/mK
t_g = 600 °C
t_w = 80 °C

Using equations 5.11 and 5.14 the rate of heat transfer per unit tube area is:

- without scale Φ/A = 1.65 MW/m^2
- with scale Φ/A = 0.89 MW/m^2

(MW = 10^6 W)

5.2.3 Unsteady or Transient Conduction

When a slab or pipe is first heated up, it can take some time to reach steady-state conditions. During this period, temperature varies with time and the temperature gradients through the body are not constant. Those near the heated face are large while those near the external wall remain small for some time after heating has started.

It is usually difficult in these circumstances to obtain analytical solutions to problems of heat conduction and generally numerical approximation methods have to be used. The time taken to reach steady-state conditions is very variable and is an important factor which has to be taken into account when heating furnace walls and loads and in designing regenerator systems. Refer to standard texts for detailed calculation methods under transient conditions.

5.3 CONVECTION

5.3.1 Introduction and General Description

Convective heat transfer is the most important mode of heat transfer associated with liquids and gases and involves movement of the fluid to carry heat with it. The movement can be natural, due to density differences arising from temperature differences, or it can be forced, say by use of a fan. The transfer of heat by convection is commonly between a fluid and a solid surface and the rate of transfer will be increased by the use of some method of forcing the flow to provide turbulence. In general, the simplest way of increasing heat transfer is to increase the velocity of the gas or liquid to reduce the thickness of the boundary layer which clings to the solid surface. Natural convection is usually small in value unless the surface is vertical and with high surface temperature. Refer to Chapter 4 for a detailed exposition of the fluid flow considerations governing convection.

There is rarely a need to increase the heat transfer of liquid-to-surface interfaces as these are inherently very high compared to that of air or gas. It should be noted that gas-to-surface heat transfer is usually the lowest and therefore the controlling factor of a system.

In the preceding Section it was shown that the basic equation of conduction can give analytical solutions of heat transfer problems for single cases where information such as surface temperatures and conductivities are accurately known. Charts and tables can be used for more complicated problems. However, the solution of the differential equations that govern the transfer of heat by convection usually present considerable mathematical difficulties. There are a number of methods of calculating rates of heat transfer due to convection and the most important aspects are described below, although detailed information should be obtained from standard textbooks.

Boundary layer theory

In boundary layer theory it is assumed that the fluid at the wall adheres to it and therefore that the heat flow at the wall is by conduction and not by convection.

Reynolds analogy

This method studies the behaviour of the boundary layer and makes use of the similarity between the mechanism of fluid friction in the boundary layer and the transfer of heat by convection. This analogy is rarely valid, but when it is, rates of heat transfer can be predicted from the measurement of the sheer stress between a fluid and a wall.

Dynamic similarity

A third method is concerned with model testing to obtain information in a form which is applicable to similar systems of any scale. It is based on the principle of dynamic similarity, and employs the method of dimensional analysis for the correlation of empirical data. There is a wealth of published data, and equations based on experimental results using dynamic similarity are widely available. As a matter of interest some of the more important dimensionless groups which enable empirical results to be applied to real problems are:

Pr (Prandtl number) is concerned with the properties of the gas or liquid, namely conductivity, specific heat and viscosity.

Re (Reynolds number) defines the conditions of flow when the flow depends on momentum forces. It depends on velocity, size of pipe or duct, and viscosity and is used for forced convection.

Gr (Grashof number) relates temperature differences and buoyancy forces for flows which depend on such forces and is therefore used for natural convection.

Nu (Nusselt number) relates the system size to heat transfer.

The basic equation for convection is the Newton equation:

$$\Phi = hA(t_g - t_s) \qquad 5.15$$

where Φ	=	heat transfer rate	W
h	=	heat transfer coefficient	$W/m^2 K$
A	=	surface area	m^2
t_g	=	gas or liquid temperature	°C
t_s	=	surface temperature	°C

The problem is finding h, the heat transfer coefficient. This is found by the use of known empirical relationships using the dimensionless groups described above.

For forced convection in turbulent flows $Nu = a(Re)^b (Pr)^c$ 5.16

where a, b and c are constants which depend on the system and its geometry.

and $Nu = hD/k$ thus $h = Nu\, k/D$ 5.17

D	=	dimension, usually diameter upon which Re was based	m
k	=	thermal conductivity of the fluid	W/mK
Re	=	$\rho Dv/\mu$	dimensionless
Pr	=	$\mu C_p/k$	dimensionless
ρ	=	gas or liquid density	kg/m^3
v	=	mean flow velocity	m/s
μ	=	dynamic viscosity of gas or liquid	Pa.s

For flow in pipes a = 0.023, b = 0.8, c = 0.3 where Re is greater than 10 000 and where temperature difference between surface and fluid is less than 5 K for liquids and 55 K for gases. Once h has been found then the heat transferred is calculated from equation 5.15.

This is just one example of how the empirical relationships can be used. Refer to standard texts for further examples for various geometries.

Heat transfer from, or to, the inside wall of a tube from, or to gas

At any particular point in a system the heat transfer from, or to, a gas flowing in a tube can be found from equations 5.16 and 5.17 by substitution:

$$h = [(8.1 \times 10^{-5}) c_p (\rho v)^{0.8}]\ D^{0.2}$$

where c_p	=	specific heat capacity of gas from tables	J/kgK
ρ	=	density of gas, from tables	kg/m^3
v	=	mean velocity	m/s
D	=	diameter or hydraulic mean diameter if not circular	m

If the duct is not circular then the hydraulic mean diameter would be used. This is 4 times the cross sectional area divided by the perimeter.

The heat transfer Φ for a length of pipe is then:

$$\Phi = hA[t_g - t_s] \quad \text{W} \qquad 5.19$$

where A = area of inside surface = πDI — m^2
t_g = temperature of gas — °C
t_s = temperature in inside surface — °C

Obviously, for a long length of pipe the value of $t_g - t_s$ would need to be a mean value for accurate calculations since both these temperatures would be changing along the length of the tube. The determination of the mean temperature can be complex. Also the property values taken from tables, i.e. density and specific heat capacity, should be evaluated at a mean of the temperature at the two ends of the tube.

5.3.2 Improving Convective Heat Transfer

The simplest way to improve the rate of convective heat transfer is to increase the velocity of the liquid or gas so that it reduces the thickness of the boundary layer that clings to the surface of the solid. Although this may not be necessary with liquid/solid interfaces, where gases are involved the rate of heat transfer is low and usually needs to be increased to reduce heat exchange surface area to an optimum level.

For given conditions of temperature, density, viscosity etc, the heat transfer coefficient varies as the fluid velocity to a power of less than one. In the case of a gas inside a tube, the relationship is:

$$h \propto v^{0.8} \qquad 5.20$$

Therefore to increase h and thus either transfer more heat or make the heat exchanger more compact, one would only have to increase the velocity. However, increasing the velocity also increases the pressure drop (Δp) by a factor proportional to v^2.

$$\Delta p \propto v^2 \qquad 5.21$$

There is therefore an economical limit to increasing velocity and other means are used to promote convective heat transfer and achieve compact heat exchangers. These are, for example, increasing the effective surface area by the use of fins or similar devices.

Jet impingement

One particularly effective method of attaining high heat transfer rates is to make a jet of fluid, particularly a hot gas, impinge onto a surface, as shown schematically in Figure 5.8. By this means values of h of around 200 $W/m^2 K$ can be achieved, while a typical value for pipe flow is about 50 $W/m^2 K$. Gas firing is particularly suitable to such methods, since the impingement of the flame on a surface does not easily form sooting, whereas serious sooting problems arise from oil burning. Indeed, it is vital to avoid any possibility of contact between a cold surface and an oil flame.

5.3.3 Boiling Heat Transfer

In the case of heat transfer across a surface to a boiling liquid the value of the heat transfer coefficient is very dependent on the temperature difference ΔT between the hot surface and the boiling

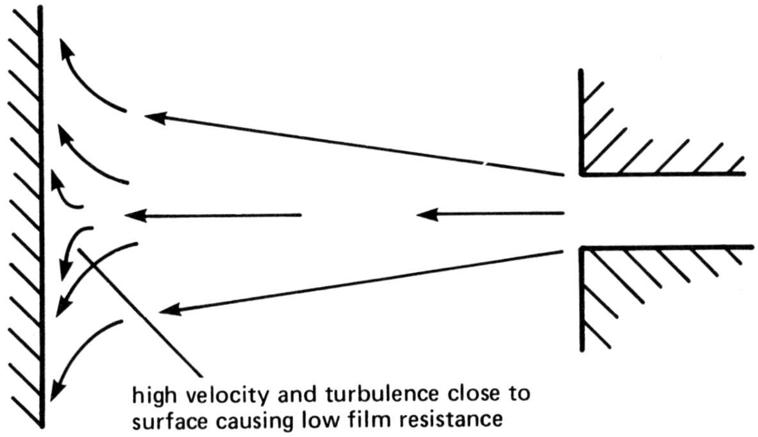

Fig. 5.8 Jet impingement on a surface

liquid. It is therefore important that in designing equipment for boiling the correct value of ΔT is obtained, in order to give the highest values of heat transfer coefficient.

Analysis of heat transfer to boiling liquids is difficult and theories are still incomplete. There are two distinct types of boiling phenomena: pool boiling and forced flow boiling.

Pool boiling

Pool boiling occurs when a large volume of liquid is heated by a submerged surface. Free convection currents stimulated by agitation of the rising vapour bubbles cause motion within the liquid. The mechanism is shown in Figure 5.9, and it is clear that there are three stages in the process.

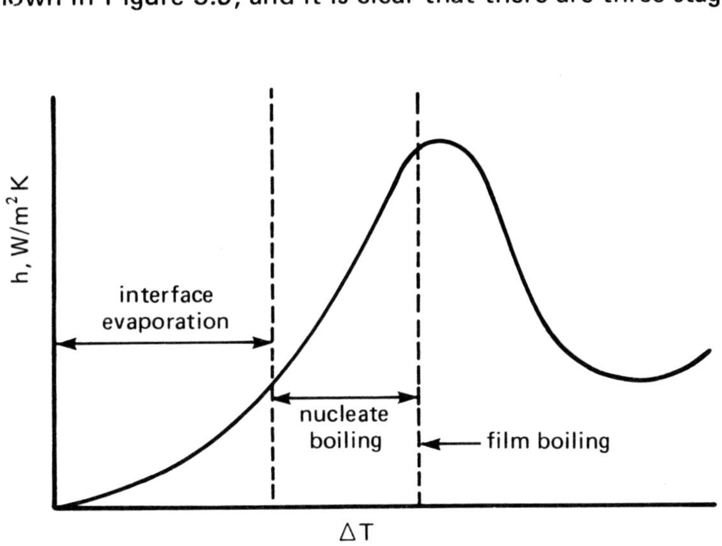

Fig. 5.9 Schematic of boiling heat transfer

Interface evaporation. The bubbles of vapour formed on the heated surface move to the liquid by natural convection, and so exert little agitation on the liquid.

Nucleate boiling. The bubbles form more rapidly and exert appreciable agitation on the liquid, so giving a steep rise in the value of the heat transfer coefficient. This is the most important region of the curve for boiling in industrial equipment.

Film boiling. If the value of ΔT is increased further, the bubbles will form so rapidly that they will combine to form a vapour blanket over the surface and so prevent fresh liquid from flowing onto that surface. This produces a considerable fall in the value of the heat transfer coefficient. For this reason film boiling should be avoided.

Forced-flow boiling

A vertical boiler tube provides an illustration of the various regimes occurring with forced flow boiling. In such a once-through tube liquid enters at the bottom and gradually evaporates until a dry or superheated vapour leaves at the top. It is assumed that there is a constant heat flux along the length of the tube.

In the first section no boiling occurs and heat flow is governed by normal forced convection.

In the next section although the bulk of the liquid is still below saturation temperature, bubbles form at the wall and collapse as they move inward.

Simple nucleate boiling occurs in the next section and when the vapour content increases further this leads on to another section of slug flow with nucleation sites in the film on the wall.

Further evaporation causes the flow to change to annular flow. Heat is transferred through the film so formed by conduction. Eventually the film will completely evaporate and there will be a sudden rise in wall temperature due to the low heat transfer coefficient at the unwetted wall.

Evaporation will continue, but as a result of radiation from the wall, together with convection of the superheated vapour surrounding any mist will eventually give rise to entirely superheated vapour.

The action of boiling liquids against surfaces can give heat transfer rates to the surface up to 200 times greater than normal forced convection. For specialised treatment refer to standard texts.

5.3.4 Condensation

In the case of a vapour condensing on a cold surface, the film of liquid which is formed on the surface acts as a barrier to the heat transfer between the vapour and the surface. It is, therefore, important to arrange the equipment so that the liquid film is rapidly removed from the surface, keeping the cold surface well exposed to fresh vapour. The heat transfer coefficient is greatly dependent on the thickness of the liquid film that is allowed to collect on the surface.

When condensation occurs in, for example, heat recovery from a drying process or in a direct contact water heater, the heat transfer rate is increased. This can result in heat transfer rates up to 40 times greater than those achieved with normal convection.

For treatment of actual heat transfer by condensation refer to standard texts. The assessment of heating plant flue losses when condensation is occurring is also detailed in Chapter 1.

5.3.5 Overall Heat Transfer

For many industrial units, for example heat exchangers, it is necessary to determine the amount of heat which will be transferred from one fluid through some dividing wall to a second fluid. This requires the application of an equation for the heat transferred from the first fluid to the wall, then for the heat transferred from the wall to the second fluid. The resistance of the wall to heat flow can usually be neglected, leaving only the convective heat transfer to be considered when it can be shown that an overall expression of the following form is applicable:

$$\Phi = K A (t_{f1} - t_{f2}) \quad 5.22$$

where t_{f1} = temperature of first fluid °C
t_{f2} = temperature of second fluid °C
A = area of heat transfer m^2
K = overall heat transfer coefficient (same concept as U value) where $1/K = (1/h_1)+(1/h_2)$ $W/m^2 K$

It can be seen that if the values of h_1, and h_2 are similar then they make a more or less equal contribution to the value of K. If, however, they are very different one value will have far more influence. The more influential one is known as the controlling factor. However much improvement is made to the non-controlling factor, the improvement to K is limited by the controlling factor. In any case where an overall heat transfer factor is made up of several component factors, attention for improvement must be paid to the controlling factors first, since these absolutely limit any improvement.

If the dividing wall is not made of metal and/or it is thick then:

$$1/K = (1/h_1) + (d/k) + (1/h_2) \quad 5.23$$

where d = thickness of wall
k = conductivity of wall

The heat transfer q per unit area is thus:

$$q = (t_{f1} - t_{f2})/[(1/h_1) + (d/k) + (1/h_2)] \quad 5.24$$

In a heat exchanger both fluid temperatures change as they pass along the exchanger. Therefore, the difference in temperature between fluids will vary. This is illustrated in Figure 5.10 for both parallel flow and counterflow heat exchangers.

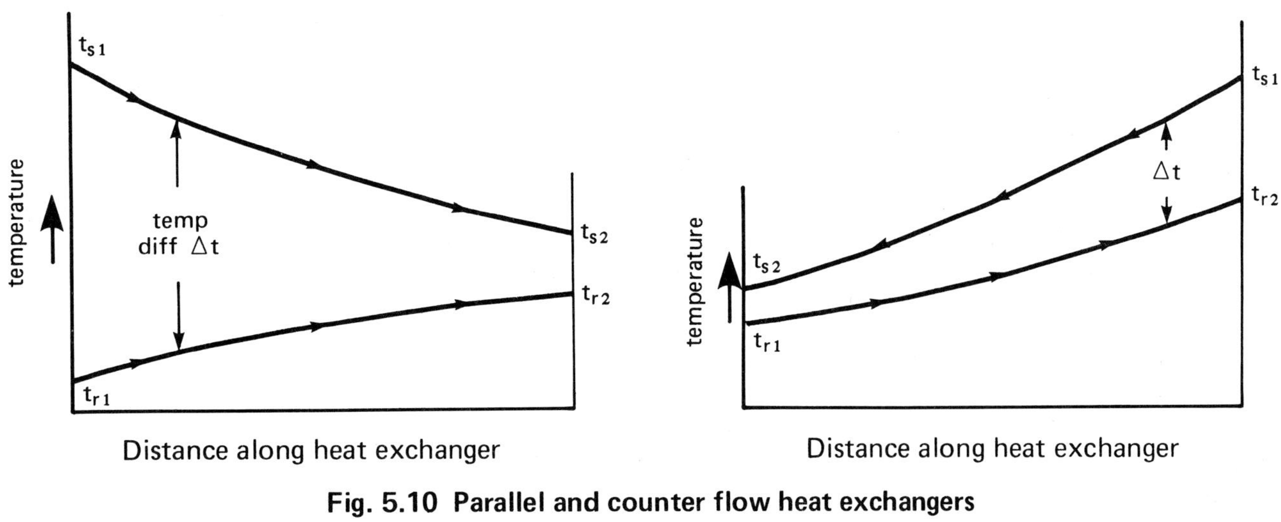

Fig. 5.10 Parallel and counter flow heat exchangers

Let the difference in fluid temperature at one end of the exchanger be Δt_1 and at the other end be Δt_2. Provided that one difference is no more than twice the other, then the mean temperature difference calculated as follows can be used to replace the temperature difference in equation 5.24.

$$\text{MTD} = (\Delta t_1 + \Delta t_2)/2 \qquad 5.25$$

If, however, the differences vary by more than a factor of two, then the logarithmic mean must be used:

$$\text{LMTD} = (\Delta t_1 + \Delta t_2)/\ln(\Delta t_1/\Delta t_2) \qquad 5.26$$

Fouling of heat exchangers

The effect of scale build-up from water in heat exchangers will give considerable resistance to heat transfer. To enable prediction to be made this must be allowed for in the calculation of K:

$$1/K = 1/h_1 + 1/h_s + 1/h_2 \qquad 5.27$$

The heat transfer coefficient for scale, h_s can be estimated using Table 5.4.

Table 5.4 Heat Transfer Coefficient of Boiler scale

Water Type	Heat Transfer Coefficient h_s $W/m^2 K$
Distilled water	11 300
Treated boiler feedwater	11 300
Treated cooling tower water	5 860
Hard water	2 090

Alternatively, if it is possible to have an actual scale layer analysed then its actual conductivity can be found from Table 5.5 and substituted in the place of $1/h_s$ i.e.:

$$1/K = 1/h_1 + d_s/k_s + 1/h_2 \qquad 5.28$$

where d_s = the thickness of the scale layer m
k_s = the conductivity of the scale layer W/mK

Table 5.5 Thermal Conductivity of Boiler Scale

Relative Density	Temperature	Conductivity k_s W/mK
1.27	30 °C	1.31
2.14	30 °C	3.23
0.05	100-300 °C	0.074 - 0.01
1.5	100-300 °C	0.36 - 0.43
2.5	100-300 °C	1.75 - 2.32
Silicate 0.29-0.97	300 °C	0.08 - 0.17
Sulphate 0.19-0.02	300 °C	0.58 - 2.32

Extensive details and practical examples relating to the design and use of heat exchangers can be found in Chapter 7.

5.3.6 Heat Transfer by Natural Convection

When a temperature difference is established between, for example, a wall and a stationary fluid such as air, then the fluid adjacent to the wall will move upward if the wall temperature is higher than the fluid. The fluid will move downwards if the temperature is lower. The movement is caused by the temperature gradient which sets up density gradients in the fluid resulting in buoyancy forces and free convection currents. The rate of heat transfer depends mainly on the fluid motion.

It is possible to derive simple formula for heat loss from hot surfaces due to natural convection in air. This is because the physical properties in the Nusselt, Grashof and Prandtl numbers can be assumed constant. Refer to standard textbooks for full derivation of the formulae. The dependence of heat transfer coefficient h on temperature can be derived as follows:

For laminar flow:

$$h \propto [(t_s - t_a)/l]^{0.25} \qquad 5.29$$

For turbulent flow:

$$h \propto (t_s - t_a)^{0.33} \qquad 5.30$$

where h	=	heat transfer coefficient	$W/m^2 K$
t_s	=	temperature of surface	°C
t_a	=	air temperature	°C
l	=	relevant dimension of surface	m

In free convection the product of the Prandtl and Grashof numbers serves as a criterion of turbulence, analogous to the Reynolds number in forced convection. The exact value where turbulence is established depends on the geometric configuration:

$$Ra = (Pr \times Gr) = 6.4(t_s - t_a)l^3 \times 10^7 \qquad 5.31$$

where l = the characteristic length for the particular geometric configuration
Ra = $(Pr \times Gr)$ is the Rayleigh number

An illustration of the basic technique for evaluating heat transfer coefficients is given below for a range of common situations and geometric configurations.

Horizontal cylinders

Characteristic dimension l is the diameter
For laminar flow, $10^4 < Ra < 10^9$ then
$h = 1.31[(t_s - t_a)/l]^{0.25}$
For turbulent flow, $10^9 < Ra < 10^{12}$ then
$h = 1.24\ (t_s - t_a)^{0.33}$

Vertical surfaces

Characteristic dimension l is the height
For laminar flow, $10^4 < Ra < 10^9$ then
$h = 1.41[(t_s - t_a)/l]^{0.25}$
For turbulent flow, $10^9 < Ra < 10^{12}$ then
$h = 1.31\ (t_s - t_a)^{0.33}$

Horizontal surfaces

Characteristic dimension l is the length of side

Hot facing upwards:
For laminar flow, $10^5 < Ra < 10^8$ then
$h = 1.31\,[(t_s - t_a)/l]^{0.25}$

Cold facing downwards:
For turbulent flow, $Ra > 10^8$ then
$h = 1.52\,(t_s - t_a)^{0.33}$

Hot facing downwards or cold facing upwards:
Flow is always laminar and
$h = 0.58\,[(t_s - t_a)/l]^{0.25}$

Published tables of heat transfer coefficients and heat emissions are available to enable estimates to be made when carrying out engineering calculations. Refer, for example, to CIBSE A3 1986. However, an example using the above equations is given to illustrate the method for ready estimation of heat transfer.

Example 5.5 A heating panel has the form of a vertical rectangle 2 m long by 0.8 m high. It convects freely from both surfaces. The surface temperature is 85 °C. Find the approximate heat transfer by natural convection. The temperature of the room is 20 °C.

$(t_s - t_a) = 65\text{ K}$
$l = 0.8\text{ m}$

Use formula 5.31 to derive Ra:
$Ra = (0.8)^3 \times 6.4 \times 10^7 \times 65 = 2.13 \times 10^9$

Therefore flow is turbulent since $Ra > 10^9$
The heat transfer coefficient h for vertical surfaces, turbulent flow, is given by

$$
\begin{aligned}
h &= 1.31\,(t_s - t_a)^{0.33} \\
&= 1.31 \times 65^{0.33} \\
&= 5.19\text{ W/m}^2\text{ K}
\end{aligned}
$$

Thus the heat transfer by natural convection, Φ_c is:

$$
\begin{aligned}
\Phi_c &= h\,A\,(t_s - t_a) \\
&= 5.19 \times 2 \times 2 \times 0.8 \times 65 \\
&= \mathbf{1\,080\ W}\ (1.08\text{ kW})
\end{aligned}
$$

Note that the above calculation takes no account of radiation. Refer to Example 5.6 for radiation heat transfer from this panel.

Where there is a variation in surface temperature with position, the method used is to divide the surface up into areas of approximately 0.15 m² and measure the temperature at the centre of each square. The heat transfer is then evaluated for groups of squares that are at 5 °C intervals. Heat dissipation by convection is the sum of the solutions for each group of areas.

5.4 RADIATION

5.4.1 Introduction and General Description

Heat transfer by radiation is the one mode which does not depend upon the presence of material solid or fluid between the source of heat, and the recipient or sink. Radiant heat will pass through air but a certain amount will be absorbed by flue gases because of the presence of carbon dioxide and water vapour. Radiation from hot combustion gases is dependent on the thickness of the gas layer. The physical laws governing the behaviour of thermal radiation bear a strong resemblance to those which govern the behaviour of light since both are of the same family. Therefore radiant heat can be emitted by a source, be received, reflected and transmitted. As with light, the degree of accomplishment of each of these depends upon properties of the radiation such as wavelength, and upon properties of the materials involved. Also the relative positions of each surface affect the rates of heat transfer.

Thermal radiation forms a small part of the electro-magnetic spectrum and is associated with part of the ultra-violet, all of the visible, and some of the infra-red radiation. Thermal radiation is energy emitted by solids, liquids and gases by virtue of their temperature above absolute zero.

A 'black body' emits radiation according to the equation:

$$M = \sigma T^4 \qquad 5.32$$

where M = radiant exitance, i.e. energy emitted in unit time from unit area, W/m^2
σ = the Stefan-Boltzmann constant (5.57×10^{-8} $W/m^2 K^4$)
T = absolute temperature K

A black body is a perfectly radiating body and emits at any given temperature the maximum possible energy at all wavelengths. The energy emitted will be less for real materials. The above equation defines the energy emission rather than the nett exchange between it and its surroundings. The emitting and absorption characteristics of surfaces and the 'view' that surfaces have of each other are factors which enter the consideration of radiation exchanges.

Emissivity

In reality few materials behave as black bodies because they emit only a proportion of the energy emitted by a black body at the same temperature. This proportion is governed by the emissivity ϵ and Equation 5.32 now becomes:

$$M = \sigma \epsilon T^4 \qquad 5.33$$

The value of ϵ for a material depends on the wavelength of the radiation, but for most radiation calculations we normally take a constant value over the spectrum and then refer to the body as a 'grey' body. The emissivity also varies from one material to another and on the surface condition of the material. Firebrick, for example, has an emissivity of 0.8, whereas polished aluminium has a value of 0.03. Values of emissivity for a range of materials are given in Table 5.6.

The emissivity of a material is also the 'absorptivity – the fraction of the radiation falling on the surface that is absorbed. Thus highly reflective surfaces, which have low absorptivity, will also be poor emitters of radiation.

Table 5.6 Emissivities of Various Surfaces

Materials	°C	Emissivity
Gold, polished	130	0.018
Gold, polished	400	0.022
Silver	20	0.02
Copper, polished	20	0.03
Copper, lightly oxidised	20	0.037
Copper, scraped	20	0.07
Copper, black oxidised	20	0.78
Copper, oxidised	130	0.76
Aluminium, bright rolled	170	0.039
Aluminium, bright rolled	500	0.05
Aluminium paint	100	0.20-0.40
Silumin, cast, polished	150	0.186
Nickel, bright matte	100	0.041
Nickel, polished	100	0.045
Manganin, bright rolled	118	0.048
Chrome, polished	150	0.058
Iron, bright etched	150	0.128
Iron, bright abrased	20	0.24
Iron, red rusted	20	0.61
Iron, hot rolled	20	0.77
Iron, hot rolled	130	0.60
Iron, hot cast	100	0.80
Iron, heavily rusted	20	0.85
Iron, heat-resistant oxidised	80	0.613
Iron, heat-resistant oxidised	200	0.639
Zinc, grey oxidised	20	0.23-0.28
Lead, grey oxidised	20	0.28
Bismuth, bright	80	0.34
Corundum, emery rough	80	0.855
Clay, fired	70	0.91
Lacquer, white	100	0.925
Red lead	100	0.93
Enamel, lacquer	20	0.85-0.95
Lacquer, black matt	80	0.97
Bakelite lacquer	80	0.935
Brick, mortar, plaster	20	0.93
Porcelain	20	0.92-0.94
Glass	90	0.94
Ice, smooth, water	0	0.966
Ice, rough crystals	0	0.985
Waterglass	20	0.96
Paper	95	0.92
Wood, beech	70	0.935
Tar paper	20	0.93

View factor

If we consider two 'black' surfaces of areas A_1 and A_2, then we can define a 'view factor' F_{12}, which equals the fraction of the total energy radiated from A_1, which is intercepted by A_2. This fraction is dependent on the relative sizes, shapes and orientations of the two surfaces. For radiation exchange between these two black surfaces:

$$\Phi_{12} = A_1.F_{12}\, \sigma\, (T_1^4 - T_2^4) \qquad 5.34$$

For 'grey' surfaces the situation is more complicated because some radiation will be reflected and hence re-radiated from the surfaces.

Although the calculation of view and configuration factors is outside the scope of this publication, special cases are considered here to show how simplified calculations can be made.

5.4.2 Radiation of an Object to its Surroundings

The special case of a body which is emitting radiation to large and distant surroundings does not require the analysis of view factor. This simplifies the mathematics and the result is a useful tool.

The basic equation for this case is:

$$\Phi = \sigma A(\epsilon_1 T_1^4 - \epsilon_2 T_2^4) \qquad \text{W} \qquad 5.35$$

where σ = 5.67×10^{-8} W/m^2K^4
T_1 = surface temperature of body K
T_2 = temperature of surroundings K
ϵ_1 = emissivity of body surface
ϵ_2 = emissivity of surroundings

Example 5.6 A heating panel at 85 °C and 2 m long by 0.8 m high is free to radiate to the room on one side. The emissivity of the panel surface is 0.6 and the temperature in the room is 20 °C. What is the radiation heat transfer to the room?

T_1 and T_2 are 358 K and 293 K respectively. Using formula 5.35 and ignoring ϵ_2, radiation heat transfer is:

$$5.67 \times 10^{-8} \times 0.6 \times 2 \times 0.8\, (358^4 - 293^4)$$

= **493 W**

If the temperature of a surface varies then summate the heat transfer for groups of squares on the surface as described in Section 5.3.6.

Radiation between two parallel surfaces

Another special case is where there are two surfaces parallel and fairly close together. Nett heat transfer by radiation is then given by:

$$\Phi = \sigma A\, (T_1^4 - T_2^4)/[(1/\epsilon_1) + (1/\epsilon_2) - 1] \qquad 5.36$$

where A = surface area m^2
T_1 = temperature of hottest surface K
T_2 = temperature of other surface K
ϵ_1 = emissivity of hottest surface
ϵ_2 = emissivity of coldest surface

Example 5.7. The heating panel in Example 5.6 is situated close to a wall. The surface temperature of the wall behind the panel is 42 °C (315 K). What is the heat loss through the wall if the emissivity of both wall and panel is 0.6? Assume no convection loss from the wall.

Using formula 5.36

$$\Phi = 5.67 \times 10^{-8} \times 2 \times 0.8\,(358^4 - 315^4)/[(1/0.6) + (1/0.6) - 1]$$

$$= \mathbf{256\ W}$$

The surface area for radiation is the area of the heating panel projected perpendicular to the room wall.

5.4.3 Radiation from Flames

The emission of radiation from flames arises from molecules and particles within the flame gases. A yellow flame, due to the presence of a high level of suspended particles, is called a luminous flame, while a blue flame, arising from the presence of the reacting gases only, is described as a non-luminous flame.

The mathematical treatment of radiation in these cases is complex and is frequently tackled by subdividing the flame or gas into volume zones, characterised by a mean beam length defined in terms of the volume to surface area of the gas, and which are each assumed to be at uniform temperature and concentration. Equations can then be established which describe the radiation transfer between one volume zone and each of the other zones in turn. This is the basis of most furnace radiation models.

Non-luminous gas radiation

Most gases, for example, hydrogen, oxygen and nitrogen do not emit or absorb thermal radiation and so do not contribute to the radiative heat transfer within enclosures. However, polyatomic gases and, in particular, carbon dioxide and water vapour are important emitters and absorbers of radiation. Thus, since flames contain significant concentrations of these gases, they play an important part in heat transfer in furnaces. Since radiation is emitted and absorbed over spectral wave bands the emissivity of furnace gases are temperature dependent. The emissivities of H_2O, CO, SO_2 and H_2O are known for a range of conditions and for mixtures of CO_2 and H_2O empirical correction factors are available to account for spectral overlaps. The effective mean beam length of the gas is a further factor which must be evaluated, using charts, so that the emissivity can be estimated. Refer to standard textbooks for detailed techniques and tabulated data.

Luminous flame radiation

The luminosity of luminous gas flames and oil flames is caused by solid particles resulting from the thermal decomposition of the hydrocarbon. These particles consist of finely divided carbon, which, if not consumed in the flame, will emerge as soot. The particles are roughly spherical in shape, very small in diameter and clustered together in long chains. Each soot particle is found to be made up of about 1 000 small graphitic crystals around two nanometres in size each having a ratio of carbon atoms to hydrogen atoms ranging from 8:1 to 25:1, dependent upon the fuel. Diffusion flames and pre-mixed slow burning flames produce larger particles whilst pre-mixed fast burning flames produce smaller particles. The rate of soot formation depends on the mixing history of the flame and upon the carbon/hydrogen ratio of the fuel as well as the structure of the fuel molecules. It is not possible to determine the properties of a luminous sooty flame from a knowledge of air/fuel input condition, as the rates of formation and combustion are not predictable. However, flame emissivities have been measured over a wide range of operating variables and reasonable estimates of practical flame emissivities may be obtained from published data.

Highly luminous flames such as oil flames or coal flames emit high intensity radiation compared to non-luminous gas flames, which have to rely on the CO_2 and H_2O concentrations in the flame. Figure 5.11 shows a comparison of the radiation from luminous and non-luminous flames. The radiation emitted by the CO_2 and H_2O in the gas flame is known as 'banded radiation' because it is emitted at fixed wavelength intervals only, whereas the radiation from soot-laden flames is emitted over the entire spectral range, and so is known as continuous radiation.

Oil and coal flames, although giving higher radiative heat transfer rates compared to gas flames, are not amenable to forced convection heating, as mentioned earlier. Thus much greater convective heat transfer is obtainable from gas flames using direct impingement heating.

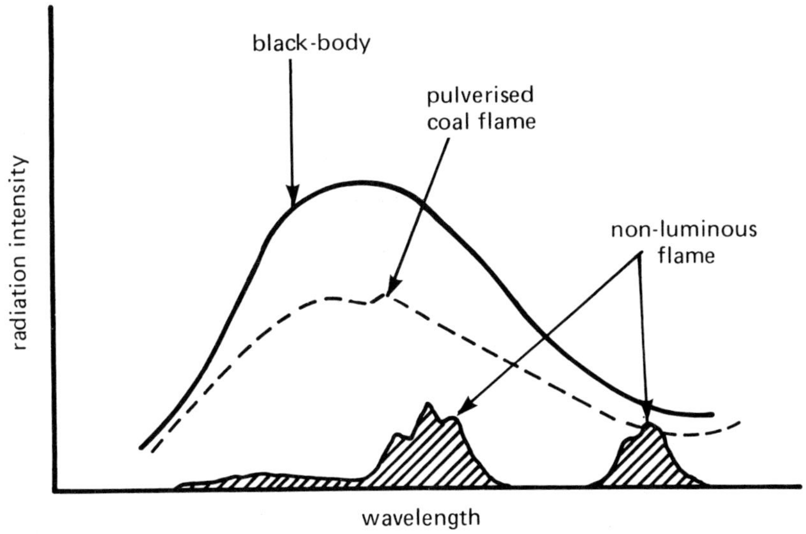

Fig. 5.11 Radiation from luminous and non-luminous flames

Calculation of radiation between a gas and a surrounding surface.

The heat exchange, Φ, between a gas and a non-reflecting surrounding surface, is given approximately by:

$$\Phi = \sigma A(\epsilon_{gas} T^4_{gas} - \alpha T^4_{surface}) \qquad 5.37$$

where σ = Stefan-Boltzmann constant
A = surface area
ϵ = emissivity obtainable from charts in standard texts
α = absorptivity of gas obtainable from charts in standard texts
T_{gas} = absolute temperature of the gas
$T_{surface}$ = absolute temperature of the surface

5.4.4 Radiation in Furnaces

Heat is transferred in open-flame furnaces mainly by radiation processes. These are made up of direct radiation from the flame and indirect re-radiation from the refractory surfaces. In high-temperature furnaces, radiation accounts for 95% of the heat transferred.

Because radiation is proportional to the fourth power of the gas temperature, the rate of heat transfer is very sensitive to this parameter. For example, if the combustion product temperature in a furnace operating at 1 600 °C drops by 100 K, there would be a 25% reduction in heat transfer. Likewise, an increase in temperature of 20 K over an initial temperature of 1 200 °C would increase heat transfer in the ratio $1\,493^4/1\,473^4$ which is equivalent to 5.5%. The intensity of radiation is, of course, influenced by the emissivities of the radiation components. Within certain limits, it is therefore generally beneficial to operate furnaces at as high a temperature as possible.

Indirect radiation results from furnace walls, roofs etc. absorbing heat from the combustion gases which is then re-radiated. The refractory surfaces radiate heat through non-luminous flames as they are semi-transparent to heat. Stock placed in the furnace therefore receives heat from both the combustion gases and the refractory surfaces.

The heat loss by direct radiation from furnace openings can be significant at high temperatures. Figure 5.12 can be used to estimate these losses by reading off the effective emissivity.

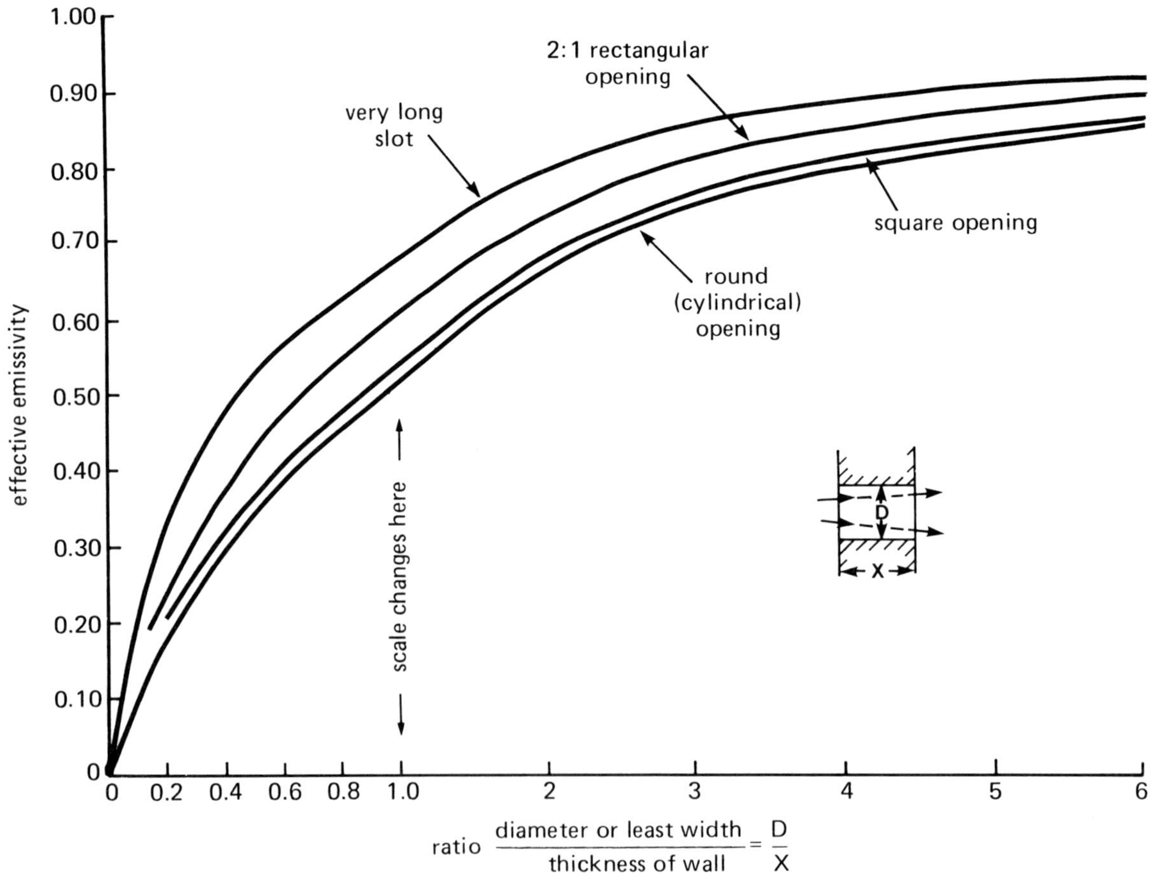

Fig. 5.12 Radiation through openings of various shapes as a fraction of the radiation from a freely exposed surface of the same area as the cross section of the opening

Example 5.8 A furnace front is bricked up with a 125 mm wall, through which an observation opening of 60 mm by 100 mm is left open. If the furnace temperature is 1 200 °C (1 473 K) how much heat escapes through the hole by radiation?

The calculated radiant exitance per m^2, M, due to a black body at 1 473 K is from equation 5.32:

$$
\begin{aligned}
M &= \sigma T^4 \\
&= 5.67 \times 10^{-8} \times 1\,473^4 \\
&= 266\,928 \text{ W/m}^2
\end{aligned}
$$

For an equivalent diaphragm area of 0.06 m x 0.10 m = 0.006 m^2 the heat loss from the freely exposed diaphragm is therefore:

$$266\,928 \times 0.006 = 1\,602 \text{ W}$$

The effective emissivity is determined from Figure 5.12:

Ratio of least width of opening to thickness of wall D/X = 60/125 = 0.48
Ratio of sides of rectangle (equivalent diaphragm) = 100/60 = 1.7

For D/X = 0.48 and interpolating between the Figure 5.12 curves for a rectangular opening in the ratio 1.7:1, the effective emissivity is thus 0.42.

The actual radiation through the opening, therefore, is:

0.42 x 1 602 = **673 W.**

5.5 HEAT TRANSFER IN PRACTICE

In almost every practical situation, heat is being transferred between bodies by more than one of the three different machanisms. In carrying out calculations we have to consider them separately in order to make the calculations simpler. However, we must also bear in mind in any design study that geometric or aerodynamic factors affect the mechanisms in different ways. For example, if convection were the only mode then we would want small volumes and fast moving gases. However, if the flame is highly luminous, this use of low volumes would suppress the radiation, and so overall we would lose efficiency. Similarly, for radiation-only systems, large volumes are better, but this reduces the convective heat transfer and also increases conduction losses. In some cases we deliberately suppress one mechanism. These aspects will be considered in this Section, covering different types of heat transfer equipment in a range of applications from low to high temperature. The examples are merely pointers and are not intended to provide a comprehensive coverage of all applications. It will also be seen that heat transfer through the temperature range affects the selection of materials and methods of construction.

5.5.1 Shell Boilers

A typical shell boiler, widely used for raising low pressure steam, is shown schematically in Figure 5.13. The boiler is divided into a number of 'passes' to transfer heat from the hot combustion products to the water on the shell side.

The first process is the combustion of the fuel and this takes place in the large diameter tube called the furnace tube or fire tube. The flame will fill about 80% of the tube, and the chemical heat release will cause the products of combustion to rise in temperature. As this process occurs heat will be transferred to the fire tube walls by radiation and convection.

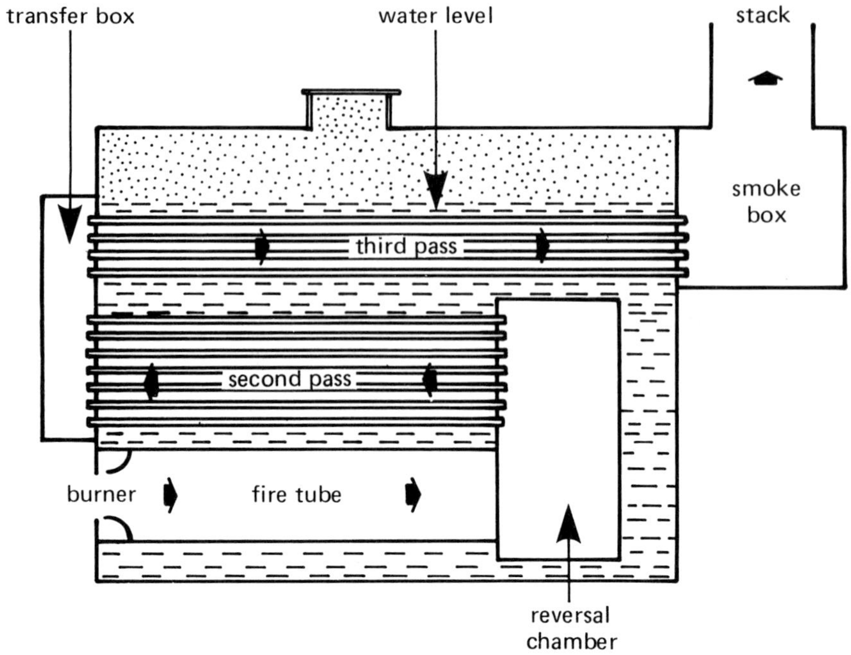

Fig. 5.13 Schematic diagram of a typical modern shell boiler

When a shell boiler initially designed for oil firing is fired on gas there is some redistribution of the heat between the passes. This is because oil burns with a luminous flame, whereas gas generally produces a non-luminous flame. A heat balance on a typical clean boiler at full rating is shown in Table 5.7.

Table 5.7 Heat Transfer in a Shell Boiler

	Heat Transferred %	
	Oil	Gas
1st Pass/Combustion Chamber	42	32
2nd Pass	33	39
3rd Pass	9	9.5
Flue Loss	16	19.5
Radiation Loss from Structure	2	2
Overall Gross Boiler Efficiency	82	78.5

The difference of 3½% in boiler efficiency is a result of the difference in C:H ratio of the fuels and thus combustion water vapour production between oil and gas. The heat transfer surfaces on an oil fired boiler quickly become fouled because soot is deposited, and so its efficiency falls off more quickly than that of a gas fired boiler doing a similar task. After a short period of operation, the efficiencies for the two fuels would, therefore, tend to be similar.

In the fire tube itself the average heat fluxes are in the order of 200 kW/m^2 for oil and 120 kW/m^2 for gas. In the case of oil, 70% of this heat is transferred by radiation and 30% by convection, whilst for gas the values are 55% by radiation and 45% by convection. The oil flame reaches a peak temperature of around 1 500 °C, with an emissivity of about 0.6, giving a peak heat flux of as high as 474 kW/m^2. The gas flame reaches a temperature of about 1 600 °C, but with an emissivity of of only 0.18, giving a peak heat flux of 170 kW/m^2. Because of these differences in heat transfer rate, the temperature of the combustion products leaving the fire tube is about 100 to 150 K higher for gas than for oil. However, the final stack temperatures come to be about equal as there is more than sufficient subsequent heat transfer area available. For the second and third passes most of the heat transfer is by convection; as the tube size is reduced towards the smoke box from typically 1 m to around 50 mm diameter, the velocity of the combustion products increases, so increasing the convection. In addition, the steady fall in temperature has a far greater effect on radiation (T^4 law) than on convection.

The conditions on the water side must not be ignored, and water treatment is vitally important to the safe and efficient use of a boiler, whether fired by gas, oil or coal. The laying down of scale on the water side because of untreated water has already been shown in Section 5.2.2 to have a marked effect on heat transfer rates. It also has the effect of increasing metal temperatures because of its poor thermal conductivity and in some cases it has resulted in key metal temperatures rising above the metallurgical limits, causing a boiler failure.

5.5.2 Tank Heating

In many processes, large vats and tanks of aqueous solutions have to be heated up to temperatures between 50 and 100 °C. There are many methods of heating these tanks, and one of the commonest is to use steam coils, where steam from the boilerhouse is condensed in a small bore tube of up to 50 mm diameter, utilizing the latent heat. The build up of condensate is controlled by a steam trap. These tanks are very often uninsulated, and so considerable heat losses can be found in practice. The main sources of heat loss from such a tank will be:

- Evaporation losses from the surface of the water:

$$\Phi_e = 1.34 \times 10^{-7} A_w \Delta H [1 + (v/1.17)] (p_s - p) \quad 5.38$$

- Natural convection losses from all exposed surfaces:

$$\Phi_c = A_c C(T_s - T_o)^{1.25} \quad 5.39$$

- Radiation losses from all exposed surfaces:

$$\Phi_r = A_r \sigma \epsilon (T_s^4 - T_o^4) \quad 5.40$$

where

A_c = convective heat transfer area (usually top and sides)
A_r = radiative heat transfer area (all exposed surfaces)
A_w = exposed water surface area
C = a constant for natural convection
ΔH = latent heat of evaporation of water at tank temperature
p = saturated steam pressure at tank temperature
p_s = partial pressure of steam at tank temperature
T_o = ambient air temperature (absolute)
T_s = surface temperature (absolute)
v = velocity of air over the surface of the tank
ϵ = emissivity of the surface(s)
σ = Stefan-Boltzmann constant

Figure 5.14 shows that as the water temperature rises, the evaporation losses increase markedly, and that they make up the bulk of the heat losses. If the tank is covered with some type of insulation, the evaporation can be suppressed, and the losses reduced. One convenient method is to use hollow plastic balls, so reducing the amount of exposed water surface by about 90%. The effect of this is also shown in Figure 5.14. A considerable reduction in losses is easily achieved, and further reduction may be achieved by insulating the sides of the tank, thus reducing the value of T_s in Equations 5.39 and 5.40.

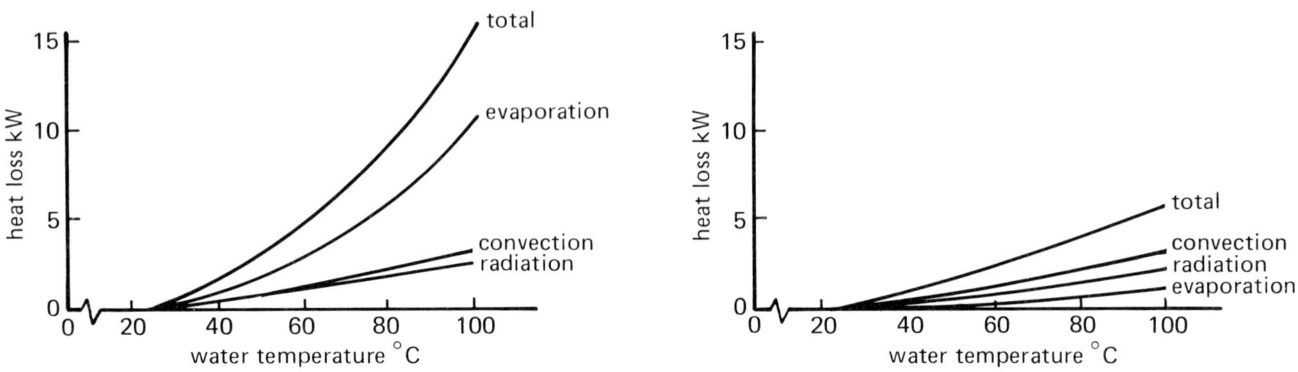

Uninsulated Water surface insulated

Fig. 5.14 Calculated heat losses from a water tank

An overall heat transfer coefficient based on the inside heat transfer coefficient can be calculated for condensing steam, the conductivity and thickness of the tube, and the natural convection from the tube to the water bath. A typical value would be 850 W/m^2 K. An alternative method of heating would be to burn gas in the tube, as this would avoid boiler and steam distribution losses. Again, an overall coefficient of heat transfer can be calculated and a typical value is about 70 W/m^2 K. The value is much lower because there is a much lower coefficient of heat transfer from combustion products to the inside of tube. However, there is also a considerable difference in the temperature driving force. Steam at 5.5 bar gauge would be at 162 °C, giving a temprature difference of 112 K, whereas combustion products would be at about 960 °C, giving a temperature difference of 910 K. This brings the value of heat transfer rate much closer together, but with steam still theoretically better. In practice, however, the situation is different. Condensing steam is held up in the system, and when the trap opens, uncondensed steam is released, causing heavy losses. For these reasons a gas fired system is more practically efficient, as shown in Figure 5.15. This advantage is increased if we also consider the efficiency of primary energy usage in generating the steam and the losses incurred in distribution and in condensate recovery.

5.5.3 High Temperature Furnaces

The previous sections considered two 'low' temperature industrial heat transfer applications, where the 'cold' surfaces are normally under 1 000 °C. This section is concerned with high temperature applications, heating objects to above 1 000 °C. including ferrous and non-ferrous stock for forming, ceramic kilns, glass tanks and other processes. In these applications radiative heat transfer plays a more important role.

A simple box type furnace as found in most forging shops is shown schematically in Figure 5.16.The furnace will probably be build of firebrick with a steel outer case, and will be fired with one or more gas or oil burners. The combustion products are flued either through a roof chimney, or just above hearth level. Billets of various shapes, sizes and temperatures are positioned on the hearth.

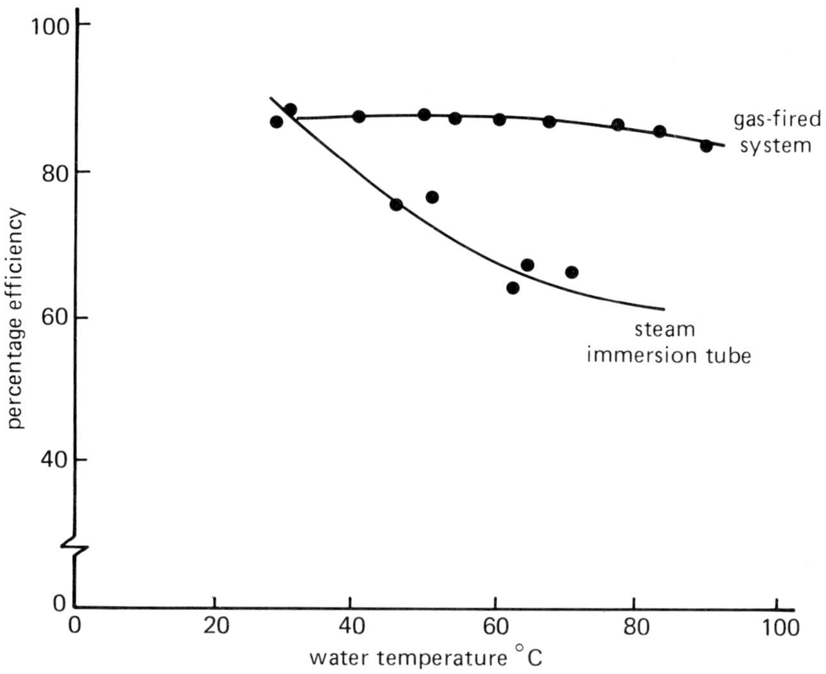

Fig. 5.15 Comparison of thermal efficiency of immersion tube systems

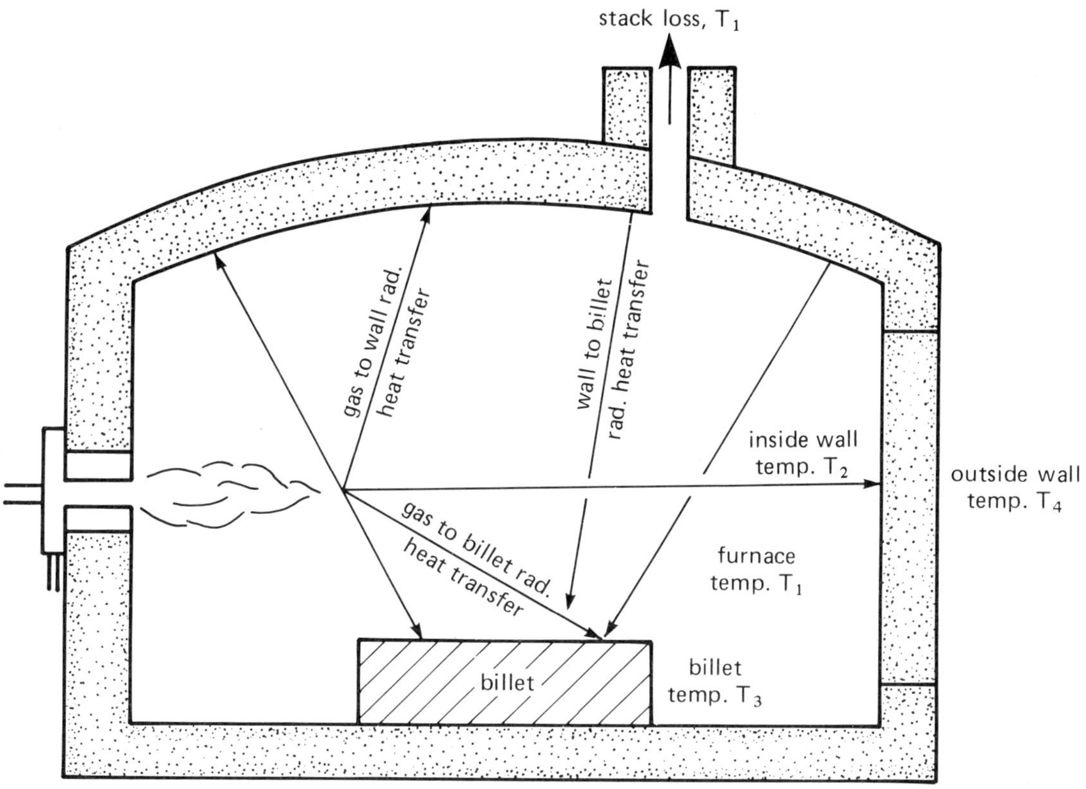

Fig. 5.16 Heat transfer in a furnace

The heat from the flame is radiated to the billet by various paths as shown. For an oil flame most of the heat is radiated directly from the flame onto the billets. For a gas flame, some heat is radiated directly onto the billets, while a considerable amount is radiated to the walls and from there re-radiated onto the billets. In both cases convective heat transfer plays a relatively small part. Typical combustion product exit temperatures will be above 1 150 °C depending on the billet and wall temperatures. The overall thermal efficiency figures will be low, with a value of 40-50% for fully loaded, well built and leak-proof furnaces. In practice a value of 20% would be considered a good overall figure. Even though the oil fired and gas fired furnaces will have differing radiative heat transfer paths, the overall efficiencies will not differ significantly. Gas flames may be slightly less efficient, but this can be overcome by small adjustments in burner position and by reducing the excess air levels. This is shown graphically in Figure 5.17.

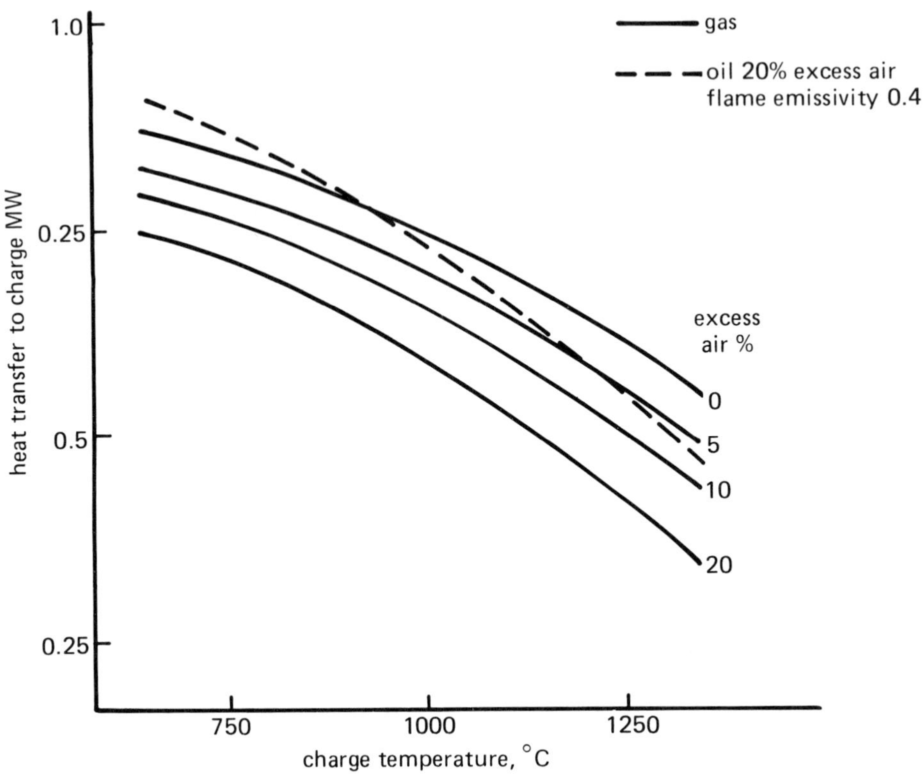

Fig. 5.17 Comparison of oil firing with gas firing in a reverberatory furnace

Recuperation

In a simple furnace the combustion products are exhausted at high temperatures, and these will therefore contain a considerable amount of useful heat. If this heat could be recovered the efficiency could therefore be increased. One of the best methods of using this heat is to preheat the incoming air. This has the added advantage of increasing the flame temperature, and hence greatly increasing radiative ttansfer. In Table 5.8 two gas fired systems are compared, one being a conventional burner without air preheat, the other a similar burner but with 350 °C air preheat, where an improvement in efficiency of over 11% is obtained. Percentage savings will be improved as air preheat temperature rise is increased. The radiative heat transfer path is from flame to wall to billet.

If the gas flow rate is maintained constant, preheating the air increases radiative heat transfer and the efficiency is increased. Alternatively, the same heat transfer rates can be achieved using less gas. Of these two alternatives, the former gives higher throughput, the latter uses less gas.

The recuperative burner, in which combustion products are drawn back through the burner to preheat the incoming air, is a convenient and cost-effective means of achieving preheat. Savings of over 20% can be readily realised, increasing up to 40% as the time that the billet is held at high temperature is increased. The effect of temperature and soaking period on fuel consumption is shown in Table 5.9.

Table 5.8 Typical Heat Fluxes and Temperatures for a Gas-Fired Forge Furnace with and without Air Preheat, Conventional Burner

Gas Rate = 66 sm^3/h (710 kW) | Excess Air = 0% (stoich.) | Outside Wall (20 cm) Temp. = 100 °C

Temperatures, °C			Heat Fluxes, kW/m^2			Heat transfer kW		Eff. increase %
Billet	Inside Wall	Combustion Product	Gas to Billet	Wall to Billet	Convective to Billet	To Billet	Through Wall	
No Air Preheat								
1030	1164	1279	17.7	49.2	0.29	224	37.7	–
Air Preheat = 350 °C rise								
1030	1193	1330*	21.6	62.4	0.4	283	38.6	11.1

*At entry to recuperator.

Table 5.9 Effect of Process Time on Fuel Savings by Replacing a Conventional Burner with a Recuperative Burner

Maximum Temperature °C	Soaking Period h	Fuel Saving %
1 100	17	23
1 350	24	36
1 200	33	43

The effect of using a recuperative burner in an intermittent kiln compared with a conventional burner is shown in Figure 5.18.

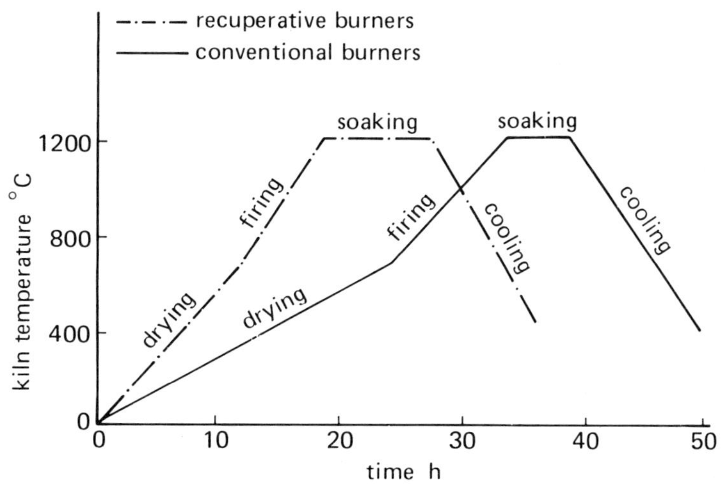

Fig. 5.18 Intermittent kiln cycles with recuperative and conventional burners

In the case of oil fired furnaces, external recuperators may be used (recuperative burners would lose effectiveness rapidly as heat transfer surfaces become sooted up). These often consist of two chequer-brick filled chambers called regenerators, the hot combustion products and the cold incoming air each flowing alternately through the chambers and the chequer bricks alternatively gaining and releasing heat. They are costly to build and maintain, often not being replaced when they deteriorate.

Regenerative burners use the regenerator principle but in a compact form in a package integral (in most cases) with the burner and can achieve greater savings than recuperative burners. Moreover, they may be successfully applied to furnaces with particulate-laden atmospheres.

Consult Chapters 2, 3 and 7 for more detailed treatment of heat recovery burners.

Rapid heating

A rather different high temperature application is the rapid heating of stock. In some cases it is important to bring billets to working temperature as quickly as possible. Reasons may include the avoidance of scale formation, preventing the furnace acting as a bottleneck in a highly automated process, or requiring only part of the billet to be heated. In some cases such as this it is useful to increase convective heat transfer and reduce radiative transfer, and it is here that gas flames are particularly suitable, since they do not contain soot particles which might be deposited on the surfaces to be heated.

There are two ways to enhance convection:

- Fire the burner directly onto the billet (direct inpingement)
- Pass the hot combustion products tangentially around or over the billet at high speed.

In both cases, high velocity burners are used and recuperators may or may not be used, depending on the design of the furnace. Comparative heating times for three different types of rapid heating furnace are given in Figure 5.19 for a range of configurations and burner types. However, there is

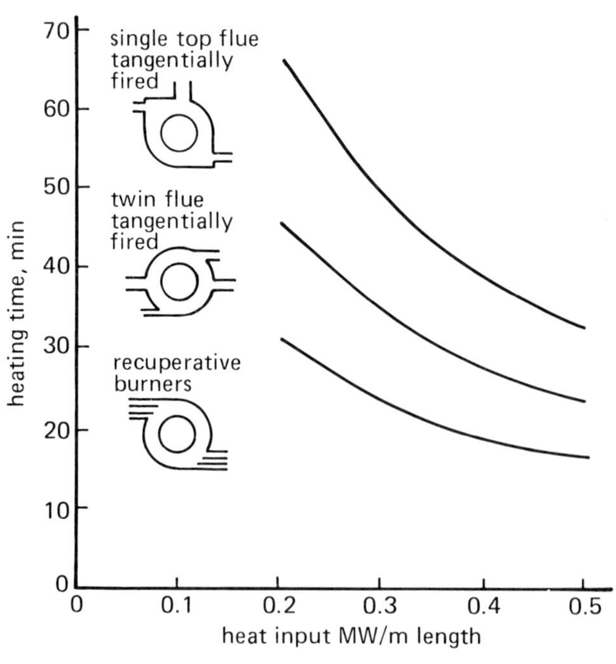

Fig. 5.19 Rapid heating furnaces: theoretical heating times for a 160 mm diameter mild steel billet in 250 mm diameter furnaces

another factor to be considered in designing these furnaces, which is the nature of the material to be heated. It is possible that the use of direct impingement furnaces for heating alloys of poor thermal conductivity, for example stainless steels, will give billets with temperatures that vary through the billet. Alloys that have high emissivity values will tend to receive most of their heat by re-radiation from the furnace walls, and here the tangential firing furnace is normally preferred.

5.5.4 Refractory Materials for Furnaces

It is important to consider the material used in the construction of the furnace, since this may affect its performance. The choice of refractory will depend on cost, strength and availability. Table 5.10 shows the effect of material selection on the performance of otherwise similar furnaces.

Table 5.10 Kiln Heat Input for Various Kiln Wall Insulated Linings

Kiln Wall Refractory	Total Heat Input - kW	
	Recuperative Burner	Non-Recuperative Burner
MPK30 Brick	4 100	6 800
MPK28 Brick	3 400	5 600
Ceramic Fibre	2 400	NA

It should also be noted that the important factor is the properties of the refractory at the service temperature rather than at lower temperatures. For example, the thermal conductivity of MPK28 is 0.29 W/mK at 200 °C and 0.45 W/mK at 1 300 °C - an increase of 50%. In general this increase will be even greater for other materials.

Consult Chapter 8 for detailed consideration of refractories.

REFERENCES

Flames: Their Structure, Radiation and Temperature. Gaydon A.G. and Wolfhard H.G. Chapman & Hall 1970.

Heat Transfer. Bayley F.J., Owen J.M. and Turner A.B. Nelson, 1972.

The Science of Flames and Furnaces. Thring M.W., Chapman & Hall 1972.

Engineering Thermodynamics, Work and Heat Transfer. Rogers F.F.C. and Mayhew Y.R. Longman, 1980.

Thermal Radiation Heat Transfer. Siegal R and Howell J.R., 2nd Edition, McGraw-Hill, 1980.

Heat Transfer. Holman J.P. 5th Edition, McGraw-Hill, 1981.

Heat Transmission. McAdams W.H., 3rd Edition, McGraw-Hill, 1983.

Heat Transfer. Chapman A.J. 4th Edition, Collier Macmillan, London, 1984.

CIBSE Guide. Chartered Institution of Building Services Engineers. In booklet form, various dates.

Gas Utilisation Equipment 6

CHAPTER 6

GAS UTILISATION EQUIPMENT

6.1 SPACE HEATING

6.1.1 Introduction

The thermal comfort of personnel may be achieved by warming their surroundings (convective heating) or by warming them by radiation where the surroundings receive heat only incidentally. A combination of convective and radiant heating is often employed.

Centralised heating is where a central heat source serves a building via a distributed medium. De-centralised heating is where a heat source serves the immediate location in which it is placed, with or without the intervention of a medium.

The types and methods of space heating industrial buildings are normally dictated by:

- Building fabric and dimensions.
- Type of environment required.
- Nature of the work carried out within the building.
- Building occupancy patterns.

In determining the type of heating to be installed in a given building, there are three main decisions to be made if installation and running costs are to be minimised. These are:

- Heating method (i.e. radiant or convective, direct or indirect).
- Fuel/Heating medium (i.e. wet or dry).
- Type of heat emitter.

The three factors are inevitably closely inter-related; discussions on one will affect the others. The order in which decisions must be taken will be largely determined by site conditions and are likely to be different for new or existing buildings, large or small sites, varying hours of operation etc.

To arrive at a cost effective solution account must be taken of projected capital and running costs, expected life and maintenance requirements.

In tall buildings, poorly insulated structures or areas having high ventilation rates, radiant heating is usually more effective and cheaper to operate than convective systems.

Smaller, heavy weight buildings, well insulated and with little or no forced ventilation may be more suited to convective heating. Where such buildings have high ventilation rates, high efficiency direct fired make-up air systems should be considered.

Generally, the cheaper, non-premium fuels (interruptible gas, coal and heavy fuel oil) are more suitable for large centralised plant to produce steam or medium temperature/high temperature hot water (m.t.h.w./h.t.h.w.). Premium fuels (firm gas, gas oil, LPG and electricity) have in the past, only been used in remote areas of large sites where a main service extension would be prohibitively expensive. It has become increasingly recognised, however, that in a large number of cases, the efficiency advantage of decentralised space and process heating over centralised plant, more than outweighs the unit cost disadvantage of firm gas. This is particularly the case where demand for steam or hot water is of a seasonal nature i.e. small process heating requirements. In this situation the overall system efficiency of centralised plant can be as low as 50% when low-load boiler efficiencies, boiler and distribution system losses, blow-down losses, flash steam and leakage losses etc are taken into account.

6.1.2 Wet Systems

Low temperature hot water (l.t.h.w.)

Domestic premises, offices and commercial areas are normally most suited to l.t.h.w. systems fed from a separate boiler or steam to water or water to water calorifiers. The majority of l.t.h.w. systems employ relatively small bore distribution pipework and circulating pumps, very few large diameter pipe, gravity circulating systems being used nowadays. System basics are shown in Figure 6.1.

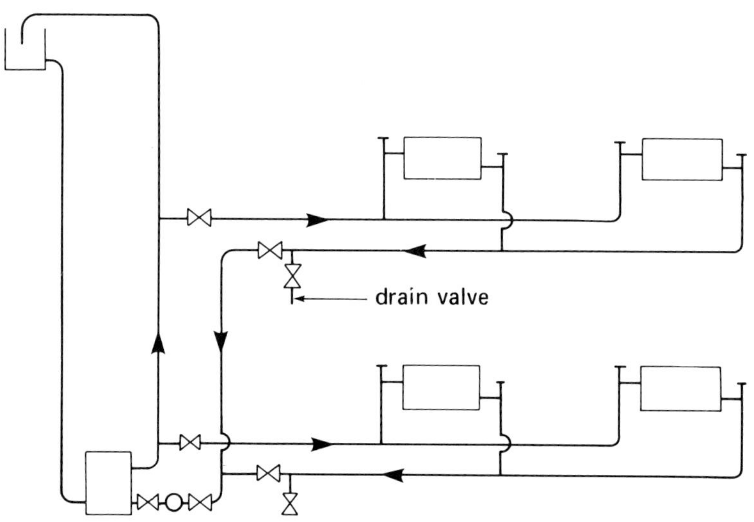

Fig. 6.1 Illustration of typical l.t.h.w. heating system

There is a wide variety of types and designs of l.t.h.w. boilers which includes:

- Non-condensing and condensing boilers.
- Conventional and balanced flue boilers.
- Cast iron sectional boilers.
- Modular boilers.
- Floor standing and wall hung steel boilers.
- Reverse flow boilers.
- Combination boilers.

Section 6.4 details the features of these boilers.

Probably the most popular form of heating for offices and small premises is the l.t.h.w. double pipe (separate flow and return) and radiator system. Radiators can be steel panel, column, or cast iron (for heavy duties) and can be wall or ceiling mounted (usually the former). Radiators, where possible, are positioned below windows to counteract downdraughts. Approximately 50% of heat transferred to the space is by radiation and 50% by convection by radiators using l.t.h.w., but these proportions vary widely with surface emissivity, air movement, temperature difference, radiator positioning etc.

Temperature control is very easy by means of pump switching thermostat and/or individual Thermostatic Radiator Valves (TRVs) and/or weather compensating control of the flow temperature, depending on the size of the installation.

Natural convectors have higher outputs than radiators and can be cabinet, continuous or skirting types. Advantages are uniform heat distribution, neat appearance and less wall space requirement. Most modern types have finned tube heat exchangers, which require regular cleaning to maintain full rated output.

Local temperature control is more difficult than with radiators and is usually via centralised pump switching thermostat and compensating system or with motorised zone valves on the larger system.

Fan assisted convectors also have higher outputs, but are more expensive and require an electrical supply to each. They are invariably of the cabinet type and have simple integral fan switching controls. Additional controls are usually necessary to prevent overheating and to prevent fan operation when water flow temperature is not hot enough.

Embedded l.t.h.w. coils or panels can be installed in floors or ceilings during construction, but are rarely used nowadays. Capital costs are high and the need for floor or roof insulation is of prime importance. Maintenance is an obvious difficulty if the need arises. The very high thermal inertia of such a system makes temperature control and intermittent operation difficult.

Heated ceilings are a variation on embedded coils and usually involve l.t.h.w. coils suspended above a perforated aluminium or a fibrous plaster ceiling. Alternatively, fan convectors can be used in the ceiling void to give a mixture of radiant and convective heating. Again, the insulation qualities of the roof are important in determining the running costs of this type of system. Costs are lower than for embedded coils, but the system suffers similar disadvantages, albeit to a lesser degree.

Fan assisted unit heaters with l.t.h.w. heater batteries are occasionally used. Problems can sometimes arise with heaters blowing 'cold' if water flow temperature falls much below 82 °C (180 °F) and their use on l.t.h.w. is not recommended.

Gas heated unvented hot water water storage systems

In 1985, the Building Regulations (England and Wales) were amended and now provide for safety requirements for unvented hot water storage. The 1986 Model Water Byelaws, which were adopted by local water undertakings in 1988, withdrew the previous ban on hot water storage of more than 15 litres from direct connection to the mains. These new regulations and byelaws have opened the way for a new generation of systems for water heating by gas.

The topic is covered in British Gas Interim Publication DM/7 August 1987.

The applications are for hot water 'for domestic purposes' as defined by the water byelaws, i.e. for the purposes of drinking, washing, cooking and sanitary use. The products described may be installed to provide 'domestic' hot water not only for dwellings but also for shops, offices, factories and other commercial applications.

Until now, as a consequence of the UK water byelaws, hot water storage vessels of more than 15 litres have incorporated a vent to atmosphere and they could not be connected directly to the mains water supply. However, almost all of the rest of the developed world do not have such a restriction.

The advantages of these systems include no cistern or pipework required in the roof space, quick to install as a unit or a package, flexibility in storage vessel location and no open cistern reducing risk of contamination.

Hot water is stored under pressure in a vessel fitted with appropriate safety and functional controls but without a vent to atmosphere. Most domestic systems are indirect when a remote gas boiler heats water in the primary circuit and coil in the cylinder. Direct systems are where the gas burner heats the surface of the storage vessel and the products of combustion are flued through the vessel.

Figure 6.2 shows an indirectly heated unvented hot water scheme. The cylinders can be either copper or steel lined with high density polyethylene to give protection against corrosion. The copper system may be designed to work at a pressure of 2 bar with a range of storage capacities from 80 litres to 250 litres. The steel system has an operating pressure of 3 bar and a range of storage capacities for 70 litres to 170 litres.

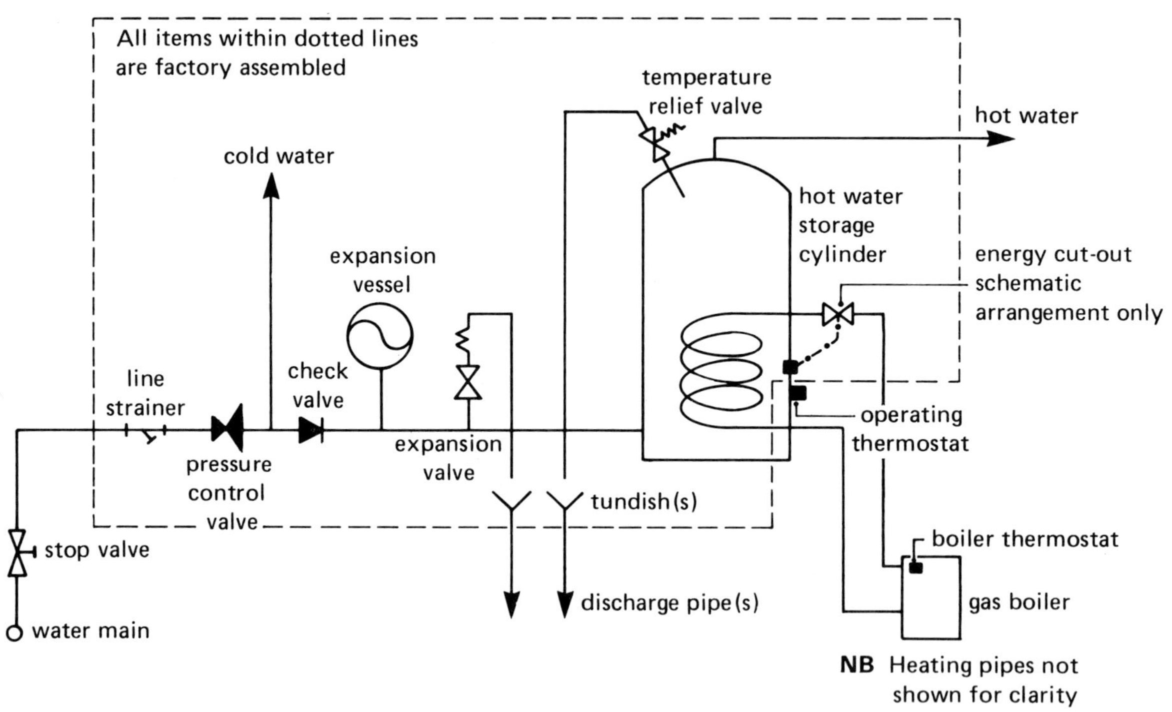

Fig. 6.2 Indirectly heated unvented hot water scheme

Fig. 6.3 shows the main components of an unvented indirect hot water storage system. Cold mains water passes through a line strainer to prevent foreign bodies passing into the system. A pressure reducing or limiting valve regulates the outlet pressure from this valve to the system to a maximum of 2 bar. The water then passes through a check valve, which is a non-return valve to prevent back-flow of water from the system into the mains water should there be a reduction of pressure in the mains. The expansion vessel is to accommodate the increase in volume of the expansion of water under normal heating conditions. An elastomeric membrane or diaphragm with air or inert gas above it to accommodate the expansion of water beneath is installed on the inlet side of the storage cylinder. An expansion valve automatically discharges water when the pressure in the system has reached a predetermined level. This protects against failure of the pressure reducing valve and failure of the expansion vessel. The temperature and pressure relief valve will protect the system against failure of the control thermostat and energy cut out. Both the expansion valve and the temperature relief valve will discharge into a tundish or air break device. This will prevent blocked discharge pipes negating the operation of both these devices.

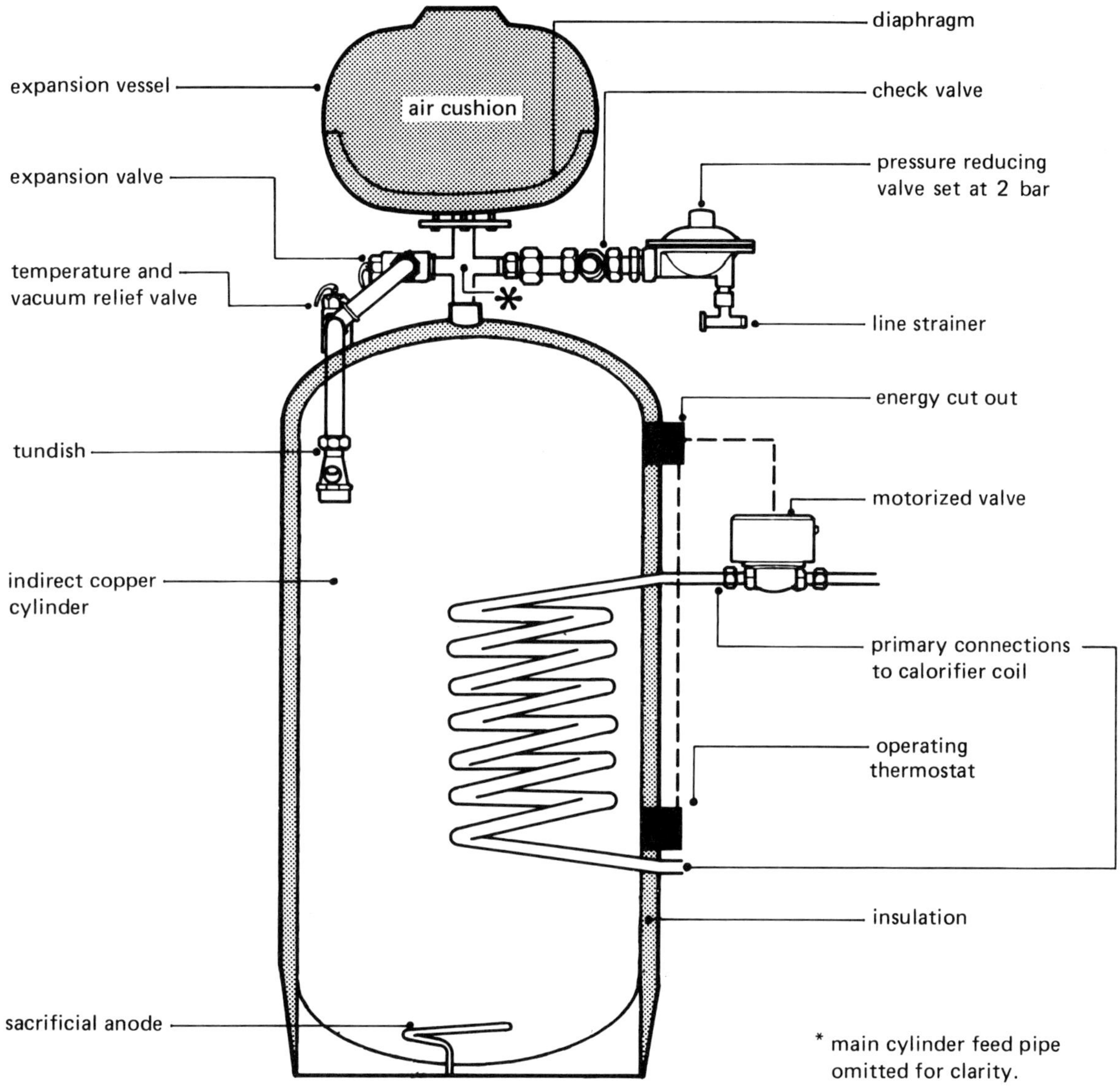

Fig. 6.3 Schematic layout of unvented indirect hot water storage copper cylinder

Medium temperature/high temperature hot water (m.t.h.w./h.t.h.w.)

Pressurised hot water systems are mainly used on large commercial/industrial sites where the predominant requirement is for space heating. Centralised shell type boilers and, to a lesser extent, water tube boilers, are the main boiler types employed. Section 6.4 discusses these boilers in detail.

In order to operate a hot water system at temperatures above atmospheric boiling point, the system must be pressurised. This can be accomplished internally by operating the boiler with a steam cushion (i.e. not fully flooded), or externally by means of nitrogen pressurisation, when the boiler is fully flooded. The latter system is preferred because it is not dependent on matching boiler pressure to the required water temperature, and is illustrated in Figure 6.4.

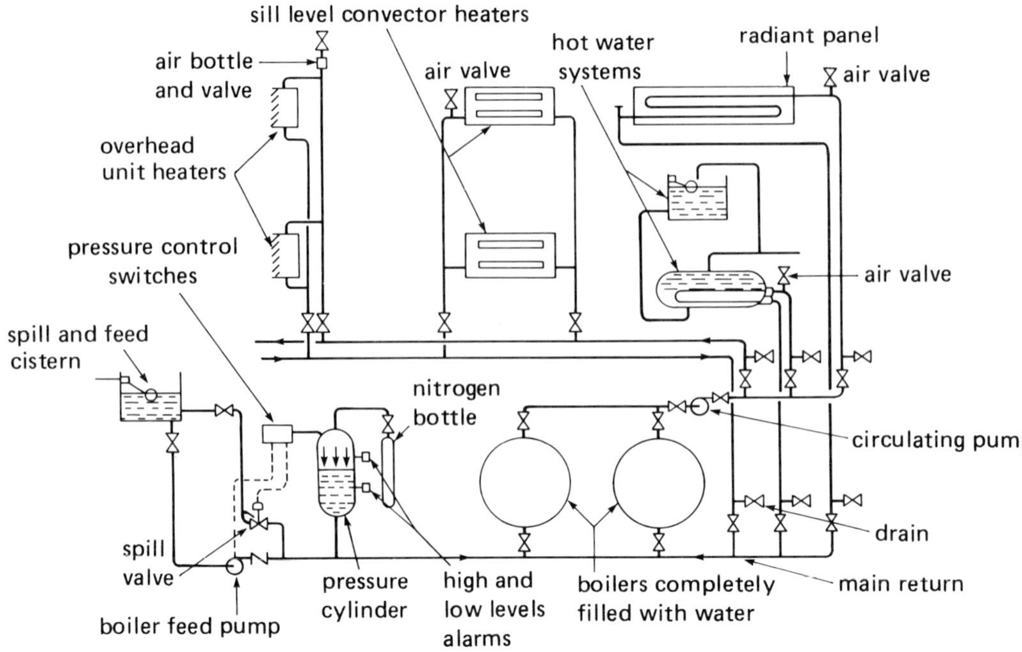

Fig. 6.4 Nitrogen pressurisation system

Operating pressures and temperatures are usually in the ranges 3 bar to 11 bar and 120 °C to 200 °C.

The main advantages of nitrogen pressurised hot water systems over steam systems for space heating are:

- The boiler/burner operation is less complicated. This is because the boilers and distribution pipework systems are fully flooded and the boilers will not shut down through overloading or through high/low water conditions. Burner operation will be controlled only by system temperature requirements.
- Reduced water treatment costs, due to the system being fully flooded and sealed. Once the system has been charged with chemicals, only minimal additional water treatment will be required.
- Water usage will be reduced as there will be no make-up requirement to replace losses from boiler blowdown, flash steam losses from feed tanks or intermediate condensate collection vessels.
- Maintenance of pipework, valves and fittings will be reduced. The major costs of maintaining steam and condensate systems are due to corrosion in the condensate collection and return systems caused by dissolved gases (CO_2 and O_2) in the condensate. Condensate with these gases present can be extremely aggressive on pipework and ancillary equipment.
- Scaling in boilers will be reduced because of the lack of suspended and dissolved solids in the boiler water.
- Although the basic pressurised water system requires greater care in design, once commissioned, the control of heating systems is generally much simpler and more accurate than with steam. This is because water distribution is by a two-pipe system which allows the use of modulating three-port valves, which can be arranged for mixing or diverting duties. These can be controlled by a single thermostat or the averaging of a multi-thermostat installation.

Steam systems

Boilers and fuel types used for centralised steam systems are identical to those for pressurised water systems.

The chief advantages of steam systems over pressurised water systems are:

- Steam systems are more flexible than h.t.h.w. in that pipework is more easily adapted.
- Steam system breakdowns can be repaired quicker due to faster cooling times and less draining of the system.
- Steam systems do not require pumping, except in some cases for condensate returns.
- Pipework is smaller for a given heat requirement, giving lower installation costs.
- A steam leak will vaporise immediately whereas an h.t.h.w. leak falls as boiling water causing a possible personnel hazard.
- Steam to l.t.h.w. heat exchangers are smaller than h.t.h.w. to l.t.h.w. exchangers reducing plant costs.
- Heat emitters, either unit heaters or radiant, have a higher rated output with steam than with h.t.h.w. e.g. a unit heater with a given heat transfer surface and an entering air temperature of 15 °C, h.t.h.w. at 150 °C having an output of 9.15 kW, on steam at 4 bar, 152 °C, would have an output of 14.94 kW. Thus, emitter costs are reduced due to fewer being required.

Types of heat emitters for centralised systems

Radiant panels with steam or h.t.h.w. as the heating medium have been effectively used in heavy workshops and tall buildings and for spot heating in otherwise unheated areas. The higher the operating temperature the greater is the amount of useful radiant heat.

Large numbers of panels are needed if areas are large, leading to high installation costs due to extensive piping and heater connections.

No secondary services are required but such systems need a lot of wall space and/or columns for supporting the panels. Warm-up from cold is rapid but local temperature control is difficult without expensive thermostatic individual or zone control valves. Panels can be arranged for one side or dual side heat emission. Little maintenance is required except on steam traps sets or control valves where appropriate.

Radiant strip heating has largely displaced the traditional radiant panel. This consists of continuous lengths of panel, generally mounted horizontally at high level. Lengths can be as long as 100 m. The basic principle is similar to radiant panels i.e. a plate heated by contact with a hot pipe. Strip can be of the single or multi-pipe design. Although pipe section is normally circular, triangular-section pipe can be used to increase radiation efficiency and heat emission per unit length. Installation costs can be lower than with panels as the strip replaces sections of distribution pipework and fewer connections are required. Radiant strip systems can be combined with fluorescent lighting systems.

When fuel prices and installation costs were low, festooned pipework heating systems were commonly used in old buildings. This is no longer the case as festooned pipework or bare heating coils are extremely wasteful, up to 75% of total output being lost via natural convection and upwards radiation, causing excessive temperature gradients.

The unit heater, which is simply a finned-tube heater battery and fan, is still extensively used in many installations on steam or pressurised water. High unit heat outputs, compared to radiant emitters mean fewer heaters and connections and hence lower installation costs. This is partly offset by the need for an electrical supply to each heater. Higher air temperatures are required than with radiant heating to give the same environmental comfort. Excessive temperature gradients can occur in high buildings and this has to be allowed for in the installed capacity depending on whether horizontal or downward discharge units are used. For improved temperature distribution a number of small units are preferred to a few large units. Unit heaters can be used to temper draughts particularly near doorways and can be used to give fresh air input by fitting an air-inlet duct to the outside. Dirty environments should be avoided as heater batteries would become blocked. Temperature control is very easy by fan switching thermostats on individual or groups of heaters.

Steam or h.t.h.w. heater batteries can be used in plenum or ducted hot air systems in industrial premises. Other systems have largely replaced this type of heating in large buildings since, apart from the usual disadvantage of temperature gradients, higher space temperatures and longer warm up periods, the system is expensive to install and maintain. Filter changes, steam trap, control valve and fan maintenance and heater battery cleaning are required. Local temperature control is a problem and the large ducts required are cumbersome and give little flexibility in the event of layout changes. The advantages of the system lie in its ability to recirculate warm air in the space heating season and to provide fresh air input in the summer, with the heater battery isolated.

Thermal oil

Thermal oil has come into widespread use for process heating in recent years because of its high temperature capacity at low working pressures. For example, an operating temperature of 260 °C requires a working pressure of about 0.7 bar, whereas saturated steam would have to be at about 48 bar. Thermal oil is rarely chosen as the medium purely for space heating. In those instances where it is used, there is normally a process requirement and it has been convenient to supply space heating via a thermal oil/steam or hot water heat exchanger. Thermal oil boilers are discussed in more detail in Section 6.4.

6.1.3 Direct Gas Fired Dry Systems

Radiant systems

Radiant plaque heaters can provide an efficient heating system in buildings with a roof or ceiling height of not less than 3.4 m. The basic unit is a gas-fired ceramic panel and the products of combustion are released directly to the atmosphere being heated. Very high surface-emission rates can be achieved, about 170 kW/m^2 and units range from 4 to 20 kW. Surface temperatures are in the region of 800 to 900 °C and the radiant efficiency is about 35 to 40%.

One of the advantages of this system is that the heaters can be mounted in positions where direct heating of persons, machinery or stock can be achieved without attempting to heat the whole volume of air. However, this concentration of heat on individuals is often limited by factors such as ventilation rate and mobility of persons.The heaters can be controlled thermostatically and/or by time switch either in total or in groups, the latter giving good flexibility.

Overhead radiant heater installations are often a little cheaper in capital cost than air-convection systems and running costs may be lower because the air temperature can be 3 °C lower than for air-heating systems for similar comfort conditions. However, a high proportion of the heat generated may be carried upwards by the combustion products. Another disadvantage is the possibility of condensation in the roof space, e.g. on steelwork. In general then they are suitable for use in buildings having a fairly high ventilation rate, or for open-air heating, e.g. in loading bays and sports stadia.

Maintenance required is panel cleaning at the start of each heating season. This may be a considerable task on large installations. Panel life is given as 10,000 h. Figure 6.5 shows a typical ceramic radiant heater burner plaque.

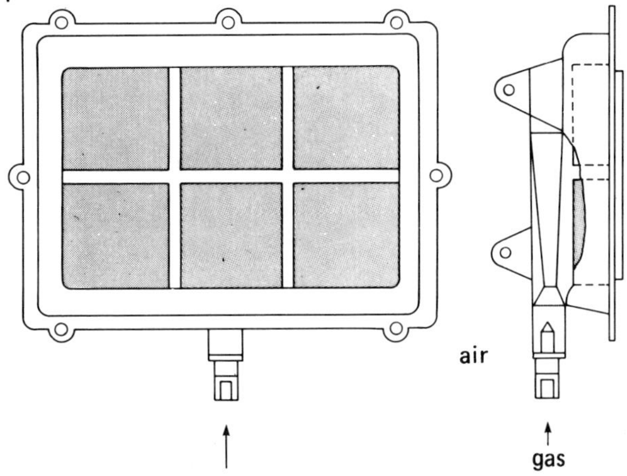

Fig. 6.5 Radiant plaque heater

Where the whole building does not necessarily require heating, but only localised areas or bays, the overhead radiant tube system should be considered. However, with sufficient numbers of tubes strategically positioned they can also just as successfully be used for heating complete buildings. This involves piping the fuel to the point of use, rather than having a centralised boiler plant with distribution losses in steam or hot water to heaters. These heaters are therefore very efficient, but also, they do not heat the air, so the air temperature is several degrees lower than would be the case with conventional heating to give the same feeling of warmth.

A typical heater is shown in Figure 6.6. It consists of a gas burner firing down the first tube of a twin tube arrangement, in which the tubes are connected at their extremity by a return U bend. The gas products are drawn through the tube by means of an extract fan which also draws air into the

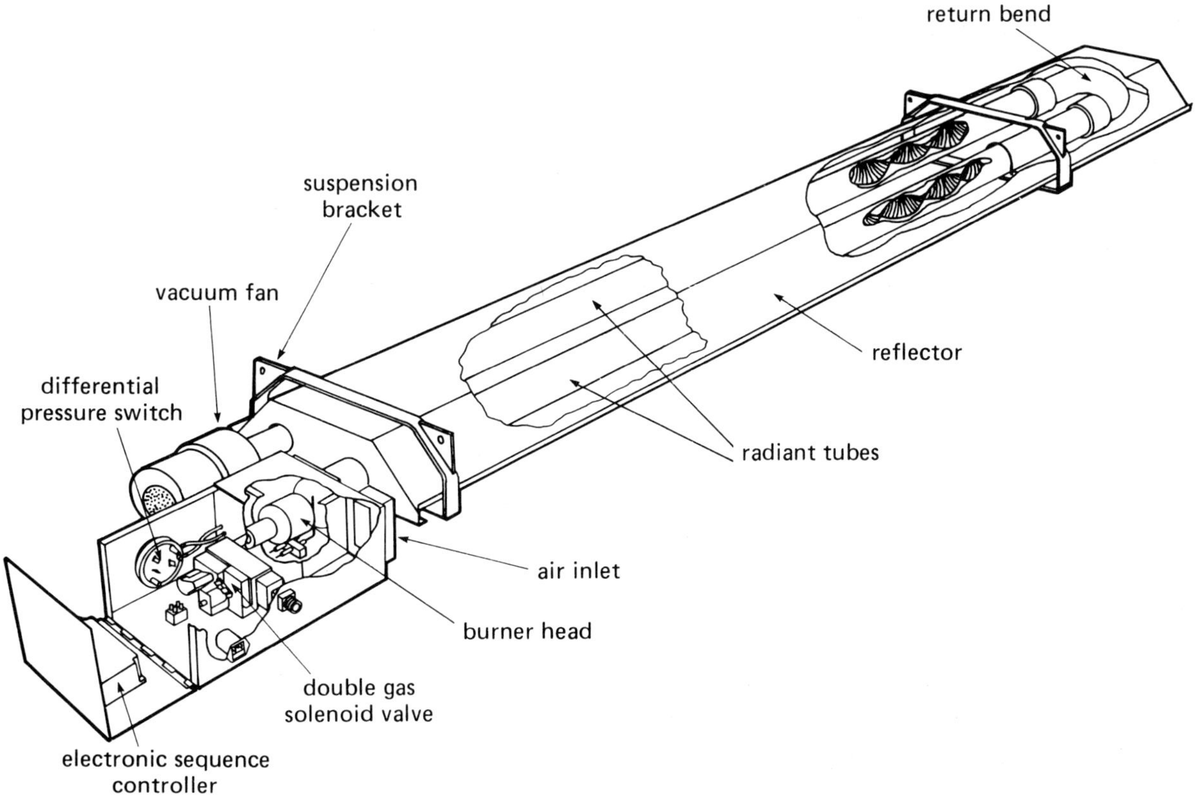

Fig 6.6 Radiant tube heater (Ambi Rad)

burner combustion chamber. Above the tube a polished aluminium or stainless steel reflector reflects the radiant heat downwards. Because the hottest and the coldest parts of the tube are adjacent to one another, there is equal temperature distribution along the length of the tube. The induced draught gas burner is automatically controlled during light up and fitted with flame rectification supervision. They should be mounted at a height of 3.6-15 m above the floor. Tube heaters may be flued individually or collectively and alternatively if there is adequate natural ventilation, as in large industrial buildings and warehouses, no flue is required. Maintenance only consists of keeping the reflector clean to maximise downward heat reflection and ensuring that the air intake to the unit is not blocked with dust.

The combustion efficiency is around 85% with a radiant efficiency of about 65%. The length of these units range from 3.5 to 6 m with outputs approximately 6 to 30 kW.

There are of course various types of radiant tube systems. If flueing is essential in the building due to poor ventilation and several heaters are to be installed, the herring-bone system can be employed. This incorporates up to about ten linear heater units and just one vacuum fan all linked together by a lightweight manifold. Only one flue to the outside of the building is required for all the heaters.

Another system employed by Nor-Ray-Vac Ltd, is with a continuous radiant tube system around the building to be heated. At regular intervals along the length of the tube, heaters boost the temperature to counterbalance the losses along the previous section of tube (Figure 6.7).

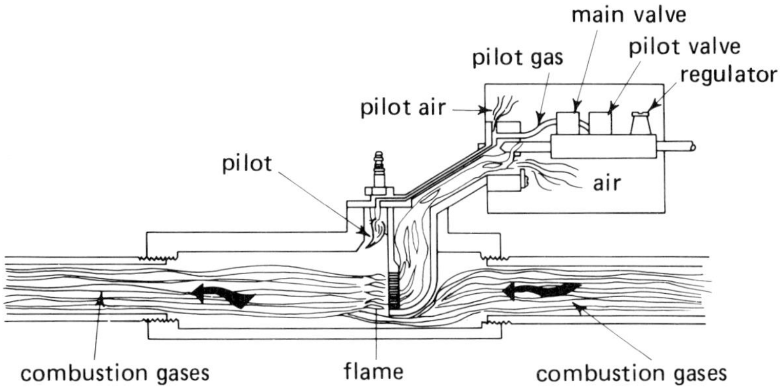

Fig. 6.7 Continuous overhead radiant tube system (cut-away view of combustion chamber)

For larger heat outputs a floor mounted gas fired combustion chamber supplies hot air to either a single tube or bank of tubes, with an exhaust fan at its extreme end to flue the products to the outside of the building. Overhead air-heated tube systems use air as the heat transfer medium. The radiating surface is made up of banks of circular metal ducts, insulated on the top and protected from cross-draughts at the sides by metal screens. A directly fired heater supplies air at a temperature to 200 °C and a centrifugal fan drives the air around the closed ducting circuit (Fig. 6.8). The average surface temperature of the tube banks is about 150 °C and the radiant efficiency is about 70%. Tube sizes range from 230 to 600 mm and the output per metre run of tube bank ranges from 0.5 to 5.0 kW. The load limit per system is about 600 kW. The fan power requirements amount to about 2% of the total energy.

The capital or first cost is less than that of a piped radiant panel system with a central boiler plant, but more than that of a gas-fired unit radiant heater installation. This system can be compared with the traditional piped radiant heating system and has the advantage over directly fired emitters in having a more uniform spread of heat and reduced maintenance because of fewer heat-generating units, but with the usual risk of centralised plant, that a heater failure means total loss of heat to the building. The heater is flued to the outside air, so there are no condensation problems.

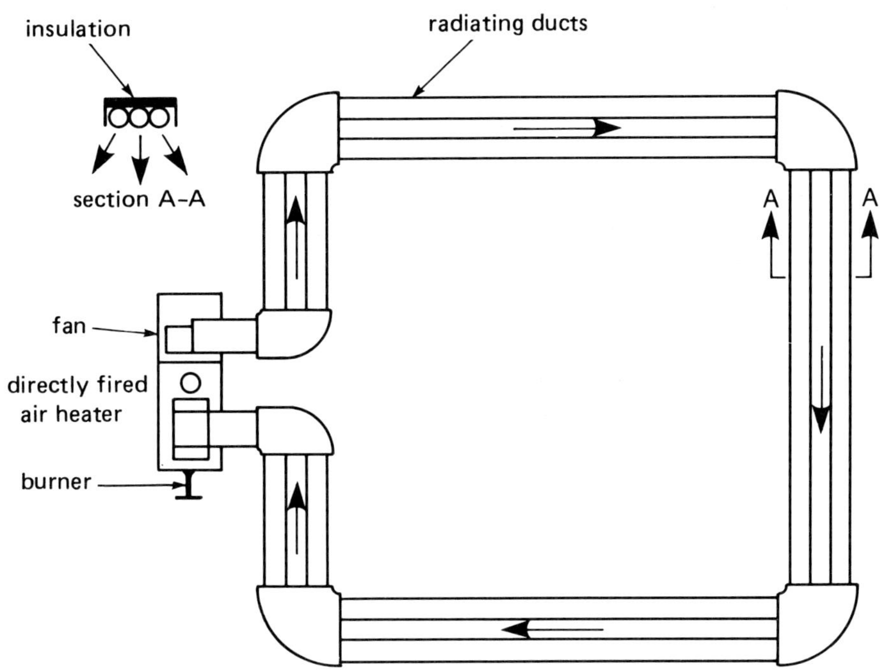

Fig. 6.8 Operating principle of overhead air-heated radiant tubes

These radiant tube systems are suitable for heating storerooms, warehouses, bus and coach stations, workshops, garages, canteens and sports halls.

Whilst gas fires are normally associated with the domestic market they can have application in some areas of commercial or industrial premises, e.g. small intermittently occupied rooms such as board-rooms, reception rooms, etc. Gas fires have a convective, as well as radiant, component and can be aesthetically pleasing and give good heating conditions at an efficiency of around 60%.

Convective systems

Direct fired convective heating is achieved by burning gas in a controlled flow of air and passing the heated air together with the products of combustion directly into the building to be heated. This is possible because natural gas contains no sulphurous gases and therefore, if correctly mixed and burned, only carbon dioxide, water vapour, nitrogen and oxygen should result. This form of heating gives a maximum thermal efficiency of 92%, as there is no flue loss.

There are various types of systems; the suspended unit heater, the floor standing units and the larger horizontal modes as shown in Figure 6.9. Here the hot air is distributed in ductwork and directed downwards into the workspace.

Another system employs a combustion chamber from which the products of combustion are directed at high temperature and pressure in large diameter ductwork at high level. These products are discharged through jets which entrain ambient air, the resulting mixture being thrust downwards via discharge nozzles (Figure 6.10 - The Casaire System).

Although direct fired heating is ideal for many applications such as warehouses and stores there are certain limitations. One is that it would generally not be suitable for steel warehouses where condensation of the water vapour would cause rusting of the steel stock. Also, there is the possible build up of products of combustion. Providing adequate excess air is supplied and the burner is properly designed, carbon monoxide, aldehydes, and the oxides of nitrogen are no problem. The threshold

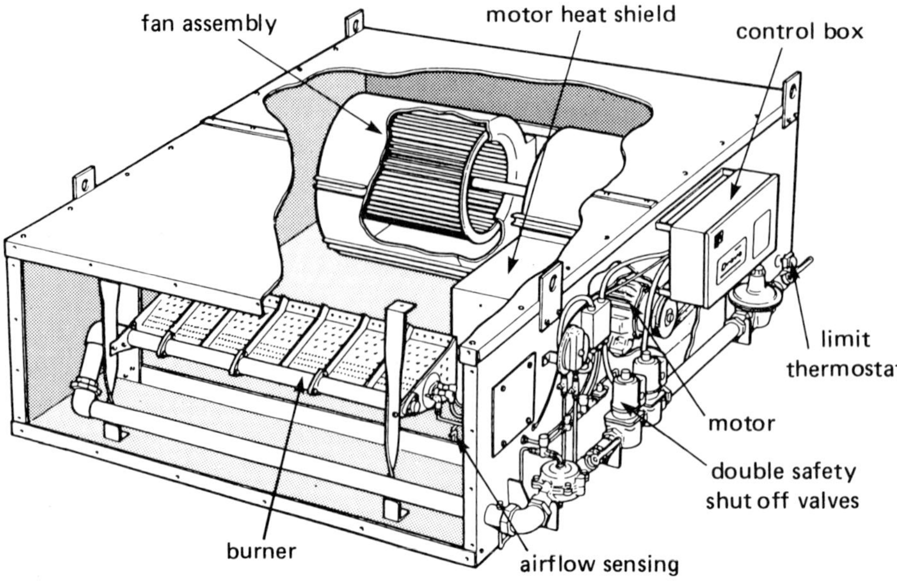

Fig. 6.9 Dravo direct gas fired air heating system

limitation for carbon dioxide is 2 800 ppm and provided this is not exceeded the carbon monoxide and oxides of nitrogen will be negligible. There are already 300 ppm of CO_2 atmospheric air, so there is a limitation in the CO_2 increase of 2 500 ppm. This will correspond to a temperature rise of 55 °C, and provided this is not exceeded the system will be satisfactory.

There are two documents concerned with the prevention of products build up:

- BS 5990 : 1981 Direct Gas Fired Forced Convection Air Heaters for Space Heating (60 kW up to 2 MW input) Safety and Performance Requirements.
- BS 6230 : 1982 Specification for Installation of Gas Fired Forced Convection Air Heaters for Commercial and Industrial Space Heating of Rated Input Exceeding 60 kW.

The first standard specifies heater performance including combustion quality, whilst the second covers installation requirements, such as ventilation and heat input to prevent products build-up. This document is the most important as far as gas installation engineers are concerned.

Provided that the system works on 100% fresh air without recirculation and the temperature rise is less than 55 °C, then the CO_2 in the discharge will not exceed a rise of 2 500 ppm and there will be no difficulty in meeting the Code of Practice BS 6230.

Examples of where this will be satisfactory are in paint spray booths, welding shops, commercial kitchens and swimming baths. These were originally called make-up air heating systems where the heat lost by ventilation was replaced by a direct fired heating system. In many instances now though, the direct fired system is the sole source of heat, so the term 'make-up' has been replaced by direct fired heating. Frequently an extract fan is linked with the direct fired heater, and so interlocked that the heater cannot work unless the extract fan is running. Air is not only drawn in through the heater, but also through windows, doors and any other apertures or crevices.

It is frequent practice to pressurise the building with the hot gases to prevent draughts etc. In this case no extraction fan is required, the pressure being vented through flaps in the walls.

Direct fired heating systems are usually fitted with modulating control, because an on/off system would cause cold draughts. Generally the turn down is 10:1 to 15:1, but some systems operate on a turndown of 20:1.

There are some high temperature direct fired heaters in which the temperature rise is greater than 55 °C and hence the resulting carbon dioxide content is greater than 2 800 ppm. In such cases, the discharge must be diluted with some fresh air before entering the workspace. One example is the Casaire System, where the temperature rise is above 130 °C, using a high pressure centrifugal fan and a duct pressure of 3.3 mbar (Figure 6.10). Turn down is about 25:1

The combustion products are discharged through nozzles where the ambient air is entrained and thrust downwards into the workspace. This type of system relies on adequate ventilation air to keep the carbon dioxide well below the threshold limit value.

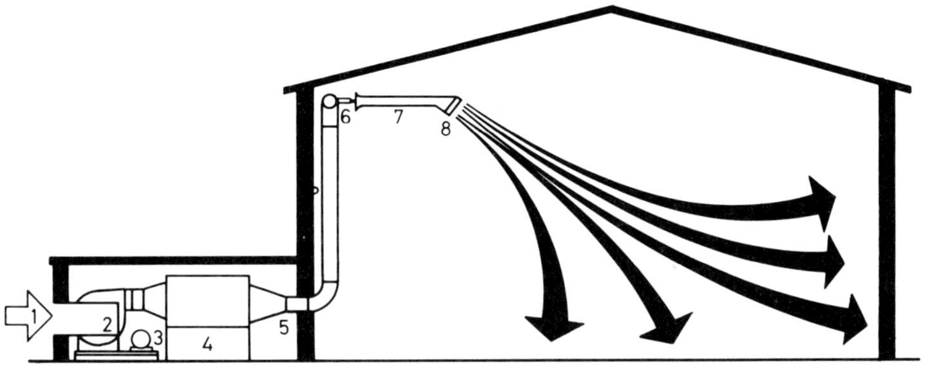

1. 100% fresh air intake
2. high efficiency centrifugal fan
3. fan motor
4. compact 'Casaire' air heater
5. small diameter insulated ducting
6. high velocity nozzles
7. lightweight aluminium venturi diffusers
8. warm air discharge with adjustable aerofoil blades

Fig. 6.10 Casaire direct gas fired heating system

For church heating the Crolla Engineering Ltd direct firing heater is often employed. The heater unit itself is located outside the church and the hot air supplied by ductwork to inlet grills on the walls. These heaters are remarkably quiet and the temperature build up is gradual which keeps the church fabric in good condition.

6.1.4 Indirect Gas Fired Dry Systems

By definition all indirect fired dry systems provide convective heating in the form of hot air.

Convector heaters

Like gas fires, conventionally flued natural convector heaters are normally to be found in domestic or small commercial premises, input ratings range from 7.5 to 14.5 kW. Where a suitable flue can not be made available, balanced flue convectors can be used, providing they can be fitted on an outside wall. Capacities are similar to natural convectors.

Where they are installed in school or waiting rooms, it is advisable to fit an internal guard. A terminal guard is also necessary on the outside of the building. Balanced flue convectors are particularly suitable for prefabricated buildings on industrial sites. Heaters can be fitted with a low voltage system which enables a multi-heater installation to operate automatically with zone time and temperature control.

Some convector heaters are available with fan assisted output which aids recirculation and improves temperature uniformity.

Suspended unit heaters

These appliances are used in industrial premises and commercial buildings and consist of a combustion chamber, a secondary heat transfer matrix and a flue output. Air is blown or drawn through the heater by a fan and passes directly into the building through louvred outlets. The louvres can be adjusted horizontally and vertically to direct the warm air where required.

Outputs range from 12 kW to 75 kW. Typical construction is shown in Figure 6.11. Normal material of construction is stainless steel.

Most models work with an air-discharge temperature of 65 °C and the combustion efficiency, on a well adjusted unit, is between 80 and 85% on natural gas.

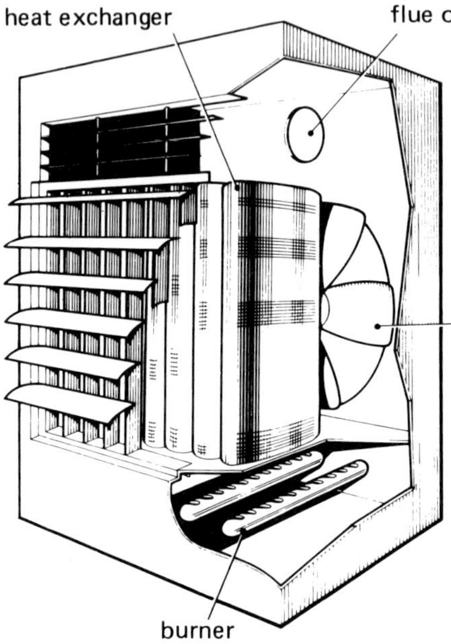

Fig. 6.11 Gas fired unit heater

Temperature distribution needs careful consideration with these units and, in general, a small number of large units are less effective than a greater number of small units. Ductwork can be attached to the discharge face to improve air and hence temperature distribution. This clearly increases installation costs. Most conventional units are fitted with quiet axial fans and correctly set outlet louvres will usually direct diffused warmth a distance of up to 15 m. If a longer air-throw is required a centrifugal fan replaces the axial type, as it does when distribution ductwork is attached to the heater.

Installation of several unit heaters makes it possible to selectively control heating in separate areas, by means of zone thermostatic and time controls. A multi-heater installation usually prevents total loss of heating which could be the case with a central plant system in the event of a single component failure. In low height buildings, they are a suitable option when considering decentralisation.

Multiple-unit installations can pose maintenance problems because of access difficulties, which could lead to a gradual loss of efficiency.

There is now a condensing version of this marketed by Reznor Ltd. The flue gas exhaust pipe and the condensation waste pipe can both be made from plastic because of the low exhaust temperature. It consists of a conventional heat exchanger unit with an additional stainless steel heat exchanger unit where the return air from the building first enters the system. This will give an extra few per cent combustion efficiency.

Floor standing unit heater

Where a higher heating demand is required and floor space is not at a premium, floor standing unit heaters, with ratings up to 1 MW, can be installed. A typical heater is shown in Figure 6.12. An automatically controlled gas burner fires into a large combustion chamber and the products of combustion are drawn downwards over secondary heat transfer tubes by means of an exhaust fan. The same motor for this fan also drives the combustion air fan. In the base of the unit, recirculated cool air is drawn in by recirculation fans which deliver the air over the heat transfer tubes, round the outside of the combustion chamber and hence out through the discharge air nozzles. These nozzles can be rotated through 360° to supply heated air to all parts of the buildings. These nozzles usually have horizontal vanes which are fully adjustable and some models can be adapted for duct discharge for heating small offices or partitional areas. For heating very large premises a multi-unit installation is advisable.

In general, these heaters suffer the usual disadvantages of convective heating from low level and should not be used in tall buildings. Large heaters need a lot of clear air space and conditions can be uncomfortable near free-blowing units. Ducting the warm air to overcome this problem increases capital costs. Time and temperature control is usually by separate time clock and room thermostat (or inlet air thermostat) with a high limit discharge air temperature cut-off.

Condensing versions of this type of heater are now marketed by Cov Rad Dravo (Condensaire) and others. In this system a modulating packaged burner and stainless steel heat exchanger with large surface area for maximum heat transfer are used. The layout of this heater is the same as the standard system.

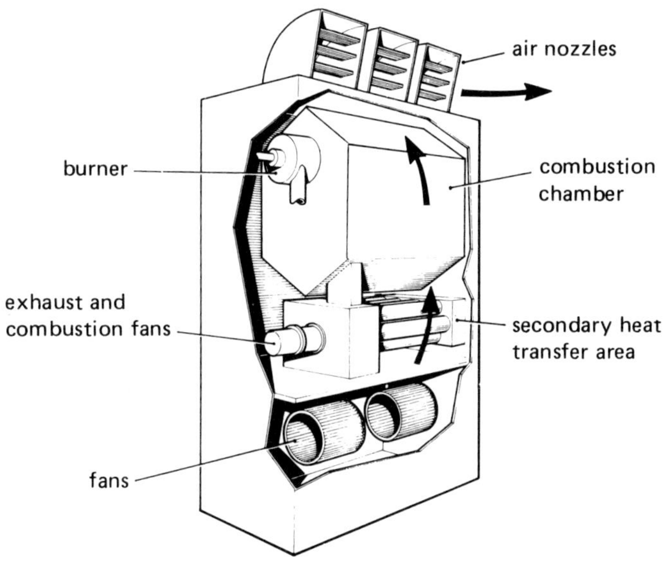

Fig. 6.12 Floor standing heater

Lennox Pulsed Combustion Floor Standing Unit Heater

The Lennox pulsed combustion up-flow gas air heaters, as shown in Figure 6.13, provide energy efficiencies of up to 96% with the output capacity almost the same as the input. The units operate on the pulse combustion principle and do not require a pilot burner, main burners, conventional flue or chimney. Compact standard size cabinet design, with side or bottom return air entry, permits installation in a basement, utility room or closet. Lennox add-on evaporator coils, electronic air cleaners and power humidifiers can easily be added to the heater for Total Comfort all season installations.

The high efficiency is achieved through a heat exchanger design which features a finned cast iron combustion chamber, temperature resistant steel tailpipe, aluminised steel exhaust decoupler section and a finned stainless steel condenser coil similar to an air conditioner coil. Moisture, in the products of combustion, is condensed in the coil, thus recovering almost all usable heat out of the gas. Since most of the combustion heat is utilised in the heat transfer from the coil, flue vent temperatures are as low as 38 to 54 °C (100 to 130 °F) allowing the use of 51 mm (2 inch) diameter PVC (polyvinyl chloride) pipe for venting. The heater can be vented through a side wall, roof or to the top of an existing chimney with up to 11 m (35 ft) of PVC pipe and four 90 degree elbows. Condensate created in the coil may be disposed of in a floor drain. The condensate (pH range from 4 to 6) is not harmful to standard household plumbing and can be drained into city sewers and septic tanks without damage.

An automotive type spark plug is used for ignition on the initial cycle only, saving gas and electrical energy. Due to the pulse combustion principle the use of atmospheric gas burners is eliminated with the combustion process confined to the heat exchanger combustion chamber. The sealed combustion system virtually eliminates the loss of conditioned air due to combustion and stack dilution. Combustion air is piped to the heater with the same type PVC pipe as used for exhaust gases.

The heater is equipped with a multi-functional gas valve, gas intake flapper valve and air intake flapper valve. Also included and factory installed are a purge blower, spark plug ignitor and flame sensor with solid-state control circuit board.

The process of pulse combustion begins as gas and air introduced into the sealed combustion chamber with the spark plug ignitor. Spark from the plug ignites the air/gas mixture, which in turn causes a positive pressure build-up that closes the gas and air inlets. This pressure relieves itself by forcing the products of combustion out of the combustion chamber through the tail-pipe into the heat exchanger exhaust decoupler and on into the heat exchanger coil. As the combustion chamber empties, its pressure becomes negative, drawing in air and gas for ignition of the next pulse of combustion. At the same instant, part of the pressure pulse is reflected back from the tailpipe at the top of the combustion chamber. The flame remnants of the previous pulse of combustion ignite the new air/gas in the chamber, continuing the cycle. Once combustion is started, it is self sustaining allowing the purge blower and spark plug ignitor to be turned off. Each pulse of air/gas mixture is ignited at a range of 60 to 70 times per second. Almost complete combustion occurs with each pulse. The force of these series of ignitions creates great turbulence which forces the products of combustion through the entire heat exchanger assembly resulting in maximum heat transfer and most efficient operation.

Sequence of operation – the room thermostat on a demand for heat will initiate purge blower operation for a pre-purge cycle (34 seconds) followed by energising of ignition and opening of the gas valve. As ignition occurs the flame sensor senses proof of ignition and de-energises the spark ignitor and purge blower. Heater blower operation is initiated 30 to 45 seconds after combustion ignition. When the thermostat is satisfied, gas valve closes and the purge blower is re-energised for a post purge cycle (34 seconds). Heater blower will remain in operation until pre-set temperature setting of 32 °C (90 °F) of fan control is reached. Should loss of flame occur before thermostat is satisfied, the flame sensor controls will initiate 3 to 5 attempts at re-ignition before locking out unit operation. Additionally, loss of either combustion intake air or flue exhaust will automatically shut the system down.

Four units are made from 11.3 kW (38,400 Btu/h) output to 27.5 kW (94,000 Btu/h) output. They are remarkably quiet in operation as the heat exchanger section is completely lined with thick foil faced fibre glass insulation. The unit should be installed on neoprene rubber mounting pads to reduce vibration and flexible duct connectors should be fitted in the supply air plenum and return air plenum.

These systems are ideal for heating small commercial premises such as offices, retail shop outlets and banks and building societies. They are suitable for fitting into small spaces such as broom cupboards or in basements.

The hot air outlets are frequently through air grills in the ceiling of the room occupied.

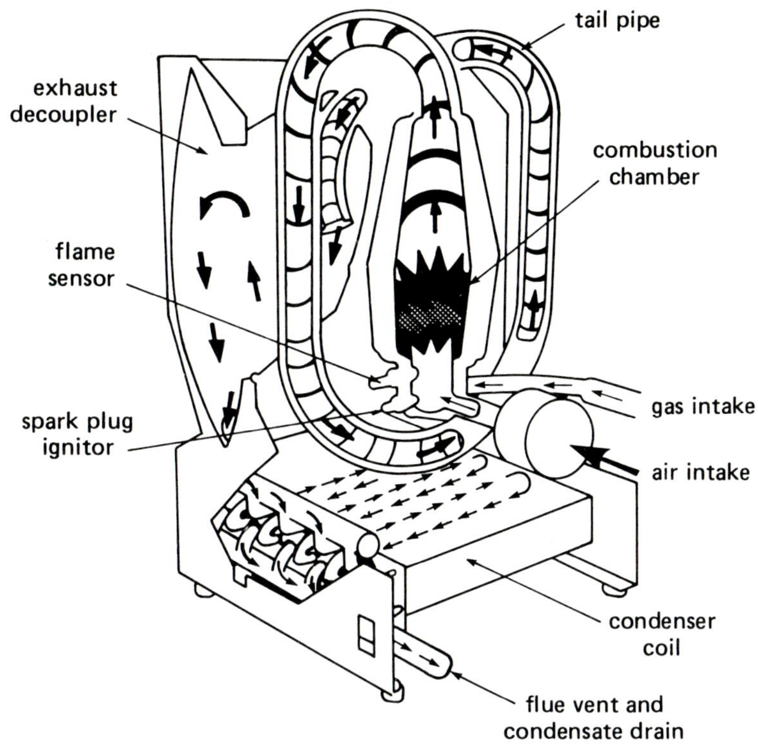

Fig. 6.13 Lennox pulsed combustion air heater

Roof Mounted Unit Heaters

As the name implies, these models are mounted in the roof and models are available with output from 18 to 180 kW. Some models have very large downward heating throws which make them an alternative to radiant heating in high buildings, large exhibition halls etc. where the obstruction caused by suspended radiant heaters may be inconvenient.

6.1.5 Controls

Need for control

The prime function is to maintain the desired comfort conditions. The secondary but very important function is to minimise energy consumption.

Energy is continually wasted in the majority of installations by:

- Maintaining too high a temperature, largely due to the misconception that one or two degrees extra is too small to matter. As, however, the average temperature difference between outside and inside over a heating season is about 11 °C, each degree of overheat raises fuel consumption by 9%. The 2 to 3 °C overheat often encountered is thus equivalent to an 18 to 27% heating fuel wastage.

- Turning on too soon. Preheat is needed to ensure premises are at the required temperature by the time occupancy starts. It is quite common for design temperatures to be attained hours before the period of use; each unnecessary hour of heating per day adds about 10% to the heating bill (day occupied premises).

- Excessive ventilation. Air within a building originates as cold outside air and a considerable proportion of the total heating requirement is needed to warm the air to the required inside temperature. Extraction over and above the acceptable minimum means that more cold air input is needed with consequent excessive fuel usage. The classic double waste situation occurs when occupants increase ventilation to reduce over-heating, instead of turning the heating down or off.

Temperature

Manual on/off control is appropriate in very few instances. It is sometimes used in rapidly responding systems (e.g. warm air or radiant plaques) where heat requirement is intermittent and spasmodic. Its application is mainly for time control rather than temperature. Automatic comfort control is much more satisfactory than manual and is achieved differently for different heating systems.

In all but domestic sized systems, wet central heating systems are required by the Building Regulations to be controlled by a weather compensator. This reduces the temperature of circulated hot water as the outside temperature increases. In addition to this, thermostatic control of the heat emitter (thermostatic radiator valves), is desirable to give individual control of each room. It should be noted, however, that problems can arise with fanned convectors using a compensated hot water circuit. These problems are due to the sensation of blowing 'cold' as the water temperature reduces. They are therefore best controlled by their own thermostat (which switches the fan on and off), and fed with non compensated hot water.

Indirect fired warm air systems are usually controlled on either a high/low or on/off basis using an air thermostat.

Direct fired warm air heating, particularly if it is make up air heating, may require analogue room temperature sensing and modulating control.

When the heat source is predominantly radiant e.g. gas fired radiant tubes and plaques, comfort is achieved by increasing the mean radiant temperature within the room rather than the air temperature. Comfort will therefore be achieved at a lower air temperature than with warm air heating. A conventional air thermostat is designed to respond to air temperature and not radiant temperature, whereas human perception of comfort depends on both. A radiant heating system should therefore be controlled by a sensor which responds to radiant heat. Such sensors are referred to as black bulb thermostats and are available from suppliers of radiant heaters.

Time

As already stated, heating systems should be operated for the minimum time necessary to maintain the required temperatures during the period of occupancy. The legal requirement is for statutory minimum room temperature to be reached not later than one hour after work commences.

The simplest control is a time clock at the heat source to switch it on in time to preheat the premises and switch it off at the end of the occupancy period. Seven day clocks, capable of independent daily setting are normally used. Less preheat time is required in warmer weather, and start-up times should be varied in accordance with outside temperature. In practice, this is rarely done and installations consuming a significant amount of fuel should be fitted with more sophisticated controls.

Depending on the building construction, occupancy pattern, etc., it can often be demonstrated that it is more economic to maintain a reduced level of heating overnight, than to shut off completely, lose the heat stored in the structure and then replace it on start-up. Night depression or set back controllers work on this principle. The overnight temperature is controlled at about 4 to 8 °C below the day time level, the amount of set back being either fixed, or varied according to outside temperature.

Optimum start/stop control is designed for use in all types of buildings which are intermittently heated, the object being to ensure optimum start-up and switch-off procedures so that the building may be at the intended temperature during normal occupancy times. The system relies on measuring the rate of heat decay in the building overnight and the outside ambient temperature; it then computes the start-up time required to heat the building in time for its occupancy and the switch-off time which will not adversely affect comfort conditions before the building is vacated. Frost protection safeguards are inbuilt; it caters for weekend or holiday closure with the resultant cold Monday morning start. There is a requirement in the Building Regulations for heating systems greater than 100 kW in intermittently occupied buildings other than dwellings to have optimum start control.

Degree days

The quantity of energy required to heat a building depends mainly on the difference between internal and external temperature. Thus, for an inside temperature of 20 °C, the amount of heat needed when it is 0 °C outside is about twice that needed for 10 °C.

Degree days are calculated each day according to the difference between actual temperature and a fixed base (usually 15.5 °C) in such a way that the figure may be used quantitatively. In other words, heat required is proportional to degree days; if degree days are, for example, 25% lower one month than the previous month, heating energy consumption should be 25% less.

The correlation between energy used and degree days depends on the pattern and nature of work. It is adversely affected by variation in working hours, casual heat gains from machinery etc. The degree day concept is nevertheless a most useful monitoring tool and there are few sites where its use is not beneficial. One of its main values is to evaluate the savings achieved by conservation measures which would otherwise be masked by weather variations.

In experienced hands, degree days can be used to yield a considerable amount of information. Where space heating is not the only part of the load on a particular boiler plant, for example, its proportion of the total load can sometimes be assessed from degree day graphs (Figure 6.14) or linear regression analysis without direct measurement. Figure 6.15 is a scattergram of fuel consump-

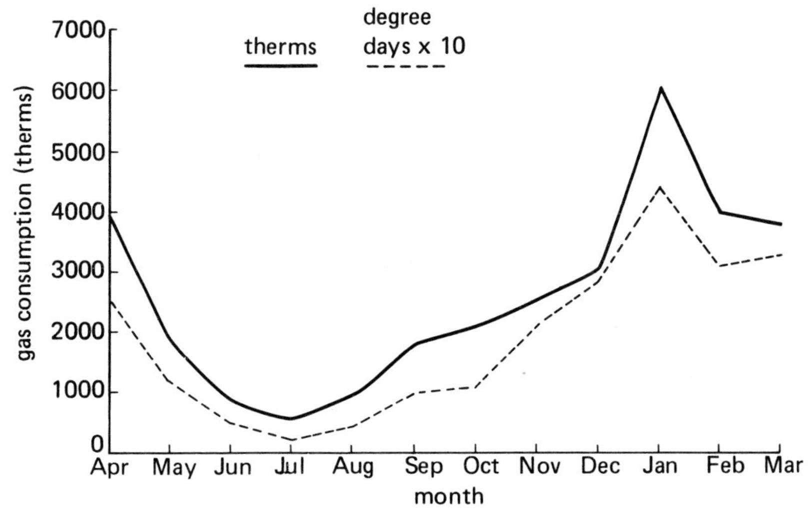

Fig. 6.14 Degree day and fuel consumption graphs

tion versus degree days, with the line of best fit, developed from regression analysis, superimposed, The lowest point of the graph of fuel consumption in Figure 6.14 and the point at which the regression line intercepts the 'Y' axis (Figure 6.15) both indicate the fuel usage which is not influenced by the weather i.e. non-space heating load. These graphs use actual data collected for a large commercial building and are used for energy monitoring and targeting.

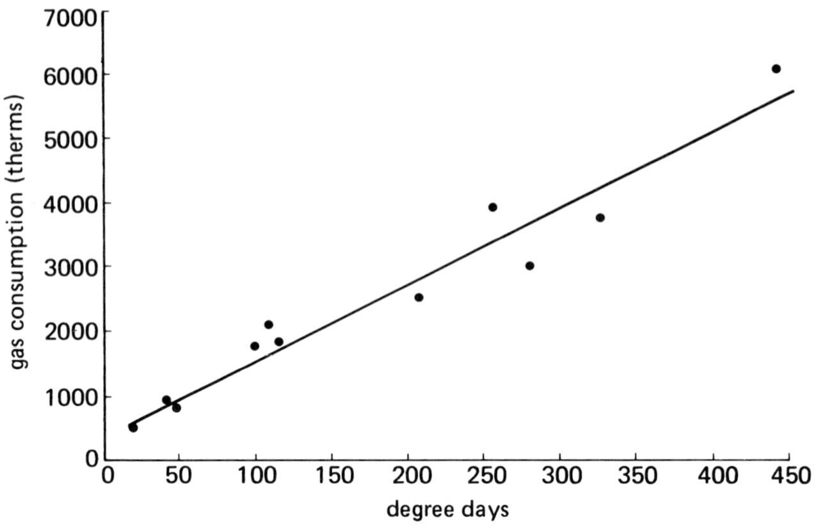

Fig. 6.15 Fuel consumption/degree day scattergram

Electronic energy management systems

Energy Management Systems (E.M.S.) are being used increasingly because they have proved themselves to be cost effective. The management systems are based on a computer which is capable of reading operational variables such as energy consumptions, temperatures, flowrates, humidity etc. and simultaneously controlling specific services within required constraints.

Electronic management systems can prove particularly cost-effective on sites where they can be used to control the environmental conditions of a number of different buildings or offices. The same system can also be used to read, log and process data, warn of plant failure and inform of planned or required maintenance.

Although an E.M.S. is sophisticated, its operation is relatively simple and it can be an excellent contributory factor in reducing energy costs. Typically an E.M.S. can save energy by minimising the unnecessary use of energy and by optimising plant efficiency. In addition, the system can be used to provide management with information which can be used to save energy by improving maintenance standards and operating procedures. Ultimately, these systems encourage the awareness of energy consumption and promote excellence in good housekeeping.

Energy Management Systems can take over the functions of stand-alone controls such as boiler sequencing control, thermostatic and weather compensating control and optimum start/stop controls. On extensive sites with multiple heating zones, control functions can include:

- Multi-set point temperature control.
- Multi-time control.

Additional savings could be anticipated by energy consumption monitoring and targeting such that departmental managers would be made aware of, and accountable for, their energy consumption.

Further savings would also be made from the reduction in personnel required to monitor and control the existing services.

A reliable and long term method of achieving and maintaining effective control over factory services could be provided by an Energy Management System. The specific site requirements need to be most carefully evaluated. A structured and disciplined approach is needed in order to be able to define accurately the control and monitoring requirements of an individual system. An Energy Management System needs to be economical, cost effective and provide the necessary flexibility for the changing standards of the modern factory environment. The system must also be capable of expansion at a later date in order to monitor and control additional areas if so required.

See Chapter 14 for details of the equipment used in these systems.

6.2 HOT WATER SERVICES

6.2.1 Introduction

Large amounts of energy are used in the UK to provide hot water services (h.w.s). Such services are defined as hot water used for personal hygiene (hand washing, showers etc.), dish-washing, food preparation, and factory/plant washdown (e.g. food industry, chemical industry etc.), but not process hot water.

The design and efficient operation of hot water services can be difficult, especially where, as in most industrial and large commercial buildings, it is required in widely separated localities. It has been traditionally produced by a centralised heat source with primary and secondary pipework distribution systems, but heat losses even from insulated lines, if very lengthy, can obviate the other benefits. The use of locally produced hot water services, with minimum distribution pipework, has become more commonplace, with natural gas as the main heat course.

The type and capacity of h.w.s. systems used depend on locations and demand volumes, heat recovery periods required and the nature of application e.g. intermittent or continuous.

Hot water systems are widely believed to be the main source of outbreaks of Legionnaires Disease. Currently, only systems that service commercial sector buildings with a large number of occupants e.g. hotels and hospitals, have been correlated with recognised outbreaks. Some sporadic cases however are suspected from colonised hot water systems in other buildings. The Building Research Establishment (BRE) and the Chartered Institution of Building Services Engineers (CIBSE) advise that no hot water systems should incorporate storage between 20 and 25 °C, as it is between these temperatures that bacteria can multiply. The exception is where water is subsequently heated to 60 °C, for delivery to the user. Generally, storage at 55 to 60 °C and delivery at 50 to 55 °C are recommended unless there are special circumstances.

6.2.2 Non-Dedicated Systems

Where there is a requirement for space heating and/or process heat in the form of low, medium or high temperature hot water, steam or thermal oil, h.w.s. production is often an integral part of the system.

The primary heating medium is generated in boilers, as discussed in Section 6.1 and described in detail in Section 6.4. The hot water is produced in heat exchangers which can range from the domestic hot water cylinder type to large storage and (to a lesser extent) heating or non-storage calorifiers or plate exchangers in industrial premises.

Non-storage calorifiers and plate heat exchangers are normally only used when demand for h.w.s. is relatively continuous and/or where fast heat recovery rates are required..

Figures 6.16 and 6.17 illustrate typical storage and non-storage calorifiers used for h.w.s. production.

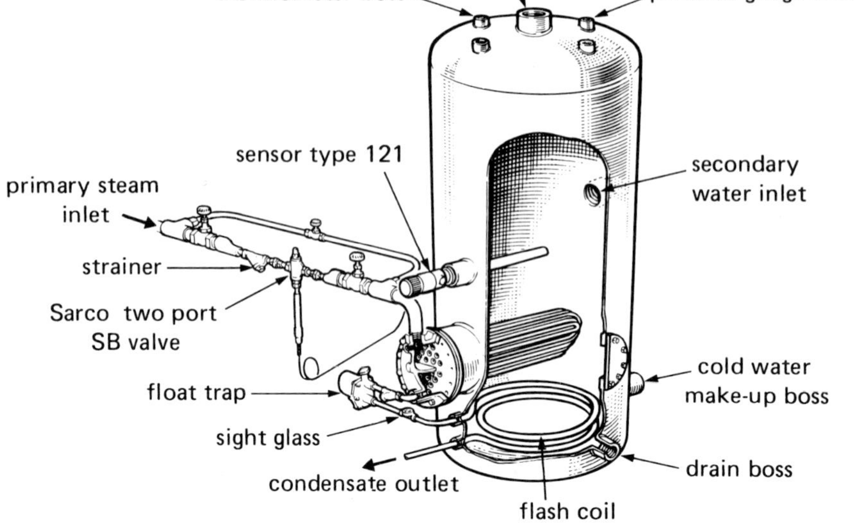

Fig. 6.16 Storage calorifier

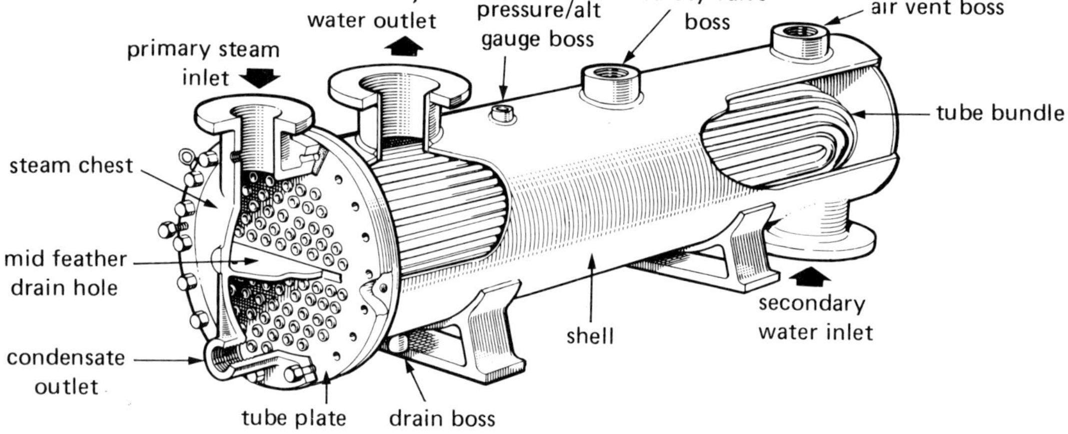

Fig. 6.17 Non-storage calorifier

The two main disadvantages of h.w.s. based on centralised boiler plant are:

- Heat and other losses from primary and secondary distribution pipework, as discussed in Section 6.1
- The adverse effect on system efficiency during the summer months if central boiler plant is primarily for space heating, i.e. boiler plant rating is then grossly oversized for h.w.s. requirement only.

These problems can be partly overcome by ensuring that space heating and h.w.s. primary distribution pipework are separate. In addition, with multi-boiler installations, careful consideration of winter and summer demand profiles may enable a combination of boiler sizes which can be operated to more closely follow the demand pattern. This will reduce low-load efficiency losses but capital costs will be increased.

Combination boilers

Losses between the boiler and calorifier and installation costs can both be reduced by the use of a combination boiler, where the combustion chamber and boiler for space heating and the calorifier modules for h.w.s. are both in the same casing. This has the following advantages:

- Reduced space for complete installation.
- The installation is easier and cheaper to install.
- Improved operating efficiency, due to reduction of heat losses from primary pipework, and faster calorifier regeneration.

An example of this system is the Hoval Boiler TKD-R (Figure 6.18) which has a range of 265 kW to 14.6 MW for combined central heating and hot water supply.

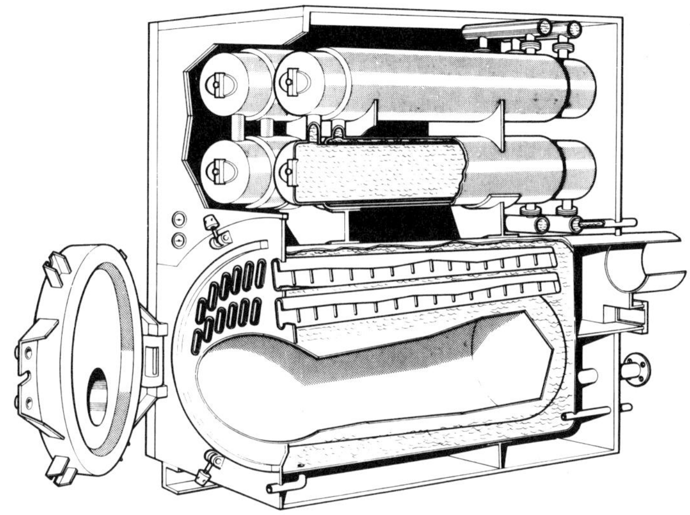

Fig. 6.18 Hoval combination boiler

The high output module calorifiers are made of special corrosion resistant stainless steel. They can be top mounted on the boiler as shown, or free standing beside the boiler or in remote calorifier chambers.

These boilers are installed in schools, hotels, hospitals etc., or in basement or rooftop boiler houses.

Direct water heating

Hot water can be produced by mixing the primary heating medium (steam or water) with cold water via mixing valves or direct injection. These can provide virtually instantaneous hot water, but suffer the disadvantage of the need to replace the primary steam/hot water so used with cold make-up water, which not only requires heating but which also requires chemical or other treatment. With steam boilers, blow-down losses will also be increased.

The application of direct mixing is nowadays limited to larger installations which use significant quantities of hot water for washing-down plant, equipment and floors.

6.2.3 Dedicated Systems

In the majority of cases, h.w.s. can be provided efficiently and cheaply by equipment dedicated to its supply.

Discrete boiler

The disadvantages of a centralised boiler system providing both space heating and h.w.s. can be partly overcome by installing a separate boiler for h.w.s. only. This will obviate the efficiency penalty during low load operation outside of the space heating period. If it is part of a central boiler installation, however, primary distribution losses could still be significant. Installation at point of use will remedy this. In extensive premises with scattered users, the installation of multiple boilers and calorifiers would not be cost effective.

Storage water heaters

Increasing use is now being made of storage water heaters which use gas directly to heat the water. Their efficiency is high and capital costs are not excessive. A typical unit is shown in Figure 6.19.

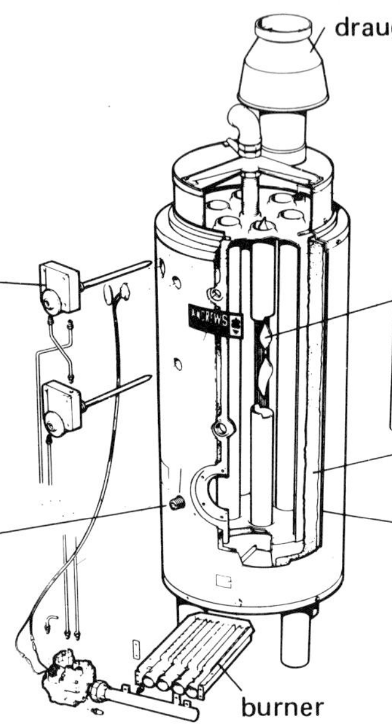

Fig. 6.19 Andrews gas fired water storage heater

They essentially consist of a gas burner firing upwards through either a single or up to six tubes. These are surrounded by a jacket of water which is insulated with mineral fibre and surrounded by a stove enamelled outer casing. The inside casing is lined with glass to retard corrosion and prevent adhesion of any scale. There is a removable sacrificial magnesium anode which will corrode in preference to the steel case itself. In each flue is a flue baffle to increase turbulence and thus heating efficiency. Typical efficiencies of these units are around 70%. Sizes range from 75 litres to 373 litres with heat recovery rates ranging from 103 litres/h to 1 620 litres/h.

They are designed for use in apartments, small hotels, schools, nursing homes, stores, churches and small industrial applications, and are reliable and easy to install and maintain.

Where large volumes of hot water are required with fast recovery rates, the Beaumont type water heater can be used (Figure 6.20). This has a range of storage from 450 to 2 250 litres, with a recovery rate of 1 350 to 4 500 litres/h, with a 55 °C temperature rise.

The construction is essentially twin tubular heat exchangers which fit horizontally into a rolled steel cylinder which has an internal polypropylene coating for protection against corrosion. Completely removable mild steel end plates provide access for cleaning and maintenance. The cylinders are constructed in sections bolted together on flanges, so the capacity of the units can, if necessary,

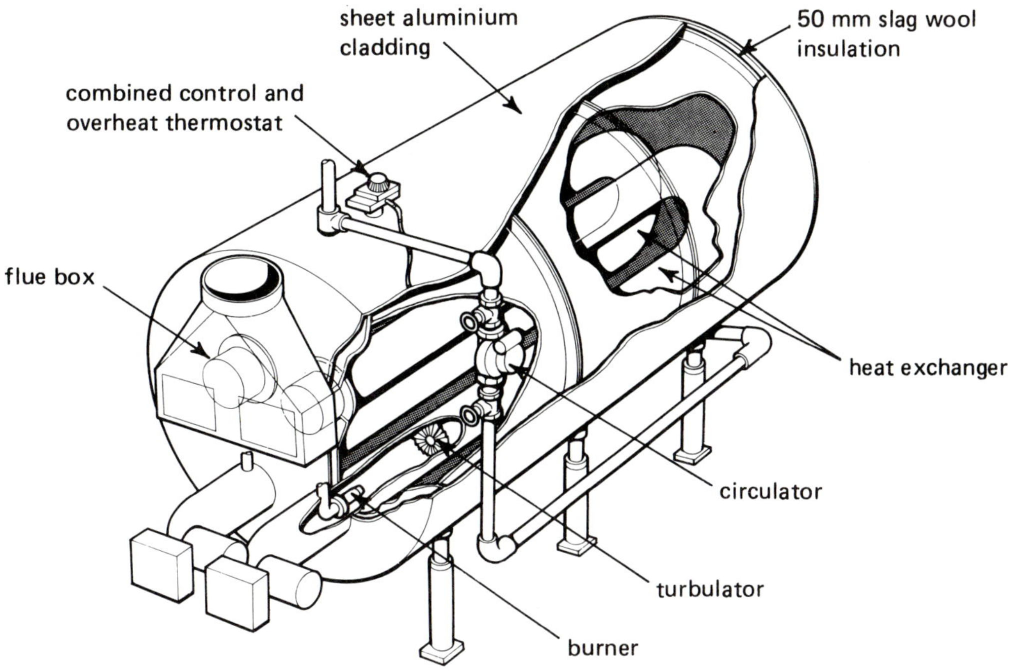

Fig. 6.20 Beaumont water heater

be increased. Two packaged burners fire one into each tube, so some heat will be maintained even if one unit fails. The temperature uniformity inside the water cylinder is improved by an external circulator and connection pipes. The cylinders are insulated by 50 mm slag wool insulation which is covered on the outside by sheet aluminium cladding. The combustion heat exchangers contain turbulators which further increase heat transfer by convection.

Efficiencies in excess of 70% are achieved with these units. Because of their construction and accessibility they lend themselves to waste heat recovery from furnaces and incinerator stacks. The continuous expansion and contraction of the unit means that any scale formed will not adhere to the inside surfaces, but fall to the bottom of the cylinder.

Condensing storage water heaters

Condensing storage water heaters offer the possibility of substantial increases in the efficiency of storage water heating. These units normally have a fan assisted conventional flue system or a balanced flue. Fanned flues permit the use of horizontal runs and allow the units to be positioned away from outside walls. It is necessary to provide an additional condensate pipe to remove flue condensate to drain (see Section 6.4 on condensing boilers).

These appliances are available from 12 to 70 kW with an output of 200 to 1 200 litres/h of water at 65 °C. A sketch of one of these units is shown in Figure 6.21.

Instantaneous water heaters

The gas fired instantaneous water heater has negligible storage and a comparatively larger heat input rate than other direct gas fired systems with similar hot water outputs. This provides an immediate supply of hot water; only the hot water that is required is heated and no storage is involved. As the name implies, the heat recovery period is virtually zero and it can be compared with a large boiler serving a non-storage calorifier.

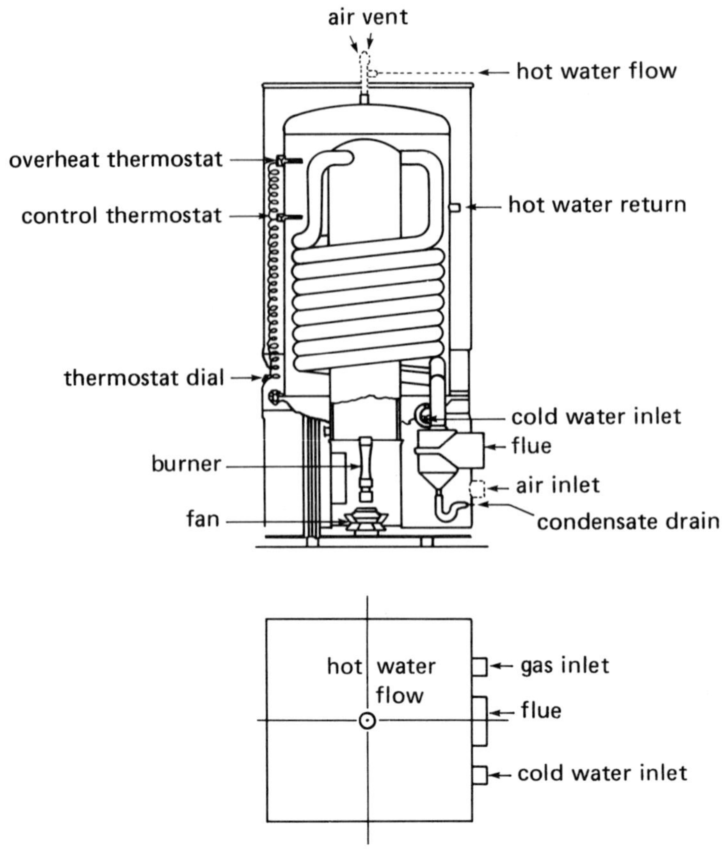

Fig. 6.21 Condensing storage water heater

Instantaneous heaters can be used to supply remote or scattered draw-off points individually, dispensing with long lengths of secondary pipework, which heavily outweighs the costs of gas service pipework. Single or multipoint units are available, most of which are of the balanced flue type.

6.2.4 Hot Water Services Design Considerations

Plant capacity

In the past, the sizing of plant in hot water systems has been based on guidelines with little evidence to substantiate them. This has usually resulted in plant capacity based on an estimate of water usage per person to which a safety margin has been added.

British Gas plc, Watson House staff investigated the h.w.s. requirement of forty commercial buildings and the results confirmed that, in many cases, boiler plant and storage calorifiers were too large. British Gas plc has published the guide to Hot Water Plant Sizing for Commercial Buildings. This Guide provides information on hot water requirement for catering and 'service' (washing, bathing and cleaning) use. It also enables plant to be sized correctly so that maximum h.w.s. demand can be met without a lot of over-capacity, and covers heat inputs/storage capacities to cater for a range from zero (instantaneous heating) to three hour heat recovery period systems.

It must be stressed that the Guide is purely for commercial premises and extrapolation to industrial buildings should only be used with caution.

Oversizing of h.w.s. plant is still a major cause of energy wastage.

Choice of System

The choice of hot water supply systems can provide a useful contribution to energy saving. Probably the most important consideration when designing a system is the separation of space heating from h.w.s. By employing applicances for each service separately, the Summer operation can be improved considerably. If heating and hot water are provided by the same boiler, then during Summer the boiler will be grossly oversized for meeting hot water requirements only. If, however, separate appliances are used, then in Summer the heating boiler will be off and the hot water boiler will work at its design load, giving higher efficiencies than in a combined system. Significant savings can be achieved if this type of design is employed when changing existing boilers.

The use of short pipe runs is also a valuable design consideration. Long dead legs will increase system losses and waste energy.These can be avoided at the initial design stage. Code of Practice CP 310 specifies maximum lengths of dead legs and this should be consulted when setting out pipe runs.

The idea of short pipe runs can also be applied to circulating pipework. If, for instance, there is a circulating system in a large building, supplying draw-off points which are a long way from the boiler, then it may be advantageous to install separate appliances at these points. This could in some instances reduce the secondary pipework and hence system losses. The distant draw-off points could be supplied by small instantaneous or storage water heaters. This design method does depend upon the type of building, its layout and the system already employed. It can, however, provide savings in running costs in many cases.

Oversizing pipework and pumps can also increase system losses as can oversizing the storage cylinder. These can be avoided by using the Guide referred to above.

6.2.5 Controls

By controlling h.w.s. systems correctly, considerable energy savings can be made. The two main questions to be considered are:

- What temperature is the water required at?
- When is the water required?

Temperature

Recommended maximum hot water storage and delivery temperatures (see Section 6.2.1) are 55 °C and 50 °C respectively. These temperatures should be adhered to as closely as possible, with the exception of water used in catering applications where a delivery temperature of 60 °C should be maintained. There are many instances where hot water is stored or delivered at temperatures 20 °C in excess of those recommended. This results in the use of 40% more energy than is necessary for h.w.s. production.

Thermostatic control of the primary heat source by means of a thermostatic valve operating under the control of a thermostat in the storage vessel or in the delivery line (instantaneous heaters) will ensure energy consumption is kept to a minimum. On dedicated boiler/calorifier systems, controlling the boiler operation from the calorifier temperature and not the boiler flow temperature will reduce boiler cycling and save fuel.

Time

The incidence of night and weekend operation of h.w.s. heating plant in day occupancy premises is high. Even when no hot water is being used, radiation losses, pump operation etc. can be a significant but avoidable waste of fuel and electricity.

Simple time clock control on the heating source and circulating pumps will effectively restrict plant operation periods to those required.

Energy Management Systems

As with space heating, h.w.s. can be brought under the dictates of an Energy Management System which not only provides time and temperature control, but can also monitor hot water temperatures and usage, if considered necessary.

6.3 INDUSTRIAL LIQUID HEATING

6.3.1 Introduction

The main industries employing process liquid heating are:

- Food and Drink
- Textiles
- Leather Tanning
- Chemicals, Paints and Dyes
- Engineering
- Detergents

Traditionally, process liquids have been heated by a distributed medium (steam, hot water, thermal oil) generated in a central boiler plant.

The overall thermal efficiency to point of use of a centralised system is much lower than that of the boiler plant itself because of inherent system losses, as discussed in Section 6.1. Notwithstanding the advantages of centralised systems (e.g. fuel flexibility), there is an increasing trend towards applications of gas fired appliances at point of use, the efficiencies of which are much higher. Most of these appliances have been developed by British Gas plc at Midland Research Station (MRS) and are manufactured under licence by a number of reputable companies.

Selection and sizing of the most suitable gas fired equipment for conversion or replacement of heated process liquid vessels is dictated by:

- Dimensions of the tank or vessel.
- Degree of insulation on the tank or vessel.
- Whether surface insulation is fitted.
- Type of liquid to be heated.
- Required liquid temperature.
- Allowable heat-up time from cold.

- Work throughput rate (e.g. on plating lines).
- Available free space to accommodate heat transfer surfaces.

6.3.2 Steam, Hot Water and Thermal Fluid Heating

With centralised systems distributing the above media to point of use, heat transfer is normally indirect via submerged coils or panels in, or by jackets surrounding, the tank or vessel containing the liquid to be heated. The exception is where direct steam injection is employed. The latter can only be used if the liquid to be heated is water or can tolerate aqueous dilution. (The same comment applies to returning condensate from pipe coils back into the tank being heated). Typical applications of distributed heating media are shown in Figure 6.22.

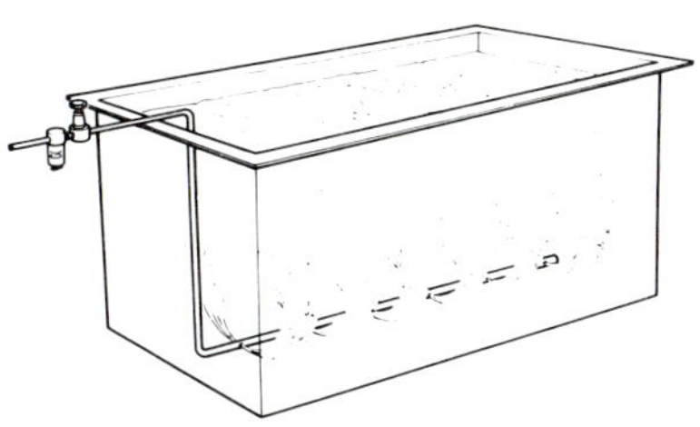

Direct steam injection

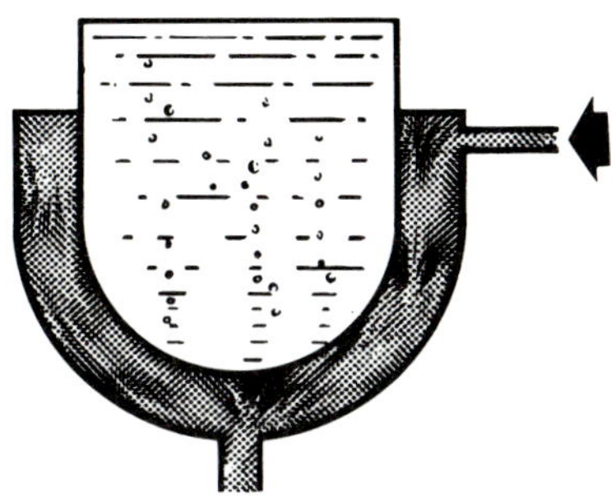

Steam jacketed vessel

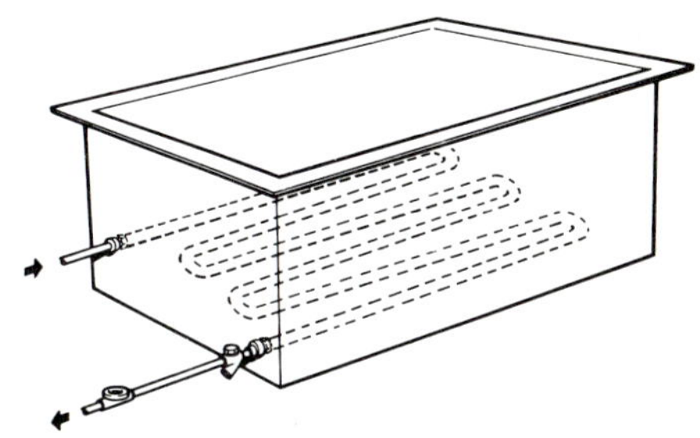

Horizontal heating coil

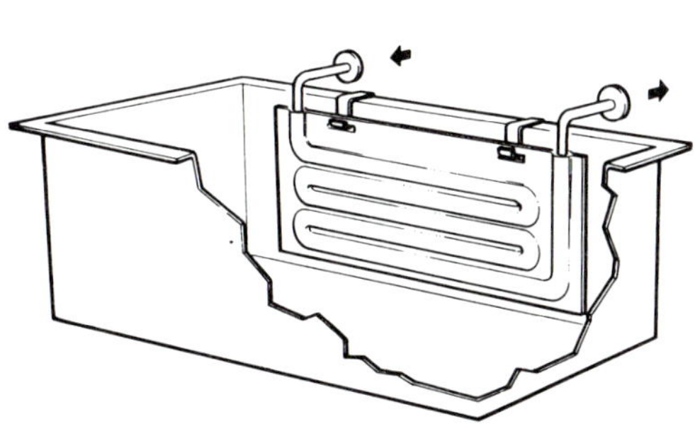

Vertical heating pannel

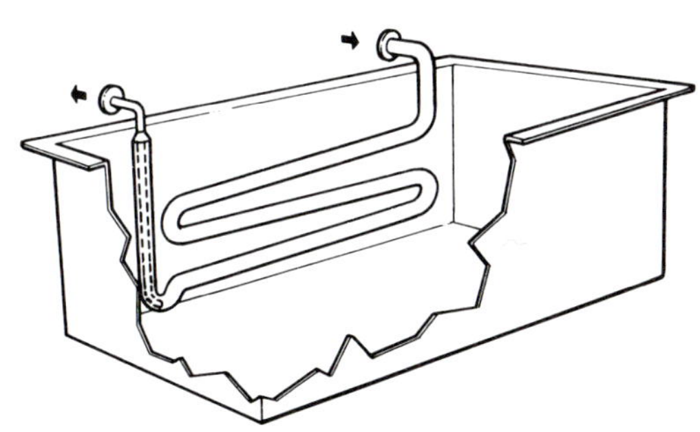

Vertical heating coil

Fig. 6.22 Application of distributed media at point of use

There are few instances where steam or hot water heated process vessels cannot be replaced with direct fired gas appliances, either by conversion or complete replacement of existing tanks etc. Such appliances are suitable for most aqueous solutions including chemical solutions, mild acids and alkalis. Selection criteria must, of course, cover corrosion and contamination potential, temperature limitations etc.

6.3.3 External Gas Heating

External heating of tanks with bar burners is less common than it was. The system is simple. easy to install and control, has low capital cost and does not obstruct the interior of the tank. Heat input is only restricted by space available. The overriding disadvantage is low efficiency, 25% being typical.

Ducting combustion products through a jacket on the outside of the tank can increase efficiency to up to 50%. Exposed flames are a potential hazard.

6.3.4 Immersion Tube Heating

Immersion tube heating gives much higher efficiencies than with bottom heated vessels.

Natural draught immersion tube

Where there are no space restrictions, immersion tubes fired by natural draught burners offer the advantages of simplicity and reliability. They have been used for many years for industrial liquid heating giving efficiencies up to 70%.

The main disadvantage is the low firing intensity (heat input per unit of cross-sectional area of tube) of about 5 W/mm^2, which in practice necessitates the use of large diameter tubes. Consequently, this restricts the heat input possible for any given vessel. Large tubes also occupy an excessive amount of tank space and can be expensive if the liquid to be heated dictates special materials of construction. Figure 6.23 illustrates some typical installations.

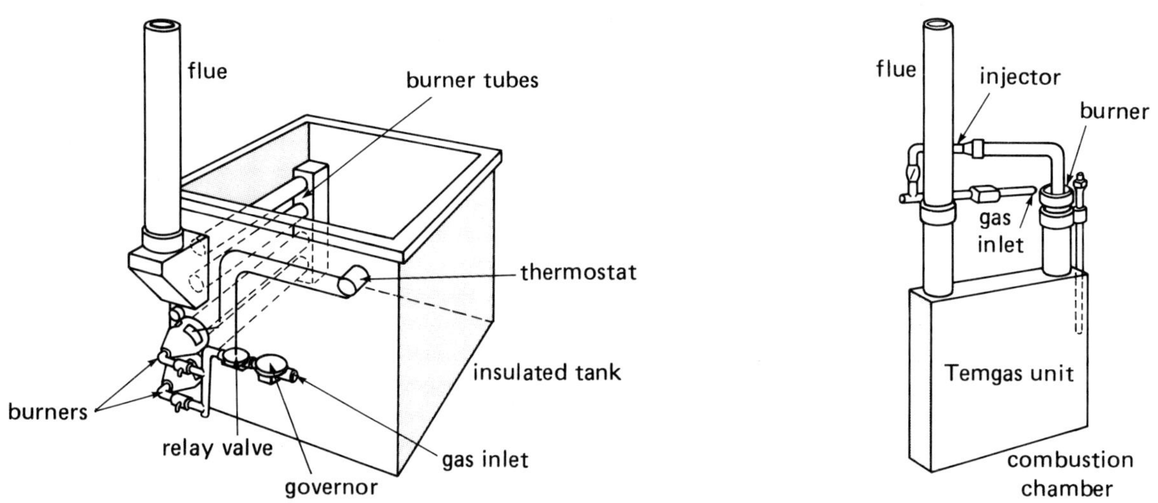

Simple natural-draught system (e.g. Selas) — Temgas immersion heater

Fig. 6.23 Natural-draught immersion-tube systems

The relationship between thermal efficiency and tube length/diameter is shown in Figure 6.24. It clearly shows the efficiency benefits of high length-to-diameter ratios.

The constraints on thermal input rates (normally 7 kW to 70 kW) and the relatively low thermal efficiencies obtainable with natural draught tubes, limit their application for industrial process liquid heating. To overcome these limitations the Midlands Research Station of British Gas plc developed forced or induced draught medium intensity and high intensity forced draught immersion tube burners.

Medium intensity immersion tube burners

These burners have been developed for the heating of smaller vats, tank and storage vessels. Firing intensities, although higher than with natural draught burners, are not high at around 14 W/mm^2 of tube cross-sectional area. Maximum thermal input rates are about 60 kW. The main advantage is high efficiency (80% plus) with length : diameter ratios of between 80 : 1 and 100 : 1. If it is only

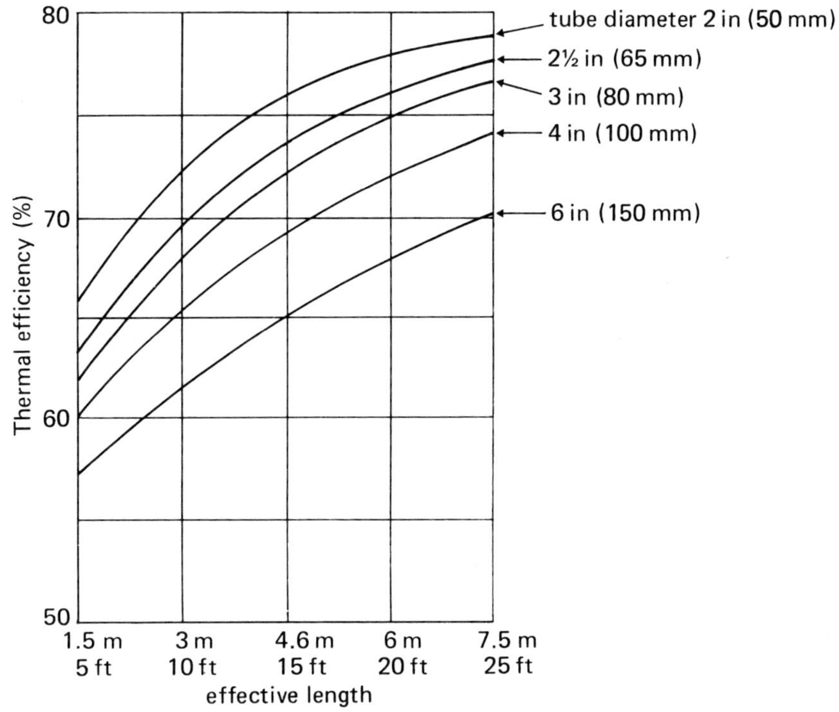

Fig. 6.24 Thermal efficiency v tube length/diameter

possible to install shorter immersion tube lengths, the fitting of turbulators will enable the high efficiencies to be achieved. Other advantages are simplicity, reliability and low capital cost. Figure 6.25 shows a schematic arrangement of a medium intensity immersion tube installation.

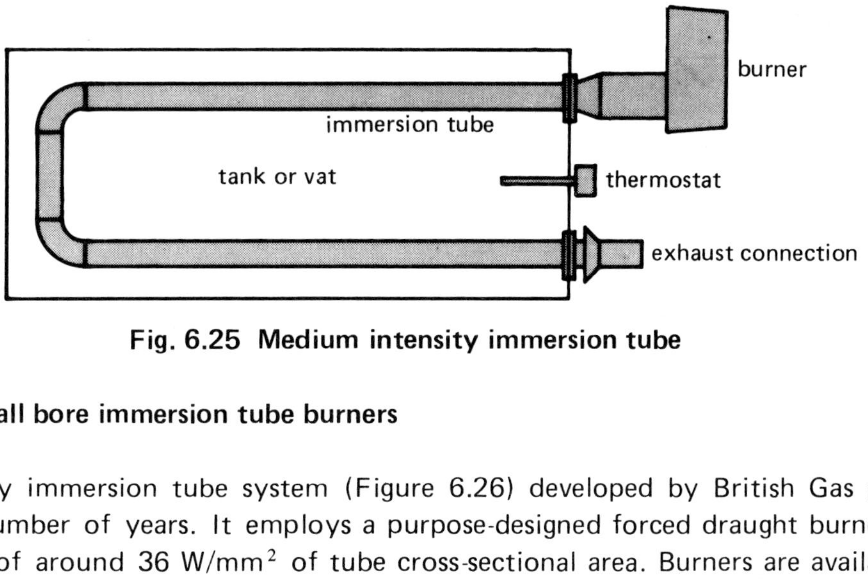

Fig. 6.25 Medium intensity immersion tube

High intensity small bore immersion tube burners

The high intensity immersion tube system (Figure 6.26) developed by British Gas plc has been available for a number of years. It employs a purpose-designed forced draught burner to achieve firing intensities of around 36 W/mm^2 of tube cross-sectional area. Burners are available for heat input rates of up to 600 kW or higher if a gas booster is used. Tube sizes range from 25 mm to 150 mm diameter.

The system incorporates a combined all metal combustion chamber and burner head. The combustion chamber is immersed in the tank solution in order to obtain maximum heat release and to give low external operating temperatures at the burner head. High intensity combustion occurs in the chamber, the hot combustion products then discharge through a converging nozzle into the immersion tube. Where it is not possible to install internal combustion chambers, external refractory lined or water cooled chambers can be fitted.

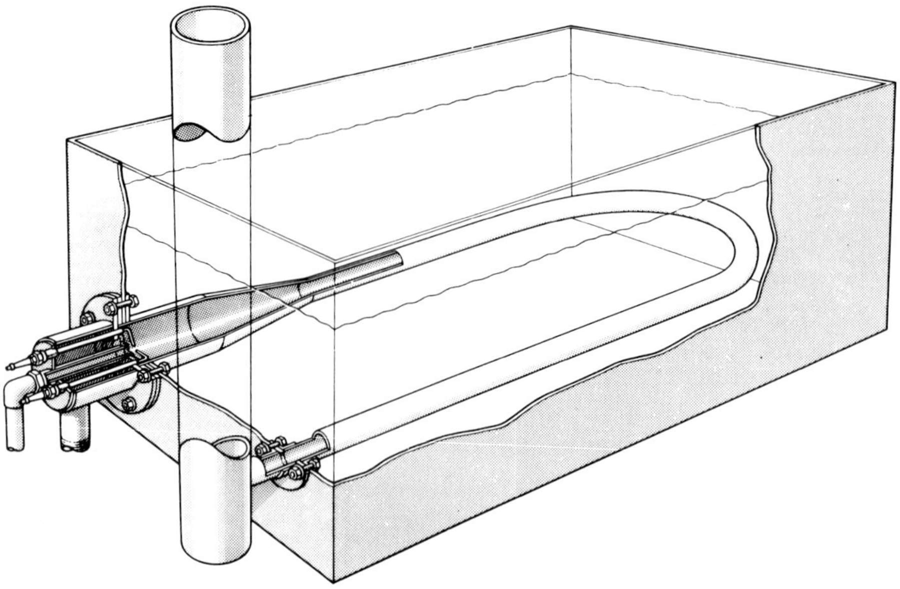

Fig. 6.26 MRS small bore immersion tube burner

Efficiencies of over 80% can be expected when immersion tubes have a length : diameter ratio of 140:1 or greater. Figure 6.27 shows the dependence of efficiency on L:D ratio.

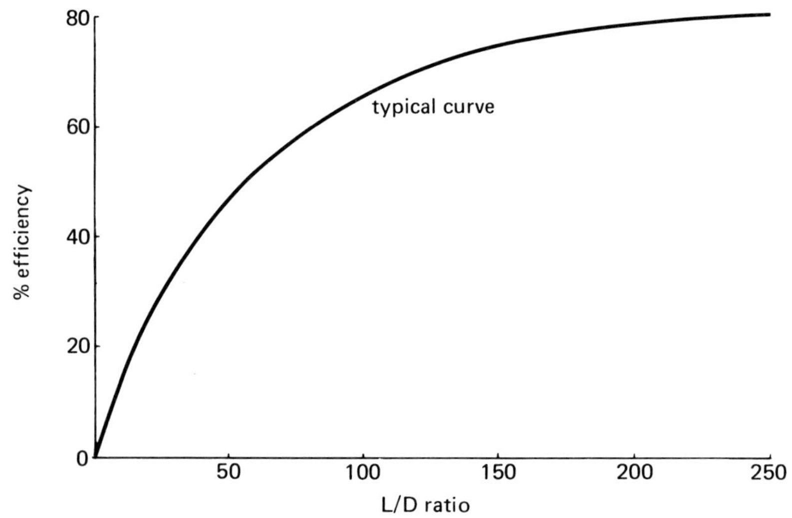

Fig. 6.27 Efficiency v length/diameter ratio

Due to high air pressure requirement, fan noise can pose problems if the unit is not carefully selected. The relatively high capital cost of the system means that it becomes more attractive as required heat input increases.

Multi-tube heat exchanger

The multi-tube heat exchanger was developed by MRS for application to small volume tanks where higher heat input rates are required or where the installation of a single immersion tube would not be practical, because of space limitations. The heat exchanger basically consists of two header tubes joined by a large number of small tubes to form a grid, as illustrated in Figure 6.28.

The combustion header is designed to ensure complete air and gas mixing to give correct combustion conditions prior to the gases discharging into the multi-tube grid and being exhausted via the exhaust manifold.

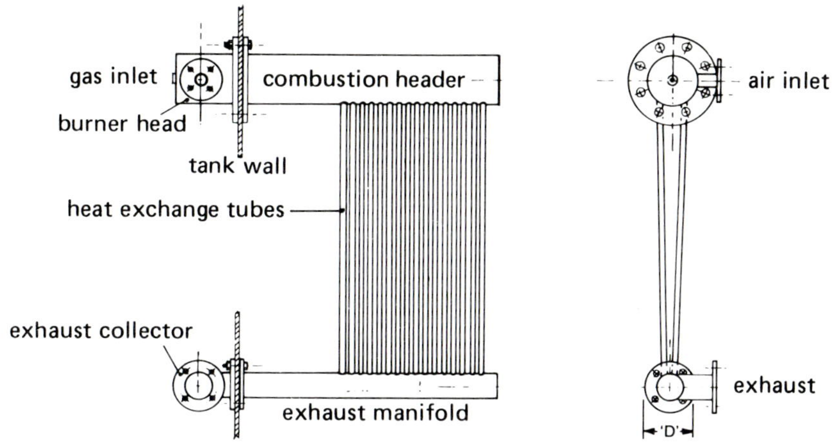

Fig. 6.28 Typical multi-tube heat exchanger

The design of the burner/heat exchanger enables it to be mounted either horizontal, vertical or at any angle between as long as the centre line of the combustion header is always above that of the exhaust manifold. Figure 6.28 above, therefore, can be regarded as showing either a plan view or side elevation. The longitudinal centres of both headers must remain parallel to the horizontal to eliminate condensate build-up.

Gross heat input rates go up to 600 kW.

Multi-tube high intensity immersion tubes

Multi-tube immersion heaters have been designed specifically as one-piece plug-in units for single hole mounting. This simplifies the installation within the vessel to be heated and also the associated gas burner controls. In many cases it is possible to replace the existing steam/hot water heating coil, in a calorifier for instance, with a multi-tube unit with little or no modification to the vessel.

The unit consists of a large diameter combustion header with a number of small tubes attached at the end remote from the burner. These tubes return along the length of the header as a concrete ring. A schematic diagram of the heater is shown in Figure 6.29.

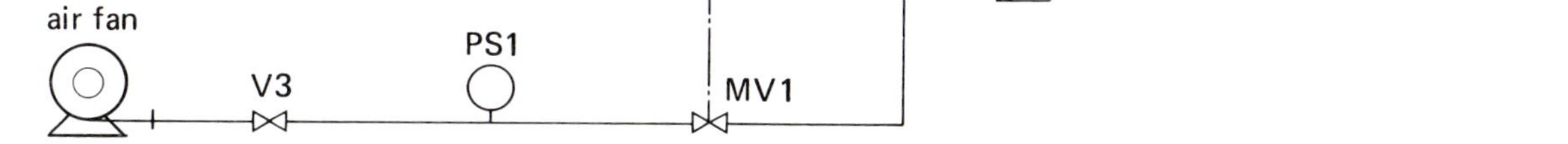

V1 = main gas isolating valves
PCV1 = pressure control governor
SV1 = main gas safety shut off valves
MV1/MV2 = linked motorised gas and air control valves
PCV2 = start gas governor
SV2 = start gas safety shut off valves
V2 = start gas isolating valve
V3 = air restrictor valve
V4 = start gas restrictor valve
PS1 = air pressure switch

Fig. 6.29 Multi-tube high intensity immersion heater

Heaters are available from 38 kW to 470 kW which are suitable for most standard storage calorifiers, tanks and vats. High combustion product velocities and hence high heat transfer rates give efficiencies of around 80% or more, based on the gross heat input.

Multi-tube in-line circulation heater

The multi-tube immersion tube unit (illustrated in Figure 6.30) can also be used as the heat exchanger for an in-line circulation heater. The high heat transfer rates can eliminate the necessity for water/ solution storage hence reducing the standby heat losses. Outputs and efficiencies are similar to those quoted above.

Fig. 6.30 Multi-tube in-line heater

The heater body is constructed of mild steel or stainless steel, depending on process requirements. Flanged inlet and outlet connections are provided in the top of the vessel as is a drain connection in the bottom. Access for cleaning and inspection is via a removable end plate.

6.3.5 Direct Contact Heating

Most systems currently employed in process liquid heating are indirect because there is usually a steel or cast iron membrane separating the liquid from the hot combustion products. Heat is transferred by conduction and convection. Heat transfer efficiency depends on the extent of scale formation on the liquid side and combustion deposits on the hot gas side, thickness of heat exchange surface, water and hot gas velocities and other variables. This normally limits the operating efficiencies to just over 80%, under favourable conditions, in a non-condensing system.

Submerged combustion

Direct contact heating/evaporation of process liquids has been practiced for many years. In a submerged combustor, fuel and air are burnt in a combustion chamber and the products directed downwards through a dip tube which projects below the liquid surface. The tube is designed to promote bubble formation and give good mixing between products and liquid by providing a large heat transfer area in a small volume. The length of the dip pipe is a compromise between gas: liquid contact area and resultant back pressure. Even so, to obtain good performance the burner supply pressures required are high: up to 0.3 bar. This will often require the use of a gas booster in addition to a high pressure air fan, adding significantly to the capital cost of an already expensive item of equipment. Figure 6.31 illustrates two types of submerged combustor.

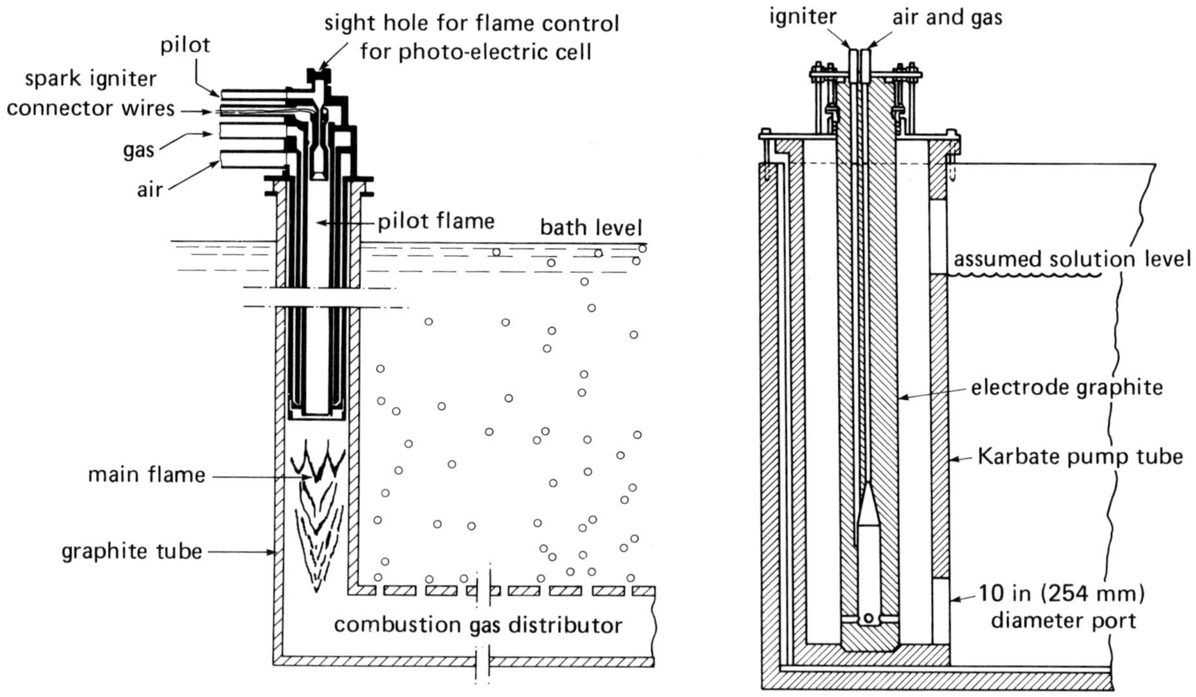

Fig. 6.31 Submerged combustors

As liquid temperature rises the water carrying capacity of the dry gases increases until all the useful heat supplied is used to evaporate water rather than raise its temperature. This occurs at a liquid temperature of 89 °C as shown in Figure 6.32. For normal liquid heating processes, the working temperature should be no higher than 70 °C where an efficiency of 80% can be expected.

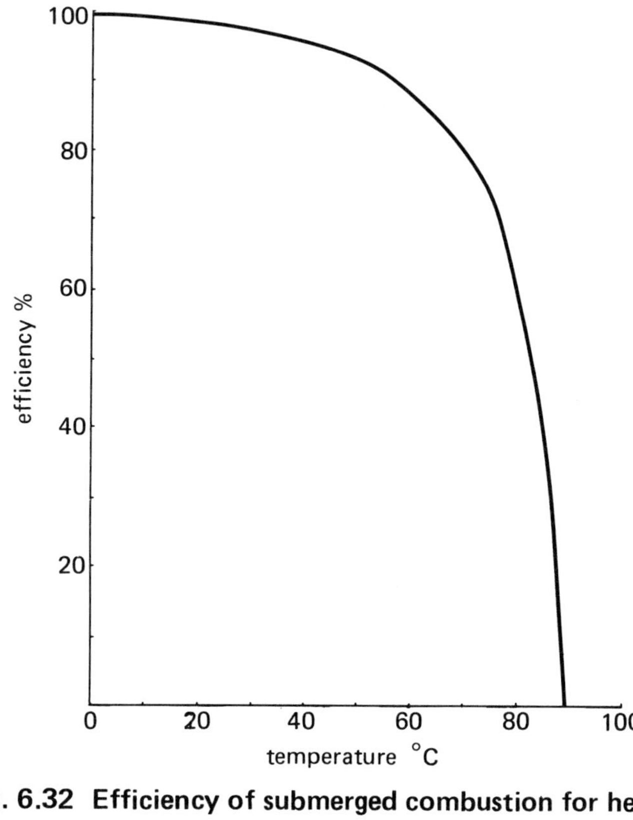

Fig. 6.32 Efficiency of submerged combustion for heating

The evaporation effect can be valuable in some cases; Figure 6.33 shows that submerged combustion is a very efficient method of evaporation.

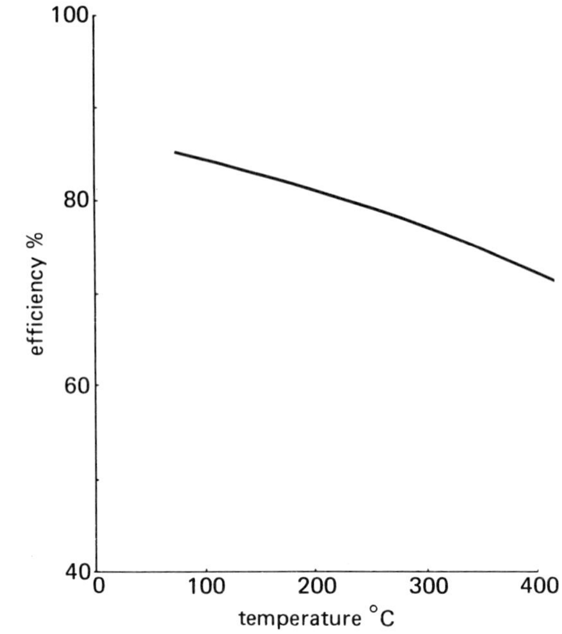

Fig. 6.33 Efficiency of submerged combustion for evaporation

Direct contact water heaters

A relatively recent successful development by British Gas plc is the Direct Contact Water Heater. This unit gives the high efficiency of direct contact heating without the disadvantages of conventional submerged combustion by combining the benefits of immersion tube heating and spray recuperation. The combination of direct contact heating in a counterflow recuperator column and indirect heating using an immersion tube at the base of the column overcome the evaporation disadvantage and enables both high water temperatures and high efficiencies to be achieved. The Nordsea 'Cascade' heater illustrated in Figure 6.34a employs these combined techniques.

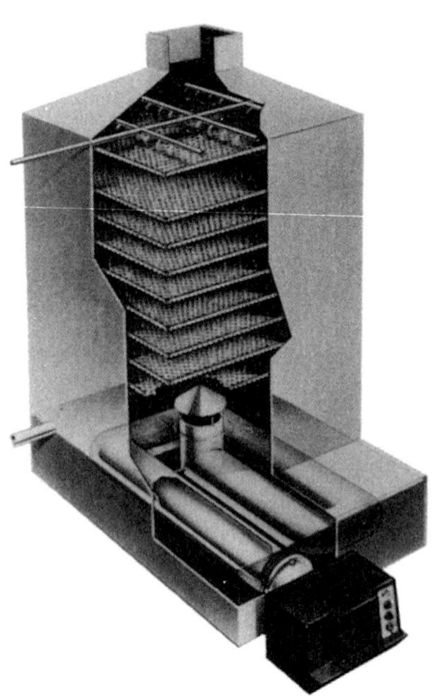

Fig. 6.34a Direct contact water heater with stainless steel baffles

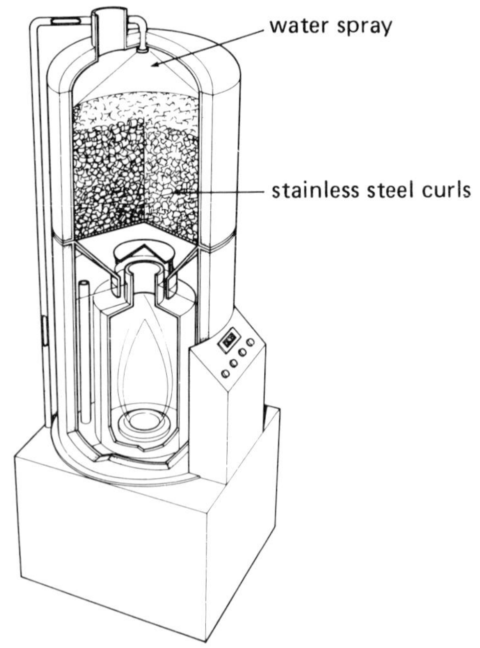

Fig. 6.34b Direct contact water heater with stainless steel curved shapes

The unit consists of a stainless steel sump tank in which is fitted a multi-pass stainless steel gas fired immersion tube. A stainless steel tower, containing a number of perforated baffle trays, sits on top of the sump tank. Cold water is fed to the top of the tower through a distribution manifold and then flows down the tower over the perforated trays, countercurrent to the products of combustion. On leaving the bottom of the tower, the water will have been heated to 50 to 60 °C. The temperature is then further raised in the sump tank, by the short length of immersion tube, to up to 95 °C if required. Approximately 50% of the total heat to the water is transferred by the immersion tube. The overall efficiency of the unit is dependent on the cold water flow rate and inlet temperature. Up to 98% efficiency is possible with cold water at 15 °C.

Because of the water's direct contact with combustion products, a slight reduction in pH level occurs, due mainly to carbonic acid formation. In most applications this can be disregarded. The latest design uses stainless steel small curved shapes to form a packing with high contact areas instead of the perforated trays. The burner fires from the base vertically upwards.

Direct contact water heaters are manufactured in sizes ranging from around 60 kW to 3 000 kW and can be of single tower/single burner or twin tower/twin burner configuration. They can be used in conjunction with hot water storage systems or as instantaneous water heaters, and have been successfully applied in, for example, the leather tanning and food industries.

6.3.6 Summary

It is clear that there is a wide range of liquid heating equipment available and the heating efficiency can vary greatly. Equipment selection can not be simply based on efficiency; process requirement and comparative capital costs have also to be considered. Each particular process must be assessed individually before the best system for that process can be selected.

6.4 HOT WATER, STEAM AND THERMAL FLUID BOILERS

6.4.1 Introduction

Before the advent of North Sea gas the British gas industry's share of the steam raising and bulk water heating market was very small. Coal and oil were the predominant fuels in this market. The availability of large quantities of natural gas has allowed British Gas plc to make appreciable inroads into the industrial boiler market.

The majority of gas fired boilers were originally designed for another fuel. In many boilers gas is the third fuel to have been used (after coal and oil). However, a growing number of new boilers have been designed specifically for gas firing and for gas/oil dual-fuel firing.

The very early boilers were spherically shaped objects which sat on brick settings with a fire lit underneath them, rather like the old kettle. Needless to say the efficiency was very low. The biggest step forward was to put the water around the flame, and then to introduce extra surface to take heat out of the gases produced from the flames. Thus the Cornish Boiler and later the Lancashire Boiler both made use of convective heat transfer. The requirement for a more compact boiler for the marine industry led to the development of the Scotch Marine Boiler where the gases from the furnace tubes were led, by means of small bore tubes, through the water instead of around the outside.

The Economic Boiler is a direct descendant of the Scotch Marine Boiler, and the modern Package Boiler is a further refinement. All these developments have resulted in compact boilers with more use being made of convective heat transfer.

The Water Tube Boiler has developed along similar lines with smaller combustion chambers, more convection passes, and the addition of Air Heaters and Economisers.

Smaller, low temperature hot water (l.t.h.w.) boilers have also progressed from the old Cast Iron Sectional (C.I.S.) boiler through welded steel boilers to finned tube, modular and condensing boilers.

The result of this progress in development is that modern boilers are inherently capable of high operating efficiencies which can be in excess of 90% with gas fired condensing boilers, under favourable load conditions.

6.4.2 Low Temperature Hot Water Boilers

Low temperature hot water boilers are extensively used in commercial and small industrial buildings to supply space heating and/or hot water service needs as discussed in Section 6.1 and 6.2.

Sectional boilers

Sectional boilers (see Figure 6.35) are widely used. Ratings normally range from 10 kW to 1 500 kW representing the domestic central heating market through to the large commercial and industrial market. Current gas-fired designs employ natural draught systems in the small sizes but use forced-draught packaged burners in large sizes permitting much higher heat-transfer rates and firing intensities. Overall heat transfer rates of 15 kW/m^2 for natural draught boilers compare with 40 kW/m^2 for a modern boiler designed for forced convection. The firing intensity can be an order of magnitude higher with the latter system i.e. 3 MW/m^3 compared with 0.3 MW/m^3 with a natural draught burner.

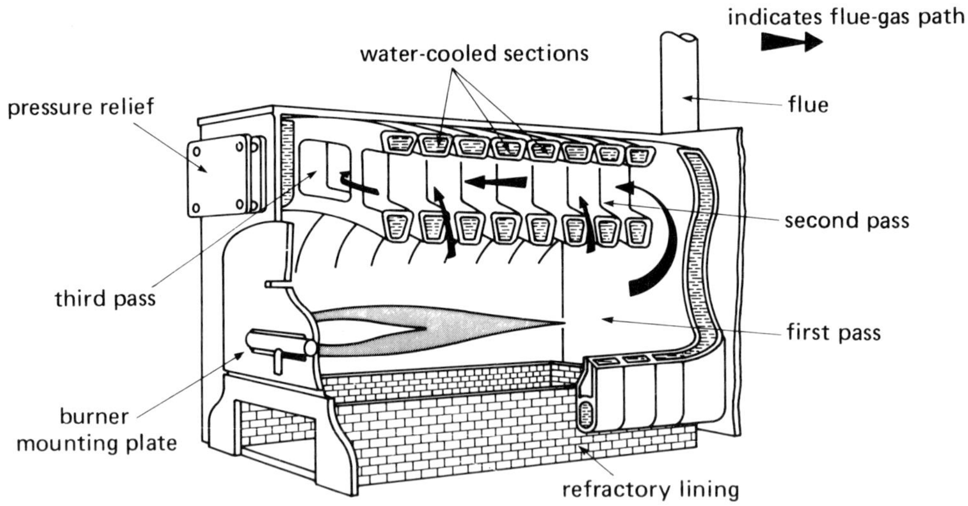

Fig. 6.35 Cast iron sectional boiler (Tasso F Series)

Modular boilers

The modular boiler system employs the principle of utilising several small boilers instead of one large boiler to meet load requirements. This gives several advantages, not least of which is improved efficiency. Low efficiencies at part load are overcome with a modular system because each module operates at full load. When more heat is called for, another module is automatically switched on by sequential thermostatic control. Another advantage is that, with the number of modules, the need to install extra capacity as standby against boiler failure is virtually eliminated. The small size allows quiet natural draught burners to be used and makes elaborate safeguards unnecessary. Installation of small modules is also simple.

The conventional modular system has cast iron sectional boilers arranged, side by side, with a sequence controller which can select which of the units take the base load, e.g. Hamworthy. This ensures that the same boilers are not always on load. A major drawback of this configuration is that water flows through all the boilers at all times. When some of the modules are not required, e.g. on reduced load, they then act as radiators, the natural draught drawing cold air across the heating surface of the boiler, unless automatic flue dampers and/or power operated valves in the water flow pipes are installed to stop the flow of air and/or water through the non-firing modules. Dampers and valves so fitted must, of course, be interlocked with the burner control to prevent firing unless they are proved to be open.

An alternative arrangement of modular boilers is now to have all the modules in the same boiler casing. One such system is the Ideal Concord boiler which can have from one to six modules in the same casing (see Fig. 6.36 for diagrammatic arrangement of the boiler and a single module). Each of the modules has an output rating of 50 kW so the maximum installed capacity of a system would be 300 kW with six modules. Each module has its own independent control system including electronic sequence controller and flame protection. Individual module thermostats are adjusted to give a sequence of operation so that the lower modules are in operation the longest time. Each heat exchange module is constructed of finned copper tubes arranged circumferentially around the gas burner, to give a low thermal mass installation and minimum heat up times. One novel design feature is that gas via a zero governor is drawn through the eye of the combustion air fan giving a stoichiometric premix of gas and air to the burner. As can be seen from the diagram there is a single flue for all modules, and the arrangement of modules is such that the products of combustion from the lower modules pass through the heating surfaces of the higher modules. The advantage here is that on low load with only the lower modules firing any heat remaining in the flue products is picked up by the higher modules.

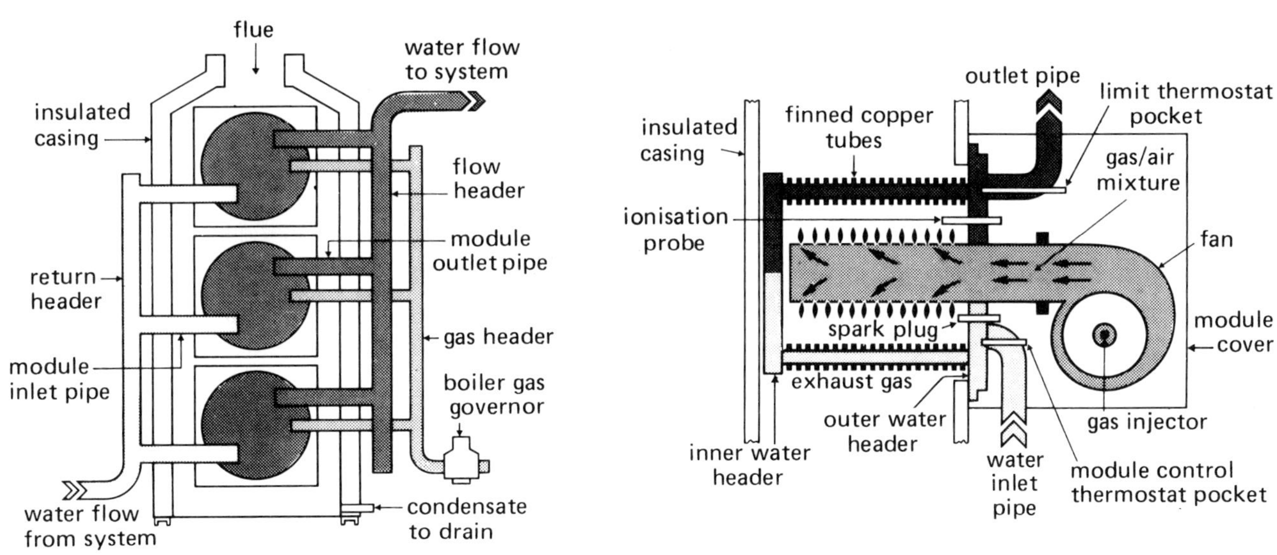

Fig. 6.36 Modular boiler (Stelrad Ideal Concord)

Another similar system with all the modules in the same casing employs natural draught burners, the construction of the modules being very similar to conventional instantaneous water heaters. An example of this system is the Chaffoteaux Flexiflame boiler which can have up to three modules in the same casing each with an output rating of 41 kW giving 123 kW with three modules.

High efficiency finned tube boilers

Efficiency on most natural draught boilers is comparatively low because there is no control over the amount of air which passes through the appliance, and in consequence the flue losses are increased as the load is reduced. In an attempt to improve the efficiency of a typical finned tube boiler, Raypack have developed an automatic air damper which modulates with the gas valve to maintain a constant air-gas ratio. The result is the Econoflame boiler which incorporates this device, and is marketed in the UK by Stokvis Ltd. This boiler is made in seventeen sizes having outputs from 104 to 1 066 kW. Modulation of the burner is controlled by a sensor detecting the water flow temperature from the boiler. A modulating motor drives a butterfly valve in the gas supply and also controls a damper which regulates the flow rate of combustion air. The air damper is a flatplate beneath the burner and parallel with the floor. It is moved upwards by an eccentric arrangement to reduce the amount of air supplied to the burner. The efficiency on the gross basis is over 30% at all loads between 25% and 100%. Figure 37a shows the Econoflame boiler, and 6.37b the characteristic curves for high efficiency finned tube boilers and traditional Cast Iron Sectional boilers.

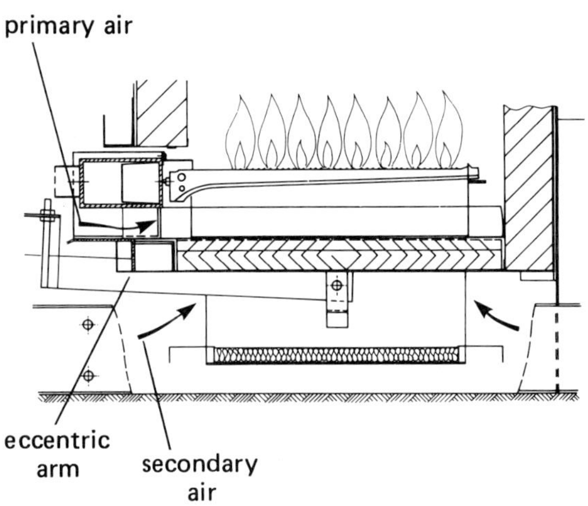

Fig. 6.37a Econoflame boiler

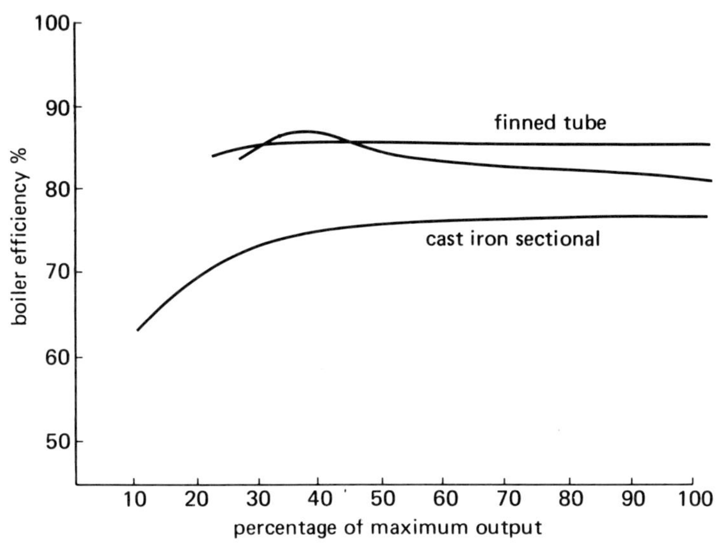

Fig. 6.37b Commercial boiler efficiency curves

Condensing boilers

When gas is burnt in a conventional l.t.h.w. boiler, the heat exchanger is designed so as not to cause condensation of the flue gases on the boiler surfaces during normal operation, although this may sometimes occur, particularly during start-up from cold. Boiler return water temperatures have also been traditionally high to prevent thermal shock and condensation. Flue gas exit temperatures are, therefore, relatively high. The latent heat of the water formed in the products of combustion with conventional boilers is lost as steam in the flue gases. To achieve an improvement in boiler efficiency, the water needs to be purposely condensed and its latent heat recovered into the boiler water.

The condensing boiler is specifically designed to abstract more heat, including a proportion of the latent heat, from the flue gases. This results in condensation forming on boiler internal surfaces, providing that the boiler return water is low enough to cool the flue gases below their dewpoint of around 55 °C.

The increased boiler surface area necessary for condensation may be in the form of an extra large heat exchanger or by an additional heat exchanger, the latter being the normal choice. The secondary heat exchange surfaces must be corrosion-resistant since the condensate is slightly acidic. Materials used are aluminium alloys or stainless steel. Arrangements must be made to collect the condensate from the secondary heat exchanger in a sump and then discharge it, via a water sealed trap, to drain.

Because the additional heat exchange surface increases the pressure drop through the boiler and because the lower flue gas temperatures reduce buoyancy draught, a forced draught or induced draught fan is invariably required.

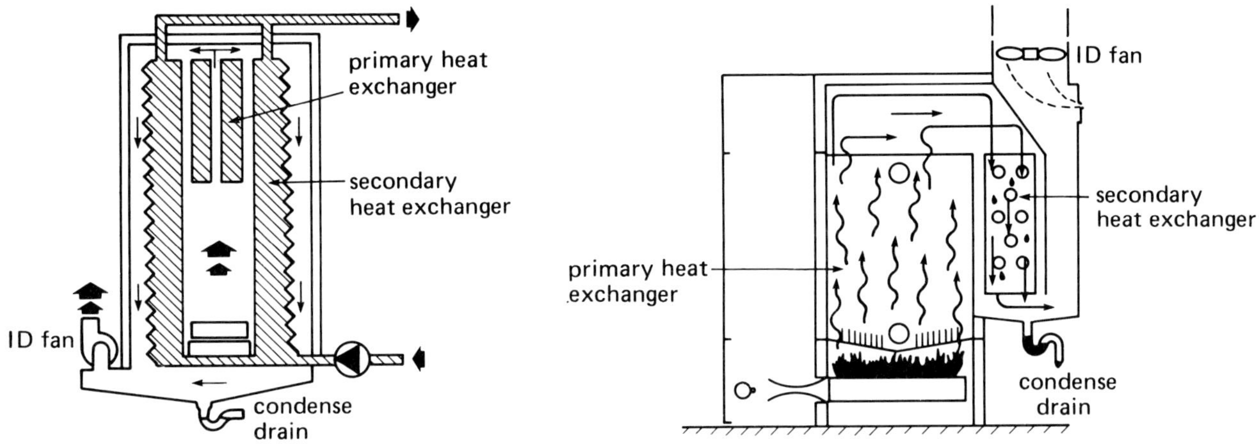

Fig. 6.38 Typical condensing boilers

Figure 6.38 illustrates two designs of condensing boilers. The earliest, installed in the UK in 1983, were from Holland and France. Whilst European manufacturers still dominate the market (at least seventeen supplying the domestic and small commercial market), in Great Britain there are nine manufacturers or importers offering condensing boilers for the commercial and small industrial market. The range of boiler outputs is 20 kW to 1 280 kW. Prices are typically double that of equivalent conventional boilers, but operating cost savings can still prove them to be cost effective.

The theoretical maximum efficiency of a conventional gas fired boiler is 85% gross. If the water vapour in the flue gases can be condensed, it is possible to raise the efficiency to around 95% (see Figure 6.39).

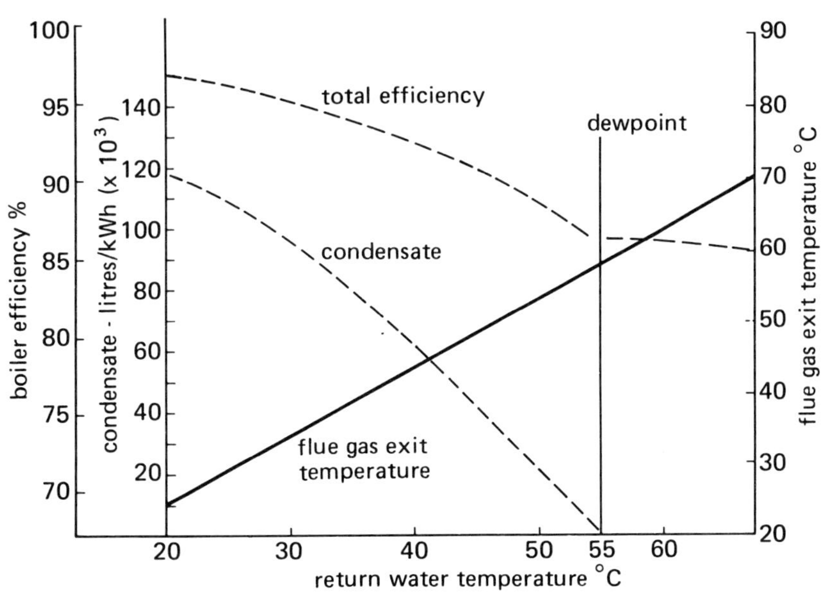

Fig. 6.39 Effect of condensing on efficiency

The ability to achieve condensing conditions depends on the boiler design, the type of heating system installed and the controls fitted. Because the efficiency varies with return water temperature and because condensate needs removing from the flue and the boiler, installation of condensing boilers requires special consideration. British Gas plc document IM/22 Installation Guide to High Efficiency Condensing Boilers contains advice on this matter.

Lower return water temperatures to the boiler improve efficiency. Figure 6.40 shows the part load efficiency curves which were measured by British Gas plc for a typical condensing boiler. Flow temperatures are shown which were about 10 °C higher than the corresponding return temperature.

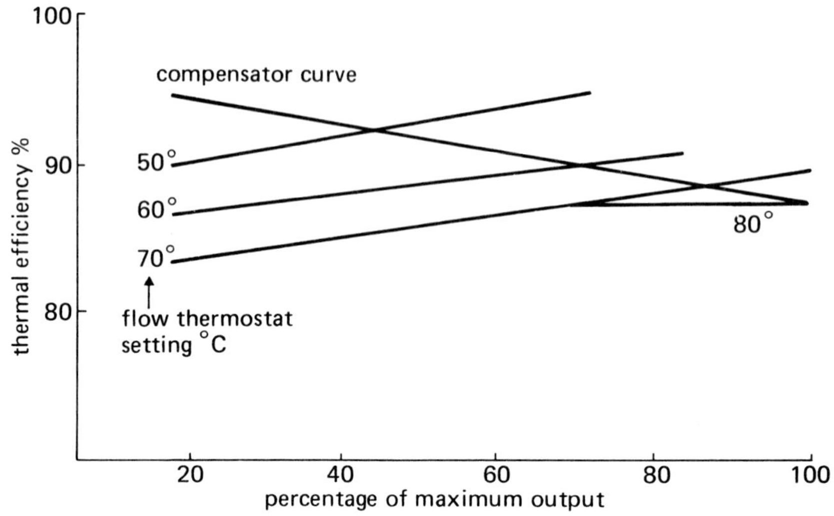

Fig. 6.40 Operating curve for a condensing boiler under direct compensator control

There are a number of ways by which return temperatures can be reduced. Oversized heat emitters would achieve this. If, for example, radiator surface area could be doubled, the required flow temperature could be lowered to 60 °C. However, the extra cost of larger emitters would not generally be worthwhile. Low temperature circuits using underfloor heating can be used but extra costs need careful assessment and such systems are not usually suitable for retrofitting.

For a simple l.t.h.w. heating system, the best way of operating a condensing boiler is to use a weather compensating controller operating directly on the boiler. Figure 6.40 above shows the effect of compensating control on the performance of a condensing boiler. Field tests have demonstrated seasonal efficiencies of up to 89% when such controls are properly installed. A simple compensated heating circuit is illustrated in Figure 6.41.

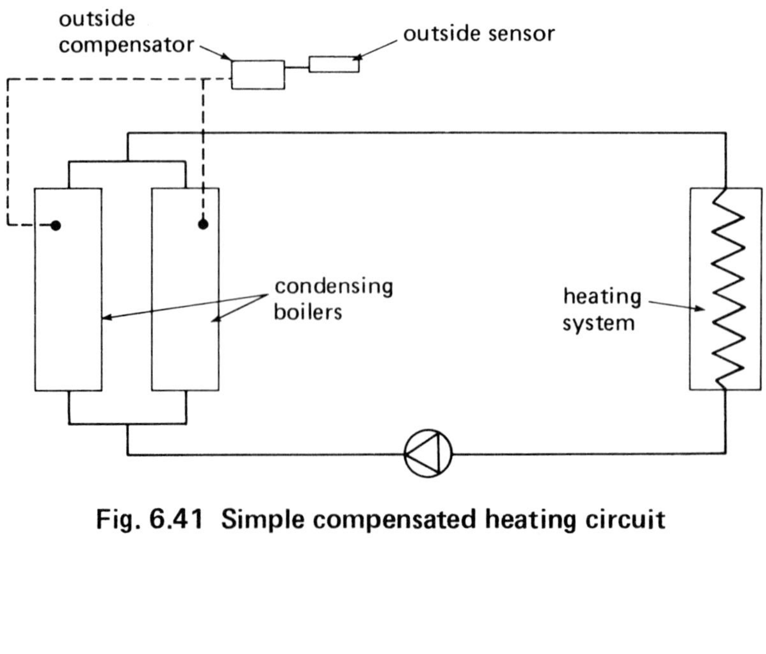

Fig. 6.41 Simple compensated heating circuit

Where there is a need to protect the primary heat exchanger, a three port valve is used to maintain temperature above the design limit (Figure 6.42). In this circuit, return water at reduced temperature will pass through the condensing heat exchanger when the three port valve is controlled by the weather compensator.

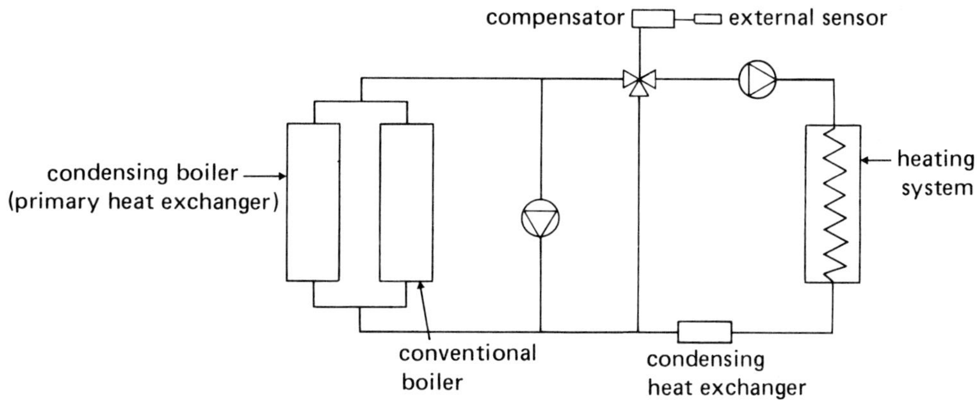

Fig. 6.42 Condensing boiler in a three port valve system

In either the simple circuit (Figure 6.41) or the three port valve circuit (Figure 6.42), the effect of reduced temperature of water passing through the condensing heat exchanger is to raise boiler efficiency. Nearly all condensing boilers can be installed in one of these two ways.

If measures to reduce building heat loss have been implemented after the heating system was installed, the original heat emitters will be larger than necessary. Therefore, unlike conventional boiler designs, the effect can be to improve the efficiency of condensing boilers.

Where thermostatic radiator valves are installed, they can lead to either a significant reduction in flow water rate, or a rise in return water temperature, both of which are better avoided. Certain types of high temperature heating circuits are also unsuitable for condensing boilers, such as circuits for fan convector emitters or where a high head of water pressure has allowed flow temperatures of 100 °C to be used.

Condensing boilers cost more than traditional plant (typically double). In a number of field trials carried out by Watson House, the measured operating cost savings have been compared with the additional capital cost for condensing boilers. Where traditional plant has come to the end of its useful working life and needs replacing, condensing boilers, or a combination of condensing and traditional plant could be used. Figure 6.43 shows the results of payback analysis on the data from the field trials. It relates additional costs per kW output and payback for various levels of boiler operating hours.

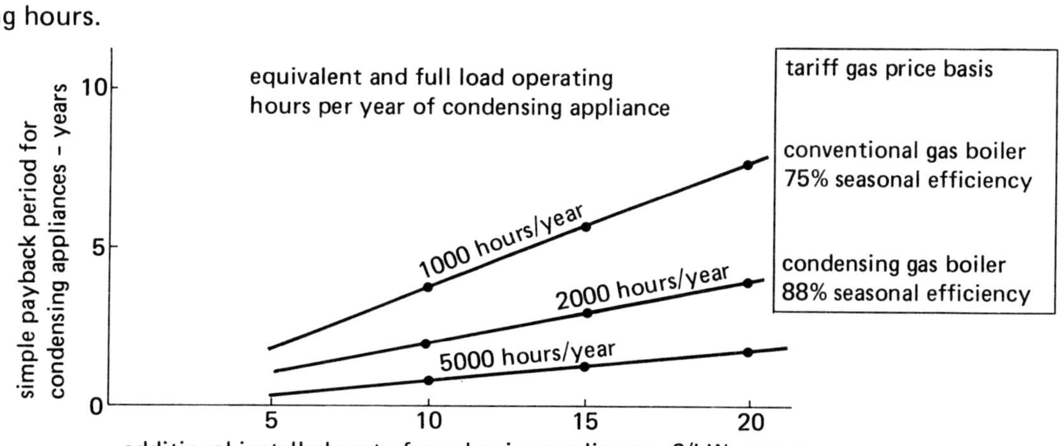

Fig. 6.43 Variation of simple payback with operating hours and installation cost of condensing boiler

6.4.3 Steam Boilers

Gas fired steam raising boilers are used extensively throughout British industry. The larger installations are on interruptible gas supplies and use dual-fuel burners to take advantage of lower prices, but there is a growing number of firm gas fired boilers where customers consider that the advantages of security of supply, ease of use and control, consistently high efficiency, cleanliness etc. outweigh the higher fuel price.

Steam is used for process and/or space heating and is transmitted by its own pressure. Its useful heat per unit mass is much greater than that for liquid media and it thus requires much smaller pipelines to transmit the same amount of energy. Steam trapping and condensate return systems entail expensive maintenance and replacement costs, and if there is no scope for flash steam recovery, the thermal transmission loss will be high.

Industrial steam boilers broadly fall into two categories:

- Fire-tube or shell boilers in which the hot gases are contained in tubes surrounded by water.
- Water-tube boilers in which water filled tubes are surrounded by combustion products.

Fire-tube or shell boilers

A shell boiler, because of its construction, is normally limited to a maximum working pressure of about 17 bar. Higher pressures would require the use of abnormally thick shell plates. Capacity is also limited by constructional features to around 18 MW.

Integral fan packaged dual-fuel or gas only burners (see Chapters 2 and 3) are now almost exclusively used for shell boiler firing.

The first fire-tube boilers to be widely used in industry were the brick-set Cornish and Lancashire boilers. Such boilers were successfully converted to natural gas firing with the advent of supplies from the North Sea. However, their use has declined to such a degree as to make them only of academic interest and no further comments are considered to be necessary.

The shell boiler market is dominated by horizontal economic and package boilers, with a few vertical boilers installed where floor space has been at a premium. Figure 6.44 illustrates two variants of the latter type.

The vertical cross-tube boiler is effectively a single pass boiler which is inherently inefficient (no more than 60% gross). The three-pass multi-tubular incorporates much more convective heat transfer surface and efficiencies between 75% and 80% can be achieved.

Two-pass dry-back economic boilers (Figure 6.45) succeeded Lancashire boilers as the work horses of British industry. Again, originally designed for coal firing, heat transfer rates and firing intensities were low. Conversion to gas firing has been successfully carried out with no adverse effect on boiler steam output. Efficiencies up to 75% are attainable with such boilers.

In the quest for higher efficiencies and lower costs, the oil fired package boiler was developed with a lot more convective heat transfer surface (more fire tubes) to take advantage of the high furnace heat release rates possible with oil firing. The size of these boilers was kept down by adding convection tube passes leading to the three and four-pass boilers of today. The first package boilers were

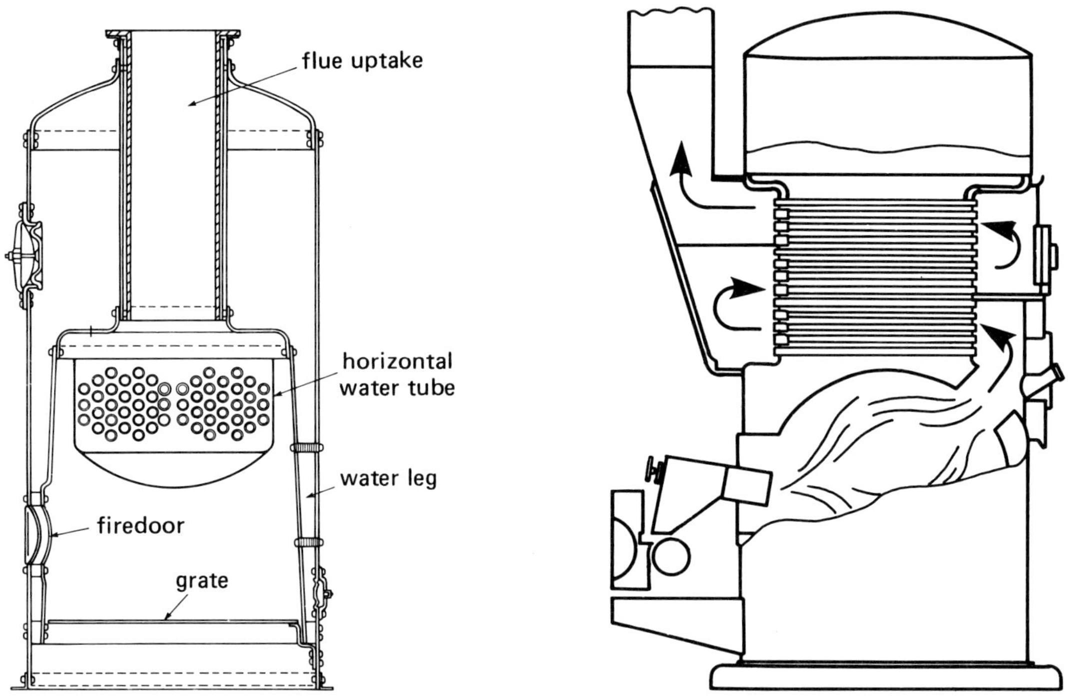

Cross-tube Multi-tubular (e.g. Cochran)

Fig. 6.44 Vertical boilers

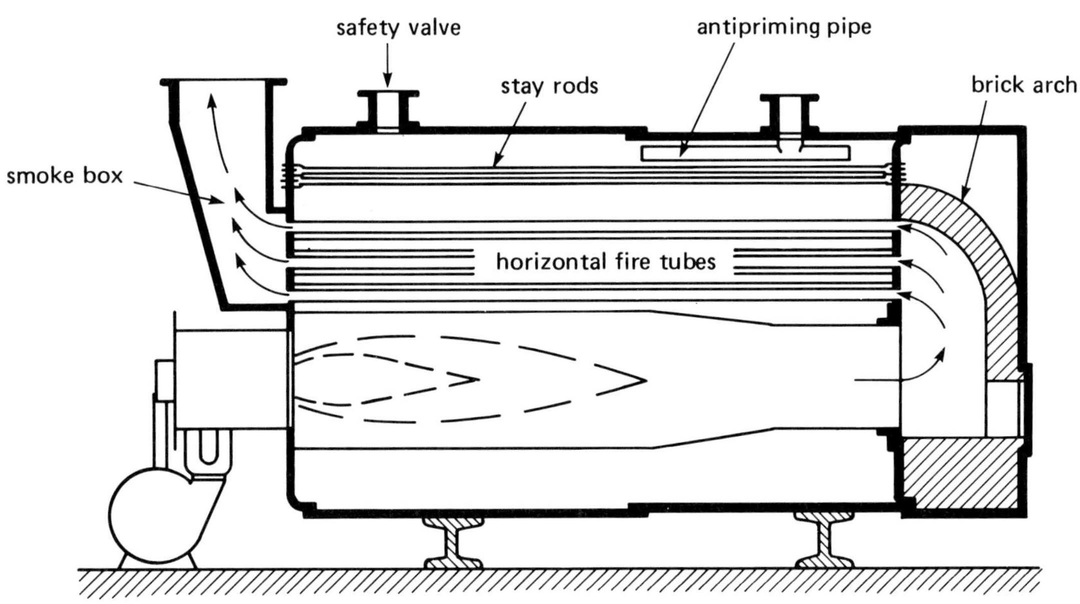

Fig. 6.45 Two-pass dry-back economic boiler

of dry-back design, but the majority of modern boilers are of wet-back construction i.e. the reversal chamber between the furnace tube (first-pass) and the first row of fire tubes (second-pass) is surrounded by water. Figure 6.46 illustrates a typical three-pass wet-back design.

The compact design of these high firing intensity boilers means that heat transfer surface in the furnace is lower than that in the older Economic boilers and the heat flux, especially across the tube plate, is increased significantly. Correct water treatment to prevent scale build up on the water side is, therefore, essential to prevent local overheating and potential metal failure. It is also essential that complete combustion takes place within the furnace tube to avoid burning off in the reversal

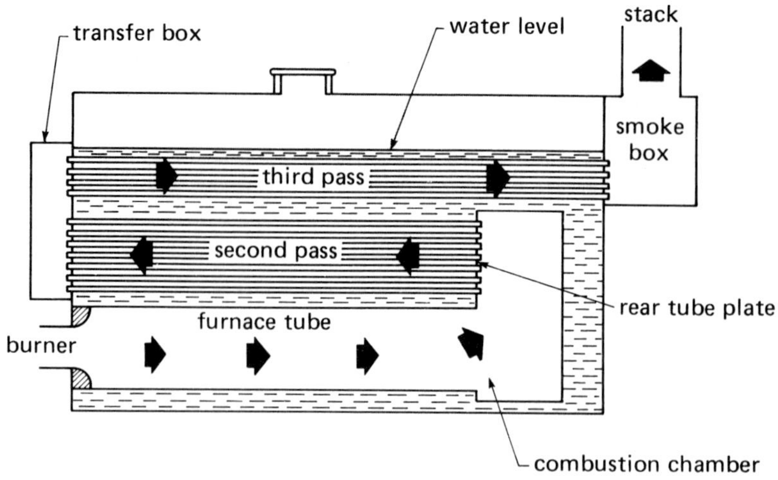

Fig. 6.46 Three-pass wet-back boiler

chamber which would lead to a higher back-end temperature. This would further exacerbate the high heat flux in this region of the boiler.

These precautions are necessary whether the fuel is oil or gas. However, a gas fired boiler is more vulnerable because radiant heat transfer in the furnace tube from a non-luminous gas flame is lower than that from a luminous oil flame leading to higher temperatures in the reversal chamber; this is only partly compensated for by the higher non-luminous emissivity. In addition the higher water vapour content of combustion products from gas firing increases specific heat capacity and thermal conductivity, thereby increasing convective heat transfer.

Tube-ends and tube-plate ligaments are the parts most vulnerable to damage. BS 2790 : 1986 does not differentiate between oil, gas or pulverised fuel firing of shell boilers in terms of permissible heat input rates, nor with regard to tube to tube plate attachment. Permissible heat input rate is dependent on furnace tube diameter (see Figure 6.47). The Standard Specification states that for tubes exposed to flame or gas temperatures exceeding 600 °C, the ends of welded tubes are to be dressed flush with the welds.

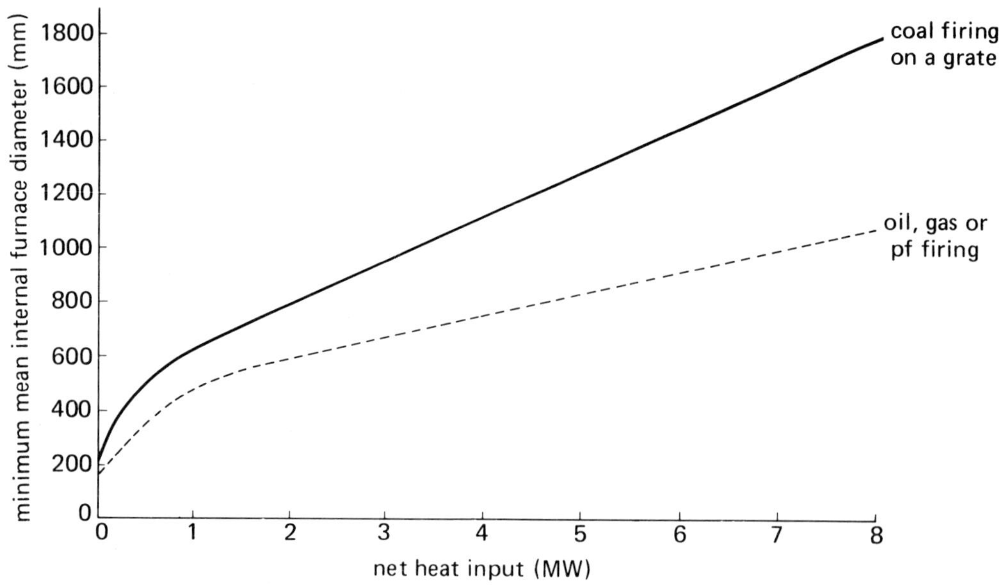

Fig. 6.47 Furnace tube diameter v nett heat input

The fact that natural gas and oil are now treated the same dispels the view previously held by a number of boiler manufacturers, that shell boilers converted to gas would need to be downrated. It was suspected at that time that the higher back-end temperature would lead to unacceptably high tube-end and tube-plate temperatures. Field trial work carried out by British Gas plc, supported by mathematical modelling techniques, had a major bearing on present-day thinking. Results of gas and metal temperature measurements at various firing rates and excess air levels for both oil and gas firing are shown graphically in Figure 6.48.

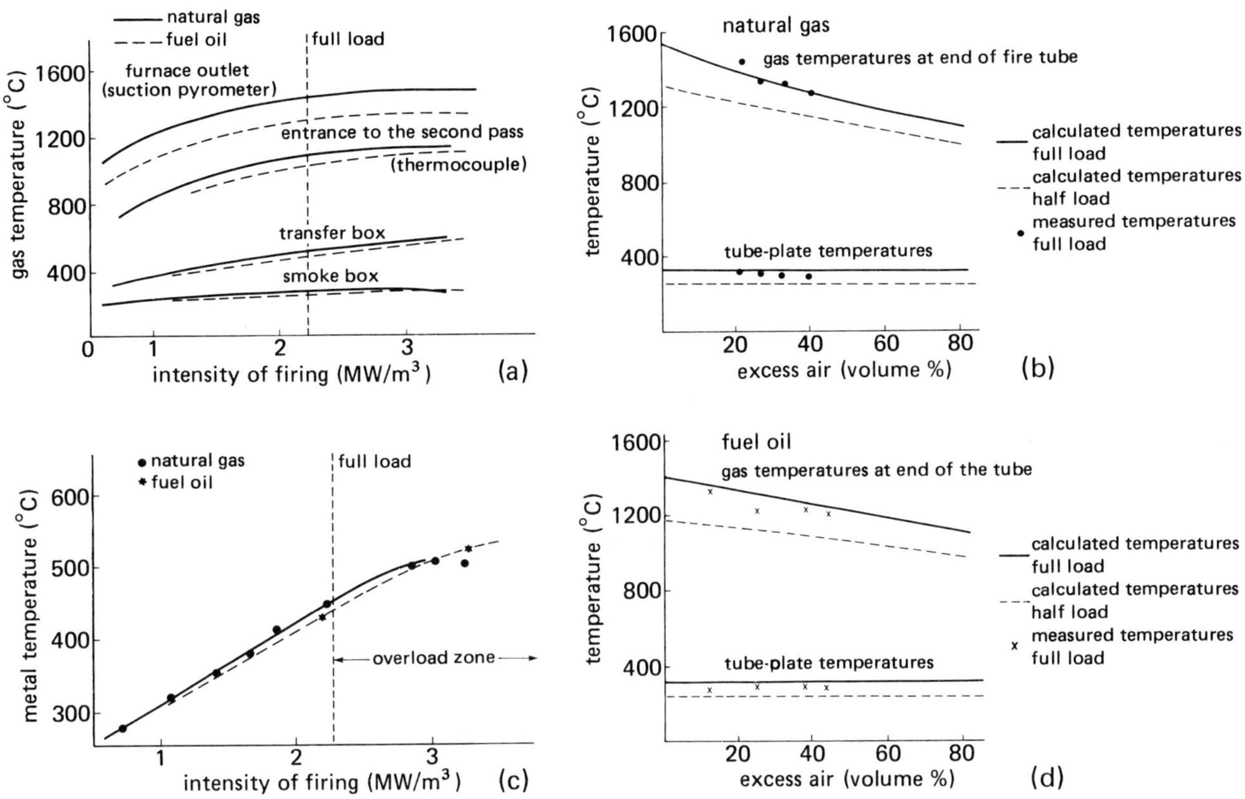

Fig. 6.48 Measured and calculated gas and metal temperatures for a packaged boiler of high firing intensity

It is clear from Figure 6.48 that, although the temperature of the combustion products is about 150 °C higher at the end of the firetube on gas firing, the corresponding metal temperatures differ by only about 20 °C even at 45% overload. Clearly on this basis there is no necessity for downrating. An increase in excess of air level modifies the heat transfer in three ways: the flame temperature is reduced, the flame emissivity is reduced by decreasing the concentration of radiating species and, since the mass flow is increased, the convective heat transfer coefficient is increased. Figure 6.48 (c) and (d) show the effect on gas metal temperatures and it is evident that the metal temperatures remain almost constant with both fuels. A few practical instances have occurred where the temperature of the furnace-tube exit gases has been substantially higher than calculated. These have been attributed either to the use of an excessively long narrow flame or to poor air distribution across the furnace-tube cross-section. In all cases modification of the burner has led to acceptable back-end temperatures.

In dry-back boilers, re-radiation from the refractory back and the reduced area of water-cooled surface tend to increase the heat flux to the tube plate and tube ends. Consequently, firing intensities for such boilers tend to be kept to fairly low levels. Even with highly rated dry-back boilers, however, British Gas plc concluded that changeover presented no cause for concern provided that unsatisfactory burners are avoided.

The development of the reverse-flow shell boiler was an elegant means of reducing the heat flux over the rear tube-plate. In effect two of the three passes take place in the furnace tube. The burner fires into the furnace tube and the combustion products turn back on themselves on reaching the stopped end of the tube and then flow, via the front reversal chamber, into the third-pass of the fire tubes. This means that the gases are cooler on entering the tubes. The method of firing produces eddy recirculation within the furnace tube which further enhances heat transfer in this part of the boiler. A typical reverse-flow boiler is shown in Figure 6.49.

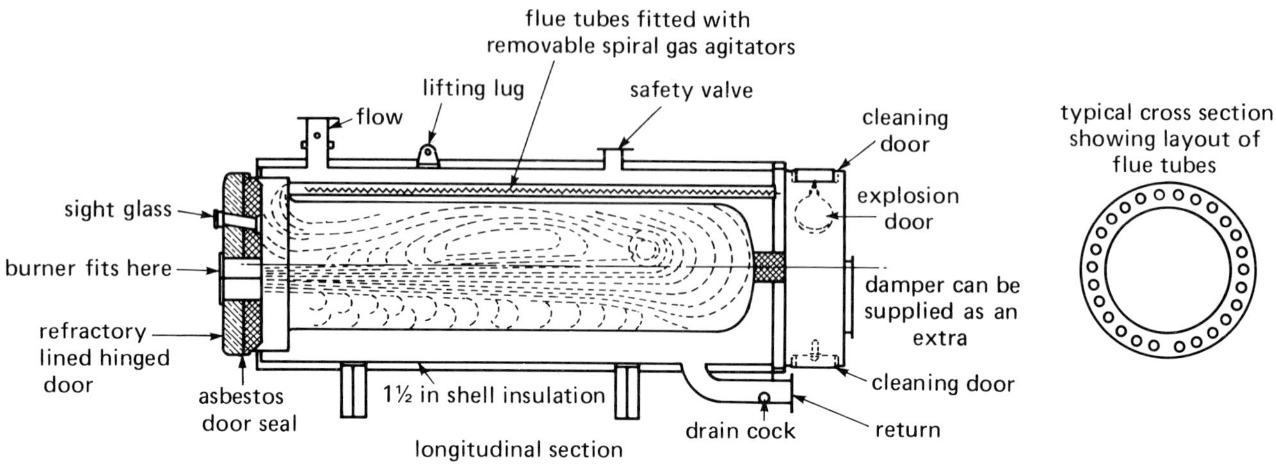

Fig. 6.49 Reverse-flow boiler

Water-tube boilers

When pressures in excess of 17 bar to 21 bar and/or outputs of over 18 MW are needed, water tube boilers are used. There are three main types. These are:

- Natural circulation where water circulation is promoted by the difference in density between the water in the fully flooded section of the boilers (e.g. downcomer tubes from the steam drums) and the steam/water mixture in the steaming tubes.
- Forced circulation. This type uses pumps to provide or assist water circulation in the boilers.
- Once-through, super-critical boilers where the operating pressure is such that water density approaches that of steam. Very high pressure feed pumps are employed.

The last two types of boilers are of limited interest only, as by far the majority of water-tube boilers used in industry are of the natural circulation type.

A water-tube boiler is essentially a series of drums and headers connected by relatively small diameter water tubes. Water tubes can be shaped to a wide variety of boiler configurations but essentially water-tube boilers consist of a furnace or combustion chamber occupying about half of the total volume, the other half being termed the convection zone and usually accommodating the superheater. The geometry of the combustion chamber differs totally from that used in shell boilers; it is roughly cuboidal as opposed to long and cylindrical. Figure 6.50 shows a typical industrial water-tube boiler. The furnace chamber may be refractory lined or, more commonly in modern practice, completely surrounded by water tubes, i.e. 'water walls'. Whilst pulverized fuel has dominated large scale power generation, oil firing replacing chain-grate stokers and subsequently gas firing (on dual fuel burners) replacing oil firing, has been the trend in industry.

Most water-tube boilers for industrial purposes are fired by register burners in the range 3 to 20 MW. The combustion air supply ducts (wind boxes) are normally incorporated into the boiler walls and

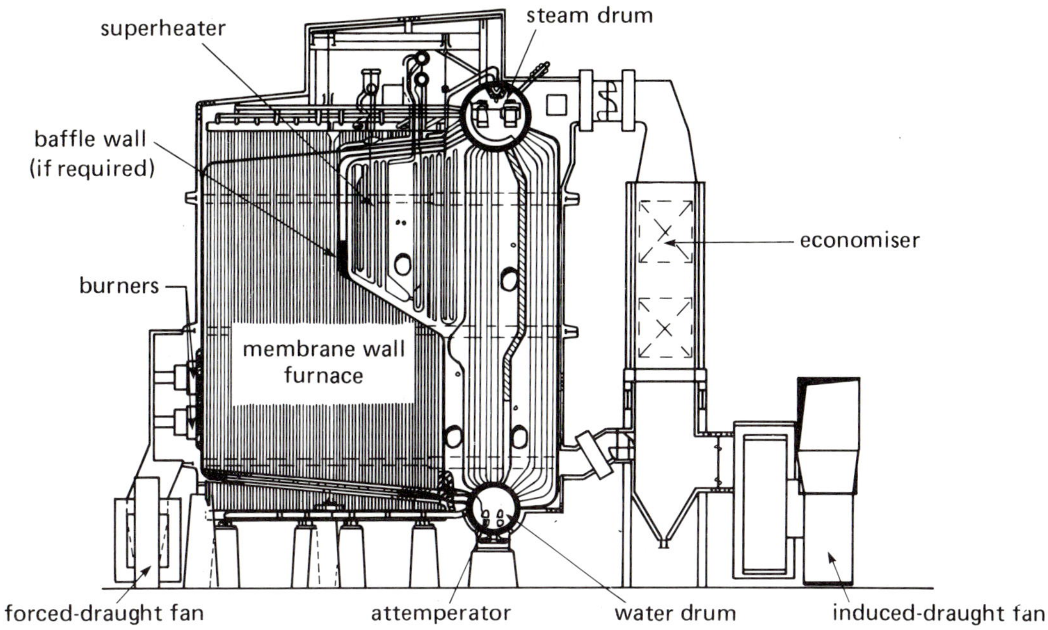

Fig. 6.50 Typical industrial water-tube boiler

are generally supplied by a single large forced draught fan. Register burners are discussed in more detail in Chapters 2 and 3.

Industrial water-tube boilers in the smaller size range of 10 to 30 MW are normally of the package type. They have the same advantages over site erected boilers as package shell boilers, i.e. they can be pre-tested and pre-set before delivery to site. Once on site, installation and commissioning is much quicker.

Coiled tube flash steam generation

These small scale boilers may also be classified as water-tube boilers. Their stored water capacity is very small and steam is available within a few minutes of lighting up. Evaporation capacity is normally in the range 130 to 2 050 kW.

The combustion chamber is formed by the cylindrical space enclosed by an inner coil; the gases then flow multipass through successive layers of coils. Gas and gas oil fired types are in use. The burner turndown must provide the low fuel input required for low steam flows and also be able to prevent widely fluctuating steam pressures resulting from burner cycling. To provide rapid starting the prepurge air flows are minimised but, otherwise, the controls are similar to those used on other packaged boiler burners. A coiled tube flash steam generator is shown in Figure 6.51.

Annular coiled tube boilers

The annular coil boiler is designed on the once-through, forced circulation principle and is suitable for medium and high pressure, saturated or superheated steam generation. Evaporative capacities range from 1.5 MW up to 30 MW at pressures up to 172 bar. Steam temperatures of up to 540 °C can be achieved when fitted with a superheater.

Figure 6.52 illustrates an M.E. Boilers Ltd. up-fired boiler. The design shown embodies integral economisers, superheater and boiler heating surfaces, arranged as a nest of concentric helical tube coils within a double air casing. All the pressure parts are of tubular. all-welded construction and are

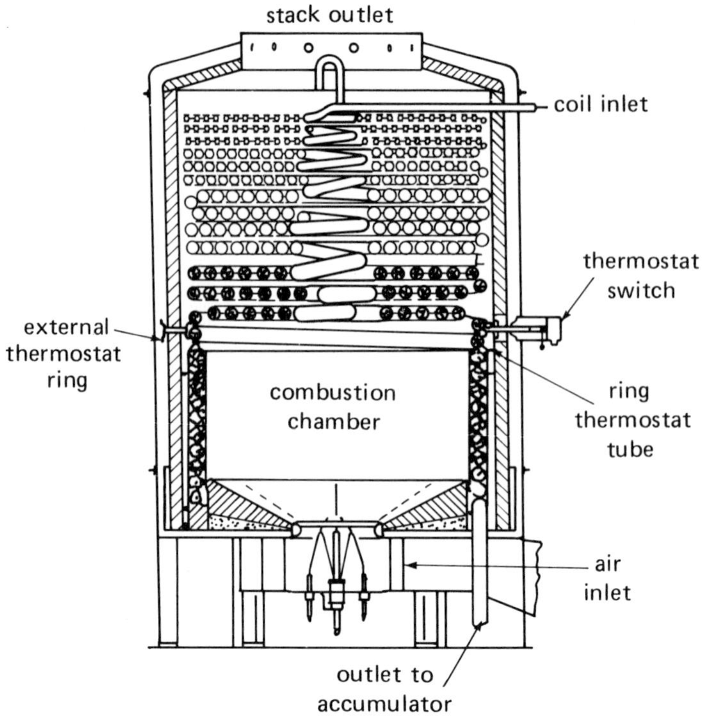

Fig. 6.51 Flash steam generator (Clayton, Stone Vapour)

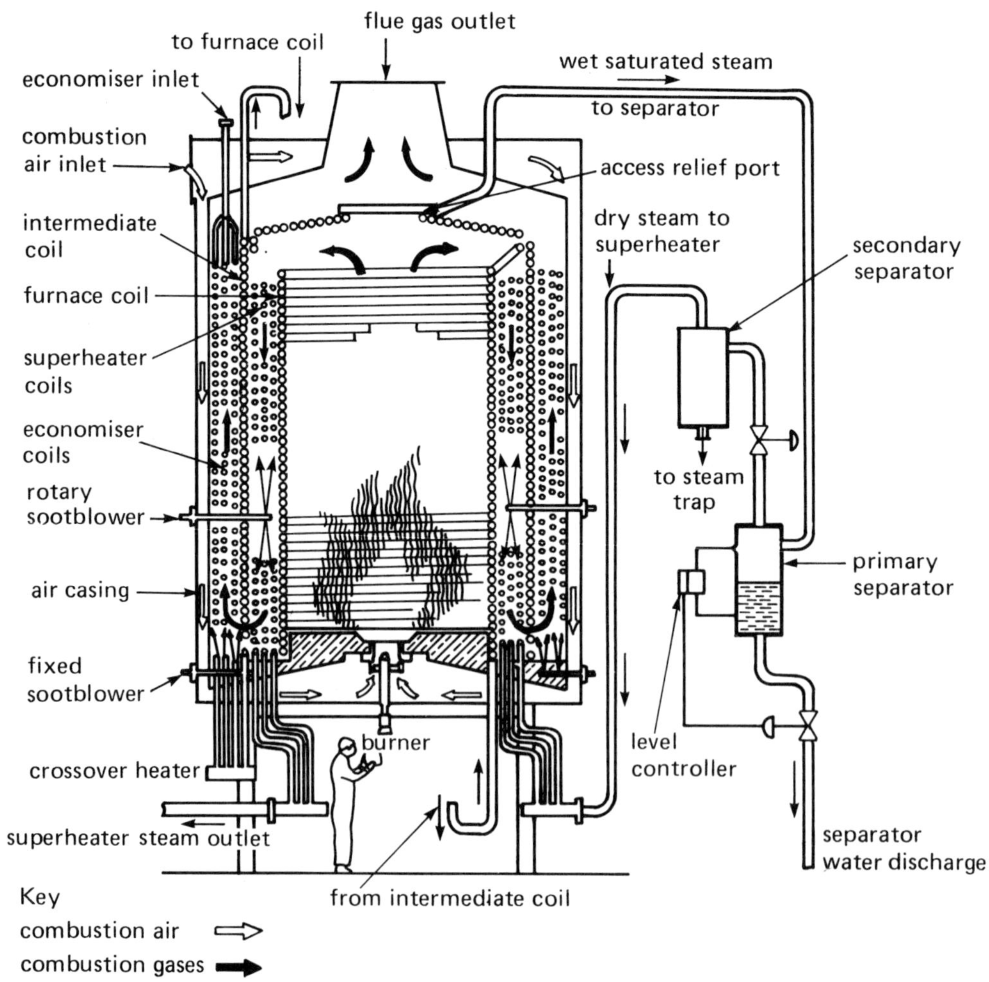

Fig. 6.52 Annular coiled-tube boiler

supported from the boiler base. This allows free cubic expansion independent of the casing. The final evaporation stage occurs in the furnace and roof coils, which are protected from drying out and dissolved salts deposition by ensuring that the exit steam condition is at least 10% wet. The wet saturated steam is fed to a steam separator system, from which dry steam flows to the superheater or process. The residual water from the separator is discharged to a flash vessel or other heat recovery system.

The boiler operates under pressurised firing with the burner normally firing vertically upwards. However, downward firing can be accommodated if necessary. Combustion air is preheated in the outer air casing and passes radially into the windbox. Combustion products from the cylindrical furnace flow radially outwards and down through the annular convection pass in which a superheater can be located if required. The gases then flow upwards over the economiser coils to the boiler flue. Design gross thermal efficiency is claimed to be 84%.

The boiler can be fired on oil or gas either independently or combined. For oil firing, steam atomisation is generally used and an angled, auxiliary burner is provided to allow onload cleaning of the main burner nozzle. Soot blowers are also fitted for cleaning the boiler, superheaters and economiser heating surface. With gas firing the auxiliary burner and soot blowers are unnecessary.

The only refractory material used in the boiler is that forming the furnace base and the burner quarl. This, together with the fact that there are no boiler drums or headers, gives low thermal inertia and permits rapid start-up.

Water-treatment

Ideally, process and heating steam is condensed in the user equipment and returned to the boiler for re-evaporation. As it is effectively distilled water and preheated too, it is an ideal feed-water. However, some processes use steam direct and there are losses through leakage and flashing condensate, therefore return to the boiler is invariably less than 100%. The balance, or make-up water is in most cases drawn from the town mains supply.

Mains water is a most unsuitable fluid for steam boilers and the consequences of using it without treatment and control range from undesirable to catastrophic. It contains dissolved minerals and gases. In the boiler, the minerals concentrate and decompose giving scale and sludge and releasing more gases which leave with the steam and make it, and the condensate, highly corrosive. As this 'soup' thickens, it foams and some can carry forward with the steam creating contamination and deposition problems.

The first essential is to minimise the make-up problem by returning as much condensate as possible to the boilers, and there are few industrial sites where condensate return is all that it could be. First, ensure that any processes on direct steam which can use it indirect are so converted. Secondly, collect any condensate running to drain. Thirdly, lag all condensate return pipework. A problem may now arise which becomes more acute as the proportion returned increases, i.e. flash steam. This is the steam which evaporates from the condensate as its pressure drops from that in the using equipment to atmospheric pressure in the feedwater tank or intermediate collection point. The higher the initial pressure, the greater the flash and for this reason, among others, steam should be used at the lowest pressure the user can accept. For example, condensate at 10 bar will lose 16% as flash vented to free atmosphere and the proportions at other pressures are:-

7 bar — 13%
5 bar — 11%
2 bar — 6%

Condensate should, therefore, be collected initially in vessels from which the flash steam can be piped to a low-pressure user and its condensate in turn returned to the boiler plant.

Having reduced the make-up water proportion to a minimum, its conversion to a suitable medium for evaporation may be tackled. Water treatment is complex and is a matter for specialists – a reputable firm should be engaged to advise on what is needed for the water in that district and the type of boiler plant in use. Thereafter, its application and control is simple and within the capacity of a trained operator with a modicum of supervision. In general terms, treatment and control comprise:-

- Softening. The scale-forming calcium and magnesium salts ('hardness') are converted into more innocuous compounds by ionic (base exchange softener) or chemical (additives) means, or inhibited (addition of sequestering agents).
- De-aeration, i.e. removal of dissolved gases from the feedwater. Maintenance of high feed-water temperatures in the hot well accomplishes this automatically which is another reason for effective condensate return. Otherwise, the addition of oxygen scavengers such as accelerated sulphites may be needed. In some cases, mechanical de-aerators are installed.
- Conditioning. Compounds are added to the feedwater to maintain the desired chemical balance within the boiler, prevent foaming etc.
- Blowdown. Water from the boiler is discharged to drain to maintain the concentration of solids (TDS – total dissolved solids) below the maximum the type of boiler will tolerate. With high proportions of make-up water, blowdown quantities are considerable and it should then be on a continuous basis with sensible heat and flash steam recovery to minimise the heat and water loss.

The standard of treatment required depends on the characteristics of the water and the type of boiler plant.

The nature of the water supplied depends ultimately on the geology of the area from which it is abstracted and ranges from very soft with low TDS to very hard with high TDS. The latter requires softening plant or better, the former often only needs additive treatment. A water authority sometimes has more than one abstraction source and it should not be assumed that mains water is consistent in quality.

The quality and purity of feedwater required rises as the design sophistication and/or working pressure of the boiler plant increase. High pressure water tube steam boilers need high purity water and it is customary to demineralise (i.e. remove all dissolved solids) and de-aerate the feedwater to very high standards.

6.4.4 Pressurised Water Boilers

Medium temperature and high temperature hot water (m.t.h.w. and h.t.h.w.) boilers are normally used where the predominant requirement is for space heating.

Boilers and combustion equipment used are almost identical to the shell and industrial steam boilers described previously. The only difference is that the distributed heating medium is hot water and the pumped circulating system is fully flooded.

Methods of system pressurisation are described in Section 6.1, which also discusses the relative merits of steam and pressurised water systems.

6.4.5 Thermal Fluid Heaters

Whilst not strictly boilers, thermal fluid heaters can be regarded as such in terms of their gas combustion system requirements.

Where an indirectly heated process requires relatively high temperatures and hot air or combustion products are unacceptable or impractical as the heating medium, thermal fluid is often used in preference to steam or hot water. The higher the temperature required the higher a steam or hot water system pressure will be. A conventional shell boiler is normally limited to a maximum working pressure of 17 bar. At this pressure, the temperature of saturated steam or pressurised hot water is 207 °C. Thermal fluids have been developed to overcome the disadvantage of the temperature/pressure relationship.

Thermal fluids

In general there are two types of thermal fluid which are:

- Mineral oil based
- Synthetic chemical compounds.

Both these types are stable in the liquid phase at temperatures up to around 320 °C at atmospheric pressure. This means that heat can be transferred to a process at this temperature, the pressure requirement being only the frictional loss in the thermal fluid distribution circuit; this will not normally be more than 2 bar. The equivalent saturated steam or h.t.h.w. pressure, at this temperature, would be over 100 bar.

Synthetic compounds like eutectic mixtures of diphenyl/diphenyl oxide, usually referred to by the trade names Thermex or Dowtherm, can be used in the liquid or vapour phase up to 400 °C at pressures well below those of equivalent temperature steam.

Thermal fluids are thermally stable (i.e. do not degrade) at the above temperatures when used in correctly designed equipment.

Types of thermal fluid heaters

Equipment design must ensure that a low heat flux and high heat transfer coefficient obtains at all times in the heater. This minimises the temperature difference across the fluid film in contact with the heat transfer surfaces. Heat distribution across the heat transfer surfaces must be uniform and flame impingement must be avoided.

The two types of heater commonly used in British industry are:

- The finned tube unit.
- The helical coil unit.

A finned tube unit (Beverley) is illustrated in Figure 6.53. The design includes a cylindrical combustion chamber with a staggered extended surface (which is about five times larger than the surface area of the combustion chamber) in contact with the thermal fluid. This arrangement allows a low heat flux into the oil and a high heat transfer coefficient caused by the turbulent flow over the extended surfaces. The convection section is mounted outside the finned surface giving high thermal efficiency.

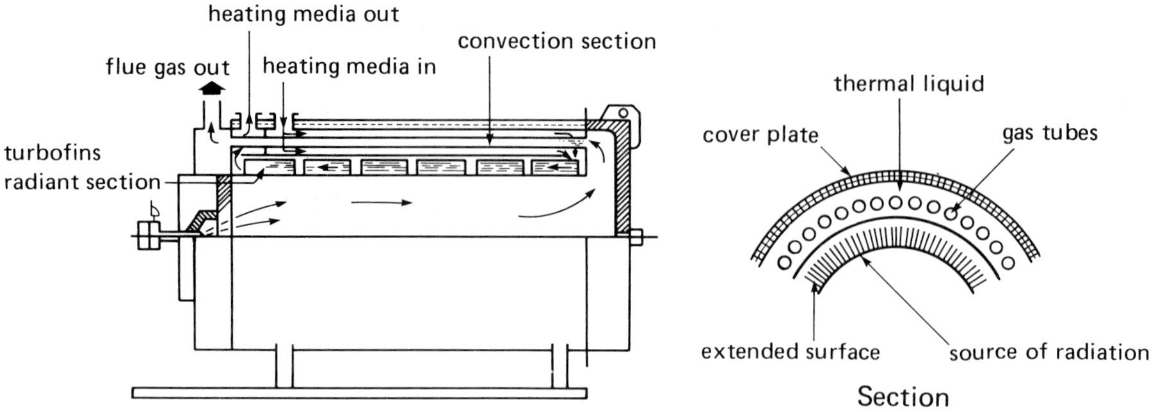

Fig. 6.53 Finned tube thermal oil heater

Helical coil heaters can be fired either horizontally or vertically and examples (Contrapol) are shown in Figure 6.54. The burner fires along the axis of the coil, transferring heat, mainly by radiation, to the inside surfaces. At the end of the combustion chamber formed by the coil the flow of combustion products reverses and the gases pass over the outside of the coil to the exit flue. Heat transfer in this section is mainly by convection. The coils are wound with adjacent turns touching which are welded together to form a firetube.

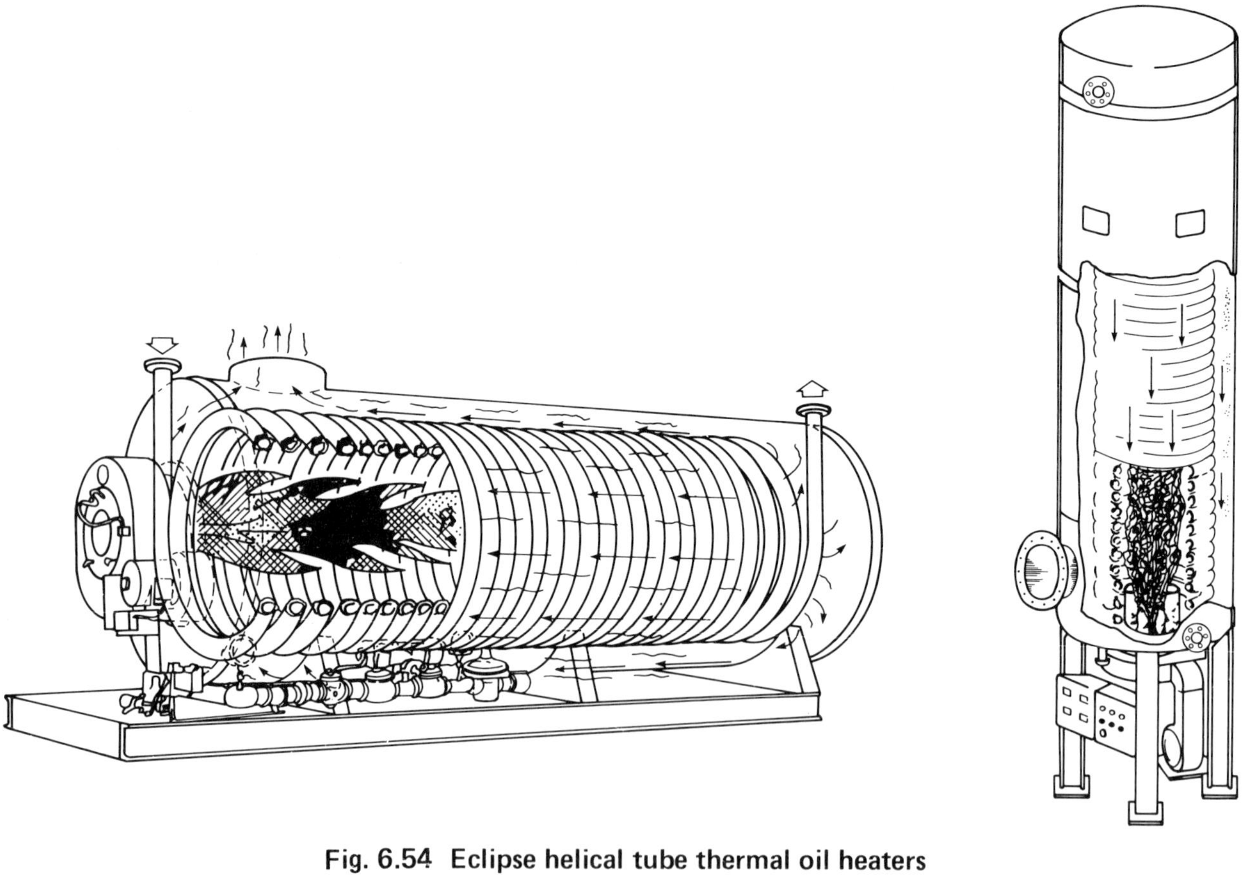

Fig. 6.54 Eclipse helical tube thermal oil heaters

The thermal efficiency of tube type heaters can be increased by the addition of an external secondary convection economiser as illustrated in Figure 6.55.

Thermex or Dowtherm thermal fluids are sometimes used in vapour systems instead of in the liquid phase. At its maximum operating temperature of 400 °C the vapour pressure is about 10 bar. At

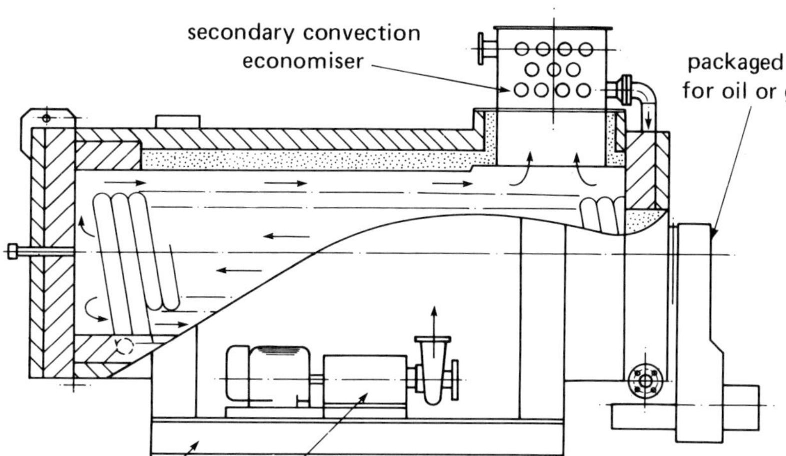

Fig. 6.55 Beverley Turbocoil thermal fluid heater

this pressure, steam would have a temperature of only 185 °C. The fluid is heated in the liquid phase and flashed off through a pressure control valve into a vapour separator. The condensed fluid passes back to the heater and the vapour passes to the process where it is condensed and returned to the fluid heater.

Medium transfer

Heat energy may be transferred from thermal oil to another medium to suit the end user, as with the steam to water calorifiers and steam to water to air batteries already discussed. The temperature gradient from primary to secondary medium is, of course, always downwards.

The high working temperature of thermal oil enables it to generate steam to quite high pressures. Where steam is a relatively small proportion of the total heat demand it is more economic to use a thermal oil steam generator, several types of which are commercially available, than a fired boiler. Similarly, thermal oil can generate hot water in calorifier-type heat exchangers.

System safeguards

Volumetric expansion of mineral oil when heated from cold up to its working temperature can be over 25%, necessitating the use of an expansion tank.

Whilst thermal fluid distribution pipework does not have to withstand high pressures, the elevated temperatures usually call for the use of fully welded seamless tube and special attention to compensation for expansion. The fluids are searching, and valve and shaft glands usually require specialist packing materials or techniques.

Venting points should be fitted at high points in the thermal fluid circuit to remove air and volatiles on start-up.

6.4.6 Control and Instruments

Reasons for use

Controls and instrumentation have the following main functions:

Operational – to provide the quality of heat output as required
– to regulate combustion

Statutory — to ensure safety.
— to prevent illegal emissions.

Information — to monitor performance.
— to promote efficiency.
— to assist maintenance and trouble-shooting.

In practice, these are overlapping requirements and one controller or instrument may have several functions. The need for adequate provision is becoming increasingly recognised, especially where it acts as a tool to reduce operating costs; a relatively small boiler plant with an average output of 5 t/h of steam over a 120 hour week, for example, will consume between £3 m and £5 m worth of fuel at 1990 prices over its working life, and instrumentation to save just a few per cent will be more than justified. In addition, the greater knowledge of what is happening focuses more attention on the plant from operating and supervising staff (boiler plant has traditionally been the most neglected aspect of a firm's activities) with attendant benefits in performance, maintenance standards etc.

Operational

Changes in demand cause the pressure or temperature of the medium to rise or fall and these variations are sensed and used to regulate the flow of fuel accordingly. Fuel regulation is linked with combustion air regulation so that air/fuel ratios are correct throughout the firing range, as described in Chapters 2 and 3.

With multi-boiler installations, controllers may be programmed in sequence. Alternatively, one or more boilers may be left fixed at an optimum firing rate, with another boiler responding to the fluctuations.

Statutory

The boiler pressure gauge is the only statutory instrument as such. Modern boiler plants required to operate automatically and unattended have to satisfy stringent requirements laid down by the insurance companies, and as it is mandatory to insure boilers, these requirements virtually have force of law. Provisions cover such aspects as boiler water level detection/control/alarm, purging and fail-safe operation of burner sequencing etc. e.g. Requirements for the Insurance of Automatically Controlled Steam Boilers (Associated Offices Technical Committee). BS 5885 : 1988 covers the requirement for the latter. Prevention of illegal emissions (e.g. smoke) is integral with the combustion control equipment.

Information

Probably the most neglected field, because it is largely optional in contrast with the previous two sections where the equipment is either integral or enforced. It covers fuel, combustion and medium. Fuel should be metered and its handling/firing parameters (pressure, temperature etc. as appropriate) measured.

Combustion efficiency is indicated by CO_2, O_2 and CO measurement and, with exit flue gas temperature, enables a continuous check on flue gas losses.

The medium should be measured, to monitor both heat demand and overall performance. With hot water, flow and return temperatures plus, in the large installations, water flow and integrated heat metering should be known. With steam boilers, make-up water, total feedwater and steam flow should be metered, together with temperatures and/or pressures as appropriate, so that the operation of the feedwater system, blowdown etc. can be accurately assessed.

Elapsed time counters fitted to the combustion appliance on each boiler record duration of operation and facilitate schedule rotation, maintenance programming etc.

6.4.7 Maintenance

Boiler plant is often not available for maintenance and repair for any length of time, except for an annual fortnight's shutdown. Even in multi-unit installations, where one unit is always off, it usually acts as standby and may be required to come on stream at short notice if a working unit breaks down. Thus, there is a strong emphasis on scheduled preventive maintenance supported by a routine checking procedure by the operator. These schedules are laid down by the equipment manufacturers and, if adhered to, give the best insurance against failure. Breakdowns may nevertheless still occur and the corollary of good scheduled maintenance is intelligent spares holdings, based on identifying critical items (i.e. those whose failure shuts down a boiler or the whole plant) and long delivery items.

Maintenance of instruments, often neglected, is of paramount importance, because they may give advance warning of a plant maintenance requirement before it becomes critical. Steadily rising exit flue gas temperature, for example, indicates heating surface fouling and a boiler outage and tube cleaning can be programmed; a rise faster than normal could indicate defective firing conditions which should be rectified immediately.

The remaining maintenance is carried out during the annual shutdown for statutory inspection, when, in addition to the statutory requirements, such items as main steam valves, reducing valves, leaking joints on the primary mains may be attended to. Temporary repairs and jury rigs made during the year should be reconstituted properly.

6.4.8 Selection of New Gas Fired Boiler Plant

Assuming that particular circumstances require a distributed heating medium system as opposed to point of use gas equipment, the main factors to be taken into account are load, type and cost. An open mind is essential; do not assume that an up-to-date replacement of the previous plant is all that is needed.

Load

Assess the load in terms of its magnitude and nature, i.e. short term, daily, weekly and seasonal variations. It should be split into logical components e.g. process and space heating, and by location if spread over a wide area.

Determine the various grades of heat required. The most demanding user will dictate the pressure or temperature of the new plant and if it is only a small proportion of the total or is geographically remote, it may be worth making separate provision for it. The aim should always be to generate at the lowest energy level – the higher the pressure or temperature of the boiler plant, the higher the initial cost, running cost and transmission losses. It is often possible to downgrade process requirements by replacing heat exchange surfaces.

Study the data obtained and evaluate heat recovery and other means of reducing the load. Schemes previously rejected as uneconomic may now be tenable because the basic energy saving is now enhanced by the capital saving of a smaller new boiler plant. Sharp intermittent peak loads originating from a single user or department should, if possible, be eliminated by work re-scheduling or plant modification; again, there will be the capital saving of a smaller boiler plant plus economies in running costs.

Include estimates of future requirements based on proposed extensions, production increases etc. These may be catered for immediately, or the new plant designed for easy future extension.

From this first stage will emerge some provisional ideas of the size of plant and whether it should be centralised or split into separate units.

Medium

Media in general use are hot water, steam and thermal oil, which have all been discussed in previous sections of this Chapter.

Unit size

Boiler efficiency drops as load decreases, markedly so below the lower turndown limit and gas burners are cycling on/off. Periodic load variations (between day and night, weekday and weekend, summer and winter) are, therefore, best met by having several boilers each rated for a proportion of the total maximum load, the number in use at any given time being only those needed to meet the load at optimum firing conditions.

The total number of units comprises those required to meet the maximum load plus standby in the event of breakdown. The latter should not be decided until unit size is settled, on the reasonable assumption that breakdown would occur on one unit only.

Consider, for example, the following three possible cases where a total boiler capacity of 100 nominal units is required:

- Load constant throughout year. Required, two boilers each rated 100 units; one working, one standby at all times.

- Load 50/50 process and space heating. Required, three boilers each rated 50 units, summer, one working, two standby; winter, two working, one standby.

  Alternatively, three boilers each rated 33⅓ units; summer, autumn, spring, two working, one standby; winter, three working, no standby and heating standard reduced if a boiler goes out of commission.

- Load space heating only. Required, five boilers each rated 25 units; number working according to load, one standby minimum under peak winter conditions.

  Alternatively, four boilers each 25 units; 25% drop in heating standards if breakdown occurs in winter.

Type

The type required will depend on unit size, medium and operating pressure. Table 6.1 gives the salient features of the range of plant commonly used today.

Table 6.1 Characteristics of various types of boiler

Type	Output range kW	Max pressure bar	Medium	Application
Cast Iron sectional	10 to 1 500	1	Water	Space heating - small scale
Welded Steel (non shell)	10 to 1 200	5	Water	Space heating - small scale
Finned Tube	17 to 1 000	10	Water	Space heating - small scale
Shell fire tube	70 to 18 000	17	Water Steam	General process and/or space heating
Coiled-tube/ annular	100 to 30 000	172	Thermal oil Steam	Process
Flash coiled-tube	300 to 3 000	80	Steam	Process
Water tube (industrial models)	4 500 to 100 000	120	Water Steam	Large scale space heating and/or process. Power generation

Final assessment and cost

The preceding sections have discussed the various aspects separately, but they are all, in practice, inter-related. The analysis of these inter-relations is the last stage of the exercise and usually results in several alternative schemes. Figure 6.56 illustrates diagrammatically seven possible options for an imaginary factory requiring three grades of heat, showing various combinations of plant and media dependent on the relative sizes of the loads, their distance apart and so on; the list is not exhaustive.

Capital costs are relatively straightforward and need no further comment beyond stating that some scheme re-adjustment and re-costing may be needed if the price of the standby provision is greater than its 'insurance' value.

Running costs must be evaluated realistically and comprehensively and related to capital costs. By far the major component is fuel cost and it is not unusual for this to equal the initial capital outlay before the end of the first year's operation. There is, therefore, a compelling economic case for spending more money initially to ensure maximum possible thermal efficiency. Typical measures would be:

- More comprehensive and precise combustion control equipment. Better designed/constructed firing equipment of equivalent standard to the control system - they are sometimes mismatched and the resultant performance will be no better than the poorest component.
- More comprehensive instrumentation to monitor performance - it is commonplace for fuel wastage to continue year in and year out simply because no-one is aware that it is occurring.
- Better condensate and water handling and treatment, blowdown heat recovery.

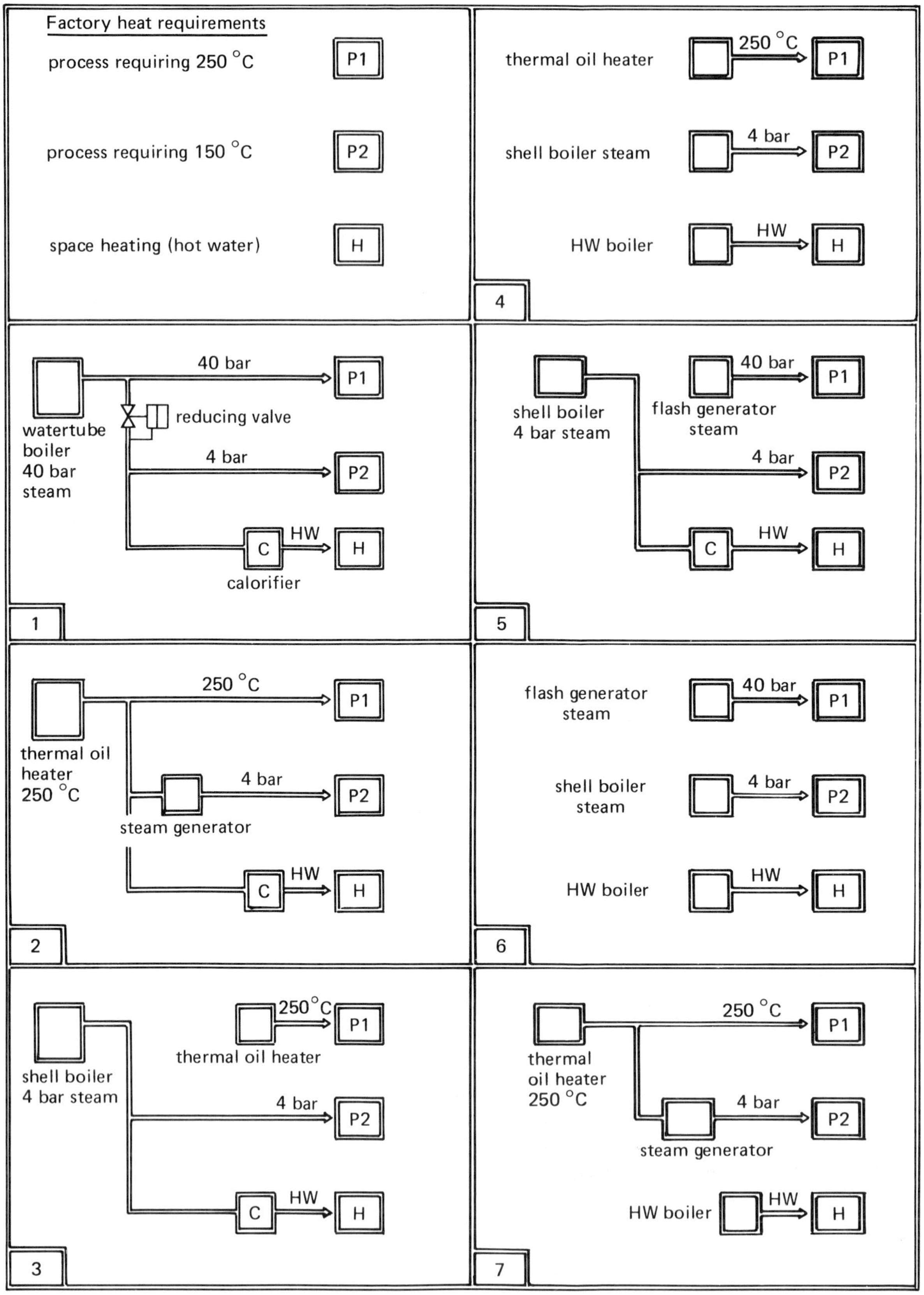

Fig. 6.56 Example of boiler plant option analysis

Other running costs are operating labour, maintenance labour and materials, electricity for fans and pumps etc., make-up water and treatment, capital charges, annual value of land occupied, general overheads.

Capital and running costs may now be correlated according to the firm's normal accounting practice (based typically on 15 to 20 years working life), the various schemes evaluated and a final decision reached.

6.4.9 Increasing the Efficiency of Gas Fired Boilers

The major heat loss with all boilers is the sensible and latent heat of flue gas constituents. The amount of heat lost varies directly with the flue gas temperature and mass flow. The latter can be minimised by ensuring that combustion control systems maintain good air/gas ratio at all points on the firing range i.e. that they optimise excess air levels.

Oxygen trim control

Oxygen trim control is one way of optimising excess air levels in boilers. There are several systems available which are dealt with in Chapter 14. In general they are only cost-effective on larger boilers where an energy saving of 1% can give a reasonable payback period. Caution should be used in determining O_2 set points, particularly under low-fire conditions as, on most burners, mixing performance deteriorates. The possibility of increased CO formation with subsequent burn-off in the reversal chamber of a shell boiler could have catastrophic effects on boiler structural stability.

Carbon monoxide trim control

CO trim is a further method of controlling excess air levels in boilers. However the problem is determining a suitable set point. The excess air v CO curve, for most burners, is as shown in Figure 6.57 and consists of a horizontal line and a near vertical line joined by a short curve. Controlling at a set point on the horizontal section results in a possible wide range of excess air, controlling on the curve can result in hunting because of the probability of rapid changes in CO levels, and controlling on the vertical portion results in too high a level of CO. Such controllers are expensive and are best suited to very large water tube boilers operating at low excess air levels. Payback times can then be small but, even so, possibly they are best used as an indication for final manual trim when load conditions are constant.

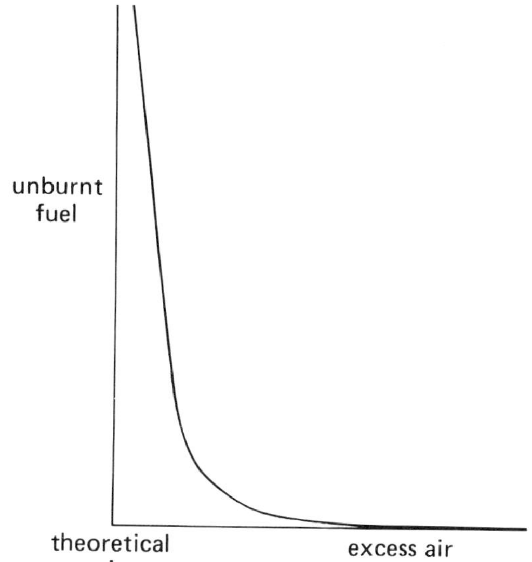

Fig. 6.57 Typical burner performance curve : excess air v CO

Reducing flue gas temperature

The lower the final flue gas temperature from a boiler, the less efficiency is dependent on excess air.

The heat flux in sectional boilers designed for solid fuel firing is very low, around one-third of that applying in boilers designed for gas firing. This is due primarily to the low combustion product velocities over the heat transfer surfaces. Dramatic increases in heat transfer and consequently thermal rating may be obtained on changeover using a forced draught gas burner and by reducing the cross sections of the second and/or third pass while still retaining the heat transfer surfaces exposed.

Inserts sized to produce increases in second-pass mean velocities (up to 450% increase in velocity) can increase the efficiency by about 5% at nominal rating or, alternatively, for the same efficiency the boiler may be operated at about double its nominal rating. Increasing the velocities in this way pressurises the combustion chamber. Increases in third-pass velocity are far less effective due to the lower gas-metal temperature differential. Since uprating leads to pressurisation of the boiler gas ways, the risk of combustion-product discharge into the boiler room must be guarded against by thorough boiler preparation. In addition it is necessary to protect the back end of the combustion chamber from flame impingement and to avoid hot-gas short circuiting by sealing the gaps between the combustion chamber and the second and third passes to maximise flue-gas travel. The effect of increasing the velocity in the second pass is shown in Figure 6.58.

Similarly, turbulators are available for shell boilers although there are drawbacks. These are, an increase in back pressure on the burner with possible downrating implications, and, with dual fuel burners, the danger of blockage from particulates when firing on oil.

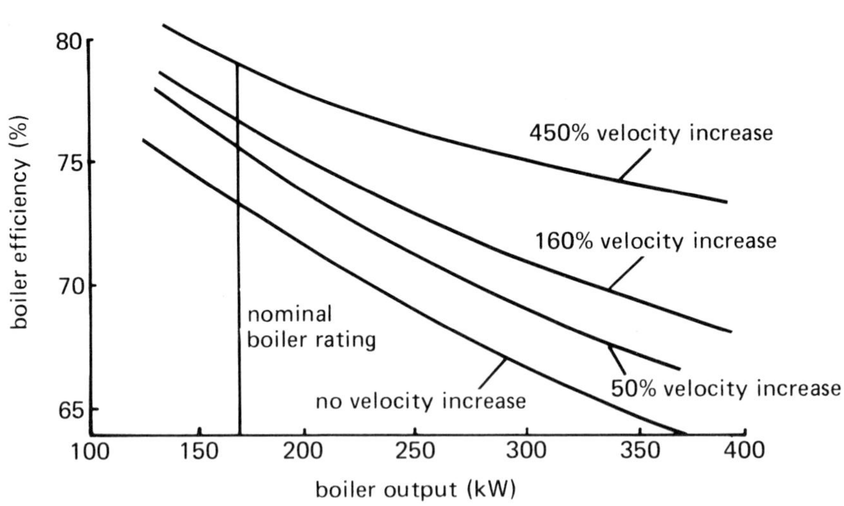

Fig. 6.58 The effect of increasing the gas velocity in the second pass of a sectional boiler

Economisers are now available in a wide range of sizes and can be applied to very nearly all boilers. They are add-on devices and either preheat the return water in the case of hot water boilers or preheat steam boiler feedwater. Typical savings from the fitting of economisers are in the range of 4% to 6% and they reduce flue gas temperatures by about 90 to 100 °C. They can result therefore in significant savings, and payback times (although dependent on the boiler annual load) are normally satisfactory.

The reduction in flue gas temperature achieved may give rise to the following operational difficulties which need consideration:

- On dual-fuel operation, when on oil, the surface temperatures within the economiser may be in the acid dewpoint range with resultant corrosion and smutting problems. The solution is to provide a by-pass on the economiser which takes the flue gases direct to the chimney when operating on oil.

- Particularly on the smaller commercial hot water boiler, there may be condensation in the chimney. This aspect needs consideration especially when the boiler is quite efficient without an economiser.

- There will be a reduction in the natural draught available and this may be of concern on natural draught boiler installations.

The effect on energy consumption of preheating combustion air by means of flue gases is well known. It is possible to do this on gas fired boilers. However, most burners used make the effective use of air preheat exceedingly difficult and air preheat is generally only used on the larger water tube boiler.

One way to use the heat in the flue gas from a boiler is to install an effective low temperature heat exchanger, and use this to heat air for another process taking care to avoid contamination of the air with products of combustion.

The fact that the products of combustion from gas are very clean enables them to be cooled to much lower temperatures than is the case with other fuels. The spray recuperator takes advantage of this fact. The recuperator is usually an add-on unit for the boiler. The flue gas passes up a tower in which are a series of baffles and thence to the chimney. Water is sprayed counter current to the flue gas cascading off the baffles. This direct contact water heating results in high overall efficiencies being achieved. Not only does the direct contact reduce the flue gas temperature but additionally, depending upon the water temperature at the inlet to the recuperator, some of the water formed by the combustion process is condensed thus recovering the latent heat. This can raise the overall efficiency of the boiler plus recuperator to over 90% (gross basis) giving fuel savings of 10 to 17%. The water is available from the bottom of the recuperator at a temperature roughly equivalent to the water dew point of the flue gas, about 55 °C. This water can be used for any process that requires, or can make use of, a low temperature heat source. Examples are preheating of boiler feed water, supply to calorifiers, supply to the first bank of air conditioning plant or directly for under-floor heating. Although the initial cost of the installation of such a unit is high, payback times are attractive.

A logical development of the above methods of reducing flue gas temperature was the condensing boiler which is discussed in detail in Section 6.4.2. Spray recuperators, economisers and other flue gas heat recovery equipment are detailed in Chapter 7.

6.5 DRYING

6.5.1 Introduction

Drying generally refers to any process in which a liquid is removed from a solid. The type of drying process of most interest to industrial gas engineers involves the application of heat to evaporate moisture (usually water) from a solid, and the removal of the resulting vapour.

Drying involves:

- Heat transfer by radiation, conduction or convection to supply the latent heat necessary to evaporate liquid from a solid material.

- Mass transfer of liquid or vapour from within the solid to the surface and the removal of vapour from the solid surface.

These two processes occur simultaneously and the factors governing the rate of each process determines the rate of drying.

The means of drying are many and varied and involve indirect or direct heating methods and intermittent or continuous processes.

6.5.2 Mechanism of Drying

When a moist material is dried the rate of moisture removal with respect to time is as shown in Figure 6.59.

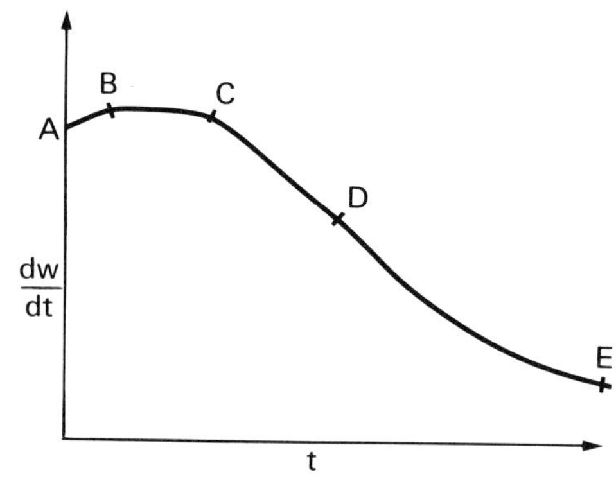

Fig. 6.59 Drying rate v time

The curve shows that the drying process is not a smooth continuous one in which a single mechanism controls throughout, but that three basic drying stages occur.

- Curve BC is the constant drying rate region, and is the evaporation of free moisture from the saturated solid surface. In this stage, the rate of liquid removal is less than or equal to the rate of liquid permeating from the solid interior. It is called the constant rate period of drying, as, for a given drying air temperature, the rate of change of moisture content of the solid with time is constant. It is controlled simply by the rate of heat transfer to the evaporaing surface. The moisture content of the solid at the end of this period is termed the Critical Moisture Content.

- Curve CD is the first of the falling rate periods, and is the evaporation of free moisture from the unsaturated solid surface. In this stage, the rate of liquid removal from the solid surface is greater than the rate of liquid permeating from the solid interior and hence the solid surface can no longer be wholly maintained saturated by moisture. As drying proceeds, the difference between the two liquid rates increases and hence drying rate progressively falls.

- Curve DE onwards is the second of the falling rate regions, and is the evaporation of moisture from the interior of the solid. In this stage the plane of evaporation has moved within the solid; again, as drying proceeds the drying rate falls off. It is this period that takes most of the drying time.

Note that AB represents the heating up period - the time involved here is usually insignificant compared with the total drying time.

These three stages do not necessarily occur in all drying processes. Clearly, if the solid is at or below its Critical Moisture Content before drying, there will be no constant rate period. Similarly, the final moisture content may be higher than the point at which the first falling rate period ends, in which case there will be no second falling rate period.

The effects on the drying process of the variables present in the governing equations may be summarised as follows:

- Drying-air temperature. The drying rate is directly proportional to the difference between the drying-air temperature and the solid-surface temperature. An increase in air temperature is often the simplest way to increase drying rates and therefore the capacity of the drier.
- Solids temperature. Faster and more efficient drying is obtained when the material is fed to the drier at temperatures above ambient.
- Air velocity. An increase in air velocity increases the heat and mass transfer coefficients and therefore increases the drying rate. This is due to increased turbulence at the solid surface reducing the resistance of the laminar film. In applications involving the passage of air through a granular bed of material, the velocity can be increased until the bed becomes fluidised. This increases the area of material available for heat and mass transfer.
- Final moisture content. An increase in the final desired moisture content decreases the time needed for drying without affecting the drying rate. It is generally uneconomic to dry material below its equilibrium moisture content in atmospheric air.
- Air distribution. The air must be brought into intimate contact with the whole drying surface to make the best use of its potential for heat and mass transfer. The air distribution system must of necessity be very carefully designed in relation to other parameters such as the material being dried, shape and size of drier, and positioning of baffles.
- Direction of air flow. The air-flow pattern through, over, under or around the material being dried depends on the nature of the material, the means of handling it and the space available.

6.5.3 Convection Drying

This is the most common method of drying used in British industry. The heat transfer occurs by direct contact between the wet solid and hot gases. The resultant vapour is then carried away by the drying medium. In the majority of convection driers the liquid to be evaporated is water and the drying medium is hot air. Typical manufacturers of all types of convection drying ovens are JLS Engineering Co. Ltd.

The hot gas can be:

- Indirectly heated air. Air is blown across a heat exchanger which is heated by combustion products, hot water or steam. The air does not come into direct contact with the heating medium, and therefore cannot be contaminated in any way. The only need for indirectly heated air occurs when the product being dried is very sensitive to contamination or gives off flammable vapour.
- Directly heated air. The drying air is heated by and mixes intimately with the combustion products from a burner. The fact that only a very few substances cannot tolerate the

combustion products of natural gas means that packaged direct gas-fired air heaters located outside the drying systems are extensively used. Direct firing gives an increase in efficiency of between 20 and 25% over indirect firing. Temperature control is also improved because, by adjustment of heat input, operating temperatures can be reached very quickly and then kept within close limits during operation. This system can also provide a cooling effect, if necessary, by shutting off the gas supply and maintaining the air flow. Two other advantages of this system are cheaper installation costs and adaptability. With a packaged air heater, all the necessary controls are included, which means that replacement costs of standard items, maintenance and labour charges can be kept to a minimum. Existing convection driers can be easily and cheaply converted to direct firing. The faster, more easily controlled drying will give lower costs, increased production and a better product quality.

- Inert gases. These are only used in very special cases where air oxidizes or otherwise damages the product being dried. In some cases the combustion products of a stoichiometric mixture of fuel and air can be regarded as an inert gas.

- Steam. Steam is still a common heating medium for use in driers. This is essentially because the process is often regarded as a medium-temperature process and the steam can provide the correct degree of heat for many drying operations. However, the need for large and expensive ancillary equipment and the limited temperatures are big drawbacks where economy, flexibility and high production rates are required.

Air is a very convenient medium for heat and mass transfer. For heat transfer, air has a low specific heat capacity and can, therefore, be raised in temperature very easily. For mass transfer, a small increase in the temperature of air can greatly increase the quantity of water it can carry.

Psychrometry

The principal objective of heating air for drying is to increase its moisture carrying capacity. Therefore the drying effect of air is governed by its 'relative humidity' (RH) or percentage saturation (see Chapter 14 for definitions).

The heat required to evaporate a given quantity of water at a range of drying temperatures can be calculated from the enthalpies of the drier inlet and exhaust air streams. The enthalpy of moist air is dictated by its temperature and humidity:

$$h = h_a + Wh_w \qquad 6.1$$

where

h	=	Enthalpy of moist air	kJ/kg dry air
h_a	=	Enthalpy of dry air	kJ/kg
h_w	=	Enthalpy of water vapour	kJ/kg
W	=	Specific humidity	kg H_2O/kg dry air

Thus to follow the progress of a drying operation, access is needed to temperature, enthalpy and humidity data. Such information is provided graphically in psychrometric charts which can be constructed to incorporate a wide range of data. The following properties of air/water vapour mixtures are generally included:

- Dry bulb temperature °C
- Moisture content kg/kg dry air
- Relative humidity %
- Specific volume of mixture m^3/kg
- Total enthalpy of mixture kJ/kg
- Wet bulb temperature °C

A given sample of air can be specified if any two of the above quantities are known, the remaining properties can be read off the chart. Figure 6.60 is an example of a low temperature psychrometric chart published by CIBSE.

At temperatures above the boiling point of water (e.g. 100 °C at atmospheric pressure), high temperature hygrometric charts are used, an example of which is illustrated in Figure 6.61. The figure is reproduced from Chapter 1 and has limited information, but charts as comprehensive as those for low temperature are available.

Micro-computer programmes have been developed at MRS which replace psychrometric charts and enable drier energy performance to be assessed.

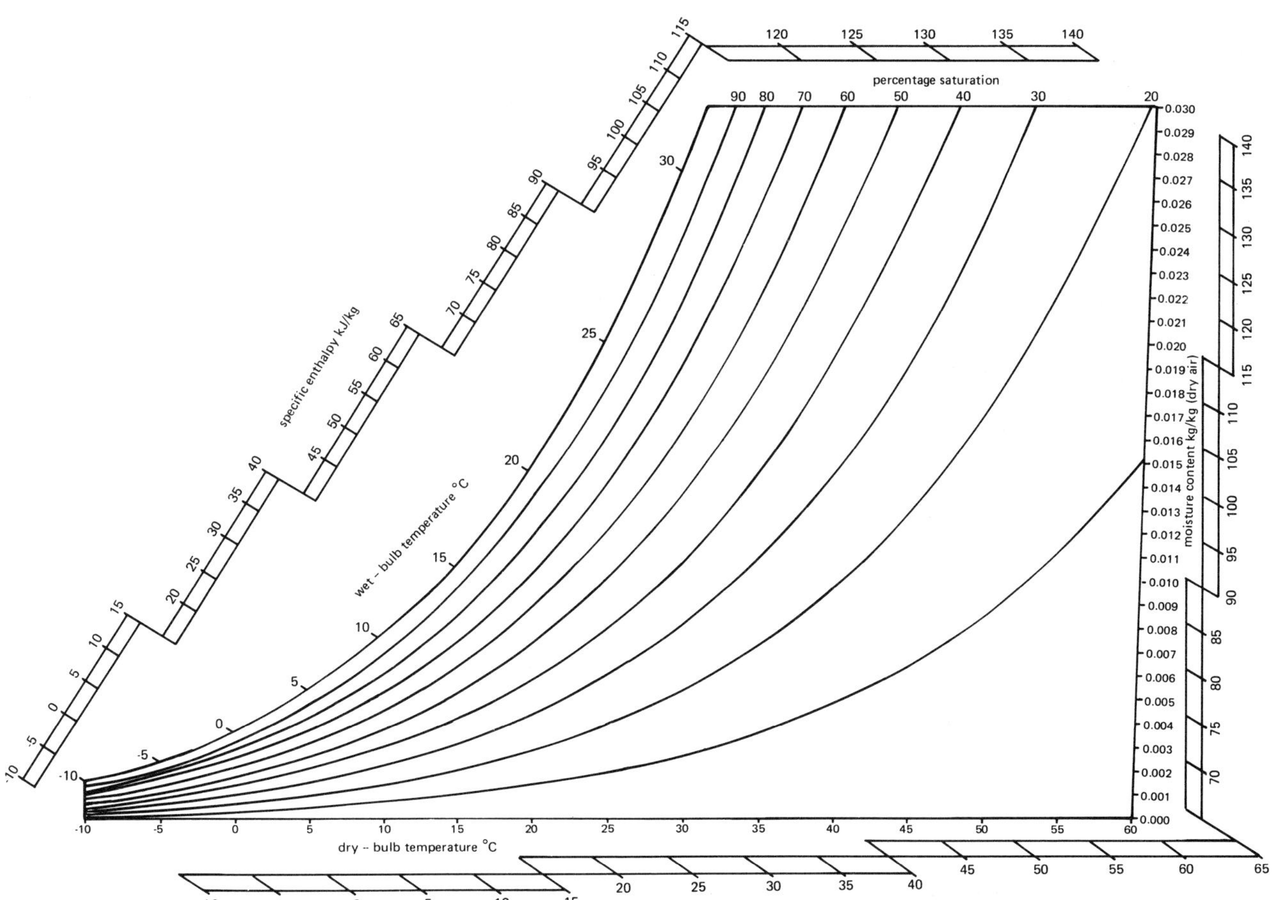

Fig. 6.60 Psychrometric Chart

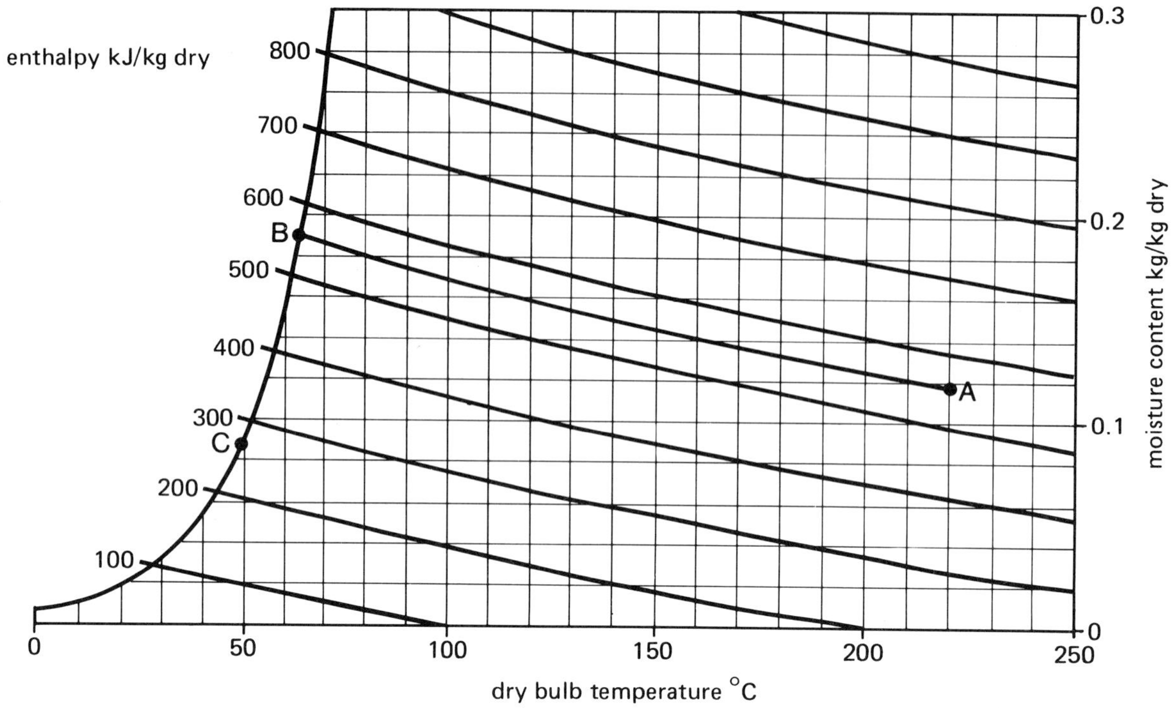

Fig. 6.61 High temperature hygrometric chart

Recirculation

One way in which the thermal efficiency of a drier may be increased and its fuel costs lowered is to reheat some of the used air and pass it back through the drier. Recirculation air also tends to give a more even temperature distribution and therefore a more uniform drying operation than a simple single-stage drier. On the other hand recirculation results in a smaller mean wet-bulb depression. This means that for the same dry-bulb temperature and velocity the drying time will be longer. A limit is reached when the recirculation rate causes the humidity in the drier to rise too quickly and so seriously retard the drying process.

The necessary degree of recirculation can be produced by hot gas fans or high velocity jet burners. The choice depends on temperature and economics.

Methods of gas-solid contacting

In the process of drying a solid by a hot gas, heat and mass transfer occur at the surface of the solid. To obtain maximum efficiency there must be:

- Maximum exposure of the solid's surface to the gas phase.
- Thorough mixing of gas and solids.

A number of possible solids bed configurations are illustrated in Figure 6.62. They comprise:

A A static solids bed, for example a tray drier. There is no relative motion between the particles, but the tray can be moving in any direction.

B A moving solids bed, in this case a rotary-type drier. In a slightly expanded bed the motion of the solids relative to each other is achieved by mechanical agitation. Another example is a tower drier where solids flow is due to gravity.

C A fluidised solids bed. In a fluidised bed drier, an expanded bed of solids exists where the solids are suspended by the drag forces caused by the gas phase passing at some critical velocity through the voids between the particles. The solids and gas phases inter-mix, behaving like a boiling fluid.

D A dilute solids bed. This is a fully expanded bed in which the solids are so widely separated that essentially they exert no influence on each other. The density is that of the gas phase alone. This type is the pneumatic conveying drier. Here the gas velocity at all points is greater than the terminal settling velocity of the solids. The particles are therefore lifted and continuously conveyed by the gas.

E Another type of dilute solids bed, the spray drier. Here a liquid feed is atomized and projected into a large drying chamber, from which it settles by gravity.

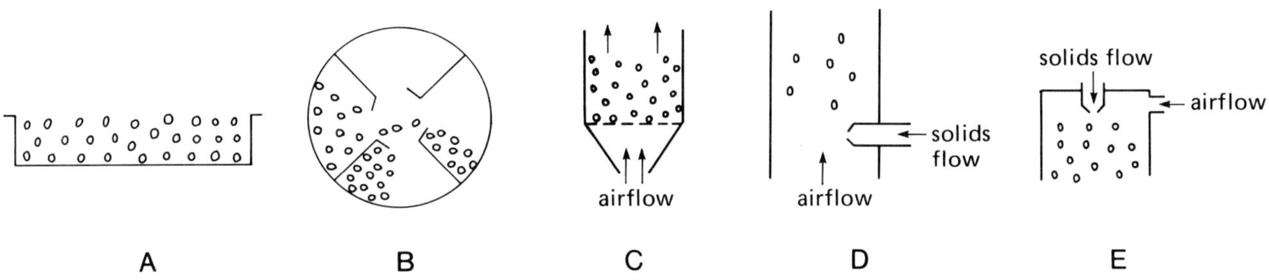

Fig. 6.62 Solids bed configurations

Examples of some of the ways in which the hot gas contacts a bed of solids are shown in Figure 6.63.

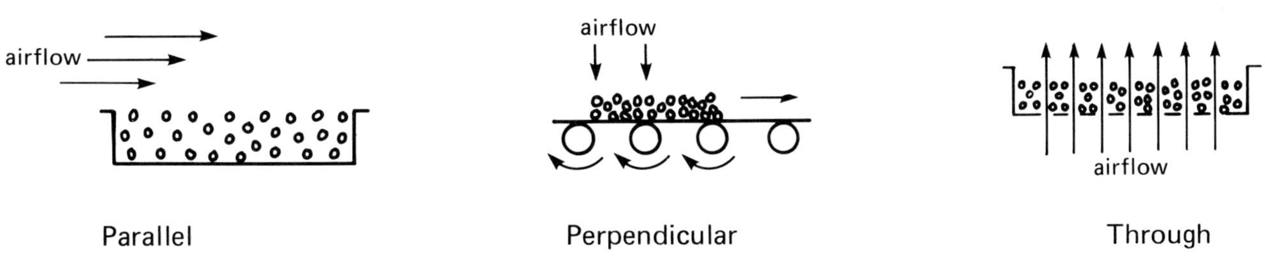

Fig. 6.63 Contact between hot gas and solids bed

Parallel flow. This type of flow occurs in a tray drier. The gas flows in a direction parallel to the solids surface. Contacting is mainly at the phase interface but with penetration of gas into the voids between the solids near to the surface.

Perpendicular flow. The gas flows in a direction normal to the phase interface, e.g. hot air impinges on the surface with some penetration into the voids between the solids being conveyed across the drying chamber.

Through flow. The gas penetrates and flows through the voids between the solids, circulating more or less freely around individual particles. The solids can be static, moving, fluidised or dilute. The Figure shows a perforated tray drier.

Figure 6.64 illustrates the three flow modes used to describe continuous drying equipment.

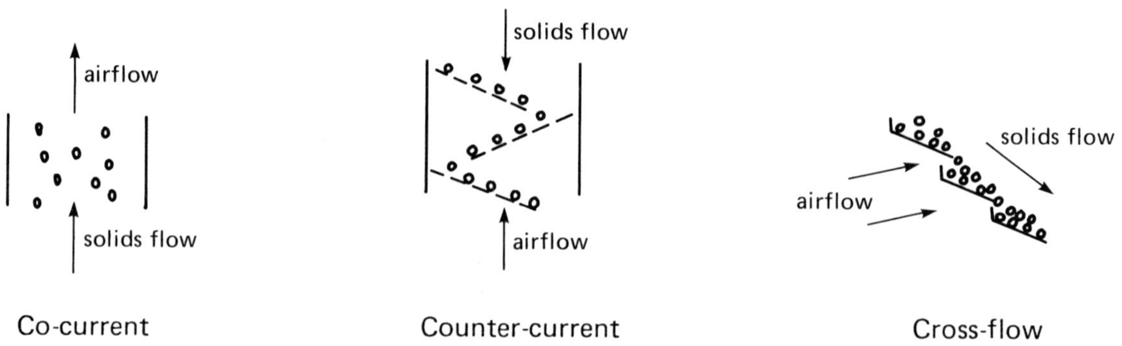

Fig. 6.64 Continuous drier flow modes

Co-current flow. Gas and solids flowing in the same direction, e.g. upwards in pneumatic conveyor.
Counter-current flow. Direction of gas flow is opposite to direction of solids.
Cross flow. Gas flows at right angles to direction of solids, e.g. Cascade type drier.

6.5.4 Radiation Drying

The effectiveness of radiation drying depends on the efficient generation and transmission of infra-red radiant energy and its absorption by the article to be dried. The materials being treated must present a flat, smooth plane, normal to the path of radiation. In practice, articles with more than one plane surface can be dried successfully by the suitable arrangement of several radiation surfaces.

Articles of irregular shape will suffer from part-shadowing if radiation is the only drying mode used. This effect can be lessened by radiation within the irregularities, conduction through the article, and natural convection currents within the irregularities.

Supplying forced warm-air convection in addition to the convector source will overcome the shadowing effect.

It is apparent from the above that radiation methods of drying are most suitable for treating coated and inked paper and board, textiles, painted metal sheets and other flat products. This range can be extended to cover thin beds of granular, powdered or fibrous products if provision is made to present all the surfaces of the products to the radiation by vibrating or raking the beds to turn the product continuously.

Radiant heaters have a number of advantages in comparison with the other heat transfer modes used in drying. They give instantaneous heat and can reach operating temperatures from cold in a few minutes, whereas convection usually requires 15 to 30 minutes. They are simple, flexible, light, compact and adaptable. They can be installed exactly as required to suit individual products. Installation charges are lower, they are easily controlled and the radiant flux density can be varied quickly to suit a variety of surface conditions. Very little of the radiant energy is lost to the surrounding air, and units can be open to allow the drying processes to be observed.

These factors make radiant driers particularly suitable for continuous operation. The ability to grade the radiation along the drier means that when the material is wet, strong heat can be applied in the initial stages of drying at the entrance, and gentle heat near the exit. Conversely, the product can be heated gently in the initial stages and more strongly in the later stages when heat treatment of the products is required.

The nature of radiant heat also gives rise to a number of disadvantages with this mode of drying as opposed to other methods. Drying rates by radiation alone are adversely affected by irregular surfaces. This effect is lessened to an extent by re-radiation, conduction and the introduction of convection. Different absorption capacities of variously coloured paint and surfaces means that dark matt surfaces absorb more radiation and therefore dry more rapidly than light polished surfaces and the flux density needs adjustment. The thickness and absorptivity of the surface will cause other problems; thus a lightly absorptive surface will give drying in depth, but low absorptivity may result in surface hardening and incomplete drying.

These factors mean that before employing radiation drying, the emissivity of the surface to be dried should be known, indicating the amount of heat which can be absorbed, reflected and transmitted. Then a radiant source is selected with the correct emissivity and wavelength to match those of the material to be dried.

Gas radiant heaters are available with a large range of emissivities which enables them to be used as heat sources in an equally large range of drying applications. High temperature ceramic plaques attain temperatures of up to 980 °C in about five minutes providing a maximum wavelength intensity of 4.7 μm and a flux density of up to 160 kW/m^2. Medium/low temperature radiant panels are made to attain temperatures of up to 700 °C with a maximum intensity of 7.2 μm and a flux density of 13 kW/m^2.

Luminous Walls

The system produces high intensity radiant energy. It operates successfully in any plane, can be constructed to any size, and can be controlled instantly e.g. Maywick, Figure 6.65a. A correctly proportioned air/gas mixture is passed through a refractory wall of uniform porosity. Combustion takes place in the refractory surface, producing incandescent heat at temperatures which can exceed 1000 °C. Such temperatures are reached within seconds of ignition and cooling on gas interruption is almost instantaneous. The immediate proximity of the flames to the burner, the high emissivity of the luminous wall and its low conductivity account for the rapid heating and cooling cycle. The luminous wall system is particularly adaptable for specialised requirements where speed and uniformity of heating are necessary. The heat input can be varied between 63 and 252 kW/m^2 (20 000 and 80 000 Btu/ft^2 h) of firing areas. Whereas radiant plaques will operate up to a process temperature of around 200 °C, a luminous wall will operate up to 1000 °C. Luminous walls have good temperature distribution with no gaps between, smooth cross ignition across the walls, a turn down of 4:1 and high durability for arduous processes.

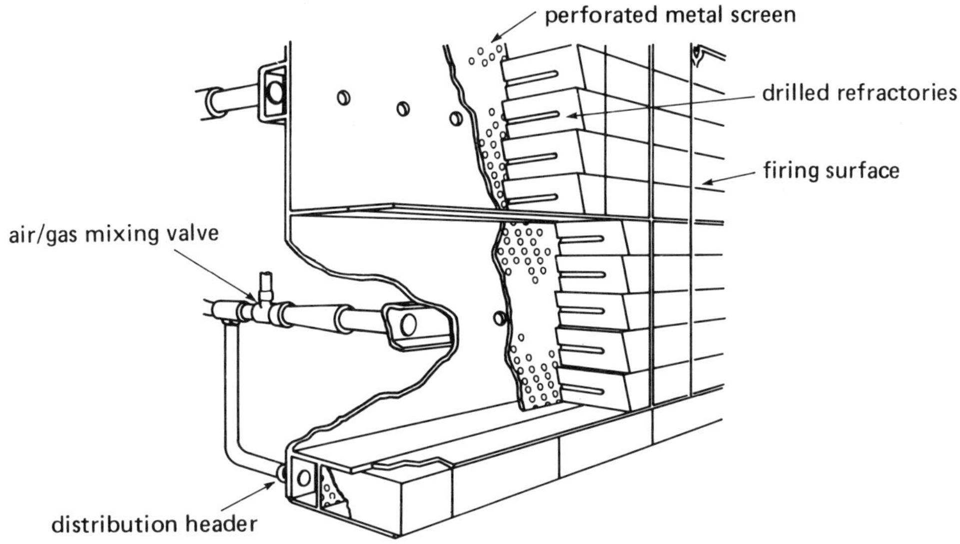

Fig. 6.65a Luminous Wall system (Maywick)

Arrangement of the radiant sources

The radiant heaters can be placed in a variety of ways to suit the particular drying operation. Figure 6.65b shows a number of possible arrangements.

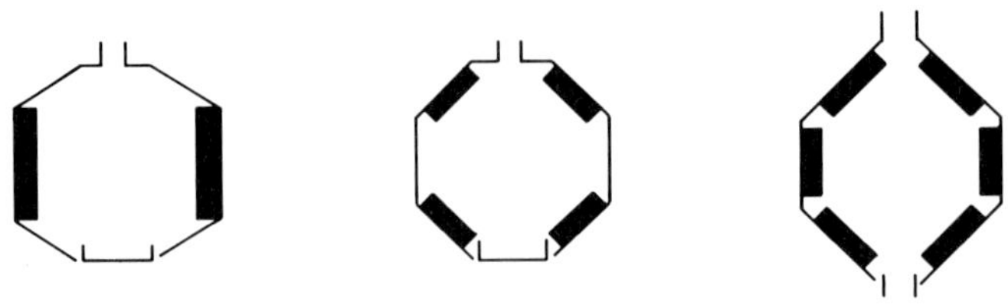

Fig. 6.65b Possible arrangements of infra-red radiant plaques in a drying tunnel (sectional views)

Usually, when flat articles moving in a horizontal plane are being dried, the radiant sources are placed in the roof of the drier. However, in certain instances the source is placed below the material to achieve some other process advantage, e.g. when drying size coatings in carpet manufacture, the steam from the size rising through the carpet helps to 'burst' the pile.

Radiation onto flat surfaces creates natural convection currents which take away evaporated vapours. This is most effective when the radiation source is mounted vertically or at an angle to the horizontal. When the radiants are mounted horizontally in the roof, the operation is improved if some air circulation is introduced to take away the vapours and combustion products.

The trend in continuous radiation driers is towards high-temperature sources and fast conveyor speeds. Metal surfaces are often suitably placed in the drier to reflect heat from the radiant heaters onto the drying surface to improve efficiency.

Curing

In most drying operations, solvent removal is unconnected with any chemical change in the product. However, in some instances solvent evaporation is connected with oxidation and hardening of oils and resins. Consequently radiation drying can overlap with curing, where the removal of solvent by drying forms a part of the curing process, increasing the strength and durability of the finished product.

6.5.5 Conduction Drying

This method of drying involves heat transfer to the wet solid by physical contact with the heated surface of the drying equipment.

The heat source can be:

- Saturated or superheated steam. The former is the most common as it gives up latent heat only in accordance with demand and it is economical, especially if use can be made of low-cost steam from some other process. Efficiencies up to 75% can be achieved. However, efficiency falls as the final moisture content required increases. A major disadvantage is that high drying temperatures can not be obtained without high steam pressure and/or superheat. This is not normally cost effective.
- Hot water or other heat transfer fluid. These give a faster heat transfer than steam.
- Combustion gases. These also give faster heat transfer.

Conduction drying methods are suitable for complete enclosure and are, therefore, adaptable to drying under vacuum or inert systems. Closed systems enable expensive solvents or gases to be recovered more easily.

Roller drying

This is the most popular conduction drying technique and is used to dry or pre-dry thin flat products e.g. paper and textiles where multiple roller systems are used extensively.

The method is useful in dehydrating liquids, pastes and free flowing slurries. The fluid is applied as a thin layer on to the clean, smooth outer surface of a continuously rotating roller, heated internally by steam or combustion products. A slurry often dries in a few seconds (less than the time for one revolution of the roller) and is scraped off by a knife as shown in Figure 6.66

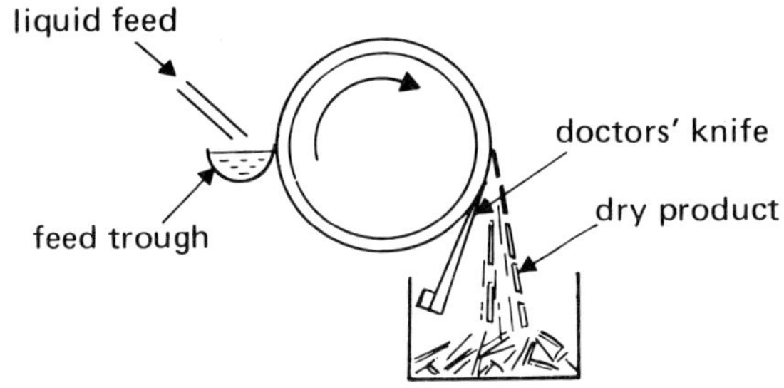

Single roller drier

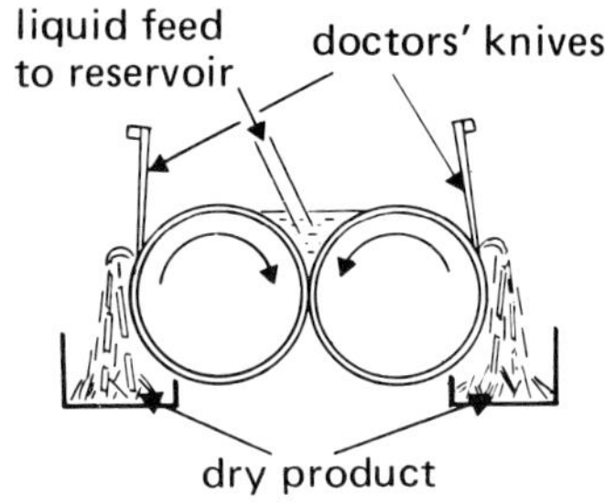

Double roller drier

Fig. 6.66 Roller driers

6.5.6 Direct Flame Drying

The application of a naked flame for drying is confined to a very few articles. The method involves radiation, conduction and convection of heat simultaneously to the article being dried and therefore control is vitally important. Regular flame shapes and constant combustion characteristics must be guaranteed. The flame itself must be soft and consistent to avoid burning or scouring the surface of the material being dried.

It is used for drying printing inks on paper surfaces, and in foundries for drying the blacking on large immovable moulds, for drying ladles and shanks and the skins of intricate box moulds.

In foundries, modern methods and automation are replacing the hit-and-miss methods of 'torch' drying of moulds and cores. Modern ladle driers no longer depend on direct flame contact with the ladle refractories. High velocity air blast burners with short, sharp flames are used to recirculate hot combustion products round the ladle, using their scrubbing action to dry and pre-heat them.

6.5.7 Selection of Drying Plant

Three main areas need addressing when choosing individual drying plant viz:

- The physical and chemical properties of the wet material and its drying characteristics (particle size, abrasiveness, corrosiveness, toxicity, flammability, initial and final moisture contents, type of moisture, permissible drying temperature, etc).
- The ultimate product quality required (bulk density, state of sub-division, shrinkage, contamination, decomposition, uniformity, etc).
- Capital and operating costs.

The physical form of the wet and dry material determines the way the material is handled in the drier; the physical and chemical properties dictate the operating conditions during drying. Drier dimensions and mechanics have to accommodate the material's physical form, and the aerodynamic and thermal design must provide satisfactory drying conditions. The drier may have to be flexible enough to cater for a range of products.

The quantity and rate of production will influence the choice of a batch or continuous process and thus affect labour and operating costs. Processes up-stream and down-stream of the drier (e.g. size reduction, dust and solvent recovery, bagging/packaging etc.) affect equipment choice and capital and running costs. Facilities available and constraints at the site (space, fuels, power, permissible environmental pollution limits, etc) also have to be taken into consideration.

An important consideration in the choice of a drier for a particular duty is the susceptibility of the material being dried to damage from high temperatures, combustion products or agitation. Materials may be broadly divided into five classes with regard to these effects. These are listed below in ascending order of vulnerability.

- Construction materials - e.g. sands, clays, gravels, bitumastics, etc. This group can withstand very high drying temperatures, contaminated atmospheres and agitation. It is common, therefore, for drying to be carried out in Rotary Driers heated with air and/or combustion products from a direct fired air heater. The only limitations on temperatures, atmosphere and agitation are those imposed by the material of the drier.

- China clay, fuller's earth, talc rock, kaolin, etc. This group can withstand high drying temperatures, a large degree of agitation, but no contaminated atmospheres. Indirect fired air heaters or radiant heat are usually employed here.

- Sawdust, wood chips, lignite, fertilizers, gypsum and various chemicals. This group can withstand agitation and contaminated atmospheres, but not high drying temperatures (greater than 200 °C) because of the dangers of chemical changes or ignition. Such products are therefore commonly dried, with agitation, by direct air heating methods, sometimes employing recirculation of the air to produce lower, even temperatures.

- Grain, grass, seeds, malt, soaps, detergents, various chemicals and pharmaceuticals, plus many food products (fruit, vegetables, meat, eggs, coffee, gelatin and starch foods). Here, agitation is usually in order, but high temperatures and contaminated atmospheres are injurious. Common drying methods are by agitation in indirectly heated warm air systems. However the clean nature of natural gas combustion products allow direct firing with gas in many cases.

- Ceramic and metal wares, products with any kind of surface treatment, large products such as paper, boards and timber, also sticky products and some food products. In these cases no agitation can be employed because of the physical structure and size of the product. Conveyorised or batch systems can be employed using hot air recirculation or radiant heat, or conducted heat.

There are no hard and fast rules and the above classifications are not always strictly adhered to. For instance, modern drying methods allow much higher drying temperatures to be used with heat-sensitive materials because of the faster drying methods available. Direct gas-fired air heating systems can often be employed where the product is susceptible to contaminating atmospheres.

6.5.8 Types of Driers

Driers may be intermittent (batch) or continuous in operation. The latter type is preferable in most cases as less labour, fuel and floor space are required and a more uniform product is produced. Conversely, batch driers have low initial cost and maintenance requirements and are very versatile in their application. Figure 6.67 shows the basic categories of drier used in industry.

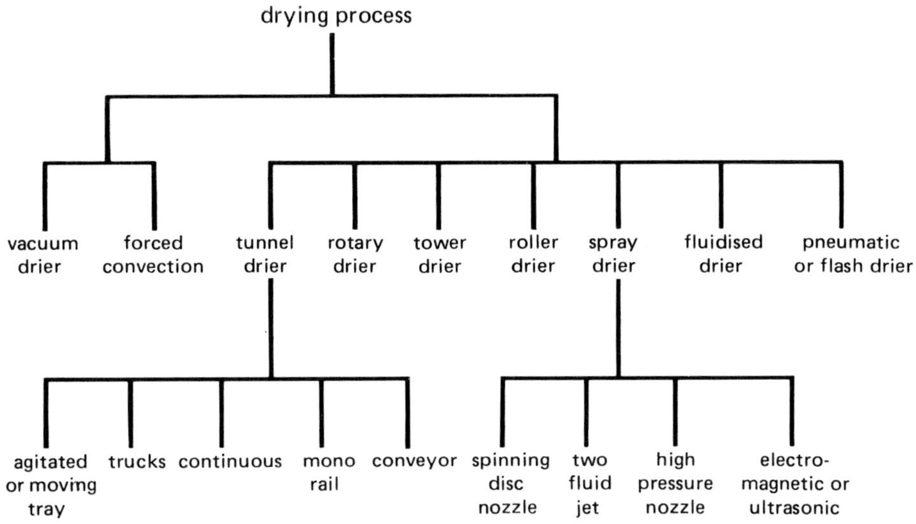

Fig. 6.67 Drier categories

Intermittent driers

Batch drying is generally limited to relatively low product throughputs. Drying times are usually longer than for continuous driers. Temperature, humidity and material moisture content conditions are constantly changing which makes precise control difficult. Since the drier structure has to be heated up during each cycle, the thermal efficiency is inherently lower than for the continuous counterpart. This disadvantage can be minimised by the use of well insulated, low thermal mass construction.

The commonest type of low temperature batch drier is the box oven. In addition to aqueous drying, they are also used in paint drying and curing where the drying process involves solvent evaporation and, in the case of oxidisable paints, the additional stage of oxidation and polymerisation. Force drying is the term used to describe paint drying at temperatures up to about 80 °C. Above this temperature the process is normally referred to as baking, stoving or curing.

Both direct and indirect fired box ovens are in use, the former called double-case ovens and the latter treble-case ovens (see Figure 6.68). The heat source can be steam or hot water coils instead of direct gas.

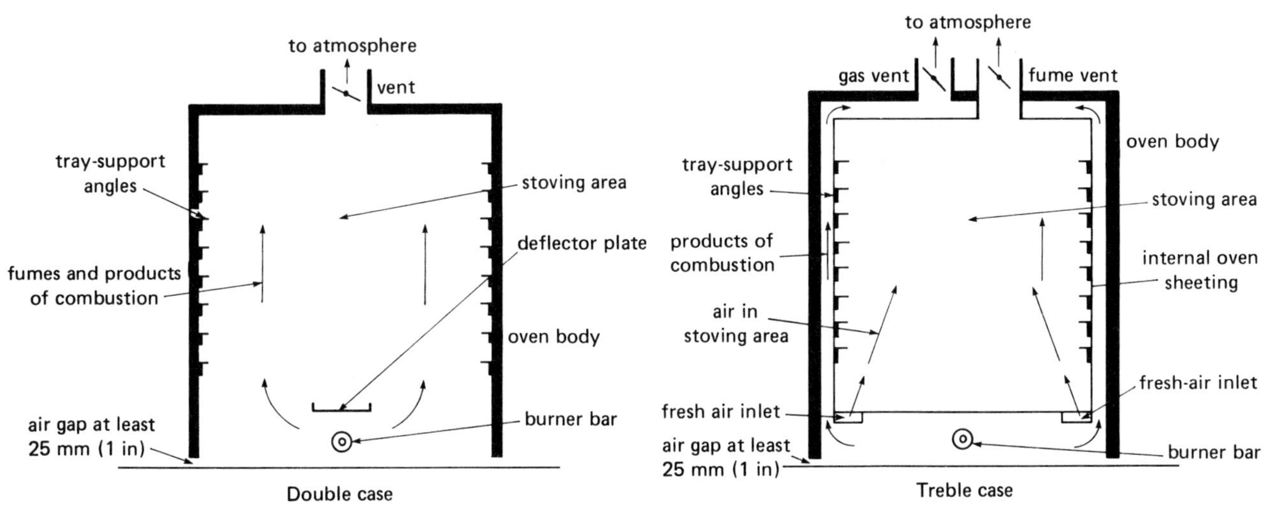

Fig. 6.68 Natural convection ovens

Gas fired forced convection ovens can make use of an integral direct fired air heater sited either upstream or downstream of the recirculation fan. Suction burners are used with the former and forced draught pre-mix or nozzle-mix burners for the latter, as illustrated in Figure 6.69.

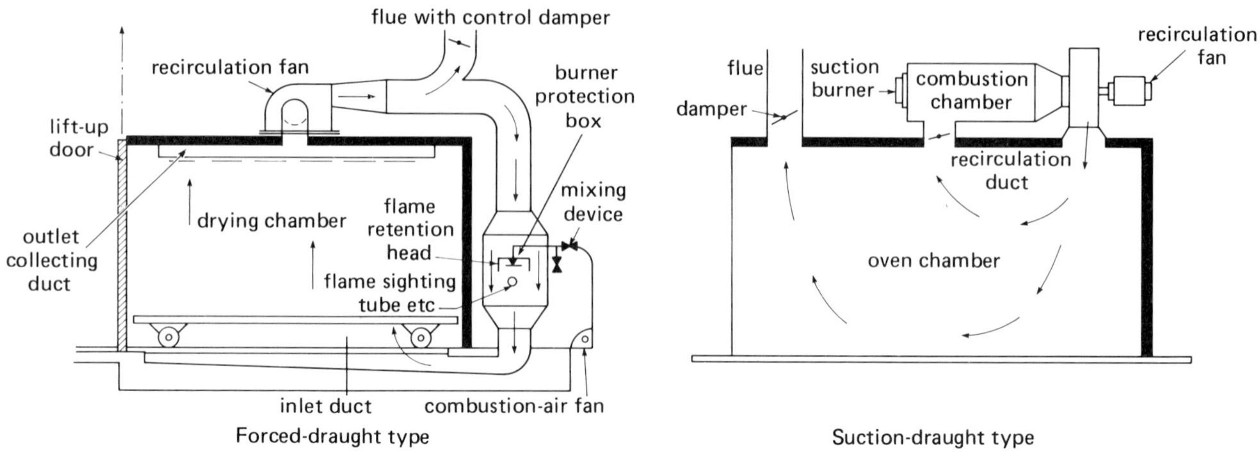

Fig. 6.69 Recirculating drying ovens

To conserve heat, most of the air is recirculated (consistent with moisture burden); fresh air can be as little as 5 to 15% of total air flow.

Drying rooms are sometimes employed for expensive products (e.g. tobacco, bacon, leather) which are hung or stacked in the room. Very low temperatures and long drying periods are normally involved, with air recirculation used to improve thermal distribution and thermal efficiency.

The thermal sensitivity of a material fixes the maximum temperature at which it may be dried. If the temperature limit is low, evaporative rates will be low and material residence time high. The use of vacuum techniques can decrease the residence time. The double cone vacuum drier is a simple batch drier and basically consists of a double skin container. The material to be dried is loaded into the inner container which is then evacuated, the heating medium being fed into the annulus formed by the double skin.

Continuous driers

These are generally more efficient, faster and more easily controlled than batch driers. Included in this category are tunnel, rotary, roller, tower, spray, fluidised bed and pneumatic (or flash) driers.

In convection tunnel driers the stock to be dried is moved progressively through the drier either co-current or counter-current with the air. They include a wide range of conveyance methods as follows:

- Agitation or moving tray; generally used for non-sticky, granular material.
- Truck or car; widely used in the heavy clay and ceramics industries (see Section 6.7).
- Overhead monorail; drying and curing of painted metal articles commonly use this method (Figure 6.70)
- Self-supporting; in the case of paper and textiles drying, the material forms its own conveyor being held in tension and moved through the tunnel by rollers.

– Conveyor; the conveyor can be a continuous belt, perforated or mat, or made from woven wire, webbing, metal sections etc.

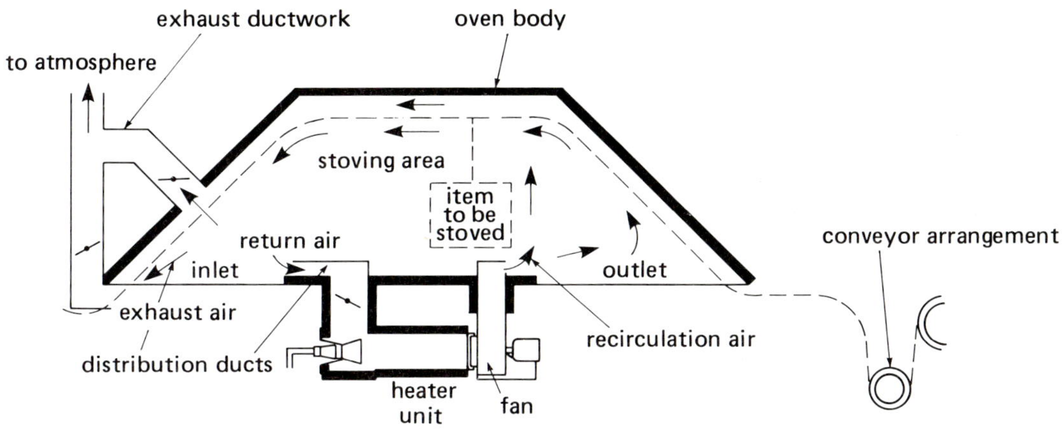

Fig. 6.70 Camel-back conveyor oven

Tunnel driers are zoned to give the correct time and temperature profile for satisfactory drying by means of air temperature (and sometimes humidity) control in sections along the drier.

Rotary driers are mainly used for granular (usually free flowing) materials like sand and some forms of chemicals, fertilizers etc. Figure 6.71 illustrates a recirculating inclined rotary drier. The material to be dried is fed into the upper end of the cylinder and is propelled through the drier by virtue of the cylinder's rotation and inclination. Internal baffles (flights) are fitted which alternatively lift and tumble the material particles through the air stream, exposing the maximum surface area. Flight geometry affects the drying rate. For wet materials, simple radial flights are normally used, whereas flights incorporating a 90° bend are used for materials that are free-flowing or nearly dry. Air flow may be counter-current or co-current to the material. When a low final moisture content is required, a counter-current flow is used to enable the hottest, driest air to contact the driest material. Co-current flow is used for heat-sensitive materials.

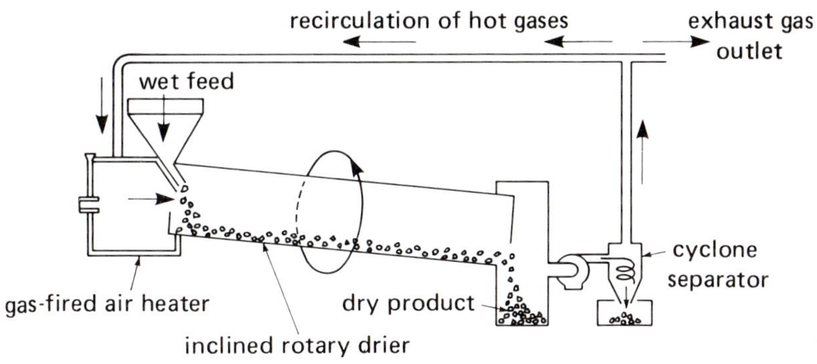

Fig. 6.71 Single-shell rotary drier with recirculation

In the double-shell rotary drier, hot air passes through the central cylinder and then back along the annulus formed with the outer shell. The material thus receives both convected heat in the cylinder and conducted heat from the inner cylinder surface.

For inlet gas temperatures of 530 to 750 °C evaporation rates are approximately 32 to 48 kg water/m^3 of cylinder volume for single shell driers and about 64 to 80 kg/m^3 for double shell driers.

Double shell driers are widely used for sand drying. Jet driven recirculation is often used as the high temperatures involved preclude the cost effective use of recirculating fans. The efficiency of a double shell drier is typically 60%.

Roller driers (see Figure 6.66) were discussed in Section 6.5.5 on Conduction Drying.

Tower driers are often used for drying grain, seed, sand and other granular free flowing materials. Figure 6.72 shows a typical application for grain drying. The grain falls through the tower counter-current to the flow of drying air. Residence time is increased by means of staggered trays or baffles.

Grain drying takes place in the falling rate period when the controlling factor is the rate of diffusion of moisture from inside the grain to its surface and cannot take place above a specific maximum rate without damaging the grain. Maximum temperature of grain in the dry state should never exceed 50 °C.

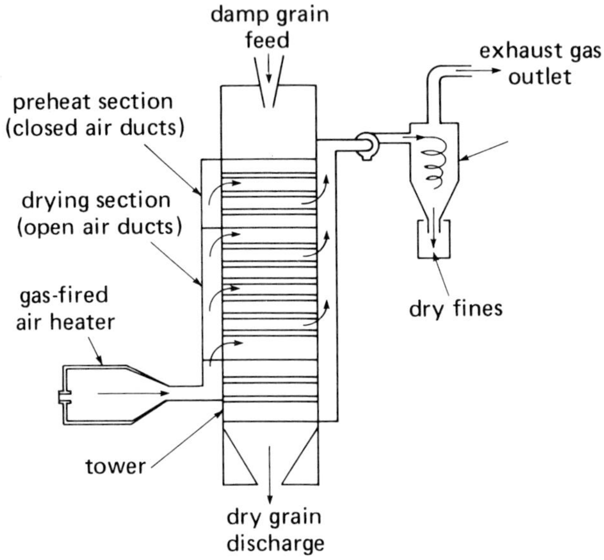

Fig. 6.72 Grain drying in a tower drier

Spray drying techniques are used extensively for materials which are sensitive to heat and/or contamination. A wide range of materials from foodstuffs through dyestuffs, pharmaceuticals, chemicals, ceramic tile dust to soaps and detergents use spray driers. In general, any material which can be pumped as a paste, solution or slurry can be spray dried. The slurry of the material to be dried enters at the top of the drier where it is atomised by means of either a spinning disc, a high pressure nozzle, a two-fluid jet or electromagnetic or ultrasonic nozzle. Atomisation produces a very high surface area to volume ratio allowing a rapid drying rate. This, together with the short residence time in the drier, enables relatively high drying air temperatures to be used without overheating the material, as the air is cooled by evaporation from the droplets and the dry product leaves the system before its temperature can be raised appreciably. Figure 6.73 shows a typical spray drier in outline.

Most of the particles (above 90%) reach the bottom of the chamber, the small proportion of fines carried over being collected in a cyclone separator. Precise spherical particles (solid or hollow) are produced where size distribution can be controlled. Maintenance costs are low and the graded marketable product often requires no subsequent treatment. The drying air may be indirectly or directly heated. Direct gas-fired air heaters are suitable for nearly all materials but other, less refined fuels are rarely suitable for direct firing. Whilst hot air is normally injected at the top of the drier as shown in Figure 6.73, in some applications, e.g. detergent drying, air is introduced at the bottom of the drying chamber, with the exhaust at the top.

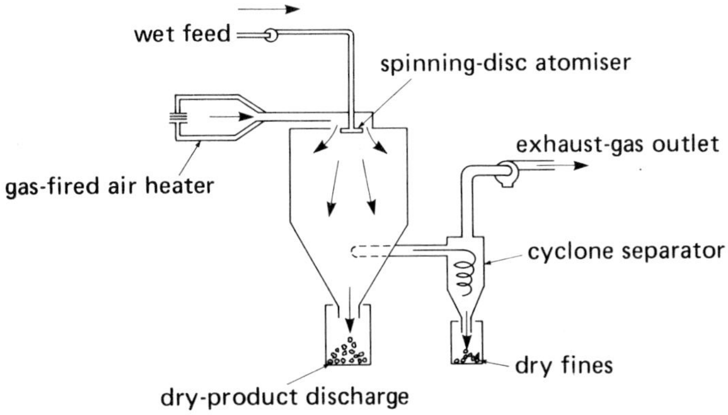

Fig. 6.73 Spray drier

Fluidised bed driers are finding increasing application, especially in large scale processes, e.g. ores and minerals, chemicals and some food products such as grain, rice, nuts, etc. Fig. 6.74 shows the general features of a fluidised bed drier. Each particle to be dried is surrounded by heated air which supports the particles and the agitation produced provides uniform temperatures in the bed. As with spray drying, direct gas-fired air heaters provide a suitable heating means for almost all materials.

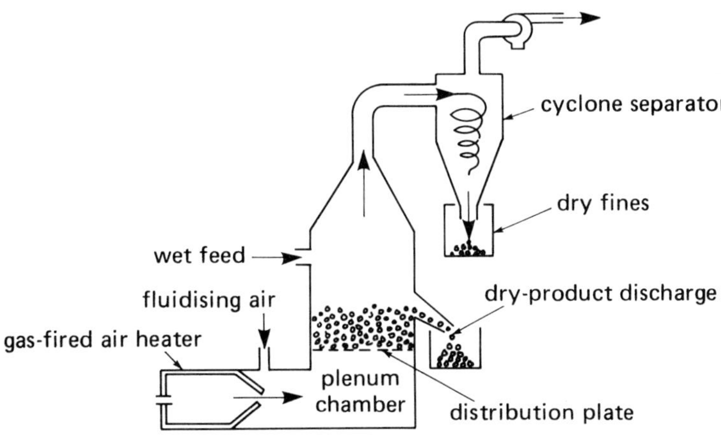

Fig. 6.74 Fluidised-bed drier

Fluidised-bed driers have the following advantages:

- Heat-transfer rates are very high, leading to compact units of high efficiency. Mass-transfer rates are also high, aided by the large transfer surface available.
- Particle-breakage rates are low since the particles are cushioned from each other by the surrounding air.
- Adaptability. Fluidised-bed driers are applicable to most materials and are particularly useful for difficult-to-handle materials. However, a wide particle size range presents problems since the fluidising velocities needed for large particles will elutriate the smaller particles.
- Drying and classifying can be done simultaneously, eliminating the need for subsequent screening and dust removal.

Flash (or pneumatic) driers, in their simplest form, consist of an heated air conveyor system. Drying is generally very fast. In fluidised bed driers, the solids are still in a dense state of aggregation whereas in flash driers they are conveyed by a high velocity air stream in a relatively dilute suspension.

Particulate solids are injected into a fast moving stream of hot air having a velocity typically between 18 and 30 m/s. The dried solids are usually separated from the gas phase in a cyclone and bag filter as shown in Figure 6.75. With materials of very high moisture content, it is difficult to achieve even the limited residence time required for drying, and systems which recycle part of the product are employed.

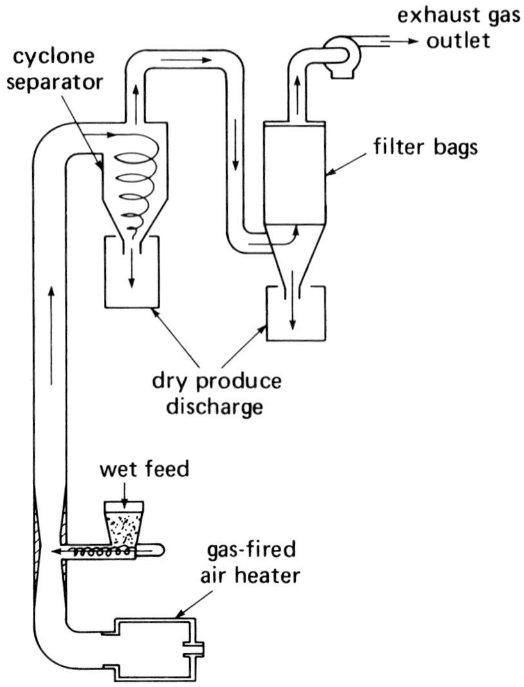

Fig. 6.75 Pneumatic conveying drier

6.5.9 Improving the Performance of Driers

Many driers, even those using the more efficient direct firing, operate at efficiencies well below their potential maximum. Studies at MRS have identified a number of improvements which can be considered, in addition to the obvious measures of better thermal insualtion and air leakage reduction. A software package entitled 'Drier Energy Evaluation Program' (DEEP) has been developed by MRS to assist in optimising performance of existing systems.

The thermal characteristics inferred by MRS of straight through driers are shown in Figure 6.76.

Calculated heat consumption is related to drier efficiency which will increase when:

- Drying air temperature is raised with no significant rise in outlet air temperature.
- Exhaust air humidity is increased.
- Exhaust stream temperature is reduced.

Increasing drying air temperature results in greater evaporative capacity giving the options of increased product throughput and/or reduced air flow, both of which will reduce specific energy consumption. Such improvements could not be obtained if product characteristics limit the drying temperature e.g. timber drying temperature limit is 95 °C.

The above specific energy consumption trends developed by MRS for an ideal straight-through drier show that significant reductions can be made by optimising the adjustment of drying parameters.

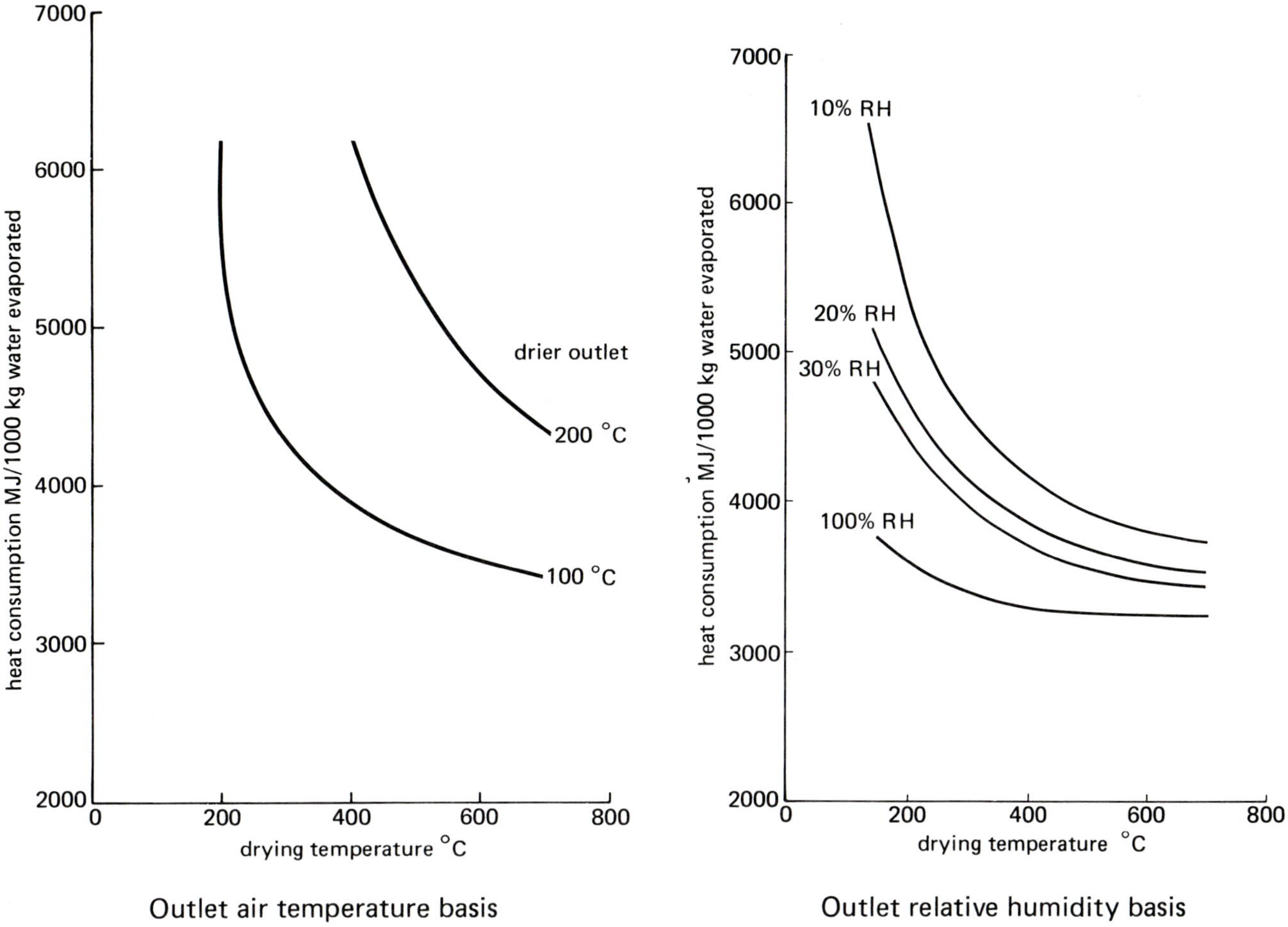

Fig. 6.76 Heat consumption curves for drying

Where drier parameters have been optimised, further improvements can be obtained by:

- Air recirculation.
- Exhaust heat recovery.
- Heat recovery with recirculation.

Driers with recirculation

Partial recirculation of exhaust gases gives higher exhaust humidities. Improvement in fuel consumption from increasing exhaust recirculation rates is illustrated in Figure 6.77. Savings of up to 35% can be achieved by incorporating a recirculating system in the drier design. The law of diminishing returns applies to recirculation ratios and the improvement in efficiency for recirculation ratios in excess of 50% is not significant when compared to 25% recirculation. With high recirculation ratios there is the possibility of saturation of the recycled air stream.

For processes involving solvent removal or heavily contaminated exhaust streams, recirculation may not be feasible because of the possibility of operating within hazardous limits or the need for expensive gas clean-up systems to avoid duct or fan blockage.

Exhaust heat recovery

Exhaust heat recovery systems should only be considered after drier efficiency has been improved by optimisation of design parameters, including the possibility of exhaust recirculation. The effective-

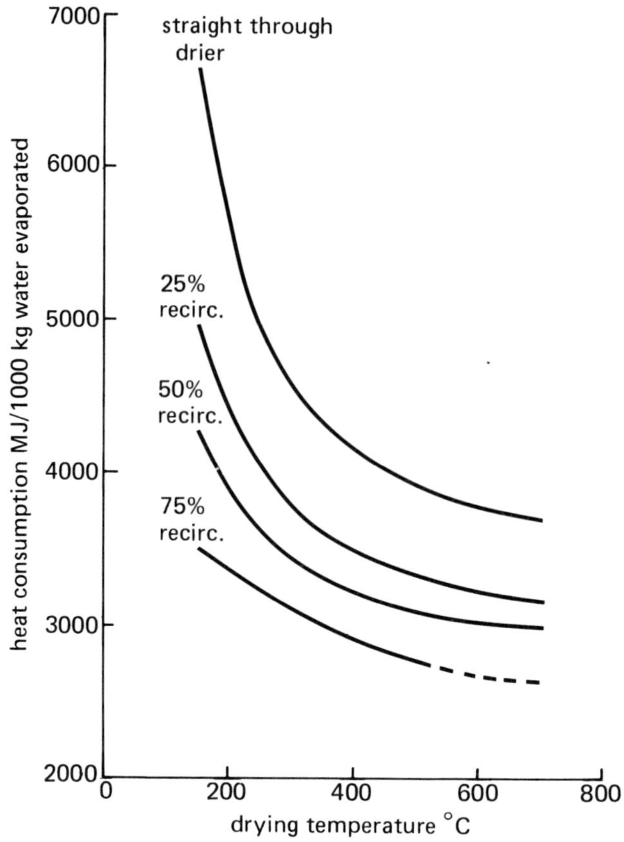

Fig. 6.77 Effect of recirculation on heat consumption

ness of exhaust heat recovery applied to a recirculation drier is only significant at temperatures below 200 °C. This reflects the fact that as drier design becomes more efficient, exhaust heat recovery is less favourable.

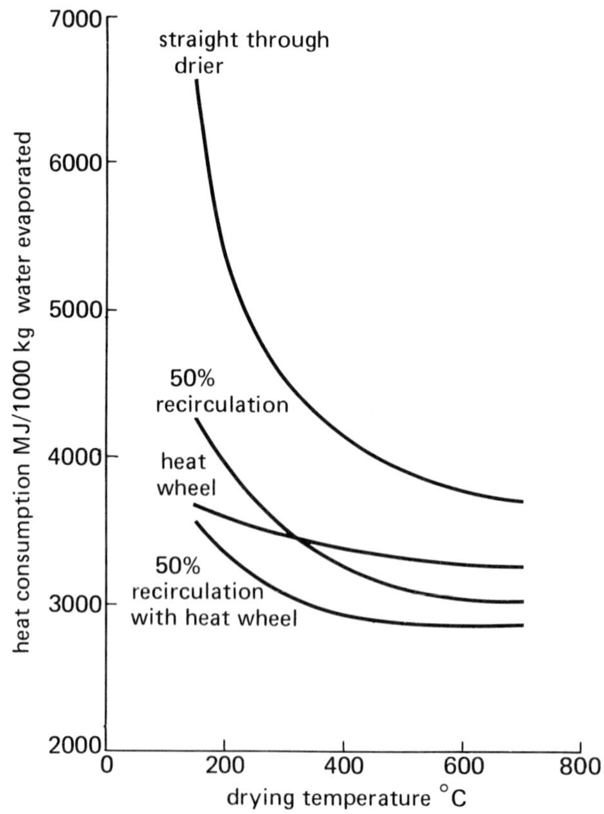

Fig. 6.78 Effects of recirculation and exhaust heat recovery on heat consumption.

The results of the MRS work, illustrated in Figure 6.78, indicate that:

- Recirculation together with exhaust heat recovery offers no significant improvement over either individual method.
- Since there are only minor differences in energy savings, the choice between recirculation and exhaust heat recovery will be dictated by the particular drying application and equipment costs.

Possible heat recovery devices together with some of their characteristics are listed in Table 6.2. For detailed treatment, consult Chapter 7.

Table 6.2 Characteristics of heat recovery equipment

Recovery device	Temperature efficiency (%)	Comments
Rotary heat exchanger	70 - 80	Suitable for exhaust streams up to 700 °C Maintenance required on rotor and moving parts. Possibility of cross-contamination.
Plate heat exchanger	60 - 70	Exhaust streams up to 400 °C. Advantage of no moving parts.
Run-around coils	50 - 60	Temperature limited by circulating fluid. Useful for distant inlet and exhaust streams.
Heat pipe	50 - 60	400 °C exhaust stream. Self-contained units with no moving parts.
Heat pump	-	State-of-art technology. Limited by low temperature exhaust requirement of less than 80 °C.
Spray recuperator	50 - 60	Large unit size. Exhaust stream up to 400 °C. Disadvantage is that the unit outputs hot water.

Radiant Heating

Recent developments in the use of ceramic fibres and porous metallic materials for radiant panels used in radiation driers should increase the potential for gas firing. These materials offer almost instantaneous heating or cooling of the radiant surface with consequent improved control over the drying process. Examples include the Stordy-Marsden radiant panel and the Maywick luminous-wall system shown in Chapter 2.

6.5.10 New Drying Techniques

Considerable increases in process efficiency are likely to result from the application of new techniques now becoming available. Heat pumps can recover both sensible and latent heat from the drier exhaust and up-grade it to working temperature. They can be driven by an electric motor or, more cost effectively, by a gas-fuelled engine, which can provide steam or hot water for use elsewhere in the process through heat recovered from the engine cooling water and exhaust. Current work aimed at improving performance and reducing capital and maintenance costs will increase the range of applications for which heat pumps are economically attractive and this trend will be accelerated by the future availability of absorption heat pumps, which are simpler in design and should be of lower cost. These units are currently under development in several countries.

Very high efficiency would result from carrying out the drying process in a well insulated chamber, with 100% recirculation of exhaust air. This is not possible unless moisture can be removed before the air is returned to the drier but should be rendered viable by the use of a desiccant material regenerated by direct gas heating. Suitable equipment, including desiccants able to withstand higher temperatures, is becoming available and offers considerable potential advantages.

Munters Ltd in conjunction with MRS have developed a gas fired dehumidifier, which should give substantial cost savings over the equivalent electrically heated system. The Munters wheel is partitioned in two zones. In the working zone moist air from the building is passed over the wheel continuously and the moisture extracted and absorbed by the desiccant. The warm air passes to the building. In the reactivation zone the process is reversed as air, heated by the direct gas firing drives off the absorbed water which is then exhausted from the building. There are few moving parts and the dehumidifier requires minimal attention.

Instead of chilling and reheating as with conventional heat pump dehumidifiers, the gas fired drier simply removes the water vapour and applies an even heat distribution to give maximum humidity control. A diagram of the wheel is shown in Figure 6.79a and also in Chapter 7, Figure 7.47.

6.5.11 Catalytic Radiant Panels

Catalytic radiant panels are infra-red emitting units heated by flameless combustion of natural gas. The radiant element of these units is made of a refractory material impregnated with catalysts that bring about catalytic oxidation of natural gas.

Catalytic radiant panels are used to bake paints and varnishes. The basic advantages are low energy consumption and suitability for use in flammable atmospheres and atmospheres containing vapours of the organic solvents commonly employed to apply paints and varnishes.

The system consists of a venturi mixing tube providing a partial premixture of air and gas, a fan to impel the secondary air, and a catalyst impregnated refractory panel forming the infra-red and radiating surface.

Uses include paint drying, textile printing, space heating, treatment of plastics, inks, varnishing, screen printing, animal and plant rearing, sports halls, display centres and workshops etc.

A venturi mixing tube is used. The partial air/gas premixture, obtained by induction of outside air, flows through the refractory panel after leaving the premixing chamber. This premixture is air-deficient (15-20% gas in air).

At the start of the catalytic combustion reaction a choke solenoid is used to enrich the mixture with gas for 20 seconds. The fan blows the secondary air drawn from the oven against the radiating surface of the panel. The secondary air diffuses through the mass of the refractory panel and meets the partial air/gas premixture, so contributing to the catalytic oxidation of the natural gas fuel. This excess air moreover makes the catalytic reaction insensitive to the unavoidable air movements occurring in enamelling stoves. It also helps keep the temperature uniform.

Catalytic combustion of the natural gas takes place in the refractory panel. This panel consists of refractory fibres in which are distributed particles of platinum-based catalysts. The catalysts initiate and accelerate the natural gas catalytic oxidising reaction which heats the refractory panel without a visible flame. The temperature obtained at the emitting surface ranges from 600 to 700 °C (1112 to 1292 °F).

The emitted infra-red energy is then absorbed directly by the organic coatings being processed, i.e. by the alkyd paints, acrylics, polyurethanes etc.

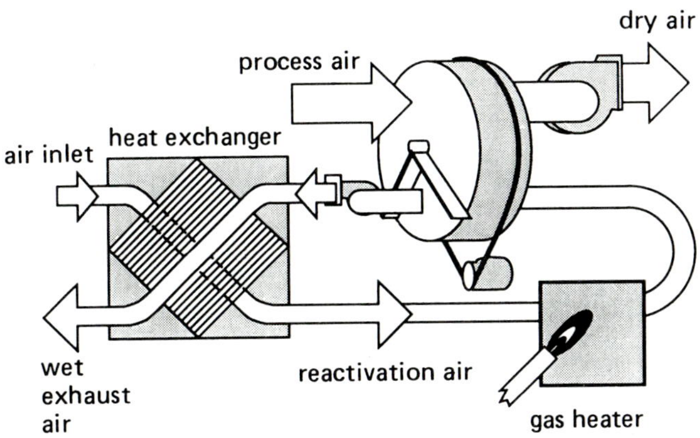

Fig. 6.79a Schematic of Munters gas fired desiccant wheel dehumidifier

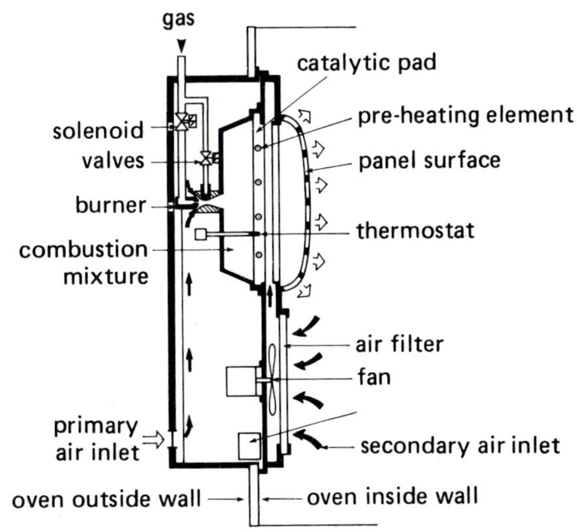

Fig. 6.79b Catalytic thermoreactor

At the start-up, a preheating resistor element heats the panel to around 300 °C (575 °F) in five minutes. Then, following the simultaneous opening of the gas solenoid valve and the choke solenoid valve (the latter remaining open for 20 seconds), the operating temperature is rapidly attained by catalytic combustion of the gas.

A thermocouple within the panel allows automatic start-up. The air/gas premixture is only allowed to the matrix when the matrix is at a certain minimum temperature.

Enamelling ovens nearly always involve emissions of flammable solvent vapours, which makes safety a prime consideration of such installations. A major advantage of catalytic radiant panels is that they will not ignite the solvent vapours commonly found in paint and varnish applications. As a result, in France, radiant panels have been exempted since 1973 from two regulatory requirements based on a 1917 law prohibiting the presence of a flame or hot points above 150 °C (302 °F) in workshops, ovens and other enclosed premises containing flammable vapours like those occurring in the treatment of paints and varnishes.

Flaming up can occur when the three following conditions appear at the same time:

- mixing of fuel and air
- concentration of the combustible somewhere between its lower and upper flammability limits
- temperature at least equal to the self ignition temperature.

If any of these three conditions is lacking there is no ignition.

The radiant panel oxidises the solvents (hydrocarbons) coming into contact with its emitting surface in a similar manner as it oxidises the natural gas fuel of the system. Thus a solvent concentration gradient is established, having a zero value at the panel surface.

Consequently, the solvent concentration throughout the area where the temperature is higher than the self-ignition temperature is less than the lower flammability limit and ignition cannot occur. With a conventional infra-red heater raised to the same temperature, on the other hand, ignition will take place if the solvent concentration exceeds the lower flammability limit.

Catalytic radiant panels effect a partial oxidation of the solvent vapours released into the oven and so reduce their concentration.

The elimination of hazards associated with flammable atmospheres and the oxidation of the solvents involved allow oven ventilation to be reduced, which leads to considerable energy saving compared with conventional processes.

6.6 GLASS

6.6.1 Introduction

Glass products cover a very wide range and are used in almost every other industry e.g. containers for food and drinks, components in electronics, in the building industry, in transport, in illumination engineering and in many domestic applications.

The main raw materials used in glass production are sand, limestone and soda ash. Manufacturing processes range from hand-forming of articles (e.g. lead crystal sector) to the highly automated processes employed in the container and flat glass sectors.

Natural gas and heavy fuel oil are the two energy forms most used in the industry and the greater part (between 70% and 80%) is used for glass melting and conditioning. Price is the main criterion in fuel selection, although legislative restrictions and social pressure relating to environmental pollution can be significant in individual cases. The industry has, therefore, adapted plant and equipment to use either fuel so that advantage can be taken of relative price movements and to obtain interruptible gas contracts.

Although differences are found when the production processes of various sectors of the industry are compared, the stages of glass manufacture common to all sectors can generally be expressed as shown in Figure 6.80.

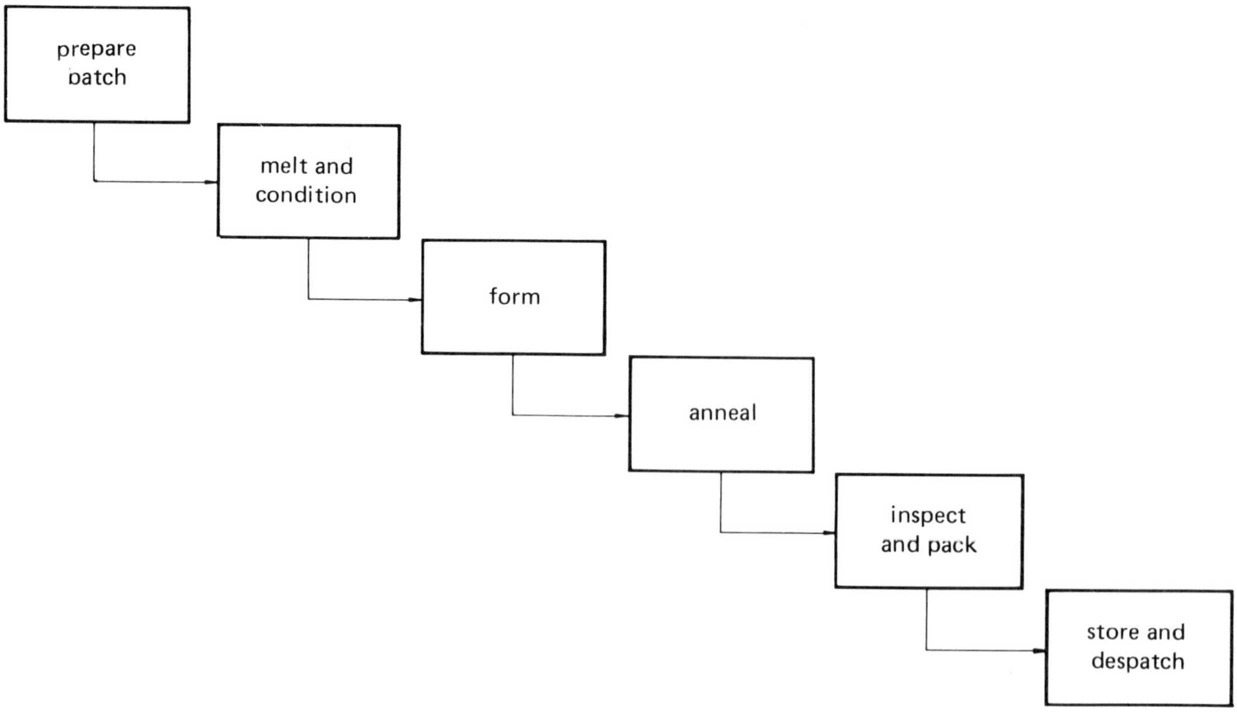

Fig. 6.80 Stages in most glass manufacturing processes

6.6.2 Batch Preparation

This includes the weighing and mixing of the raw materials and delivery of the mixed batch to the furnace. Sand, limestone and soda ash are the main raw materials for container glass and, with a relatively small amount of dolomite, also for flat glass. These two sectors represent almost 90% of glass production. Small quantities of other additives are used to give different characteristics to individual glasses but these are usually a very small percentage of the total raw materials.

The raw materials introduce different oxides into the glass which can be classified according to the part played in forming the glass structure as follows:

- Glass formers. Sand consists of almost 100% silica, which is the actual glass former.
- Stabilisers. Calcium oxide, magnesium oxide and alumina are the main stabilising oxides and are introduced by limestone and dolomite.
- Fluxes. Soda ash is a fluxing agent which decomposes to sodium oxide. Sodium sulphate helps in refining and homogenising the glass during the melt.

Cullet (scrap glass) is added to the mixture, not only for economy, but also to improve the rate of melting. The batch can include up to 35% of cullet.

In a factory with a small output, the batch mixing process could be by a simple rotary mixer, all materials being manually weighed. A batch plant at a large works would be fully automated as in Figure 6.81 which shows the way in which batch constituents are fed into a reception hopper and distributed to their respective storage silos. The individual materials are then drawn off and accurately weighed. Correct weighing and proportioning are critical.

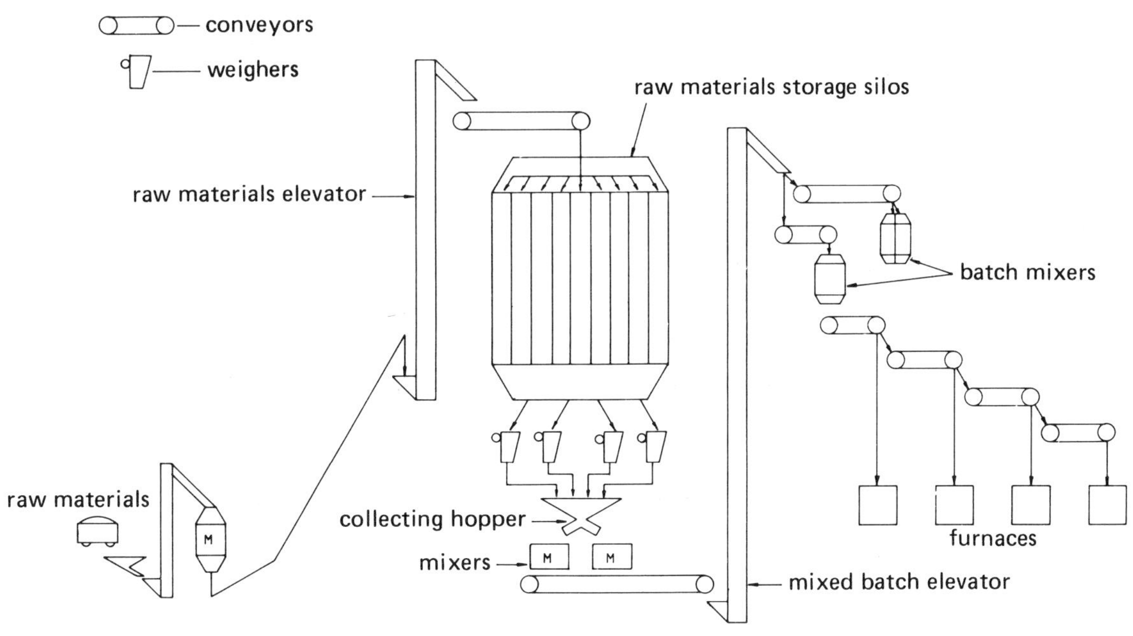

Fig. 6.81 Salient features of automatic batch preparation

The mixed batch is mechanically conveyed and charged into the melting chamber of the furnace. Charging needs to be as continuous as possible to avoid disturbance to the melting process which could adversely affect glass quality. Dust production should be minimised since alkali compounds in the batch can attack furnace refractories. They also may be carried into the firing ports and regenerators by the flame, resulting in deposition in chequer work channels and reducing regeneration efficiency. Screw feeders, finned rollers or pusher bars are used for continuous charging to give a constant flow of materials into the glass tank.

6.6.3 Glass Melting and Conditioning

Melting and conditioning employs the largest proportion of total fuel used and involves:

- Heating of the charge until it is fused.
- Continuation of heating until all reactions are complete and the mixture is homogeneous and free from inclusions such as bubbles.
- Maintaining the melt until required for further processing.

Table 6.3 indicates energy used for melting and conditioning in the various sectors of the glass industry.

Table 6.3 Energy used for glass melting and conditioning

Product	% Total Energy	Specific energy consumption (GJ/t)	
		Average	Range
Glass containers	79.5	10.9	8.5 - 14.2
Flat glass (Melting and annealing)	87.5	10.4	10.4
Glass fibre insulation (Melting and conditioning)	35.1	14.6	14.6
Reinforced glass fibre 49.1% (15 GJ/t) for melting and 28.6% (8.7 GJ/t) for conditioning.	77.1	23.7	23.7
Lead crystal (Wide specific energy range due to varying utilisation factors of pot furnaces).	88.7	59.5	38.7 - 80.3
Domestic ware	67.6	15.7	15.7
Tubing	55.9	12.2	12.2

There are two general types of glass melting furnaces viz. pot furnaces and tank furnaces. In the UK approximately 5% of total glass produced is melted in pot furnaces, the remainder being melted in tank furnaces. Tank furnaces are capable of melting up to 5 000 t/week. Single pot furnaces may melt as little as 2 t/week.

6.6.4 Pot Furnaces

Pot furnaces are the oldest form in use and represent intermittent or batch operation. The pots are made of fireclay or fireclay and sillimanite and range in capacity from around 70 kg to 2.7 t. The most common size of pot is approximately 700 kg.

Originally the pots were heated externally by a fire beneath the pot, the exhaust gases passing around the pot which was housed in a brick enclosure. As the quantities of glass required increased, the size and complexity of the pots also grew. Eventually it became uneconomic to fire pots from cold in the main furnace and the 'pot arch' furnace was introduced for pre-heating purposes. The pot arch furnace, shown in Figure 6.82, is used to bring unfired pots from room temperature up to the ceramic bonding temperature. The pot can then be transferred to the main furnace with little danger of failure due to thermal shock, mechanical damage or distortion.

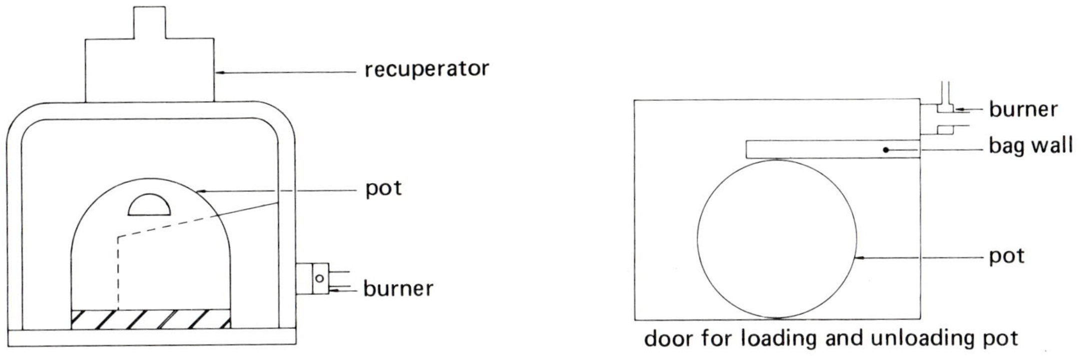

Fig. 6.82 Pot arch furnace

Pot arches are simple enclosures, ideally with internal air recirculation facilities to improve temperature uniformity. Gas firing is common and recuperators are normally fitted to increase thermal efficiency. Pot furnaces are used to make lead crystal or coloured glass and may contain only one or as many as twenty pots.

Heat recovery

Heat recovery from pot furnaces can be by recuperators or regeneration. Single pot furnaces normally have a small metallic recuperator, either alone or in conjunction with a ceramic recuperator. They are usually fired by one or two low pressure neat gas or nozzle mixing burners. Figure 6.83 shows the latter, where combustion air is supplied by a fan and is preheated in a metallic recuperator. Often the recuperator is of the radiation type design. Pot furnace operating temperatures range from 1 320 to 1 440 °C with metallic recuperators mounted on the top of the furnace.

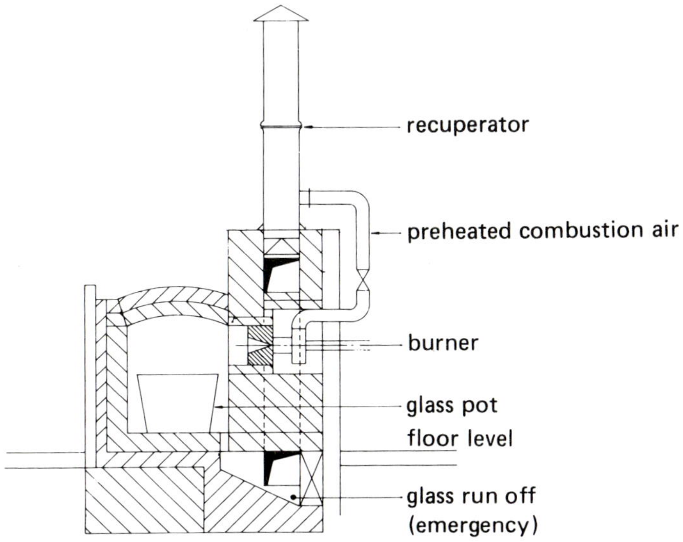

Fig. 6.83 Single pot furnace

Multiple-pot furnaces (Figure 6.84) may have the pots arranged either in rows or in a circular pattern. For regenerative operation, two rows of pots are normally found, with burners and regenerators at each end of the rows. Recuperative multi-pot furnaces normally are fired by a single burner and use a round configuration, with the ceramic recuperator beneath the circular chamber. The pots are clustered around the central burner port. The multi-recuperator tube banks are located below ground level and products of combustion flow up from the eye of the furnace over the pots and down through flue openings at the periphery of the hearth into the tube banks prior to going to the stack. Air preheat temperatures achieved in ceramic recuperators are 800 to 1 000 °C.

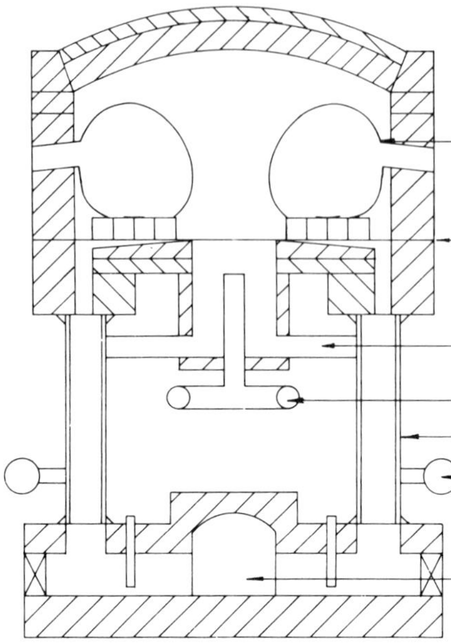

Fig. 6.84 Multi-pot furnace

New developments

The application of regenerative burners to pot furnaces has been demonstrated on a soda lime glass open-pot furnace used in the production of lamp shades (Figure 6.85). Manual charging through the hand gathering port is performed once per day with the furnace at 1 150 °C. The furnace is heated to 1 400 °C for melting and refining and the temperature is then returned to 1 150 °C for gathering and working of the glass.

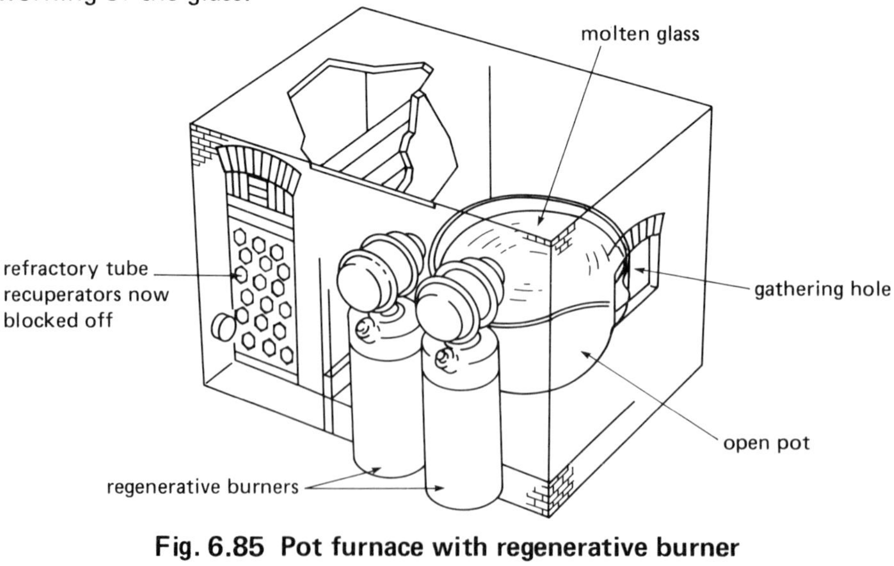

Fig. 6.85 Pot furnace with regenerative burner

The furnace was purposely chosen to provide a practical demonstration of regenerative burners under conditions of continuous high temperature operation and 'dirty' exhaust gases. Typical air preheat temperatures of 1 300 °C are achieved when the furnace is at 1 400 °C. This compares with an air preheat of around 700 °C with the previous refractory tube recuperator i.e. the regenerative system savings are 20%. The initial trial confirmed the need for regular bed cleaning to remove material deposited in the regenerators. The deposits are reported to be generally of a dusty nature and the packing can be easily removed and cleaned. Cleaning frequency is typically six weeks and takes approximately one hour to complete. The furnace does not need to be taken out of service during cleaning.

6.6.5 Tank Furnaces

A tank furnace is basically a bath, the bottom and sides of which are constructed of large blocks of refractory material. The glass is melted by direct open-flame firing across the glass surface and furnaces can have melting capacities of up to 5 000 t/week. Glass tanks are essentially reverberatory furnaces.

The most commonly used tank furnaces in the UK are:

- Recuperative continuous tanks.
- Regenerative continuous tanks.

Other types of furnace e.g. day tanks, unit melters, double-crown furnaces etc. are rarely found. No futher remarks on these are made.

The majority of furnaces in the container and flat glass sectors of the industry are of the regenerative type; recuperative furnaces are used in the production of glass fibre insulation and account for a minority of total glass production.

Continuous tank furnaces, operating 24 h/day will remain in operation for between 5 and 8 years. The necessity for long periods ('campaigns') between repairs/rebuilds demands careful furnace and burner design and application. Combustion must be established within a specific space in the furnace, near to the glass surface. Injection of fuel into this space is dictated by the burner design and its position at the firing port. Air flow patterns are determined by the aerodynamics of the combustion chamber and are established at the design stage.

6.6.6 Recuperative Continuous Tank Furnaces

The simplest applications of recuperators to glass furnaces have involved the fitting of suitable heat exchange devices to previously cold air, direct fired furnaces. The device is fitted at the exhaust port, and preheated combustion air is taken in an insulated duct to the burner port. Thus a single burner furnace can be converted simply by the application of a metallic or ceramic recuperator. With multiple flame furnaces, there can be separate recuperators for each burner, or else a manifold system for the hot combustion air, which is distributed from a single heat exchanger to the burners. (Fig. 6.86).

An example of a multiple flame furnace with separate recuperators is the Uniflow furnace made by Teisen Furnaces Limited. This is shown in Figure 6.87. The uniflow tank is an end fired glass melting tank with a multipass horizontal, hexagonal refractory recuperator which the company had designed and developed themselves. In their basic 'Uniflow' design, the recuperative principle had been mainly exploited by positioning the whole of the recuperator underneath the tank, thus saving valuable space.

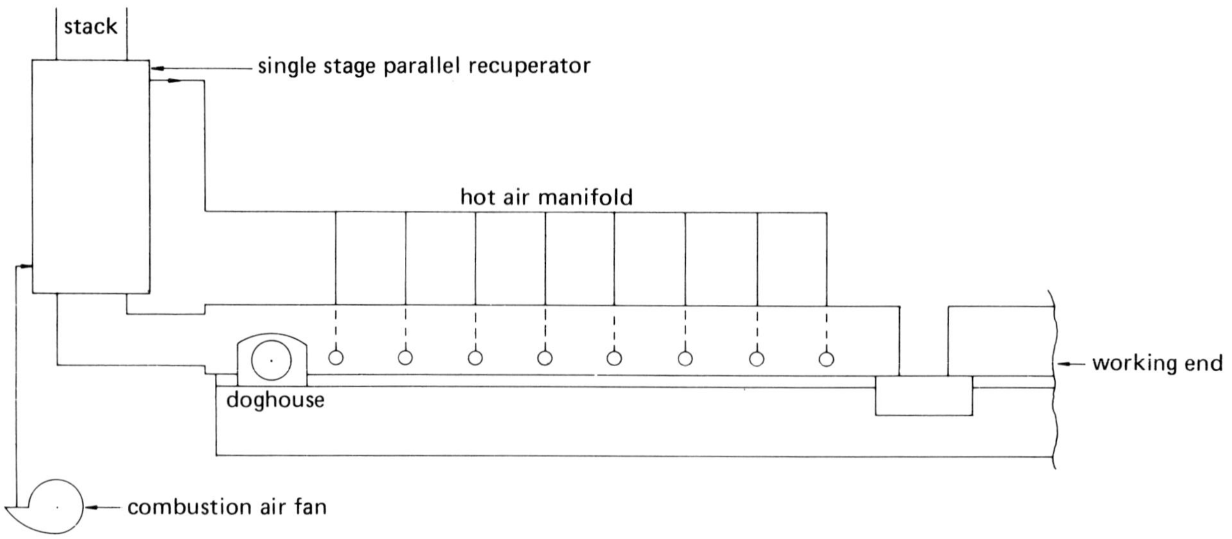

Fig. 6.86 Typical arrangement for a recuperative side-fired furnace

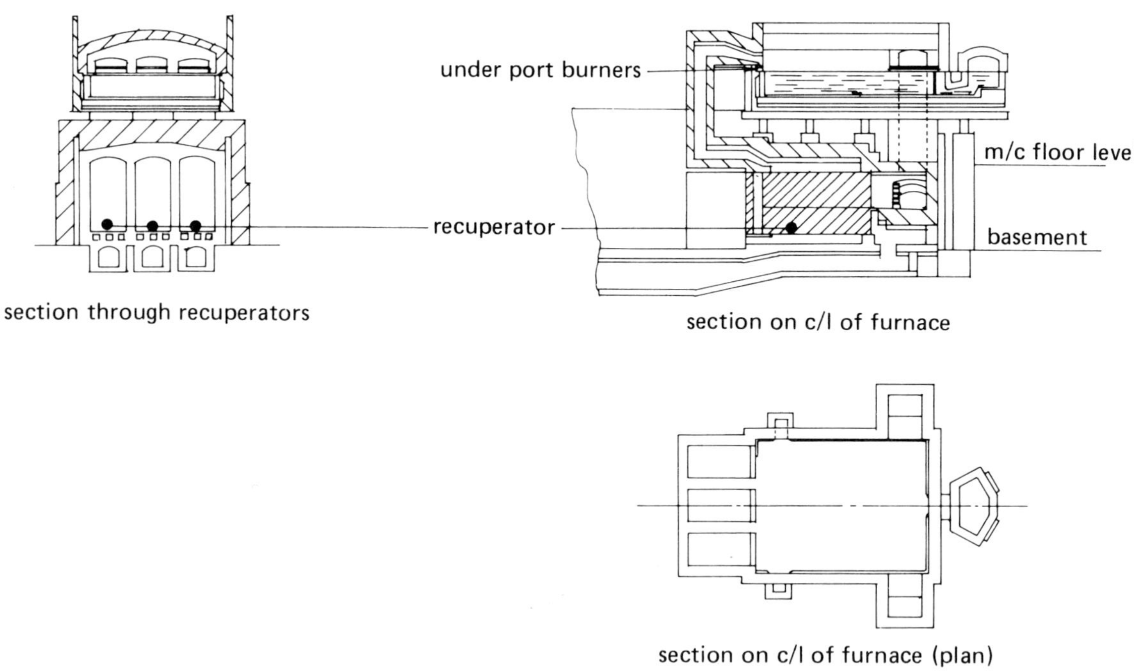

Fig. 6.87 Typical design of end-fired recuperative furnace

With the increase of melting temperatures which has generally taken place in the industry, pulls of over 100 tonnes are now established for this type of furnace. Melting areas of such furnaces are up to 45 m^2. A performance curve for this size of furnace is shown in Figure 6.88.

To avoid long manifolds or ducts, furnaces have been designed with adjacent burner and exhaust ports. An example is the twin burner, single exhaust port furnace as shown in Figure 6.89. The flame path represents a double-horseshoe.

The advantages of recuperative furnaces include the steady level of preheat attained compared to the regenerative furnace, albeit at a lower average temperature. Also recuperators are usually located above or below the melter chamber, and floor space requirements are therefore less. Basements for regenerators, flues and reversing valves are also eliminated, and surface areas for heat loss are usually much lower. In small furnaces, this reduction of heat loss may mean that a recuperative

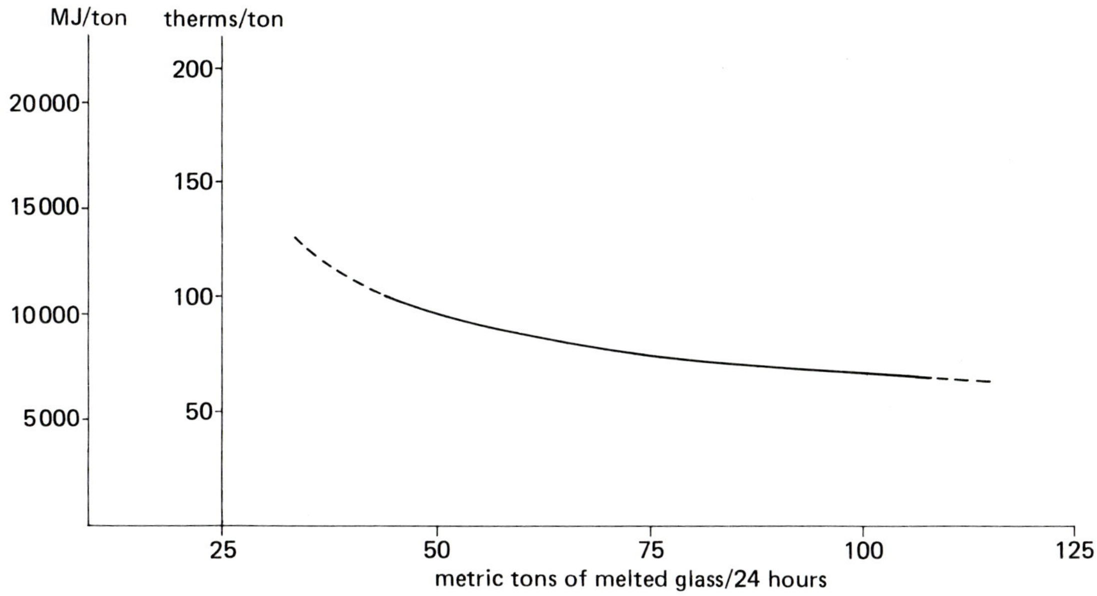

Fig. 6.88 Performance curve for a 45 m² recuperative tank

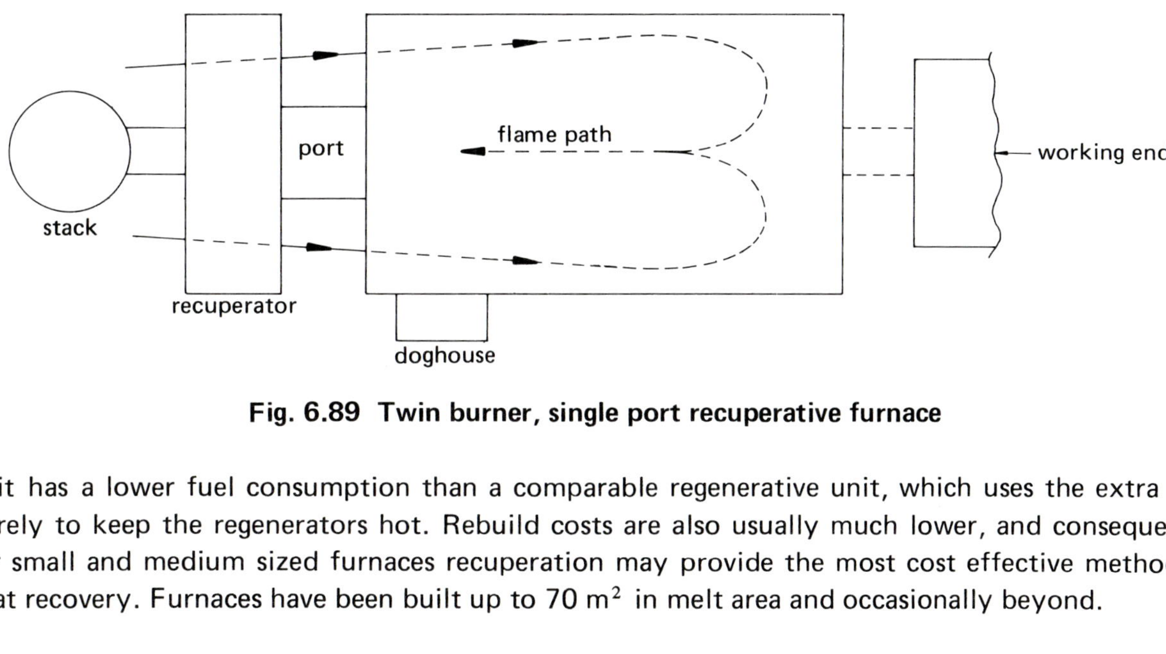

Fig. 6.89 Twin burner, single port recuperative furnace

unit has a lower fuel consumption than a comparable regenerative unit, which uses the extra fuel purely to keep the regenerators hot. Rebuild costs are also usually much lower, and consequently for small and medium sized furnaces recuperation may provide the most cost effective method of heat recovery. Furnaces have been built up to 70 m^2 in melt area and occasionally beyond.

6.6.7 Regenerative Continuous Tank Furnaces

As with the recuperative tanks, regenerative furnaces can be end-fired or cross-fired.

End-port tanks

Uneven width-wise temperature distribution can be a problem with some end-port recuperative furnaces. An obvious way to overcome this is to change the direction of firing at intervals. This is the principle of the end-port regenerative furnace (Figure 6.90).

The heat release patterns for both gas and oil flames give a reasonable temperature distribution in this type of furnace, provided that the length to width ratio is not too high and that the flame length is correct. Obviously impingement on the front wall of the furnace as the flame turns around can be a

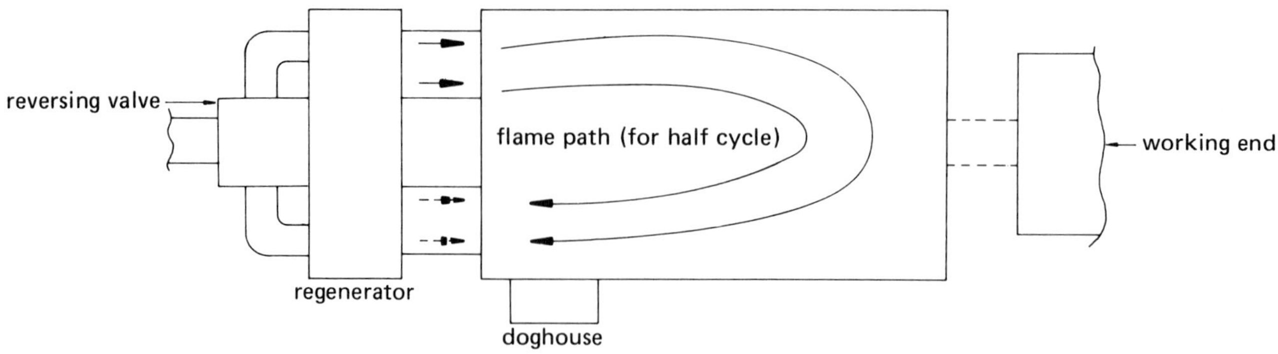

Fig. 6.90 Plan of an end-port regenerative furnace

source of heavy wear, and the portion of the back wall between the port openings is also a vulnerable area. Careful design and choice of refractories are therefore essential. The regenerators of an end-port furnace can have either vertical or horizontal flues, and can be single or double pass. They can be located behind the furnace or beneath it. The reversing valve will be interposed between the regenerators and the chimney, which may be individual to the furnace or common to several. See Figure 6.91.

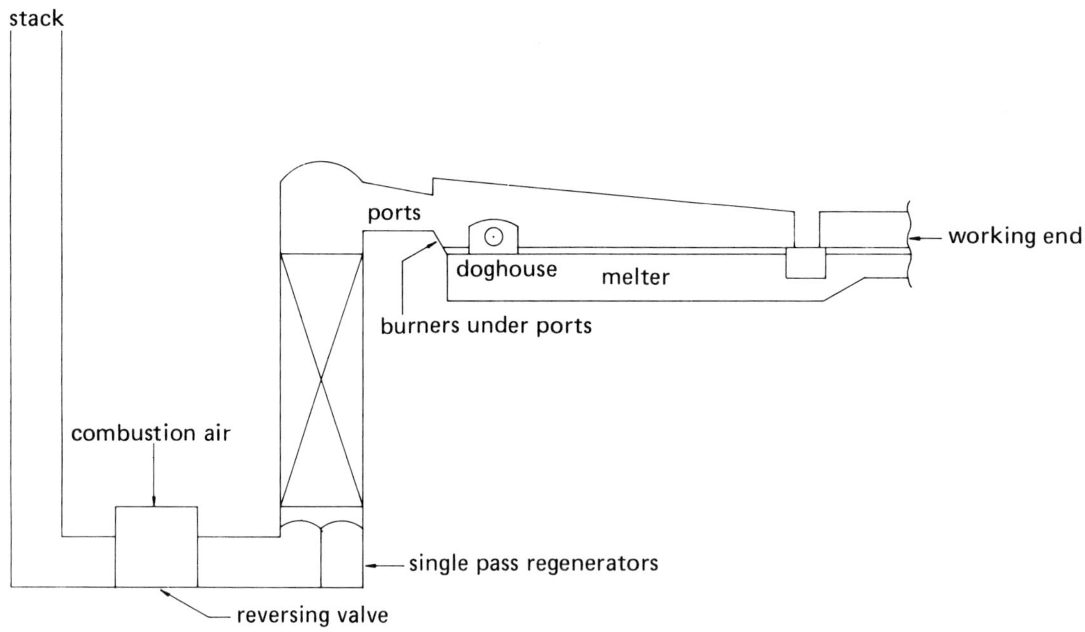

Fig. 6.91 End-port regenerative furnace with single pass regenerators

End-port furnaces normally range up to 75 m^2 in melt area with outputs up to around 200 t/day. The unobstructed side and narrowness of the layout make them particularly attractive for installations with only a few forming lines connected. Batch feeding is usually via a 'dog house' on the furnace side near the back wall. The batch should flow along the back wall, turn forward and then be held back by the reverse current, so causing it to recross the melt surface and provide a complete covering of the glass surface in the back portion of the melter chamber.

Performance data for a typical end-port regenerative furnace (single port; 3 burners per port) are as follows:

Glass temperature	1 245 °C (at distribution point)
Melting chamber area	62 m^2
Maximum thermal rating	45 GJ/h
Maximum glass production	170 t/day
Normal glass production	145 t/day
Air preheat	1 460 °C
Combustion flow	12 640 m^3/h
Specific fuel consumption	5.3 GJ/t

Cross-fired tanks

Ideally, in order to get the required temperature profile in a regenerative tank furnace, fuel input should be varied along its length. This can be achieved by means of a large number of burners distributed along the furnace side wall. A cross-fired recuperative furnace can approach this ideal. However, for larger tanks, where regenerative heat recovery is desirable, the need for symmetry on waste heat and firing sides, to facilitate reversing operation, means that exhaust ports have to be located opposite to burners.

Achieving similar flows through the regenerators in both directions is difficult. The compromise solution is to have a relatively small number of ports. Simple furnaces may have one port at each end, but it is more usual for there to be between three and eight ports as illustrated in Figure 6.92.

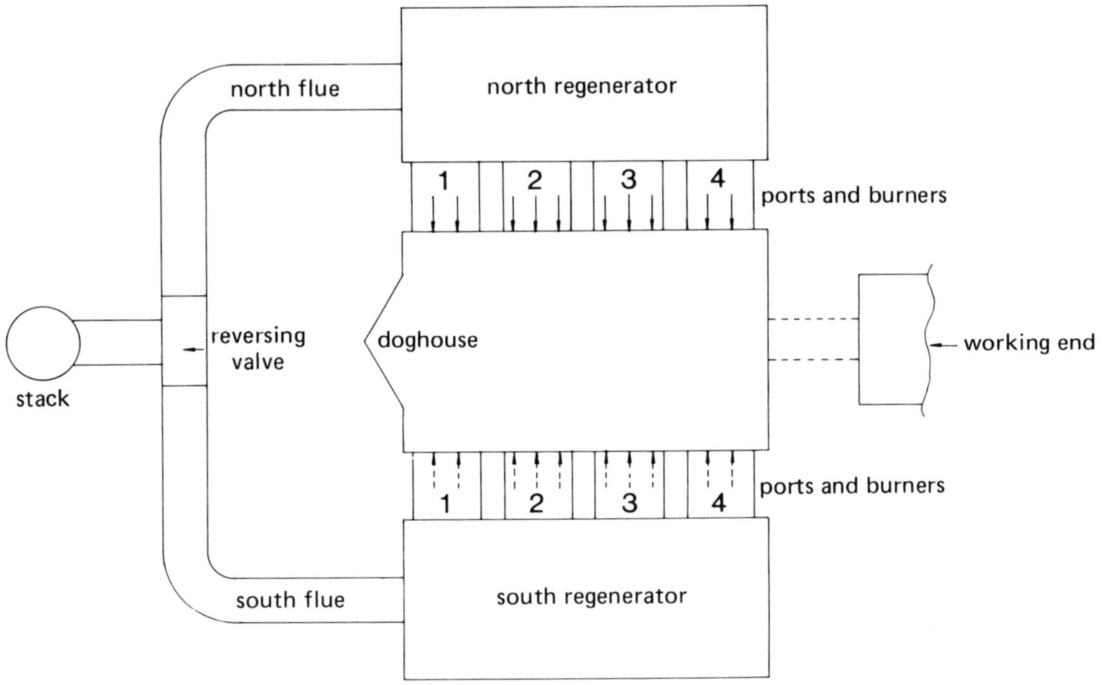

Fig. 6.92 Plan of typical four-port cross-fired furnace

The actual number of ports depends on the furnace size, the fuel used and the type of glass produced. For example, in the manufacture of float glass a cross-fired regenerative furnace with six to eight ports on each side is used as shown in Figure 6.93.

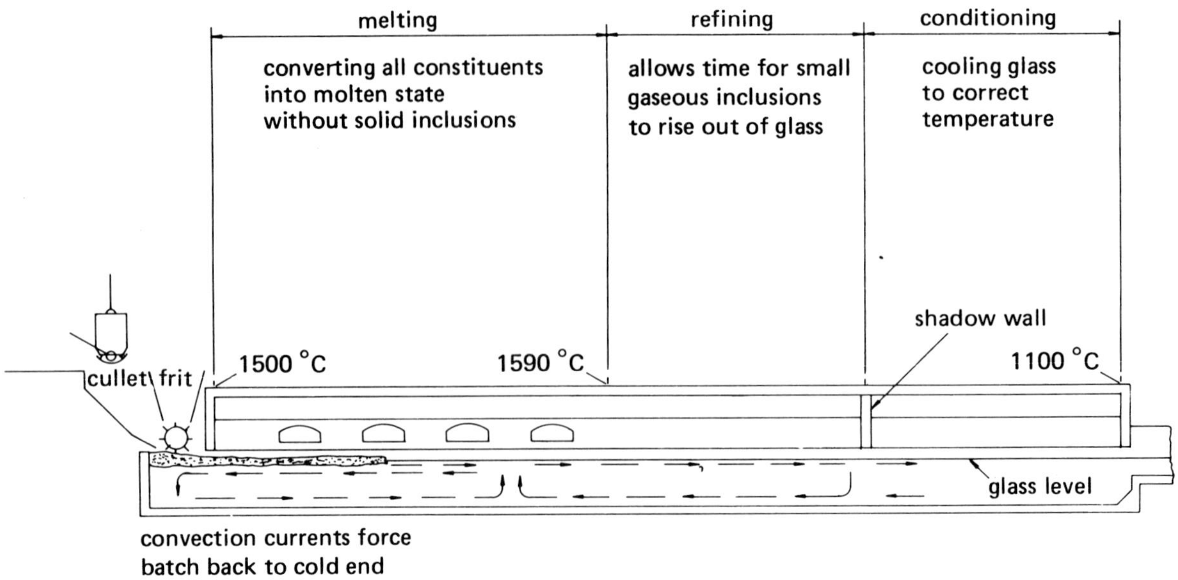

Fig. 6.93 Diagrammatic section through a float glass tank

Performance figures for cross-fired regenerative continuous tanks will vary according to the type of glass made. Typical figures for a medium sized container glass furnace having two ports per side and four burners per port would be:

Glass temperature	1 320 °C
Maximum thermal rating	32 GJ/h
Maximum glass production	3.75 t/h
Air preheat	1 150 °C
Specific fuel consumption	8.4 GJ/t

Figure 6.94 is a Sankey diagram illustrating the heat balance of such a furnace.

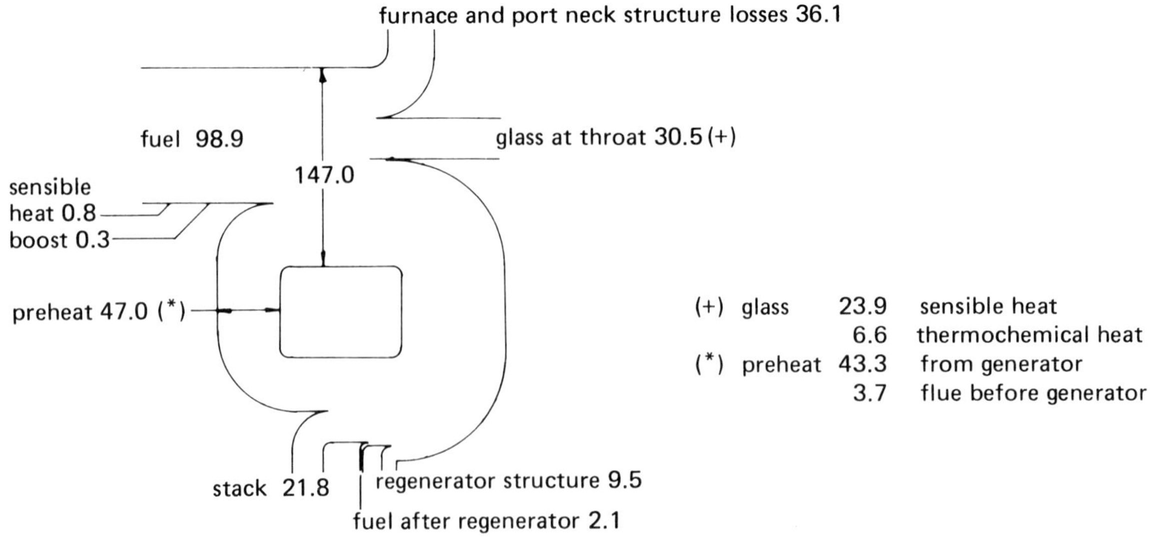

Fig. 6.94 Sankey diagram for a typical medium-sized container furnace

6.6.8 Firing systems for Regenerative Tank Furnaces

Burner configurations

The method of firing varies with furnace design and type of fuel, but the method of combustion air introduction is common. Combustion air is heated as it passes through the regenerator and then flows into the furnace through the ports in the side of the furnace superstructure. The air passes over the burner, mixing with the fuel, and the products of combustion traverse the furnace and exit through a similar port and regenerator system (see Figure 6.95).

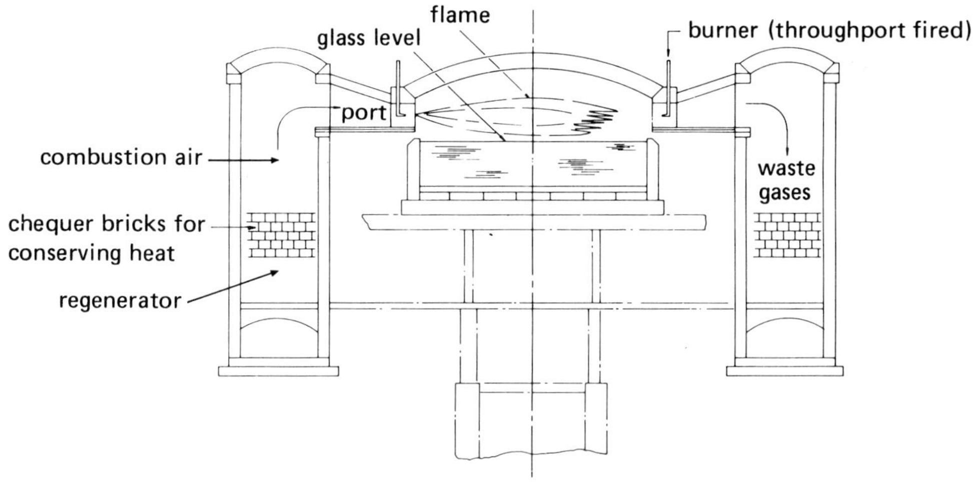

Fig. 6.95 Cross-section through tank and regenerators

The most common methods of burner installations are:

- Overport
- Underport
- Throughport
- Sideport

These methods are illustrated in Figure 6.96. The method of application is dictated by practical considerations of convenience, accessibility and combustion governing the propagation of the flame and the gas flows within the furnace.

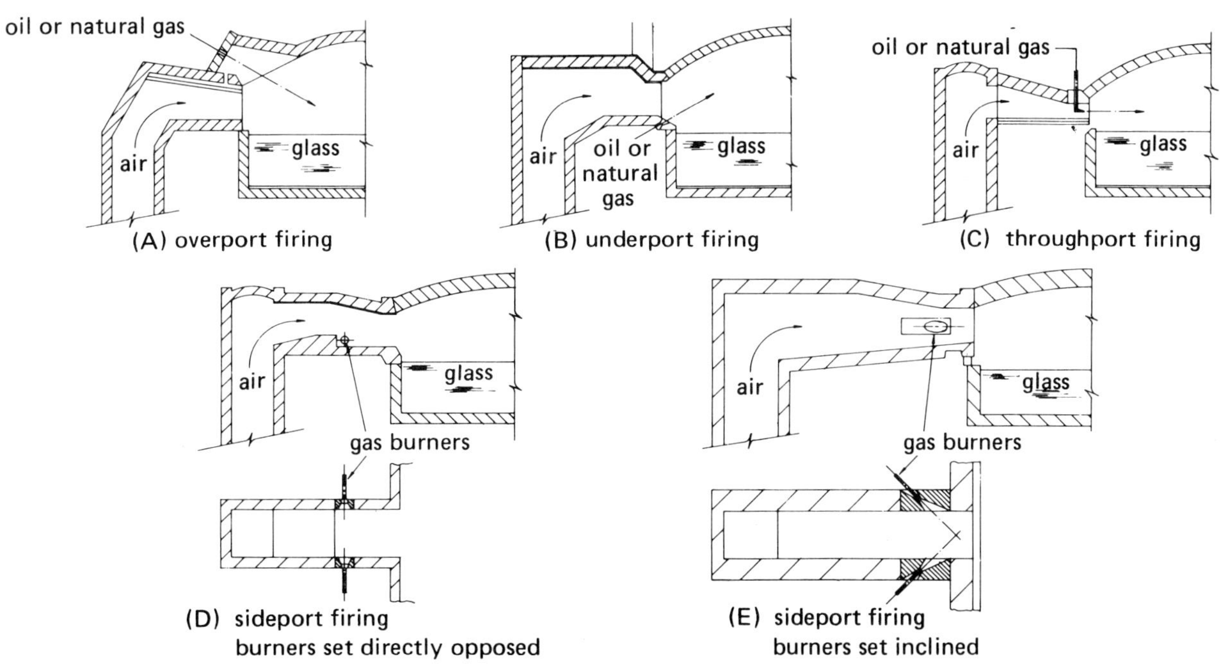

Fig. 6.96 Burner configurations for regenerative tank firing

Combustion and radiation characteristics of oil and gas flames

Oil is atomised mainly by compressed air atomisers or combined pressure oil/compressed air atomisers and when introduced into the furnace produces oil droplets of diameter from 30 to 100 μm depending upon the atomiser. Initially, the more volatile hydrocarbons are driven off from the oil as the droplets become heated and can burn rapidly with oxygen. A solid carbon skeleton is left behind which takes the form of incandescent soot particles giving the flame its luminosity. The carbon particles take considerably longer to burn than the hydrocarbons.

With the combustion of natural gas, diffusion of gas and air takes place on a molecular level and the chemical reaction time is extremely short. The rate of combustion of a natural gas flame is therefore controlled by the physical mixing of gas with air.

Depending on the method of application, natural gas flames may be either mainly luminous or non-luminous whereas in the case of oil, the flame is always luminous. In a comparison of the two fuels it is necessary to study the radiation from both luminous and non-luminous flames and how the radiation is transmitted to the glass bath in the confined space bounded by the walls and crown of the furnace, which must be also taken into consideration.

The luminous oil flame has a high emissivity and radiates strongly at all wavelengths both to the glass and to the furnace crown. Most of the radiation to the crown is re-radiated by the refractory and re-absorbed by the luminous flame, because as it emits heat at all wavelengths, the flame will similarly absorb heat at all wavelengths. Thus a much smaller proportion of the crown radiation penetrates the flame to reach the glass bath. It has been reported that in the case of oil, the energy transmitted by the flame is in the order of 60%, the rest being provided by the crown. A luminous oil flame therefore contains a larger concentration of solid carbon particles which act as grey bodies radiating over a wide spectral range.

A natural gas flame contains no soot and it radiates from combustion products within limited narrow spectral bands which lie outside the visible spectrum. The flame can be made luminous (at least in its early stages) by cracking the methane at about 1 000 °C in the absence of oxygen. The mode of heat transfer with a non-luminous gas flame is entirely different from that with a luminous flame. The furnace crown plays a major part in transferring heat to the glass bath, because this type of flame is mostly transparent to radiation. Most of the heat radiated from the flame will be absorbed by the crown and only a small amount reflected. The absorbed energy is then re-radiated over a much wider sprectral band; in other words, the crown converts the highly selective gas radiation into the normal continuous emission of a solid body. The directly reflected radiation, having the same wavelength as that originally emitted by the gas, is mainly re-absorbed by the gas. Approximately 75% of the energy radiated from the crown is capable of reaching the glass surface through the non-luminous flame.

6.6.9 Natural Gas Firing of Regenerative Furnaces

Glass melting is very energy intensive and glass manufacturers require fuel at a non-premium price. They also require the flexibility in combustion systems to allow them to take advantage of relative fuel price movements. This means that interruptible gas contracts and dual-fuel gas and heavy fuel oil facilities are the norm.

In the early 1970s many changeovers of glass melting tanks from heavy fuel oil to natural gas tried to reproduce the luminosity of oil flames. This was done by delaying mixing in an attempt to 'crack' the gas to give luminous particles. The technique proved to be unsuccessful. British Gas plc has undertaken a programme of work to improve the performance of gas-fired glass tanks. This programme includes mathematical and physical modelling, burner design and testing as well as field trials. Two important parameters identified for natural gas firing are:

- Heat release profile: this essentially depends on the flame length.
- Recirculation ratio: this relates the volume flow of combustion products recirculating in the tank, relative to the volume flow of flame which is mainly in a forward direction.

The length of the flame is determined by the thermal input and the supply pressure. A shorter flame is produced by a higher pressure due to the additional energy available for mixing. However, this higher thrust promotes recirculation of hot combustion products within the tank, which then get drawn into the root of the flame along with the combustion air. Mixing takes longer, which retards the completion of combustion and results in a longer flame. The actual flame length is, therefore, a resultant of these two opposing factors.

Conclusions from the work carried out were that a short flame is required, which has a high heat release in its early stages, and which causes little additional recirculation of the combustion products in the tank. Additional luminosity is always valuable as long as it does not result in a long flame which requires high excess air levels. Also, from a practical stand-point, the burners should be capable of operating at low pressure to match the pressure available to customers.

The simplest way of achieving short gas flames is to use high velocity jets issuing from single hole nozzles as shown in Figure 6.97. Smaller, high velocity jets produce shorter flames. Flame length can also be reduced by using several nozzles in the burner i.e. reducing the size of each nozzle. However, there is a limit to the degree of flame shortening that can be achieved in this way, since adjacent jets ultimately interfere with one another.

On an existing furnace, it is often not possible to alter the firing arrangement. Calculations show that, in order to achieve flame lengths of about half the available port to port distance, which mathematical modelling suggested would be effective, relatively high gas pressures are required of around 1.4 bar. Such pressures are not normally available to British Gas plc customers. In addition

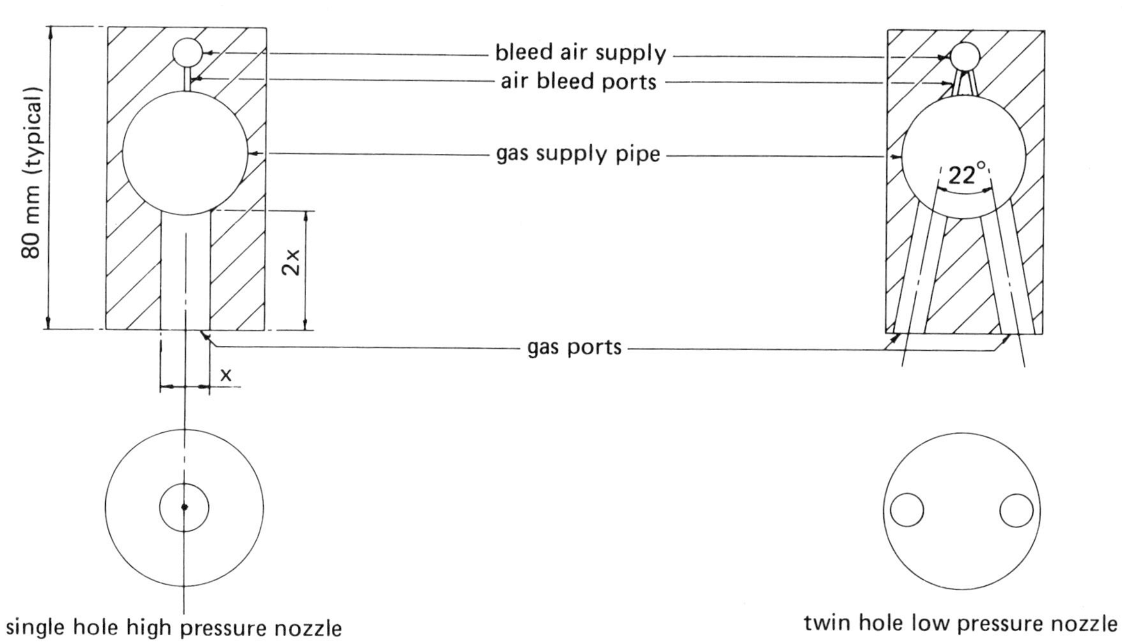

Fig. 6.97 Burner nozzles

the use of very high velocity jets brings the risk of flame impingement on the crown or of increased batch carryover. They can also create a massive entrainment of cold air through the quarls of under-port burners. An alternative nozzle design was therefore sought, which would achieve the required flame length without the need for such high gas pressures. After a programme of tests using the acid-alkali water modelling technique, a suitable gas nozzle using two holes angled at 22° was identified (Figure 6.97). Full scale tests at the Midlands Research Station Coleshill Burner Laboratory confirmed the performance of this type of nozzle and indicated that there was little benefit to be gained from increasing the number of holes beyond two. The angle of 22° is the optimum value to prevent the two jets from recombining, while avoiding mutual interference between neighbouring burners. With nozzles such as this, acceptable flame lengths may be achieved with a gas pressure of 0.2 bar.

A burner design is being developed at MRS for tanks where only low gas pressures (several mbar) are available. By using compressed air as the driving fluid, the pressure at which the gas is used can be increased. The rate of air consumption is similar to that which would be used for atomisation by an equivalently rated oil burner. A section through a typical burner is shown in Figure 6.98 together with its performance.

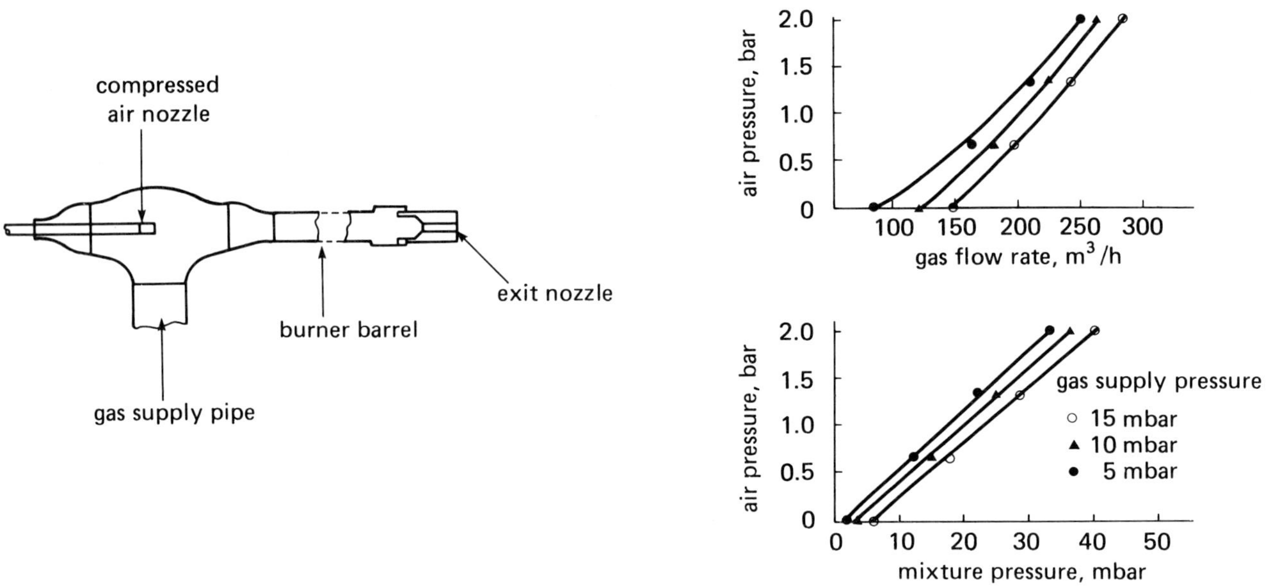

Fig. 6.98 Compressed-air-assisted burner

The above design still results in a low mixture pressure and is unlikely to be suitable for narrow cross-fired tanks. There are, however, potential applications in firing gas on end-fired tanks.

Where higher gas pressures are available there is more scope for using different burner systems. Port loadings, bath dimensions and the number of burner nozzles to be used must be considered before a suitable operating pressure can be determined. The burner system is then usually a compromise design based on the tank thermal requirements and the flame characteristics.

Throughport firing

Throughport firing of glass tanks necessitates mounting the gas nozzles in a water cooled burner body (Figure 6.99) inserted through a hole in the bottom of the ports.

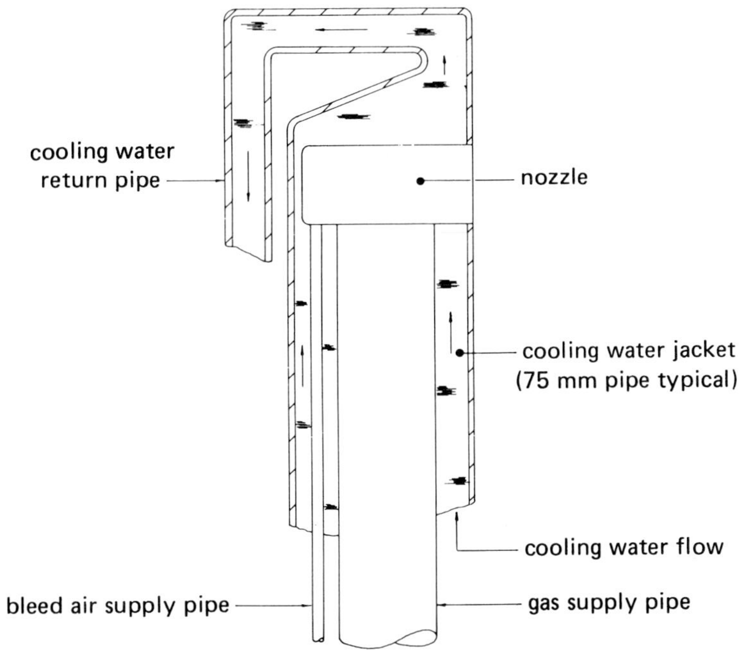

Fig. 6.99 Part section of throughport water-cooled burner

Multiple burners with twin hole nozzles at an angle of 22° minimise interference between neighbouring burners and prevent flame impingement on the glass surface or the furnace crown. The arrangement is illustrated in Figure 6.100. Using this type of burner it has been possible to fire glass tanks on gas and heavy fuel oil with similar net efficiencies.

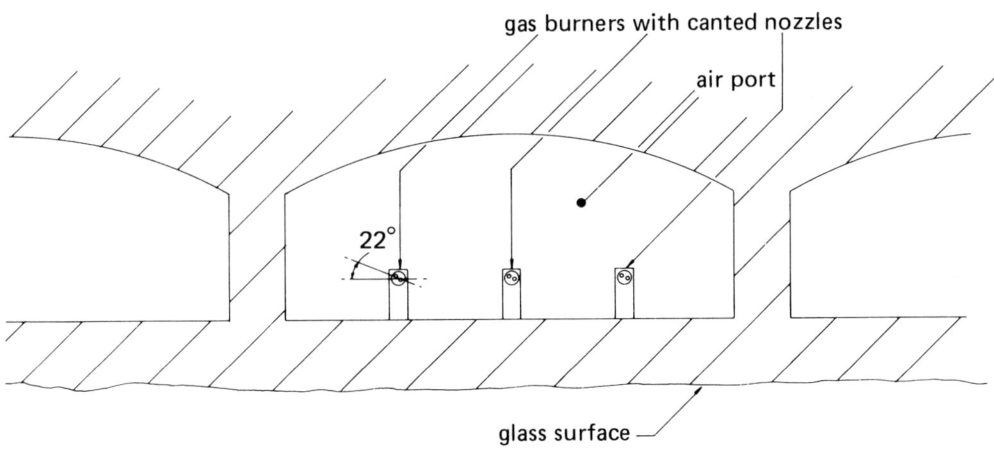

Fig. 6.100 Disposition of canted nozzles on multiple burner

The use of water-cooling with its attendant cooling loss is a disadvantage with this type of firing system. The loss, however, is not more than 1% with the burner design illustrated. An air purge jet is incorporated in the gas nozzles to keep the nozzles clear during the non-firing part of the regenerative cycle.

Underport/overport firing

Overport firing is not common as it has many disadvantages. The method of location can cause failure of the refractory brickwork because the burners are positioned where the ports are made up to the main crown. In addition, the burners are difficult to adjust and replace as personnel can be exposed to intense heat.

Similar principles to throughport firing apply to underport firing. The angle at which the burner is installed to the horizontal, however, is critical, especially when using high pressure jets. If the angle is too low, the gas remains below the bulk of the combustion air giving a slow-mixing flame and poor performance. Too high an angle causes flame impingement on the crown and the risk of crown failure.

Underport firing is suitable for both end-fired and cross-fired furnaces, for both container glass and flat glass production. Its major advantage is the ability to change over from oil to gas, or vice versa, simply by exchanging the burner lance without changing the burner block. A particular feature of underport firing is the use of relatively large burner quarls, through which the burners fire into the furnace chamber. It has been found that great quantities of cold air are induced into the furnace through these quarls, with up to 20% of the stoichiometric air requirement entering the furnace (and hence bypassing the regenerators) under some circumstances. Lower pressure burners, smaller quarls or an increase in the size of the burner body alleviate this problem, which otherwise causes a noticeable increase in fuel consumption.

Although the underport burners are mounted outside the furnace, they are nevertheless subject to considerable thermal radiation as well as to 'sting out' when not firing. A small purge of cooling air passed down the nozzle as shown in Figure 6.97 has been found to prevent the nozzles overheating and hence to reduce maintenance. The twin hole nozzle can be applied to underport or overport fired furnaces. The Körting jet type burner operates on the ejector principle i.e. the available natural gas pressure is utilised to induce atmospheric air into the burner as shown in Figure 6.101.

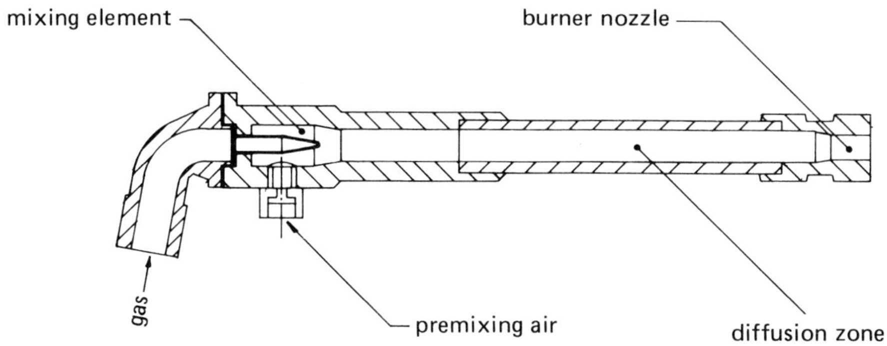

Fig. 6.101 Körting natural gas burner

The gas is introduced into the burner through the gas jet causing a negative pressure in the burner body, thus allowing atmospheric air to be drawn into the burner. The gas and air are mixed in the diffusion zone and enter the hot air flow through the furnace port. The amount of premixing air induced is between 1% and 3% of the total combustion air requirement. By regulating the premixing air, the flame length can be adjusted to 2/3 to 3/4 of the furnace chamber width/length, thereby preventing after-burning in the regenerators.

Sideport firing is a particularly attractive option for gas. However, British Gas plc involvement has normally centred around changeover of oil fired tanks fitted for underport or throughport operation with changeover occuring during the furnace campaign. It has normally been impossible to accommodate a sideport firing system in such cases, or to make the necessary air port alternations.

Körting gas burners have been used in sideport firing. The burner nozzles are retracted by approximately 800 to 1 000 mm (measured from the melting tank border) in order to attain good mixing of the gas and air in the port (see Figure 6.102).

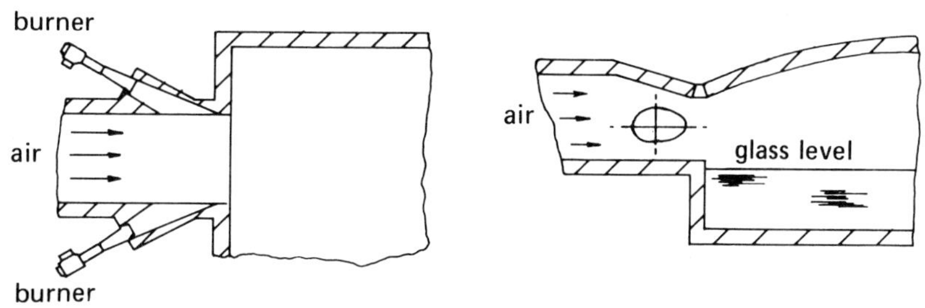

Fig. 6.102 Körting angle sideport arrangement

The angle of incidence of the nozzles is about 45° for sideport arrangement. Port width occasionally presents difficulty on changeover with angle firing, because some port designs for oil firing can be wide. This makes uniform gas distribution across the ports difficult to achieve. In these cases, it is sometimes possible to use right-angle sideport firing. An example is that used by Laidlaw Drew as illustrated in Figure 6.103. Sideport firing techniques should achieve the benefits of throughports firing whilst reducing the problem of water cooling

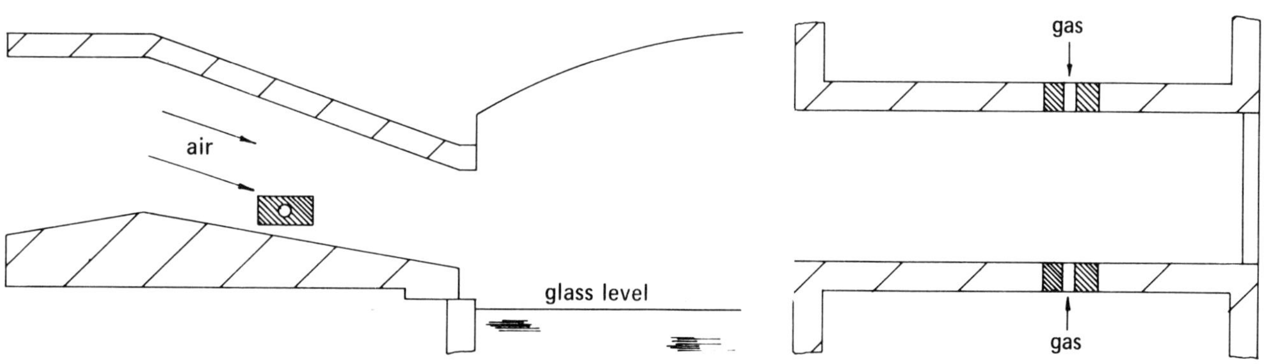

Fig. 6.103 Laidlaw Drew right-angled sideport arrangement

6.6.10 Furnace Preheating

With a newly built furnace, gas is used to heat up the furnace structure before the main burners are applied. Raising the temperature of a structure as complex as a modern glass furnace requires extreme care and ingenuity. A rigid time/temperature programme must be followed to ensure uniform expansion of the whole furnace and since the crown and casement walls are generally of silica brick, extreme care is required whilst heating through the two phase changes. Other refractories also have their critical temperature points.

Furnace preheating is almost exclusively performed with natural gas, in various ways. The system which has gained increasing acceptance during recent years consists of a rapid recirculation of hot gases round the sidewalls of the furnace to heat the structure by convection alone, and the technique of jet burner recirculation has developed. Such a system consists of a number of jet burners disposed around the furnace in a manner where they are most likely to set up a single major recirculation of gases round the side walls. Figure 6.104 shows possible arrangements of burners. Temperature readings are obtained from various thermocouples positioned in the crown, and temperatures are indicated at a central point.

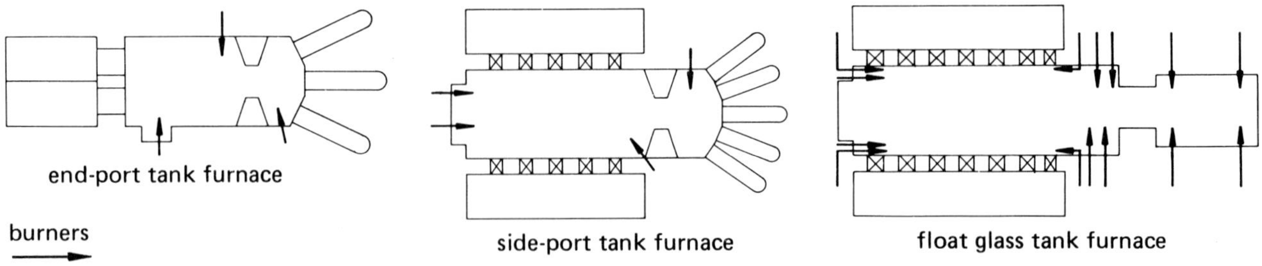

Fig. 6.104 Typical glass tank settings preheated by jet burner recirculation

The burners are designed so that the jet entrains a large proportion of waste gases and this enables both uniform temperature and high overall heat transfer to be obtained. The quantity of air and fuel can be varied over a wide range and during initial drying out the burners are set to deliver high volumes of air at low temperature. Figure 6.105 shows a typical warm-up schedule.

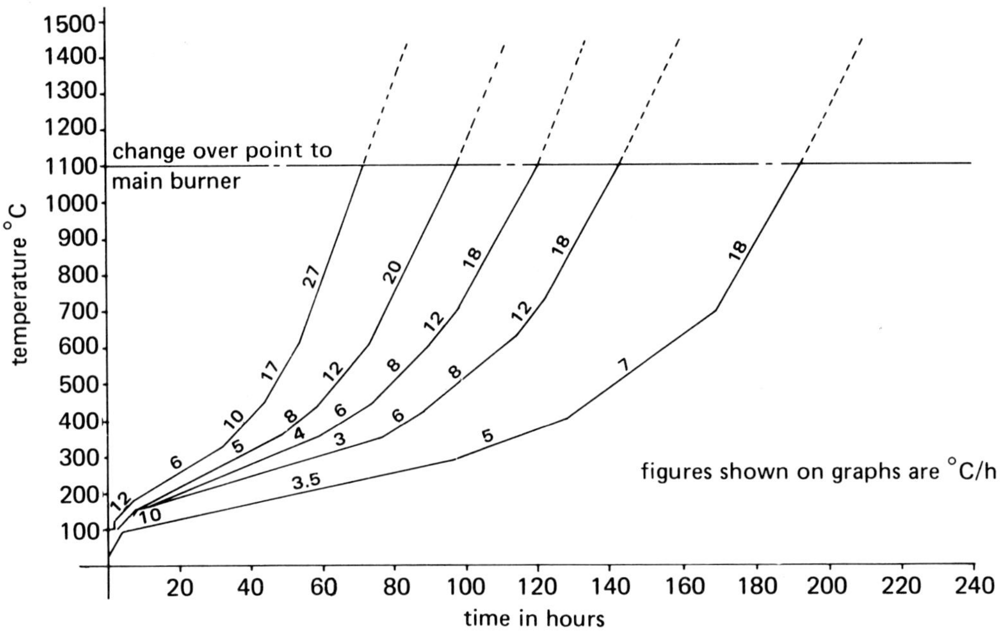

Fig. 6.105 Typical heat-up curves

6.6.11 Hot Repairs

During a furnace campaign, which can vary in duration between say three and eight years, it is necessary to make periodic repairs to the structure of the furnace. Such repairs may be made while the furnace is still hot and without completely shutting down. These are known as 'hot repairs'.

The most extensive of these repairs is the renewal of brickwork in the regenerators which generally suffer more erosion than any other furnace part. During such a repair, the tank temperature is reduced to about 800 to 850 °C, this being just high enough to prevent spalling of the refractories. The regenerators are then isolated from the rest of the furnace by either refractory screens or water cooled dampers built in the neck of the regenerator ports. Repairs will commence immediately after the regenerators are sufficiently cool. The temperature of the furnace is maintained over this period by a number of air blast burners firing through holes in the furnace sidewalls.

6.6.12 Glass Furnace Energy Saving Opportunities

Thermal efficiency of glass furnaces

Specific consumption trends for the larger tank furnaces in the container glass sector provide a good illustration of the progress that has been made and is continuing. The 8 GJ/t considered to be good practice in the 1960s has been reduced to 5.25 GJ/t in the 1980s; 4.75 GJ/t is predicted for the 1990s and should ensue from better regenerator designs and higher standards of thermal insulation.

The main factors influencing thermal efficiency are:

- The greater the production rate of glass the less fuel will be required (Figure 6.106). Large tank furnaces will be more efficient than small tank furnaces. Small tank furnaces will be more efficient than single pot furnaces.
- The lower the temperature of glass required, the less fuel will be used. Thus the amount of fuel to produce boro-silicate glass will be more than to produce the same amount of soda lime glass (all other things being equal). At a furnace temperature of 1 400 °C, the products of combustion leaving the furnace contain 70 to 80% of all the heat in the fuel. At 1 600 °C they contain 80 to 90%.
- Regenerators recover more heat than recuperators. Hence furnaces fitted with regenerators should be more efficient than those fitted with recuperators.
- The furnace at the end of its life or campaign will be less efficient than at the beginning due to deterioration and blocking of regenerators etc. The increase in energy consumption over a campaign life will be 15 to 25%.
- The design of the tank is very important. Figure 6.106 shows that end-fired tanks are more efficient than cross-fired tanks, which in turn are more efficient than unit melters. Under otherwise equal conditions, cross-fired tanks have a heat consumption about 15% higher than a similarly loaded horseshoe flame tank.
- The make-up of the batch is also important. The heat requirement for cullet is less than that required for batch. A figure of 2½% fuel saving for every 10% cullet added has been stated.

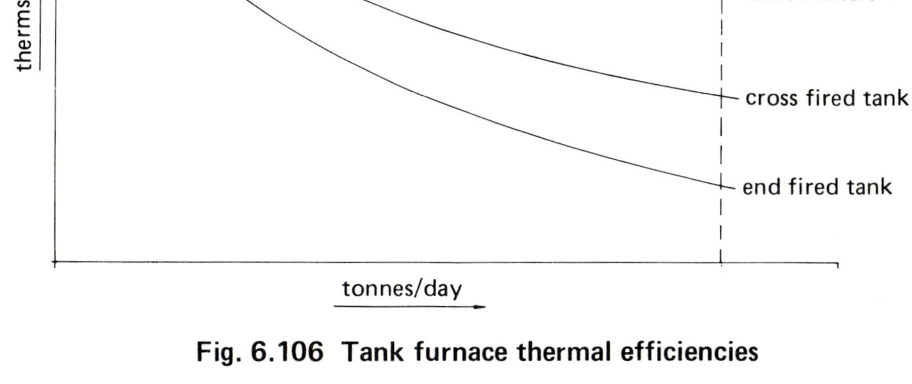

Fig. 6.106 Tank furnace thermal efficiencies

Stability in furnace control is equally, if not more, important than combustion efficiency control. The production of good quality glass depends on consistency of furnace operation. A reduced rejection rate equates to a saving in the energy required to manufacture a given saleable output.

Temperature monitoring in the glass at various points in the furnace is used to aid control of conventional furnaces. Measurement of exhaust gas content is now commonly performed to monitor and control excess air levels. Minimising excess air can produce significant savings, e.g. reducing O_2 content by 1% in the waste gases from a 200 t/day furnace melting soda-lime glass, will save approximately 1% fuel. Excess air levels should not be reduced to such a point that carbon monoxide is produced as this can lead to burn-out in the regenerators and structural damage.

Oxygen enrichment

Oxygen enrichment of combustion air is an established technique, pioneered in the steel industry, subsequently taken up by the glass industry. Originally it was used mainly to boost output, but its ability to produce fuel savings is now recognised.

The majority of combustion processes use ambient air as the source of oxygen. Air contains roughly 21% oxygen and 78% nitrogen, with the balance being small quantities of inert gases. If the proportion of oxygen is increased, the characteristics of the flame change. Whatever the level to which the oxygen in the combustion air is raised, the following effects will always be noted to an extent dependent upon the degree of oxygen addition:

- Increased flame temperature.
- Increased burning velocity.
- Reduced ignition temperature.

Combustion occurs as a result of high energy collision between molecules of fuel and oxygen. In a conventional air fuel system, this mechanism is retarded by the presence of nitrogen. The nitrogen interferes in two ways. Firstly, it conveys heat away from the reaction and secondly, it reduces the opportunity for oxygen and fuel molecules to collide. Thus, a reduction in the proportion of nitrogen present increases the kinetics of the combustion reaction and leads to the three effects listed above.

Flame temperatures are increased dramatically by increasing the proportion of oxygen in the combustion air. The higher temperatures resulting from oxygen addition create higher heat transfer rates by radiation, conduction and convection. Which of these three modes of heat transfer is affected the most depends on such factors as flame type and shape, degree of oxygen addition, furnace environment etc. However, with the heat transfer rate proportional to the fourth power of the temperature difference, radiation is likely to be dramatically affected. Heat radiated from the flame does not truly follow the Stefan-Boltzmann Law but the heat radiated from the furnace structure does.

Bearing in mind possible future regulations limiting nitrogen oxides emission, increasing flame temperatures should be approached with caution (See Section 6.6.14).

Oxygen enrichment and its application are covered in Chapters 1 and 2.

Higher cullet ratios

In practice a saving of approximately 0.2% in energy for glass melting can be achieved with a 1% increase in the cullet ratio. The main restrictions are found in the limited availability of cullet of the appropriate quality and the increasingly stringent demands for quality of ware from packers and consumers alike, with the resultant need to ensure closer control of glass homogeneity. The cost of collecting cullet, sorting, transport and processing for re-use is also a factor. Environmental, rather than technical considerations are likely to influence the amount of cullet recycled.

6.6.13 Glass Finishing Processes

Finishing processes generally require premium fuels and consume about 20 to 30% of the energy used in the glass industry. The main processes are conditioning (forehearth heating), annealing and decorating.

Forehearths/feeders/canals

After the glass has been made in the tank furnace it passes to the forehearth or canal. Forehearths take different forms according to the product, but their function is always the same, i.e. to deliver the glass in the correct condition for use. The words 'correct condition' imply mainly correct viscosity, which is a function of temperature, viscosity varying very rapidly with temperature in the normal working range. Homogeneity of temperature within the glass is also essential.

In the container and tableware industry, after conditioning in the forehearth, the glass flows to the nose cup where it is in effect pumped to the forming machine, the 'pump' mechanism being called a 'feeder'. There is often confusion between the words feeder and forehearth since in some sections of the glass industry, the forehearth itself is sometimes called a feeder. Forehearths are used also in the production of glass fibre products and in the production of wired and patterned glass in the flat glass industry. Here they are often referred to as canals.

Figure 6.107 shows a layout of forehearths for the production of tableware. Each forehearth is a self-contained production unit with a length varying according to the plant layout and production required. The length of the forehearth depends upon the space available for positioning the automatic machines and on the length of the forehearth considered necessary to gain control over the temperature of the glass as it leaves the tank furnace.

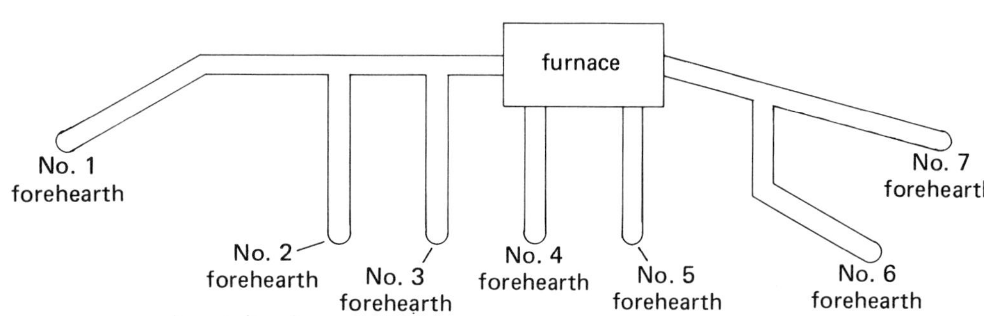

Product
No. 1 forehearth — pressware (bowls/jugs)
Nos. 2, 3 & 6 forehearth — stemware
Nos. 4, 5 & 7 forehearth — tumblers

Fig. 6.107 Typical layout of forehearths (tableware industry)

Figure 6.108 shows a cross-sectional view of a typical forehearth used in the container industry. The forehearth consists of several zones along its length, each zone being heated by a row of premixed air blast burners manifolded at close centres (150 mm) along each side wall. With the exception of the delivery or nose end of the forehearths, all burners are automatically temperature controlled by thermocouples positioned at the exit of each zone. At the top of each zone is a large opening covered with a damper tile, which may be manually adjusted.

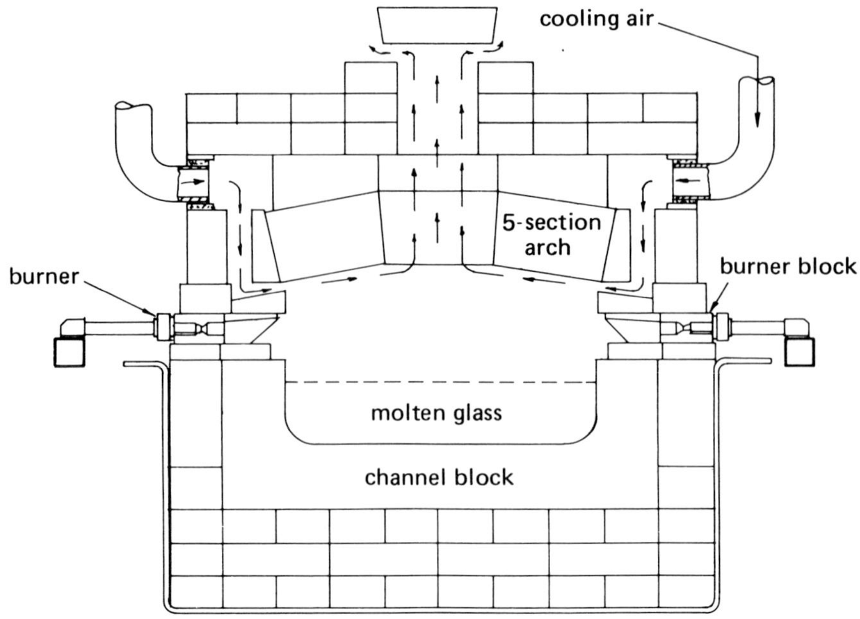

Fig. 6.108 Container industry forehearth

Controlled introduction of cooling air is required for container forehearths because they are short, glass throughput is high and gob temperature is relatively low. Forehearth dimensions are typically 1.2 m to 10 m long and 410 mm wide. Glass depth is between 150 and 250 mm. Thermal performance figures of 34 GJ/m^2 of glass surface area per week are usually acceptable. Some container forehearths can use as little as 23 GJ/m^2 per week.

Forehearths used in the tableware industry tend to be longer. Glass throughput is not normally as high as with container forehearths and gob temperatures are higher. Structural losses are therefore usually higher and cooling air is not normally required. A typical performance figure is 40 GJ/m^2 per week. Comparison with container forehearth figures should not be made because of the different manufacturing requirements.

The function of a forehearth is normally the controlled cooling of the glass from inlet to outlet. The heat loss from the structure normally exceeds the heat to be removed from the glass. In general, high throughput forehearths use a higher rate of cooling air. Higher gas input rates are therefore needed to compensate for the large amount of heat lost to the cooling air. Since the forehearth may be up to 1.2 m wide, at high glass throughput rates, heat needs to be lost from the centre but edge heat may be required to maintain temperature uniformity and glass quality. High levels of side heat are required for low throughput rates. Figure 6.109 shows how specific gas consumption varies with glass throughput rates for short and long forehearths.

Energy conservation is of secondary importance to the prime objective of achieving the correct glass temperature and viscosity.

Glass annealing

When glass has been formed, the stress induced by the forming operation must be reduced to an acceptable level by subjecting the glass to a controlled heating and cooling process, ultimately bringing the glass temperature down to ambient. It is done by conveying the product through a lehr (tunnel) in which the necessary temperature profile is established.

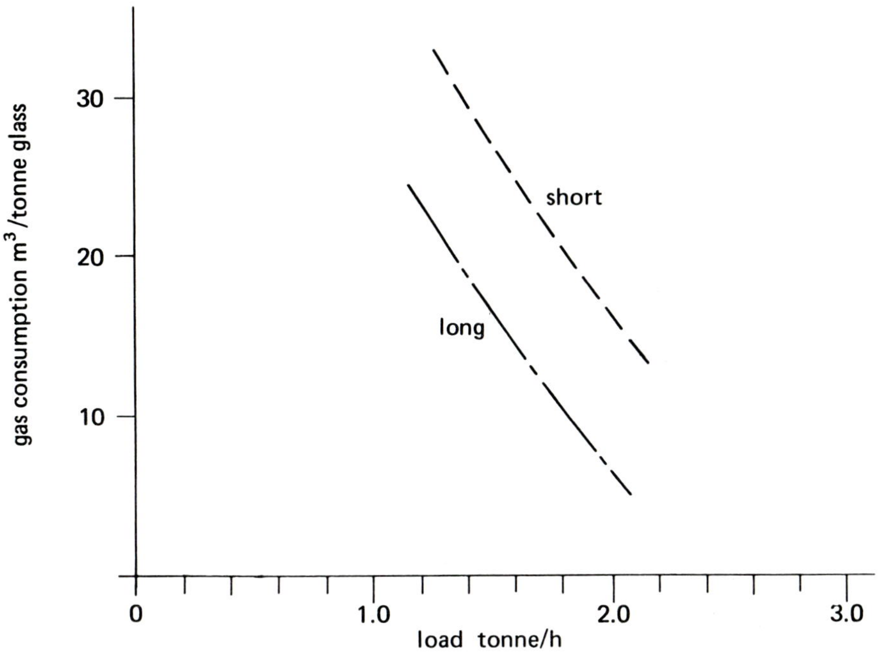

Fig. 6.109 Forehearth efficiency

The initial temperature for annealing glass will lie between 430 °C and 700 °C, depending upon the glass composition and the wall thickness of the product. The wall thickness and mass of the produce will also greatly affect the time required to achieve acceptable levels of stress. Once the glass has been cooled gradually to below 400 °C the rate of cooling may be increased quite considerably, until ambient temperature is reached.

However, the annealing and tempering (toughening) of glass is an extremely complex subject about which only a very superficial treatment can be given here, but, briefly, the process consists of four distinct stages:

- Bringing to the required temperature.
- Stabilisation at this predetermined temperature for sufficient time to permit relaxation of the stresses present.
- Slow and controlled cooling through the annealing range where the glass passes through distinct stages of equilibrium.
- Rapid cooling to ambient temperature, the cooling rate being so designed that any potential weakness with respect to thermal shock will become evident.

In some processes, for example flat glass manufacture, the first stage is omitted because the glass at that point of manufacture is already above the required annealing temperature and minimal heat is required in the lehr for the controlled cooling.

Figure 6.110 shows a typical annealing lehr for container glass with an approximate time/temperature curve.

There are two broad categories of lehr – continuous and batch. The batch type lehr tends to be used for the annealing of specialist products and is uncommon. Mass-produced articles are annealed in continuous lehrs. Lehrs are commonly used for annealing, decorating, tempering and bending in

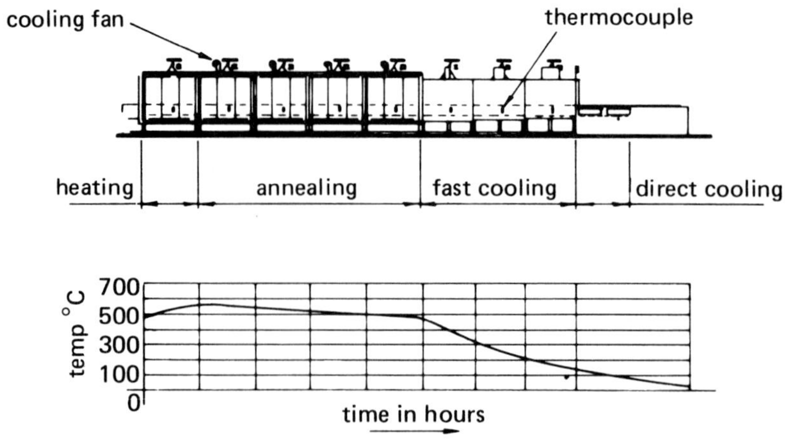

Fig. 6.110 Glass annealing

the container, flat glass and tableware industries, as well as for annealing in the TV tube and lighting industries.

The older design of lehr is indirectly fired due to the fact that the products of combustion from towns gas contained sulphur which deposited a white sodium sulphate 'bloom' on the glass. One typical design (shown in Figure 6.111) utilises radiant tube heating systems and incorporates a paddle type fan in the roof to aid convection heating.

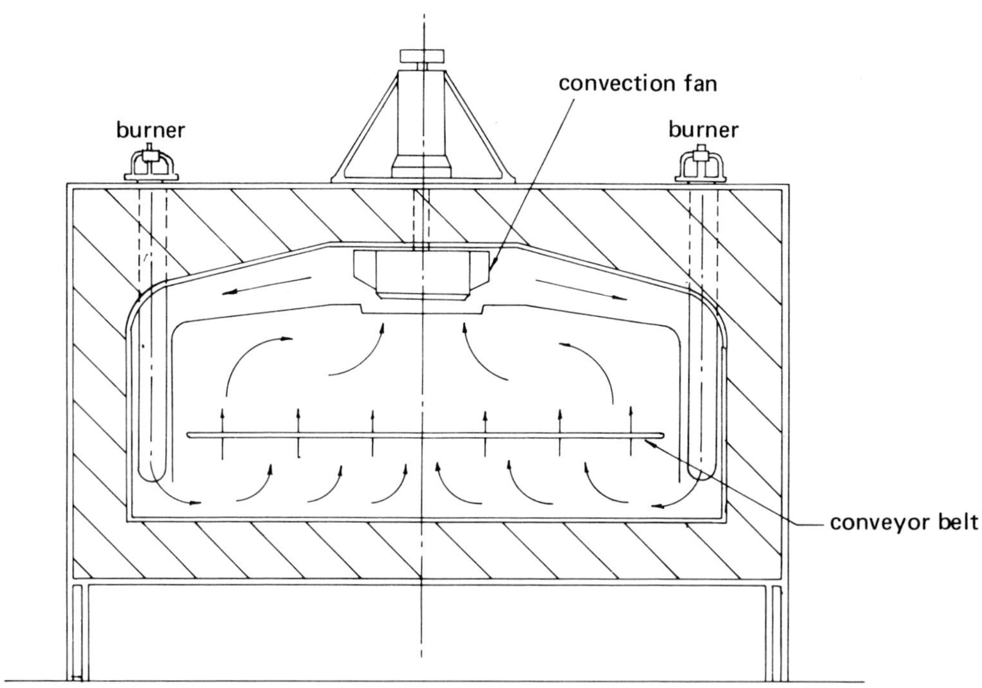

Fig. 6.111 Vertical radiant tube annealing lehr

With the introduction of natural gas and effectively zero levels of sulphur in the products of combustion, direct firing became acceptable. This is obviously more efficient on a thermal basis. Newer designs of lehr are also generally of lighter construction, more heavily insulated and utilise woven heat resisting mesh conveyor belting. A typical design is shown in Figure 6.112. Cooling is normally carried out on a convection-heated lehr by recirculating the gases in a similar way to that for the heating zones, that is the recirculated gases are passed through the ware and the temperature is controlled in zones either by means of diluting with cold air to give direct cooling, or by the use of heat transfer panels arranged in the recirculating ducts to provide indirect cooling of the ware.

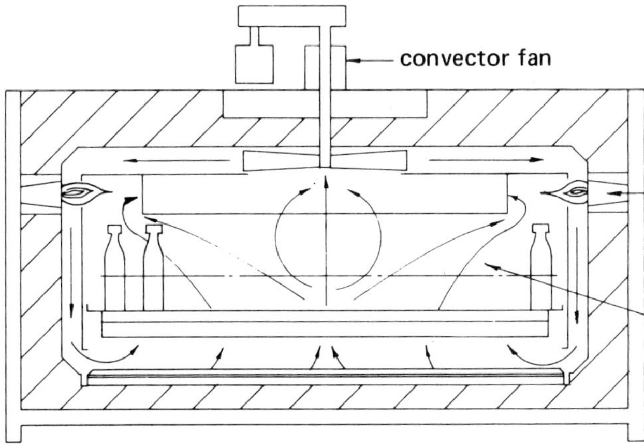

Fig. 6.112 Direct fired lehr

Flat glass annealing lehrs require special consideration because of the shape of the glass being processed. It is usually necessary to heat the edges of the ribbon in preference to the centre to avoid the large stress levels which would arise from uneven cooling across the ribbon width. Ribbon widths may be from 1 m to 3.7 m. For float glass and general window glasses the annealing temperature is typically about 550 °C.

Flat glass lehrs are between 76 m and 122 m long. The natural gas consumption on such a lehr would be 85 to 200 m^3 per hour.

Traditionally, firing is accomplished indirectly through a series of tunnels in the brickwork which run longitudinally down the lehr. Flue offtakes positioned along the length of the flues, pass to a main stack with an extractor fan providing constant conditions. Figure 6.113 shows this design. Burners are positioned in the tunnels at appropriate intervals with a greater concentration at the hot end.

In float glass manufacture, the trend is now towards multi-production short runs of varying dimensions to suit warehouse requirements, and this calls for changing conditions and faster temperature responses.

Modern lehrs for flat glass production are of lightweight, insulated design and are directly heated. Those for 'rolled plate' production may incorporate nothing more than edge burners and even be without the flue systems mentioned above.

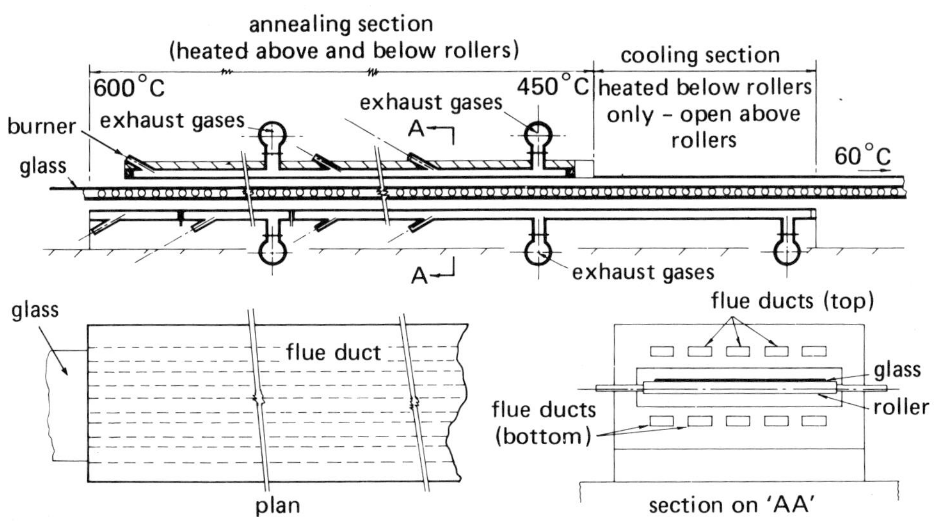

Fig. 6.113 Flat glass annealing lehr

Lehrs are also used for decorating glassware in which enamel designs are fused on to the glass surface. The designs may be transmitted on to the glass either by the application of transfers or by a silk screen process. When printing is completed, the enamel is 'fired in' at temperatures between 480 °C and 750 °C, depending upon the articles being processed and the enamel to be fired, in either a direct-fired lehr as previously described or an indirectly-fired lehr, often of the radiant tube type. Here the firing zone occupies the initial sections of the lehr and the radiant tubes may be placed either beneath the belt or in the side walls of the lehr. Air-circulating fans are attached to the top of the lehr in the firing zone to improve temperature uniformity. After the firing zone, there is a cooling zone, and then the standard anneal previously described takes place along the remainder of the lehr.

Opportunities for saving energy on lehrs

The following opportunities for saving energy should be considered:

- Glass composition and container design developments may eventually lead to lower maximum temperature requirements and shorter cycle times in the processes considered.
- In a lot of cases, to save capital expense, cooling zones have been uninsulated. Much heat is then lost to the environment from standing losses. An improvement in the lehr construction to reduce these losses, thus releasing more heat to be recovered from within the lehr for export to shop space heating, or to a preheater zone in any of the cold-loaded processes, may be viable.
- The belt specification should be carefully examined on existing lehrs when replacement is contemplated, bearing in mind that lightweight stainless steel weave types are now available. Internal belt return is only a means of reducing the heat required to reheat the belt in each cycle of its operation and a reduction in belt weight without sacrificing strength and life, thus achieving a further and permanent reduction in reheat, should be carefully considered.
- Front and back adjustable doors should be serviceable, since any uncontrolled air ingress into the lehr implies additional heating costs.
- Minimum annealing and decorating cycles, in terms of both time and temperature, should be established consistent with good performance for acceptable annealing, heat input constraints and drive power requirements.
- The best and most efficient heat recovery systems are those which recover heat in the process itself and thus reduce the lehr running cost directly. Such internal heat recovery systems, in their simplest forms, are automatic tunnel pressure controllers and charge door air curtains, to exclude cold air ingress. The most sophisticated systems for cold-loaded processes are recirculated hot gas ductwork and fan arrangements between heating and cooling sections.
- Older lehrs which were originally designed for indirect firing on towns gas can be changed to direct natural gas firing. Potential fuel savings are up to 30%.

6.6.14 Glass Product Manufacturing Processes

Some of the more common glass product manufacturing processes are described below.

Container manufacture

The processes in container manufacture are shown in Figure 6.114. Containers may be made by semi-automatic or fully automatic methods, dependent on the products. The essential difference is the means of glass delivery to the mould and the method of transferring the container from the blank mould to the finishing mould.

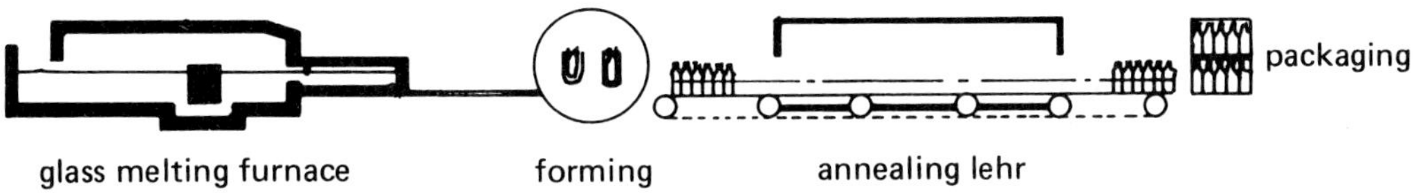

Fig. 6.114 Stages in container manufacture

In the semi-automatic process the gob is gathered manually, and is dropped into a preliminary mould. Compressed air is introduced to form the neck of the article. This initial shape (parison) is then transferred to the finishing mould in which the final shape is blown. Figure 6.115 illustrates the process.

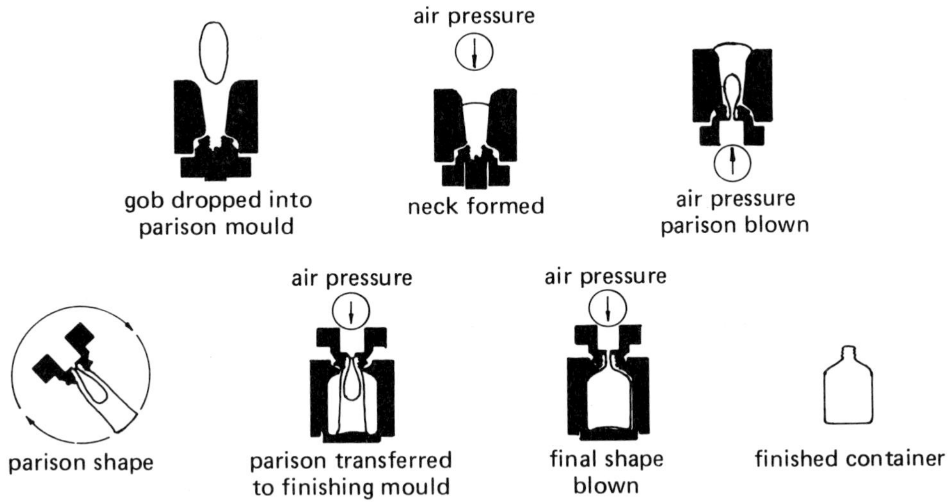

Fig. 6.115 Semi-automatic container manufacture

In the automatic process, molten glass is fed into a machine where it is automatically blown. First a parison shape is blown which is transferred to a second mould where the container is blown to its final form. Narrow-neck ware is usually made by the 'blow and blow' process. The alternative method for wide-necked containers is the 'press and blow' process. Both are depicted in Figure 6.116.

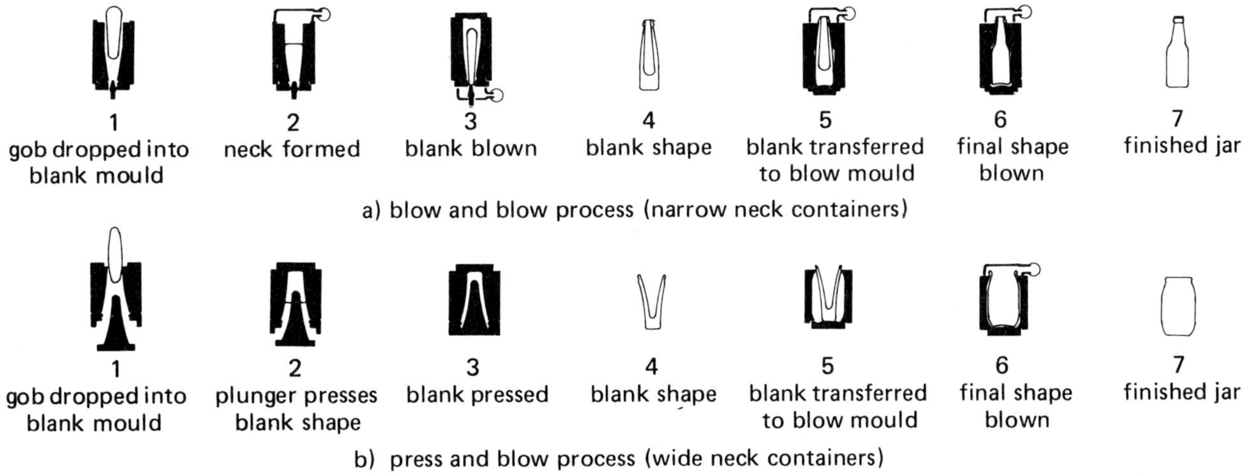

Fig. 6.116 Automatic container manufacture

Float glass manufacture

The float process was conceived as a means of combining the brilliant 'fire-finish' of *sheet* glass with the freedom from optical distortion of *plate* glass. Since its introduction by Pilkington in 1959 it has largely replaced the sheet and plate processes world-wide.

The glass obtains its lustrous finish and perfect flatness by floating on a bath of molten tin in a controlled reducing atmosphere. The glass ribbon is then cooled, as it advances, until the surfaces are hard enough to leave the batch without the annealing lehr rollers marking the surface. Float glass is a large scale process which requires melting furnaces of 1 500 to 5 000 t/week capacity. A flow diagram for a float glass line is shown in Figure 6.117.

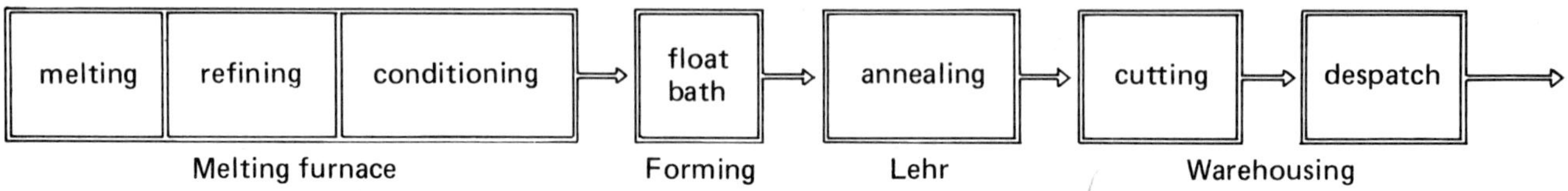

Fig. 6.117 Flow diagram of a float glass line

The glass passes through the canal and enters the float bath to form a continuous ribbon. It is at this point that the float glass process departs from the traditional methods of flat glass manufacture. Figure 6.118 shows the principle of construction and operation of the float bath.

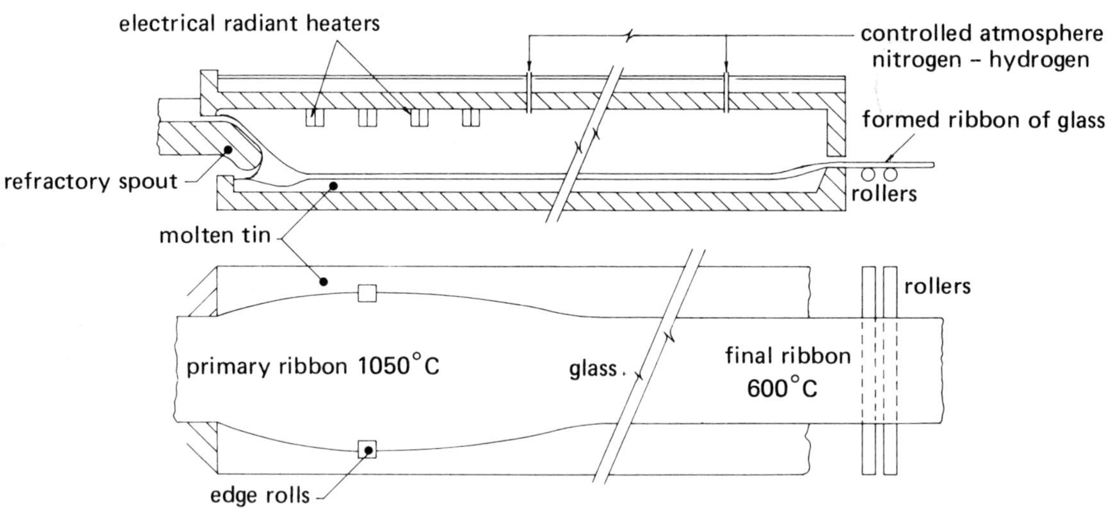

Fig. 6.118 Float bath

The float chamber consists of a shallow refractory bath in a steel casing with a suspended refractory roof. It is 52 m long by 7 m wide and the bottom is covered by several cm of molten tin. The molten glass passes over a refractory lip into the float bath and floats in a continuous ribbon on the surface of the molten tin. Above the ribbon is a reducing atmosphere of hydrogen and nitrogen (ratio about 1:10) to prevent oxidation of the tin, and the ribbon is heated from above at a strictly controlled temperature. As it passes over the first part of the bath, the ribbon is held at a temperature high enough and for a long enough period for all the irregularities to be melted out and for the surface to become flat and parallel. Because the surface of the molten tin is flat, the bottom surface of the glass will also be flat.

The ribbon of glass is then cooled down while still advancing across the molten tin until the surfaces are hard enough to be taken out of the float bath into the annealing lehr, without the lehr rollers marking the bottom surface. The glass is produced without touching a solid surface before it has hardened, resulting in a fire-polished finish needing neither grinding nor polishing.

When the glass is left to spread freely over the surface of the molten tin, it will attain its own equilibrium thickness (close to 6 mm) depending on its temperature, viscosity and surface tension. The desired substance (thickness) of the finished glass is achieved by imposing physical restraints to the ribbon and by adjusting heating, cooling and the rate of input of glass and speed of draw off.

The temperature of the bath is carefully controlled by banks of electrical radiant heaters which are suspended vertically through the roof. The electricity consumed varies considerably with the substance (glass thickness) and the load being produced.

Notwithstanding the electricity, there are numerous intermittent applications for gas. Natural gas burners are used to maintain bath temperatures during breakdowns and interruptions to production. One such application is to provide local heat at the point where the glass leaves the bath and is fed onto the rollers of the annealing lehr. Any reduction in the normal operating temperatures would result in cracking and breaking up of the glass when production was re-started. Another use of natural gas is to supplement the electrical heaters during the initial warming up operation. The gas burners used in all these applications must be set gas rich to ensure the exclusion of oxygen from the bath atmosphere.

Glass fibre manufacture

Glass fibres are formed when drops of molten glass, held in the viscosity range of 500 to 1 000 poise, are attenuated, that is made longer and thinner. At lower viscosities the fibre tension becomes too high and the fibre breaks under attentuation. The optimum temperature for fibre forming depends on the particular properties of the glass itself, such as viscosity, surface tension and infra-red emission, which are functions of composition and temperature.

Glass fibres are produced in two main forms: discontinuous filament or wool for insulation, and continuous filament for reinforcement and textiles. Insulation fibre is manufactured by a rotary process in which the fibres are attenuated by the actions of centrifugal force and hot gases, sprayed with binder, and formed into a mat which is shaped and cured in an oven. Continuous filament is formed by mechanical attenuation from orifices in the base of a platinum bush; the filaments are gathered into a strand, coated with an organic size, and wound on to spools for further processing into finished products – mats, etc.

Figure 6.119 illustrates the manufacture of insulation glass fibre.

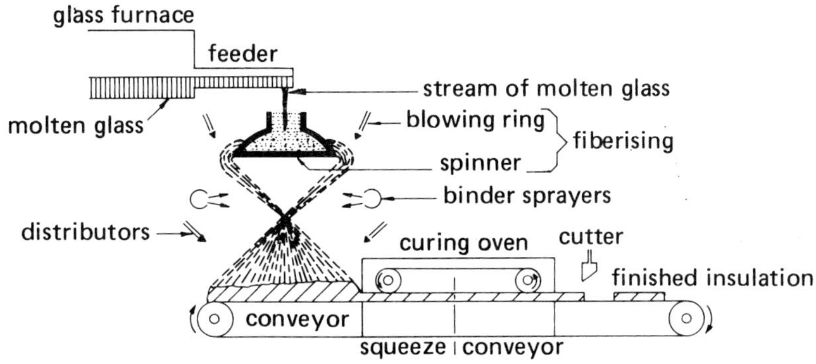

Fig. 6.119 Production of insulation glass fibre

Conditioned glass, usually a soda-borosilicate, from the forehearth is fed into a rotating cylinder which is rotating horizontally at about 3 000 rev/min. The cylindrical sides of the inner part of the spinner contain a multiplicity of holes through which the glass passes and is projected against the inner wall of the outer part of the spinner, where the glass forms a thin layer. A band 30 to 50 mm wide in the outer part of the spinner, and about 200 mm in diameter is provided with about 4 000 holes through which the glass is projected horizontally as a result of centrifugal force. Immediately outside the rotating spinner and surrounding it in an annular configuration is a series of high temperature gas burners which deflect the spun glass vertically downwards and, at the same time, attenuate the fibres.

Fibres formed in this fashion have a diameter of 6 to 8 μm. The spinner is constructed from high-temperature alloys and has to be carefully manufactured and dynamically balanced. It manufactures up to 5 tonne per day of fibre.

Immediately below the spinner the moving fibres are sprayed with a suitable binder, usually a phenolic emulsion, and are then deposited on a moving conveyor belt under suction. From there the mat of fibres moves into a tunnel oven for drying and curing at about 115 °C. The resulting insulation mat can be used as such or forms the raw material for other insulation products, e.g. preformed sections of pipe insulation.

Compared with other types of glass production, differences can be seen. For example, the melting and conditioning process uses a much lower proportion of the total energy than in the other sections of the industry. Normally melting accounts for between 60 and 80% of the total whereas in the case of insulation glass fibre the proportion is approximately 35%. The other area in glass-fibre production which uses a large amount of energy, fiberising, accounts for about 41%.

Figure 6.120 illustrates the principles of reinforcement fibre manufacture. The manufacture of continuous filament glass fibre requires a supply of glass of very high quality and at a uniform and suitable viscosity which, in practice, means temperature. The most common glass composition used is known as E glass an alkali-free calcium-magnesium-alumino-borosilicate glass.

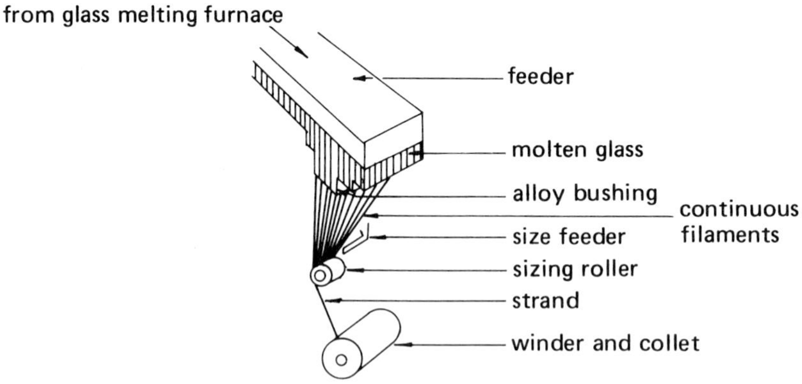

Fig. 6.120 Production of reinforcement glass fibre

The glass flows from an extensive forehearth system and passes via slots in the floor refractories into the fibre-forming furnaces or units, usually called bushings. The bushings are made of platinum with several hundred orifices in the base through which individual filaments are drawn. The drawing temperature must be precisely controlled and the main forehearth heating may be supplemented by electrical heat input to the bushing itself to improve temperature control at this point. Fibres, formed by drawing together one hundred or so individual filaments, are coated with an organic size to protect the surface and are wound onto reels. Full reels are dried in continuous tunnel ovens at a temperature of about 115 °C.

To produce chopped strand mat, the continuous fibre is cut into short lengths which are coated with a bonding agent, pressed into a thin mat and reheated to cure the organic material.

Manufacture of lead crystal glassware

Lead crystal glassware is the product of the more traditional sector of the glass industry. Its high refractive index gives a characteristic sparkle and its comparative softness facilitates cutting or engraving for decorative purposes. In contrast with most other sections of the industry, the forming and decorating processes are performed mainly by hand but even so they contribute relatively little to energy consumption, because of the large proportion of total energy used in melting (about 80%).

Melting is mostly accomplished in refractory pots which are heated in single or multi-pot furnaces.

The method of forming is to make a 'gather' of glass on a long tubular piece of metal (blowing iron) through the middle of which the operator can blow, causing a shape to be formed. A mould is used in some cases to determine the final shape. In the manufacture of wine glasses, stems and feet have to be attached and shaped at a temperature which allows two pieces of glass to be welded together. The same applies for jugs with handles, and other similar objects.

As in other sections of the industry the glass then has to be annealed. The articles are commonly blown from what is to be the open end of the article which means that this edge will require to be 'finished'. The excess glass is removed by the application of a hard metal (tungsten carbide) point and a gas jet (for heavier pieces a diamond saw is often used). The resulting edge is ground and polished, or it can be 'fire-finished', a process which melts the edge so that it flows into a smooth safe surface. The articles to be decorated by cutting are then marked out in such a way that a guide is given to the cutter. Then the glass is cut in one of three ways, the heavy cutting being done over a carborundum or diamond-impregnated wheel, the lighter cutting under a smaller finer wheel and engraving done on a series of much smaller copper wheels.

After cutting, the glass is acid polished in a heated acid bath. This brings the surface to the highly reflective appearance which is a feature of this type of glass. Figure 6.121 shows twelve steps which may be involved in hand making a crystal glass article.

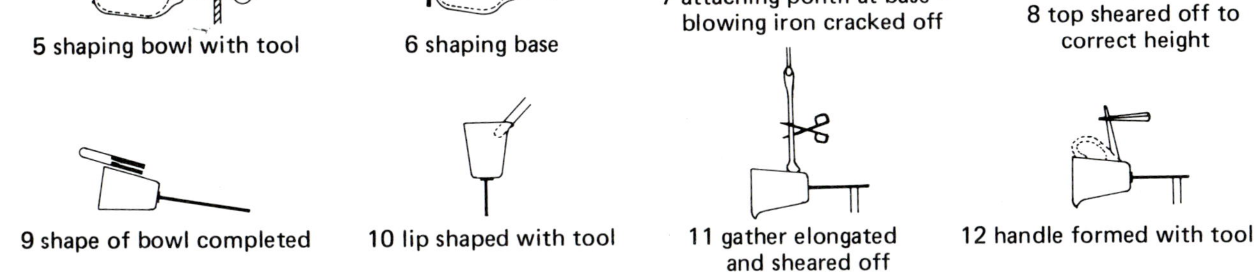

Fig. 6.121 Manufacture of lead crystal

Electric lamp manufacture

Soda-lime, lead and hard glasses are among the different types of glass used for lamp making and in recent years quartz has also become more popular. Figure 6.122 shows the manufacture of electric lamps. A ribbon of glass flows continuously from the forehearth of the furnace between water cooled rollers. The ribbon so formed is carried forward on orifice plates. Blowheads on a conveyor system blow the glass through the holes in the plates and moulds from below the glass meet and close around these 'blisters'. The moulds fall away revealing the formed lamps which are cooled by air jets and lopped off the ribbon. They fall into scoops on a rotary turntable which tip them on to the conveyor belt to the annealing lehr. Up to 2 000 per minute can be produced on such a machine.

This glass envelope is then supplied with an annealed glass stem containing tungsten filament wires. Following this mounting the lamp is sealed, using gas/air/oxygen burners and is then evacuated prior to filling with an inert gas.

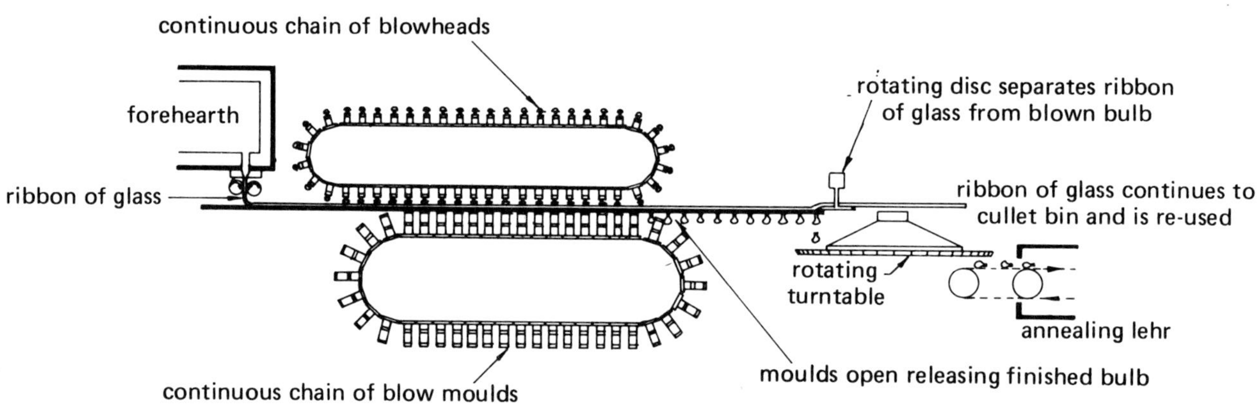

Fig. 6.122 Electric lamp manufacture

Flame working of glass

The use of working flames is largely confined to the domestic, industrial and scientific sectors of the glass industry. The widely varying applications of working flames have the same basic requirement i.e. a clean, controllable, stable flame. The choice, therefore, can be natural gas, LPG or hydrogen. Natural gas is normally preferred because of its cleanliness and controllability. Hydrogen is used for certain applications because of its combustion characteristics i.e. it burns with a hotter flame giving a high level of heat intensity when the flame impinges on the glass.

Working flame burners can be fixtures on machinery, lathes etc., or can be hand held torches. Air/gas burners are generally used where high temperatures or high production rates are required. Oxygen/gas burners can be used for the latter and also for cutting and sealing. Working flame burners are treated in more detail in Chapter 2.

6.6.15 Air Pollution Control

The 1987 UK and EEC proposals for air pollution control legislation are of great concern to the glass industry. High furnace temperature requirements, dependence on fossil fuels and the use of sulphates in the raw materials mean that the three principal polluting emissions are sulphur oxides, sulphate particulates and nitrogen oxides. The former two result from fuel combustion and raw materials volatilization; nitrogen oxides are mainly formed from the oxidation of atmospheric nitrogen in the high temperature flames.

Emissions from glass furnaces can be controlled either internally or externally.

In-furnace control of sulphur oxides (SO_X)

To achieve a reduction in SO_X emissions would need a reduction in the sulphur content of the fuel, a decrease in the raw materials sulphate ratio and an increase in the proportion of waste glass or cullet used.

Sulphate ratios have been reduced considerably in recent years, for economic reasons, and most furnaces already operate at optimum sulphate levels consistent with the satisfactory removal of bubbles and imperfections in the glass. Compliance with proposed SO_X limits can be achieved with natural gas firing but not with heavy fuel oil firing. However, even with gas firing, the industry is of the opinion that particulate emissions (mainly sulphates) could not be met. The consensus is that in-furnace control of SO_X and related particulate emissions is insufficient to bring glass furnaces into compliance with existing and proposed European legislation.

External control of SO_X

SO_X and particulate emissions can be controlled by add-on facilities i.e. flue gas desulphurisation (FGD). Capital and operating costs of such systems are very high and would lead to a significant increase in glass production costs.

In-furnace control of nitrogen oxides (NO_X)

Nitrogen oxides are generated in glass furnaces by the direct oxidation of atmospheric nitrogen in high temperature flames; the nitrogen content of the fuel makes little contribution.

Figure 6.123 is a graph of the rate of formation of nitrogen oxides from its elements against temperature. Large glass making furnaces usually operate at a temperature in the region of the curve with the steepest gradient.

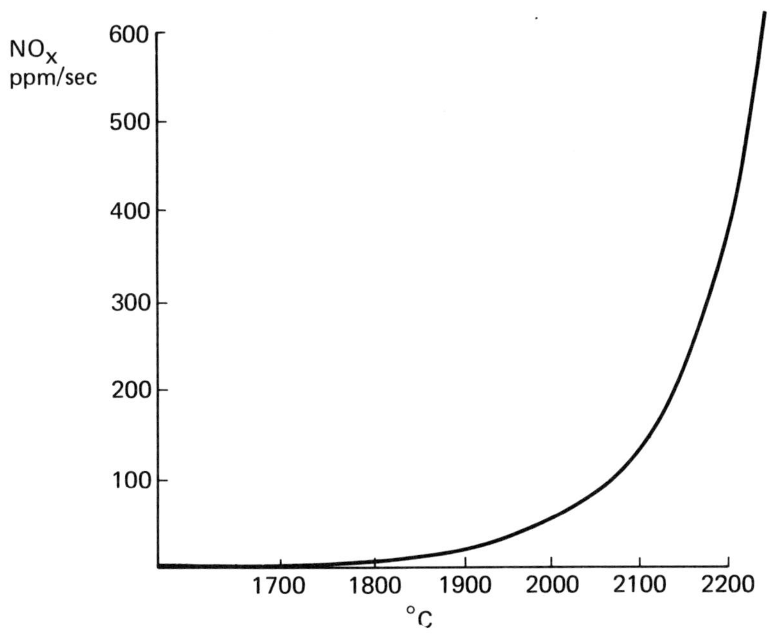

Fig. 6.123 Rate of formation of NO_X from elements

There are three ways in which the rate of formation of NO_x can be reduced:

- Reduce flame temperature
- Reduce excess air
- Reduce gas residence time.

Reducing the flame temperature reduces the heat transfer from the flame to the glass and to the furnace crown (and re-radiation from the crown) according to the fourth power law. This would adversely affect melting efficiency and product quality. Gas fired furnaces normally require a higher flame temperature compared with oil firing to achieve the requisite total heat transfer from the flame and furnace crown. Whilst the increased heat losses in the flue gases can be compensated for by higher air preheat temperatures, increased NO_x formation can not. A change from oil to gas firing is advantageous for SO_x and particulate control, but not for NO_x control.

Minimising excess air levels has a beneficial effect on both furnace efficiency and control of NO_x emissions. It is doubtful, however, that further improvement in air/gas ratio control would enable the proposed NO_x limits for large combustion plant to be achieved.

Reducing the time in which nitrogen and oxygen are in contact will reduce the amount of NO_x produced. Burner, port and furnace design to this end will help. It has been suggested that phased combustion along the length of the flame may be possible on end-fired furnaces. It is unlikely that flame length on cross-fired furnaces is long enough to achieve this.

External control of NO_x

Two flue gas denitrification processes are currently being studied, neither of which is technically or economically proven yet. They involve the reduction of nitrogen oxides to nitrogen and water; one by means of ammonia injection in the presence of a catalyst; the other by ammonia or an ammonia/hydrogen mixture injection without a catalyst. Gas clean-up requirement and reaction temperatures needed are the main stumbling blocks.

6.7 CERAMICS AND CLAY

6.7.1 Introduction

In a relatively short period of time, natural gas has ousted coal and oil to become the predominant fuel in the pottery, heavy clay and refractory products industries.

The term 'ceramic' applies to a material formed from clay or a similar substance in the plastic state which is then dried and fired at a temperature high enough to provide the necessary strength and durability. Ceramic products can be broadly divided into:

- Pottery, e.g. tableware, sanitary ware, wall tiles etc.
- Heavy clay products, e.g. bricks, drainpipes, refractories etc.

With the exception of some special ceramics and refractories, the basic raw material of ceramic products is clay, which is used in a variety of forms both naturally occurring and artificially blended. The major constituents are kaolin, feldspar, silica, lime, magnesia and iron.

The structure of clay is variable and complex. The relative proportions of one or more constituents will largely determine the final properties of a product, and also the measures needed to achieve the desired result. Mixtures of clays are used in many cases. The general process by which a clay body is prepared can be summarised as:

- Size reduction, with calcination when required.
- Separation and removal of contaminants.
- Blending.
- Removal of excess water.
- Shaping.

With the exception of calcination, these are mechanical processes to give a uniform body which is mechanically suitable for further shaping, if required.

The subsequent processes of drying and firing are discussed under product headings below.

6.7.2 Pottery Industry

For products such as earthenware tableware, wall tiles and electrical ware, other materials such as various forms of silica and fluxes are added to the clays. Silica may be regarded as a cheap white filler with a suitably high thermal expansion, and a flux is a material used to promote vitrification or liquid-phase sintering at firing temperatures of about 1 100 °C to 1 250 °C. Bone china contains clay and flux but is unusual in that it also contains about 50 per cent of animal bone, which functions principally as a filler.

Cornish stone, feldspar and nepheline syenite are the main fluxes, and like silica and bone they must be crushed and ground very finely before they are mixed with the clay slurry to form a suspension known as slip. Before crushing and grinding, the bone and silica (as flint) must be calcined.

After as many impurities as possible have been removed from the slip by sieving and magnetic separation, the water content is reduced by filter pressing to make a plastic body. The filter-press cakes so produced are processed in a pug-mill to remove air and to consolidate the body, which is then extruded through a nozzle as a continuous column.

From this, many articles are made by jiggering or jolleying on, or in rotating plaster-of-Paris moulds; others, such as hollow articles of complex shape, are made by slip casting in moulds. Once made, both the clayware articles and the moulds are dried. It is normal practice in the industry to recover any unfired scrap material and after reprocessing to incorporate it in the slip. Tiles are usually pressed from dusts of suitable moisture content prepared by spray drying the slip, although some floor-tile dusts are still made from filter-pressed or drum-dried slips. Again, after making, the tiles are dried before firing.

In the manufacture of pottery the ware may undergo three firing processes. Tableware and wall tiles, for example, usually receive a biscuit firing at a temperature of about 1 150 °C to 1 250 °C to convert the clay-ware into hard, strong, sometimes non-porous and, in the case of bone china and porcelain tableware, translucent articles. Subsequently a suspension of powdered glaze in water is applied. Drying may be necessary, before glost firing at a temperature of about 1 050 °C to 1 100 °C. The main purpose of a glaze is to provide a smooth, impervious and often glossy finish to the ware. Fired scrap is not usually recovered, but, in the case of wall tiles, both biscuit and glost scrap may be ground and incorporated in the body.

Decoration may be applied either under or on the glaze. If the decoration is under-glaze, it is applied to the biscuit ware, and a further firing (called 'hardening-on') may be necessary before glazing with a transparent glaze. On-glaze decoration is applied in many ways; in the case of the more expensive tableware, several methods of application may be used on a single article, with a separate firing after each application. On-glaze decorated ware is fired to about 800 °C to 'weld' the decoration securely to the underlying glaze.

Products such as electrical ware and sanitaryware are glazed in the clayware state and fired once to about 1 200 °C.

Most of the pottery produced in the UK is fired in an oxidising atmosphere, although the production of certain colours used for decoration does demand fairly carefully controlled reducing atmospheres.

Drying

Drying principles and types of driers (intermittent and continuous) are covered in Section 6.5 of this chapter. Convection drying is almost always used.

In small scale production, it is customary to find batch or intermittent operation driers in use. By various means the drying gas is brought into as uniform as possible contact with the stationary ware. The heat required for drying is abstracted from the gas, during its passage over the ware, and the gases leaving the driers are usually recirculated through a heater to restore them to their original temperature. Naturally water absorbed into the gas is also recirculated, and by adjusting the relative intake of fresh air, a degree of humidity control can be achieved in the drier. Recirculation not only increases the thermal efficiency of the drier, but also makes a considerable contribution to the uniformity of drying. On the other hand, recirculation produces a smaller wet-bulb temperature depression, and consequently prolongs drying time.

In a tunnel drier, intended for continuous production of identical (or similar) shapes, the ware is stacked on multi-deck trolley frames. Figure 6.124 shows both counterflow and parallel flow of ware and air, in different parts of the drying sequence.

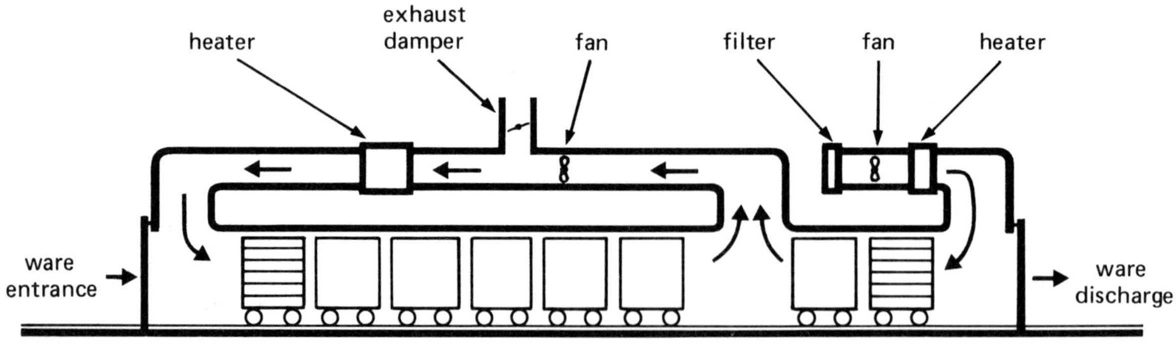

Fig. 6.124 Tunnel drier with recirculation

Drying temperatures are comparatively low, therefore low grade heat from the firing process or other process can be utilised. Air from the cooling zone of associated tunnel kilns can often serve the heat requirements of the drier, with or without supplementary direct gas firing. Where it is not possible to match the availability of heat from another process with drier demand (e.g. on some intermittent drying/firing process), direct gas fired heating is frequently used. The use of steam heat exchangers for drier air heating is becoming less frequent.

Firing

Gas fired tunnel and intermittent kilns are the two main types used in the pottery industry. Either type can be directly or indirectly fired.

The main developments in firing methods for pottery since the 1950s have been:

- A trend towards once-firing certain types of glazed ware, e.g. vitreous sanitary ware.
- A changeover in the type of fuel used. Coal in intermittent kilns has given way to natural gas, LPG, electricity and oil.
- The replacement of saggar and muffle-furnace firing by open-firing methods using gas and yielding high efficiencies.
- The introduction of new kiln types, e.g. modern car tunnel kilns, multipassage kilns and continuous kilns of small cross-section. Also the development of modern intermittent kilns fired by gas.
- The attainment of faster firing cycles, e.g. by firing a stream of single articles.

Intermittent kilns

The pottery industry adopted the car-bogie tunnel kiln in large quantities in the 1950s as an efficient replacement for the old solid-fuel-fired intermittent bottle kilns. Large tunnel kilns must, however, be run with nearly maximum payloads to ensure efficient operation. Market conditions and/or small production rates may make this impossible and so the flexibility of intermittent kilns has led to their reintroduction in a convenient and efficient form.

Gas-fired intermittent kilns may be of the box, bogie-hearth or portable-cover-types. Figure 6.125 shows a typical portable-cover intermittent kiln fired by vertical tunnel burners along the side walls and flueing centrally through the perforated hearth. Portable-cover furnaces possess two main advantages: the ware is not disturbed after placing and one cover may be used for two bases thereby shortening heating and cooling times and making a small contribution to fuel efficiency.

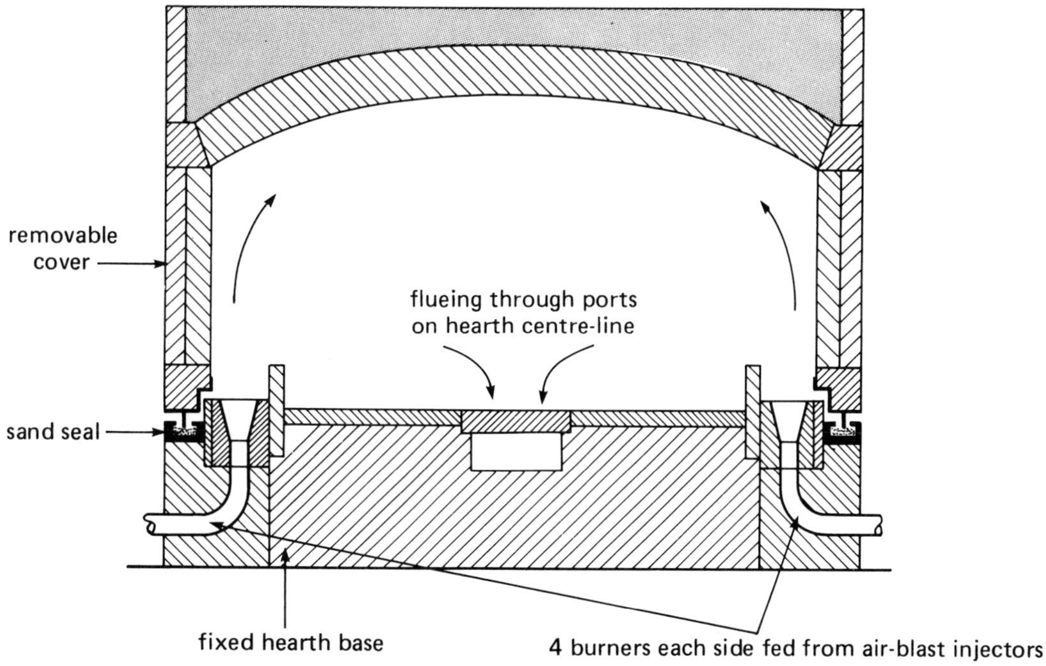

Fig. 6.125 Portable-cover pottery kiln

To obtain consistent ware quality it is vital that each piece of ware has a similar thermal history. Thus an essential feature of any satisfactory firing system is even temperature distribution. Forced recirculation using high-velocity burners is particularly useful in providing temperature uniformity in pottery kilns since the ware is normally loaded in a fairly elaborate stack such that the central pieces are shaded from direct radiation by their neighbours and by the kiln furniture. Compared with continuous kilns, in which extensive load preheating is carried out, intermittent kilns are inefficient devices unless air recuperation is practised. A relatively cheap and convenient means of combining the virtues of forced recirculation and recuperation is the use of recuperative burners. The application of the recuperative burner to an existing single-truck intermittent kiln is shown in Figure 6.126. At the end of the predetermined heating cycle, the burner switches off and the air-flow controls open to provide rapid cooling. Additional advantages are that, due to the high mass flow rate, the drying period of the cycle is reduced and the greater turn-down of the combustion system results in the optimum firing cycle.

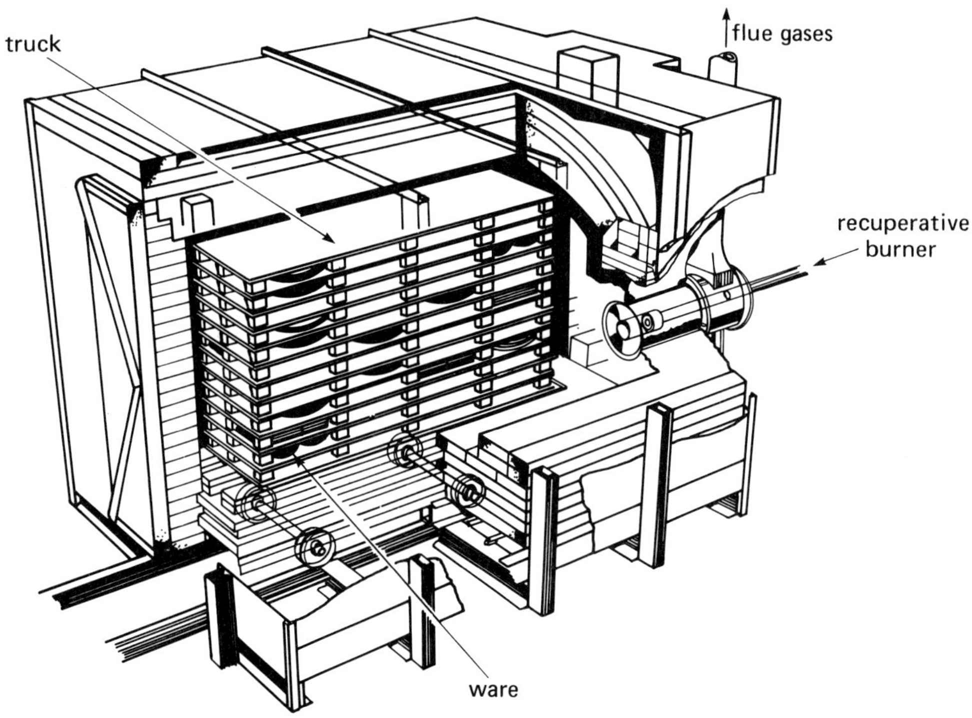

Fig. 6.126 Application of recuperative burner to single-truck intermittent kiln

Intermittent kilns have to be used where the articles to be fired are too large for tunnel kilns (e.g. in the electrical ceramics sectors). An example is illustrated in Figure 6.127.

Direct fired intermittent kilns are also popular for firing tableware and ornamental ware as in Figure 6.128.

Electric firing, which has been almost totally overtaken by gas firing, is presently limited to glost and decorating firing of tableware in small-scale applications.

Continuous kilns

Larger production rates are catered for by continuously operated tunnel kilns. The majority used in pottery firing are car-bogie tunnel kilns. These may be muffle or open fired. With the latter, the original practice was to place the ware in saggars to protect it from direct flame impingement and from the effects of combustion products. Direct gas firing has enabled the saggars to be eliminated

Fig. 6.127 Intermittent kiln for large articles

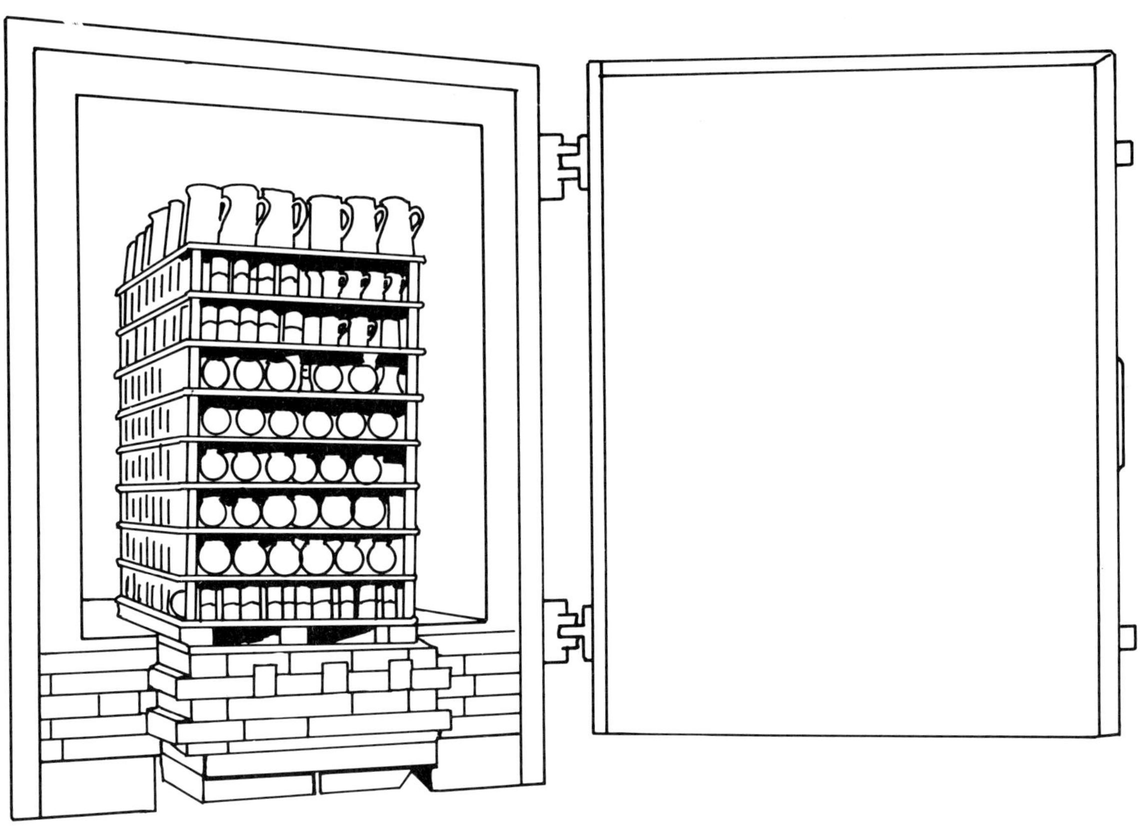

Fig. 6.128 Car-type intermittent kiln

and has been increasingly adopted in tunnel kilns for biscuit firing and, in many instances, for glost firing.

Figure 6.129 is a diagrammatic representation of a direct-fired tunnel kiln viewed in cross-section. Horizontal gas burners are directed under the load and other burners, which may be directed either horizontally or vertically, are used to heat the upper part of the load.

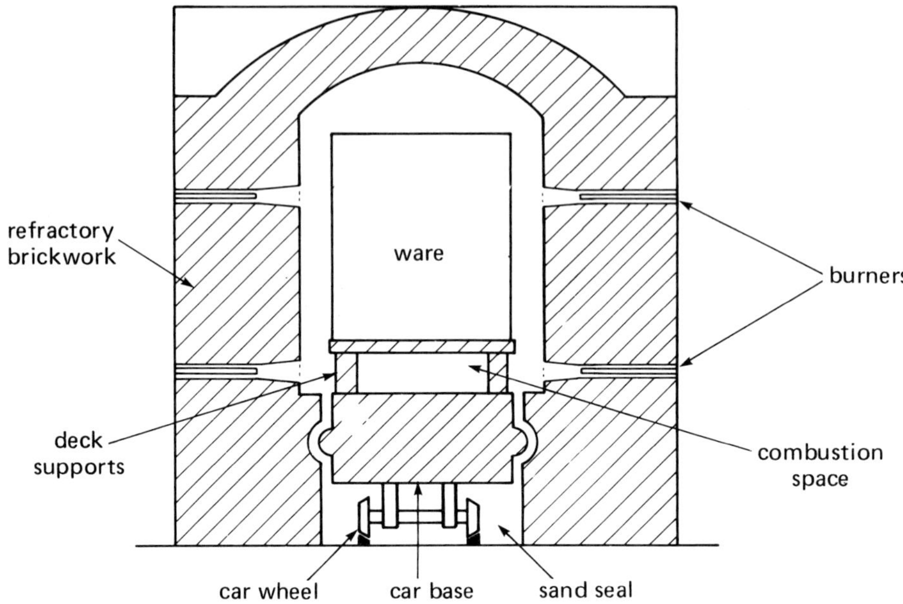

Fig. 6.129 Cross-sectional diagram of direct-fired tunnel kiln

Figure 6.130 is a diagrammatic side view of the same type of kiln and shows the loaded cars moving in one direction and the air and hot combustion gases mainly in a counterwise direction. One of the advantages of this arrangement is that air blown into the kiln from the end at which the cars leave helps to cool the ware and cars, and then enters the firing zone as preheated secondary air. Hot combustion gases from the firing zone also flow counterwise to the cars and unfired ware in the preheating zone, and give up much of their heat before leaving the kiln through exhaust ports near to the entrance. An idealised temperature distribution curve is also shown in Figure 6.130. In practice the maximum firing temperature is not maintained throughout the firing zone.

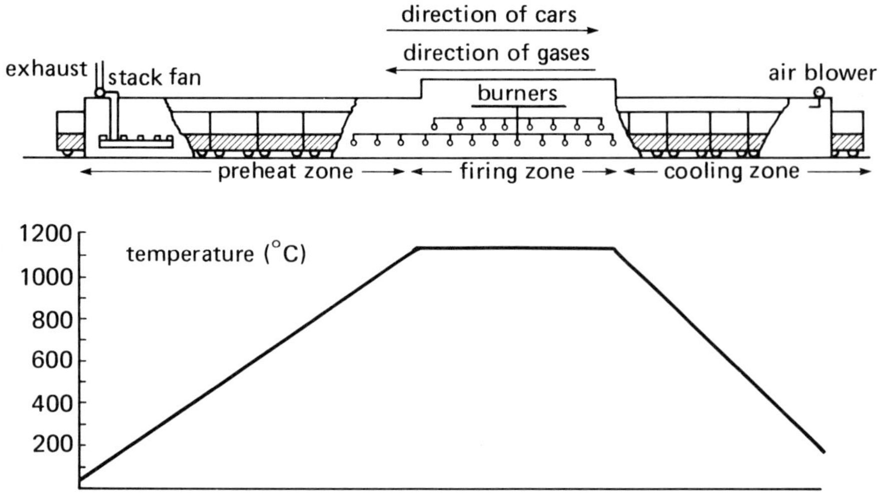

Fig. 6.130 Side elevation of direct-fired tunnel kiln and idealised firing curve

The placing of the ware on bats supported by two rows of bricks forms a flueway beneath the ware between the bricks. This avoids the cold-base problem inherent in older tunnel kilns.

Indirect or muffle tunnel kilns have combustion chambers on both sides of the kiln which prevent flames and combustion products coming into contact with the ware. The combustion gases again flow counterwise to the ware but the temperature at which they leave the kiln is much higher than is the case in a direct-fired kiln. Many present-day muffle kilns were designed to operate on producer gas and fuel oil, the muffle construction being essential to prevent ash and sulphur oxides coming into contact with the ware.

Various forms of muffle kiln are used, one of which is illustrated in simplified form in Figure 6.131. Clearly, heat transfer is mainly by radiation from the muffle walls.

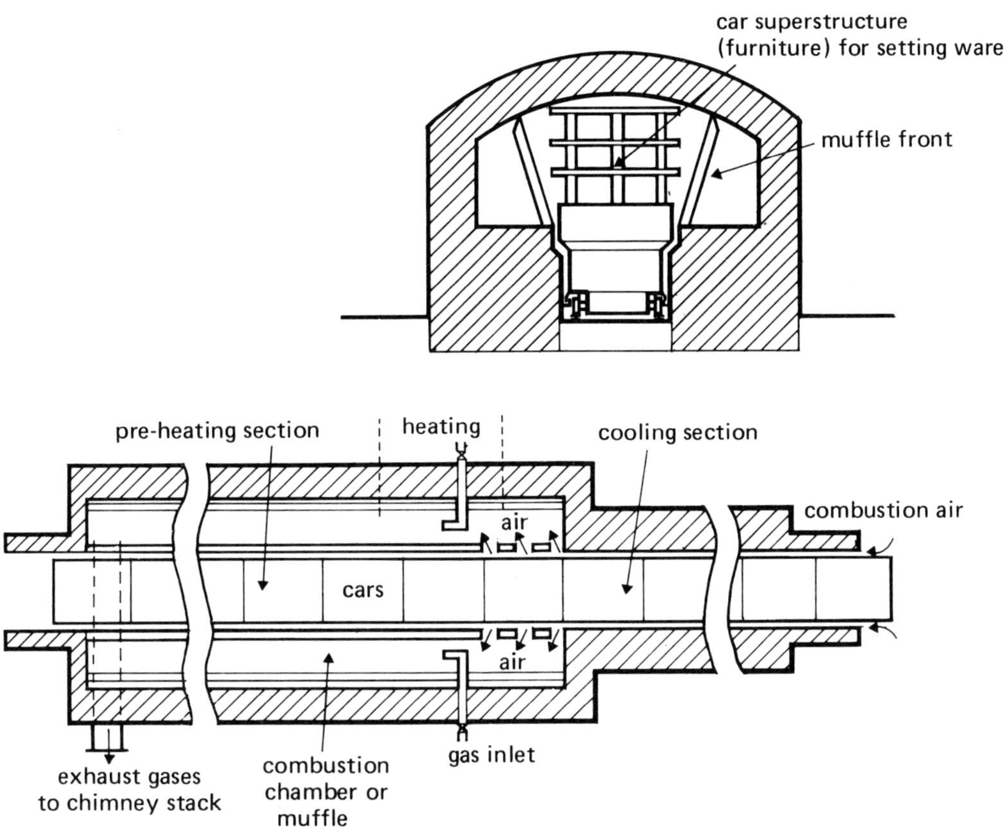

Fig. 6.131 Muffle fired tunnel kiln

In older tunnel kilns the ratio of dead load (kiln cars and furniture) to payload (ware) was very high indeed. Many advances have been made in reducing the mass of both by the use of new construction materials and the use of ceramic fibre insulation for the cars. In recent times the concept of fast and once firing of ware has been resurrected and a number of demonstration projects have been sponsored by the Energy Efficiency Office. Fast firing is the technique of placing ceramic ware in a single setting as opposed to a conventional densely packed, multiple layer setting. The objective is to reduce the firing times and minimise or even dispense with kiln furniture.

British Ceramic Research Ltd. proved that fast firing was technically feasible in the early 1970s. As a result, a number of large tableware manufacturers installed fast firing kilns. However, the kilns were not designed for good thermal efficiency, neither was advantage taken of the automatic product handling feasible with single layer settings.

ETSU demonstration projects have covered fast and once firing of ceramic tiles, biscuit earthenware and hollow-ware and vitreous tableware. Whilst significant savings potential has been demonstrated, there are still some technical problems to be overcome, especially with flatware and wall tiles. Energy savings of over 50% and labour savings (by automating handling of ware into and out of the kiln) have been reported. The development of a gas fired ceramic radiant tube kiln for the fast firing of glost and decorated tableware is regarded as having widespread application in the pottery industry.

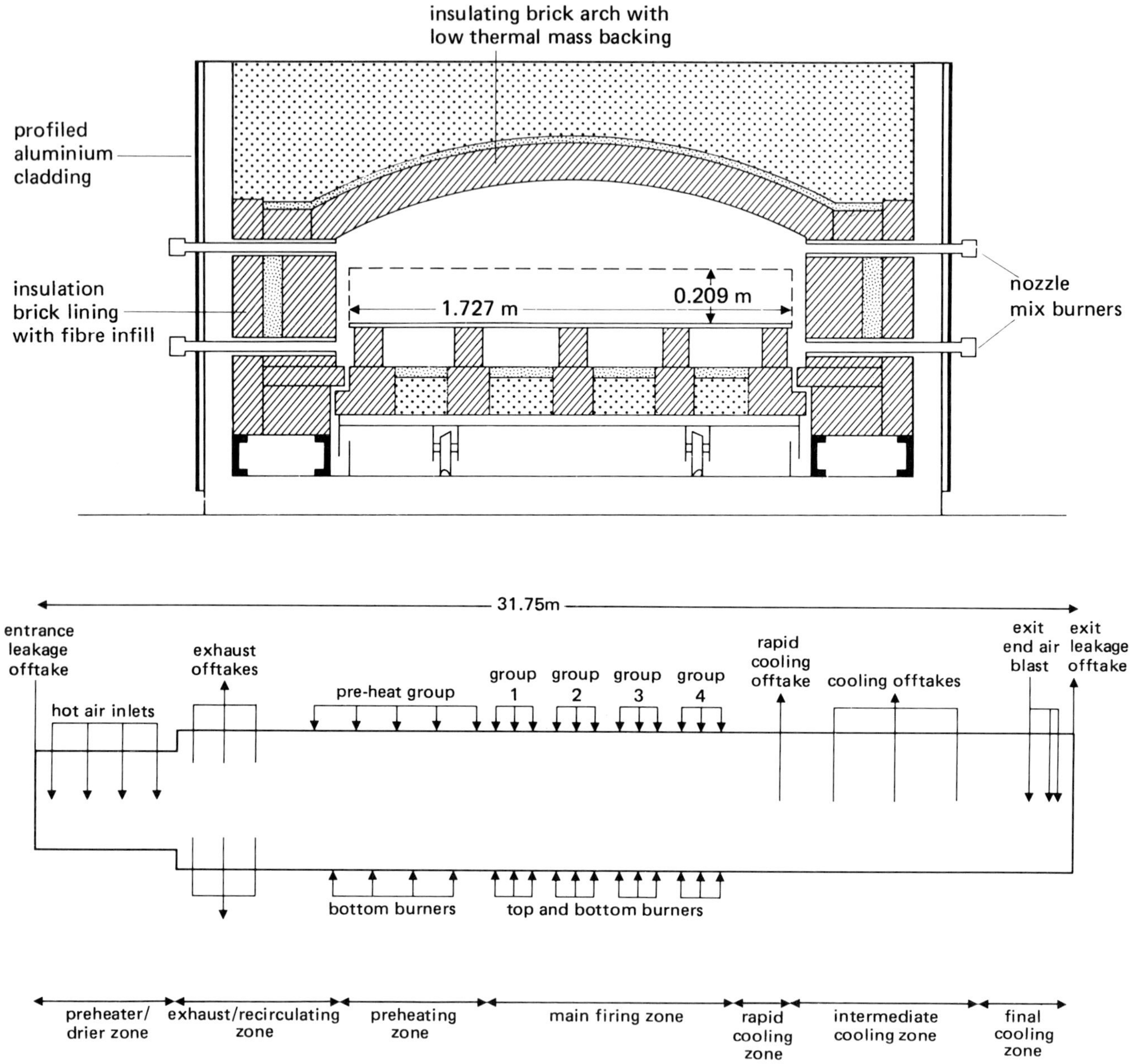

Fig. 6.132 Single-layer, fast-fire car tunnel kiln

Figure 6.132 shows the layout and cross-section of a kiln used in the fast, once firing of glazed earthenware mugs, one of the projects supported under the Energy Efficiency Demonstration Scheme (EEDS). The report on this project concluded that the installation of the kiln has been successful from both production and financial viewpoints, the main benefits being:

- Energy cost savings by once firing.
- Labour cost savings accruing from week-day only operation and elimination of biscuit firing.

- Simpler placing and emptying operations because kiln cars are single rather than multi-deck.
- Decrease in faulty ware.
- Elimination of sorting and warehousing of biscuit cups.

The importance of fast firing cycles depends largely on output rates per unit investment and running costs (including labour) compared with those for more conventional methods.

6.7.3 Building Brick Industry

The majority of British houses are built with brick walls. This is the principal market of the brick industry. There are three main categories of brick which are:

- Facing bricks used for the outer walls of buildings where appearance is important.
- Common bricks used where appearance is not important. (Other properties can be as good as facing bricks).
- Engineering bricks used where there are adverse environmental conditions. Corrosion and mechanical properties are better than the other two types.

Common bricks used to be the major product, but have been overtaken by facing bricks as light-weight concrete and other blocks have replaced common bricks for the load-bearing inner walls of houses. Engineering bricks account for less than 10% of total production.

A variety of raw materials is used for brick manufacture. Approximately 40% use Oxford Clay to produce Fletton bricks. Because of the high carbonaceous content of Oxford Clay, Fletton bricks require only about one third the average energy used in the remainder of the industry.

The specific energy requirement (energy consumption per unit of production) varies widely within the non-Fletton sector, depending on raw material properties, age of plant and drying and firing procedures. Figure 6.133 shows a process flow diagram for a typical brickworks.

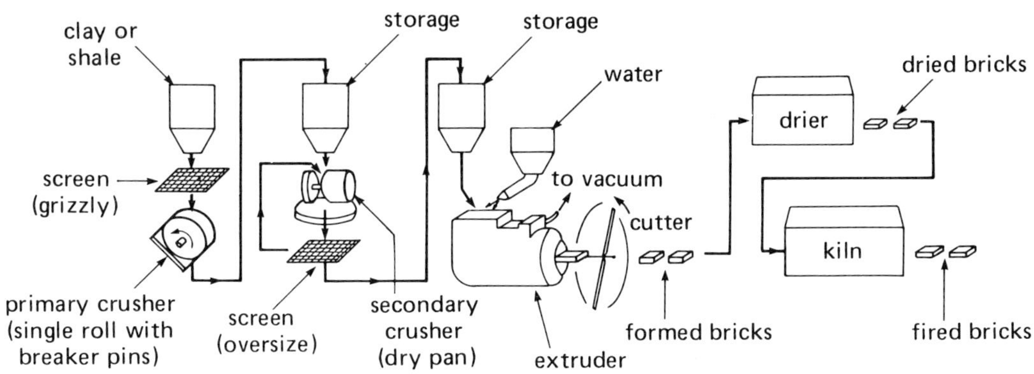

Fig. 6.133 Process flow diagram of a brickworks

Electricity is the main energy form used in the process prior to drying and firing.

Drying

Brick shaping processes require various proportions of water mixed with clay. Removal of some of this water is carried out in driers before placing the bricks in a kiln. Drying gives the bricks rigidity to help them withstand handling and setting in the kiln. Normally bricks are dried beyond the stage at which most of the shrinkage caused by drying is complete. The amount of shrinkage depends on the raw materials used and drying must be carefully controlled if percentage shrinkage is high.

Bricks made by stiff-plastic and semi-dry processes, which contain less water, can sometimes be set in the kiln without pre-drying. An overall saving in primary energy is claimed, in these instances, even though electrical power intensive machines are used.

The three principal drying methods employed in the UK are:

- Hot Floor Drying. Thermal efficiency is very low and hot floor drying methods are becoming rarer.
- Tunnel Drying. As described in the section on the pottery industry and in Section 6.5, tunnel drying can be an attractive method when used in conjunction with a tunnel kiln. The use of kiln cooling air for the drier aids overall thermal efficiency and if the kiln car is used in the drier, substantial labour savings can be made.
- Chamber Drying. A group of driers are normally built together to form a battery. Heat is supplied by steam coils/radiators, cooling air from kilns or by direct firing. They can be operated as 'intermittent air-flow driers' to shorten the drying time. This involves passing hot drying air of controlled humidity over the bricks to create an excessive drying rate i.e. if continued indefinitely the bricks would crack. To avoid this the air flow is stopped after a short time to allow relaxation of the shrinkage stresses and then the cycle is repeated.

Refer to Section 6.5 for a more detailed discussion on drying.

Firing

Nearly all facing and common bricks are fired in continuous kilns, with intermittent kilns being sometimes used for engineering bricks and quarry tiles. Continuous kilns can be subdivided into:

- Annular or continuous chamber kilns in which the ware is stationary and the firing zone moves.
- Car tunnel kilns in which the firing zone is stationary and the ware moves.

Annular kilns

This type can be further subdivided into longitudinal arch (e.g. Hoffman) and transverse arch (e.g. Staffordshire) kilns.

Figure 6.134 illustrates a 16 chamber Hoffman kiln, the operation of which is typical of continuous chamber kilns. Three chambers are shown firing, the combustion products from which are induced by stack draught or by ID fans through the four preheating chambers to the central exhaust duct thence to the stack. Two drying chambers isolated by paper screens are fed by warm air from the cooling chambers via the hot-air duct. Next to the drying zone is a chamber being 'set' (loaded), the next one is empty and the remaining chamber is being 'drawn' (emptied). Air for combustion and drying is induced through the open 'wickets' (doorways) of the last-mentioned three chambers and becomes preheated as it cools the ware in the cooling zone, some of it passing to the drying chambers and the remainder flowing to the firing zone. Each chamber is essentially identical and at regular intervals (say 20 hours) the process is advanced one chamber. The paper division isolating the drying chamber burns out as the preheating zone advances and this chamber becomes part of the preheating zone. The setting chamber then becomes the second drying chamber and so on. The chamber capacity of Hoffman kilns is typically 15 000 to 25 000 bricks although capacities up to 45 000 are not unknown.

These kilns are suited to rapid firing where control requirements are not too stringent. The traditional Hoffman kiln was limited in size because of its roof structure. Modifications to traditional design, e.g. the Suspended Ceiling Hoffman, have enabled capacity to be increased.

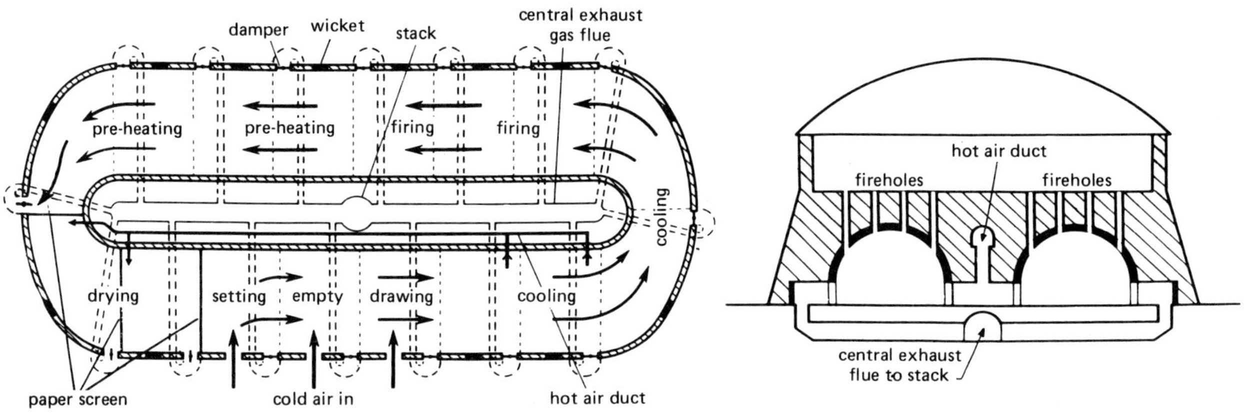

Fig. 6.134 Principles of the longitudinal arch kiln

Transverse-arch kilns operate on a similar principle to the longitudinal arch type. They are divided into a series of chambers by permanent walls, usually with an equal number of chambers placed back-to-back. Individual chambers are connected by holes through the inter-chamber walls, and the end chambers by flues through the main central wall and/or through the end wall of the kiln. The arched roof of each chamber is sprung from the inter-chamber walls and chamber capacities vary from 8 500 to 70 000 bricks. Typical capacity is between 20 000 and 40 000 bricks. Figure 6.135 shows a typical layout of an 18 chamber kiln.

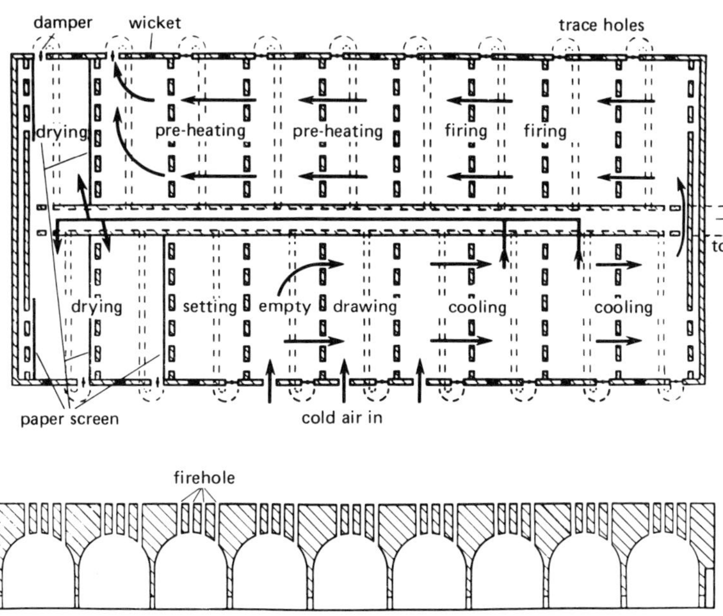

Fig. 6.135 Principle of transverse arch kiln

The method and sequence of firing is similar to that of the longitudinal arch kiln. Products of combustion and cooling air pass from one chamber to another via the holes in the inter-chamber walls. When a chamber has been set the holes are papered over, or sealed by adjustable metal dampers. A typical firing schedule is shown in Figure 6.136.

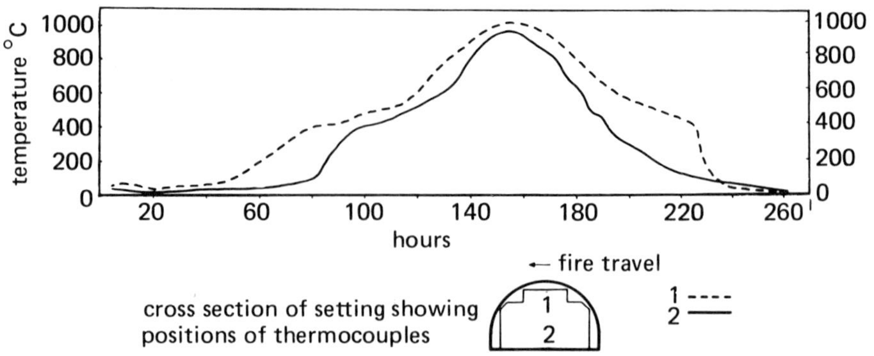

Fig. 6.136 Firing schedule for facing bricks, transverse arch kiln

It is claimed that as well as having bigger chamber capacities than Hoffman kilns, transverse-arch kilns provide a more uniform control, giving higher efficiencies or higher firing temperatures.

Tunnel kilns

Tunnel kilns for brick firing employ the identical principle to those for pottery kilns described in Section 6.7.2.

A diagram of a typical car tunnel kiln is shown in Figure 6.137. Tunnel kilns are generally displacing annular kilns as and when new kilns are required for brick making, mainly because of the labour saving effected.

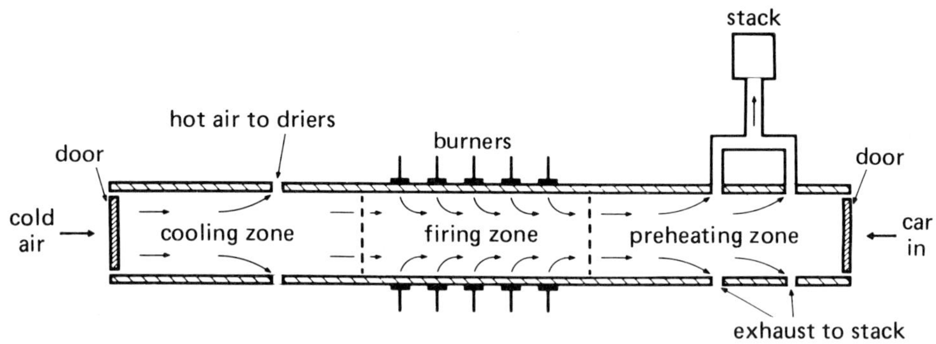

Fig. 6.137 Diagram of a car tunnel kiln

The temperatures of the exhaust gas and cooling air leaving the kiln are in the ranges 100 to 450 °C and 50 to 200 °C respectively. The flow of gases through the tunnel can be regulated by the use of fans which can either recirculate air in the preheating zone to equalise vertical temperature gradients or extract gas from the cooler part of the preheater section and inject it into a hotter part. This increases flow velocity through the bricks and thus speeds up heat transfer.

Intermittent kilns

These are kilns with a single chamber which fire bricks in batches. Intermittent kilns may be circular or rectangular in shape and may be updraught, horizontal-draught or downdraught. Downdraught kilns are the most widely used in Britain being more efficient than the other types. Figure 6.138 shows a typical downdraught intermittent kiln.

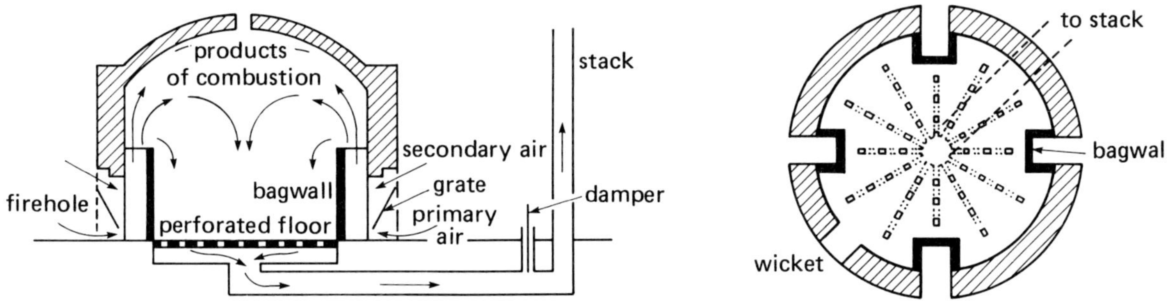

Fig. 6.138 Intermittent downdraught kiln

Downdraught kilns are used for most types of heavy clay product. Sizes range from 10 000 to 100 000 bricks with a typical capacity in the range 30 000 to 50 000 bricks. Fuel is burned at the 'fireholes' around the periphery, the ware being protected from flame impingement by 'bagwalls' which direct the combustion products to the kiln crown from where they are drawn down through the setting to radial flues in the hearth. Some variants of the downdraught kiln exist, including designs with a central interior flueway and others containing flueways in the walls.

Most intermittent kilns in Britain (beehive and rectangular) have been converted to natural gas firing by direct substitution of gas burners for the solid fuel grates. In some instances bagwalls have been retained, in others they have been removed, to provide an increase in kiln capacity.

As in the production of ceramic ware, heavy clay products are generally fired in oxidising atmospheres but the manufacture of some special products (some kinds of engineering bricks and floor quarry tiles for example) requires firing to take place under reducing conditions, at least for part of firing cycle.

The great advantages of gas firing kilns of this type are the ease with which reducing atmospheres may be produced (by adjustment of air/gas ratios at the burners) and the much improved degree of control over the composition of the kiln atmosphere which results. The impact upon the environment is also much reduced by the avoidance of the production of dark smoke.

Modern intermittent kilns are replacing the conventional kiln. An example is the shuttle kiln (Figure 6.139) in which the thick, dense, traditional firebrick walls are replaced by light, thin walls made of insulating brick and/or ceramic fibre supported externally by a steel structure. This decreases the thermal mass of the kiln and creates more space. The bricks are loaded on cars which can be rolled in and out of the kiln allowing bricks to be set and removed outside of the kiln.

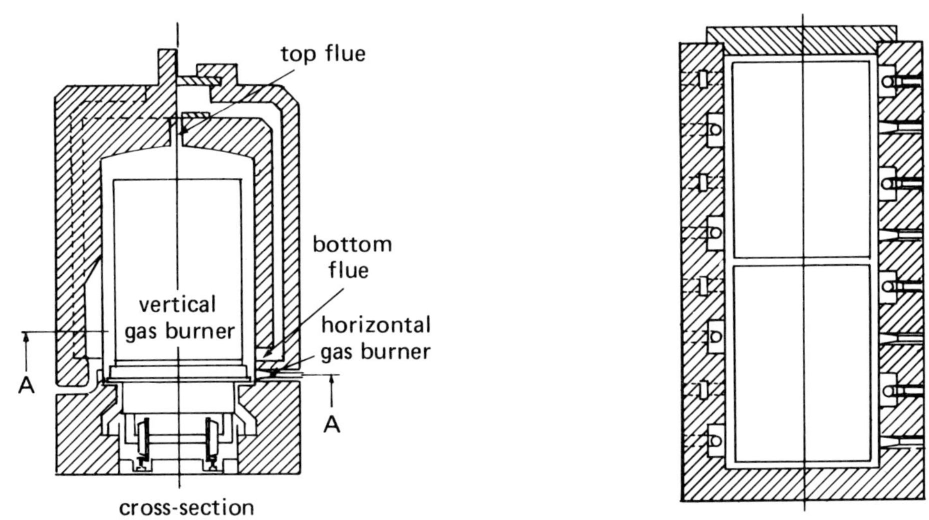

Fig. 6.139 Car bogie intermittent kiln

6.7.4 Refractories Industry

Refractories are materials that retain their shape and chemical stability for an acceptable period when subjected to high temperatures. A broad definition of a refractory material would be a material that does not soften below 1 500 °C. The properties and applications of refractories are detailed in Chapter 8. Refractory products can be in shaped or in unshaped form. Shapes include bricks and hollow-ware. Cements, castables, moudables and ramming materials are the commonest unshaped products. Figure 6.140 indicates the general processes involved in refractory production.

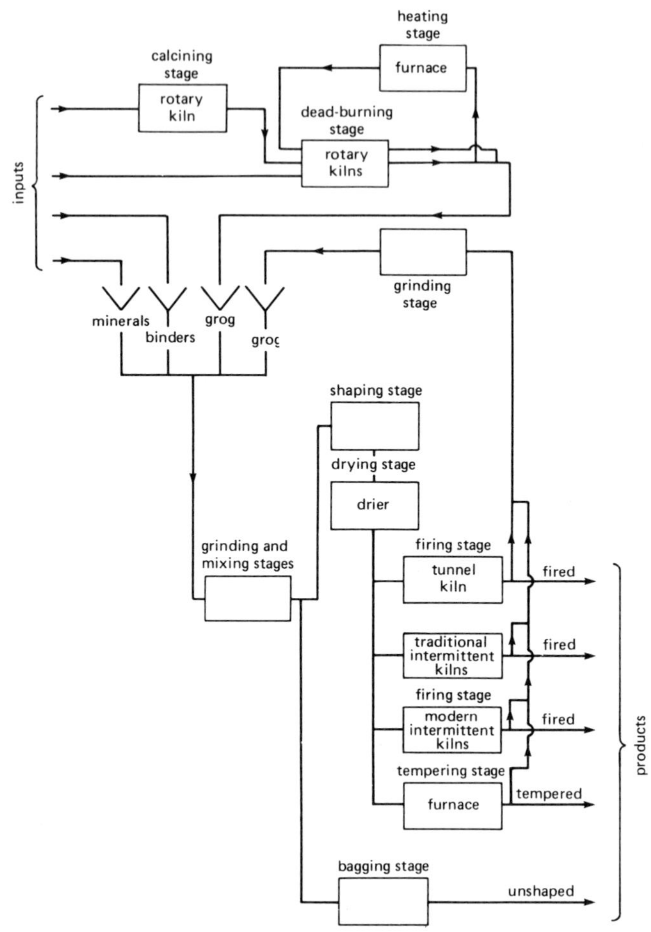

Fig. 6.140 Bulk refractories production

As with the other sectors of the ceramic industry, kilns are the major energy consumers. Driers often use warm air from kilns as described in previous sections. Continuous (tunnel, annular and rotary) and intermittent kilns are in widespread use.

Whilst intermittent kilns are inherently less efficient than continuous kilns, they have advantages which will ensure their continued use in the following situations.

- Where articles require extended firing cycles (e.g. silica, which needs at least eight days), tunnel kilns would have to be very long.

- Small scale production of particular products requiring specific firing temperature profiles would preclude tunnel kiln operation on efficiency grounds.

Continuous annular and tunnel kilns have been discussed in detail in previous Sections of this Chapter and no further comments are required.

6.8 METAL HEATING

6.8.1 Introduction

This Section deals with the furnaces and methods employed for heating metal in its final or near-final form i.e. excludes metal melting or re-heating (prior to hot working) which are covered in subsequent Sections of this Chapter. However, a number of furnaces are common to both heat-treatment and re-heating processes.

The objective of heat-treating metals is to produce mechanical or physical properties suited to their ultimate use. In its broadest sense this includes any process involving heating and cooling the solid metal by which its properties are altered. All heat-treating operations consist of subjecting a metal to a definite time-temperature cycle, which falls into three parts:

- Heating.
- Holding at temperature (soaking).
- Cooling.

The treatment of non-ferrous metals is similar in principle to that for ferrous metals, except that it is generally performed at lower temperatures.

6.8.2 Metallurgical Requirements

Heat-treatment of metals can impart or enhance properties such as hardness, ductility, machinability, toughness etc. or can remove stresses induced by previous forming or fabrication procedures.

Heating-up rates are not particularly important except in the case of some chromium steels and where workpieces are in a highly stressed condition as a result of previous cold working or hardening. The length of the soaking period is governed by the nature of the heat-treatment process and by the thickness of the section to be heated. The objective is to achieve uniformity of temperature throughout the volume of the piece.

A complete description of the metallurgy of heat-treatment is outside the scope of this book. The sections which follow will, however, introduce the requirements of the common heat-treatment processes and their terminology.

Iron-carbon equilibrium diagram

This diagram (Figure 6.141) is basically a map of the temperatures at which the various crystal structures or phase change, on very slow heating or cooling, in relation to carbon content.

Pure iron (ferrite), for example, exists from room temperature up to 910 °C (point G on the diagram). As the carbon content increases, its temperature limit reduces to around 723 °C at about 0.8% carbon. Above these temperatures it converts to austenite. The temperatures at which such transformations occur are called critical temperatures or critical points. It must be remembered that the equilibrium diagram represents behaviour under very slow heating and cooling conditions and delineates areas of composition and temperature in which the various constituents are stable. No provision is made for the situation where, by rapid cooling, constituents exist on the 'wrong' side of the critical line. Steel can also exist as martensite, an important phase in the hardening of steels, which does not appear at all in the phase diagram. Full annealing is the only type of heat-treatment which the diagram describes accurately. None of the hardening and strengthening treatments obtained by rapid cooling are fully described by the diagram.

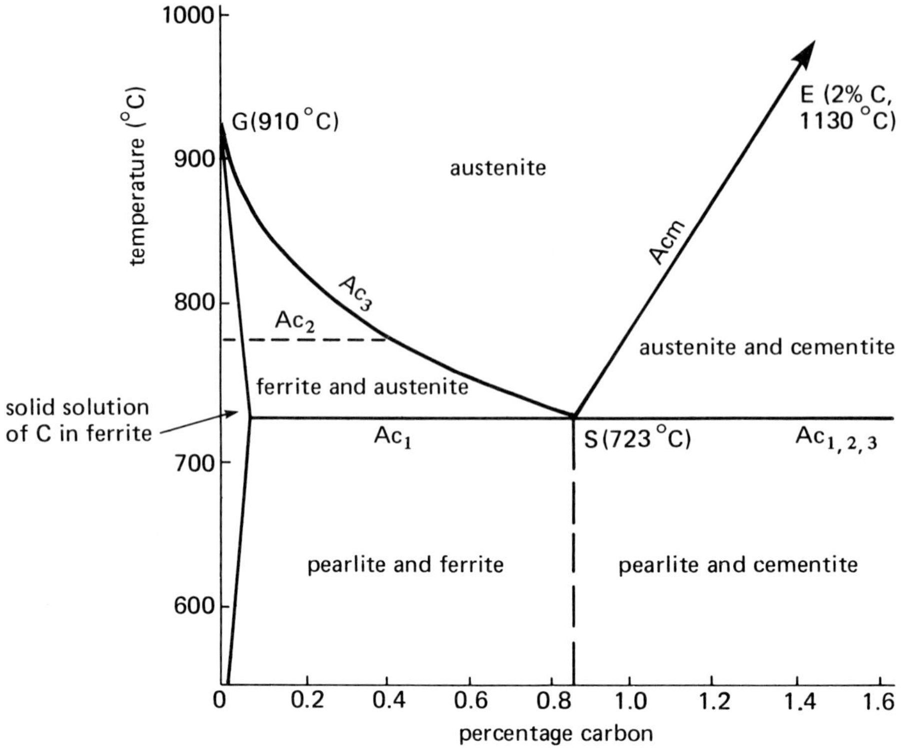

Fig. 6.141 Iron-carbon equilibrium diagram

There are two constituents in cold (annealed or normalised) steel, α-iron (Fe) and cementite (Fe_3C). If the steel is heated to a sufficiently high temperature there is only one constituent, austenite which is a solid solution of cementite in γ-iron. The diagram indicates that austenite does not exist at temperatures below 723 °C but in some alloy steels austenite is stable at room temperature. Austenite can contain various amounts of Fe_3C, the amount increasing with temperature, but the amount is limited and in the field to the right of line SE there are two constituents, austenite and cementite, the point E at 2% C representing the boundary between steels and cast irons.

Point S is of particular interest. A carbon-rich austenite, say, of 1.5% C cooling from 1 100 °C remains the same until it reaches 1 050 °C at which point the solid-solubility limit is reached and cementite comes out of solution. The carbon-depleted austenite continues to deposit more and more until the temperature approaches 723 °C. At this point iron carbide corresponding to about 0.67% C of the original 1.5% has come out of solution leaving 0.83%. At point S the iron in the austenite changes to the α-form in which the carbon solubility is low (maximum 0.025% C) and so the remaining 0.8% or so comes out of solution forming (under slow-cooling conditions) interleaved plates of cementite and ferrite, this composite structure being termed pearlite.

If on the other hand austenite of a lower carbon content, say 0.4%, is cooled it remains unchanged down to about 790 °C when the ferrite separates leaving the austenite relatively richer in carbon. The process continues down the line GS until the carbon content of the austenite has increased to 0.83% C and the temperature has fallen to 723 °C, i.e. point S. Further cooling causes the austenite to change to pearlite.

Thus in the first case the steel comprises pearlite and cementite and in the second case pearlite and ferrite.

Point S is the lowest point at which α-iron can exist and is termed the eutectoid as the solid-solution analogy with the eutectic point of liquid-solid transitions. Thus steel of around 0.83% C is commonly termed eutectoid steel, steel with $<$0.83% C is hypoeutectoid steel and steel with $>$0.83% C is hypereutectoid steel. In commercial steels containing alloying ingredients the eutectoid occurs over a band of temperatures rather than at 723 °C.

Referring again to the diagram, the line GS is designated Ac_3, the line SE, Acm, etc. This terminology requires some explanation. If a steel sample is heated and a time-temperature record is made it will show a break at 723 °C and this distortion of the plot will persist up to the line GS or SE. This 'arrest' indicates that heat is required for the α- to γ-iron transformation and for the solution of carbon in iron. The arrest on heating is designated Ac (French arrete chauffage) and further designated Ac_1 etc, to indicate the change involved. In the case of hypereutectoid steels the Acm line denotes the cementite-solution arrest points.

As previously stated the equilibrium diagram relates strictly only to very slow heating and cooling. In practice, heating and more importantly cooling, generally take place at a much faster rate. This allows austenite to be retained below the temperature indicated on the equilibrium diagram and this in turn transforms to martensite, which is very hard, and pearlite. Thus a steel of say 0.45% C quenched from 715 °C would contain ferrite, pearlite and a small amount of martensite. If it were quenched from say 750 °C its hardness would increase markedly and the microstructure would indicate that the pearlite had been entirely replaced by martensite which moreover had encroached on the ferrite areas since at sufficiently high temperatures the austenite dissolves the ferrite. Still higher temperatures render the steel wholly austenitic and consequently wholly martensitic after quenching. This process is reversed by heating and subsequent slow cooling. Martensite being brittle and hard does not itself confer very widely useful properties on the steel but it is the constituent which can readily be converted into other microstructures providing a valuable combination of hardness and strength.

Annealing

An annealing process is, in general terms, one which renders a product soft. Steel annealing may be subdivided into full annealing i.e. heating the steel above its critical temperature range and cooling it slowly, and subcritical annealing in which the steel is heated to just below its critical temperature. Full annealing results in a structure which is pearlite together with free ferrite or free carbide depending on the carbon content and has an additional effect generally described as grain-size refinement. This latter effect improves the ductility of the softened steel or, if it is to be subsequently hardened, improves the toughness of the steel as hardened. It is necessary to exceed Ac_3 to transform the ferrite wholly to austenite. However, in steels without additions for grain-size control the grain starts to coarsen at temperatures just above Ac_3 and the annealing temperature should be kept as little as possible above Ac_3. The necessary slow cooling through the critical range is usually provided by cooling in the furnace. Directly fired furnaces may be employed but increasingly bright annealing is practised using controlled atmospheres.

Subcritical annealing has as its chief aim the removal of internal stress, usually in low-carbon steels (up to about 0.25% C) which have been cold-worked. The process temperature is below the Ac_1 line. No major structural changes are promoted but grain distortion is corrected. If the steel is not to be further cold-worked, temperatures well below the critical range may be employed, e.g. 550 °C. Stress relief is commonly carried out on welded structures to render the welding stress distribution uniform. Temperatures are generally in the range of 500 to 650 °C. British Standard requirements exist for the heat treatment of welded pressure vessels. Typically for straight carbon steels stress relief at 580 to 650 °C for one hour per 25 mm thickness is required increasing to 740 °C for some alloy steels (e.g. Cr, Mo).

Normalising

Normalising consists in principle of heating steel to above its critical temperature (Ac_3 or Acm) followed by cooling in air. This may of course cover a wide range of cooling conditions. A small single component will cool much more quickly in air than will a number of components piled together leading to rather different properties in the finished product.

Normalising generally involves rather higher temperatures than are involved in full annealing with a view to obtaining full absorption of excess ferrite and equalisation of the carbon content throughout the austenite. To avoid the coarse grain structure which would result from the slow cooling of full annealing from this high temperature, the much faster air cooling is adopted.

Spheroidising

Spheroidising is essentially a type of tempering. If steels are subjected to prolonged heating close to but below Ac_1 so that no austenite is produced, the carbide which is normally present as plates or needles interspersed in the ferrite or pearlite, bainite or martensite tends to contract into spheroids. A spheroidised structure provides machinability and improves uniformity in hardening with some grades of steel.

Hardening

Hardening occurs when steels are heated above Ac_3 to develop an austenitic structure followed by quenching to develop martensite. Two factors determine the hardness after quenching: first the 'hardenability' of steel, which is a function of carbon content for straight carbon steels but which is strongly influenced by the presence of alloying elements; second the speed of quenching. Water is used for fast quenching, oil for slower quenching. Hardening generally produces a steel which is too hard and brittle for most engineering purposes and tempering is required after hardening.

Tempering

Tempering or 'drawing' involves heating martensitic steels to below Ac_1, generally in the range 120 to 675 °C, the higher temperatures inducing greater toughness. The changes occurring during tempering are complex but consist mainly of carbide precipitation and agglomeration. Short tempering times are generally undesirable. At least ½ h and preferably 1 to 2 h are required at tempering temperatures for most steels. Tempering usually follows hardening before the hardened steel has cooled to room temperature to avoid stress-cracking. The term tempering is also applied to a post-normalising stress-relief carried out at 650 to 750 °C on some classes of steel.

Patenting

Patenting is carried out on work-hardened steel wire which requires further cold-drawing. Temperatures may be as high as 1 000 to 1 050 °C but the cooling is such that a fine pearlite structure is obtained. Air cooling may be employed but in steels with high hardenability a lead-bath quench (480 to 620 °C) is used to control the cooling rate.

Austempering

In this process austenite is transformed to bainite without the intermediate production of martensite. It consists of heating steel above Ac_3 followed by quenching in a hot bath held at a temperature below that of fine-pearlite formation.

Martempering

Martempering is a quench-hardening process in which the final stages of the quench are carried out at a much slower rate than in normal quenching, i.e. in air, to minimise stresses. The steel is quenched first in a molten-salt bath which cools it rapidly to slightly above the M_S temperature (i.e. the temperature at which martensite starts to form on cooling) then cooled in air. The finished product is fully martensitic and may be subsequently tempered.

Figure 6.142 indicates temperature ranges for the more common heat treatment processes for plain carbon steels.

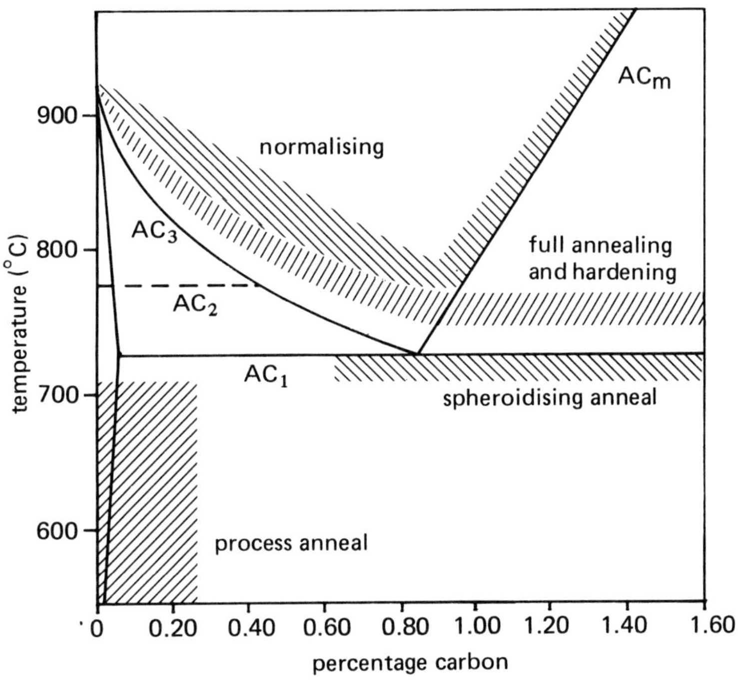

Fig. 6.142 Recommended temperature ranges for heat treating plain carbon steels

Case hardening

Case hardening is a process of hardening a ferrous alloy so that the surface layer or case is made substantially harder than the interior or core. The chemical composition of the surface layer is altered during the treatment by the addition of carbon, nitrogen, or both. The most frequently used case-hardening processes are carburising, cyaniding, carbonitriding and nitriding.

Carburising

Carburising is a process that introduces carbon into a solid ferrous alloy by heating the metal in contact with a carbonaceous material to a temperature above the Ac_3 of the steel and holding at that temperature. The depth of penetration of carbon is dependent on temperature, time at temperature, and the composition of the carburising agent. As a rough indication a carburised depth of about 0.75 to 1.25 mm can be obtained in about 4 h at 925 °C depending upon the type of carburising agent, which may be a solid, liquid or gas (see Section 6.12).

Since the primary object of carburising is to secure a hard case and a relatively soft, tough core, only low-carbon steels (up to a maximum of about 0.25% of carbon), either with or without alloying elements (nickel, chromium, manganese, molybdenum), are normally used. After carburising, the steel will have a high-carbon case graduating into the low-carbon core.

A variety of heat treatments may be used subsequent to carburising, but all of them involve quenching the steel to harden the carburising surface layer. The most simple treatment consists of quenching the steel directly from the carburising temperature; this treatment hardens both the case and core (insofar as the core is capable of being hardened). The plain carbon steels are almost always quenched in water or brine; the alloy steels are usually quenched in oil.

Cyaniding

A hard, superficial case can be obtained rapidly on low-carbon steels by cyaniding. This process involves the introduction of both carbon and nitrogen into the surface layers of the steel. Steels to be cyanided normally are heated in a molten bath of cyanide-carbonate-chloride salts (usually containing 30 to 95% of sodium cyanide) and then quenched in brine, water, or mineral oil; the temperature of operation is generally within the range of 845 °C to 875 °C. The depth of case is a function of time, temperature, and composition of the cyanide bath. The time of immersion is quite short as compared with carburising, usually varying from about 15 min to 2 h. The maximum case depth is rarely more than about 0.5 mm and the average depth is considerably less.

Carbo-nitriding

Carbo-nitriding is a process for case hardening a steel part in a gas-carburising atmosphere that contains ammonia in controlled percentages (see Section 6.12). Carbo-nitriding is used mainly as a low-cost substitute for cyaniding and, as in cyaniding, both carbon and nitrogen are added to the steel. The process is carried on above the Ac_1 temperature of the steel, and is practically up to 925 °C. Quenching in oil is sufficiently fast to attain maximum surface hardness; this moderate rate of cooling tends to minimise distortion. The depth to which carbon and nitrogen penetrates varies with temperature and time.

Nitriding

The nitriding process consists of subjecting machined and heat-treated steel, free from surface decarburisation, to the action of a nitrogenous medium, usually ammonia gas, at a temperature of about 500 °C to 540 °C, whereby a very hard surface is obtained. The surface-hardening effect is due to the absorption of nitrogen, and subsequent heat treatment of the steel is unnecessary. The time required is relatively long; normally between 1 and 2 days. The case, even after 2 days of nitriding, is generally less than 0.5 mm and the highest hardness exists in the surface layers to a depth of about 100 μm only.

Special low-alloy steels have been developed for nitriding. These steels contain elements that readily combine with nitrogen to form nitrides, the most favourable being aluminium, chromium and vanadium. Molybdenum and nickel are used in these steels to add strength and toughness. The carbon content is usually about 0.2 to 0.5%, although in some steels where high core hardness is essential, it may be as high as 1.3%. Stainless steels also can be nitrided.

Surface hardening

It is frequently desirable to harden only the surface of steels without altering the chemical composition of the surface layers. If a steel contains sufficient carbon to respond to hardening, it is possible to harden the surface layers only by very rapid heating for a short period, thus conditioning the surface for hardening by quenching.

Non-ferrous metals

The commonest non-ferrous metals used in industry are aluminium and its alloys, copper, copper alloys, lead, lead alloys, tin and zinc.

Figure 6.143 shows a generalised phase diagram for age-hardening aluminium alloys, upon which are superimposed the various heat treatment ranges.

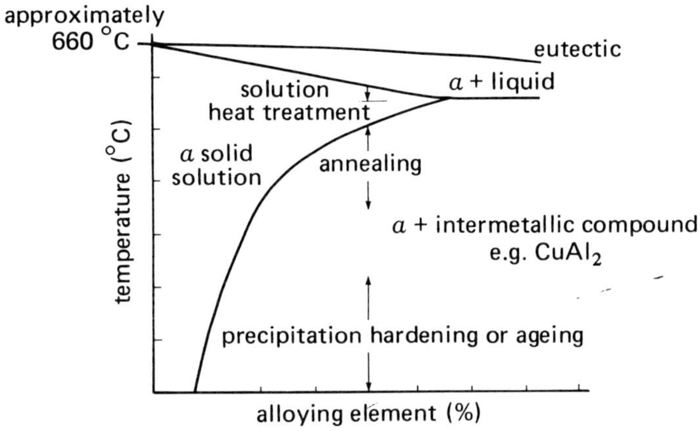

Fig. 6.143 Typical phase diagram for heat-treatable (age hardening) aluminium alloys

The main heat-treatment processes used with aluminium alloys are:

- Solution treatment. This consists of heating the alloy in the α solid-solution field for sufficiently long to take all the phases into solid solution and to anneal the metal thoroughly.
- Quenching. If the alloy is cooled rapidly it becomes supersaturated and the alloying elements are precipitated out as an aluminium-alloying-element intermetallic. The size and distribution of the precipitate determine the full properties of the alloy. Quenching is done as rapidly as possible, usually in water.
- Ageing. This is carried out at low temperatures, always below the α-solid-solution boundary. Maximum hardness is obtained with lower ageing temperatures and long times.

Solution-treatment temperatures are taken as high as possible above the α + intermediate compound boundary but at such levels as to avoid the possibility of eutectic melting. Good furnace temperature uniformity and control are essential for this process. Ageing is carried out in ovens, generally in the temperature range 130 to 240 °C for periods ranging from 1 to 50 hours.

Selection of appropriate heat-treatment atmospheres are most important in the heat-treatment of copper and its alloys. Brighter annealing of copper in the range 400 to 600 °C is comparatively easy, oxidation being prevented by the presence of small amounts of reducing gases. Apart from free oxygen and some sulphur-containing gases, combustion products are inert to copper. Lean, exothermic atmospheres, slightly on the rich side of stoichiometric, are satisfactory for pure-copper heat-treatment.

Hydrogen is inert to pure copper but may cause embrittlement in contact with copper containing copper oxide in solution. Hydrogen is capable of diffusing into the metal where it reacts with the oxide inclusions to produce water vapour at a sufficient pressure to produce blisters and embrittlement. Copper of this type (high-conductivity or tough-pitch copper) must be heat treated in a lean exothermic atmosphere containing a controlled minimum of hydrogen (<0.5 to 1.0%). Wet CO also causes embrittlement, the water-gas reaction liberating sufficient H_2 to cause embrittlement.

Hydrogen sulphide attacks copper and complete absence of this gas is a pre-requisite of bright annealing. Sulphur and organic sulphur compounds on the other hand are practically inert to copper. Annealing of copper wire is normally carried out under vacuum conditions. Even under moderate vacuum conditions the concentration of active constituents is very low, for example if the pressure in a vessel containing air is reduced to 1.3 pascals, the oxygen content is reduced to 1.3 ppm. This order of vacuum is easily obtained using mechanical pumps. There are two main reasons for vacuum annealing being used for copper wire:

- The surface of the wire is often contaminated with die lubricants having low vapour pressures and these are removed more easily when pressure is reduced.
- Vacuum annealing prevents the tendency of successive coils of wire to stick together so that on unwinding the wire breaks or kinks. Sticking of this kind is caused by the presence of a layer of copper oxide which normally reduces to the metal in prepared atmospheres giving rise to local welding. This difficulty is overcome in vacuum annealing.

The atmosphere requirements for copper alloys containing nickel, silver, or up to about 15% zinc or tin are similar to those for pure copper. Brasses containing more than about 15% zinc are difficult to heat treat without discolouration. Brasses oxidise readily in gases containing CO_2 and H_2O forming a film mainly of zinc oxide which provides some protection against further oxidation up to 600 °C. Brasses are also susceptible to staining by controlled atmospheres and lubricants containing sulphur. The major difficulty with brass is the significant loss by volatilization which may occur at temperatures above 450 °C. Volatilization is a function of the density of the controlled atmosphere and is particularly serious under vacuum conditions.

The requirements for nickel are similar to those for copper but nickel is slightly more resistant than copper to oxidation and attack by sulphur gases. Nickel alloys with metals which, like nickel, have oxides of high dissociation pressure can be bright annealed in exothermic atmospheres. On the other hand alloys containing chromium, for example, require very pure atmospheres for bright heat treatment.

Section 6.12 elaborates on atmosphere generation and applications.

6.8.3 Factors Affecting Equipment Selection

From the previous section, it is clear that a wide range of furnace and auxillary equipment is needed to cover the temperature ranges involved in metal heat-treatment (130 °C to around 1 000 °C). The equipment must be capable of control within close margins of temperature. In those cases where rapid quenching is employed, the ancillary materials-handling equipment can be as important as the furnace.

The effectiveness of a furnace or other system in heating the charge will depend mainly on the modes of heat transfer used. Radiation and convection are the modes principally concerned with heat transfer to the charge and the relative effects of each will depend largely on the operating temperature. At higher temperatures radiation will predominate, whilst at lower temperatures heat transfer by convection will be more important. Conduction heat transfer dictates the flow of heat within the charge and consequently the permissible charge surface temperatures and the temperature gradient through the charge. Radiation begins to predominate only above about 650 to 700 °C (see Figure 6. 144).

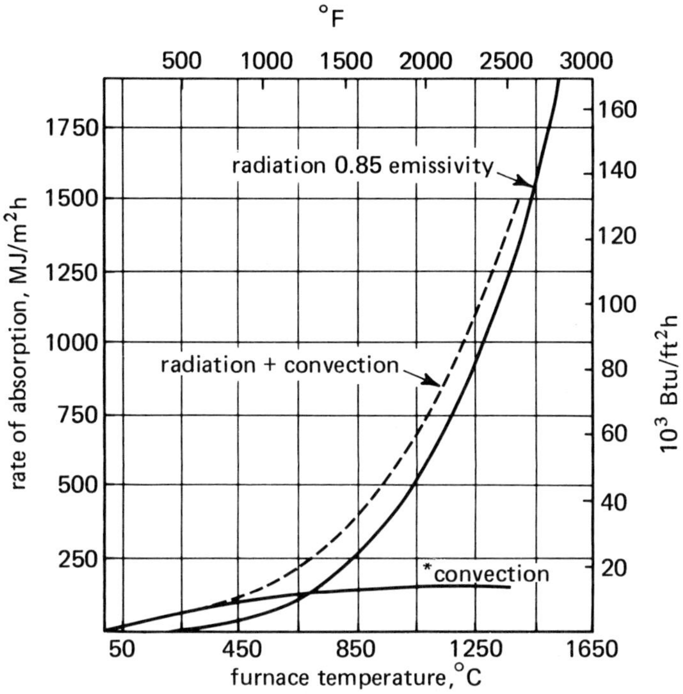

*convection curve based on 50 m/s with flow parallel to a smooth flat surface and air as the medium

Fig. 6.144 Effect of furnace temperature on rate of heat absorption

Figure 6.144 accounts for radiation from a furnace lining and the furnace gases as if they constituted a uniform source. However, in direct fired furnaces, the work chamber is frequently also the combustion chamber. When this is the case, the flame itself is often the source of temperatures well in excess of those required. Localised flames in 'sight' of the charge can not be tolerated at lower working temperatures as the required temperature uniformity of the charge would be adversely affected. The effect on the evenness of charge heating is especially important where several articles constitute one furnace load, all of which ideally need to complete their heating cycles simultaneously, to avoid fuel wastage. Uniform heat distribution throughout the charge is required. It is necessary, therefore, in those low-temperature processes to promote convective heat transfer by the use of fan- or jet-driven recirculation, often using an external combustion chamber to avoid the charge 'seeing' the flame. This is particularly important in the case of light-alloy treatment, especially solution treatment where, as indicated previously, the temperature tolerances are very small.

Recirculation at higher temperatures is also encouraged, particularly in intermittent furnaces where the natural tendency is for gases to stratify due to their buoyancy, and so lead to difficulty in keeping temperature distribution in the vertical direction uniform. Where the use of hot gas recirculation fans is difficult to justify, e.g. because of its first cost, successful use has been made of the momentum in the high speed jets from direct firing burners, performing a limited amount of entrainment and recirculation This is especially effective when an induction throat can be placed concentrically with a single jet. The objective in both cases is to encourage motion of the gases passing through the furnace, so as to overcome buoyancy effects, particularly where the furnace crown is relatively high, and to improve top-to-bottom temperature uniformity.

Factors influencing furnace design are:

- Liberation of heat by combustion.
- Transmission of heat to the charge.
- Heat losses through the furnace structure.
- Heat losses in combustion products.
- Aerodynamics of furnace gases.
- Effect of furnace atmosphere on charge.

These aspects are covered individually in other Chapters of this book. It is clear that different temperature cycles will require different furnace configurations, atmospheres and combustion systems.

Increasing labour charges and the continued growth of automation in industry has encouraged the development of increased automation in furnace operation. The trend is towards continuous furnaces with fully automatic stock transfer (into, through and out of the furnace) and combustion control.

6.8.4 Continuous Furnaces

Continuously operating furnaces may be classified into a number of types according to the method of progressing stock through the work chamber.

The seven main groups are:

- Conveyor furnaces.
- Roller hearth furnaces
- Revolving retort furnaces.
- Pusher furnaces.
- Walking beam and notched hearth furnaces.
- Rotary conveyor furnaces.
- Charge conveyor furnaces.

Pusher furnaces, walking beam and notched hearth furnaces are mainly used as re-heating furnaces and will be discussed in Section 6.9.

The operating temperature of a furnace will largely determine the type of burner system chosen.

Conveyor furnaces.

Conveyor furnaces have a hearth consisting of an endless chain or mesh belt kept moving by a drive motor such that articles loaded onto the belt at one end of the structure are conveyed, on the upper taut surface of the belt, through a long heated tunnel. The heat-treated components are then off-loaded at the exit end. Typical furnaces are shown in Figures 6.145 and 6.146.

Although conveyor furnaces may be used at temperatures up to 750 °C, or even higher, their use is often restricted to processes operating at much lower temperatures for which the operating tolerance

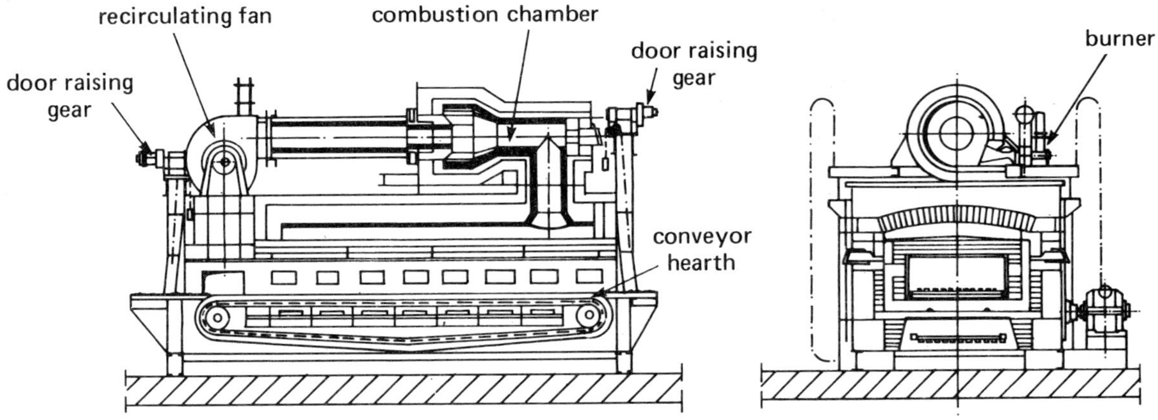

Fig. 6.145 Conveyor furnace with air-heat burner

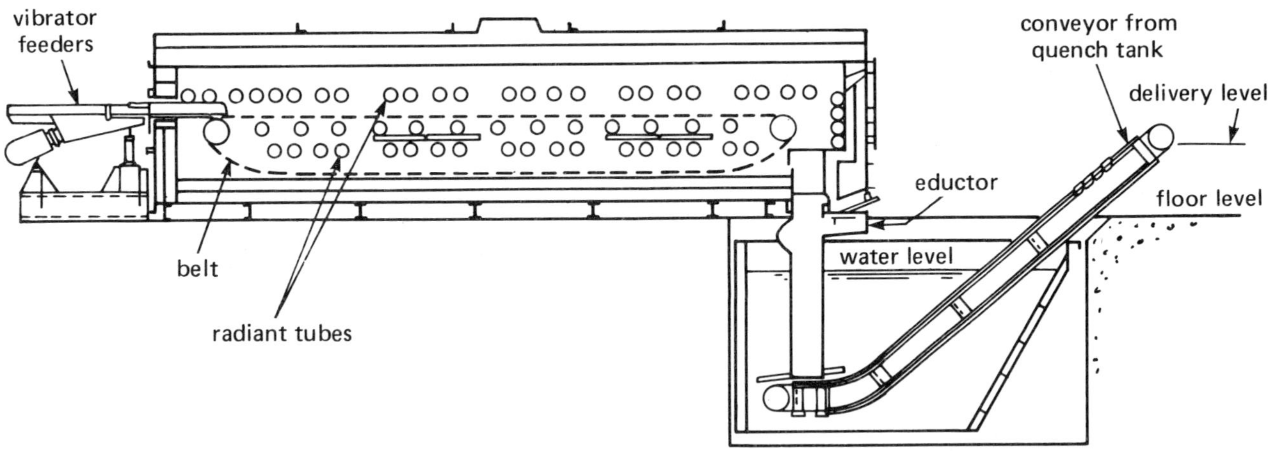

Fig. 6.146 Conveyor furnace with radiant tube burners

will be narrow and uniformity of temperature distribution of paramount importance. A typical application of a conveyor furnace would be for tempering small steel components.

The return pass of the belt may travel back through the furnace, or it may be taken beneath the furnace structure. However the belt travel is arranged, the belt will be subject to cyclical heating and cooling which in time leads to the formation of cracks. Also the maintenance of correct tension on the belt is important if belt life is to be preserved. This aspect, too, is affected by temperature uniformity, or lack of it, throughout the furnace cross-section. If parallel strands of chain are subjected to marked differences of temperature unequal expansion arises causing unequal distribution of stresses and premature failure.

The size of conveyor furnaces varies considerably, some being only 15 to 18 cm in width, whilst others may be a metre or more wide. In some the belt may carry fitments onto which the articles to be heated are placed or attached. In the case of furnaces having carrier attachments, the belt may be arranged to be wholly outside the furnace, the carrier arms projecting upwards or sideways, and travelling through a longitudinal slot in the furnace bottom or side. This arrangement is common for furnaces operating at around the temperature of red heat.

Conveyor furnaces may have several zones, each being controlled to a different temperature, so that articles in their progression through the furnace may be subjected to a prescribed heating and cooling cycle.

The burner arrangements for conveyor furnaces are numerous and may range from natural draught tunnel burners, to radiant plaques, radiant tube burners, and forced draught nozzle-mixing burners.

For furnaces operating towards the lower end of the temperature range, air-heat burners, firing into a stream of air greatly in excess of that required for complete combustion, are not uncommon. There would be a single burner and air supply for each zone of the furnace. The advantages of this arrangement are the ease of control and the improved temperature uniformity which results from moving a large body of hot gas at a temperature very close to the process temperature through the furnace (or each individual zone of the furnace). Circulation of the hot gases is by means of powerful fans which draw the gases in counterflow to the work and which also re-circulate some of the gases. The use of counterflow and recirculation techniques improve the economy of operation.

An important aspect of the design and performance of conveyor furnaces is the quantity of heat carried away in the belt, which in the case of furnaces in which the belt returns beneath the work chamber, may be considerable.

Roller hearth furnaces

The use of roller hearths offer some satisfactory solutions to a number of problems in furnace transportation. They are particularly useful in conveying plate or coiled strip materials, or sometimes bar stock. They are available for use at temperatures up to 1 200 °C.

In its simplest form the roller hearth consists of a number of horizontal, revolving rods driven from outside the furnace. The drive is frequently from a long lay shaft with worm or bevel gear transmission to the rollers. Where especially heavy stock is being treated there may be individual drive motors to several rollers, each motor turning a relatively short lay shaft serving a few rolls only. The rollers may be constructed entirely of metal or they may be refractory-covered. For the conveyance of sheet or plate, rolls which consist of a number of discs mounted on cross shafts are frequently used, the discs being staggered on successive shafts. Small components may be loaded into trays, or into racks. The shafts of the rollers, of whatever construction, are almost invariably tubular and water-cooled, a suitable rotatable gland at each end accommodating the water inlet and outlet connections. The water cooling can represent a considerable loss of heat from furnaces of this kind.

Suitable allowance should be made for both longitudinal and radial expansion of the rolls as they achieve working temperature. Rolls must also be kept rotating, irrespective of whether or not the furnace is charged, to prevent them sagging. The rotational speed must be sufficient to ensure temperature uniformity around the circumference of each roller and prevent warping.

One of the main advantages of the roller hearth furnace over conveyor furnaces is that there is less creep in the rolls than there is in the chains or mattresses of conveyor furnaces. The segments of each roll are alternatively under tension and compression during rotation whereas the belt of a conveyor is always under deliberately induced tension. Roller hearths are frequently designed to accommodate about 1% creep in 10 000 hours of use.

As far as the firing arrangements for roller hearth furnaces are concerned much depends upon the intended temperature of operation, but probably the most common is the use of nozzle-mixing gas burners. Nozzle-mixing burners offer the option of using preheated combustion air, the preheat coming from the waste gases via a recuperator. The waste gas flow in the furnace and direction of travel of the stock should, as in all continuous furnaces, be arranged in counterflow to assist in optimising the economics of operation. For furnaces operating above about 700 °C, recuperative burners can be used.

Roller hearth furnaces are quite frequently used with protective atmospheres in which case the following points should be considered:

- The burner system will almost certainly require isolation from the work chamber and the use of single ended radiant tubes is suggested, particularly since these may also incorporate recuperation.
- Where the roll axles pass through the sidewalls, the construction must be such as to minimise the escape of the protective gases.
- There must be an effective door or curtain seal where the work enters and leaves the furnace to minimise the escape of protective gases.

Figures 6.147 and 6.148 show respectively the construction of a refractory-sheathed roller and an interior view of a furnace with disc shafts. Many of these furnaces are constructed with lift-off roofs to give easier access to the rolls for maintenance purposes. Many of them are very large with up to

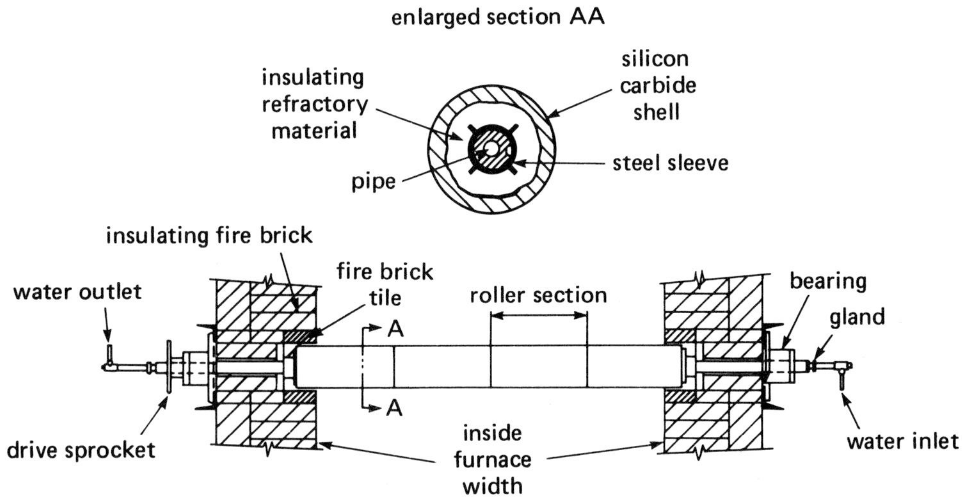

Fig. 6.147 Furnace roller with shell of silicon carbide

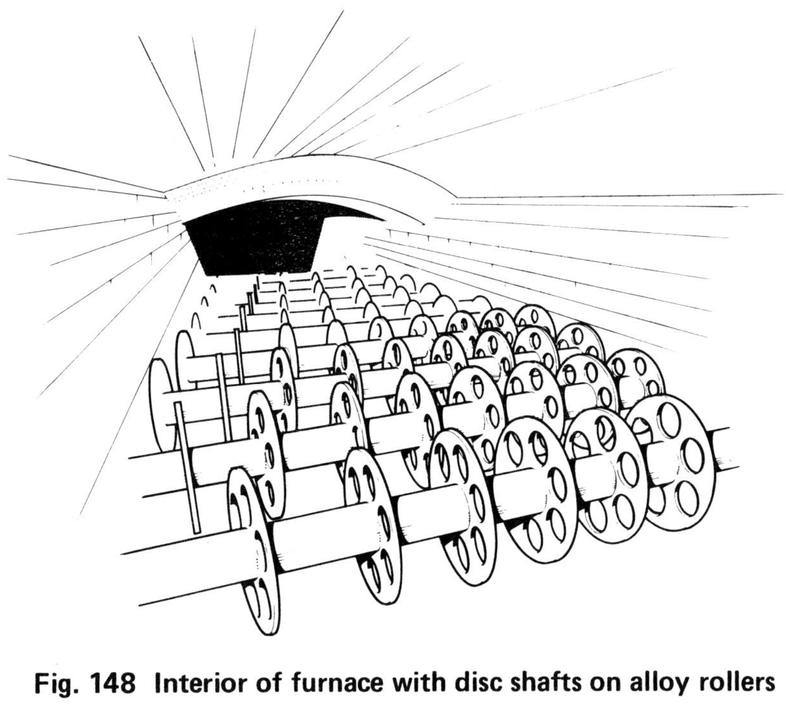

Fig. 148 Interior of furnace with disc shafts on alloy rollers

perhaps, twenty separately controlled firing zones. A furnace of that size would have some eighty to a hundred burners of about 35 to 40 kW rating each. The burners would typically be arranged in equal numbers above and below the hearth to produce temperature uniformity in the load. The speed of travel of the load through the furnace depends upon the range of the material being treated and the highest temperature required.

Revolving retort furnaces

For the heating of small articles the stock may be conveyed through a furnace, which is drum shaped, by a helical motion. The material is charged at one end of the drum and rotation of the drum carries the stock up one side in a plane at right angles to the axis of rotation. The length of arc by which the stock is lifted depends upon the coefficient of friction between the surface of the drum and the material being heated, but, at a certain height the article detaches from the side of the drum and falls back to the lowest point of the arc. If the axis of the drum is tilted with respect to the horizontal the stock will travel from the input to output end in a zig-zag motion. The rate at which stock traverses the furnace is dependent upon the degree of inclination of the drum and upon its speed of rotation. In some versions of this kind of furnace both the angle of tilt and speed of rotation may be adjustable.

Revolving retort furnaces are used for the annealing, hardening and tempering of small workpieces of fairly regular dimensions such as coil bianks, eyelets, ferrules, buttons, saw teeth, tacks, rivets, screws, springs etc., in fact any object which will pass freely through the drum.

One class of workpiece which, somewhat surprisingly, is not amenable to the inclined drum furnace is ball bearings. Balls and rollers pass too freely through the drum and to heat treat these a horizontal drum with an internal Archimedean screw is used. The rate of progression of the work through the drum is then a precise function of the pitch of the screw and the rate of revolution of the drum.

The operation of revolving retort furnaces can be made completely automatic by the use of mechanical feed and extraction devices. However, this often prevents the use of protective atmospheres as the presence of the feeding and extraction mechanism makes effective sealing of the furnace chamber difficult.

The firing arrangements for revolving retort furnaces often present some problems. Small furnaces may be fired with a single burner through one end of the drum; larger furnaces may be fired externally to the drum with the products of combustion passing round the outside of the drum before passing through it via gas ports in the drum ends.

A horizontal revolving drum with internal screw is shown in Figure 6.149

Rotary hearth furnace

For many purposes circular motion is more convenient than linear motion and in the case of rotary hearth furnaces, in which the whole hearth bed rotates, there are several other advantages. There is no relative movement between the stock and the hearth, so that hearth wear is reduced; there is no need for water-cooled rails or supports, so that heat loss is reduced; and no mechanical transport mechanisms need project into the furnace chamber, as is the case in roller-hearth, conveyor and walking-beam furnaces. The mechanism for rotating the hearth may be completely protected from the effects of high temperature by the mass of hearth structure above it., The one problem which requires care in the furnace construction is the seal between the rotating hearth and the side walls;

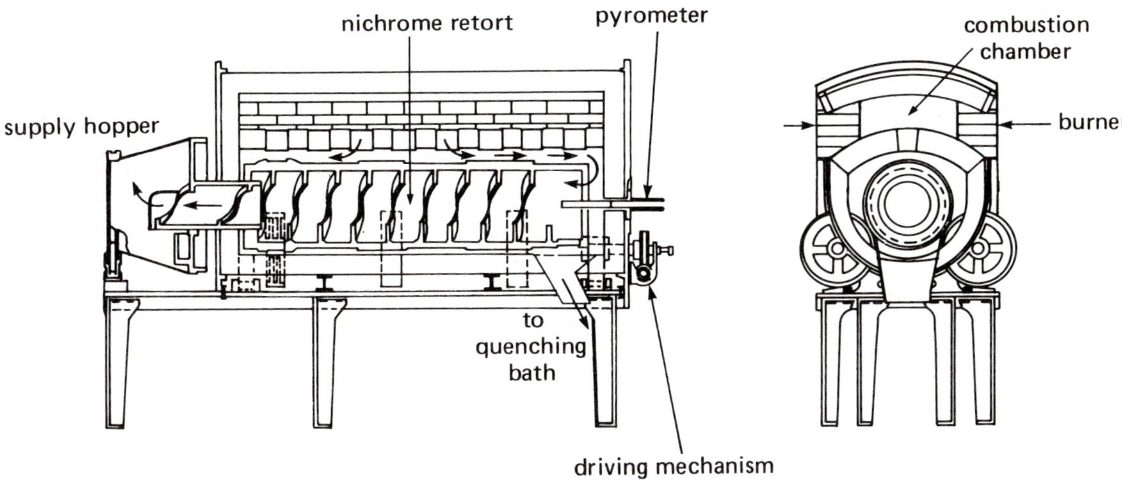

Fig. 6.149 Revolving drum furnace

it is equally undesirable for there to be an escape of hot gas from the furnace, or an ingress of cold air. If a protective atmosphere is used with the furnace, these contingencies become even more undesirable. There is usually some kind of baffle construction at the joint of hearth with side-wall, sometimes combined with a sand seal.

Like other forms of continuous furnace, rotary hearth furnaces vary greatly in size, temperature of operation, and type of work-piece treated. Their diameter may vary from a metre up to about 10 metres. They may be used for the whole range of heat treatments from tempering up to forging temperatures and they will handle work of considerable size ranging from small components to ingots weighing many tonnes. They are particularly adaptable for irregularly shaped objects which would be difficult to push, or transport otherwise, through linear furnaces. Typical furnaces are shown in Figure 6.150.

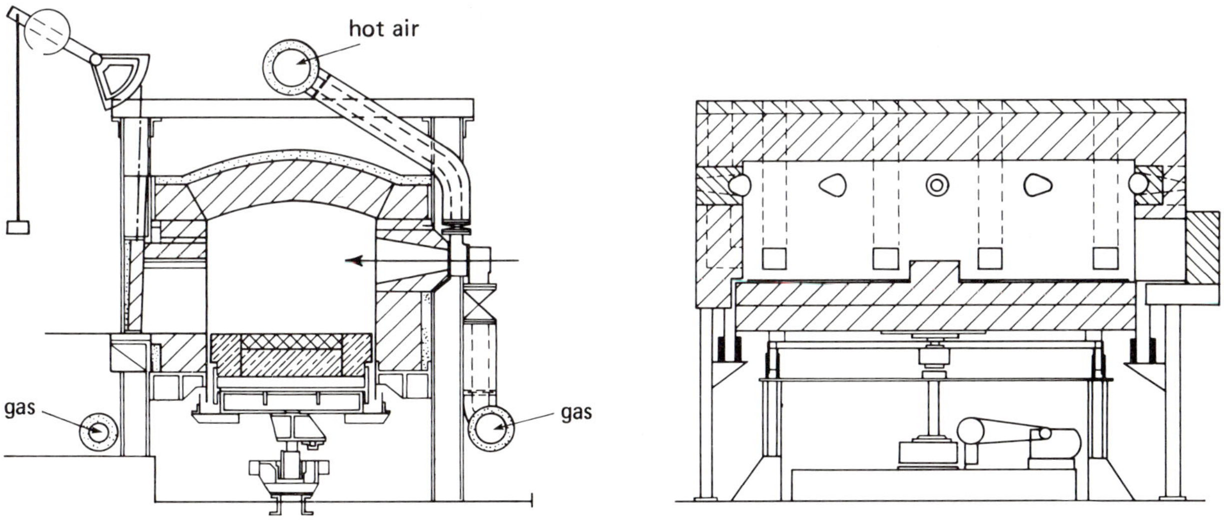

Fig. 6.150 Rotary hearth furnace

Rotary hearth furnaces may be equipped with only one door, or with two doors for loading and unloading, but an important part of the furnace structure is the mechanism for off-loading heated work and on-loading new cold work. Where pieces of some considerable size and weight are being worked the loading and unloading machinery may itself be massive and very rugged in its construction. Also, where large work is being processed, instead of having a door on the furnace, a whole segment of the furnace side-wall and roof may be removable, to facilitate access to the hearth. 'Trouble' doors are seldom necessary since there is no tendency for climbing to occur and scale may be removed via the charging door. The burners on rotary hearth furnaces are normally arranged to fire tangentially to the side wall. Where in the past numbers of air-blast or nozzle-mixing burners may have been used, distributed around the circumference, there is a growing tendency to use recuperative burners.

The higher efflux velocity of hot gases from the recuperative burner brings improved temperature distribution within the furnace, making it feasible to use fewer burners. The recovery of heat from the outgoing products clearly has benefits on the economy of operation.

Charge conveyor furnace

Wires, chains and strip metals may be pulled through furnaces under tension and thus made to act as their own conveyors. In the wire industry in particular, this method of conveying has been in use for a very long time. The stock is heated whilst being pulled through the furnace, which may be an open furnace, i.e. the stock is subject to contact with the products of combustion, or a muffle furnace in which the stock is protected from the products of combustion. The interior of the muffle may be filled with a protective atmosphere if the metallurgy of the process necessitates.

Processing of the stock may be completed by continuously drawing it through a lead bath, or salt bath, or quench tank, as necessary. For wire processing, several strands may be drawn through the furnace side by side, or each strand may be drawn through its own tube, which then acts as a muffle for that strand. To be truly continuous, furnaces for heating wire and strip must be equipped at both ends with loops which contain sufficient material to maintain the flow through the furnace whilst a new coil is put into place and whilst a processed coil is removed. Furnaces of the sort described take up a great deal of floor space if arranged horizontally, so there is a tendency to build this kind of furnace as a vertical tower or to loop the wire or strip in zig-zag fashion. Both horizontal and vertical types are shown in Figure 6.151.

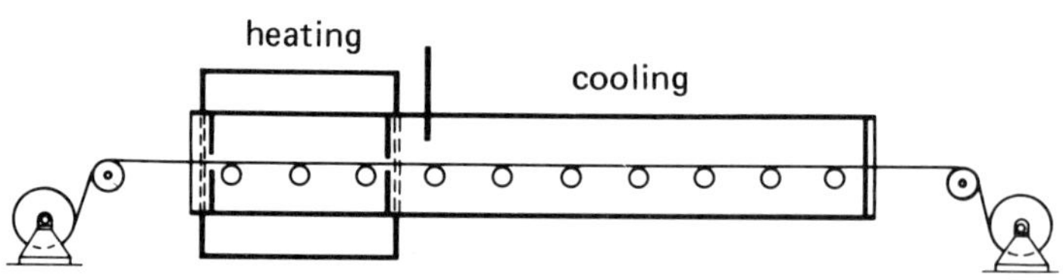

continuous furnace for heating strip steel - the strip is pulled through the heating and cooling sections

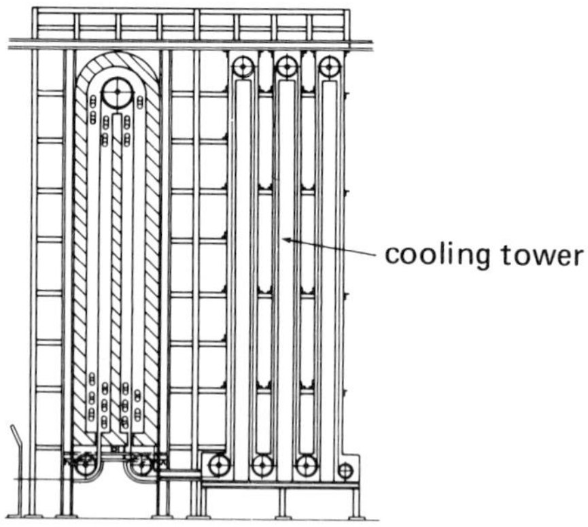

tower type furnace for annealing strip steel - in the heating tower (left) radiant tubes are provided all the way up

Fig. 6. 151 Charge conveyor furnaces for heating wire or strip

Burner systems

Remarks have already been made concerning the selection of burners for continuous furnaces but some further comment about overall burner systems is required. In gas fired furnaces the burner systems must, of course, now comply in their construction and methods of operation with British Gas plc IM/12 : Code of Practice for the Use of Gas in High Temperature Plant, according to whether the operating temperature of the plant is below or above 750 °C at the walls of the work chamber.

Where plant is expected to comply with IM/18 (Low Temperature Code), then flame detection systems will be expected and only in special circumstances, as set down in IM/18, Section 5.2.5 for Multi Burner Plant, will the omission of flame safeguards be acceptable.

IM/12 states a strong preference for the use of flame safeguards but recognises that where the plant spends long periods of time at temperatures above 750 °C, with infrequent lighting-up, such safeguards may have limited value and may be dispensed with, providing certain other provisions are met. Reference should be made to Section 5 of IM/12.

One of the objections levelled at the use of flame safeguards in the past has been the complexity and ensuing cost of equipping a large number of burners with safeguards. In this respect the use of recuperative burners frequently brings benefits in that it is usually possible to fire any given furnace with a smaller number of recuperative burners than with burners of more conventional design, so that the cost of providing flame protection should be correspondingly reduced.

Another aspect of the burner system which requires careful consideration is air/gas ratio control. Energy conservation demands that the furnace is operated at an air/gas ratio which is as close to stoichiometric as is reasonably practicable, although the metallurgical requirements of the process may have an overriding influence. The process may require changes in the air/gas ratio at certain times in the process cycle, or when certain temperatures are achieved.

Refer to Chapters 2, 9 and 10 for details of burner systems and controls.

6.8.5 Batch Furnaces

A wide variety of direct and indirect gas fired batch or intermittent furnaces are in use. They are often categorised on the basis of the charging and discharging methods employed and partly on the size of stock they can accommodate.

Many batch furnaces, designed to cater for a number of different sizes of components and types of treatment, have large hearth areas and high crowns. For a large proportion of their operating periods they can be used for small components giving a low hearth area loading leading to inefficiency. Whilst flexibility can be desirable (e.g. in some large forges, furnaces are used for re-heating as well as heat treatment) such flexibility is normally a trade-off against efficiency.

Heat treatment processes frequently employ indirect firing. This is generally the case with annealing processes for non-ferrous metals. Many steel heat treatment processes also require indirect firing, i.e. for stainless steels, nickel alloys and various annealing processes. Here, the radiant tube burner can frequently be used. The nickel-chrome alloy radiant tube can be used for various metallurgical heat treatments including vitreous enamelling, bright annealing, carburising, hardening and tempering. For higher temperature work such as sintering of ferrous alloys and bright annealing of stainless steel the ceramic radiant tube can be used.

Semi muffle furnace

In many heat treatment operations the products of combustion (certainly with gas fired plant) will have no serious effects on the furnace load. However, direct flame impingement on the stock should be avoided to prevent decarburisation and other deleterious effects. This is provided for by the inclusion of an inner brickwork structure which supports the load but allows only the products of combustion to circulate round the stock, as shown in Figure 6.152. In the diagram the burners fire under a load supporting tile, whilst in larger furnaces the tile or inner wall is higher inside the furnace to permit over-fired burners as well. In the brick industry the same purpose is served by the bag wall.

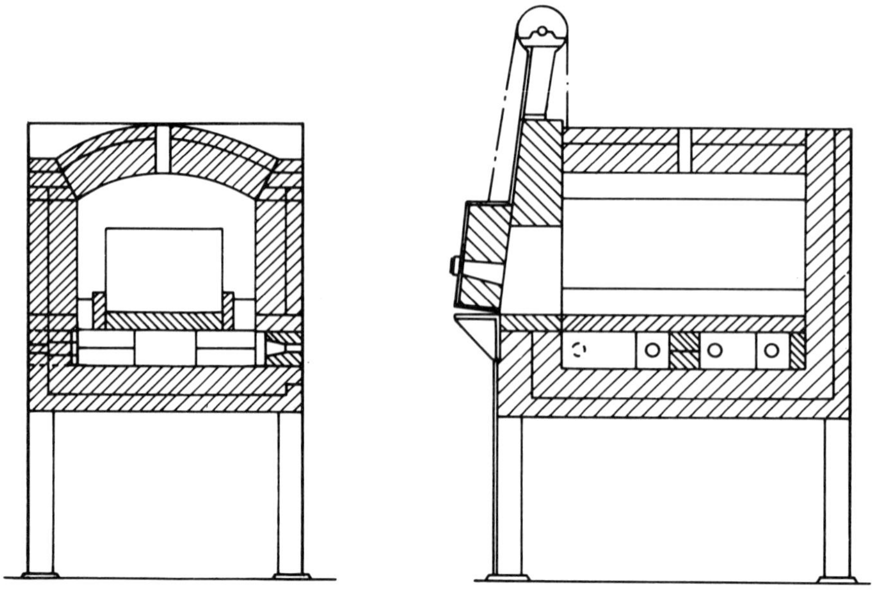

Fig. 6.152 Small batch-type, under-fired, semi muffle heat treatment furnace

Full muffle furnace

An extension of the semi muffle furnace is the full muffle system. This is necessary where not only the flame but the products of combustion, even with natural gas firing, should be excluded from the stock. An inner brickwork structure completely surrounds the stock, so that heat transfer is by radiation only from the inner brickwork refractory, as shown in Figure 6.153. This system is generally limited to small furnaces because of the extra cost of the inner brickwork. For heat treatment of larger articles other techniques are used.

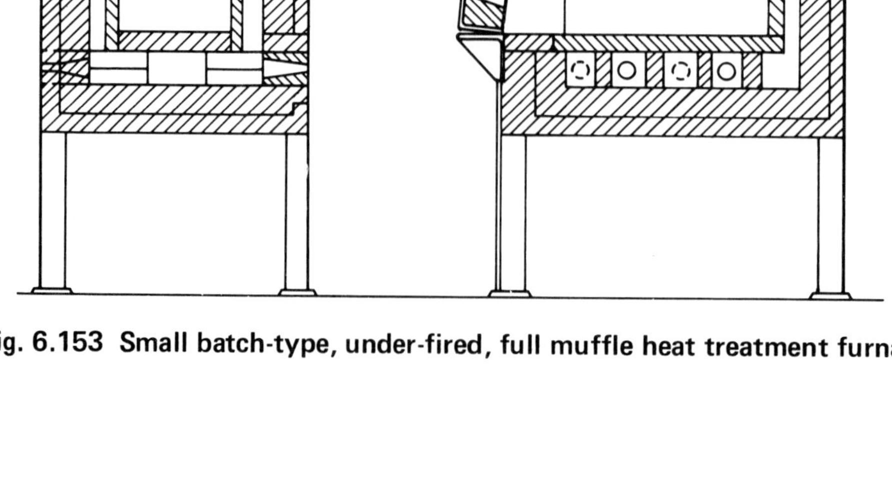

Fig. 6.153 Small batch-type, under-fired, full muffle heat treatment furnace

A modification of this technique is the portable cover furnace fired by direct burners as shown in Figure 6.154. In this type of batch annealer, radiation from the burners is indirect because the inner cover is interposed; heat transfer is from the cover to the work, assisted by convection set up by the fan in the stand base. Here the burner gases themselves activate recirculation outside the inner cover. The life of the ring covers has a pronounced bearing on process economy, and the disposition of direct firing burners will have a distinct effect on deterioration rates of the inner cover material.

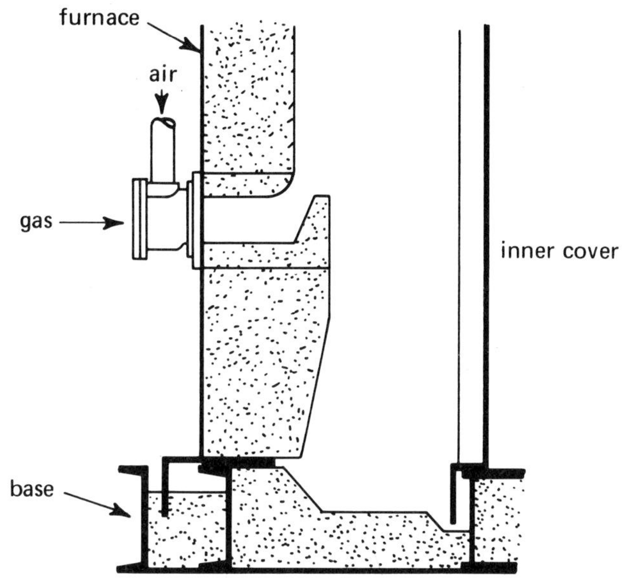

Fig. 6.154 Direct-fired burner on batch annealer portable cover furnace

Radiant tube furnaces

As an alternative to the muffle furnace, radiant tubes can be used where the products of combustion must not come into contact with the stock. The tubes are of either non-metallic or metallic construction depending on the process temperature.

The use of radiant tubes has the following advantages:

- A radiant tube gives up heat uniformly along its length, whether parallel-mixing post aerated burners, or intensive recuperative burners are used.
- Since the combustion is carried out in a space completely separated from the work chamber, there is no contact between work and combustion gases. This enables expensive muffle covers to be dispensed with, and the radiant tubes can be installed in the direct vicinity of the charge. Similarly, the use of protective furnace atmospheres is facilitated.
- Heat transfer rates are high.
- Radiant tubes can be applied to all types of furnace lay-out.
- Temperature control is simple and accurate, and temperature fluctuations of only $\pm 3\,^{\circ}C$ can be expected.

These tubes are used extensively in the annealing of non-ferrous metals where protective atmospheres are distributed through the load, and also in the heat treatment of special steels and certain enamelling processes.

Portable cover annealing furnaces

In batch annealing, steelstrip, coiled after rolling, is annealed either as close coils or opened coils, in single- or multiple-base furnaces. A distinct advantage is conferred in the ability to process mixed orders and to include small orders amongst larger ones, but disadvantages also arise, especially in non-uniformity of temperature-time history between the inner and outer wraps of the coils. The equipment used comprises furnace bases, inner covers for enclosing the charge and the cover furnace. The cover contains the burners, supply piping, burner controls and safety equipment. One such cover serves several bases, so that coils can be loaded onto other bases as the furnace is firing. On some designs the base itself is sunk into a pit, so that the cover can be reduced in size. The inner cover is a radiation shield, preventing severe temperature non-uniformity near to the source of heat, and is also sealed off from the furnace cover, using a sand seal, so that the atmosphere within the inner cover can be selected to suit the charge. The box contains a hearth, air convection guides, and a motor driven fan (see Figure 6.155). A single base can usually accept up to 7.5 t of steel, but recent tendencies to wider strip will involve furnaces accepting up to 10 t per base.

Conventionally, the majority of batch strip annealers are fired by alloy radiant tubes, the thermal energy of combustion within the tube being transmitted by a combination of radiation and convection. The radiant tube may be either installed vertically, as shown, or in the form of hair pin or thimble loops. Some tubes are now being fitted with recuperative burners to maximise combustion economy.

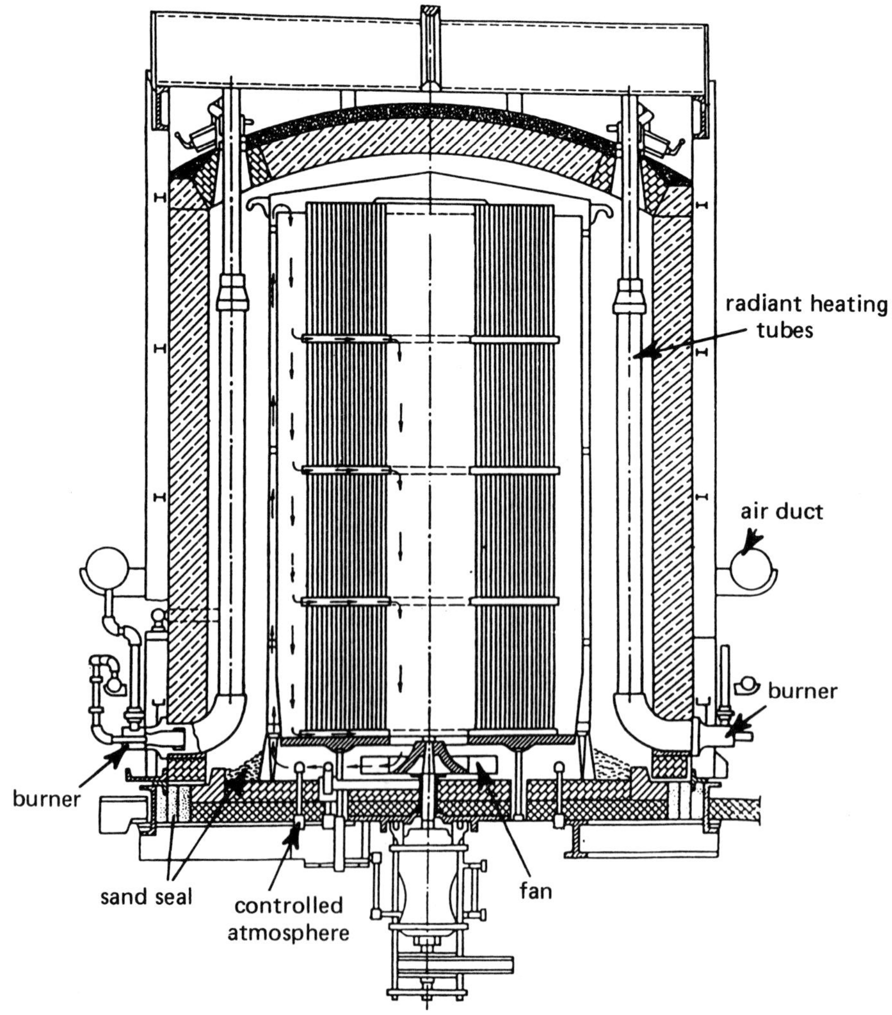

Fig. 6.155 Cover annealing furnace for steel strip

A modification of the type of furnace for annealing of very large steel bars are the lift-off cover furnaces and split-hood furnaces. These covers contain all the burners, controls, pipework etc. and are invariably lined with stack-bonded ceramic fibre insulation for lightness and efficiency. As these furnaces are fuelled by natural gas they are direct fired without any inner cover. Because of the lightness and high temperature capability of stack-bonded ceramic fibres, many direct lift-off cover furnaces are now being constructed in prefabricated modules for both reheat and annealing operations.

Bogie hearth and fixed hearth furnaces

These furnaces used for heat treatment are very similar to the furnaces used for reheating prior to rolling and forging. Indeed many furnaces are multi-purpose and can be used for either reheating or heat treatment depending on the temperature setting.

A slight variation is the car-bottom type furnace shown in Figure 6.156 where the door is integral with the moving hearth. This furnace may be under- or over-fired, or both as shown, and is used for stress relieving, annealing, normalising or carburising in boxes.

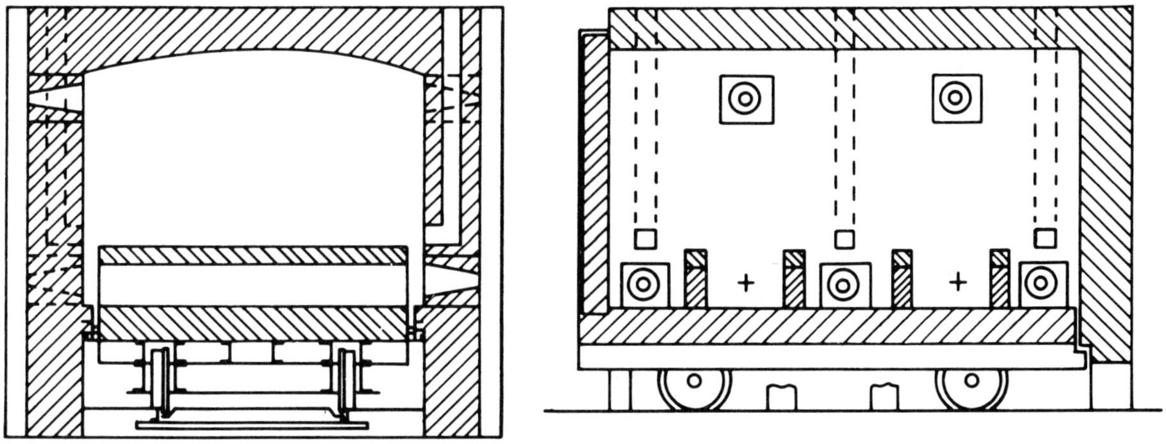

Fig. 6.156 Car-bottom type furnace

Fixed hearth furnaces are frequently used for the heat treatment of small to medium sized components. For the hardening of circular steel saw plates at 900 °C one such furnace was equipped with twelve flat flame burners in the furnace roof. These gave good temperature distribution without flame impingement. The furnace was lined throughout with ceramic fibre insulation. After 45 minutes at this temperature the plates are quenched, and then tempered at 600 °C in another fixed hearth furnace, fired by high velocity burners situated at each corner to give close temperature uniformity.

An example of a bogie hearth heat treatment furnace employing modular panels has furnace dimensions 13.7 m x 3.7 m x 2.4 m high of 50 000 kg capacity for marine propeller shafts. It is fired by two high velocity burners having a total maximum rating of 1 172 kW. They are positioned to fire parallel with the side walls from opposite ends of the furnace, which is octagonal in shape to allow adequate combustion space and prevent flame impingement on the load (see Figure 6.157).

Recuperators are mounted at each end of the furnace to preheat the combustion air. The firing configuration enables the burner to produce a temperature uniformity of ±8 °C at 650 °C and above. Operating temperatures range from 1 000 °C maximum to 500 °C minimum, the normal design temperature being 1 050 °C. A maximum combustion air preheat temperature of 650 °C is attained at temperatures above 800 °C. The novel feature of the furnace is that it is pivoted at one end like crocodile jaws and lifts at the other end to allow entry and removal of the bogie. In order to accommodate occasional extra long loads, a hinged self sealing door is incorporated in the end wall.

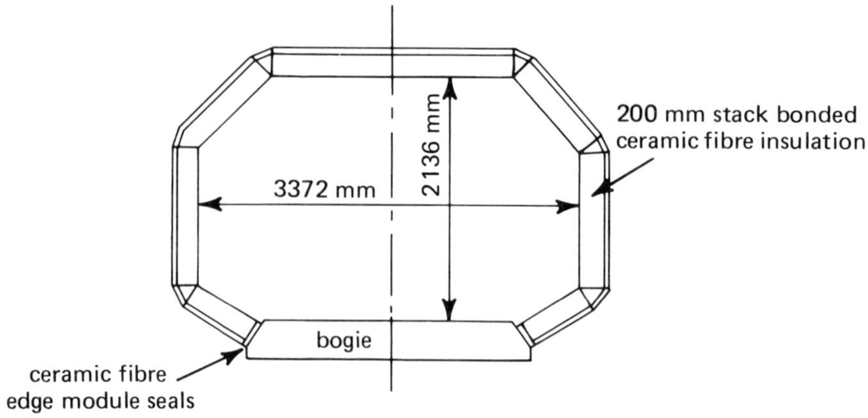

Fig. 6.157 Cross-section of 'crocodile' bogie hearth furnace

The body is constructed from standard modules having a 200 mm thickness of stack-bonded ceramic fibre insulation with a 50 mm layer of fibre held back to the casing by means of stainless steel anchors and Expamet squares. Further layers of fibre 100 mm and 50 mm respectively are then cemented into position, ensuring that all joints are adequately covered. The stack-bonded system gives a tight surface which is highly resistant to erosion. The furnace gave a 75% fuel saving on the furnace it replaced.

Aluminium coil bogie hearth annealing furnaces

After aluminium slabs from the soaking pits have been successively rolled in a breaking down mill and further in a finishing mill, the sheet has become progressively hardened. By this time it has been side trimmed and cropped either end and rolled into a coil weighing about 7 t. It must be softened by annealing before further fabrication work can be performed. A full anneal to make the sheet fully soft is carried out at temperatures between 350 and 450 °C for about 12 hours. If partial annealing is required, for example for tempering, the process temperature is between 200 and 300 °C for about 16 hours.

A typical batch furnace for this is shown in Figure 6.158. Heating is by vertical U-type radiant tubes with a controlled atmosphere of nitrogen being fed underneath the load. Coils are loaded onto a bogie which is then pushed into the furnace by a charging machine.

Lead and salt baths

Before the rapid development of the use of gaseous inert atmospheres, salt and lead baths offered the best, if not the only means, for heating with perfect freedom from scaling or decarburisation. These baths are a variant of the muffle furnace, as they surround the charge with a liquid protective atmosphere (see Figure 6.159).

The working temperatures of salt baths vary between 175 °C and 1 280 °C depending on the salt used. Advantages claimed for salt baths are:

- Faster heating by conduction; salt baths bring work up to temperature much faster than radiation or convection type furnaces in the medium temperature range such as 815 °C to 870 °C.

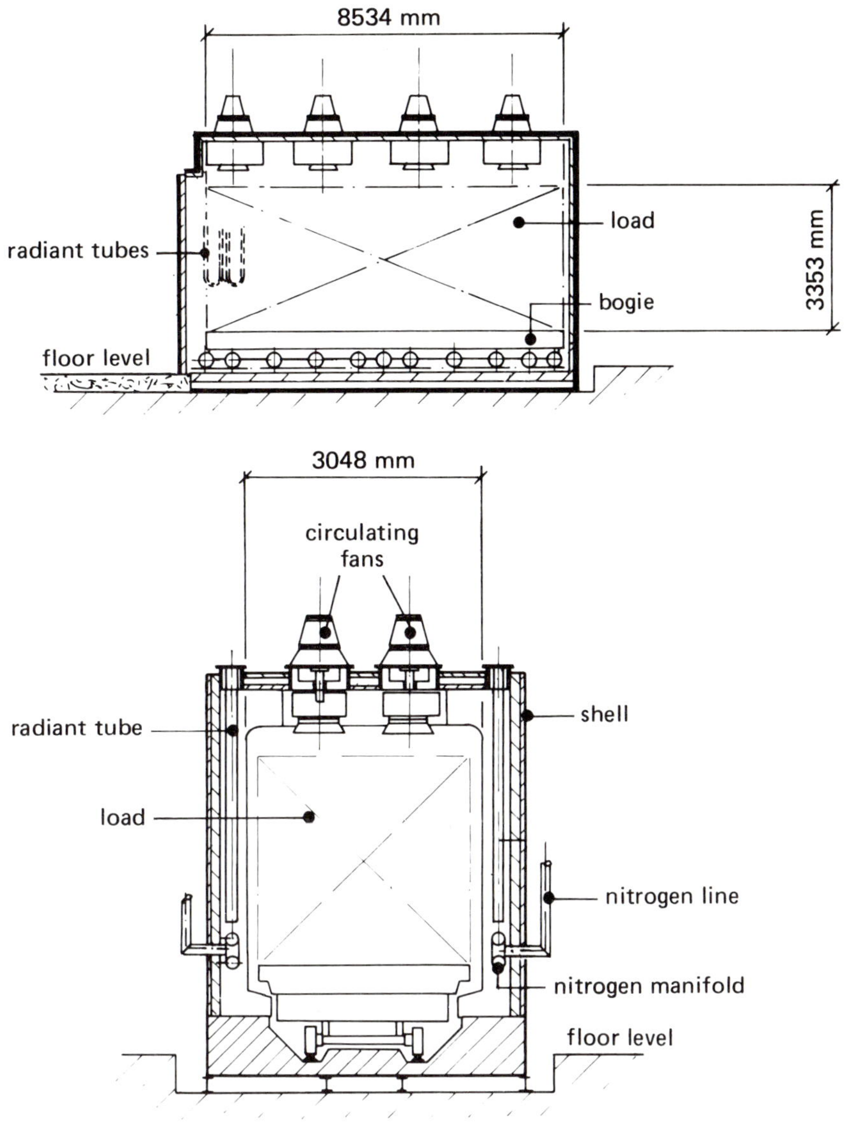

Fig. 6.158 Aluminium coil annealing furnace

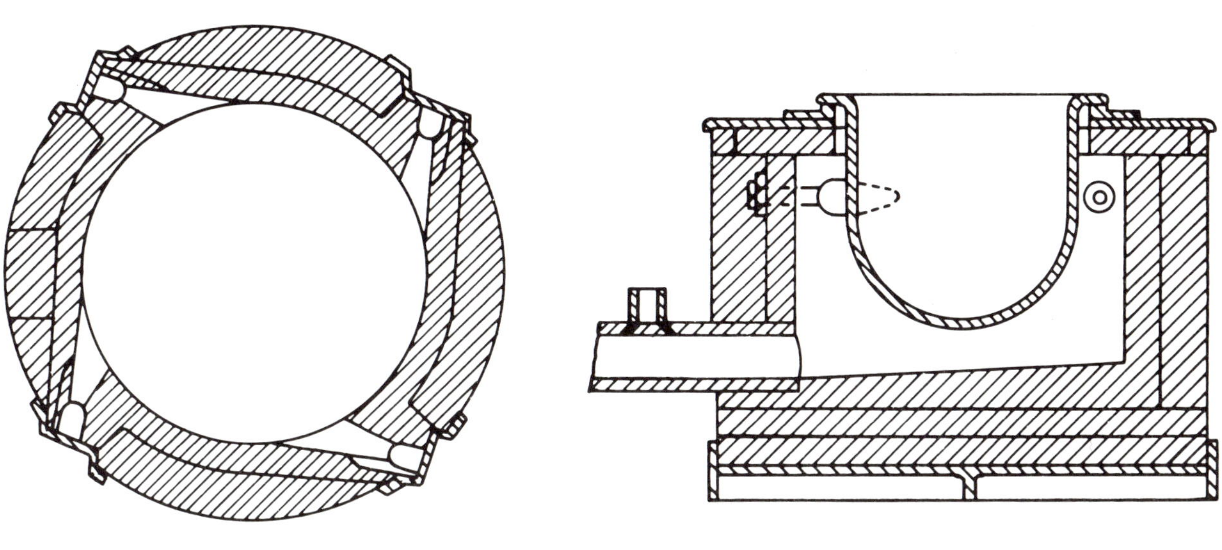

Fig. 6.159 Circular pot furnace for salt bath hardening

- Close control of temperature uniformity which results in uniform heating and avoidance of overheating.
- Selective heating is easy by immersing that portion only of the work to be treated. The fast heating inherent in salt gives a sharp line of demarcation between the heated and unheated portions of the work.
- Reduced distortion is possible due to several factors. Work when immersed in salt displaces its own volume and hence is supported to a degree and weighs 25% to 35% less than in air.
- Eliminates atmosphere problems since no air contacts the parts; scaling, oxidisation and decarburisation are eliminated.

The disadvantages of salt baths include:

- Molten salts are potentially hazardous and require special precautions; operators must be protected from splashing or dropping of the salt.
- Salt must never be allowed to come in contact with combustibles or reducing agents such as magnesium or cyanide.
- Parts entering the bath must be clean and dry. They must also be free from pockets or cavities that contain air or other gases as the possibility of explosion exists.

Maintenance on pots is considerable. Despite these apparent disadvantages salt baths continue to be extensively used.

Lead baths are also used for hardening small tools and used to a considerable degree in the wire trade for annealing, cooling and/or tempering of wire in continuous furnaces. There are certain disadvantages in the use of lead. Steel is lighter than lead and various jigs and fixtures have to be used to keep the steel immersed. It is readily oxidised and begins to vapourise at about 650 °C and volatilizes more and more rapidly as it is heated above that point. For that reason lead pots should be equipped with hoods for carrying away poisonous fumes. Sometimes lead baths are covered with charcoal to prevent evaporation.

The baths referred to so far are intended to transmit heat to the material immersed in them, and not to have any chemical action. Cyanide salt baths are an exception as these are used for producing a thin case of hard material on soft steel. As with other hardening agents the depth of case is a function of time and temperature. Cyanide vapours are extremely poisonous and hoods must be provided for carrying away fumes.

Barium salts are also used for hardening special alloys. Hardening of articles in the range 700 to 900 °C is achieved with one type of barium salt, and a delay quench by immersing in another salt is achieved in the range 200 to 500 °C.

Fluidised bed furnaces

The fluidised bed technique is increasingly being used for the heat treatment of small to medium sized metallic components. Components are submerged in a bed of finely divided sand particles which behaves like a liquid when hot air or combustion products are passed through it.

Fluidised bed heating in the past was achieved by a heat source external to the bed e.g. an air-heater. In an externally electric heated bed, the time to reach operating temperatures could be excessive. Bed heating by internal combustion of an air/gas mixture ignited in the bed, where the heat source and the fluidising medium are one and the same, increases heat transfer to the bed significantly. On start-up the air-gas mixture is ignited above the bed and quickly imparts its heat to the rapidly moving bed particles. After a period of time, combustion takes place within the bed, normally when the bed temperature has reached between 600 and 800 °C. A typical furnace is illustrated by Figure 6.160.

The disadvantages of this system are:

- For optimum control, the heat input must be independently controlled from the total fluidising medium. Fluidising velocity must be kept within the limits of that required to promote fluidisation and that which causes bed particles to be elutriated.

- High temperatures can be generated in the bed immediately above the distribution plate, which could cause damage to the distributor.

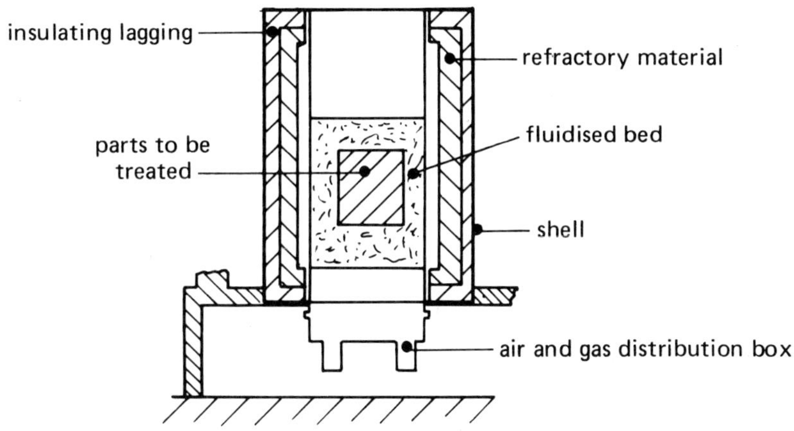

Fig. 6.160 Fluidised bed furnace

The potential heat treatment uses of fluidised beds are numerous, but each application must be decided on its own merit to obtain the best economic advantage in comparison with conventional furnaces. They have been used for the heat treatment of tool and die steels at temperatures between 700 and 1 025 °C. Another example is the hardening and tempering of constructional steels in a furnace 500 mm diameter x 600 mm deep with an output of 160 kg/h. They are also used in various carburising operations.

Fluidised bed furnaces offer the heat treatment industry significant improvements in performance and properties. The high heat transfer coefficients obtained enable products to be heated or cooled at speeds very close to those obtained in salt or lead baths. Temperature uniformity within fluidised bed furnaces is very good, being within ± 3 °C at operating temperatures. Another advantage is the recovery rates that can be achieved in relatively small work zones.

The characteristics of high heat transfer rates, fuel economy, rapid carburising and environmental control provided by fluidised bed furnaces will ensure their increased use for heat treatment.

On-site stress relieving

Welded pressure vessels are sometimes so large that stress relief can not be carried out in a conventional furnace. The technique of on-site stress relief of such vessels by using the vessel itself (suitably insulated) as the furnace, in which a high velocity burner(s) promotes recirculation of combustion products around a temporary heat-distribution system, is now well established. Temperature uniformity of ±10 °C can be achieved and fuel consumption is low, e.g. 530 MJ/t at 650 °C. Figure 6.161 illustrates the system applied to a 21 m high column.

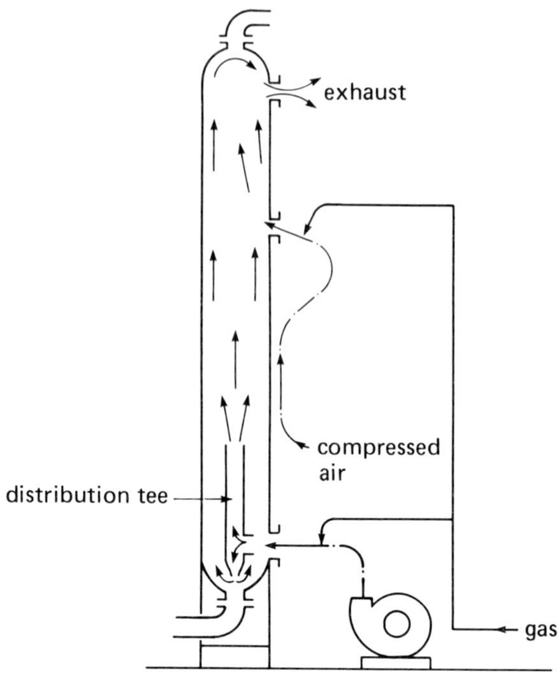

Fig. 6.161 On-site stress relief of tall columns

6.8.6 Regenerative Burner Applications

Regenerative burners (discussed in Chapter 2) have been proven and successfully demonstrated in a number of furnaces and the range of applications is increasing. Most of the installations carried out have been on fairly large furnace plant. It is likely that many of the future applications will involve smaller furnace units and the need is recognised for the development of cost effective regenerative burners of smaller sizes.

Batch furnaces

Examples of applications of regenerative burners to batch heat treatment furnaces are at British Steel's Llanwern works which specialises in the production of steel sheets. This process involves an annealing operation in which coils of sheet are heated in a protective atmosphere of nitrogen and hydrogen to 700 °C and held at this temperature for up to 25 hours. The annealing furnaces are large lift-off cover units in which the coils of steel are arranged in four stacks, each stack being fitted with an individual cover made of heat resistant steel inside which the protective atmosphere is supplied. A thermocouple in the furnace cover monitors the temperature of the enclosure outside the muffle and this temperature is raised to 900 °C over a period of between 6 and 10 hours. This is then held until the temperature of the recirculating atmosphere gas inside one of the muffles reaches the 700 °C annealing temperature. The burners are then controlled from the muffle to maintain the temperature at 700 °C for an appropriate soaking time dependent upon the material being annealed.

In common with other heat treatment operations the requirements for accurate temperature control and for provision of temperature uniformity within the furnace are of prime importance. Previously this was achieved using a large number of small burners situated along each side of the furnace. Two pairs of 1.4 MW regenerative burners were specified for this installation with the pairs situated in each end of the rectangular furnace as shown in Figure 6.162.

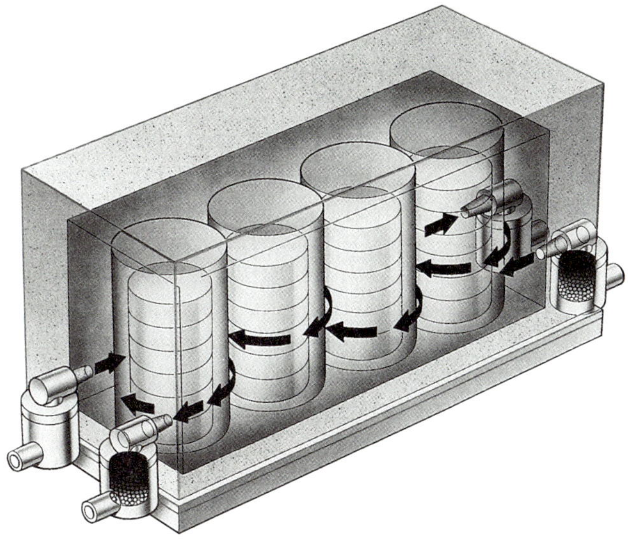

Fig. 6.162 Coil annealing furnace with two pairs of regenerative burners

The reversing controls are arranged so that the burners in diagonally opposite corners fire in pairs in order to provide a recirculating and reversing flow of gases within the furnace to aid temperature uniformity. The system has proved a most effective means of heating the furnace. Temperature equalisation within the furnace also now appears to be achieved more rapidly and this can be exploited by shortening the heating cycle and thereby increasing productivity. Fuel savings, even at this relatively modest furnace temperature, are around 40% over the previous firing system and have been sufficient to recover the entire cost of the conversion in less than 12 months. Following the successful pilot scheme a further 8 furnaces were converted on this site.

Continuous furnaces

The continuous heat treatment of steel is often carried out in long furnaces which may have a number of heating zones operated at different temperatures. In such applications the number of burners used are often larger than in a batch process in order to provide the required heating profile. Applications of regenerative burners have been carried out on furnaces used for the continuous heat treatment of stainless steel plate and strip at British Steel Stainless Division. In both cases, firing from only one side of the furnace was specified in order to simplify installation, reduce the pipework and ease maintenance. As part of the feasibility studies, physical and mathematical modelling were carried out at British Steel, Swinden Laboratories to predict flow patterns in the furnace and to determine the required heat input.

The continuous annealing of plates is carried out in a roller hearth furnace which is 26 m long. The plates enter the furnace directly from the rolling mill at around 600 °C and are then heated and soaked at a temperature of between 1 050 and 1 080 °C. The furnace was previously fired with 32 nozzle-mixing burners spaced along each side of the furnace, giving a total thermal input of around 6 MW (200 therm/h). The exhaust gases were passed through a recuperator which provided an air preheat of around 250 °C.

The design studies indicated that nine pairs of regenerative burners with gas inputs ranging from 300 kW to 900 kW would provide both a suitable flow pattern and heat input. The burners were mounted along one side to fire above the plates with space beneath for the gases to recirculate before discharging through an adjacent burner. The physical modelling showed that with this arrangement, even with the largest stock in the furnace, 50% of the flow passed beneath the plates and this was considered to be satisfactory for heating. In addition, following reversal, the flow pattern was fully re-established in less than 5 seconds. Hence, since the average time between reversals would be around 120 seconds the effects could be neglected. The practical performance confirmed the findings of the design studies. Throughput rates have increased dramatically and uniform heating has been maintained at the higher throughputs. Air preheats, measured at the regenerator outlets, are in excess of 900 °C and the exit flue gas temperatures are controlled at less than 150 °C. Fuel savings (including those from other furnace improvements) are over 60%. The furnace before and after modification is shown in diagrammatic form in Figure 6.163.

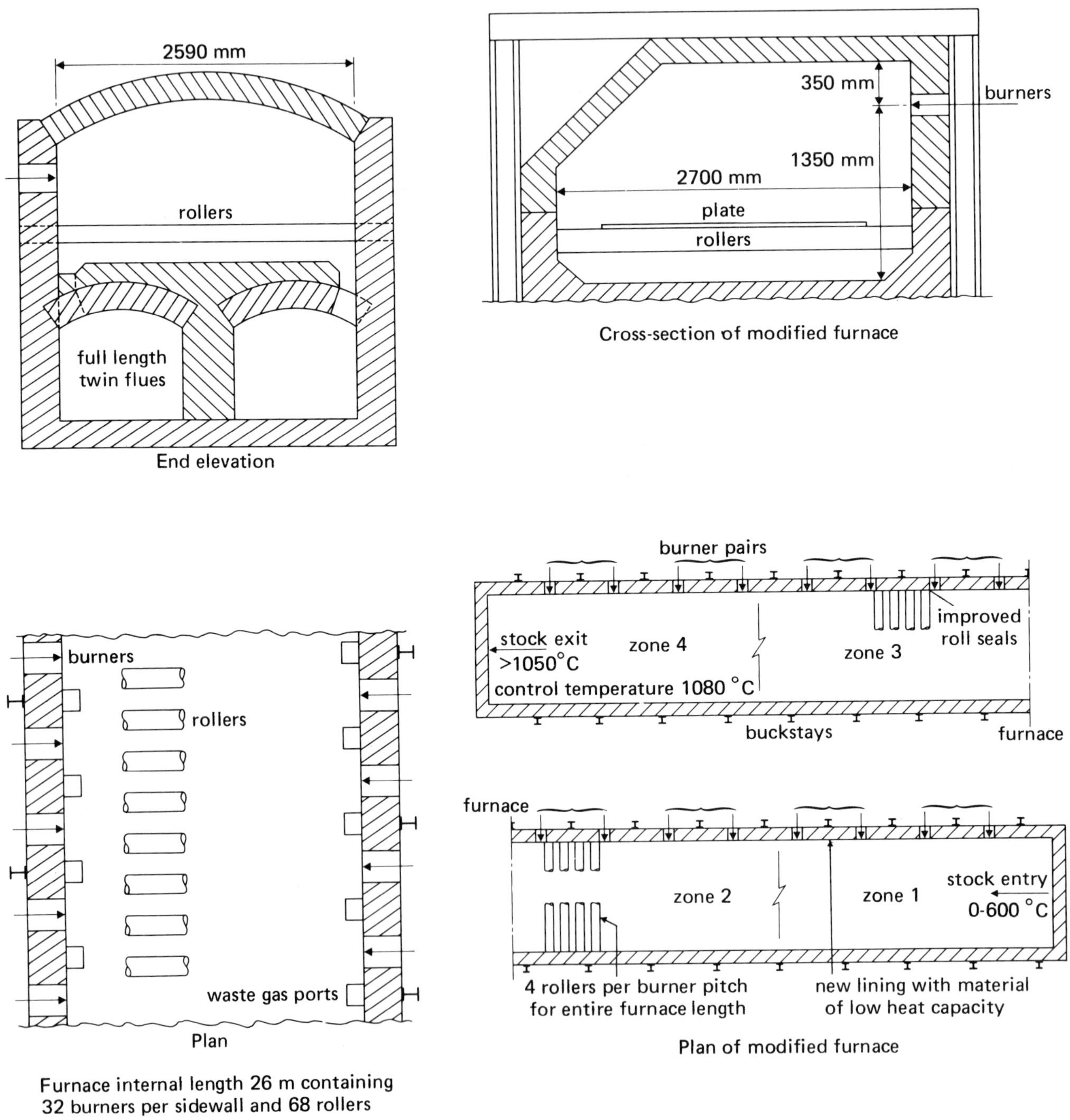

Fig. 6.163 Continuous plate heat treatment furnace

6.8.7 Sequential burner control systems

Sequential firing is the name given to a system of control developed for use with medium or high velocity burners, replacing traditional high/low or modulating control options normally associated with industrial heat treatment furnace applications and brick kilns. The main benefits are increased rate of convective heat transfer and maintenance of a uniform temperature distribution within the furnace. Over recent years users have wanted more flexibility and very often require different processes through the same furnace from stress relieving to general heat treatment applications. This range of operating temperatures makes traditional combustion systems very difficult to size, set and to obtain close temperature uniformity.

In heat treatment there are some burners working with 1000 per cent excess air on applications starting from cold with a set temperature climb, the excess being used to give circulation of mass flow within a furnace to maintain temperature uniformity rather than close ratio firing. At a temperature of 800 °C a burner working on perfect combustion would give 58% gross fuel input but at 100% excess air this would be reduced to 34%. To operate a successful sequential fired system it is essential that the correct type of burner is used with matching air and gas control system, both allied to the impulse sequential controller.

Consideration should be given to the width of furnace when side firing as too high a velocity on a narrow furnace will give hot spots on the opposing walls. The effectiveness of velocity also depends on the hearth construction and circulation space between loadings, normally burners are fitted on the side of the furnace and widths up to 8 m can be accommodated.

It is clear from the above that the most efficient operation is achieved with a burner operating at full rated heat output using a stoichiometric air/fuel mixture. The commonly used practice of employing high excess air levels to maintain the stirring action in the furnace as the burner heat input is reduced as the control mode is bad practice. It is inefficient both in providing high momentum in the furnace and in fuel usage. There are very few applications which specifically require excess air levels hence the use of sequential control firing. Typically 15% fuel savings can be made against a continuously proportional controlled system.

It has been found that the best method of achieving the criteria laid down is to pulse fire the burners in sequential rotation. In the U.K. it is necessary to pulse the burner to high fire from a low fire state and a 20:1 or better turndown is required. In the rest of Europe this low fire state can in fact be totally off. The burner is pulsed from low fire to high fire in approximately 1 second and the time that the burner fires in the sequential control mode is between 5 and 10 seconds. This method of combustion control eliminates the need for excess air by operating the burner at its full rated heat output. Turndown is achieved by maintaining a pre-set burner on-time and varying the duration of the off-time. Pulsing allows the fuel air mixture to be stoichiometric at all times.

A signal from a conventional temperature controller is fed to an analogue to digital converter. The pulses produced are counted and timed and an on/off signal produced and applied to the burners. Pulse frequency is proportional to the control signal and determines the number of burners firing. The on time may be set on commissioning, as may be sequence of burner firing. A pulse control unit designed for high and medium velocity burners and radiant tube burners, sequentially pulsed fired up to 16 burners from one control unit (see Figure 6.164). The on pulses are of brief duration and can be set from 1 to 150 seconds.

Additionally it is possible to connect each zone to a heating/cooling cycle. Controlled cooling from set temperature is sometimes required and the sequential control system accommodates controlled cooling via the burners in the same impulse sequence as would be used on heat up. Therefore the impulse system works for both heating and controlling the cooling temperature.

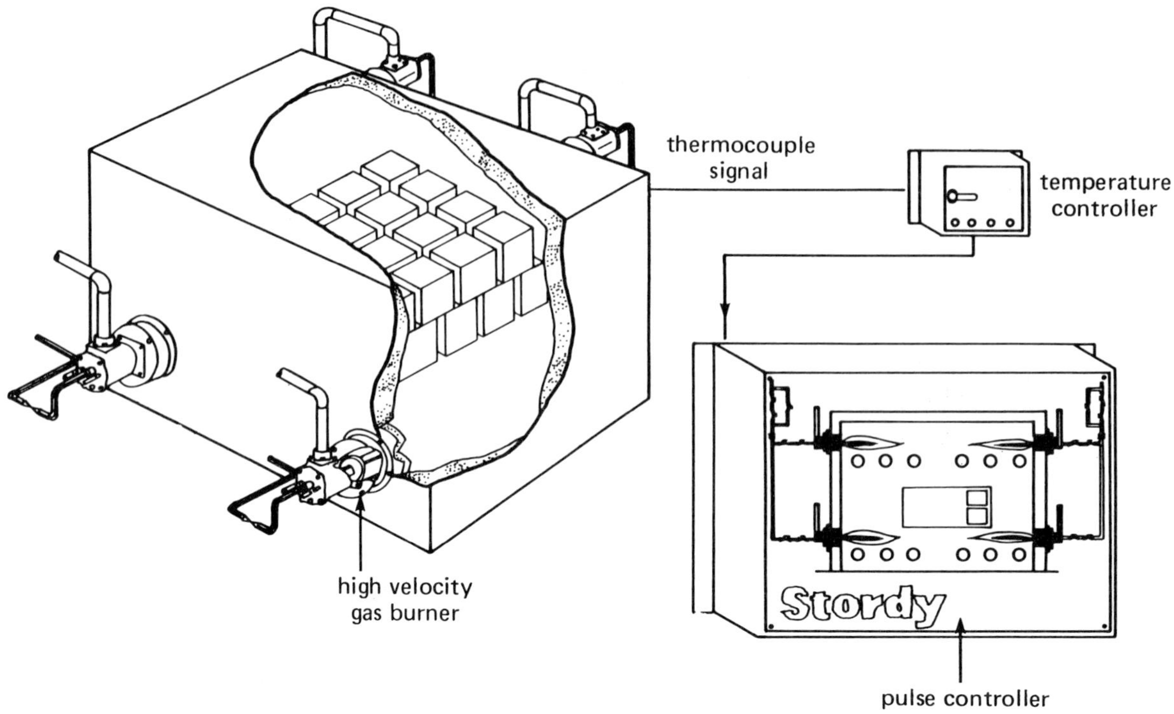

Fig. 6.164 Stordy electronic pulse controller for sequential burner firing

6.9 METAL RE–HEATING

6.9.1 Introduction

This section deals with furnaces for heating metals to temperatures suitable for hot-working processes such as forging, rolling, pressing, extruding etc. Hot-working temperatures are generally in the range of 1 000 to 1 400 °C for steel and 480 to 550 °C for aluminium alloys.

Re-heating is required at various stages of the steel-forming process, involving the use of both intermittent and continuous furnaces. Intermittent types include pit furnaces (soaking pits), bogie hearth furnaces and fixed hearth furnaces. Continuous furnaces are predominantly of the pusher type but rotary hearth, walking beam and notched hearth furnaces are also used.

In large aluminium rolling plants the pre-heating furnace must accept blocks weighing up to 10 t and be capable of heating to a closely controlled temperature. Sometimes material has to be soaked for homogenisation of composition and structure. Pit furnaces are commonly used. In smaller installations, continuous horizontal furnaces may be used e.g. conveyor or walking beam furnaces.

Both zinc and lead are hot-worked at temperatures which are low compared with other metals. Zinc slabs are heated to a working temperature of 230 °C, normally in reverberatory furnaces. It can also be continuously cast as strip, which can be directly cold rolled. Lead is typically poured molten into the chamber of an extrusion press where it solidifies and is extruded. There is, therefore, no pre-heating required.

Rapid heating techniques have resulted in the development of batch and continuous rapid heating furnaces for steel and aluminium for both total and localised heating.

The factors influencing equipment selection are similar to those for the metal heating processes discussed in Section 6.8.

6.9.2 Intermittent Furnaces

The batch type re-heating furnace is favoured for forge, press and dry stampwork where furnace dimensions are usually dictated by the character and dimensions of the stock. The stock may be in the form of ingots or blooms; charging is usually by special purpose machines or fork-lift trucks. These furnaces generally operate in the range 1 200 to 1 400 °C (steel re-heating) in which the dominant mode of heat transfer is radiation. The general requirements are:

- Furnace chamber temperature must be maintained high enough to provide work on demand at the required temperature.
- Heating of work to forging temperature should be as rapid as possible with the lowest achievable specific fuel consumption.
- Temperature uniformity throughout the work piece.
- Overheating should be avoided to prevent rapid scaling of the stock.

Fixed hearth furnaces

Where the work weight is relatively low and the possible variation in shape is wide this type of furnace will be encountered (see Figure 6.165). Here the hearth is permanently attached to the

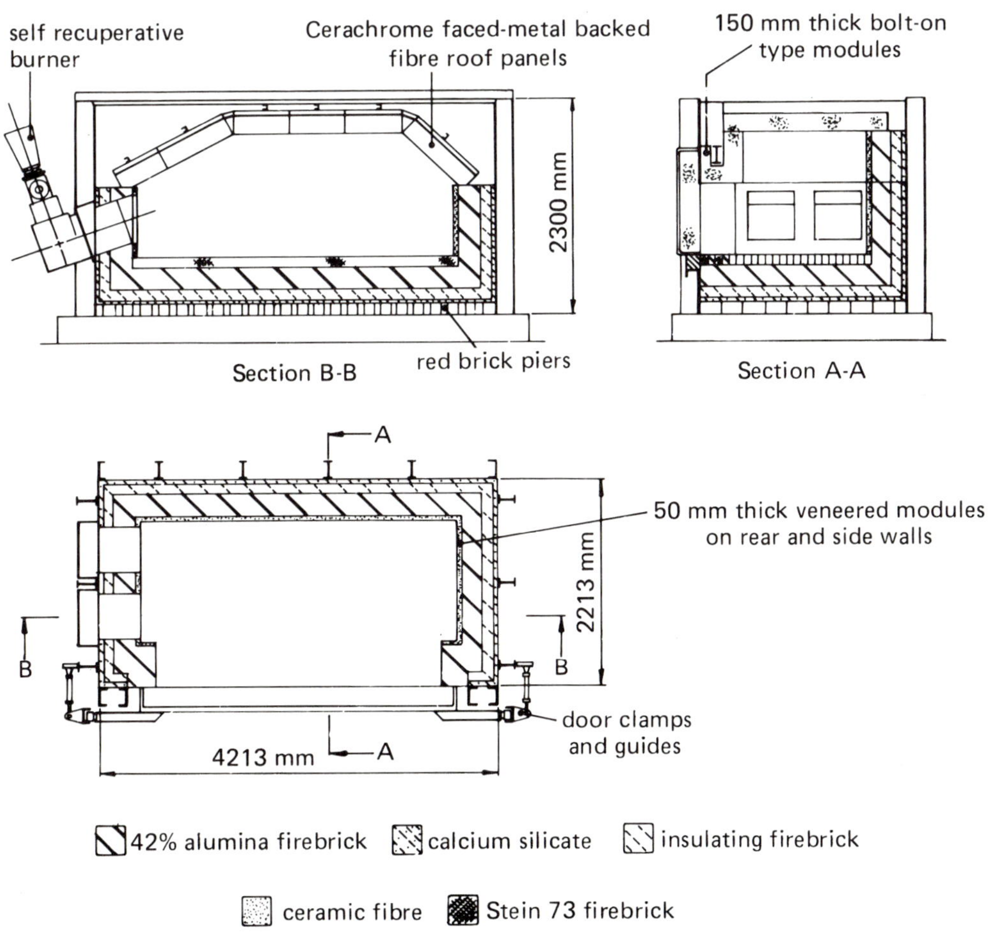

Fig. 6.165 Solid hearth forge furnace

furnace structure and is constructed of hard firebricks to withstand the weight of the charge and normal wear and tear of loading and unloading.

Depending on the size of furnace and workpiece, loading and unloading may be by hand tongs, fork-lift truck or charging machine. Generally, as a hot workpiece is removed for hot working, another cold component is loaded into the furnace and kept there until the required temperature is reached. This form of furnace is frequently in use for bar-end heating and reheating, or for smaller drop-forgings. Recuperation, certainly on the smaller furnaces, is still not a common feature but on the larger furnaces, recuperative burners and other air preheat devices are increasingly being installed. For a relatively small furnace with a hearth area of only 0.5 m^2 the specific fuel consumption will be in the region of 3.5 MJ/kg without recuperation. Generally, burners are of the nozzle-mix variety with good turn down capabilities, but air blast premix burners are also used.

Bogie hearth furnaces

These furnaces are normally very large structures, where the stock to be heated is loaded on a bogie outside the furnace itself. This bogie is sometimes sunk into a pit in the floor so that the load is at floor level, for ease of loading. Alternatively the bogie itself will run in rails at floor level. In either case the loaded bogie will be pushed into the furnace by a charging machine or tractor. The furnaces are extremely flexible and, because of their large internal dimensions, can accept large vessels, welded fabrications and castings as well as smaller components for stress relieving in the range 600 °C to 900 °C. They are used in the upper temperature range to 1 300 °C for re-heating prior to forging. Because of this size and temperature range the thermal efficiency is generally relatively low. Two such furnaces are shown in Figure 6.166. The moving parts are protected from the hot gases by longitudinal sand seals.

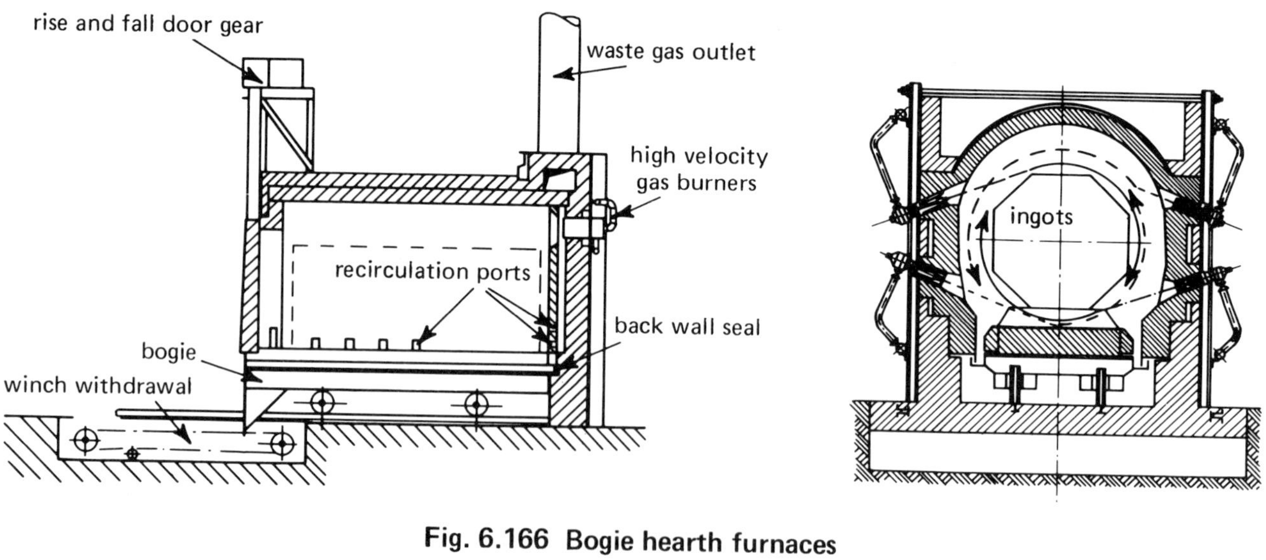

Fig. 6.166 Bogie hearth furnaces

Where the furnace is utilised for stress relieving, good temperature distribution is essential. In this case, in the temperature range 600 °C to 900 °C, recirculation of combustion products must be employed so that full use of convective and radiative heat transfer is obtained. At medium flow temperature, fan recirculation should be used, but at high temperatures the only practical method is to employ high velocity burners for good temperature distribution.

To improve overall thermal efficiency of these flexible furnaces several options are available when a furnace rebuild becomes necessary. If the brickwork is particularly poor it should be replaced completely with ceramic fibre of stack-bonded construction. When the brickwork is reasonably satisfactory but just cracked in several places, it may be more economical to apply the low thermal mass lining directly to the existing brickwork in the form of a veneer. This consists of gluing fibre panels normally 250 mm square x 35 to 55 mm thick to the brickwork with a suitable cement. The former method will give the best results because less heat will be absorbed during each heat up cycle. However, even adopting the latter system will mean that the furnace will reach temperature more quickly than before, and high heat transfer by radiation will occur earlier in the heating cycle, resulting in a shorter cycle time and higher furnace efficiency.

High velocity burners have already been discussed to promote good temperature distribution, but even so, when the furnace is operating at high temperature, a large proportion of the heat input is lost up the flue. Separate recuperators external to larger furnaces are now being installed, but recuperative burners are often more convenient. These are generally designed to operate as high velocity burners, and if the furnace is in regular use, reasonably short payback periods can be achieved. An additional benefit of these furnaces is that flame protection can economically be applied especially if one or two strategically placed recuperative burners replace several conventional burners. This is even more important where, on some processes, the furnace may be operating below 750 °C when flame protection will be essential.

Slot forge furnaces

This type of furnace is typically one of the most inefficient operating in the metal working industry today. It is constructed generally as shown in Figure 6.167.

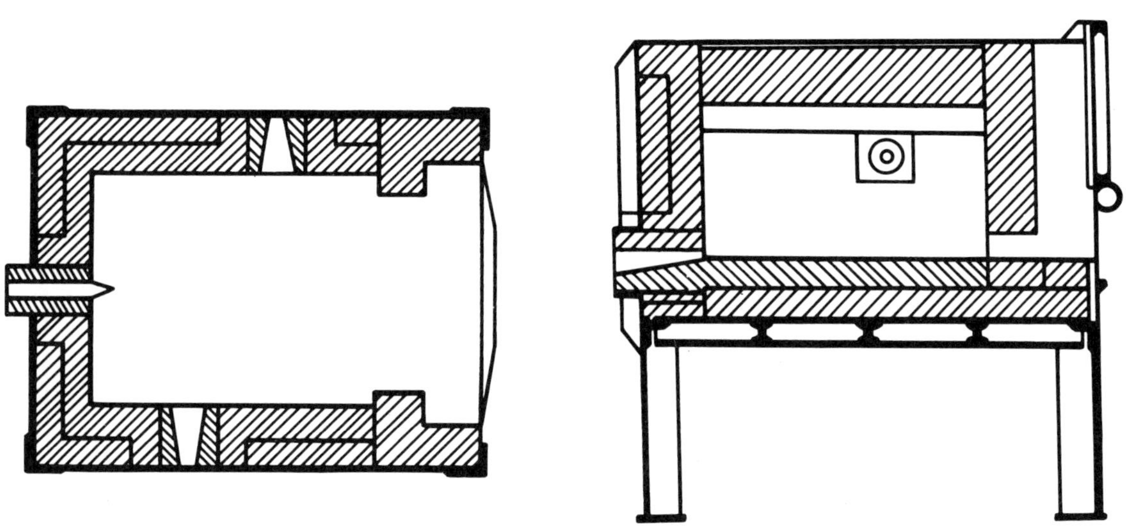

Fig. 6.167 Slot-type forge furnace, top-fired from both sides

These furnaces generally work with a high temperature head, and many have been traditionally oil fired although gas firing is now more common. A working temperature range of 1 200 to 1 500 °C is typical. The stock is generally relatively small rounds or squares 25 to 75 mm or flat steel strip. This type of furnace is used where one end only of the bar requires forging. Non-blow-off tips, gap fired from either side of the furnace are commonly used. Flat flame burners are also used to prevent flame impingement on the stock.

Pin slot forge furnaces are worked in a similar way and are of similar construction. Holes slightly larger in diameter than the pins to be heated replace the continuous slot. A relatively new design of pin slot forge furnace for small hand tools uses a recuperative radiant tube burner fired from side to side of the furnace with one end spring loaded. It has a high temperature insulating brick hearth and ceramic fibre lined walls and roof. An ordinary recuperative burner is not used because the products would be lost through the slots, and an oxidising atmosphere should be avoided so that no tenacious scale is formed on the tools.

Attempts at improving the efficiency of slot forge furnaces have centred on rapid heating techniques using high intensity gas burners and customised slot gaps. These are discussed in the section on Rapid Heating Techniques (Section 6.9.4). Another development is the regenerative slot forge furnace shown in Figure 6.168.

It comprises a rectangular chamber with walls and roof made from ceramic fibre modules to provide both a high degree of insulation and low thermal mass for rapid warm up. The furnace roof can be lifted to facilitate easy maintenance. The chamber is supported by a leg at each end and in each of these is contained a rectangular regenerator bed. Gas is introduced directly into the furnace chamber through a simple burner arrangement situated immediately above the regenerator. The control system has been made as simple as possible whilst remaining compatible with safe operation, and the pipework and reversal valve are located neatly beneath the furnace. Further cost savings have been made by the use of a single fan to provide both the combustion air supply and flue gas extraction through the use of two ejectors.

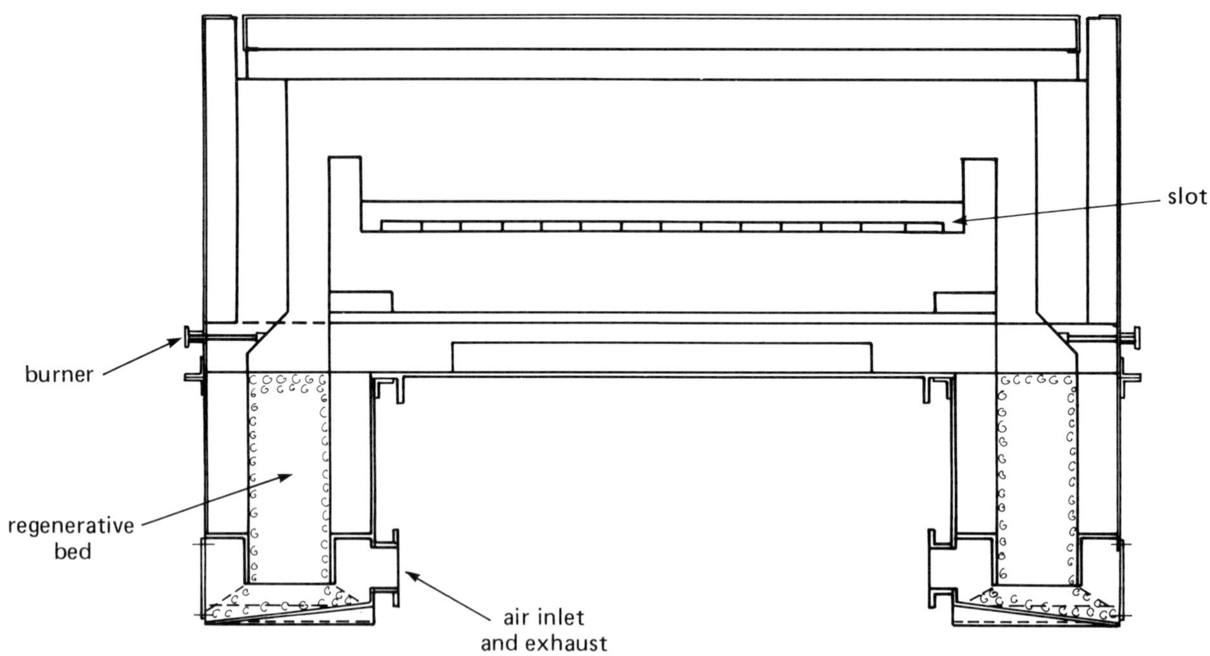

Fig. 6.168 Regenerative slot forge furnace with regenerators in furnace legs

The furnace is very quiet in operation due to the low velocity mixing of air and gas and also gives a very uniform temperature distribution across the slot. The control of furnace pressure provided by the regenerative system ensures that there is minimum discharge of gases through the slot. Operator comfort is thus considerably improved and the need for air curtains, frequently employed on traditional furnaces of this type, is eliminated. Efficiencies well in excess of 50% have been recorded and represent fuel savings of some 80% over conventional practice.

The compact regenerator construction is such that it is well suited for incorporation into the structure of a furnace and further developments along this line are anticipated. These include the extension of the slot forge arrangement into a long furnace suitable for the continuous re-heating of steel and other materials.

Soaking pit furnaces

Soaking pits are used for re-heating steel ingots and aluminium slabs prior to rolling into sheet. They are, therefore, the link between the melting shop and the rolling mills. The furnaces are sometimes sunk into the floor, but are now generally built above ground. The brickwork construction and insulation differ for ferrous and non-ferrous metals because of the different process working temperature.

These furnaces are very large structures because of the size of the stock involved. The mode of operation is similar for both steel and aluminium. The removable lids on top of the furnace uncover that part of the furnace which is being loaded or unloaded and are removed by an overhead gantry crane system which runs over all the pits. The ingots or slabs are picked up by a mechanical grab attached to the overhead crane. These furnaces have traditionally been fired by heavy fuel oil in both industries but natural gas is now common. Initially, gas firing was by large nozzle-mixing burners, but high velocity burners are now more popular. Regenerative burners are expected to find increasing application.

Steel soaking pits

Pit internal dimensions are typically approximately 10 m x 2.5 m x 4 m high and furnaces operate in the temperature range 1 250 to 1 350 °C (see Figure 6.169). They have massive brickwork proportions of around 700 mm hot face refractory brickwork and 230 mm high temperature insulation bricks. This is because the furnaces generally operate 24 h/day and the inside temperature very seldom drops. Therefore, furnace construction is similar to that of a continuous furnace, because the heat is absorbed by the brickwork once only per week. The lids on the majority of pits are still firebricks 225 to 300 mm thick.

The system normally operates on a batch system, where the ingots, after being cast, are loaded into the soaking pit which is then progressively emptied until the next batch is loaded. The ingots generally weigh between 4.5 and 6 t and the pits can accommodate up to 20 ingots which, when loaded, rest against the sides of the pit. For this reason the furnace is constructed from firebrick which has reasonable resistance to mechanical abrasion.

External recuperators are currently more popular than the old regenerator type combustion air preheaters. Fuel consumption varies between wide limits depending on:

- Percentage of cold ingots charged.
- Dwell-time in the soaking pit, imposed by the casting programme.
- Time between casting and charging into the pit.

In the worst case, with all ingots charged from cold, a specific consumption of 1.65 MJ/kg would be typical. The heating capacities, based on hearth area, are around 500 $kg/m^2 h$. The soak time in the furnace is approximately 1 hour for a hot charge straight from casting, up to about 3 hours for a cold charge.

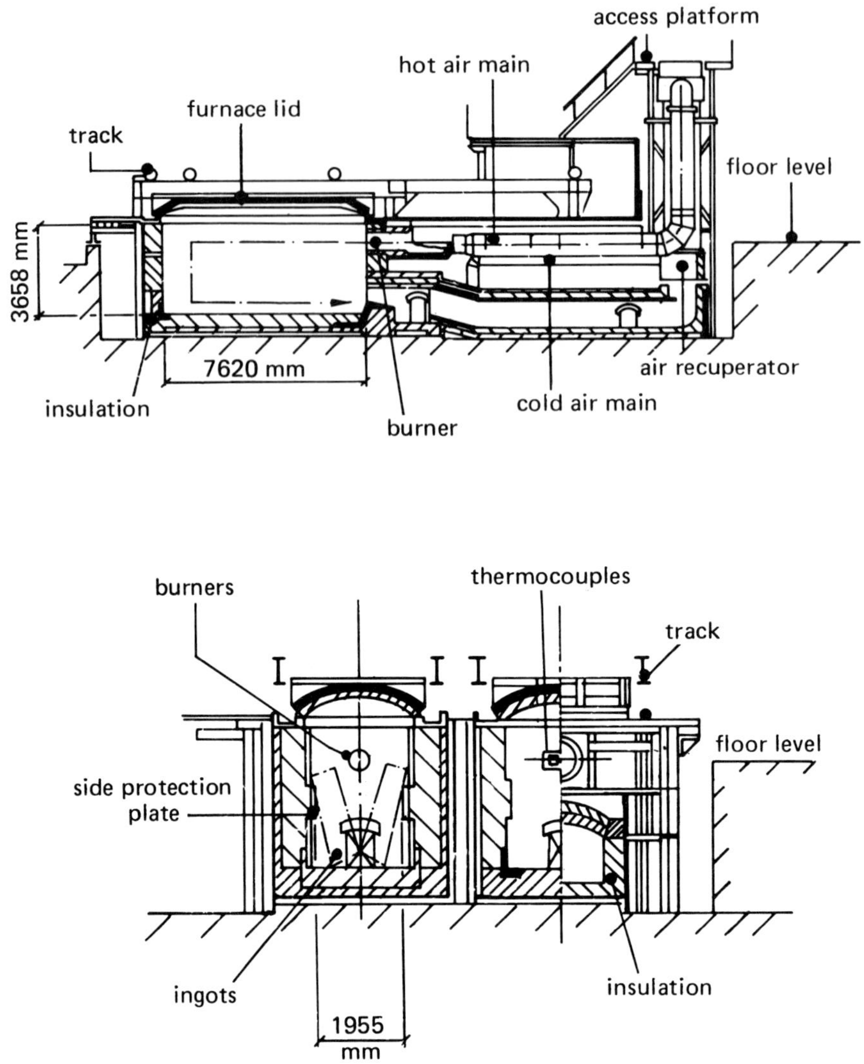

Fig. 6.169 Twin steel soaking pit

Burner control is usually fully modulating using a three term control system with air-lead. Oxygen addition has been successfully used on soaking pits giving fuel savings of between 10 and 20%, and increased production rates.

Aluminium soaking pits

These furnaces operate at a temperature of 550 °C, and therefore the standard of insulation required is not as high as with steel soaking pits. The mode of operation is that as one continously cast slab of aluminium is removed from the furnace another one is loaded into the space vacated and so the furnace is always full. A typical furnace pit as shown in Figure 6.170 is approximately 5 m x 4.5 m x 5 m deep. It will hold about fourteen cast slabs of aluminium each weighing about 8 t. The soak time is about 12 hours to reach a uniform temperature of 550 °C before transfer to the breaking down mill to be rolled into sheet.

On the furnace shown there are eight nozzle-mixing burners: two burners to each of the four zones. The two burners in each zone are situated one above the other at each corner of the pit. Twin hot gas recirculating fan units are positioned at each end of the pit chamber, arranged to discharge hot recirculated gases into a plenum chamber under the distribution plate and load stools. After passing

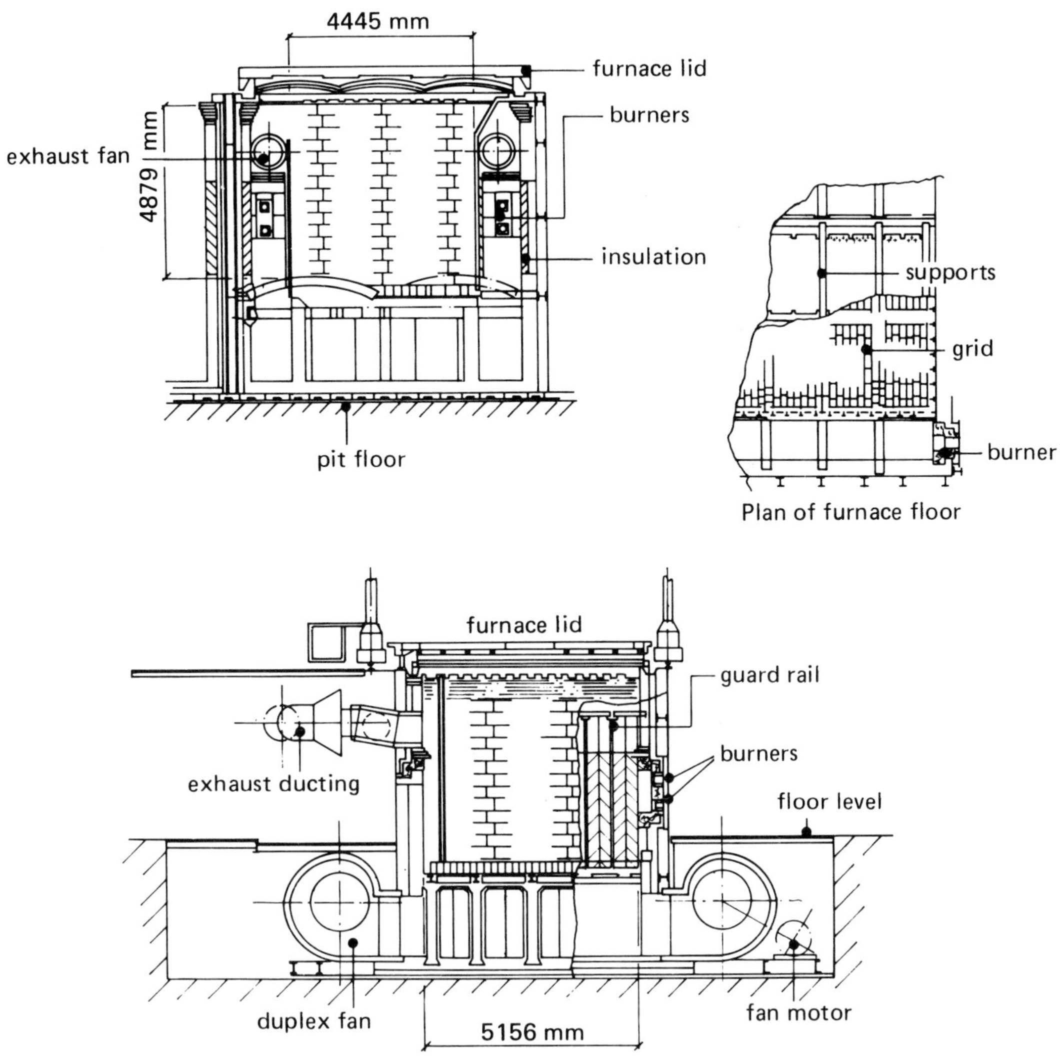

Fig. 6.170 Aluminium soaking pit

up through the load the gases are drawn to each side of the pit and over longitudinal baffle plates. The direction of the flow of recirculating gases is then directly downwards through the chamber formed between the outer walls and the pit and the baffle plates. During their downward path, the recirculating gases mix with further hot products from the burners. The gases then pass into the recirculating fan connecting ducts and commence a further cycle. Eventually the hot combustion products pass into short, damper-controlled ducts which control the pressure inside the pits before passing into an air recuperator then to the exhaust stack.

The removable cover of each soaking pit is formed from RSJs (Rolled Steel Joists) from which a steel plate top is suspended with a total of 300 mm insulation. This is composed of 50 mm needled ceramic fibre blanket, 100 mm low temperature ceramic fibre blanket backed by 150 mm standard loose fibre insulation. The whole structure is held in place by stainless steel pins (see Figure 6.171). The lid seal is achieved by a square box section underneath the periphery of the lid. This embeds itself into a channel around the periphery of each pit into which is formed three layers of 25 mm thick Kaowool pinned into place. The covers are removed by a travelling overhead gantry crane fitted with a mechanical grab, which travels over several pits in the same line. Ingots with sawn bottoms are placed in the pit chamber standing freely on the load grid. The temperature control thermocouple is positioned at the inlet to the recirculating fans which in turn controls a butterfly valve in the air supply line.

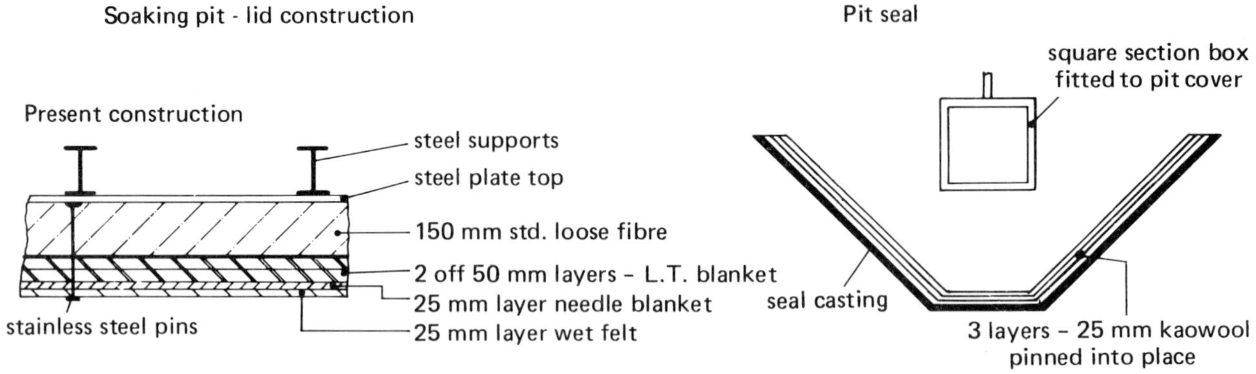

Fig. 171 Soaking pit - lid construction and pit seal

The external walls of the pits are about 380 mm thick, consisting of an inner 230 mm lining of firebricks backed with 150 mm thickness of slab type insulation. The refractory pier side walls supporting the load stools are also constructed of firebrick materials. The floors of the pit chambers are approximately 270 mm thick consisting of a 150 mm layer of insulating bricks laid on the pit base plate with a top 120 mm thick paving of firebrick. Insulated panels are fitted on the side baffle plate adjacent to the burners, to prevent radiation from the burner flame onto the load.

Typical fuel consumption of aluminium soaking pits is 2 to 2.5 MJ/kg, according to type of load.

Lift-off cover furnaces

In large re-heating, annealing operations in the forging industry it is often more convenient to load the large forging onto a purpose made base at floor level, and lift the furnace onto its base afterwards. The lift-off cover therefore contains all the refractory and insulating brickwork, the burners, the pipework and all control equipment. The flue is generally located in the centre of the base.

An example of a new furnace of this type for heating and annealing of large marine crankshafts has a lift-off cover with external dimensions of approximately 12 m x 9 m. To keep the structure as light as possible it is insulated throughout with stack-bonded ceramic fibre insulation about 305 mm thick held in position with stainless steel pins. Four high velocity burners are strategically placed in the side walls to give tangential firing, for good temperature distribution and uniformity. It can be used for annealing in the range 600 to 700 °C and is also suitable for re-heating operations for forging at higher temperatures.

Split-hood furnaces

Where very long bars of varying lengths for different batches are heated, split-hood furnaces are used. These consist of two hoods which slide together for very long bars, and contain all the necessary controls, burners, insulation etc. (see Figure 6.172). It forms, in effect, a completely sealed furnace.

Where shorter bars are to be treated, only one of the hoods may be necessary. The end is then blanked off by an end wall which is lifted into position. This technique saves fuel as only half the number of burners are utilised when treating shorter bars. When two batches of short bars are to be treated simultaneously they can be independently heated and controlled by having the end wall separating the two hood ends, which can fire simultaneously. This type of furnace is, therefore, very flexible.

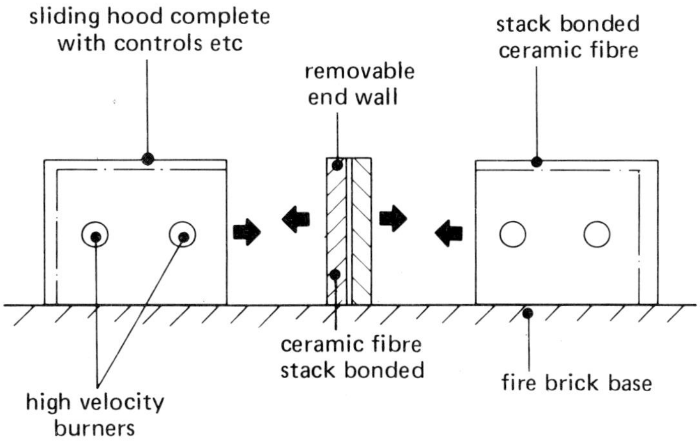

Fig. 6.172 Split-hood furnace for large bars

A modern furnace of this type has the split hoods lined throughout with stack-bonded ceramic fibre about 300 mm thick and held in position with stainless steel pins. The removable end wall is also lined on both sides with stack-bonded ceramic fibre as above. High velocity burners are fitted to give good temperature distribution.

6.9.3 Continuous Furnaces

The majority of continuous re-heating furnaces for slabs, blooms or billets are of the counter-flow, pusher type. Walking-beam and notched-hearth furnaces are also used for metal re-heating, as well as for heat-treatment purposes.

Pusher furnaces

An effective means of moving material through a furnace is to push it over a hearth. When it is required to heat steel to rolling or forging temperatures (1 100 °C to perhaps 1 400 °C) the pusher type furnace is the most practicable and widely used continuous furnace with straight line motion. The pusher type of furnace is adaptable for processing a large variety of stock sizes, but the stock is usually (although not invariably) of square or rectangular bar section, or billets, in a variety of lengths. The furnaces differ in size, method of discharge of the heated stock, and the construction of the hearth.

Pusher furnaces may be either end-discharge or side-discharge. With end-discharge the stock may be allowed to fall or roll by gravity on to a conveyor to carry the heated stock to the forge or to the rolling mill, or to whatever the production process may be. Several end-discharge furnaces may be placed side by side to discharge on to the same conveyor, if production so demands. Side-discharge furnaces can be used when long bars of comparatively small section have to be heated for rolling in continuous mills. The side extraction arrangements may be manual or, preferably, by hydraulic or pneumatic ram.

Pusher furnaces work best with bars or billets of uniform shape and square edges. Any roundness of the edges or lack of uniformity of shape creates the risk of pieces climbing on top of each other, and for this reason pusher furnaces may be equipped with 'trouble doors' to permit blockages to be cleared. Plates and sheet material are not pushed through furnaces because of their tendency to slide over one another or to buckle. The danger of piling up is reduced by inclining the hearth downwards towards the exit end. A slope of 10 degrees is sufficient.

If the material to be heated is pushed along the flat hearth of the furnace considerable force is required to overcome the friction of several workpieces resting on the hearth, the hearth suffers wear and one side of the workpiece fails to achieve full temperature. It is usual, therefore, for the stock to be carried on skid rails and sometimes for there to be facilities for turning each piece over once during its passage through the furnace. Skid rails may be totally of refractory material, or they may be water-cooled pipes with a refractory skin. Water-cooled skid rails give a longer service life but may be the cause of much heat loss from the furnace and they tend to worsen the effect of cold spots on the sides of the billets in contact with the skids, despite the refractory skin on the rails.

The great variety in size and shape of stock heated in pusher furnaces creates similar variety in the design of the furnaces and in their firing arrangements. Furnaces for heating large blooms may incorporate multiple zones, and in the past have typically been liquid fuel fired, having a single or perhaps two burners at the end of each zone, firing counterflow to the movement of the stock (see Figure 6.173).

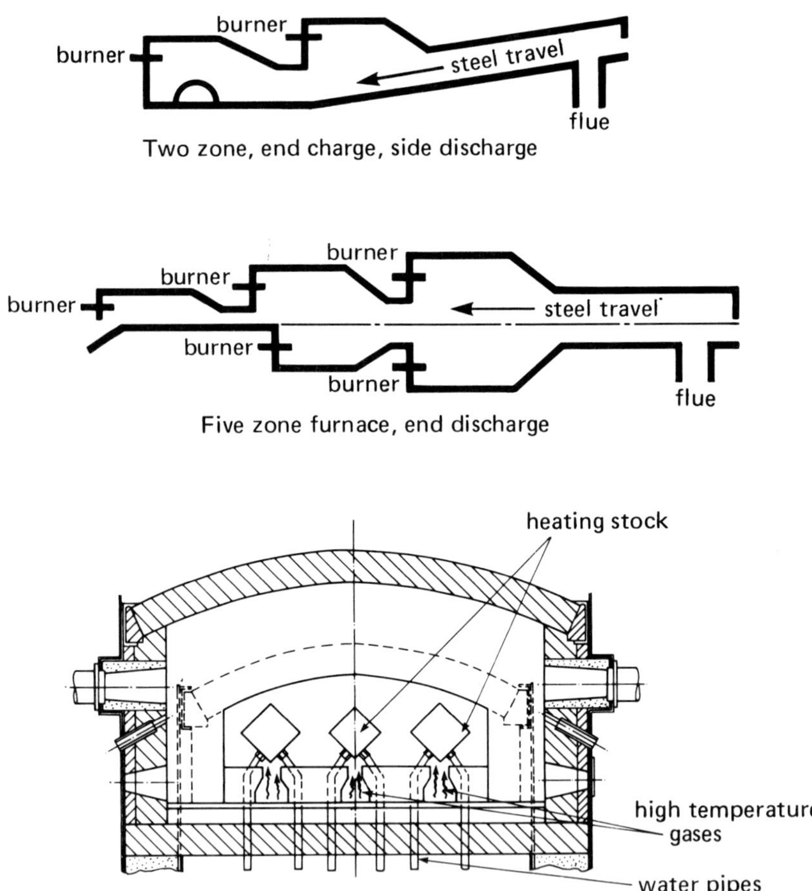

Fig. 6.173 Some examples of pusher type continuous, billet re-heating furnaces

Furnaces for heating bars and rounds of small cross-section have generally been gas fired and have, in recent years, been the subject of much development work designed to improve their thermal performance. These developments have followed a number of well defined paths:

- The design of furnaces of a size which will accommodate just the size of material to be heated.
- The use of materials of construction of improved thermal performance.
- The use of high velocity burners to improve heat transfer to the stock.

– The use of recuperative techniques to recover heat discharged in the products of combustion. Recuperation may be by the use of a load pre-heating section, the direction of flow of the products of combustion being such as to preheat the ingoing stock, or by transfer of waste heat into the combustion air, or by a combination of both techniques.

In gas fired pusher furnaces, almost every type of gas burner has found application but the need to incorporate recuperation is now causing a concentration on two main types i.e. nozzle-mixing forced draught burners and recuperative burners. Nozzle-mixing burners permit the use of external recuperators. High velocity, nozzle-mixing burners promote improved temperature uniformity and improved heat transfer. Recuperative burners have some measure of high velocity characteristics and their self-contained design makes the provision of waste heat recovery simpler and more convenient. However, there are limitations on the maximum temperature at which the integral heat exchangers may be operated. The normal maximum furnace temperature at which recuperative burners may be safely applied is about 1 400 °C. In some specialised applications they may be used at slightly higher temperatures.

Walking-beam and notched-hearth furnaces

Next to roller hearth furnaces, walking-beam furnaces are probably one of the most popular designs of continuous furnace. Notched-hearth furnaces, although less common, have been grouped with walking-beam installations for present purposes, since both machines cause the stock to move forward in small, integral steps.

In the walking-beam furnace (Figure 6.174) two or more longitudinal beams rest in slots running the length of the furnace hearth. The beams are linked to a mechanism which at intervals causes the beams to: rise beneath the stock resting on the hearth, lifting the stock off the hearth; move forward by a small distance; lower into the slots again, depositing the stock back onto the hearth; and move backward in the slots to the original position. Thus the work to be heated is moved forward in regular, discrete increments. The furnace hearth usually consists of a number of raised refractory ribs on which the stock rests, so that hot products of combustion may pass on all sides of the stock,

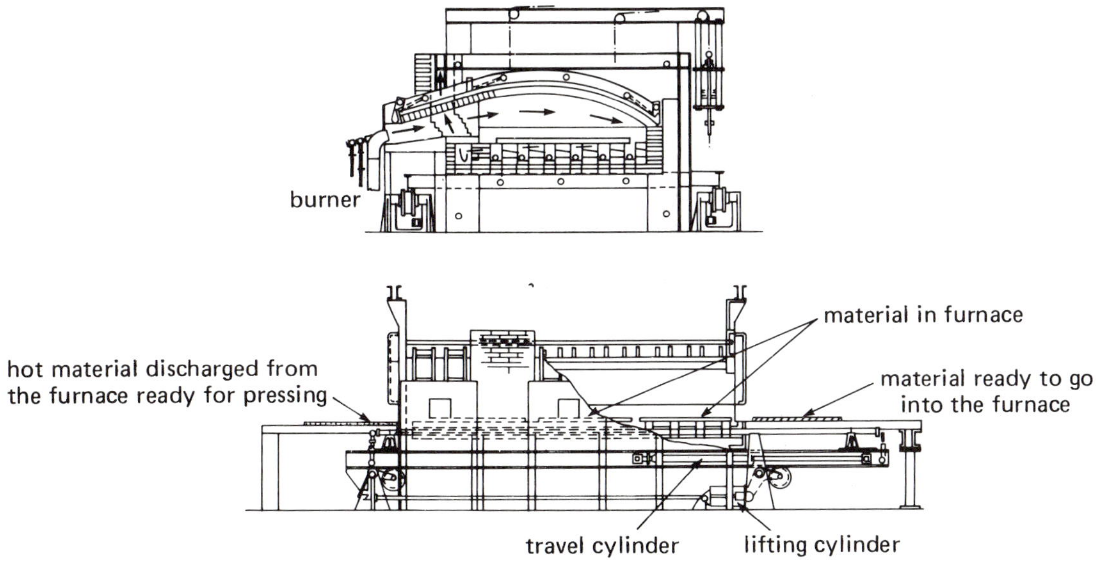

The rocker bars travel through a rectangular path and are operated by two hydraulic cylinders

Fig. 6.174 Walking beam furnace for plate heating

with minimal areas of contact with refractory to cause cold spots. Sometimes, instead of the combination of rocker-bars and fixed hearth, two sets of rocker-bars are used so that the load is alternatively carried on one set of bars whilst the other is returning to its starting position; the motion of the load with this arrangement is more nearly continuous.

The design and construction of walking-beam furnaces, if the mechanism is to have a useful life-span, is a compromise. The temperatures at which they are worked must also be limited if the operation is to be reasonably trouble free. The compromise arises from the need to protect the rocker bars from excessive heat and yet allow sufficient clearances to permit free operation. Problems of operation result from scale formation. Loose scale finds its way into the slots in which the rockers operate and quickly fouls up the mechanism. It is for reasons of limiting scale formation that the temperatures of operation are restricted.

Notched-hearth furnaces (Figure 6.175) are used for heating bars, usually of square section and of a length greater than the cross-sectional dimensions. The bars lie transversely in the furnace in raised 'V' notches. A series of rams below hearth level are worked in sequence to push each bar from the notch in which it lies to the next notch down the furnace, and subsequently into the discharge chute at the exit of the furnace.

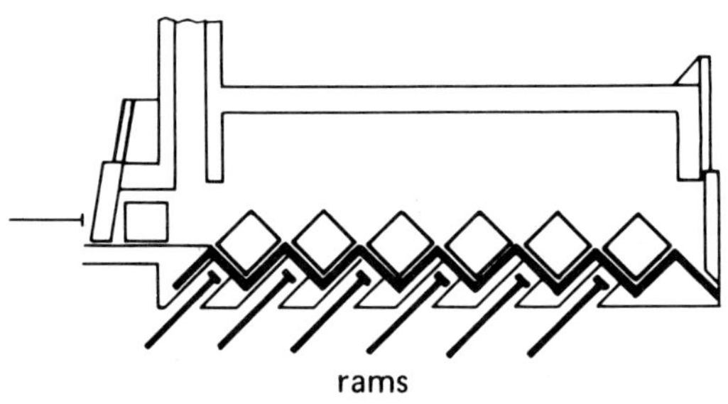

Fig. 6.175 Notched-hearth furnace

The remarks made about the mechanical problems, scaling and temperature limitations of walking-beam furnaces apply equally well to notched-hearth furnaces. Also, the remarks made about burners, firing arrangements, the need for recuperative techniques to ensure economic operation etc. concerning roller-hearth furnaces and conveyor furnaces may be applied equally well to walking-beam and notched-hearth furnaces.

6.9.4 Rapid Heating Techniques

The re-heating process influences both the cost and quality of the final product. In many applications, rapid heating techniques can provide significant benefits. Any method of increasing the rate at which material can be heated compared to that achieved in conventional furnaces can be described as rapid heating. In conventional furnaces, raising the stock above the hearth to increase the surface area available for radiant heat transfer, or raising the furnace temperature to increase the 'temperature head', can increase heat transfer rate.

With natural gas, increased convective heat transfer to the stock can be obtained by causing hot combustion products to flow across the surface of the material at relatively high velocities. Heat is transferred directly from the combustion products at a rate proportional to the gas velocity and hence the burner throughput. This, together with the low thermal mass of the gases, results in a very low thermal inertia system which enables the convective heat flux to the stock and the stock temperature to be easily controlled by burner throughput adjustment. Consequently, full advantage can be taken of the very large temperature difference between the combustion products of natural gas (adiabatic flame temperature 1 950 °C) and the metal to be heated.

Because of this large temperature difference, very high heat fluxes can be obtained over the full range of surface temperatures encountered in re-heating processes. Coupled with the low thermal inertia of the system, which enables stock temperature (as opposed to furnace temperature) to be accurately controlled, this ensures that the resulting high heating rates can be maintained until the required stock temperature is reached. In practice, heating rates of 400 °C/min are possible for 50 mm square section steel and low stock temperatures, and over 100 °C/min can be maintained for the whole range of surface temperatures encountered in steel re-heating.

The high velocity flows required are achieved by one or both of the following:

- Using burners where combustion is complete within the burner tunnel and directing the gases on to the surface of the material being heated.
- Matching the shape of the furnace chamber enclosing the stock to the shape of the material being heated, enabling high velocity flow over a large proportion of the material surface.

Most of the designs of rapid heaters currently in use incorporate a combination of these methods.

Benefits of rapid heating

The high heating rates and accurate stock temperature control offered by rapid heating techniques result in a number of benefits.

Energy consumption is reduced because of the faster start-up times from 'cold' and the ability to turn a rapid heater down or off during short breaks in production. It is not necessary to remove the hot material from the furnace during production breaks or at the end of a working period.

Consistently accurate stock discharge temperatures can be achieved, irrespective of working rate or hold-ups. Furnace scheduling can be simplified, as different set-point temperatures can be achieved in a matter of minutes, if required.

The ability to control stock temperatures whilst maintaining high heating rates right up to set point, without the risk of overheating, reduces the time stock spends at high temperature. This reduces metal loss from surface oxidation and results in low levels of decarburisation. Most stringent specifications can, therefore, be met by rapid heating.

Automation of rapid heaters is relatively easy and they can be incorporated into fully automatic production schemes leading to reduced labour costs. The size of rapid heaters, compared with conventional furnaces, is greatly reduced because of the high heat fluxes obtained. This means that capital cost and floor space requirements are significantly lower. Maintenance costs are also reduced.

Types of rapid heaters

For heating single small steel or aluminium billets up to 225 mm diameter and 835 mm long for extrusion purposes, the single cell high speed heating furnace is sometimes used (Figure 6.176).

The billet rests on a bogie hearth which can be extracted for loading and unloading and is positioned in the centre of the furnace. Gas and air are mixed in air blast injectors and the mixture passes, via a manifold on each side of the furnace, to the burners. Combustion products are fired

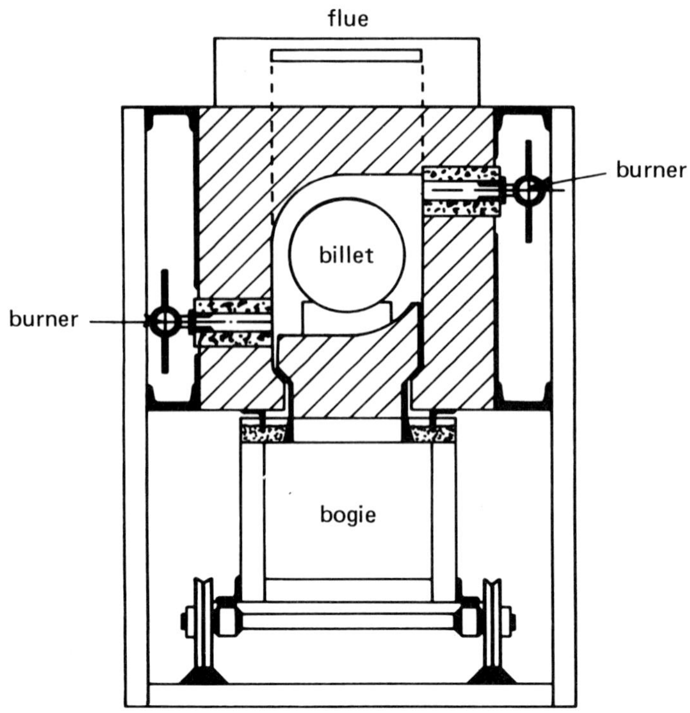

Fig. 6.176 Single cell rapid billet heating furnace

at high velocity tangentially around the billet in the space between the furnace and the billet. The shape of the furnace promotes good circumferential temperature uniformity, with heat being transferred by convection and radiation. The furnace chamber is constructed of lightweight castable refractory to give fast heat-up time. However, it is much more common to encounter the technique as a continuous furnace where the billets are pushed through by a hydraulic ram.

For aluminium billet reheating prior to extrusion, the billets are heated to 550 °C with burners arranged in zones along each side of the furnace. An example of this is the furnace manufactured by Granco Ltd, and marketed in this country by Mechatherm Ltd.

Figure 6.177 shows, in diagrammatic form, two types of part-bar rapid heaters in which gas and air are premixed in a manifold and then passed at high velocity through a number of ports into a refractory lined tunnel where combustion takes place. The stock is heated by convection from the hot gases and by radiation from the refractory surfaces.

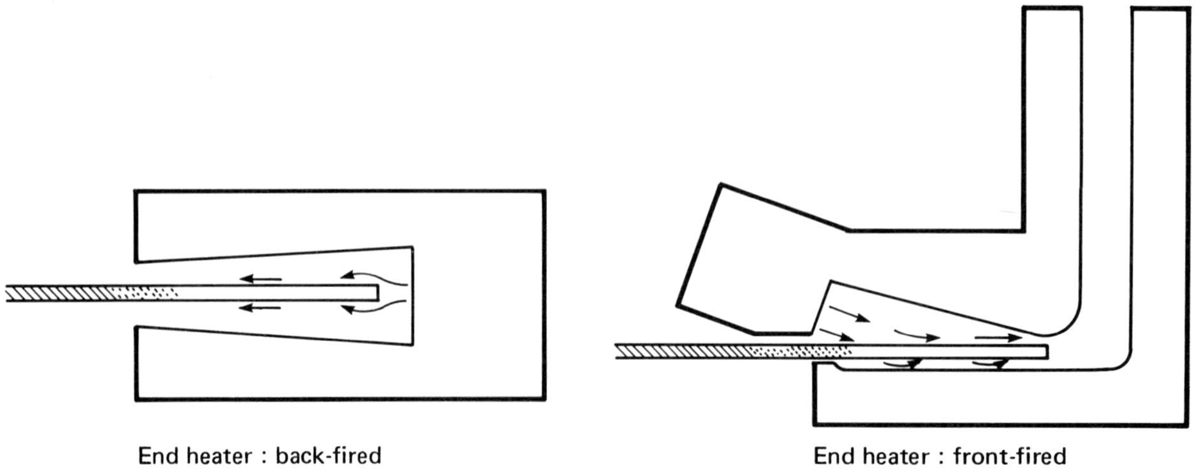

Fig. 6.177 Examples of part-bar rapid heaters

Where only small areas of a piece of stock remote from either end of the piece need heating, a heater developed by MRS, which relies on direct impingement of hot gas jets from high velocity burners onto the stock surface, can be used (Figure 6.178).

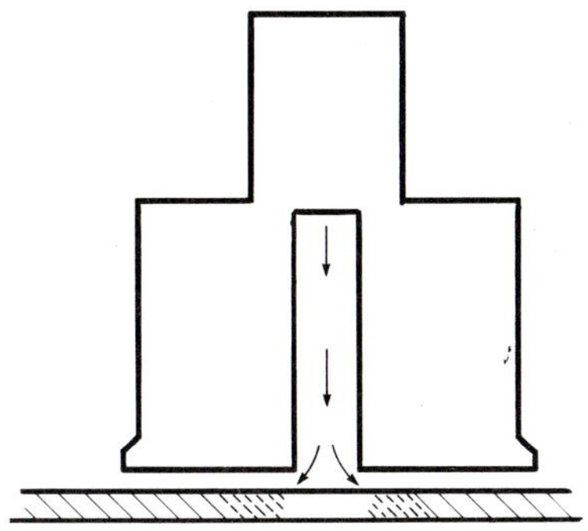

Fig. 6.178 Jet impingement heater

Heaters based on the principles of heating the stock within the tunnel of a high velocity burner or by direct impingement have been successfully applied and achieve the benefits of rapid start-up from cold (typically 10 minutes) and very high heating rates. Whilst thermal efficiencies are often good in comparison with the furnaces such rapid heaters replace, efficiencies are lower than those available from conventional continuous furnaces. This can be remedied by the adoption of waste heat recovery, either by air preheat using recuperative burners, or through the use of a load preheat zone based on the counterflow heat exchange principle. These composite designs retain the benefits of rapid start-up and high heating rates, with improved thermal efficiency being achieved through the promotion of convective heating in the preheat zone as well as in the rapid heating zone. Thermal efficiencies of up to 50% can be obtained with heaters having load recuperation. Figure 6.179 illustrates such a furnace manufactured by Fairbank Brierley to an MRS design.

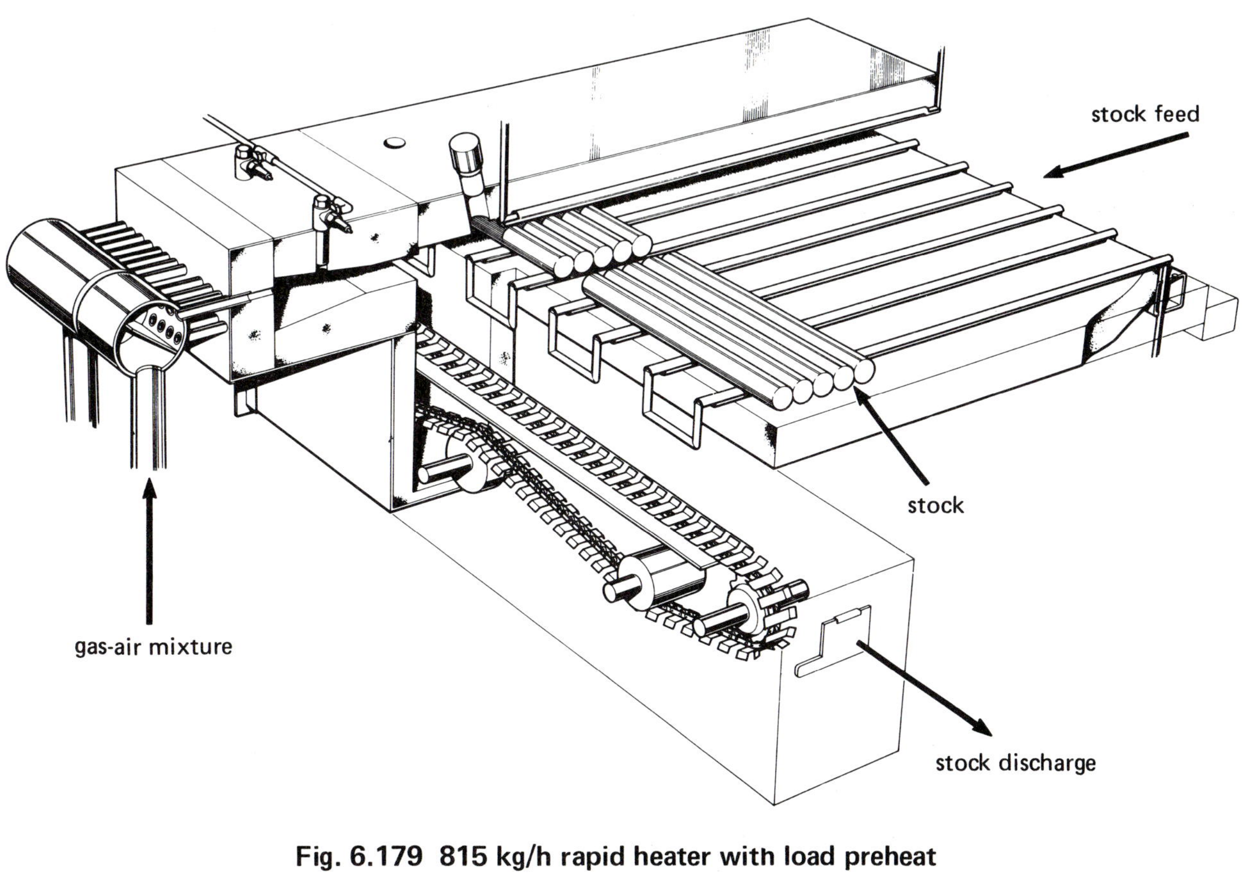

Fig. 6.179 815 kg/h rapid heater with load preheat

Rapid heating applications

Most of the rapid heaters used currently are engaged in fairly light duty applications, with stock throughputs of less than 2 t/h. However furnaces are available with capacities of over 30 t/h of steel at hot rolling temperatures. Rapid heating can be applied to most ferrous and non-ferrous metal re-heating processes, e.g. forging, pressing, hot forming, rolling, extruding and cropping. The technique is particularly attractive in situations where the production pattern is intermittent or vulnerable to stoppages elsewhere in the plant. In such circumstances the use of conventional furnaces causes excessive fuel consumption and loss of control of metal quality.

Gas fired rapid heating provides a comparatively low cost means of automation and its use is being extended into heat treatment applications and lower temperature processes. For further information on rapid heating techniques, reference should be made to British Gas MRS Publication E347, Rapid Heating in Perspective, and Publication E442 Gas Fired Rapid Heating Ten Years On.

6.10 METAL MELTING

6.10.1 Introduction

With the exception of injection in blast furnaces for the production of pig iron, natural gas usage is not significant in primary metal extraction processes. It is mainly used in secondary melting and refining processes in the iron casting and non-ferrous metal industries.

6.10.2 Iron and Steel Making

The production route for the manufacture of steel both from iron ore and scrap is shown in the simplified flow diagram, Figure 6.180.

At the beginning of the process, coal is converted to coke in coke ovens. Here, coke oven gas and other by-products are formed in addition to coke. The bulk of the coke passes to the blast furnaces, where it possesses a triple role. It acts firstly as a fuel which in burning raises the temperature in the furnace, secondly as a reducing agent to reduce iron oxide to iron, and thirdly as a physical support for the 'burden' while at the same time being porous enough to allow the hot gases to permeate to the top of the furnace.

The ore is usually charged to the blast furnace in the form of sinter which is basically an agglomeration of ore, coke breeze and limestone roasted together in the sinter plant to form a clinker, which not only has good physical characteristics to support the burden in the furnace, but has lost a proportion of unwanted volatile matter in the roasting process.

Most modern blast furnaces rely primarily on sinter and coke as their burden, whilst others have a variety of burden materials. Alternatively some ores are prepared as pellets by bonding very finely ground ore. The limestone in the charge combines with impurities in the iron ore, forming a liquid slag, which being lighter than the metal, floats on top of it and is removed separately from the furnace.

Both the coke ovens and the blast furnaces produce by-product gases (of 19.9 MJ/m^3 and 3.2 MJ/m^3 specific enthalpies respectively) which are used, as far as possible, totally in other departments of integrated steelworks. Coke oven gas is used in power station boilers, in annealing furnaces and in soaking pits and reheating furnaces, neat or mixed with blast furnace gas. Blast furnace gas is used directly in power station boilers and in stoves with or without coke oven gas or natural gas for heating the air to the blast furnaces

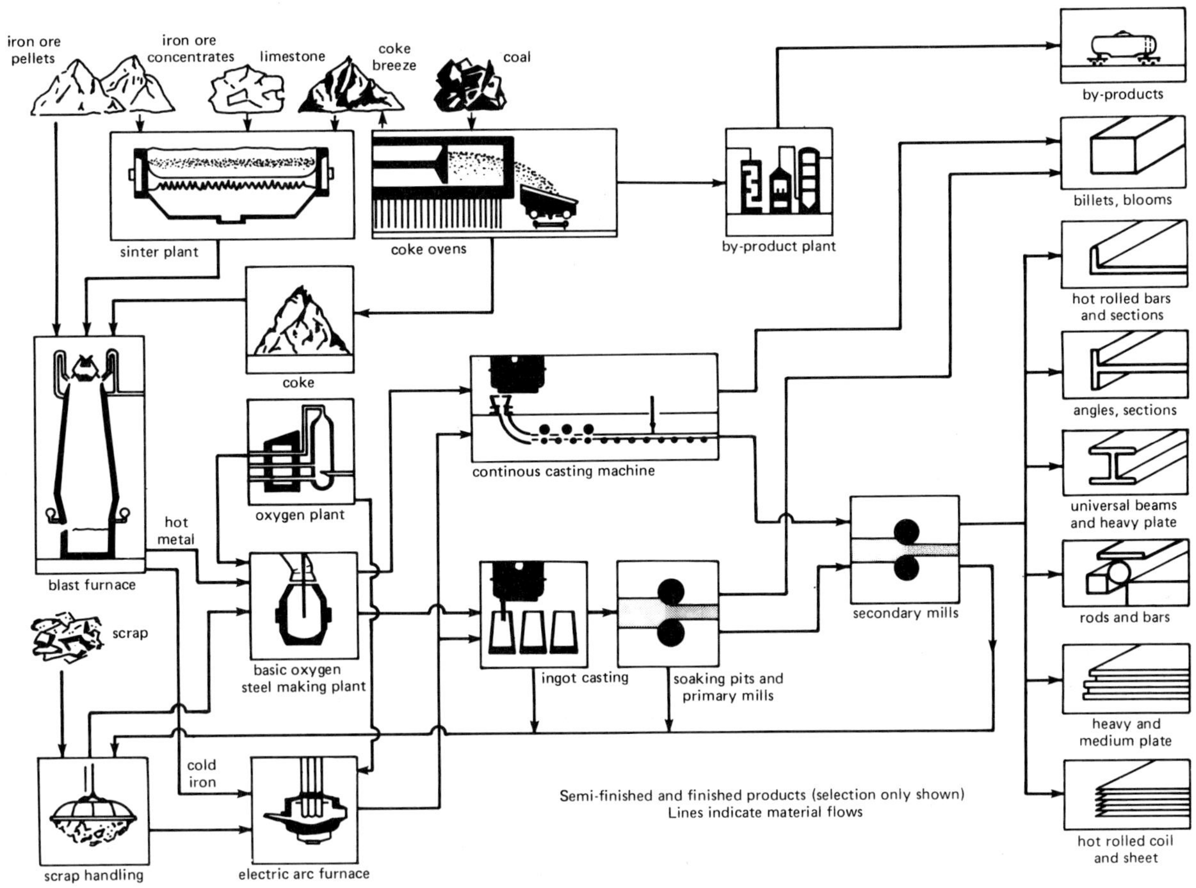

Fig. 6.180 Simplified iron and steel manufacturing process flow diagram

The iron produced by the blast furnace is termed 'hot-metal' and is mainly converted into steel by charging to a steelmaking furnace. In this process the 'hot metal' reacts with oxygen or iron oxide to remove excess carbon and thus becomes steel. Some 'hot metal' is used cold, for example in the production of iron castings and in the electric arc furnace.

Blast furnaces

Figure 6.181 is a diagrammatic representation of a blast furnace.

Coke, recirculated blast furnace gas to the stoves and other fuels, injected at the tuyeres, are used to smelt iron ore, producing iron for feeding to the steelmaking plant.

The gas from the top of the furnace is passed through a gas cleaning plant and is used for a number of heating purposes on the works. A large proportion of the gas is recycled to heat the stoves of the blast furnace system. The latter are for the production of 'hot blast' to the furnace which is admitted, through tuyeres near its base, at a temperature of around 1 000 °C.

Fuel oil and natural gas are injected through the tuyeres to reduce coke requirements, providing that it reduces the overall costs; the economics are determined by the relative costs of coke and the injectants at the point of use. Maximum coke replacement rates are approximately 20% for oil and 12% for natural gas, the replacement ratio for natural gas being about 600 m^3/tonne of coke replaced. As injectants have a cooling effect, steam injection at the tuyeres to control flame temperature would not normally be used when other injectants are employed.

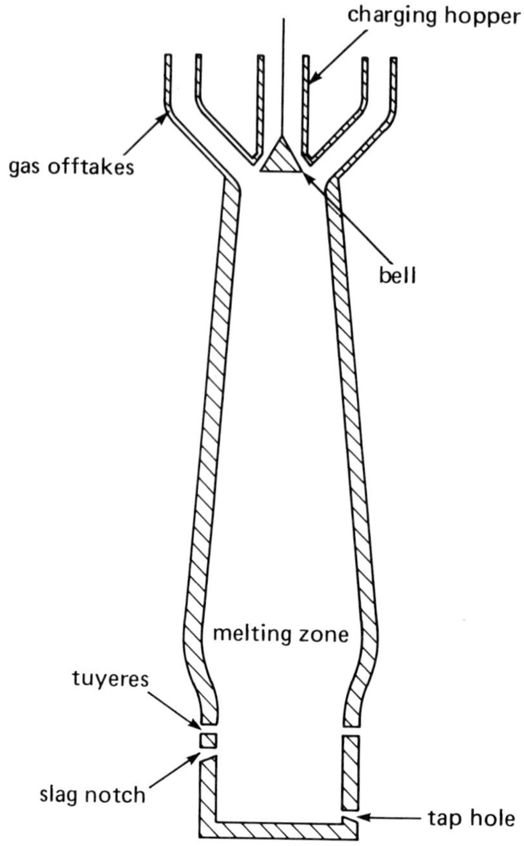

Fig. 6.181 Blast furnaces

Oxygen is sometimes introduced with the blast air to reduce the quantity of inert nitrogen flowing through the furnace.

Sintering plant

The performance of blast furnaces is increased by charging iron ore in the form of sinter rather than directly as ore. The sintering process consists of mixing iron ore with coke breeze followed by a combustion source to promote surface melting of the ore particles. Between 7 and 10% coke breeze is used and the mixture is spread on a travelling grate and ignited by a bank of gas burners firing downwards from an ignition hood. The heat requirement is about 2 000 MJ/t of finished sinter of which coke supplies 80% and the ignition system 6%, the balance being supplied by recycled flue dust. In general, sintering is carried out at integrated works and natural gas would generally be in competition with site-produced coke-oven gas.

Steel making

Currently there are two types of steel making furnaces used in the UK. These are:

- Basic Oxygen Steel Furnaces (BOS).
- Electric Arc Furnaces (EAF).

Open hearth furnaces ceased to be used in the UK in 1980.

The BOS process is the major method of making steel and consists of blowing a high velocity jet of oxygen on to the 'hot metal'. Steel scrap is also charged to the furnace to absorb excess heat as the removal of impurities by oxidation is highly exothermic.

Electric arc furnaces (EAF) use only cold scrap metal and melting is achieved by means of striking an electric arc between graphite electrodes and the charge. The metal charge is internally recycled scrap together with externally purchased scrap. The essential features of an electric arc steel-melting furnace are shown in Figure 6.182. Carbon electrodes are located in the roof of the furnace and these may be lowered or retracted as the melt proceeds. Arc furnaces range in capacity from around 10 t up to 130 t.

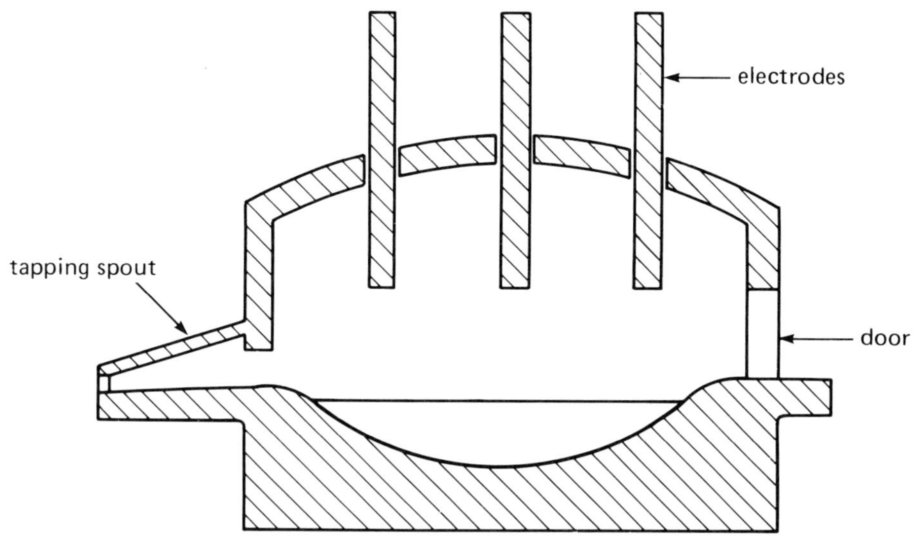

Fig. 6.182 Electric arc furnace

Considerable development work has been carried out on assisted melting in electric arc melters using oxygen-gas burners and, to a lesser extent, air-gas burners. These have been successful in overcoming electricity-supply restrictions and increasing production together with reducing energy costs in some instances. For their most effective application, the combustion products should be directed at the base of the scrap charge and allowed to permeate through the charge before being exhausted. When melting is under way and the bath has become flat, the burner effectiveness, i.e. ratio of electrical power saved to burner input, is much reduced.

The location of the burner is largely a matter of convenience and is, to some extent, governed by the availability of space around the furnace. Wall-mounting, roof-firing and door-firing arrangements have all been employed, the latter having the advantage that some degree of manipulation is possible. Oxidation and decarburisation are difficult to avoid when the burner is operated at stoichiometric oxygen-gas ratio because the oxygen-gas flame contains a large proportion of dissociated combustion products including free oxygen. Consequently gas-rich flames with oxygen-gas ratios between 75 and 100% stoichiometric are employed.

Air-gas burners have been applied to a limited extent but the reduced heat-transfer rate requires a greater heat input for an equivalent effect and the lower flame temperature restricts their use to the earlier stages of melting.

In general, assisted melting has proved successful particularly in small under-powered arc furnaces where significant reduction in tap-to-tap times have been achieved. The burner is least effective when the bath is flat and optimum savings are achieved in energy costs if the burner is used for less than 50% of the time.

6.10.3 Cast Iron

Cast iron is an alloy of iron with carbon that occurs naturally if iron is melted in a carbonaceous environment. The addition of carbon to iron lowers its melting point. The melting point decreases from 1 538 °C for pure iron to a minimum of 1 150 °C at 4.3% carbon. This decrease in temperature simplifies and reduces the cost of the casting process. The carbon content of cast iron usually lies between 1.7 and 4.5%, the actual value being determined by the required material properties.

Cast iron does not describe a single product but a wide range of materials which can be classified as follows:

- Grey iron.
- White iron.
- Malleable iron.
- Spheroidal graphite iron (s.g. - also termed ductile or modular iron).
- Alloy cast iron.

The cupola is the most popular melting furnace used in iron foundries and resembles a small blast furnace. The main difference is that the metallic charge is not iron ore but a mixture of some or all of 'foundry returns', iron scrap, steel scrap and pig iron. The proportions of these materials is dictated by the grade of iron required and their relative cost. The charge is introduced through an opening in the stack (charging door) in alternating layers of metal and coke together with small additions of limestone. Most cupolas use a cold air blast but hot-blast units are also in use, particularly where continuous supplies of molten iron are needed.

Partial replacement of coke by natural gas is possible. Figure 6.183 shows a cupola modified to incorporate four nozzle-mix burners firing into refractory lined combustion tubes.

The Cokeless Cupola

The cokeless cupola developed by the Cokeless Cupola Company Ltd, completely eliminates the use of coke for melting iron with substitution by natural gas, propane, diesel oil or other suitable fuels. Existing cupolas can be readily converted or new ones designed for specific applications.

The main features shown in Fig. 6.184 include the following. A water cooled grate consisting of steel tubes which may be coated with refractory for insulation on smaller cupolas. This supports the specially developed refractory spheres about 150 mm diameter, which act as a heat exchanger. Below the grate are the burners which are operated to give partially reducing conditions inside the cupola to reduce oxidation losses. The hot gases from the burners maintain the bed of spheres at high temperature and also pre-heat and melt the metal in the shaft. The metal is superheated in passing through the bed and is then collected in the well prior to tapping out in the usual way. A suitable recarburiser such as graphite powder is continually injected into the well to give the correct carbon analysis. A typical installation taps about 6 tonnes per hour high quality grey iron.

Some of the main advantages include the following. Because there is no coke used, depending on the relative costs of coke and other fuels, this could be a major advantage. There is no visible emission from the stack and the actual solids emitted are very low. Only a simple form of fume control is necessary and a simple wet cap will enable the cupola to meet all known environmental

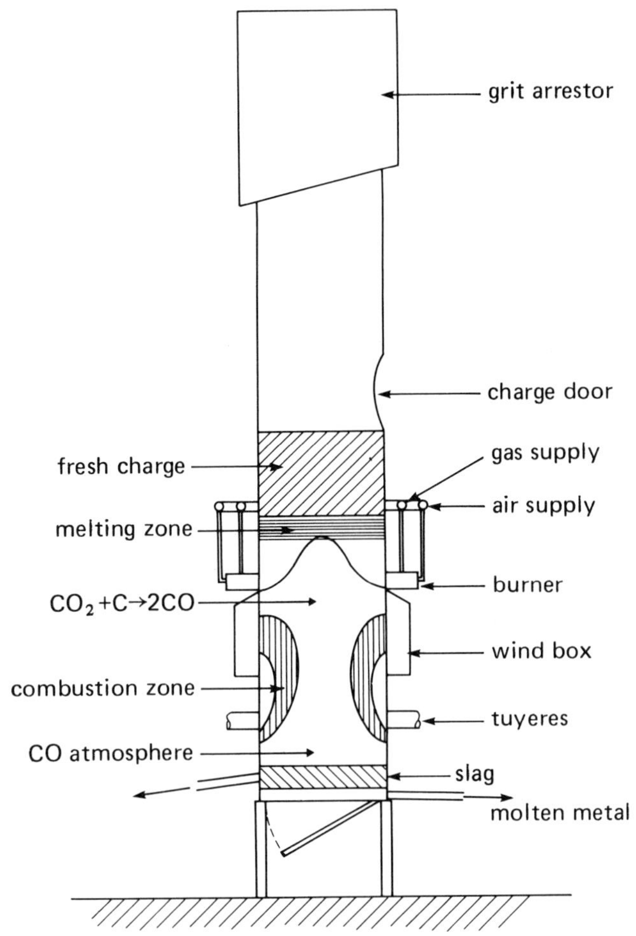

Fig. 6.183 Cupola melting furnace

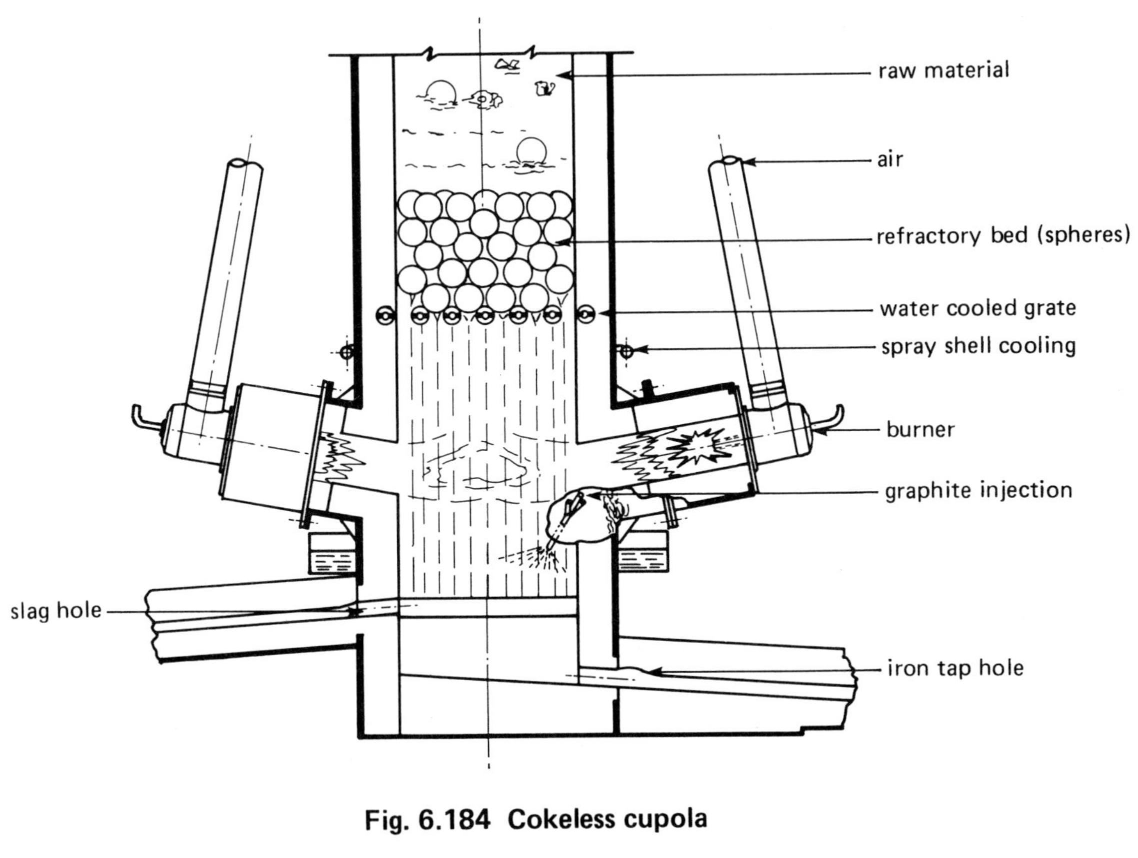

Fig. 6.184 Cokeless cupola

standards. Indeed, this cleanliness may well be the main selling point in locations where the local authority has demanded a reduction in stack emission. The sulphur content of the tapped iron is below 0.01% making it ideal for the production of ductile iron. There is less refractory lining wear and longer campaigns are possible, and less slag is made and hence there are less disposal problems. The iron temperature and iron analysis are far more consistent because of the control of injection of the graphite. It makes possible the production of high quality metal much easier to produce good castings.

There are two main ways of using the Cokeless Cupola. Firstly as a unit melter where the metal is poured directly into moulds. This is ideal in countries where coke is expensive for producing iron with consistent quality above 1450 °C. Secondly as a prime melter where the best economics are achieved. Here the metal is tapped at a lower temperature (about 1340 °C) and hence the ceramic balls last longer. The metal is then transferred to an electric furnace for reheating and recarburising.

6.10.4 Aluminium Melting Furnaces

The major use of gas for melting in the aluminium industry is the recovery of scrap rejected during the manufacture of aluminium products or from scrap products after their normal service life. Recovery is achieved by melting and blending the scrap to provide metal suitable for re-use, mainly by the foundry industry, thus reducing the demand for primary aluminium.

Melting aluminium presents two main problems. First, it is highly reactive to oxidising agents, the oxidation rate increasing sharply above the melting point (see Figure 6.185).

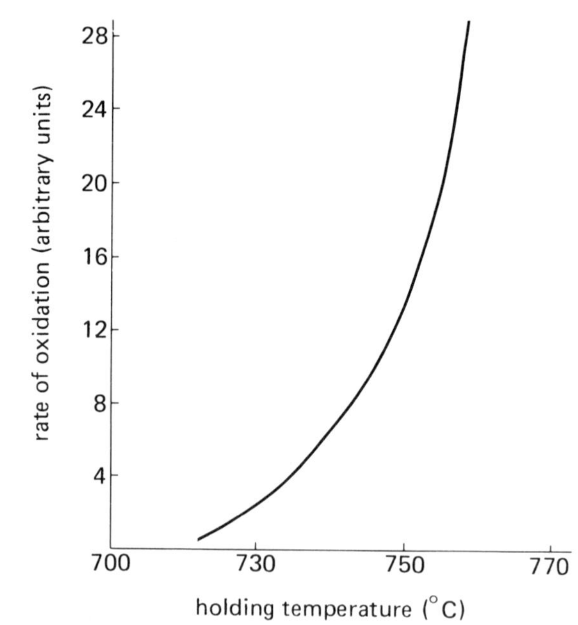

Fig. 6.185 Oxidation rate of aluminium

If the oxide film remains undisturbed, further reaction is minimised, but this is rarely possible in practice. Agitation of the bath surface has the added disadvantage of mixing the surface oxide skin into the molten metal, and since the metal and oxide have similar densities, oxide particles are likely to be carried into the castings. Oxides forming on the surface of molten metal can be removed by surface-cleaning fluxes. The oxides separate to form a dry powdery dross which is removed by skimming.

The other major problem is hydrogen pick-up. Hydrogen derived from moisture and dirt on the charge and from combustion products, dissolves readily in molten aluminium. However, solubility reduces drastically on solidification causing hydrogen in excess of the solid solubility to come out of solution and resulting in 'pinhole porosity'. It is often necessary to 'de-gas' the metal. Gaseous agents bubbled through the melt are chlorine or chlorine-nitrogen mixtures. Pelleted solids, e.g. hexachloroethane, can also be used.

Reverberatory furnaces

Reverberatory furnaces are generally used to melt large quantities of metal to supply holding furnaces or to melt scrap on a large scale. Capacities normally range from 1 t to 50 t, but there are a few instances up to 90 t. These furnaces are directly fired and the metal is unprotected from combustion products.

The metal is contained in a shallow hearth within a steel cased refractory box with a stack at one end and gas burners firing more or less horizontally across the metal surface at the other end. The combustion space is large and the charging door is big to enable bulky pieces of metal to be charged. Smaller doors give access for stirring metal, removing drosses and fluxes and cleaning the refractory walls. Figure 6.186 illustrates a typical reverberatory furnace.

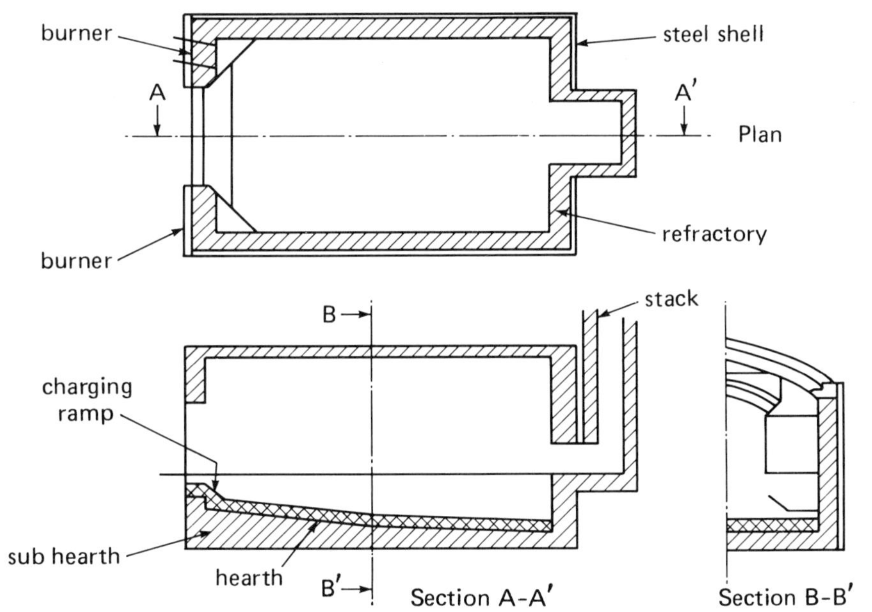

Fig. 6.186 Reverberatory furnace

There are two basic types of reverberatory furnace for bulk aluminium melting, viz wet-hearth and dry-hearth. In the former there is an open forehearth or sidewell in which the metal charge is immersed in an already molten bath without contact with combustion gases. This minimises metal loss. In dry-hearth furnaces there is a raised sloping platform near the stack on to which the charge is placed and which is enveloped by combustion products. The solid charge melts and runs down the hearth into the holding chamber. Because of the much larger exposed heat transfer area, dry-hearth furnaces have higher melting rates but, for the same reason, have higher melting losses. The dry-hearth needs to be kept free from dross build-up. This type of furnace is used to melt mixed scrap where the aluminium melts and runs into the furnace, leaving steel parts which can be removed.

For rapid and flexible melting, compact, tilting reverberatory type furnaces are commonly used. Another variant is the rotary reverberatory furnace which partially rotates, the hot walls of the furnace being continually in contact with the molten metal. Low grade scrap, e.g. swarf, can be successfully melted in this furnace due to the mixing action resulting from rotation. Burners for reverberatory furnaces are normally conventional nozzle-mix designs. Furnace efficiencies are not high and there is considerable scope for heat recovery by load preheating or combustion air preheating., The contaminated nature of waste gases has, in the past, prevented the use of the latter. However, the development of the regenerative burner and the energy savings achieved on 'clean' installations suggested that the same technique could be used on contaminated waste gases generated in secondary aluminium melting. One such installation is a 40 t forewell furnace previously fired by two nozzle-mix burners, replaced with a regenerative burner system as shown in Figure 6.187.

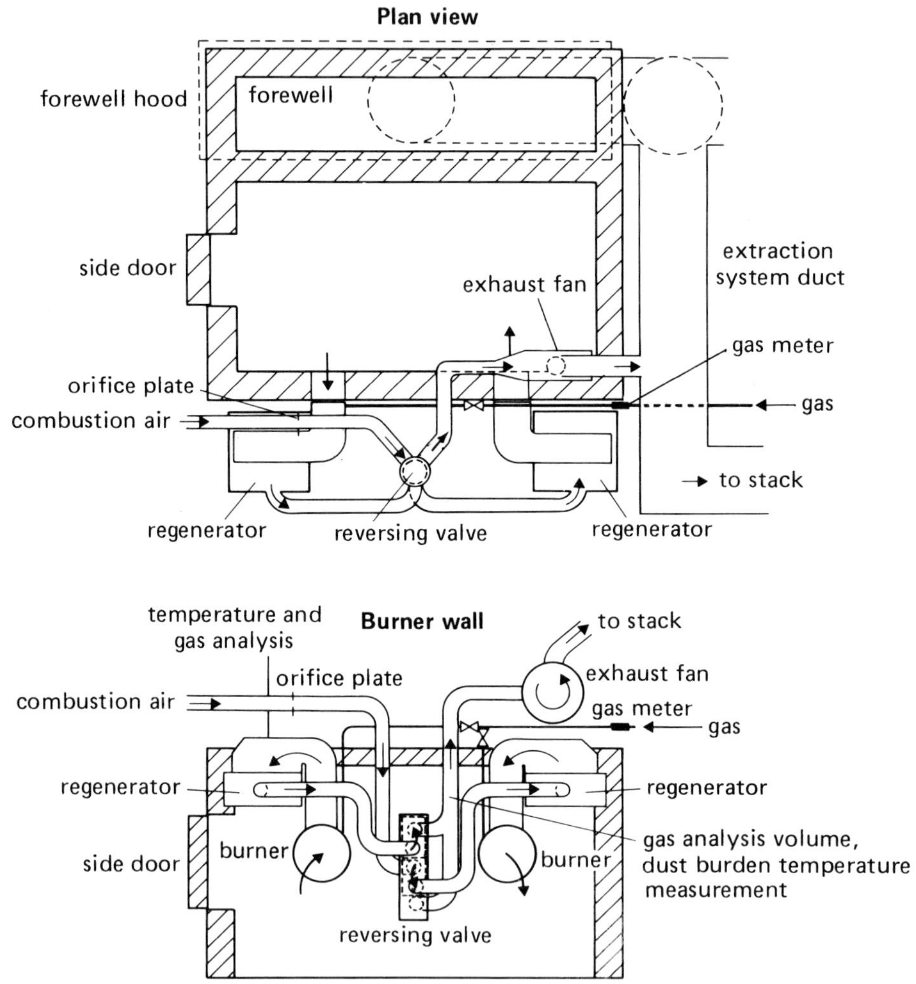

Fig. 6.187 Aluminium melting furnace with regenerative burner.

Average fuel savings of nearly 30% have been demonstrated on this furnace, the actual savings on individual casts being between 24% and 40%, depending on melting rate. The regenerative burner system has proved able to cope with the majority of contaminated waste gases generated, although some change of practice has been necessary. This has mainly involved discontinuing the addition of material containing a salt flux to the furnace, as it causes blocking of the regenerators. Regular maintenance is needed to keep the regenerator bed free from blockage by dust. Cleaning is normally carried out every two weeks, but when a large number of casts are melted or scrap is abnormally dirty, weekly cleaning may be necessary. The contamination is easily removed by washing. The regenerator beds remove 95% of the dust discharged from the furnace.

Crucible furnaces

The crucible furnace is common in aluminium foundries which need small amounts of molten metal frequently, or for holding molten metal. Crucible furnaces are generally classified on the basis of the means of removal of the molten metal. In pit-type or lift-out crucible furnaces, the crucible is removed by manual or mechanical tongs for pouring. Tilting furnaces eliminate the need for crucible removal and are suitable for larger capacities. The whole furnace tilts about axes which may be on the centre line or at the pouring lip. Bale-out furnaces are mainly used for die-casting and may be melting/holding furnaces or simply holding furnaces supplied from a bulk melter. Capacities of crucible furnaces range from 150 kg to 1 000 kg. Figure 6.188 shows an air blast burner firing tangentially into the base of the annulus between the lining and the crucible. This has, in the past, constituted the normal firing method. Nozzle-mix and packaged burners are also used.

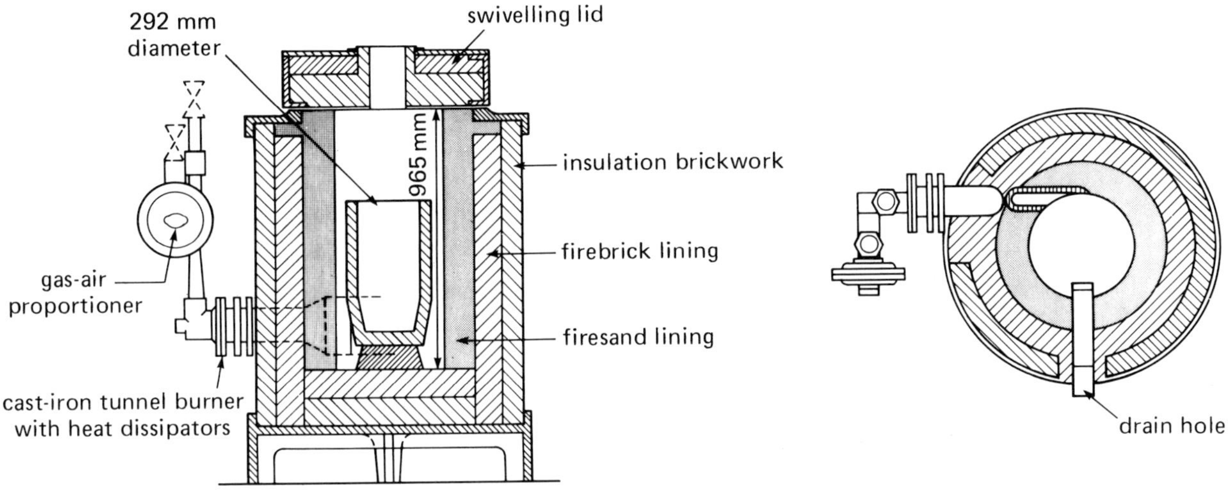

Fig. 6.188 Tangentially fired crucible furnaces

Melting via crucible furnaces is less efficient than direct firing, but the application of recuperative burners has led to significant improvement in this respect. Crucible furnaces lend themselves to recuperative burner firing, the burner momentum producing recirculation in the annular space. Savings of between 25% and 60% can be achieved over traditional burners depending on the application.

Fuel savings of between 25 and 50% have been obtained when recuperative burners are fitted to baleout crucible furnaces used for melting and holding non-ferrous metals such as aluminium and bronze (Figure 6.189). Physical modelling of early applications demonstrated good gas recirculation both in the vertical and horizontal plane, thus giving uniform crucible temperatures under full rate and turndown conditions. On new designs of furnaces, mathematical modelling has been carried out to optimise the recirculation space around the crucible for maximum heat transfer.

Crucible furnaces (especially those used intermittently) benefit from hot-face insulation of low thermal mass e.g. ceramic fibre castables (See Chapter 7).

Another form of furnace using a crucible for heat transfer to the metal is the immersed crucible furnace. Figure 6.190 illustrates such a furnace fitted with an external recuperator. The gas burner fires into a ceramic crucible immersed in the metal. Combustion gases, leaving through a port in the burner cover, pass over the charge before being extracted from the furnace through the recuperator by an eductor in the flue. Ingots and scrap are charged into the exhaust offtake and are preheated by the exhaust gases leaving the furnace. The use of a remote recuperator rather than a recuperative burner retains the benefit of load preheating as well as combustion air preheat.

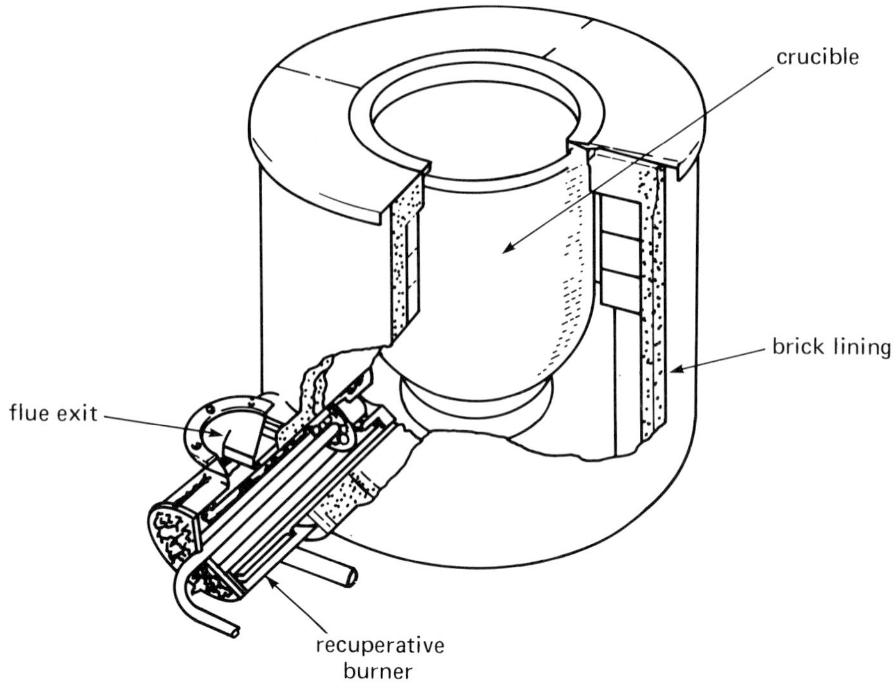

Fig. 6.189 Aluminium bale-out furnace with recuperative burner

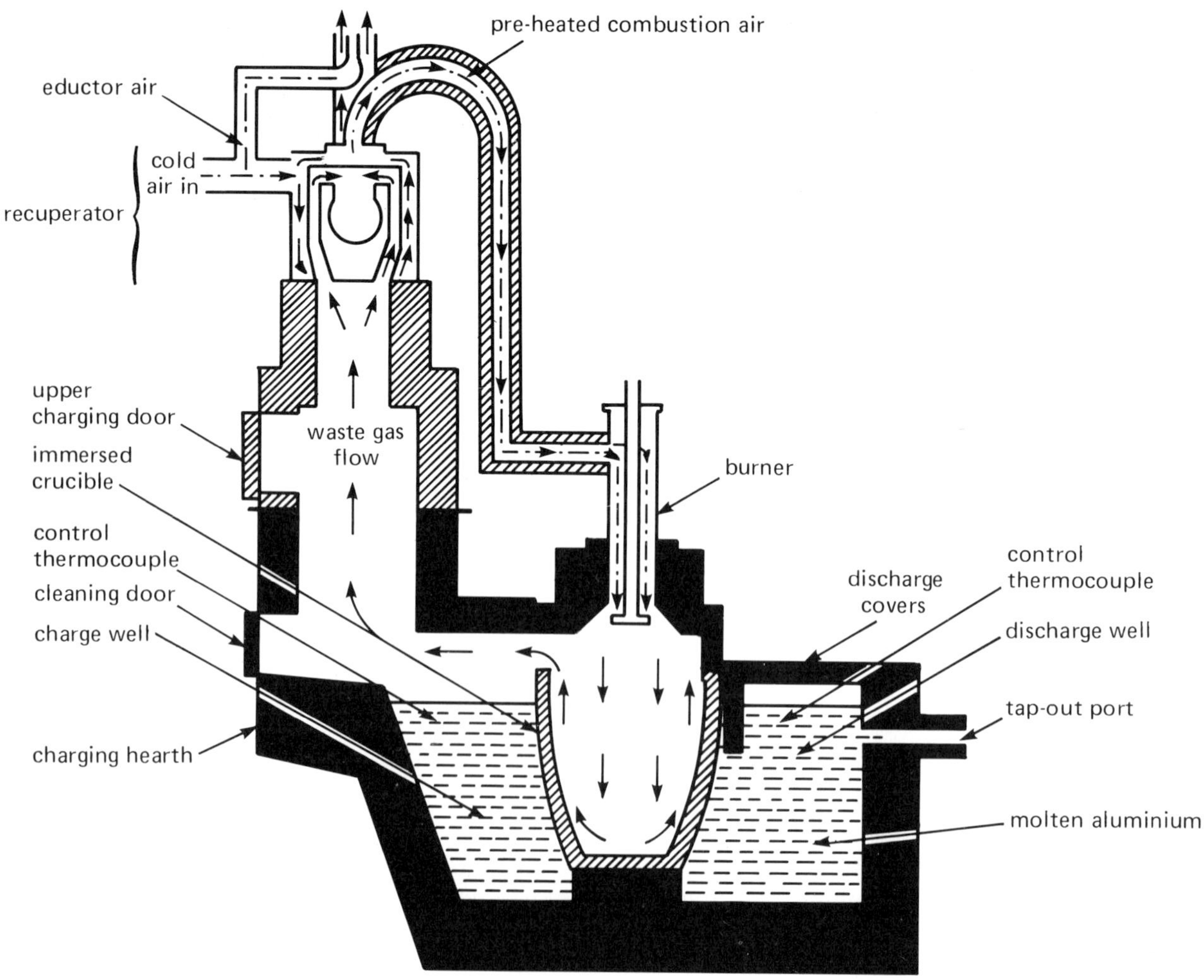

Fig. 6.190 Immersed crucible furnace with recuperator

Ceramic immersion tubes

The development of ceramic immersion tubes for melting and holding non-ferrous metals is now playing a significant role in saving fuel and improving temperature uniformity and metal quality. The technique has already been illustrated (Section 6.3) and utilises many of the features developed for the ceramic radiant tube, including a built-in recuperator for high thermal efficiency.

An aluminium holding furnace using a ceramic immersion tube, developed jointly by MRS and British Non-Ferrous Metals Technology Centre, achieves both high fuel savings and low metal loss (see Figure 6.191).

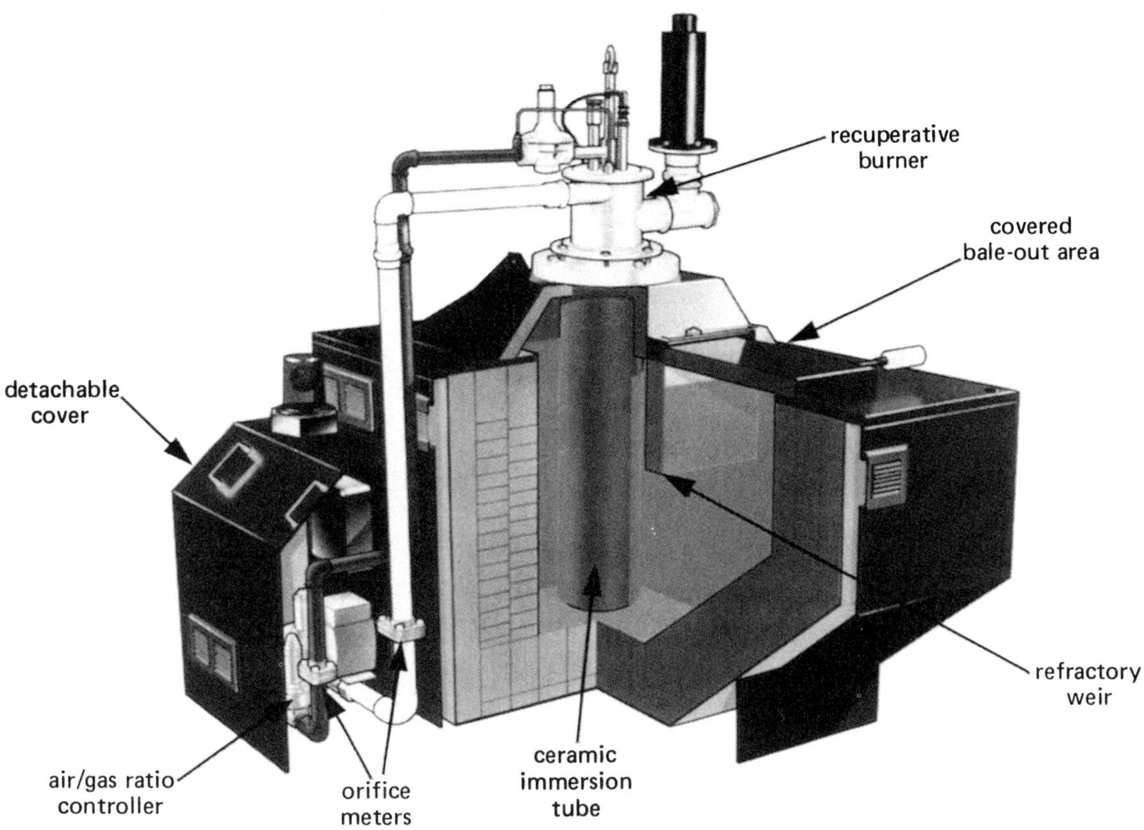

Fig. 6.191 Aluminium bale-out bath with ceramic immersion tube

Tower melting furnaces

These furnaces are equipped with separate melting and holding burners and chambers. The burners for melting and holding are controlled separately. A furnace of this type is manufactured by Striko Ltd. and shown in Fig. 6.192. The furnace can be charged from overhead manually into the charging flue chamber as shown in the diagram, or by automatic charging equipment. The waste gases from the melting and holding burners preheat the charged material which can be aluminium ingots or scrap. Contaminants are burnt off and therefore do not come into contact with molten aluminium thus ensuring improved metal quality. Automatic temperature control regulates the temperature of the waste gas, including the charging cycle, melting time and melting capacity.

This furnace can be used whenever molten aluminium has to be available continuously i.e. in high pressure die casting, gravity die casting and sand casting foundries. The advantages include reduced energy costs through high thermal efficiency and optimum insulation. Thermal efficiencies on melting of around 60% are attainable with waste gas temperatures under 300 °C.

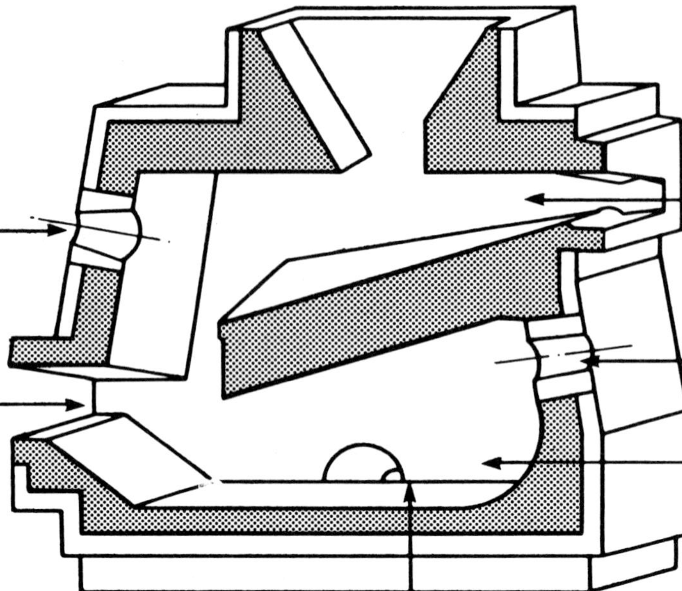

Fig. 6.192 The Striko furnace

6.10.5 Zinc and Lead Melting Furnaces

Furnaces used in the zinc and lead industries are primarily for melting the metal and for holding it in the molten state, often to bring about a desired reaction. The two main types of furnaces are:

- The blast furnace. Like the steel blast furnace, this is a vertical shaft furnace used mainly in the primary production of lead and zinc, but which is also used in secondary lead production.
- The crucible furnace. There are two types with differing functions. One is made of steel (a 'kettle') used in lead refining and having a capacity of over 100 t. The second type is much smaller and is used for holding zinc or lead prior to casting.

Rotary and reverberatory furnaces are also used in the refining of lead.

The blast furnace used in primary zinc-lead extraction is a cylindrical shaft furnace lined with refractory and cased in steel (Figure 6.193).

Hot air is fed into the bottom of the furnace and a charge consisting of sinter and coke, dried and preheated by combustible gas, is fed into the top. Reactions taking place within the furnace permit removal of non-volatile lead at the base of the furnace while zinc vapour, together with combustible gas, passes out at the top, the zinc being subsequently condensed out. Energy is subsequently recovered from the combustible gas by burning it in a Cowper stove through which colder air is drawn in order to provide preheated air to the tuyeres.

Cupola

The cupola, working on similar principles to the blast furnace for the primary extraction of lead, is frequently employed in secondary lead extraction (Figure 6.194). In this the shell, which may be water cooled, is top charged with coke and lead-bearing scrap, together with slag-making materials.

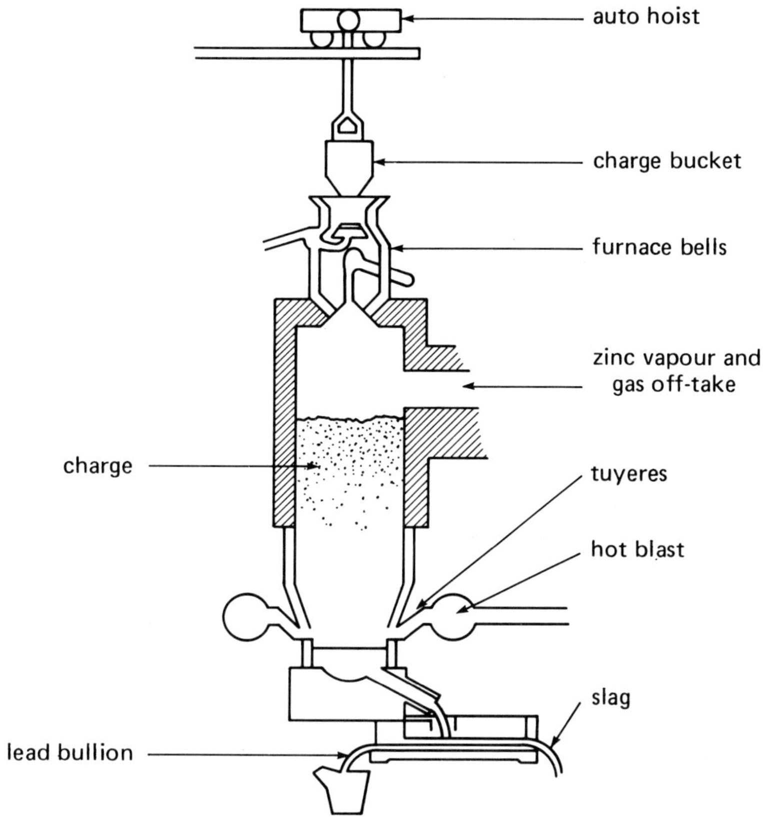

Fig. 6.193 Zinc-lead blast furnace

Air introduced at the bottom of the furnace develops a gaseous reducing atmosphere in which the lead is melted, entrained lead oxide and sulphates are reduced to lead, and residuals are converted to a slag which can be removed.

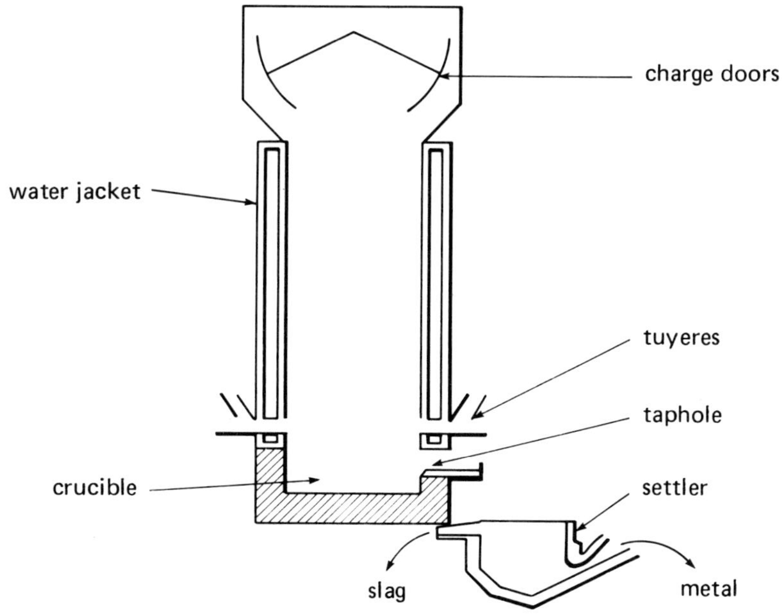

Fig. 6.194 Lead melting cupola

Crucible furnaces

The large type of crucible furnace is used in lead refining, typical capacities being 100 to 200 tonnes (Figure 6.195).

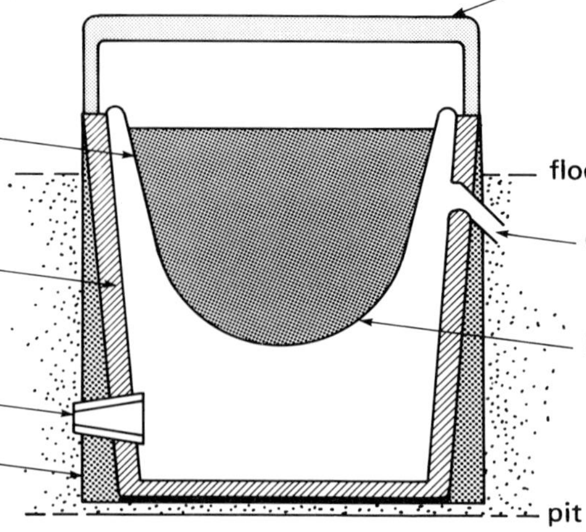

Fig. 6.195 Lead melting crucible furnace

The furnaces are used for melting lead bullion and holding it in the molten state to remove impurities, mainly silver. The crucible or kettle is made of steel and is heated by burners firing tangentially into the annular space between the steel crucible and a refractory insulating surround. Apart from reducing heat losses there is an environmental advantage in using a lid to cover them for fume containment. Recuperative burners also can be used to effect on lead kettles, as can low thermal mass hot-face insulation.

Small (about 200 kg) crucible furnaces are used to hold zinc and lead in the molten state. These furnaces are operated in a similar way to those described in the previous section on aluminium melting.

Reverberatory furnaces

Reverberatory furnaces are frequently used in the initial stages of refining lead, since vertical shaft furnaces and crucible furnaces are limited in size and blast furnaces cannot accept bulky charges. They are often preferred because the oxidising atmosphere used in refining can be controlled and hence the rate of slag formation and its composition. In addition, reverberatory furnaces are convenient for adding molten metal which may be needed for blending.

Immersion tube melting furnaces

As with aluminium melting, ceramic immersion tubes are used to great effect in the melting of zinc prior to refining. An example is illustrated schematically in Figure 6.196.

The furnace bath is fitted with two immersion tubes to melt 4 t/day of ingots. The molten metal is then fed through an offtake launder to a distillation column for separation of cadmium and zinc. This bath replaced a reverberatory furnace and fuel savings of almost 70% were achieved thereby. Metal loss was also substantially reduced.

Galvanising baths

Temperature uniformity is of prime importance in kettle type galvanising baths to avoid hot spots and increase the working life of the kettle. This was achieved in the past by the use of a large number of small burners in a variety of configurations.

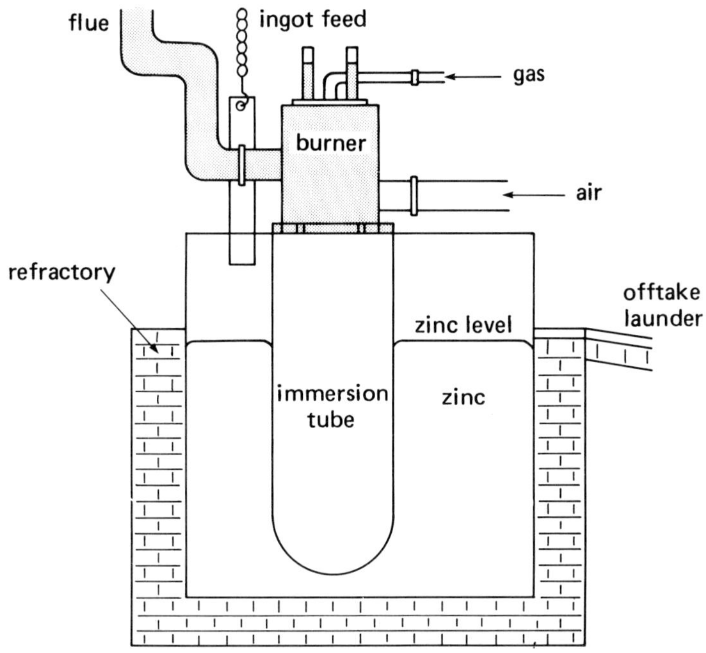

Fig. 6.196 Zinc melting furnace

With the development of high velocity burners, jet-driven recirculation techniques were employed. Figure 6.197 shows a horizontal cross-section through a galvanising bath fired by a single high-velocity burner. Combustion products from the burner flow along the low-level heating ducts along

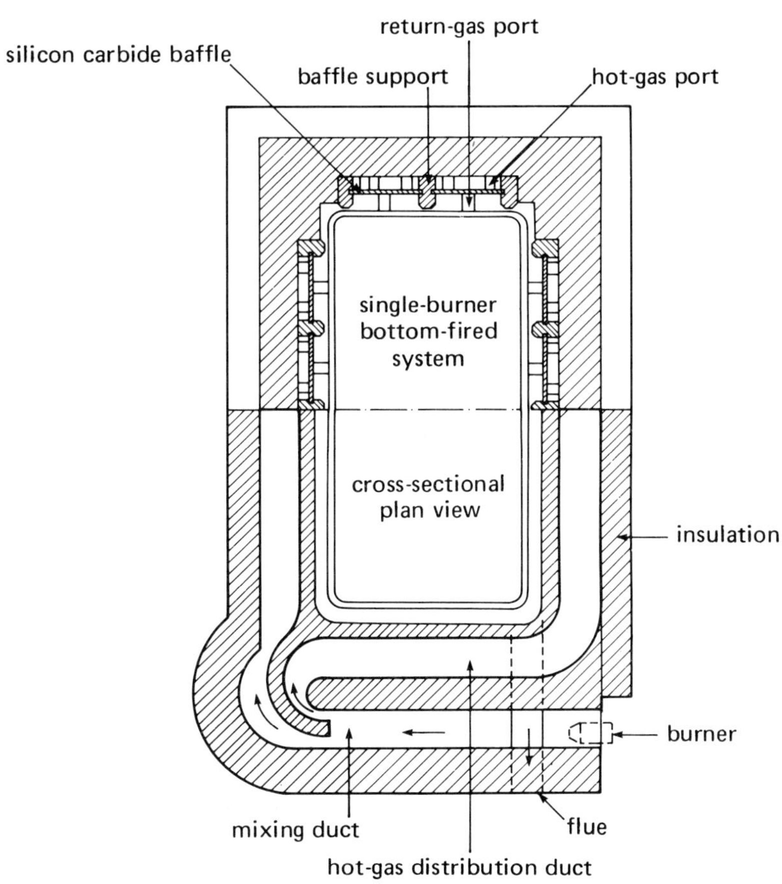

Fig. 6.197 Galvanising setting with single-burner recirculating firing

the bath sides, are discharged through ports and pass upward between the setting and the bath, over a baffle and down over the bath sides to the collecting duct. The gases then flow to the mixing duct where a proportion is recirculated. Because of the high mass flow and low temperature head, heat transfer is mainly by convection, supplemented by low-temperature radiation from the baffle walls. As well as giving temperature uniformity, the reduced number of burners enables automatic controls to be installed relatively cheaply.

Another system which employs high-velocity burners is the Top Heat System in which the surface of the zinc, contained in a ceramic lined bath, is heated within a reverberatory chamber or canopy which sits over the bath leaving a proportion of the zinc surface free for dipping the steel to be galvanised (see Figure 6.198). A top heated bath has an overall width about 2½ times that needed for dipping purposes and this necessitates a large volume of zinc being stored in the bath. The typical size/output ratio (the weight of zinc in a bath to the hourly output of galvanised steel) for a top heated bath is 40:1. Temperature variation from the top to the bottom of the bath can be large, e.g. 60 °C over 1 metre depth, with the bottom being cooler. Operation relies upon some stirring of the zinc by the stock.

High temperatures are generated in the canopy to transfer heat by radiation and convection to the zinc, which has a low absorptivity value. This results in high exhaust gas temperatures (1 000 °C plus) and hence low thermal efficiency (40 to 50%). Although the specific fuel consumption of top heat baths is normally higher than for a kettle, they have the advantages of very long bath life (20 to 30 years) and the ability to cope with temperatures higher than 460 °C. This makes them suitable for high temperature galvanising at 550 °C.

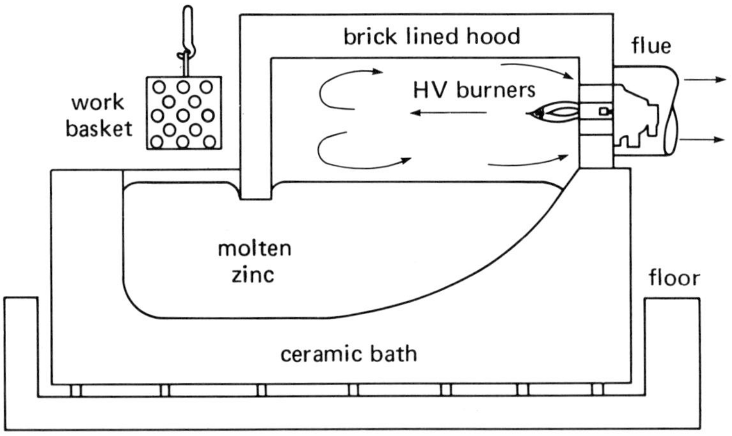

Fig. 6.198 Top heated ceramic lined galvanising bath

A ceramic immersion tube system has been developed by British Gas plc for use in ceramic lined baths rather than steel kettles. Its advantages are:

- Less dross formation.
- High output per volume of zinc.
- Uniform temperature.
- Lower fuel consumption.
- High temperature capability

The tubes provide a compact and efficient means of heating galvanising baths and may be mounted along one side, giving the operators access to the other side. Heavy refractory lined blocks are used to mount the tubes in the vertical position. These blocks are 0.4 m square so that the bath is only

0.4 m wider internally than the width needed for dipping. This represents a considerable saving in space compared with top heat.

Heat transmission is up to 250 kW/m^2 of bath surface occupied. Consequently, the bath may be shorter than a kettle, in cases where the length is not the limiting factor, e.g. wire galvanising. The size/output ratio may be as low as 8:1 compared with a typical 40:1 for top heated baths.

The control system is designed such that each tube has its own air/gas ratio controller, ignition transformer, flame detector, sequence programmer and safety shut-off valves. This gives completely independent operation for each tube, with proportional control over a 4:1 turndown ratio and the option of switching some tubes off during holding periods.

Tubes require periodic replacement, which is straightforward and can be done without stopping production. A spare tube, with its burner, is preheated for about 30 minutes by the side of the bath. The old tube is disconnected from the gas, air and electrical services and is then hoisted out of the bath. The new tube is lowered into position and reconnected.

The ceramic immersion tube system has been applied to ceramic lined baths for both wire and general spin galvanising. Figure 6.199 illustrates a purpose-built bath heated by 6 tubes which replaced a fairly modern steel kettle system. The energy saving was in the range 20 to 30% and output can be up to double that of the similarly sized kettle system.

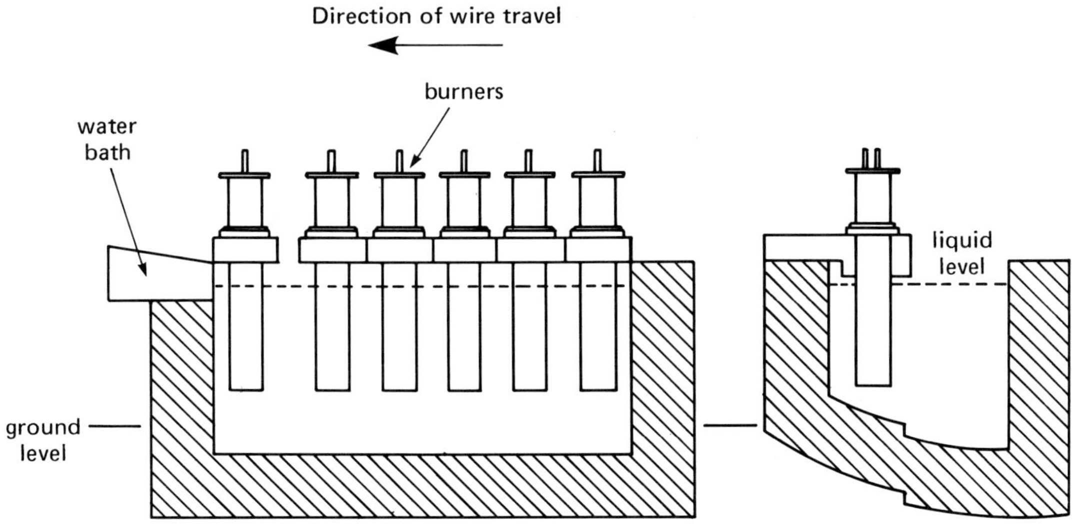

Fig. 6.199 Ceramic radiant tube galvanising bath

6.11 INCINERATION

6.11.1 Introduction

Incineration is defined as the process of igniting and burning a solid, semi-solid, liquid or gaseous combustible waste.

Basically there are two general types of waste which are incinerated:

- Material which is inoffensive but which, if allowed to accumulate increases the risk of fire.
- Material which is organic or putrescent and offensive or dangerous to health.

British Standard BS 3316 categorises the more general types of waste for incineration into three types as follows:

- Rubbish: including combustible waste such as mixed papers, packaging materials, foliage and combustible floor sweepings, from domestic, commercial and industrial activities. This waste can be expected to contain up to 25% moisture, 10% of incombustible solids and have a specific enthalpy of 15 000 kJ/kg. It should require no auxiliary fuel input.

- Garbage: consisting of animal and vegetable wastes from restaurants, hotels, markets and similar premises. This type of waste can be expected to contain up to 70% moisture, up to 5% incombustible solids and to have a specific enthalpy of 5 800 kJ/kg. An auxiliary fuel input of more than 1 950 W/kg of waste incinerated will be needed with garbage.

- Refuse: consisting of an approximately even mixture of rubbish and garbage, commonly from residential sources. A moisture content of up to 50%, an incombustible solids content of 10% and a specific enthalpy of 10 000 kJ/kg is common for this type of waste. Auxiliary fuel input, which may not be required for the whole of the operating period, is approximately 975 W/kg of waste.

In addition to the above types of waste there are others which are burned in large quantities which require specifically designed incineration plant. These include rubber, plastics, wood waste, liquid, semi-liquid or gaseous wastes (e.g. paints, solvents, sludge or fumes from industrial operations) animal carcasses and human remains.

6.11.2 Factors Influencing Incinerator Selection

Natural gas is the preferred fuel for incinerators but it is clear from the above that the opportunity for its use depends on the nature of the waste, its autothermic ability and its propensity for producing gaseous pollutants. Gas may be required for one or more of the following functions:

- To ignite the waste by means of ignition burners.
- To maintain ignition and burning of the waste by means of auxiliary fuel burners.
- To burn out combustible particulates and/or noxious gaseous pollutants in the exhaust gases by means of afterburners.

General incinerators are required to deal with a variety of refuse, which is in the main relatively dry i.e. about 25% moisture content. Occasionally and unpredictably, moisture content will rise if the refuse sometimes contains more wet material e.g. canteen waste. Incinerator selection must be based on as accurate as possible analysis of the waste and possible range of moisture content etc. For instance, in the above case, for the majority of the time no auxiliary fuel would be needed. However, on those occasions when high moisture waste has to be dealt with, the provision of auxiliary burners is essential.

The quantity, composition, variability and characteristics of the waste to be disposed of, therefore, need to be taken into account.

6.11.3 Types of Incinerator

Incinerators used for waste having a relatively high calorific value which can sustain combustion without auxiliary fuel input and which do not require afterburners, clearly do not provide a market for gas. Large scale municipal refuse incinerators are an example.

Fume incinerators

There are a number of industrial processes (e.g. metal coating ovens) which produce combustible/ toxic products which have to be removed. If filters or scrubbers are inadequate for the task, incineration is called for. This could be purely for pollution control or also to recover energy. If the specific enthalpy and the quantity of gaseous pollutants is high enough it is possible to recover more energy than is used in the incineration process. Almost all cost-viable process incineration systems involve heat recovery integral with the process producing the pollutant. However heat recovery into process air, steam, hot water etc. for use in external processes should not be ruled out. Figure 6.200 shows a Contrapol TV-A fume incinerator.

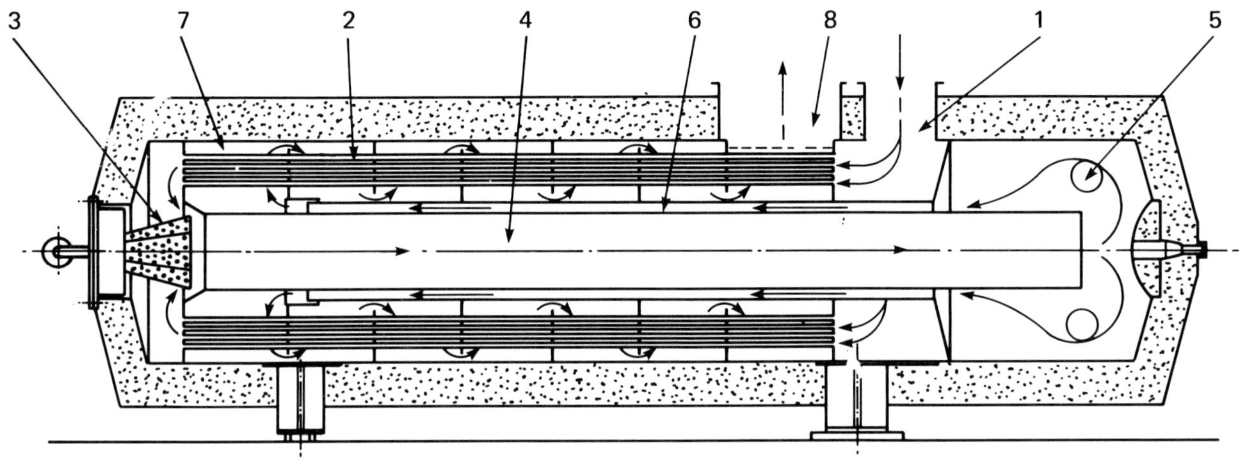

Fig. 6.200 Fume incinerator

With this unit the contaminated exhaust gases enter the pre-evaporation chamber (1) via a tangential inlet. Any condensibles are deposited on the chamber walls and evaporated. The exhaust gases pass into the tubes of a heat exchanger (2) and are preheated by the hot exhaust gases. The auxiliary fuel required to achieve the reaction temperature of around 700 to 750 °C is provided by the gas burner (3). A high velocity venturi flame tube (4) in which the hot gases are mixed gives a uniform temperature across the flow section. To ensure complete burn-out, the gases are held in an integral residence chamber (5) under turbulent conditions. To minimise temperature gradient during incineration the hot, clean gases flow around the high velocity flame tubes (6), pass over the tubes of the fume preheater (7) and then to atmosphere or secondary heat recovery systems (8).

All-purpose waste incinerators

Where waste quality is very variable i.e. from high-enthalpy relatively dry waste through very wet, low-enthalpy waste and animal remains or pathological waste, an incinerator must be capable of dealing with all circumstances. In such cases the incinerator is usually made up in modular form as follows:

- A primary chamber for 'good' waste, which is charged onto a grate.
- A secondary chamber, for the disposal of wet waste or to provide a crematory facility, where the waste is charged onto a solid hearth.
- An afterburner chamber for the removal of smoke and other exhaust gas pollutants.
- A dust settling chamber incorporating water sprays or a grit arrestor to remove particulate emissions.

Each of the incineration chambers (grate, hearth and afterburner) have their own independently controlled gas burner system. Figure 6.201 illustrates a general purpose incinerator made by Hodgkinson Bennis.

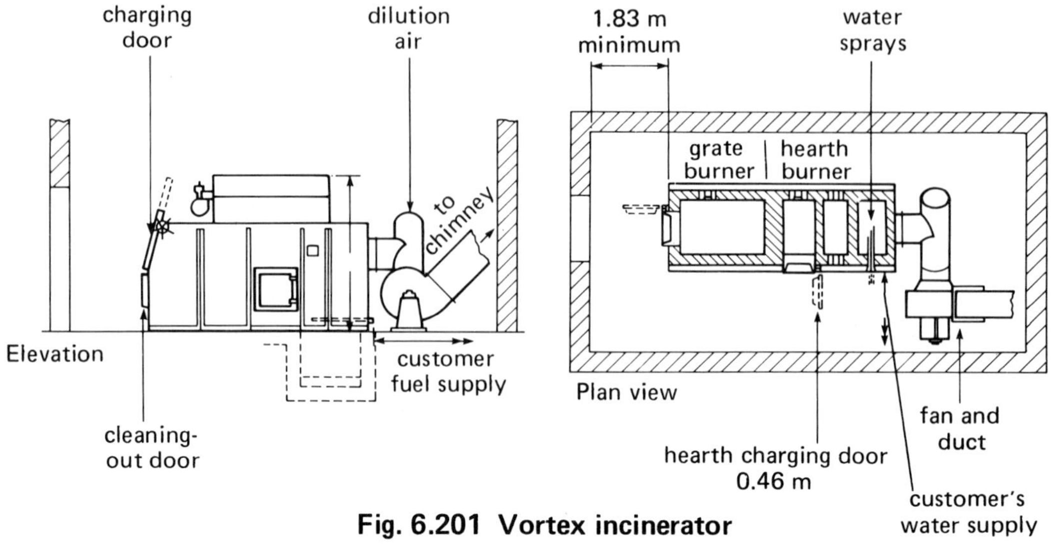

Fig. 6.201 Vortex incinerator

Starved air incinerators

The previous incinerators described all induce active combustion, with the introduction of large volumes of air creating turbulent conditions and thus carryover of particulate material which has to be removed. Starved air incinerators use an initial pyrolysis stage followed by the burning of the partial combustion products in a second afterburning stage.

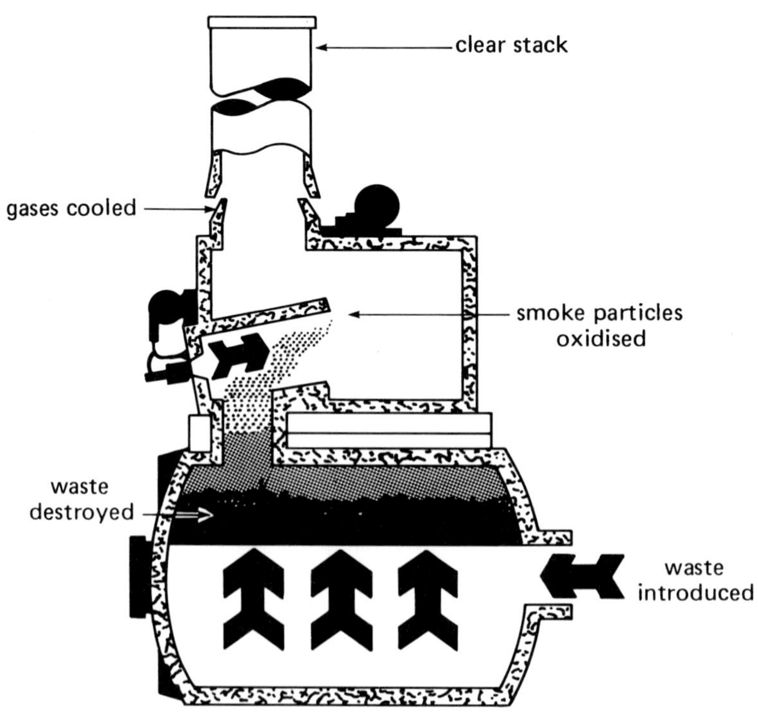

Fig. 6.202 Starved air incinerator

Figure 6.202 is a diagram of a Consumat incinerator which employs these principles. The equipment consists of a refractory lined combustion chamber into which waste is charged. A small amount of air is admitted and the waste is heated by small auxiliary burners, is pyrolised and burnt under quiescent conditions. Carry-over of particulate material is claimed, therefore, to be minimised. The products from the partial combustion passed into an afterburner chamber, are mixed with additional air and increased in temperature to between 1 000 °C and 1 200 °C to ensure burn out of smoke and pollutants. The exhaust gases are cooled by dilution with ambient air before discharge to atmosphere. Alternatively, the hot gases can be directed to a waste heat boiler or other recovery device.

6.12 ATMOSPHERE GENERATION

6.12.1 Introduction

Fuel gases and products of their complete or incomplete combustion are used industrially for preparing atmospheres to surround, protect or react with a product in various processing operations.

Protective gases are used to minimise or prevent undesirable reactions such as oxidation or decarburisation, e.g. in annealing, brazing, hardening, normalising and sintering. Reactive gases chemically change a product, usually its surface as in carburising or nitriding steels. Atmospheres may be reactive in some processes and protective in others. For example, water vapour is undesirable when treating steel but may be used in annealing copper.

The atmospheres that can be obtained from natural gas are lean exothermic, rich exothermic, endothermic prepared natural gas and natural gas itself. Each of these types of atmosphere can be used in a protective or active role. The characteristics of each constituent, separately or mixed with others, must be appreciated. For low carbon steels at normal hardening temperatures:

- Carbon dioxide, water vapour and oxygen are oxidising and decarburising.
- Hydrogen is reducing.
- Carbon monoxide and hydrocarbons are both carburising and reducing.
- Nitrogen is inert for most practical purposes.

It is clear that, for all protective applications, pure nitrogen should give the best results. The cost of pure nitrogen is high, however, and for the majority of processes limited contamination of the nitrogen is permitted. For active atmospheres, the relative proportions of atmosphere constituents must be determined from a knowledge of the process and material involved. Prepared atmospheres are extensively used in heat treatment furnaces, saving expensive metals by reducing scale and often eliminating a machining operation.

Refer to Chapter 1 for a more exhaustive exposition of the reaction involved. Reference should also be made to British Gas Publication IM/9, Code of Practice for the use of Gas in Atmosphere Gas Generators and Associated Plant, Part 1 Exothermic Atmosphere Gas Generators, Part 2 Endothermic Atmospheric Gas Generators, Part 3 Use of Generated Atmosphere Gas in Associated Plant.

6.12.2 Production and Properties of Exothermic Atmospheres

Exothermic atmosphere generators consist essentially of a burner, combustion chamber and atmosphere cooler. Premix burners are normally used, the most common arrangement being a machine-premix system which combines accurate air/gas ratio control with a high mixture pressure to overcome the high system resistance. Figure 6.203 is a diagram of a typical exothermic atmosphere generator.

A rich exothermic atmosphere is produced by burning air/natural gas mixtures having an air/gas ratio of between 6:1 and 5:1. Lean atmospheres are obtained with combustion around stoichiometric conditions; generally slightly below to produce a gas consisting of CO_2, N_2, and H_2O with a small excess of combustibles.

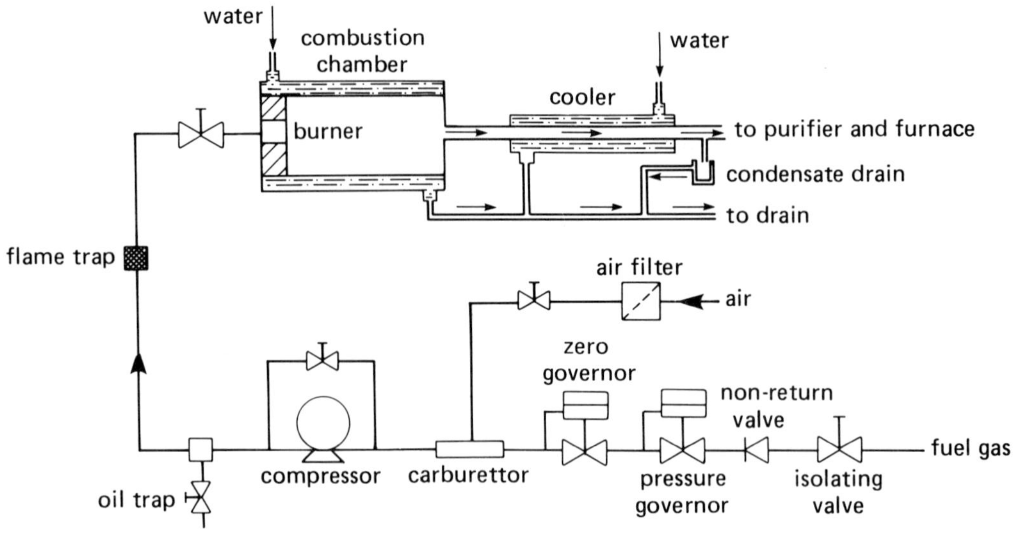

Fig 6.203 Exothermic atmosphere generator

The analysis of rich exothermic atmospheres varies, but for a natural gas generator, 15% H_2, 10% CO, 5% CO_2 and 1% CH_4 (dry basis) would be typical. Rich exothermic gas may be used 'raw' or 'stripped', i.e. after removal of CO_2 and water vapour. These can be removed by one of three methods:

- Compression over water.
- The use of molecular sieves (pressure swing adsorption).
- The use of organic solvents.

Stripped rich exothermic gas is very reducing and may be used as an alternative to endothermic gases. Near-pure nitrogen can be produced by stripping CO_2 and water from lean exothermic atmospheres.

6.12.3 Production and Properties of Endothermic Atmospheres

Endothermic atmospheres can not be produced by combustion in the usual way. They are produced by the catalyst reforming of natural gas with a small amount of air. Air/gas mixture ratio is about 2.4:1. A machine-premix unit feeds the air/gas mixture into a nickel-chrome steel or refractory retort containing nickel impregnated refractory as a catalyst, which is heated externally to a temperature normally between 1 050 °C and 1 100 °C. The objective is to produce near-exact conversion of methane to H_2, CO and N_2. A typical composition of an endothermic atmosphere generated from natural gas is: 0.2% CO_2, 19.8% CO, 39.4% H_2, 0.2% CH_4, balance N_2 and traces of H_2O and other hydrocarbons. The quantities of CO_2 and H_2O are very sensitive to changes in air/gas ratio. Since even a small increase in these decarburising agents may be significant, accurate ratio control is required. Natural gas is a particularly suitable feedstock because of its consistent chemical composition.

For gas carburising purposes, it is usually necessary to enrich endothermic gas by the addition of a few per cent of natural gas to increase its carbon potential.

6.12.4 Non-Fuel-Gas Based Atmospheres

The alternative source of industrial, controlled atmospheres to fuel gases is single, relatively pure gases in cylinders.

Pure hydrogen is used when extreme reducing conditions are required and argon for processes in which nitrogen is not sufficiently inert. Ammonia, however, is the only cylinder gas which is widely used.

Ammonia is used as the raw gas, undiluted for nitriding purposes and as an addition in the carbo-nitriding process. Nitriding is a simple but lengthy process carried out in ammonia atmospheres at about 500 °C, special alloy steels containing aluminium probably being treated. The process depends upon the partial dissociation of the ammonia on the surface of the work allowing the nascent nitrogen produced to combine with the alloying elements of the steel to form nitrides. Atmosphere control is normally achieved by using sufficient flow to prevent the undissociated ammonia content falling below about 70% where possible, particularly during the early stages of the treatment.

The main use of ammonia, however, is in the following forms:

- As dissociated or 'cracked' ammonia giving a highly reducing atmosphere.
- As 'burned' ammonia, ranging from moderately reducing nitrogen/hydrogen mixtures containing 25% hydrogen to pure nitrogen.

Cracked ammonia is produced when ammonia is dissociated into its elements by passing it over a heated catalyst, the result giving a gas having a composition of 75% hydrogen and 25% nitrogen. The dewpoint of the gas is normally between 15 and 40 °C, depending on the grade of ammonia used and the contamination occurring during the vapourising and cracking process. Burned ammonia is now prepared using a single unit which converts the ammonia/air mixture to nitrogen, hydrogen and water directly over a catalyst at about 850 °C.

6.12.5 Application of Controlled Atmospheres

Table 6.4 lists the characteristics and applications of protective atmospheres.

Carbon steels

In general, decarburising is unimportant with low-carbon steels and the main concern is to prevent oxidation. This can be achieved using a rich exothermic gas at normal heat treatment temperatures. Many low-temperature stress-relieving processes are carried out on mild-steel fabrications in directly-fired furnaces using stoichiometric combustion. The limited degree of oxidation occurring is regarded as acceptable.

Similar considerations regarding the prevention of oxidation apply to high carbon steels as with mild steel, although the temperatures involved are normally higher. Minimising decarburisation, however, is the main requirement. As indicated previously, an atmosphere which is non-oxidising may be quite strongly decarburising and much lower CO_2 and H_2O levels must be used. For short, low-temperature heat treatment processes, rich exothermic atmospheres may be used, accepting some small carbon loss. At high temperatures, the choice is limited to endothermic or stripped rich exothermic gases.

Alloy steels

Prevention of oxidation and decarburisation of low-alloy steels is relatively easy. Similar atmospheres to those used for low-carbon steels may be utilised if the combined concentrations of chromium, manganese and silicon total less than 1.5%.

High-alloy steels are very prone to oxidation and may require bright heat treatment in dissociated ammonia.

Table 6.4 Furnace Atmospheres

Atmosphere	Source	Approximate composition % DP = dewpoint	Process	Metals	Advantages	Disadvantages
Rich exothermic (1)	Partial combustion of fuel gases	9 - 12 CO 11 - 15 H_2 7 - 5 CO_2 2 - 3 H_2O 1 - 2 CH_4 Bal N_2	Normalising, copper brazing and sintering	Ferrous	Cheap	Can cause decarburisation and sooting
Lean exothermic (2)	As (1)	0 - 3 CO 0 - 4 H_2 13 - 10 CO_2 2 - 3 H_2O Bal N_2	Bright and clean annealing	Copper Nickel Brasses Aluminium	Cheap; non-explosive	Sulphur removal sometimes necessary; Traces of O_2 near full combustion
Rich stripped exothermic (3)	As (1) but carbon dioxide and water removed	10 - 13 CO 12 - 15 H_2 Bal N_2	Substitute for (6) for most purposes		Relatively cheap; non-decarburising	Can cause sooting; Increased plant size and sometimes increased maintenance over (1)
Lean stripped exothermic (4)	As (3)	0 - 3 CO 0 - 4 H_2 Bal N_2	Annealing; carrier gas for carbon restoration	Ferrous metals	As (3); also non-explosive and non-carburising	As (3); traces of O_2 near full combustion
Modified stripped exothermic (5)	As (1) but CO_2 removed, water gas shift, CO_2 and H_2O removed	3 - 12 H_2 Bal N_2	Long cycle annealing	Low-carbon and mild steel	Non-sooting; cheaper than (8)	. . .
Endothermic (6)	Catalytic reaction of fuel gas and air	20 - 25 CO 30 - 45 H_2 0.5 - 1.0 CH_4 -15 to +15°C DP	Hardening: brazing and sintering: carrier gas for carburising and carbonitriding	Ferrous materials	Non-decarburising	Explosive; can cause sooting; good maintenance and control required
Cracked ammonia (7)	Catalytic cracking of ammonia	75 H_2 25 N_2 -15 to -70°C DP*	Bright and clean annealing; sintering and brazing	Stainless and other steels; certain copper & nickel alloys	Pure; cheaper than (10)	Explosive; can cause nitriding effects
Partially burnt ammonia (8)	Catalytic oxidation of ammonia	0.5 - 25 H_2 Bal N_2 +5 to -70°C DP	As (7)	As (7)	Pure; cheaper and less explosive than (7)	. . .
Fully burnt ammonia (9)	As (8)	0 to 0.5 H_2 Bal N_2 +5 to -70°C DP	Special annealing hardening and sintering	Special non-ferrous metals and steels	Non-explosive; purer than (4)	. . .
Hydrogen† (purer grades) (10)	Electrolysis of water Diffusion purification of (7) In large bottles	99.99 + H_2 99.999 + H_2 99.99 + H_2	As (7) but additional applications where nitriding must be avoided	As (7)	Very pure; no nitriding effects; bottles require little attention and capital	Expensive; explosive
Nitrogen† (purer grades) (11)	In large bottles as liquid	99.99 + N_2	As (4) and (9) but generally more exacting applications	As (4) and (9)	Pure; non-explosive; inert to most metals; little attention and capital needed	Expensive
Argon (purer grades) (12)	In large bottles as liquid	99.99 + A	Bright annealing and other heat treatments on special components	Titanium; special steels Nimonic alloys	Non-explosive; inert; little attention and capital needed	Very expensive
Steam (13)	Electric boiler	99.9 + H_2O	Tempering and blueing; bright-annealing	Steel; copper	Very cheap; non-explosive, little attention needed	Condensate difficulties

*Variable according to grade and type of raw material, consumption rate etc. †Mixtures of these gases can be used instead of (7) (8) and (9)

Carburising

Gas carburising is the most widely used method as it is quick, clean and controllable. Either endothermic gas or stripped rich exothermic gas can be used as a carrier gas, the carbon potential being increased by the controlled addition of natural gas.

Carbo-nitriding, in which both carbon and nitrogen enter the steel surface, also uses endothermic gas, with the addition of controlled amounts of ammonia. The process is carried out between 850 °C and 925 °C.

Non-ferrous metals

The bright heat treatment of non-ferrous metals and their alloys usually present fewer problems than the treatment of carbon steels, since the question of balanced oxidation-reduction reaction is less critical.

Non-ferrous metals can be broadly split with the following categories:

- Those forming oxides with high dissociation pressures permitting the use of exothermic atmospheres (e.g. nickel, cobalt, copper, molybdenum).
- Those forming oxides with low dissociation pressures with which oxidation will take place even in an endothermic atmosphere (e.g. chromium and manganese). These require a pure atmosphere such as pure hydrogen or dissociated ammonia.

Copper is easily bright annealed, oxidation being prevented by small amounts of reducing gases. Lean exothermic atmospheres, slightly on the rich side of stoichiometric, are satisfactory for pure copper heat treatment.

The requirements for nickel are similar to those for copper, but nickel is slightly more resistant to oxidation. Nickel alloys with metals have oxides of high dissociation pressure (like nickel) and can be bright annealed in exothermic atmospheres. Conversely, nickel alloys containing chromium need very pure atmospheres for bright heat treatment.

6.13 CATERING

6.13.1 Introduction

Energy is consumed in the storage, preparation, cooking and service of food, and in dishwashing and environmental services. In meal production most of the energy is used in the cooking process, with relatively little used in preparation.

Gas is the primary fuel used in cooking by the majority of the catering industry. However, energy costs are a small component of the overall costs of catering operations and have tended not to receive the attention due.

6.13.2 Catering Equipment

Full approval testing of catering appliances by Watson House was discontinued and the scheme now in operation is 'Tested for Safety' only. Steps currently being taken to harmonise European standards through the Comite European de Normalisation (CEN) are expected to result in the replacement of BS 5314 with EN 203.

Commercial cooking equipment is generally larger, more specialised and self contained than domestic equipment. The size and durability are broadly classified in the categories light, medium and heavy duty, with the following applications:

- Light duty: snack bars, cafes, small hotels, guest houses, back bars in public houses.
- Medium duty: coffee shops, commercial restaurants and most hotels.
- Heavy duty: industrial and welfare catering and large hotels.

The following illustrations, descriptions and applications of the main types of equipment used in commercial catering establishments are taken from the British Gas plc publication 'Catering Appliances Tested for Safety'.

Ranges

A range is a composite unit made up of a boiling table (hotplate) and a general purpose oven, occasionally with the addition of a grill fitted in the pot rack at eye level. Medium duty ranges (Figure 6.204) usually have open-ring burners, and sometimes include a grill. Solid tops are also available. Ovens are normally internally heated with side hinged doors, single or double according to width. Basically, these units are larger and stronger versions of the domestic cooker.

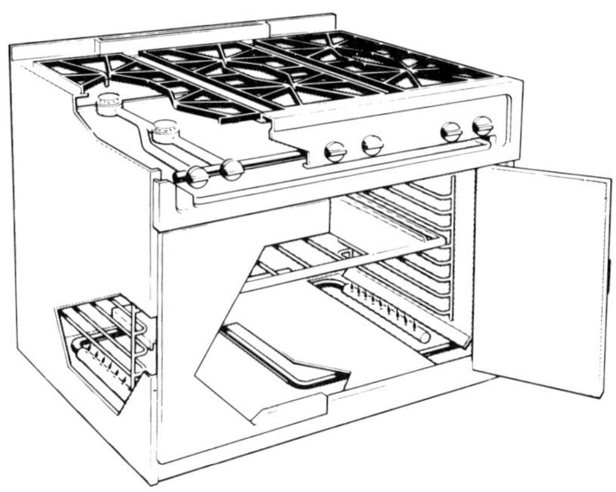

Fig. 6.204 Medium duty range

Heavy duty ranges (Fig. 6.205) may have a solid top, open-ring burners, griddle plate or a combination of the last two with a semi-externally heated oven with drop-down doors.

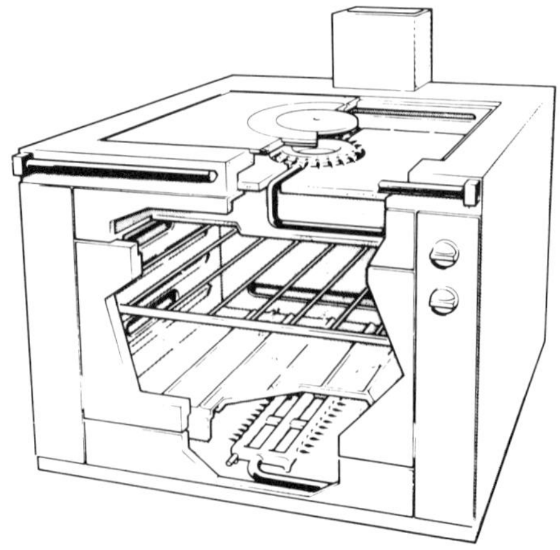

Fig. 6.205 Heavy duty range

A range is found in nearly every kitchen and can be used for all cooking purposes. Because of their comparatively light construction, medium-duty ranges are more suitable for small establishments such as cafes, public house kitchens and schools and in similar situations where use is moderate or intermittent. Heavy-duty appliances are designed for continuous use under severe conditions with weighty utensils such as in the kitchen of a hotel, large restaurant or hospital.

A single-oven range will normally cater for up to fifty persons and a double-oven unit from fifty to a hundred. The increase in size above this number is not pro-rata as it is usual to introduce other equipment in the interests of efficiency. However, much will depend on the scope of the menu. Where only one oven is in use the internally-heated type may be preferred as the temperature gradient will permit the cooking of different dishes at the same time.

Boiling tables and stockpot stoves

Boiling tables are available in two types:

- Open-top with a series of ring burners.
- Solid-top heated by single or multiple ring burners or jet burner.

Stockpot stoves (Figure 6.206a) are low-level boiling tables, generally 610 mm high, usually with an open-ring burner although a solid-top version is available.

Fig. 6.206a Stockpot stove

Fig. 6.206b Open top boiling table

Boiling tables can be used for all boiling top operations; stockpot stoves by virtue of their low height are more suitable for larger utensils. The choice between the open or solid type will depend on the type and amount of use. The open-top table (Figure 6.206b) is more commonly used, with the solid type for continuous heavy-duty catering.

General purpose and roasting ovens

There are two principal methods of heating this type of oven:

- Internal heating, with visible flames inside the oven, similar to the domestic cooker, and with a temperature gradient rising from the bottom to the top of the oven. Roasting ovens are usually of this type.

– Semi-external heat, with the burner or burners placed under a metal plate called the sole. The hot gases enter the oven round the edges of the sole plate or through vents in and over the tops of the inner walls of the evenly heated oven. Figure 6.207 illustrates such an oven.

Fig. 6.207 General purpose oven

Heavy duty ovens are usually fitted with drop-down doors, and medium duty and roasting ovens have side-hinged single or double doors according to the width. Both types of oven can be mounted in tiers one above the other on a stand. This type is suitable for all oven work (roasting and baking). Internally-heated ovens with their fast heat-up are suitable for small establishments using only one or two units and for intermittent use. Semi-externally heated ovens are better for large kitchens where several ovens are installed, each being used for only one kind of dish at a time. At least one oven should be wide enough to accommodate the largest joint likely to be required. Roasting ovens are designed for cooking very large joints and bulk roasting and can be supplied as single units or with a double compartment.

Forced convection ovens

These may be externally or semi-externally heated (Figure 6.208), the hot air being circulated by means of an electrically operated fan. The required heat is reached in a comparatively short time and an even temperature is maintained throughout the oven. A forced convection oven is suitable for all normal roasting and baking. With its even temperature distribution the full capacity of the oven may be utilised and cooking times reduced. The unit is eminently suitable for rapid re-heating and end-cooking of frozen or chilled foods.

Steaming ovens

There are two types of steaming oven available: atmospheric and pressure, as shown in Figure 6.209. The atmospheric type is available in light or heavy duty models. The light duty atmospheric model consists simply of a separate oven and burner stand. The water supply is fed automatically through

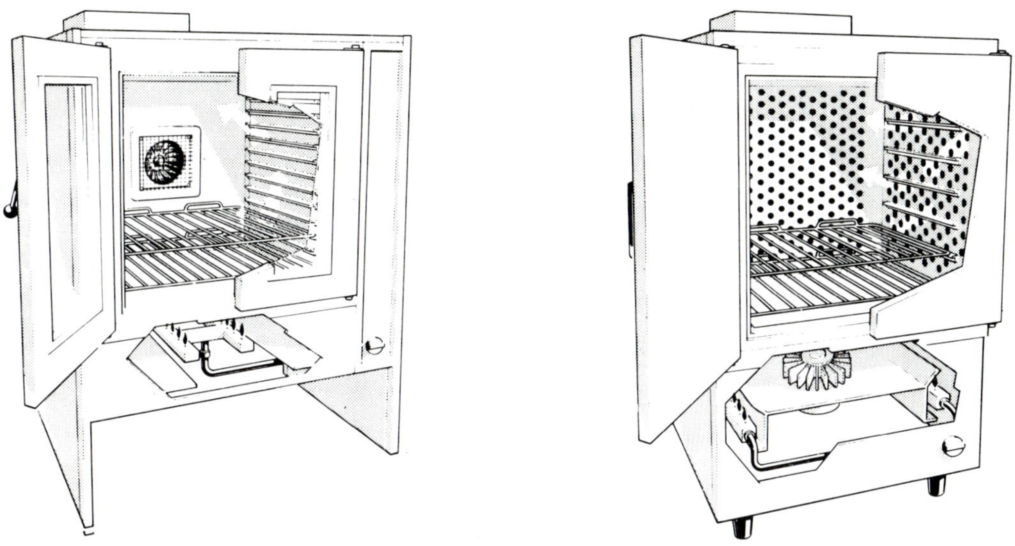

Externally heated   Semi-externally heated

Fig. 6.208 Forced convection ovens

a cistern. The heavy duty atmospheric type is usually of heavier construction and the water level in the pan is automatically maintained by a feed cistern.

The pressure steamer generates steam in a cast aluminium cooking chamber and is fitted with an integral water feed tank and high pressure ball valve.

Once the cooking cycle has commenced automatic controls maintain the oven pressure at between 0.17 bar to 0.55 bar depending on the requirements of the food being cooked. Pressure steamers are fitted with safety doors that are impossible to operate whilst the unit is under pressure.

High pressure steamers because of their higher operating temperatures shorten cooking times.

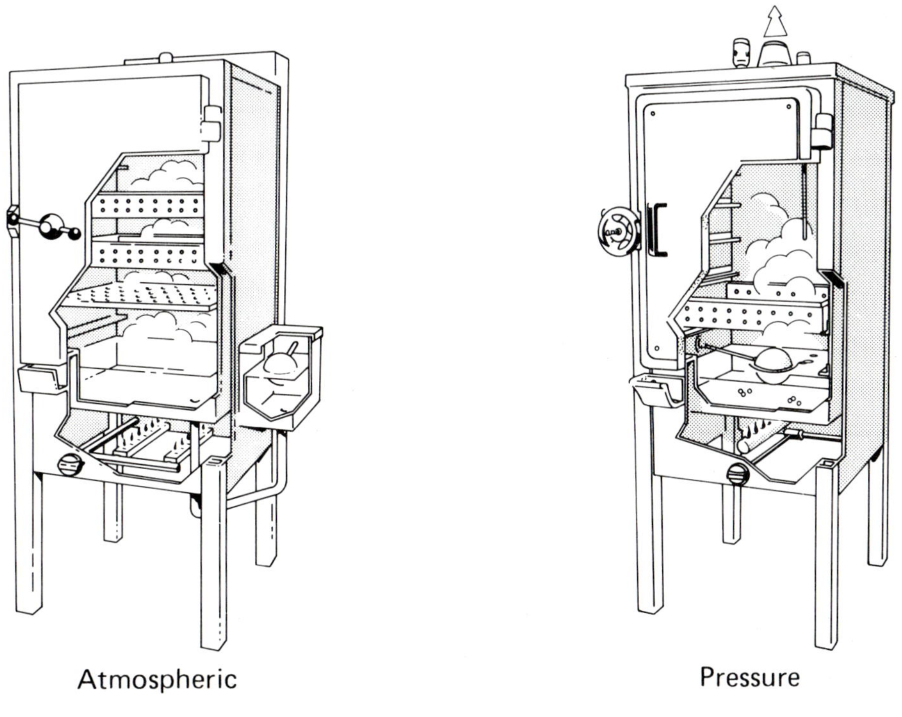

Atmospheric   Pressure

Fig. 6.209 Steaming ovens

Combination Ovens

Sometimes known as combi ovens, these units combine the function of an atmospheric steamer and a forced air convection oven. Figure 6.210 shows the Bartlett Harmony Combination Oven. This means that, for example, meats can be cooked by firstly steaming and finished off by dry forced convected heat to provide a roasted appearance and flavour but with a greatly reduced weight loss. The dual function of these units means that they are suitable for cooking almost every type of food. These ovens are also fitted with controls which make the operation fully automatic.

In steam mode steam is rapidly generated. The convection fan draws the steam into its centre then circulates it around the oven. Steam builds up in layers until forced down the drain tube over the steam control thermostat phial which cuts off and re-engages the heat source to keep the oven steam saturated.

In convection mode, the fan draws air into the heat exchanger chamber to be heated rapidly and re-circulated. Oven temperature is then monitored by thermostatic control.

In the combination mode, steam and heated air are controlled by their individual thermostats. The fan distributes both to ensure oven efficiency at the required temperature.

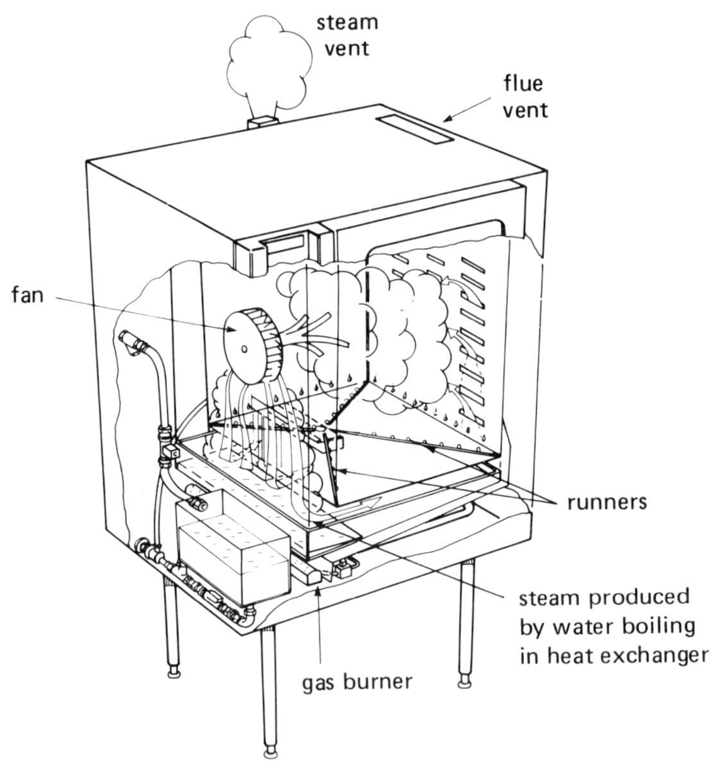

Fig. 6.210 Bartlett Harmony combination oven in the steam mode

Deep fat fryers

Essentially, a deep-fat fryer is a heated pan containing oil or fat. There are two main types: one with a V-shaped pan heated above the bottom of the V; the other fitted with burners firing into tubes immersed in the frying medium. Figure 6.211A illustrates the former type.

In both there is a zone of relatively cool oil below the source of heat into which food particles can sink and remain without charring. There is also a large-capacity flat-bottomed type (Figure 6.211B) used in fish and chip shops.

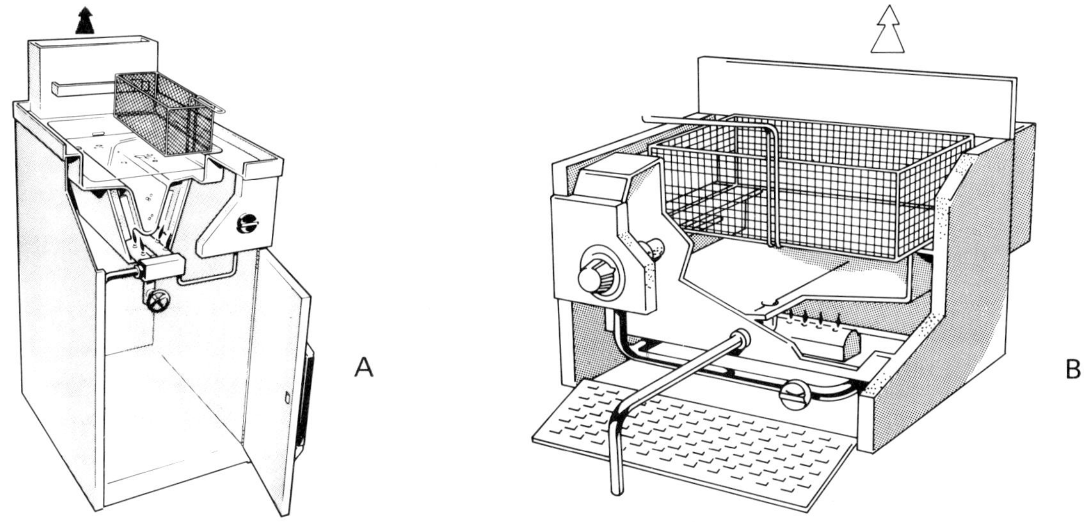

Fig. 6.211 Fryers

They can be used for cooking all fried foods (chips, fish, fritters etc) that are immersed into a heating medium, i.e. the oil or fat. The type of fryer chosen will depend on the type of use required. Foods can be fried directly in most flat-bottomed fryers, but baskets are frequently used to contain chips, scampi, etc.

Fryers are also available as pressure vessels which has the effect of speeding the cooking operation due to the higher temperatures obtainable in the cooking medium.

Grillers

The two main types of grillers (Figure 6.212) are:

- Underfired: sometimes known as flare grills with the source of heat below the meat.
  Burners are located underneath refractory bricks or lava rock. The liquid or fat falls on the heated material, flares, and the smoke and flare impart a charcoal-cooked flavour and appearance to the meat.
- Overfired (Salamanders): the source of heat is from refractory bricks or a metal fret placed above the food.

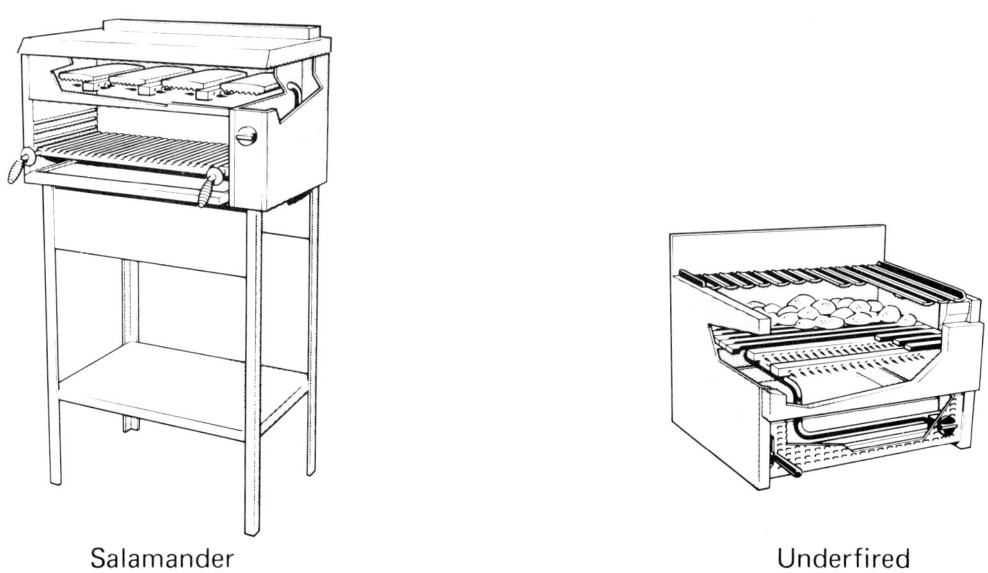

Fig. 6.212 Grillers

Grillers are used to cook food by radiant heat. Underfired grillers are mainly used for chops and steaks, but overfired grillers can also be used for making toast and for quick heating of dishes before service, and are thus more versatile. Underfired grillers with their flare effect are often used when cooking is carried out in view of the public.

Griddle plate

Griddle plates (or dry-fry plates) are solid plates, usually provided with a drain channel, and heated from below by a bar burner. Figure 6.213 shows a typical griddle plate. They are used for fast or continuous frying of eggs, bacon, liver, steak, hamburgers etc, and are common in snack bars and grill rooms for call-order cooking.

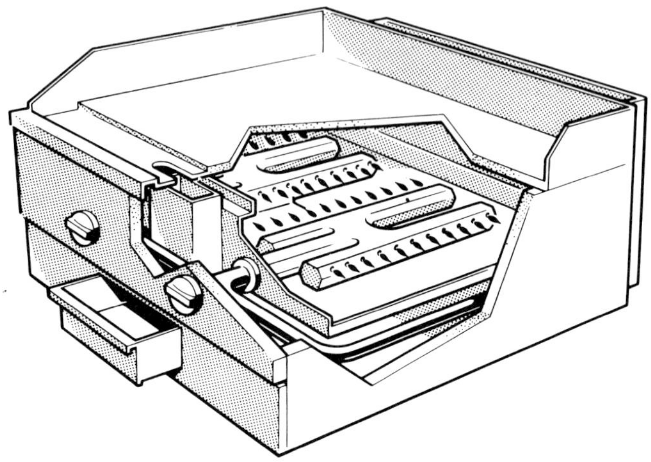

Fig. 6.213 Griddle plate

Bains-marie

Bains-marie are used to keep cooked foods hot prior to, or during service. They can be the open type, in which food containers are placed in hot water to retain their heat (Figure 6.214), or fitted with filler plates and containers which can be either directly heated by gas or indirectly by hot water or steam.

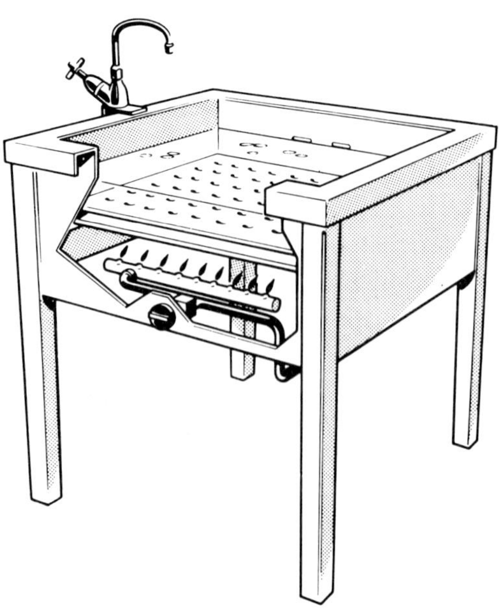

Fig. 6.214 Open top bains-marie

Hot-cupboards

These cabinets which may be directly heated either by a burner sited below a baffle plate, or indirectly by hot gases conducted round the cabinet through channels, or by steam generated in a well in the base. They are used to heat plates and to keep cooked food hot during the delay between cooking and serving.

The directly-heated hot-cupboard (Figure 2.215) is most suitable for plate warming at about 60 °C, and the indirectly-heated hot-cupboard for keeping food hot at about 82 °C. Hot-cupboards may be thermostatically controlled to pre-set temperarures. There are free-standing hot-cupboards or wall-fitted models to save floor space and tiered models to fit narrow spaces. Counter display units with glass fronts are also available. Doors may be hinged or sliding, or roller-shutter doors can be fitted to give full width opening to facilitate handling of dishes. Most models can be obtained with doors both sides if required. Hot-cupboards can be obtained with fitted or open bains-marie.

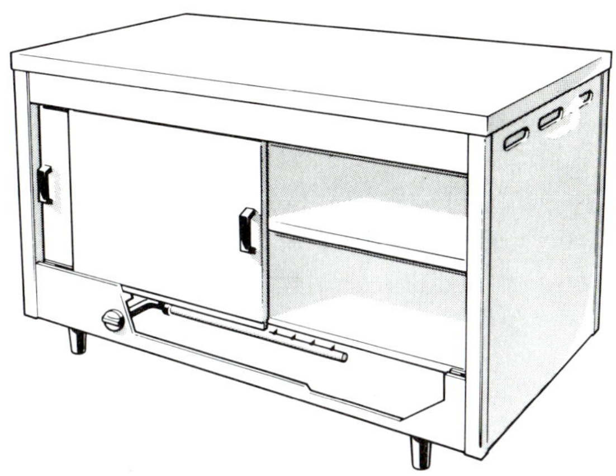

Fig. 6.215 Hot-cupboards

Fig. 6.216 Combined microwave and gas fired forced air convection oven

Combined microwave and gas fired forced air convection oven

These units offer the speed of microwave cooking with the ability of forced air convection to cook pastry items, roast meats and to impart a browned appearance to the food (eg Merrychef) Fig. 6.216.

The units are basically small - a typical oven cavity is 350 mm cubed which makes them suitable for counter top use in pubs, wine bars and small restaurants.

As a comparison with conventional cooking methods, these units can roast a 1.6 kg fresh chicken in 15 minutes and bake 1.8 kg of jacket potatoes in 17 minutes.

Boiling pans

There are two main types of boiling pan (Figure 6.217):

- Directly heated by the gas burner.
- Jacketed - a type of double saucepan in which the heat is transmitted from the burner to the contents of the inner pan through steam or water, thus eliminating hot spots.

The single pan is used for cooking root vegetables, boiling meat and making soups. The jacketed pan is used for custards, milk puddings, porridge and other viscous liquids that burn easily. Boiling pans are available in sizes from 45 to 180 litres capacity, but it is better to install several small pans rather than one large unit in order to achieve greater flexibility. This is because the smaller pans give a faster temperature recovery and hold smaller quantities of vegetables.

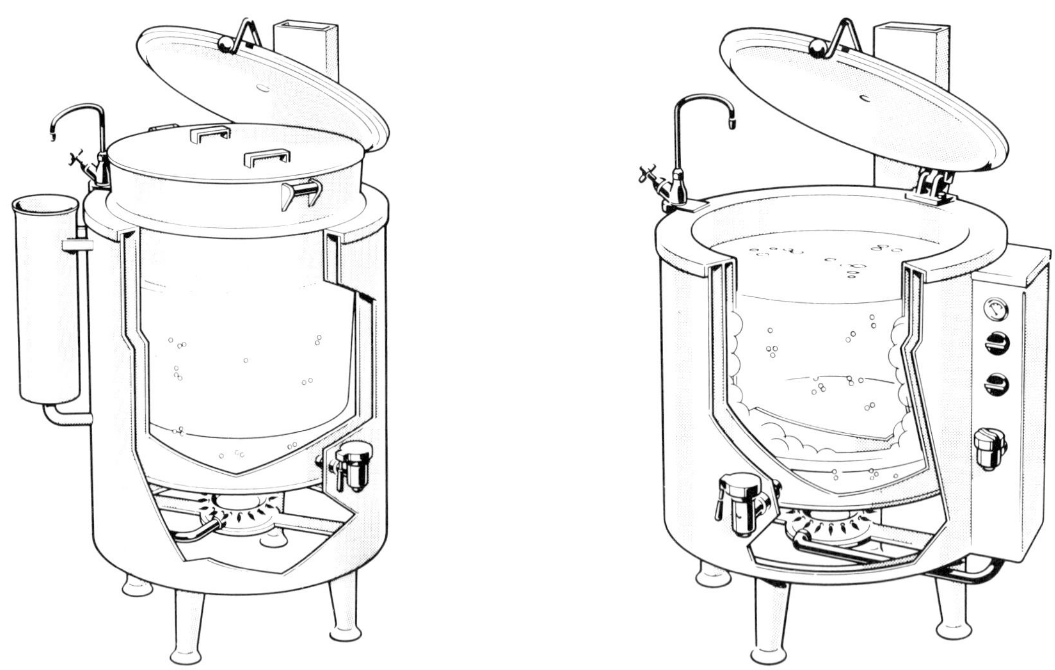

Fig. 6.217 Boiling pans

Water boilers

There are a number of types of small water boilers used mainly to provide hot water or hot milk for making beverages. Bulk water boilers (urns) are used when a known quantity of water is required. An example is shown in Figure 6.218. The container can be copper or stainless steel.

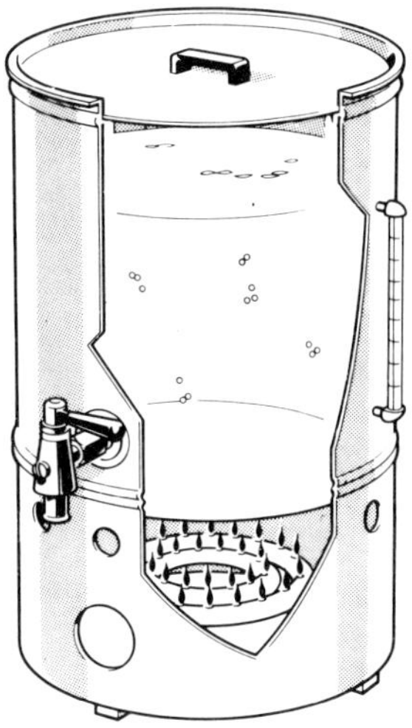

Fig. 6.218 Bulk water boiler

Cafe boilers used to supply instant boiling water on demand are available in two types (Figure 6.219):

- Expansion or overflow boilers.
- Pressure boilers.

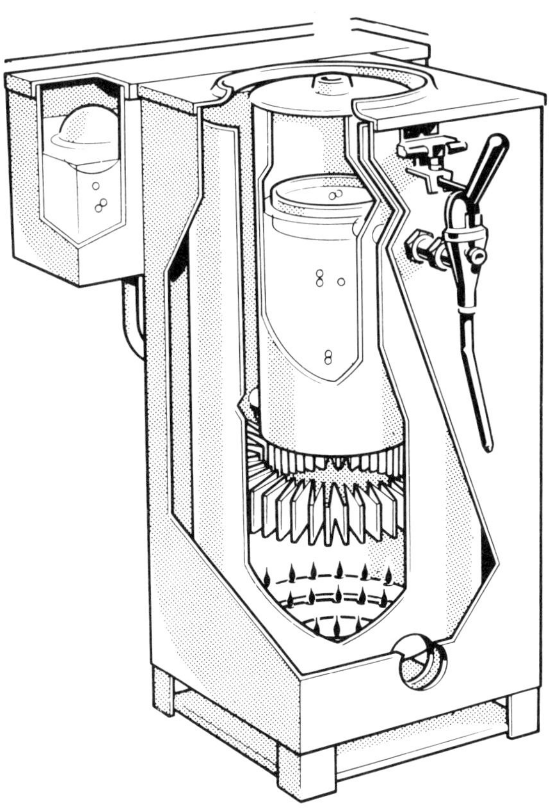

Expansion

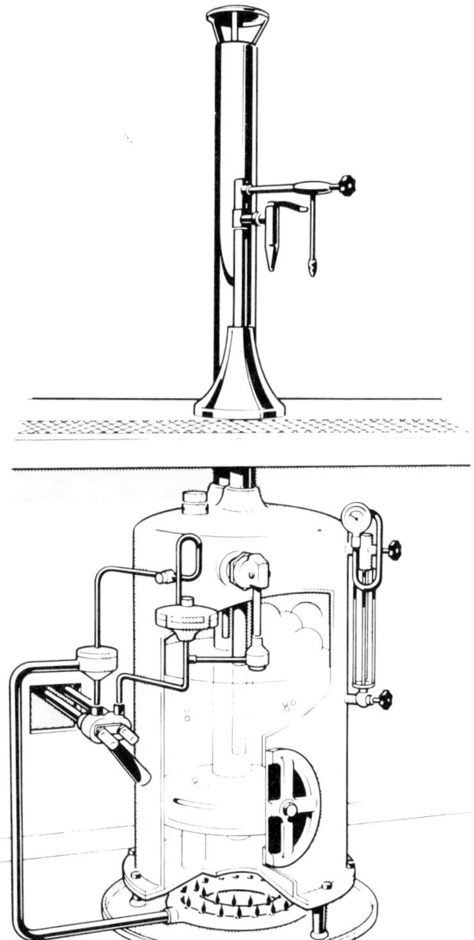

Pressure

Fig. 6.219 Instant boilers

The operation of the former is based on the principle that water expands when heated. Some models incorporate a compartment for storing a reserve of hot water. Pressure boilers are housed under the serving counter and heat and store water at around 105 °C at a pressure of up to 500 mbar. Steam injectors for heating soups, etc., and extended arms for infusers can be supplied with pressure boilers.

Dishwashers

There are various styles and sizes of dishwasher. The three main types are:

- Under-counter models, loaded like an oven.
- Manually operated or semi-automatic machines into which racks are pushed after loading.
- Conveyor or fully automatic machines.

Washers can have a timed washing cycle and provision for the automatic addition of detergents and drying agents.

Most commercial dishwashers are supplied with hot water at between 50 °C and 55 °C. Supplementary heating in the form of electric immersion elements or steam/hot water coils in the tanks of the wash and rinse sections maintain/boost temperatures of 55 °C and up to 85 °C respectively, in conveyor machines. In the larger dishwashers there can be a pre-wash and pre-rinse. Hot, relatively clean rinse water is often diverted to the wash tank, thus saving both heat and water. Figure 6.220 shows a modern automatic conveyor washer.

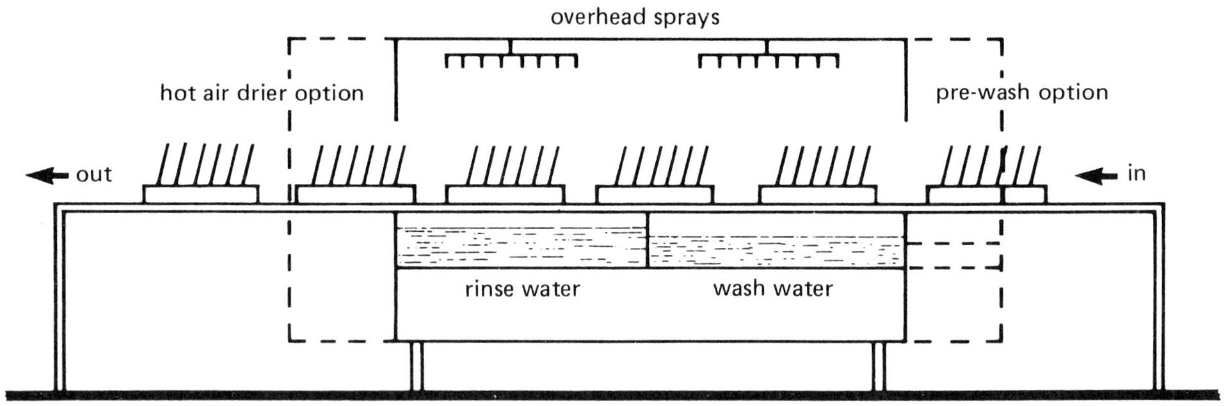

Fig. 6.220 Automatic conveyor dishwasher

Drying tunnels can be incorporated into conveyor machine design. Hot air at about 95 °C is the drying medium, the heat source being either steam or electricity via heat exchangers. Most drying tunnels incorporate hot air recirculation to improve energy efficiency. Heat can be recovered from the moisture laden exhaust air via an air/cold water heat exchanger. A temperature rise from 10 °C to 55 °C is claimed by some manufacturers.

Heat can also be recovered from hot drain water on continuous dishwashers. This is especially valuable where the water feed is cold. Water temperature can be increased by as much as 35 °C.

6.13.3 Catering operations

Catering operations cover a wide range of requirements. Restaurants in large hotels have normally to be available for the three main meal periods and coffee shops may serve meals up to 18 hours per day. Some commercial restaurants, including fast food outlets, operate over a 15 hour day.

Residential and institutional (e.g. hospital) services provide full meal services throughout the day. Staff canteens are also available over long periods in these premises, especially where services are required for night-shift work. At the other extreme, canteens for employee or welfare catering may only operate during the mid-day meal period or with a limited snack service at other times of the day.

There are normally two to four hours of concentrated activity before and during each meal period, in preparing, cooking and serving food. Some of the preparation may be carried out in advance or food may be supplied ready-prepared or pre-cooked and frozen. The latter requires end cooking or regeneration, which requires less time and energy.

Cooking is usually performed in batches and production is such that the supply of suitable quantities of freshly cooked food matches customer demand.

6.13.4 Energy Usage

Intermittent periods of high energy demand are caused by cooking equipment. Most commercial cooking appliances are highly rated and, apart from ovens and enclosed boilers/steamers, this energy is discharged as heat continuously into the kitchen environment and needs to be removed to moderate room temperature.

Food cooking

The amount of energy required to cook a unit quantity of food depends on the heat transfer process, as well as on the type and capacity of the equipment used. For instance, frying and grilling require relatively high energy inputs but forced convection ovens are comparatively efficient, if well insulated.

A high proportion of energy is absorbed in preheating the equipment from cold to operating temperatures. Overall, this accounts for about 25% of the energy used. Large preheat losses occur in fryers and boiling pans due to the relatively large mass of oil or water. Solid top ranges and other heavy equipment also have high pre-heat requirements. The pre-heat losses in boiling tops are mainly due to the transfer of heat to the utensils and thence to the boiling or frying medium. Pre-heat losses are significantly reduced if the equipment is continuously used or has a well insulated casing. Even when equipment is fully utilised, the actual energy absorbed by the food is only a small percentage of the total energy input, ranging from 35% in a convection oven to less than 10% with a boiling pan.

Energy consumption per meal depends on the nature of the operation, scale of production, installed capacity and type of equipment. The normal range is 3.3 MJ to 8.7 MJ per meal (excluding dishwashing, direct hot water and environmental services), but consumption can be as high as 45 MJ/meal in high class a la carte restaurants.

Energy monitoring carried out by the University of Surrey indicated the specific energy consumption figures for different operations shown in Table 6.5. The figures relate only to the energy used for food preparation, cooking and service.

Table 6.5 Energy Consumptions per Meal

Type of operation	MJ/meal
Employee catering — large scale	3.5 - 4.5
— medium scale	4.8 - 6.6
University and school catering	8.8 - 9.6
Hospital catering	7.7 - 9.0
Snack bars, pub counter meals	3.0 - 4.5
Store restaurants, coffee shops, cafes	4.0 - 5.6
Steak houses, pub meals	7.5 - 21.0
Traditional English restaurants	8.8 - 10.0
High class restaurants	33.0 - 50.0

The high level of energy consumption in haute cuisine establishments is due to the wide menu choice and the inherent wastage in the operation.

Dishwashing and hot water supplies

Table 6.6 following shows the average quantities of hot water used per meal in catering services developed from work carried out by British Gas plc, and the energy requirement based on a water supply temperature of 65 °C and 70% boiler efficiency.

Table 6.6 Water and Energy Usage per Meal

Premises	Water litres/meal	Energy MJ/meal
Schools	5.5	2.01
Restaurants	8.2	2.99
Stores	8.0	2.92
Offices	9.7	3.54
Hotels	13.7	5.00

Dishwashing requirements depend on the type of machine and the extent of boost heat installed. The amount of water used generally increases with size from 3 litres/meal to 4.7 litres/meal or more. With a connected hot water supply, the additional energy required for temperature boosting to 82 °C varies between 0.3 and 0.5 MJ/meal.

Energy demands of dishwashing machines are high. A small intermittent dishwasher needs about 16 kW and large conveyor type equipment (1 500 to 5 000 plates/hour) uses between 47 and 87 kW. Allowing for overall energy consumption and 50% machine utilisation, dishwashing accounts for between 2.8 and 4.4 MJ/meal. In practice much greater water and energy usage is often involved because equipment capacities are normally related to peak requirements resulting in very low utilisation factors.

With the large conveyor type machines, facilities for variable conveyor speed, hot air drying and heat recovery are available. Heat recovery equipment can involve the use of a vapour condenser, condenser-economiser or condenser combined with an air to water heat pump.

Energy savings opportunities

Many kitchens have an excess of cooking equipment or capacity, the higher installed capacity affecting space, capital and operating costs. Equipment utilisation is often found to be very low leading to low energy efficiency. Wide fluctuations in energy use per meal can be due to equipment being switched on in advance of use or left on continuously. More precise scheduling of meal production can contribute to rationalisation in equipment capacity and extent of use.

Energy savings can be achieved by improvements in good housekeeping and/or maintenance where:

- Equipment is left full on for only occasional use; particularly boiling tops.
- Lids of pans are left off.
- Automatic controls and times are manually overridden.
- Heat transfer surfaces are distorted or corroded or fouled with dirt or scale, reducing emission and contact.
- Gas jets are worn and corroded, badly adjusted or malfunctioning.
- Steam or hot air is leaking from distorted or damaged doors and fittings.
- Dishwashers run with only part loads and with jet and cycle control defects.

Heat recovery potential exists from:

- Dishwasher liquid effluent and vapour laden exhausts.
- Kitchen extract systems.

The economic feasibility of heat recovery will depend on the circumstances and the extent to which heat transfer surface fouling by grease, soap and scum can be minimised. It can be argued that with modern equipment design having higher efficiency, better programme control and the retention and recycling of heat within the equipment, the potential for external heat recovery will reduce.

6.14 BAKING OVENS

6.14.1 Introduction

Baking is concerned with the range of static and continuous ovens which are used for the final cooking of bread, cakes, pies and biscuits. These ovens go back to the times when they were mainly of batch type and originally were solid fuel fired. The present day range of ovens cover small and large batch ovens, and, for mass production, the travelling or continuous conveyor ovens. Heating methods are varied to suit the range of sizes and also the age of the equipment; some of these ovens are in fact very old and date from a time when design was based on fuels other than gas.

The preparation of the food prior to baking is of interest, but comment will be restricted to those areas concerned with heat. A typical bread line would consist of a proving oven, heated by steam coils, with live steam and extract fans to control the humidity; here the dough rises in the individual tins before passing into the main baking oven.

Ovens bake a very wide range of products and the actual heating requirement will vary tremendously.

There are therefore a wide array of oven types some batch, others continuous, some very old and others comparatively new designs. Most of the types are listed below:

a) Steam tubes or draw plate batch type

b) Peel ovens

c) Hot air and forced circulation

d) Reel ovens

e) Rack ovens

f) Travelling or continuous conveyor ovens.

6.14.2 Steam Tube Ovens

This method uses rows of small-bore wrought iron tubes, each one containing a fixed quantity of distilled water which turns into steam when the tube ends are heated. While the tube ends are heated by radiation and flue gases in the case of solid fuel or oil fired examples, gas versions have a small gas jet from a bar burner under each end of the tube. A development in steam tube ovens is to use high velocity single burner systems to heat a combustion chamber which contains part of the individual steam tubes or part of an annular steam tube. This ensures no flame impingement on the tube and high rates of heat transfer are achieved.

The steam tube principle lends itself very well to multi-deck construction. For example, a four deck oven will have five rows of tubes, the intermediate ones supplying top heat to one deck and bottom heat to the next. Gas firing enables the correct amount of heat to be supplied to each tube so that no one tube is likely to get overheated. The steam pressures within the tubes must be high to achieve at least 301 °C (575 °F) to maintain a baking temperature of 232 °C (450 °F), which means that the steam pressure will be about 75 bar (1 100 lb/in^2).

Figure 6.221 shows a diagrammatic arrangement of a three deck steam tube oven. In fact the tubes slope downwards to the burner to enable the water in the cold tubes to be at the best position to start up. The size of these ovens is 3 m x 2.5 m (10 ft x 8 ft) approximately.

To prevent the steam tube exploding, the overheat thermostat protection system was developed in 1972. This does not cater for full fail safe protection. Flame safeguards should be fitted when possible instead of low pressure cut off valves. The system shown here only provides for the prevention of a steam tube exploding due to overheating.

6.14.3 Peel Ovens

A peel oven is a deck oven in which goods to be baked are set and after baking are withdrawn by means of a peel which is a flat wooden blade on the end of a long wooden handle. The oven may be of brick construction or of sheet metal. A variety of methods can be employed to heat a peel oven - side flame, steam tube, hot air and hot air forced convection.

A peel oven is still the backbone of the baking industry. It gives an excellent bake, but requires a skilled operator to load and unload particularly the larger sizes. Average size 2.5 m x 2 m.

A side flue oven will frequently have a flame retention head burner arranged so that the flame will project through a hole in the oven. The main gas supply to the burner and the auxiliary supply to the by-pass are led through an interlocking cock. The flue damper is connected by cable to the

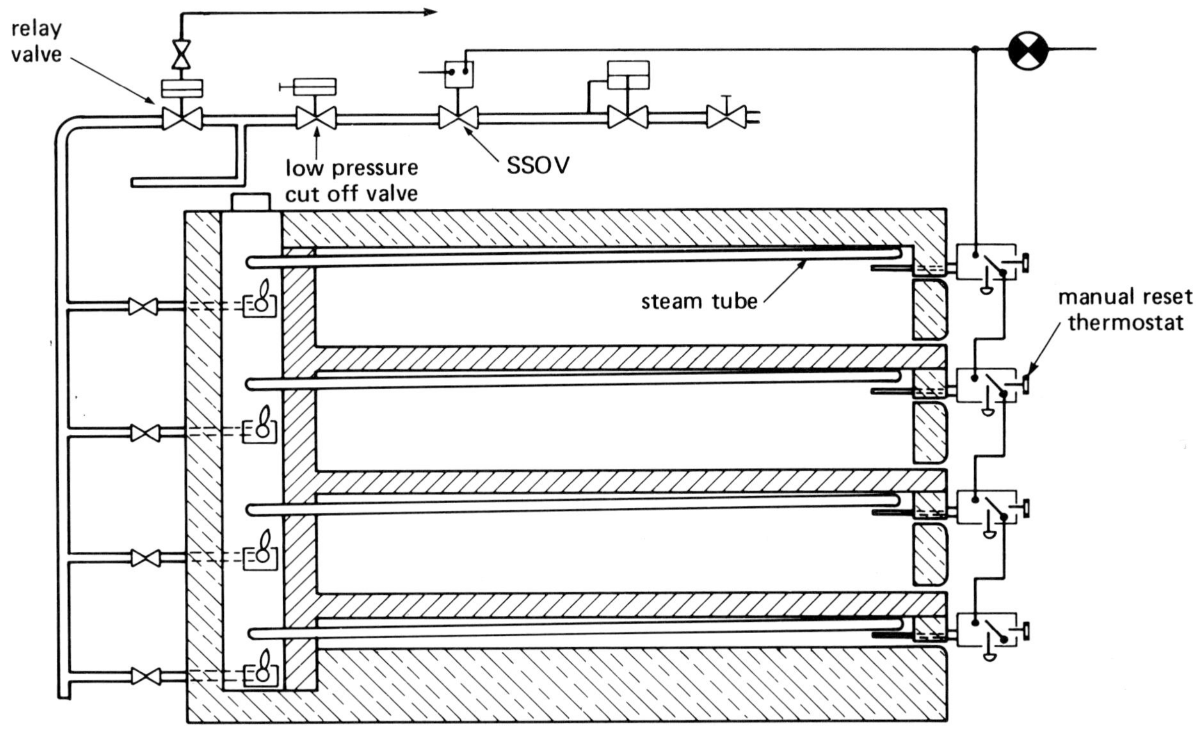

Fig. 6.221 Three deck steam tube oven with overheat thermostatic protection

interlocking cock ensuring that the main gas cannot be turned on unless the flue damper has been fully withdrawn. Existing versions of this oven would ideally have electronic flame protection and the damper either cut away or interlocked with the starting procedure.

6.14.4 Hot Air Ovens

Figure 6.222 shows a diagrammatic arrangement of a hot air oven. The products of combustion are conveyed around the outside of the baking chambers by flue ways. The best and most efficient example of this method was the heat trap oven, regarded by many as the Rolls Royce of peel ovens.

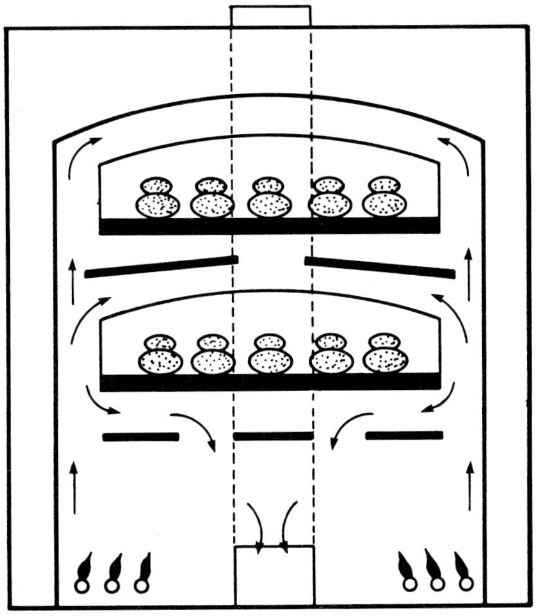

Fig. 6.222 Hot air oven

As with steam tube ovens, depending on the manufacturer, decks can be heated and controlled collectively or individually. Hot air ovens are very flexible in use and can be employed for bread or confectionery baking in the higher or lower temperature ranges.

Hot air forced convection ovens are similar to above but the products are circulated at high speed giving quicker heat up and more even heat distribution. This type of oven is very popular on the Continent. It offers self generation of steam with powered extraction individually from each deck. Again, ideally suited for multi-deck construction. Figure 6.223 shows a hot air forced convection oven.

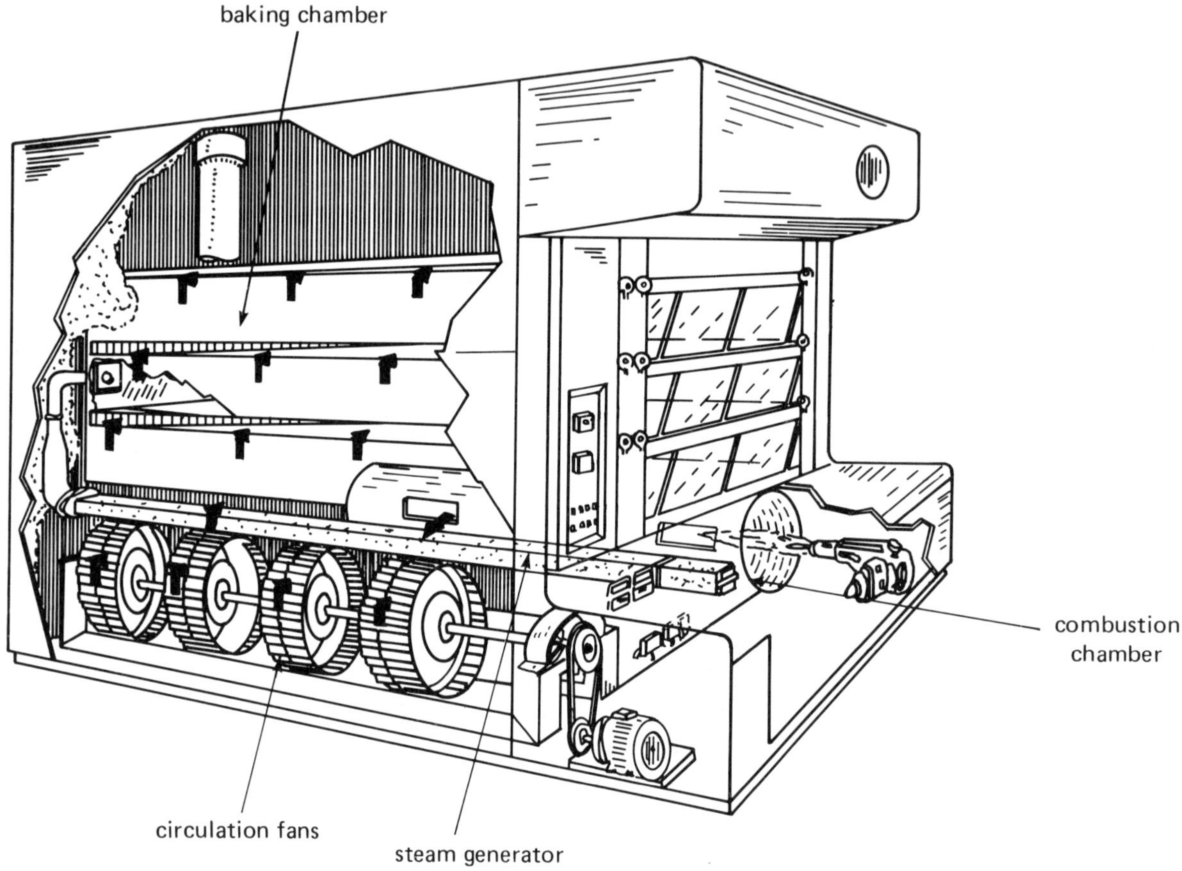

Fig. 6. 223 Hot air forced convection oven

6.14.5 Reel Ovens

This is essentially a large box containing a reel rather like a fairground Ferris wheel; Fig. 6.224 shows a diagrammatic arrangement. The carriers on which the goods are baked are evenly spaced around the outside of the reel and the drive to the reel is usually geared so that each revolution takes approximately 2.5 minutes. Variable speed and or reversing reels can be fitted but variable speed is seldom considered to be worth its cost. The great majority of reel ovens are direct gas fired as standard. The indirectly fired reel oven takes twice as long to heat up, is less flexible and is marginally more expensive to run.

The reel oven is probably the most versatile oven in use today, quick to heat up to temperature (30-40 minutes to 260 °C [599 deg F]), can be moved up or down quickly, will bake virtually anything to a high standard and does not need skilled operators. One outstanding virtue of this

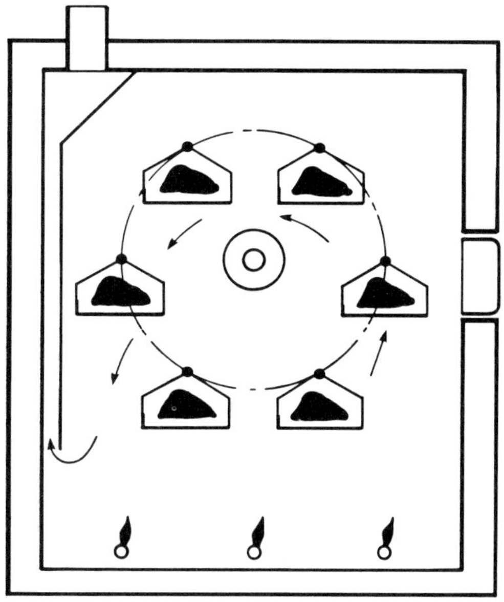

Fig. 6.224 Reel Oven

type of unit is its tolerance in taking widely differing articles requiring different temperatures and baking times at one and the same time to produce a perfect result with them all. This is due to the very mellow heat and the fact that each article returns to the oven door every 2.5 minutes and can be removed when desired.

There is an alternative method of heating the batch type ovens which consists of a central burner, which may be of the packaged type. This is fired into a combustion chamber from which the hot gases are delivered with recirculation air into the oven cooking space. In this way temperature uniformity is achieved by the recirculation of the gas by a hot gas fan.

6.14.6 Rack Ovens

This is a forced convection oven and the term 'rack' is derived from the method of loading whereby the trays of goods are placed in a rack which is then pushed into the oven. Most rack ovens are indirect, the heat source being a heat exchanger over which air is drawn and passed through the baking chamber before being returned to the heat exchanger. There are several different methods of applying the same basic principle. Each rack - or racks, depending on the size of the baking chamber - can be revolved on its own axis in the chamber. Alternatively the racks can be carried round on a turntable. Another method is to move the rack forwards and backwards, each change of direction being accomplished by a reversal in air-flow direction. Size for size, both physically and in terms of capacity, it occupies only a quarter of the floor space of its competitors. It costs less to buy than equivalent capacity conventional ovens, is cheaper to run (for example, up to 232 °C (450 deg F) in 10-15 minutes) and, very important in present circumstances, shows spectacular savings in labour requirements. For instance, on a twin four rack installation, one man can comfortably handle 8 000 140 g (5 oz) meat pies per hour. Most rack ovens can be supplied complete with their own steam generation system and provers. Again they will bake virtually anything. Although not quite as versatile as the reel, for batch production they are probably without equal.

Most turntable rack ovens are motor driven from beneath the turntable. The trays are loaded outside the oven onto a multi-deck trolley, and the trolleys then pushed onto the turntable. On some newer systems, the turntable is suspended and driven from the roof of the oven. With the suspended rack ovens, when the door of the oven closes, the racks are lifted off the floor, and the turntable will start to rotate.

6.14.7 Travelling or Conveyor Type Ovens

The evolution of ovens naturally led to the mechanisation of 'straight line' production for bread, biscuits and other products in the plant bakeries. The items to be baked are put on one end of a conveyor and transported continuously through the plant to be discharged at the other. There are many types of travelling oven with different lengths, widths, band design and burner arrangements. The length of the travelling oven can be within the range of about 12–150 m, and may have band widths of 800 mm, 1 000 mm or 1 200 mm. Figure 6.225 shows a direct gas-fired oven with heat being generated in the chamber by means of a number of individual ribbon-type gas burners, the burners being grouped above and below the baking band into independently controlled heating zones along the length of the oven. Steam and the products of combustion are exhausted from the oven chamber by a separately controlled extraction system for each zone.

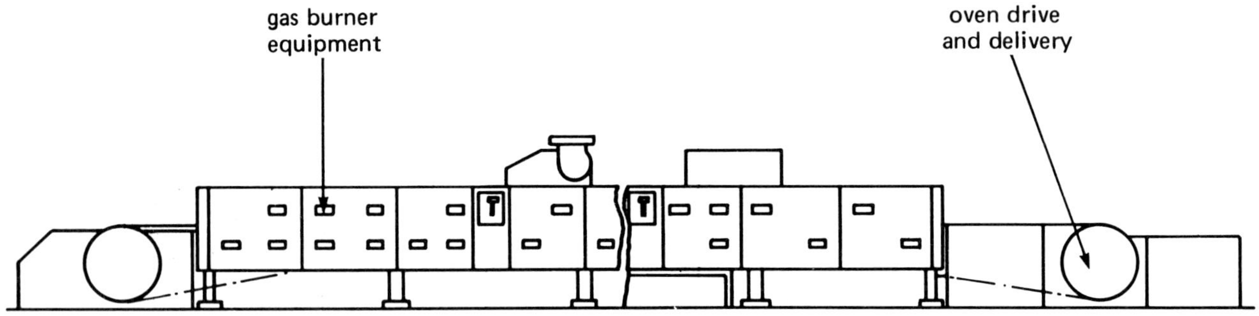

Fig. 6.225 Direct fired travelling band oven

The burners are supplied with a mixture of air and gas from air blast injectors which use pressure air from a fan and the pressure air induces gas governed at zero pressure. This system is stable and adjustable to give optimum combustion conditions.

The thermal rating for biscuit production can be from 19 kW/m^2 (6 250 $Btu/ft^2 h$) of oven band up to 22 kW/m^2 (7 100 $Btu/ft^2 h$) of oven band. For crackers this can be increased to 28 kW/m^2 (9 000 $Btu/ft^2 h$).

Travelling ovens can also be indirectly fired and can have as the principal source of heat transfer either radiation or forced convection. Figure 2.226 shows a radiation system of heating, the products of combustion recirculating through 100 mm diameter tubes above and below the band The tubes are at negative pressure to ensure that the products of combustion do not contact the goods in the event of leakage.

Figure 6.227 shows an indirect system designed to transfer heat principally by forced convection. The system claims to give uniformity of air distribution resulting in an improved rate of heat transfer, accurate control with rapid response, fuel economy and improved outputs. The hot air is distributed above and below the band directing the hot air streams at the product. Steam can be exhausted from each oven zone by separately controlled extraction systems. The burner fires into a chamber below the oven band and hot gases mix with recirculated bake chamber air which is provided by a specially designed double inlet fan. The maximum standard heat rating is 50 kW/m^2 (16 000 $Btu/ft^2 h$) of oven band.

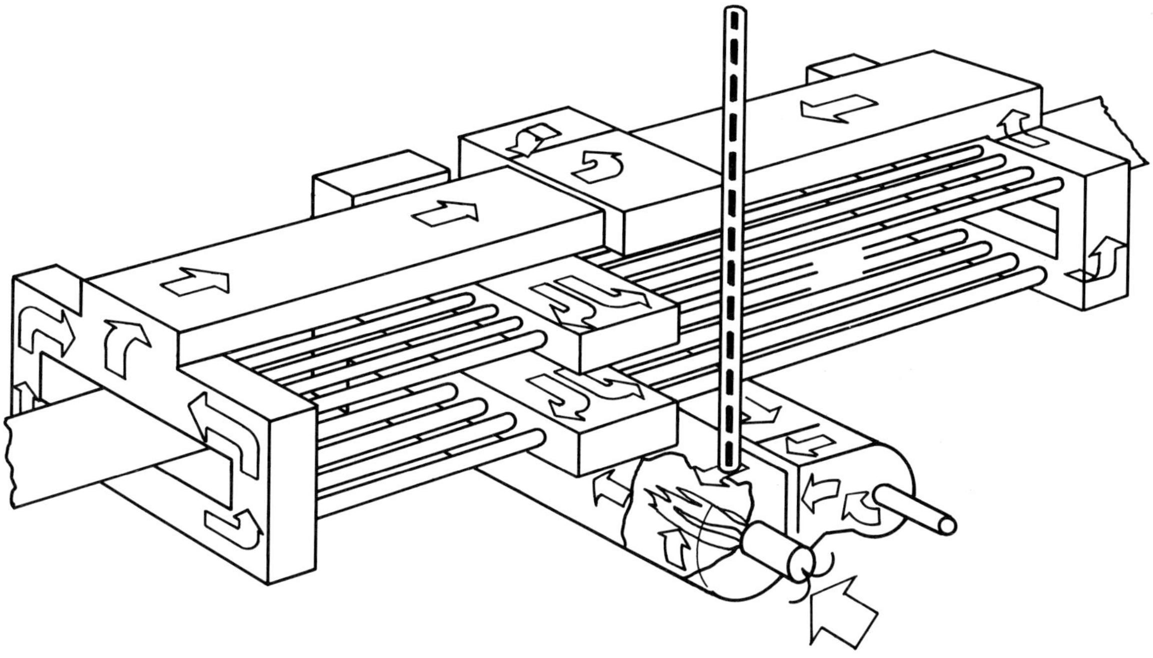

Fig. 6.226 Radiation heated travelling band oven

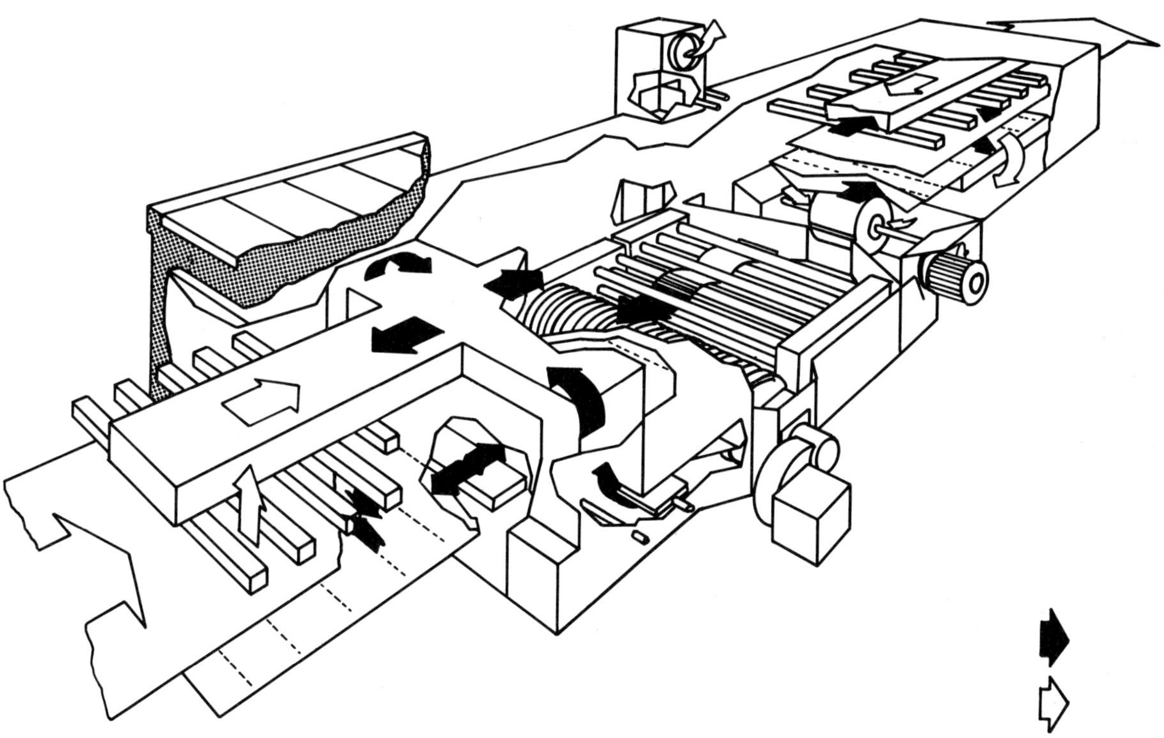

Fig. 6.227 Indirect fired convection oven

Another type of travelling oven is one where the burner system is located on top of the oven and the hot products of combustion pass through a hot gas fan into plenum chambers above and below the band. A proportion of these products is then recirculated back through the burner. This is an example of a direct-fired forced convection oven.

An example of production capacities of travelling ovens is a unit fired by three forced draught packaged burners each with a rating of 465 kW (1 575 000 Btu/h) The baking chamber is 28 m long x 3 m wide and is heated by a combination of radiant heat and controlled turbulence. There are three zones with turbulence being provided on the second and third zones by means of a floor-mounted recirculating fan supplying delivery ducts above the top radiator and below the bottom ones with nozzles directed towards the baking sole, extraction being achieved through the ducting from openings in the baking chamber. Dampers in the delivery ducts enable the operators to control the amount of turbulence.

One of the modern innovations in recent years has been the introduction of a very large oven, of American origins, for bread baking. The production rating is about 8 000 standard loaves per hour. The oven is charged from the top in a spiral fashion and has multiple pre-mixed ribbon burners as the heat source. The energy saving in one factory with this oven is projected as 380 kW (about 100 000 therms/annum).

6.14.8 Summary

Having earlier considered the various types of oven currently available, experience shows that despite a considerable degree of difference in heating techniques and control sophistication, the baking industry is markedly predictable in its choice of ovens. The majority of small bakers, for instance, will use a peel oven, mainly hot air, but a substantial number will opt for steam-tube heating. It is in fact still possible to find steam-tube drawplate ovens in use. The advent of modern cantilevered drawplates has given this concept a new lease of life. Substantial numbers of small bakeries also use reel ovens, thereby obtaining a very wide degree of production flexibility. In some views the reel oven is arguably the most versatile oven available and its capabilities somewhat underrated. The arrival of the compact rack oven has led to an increasing number of these units being installed in smaller bakeries.

It is probably in the medium size baking section where the rack oven has made the greatest inroads. Its outstanding advantages of high production capacity, minimal floor space, coupled with spectacular reductions in labour requirement and, last but not least, its extremely low running costs have made it the basic tool of this section of the industry. Even so, there are still a large number of reel ovens in use and indeed some peel ovens.

In the plant bakeries highly sophisticated production processes mean there is little alternative to the travelling oven and indeed it is difficult to envisage a more effective unit than the modern traveller. There are some interesting variations on the theme as instanced by several Continental multi-deck designs and in particular the new large American oven. Even so, many plant bakeries have taken advantage of rack ovens to isolate a quite substantial production requirement to a corner of the bakery.

Despite the foregoing there are still examples of plant bakeries using banks of steam tube ovens to produce, for example, slab cake. The mellow heat of the steam tube oven is still considered by many to be the ideal medium for this type of cake. So, notwithstanding the great advances in oven technology, it is probably safe to say that tradition dies hard.

6.15 AIR CONDITIONING

6.15.1 Introduction

The principle of the absorption cycle for gas fired chilling and cooling is now well established. Air conditioning by the method is now being used in several public buildings, executive offices and manufacturing organisations in the UK.

The use of ammonia absorption refrigeration for air conditioning offers the following advantages over other methods of chilling, namely reliability, low maintenance costs, environmentally friendly, low noise and vibration levels, no major moving parts to breakdown and the option of combined chilling and heating.

6.15.2 The Absorption Cycle

Heat from the gas burner is applied to the generator causing the liquid inside to boil and a high percentage of refrigerant vapour to rise to the top of the generator, the remaining liquid gravitating to the bottom. The vapour then enters the levelling chamber where it loses some of the carry-over water and hence to the rectifier. Here any remaining water vapour condenses out onto the cooler coil which passes through it on its way back to the generator. The ammonia vapour leaves the rectifier at high temperature and pressure and enters the U-shaped condenser coils. The condenser fan moves ambient air across the outside of these coils which removes heat from the refrigerant vapour condensing it to a liquid. The liquid ammonia passes through a heat exchanger and into the evaporator coil, where water flowing over the outside of this contains heat removed from the conditioned space.

The refrigerant then boils at low temperature and the water is chilled as it drains into the bottom of the chiller tank to be recirculated to the conditioned space. The vapour then enters the solution cooled absorber where it is reunited with the weak solution. The hot solution now passes through the air cooled absorber giving up heat to the cooler ambient air being drawn across the coils. The strong solution then enters the solution pump where it is pumped through the inside coil of the rectifier, then the inside coil of the solution cooled absorber before returning to the generator.

Gas fired absorption chillers and chiller/heaters are available in a range of sizes with outputs ranging from 10.6 kW to 17.6 kW for cooling and 26.4 kW to 36.6 kW for heating. Electrolux-Servel chillers are an example of the equipment marketed by Elstree Air Conditoning Ltd.

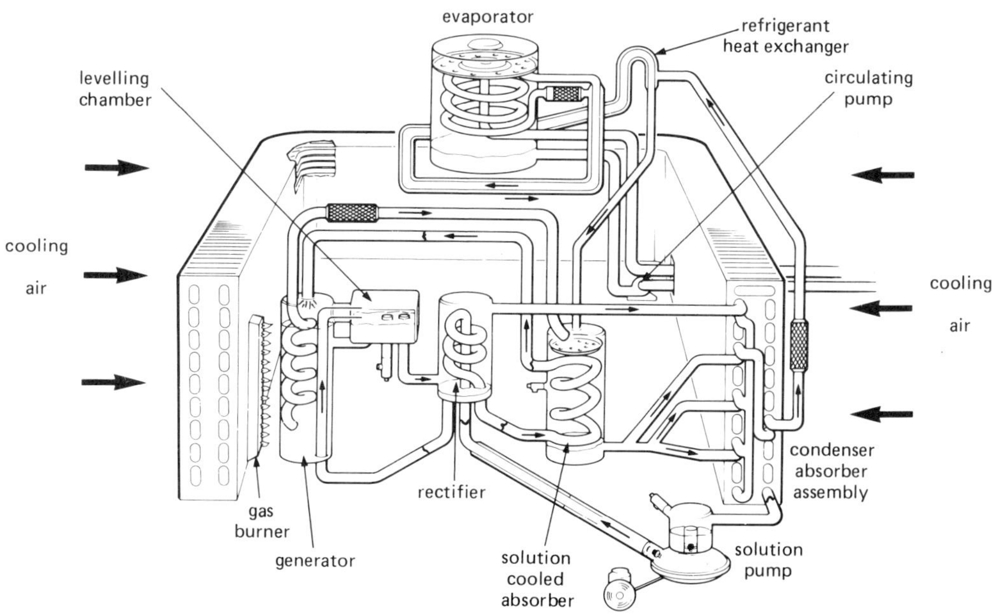

Fig. 6.228 The absorption cycle

REFERENCES

BRITISH GAS plc PUBLICATIONS

IM/9 1st Edition 1977 (Parts 1, 2 and 3): Code of Practice for the Use of Gas in Atmosphere Gas Generators and Associated Plant

IM/10 2nd Edition 1985: Technical Notes on Changeover to Gas of Central Heating and Hot Water Boilers for Non-Domestic Applications.

IM/11 1st Edition 1979 with 1982 Amendments; Flues for Commercial and Industrial Gas Fired Boilers and Air Heaters

IM/12 2nd Edition 1989: Code of Practice for the Use of Gas in High Temperature Plant.

IM/18 1st Edition 1982: Code of Practice for the Use of Gas in Low Temperature Plant.

IM/19 1st Edition 1982 Parts 1 and 2: Automatic Flue Dampers for Use with Gas Fired Space Heating and Water Heating Appliances.

IM/22 1st Edition 1986: Installation Guide to High Efficiency (Condensing) Boilers – Industrial and Commercial Applications.

DM/7 1st Edition 1987: A Guide to the Selection, Installation and Maintenance of Gas Heated, Unvented Hot Water Storage Systems.

The British Gas Guide to Hot Water Plant Sizing for Commercial Buildings, 1983

Catering Appliances tested for Safety, 1987.

MRS : E347 Rapid Heating in Perspective.

MRS : E442 Gas Fired Rapid Heating Ten Years On.

BRITISH STANDARDS

BS 845 : 1987 Assessing Thermal Performance of Boilers for Steam, Hot Water and High Temperature Heat Transfer Fluids.

BS 2486 : 1978 Treatment of Water for Land Boilers.

BS 2790 : 1986 Design and Manufacture of Shell Boilers of Welded Construction.

BS 3316 : 1987 Requirement for Incineration of Hospital Waste.

BS 5885 : 1988 Specification for Industrial Gas Burners of Input Rating 60 kW and Above.

BS 5978 : 1983 Safety and Performance of Gas Fired Hot Water Boilers (60 kW to 2 MW Input).

BS 6644 : 1986 Specification for Gas Fired Hot Water Boilers of Rated Input Between 60 kW and 2 MW.

BS 5990 : 1981 Direct Gas Fired Forced Convection Air Heaters for Space Heating (60 kW up to 2 MW Input).

BS 6230 : 1982 Specification for Installation of Gas Fired Forced Convection Air Heaters for Commerial and Industrial Space Heaing of Rated Input Exceeding 60 kW.

DEPARTMENT OF ENERGY PUBLICATIONS

Energy Audit Series, issued jointly with DTI, covering energy consumption and conservation in 21 industries/processes.

Energy Efficiency Demonstration Scheme (EEDS) Project Profiles, Extended Project Profiles and Monitoring Reports. Energy Technology Support Unit (ETSU).

OTHER PUBLICATIONS

Chemical Engineers' Handbook. Perry R.H., Chilton C.H. (editors), 5th Edition, McGraw-Hill, 1981.

CIBSE Guide. Chartered Institution of Building Services Engineers. In booklet form, various dates.

The Efficient Use of Energy, Dryden I.G.C. (Ed). IPC Science and Technology Press. 1975.

The Efficient Use of Steam. Goodall P.M. (Ed). IPC Science and Technology Press. 1980.

Heating, Ventilating and Air conditioning. Hall F. The Construction Press. 1980,

Industrial Gas Utilisation: Engineering Principles and Practices. Pritchard R., Guy J.J., Connor N.E. Bowker. 1977.

The Science of Flames and Furnaces. Thring M.W. Chapman and Hall 1972.

UK Steam Tables in S.I. Units. UK Committee on the Properties of Steam. Edward Arnold. 1972.

Waste Heat Recovery 7

Acknowledgement

The following diagrams are published by kind permission of the Department of Energy via the Energy Technology Support Unit.

Figures 7.71 to 7.96 inclusive.

CHAPTER 7

WASTE HEAT RECOVERY

7.1 INTRODUCTION

It is an inevitable theoretical and practical consequence of using energy that some will be wasted as a result.

The potential for recovering this waste heat and putting it to some useful purpose has always been present. In recent years, however, rising fuel prices and technological developments have enabled a greater number of techniques to become both economically and practically viable.

All heat recovery schemes can be categorised by three parameters; the heat source, the heat recipient and the method by which the heat is transferred. The combinations and permutations are legion and range from long-established systems such as boiler economisers, whereby waste heat in the flue gas is used to preheat the boiler feedwater via a heat exchanger, through to newer designs such as heat pump de-humidifiers, whereby the latent heat contained in moist ventilation exhaust air is recovered and used to preheat the corresponding incoming air.

Later in this Chapter the wide range of available techniques are discussed individually, although for convenience they have been grouped together as follows:

- Those which use the waste heat *directly* (e.g. by recirculation).
- Those which *transfer* the heat from a *liquid* to another *liquid* at a lower temperature.
- Those which transfer heat from a *gas* to another *gas* at a lower temperature.
- Those which transfer heat from a *gas* to a *liquid* at a lower temperature.
- Those which transfer heat from a source to a recipient at a *higher* temperature.

Following discussion of the individual techniques the final section in this Chapter presents a number of successful case histories, drawn from a range of processes, which illustrate how the techniques can be used, sometimes in combination with each other.

The first stage, however, is to discuss the practical design considerations which will affect the choice of any waste heat recovery scheme. This is followed by a brief discussion of heat transfer theory as applicable to waste heat recovery techniques.

7.2 GENERAL DESIGN CONSIDERATIONS

For the waste heat recovery scheme to come to fruition three fundamental conditions must exist, namely:

- There must be a suitable waste heat source.
- There must be a corresponding potential heat recipient.
- The engineering implications of transferring heat from the source to the recipient must be neither technically nor economically prohibitive.

7.2.1 Waste Heat Source

Waste heat is available from gaseous, vapour, liquid or solid sources. Examples of gaseous/vapour sources include combustion products and extract air from buildings or processes such as drying plant. These sources will contain sensible heat and often a significant quantity of latent heat from associated vapours. The latent heat content can significantly increase the potential energy of the source. In liquid form, heat is available from process effluents e.g. from washing processes, condensed liquids in chemical processes, cooling liquids or condensates from steam systems. Waste heat from a solid source may be the heat contained in a heated workload which is allowed to cool after its process cycle, the heat being lost to atmosphere representing a potential source for heat recovery.

Whatever the source of the waste heat, consideration of three key parameters will help define its suitability for subsequent recovery; these are its quantity, its grade and its quality.

Quantity

The heat content of a waste source can often be calculated from direct measurement of its flowrate and temperature, combined with a knowledge of its composition.

This form of assessment is, of course, invaluable but should, wherever possible, be incorporated into full mass and heat balances for the process producing the waste heat concerned. An example of a heat balance is given in Figure 7.1 and represents a complete account of all heat flows into and out of a process.

The heat balance can serve two useful purposes. Firstly, it should confirm the accuracy of the initial quantification of the waste heat availability. Secondly, it will enable the overall thermal efficiency of the process to be established. This is an important point as inefficient processes are often the first to be considered for waste heat recovery. Before embarking on a recovery project, however, it is vital to examine any options for improving the thermal efficiency of the process itself. It is frequently more economically attractive to reduce the quantity of waste heat generated, for example by eliminating unwanted air infiltration into a furnace, rather than to recover the much larger quantity of waste heat generated by an inefficient operation.

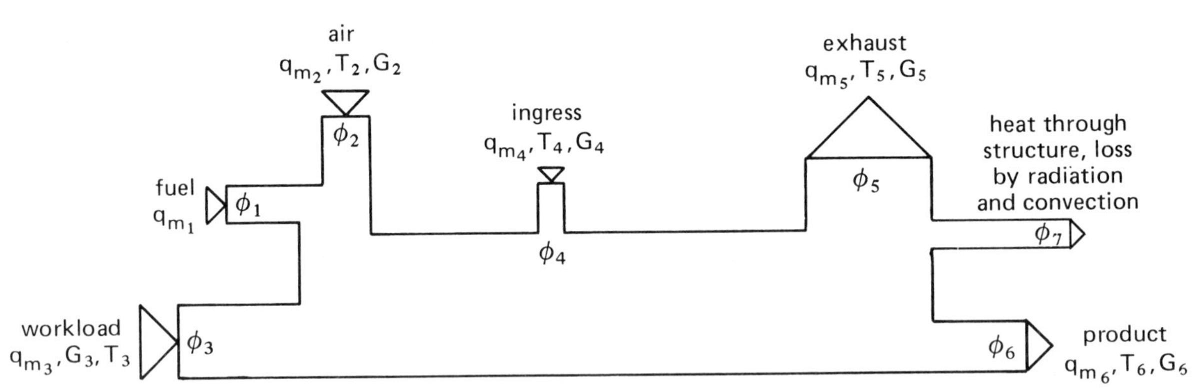

Mass balance

$$q_{m1} + q_{m2} + q_{m3} + q_{m4} = q_{m5} + q_{m6}$$

Heat balance

$$\phi_1 + \phi_2 + \phi_3 + \phi_4 = \phi_5 + \phi_6 + \phi_7$$

q_m mass flow
G moisture content
T temperature
ϕ heat flow rate

Fig. 7.1 Typical heat balance (at steady state)

Detailed guidance of how to quantify heat contents and construct mass and heat balances is outside the scope of this Chapter: interested readers should therefore consult References.

One word of caution will be sounded, however. Heat contents are always quoted relative to a base set of conditions which may include temperature, pressure and phase (e.g. whether water is liquid or gaseous). It is imperative that all heat contents used to compile a balance are quoted to the same set of base conditions. Confusion over this fundamental requirement is the most common cause of erroneous heat balances.

Grade

In the context of waste heat recovery the term 'grade' is really synonymous with the word 'temperature'. A waste heat source is potentially more useful the higher its temperature as it allows a greater scope for transfer to cooler recipients.

As a general rule heat is usually referred to as 'low grade' at temperatures below 350 °C. Above this temperature the term 'high grade' is used.

Quality

The quality of the waste heat is the greatest problem for most waste heat recovery schemes. Quality considerations principally encompass any contaminants present in the waste stream along with the profile of heat availability. Unfortunately many waste heat sources are exhaust or effluent streams and hence, almost by definition, are contaminated either physically or chemically.

Physical contamination with particulate matter can lead to erosion or fouling problems in the heat exchangers. Chemical contamination can give rise to both corrosion and fouling problems. Certain contamination, for example by toxic substances, may demand that extra precautions are taken to ensure that no physical cross-contamination can occur between heat source and recipient.

Forewarned, most contamination problems can be taken into account at the design stage of a recovery project. The implications may be that the scheme becomes too expensive to be economically viable. Better this, however, than to install a system which develops prohibitive maintenance costs or even fails to work at all.

The profiles of heat availability can be very variable, particularly those emanating from batch processes. Recovering heat from a steady flow is usually easier than recovering the same quantity of heat in an intermittent form. The latter requires larger, more expensive, heat recovery equipment to handle the greater maximum flow rates, and can give problems when trying to match heat demand to its availability.

7.2.2 Heat Recipient

Having a suitable heat recipient is equally as important as having an adequate heat source.

Quantity and grade

The quantity of heat that can be recovered from any particular heat source will, in most cases, vary inversely with the grade of heat that the recipient requires. Put another way, the smaller the temperature differential between source and recipient the less heat that can be transferred between them.

In situations where the waste heat has a high latent heat component, for example humid drier exhaust gases, the temperature of the recipient can be critical in ensuring significant heat recovery, i.e. it must be low enough to accomplish condensation.

Quality

The major quality problem associated with heat recipients is that of heat demand profile. It is always desirable for this to match, as closely as possible, that of the heat available. For this reason recovery schemes which re-use waste heat within the same process that has generated it are very attractive. In most other cases, close matching is rare and heat recovery has to be configured accordingly.

7.2.3 System Configuration

The first essential is to produce load profiles of both source and recipient heat loads, and compare them. Variations and mis-match may be short-term (hourly or less), medium term (day/night) or weekday/weekend) or long-term (seasonal); frequently in combination.

If availability is always considerably greater than demand, all that is required is to partially by-pass or shunt the unrequired heat from the heat recovery equipment and the demand will always be satisfied. Conversely, the recipient is similarly shunted and all the available heat will be absorbed, but in this case a supplementary heat source must be provided to meet the balance of recipient demand. These are relatively straightforward cases, where the load profile of one is always fully contained within the other. If, in either case, there are periods of zero availability but continuing demand, the latter may need full standby provision.

Where the heat quantities are more nearly equal but the profiles differ, comparison will show the degree of coincidence, i.e. the amount of recipient requirement that can be achieved by heat exchange as a percentage of the total demand. Determining the coincidence can be laborious, necessitating in some instances comparison of hourly, daily, weekly and seasonal profiles, but is essential to the technical and financial success of heat recovery. If the coincidence is high, simple heat recovery may well suffice, accepting that sometimes recoverable heat will be lost and at other times the recipient will be undersupplied. Heat storage, to absorb available heat when in surplus and satisfy recipient heat when higher than availability, can be sized by analysing the non-coincident profiles; its added cost must then be assessed against the marginal benefit gained.

Heat storage in this context is invariably low grade, usually in the form of hot water. To minimise tank capacity and hence costs, it should be stored at the highest temperature it can be recovered at and subsequently blended with cold water if required at a lower temperature. Considerable Research and Development work is being devoted to finding more compact and higher grade means of heat storage. One development is shown in Figure 7.2. The vessel is filled with spherical polypropylene nodules containing eutectic salts and hydrates. Heat is stored by melting the salts and

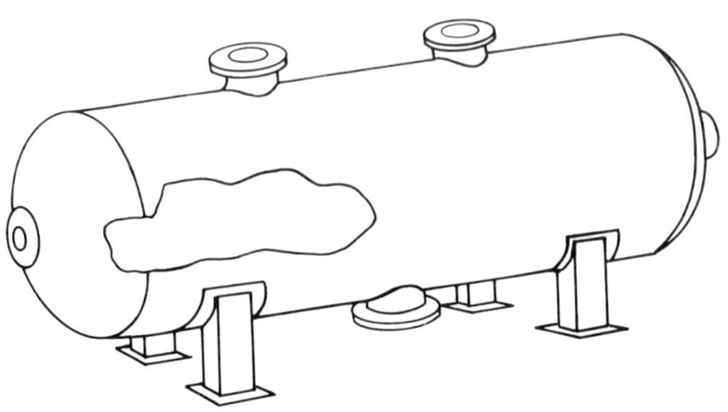

Fig. 7.2 Heat storage vessel

recovered by their re-crystallisation. Operating temperatures of between – 40 °C and +95 °C can be specified.

It will be evident that the configuration of a heat recovery scheme can be relatively complex even for an ostensibly simple application. Complexity increases considerably when there are several heat sources and recipients which often require partial or total integration. Typical examples are the textiles and food industries, where an integrated scheme may comprise heat exchangers of various types, source shunting, recipient shunting, buffer hot water storage, supplementary heat supply and the associated piping systems, pumps, fans, instruments and controls.

It will also be clear that heat exchangers, sized normally for the maximum load, may often be operating under capacity. It is an important part of the design task to assess load factors and review the part-load performance of the equipment under consideration.

7.2.4 Engineering Implications

The engineering implications of waste heat recovery are diverse, particularly if it is being considered as a retrofit on an existing plant.

Impact on existing plant

Considerations will range from the obvious, such as whether there is enough space and whether existing structures are strong enough to support additional equipment, through to more subtle issues. For example, if a heat exchanger is used it will impose an additional pressure drop which may require a more powerful fan or pump to be installed. Similarly the exchanger may upset the pressure balance within existing furnaces etc.

Most heat recovery systems also require the provision of additional services, for example compressed air or electricity to drive fan motors. If you are unfortunate enough to have your existing services running at maximum capacity then the additional load could have far reaching consequences.

Instrumentation

Adequate instrumentation is essential for all heat recovery systems in order to confirm the initial evaluation, monitor performance, define optimum maintenance requirements (e.g. heat exchanger cleaning frequency) and assess the impact of any subsequent production/space heating changes which alter the heat and mass flows. Its nature, extent and sophistication are governed by the type and size of the system and can only be considered here in general terms. The minimum requirement is inlet and outlet temperatures of source and recipient, usually simple and inexpensive to measure directly except for solids (e.g. furnace load recuperation) where inferred methods usually have to be employed. Pressure and/or pressure drop measurement is desirable and may be essential where there is a fouling potential. Less frequent requirements, but justifiable by the size of the system or in some applications crucial, are flow measurement and the analysis of one or more critical constituents of the fluids handled. The choice, positioning and maintenance of in-line sensors, samplers etc. must take into account any likelihood of fouling, blockage or corrosion.

Control

All installations require some form of control ranging from manual to fully computerised automated systems. Its importance cannot be over-emphasised; its absence or inadequacy can lead to heat recovery being a burden instead of a benefit, and consequently abandoned. As with instrumentation, the degree of control to be fitted is specific to each heat recovery application. The whole field of control is involved and only a brief review can be given here.

Isolation. Essential for every installation in at least its simplest form of manually-operated valves/dampers and by-passes, to enable off-line cleaning and other maintenance while base plant continues operating. Production (e.g. if back-pressure rises to unacceptable level) or safety considerations may dictate automatic isolation.

Temperature. In many schemes this is the only control required and comprises a modulating valve/damper regulated by recipient flow outlet temperature. Temperature control may be required to protect susceptible materials of construction e.g. to prevent source fluid temperatures falling below corrosion dewpoint, or in times of low recipient demand rising above the maximum temperature that subsequent ducting or chimney can tolerate.

Pressure. Controls impulsed by pressure or pressure drop may be used to operate on-line cleaning systems, continuous-roll dust filters etc. Pressure reliefs ranging from explosion panels to safety valves may be needed for protection against over-pressure.

Time. Where the accumulation of deposits can be predicted or learnt from experience, pressure controls (apart from over-riders) are replaced with frequency/duration timers which operate on-line cleaning, shaking or reverse flow on bag filters etc.

Interlocks. Some applications may require heat recovery plant to be by-passed until base plant has reached operational conditions from a cold start, or while it is temporarily stopped or slowed, e.g. by breakdown. This is achievable by interlocking main switches, or using a critical base plant condition to start or stop heat recovery, in some cases incorporating timers and delays.

Interface. Where there is an imbalance between source and heat recipient heat flows, additional sub-systems are often incorporated, requiring their own controls and interfacing with the heat exchange equipment controls. Sub-systems typically comprise source and recipient shunts, temperature-actuated modulating flow controllers; supplementary or standby heat input units, with their own control systems but set only to operate when and to the extend needed; buffer heat storage. An integrated control system is imperative if the full potential is to be realised.

The advent of micro-processors brings excellent control facilities within the reach of all but the smallest projects. Furthermore, the increasing number of sites already fitted with electronic energy management systems enables at least monitoring, and often integrated control, to be incorporated into a heat recovery scheme for little added cost.

Maintenance

Escalating maintenance costs have been the downfall of many waste heat recovery schemes. Maintenance requirements must be carefully considered at the design stage, and adequate provision made for cleaning, heat exchanger tube replacement etc. Similarly, in-situ cleaning apparatus may be specified.

As already mentioned, provision should be made to allow the heat recovery system to be by-passed thus allowing the heat source process to continue operating without interruption.

By-pass and flow regulation valves/dampers should be routinely inspected. Manual regulators should be checked for position and, if normally closed, for tightness of seal. Automatic regulators should be checked over the operating range to ensure correct response to the actuating signal.

7.3 HEAT EXCHANGE

Heat exchangers feature in virtually all waste heat recovery systems. A knowledge of the parameters governing their performance can therefore be useful. A more fundamental treatment of heat transfer is given in Chapter 5.

7.3.1 Flow Configurations

On liquid heating applications two basic flow configurations are most frequently used, namely 'counter flow' where the source and recipient fluids pass through the heat exchanger in opposite directions, and 'parallel flow' where they flow in the same direction. These are shown diagrammatically in Figures 7.3 and 7.4 respectively, along with their characteristic temperature profiles.

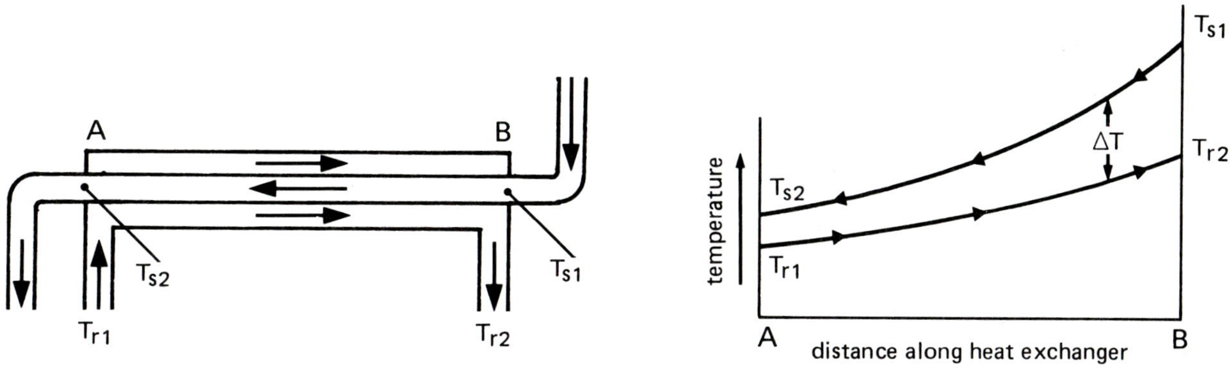

Fig. 7.3 Counter flow heat exchange

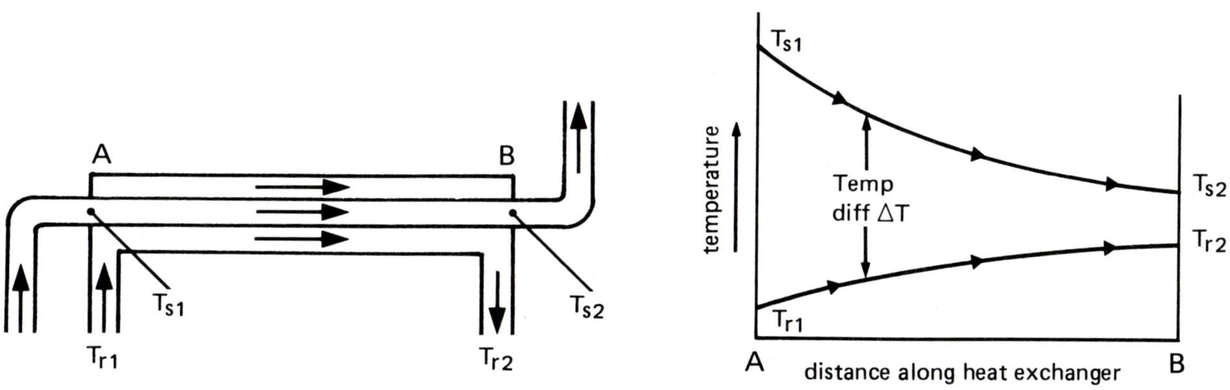

Fig. 7.4 Parallel flow heat exchange

It is important to note that in the counter flow situation the exit temperature of the recipient fluid (T_{r2}) can be higher than the exit temperature of the source fluid (T_{s2}). It is impossible to achieve this desirable result using a parallel flow heat exchanger.

In gas-to-gas heat transfer applications a third configuration, namely cross flow, finds wide acceptance, Figure 7.5 shows this diagrammatically along with its characteristic temperature profile.

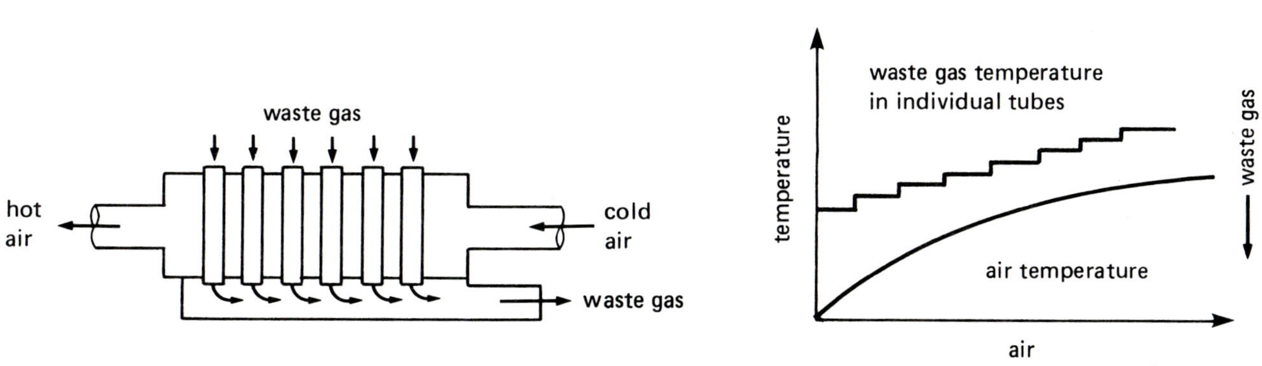

Fig. 7.5 Cross flow heat exchange

7.3.2 Equations of Heat Transfer

Let us consider two fluids flowing through a heat exchanger. Heat will flow from the hotter (source) fluid to the colder (recipient) one. We shall define the following where subscripts 's' and 'r' refer to source and recipient respectively:

Φ	=	heat transfer rate, W
q_{ms}, q_{mr}	=	fluid mass flowrate, kg/s
c_s, c_r	=	fluid mean specific heat capacity, J/(kg K)
T_{s1}, T_{r1}	=	fluid inlet temperature, K
T_{s2}, T_{r2}	=	fluid outlet temperature K
$\Delta H_s, \Delta H_r$	=	release/gain of fluid latent heat, W

The rate of heat loss from the source fluid is:

$$\Phi = q_{ms}c_s(T_{s1} - T_{s2}) + \Delta H_s \tag{7.1}$$

which, under steady state conditions and ignoring losses must equal the rate of heat gain by the recipient fluid:

$$\Phi = q_{mr}c_r(T_{r2} - T_{r1}) + \Delta H_r \tag{7.2}$$

Thus:

$$q_{ms}c_s(T_{s1} - T_{s2}) + \Delta H_s = q_{mr}c_r(T_{r2} - T_{r1}) + \Delta H_r \tag{7.3}$$

This equality enables flow rates to be balanced or exit temperatures and heat transfer rates to be calculated given a set of base design data.

Example 7.1

A dyehouse liquid effluent stream's flowrate is 1.5 kg/s and is at a temperature of 60 °C (333 K). Its specific heat capacity is 4 190 J/(kg K). It is desired to recover heat from this effluent to pre-heat boiler make-up water. The make-up water flowrate is 1.0 kg/s and is at 5 °C (278 K). The specific heat capacity of water is taken as 4 190 J/(kg K). and a make-up water exit temperature of 50 °C (323 K) is required. There are no phase changes in either fluid. Assume no heat losses from the exchanger. What will the effluent exit temperature be? What will the heat transfer rate be?

As there is no change of fluid state, H can be omitted from equation 7.3:

$$q_{ms}c_s(T_{s1} - T_{s2}) = q_{mr}c_r(T_{r2} - T_{r1})$$

i.e.

$$1.5 \times 4190 \times (333 - T_{s2}) = 1.0 \times 4190 \times (323 - 278)$$

$$T_{s2} = \underline{303 \text{ K or } \mathbf{30\ °C}}$$

Using equation 7.2:

$$\Phi = 1.0 \times 4\,190 \times (323 - 278)$$

$$= 188\,550 \text{ W or } \underline{\mathbf{188.55\ kW}}$$

(Equation 7.1 with the calculated value of T_{s2} could have been used to give the same answer.)

The next stage is to calculate the area of heat exchanger required. At any point within the exchanger the rate of heat transfer will be given by:

$$\Phi = kA\,\Delta T \qquad (7.4)$$

where k = overall heat transfer coefficient, W/(m^2 K)

A = heat transfer area, m^2

ΔT = temperature difference between source and recipient fluids, K

When considering the overall performance of a heat exchanger, however, two of these terms require special attention.

Firstly, the overall heat transfer coefficient k, which is a measure of the ease with which heat can be transferred between the two fluids. The higher the coefficient, the better. Heat transfer can be considered as taking place in a series of sequential steps. For example, first heat is transferred from the source fluid to the exchanger wall, then from one side of the exchanger wall to the other and finally from the exchanger wall to the recipient fluid. Each of these stages has its own heat transfer coefficient, given the symbol 'h'. The overall coefficient is given by:

$$1/k = 1/h_1 + 1/h_2 + 1/h_3 \qquad (7.5)$$

where h_1, h_2 and h_3 are the individual coefficients.

Whichever stage has the smallest coefficient will act as the limit to the overall heat transfer process. Almost always transfer to and from the fluids is the limiting step with transfer across the exchanger wall being much easier. Table 7.1 shows the range of individual heat transfer coefficients that are normally found.

Table 7.1 Typical Heat Transfer Coefficients

Fluid	Heat Transfer Coefficient Range W/(m^2 K)
Condensing aqueous vapours	3 000 – 6 000
Boiling aqueous solutions	2 000 – 3 100
Dilute aqueous solutions	1 500 – 2 000
Boiling and condensing organics	900 – 1 800
Molten salts	500 – 800
Dopes	100 – 500
Air (high pressure)	100 – 400
Air (low pressure)	10 – 100

The second parameter which needs attention is that of the temperature difference between the source and recipient fluids. Reference back to Figures 7.3 and 7.4 will show that this temperature difference is not constant throughout the exchanger. It is necessary therefore to define the average (or mean) temperature difference.

The average figure used is not the simple arithmetic mean but rather the 'logarithmic mean'. The logarithmic mean temperature difference (LMTD) for counter flow and parallel flow heat exchangers is given by:

$$\text{LMTD} = (\Delta T_A - \Delta T_B)/\ln(\Delta T_A/\Delta T_B) \tag{7.6}$$

Where ΔT_A = temperature difference between source and recipient fluids at one end of the exchanger, K

ΔT_B = temperature difference between source and recipient fluids at the other end of the exchanger, K

ln = natural logarithm

It does not matter which ends are designated A and B; any negative values cancel out.

Example 7.2

For the dyehouse effluent heat recovery scheme considered in Example 7.1, what area will be required for the heat exchanger given an overall heat transfer coefficient of 800 W/(m² K)?

Re-stating the necessary data:

Φ = 188 550 W
Effluent inlet temperature = 333 K
Effluent outlet temperature = 303 K
Water inlet temperature = 278 K
Water outlet temperature = 323 K

Because the water outlet temperature is to be above that of the effluent, a counter-flow exchanger is required and the temperature differences are therefore:

$$\Delta T_A = 333 - 323 = 10 \text{ K}$$
$$\Delta T_B = 303 - 278 = 25 \text{ K}$$

Using equation 7.6:

$$\text{LMTD} = (10-25)/\ln(10/25) = 16.4 \text{ K}$$

From equation 7.4:

$$A = \Phi/(k \text{ LMTD}) = 188\,550/(800 \times 16.4) = \underline{\mathbf{14.4\ m^2}}$$

Determination of the LMTD for cross flow exchangers requires the use of correction factors as outlet temperatures are not uniform (see Figure 7.5).

7.3.3 Fouling Factors

So far in this Section it has been assumed that all heat transfer surfaces are clean. In practice, of course, surfaces will become fouled with use. This can be a very serious problem with some waste streams.

Two approaches are possible to maintain thermal performance when handling dirty fluids. Firstly, provision can be made for regular surface cleaning either by shut-down, in-situ cleaning devices or a self-cleaning exchanger design. Secondly, an oversized exchanger can be specified so that even when partially fouled its performance can still meet design requirements. In practice both approaches are used and a compromise reached to ensure minimised capital and maintenance costs.

The degree of over-sizing required can be estimated using fouling factors which can be used to calculate a 'dirty' overall heat transfer coefficient as follows:

$$1/U_{dirty} = 1/U_{clean} + \text{Fouling factor} \qquad (7.7)$$

The 'dirty' overall coefficient will inevitably be smaller than the equivalent 'clean' one dictating a larger exchanger area for the same heat transfer duty. Lists of fouling factors are published based on operating experience with a range of fluids.

Example 7.3

For the dyehouse effluent heat recovery scheme considered in Examples 7.1 and 7.2, what surface area of heat exchanger will be required if we assume a fouling factor of 0.0001 m^2 K/W?

Using equation 7.7:

$$1/U_{dirty} = 1/800 + 0.0001$$

$$\therefore U_{dirty} = 741$$

$$A = 188\,550/(741 \times 16.4) = \underline{\mathbf{15.5\ m^2}}$$

(compared with 14.4 m^2 for clean surfaces)

7.3.4 Effectiveness

Heat exchanger effectiveness is a measure of the actual heat transfer as compared with the maximum available.

$$\text{Effectiveness} = \Phi/\Phi\ \text{max}$$

The effectiveness of a heat exchanger is often given in terms of temperature effectiveness and should be derived for equal flows. It does, however, only account for sensible heat transfer making no allowance for any latent heat transfer.

$$\text{Temperature effectiveness} = (T_{r2} - T_{r1})/(T_{s1} - T_{r1})$$

It is frequently expressed as a percentage and designated (misleadingly) as 'heat exchange efficiency'.

7.4 DIRECT HEAT RECOVERY

Re-use of heat directly in a building or by return to its original process using recirculation, load preheat or load recuperation are primary methods which have been common practice in some industries for many years.

7.4.1 Recirculation

Many plants, processes or buildings can make significant energy savings by recirculation of exhaust air. In so doing a proportion of the heat which would otherwise be wasted is re-used to supplement new hot air supplied to the process etc., provided that its grade and quality are suitable. Some appliances feature specially constructed internal recirculation paths. Figure 7.6 shows one such design - an industrial oven featuring a double case construction.

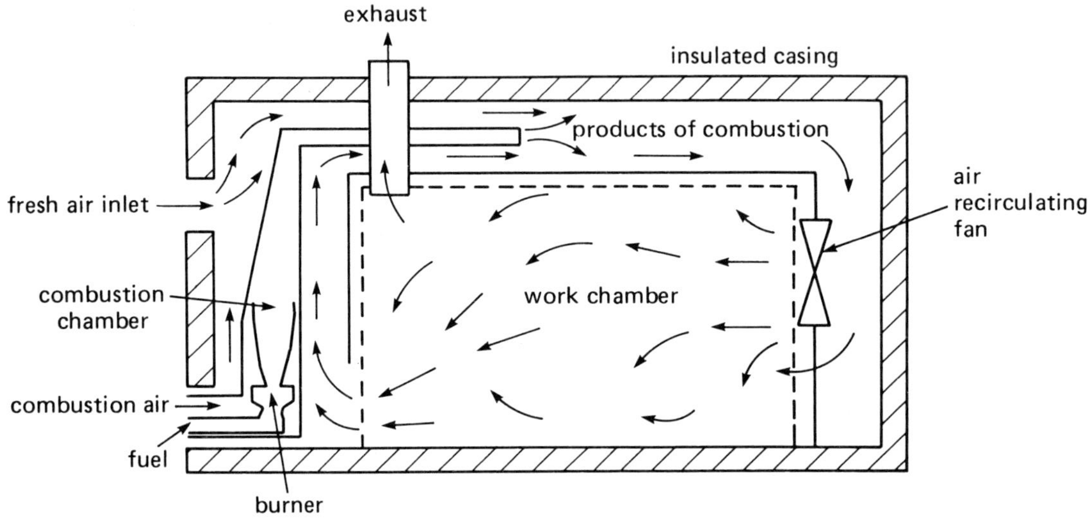

Fig. 7.6 Recirculation oven

In other cases (Figure 7.7) separate ducting can be installed to direct exhaust gases back to the inlet of a process which may be spread out in terms of size and yet where the use of insulated return ducting can make considerable fuel savings.

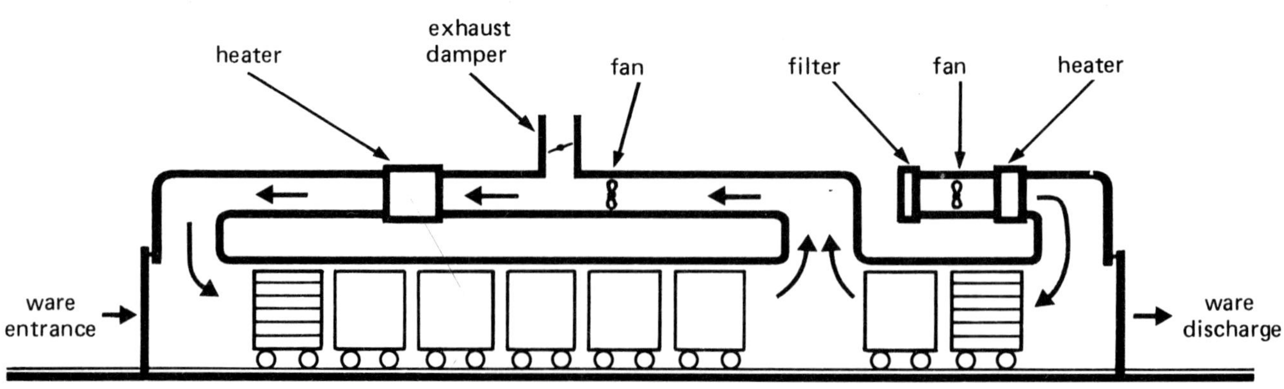

Fig. 7.7 Tunnel drier with recirculation

Other areas for recirculation of exhaust air occur in building heating and ventilation schemes. One simple form of recirculation heat recovery is the use of fans suspended in the apex of high ceilings as shown in Figure 7.8. These break up stratification and direct warm air downwards into the workspace hence recovering heat that would otherwise have been lost by conduction through the ceiling. Variants on this technique incorporate a flexible duct from high to floor level to improve redistribution.

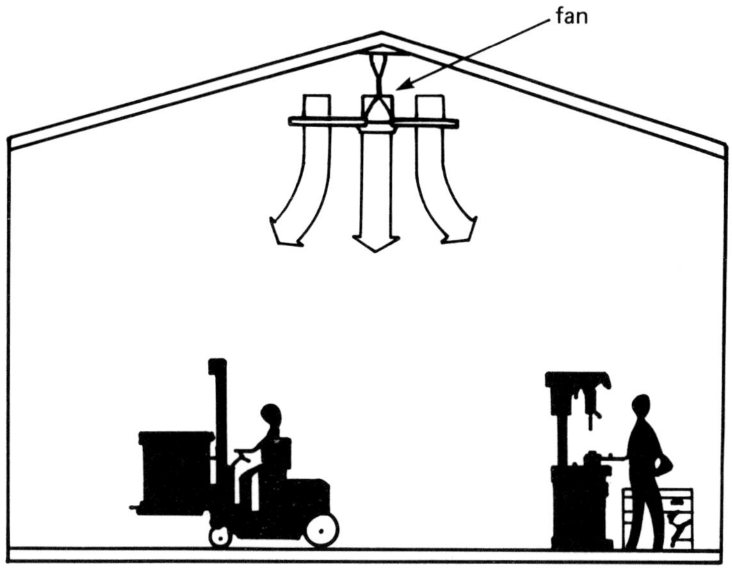

Fig. 7.8 Ceiling mounted recirculation fan

Compared with heat exchanger systems, recirculation can offer a more economically attractive proposition when technically viable. The scope for recirculation may, however, be reduced by the degree of contamination (including humidity) of the exhaust stream and the corresponding tolerance of the process to these contaminants. In some applications it may be possible to clean the exhaust stream prior to re-use, for example using activated carbon filters to clean exhaust ventilation air from garages and workshops.

7.4.2 Load Preheat/Recuperation

The direction of combustion gases towards the workload entrance of a furnace or kiln provides a means of direct heat recovery without the use of additional heat exchange units. The exhaust combustion gases come into contact with the incoming 'cold' workload and heat which would otherwise be lost to atmosphere is recovered. The workload gradually preheats from the combustion gases providing a controlled heat-up of the workload which can lead to an improvement in product quality. Potential heat from the combustion products is therefore utilised as they progressively cool. An example of this is given in Figure 7.9 which shows how billet preheating is achieved in a continuous metal reheating furnace.

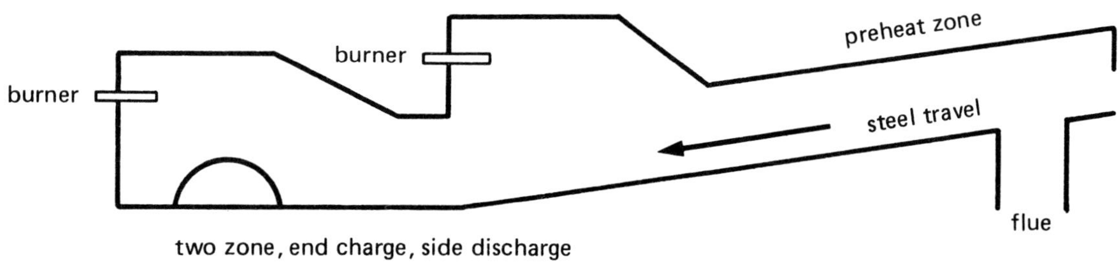

Fig. 7.9 Load preheating in continuous counterflow furnace

Similarly, heat can be recovered from a hot workload as it cools. Figure 7.10 shows such a scheme on a glass bottle making machine where hot bottles in transit to the annealing lehr are cooled by ambient air. The warmed air is collected and ducted to adjacent areas for space heating.

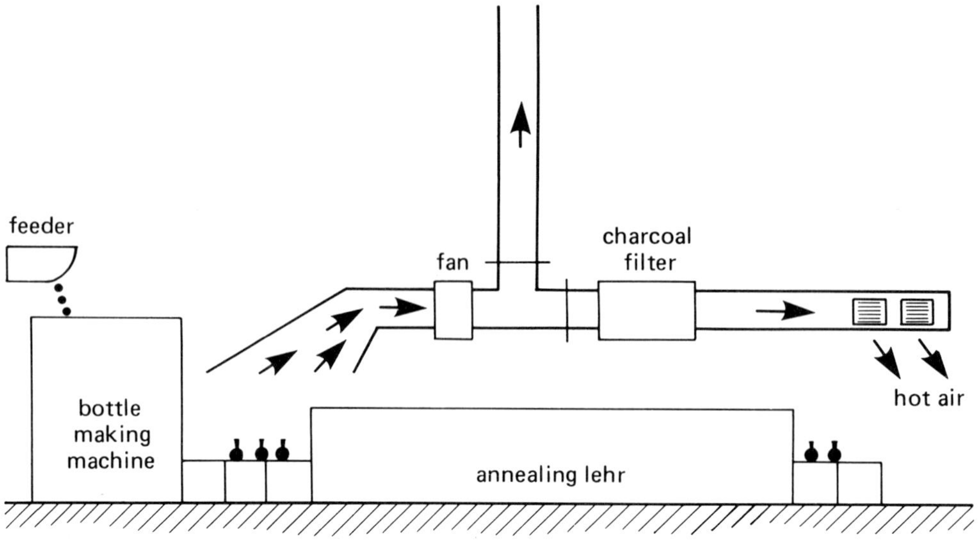

Fig. 7.10 Load recuperation in Lehr

Both load preheating and recuperation can be combined in a single process as illustrated by the continuous tunnel kiln found widely in the ceramics industry and shown in Figure 7.11.

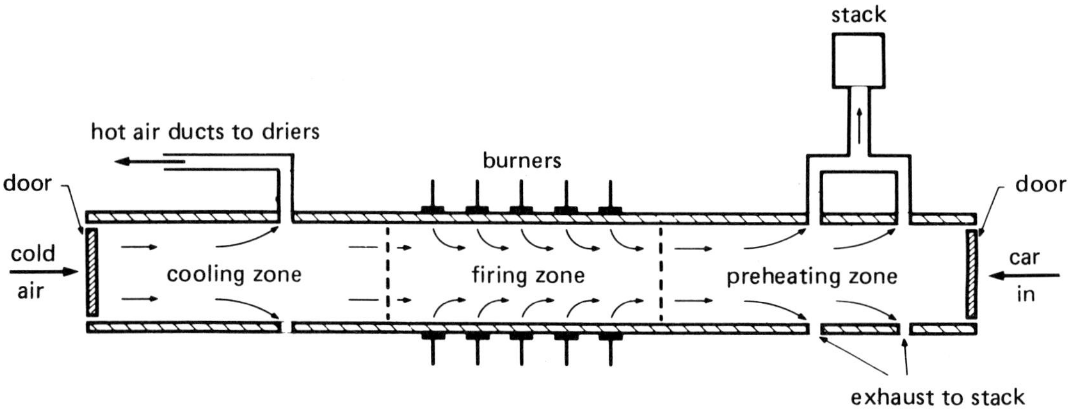

Fig . 7.11 Load preheat and recuperation continuous tunnel kiln

Each new car load of ware is preheated by contact with the cooling combustion gases passing out from the high temperature firing zone. Load recuperation is also incorporated as part of the process so avoiding thermal shock to the fired ware before it is removed from the kiln. As a consequence the air drawn in to cool the ware becomes heated and is available for other heating duties, commonly being ducted to an adjactent drier or drying room.

Whilst many load preheat/recuperation techniques have been applied to large plant, smaller purpose designed automatic furnaces are also available. Examples are rapid billet and bar end heaters which enjoy the benefits of lightweight ceramic fibre insulation and load recuperation.

7.5 LIQUID TO LIQUID HEAT EXCHANGE

A variety of liquid to liquid heat exchanger designs are available to cater for a range of temperature, pressure, space and fouling constraints.

7.5.1 Shell and Tube Heat Exchangers

Shell and tube heat exchangers are one of the oldest designs and are found widely in heat recovery applications. They comprise of a bundle of separated tubes inside a cylindrical shell, as shown in principle in Figure 7.12.

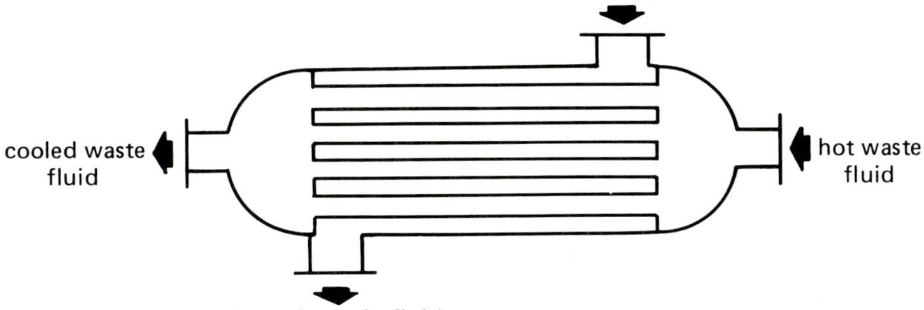

Fig. 7.12 Shell and tube heat exchanger

The two fluids, one flowing through the tubes and the other in the shell, exchange heat by conduction through the tube walls. On liquid applications, it is usual to use either plain or indented tubes. Carbon steel and stainless steel are the most commonly used materials although more exotic metals or plastics may be used to handle highly corrosive liquids. The shell and tube heat exchanger's principal benefits are that it can be constructed to withstand high pressure and temperatures. Against this is is a bulky design and is neither particularly resistant to fouling nor easy to clean.

The exchanger shown in figure 7.12 is a single pass design (i.e. the tubes pass only once through the shell). In practice multi-pass designs are frequently employed. Figures 7.13 to 7.15 illustrate some of the patterns of shell and tube exchanger which are available and all those shown are two-pass designs. Figures 7.13 to 7.15 also show the use of shell-side baffles which are invariably fitted to direct liquid flow across the tube bundles. They also provide mechanical support for the tubes. Approximately 25% of the baffle disc is cut away and arranged to establish an 'up and down' or 'side to side' flow through the shell, increasing shell-side fluid flow velocity and turbulence and thus improving heat transfer.

Fixed Tube

A fixed tube unit (Figure 7.13) consists of a bundle of straight tubes in two tube end plates, which are bolted or welded into the shell. Tube interior cleaning can readily be accomplished but shell side cleaning is more difficult.

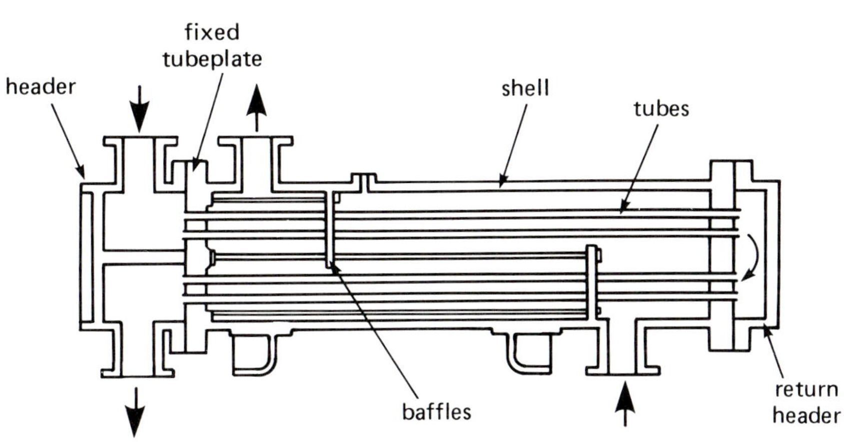

Fig. 7.13 Fixed tube shell and tube heat exchanger

U-Tube

A U-tube bundle of tubes (Figure 7.14) eliminates one of the tube end plates. The hairpin shaped tubes are connected to a header which separates the inlet and outlet flows of one of the fluids. It is available for higher temperature differentials than the fixed tube, the U arrangement allowing for expansion and contraction. The tube bundle can be removed for shell side cleaning but the U arrangement forces chemical cleaning of tube interiors.

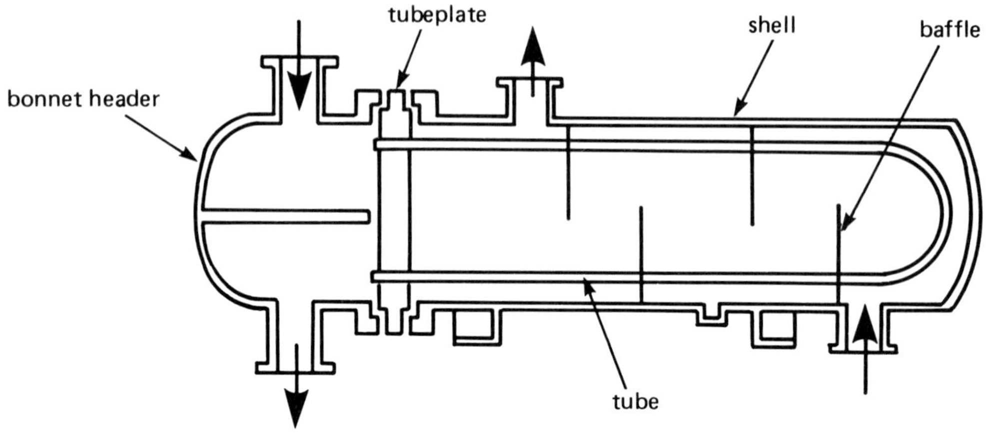

Fig. 7.14 U-tube shell and tube heat exchanger

Floating Head

The floating head type (Figure 7.15) has straight tubes connected to two end plates, one fixed and the other free to move inside the shell. Again this accommodates expansion and contraction and allows for ease of removal of tubes for maintenance.

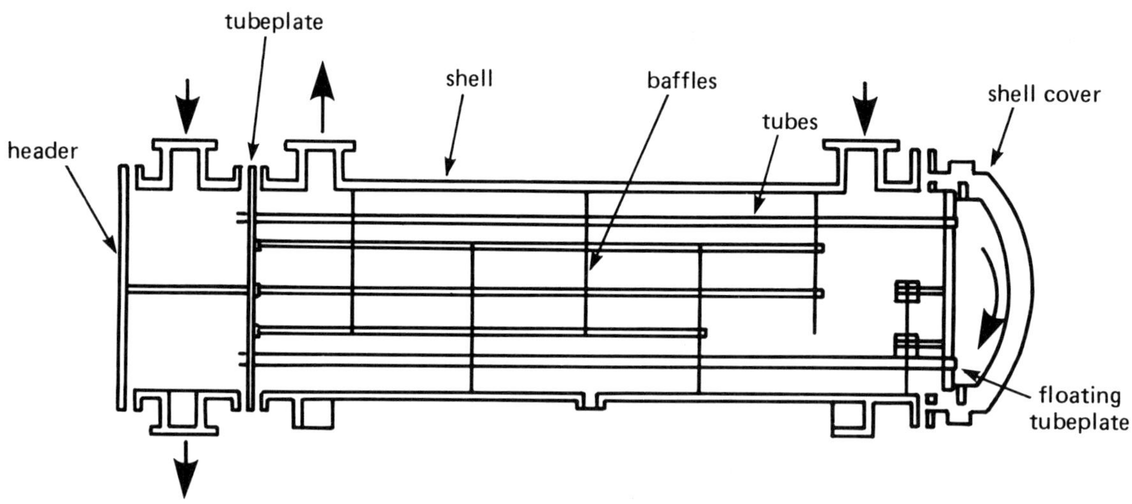

Fig. 7.15 Floating head shell and tube heat exchanger

Calorifiers

Standard shell and tube heat exchangers are normally employed on applications of a continuous nature. Where the supply or demand pattern is intermittent, storage heat exchangers or calorifiers are available whereby U-tubes are immersed in a large storage tank. This enables a reservoir of hot liquid to be built up at times of low demand ready to meet peak demands or periods of low waste heat availability.

Applications

Applications are wide ranging and one example is shown in Figure 7.16.

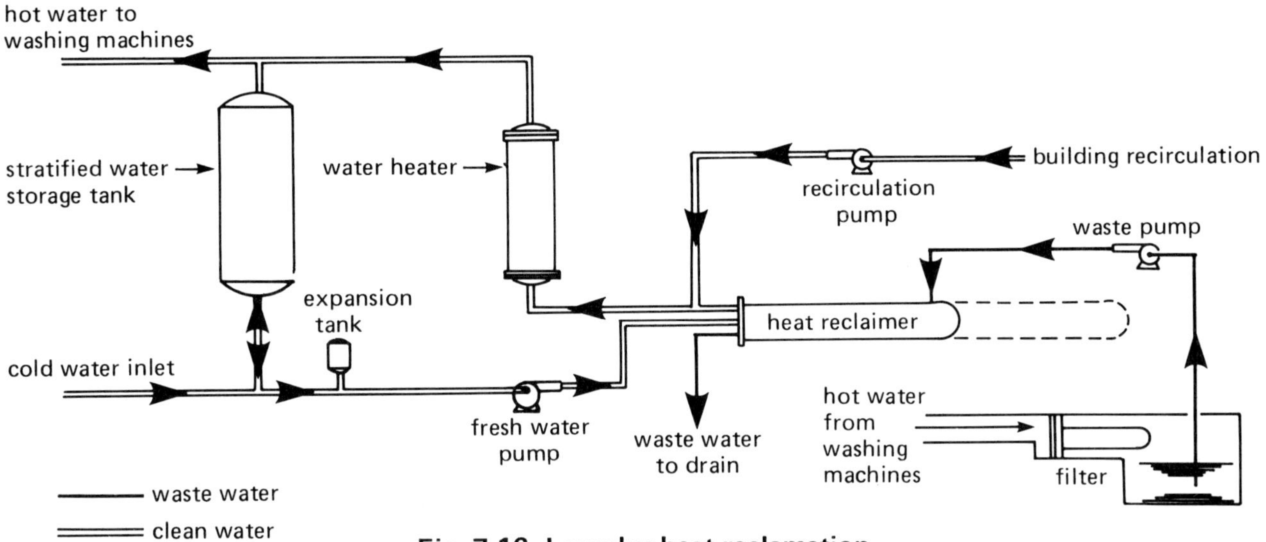

Fig. 7.16 Laundry heat reclamation

In this system, installed in a laundry, heat is recovered from waste wash and rinse water and used to preheat fresh water supplies. A U-tube exchanger is used to enable the shell to be removed periodically thus allowing the outside of the tube bundle to be cleaned.

7.5.2 Lamella Heat Exchangers

Similar to the shell and tube heat exchanger in format, the lamella heat exchanger consists of a bundle of flattened plates in an outer shell. The lamellas are formed by welding together two thin metal strips in pairs at their edges. This produces a straight channel or lamella between each pair of plates, see Figure 7.17. Each lamella incorporates a series of indentations produced by spot welding the plates together. This will impart turbulence to the fluid in the lamella to optimise heat transfer. One end of the lamella bundle is fixed in the shell, the other is supported by a floating end box to allow for expansion and enable the bundle to be removed from the shell.

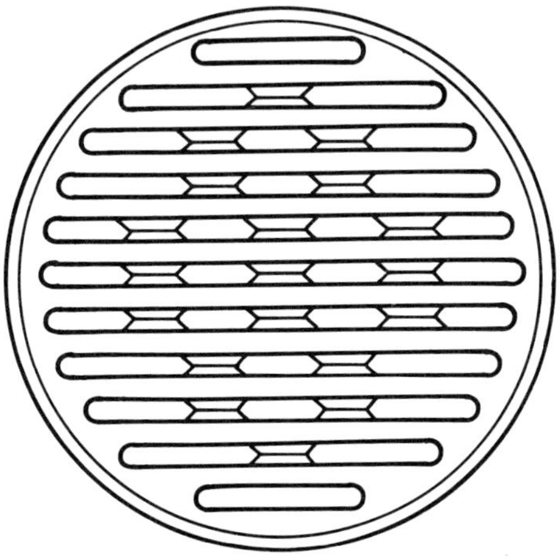

Fig. 7.17 Lamella bundle

In operation, one fluid flows through the lamellas countercurrent to the other flowing through the shell, see Figure 7.18. High turbulence, uniform flow and smooth surfaces minimise fouling in lamella heat exchangers. Contaminants can be cleaned off by suitable chemical cleaners. Examples of shell and lamella materials of construction are stainless steel, Incoloy, Hastelloy and titanium.

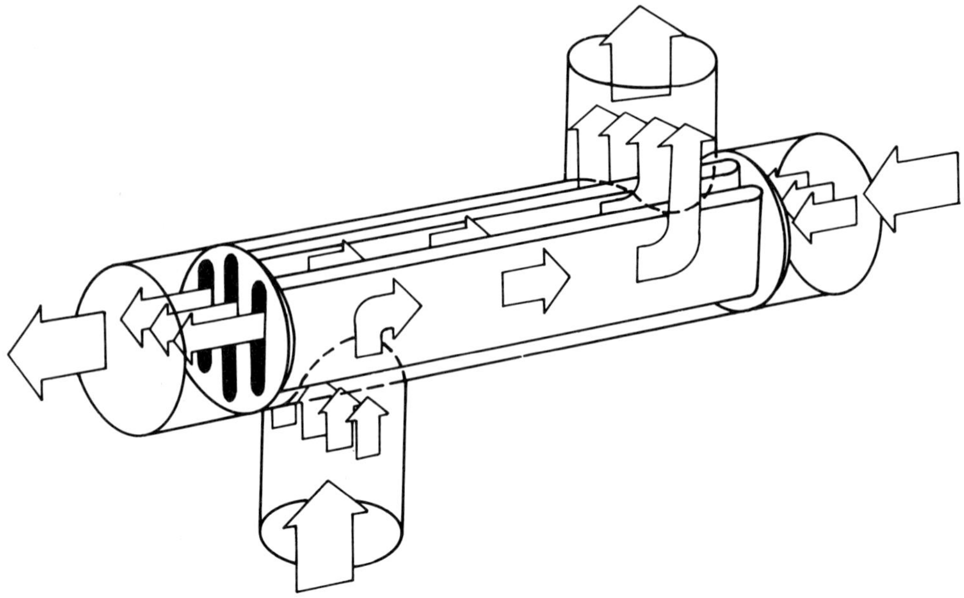

Fig. 7.18 Flow pattern in a lamella heat exchanger

Standard designs of lamella heat exchangers are suitable for pressures up to 20 bar. Maximum temperature is 220 °C with Teflon gaskets or up to 500 °C with asbestos gaskets in a stainless steel shell. They are thus generally used for high pressure, high temperature applications in the pulp, paper and chemical industries.

7.5.3 Plate Heat Exchangers

The plate heat exchanger consists of an arrangement of plates and sealing gaskets clamped together in a metal frame as shown in Figure 7.19.

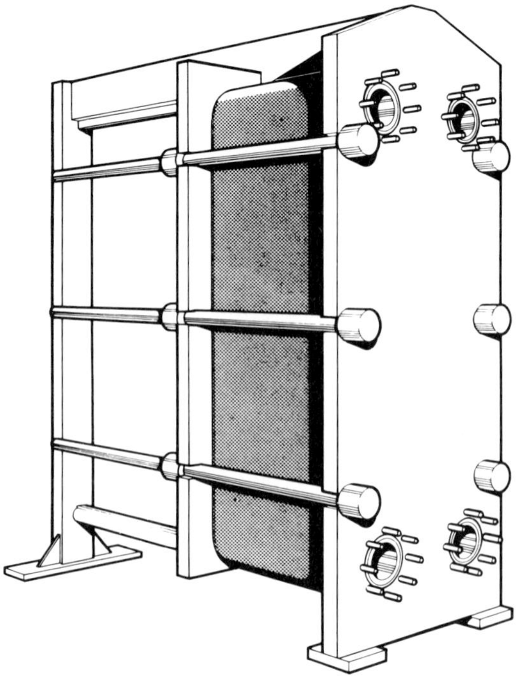

Fig. 7.19 Plate heat exchanger

A series of thin passages, through which the two media flow, are created between plates the outer edges of which are sealed with gaskets. Gasket materials are chosen for the particular application and may be of elastomeric compound, compressed asbestos fibre etc. dependent upon duty. Ports at the plate corners, in association with the gaskets, separate, extract and supply fluids to alternate passages with no intermixing, as shown in Figure 7.20.

Fluids flow countercurrent, with heat transfer across the plates being maximised by the design of plate surface. Pressed troughs or corrugations increase surface area, promote turbulence and impart rigidity to the thin plates. Stainless steels are common for plate construction although special plates of e.g. titanium, nickel or special alloys are available if required for a particular corrosive or aggressive duty.

Maximum operating temperatures substantially depend on the gasket material, of the order of 130 °C for rubber gaskets and around 200 °C with compressed asbestos fibre in heat exchangers carrying steam. Liquid pressures up to 20.6 bar can be accommodated by choice of exchanger construction.

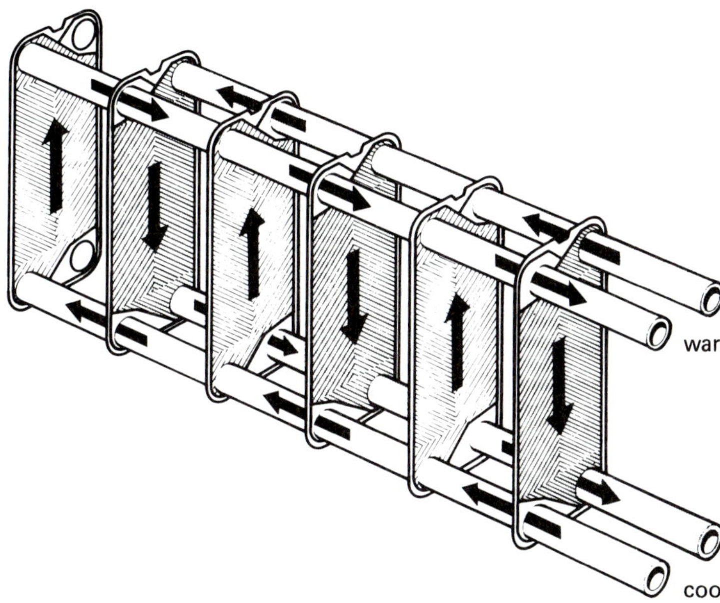

Fig. 7.20 Flow pattern in a plate heat exchanger

Plate heat exchangers are extremely compact and the arrangement of the heat exchanger in a framework allows for ready expansion or reduction of the heat exchanger capacity if requirements change. In addition, they allow easy access for cleaning following removal of the clamping plates. They are, however, comparatively little affected by fouling due to good fluid distribution and high turbulence. Faulty gaskets or heat exchange plates are similarly straightforward to replace.

Applications have included recovery from textile washing plant, milk pasteurisation, car washing plant, chemical process plant etc. They have also been used in conjunction with economisers to recover heat from boiler flue products, the plate heat exchanger transferring heat from the economiser water to the process fluid.

7.5.4 Spiral Heat Exchangers

The spiral plate heat exchanger is used where high temperature and pressures are encountered. It consists of two parallel channels formed on a central mandrel by spirally winding two long strips of metal with integral spacer pins, see Figure 7.21. The spiral plates are welded alternately along their edges to completely separate the fluids involved and prevent intermixing. The fluids pass in a spiral motion through the exchanger with heat being exchanged through the coiled metal walls.

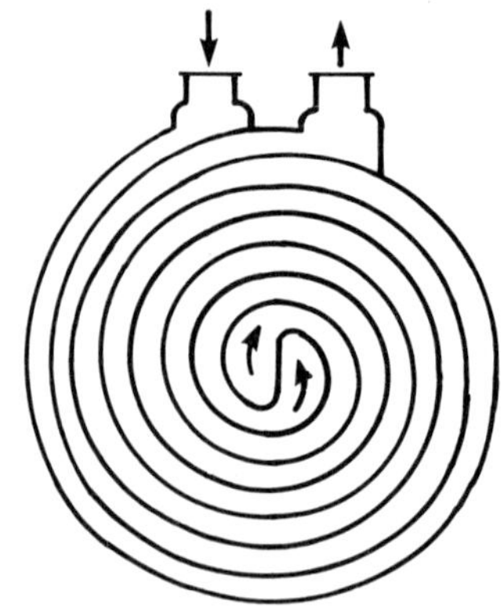

Fig. 7.21 Flow pattern in a spiral exchanger

Three patterns are available, countercurrent, spiral flow/cross flow and distributive vapour spiral flow, operating at temperatures and pressures up to 400 °C and 10 bar respectively.

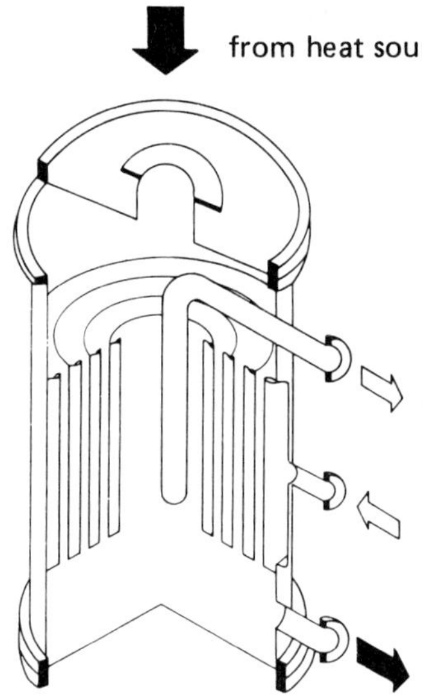

Fig. 7.22 Countercurrent spiral heat exchanger

In the countercurrent type, shown in Figure 7.22, both fluids flow spirally, the hot fluid entering at the centre and flowing out towards the periphery. The cold fluid enters at the periphery and flows countercurrent towards the centre. It is used for heat exchange where there is no phase change (e.g. liquid to liquid, gas to liquid or gas to gas).

The spiral exchanger's principal benefit is that it has inherent self-cleaning properties. This is because there is only a single passage for each fluid to travel in, whereas in many other designs a number of parallel paths exist which can lead to partial blocking and short-circuiting.

7.5.5 Rotating Element Heat Exchangers

In its simplest form the rotating element exchanger can consist of a single finned tube suspended and rotated in a liquid trough. A more sophisticated design is shown in Figure 7.23.

Here the rotating element is a series of baffled shells mounted on a hollow shaft. Clean liquid, typically water, passes through the inside of the element while a dirty stream passes through the trough. The rotation of the element promotes heat transfer and, more significantly, prevents settling of contaminants in the trough and gives the exchanger extremely good self-cleaning properties.

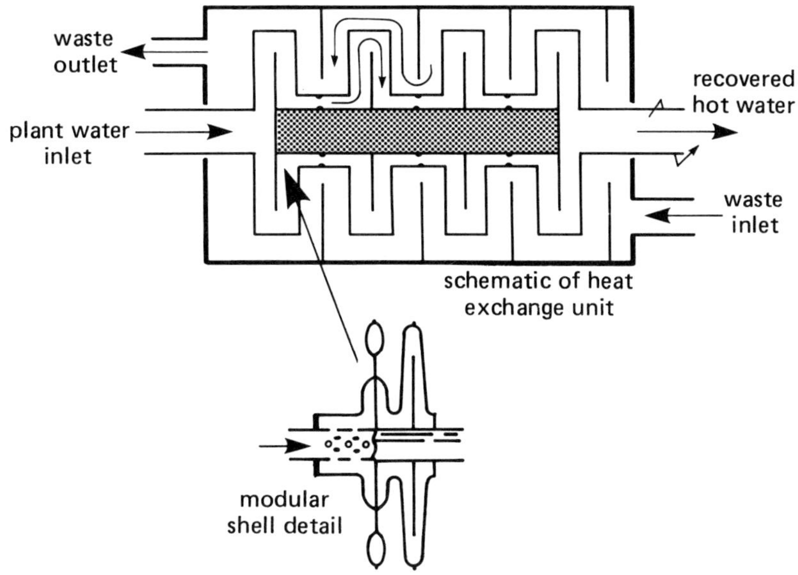

Fig. 7.23 Rotating element heat exchanger (Pozzi)

Although not particularly thermally efficient these exchangers are finding applications where effluent streams are very dirty, for example in the pulp and textile industries. Figure 7.24 shows a typical installation handling dyehouse effluent.

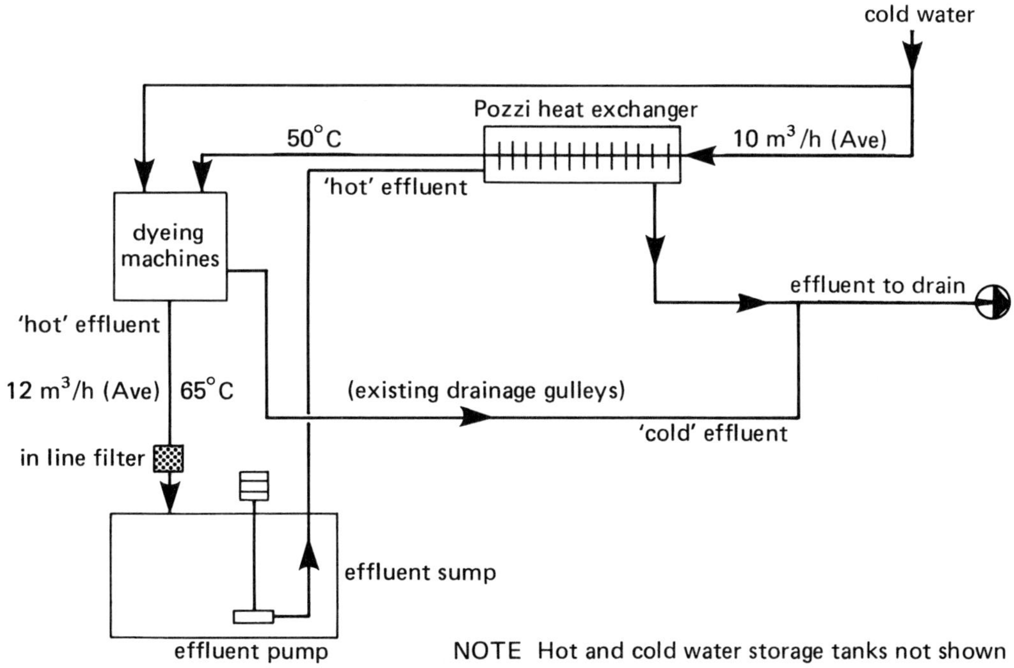

Fig. 7.24 Dyehouse waste heat recovery

7.6 GAS TO GAS HEAT EXCHANGE

Gas to gas heat exchange embodies a wider range of operating temperatures than either liquid to liquid or gas to liquid. As a consequence the technologies involved vary widely. Operations up to around 350 °C are termed 'low temperature' heat recovery and include reclaim from ventilation air and drier exhausts. Above 350 °C, heat recovery is called 'high temperature' although a significant subdivision occurs within this category, depending upon whether the predominant heat transfer mechanism is convection or radiation.

At temperatures below about 1 000 °C convection largely predominates and the velocity of the waste gas stream determines the degree of heat transfer. As velocity increases so the convective heat transfer coefficient will significantly increase. In comparison radiative heat recovery devices operate at up to 1 500 °C with heat transfer by radiation predominant. The principal constituents of combustion gases which significantly emit energy in the form of non-luminous radiation are carbon dioxide and water vapour. Their concentration, temperature and layer thickness will determine the degree of transfer of radiant heat. In high temperature plant with high mass flows the influence of carbon dioxide and water vapour in flue products from refined fuels will be high and hence above 800 °C radiative heat transfer will rapidly begin to dominate.

The potential savings from high temperature heat recovery are relatively large and so it is widespread on furnaces throughout the metal processing, ceramics and glass industries. A major end use for the recovered heat is that of combustion air preheating. This can yield significant fuel savings as illustrated by Figure 7.25.

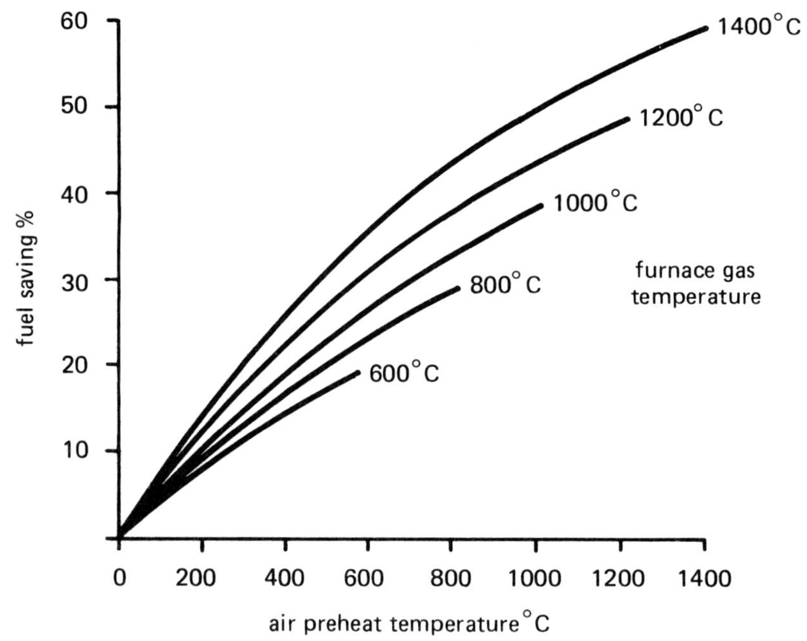

Fig. 7.25 Fuel saving v. air preheat temperature (Stoichiometric conditions)

Combustion air preheat has the added advantages of ensuring that recovered heat demand profiles exactly match those of waste heat availability and of requiring only minimal transportation of the heat from source to recipient.

Elevating combustion air temperatures (and indeed possibly gas fuel temperatures) will increase the resultant flame temperature, as illustrated by Figure 7.26. This may be of advantage in increasing product quality and/or throughput but caution must be exercised if retrofitting such schemes or local furnace overheating may occur.

The available techniques for gas to gas heat exchange (both high and low temperature) are described in the following paragraphs.

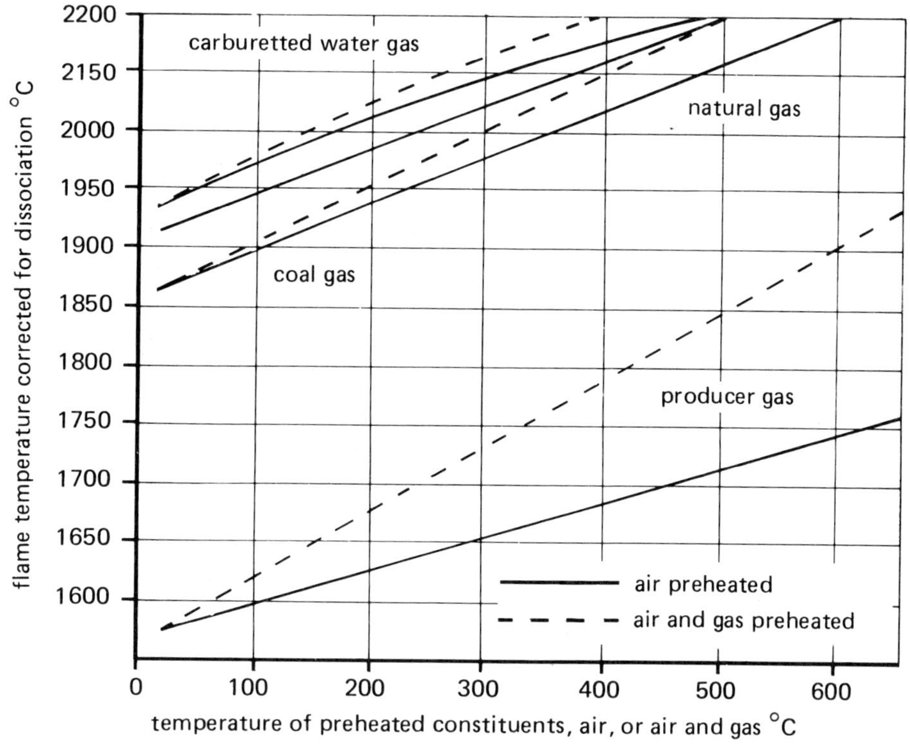

Fig. 7.26 Effect of air preheat on flame temperature

7.6.1 Tubed Recuperators

The term 'recuperator' describes a heat exchanger used for high temperature gas to gas heat exchange. Tubed heat exchangers are rarely used on low temperature gas to gas duties, except where special circumstances (e.g. highly corrosive conditions) exist.

Metallic Recuperators

Steel tube recuperators take many forms designed to allow for expansion. This may be achieved by using 'kinked' tubes, expansion joints or by fixing the tube bundle at only one end and allowing the other to 'float'.

Convective recuperators may utilise either cast or drawn tubes. By casting tubes it is possible to obtain fins on both the inside and outside in order to improve heat transfer. Figure 7.27 shows the construction of a typical 'needle tube'.

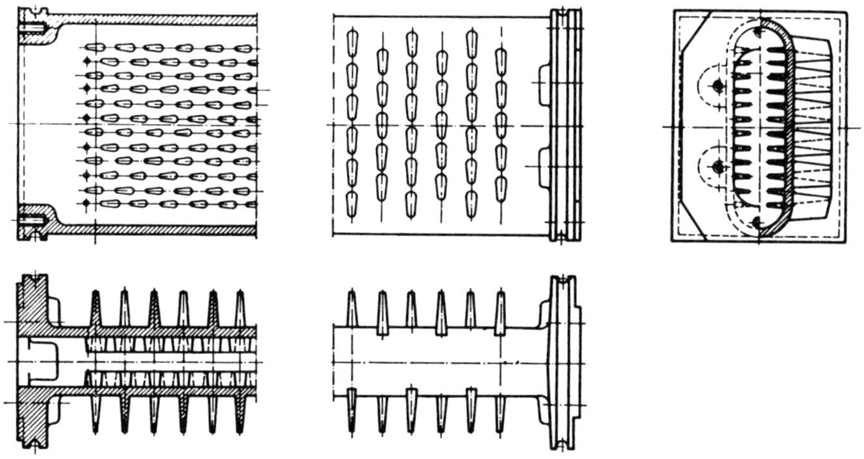

Fig. 7.27 Needle tube elements

Needle tube recuperators are used in the steel industry, the tubes being mounted vertically to prevent sagging at elevated temperatures, as shown in Figure 7.28.

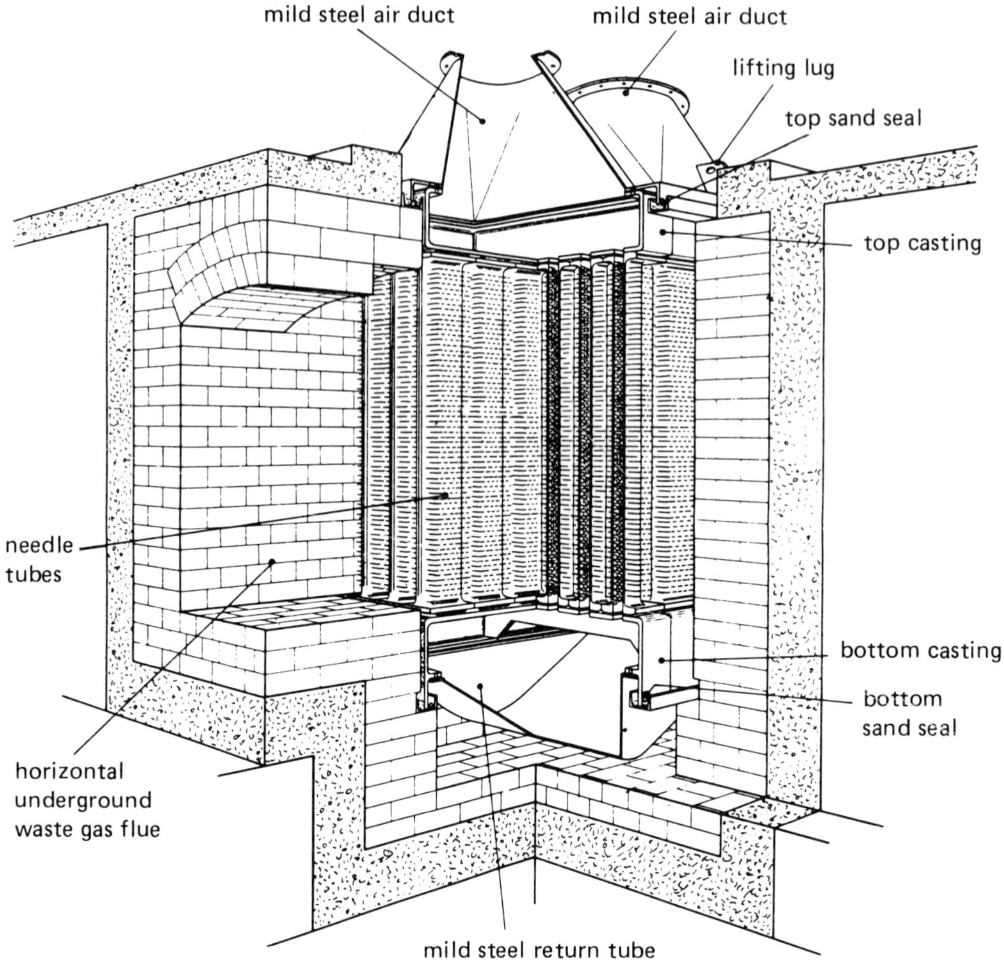

Fig. 7.28 Needle tube recuperator

Although transferring less heat per unit length than cast tubes, plain drawn tubes do find applications where the exhaust gases are heavily contaminated with particulate matter. Figure 7.29 shows a retrofit heat recovery scheme using a plain tubed convective recuperator applied to a plant manufacturing zinc oxide. In this case cleaning of the outside of the tubes is accomplished using on-line equipment.

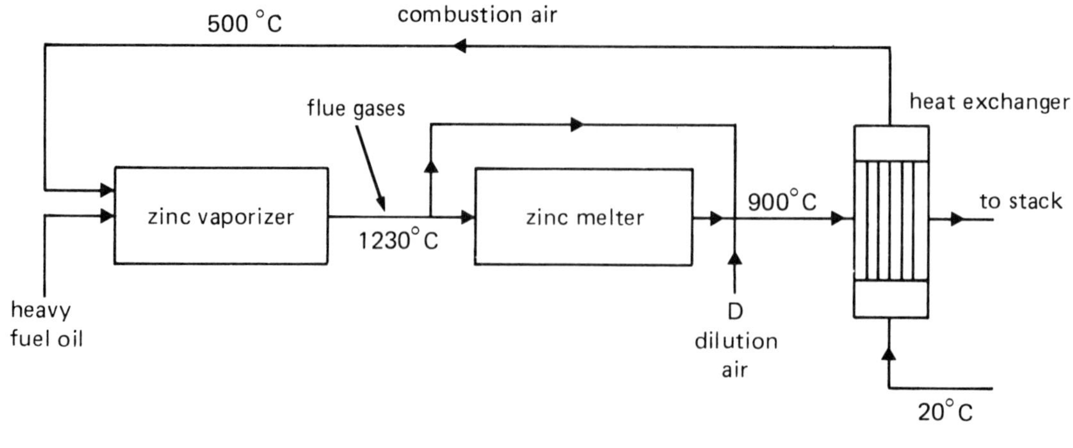

Fig. 7.29 Plain tube convective recuperator applied to a zinc oxide process

In applications where pressure losses cannot be tolerated in the exhaust gas stream an eductor arrangement can be installed to induce the gases over the heat exchange surface, as in Figure 7.30.

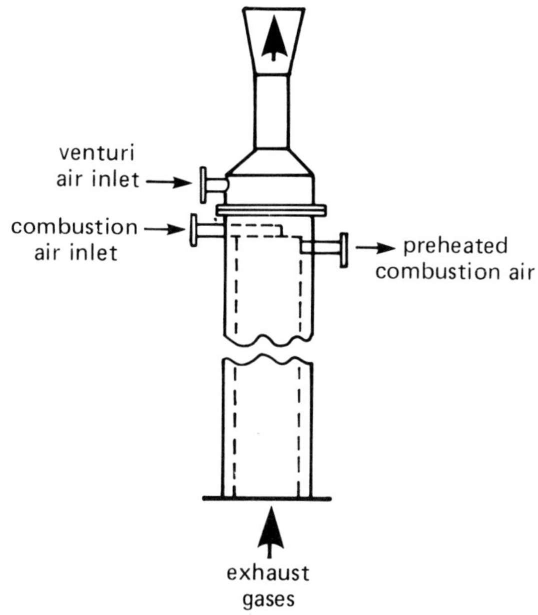

Fig. 7.30 Convective recuperator with eductor

The eductor enables higher flue gas velocities to be used improving the convective efficiency of the recuperator. The waste gases generally pass once through a convective recuperator with the supply air flowing through the tubes usually in a multipass configuration. The relatively large bore tube facilitates cleaning of contaminants from the heat exchange surface.

Above 1 000 °C the design of metallic recuperators accommodates radiative heat exchange. Larger diameter tubes are used, optimised to increase the thickness of the radiating layer which improves the radiation heat transfer coefficient. The technique offers reduced resistance to flow compared with the convective tubed recuperator and is less susceptible to fouling.

Radiative units are available as 'in-wall' recuperators (Figure 7.31) which allow insertion into furnace walls or roofs up to a depth of 300 mm. An annulus is formed by two concentric tubes which contain the combustion air flow and heat transfer takes place between the combustion gases in the annulus and the air. Their operating temperature range is up to 1 250 °C, providing air preheats of up to 600 °C.

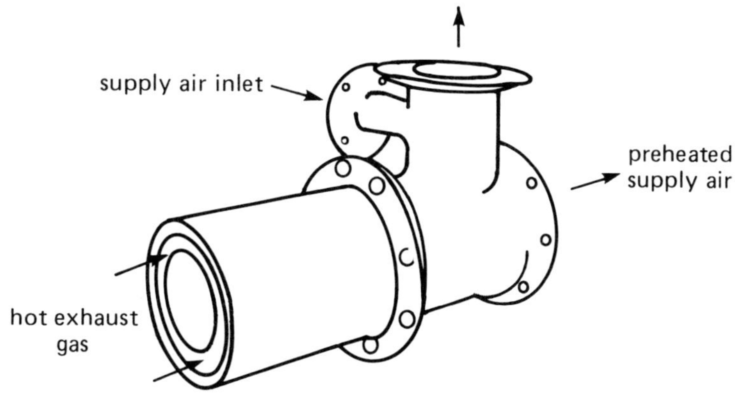

Fig. 7.31 In wall metallic recuperator

Many furnaces feature metallic recuperators in their original design as illustrated in Figure 7.32. This shows an aluminium bale out furnace which operates with combustion air temperatures of 350 °C - 400 °C.

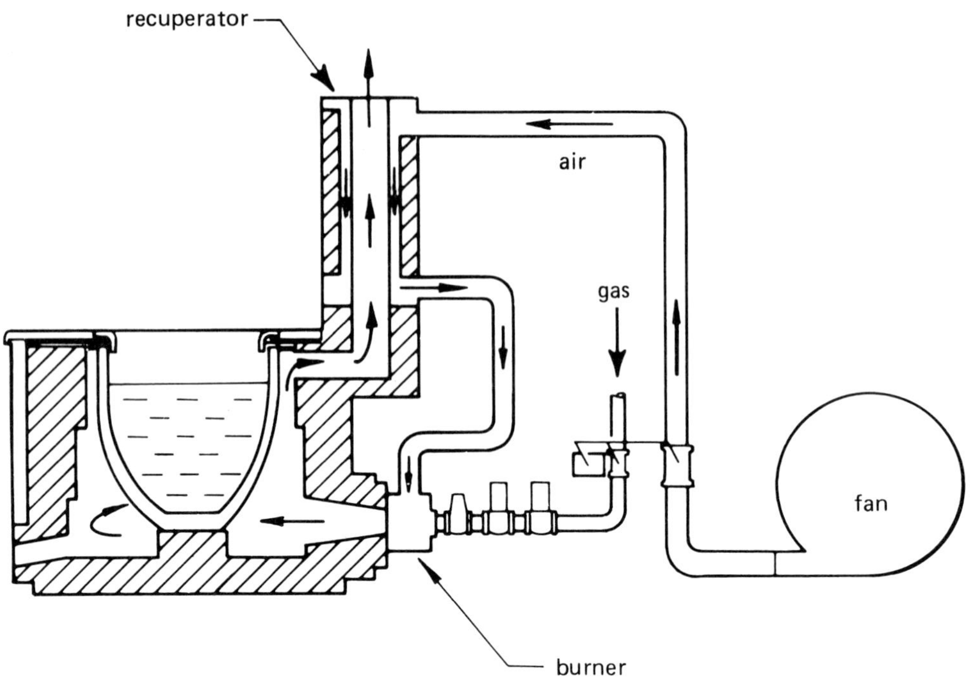

Fig. 7.32 Bale out furnace incorporating a metallic recuperator

Alternatively, much larger 'double skin' or 'shell' recuperators are available as stand-alone units, see Figure 7.33.

Whilst radiative recuperators operate at higher temperatures than the convective types, under turn-down conditions they may be subject to overheating due to inadequate air flows. This can be minimised by controlled cooling of the waste gases by air dilution or water sprays. Excess quantities of supply air may alternatively be passed through the annulus of the recuperator with the additional air being bled off before the burners. The use of manufacturers' specified minimum air flows will assist prevention of overheating and prolong the life of the recuperator.

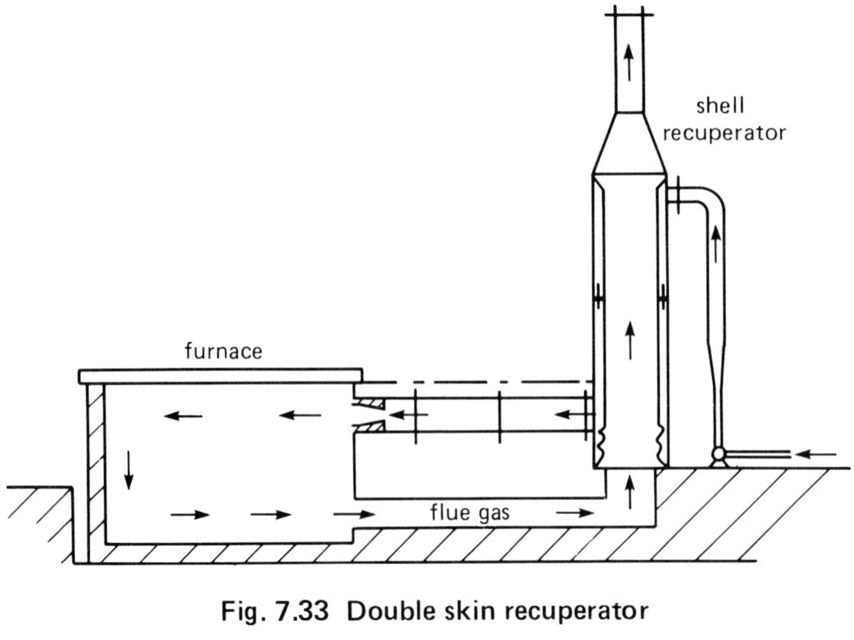

Fig. 7.33 Double skin recuperator

Ceramic Recuperators

Ceramic recuperators are used to overcome the temperature limitations of metallic recuperators. Historically, however, they have been subject to leakage problems. Even at low air pressure the leakage of supply air into the waste gas stream under suction conditions has reduced effectiveness. The British Steel Corporation development of a ceramic recuperator (Figure 7.34) using silicon carbide tubes sealed with flexible ceramic fibre wall seals is reported to have significantly reduced leakage and has achieved air preheat of 650 °C. Other manufacturers have quoted 800 °C and up to 1 200 °C, but such units are not in general use and development work continues.

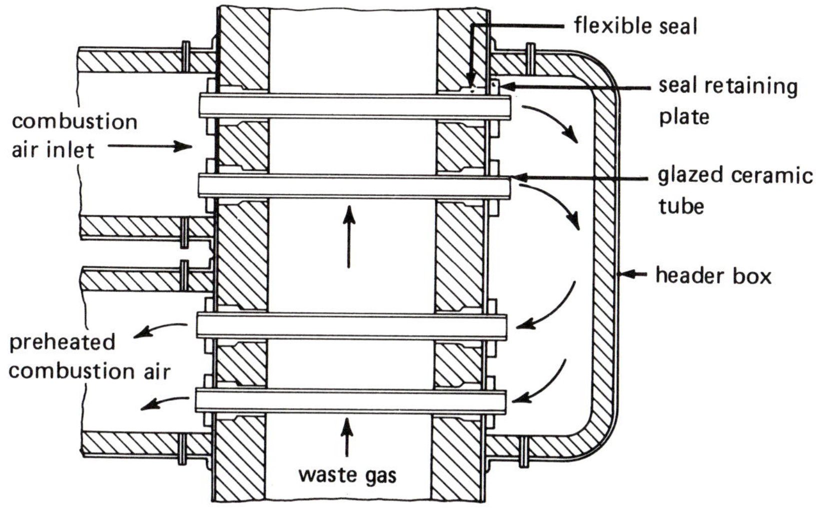

7.34 Cross section of ceramic recuperator

Long silicon carbide tubes, chosen for high strength and slag resistant characteristics, span the exhaust duct, in the case of B.S.C. from a soaking pit furnace. The tubes are sealed in a counter/cross flow arrangement with combustion air flowing through the tubes, see Figure 7.35.

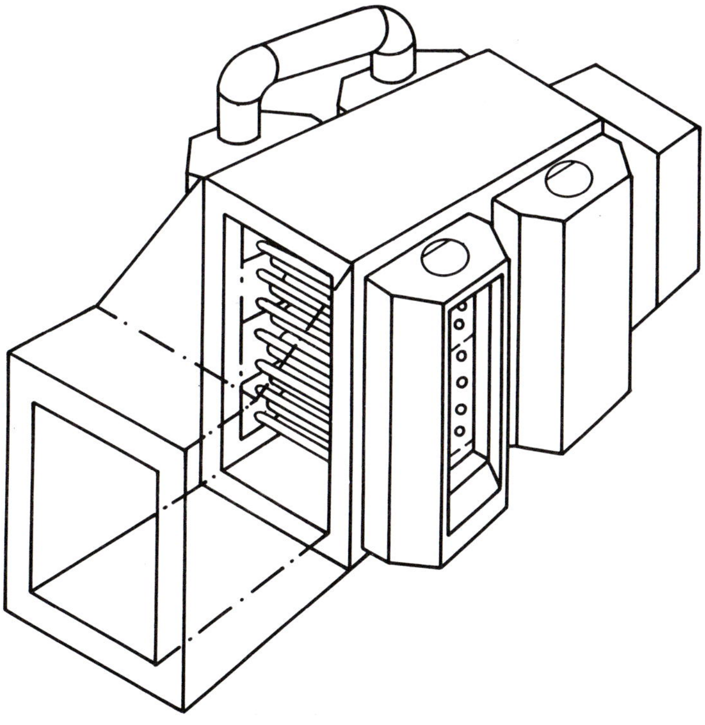

Fig. 7.35 Steel industry ceramic recuperator

A compact ceramic recuperator of similar design has been developed for smaller applications. One has been successfully demonstrated on a continuous muffle sanitary ware kiln, where the flue gases are contaminated with glaze volatiles. Glazed alumina porcelain tubes are used with fresh air passed through the tubes where it is heated prior to use in a mould drying process. Sticky, acidic deposits occur on the outside of the tubes but the glazed finish has proved to be both chemically resistant and easily cleaned.

Other ceramic recuperators, which similarly can be retrofitted into existing plant or designed into new plant, are available as laminated cubes of ceramic material, see Figure 7.36. In one design the cube is built of extruded plates of cordierite, a material said to be chemically inert and having very low thermal expansion, imparting properties of corrosion and thermal shock resistance. The ceramic cube is installed in a metal housing which directs the exhaust gases in cross-flow to the supply air passing through alternate spaces in the laminated structure. Whilst the exhaust gases are restricted to a single pass, the number of air passes in the exchanger is dependent upon available air pressure and the temperature and flow required.

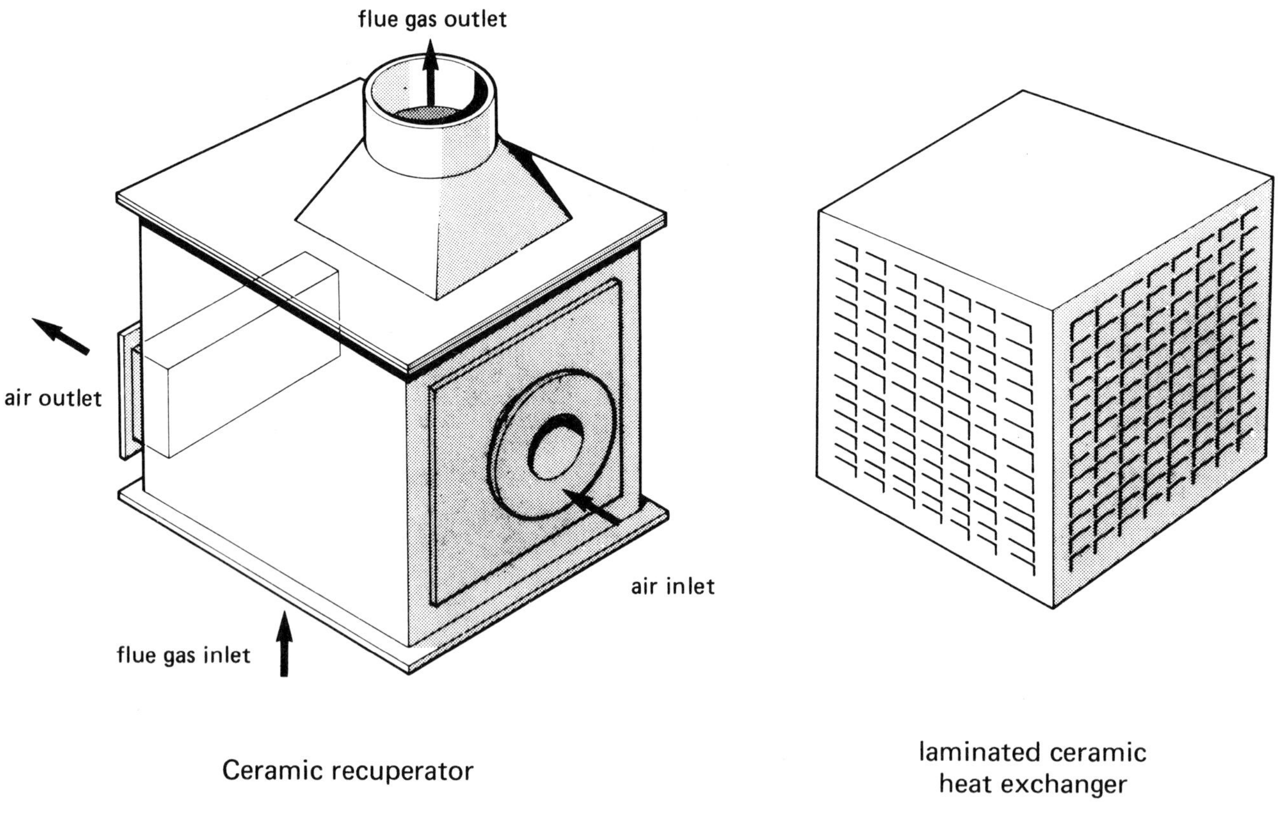

Fig. 7.36 Laminated ceramic recuperator casing and element

Glass Tubed Heat Exchangers

The glass tubed heat exchanger provides a means of gas to gas heat recovery which is resistant to corrosion and has reduced fouling characteristics compared with other means of heat recovery. It represents one of the few applications of tubed exchangers to low temperature gas to gas heat exchange.

The heat exchanger comprises a module of borosilicate glass tubes mounted and sealed by an elastomeric compound in stainless steel end plates (Figure 7.37). It is installed so that exhaust gases of up to 250 °C flow over the bank of tubes transferring heat through the tube walls to the supply air entering in cross flow. The preheated supply air is then ducted directly to the process.

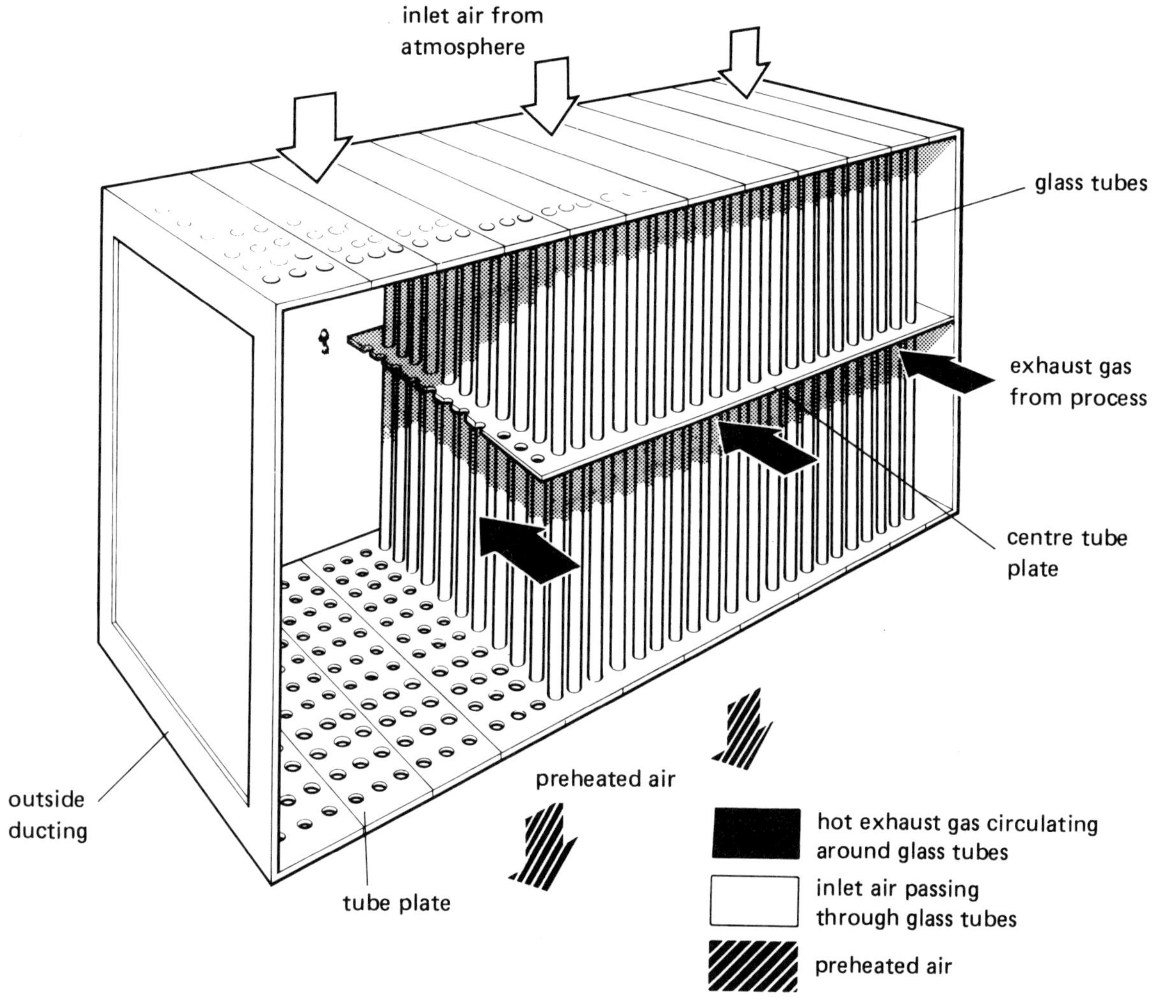

Fig. 7.37 Glass tubed heat exchanger

The corrosive nature of many exhaust gases is increased as condensation occurs upon cooling. This is often incompatible with efficient waste heat recovery which should aim to reclaim the latent heat content of gaseous waste streams. The glass tubed heat exchanger has excellent corrosion resistent properties and therefore has a particular application in such circumstances. Vapours contained in the exhaust gases condense out on the outside of the tubes and are drained from the lower tube plate.

Glass tubes have a very smooth surface minimising the attraction of particulate matter. Tubes can be washed in place by the action of mild detergent washes from spray nozzles and some units are self cleaning due to condensate in a subsequent cycle washing away deposits from a previous cycle.

Figure 7.38 shows a typical installation of a glass tube recuperator recovering heat from a brick kiln exhaust. The gases are contaminated by significant amounts of fluorides, chlorides and sulphates. The borosilicate glass tubed exchanger provides warm air which is fed to brick drying chambers.

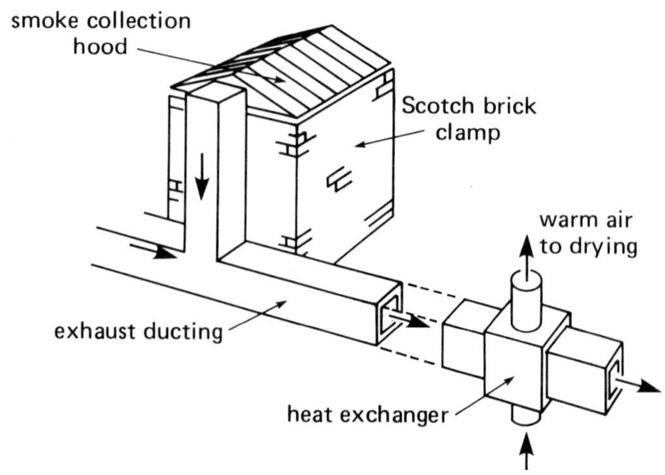

Fig. 7.38 Glass tubed heat exchanger installation

7.6.2 Recuperative Burners

The installation of conventional recuperators brings with it a number of disadvantages, for example heat resisting ducts are required to convey the hot gases to the recuperator and also ducts and manifolds to convey the hot air to the burners. This means that any burner adjustments are made on the hot side of the recuperator and there are pressure losses to be overcome in the ducts to and from the heat exchanger. A large external recuperator with its associated pipework and ducting also tends to have thermal inertia, which can be a drawback particularly with batch-type furnaces.

A heat exchanger integral with a burner would overcome many of these disadvantages and this situation has been achieved with the development of the recuperative burner. Figure 7.39 shows the design of a burner developed by the Midland Research Station of British Gas plc.

Essentially, this unit consists of a nozzle-mixing burner the air supply to which passes through a heat exchanger in counterflow to the outgoing flue gases. A plain, smooth tube, known as the air tube, separates the air and flue gases. Both fluids pass through narrow annuli, air flowing on the inside. The recuperator is integral with and positioned immediately behind the burner nozzle and

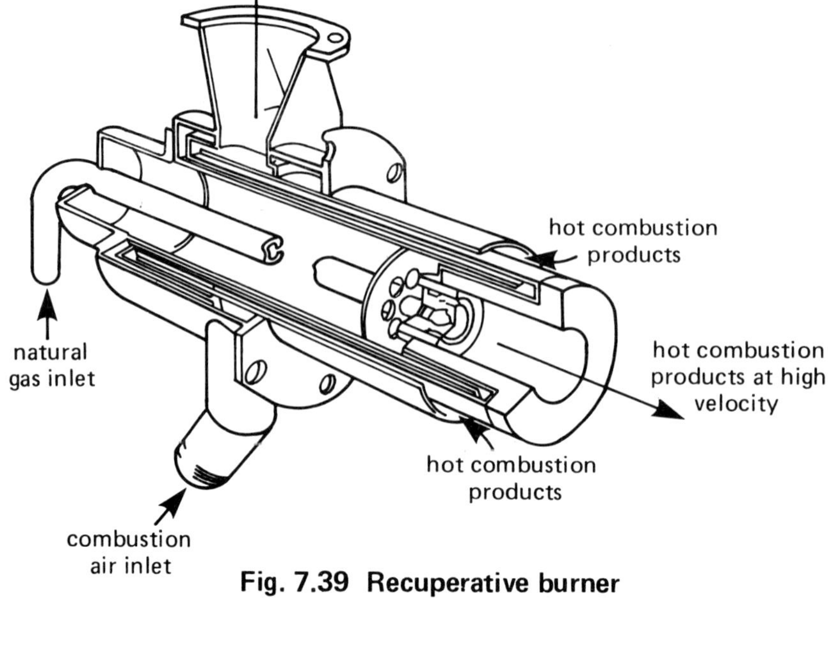

Fig. 7.39 Recuperative burner

tunnel, the front end of the air tube enclosing the refractory burner tunnel. The recuperative burner is fitted in a hole in the furnace wall, which is slightly larger in diameter than the air tube. In this way, the hottest part of the recuperator is located within the furnace wall. Combustion product temperatures of up to 1 400 °C can be accommodated. Air and gas inlets and the flue are all situated near to each other outside the furnace wall. The disposal of the flue gases is simplified since they are cooled before reaching the flue.

Recuperative burners are available in a variety of sizes and can be used as direct replacements for conventional burners. They have been applied to a number of items of high temperature plant. Examples on batch process plant are intermittent kilns, forge furnaces, heat treatment and glass and crucible furnaces.

Each compact recuperative burner is installed in the plant wall in a position to suit the heating requirement of the process. The exit velocity of the combustion products often improves heating efficiency and the recirculation flow path generated by careful siting of burners can often reduce their number from that originally required on the plant. An example of the before and after firing pattern with an intermittent kiln is shown in plan in Figure 7.40.

They are also being increasingly used on large continuous plant e.g. pusher furnaces, bogie hearth heat treatment furnaces and ingot reheating furnaces.

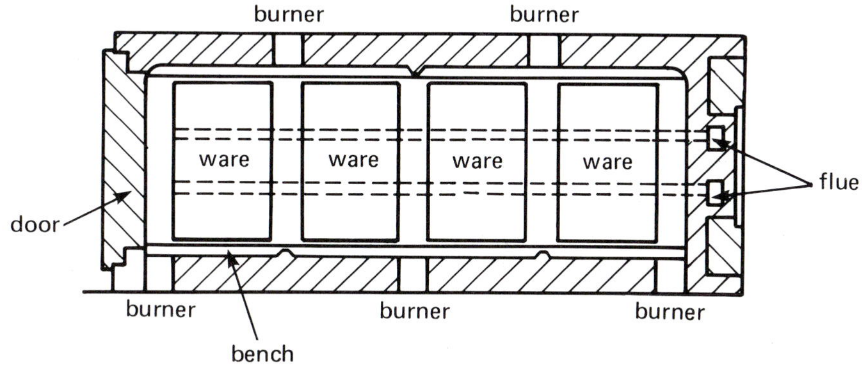

Before modification — fired by 5 conventional burners

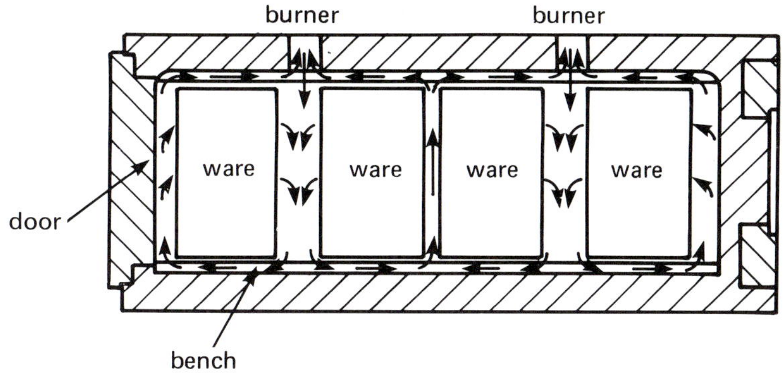

After modification - fired by 2 recuperative burners

Fig. 7.40 Recuperative burner retrofit on twin truck intermittent kiln

Much of the development in recuperative burners has centred on the reduction of manufacturing costs and on producing a design which facilitates ease of inspection and maintenance. The experience gained by Midland Research Station of British Gas plc, has aided the development of such a design which does not require the fine tolerances of the original design. It includes a conical recuperator assembly to provide the necessary heat transfer surface area, see Figure 7.41. At the same time a new low cost recuperator alloy was found to be acceptable leading to a combined reduction in manufacturing costs of the order of 50%. The various licensees now have an individual design based on this concept. The reduced costs offer increased scope by reductions of payback periods.

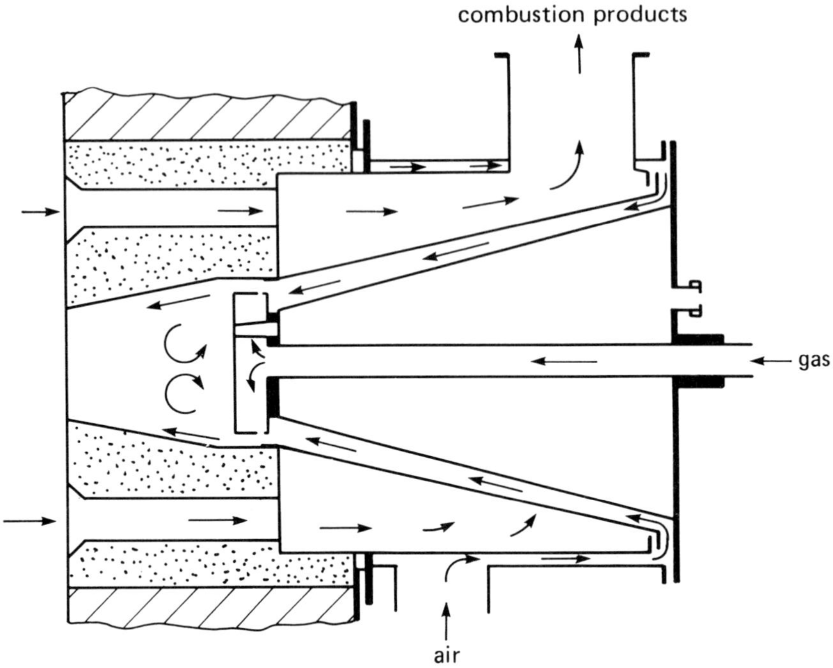

Fig. 7.41 Improved recuperative burner

A further development by one licensee is the use of their packaged system as a split burner/ recuperator. This arrangement offers flexibility of application and enables the benefits of heat recovery to be obtained on furnaces where space limitations have precluded the use of recuperative burners. The arrangement has been incorporated on an intermittent kiln used for firing refractory ware which was previously top fired and bottom flued (Figure 7.42).

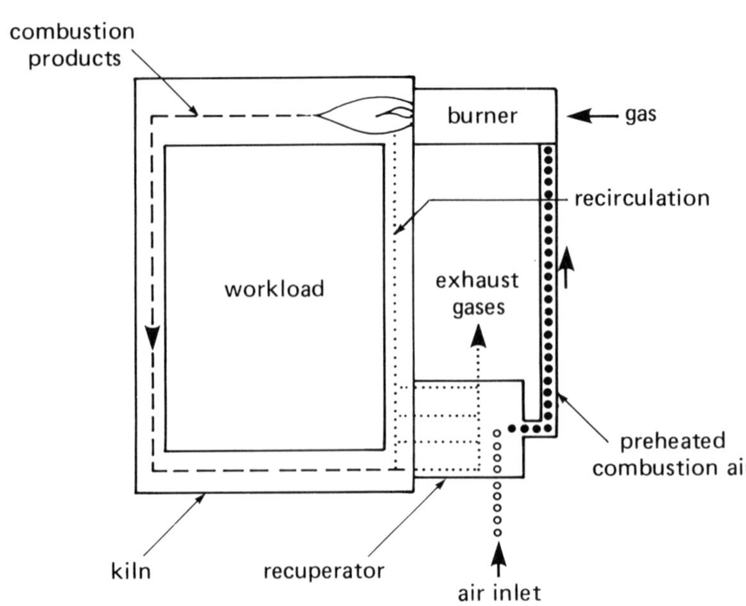

Fig. 7.42 Split burner/recuperator system

The use of conventional recuperative burners as a direct replacement for the existing burner would have resulted in short circuiting of the combustion products and poor temperature uniformity. A major loss of production capacity would also have resulted if packaged recuperative burners had been fitted in the more usual low level position since passages for recirculation would have been required around the load. The four burners were subsequently fitted at high level with individual recuperators below each burner at low level. Substantial fuel savings and temperature uniformity were achieved. Applications in heat treatment where it is necessary to top fire to avoid impingement where load size is variable, and to bottom fire to give temperature uniformity, also offer scope for this development.

In addition to the direct fired applications presented above, recuperative burners are also successfully employed on indirect systems such as radiant tube and immersion tube heaters. Applications for metallic recuperative radiant tubes include bright annealing, carburising, hardening, sintering of copper alloys and annealing of aluminium in air. Recuperative ceramic tube heaters have been developed for galvanising and aluminium holding furnaces and include some immersion applications like the aluminium bale out furnace shown in Figure 7.43.

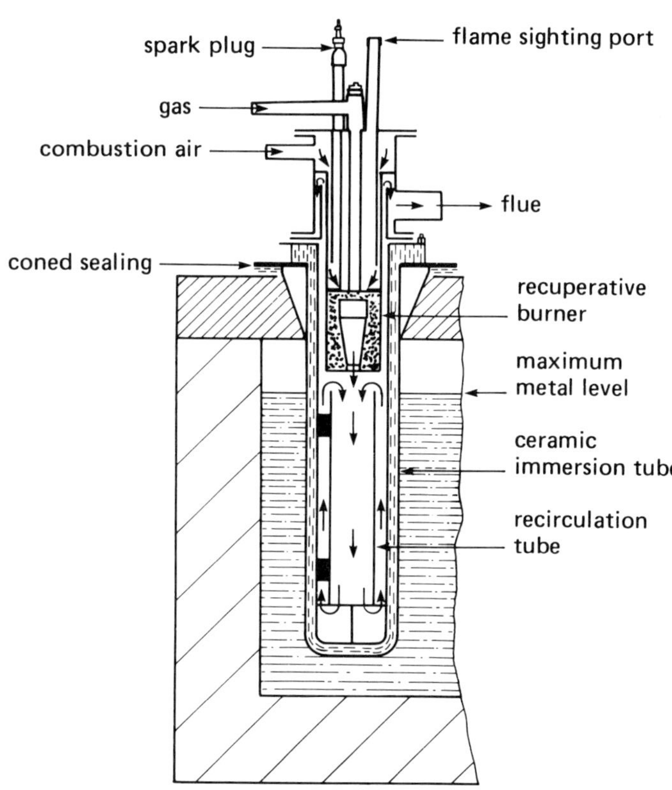

Fig. 7.43 Recuperative immersion tube heater

7.6.3 Static Regenerators

Heat recovery from very high temperature gases using a recuperator demands the use of ceramic materials. However, high temperature ceramics can be difficult to fabricate into a suitable recuperator construction because of the complex shapes necessary for a compact, effective unit. In addition they are often susceptible to failure under low tensile stresses and can be difficult to seal between the two gas streams.

An alternative approach to increased performance is the regenerator. Regenerators have traditionally been used in large continuously operating high temperature processes such as glass melting furnaces and open hearth furnaces. They have largely been incorporated into plant during its original construction and their size does not easily lend them to the retrofit market. In essence they consist of two regenerator chambers located below and to each side of the furnace, see Figure 7.44.

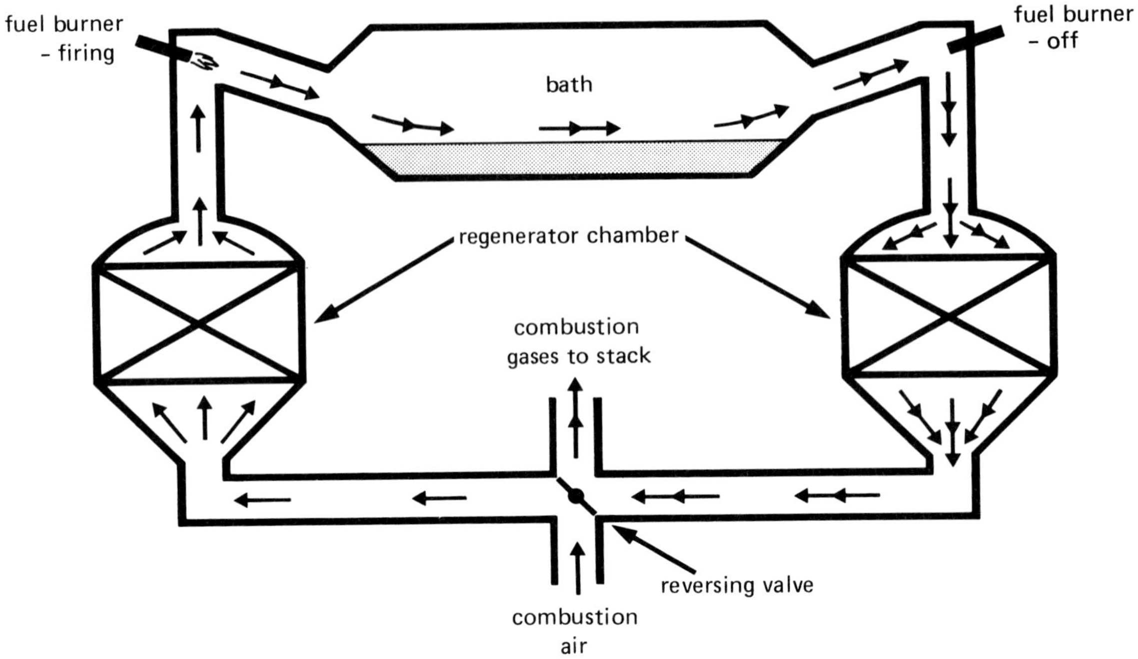

Fig. 7.44 Static regeneration on glass melting tank

They operate on a cyclic schedule with one chamber supplying combustion air whilst the other is receiving exhaust combustion products. Each chamber incorporates an internal chequer brick or rubble bed heat transfer matrix which has a large surface area to aid heat transfer. One matrix is heated by the combustion gases at around 1 400 – 1 450 °C as they pass to the flue. In the alternative matrix the combustion air supply preheats from the matrix heated during a previous cycle. A reversing valve directs the flow and when the matrix preheating the combustion air has given up sufficient heat the valve is reversed to allow the combustion air to preheat from the other chamber. Reversal times, varying bewteen 5 and 30 minutes, are arranged to give maximum fuel economy, with maximum preheat temperatures of up to 1 000 °C often possible, although these can vary significantly throughout the cycle. Optimum heat transfer through chequer brick regenerators is obtained with gas and air flows through vertical chambers. In addition careful design of the entrances to the chambers will achieve even flows through the matrices. The use of horizontal flow chambers leads to reduced effectiveness as density differences due to stratification will cause the hotter gases to travel along the top of the chamber.

Static regenerators have not been universally applied to high temperature plant because of their complexity of operation and large size. This has led to the development of alternative regenerative methods, namely regenerative burners and rotary regenerators.

7.6.4 Regenerative Burners

The regenerative burner system was developed by the Midland Research Station and comprises two burners, each of which has a regenerator chamber containing a packed bed of ceramic shapes of large surface area to provide the heat store (see Figure 7.45). The burners fire alternately for a few minutes each with the flue gases from the one firing leaving the furnace through the burner tunnel of the other. The hot gases flow through the packed bed heating it almost to the furnace exit gas temperature. On reversal, cold combustion air passes through the packed bed and preheats before entering its associated burner. The cycle is continuous with each burner firing at either full rate or turned down to a minimal firing rate. The heat recovery 'effectiveness' is high at 90% and the regenerator takes up only 2% of the space required for equivalently sized conventional units. The extensive surface area, short reversal periods and high convective heat transfer coefficients of the ceramic shapes provides an effective, compact design.

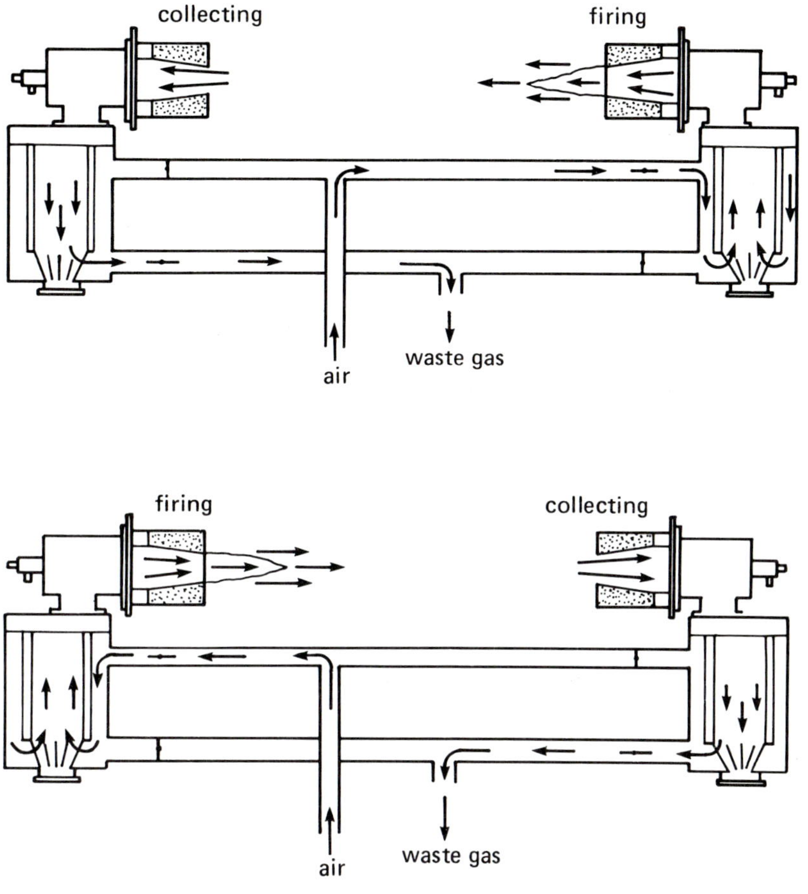

Fig. 7.45 Regenerative burner system

The system has been applied to a pot glass melting furnace operating at temperatures up to 1400 °C. The installation has proved reliable and effective and has demonstrated a further advantage of this regenerator system. It offers the facility for operation with dirty atmospheres, by ready removal of the ceramic heat exchange shapes for renewal or cleaning, in contrast to conventional metallic recuperators and recuperative burners which are susceptible to corrosion and fouling. This system has now been used on a wide range of processes such as bulk aluminium melting, steel heat treatment (batch and continuous), steel reheating (batch and continuous), barrel furnaces and the slot forge furnace.

7.6.5 Rotary Regenerators

The rotary heat regenerator provides a method of gas to gas heat exchange between two adjacent ducts, one carrying exhaust and the other incoming supply air. The device (commonly called a thermal wheel) of diameter 0.6 to 4 m is suspended between the two counter-current air streams, see Figure 7.46. Half of the wheel spans the exhaust stream and the remainder spans the supply stream. It rotates, driven by an electric motor, at slow speed absorbing heat from the exhaust flow into its matrix. The absorbed heat is transferred to the supply air as the heated matrix area progresses through the intake ducting.

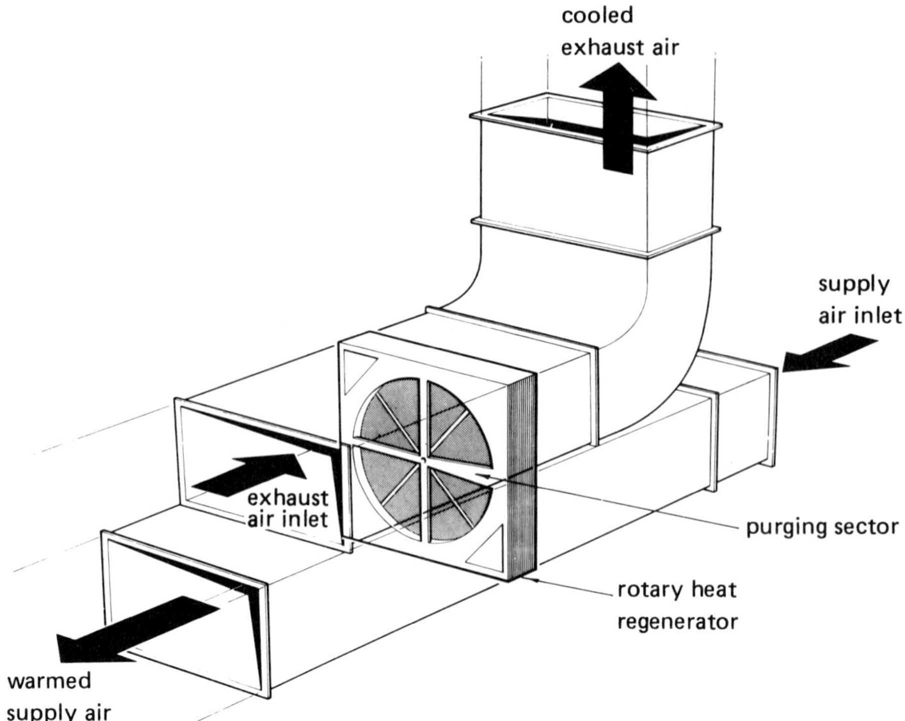

Fig. 7.46 Rotary regenerator

Speed of rotation, typically around 10 revolutions per minute, and the relative velocities of the gas streams govern the effectiveness and efficiency respectively of heat transfer. The amount of heat recovered is a function of the temperature and mass flow rate of each gas stream. Hence for balanced flows, control of the speed of rotation will regulate the heat and thus temperature available to the recipient stream. On some installations a bypass around the regenerator may be used to compensate for unequal flow rates and a similar system can be used to regulate heat recovered for varying demands.

Rotary regenerators capable of operating at high temperatures have been developed using stainless steel and ceramic matrices and are used in the glass, steel and aluminium industries. The major area of application for rotary regenerators, however, is in low temperature duties up to 200 °C.

Installations of thermal wheels in commercial applications include air conditioned office blocks, hospitals, libraries, bus garages, airport buildings, leisure centres, swimming pool halls, theatres, supermarkets, restaurants, university laboratories, film processing laboratories and shopping centres. In industry, they have been installed on a great variety of process plant including spray booths, ovens, paint plant and welding shops. Rotary heat regenerators are available for the recovery of sensible or total heat (i.e. sensible heat plus latent heat).

Those designed to recover sensible heat only consist of a heat exchange matrix of knitted aluminium, mild or stainless steel wire mesh of large surface area aiding heat transfer. The more common development, laminar flow wheels, use aluminium foil or thin stainless steel sheet rolled up to form a mass of small axial flutes or air passages. The latter are reputed to have improved pressure drop and fouling characteristics and to be easier to clean compared with the mesh matrix type.

The hygroscopic thermal wheel is used for the recovery of latent heat in addition to sensible heat. Latent heat is transferred by absorption of moisture in the exhaust air stream on to the matrix of the wheel. The matrix material can take various forms. In the Munters wheel an inorganic fibrous material, of structure resembling a roll of corrugated cardboard, is coated with a bacteriostatic dessicant salt, lithium chloride. Aluminium matrices can be coated with a dessicant layer of aluminium oxide. The lithium chloride and aluminium oxide absorb moisture which is collected by the supply air stream.

Cross contamination in thermal wheels is minimised by radial and circumferential seals on the rotor and stationary sealing strips on the casing. However, as the exhaust air stream passes through the matrix of the rotating exchanger a proportion of the air becomes entrained in the flutes. Due to the wheel's rotation this will be carried towards the incoming fresh air stream where it would mix with the supply air. The carryover of the exhaust air entrained would be directly proportional to the volume of the matrix and the speed of rotation. The inclusion of a purge sector in the unit whereby every cell is flushed out with a small amount of fresh air reduces carry-over to a negligible amount.

An integral automatic purge section (Figure 7.47) is included in the Munters wheel to reduce contamination to a quoted value of less than 0.04%. The purge unit is a segment of ducting which allows a proportion of the supply air to flush any dust, vapours or other contaminants from the matrix as it enters the supply stream and return them to the exhaust stream.

Additionally the laminar flow pattern parallel to the axes of the passages is designed to assist any particle entering a passage to flow in a straight line down the space to emerge at the other end without touching the side walls. The reversal of the air flow as the wheel rotates will assist any large particle entering the matrix to be flushed out.

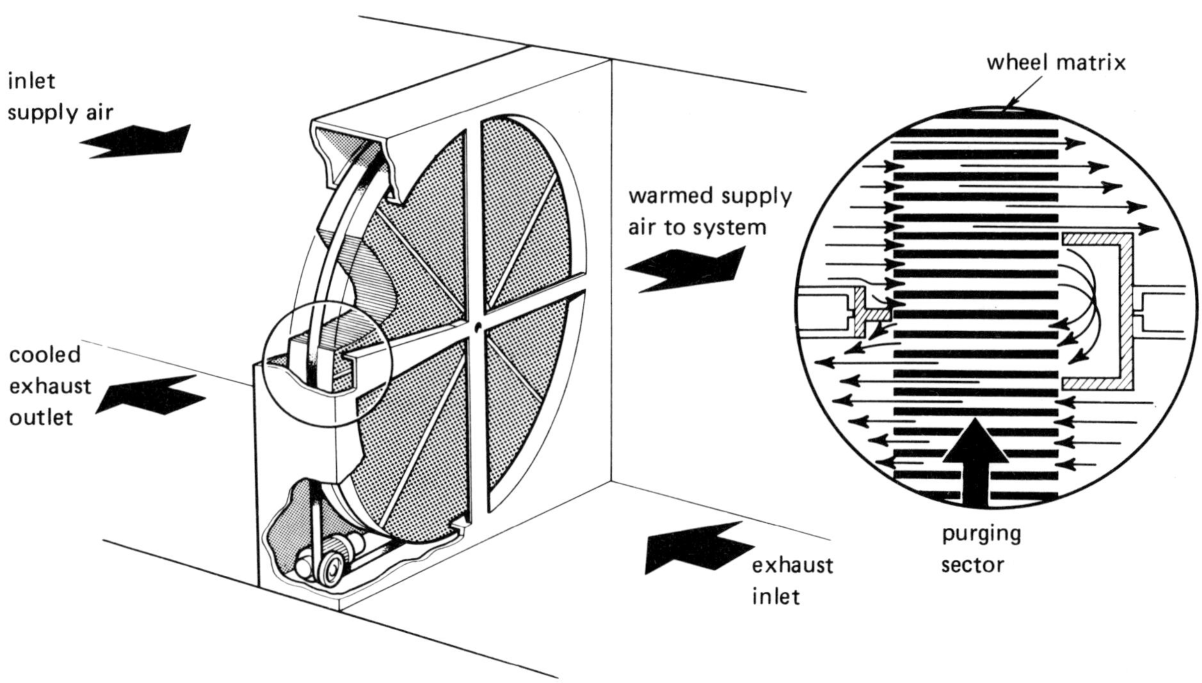

Fig. 7.47 Rotary regenerator showing purging sector

7.6.6 Plate Recuperators

Plate recuperators are a convenient modular means of heat recovery between air or gas streams. The heat exchanger consists of a series of plates separating hot and cold areas, sandwiched in a box-like structure. The incoming supply air and the outgoing exhaust pass between the plates. The channels are sealed at joints between each of the plates eliminating intermixing of the two fluids and hence contamination of the supply air. The plate recuperator is crossflow, but there are two possible ducting configurations which may be termed counter flow and crossflow.

In the counterflow type (Figure 7.48) the two ducted air streams are brought together at the exchanger with the two fluids entering and flowing from opposite directions.

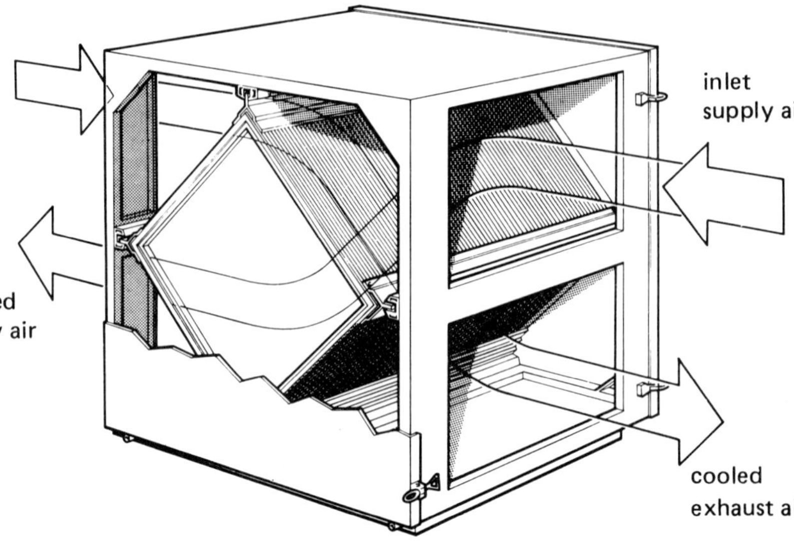

Fig. 7.48 Counterflow plate recuperator

The incoming air stream in the crossflow type (Figure 7.49) enters at 90° to the exhaust stream which can lead to more convenient ducting arrangements in some circumstances, for example where space is at a premium, although thermal effectiveness is reduced.

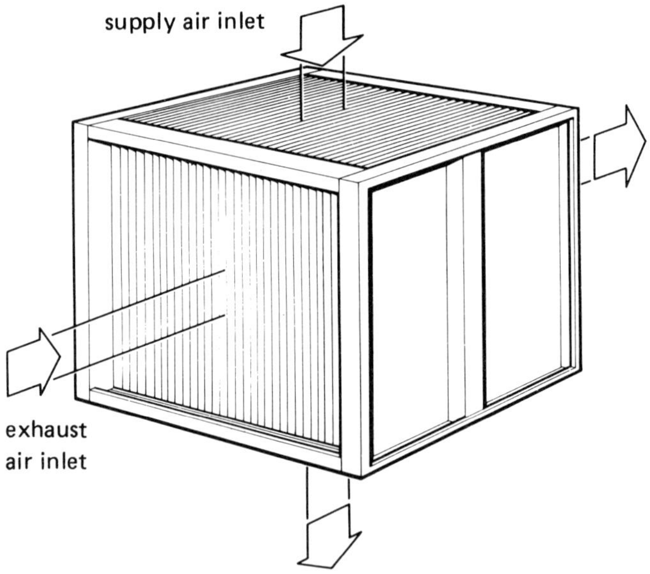

Fig. 7.49 Crossflow plate recuperator

Plate material is commonly thin aluminium sheets having structural space protrusions or corrugations which strengthen the plates and increase turbulence to assist heat transfer. An acrylic or plastic film coating can be applied to aluminium plates if corrosion is a possibility. Other plate materials e.g. aluminised steel, stainless steel or glass are also used.

Low temperature applications include spray booths, garage exhausts, hospitals, swimming pool halls, chemical and process plant, bakeries, restaurants, kitchens and laundries.

High temperature versions are usually welded and can operate up to 800 °C, for example on drying ovens and incinerators, but their use is less widespread. Designs incorporating a gasket can be used up to 300 °C.

7.6.7 Run-around Coils

Situations exist where heat can be recovered from a hot exhaust fluid stream to preheat a cool supply fluid stream but where the two streams are some distance from each other, for example where the air inlet and extraction systems are arranged remote from one another in a building. The liquid coupled run-around coil system provides a simple and relatively cheap method for effective heat recovery in such circumstances, where a single heat exchanger would entail long runs of ducting. It may be similarly applied on gas to gas or liquid to liquid systems.

In the gas to gas system, heat from the exhaust products is transferred to a heat transfer liquid by means of multi-finned tube coils mounted in the ducting. The tube fins provide a large surface area to maximise heat transfer. The optimum fin spacing is related to the cleanliness of the air streams. Commonly aluminium fins on copper headers are used for low temperature duties. The heated liquid is pumped through a pipework system (Figure 7.50) to similar coils which act as the heat sink in the supply air ducting. The liquid, usually water, or a water/glycol solution if freezing is a likely problem, releases heat through a second heat exchanger coil to raise the temperature of the supply air. It then returns to the coils in the heat source through the return loop pipework. The pipework system incorporates an expansion tank to compensate for volumetric expansion. The rate of heat transfer can be varied by the inclusion of a bypass in conjunction with a motorised three port valve controlled by a thermostat sensing the temperature of the outlet supply air.

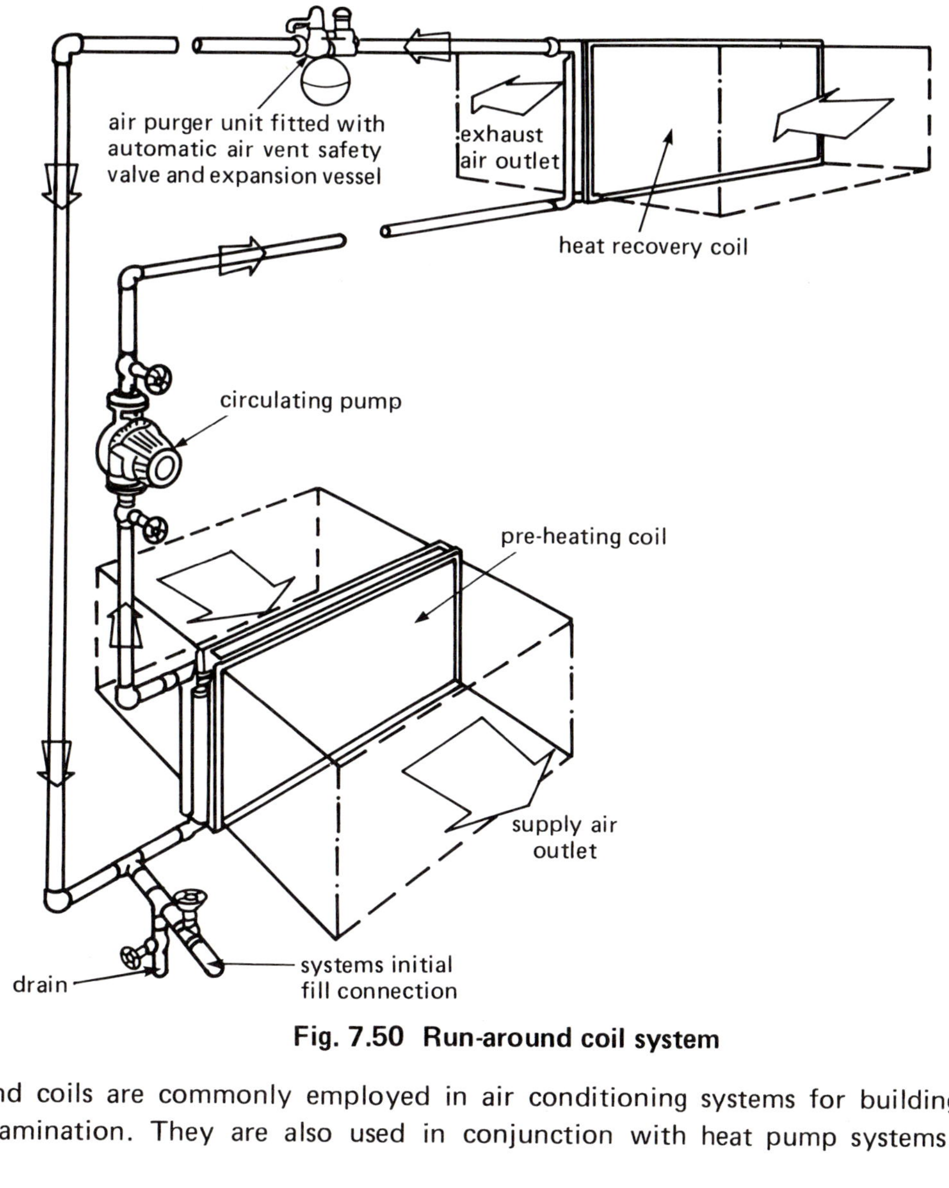

Fig. 7.50 Run-around coil system

Run-around coils are commonly employed in air conditioning systems for buildings, eliminating cross-contamination. They are also used in conjunction with heat pump systems. The concept allows waste heat to be used to heat the supply air to remote office buildings or to provide simultaneous heat recovery between multiple supply and exhaust air streams, although their effectiveness is frequently less than alternative methods.

An alternative form of run-around system, where corrosive gases are involved, is to use a glass tubed heat exchanger as one of the heat transfer elements. A modular glass tubed unit, as shown in Figure 7.51, is mounted in the corrosive vapour stream and by means of a glycol/water solution heat is transferred to the incoming air stream using a copper tube/aluminium fin type exchanger.

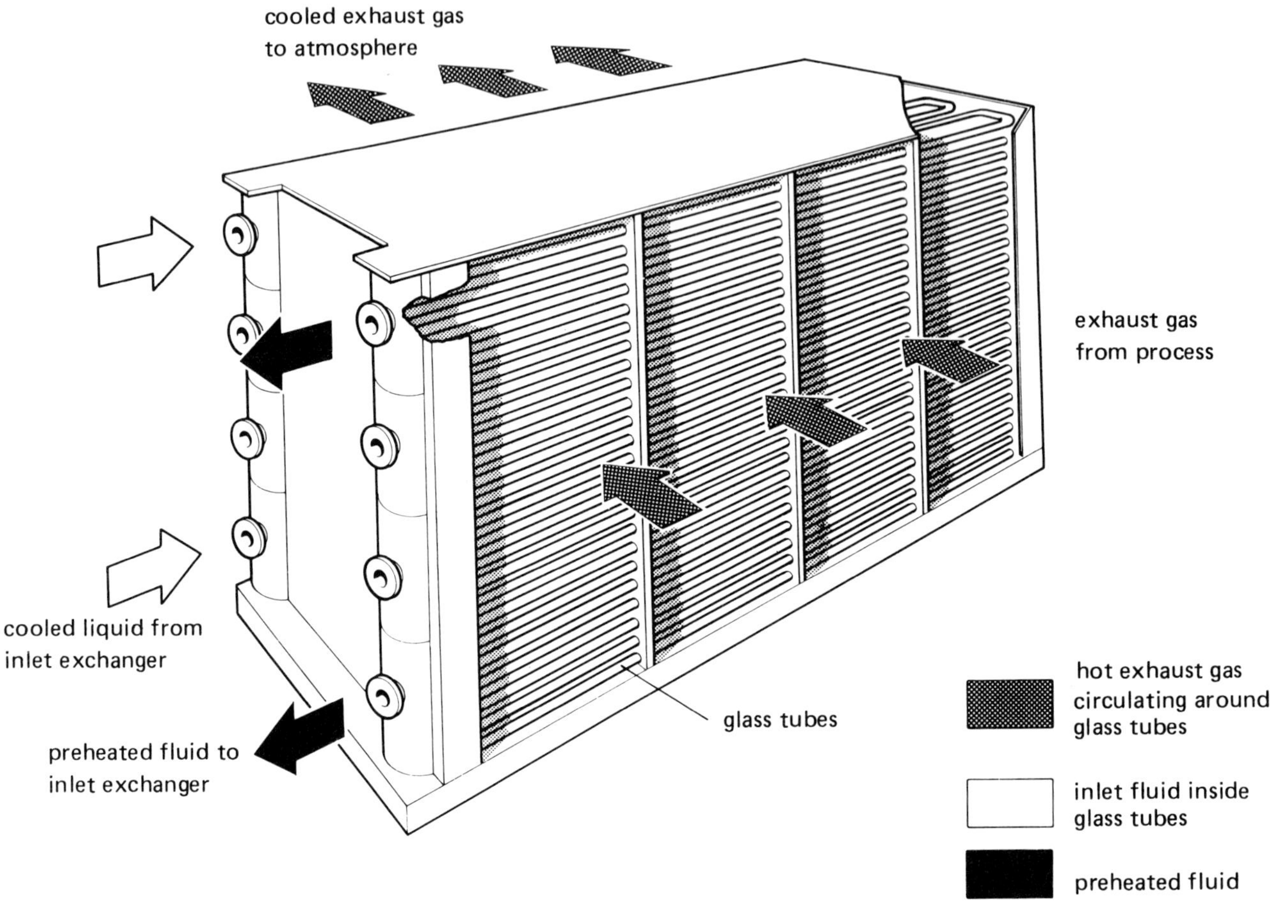

Fig. 7.51 Glass tubed heat exchanger for run-around system

In the process industries, particularly petrochemicals, a form of high temperature run-around coil system is used for heat recovery from fired process heaters. Two liquid coupled finned coils are situated, one at the exit of the process heater and the other in the combustion air supply to the burners. Commonly an organic heat transfer fluid of operating range up to 400 °C is used in the system, e.g. diphenyl-diphenyl oxide eutectic (thermax) which has a boiling point of 260 °C, or in some instances very high pressure water. In the closed loop system (Figure 7.52) the fluid is circulated as in the run-around coil system and transfers heat from the flue gases to preheat the combustion air. Included in the closed loop system is an expansion/blowdown receiver and circulation pumps; forced or induced draught fans may be necessary to overcome the resistance of the coils. Several process heaters may be linked to the same heat recovery system.

An alternative approach is the slipstream system (Figure 7.53) which diverts a proportion of the process fluid through the combustion air preheat coil. Having preheated the air the fluid is passed through the recovery coil where it captures residual heat from the flue gases before being returned, reheated, to the main process stream.

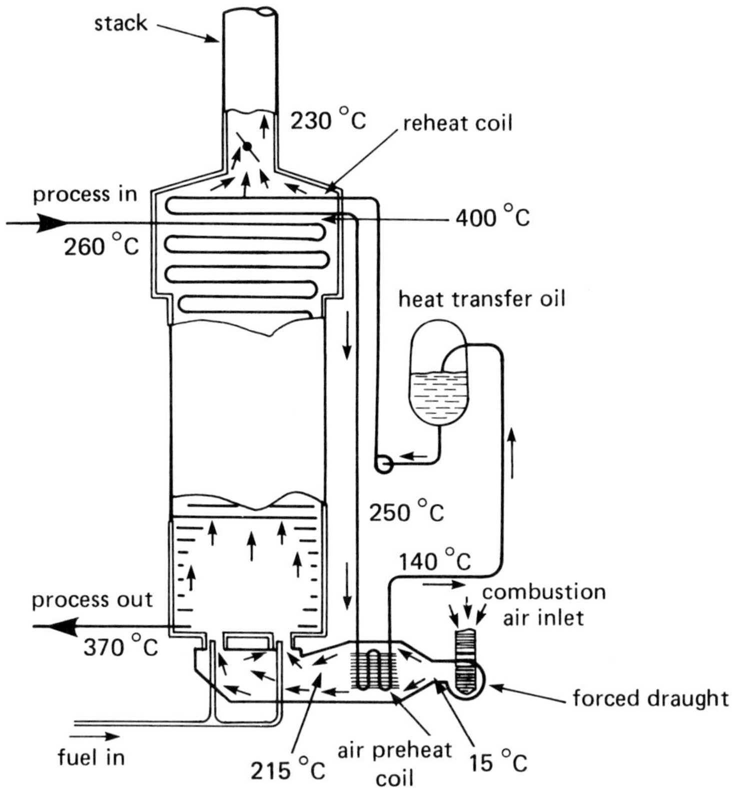

Fig. 7.52 Closed loop, liquid coupled air preheat system

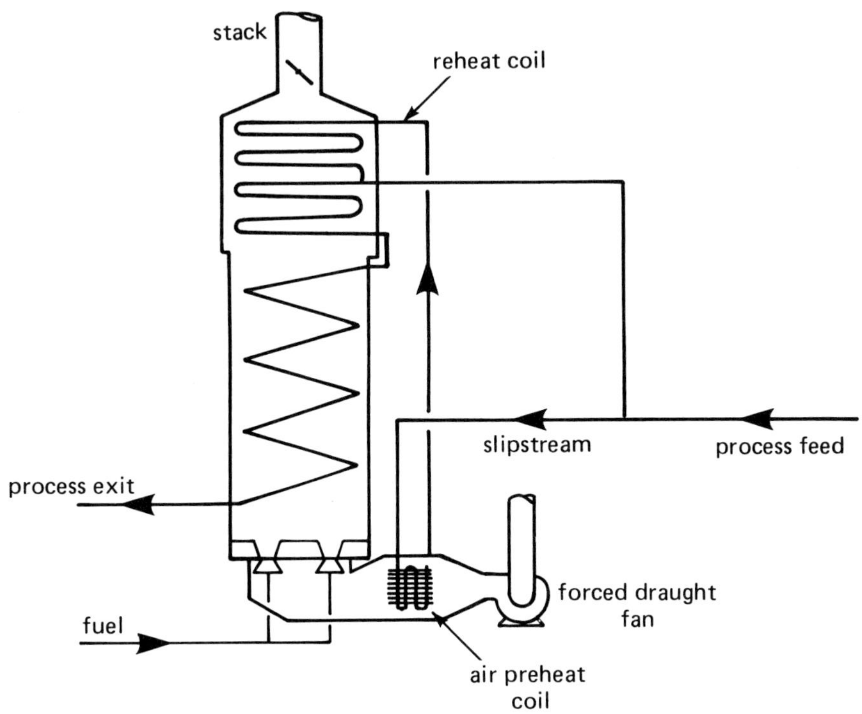

Fig. 7.53 Slip stream, liquid coupled air preheat system

7.6.8 Heat Pipes

The heat pipe is a self-contained unit with no moving mechanical parts which provides heat exchange by a reversible evaporation/condensation cycle. It consists of a sealed tube containing a small volume of working fluid (Figure 7.54).

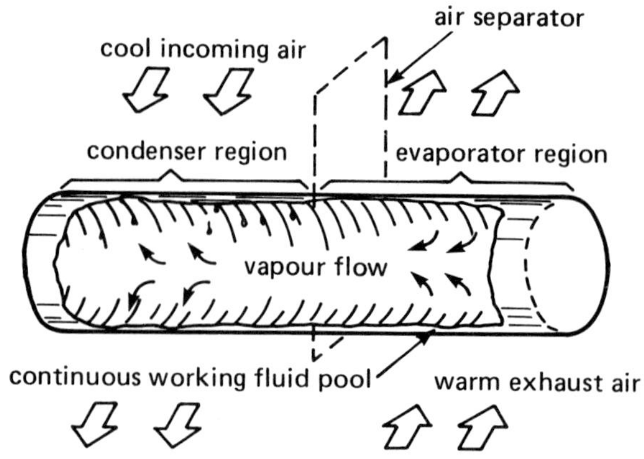

Fig. 7.54 Principle of the heat pipe

One end is situated in the 'heating' stream and the other end in the 'cooling' stream. The heating process causes the fluid to evaporate and the vapour flows to the 'cold' end of the tube where it condenses releasing its latent heat of condensation. The condensed fluid returns to the evaporation end, assisted by gravity if the tube is vertical or inclined.

Heat pipes need not rely on the assistance of gravity for return of the condensate. A wick arrangement on the inner surface of the heat pipe assists return of the condensed working fluid by capillary action. The vapourisation of the working fluid at the evaporator end, using the exhaust heat, leads to an increase of vapour pressure in the tube causing the vapour to flow to the condenser end of the pipe. At this point condensation occurs with a release of thermal energy to the supply air, before the fluid returns via capillary action aided by the wick to the evaporation zone. The wick may be of metal wire mesh, a fine gauze or may be a grooved channelled structure in the heat pipe interior wall. The choice of wick is related to the characteristics of the working fluid. The capillary action in conjunction with the wick allows the heat pipe to be used in any orientation. Heat pipes having evaporator below condenser, however, offer a heat transport somewhat in excess of heat pipes with condenser below evaporator. The heat pipe thus provides a more rapid means of transfer of thermal energy than would be possible by conduction along, for example, a solid copper bar of equivalent size.

The tube of the heat pipe can be manufactured from a range of materials dependent upon factors such as temperature of application, compatibility with working fluid and environment, thermal conductivity and ease of fabrication. Copper tubes are commonly used between 0 and 200 °C. Aluminium and stainless steel tubes are also widely used.

A range of working fluids is available to meet whatever temperature range is required. Fluorocarbons and water are for example used for the lower temperature ranges and liquid metals for special high temperature applications. A good thermal stability, high latent heat and thermal conductivity and an acceptable vapour pressure range are among properties influencing choice of fluid, as is a high surface tension to assist capillary action.

In practice, to improve heat transfer and to transport large quantities of heat several heat pipes are used in parallel with extended surface areas (Fig. 7.55). Fins, generally of the same material as the tube, (aluminium fins are used on some copper heat pipes) are fixed at the evaporator and condenser ends. The practical spacing of the fins will largely depend on the cleanliness of the gas streams involved.

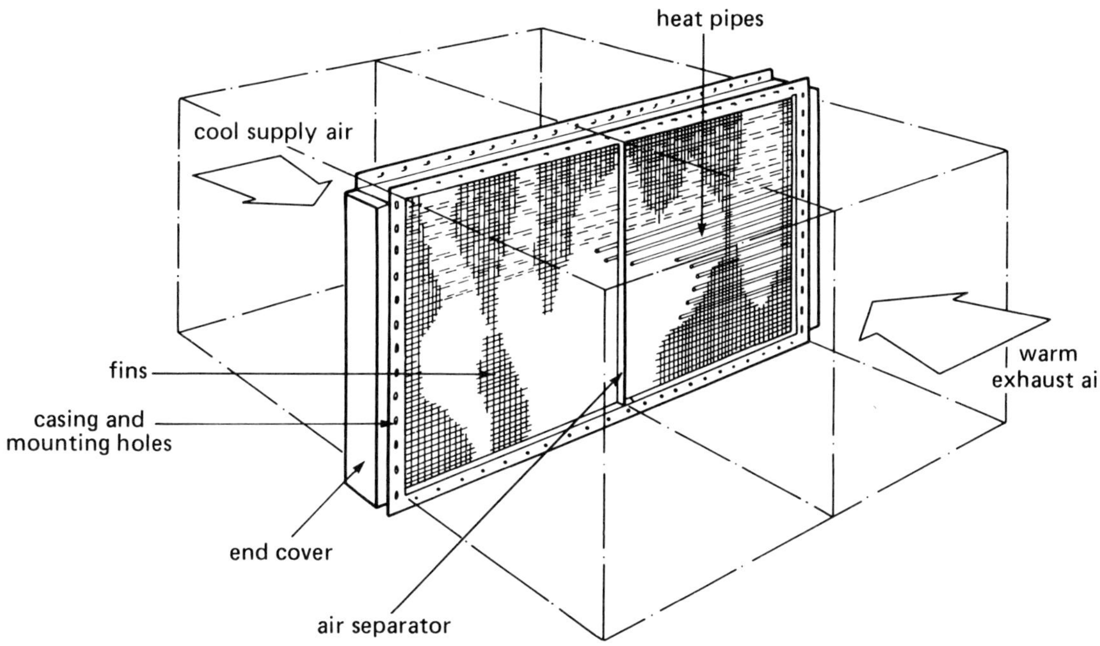

Fig. 7.55 Heat pipe recovery unit

Most gas to gas heat pipe heat exchangers are arranged with the evaporator section spanning the hot exhaust stream and the condenser section the cool supply stream. They are located between the two ducts with pipes horizontal or tilted as required with the evaporator below the condenser. The streams are separated in the exchanger by a splitter plate which both supports the heat pipes and has the advantage of preventing cross-contamination between streams. Additionally, tube ends are protected by end covers.

Heat pipes are used for heat recovery in space heating, ventilation and air conditioning applications where operating temperatures are relatively low. Process applications which have been developed include recuperation from exhaust air from dryers and ovens, kilns, laundries and welding shops. Generally, applications have been limited to exhaust temperatures below 400 °C. The operating temperature will largely be governed by the combination of tube and wick material and working fluid characteristics. Where higher exhaust temperatures are involved considerably higher mass flow of air over the condenser is required relative to the flow over the evaporation section to limit the vapour temperature within the heat pipe.

7.7 GAS TO LIQUID HEAT EXCHANGE

7.7.1 Tubular Heat Exchangers

Tubular gas to liquid heat exchangers find a range of application in waste heat recovery systems. In all cases, however, their construction reflects the disparate heat transfer characteristics of gases and liquids. Heat transfer coefficients to and from gases are much smaller than those to and from liquids. For this reason gas/liquid heat exchangers normally feature finned tubes in order to obtain a larger heat transfer area on the 'gas side'. On low temperature duties, for example as part of a run-around coil system handling ventilation extract air, aluminium finned copper tubes are usually used. The most significant applications of tubular heat exchangers, however, are as 'economisers' recovering sensible heat from boiler flue gases by preheating the feedwater. Various economiser constructions are available dependent on the fuel from which the combustion gases emanate. They mostly consist of welded steel tubes finned (Figure 7.56) to improve the heat transfer between the gases and the tubes. The optimum fin pattern and spacing will be dependent on the fuel, the relationships between cost and the efficiency required and also the tolerable back-pressure which can be imposed on the flue.

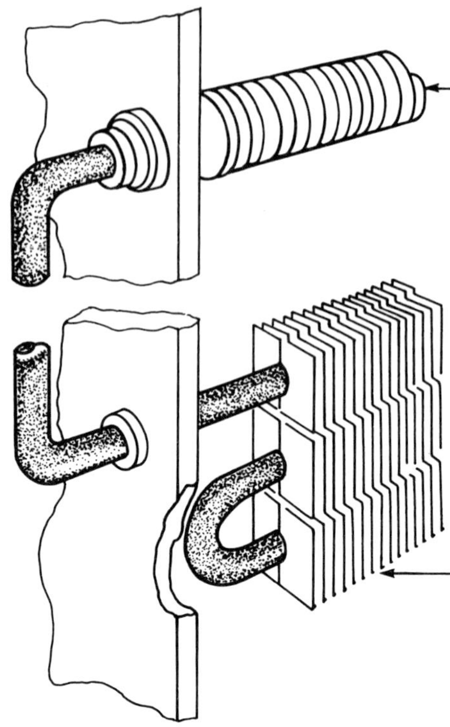

Fig. 7.56 Economiser fin patterns

With the clean combustion gases produced from natural gas fired boilers helical continuously welded finned tubes are used to produce a compact efficient unit. Where resistance through the economiser has to be minimised steel parallel patterned finned tubes are available but the resultant efficiency will be somewhat less. Parallel fin material of cast iron or cast iron sleeves can be used for flues where acid corrosion is a likely problem. On these steam, mechanical or ultrasonic cleaning devices will usually be installed to periodically clean deposits from the surfaces.

Many helical finned economisers suited to natural gas systems are produced as insulated packaged units complete with all controls. All economisers are designed and constructed to pressure vessel codes and accordingly incorporate safety relief valves. Generally helical finned type batteries are rectangular in shape but circular designs incorporating integral dampers are also available.

Induced draught fans can be installed to overcome resistance imposed by economisers. One proprietary type (Figure 7.57) is supplied as a packaged unit complete with an induced draught fan. This system is designed for a pressure drop of 2.5 mbar through the gas side of the economiser but the use of the integral fan produces a compact space-saving unit for natural gas systems.

Installed as a part of a new boiler system or, as in most cases, retrofitted to existing plant, economisers play a significant heat recovery role. In addition to the fuel savings which are available economisers can assist the boiler to provide more rapid reaction to peak loads, can extend boiler life due to reduced thermal shock, and also can increase steam production capacity. Economisers are not only used on boiler installations but can also be applied to furnaces and drying plant to produce hot water supplies. In some situations where condensate vapours are present heat transfer

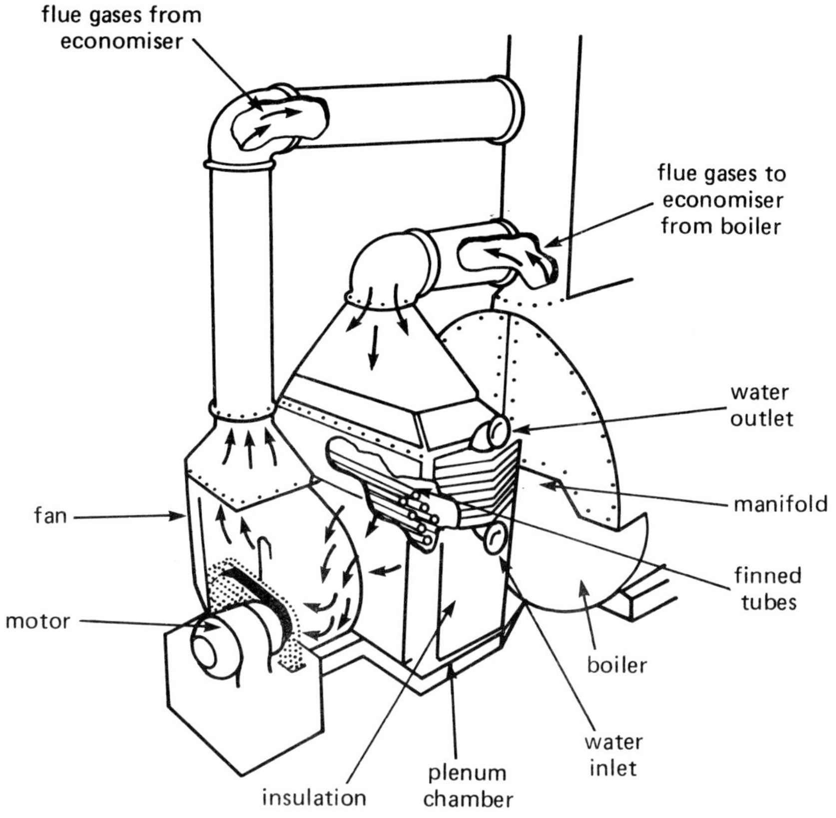

Fig. 7.57 Fan assisted economiser

can be effective. Figure 7.58 shows one such scheme which condenses a steam/petrol vapour mixture produced during the regeneration of an activated carbon adsorber. Where significant condensation occurs finning is not necessary and the stainless steel heat exchanger used in this application has plain tubes which also help to resist fouling.

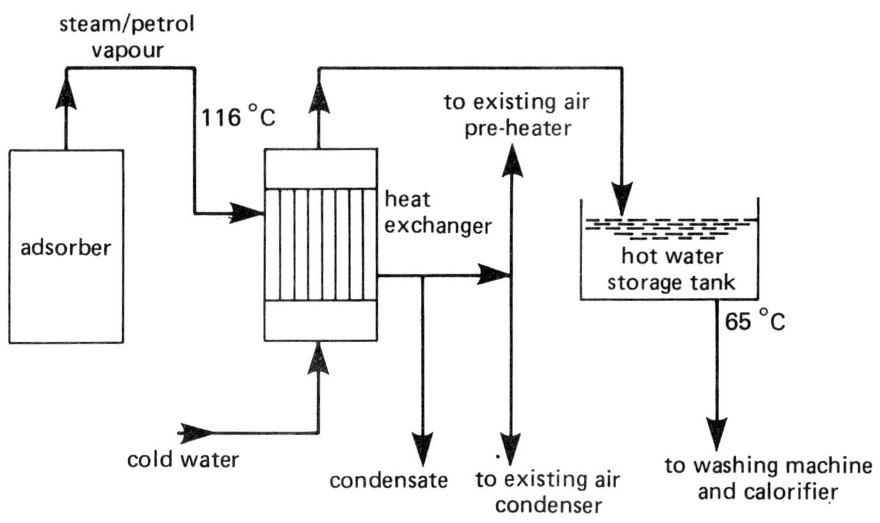

Fig. 7.58 Heat recovery by vapour condensation

On a smaller scale, proprietary waste flue heat exchangers are available for installation in the flues for forced draught boilers, ovens or air heaters. One example (Fig. 7.59), comprises a cylindrical welded steel outer cover such that an annulus of about 50 mm is formed between the components.

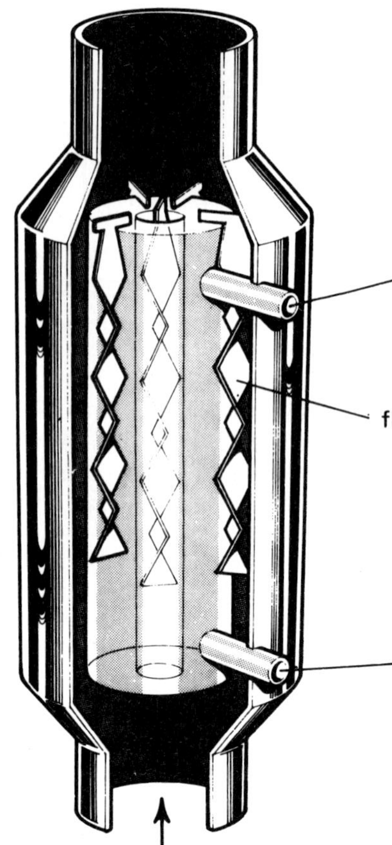

Fig. 7.59 Waste flue water heater

Flue gases pass up the annulus and baffles assist heat transfer. Other similar devices incorporate tubular water passages through the heat exchanger. The resultant hot water from such units is available for domestic hot water, space heating or for return to the boiler.

7.7.2 Fluidised Bed Heat Exchangers

A range of fluidised bed heat exchangers has been developed with liquid-carrying tubes immersed within the bed. The gas side heat transfer coefficient is very high within the bed allowing plain tubes to be used. The abrasive action of the bed particles imparts self-cleaning properties to the outside of the tubes, although in situ cleaning devices may be necessary on the underside of the gas distribution plate.

Figure 7.60 shows one heat exchanger design. In general, though, fluidised bed heat exchangers have not found widespread application.

7.7.3 Spray Recuperators

The exhaust from the combustion process in boilers or from some processes e.g. papermaking machines, contains water vapour which is normally discharged to atmosphere with a loss of latent heat. Spray recuperation is a gas to liquid heat recovery technique which recovers much of this latent heat in addition to the associated sensible heat in the exhaust. The combustion gases from natural gas fired boilers are particularly suitable for this treatment and their cleanliness does not generate acid corrosion problems other than the relatively mild effects of dissolved oxygen and carbon dioxide.

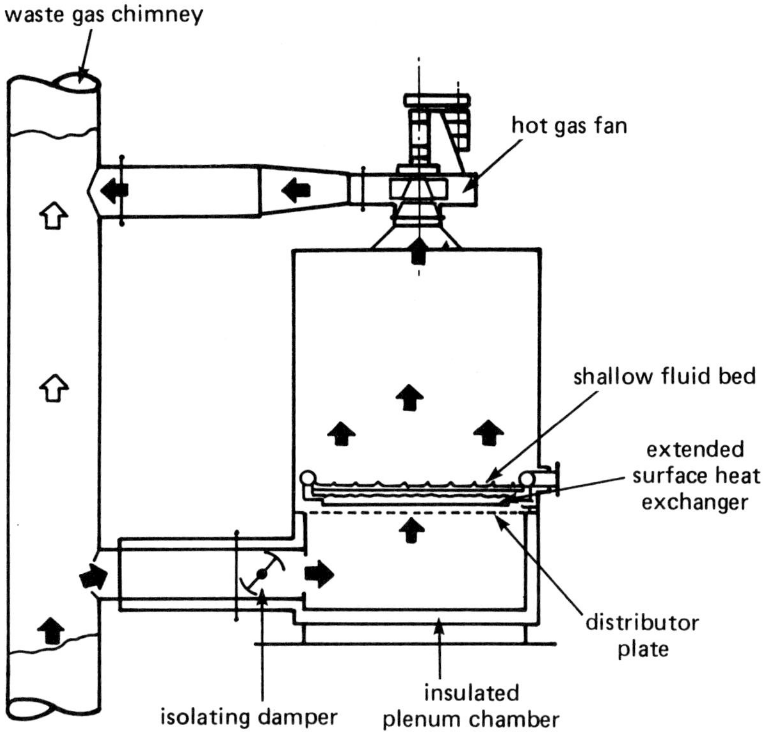

Fig. 7.60 Fluidised bed heat exchanger

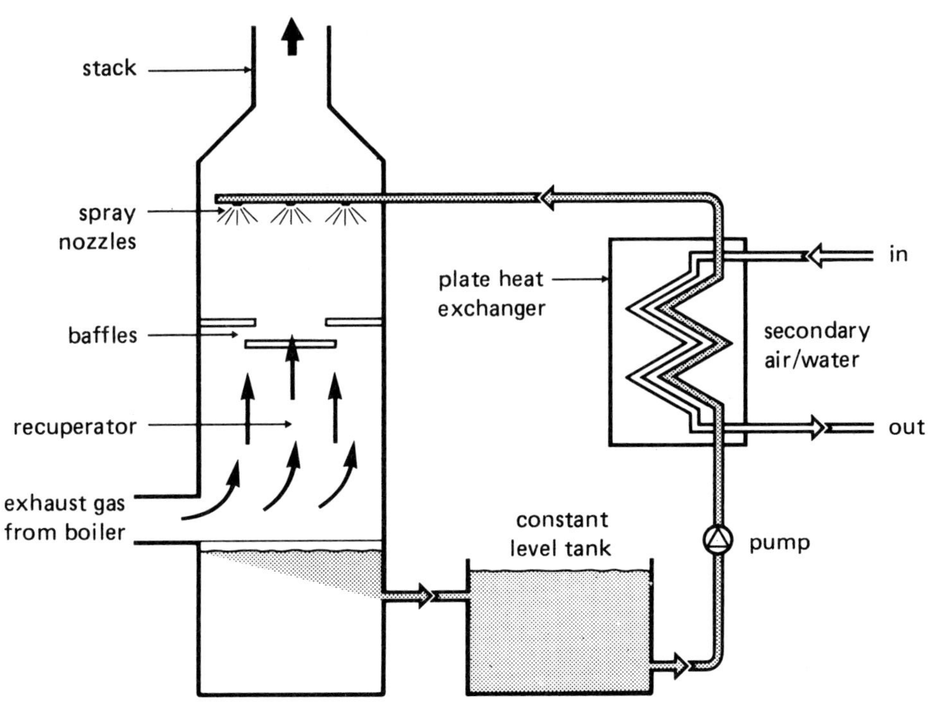

Fig. 7.61 Spray recuperation

The spray recuperator or condenser (Figure 7.61) is basically a vertical tower usually of stainless steel construction, which acts as the chimney for the installation. Integral components are spray nozzles, a liquid pipework system and inclined baffles. On a boiler system, the exhaust gases are admitted to the tower at a low level and rise towards the outlet coming into contact with a counter-current cascade of water which cools them to below their dewpoint. The sensible and much of the latent heat contents are transferred to the falling water, which collects as warmed water in a constant level tank. The tank may contain filter elements to trap any solids which may be present to prevent blockage of the system. Compensation for increase in the water content of the system due to the condensed water vapours is accommodated by an integral overflow device.

The warmed water is pumped either through a secondary heat exchange unit, often a liquid/liquid plate heat exchanger, or in some circumstances is used directly. On completion of its heating task, the water from the exchanger is returned to the spray nozzles at the top of the recuperator where another cycle commences. Where the water is used directly the increase in water content due to the condensation process in the recuperator can reduce the overall water consumption of the premises, although additional water treatment may be required.

The degree of heat recovery using a spray recuperator depends on the properties of the exhaust gas and the utilisation temperatures required from the secondary heat transfer system. The lower the utilisation temperature the greater the potential for recovery. Spray recuperators can supply water to the secondary heat transfer unit at temperatures up to 55 °C. They have been used on a range of applications in conjunction with gas fired boilers, for example boiler feedwater preheating, air heating, underfloor heating and swimming pool water heating. In dairies the low grade heat can be elevated by heat pump to provide hot water for bottle washing.

7.7.4 Waste Heat Boilers

Waste heat boilers are used to recover exhaust heat from metallurgical and chemical processes, incinerators, gas turbines and other prime movers. They are situated at the exhaust gas outlet from the plant and generally produce steam for process, space heating or power generation. In most cases their sole energy supply is in the form of waste gases at high temperatures up to 1 200 °C but in some instances a supplementary firing facility is installed. This will compensate for periods of reduction in waste heat availability. Figure 7.62 illustrates the use of a twin-tube boiler in incineration plant with one of the tubes fitted with a fuel fired burner which is used during troughs in the availability of waste for incineration.

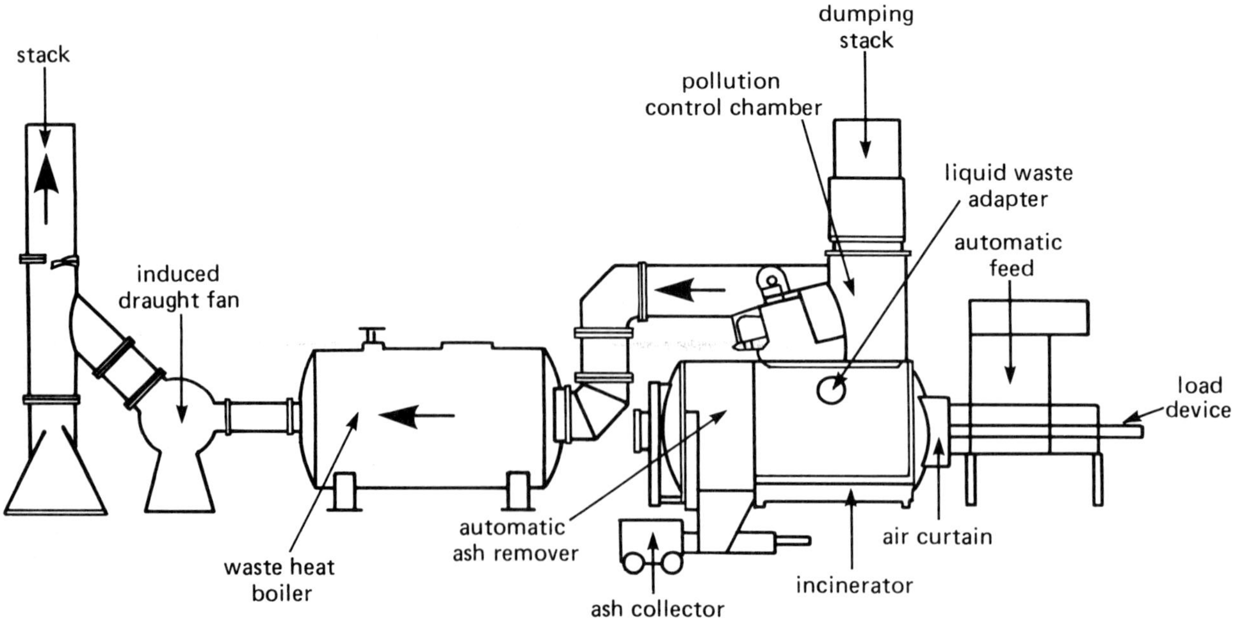

Fig. 7.62 Waste heat boiler attached to an incinerator

Waste heat boilers may be installed as the primary recovery technique or may be used in a secondary role for example accepting the exhaust from a static regenerator. Similarly there is often scope for further heat recovery after the boiler. This can take the form of an economiser or an air preheater to reduce the flue gas temperature further. The main conditions which limit the efficiency that can be achieved by a waste heat boiler are the acid dew point (where the waste gases contain sulphur oxides or other acidic constituents), and the steam or hot water temperature required. Two principal waste heat boiler designs are available, namely water tube and fire tube (or shell).

Water tube boilers

Water tube boilers are used when high steam pressures and evaporation rates are required and also when a low gas-side pressure drop is necessary. They may be fabricated with extended tubes to reduce the gas side heat transfer resistance where gas velocities are limited.

Natural and forced water circulation units are available, typically including superheater and economiser sections. A low pressure steam coil may also be fitted after the ecnomiser to provide steam for de-aeration plant or for feedwater preheating.

Natural circulation boilers (Figure 7.63) operate with vertical flows. The motive force is provided by the density difference between the water in the downcomer(s) and the hotter less dense steam/water mixture in the risers. Circulation is self-regulating and high rates can be achieved. Between one and five gas passes can be incorporated by use of baffles. Their simplicity of design allows for access for inspection and tube replacement and sootblowers or vibrators can be fitted to remove deposits where necessary.

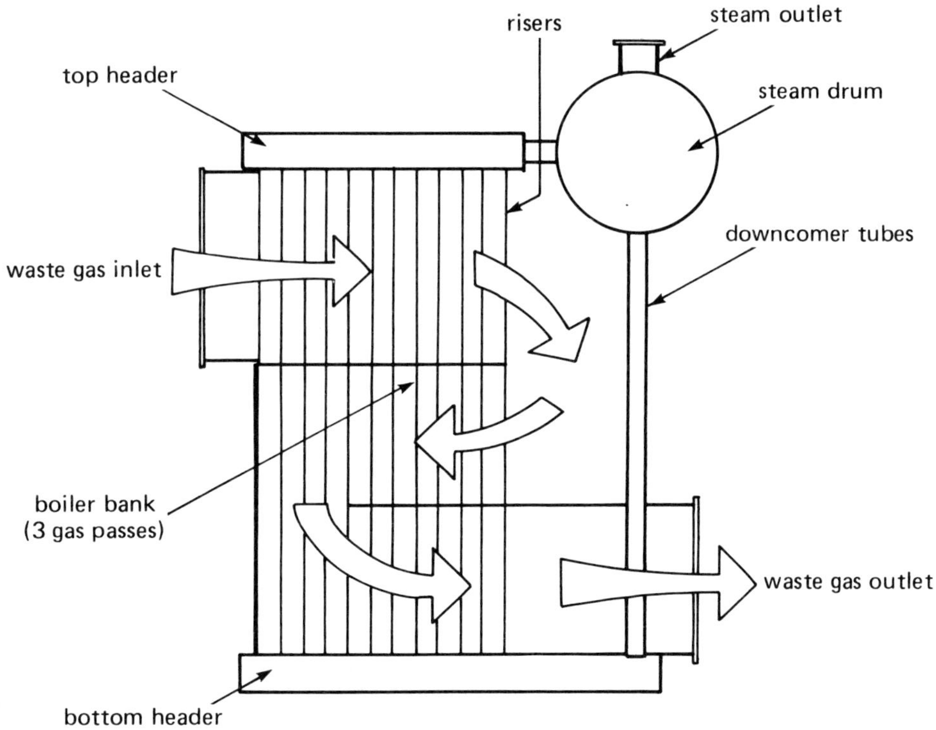

Fig. 7.63 Natural circulation water tube waste heat boiler

Forced circulation boilers (Figure 7.64) allow higher water side heat transmission rates to be achieved. This results in an arrangement of smaller diameter, lighter tubes usually of carbon steel which can be horizontally mounted. The water content is lower than for natural circulation types allowing more rapid start-up times. The plant layout will also influence the choice between natural and forced circulation boilers.

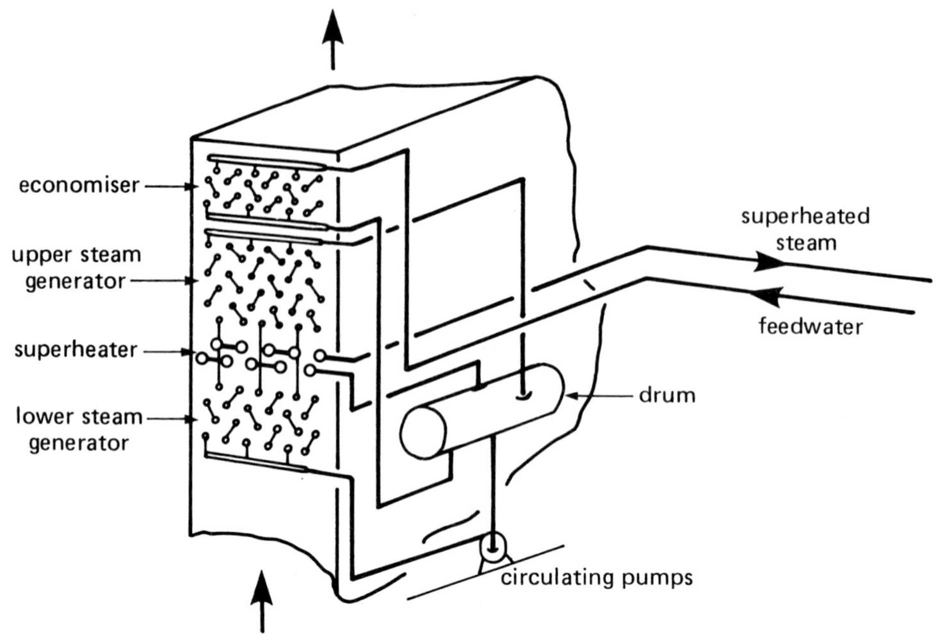

Fig. 7.64 Forced circulation water tube boiler

Fire tube waste heat boilers consist of a shell containing a bundle of steel tubes, sometimes stainless, welded into tube end plates (Figure 7.65). The hot exhaust gases pass through the tubes transferring heat to the surrounding water to generate steam or hot water supplies. The units are usually mounted horizontally in a single pass arrangement although twin pass horizontal and single pass vertical types are available. Compared with water tube units, they are less susceptible to deposition and blockage, and are more easily cleaned.

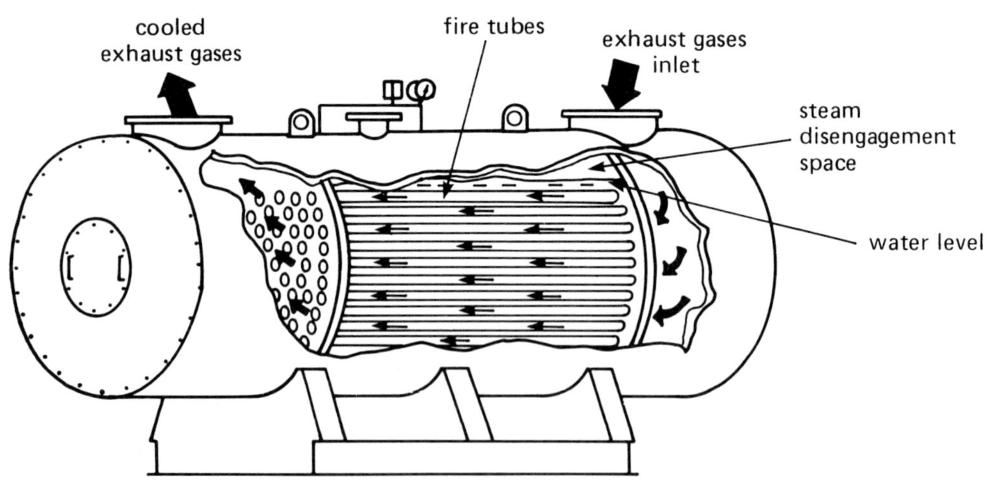

Fig. 7.65 Fire tube waste heat boiler

The steam produced can be collected in a steam drum for distribution and also to allow variations in the heat load during start-up to be absorbed. Where the quality of the steam is not critical the steam drum can be omitted and a steam disengagement space above the tube bundle incorporated.

Fire tube boilers are generally limited to gas inlet temperatures of 1 000 °C and 30 bar steam pressures. Special designs can be obtained for higher pressures. They are generally cheaper than water tube boilers for steam rates of up to 30 000 kg per hour.

The use of waste heat boilers can affect the draught conditions through the primary heating plant if careful consideration is not given to the resistance imposed by the boiler. The use of forced or more commonly induced draught fans will overcome resistance problems and also assist heat transfer. The latter will be increased by the higher gas velocities and turbulence generated which will minimise gas film thickness on the tubes.

7.8 METHODS WHICH UPGRADE RECOVERED HEAT

Waste heat is often shed at grades too low for effective re-use through conventional heat recovery systems. Some methods are available, however, which enable the recovered heat to be upgraded and hence re-used at a higher temperature than that at which it was generated. The 'catch' with these methods is that they require the expenditure of a small amount of high grade heat; this when combined with the low grade waste heat, produces a recovered heat stream of an intermediate grade.

7.8.1 Heat Pumps

The most common heat pump operating cycle employs vapour compression which uses the mechanical energy for temperature elevation. There are alternative cycles such as absorption involving a physical chemistry process for temperature elevation, together with modified thermodynamic cycles and thermo-electric processes which are only at the development stage. Only the vapour compression heat pump is considered here because units of this type are more readily available and are likely to be no more costly than the equivalent absorption units.

Vapour compression heat pumps have been designed to operate according to a number of thermodynamic cycles such as those due to Rankine, Otto, Brayton or Stirling. The ideal Carnot efficiency is most closely approached by the Rankine cycle and hence this offers the best performance prospects. Consequently, currently available electric motor and gas engine driven heat pumps employ this cycle. The components used are readily available from air conditioning and refrigeration practice.

The vapour compression cycle

The vapour compression cycle has the advantage that it is proven technology, offering a relatively high efficiency over a reasonable range of temperatures. The basic system is shown in Figure 7.66 and consists of a compressor, expansion device and two heat exchangers. The cycle is closed and the working fluid is commonly a refrigerant such as Freon operating in two phases. A description of the operation, starting with the evaporator, is as follows:

- The temperature of the working fluid in the evaporator is lower than the temperature of the source. Heat is therefore transferred to the working fluid causing it to evaporate.
- The evaporated vapour is then drawn into the compressor and, by the input of shaft power, is compressed to a higher pressure and temperature.
- The vapour leaving the compressor and entering the condenser is at a temperature higher than that of the load to be heated. During its passage through the condenser, therefore, heat is rejected to the load causing the working fluid to condense.
- The high pressure liquid then passes through an expansion device where its pressure and hence temperature are lowered. Low temperature fluid then enters the evaporator to complete the cycle.

A range of refrigerant working fluids are available commercially such as R22, R12 and R114 for low, medium and high temperature applications respectively. Although some heat pumps have operated at temperatures in excess of 120 °C, most are at present limited to less than about 70 °C at the condenser.

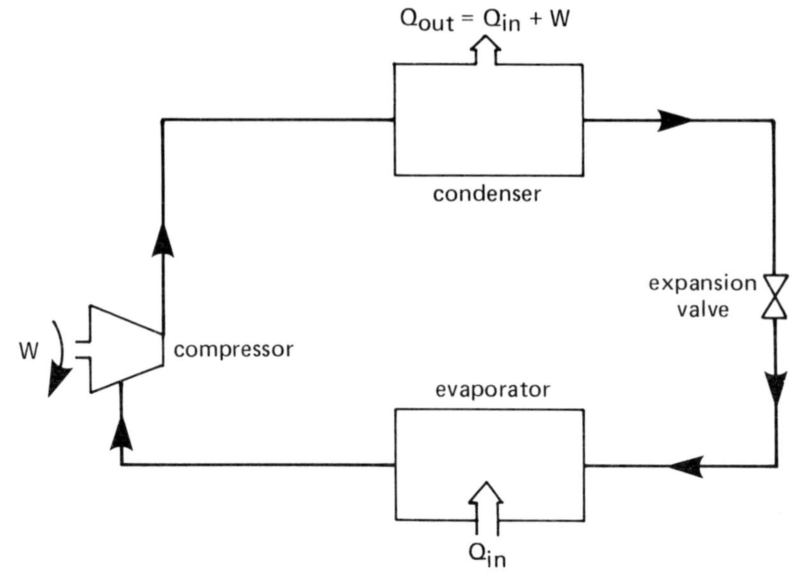

Fig. 7.66 Components used in the vapour compression heat pump

In practice, a heat pump contains more components than are shown in the diagram although these do not greatly affect the basic heat pump cycle. A number of compressor types are available such as reciprocating, sliding vane, screw and centrifugal. Reciprocating compressors are currently the most common but other types are becoming more popular for heat pump applications.

While heat pumps are commonly driven by electric motors, an alternative is to use reciprocating, spark ignition, internal combustion engines operating on natural gas. Gas engines can offer advantages over electric motors, particularly if waste heat is recovered from the engine cooling jacket and exhaust system (refer to Chapter 13 for further details).

Heat pump performance

Heat pump performance has historically been expressed as the Coefficient of Performance (COP) which is defined as the heat output from the condenser divided by the power required to drive the compressor. COP is of limited value when considering the performance of a heat pump installation since it does not take into account the efficiency of the compressor drive, power for ancilliaries or heat supplied by the engine. For this reason, another term, the Performance Effectiveness Ratio (PER) is used and this is defined as:

$$\text{PER} = \frac{\text{useful heat output from the complete installation}}{\text{energy input to the installation}}$$

The values of COP and PER for both electric motor heat pumps and gas engine heat pumps vary with the temperature difference between the load and source. PER has the higher value for the gas engine type since part of the temperature lift is being provided by the engine heat recovery. COP is the greater of the two ratios for the electric motor type but it would need to be three or four times as great, as is the typical cost of electricity in relation to that of gas, to achieve the same running costs.

Applications

Due to the relatively high capital cost of heat pump installations, they are only economically attractive in applications where long hours of operation and moderate temperature differences can be assured.

Heat pumps have been applied as a means of heat recovery where a low grade source waste heat is at a temperature below that required for re-use in part of a process. As discussed, the heat pump is used to elevate temperatures, for example in a dairy where milk cooling yields low temperature water which can be upgraded for bottle washing. The maximum load temperature from the condenser for a heat pump is around 70 °C when using ambient air at -1 °C as the source. However, where warm effluent is available at say 70 °C then condensing temperatures of 120 °C are achievable. Heat pump recovery systems have also been installed on heating and dehumidification applications, for example in drying processes and swimming pools.

A specific example is a malt drying kiln (Figure 7.67) where the source and recipient may be some distance apart and where the heat extracted by the evaporator is principally latent heat. The heat collected at the evaporator is elevated by the compressor to a temperature of 60 °C at the condenser. The resultant warmed supply air temperature to the kiln is subsequently raised to 72 °C by exhaust and water jacket heat recovery.

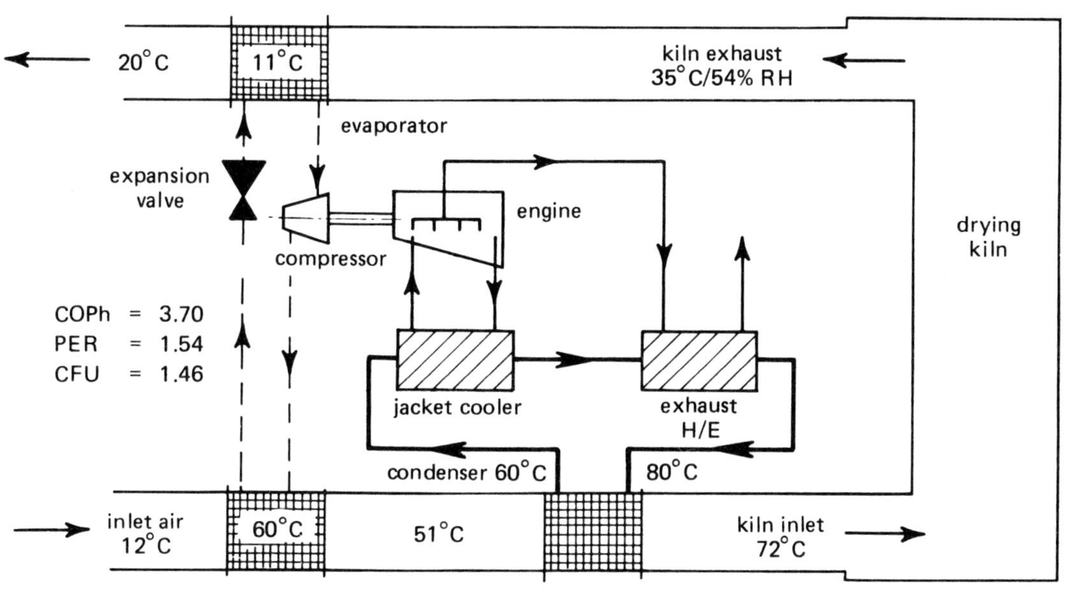

Fig. 7.67 Heat recovery on drying kiln using a gas engine driven heat pump

Heat pump applications are often used in conjunction with other heat recovery techniques (e.g. run-around coils, plate heat exchangers) to increase the overall effectiveness of the energy system.

7.8.2 Vapour Recompression

Drying, evaporation and many other industrial processes reject waste vapour, frequently steam, either to atmosphere or a condenser and for maximum energy efficiency it is important to re-use this heat wherever possible. Vapour recompression is a method of taking waste vapour and upgrading it to a useful level for process heating. Plant should be fully assessed and other methods of heat recovery such as heat exchange should also be considered in order to determine the most appropriate heat recovery system.

Thermocompression

Thermocompression is achieved by the use of high pressure steam in a venturi to compress waste vapour. The resultant combined flow will have a pressure (and hence saturation temperature) higher than that of the waste vapour and hence will be more suitable as a heat source. The heat content of the high pressure 'driving' steam is not lost as it forms a part of the combined exit stream. Figure 7.68 shows a typical installation where thermocompression is used on the Machine Glaze (M.G.) cylinder of a paper tissue machine.

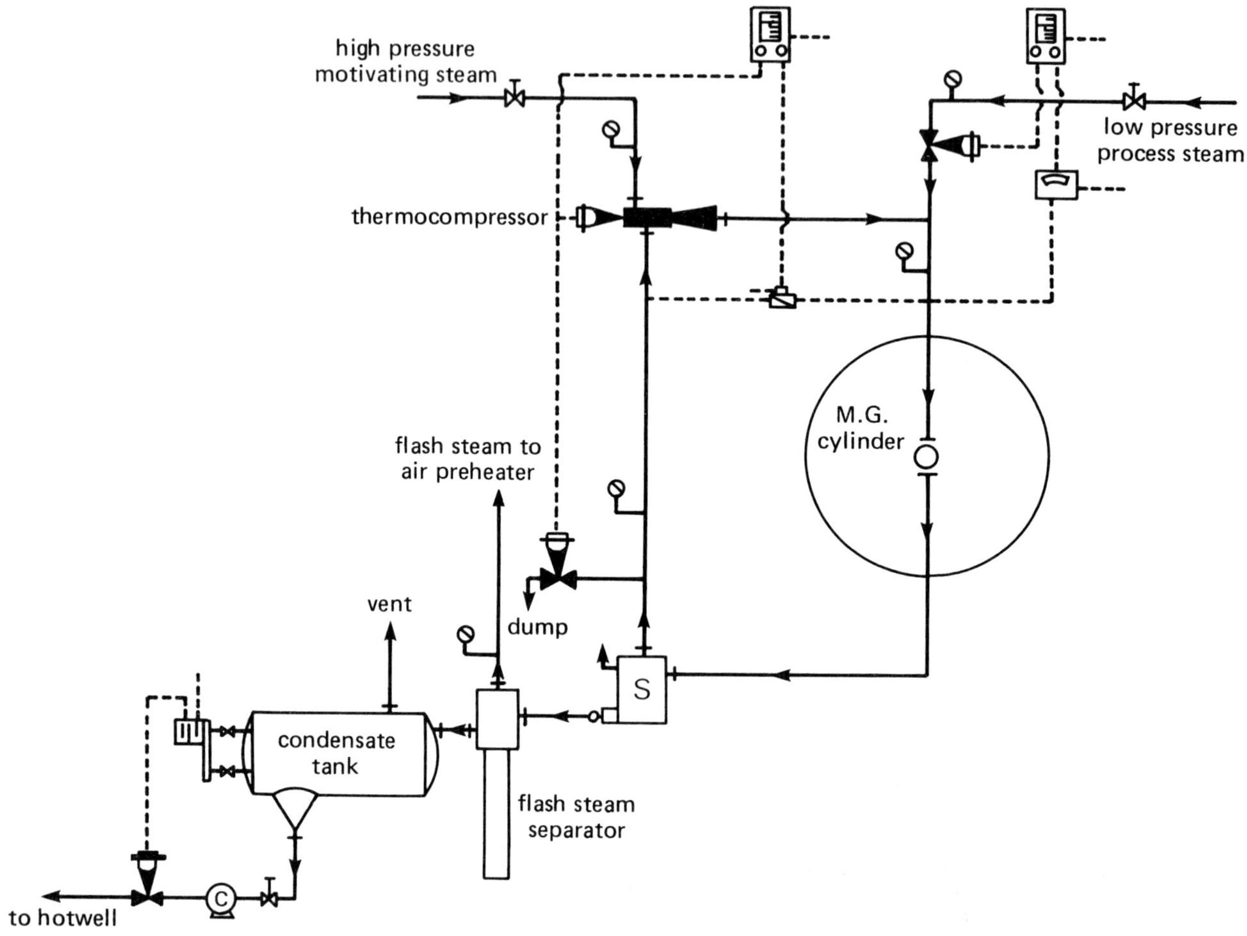

Fig. 7.68 Thermocompression on paper machine cyclinder

Exhaust steam from the cylinder passes via a condensate separator to the thermocompressor, is raised to the inlet pressure and re-introduced into the cylinder. Thermocompression is widely used in the paper industry on both single and multi-cylinder machines.

Mechanical vapour recompression (MVR)

As an alternative to thermocompression, MVR uses a mechanical compressor to boost the pressure (and hence saturation temperature) of the waste vapours. The compressor may be driven by a variety of engines (e.g. electric, gas internal combustion, steam turbine) although all require a high grade heat input as is demanded by this form of heat recovery.

The two basic forms of mechanical vapour recompression system are the open or direct cycle and the closed or indirect cycle, which are shown schematically in Figures 7.69 and 7.70 respectively.

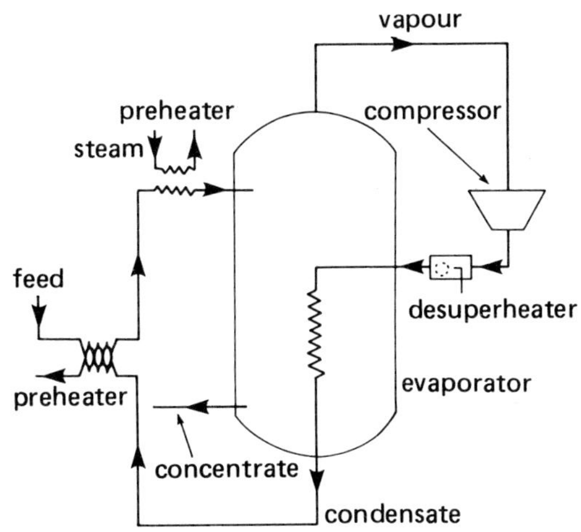

Fig. 7.69 MVR direct cycle

In the open cycle system the waste vapour is taken directly to the compressor, the resulting higher pressure vapour is then used in the heating tubes.

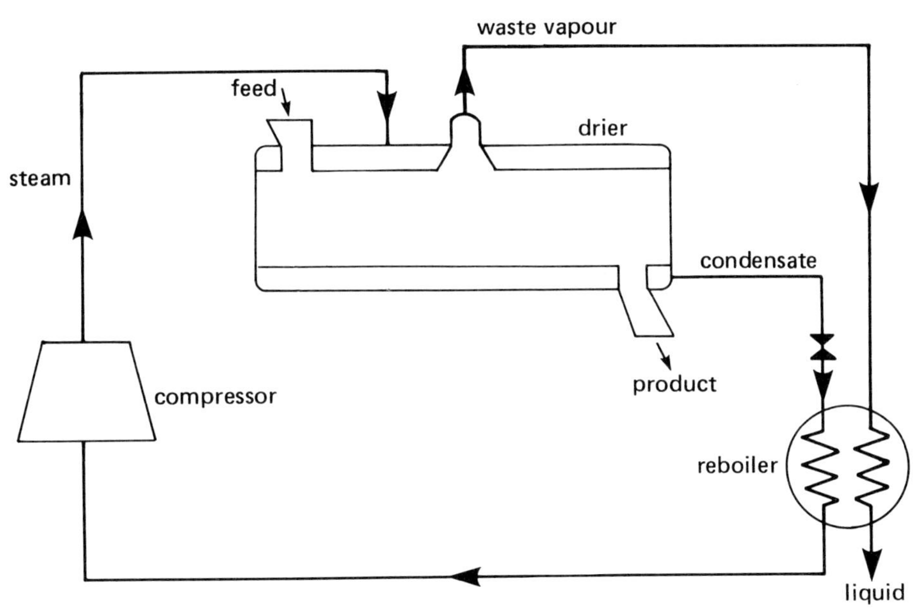

Fig. 7.70 MVR indirect cycle

The closed cycle system makes use of a heat exchanger (re-boiler) designed to accept foul or non-condensible streams such as air. Waste vapour from the process is condensed in the heat exchanger to produce clean steam at slightly sub-atmospheric pressure from the condensate from the process. This reduced pressure steam is then compressed.

Most work on MVR systems was originally carried out by the then Electricity Council but there is clearly potential for gas engine driven compressor systems particularly since costs could be reduced further by utilising the heat generated by the engine. Engine waste heat could be used to preheat condensate or to provide supplementary low pressure steam. As with thermocompression, MVR has found a range of applications in the whisky, dairy and food industries, principally on evaporation/distillation plant.

7.9 SYSTEM APPLICATIONS

The preceding sections of this Chapter have discussed the technical and operating features of a range of heat recovery techniques, presenting some examples of typical applications. The engineer is normally presented with the reverse situation, however, in that he has to select a heat recovery technique to fit an existing process or piece of equipment.

This section presents a range of common processes and describes briefly some of the *retrofit* heat recovery techniques which have been applied to them, in most cases with a reference back to the appropriate section in which the technique was described.

7.9.1 Metallurgical Furnaces

Being high temperature processes, metallurgical furnaces can be a source of high grade waste heat. Historically, static regenerators and large metallic recuperators have been used as part of the original installations. These can frequently benefit from the retrofit of secondary heat recovery plant (for example, rotary regenerators) to further reduce flue gas temperature.

A range of retrofit primary heat recovery techniques can also be applied as illustrated by the following case histories.

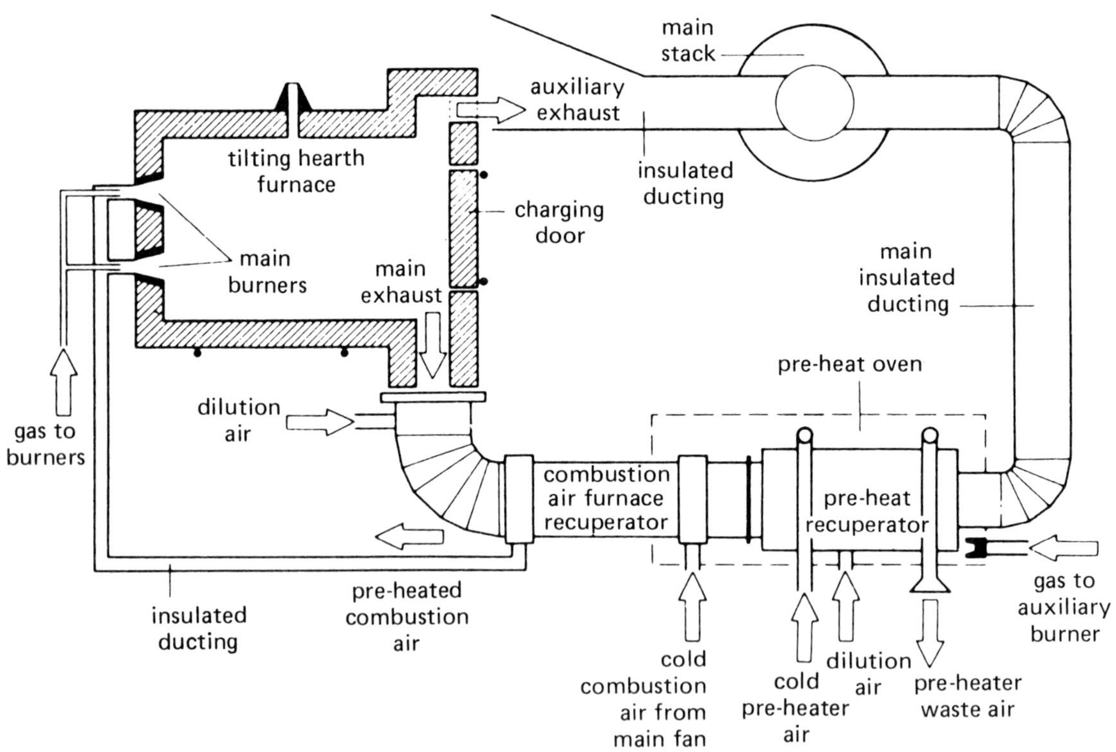

Fig. 7.71 Metallic recuperators applied to aluminium melting furnace - ref. 7.6.1

Heat source:	Exhaust gases at 1 100 °C
Technique:	Primary recovery – nickel chromium radiative recuperator
	Secondary recovery – mild steel convective recuperator
Heat recipient:	Primary recovery – combustion air at 450 °C
	Secondary recovery – preheat oven air at 300 °C
Energy saving:	15 190 GJ/a
Payback period:	18 months.

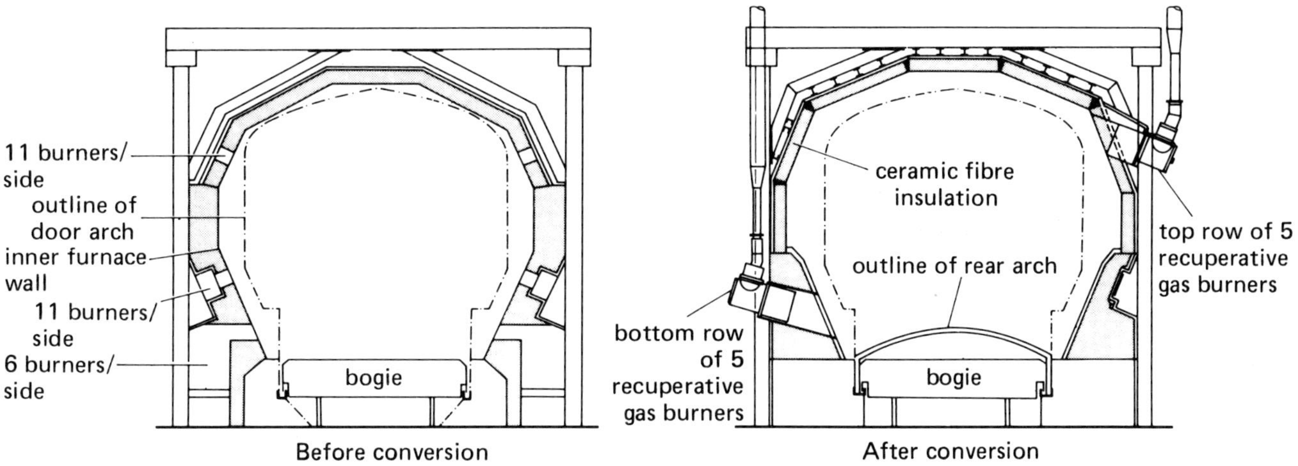

Fig. 7.72 Recuperative burners applied to a heating forge furnace - ref. 7.6.2

Heat source:	Flue gases at 1 300 °C
Technique:	Recuperative burners
Heat recipient:	Combustion air
Energy saving	38 440 GJ/a
Payback period:	20 months

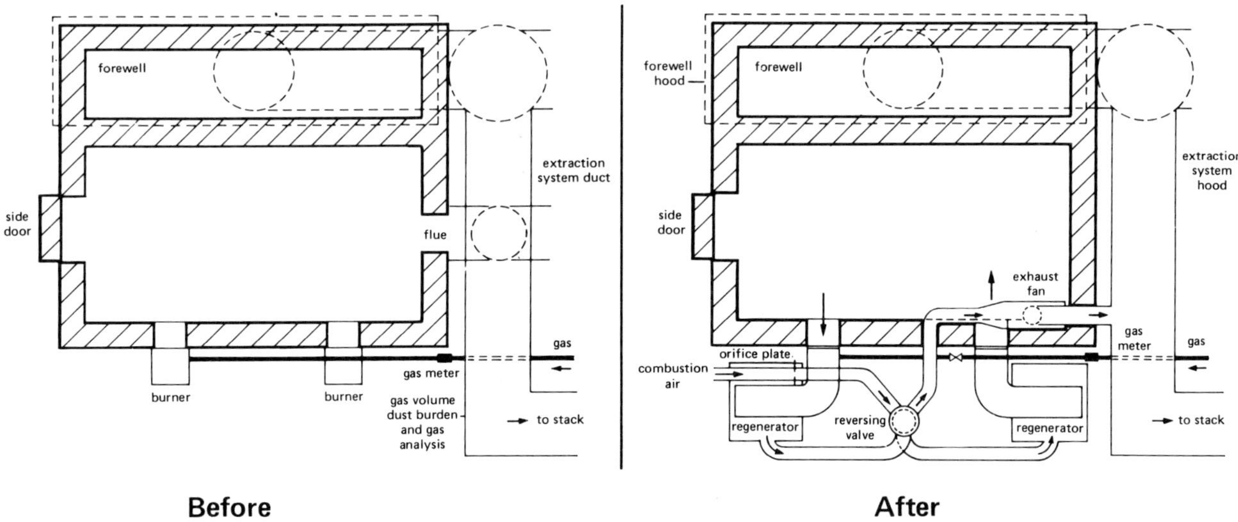

Fig. 7.73 Regenerative burners applied to an aluminium bulk melting furnace - ref. 7.6.4

Heat source:	Exhaust gases
Technique:	Regenerative ceramic burners
Heat recipient:	Combustion air
Energy saving:	12 900 GJ/a
Payback period:	1.1 years

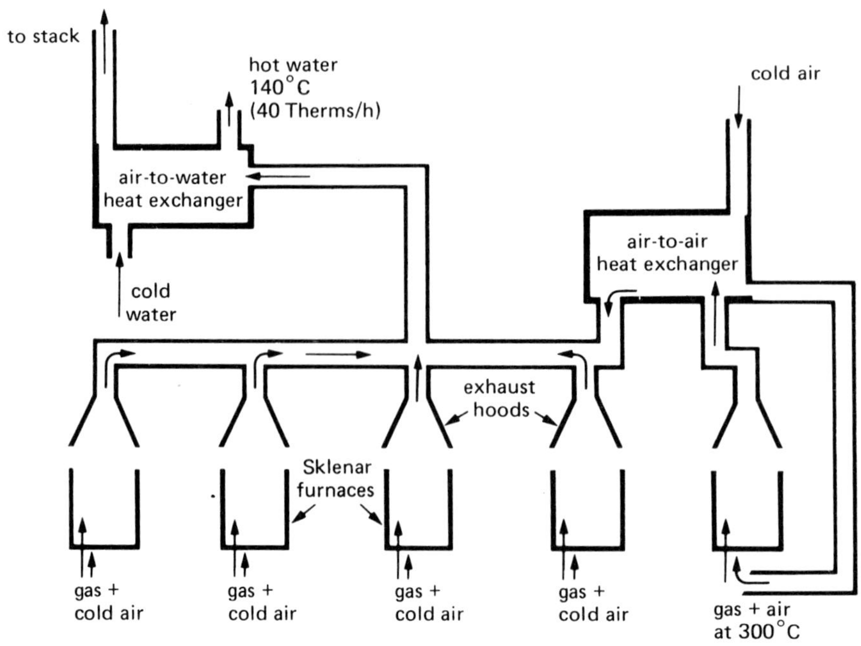

Fig. 7.74 Heat exchangers applied to copper melting furnaces - ref. 7.6.8 and 7.7.1

Heat source:	Exhaust gases at 400 °C	
Technique:	Primary	– heat pipe exchanger
	Secondary	– finned tube exchanger
Heat recipient:	Primary	– combustion air at 300 °C
	Secondary	– Hot water at 140 °C
Energy saving:	Primary	– 12 000 GJ/a
	Secondary	– 11 850 GJ/a
Payback period:	Primary	– 32 weeks
	Secondary	– 2.9 years

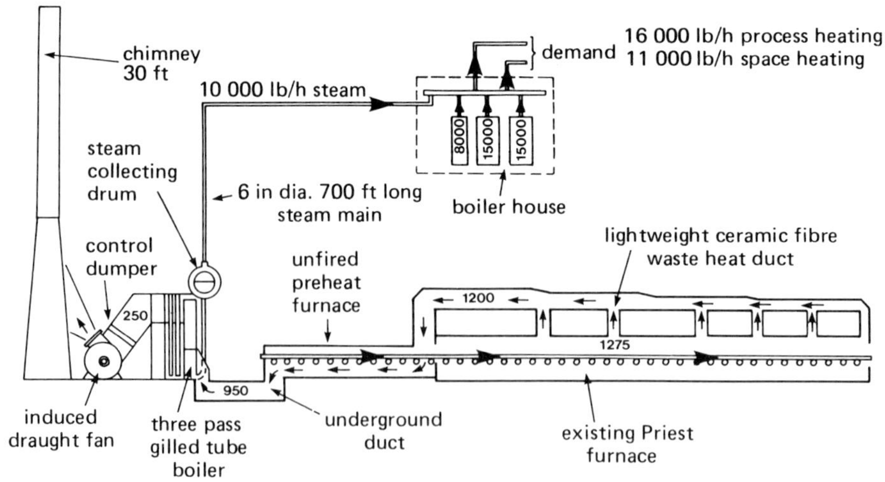

Fig. 7.75 Waste heat boiler applied to a planetary mill furnace - ref. 7.7.4

Heat source:	Exhaust gases	
Technique:	Primary	– load preheat
	Secondary	– water tube waste heat boiler
Heat recipient:	Primary	– steel slabs
	Secondary	– steam generation
Energy saving:	79 500 GJ/a	
Payback period:	1.5 years	

7.9.2 Glass Furnaces and Lehrs

As in the metals industries, glass melting furnaces have for a long time been built with static regenerators. Increasingly, however, secondary heat transfer is being practised usually via a waste heat boiler producing steam or high pressure hot water for power generation or space heating respectively.

Recent retrofits of primary heat recovery techniques have included ceramic recuperators for combustion air preheating and also the use of recuperative burners. The latter have, in some instances, achieved payback periods of under 6 months.

Annealing lehrs, whilst operating at lower temperatures than melting furnaces, still offer scope for waste heat recovery. One approach adopted, which has given a 2 year payback, is to extract hot air from the lehr and duct it directly to an adjacent warehouse for space heating purposes, as shown in Figure 7.76.

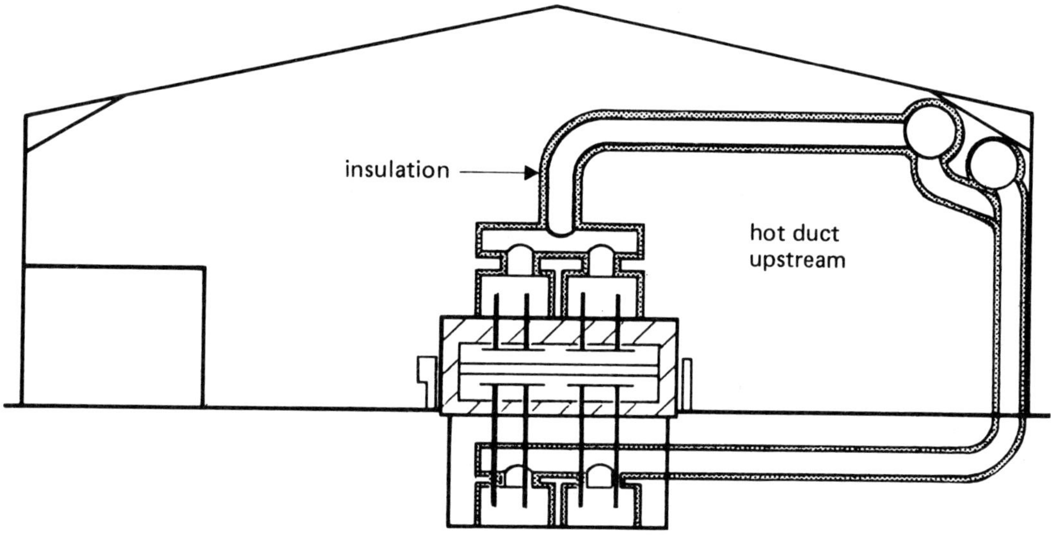

Fig. 7.76 Lehr heat recovery system

7.9.3 Ceramic Industry Kilns

Historically the major waste heat recovery techniques employed by the ceramics industry have been load preheating and recuperation, normally on continuous and batch kilns respectively.

Many attempts have been made to recover waste heat from flue gases using waste heat boilers, metallic recuperators or rotary regenerators. Most have been unsuccessful due to a variety of causes, such as fouling both from products of combustion and volatiles in the ware resulting in low temperature corrosion. More recently, however, recuperative burners have started to find acceptance and are particularly useful on batch kilns. Ceramic recuperators have also shown themselves to be suitable for use in aggressive environments and an application is illustrated in Figure 7.77.

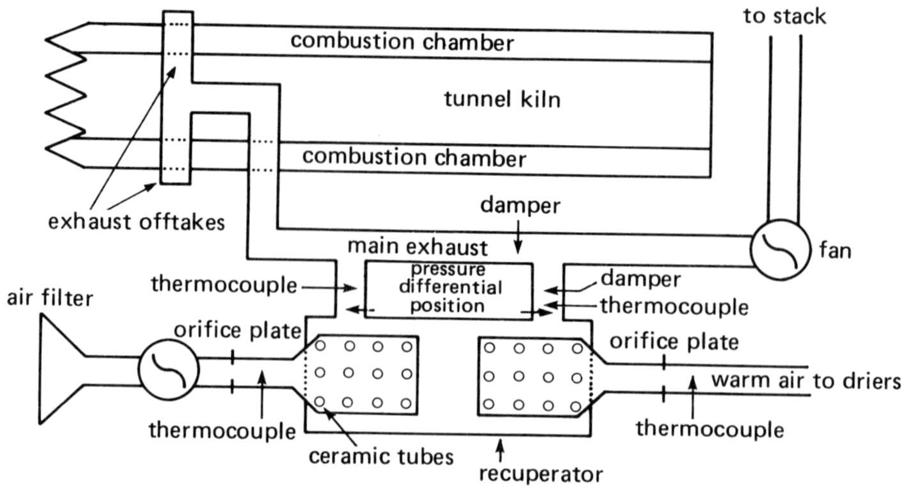

Fig. 7.77 Ceramic recuperator applied to a sanitaryware muffle kiln - ref. 7.6.1

Heat source:	Exhaust gases at 400 °C
Technique:	Glazed tube ceramic recuperator
Heat recipient:	Air supply to driers at 85 °C

7.9.4 Driers and Ovens

Driers and ovens are found in a wide variety of industries and their exhausts are characterised by a comparatively high latent heat content. A range of heat recovery techniques have been successfully applied to drier exhaust streams and a selection follows:

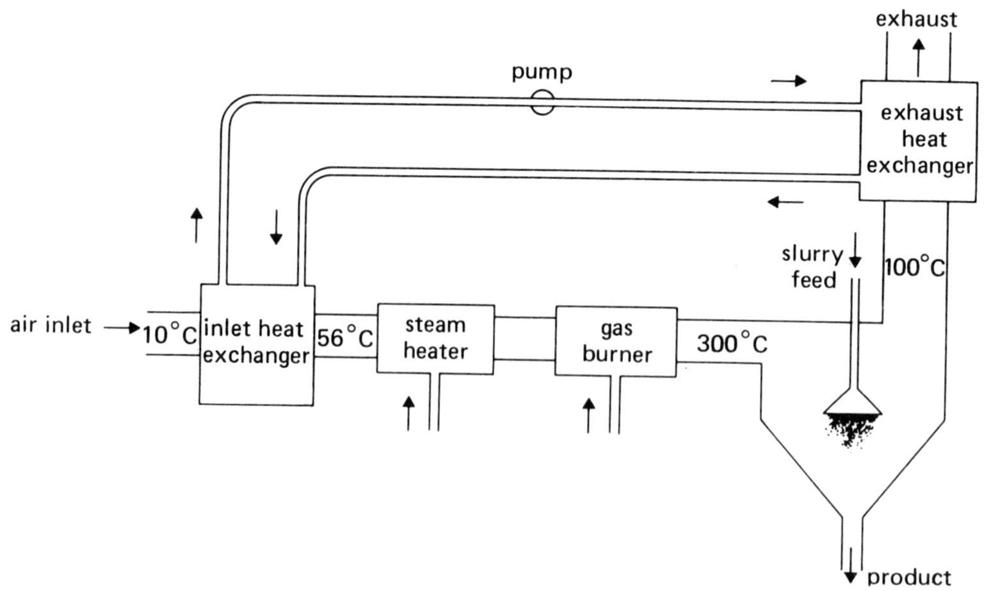

Fig. 7.78 Run-around coil applied to a dyestuff spray drier - ref. 7.6.7

Heat source:	Exhaust gases at 100 °C	Energy saving (predicted)	4 505 GJ/a
Technique:	Run-around coils	Payback period (predicted)	2 years
Heat recipient:	Combustion air at 56 °C		

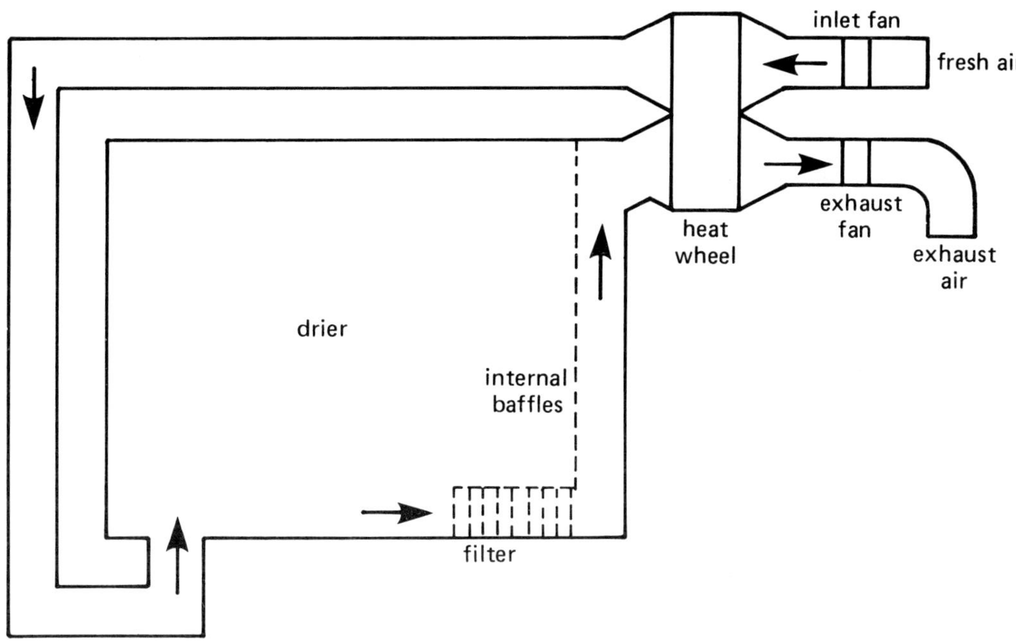

Fig. 7.79 Rotary regenerator applied to a leather drier - ref. 7.6.5

Heat source:	Exhaust gases
Technique:	Rotary regenerator
Heat recipient:	Drier supply air
Energy saving (predicted):	1 325 GJ/a
Payback period (predicted):	2.6 years

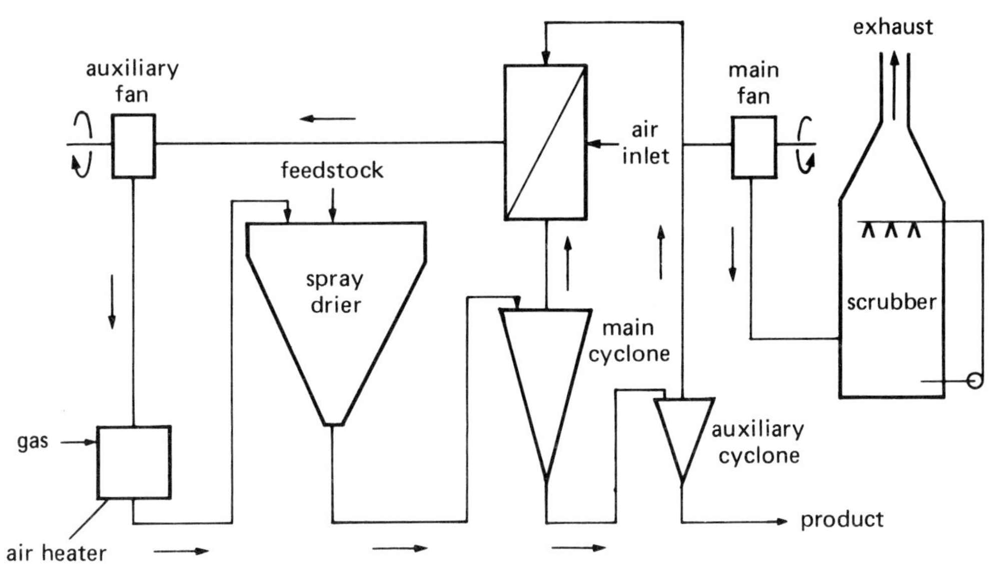

Fig. 7.80 Glass heat exchanger applied to a chemicals spray drier - ref. 7.6.1

Heat source:	Exhaust gases
Technique:	Glass tubed heat exchanger
Heat recipient:	Combustion air
Energy saving (predicted):	3 000 GJ/a
Payback period (predicted):	3.0 years

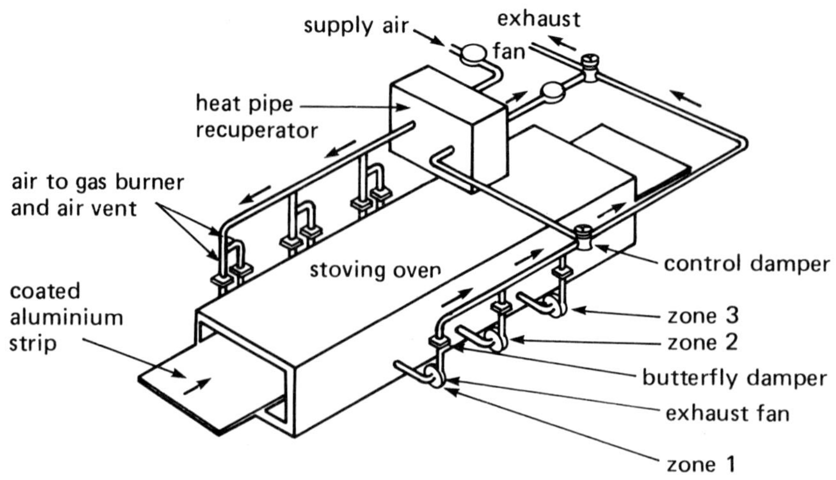

Fig. 7.81 Heat pipe recuperators applied to a paint stoving oven - ref. 7.6.8

Heat source:	Exhaust gases at 240 °C
Technique:	Heat pipe recuperator
Heat recipient:	Combustion air
Energy saving (predicted):	7 155 GJ/a
Payback period (predicted):	1.7 years.

7.9.5 Chemical Process Plant

The chemical processing industry generates a wide range of gaseous and liquid effluent streams which have the potential for waste heat recovery. So complex are the energy requirements of a large chemicals site that a new design technology has recently been developed called 'process integration'. This seeks to maximise the re-use of waste heat by the careful selection of process plant operating conditions. The technique has been pioneered, to good effect, by ICI. Whilst a detailed discussion of process integration methods is outside the scope of this chapter, there are plenty of more conventional retrofit heat recovery opportunities, two of which are illustrated below.

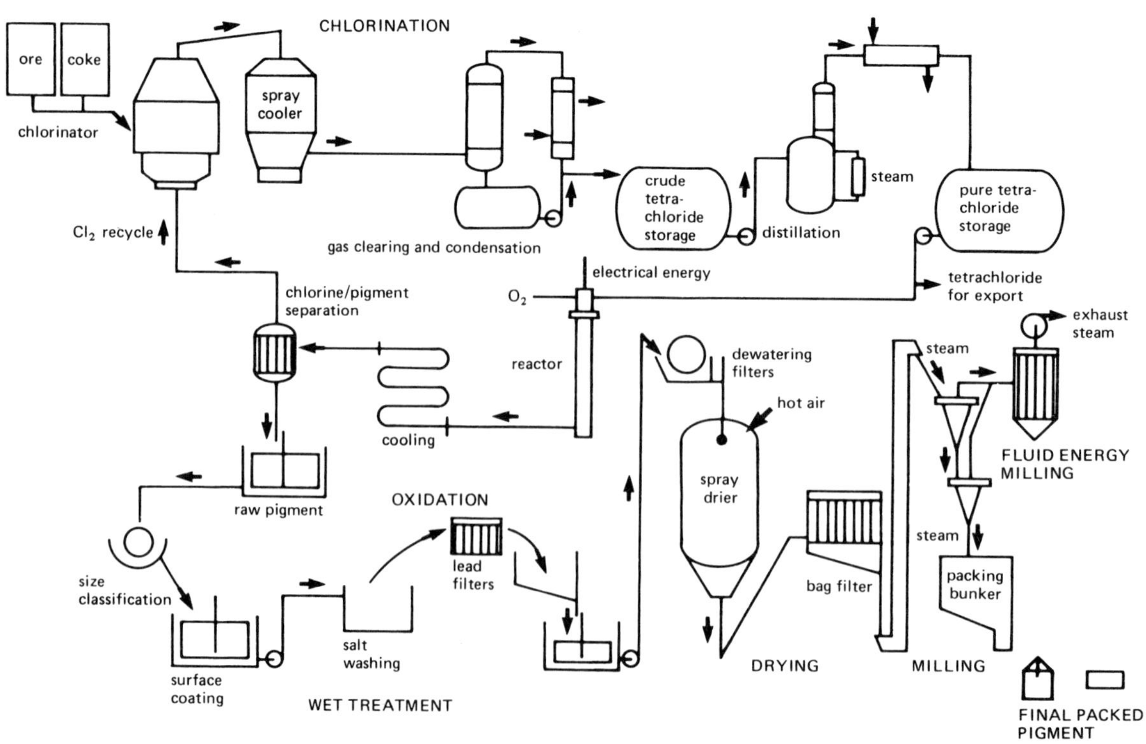

Fig. 7.82 Heat recovery from a heavily contaminated vapour stream - ref. 7.7.3

The exhaust from fluid energy mills used for grinding titanium oxide is a steam/hot air mixture contaminated with pigment particles. Conventional gas heat exchangers would become rapidly fouled. A novel system has therefore been devised whereby the stream is condensed in a spray condenser. The resultant dilute slurry is then passed through heat exchangers to preheat air and water.

Energy saving (predicted):	93 120 GJ/a
Payback period (predicted):	9 months

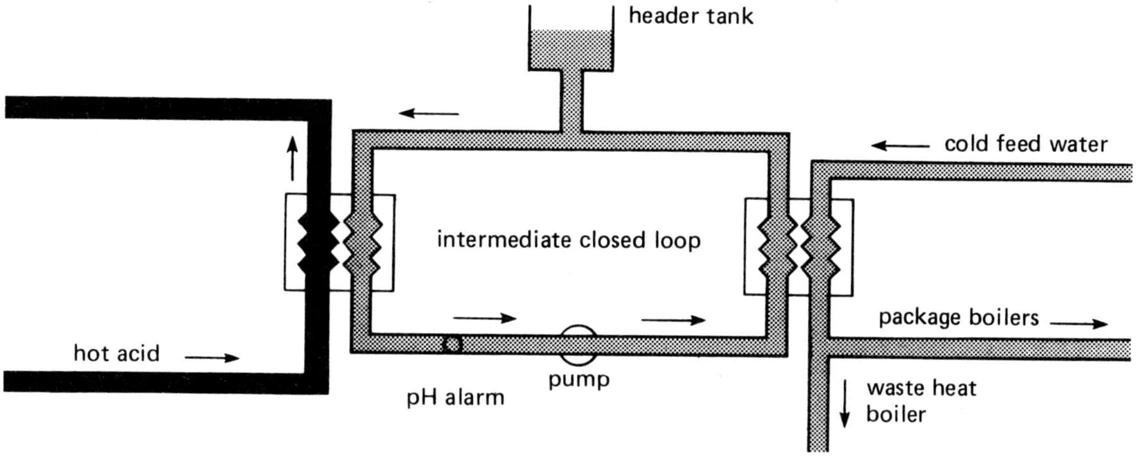

Fig. 7.83 Run-around coil applied to a liquid/liquid duty - ref. 7.6.7

Heat source:	Sulphuric acid at 90 °C
Technique:	Run-around coil (to reduce risk of cross-contamination) utilising Hastelloy plate heat exchangers
Heat recipient:	Boiler feedwater
Energy saving (predicted):	33 125 GJ/a
Payback period (predicted):	1.4 years.

7.9.6 Evaporators and Distillation Plant

Major users of distillation and evaporation processes include the chemicals and whisky industries. In both, there are opportunities for 'process integration' techniques, an example of which is now given.

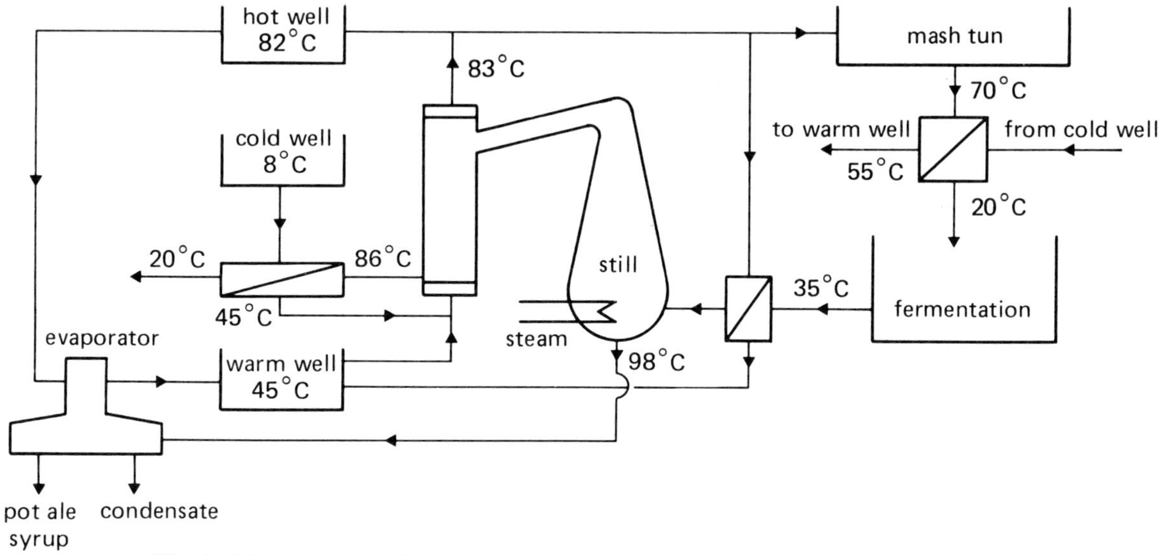

Fig. 7.84 Integrated heat recovery system in a malt whisky distillery

The process heat sources and recipients are fully integrated including the incorporation of a waste heat driven pot ale evaporator. The evaporator produces its own low grade heat which is returned to the process. The batch nature of the distillation process, coupled with the continuous operation of the evaporator, means that carefully sized and operated thermal stores (hot and warm wells) are needed.

Energy saving (predicted): 37 100 GJ/a
Payback period (predicted): 1.6 years

Vapour Recompression

Vapour recompression is finding many applications in the recycling of vapours from stills and evaporators. This example features thermocompression but mechanical vapour recompression is also beginning to find favour.

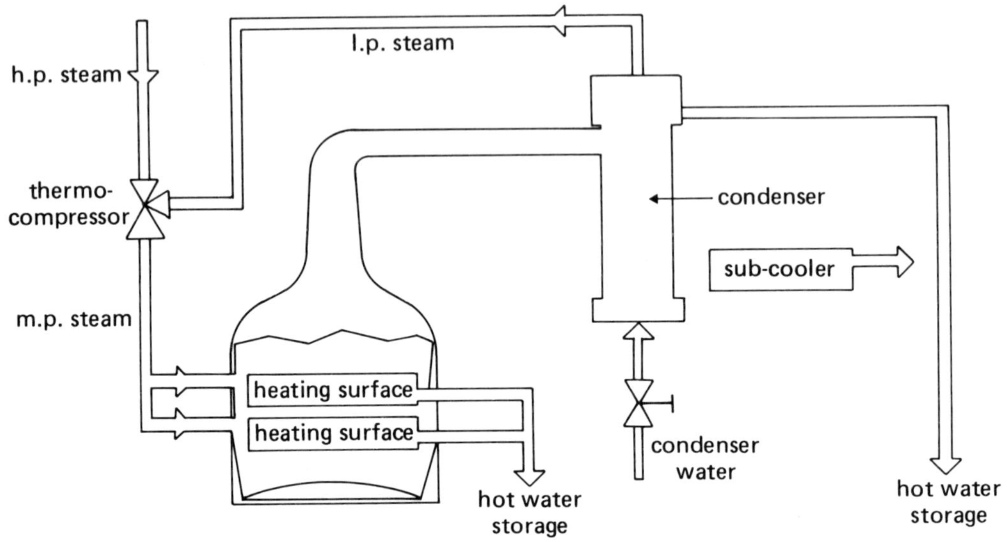

Fig. 7.85 Thermocompression applied to a whisky still - ref. 7.8.2

Energy saving (predicted): 26 500 GJ/a
Payback period (predicted) 2.6 years.

7.9.7 Paper and Board Machines

In the paper industry, large quantities of warm moist air are extracted from hoods over the drying sections of paper machines. This heat has been recovered to warm water using spray recuperators or can be used to generate preheated air for drying as shown below.

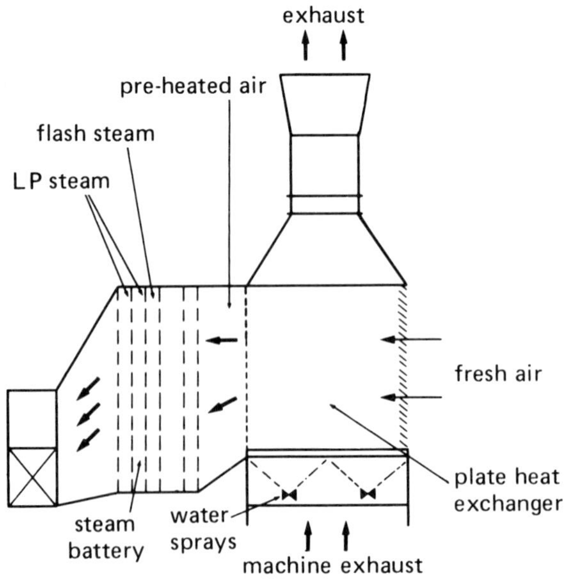

Fig. 7.86 Gas to gas plate heat exchanger applied to a paper machine - ref. 7.6.6

An unusual feature of this application is the use of water sprays to saturate the exhaust stream prior to its entry into the exchanger. This causes constant condensation within the exchanger and prevents fouling build-up of fibrous material.

Energy saving (predicted): 31 800 GJ/a
Payback period (predicted): 9 months

7.9.8 Washing plant

Washing and sterilising plants invariably produce warm liquid effluent streams even though their magnitude may be limited by careful design (i.e. making use of counterflow and recycle techniques). Heat may be recovered from the effluent using heat exchangers (if the temperature is high enough) or heat pumps if it is necessary to upgrade the waste heat.

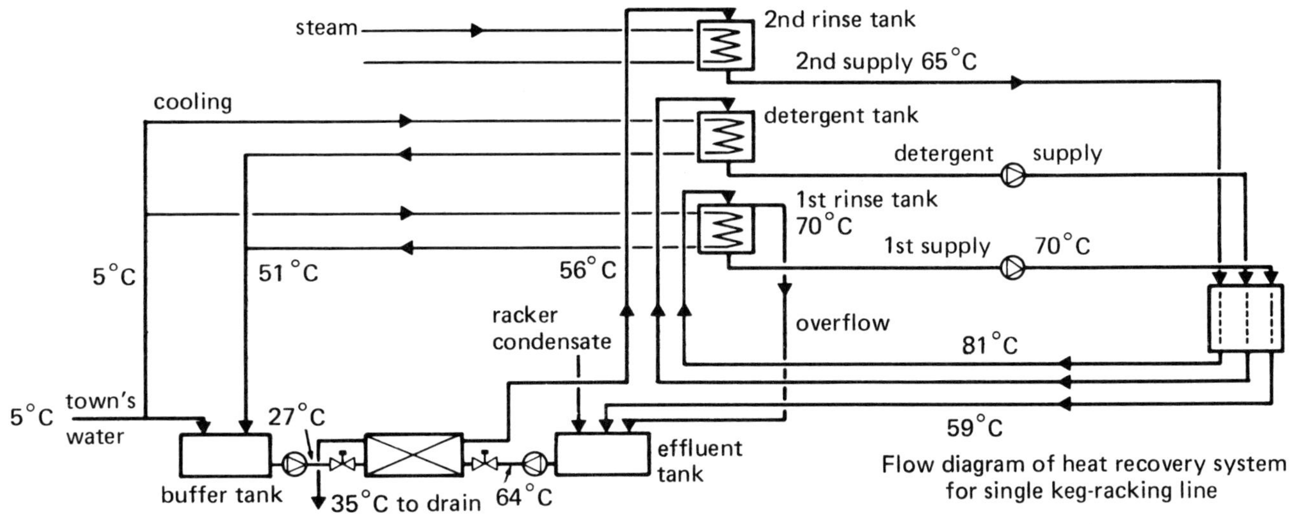

Fig. 7.87 Liquid to liquid plate heat exchanger applied to a keg washing machine - ref. 7.5.3

Heat source: Liquid effluent at 64 °C
Technique: Liquid to liquid plate heat exchanger
Heat recipient: Rinse water at 56 °C
Energy saving (predicted): 6 860 GJ/a
Payback period (predicted): 2.5 years

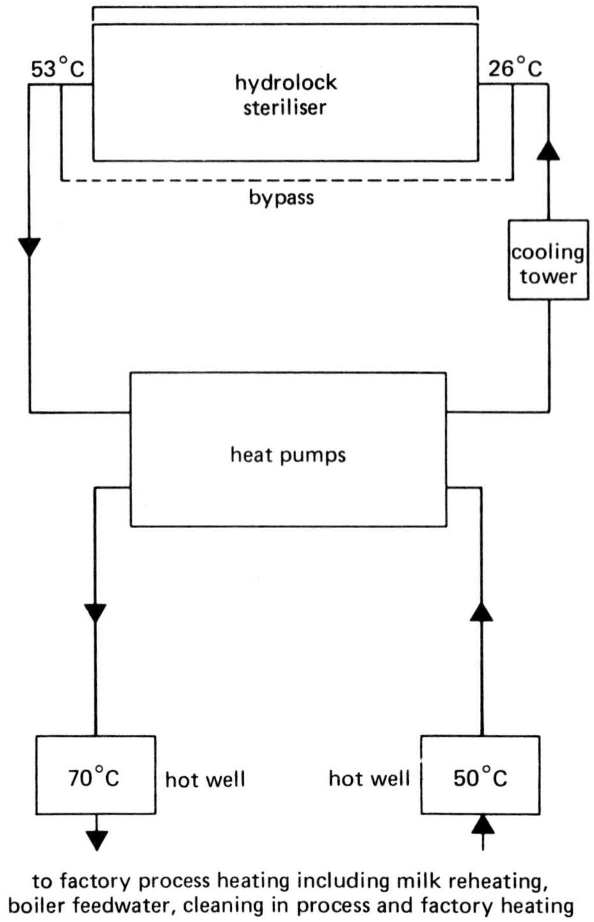

Fig. 7.88 Heat pump applied to a milk steriliser - 7.8.1

Heat source:	Cooling water at 53 °C
Technique:	Electrically driven heat pump
Heat recipient:	Process hot water at 70 °C
Energy saving (predicted):	6 095 GJ/a
Payback period (predicted):	2 years

7.9.9 Boiler and Steam Systems

Three primary areas for waste heat recovery exist with boiler plant and steam systems, namely:

- Economisers or spray condensers, to recover heat from the flue gases.
- Condensate and flash steam recovery
- Blowdown heat recovery.

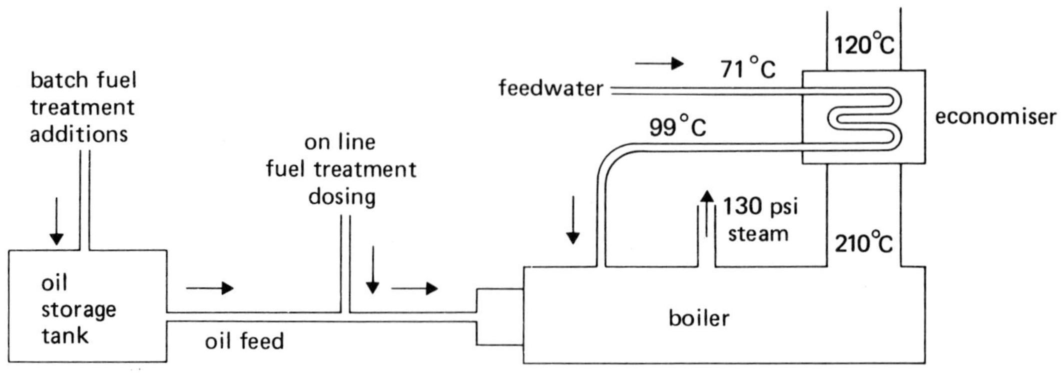

Fig. 7.89 Economiser applied to oil fired boiler - ref. 7.7.1

Economisers are widely used on a range of gas, oil and coal fired boilers. When burning sulphur bearing fuels the practical limit for heat recovery is normally defined by the acid dew point at which extremely corrosive condensation occurs.

The above example shows the use of a fuel oil treatment programme which lowers the acid dewpoint and corrosive activity. The result is that the flue gases can be cooled to 120 °C (instead of 152 °C) whilst still using a conventional cast iron economiser.

Energy saving (predicted): 4 346 GJ/a
Payback period (predicted): 2.7 years

Condensate and flash steam recovery

The condensate resulting from the indirect use of steam will still contain a significant quantity of heat. Wherever practical this condensate should be returned to the boiler feed tank so elevating the feedwater temperature and reducing fresh water treatment costs. For every 6 °C rise in feedwater temperature, approximately 1% fuel saving will be achieved.

On extensive sites the capital costs required to return condensate may not be justified. In those situations it can often be advantageous to use the hot condensate (and the flash steam which is generated when its pressure is reduced) for nearby heat users. Figure 7.90 shows one method of re-utilising condensate derived flash steam.

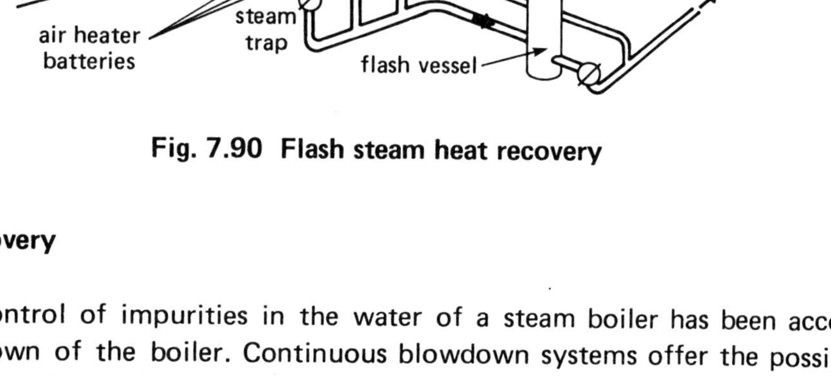

Fig. 7.90 Flash steam heat recovery

Blowdown heat recovery

Traditionally the control of impurities in the water of a steam boiler has been accomplished by intermittent blowdown of the boiler. Continuous blowdown systems offer the possibility of heat recovery from the condensate/flash steam generated to preheat the boiler feedwater, or in some instances combustion air or fuel oils. One method (Figure 7.91) directs the blowdown into a vessel producing flash steam which can be directly injected into the feedwater recovering both heat and water content. If the feedwater is already at a sufficient temperature then the flash steam can be used for low pressure heating duties. Other methods utilise heat exchanger(s) in conjunction with the flash vessel.

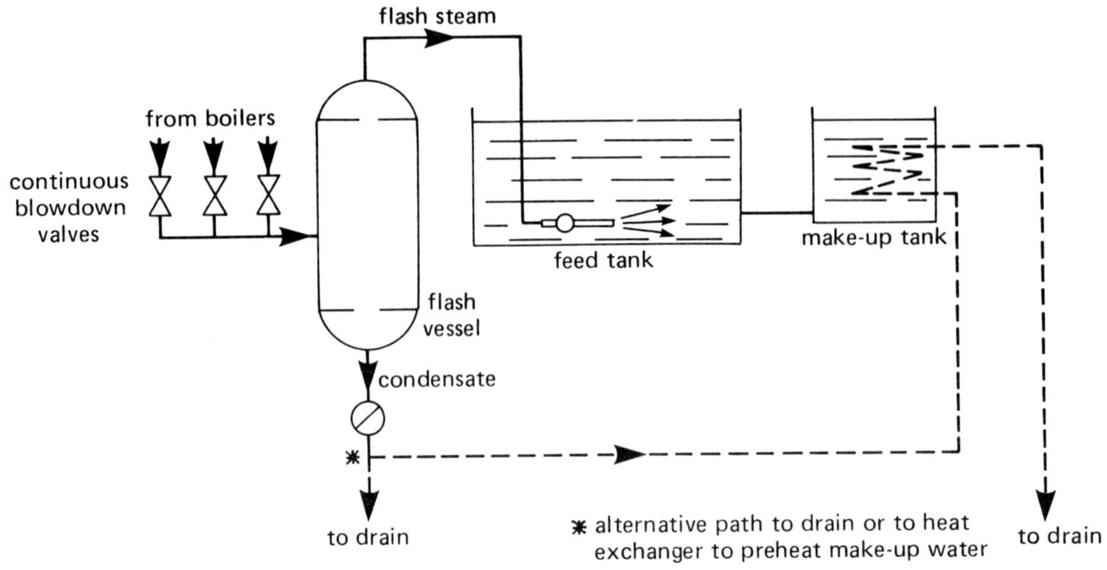

Fig. 7.91 Blowdown flash steam and sensible heat recovery

7.9.10 Prime Movers

Where prime movers such as gas turbines or internal combustion engines are used, scope may exist for waste heat recovery from the cooling systems and exhaust streams. When used to drive electricity generators, heat recovery from prime movers enables the construction of Combined Heat and Power schemes. These are discussed in some detail in Chapter 13. Figure 7.92 shows one large scale retrofit scheme whereby the exhaust from a gas turbine used for power generation is used with auxiliary fuel input to fire process heaters at an oil refinery.

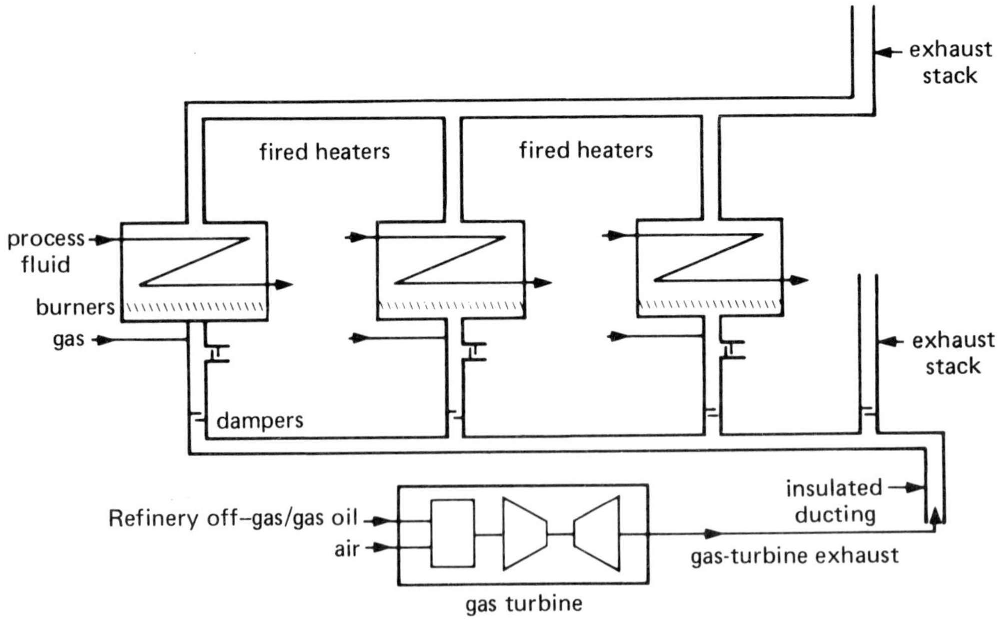

Fig. 7.92 Gas turbine and fired heaters

Energy saved:	808 000 GJ/a
Payback period:	1.6 years

7.9.11 Air Compressors

The act of compression requires power input and the heat incidentally generated results in a number of potential recovery options.

Compressor cooling

Compressors will be either air, water or oil cooled. Where air cooling is used it is often a simple matter to duct the resultant warm air to adjacent areas for space heating purposes. Water cooled compressors usually offer less scope for waste heat recovery as the compressors normally require a maximum water outlet temperature which is too low for re-use other than via a heat pump. Oil injected rotary compressors offer the greatest scope for heat recovery as oil exit temperatures are high. It is possible to generate hot water at up to 80 °C by using a water cooled oil cooler.

Air cooling

It is usual to cool compressed air prior to distribution in order to reduce condensation problems at the end users. Figure 7.93 shows one scheme where careful design has allowed hot water temperatures approaching 90 °C to be achieved using a shell and tube compressed air aftercooler taking air at 165 °C. The payback period was around two years.

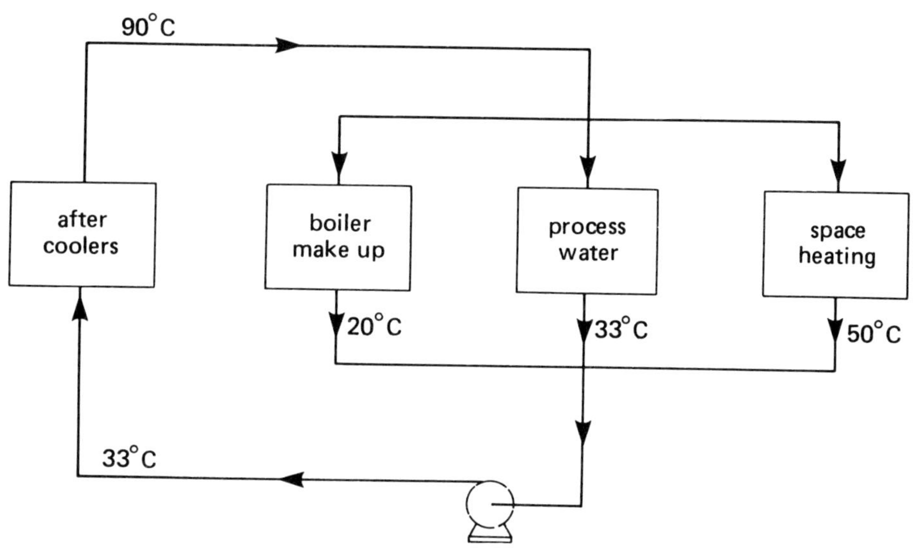

Fig. 7.93 Compressed air aftercooler - ref. 7.5.1

7.9.12 Refrigeration Plant

A considerable amount of energy is used to chill foods and liquids and refrigerate storage both in manufacturing industry and in food stores with chilled display cabinets. The refrigeration systems used provide a significant heat source which can be used for other purposes. Refrigeration technology is similar to that in heat pumps. Development of heat recovery from the compressors in refrigeration systems has increased as energy prices have risen. The chilling unit, e.g. chilled display cabinet, is the evaporator for the system. Domestic hot water can be produced from the heat available in the superheated refrigerant vapour leaving the compressor. Additional heat can be applied to the water, if necessary, to raise its temperature for other processes.

The subsequent cooling of the refrigerant vapour at the condenser in the refrigeration system is performed by air cooling. The air fans reduce the temperature of the refrigerant and the heat removed is very often dumped into the atmosphere. However, this offers potential for heat recovery providing a heat source for the the space heating system. In some food stores return air from warehouse and store areas is taken over compressors to provide surface cooling. The air gains temperature and is then ducted through the condenser coils of the refrigeration system where it further preheats. It thus offers a notable reduction in the heating requirements of the premises, as when 'top-up' heating is provided the air is used directly for store and warehouse heating. Figure 7.94 shows a total reclamation system.

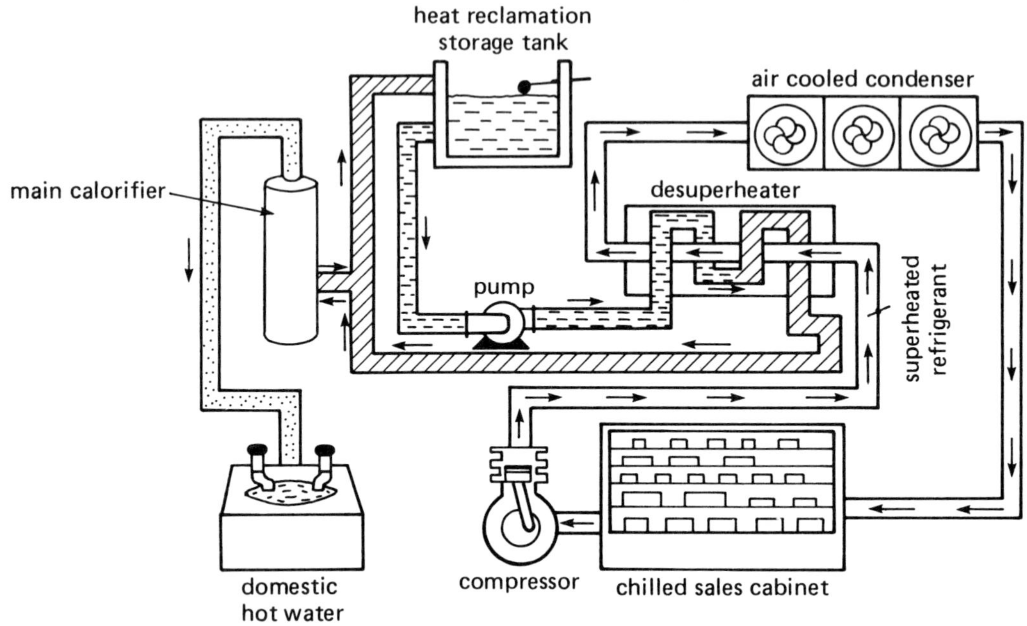

Fig. 7.94 Heat reclamation from refrigeration

7.9.13 Space Heating and Ventiliation Plant

In locations where there are significant heat gains, for example canteens or production areas, ventilation exhaust streams may be of a sufficient temperature to allow heat recovery via heat transfer (e.g. plate heat exchangers or run-around coils) to the incoming air flow. Even if temperatures are too low for these techniques recycle or heat pump methods may be used. The latter are particularly attractive where the exhaust flow has a high moisture (and hence latent heat) content. Several installations have been completed on swimming pool ventilation systems and one is shown in Figure 7.95.

7.9.14 Lighting

Lighting systems in retail and commercial buildings generally consist of banks of fluorescent tubes in the form of luminaires which by their nature contribute a significant internal heat gain to the building. Fluorescent tubes, however, provide their peak performance in terms of light output at around 40 °C. Thus to achieve optimum output cooling of the tubes is often employed. The most common cooling system used is by air flow through an air handling luminaire (Figure 7.96). This is a key component of integrated services in ceiling voids in buildings and is designed to extract air from the room into the void or exhaust ducting system. The air entering through slots in the luminaire is directed over the lighting tubes and control gear. A 10-15% increase in light output and increased control gear life results when compared with luminaires without cooling.

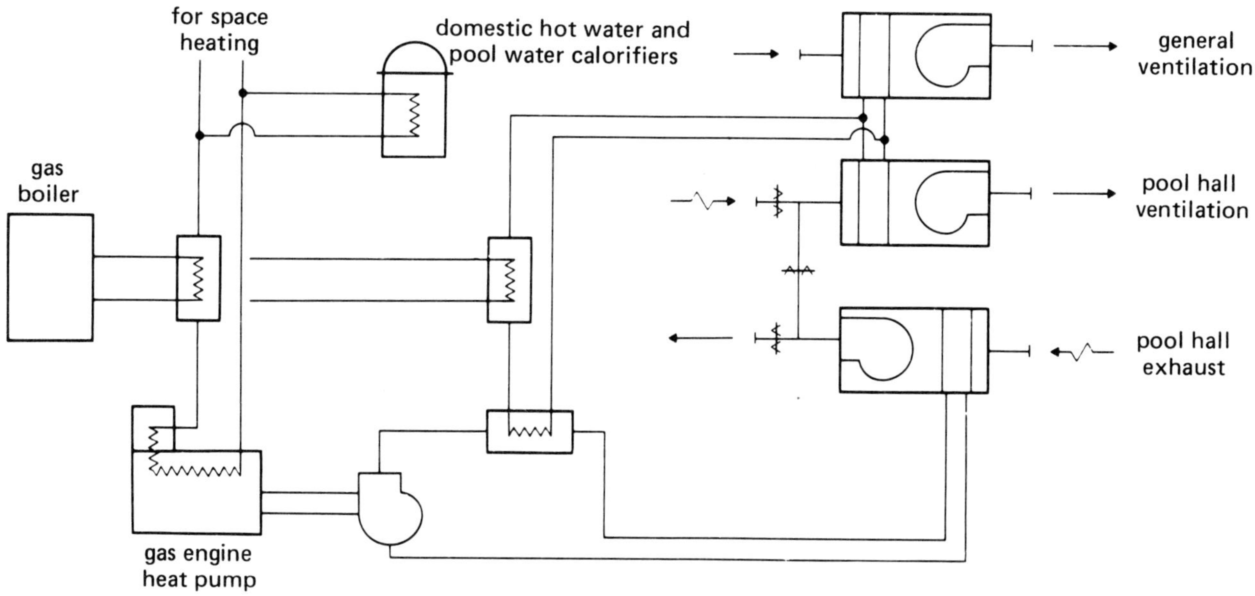

Fig. 7.95 Heat pump applied to swimming pool ventilation - 7.8.1

Heat source:	Pool ventilation air
Technique:	Gas engine driven heat pump with engine heat recovery
Heat recipient:	Fresh ventilation air and domestic hot water
Energy saving (predicted)	5 380 GJ/a
Payback period (predicted)	6 years

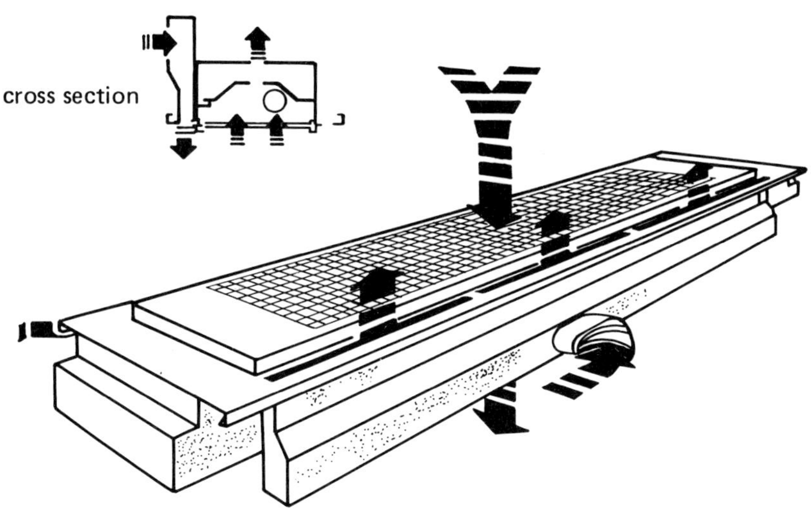

Fig. 7.96 Prismatic diffuser luminaire

Integration of the lighting system with heating and ventilation systems can lead to the heat removed from air handling luminaires being recovered for other heating duties. Up to 70% of the energy input to the lighting unit can be recovered in this way producing a supply of low grade heat, typically up to 30 °C. It is thus mainly of use for heating storerooms, cloakrooms, corridors and stairways but makes a worthwhile contribution to the building heat requirements without the need for additional heating.

REFERENCES

Building for Energy Conservation, P.W. O'Callaghan, Pergamon Press.

Chemical Engineers' Handbook, R.H. Perry & C. H. Chilton, McGraw Hill.

The Efficient Use of Energy, ed. I.G.C. Dryden, IPC Science & Technology Press

The Efficient Use of Steam, ed. P.M. Goodall, IPC Science & Technology Press

Energy for Industry, ed. P.W. O'Callaghan, Pergamon Press

Energy Management, W.R. Murphy & G. McKay, Butterworths

Energy Manager's Workbook, Energy Publications

Heat Pipes, P.D. Dunn & D.A. Reay, Pergamon Press.

Heat Pumps, R.D. Heap, E. & F.N. Spon

Heat Pumps - Design and Applications, D.A. Reay & D.B.A. MacMichael, Pergamon Press

Heat Pump Technology, H. Ludwig von Cube & Fritz Steimle, ed. E.G.A. Goodall, Butterworths

Heat Recovery Systems - A Directory of Equipment and Techniques, D.A. Reay, E. & F. N. Spon

Industrial Energy Conservation - A Handbook for Engineers and Managers (2nd ed.), D.A. Reay, Pergamon Press

Journal of Heat Recovery Systems, Pergamon Journals Ltd (Sciences)

Thermal Energy Recovery, 2nd ed. John L. Boyen, Wiley Interscience

Waste Heat Recovery from Industrial Furnaces, a series of papers presented to the Institute of Fuel, Chapman Hall.

Papers and articles on heat recovery techniques, case histories etc. appear from time to time in journals, periodicals and occasional publications such as:

Energy World/Journal, both published by the Institute of Energy

Gas Engineering and Management, Institution of Gas Engineers

Relay, Midlands Research Station, British Gas plc

Monitor, British Gas plc

Project Profiles, Extended Project Profiles, Reports, Energy Efficiency Demonstration Project Schemes, Energy Technnology Support Unit

Energy Management, Department of Energy

Energy Manager, MCM Publishing

Refractory and Insulating Materials in Furnace Construction 8

CHAPTER EIGHT

REFRACTORY & INSULATING MATERIALS IN FURNACE CONSTRUCTION

8.1 INTRODUCTION

The importance of selecting the correct refractories for the application, and using the constructional method appropriate to their properties and the required furnace duty, can scarcely be overemphasised. If correctly followed, the results will be evident in product quality, optimum energy efficiency and minimised maintenance costs. These considerations may also often be applied to existing furnaces (especially when due for major overhaul and re-bricking) to improve their performance.

This Chapter considers firstly the properties and applications of refractories, and secondly the various components of furnace construction. For information on heat transfer and the various types of furnaces used in industry, the reader should refer to Chapters 5 and 6 respectively.

8.2 FURNACE REFRACTORIES

8.2.1 Definitions of Terms

The function of refractories is to isolate or form a heat barrier by means of resisting materials in the form of a lining within the structure, in such a way that the maximum possible amount of the initial heat input will be contained and passed only to the workload.

The refractory lining surrounding a flame used for heating stock and contained within the steel structure, constitutes the furnace proper.

The term 'refractory materials' is understood to refer to those materials, or the manufactured products prepared from them, which exhibit mechanical strength and chemical stability at high temperatures, together with the required insulating properties. They must also withstand the effects of thermal shock (i.e. sudden temperature change), high resistance to wear, chemical resistance to slags and be of uniform and constant composition. 'Insulating materials' are a sub-group of refractories whose main property is resistance to heat flow, i.e. low thermal conductivity. There is in practice varying degrees of overlap between these two categories.

8.2.2 Description of Materials

Materials may be broadly classified in the five following groups, bearing in mind that insulation for the higher temperature ranges must in general be refractory in character:

- General refractories
- Insulating refractories
- Low temperature insulation
- Monolithic refractories
- Ceramic fibre insulation.

General refractory materials

These are produced mainly in the form of standard firebricks and are required for their strength and abrasion-resisting qualities on hearths, the lower areas of side walls and doors. Because of their poor insulating properties, they are confined mainly to the inner courses. Their properties are listed in Table 8.1.

Table 8.1 Properties of Refractory Brick

	Firebrick	Aluminous firebrick	Siliceous or semi-silica	Silica	Sillimanite	Chrome-Magnesite
Refractoriness Cone °C	28-31 1630-1700	32-34 1710-1750	29-31 1650-1690	31-33 1690-1730	35-36 1770-1790	38 1850
Refractoriness under load 193 kN/m^2 °C	1440-1520	1550-1620	1550-1600	1660-1700	1700	1500-1700
Bulk Density kg/m^3	1920-2080	1920-2080	1840-2000	1680-1840	2000-2160	2800-3360
Cold crushing strength MN/m^2	17.2-37.9			13.8-37.9	34.5-48.3	20.7-34.5
Thermal conductivity W/(mK)	1.1	1.1		1.2	2.4	3.6
Typical compositions SiO_2	57.8	51.3	85.0	95.7	34.0	4.5
Al_2O_3	28.6	43.1	12.5	1.1	61.4	10.3
Fe_2O_3	3.5	2.1	0.8	0.7	0.7	12.7
Cr_2O_3	–	–	–	–	–	30.2
TiO_2	0.9	1.9	0.5	0.1	1.9	-
CaO	0.5	0.5	0.2	2.1	0.9	1.0
MgO	0.7	0.3	0.2	0.1	0.4	41.0
Alkalis	1.8	1.0	0.6	0.2	0.7	-

Insulating refractory materials

These consist of insulating firebrick on high porosity. They are sufficiently refractory and tough to withstand high temperatures under some load but are light weight enough to have a low thermal capacity (Table 8.2).

They are commonly used as inside face linings of furnaces and as backing insulation to firebrick and are known as 'hot face' insulation. These bricks are made from a mixture of alumina (Al_2O_3) and silica (SiO_2) which have been mixed with combustible matter then heated sufficiently to burn away the added matter, thereby becoming porous. There are a variety of qualities available, which include those more suitable for the lower range of temperatures occurring at the outer courses of the refractory lining.

Furnace linings generally comprise such insulation refractories in the inner course where abrasion does not occur, in order both to conserve heat and to accelerate the rate at which the furnace temperature can be increased. In general for any given chemical or mineral composition, the higher the porosity the lower the bulk density and the greater the insulating ability of the material. As a result of this, however, the mechanical strength of hot face insulating refractory bricks is not great and care must be exercised in building them into walls where they may be subject to mechanical damage. They are also susceptible to the attack of ferrous slag and have a low resistance to abrasion.

Table 8.2 Properties of Insulating Bricks (Moler)

MPK Grade	Classification Temp. °C	Bulk Density kg/m^3	Cold Crushing Strength MN/m^2	* Permanent Linear Change %	** Thermal Conductivity W/m K at mean temperature °C:						Chemical Analysis % w/w:			
					200	400	600	800	1000	1200	Al_2O_3	Fe_2O_3	TiO_2	Alk
Porous	870	540	1.4	-0.5	0.117	0.145	0.179	-	-	-	10.2	5.6	0.6	1.7
Solid	870	720	4.8	-0.7	0.133	0.157	0.183	-	-	-	10.2	5.6	0.6	1.7
Rotol	920	770	9.8	-1.0	0.181	0.187	0.204	-	-	-	11.0	4.0	1.0	0.5
18/25	980	400	1.3	-0.1	0.136	0.159	0.188	0.218	-	-	24.1	4.0	1.3	2.7
20/26	1100	420	1.6	-0.5	0.165	0.186	0.216	0.252	-	-	24.1	4.0	1.3	2.7
20/36	1100	580	2.9	-0.5	0.204	0.234	0.266	0.296	-	-	26.9	3.2	1.3	2.5
20/43	1100	690	3.4	-0.4	0.273	0.291	0.319	0.361	-	-	29.8	2.5	1.2	2.4
23/42	1260	680	2.4	-0.3	0.174	0.197	0.215	0.236	0.258	-	35.1	1.0	0.2	2.2
23/48	1260	770	3.4	-0.5	0.242	0.271	0.307	0.348	0.389	-	35.2	1.5	0.9	3.0
23/60	1260	960	5.5	-0.5	0.294	0.319	0.351	0.387	0.432	-	36.6	1.4	0.7	2.7
23/70	1260	1120	9.0	-0.6	0.377	0.424	0.470	0.517	0.563	-	38.3	1.2	0.5	2.5
23/80	1260	1280	14.5	-0.6	0.534	0.562	0.595	0.633	0.670	-	38.4	1.8	0.1	2.4
25/44	1370	710	2.1	-0.4	0.194	0.214	0.238	0.268	0.304	-	40.2	0.4	0.9	0.9
25/54	1370	860	4.8	-0.4	0.274	0.307	0.338	0.367	0.390	-	41.3	1.3	0.6	2.6
26/48	1425	770	2.7	-0.4	0.291	0.311	0.335	0.360	0.399	-	47.0	0.5	1.6	1.5
26/57	1425	920	6.2	-0.5	0.286	0.310	0.335	0.361	0.399	-	41.3	1.3	0.6	2.6
28/55	1540	880	3.8	-0.3	0.330	0.343	0.355	0.370	0.385	0.403	61.0	0.3	1.4	1.0
30/67	1650	1070	4.8	-0.4	0.412	0.423	0.437	0.452	0.470	0.490	73.0	0.3	0.7	0.8
30/78	1650	1240	10.8	-0.6	0.600	0.566	0.566	0.587	0.610	0.641	68.6	0.3	0.7	0.8
33/80	1815	1280	11.1	-0.8	-	-	0.957	0.955	1.002	1.095	93.9	0.08	0.01	0.17
34/90	1870	1450	20.7	-0.8	-	-	1.295	1.188	1.118	1.095	99.3	0.08	0.01	0.13
34/110	1870	1760	17.5	-0.6	-	-	1.710	1.570	1.475	1.450	99.6	0.08	0.01	0.13

* After 24 hours at 28 °C below classification temperature
** Measured through 144 mm dimension

Low temperature insulating materials

These are light in weight, have a low thermal capacity and a low crushing strength. This group would also cover low temperature applications such as lagging, being available in various forms of material e.g. magnesia compositions, refractory wool, spun glass, aluminium foil and microporous silica.

Monolithic refractories

This description is given to a very wide and flexible range of refractory concrete materials which are suitable for the production of shapes, especially in situ in new or existing structures and in burner manufacture. Replacement burner quarls are frequently produced on site to be formed of suitable mix in a temporary mould installed in the furnace walls.

Three basic forms are usually distinguished:

- Castables
- Mouldables
- Ramming Mixtures.

Traditional refractories have made use of a ceramic bond developed as the temperature increases. However, chemically bonded castables are now available which have improved properties at reasonable cost. As this technology develops other refractory products are likely to become generally available using this technique.

In addition, stainless steel fibres can be added to monolithic refractory to improve its mechanical properties.

Castables consist of a mixture of aluminous cement and refractory aggregate, based on Diatomite and similar materials. Use is made of the hydraulic setting property of a suitable cement, with which is mixed a refractory aggregate. They are incorporated in the furnace construction, either after pre-moulding and setting, or cast in position. The hydraulic bond is destroyed by heat and is replaced by a ceramic bond, much as happens when green clay masses are kiln fired. The importance of selection of materials such that a ceramic bond will, in fact, develop during service should be emphasised. Aluminous cements are usually employed for this purpose, as typified by Ciment Fondu, well known for its rapid setting property, and by Secar grades which are basically calcium aluminates and are serviceable at higher temperatures. Typical aggregates used in British practice are Alag (sintered Ciment Fondu); calcined 42/45% alumina clays; bauxite, which is a high-alumina material, used either raw or after calcination; molochite, which is a calcined china clay; sillimanite, a calcined mineral; and various grades of fused alumina. For insulation purposes, castable mixes will contain other, lightweight aggregates based on diatomite, perlite, aglite, vermiculite, calcined porous clay aggregates, or bubble alumina which is a fused alumina expanded into bubble form.

Chemically bonded castables differ in composition by having a low cement content and a chemical binder. The bond is present upon casting and it is not necessary to raise the refractory temperature to gain strength. Moreover, the resulting refractory has increased hot strength and abrasion resistance. An increased material cost of approx 10-20% over cement bonding is entailed. Castable compositions covering various service temperature ranges and conditions are listed in Table 8.3.

Table 8.3 Compositions and Service Temperatures of Castables

Temperature Range °C	Type	Remarks
1 200	Alag + Ciment Fondu	High strength, high abrasion resistance
1 250	Calcined 42/45% alumina clay + Ciment Fondu	High strength
1 300-1 400	The same, but with leaner cement mix	General purpose
1 500	Calcined 42/45% alumina clay + Secar 250	
1 500-1 600	Bauxite + Ciment Fondu Impure fused alumina + Ciment Fondu	
	Molochite + Secar	Low ferric oxide for reducing conditions
	Sillimanite + Secar	
1 650-1 700	Brown fused alumina + Secar	
1 750-1 800	Pure white fused alumina + Secar	

Mouldables consist of refractory grogs or aggregates, bonded as a mixture with clay and water and often available in a workable condition, including a cold setting bond. A ceramic bond develops as the temperature is increased. Care must be taken in this initial firing to ensure that the temperature is such as to achieve this bond. Mouldable compositions are usually sold within the range of 40%, 60% and 80% alumina.

Ramming mixtures are similar to the moulding compositions, being differentiated by finer grain sizes to give a closer texture finished product. They are much used in the repair of damaged refractory installations and can be produced in similar compositions to mouldable and castable refractories.

The improved mechanical properties achieved by incorporating stainless steel fibres enable a hot face temperature up to 1 650°C. The fibres, 25–35 mm long, are usually added to a level of 3-5% by weight to increase the spalling and erosion resistance of the refractory. The increase in refractory life is dependent on the working environment and should be assessed on an individual basis. Obvious applications for this material are furnace doors and lintels which are subject to spalling.

Ceramic fibre insulation

Ceramic fibre material has now been marketed for several years but, initially, was perhaps slow in gaining acceptance, possibly by reason of conventional and prejudiced thinking in terms of the solid type refractories.

Because of the increasing current use of this form of furnace lining the following list of properties and capabilities should prove helpful:

- Very low thermal conductivity, considerably lower than insulating firebrick.
- Good resistance to thermal shock, no matter the rate of change when heated or cooled and with no occurrance of cracking or spalling.
- Low density (65–192 kg/m^3) compared to that of insulating firebrick.
- Low thermal mass due to its low density, resulting in faster cycle times and fuel savings when used in furnace applications, particularly with batch operation.
- High chemical purity with very low chloride content.
- Good chemical stability, unaffected by oil, water and steam and resistant to most acids (except hydrofluoric) and many alkalis.
- Available in three main grades (1 250°C, 1 400°C and 1 600°C) representing most service temperatures in intermittent use. The continuous service rating of these material grades are 1 150°C, 1 250°C and 1 400°C respectively. ICI Saffil fibres, 95% alumina composition, are suitable for use at continuous operating temperatures up to 1 600°C.
- Other characteristics of interest include incombustibility, ease of installation and good acoustic qualities.

Ceramic fibre is available in several forms:

- Fibre blanket.
- Stackbonded fibre modules.
- Rigid board.
- Paper.
- Bulk fibre wood.
- Vacuum formed moulded shapes.

Vacuum pre-formed ceramic fibre shapes are available and often simplify an awkward insulation job. These are self-supporting and take the form of simple tubes, cones or domes, pre-formed muffles, launder linings, pouring nozzle insulation, and many other useful configurations.

In conjunction with furnace linings, ceramic fibre burner quarls (Figure 8.1) prove a logical alternative to the conventional quarls and their inherent service problems. Because they are not prone to spalling they do not require as much maintenance as conventional units and a longer service life can be expected. Being 90% lighter, installation is simpler and quicker, as they can be easily manhandled without recourse to lifting gear. Figures for heat storage show that the ceramic fibre unit stores only 6% that of the firebrick quarl and 7% of a castable quarl.

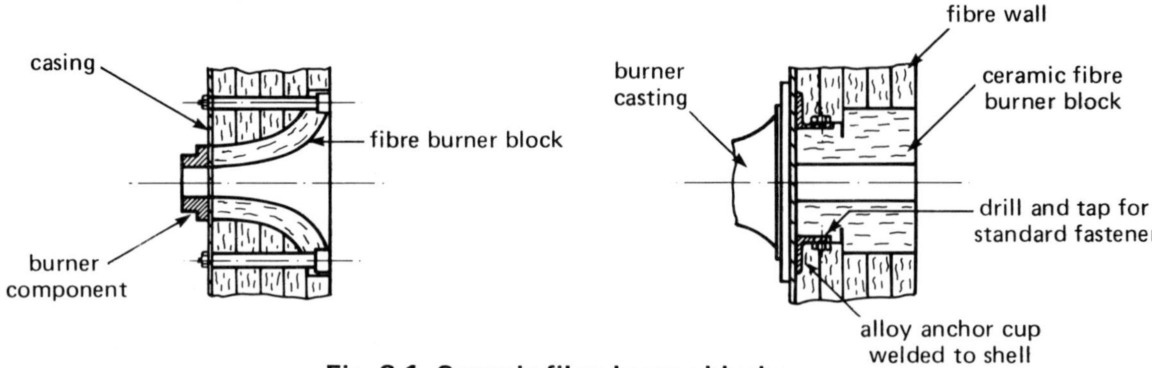

Fig. 8.1 Ceramic fibre burner blocks

By reason of its compressibility, ceramic fibre has found many useful applications as a sealing medium, replacing other jointing material. It may also be used in door seals and can replace older type sand seals.

Total fibre lined furnaces, leaving only the hearth confined to traditional refractories for their load-bearing characteristics, are now proving successful. The ceramic fibre is applied to the inside walls of the furnace shell by means of a refractory cement and expanded metal to ensure complete adhesion.

It is accepted that ceramic fibre has application limitations and may not be the answer to everyone's wish to conserve fuel. For example, in a furnace that is truly continuous in operation, such as a tunnel kiln which may only shut down once every 10 years, low thermal mass material is of little use as a lining as it only heats up once during a campaign. Even in such furnaces, however, improvements to reduce maintenance costs may well be made by the use of ceramic fibre seals and joints.

8.2.3 Choosing a Refractory

There are many factors to consider when designing a furnace lining and all designs are in some way a compromise. Specialist advice may need to be sought before a final choice is made.

It is common practice to use more than one type of material in a design to take advantage of each material's properties and economic cost. Moreover, economic cost is dependent on whether you are an equipment supplier or an end user and careful analysis is required before making a choice.

A list of the typical properties or problems that may be considered is:

- Service temperature.
- Strength.
- Resistance to physical or chemical damage.
- Thermal shock resistance.
- Weight.
- Available space.
- Heat storage capacity/heat-up cycle time.

- Cost of material.
- Cost of installation.
- Cost and speed of repair.

Service temperature is fundamental to choice of material. Figure 8.2 shows a comparison of common materials that are available.

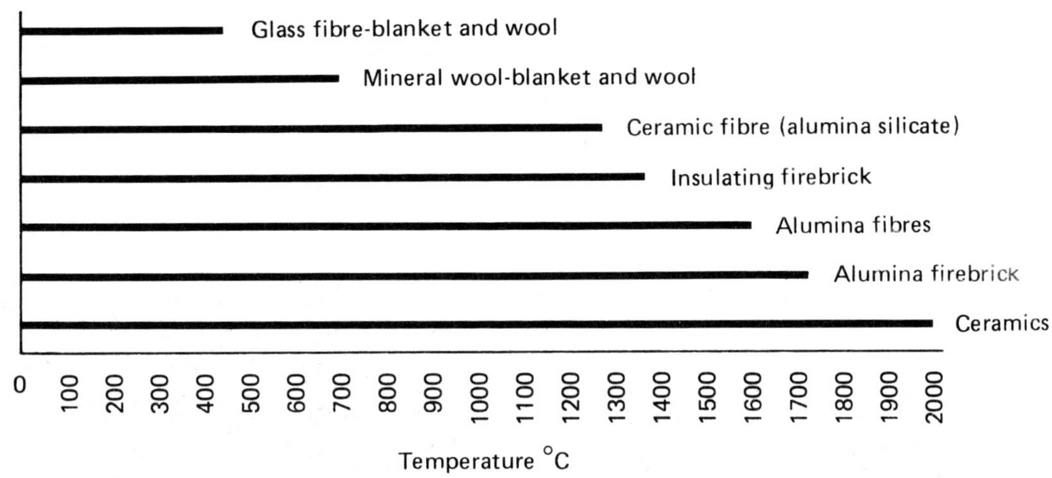

Fig. 8.2 Optimum use temperature for common refractory materials

The thermal conductivity of the material determines the thickness and quantity of material required by the furnace. The thermal conductivities shown in Figure 8.3 indicate the wide variation between different materials.

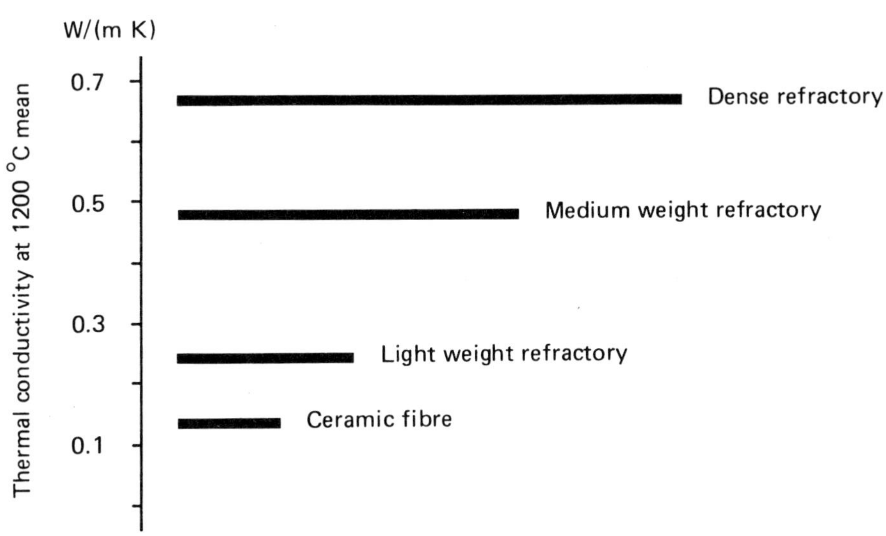

Fig. 8.3 Thermal conductivity values of common refractory materials

The diversity of refractory materials commercially available is due in large part to the variation in economic cost of using them. Initially, the cost of supply and installation must be considered and this cost will vary greatly with quantity, availability and local skills to install the material. Figure 8.4 gives an approximate price comparison for common materials achieving the same thermal duty.

The installed cost of the material is only part of the total cost consideration. For example, the cost of heating up a furnace to operating temperature increases with the thickness of the lining, thus wasting valuable fuel. If the furnace is reheated frequently then low density/high cost materials should be considered to reduce operating costs.

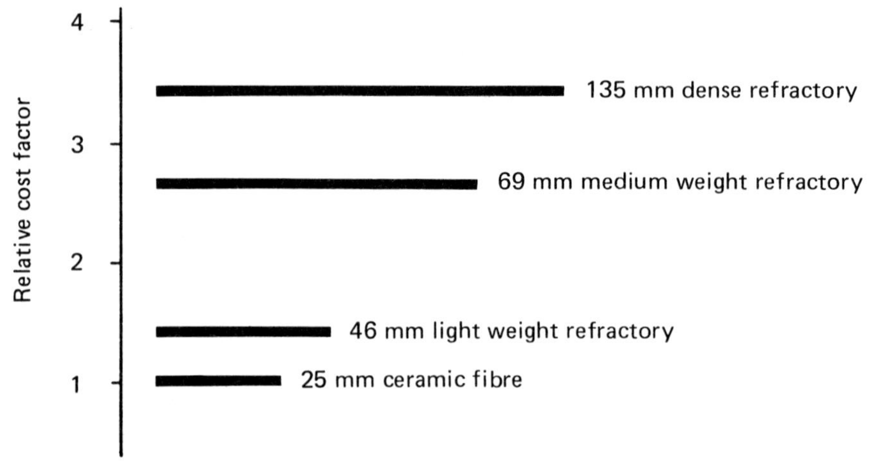

Fig. 8.4 Approximate relative cost of installed refractory materials to achieve the same thermal duty

A more intangible commercial consideration is the relative replacement cost at the end of the refractory lining life cycle. Frequent repairs may be avoided by choosing a different material to avoid spalling or mechanical damage.

8.2.4 Application of Refractories

Walls

Traditional multi-layer walls have a working face of dense strong refractory backed by one or more layers of insulating refractory, chosen to withstand the interface temperatures predicted by the solution of the heat conduction equations. A typical combination would be an alumina brick course, backed by an insulating refractory and a diatomaceous brick course respectively. This type of construction, producing furnaces of high thermal mass, although with good structural properties and maintaining equilibrium conditions, is not very energy efficient. The next step was the introduction of hot face insulating refractories for the complete inner lining, in various thicknesses and backed as necessary by low temperature insulation according to temperature requirements. Ceramic fibre linings more recently introduced are providing yet another stage in the improvement of furnace lining efficiency.

Multi-layer wall linings, particularly in high temperature applications, require to be tied back at intervals to the main furnace shell casing, as there is a tendency for the walls to fall in towards the centre of the setting at the top, especially where the arch roof is independently supported.

Hearths

Where mechanical strength is important, such as in hearth construction (the hearth itself being the most ill-treated component in any furnace) firebrick and monolithic refractories are used, due to their load bearing properties and abrasion resistance. The hearth must accept the weights of charge placed on it, withstand the resultant wear and tear associated with loading and unloading and also resist any chemical effects of the charge being processed. Monolithic hearths are, as a rule, rammed in situ, usually over a firebrick bed lining. Chrome magnesite or chromite compositions are used for this purpose with advantage, since when carefully 'burned in' they provide surfaces which are resistant to abrasion, scale attack and thermal spalling.

The depth of material used in hearth construction must be sufficiently thick to minimise heat penetration to the furnace foundations and, depending on load requirements, insulating materials are judiciously interposed. Wherever possible, of course, hearths should be supported on structural steelwork, thus providing air ventilation beneath the furnace structure.

Roof

The two main methods of roof construction are that of the masonry arch and what are known as suspended roofs, the latter built up of special shapes, usually cast, and suspended from steel girders (Figure 8.5).

Fig. 8.5 Detrick construction of flat suspended roof

The arch, however, is the common form of roof construction employing specially shaped (taper) bricks using High Temperature Insulation (HTI), firebrick or a combination of both. It is built either in rings or is of bonded construction, the outer rows often being backed by low temperature insulation. By reason of its excellent crushing strength, firebrick is used for skewbacks (Figure 8.6).

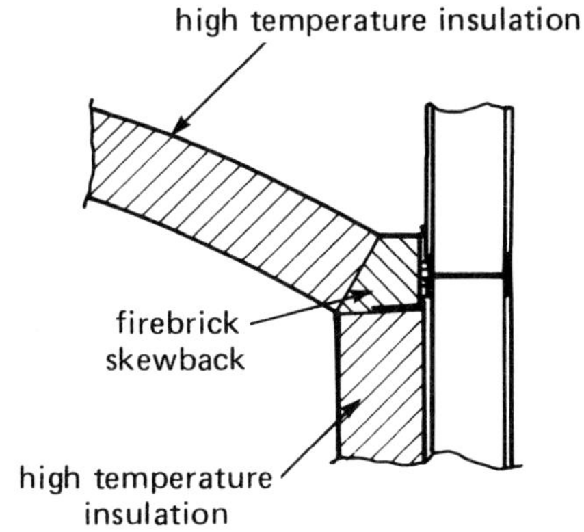

Fig. 8.6 Independent arch roof support arrangement

The arch rests with its skewback on the furnace side walls, or sometimes independently supported at the outer steelwork, upon which the arch thrust is taken as previously discussed. Spring loading for expansion allowance, whereby the abutments or skewbacks were pushed apart, was once quite often incorporated, but more commonly today the skewbacks are fixed (Figure 8.7).

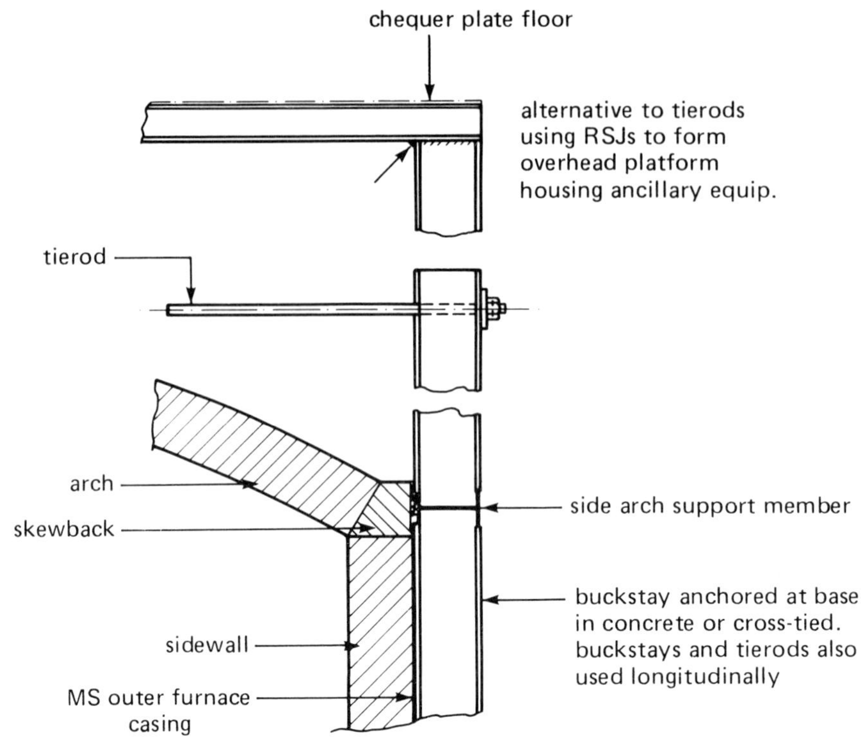

Fig. 8.7 Tie rod and side arch support

It is probable that ceramic fibre will become a common material for furnace roofs up to a temperature of 1 600°C. Ceramic fibre has the advantage of being considerably lighter in weight than traditional refractory and requires less supporting steelwork.

Doors

The refractory lining of doors initially consisted of firebrick housed in and anchored to the casing. This subsequently was superseded by the use of insulating bricks or insulating refractory castables, again suitably anchored. Castable refractory can be reinforced by adding metal fibres to increase spalling resistance and chemically bonded castable is also suitable for this type of application.

Ceramic fibre lining is currently being used, the door casing designed to incorporate a peripheral flanged edge away from the heat to form a sliding surface and eliminate abrasion.

Expansion allowance

Spalling resistance is of great importance to the correct use of any refractory brick. Dimensions will naturally change with temperature giving rise to temporary and permanent volume changes.

The percentage linear expansion of some refractories between 0°C and 1 300°C are:

Silicon carbide	up to 0.6%
Firebrick	0.6% to 0.8%
Corundum	0.8% to 1.2%
(according to corundum content)	
Silica	1.2% to 1.5%
Magnesite	1.6% to 1.8%

Thermal expansion must take place and, if suitable provision is not made for it, enormous stresses are built up and serious damage may result. Expansion during use should be allowed for by the incorporation of regularly spaced expansion joints, sized in accordance with the expansion coefficients of the material used. Experience indicates that the individual bricks do not slide over each other but that the wall expands 'en masse' as a single unit.

Ceramic fibre in bulk, strip or blanket form may be used as an infill packed into expansion joints in brick furnace structures. This prevents dust or slag collection, also prevents the formation of hot spots behind the expansion joint and at the same time provides an effective gas seal.

Joints should always be kept as thin as possible, no matter what method of bond is adopted. The strength of these joints depends on the dimensional accuracy of the bricks or shapes and in the quality of the mortar, which may be heat setting or air setting. With the first, the bond develops at temperatures of the order of 1 100°C to 1 200°C. The second type develops a strong bond on air drying, the strength being maintained throughout the working temperature range. It is important that the correct mortar is used to ensure compatability between the brick and the mortar itself.

8.2.5 Application of Monolithics

The method of application depends largely upon the shape of the furnace to be lined, circular and semi-circular settings being mostly involved. The material is normally rammed into position and replaces specially-dimensioned shapes and sizes which otherwise might prove difficult to obtain or fabricate.

Equally attractive is the reduction in the number of joints required and the savings of time and labour, although due regard must be observed to provide adequate resistance to the thermal stress which will be set up in these bigger shapes which, if the resistance is inadequate, would result in uncontrolled cracking and possible furnace atmosphere leakage.

The behaviour of refractory materials on sudden temperature change depends on their structure, thermal conductivity and, above all, on their thermal expansion coefficient.

8.2.6 Application of Ceramic Fibre

Most manufacturers provide and recommend their individual systems of fixing or anchoring the fibre applications and these have much in common, although in the light of cumulative experience and other types of fibre becoming available, fixing methods have been improved and continue to be developed.

In earlier days ceramic fibre in blanket form was more commonly promoted, the blanket being secured by uniformly spaced anchors tack-welded to the main furnace casing. With the introduction of stackbonded fibres, modular construction and the veneering of existing refractory work, different methods of fixing have been adopted.

The four main methods of ceramic installation to be considered are:

- Fibre blanket.
- Fibre modules.
- Fibre veneering.
- Fibre lined module construction.

Fibre blanket

In this method the furnace casing, either new or an existing one with refractories removed, is lined with ceramic fibre blanket of required thickness. The blanket is fixed by anchors suitably spaced and allowing for slight overlap at joints, and some typical fixing devices are shown in Fig. 8.8.

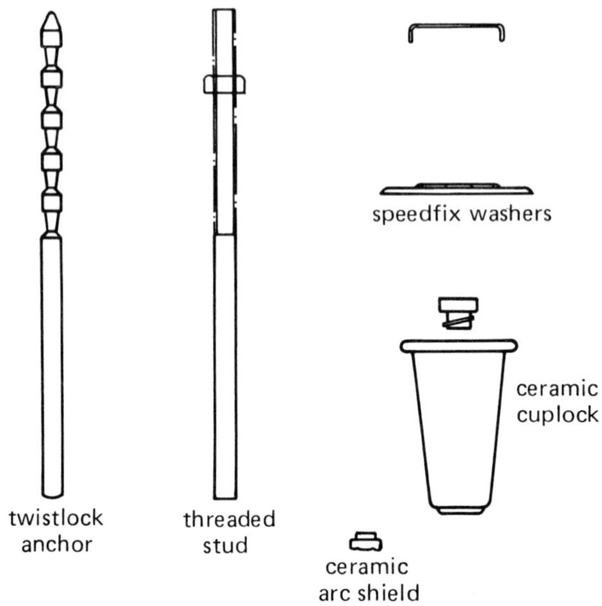

Fig. 8.8 Fixing systems for ceramic linings

Normally the anchors are of heat resistant alloy, which can withstand temperatures up to 1 250°C. For temperatures beyond this and for certain carbon rich atmospheres, a ceramic anchoring system would be required. With respect to the firebrick arch, this would be replaced with a flat blanket lined roof, supported by light steelwork. It is imperative that all anchors should have a firm layout to ensure proper anchoring of the blanket edges and carrying of the total insulation weight.

Fibre module

As an alternative to the layer construction method, refractory fibre modules have been developed which incorporate reinforcing material. The modules are normally of stackbonded material and are so arranged as to provide a variable pattern of fibre direction and thus minimise shrinkage (Figure 8.9). Approximately 0.1 m^2 of lining is attached to the shell by a single anchor which interfaces with the hardware of the module.

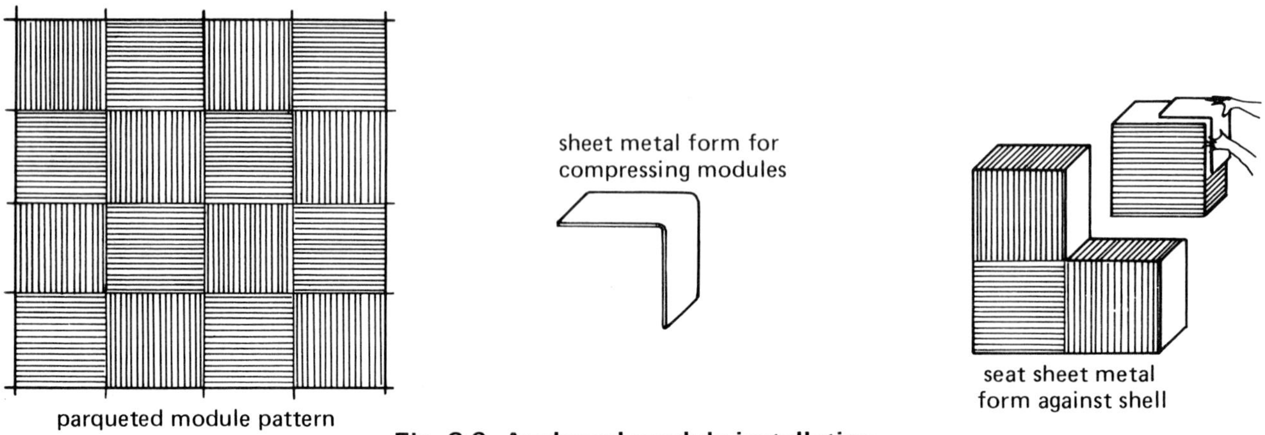

Fig. 8.9 Anchored module installation

The intricate anchor layout of the blanket method is eliminated since each module is pressed snugly against another and anchored in place. The two steps of anchor welding and subsequent impaling of each layer are replaced by one fastening step. The modules are lightweight, easy to handle and when installed have a concealed anchoring system (Figure 8.10). Alternatively they can be cemented to the casing.

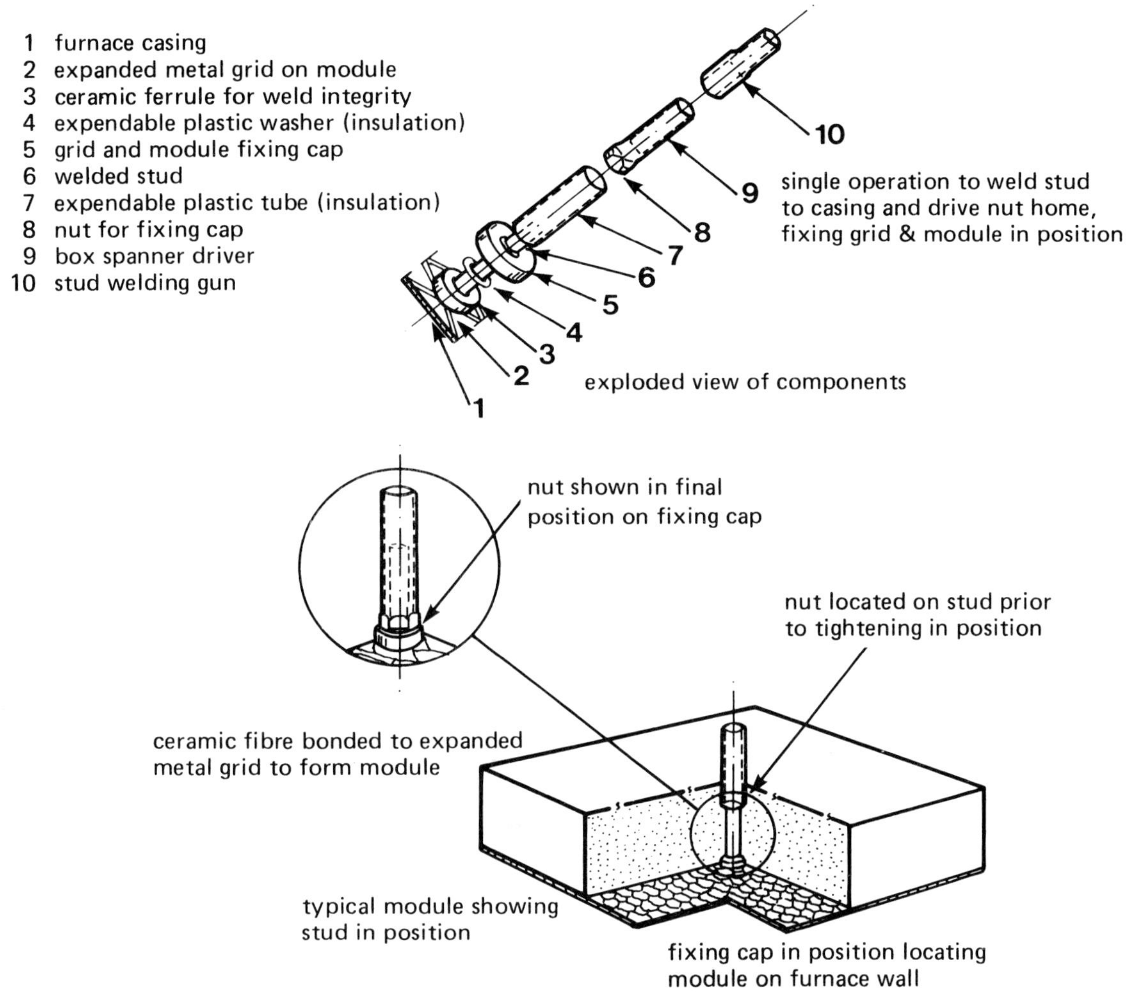

Fig. 8.10 Sauder patented 'Pyro-Block' (one shot) stud welded module

Veneering

This is the term used to describe the ceramic fibre lining of existing refractory work, which should be in reasonable condition to accept it. Any of the three basic forms of fibre may be applied.

The use of a layer of fibre applied as a veneer achieves energy savings by providing more insulation on the hot face and reducing the heat loss from cracks and open joints. Veneering also greatly reduces the amount of heat stored in the original lining, thereby providing faster cycle times.

As to the application, cement is applied to both fibre modules and refractory surface and the module presented by means of a flat board and held in position with a light applied pressure for about 5-20 seconds; this is repeated for subsequent modules, taking care to butt each one firmly against the other.

Ceramic fibre veneering has proved to be very successful, providing adequate care is taken during the installation stage. If the refractory lining to be veneered is not too severely damaged and cracked, a little additional work to provide a relatively smooth surface will enable a satisfactory lining to be achieved for almost any existing refractory surface. The main advantages of veneering may be summarised as follows:

- It is approximately 50% of the cost of a new refractory lining (depending on thicknesses)
- It takes less time to install than a complete lining replacement.
- It generates fuel savings varying from 10% to 30% depending on temperature, the original lining and other factors such as cycle times.
- Maintenance costs can be cut because fibre veneer can outlast conventional refractories that suffer from thermal shock.
- Veneering increases available production time.

Fibre lined module construction

This is a complete furnace building system (Figure 8.11) whereby the fibre modules and reinforcing material are housed in a mild steel flanged casing drilled for bolting together to form a rigid gas tight and lightweight ceramic fibre lined structure, or part structure such as a door or roof. This very successful application of modular fibre lined panels is a technique which has been largely adopted for example for the construction of large portable cover furnaces used in conjunction with the annealing of pressure vessels.

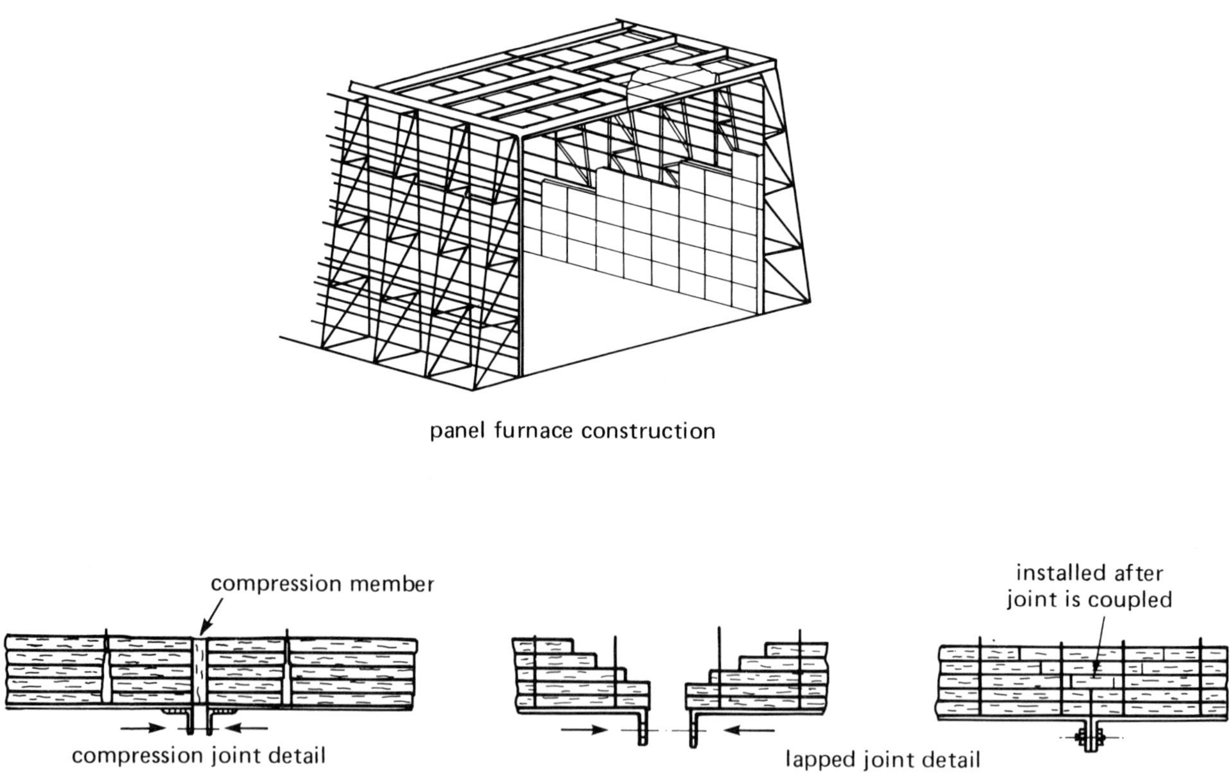

Fig. 8.11 Panel furnace construction

8.3 FURNACE CONSTRUCTION

8.3.1 Steelwork

The furnace's structure should be designed to support the chosen refractory or insulation design and this will vary greatly according to the choice of materials. Furnaces incorporating a traditional dense refractory material require a substantial steel framework for support. Ceramic fibre insulation requires a less substantial support structure due to its reduced weight for a given level of insulation.

Most furnaces basically consist of a fabricated, reinforced steel casing, refractory lined with arch roof, hearth, adjustable door, the enclosed chamber being suitably dimensioned to accommodate the stock to be heated. The side walls carry the roof arch of bonded or ring construction. The hearth is suitably designed to support the stock or charge to be heated, and is of fixed construction but may, depending on the type of stock, be in the form of a retractable bogie. The refractory arch roof rests with its skewbacks and abutments on the side walls, and particularly when up to temperature, exerts a thrust which is taken up in the furnace by the steel casing and outer reinforcing steel sections, the latter being referred to as buckstays, suitably disposed along the side walls and ends of the furnace. The buckstays are normally anchored in concrete, but may be cross tied at the base. At the top they are held by tie rods, both longitudinally and transversely, these being generally referred to as the furnace binding (Figure 8.7).

It will be evident from the general description that design principles have to be applied, particularly at certain areas in the furnace subjected to varying stress conditions, such as the arch and the various openings in the refractory lining, together with the hearth which must withstand the effect of the load and the method of loading.

8.3.2 Arch

The traditional roof arch is a critical element of furnace construction and although rapidly being superseded by low thermal mass suspended flat roofs where applicable, is still in widespread use.

The forces set up in a furnace structure are mainly:

- Lateral thrust exerted by the arch.
- Forces imposed on the steelwork binding by the expansion of the refractory walls.

The overall construction is based on sound engineering principles and designed to withstand these forces.

The arch thrust must be taken up by the binding without the yield-point stress of the binding being exceeded, using buckstays and tie rods as previously described. This also means that the thrust of the arch must be carried by a beam or support between the buckstays. The arch itself may be independently supported by the steelwork (Figure 8.7), so that none of it rests on the side wall. The forces which affect the binding of furnaces with refractory suspended roofs, and also the longitudinal binding of furnaces with sprung arches, are very much less than the forces which the cross binding of furnaces with sprung arches must withstand.

The skewback or abutments are reckoned to be fixed, and, assuming that the bricks are incompressible, the arch will rise at the crown, in which case the arch blocks will tend to gape at their outer edge and be in compression at the inner edge on the lower surfaces, thus rendering a degree of instability. In practice, however, taking into account temperature, compactness of the brick surfaces and the plastic properties of the materials used including mortars, this does not pose major problems.

The top of the arch is termed the crown and the distance between the crown and the chord forming the base line of the skewbacks or springers is the rise of the arch. The smallest safe thickness of an arch is a function of the span and the rise (Figure 8.12). It is common to make the radius of curvature of the inside of the arch equal to the span, which is equivalent to a rise of 13.4% of the span.

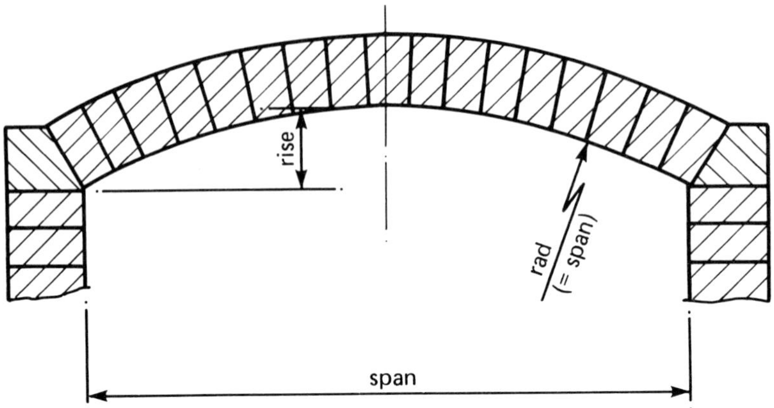

Fig. 8.12 Standard 60° arch

For high temperature furnaces a rise of 200 mm per metre of span is sometimes allowed. With regard to the relationship between the span and the thickness of brickwork, the following is given as a general guide for firebrick material only. For lightweight insulating brick, thinner arches are frequently used.

Span less than:	1.2 m	3.7 m	5.5 m	7.4 m
Brick thickness:	115 mm	230 mm	343 mm	460 mm

It is sometimes desirable to make the bottom of the arch flat instead of curved, which is done by using longer bricks in the middle of the span (Figure 8.13). This construction is known as a Jack Arch and the forces acting remain unchanged, the only effect being that of further load at the centre. The other feature associated with this type of arch is that it cannot 'rise' as easily as the sprung arch when heated.

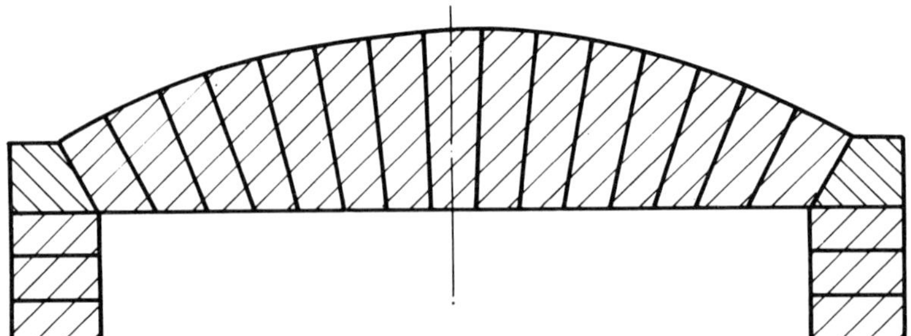

Fig. 8.13 Jack arch

For spans greater than about 4.5 m, however, flat suspended roofs are frequently used, being constructed of special shapes and usually cast and suspended from steel girders (e.g. Figure 8.5).

Ceramic fibre is eminently suitable for use as an arch material subject to the limitations of its physical properties (see section 8.2.6).

8.3.3. Door

Most doors consist of a cast or fabricated steel casing with a suitable refractory lining, and are operated by means of lifting gear which is usually integral with the furnace casing. In the case of small and medium sized furnaces, cast iron is utilised because of its durability and resistance to oxidation. Hinged doors are not recommended as they are liable to be slammed shut with consequent damage to the refractories.

Average-sized doors and their linings tend to be subject to distortion and spalling due to adverse operating conditions and temperature variation respectively.

For larger units, the door casing is designed with bolted or ribbed sections and guides to give overall strength and to offset distortion. Many larger doors are mechanically operated and counter-balanced, keeping the dead-weight to a minimum (particularly if fitted with ceramic linings).

Doors for very large heat treatment furnaces, where problems of operation can arise, may be supplied integral with the bogie hearth.

Door seating and good sealing in smaller furnaces can be effected by inclining the door and furnace front plates (Figure 8.14) and for larger furnaces by a suitable wedge or clamping arrangement, manually or pneumatically operated.

Ceramic fibre lined doors have the advantage of low weight and good insulation coupled with durability and good sealing properties, but attention must be paid to the design of the door closing mechanism to prevent the fibre from being abraded as the door opens and closes.

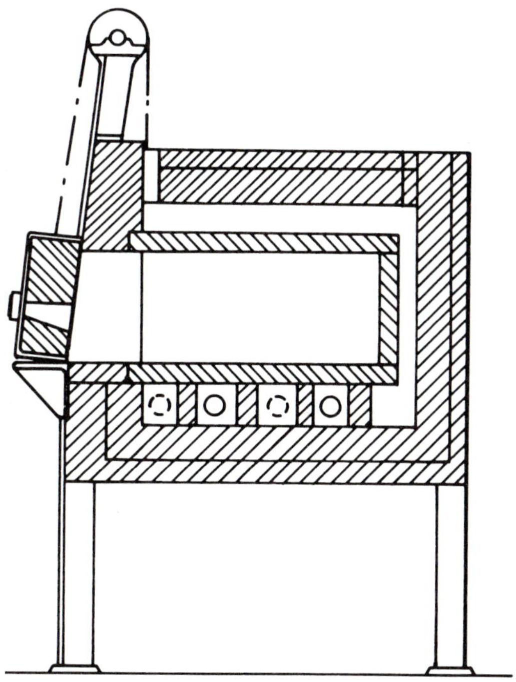

Fig. 8.14 Self sealing door

8.3.4 Openings

Openings are provided in a furnace for flues, doors, inspection ports and burner quarl mountings. The arch, with its strength and loadbearing capabilities, is the preferred method for spanning these openings.

Where openings are susceptible to mechanical damage, or are in the path of hot gases (e.g. flue ports), a double arch (Figure 8.15) is often used whereby the lower arch takes the impact of spalling etc. whilst the upper arch remains unimpaired. Door openings in particular, are most prone to mechanical damage, both in the handling of stock and careless door operation.

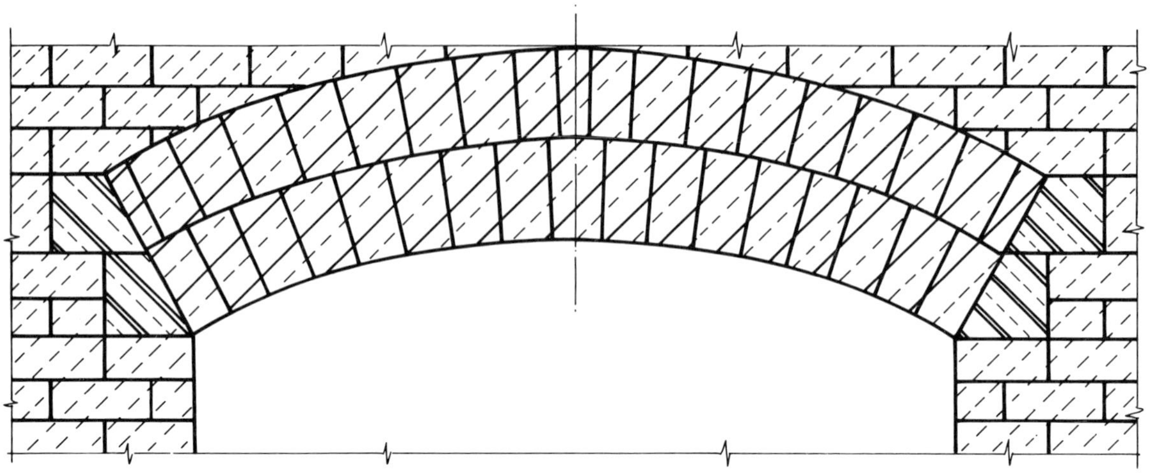

Fig. 8.15 Independent double arch

Smaller openings may be spanned by the use of refractory tiles but these, other than for burner quarl openings, tend to sag under load and crack, proving somewhat unsatisfactory in the long term (Figure 8.16). Castable refractory containing metal fibres has been used for this duty and given increased operational life.

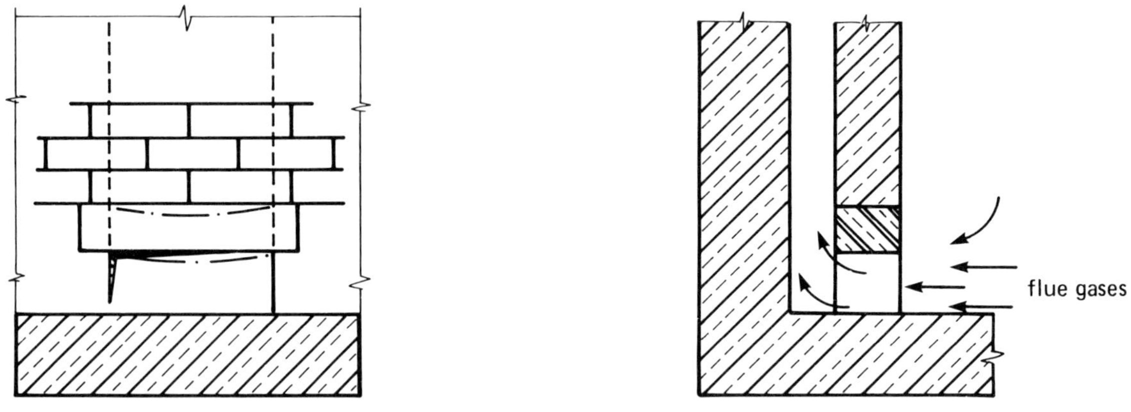

Fig. 8.16 Heat distortion of support lintel

8.3.5 Hearth

The hearth is almost invariably the most ill-treated component of the furnace. In melting furnaces, it may be subject to chemical attack from the molten charge or its slag, or both. In solids-handling furnaces, the main problem is mechanical damage; fixed hearths in particular are subject to impact and skidding forces as the charge is loaded, unloaded or pushed, but moving hearths such as bogies and kiln cars are by no means immune.

Hearth design, whilst of course subject to those basic factors (e.g. temperature suitability, expansion, thermal storage) influencing all furnace components, is thus more critically dependent on the furnace type and duty. This has resulted in a wider variety of hearth than for any other part of the furnace and it is only possible to give general guidelines in this Section. In most cases, the result in practice is a compromise which ideally should be based on cost-effectiveness, i.e. the acceptable balance between capital outlay and maintenance costs.

Choice of refractory and constructional mode must first satisfy (as applicable) the following requirements:

- Structural strength, to be able to accept the weight of the charge and in some cases the weight of the loading and/or load conveying equipment.
- Resistance to impact (often point-loading) and abrasion.
- Resistance to chemical attack from molten charge, slag or hostile atmosphere.

The other general criteria can then be applied.

In relatively non-aggressive situations, the traditional refractory/insulation multi-layer brick construction is still commonplace, although modification to minimise thermal mass is often quite feasible. Joints are, however, an even greater problem in hearths than elsewhere and as the degree of severity rises monolithic construction is favoured (usually rammed in situ).

Moving-car furnaces and kilns are a specific category which has received considerable design attention in recent years and low thermal mass cars are becoming standard in many applications. On existing installations, it is not usually economic to convert cars until they fall due for major overhaul.

REFERENCES

BRITISH STANDARDS

BS 2972 : Methods of Test for Thermal Insulation Materials.
BS 2973 : Classification and Methods of Sampling and Testing of Insulating Refractory Bricks.
RPE/9 : Draft for Development: Aluminosilicate Fibres (Type 1250)

OTHER PUBLICATIONS

The Thermal Insulation of Furnaces and Kilns, A.E. Hubbard, Britain's Fuel Problems, pub. The Fuel Economist, 1927.

Steelplant Refractories, H.G. Chesters, The United Steel Companies Ltd., 1963.

Technical Data on Fuel, J.W. Rose & J.R. Cooper, The British National Committee World Energy Conference, 1977.

Insulation of Kilns in Building Brick Production, A.E. Aldersley, Ceram. Techn. Note 274, 1978

High Temperature Thermal insulation in the UK Process Industries, R.W. Barnfield, Eng. Proj. Division, AERE, Harwell, 1981.

Codes of Practice – Ceramic Fibre Linings, The European Ceramic Fibres Industry Association, 1985.

Refractories for Iron and Steel Making, J.H. Chesters, The Metals Society.
Refractories Production and Properties, J.H. Chesters, The Metals Society.

Refractories, Norton, McGraw Hill.

Introduction to Technical Ceramics, Bewaye Maclaren.

Fibrous Insulation Heat Transfer Model/Heat Transfer in Refractory Fibre Insulation, both by R P. Tye, ASTM. STP660.

Journals:
- Thermal Insulation.
- Ceramic Review

Pottery Ovens, Fuels and Firing, S.R. Hind, The British Pottery Manufacturers' Federation.

Gas and Air Supply Controls 9

9.7.4 (cont.)

CHAPTER 9

GAS AND AIR SUPPLY CONTROLS

9.1 INTRODUCTION

For the efficient combustion and utilisation of gas, it must be controlled safely and accurately. This should be as far as possible by automatic means and not rely on the vigilance and reaction of the operator.

Many more industrial processes are now using natural gas and with escalating labour costs many processes are fully automatic. Hence all control systems must be fail safe, which has led to more reliable valves with associated electric control equipment. Also, with ever increasing fuel prices, the combustion processes should be as efficient as possible. This has caused the development of sophisticated air/fuel ratio control equipment with monitoring and adjustment over the turndown range.

This chapter will therefore examine the manual and automatic valves available; governors to give constant pressure at the burner irrespective of flow; air fuel ratio control systems to give optimum fuel efficiency by keeping combustion at or near stoichiometric, and valve proving systems to comply with the larger burner systems over 3 MW.

9.1.1 Introduction to Valve Symbols

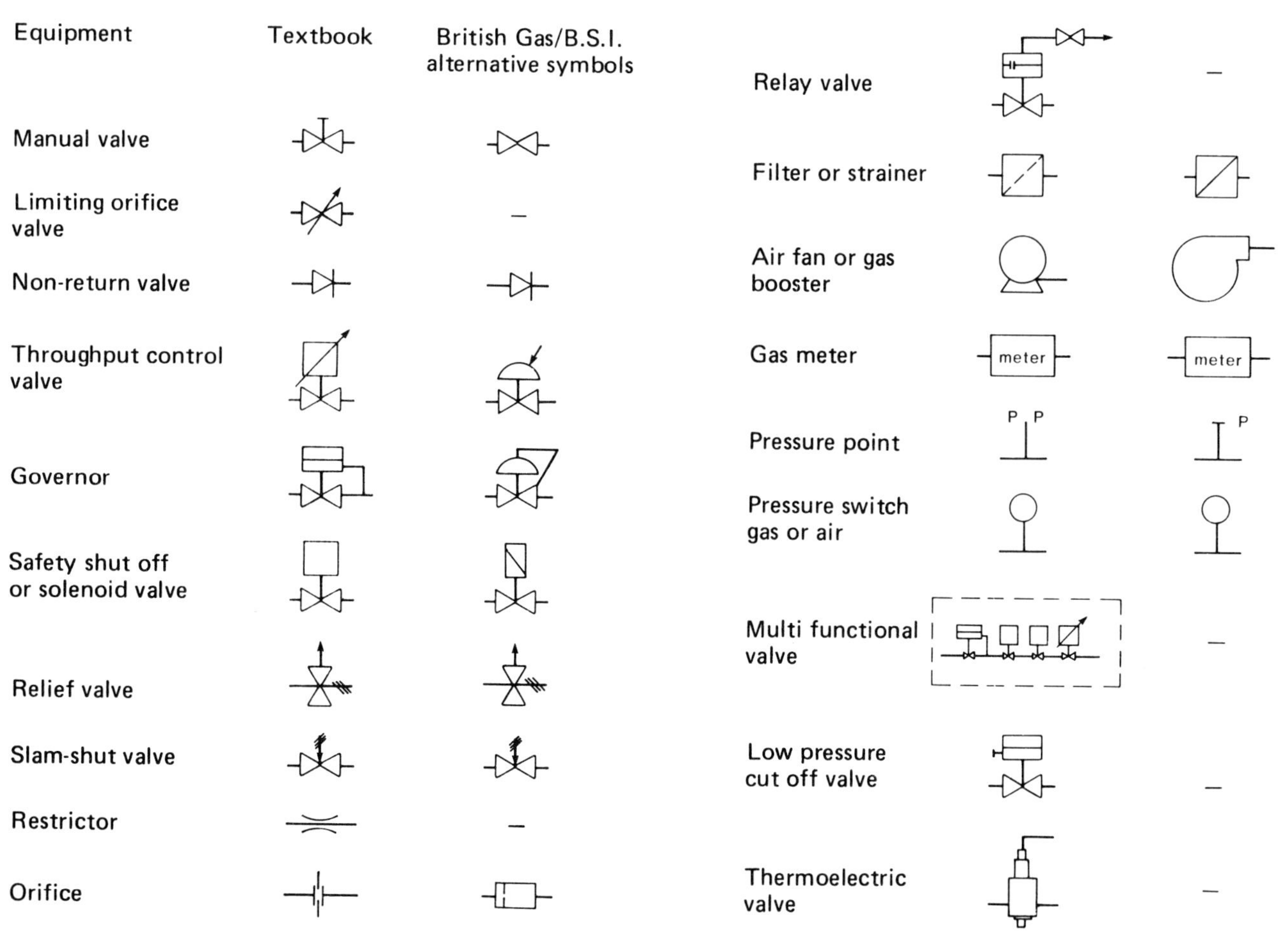

Fig. 9.1 Symbols used for indusrrial/commercial pipework equipment

The symbols shown in this book are those which have been used extensively by the Industrial/ Commercial Marketing Department of British Gas for several years and also by equipment manufacturers and customers. There are other symbols for example as used by British Gas, Gas Business Engineering, and the British Standards Institution, which may be found in other publications. These are shown alongside the symbols used in this publication in Figure 9.1.

9.2 GOVERNORS

Gas governors are basically pressure reducing valves. Their most common role is as a constant pressure device to give a constant outlet pressure from a higher and possibly varying inlet pressure, over a range of throughputs. Low pressure governors normally handle low inlet pressures up to about 0.34 bar (5 psi), and give a range of outlet pressures from about 3 to 80 mbar (1.5 - 32 in wg). High-low governors, high pressure governors for use on district governors, and zero (atmospheric) pressure governors are increasingly being used.

When used as a constant volume governor, a constant differential pressure is maintained across an orifice. This type of technique is used on recuperative burners to give constant gas flow as the burner heats up. Fig. 9.2 shows the layout of a constant volume system.

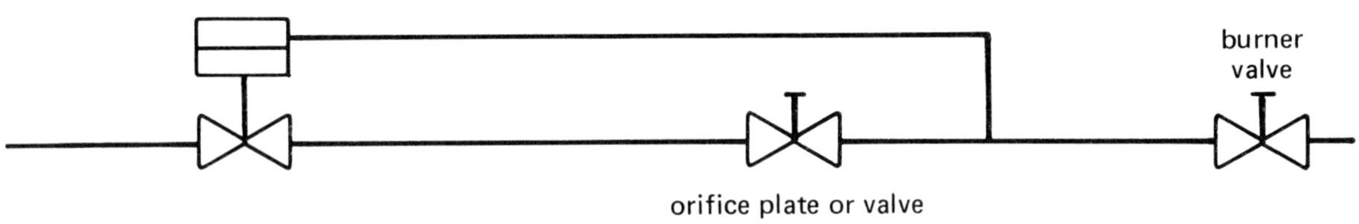

Fig. 9.2 Volumetric governing

9.2.1 Low pressure governors

A low pressure governor normally takes the form of a horizontal, spring loaded, flexible diaphragm which is exposed to the outlet gas pressure on its underside, and to atmospheric pressure on its upper side (through a vent). Any increase in gas pressure on the outlet side raises the diaphragm, partially closing the valve, and tending to restore the pressure on the outlet side to its initial value. The diaphragm reacts only to the outlet gas pressure, and the controlled outlet pressure is, of course, always lower than the inlet pressure.

A second subsidiary diaphragm is used in compensated governors. This diaphragm has the same effective area as the valve, and so compensates the effect of varying inlet pressure on the valve. With small governors such as used for pilot lines, a degree of compensation is achieved by the valve having the same area as the single diaphragm.

Governors of this type are manufactured by companies such as Jeavons Engineering and Bryan Donkin Ltd. Fig. 9.3 shows a Jeavons J48 Low Pressure Governor.

A governor will be necessary on all main and pilot gas burners encountered on industrial and commercial equipment. The pilot governor is a smaller version of a low pressure governor but without the compensating diaphragm.

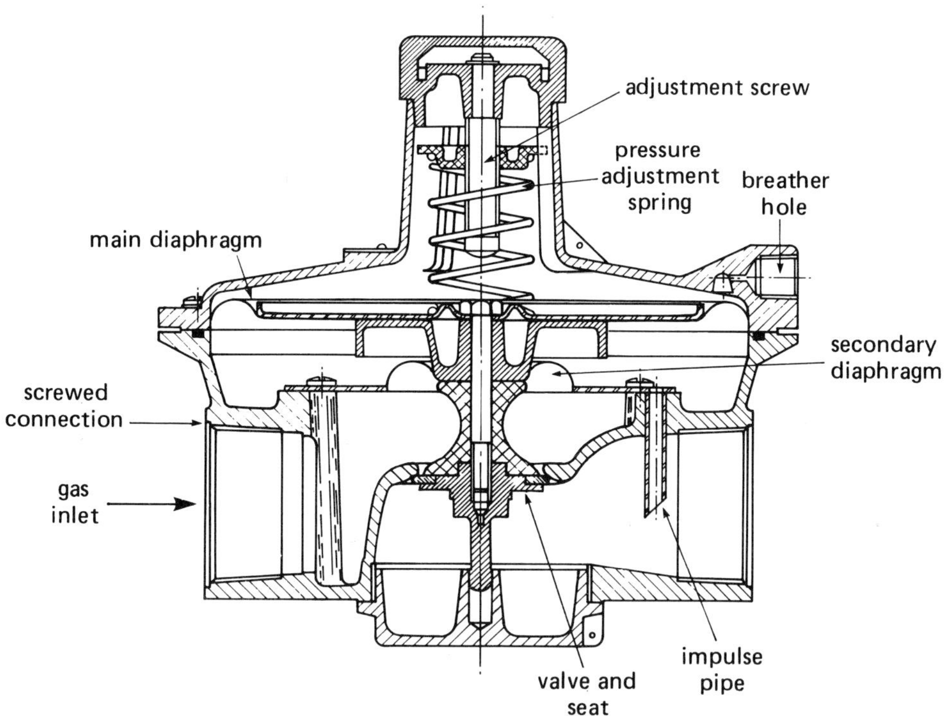

Fig. 9.3 Jeavons J48 spring loaded governor

9.2.2 High pressure and high-low pressure governors

High pressure and high-low pressure governors often work on the lever principle. As can be seen in Fig. 9.4, gas enters through a small orifice to act on a very small valve surface. The low pressure resulting from the pressure drop across this restriction, is free to act on the relatively large flexible diaphragm, which is loaded to the required outlet pressure. There results a considerable ratio between the relative forces on the diaphragm and the valve. The ratio is still further magnified by the mechanical advantage of the lever connecting the two moving components.

An internal relief valve is frequently incorporated in high pressure governors such that if the outlet pressure rises above a predetermined level, gas is allowed to escape through a vent to atmosphere. The relief valve will come into operation at approximately 22 mbar (9 in) above the outlet pressure setting of the governor. Fig. 9.4 is a governor of this type. The correct sizing and termination of the vent is important.

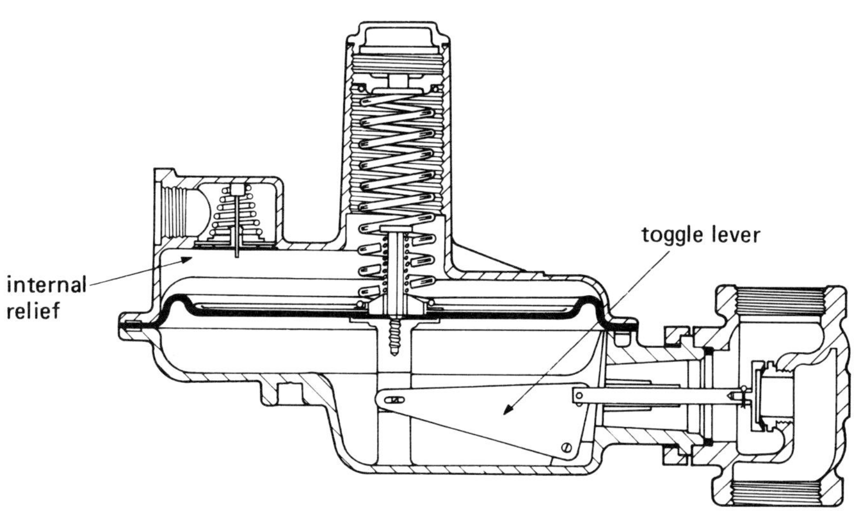

Fig. 9.4 High low pressure governor

One example of the high-low pressure governor is the Jeavons J125. This is a standard type lever operated system with a spring loaded vent valve above the main diaphragm chamber. It will accept inlet pressures of 70 mbar – 8.6 bar (1–125 psig) to give outlet pressures from 5 mbar to 760 mbar (2 in wg - 11 psig).

Options on the governor are either a full or limited capacity relief valve situated on the top of the diaphragm and over pressure and under pressure slam shuts. Figure 9.5 shows the over pressure and under pressure slam shut assembly which is attached to the governor casing by Allen screws. The diaphragm case can be fully rotated, and during inspection and servicing the case can be removed without disturbing the pipework.

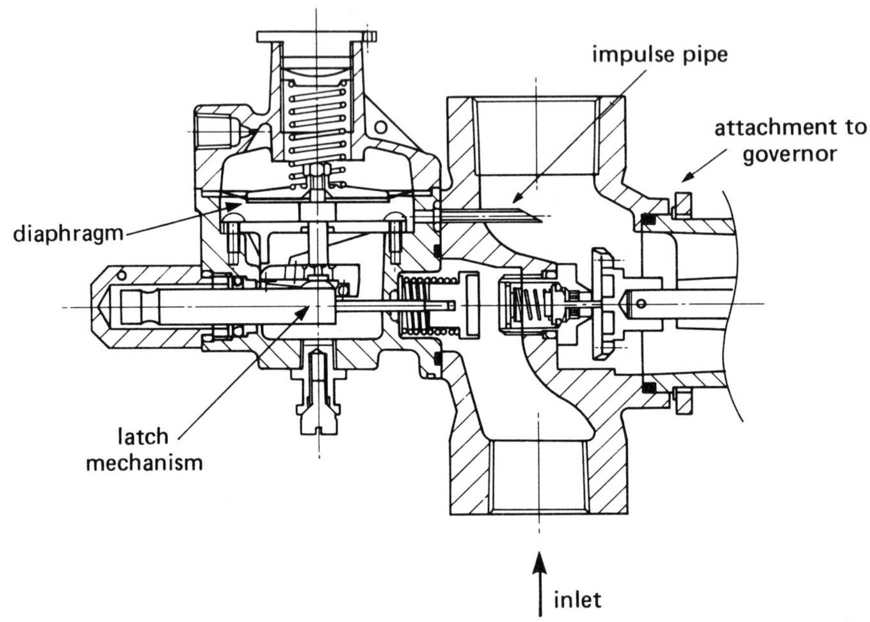

Fig. 9.5 Jeavons J125 - Over pressure and under pressure assembly

If the outlet pressure rises above a preset value as set by an adjustment screw, the diaphragm on the slam-shut valve will be lifted by gas from an impulse pipe in the gas outlet. This will trigger a latch mechanism to close the inlet to the valve, which can only be reset by an external pull shaft.

If the outlet pressure drops below a preset value the diaphragm on the slam-shut valve will be depressed by the spring thus tripping the latch and closing the inlet to the governor.

9.2.3 Zero Pressure Governors

Zero pressure (or atmosphere) governors are used principally in pressure air injection systems such as the air blast system. They are also used in the backloaded mode with pressure divider type of ratio control system. The 226Z governor manufactured by Donkin is an example of this type, Figure 9.6. This zero governor takes the form of a horizontal, flexible diaphragm exposed to atmospheric pressure on its upper side (through a vent), and to outlet pressure on its underside (through an impulse pipe). It is very similar to the low pressure governor described previously except that the weight of the diaphragm and valve parts is supported by a low-rate tension spring to which slight adjustments may be made in order to obtain slight changes in outlet gas pressure.

Any increase in outlet pressure raises the diaphragm and partially closes the valve, so tending to restore the outlet pressure to zero gauge pressure (atmospheric). Conversely, any suction at the out-

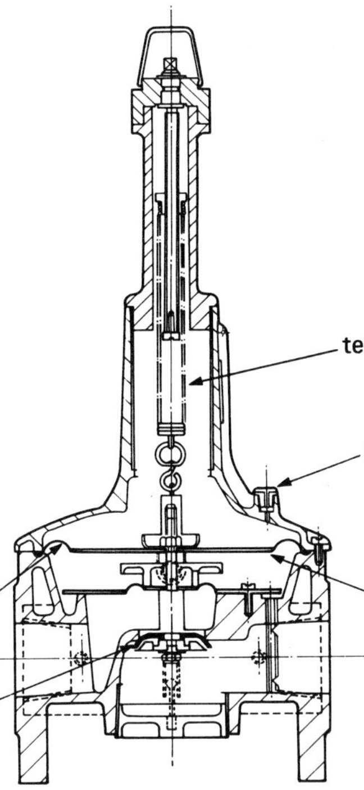

Fig. 9.6 Donkin 226 Z zero governor

let partially opens the valve and again restores equilibrium between the atmospheric and the outlet gas pressures.

For optimum performance zero governors should operate with a small pressure differential, a maximum of 5 mbar (2 in wg). A zero governor should be installed downstream of a constant pressure governor set to 5 mbar (2 in wg).

9.2.4 Pilot operated governing

The advantage of this system is that it is possible to obtain quite high outlet pressures whilst utilising full size diaphragm case assemblies resulting in maximum sensitivity. In addition, upon inlet pressure failure or in the unlikely event of main diaphragm failure, the governor will close (see Figure 9.7).

The system involves utilising a controlled pressure from a pilot governor to control the main governor. The pressure from the pilot governor would be adjusted to throw a pressure equal to the summation of the final outlet pressure required plus the spring loading of the main governor.

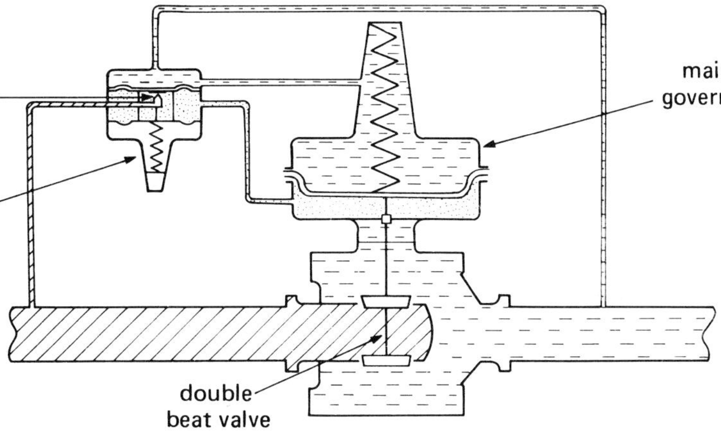

Fig. 9.7 Pilot controlled governor (schematic)

The mode of operation is that gas passes between the two diaphragms of the pilot governor and on to the underside of the main governor which will lift the seats off the valve against the compression spring. As the pressure increases on the downstream side of the main governor, this will be transmitted to the top side of the diaphragm which will move the valve nearer to its seats until the valve reaches an equilibrium position.

At the same time the downstream pressure will keep the top diaphragm of the pilot governor down tight on the jet inlet to the space between the two diaphragms, thus allowing no gas through to lift up the main governor diaphragm. Any increase in the outlet pressure will move the valve still further to the seat, restoring the pressure to the original setting. Conversely any decrease in output pressure will take pressure off the top diaphragm of the pilot governor, opening the inlet jet which will in turn lift the main governor diaphragm and open the main valve, raising the pressure to the original figure.

Upon inlet pressure or diaphragm failure, the pressure on both sides of the diaphragm will equalise and the spring will move the main governor to the shut position.

9.2.5 Pressure reducing valves

Pressure reducing valves (regulators) reduce high pressure supplies of fluid to a constant usable working pressure.

The working elements of a pressure reducing valve, shown in Fig. 9.8, consist mainly of a flexible diaphragm which controls a valve through an interconnecting valve pin, and an adjusting spring which is loaded by means of an adjusting screw.

The pressure side of the diaphragm is connected to the outlet part of the reducing valve so that regulated pressure is exerted on the diaphragm. The operation of the pressure reducing valve is essentially identical to a low pressure governor without a compensating diaphragm.

When the adjusting screw is retracted so that no load is applied to the spring, the regulator valve is closed. As the adjusting screw is turned in, it applies a load to the spring which is transmitted to the valve through the diaphragm and the valve pin, thus opening the valve. If the regulated pressure

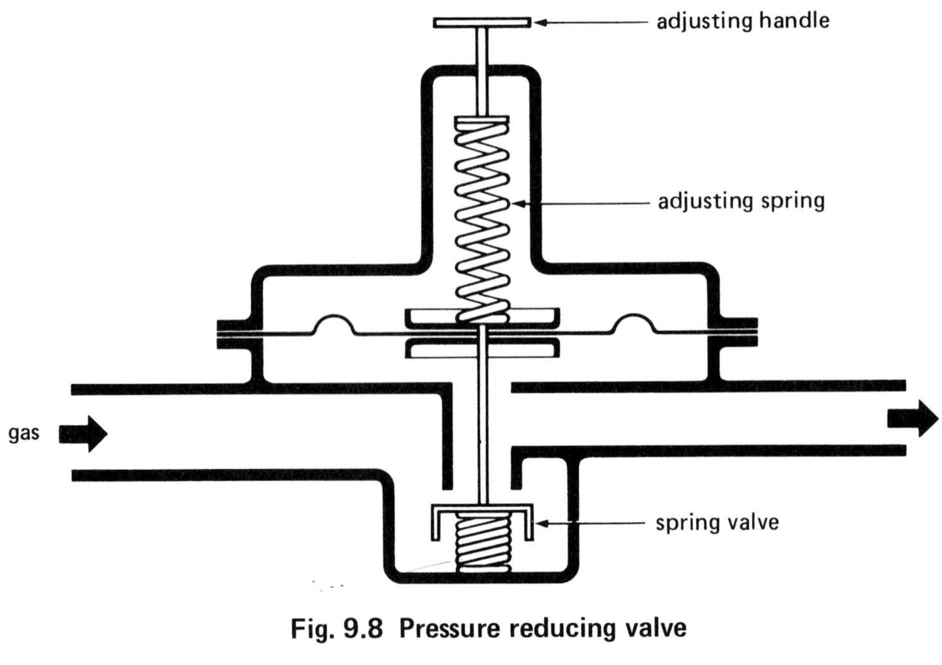

Fig. 9.8 Pressure reducing valve

tends to increase, the pressure against the diaphragm increases, forcing the diaphragm to compress the spring until the force exerted by the spring is equal to that exerted by the regulated pressure. This closes the valve just the amount necessary to compensate for the change in demand thus maintaining the desired regulated pressure.

These types of regulator are usually made to accommodate inlet pressures up to 17.2 bar (250 psi), and fitted with outlet pressure gauges. Special considerations must be made when oxygen regulators are required, to accommodate the higher pressures involved (around 138 bar (2000 psi) and to avoid the explosion hazard from contact between oil and pressurised oxygen. Reference to the oxygen suppliers is recommended.

9.2.6 Installation and Servicing of Governors

Governors should be installed in easily accessible positions, not exposed to excessive heat or freezing conditions.

(i) Remove plugs from inlet, outlet and vent connections.

(ii) The service pipe must be well blown out and cleaned so that no dust or metal turnings from newly cut threads lodge in the governor valve and prevent it closing effectively.

(iii) Install the governor in a level position or the position which the manufacturers suggest, making sure the gas will flow in the correct direction.

(iv) Remove any packing pieces and reassemble the governor.

(v) If the governor is supplied with relief valves, adjust these as the manufacturer suggests to 'blow' at the desired pressure. The vent for the relief valve(s) should be connected by a pipe to discharge any gases at a high level outside the building. The end of the vent pipe should be protected at the top by a close return bend to prevent the ingress of water.

(vi) Turn on the gas slowly.

(vii) Adjust the spring to obtain the desired outlet pressure if this is a pressure different from that for which the valve is already set.

(viii) Ensure that the top cover is properly seated to avoid 'chattering' and that the vent hole is clear.

Servicing

Periodically the downstream pressure from the governor should be checked, and the spring adjusted accordingly to give the required pressure.

Once every 10 years the manufacturers recommend that the diaphragms, spring and valve seal are replaced. These are obtained as a governor service kit from the manufacturers.

9.3 Manually Operated Valves

There are a wide selection of manual valves available to the Industrial/Commercial gas engineer. It is preferred that these be quick acting, i.e 90° turn on the valve handle from the fully open position to the fully closed. This would indicate the use of a ball, plug or soft seat butterfly valve. However, other valves such as gate valves are acceptable provided that they are frequently pressure tested and serviced to ensure closure in any emergency.

Operating levers on valves should be secured to the valve to ensure that when the valve is closed the lever is at right angles to the pipe, and when the valve is open the lever is in line with the pipe. Valves on vertical gas pipes require the levers to move upwards from the horizontal to the vertical position to open the valve.

The valves encountered in industrial and commercial premises for gas isolation are listed below. Further information on selection of valves for installation into these premises can be obtained from IM/15, Manual Valves – A Guide to Selection for Industrial and Commercial Gas Installations.

a) Plug Valves (non-lubricated and lubricated)
b) Ball Valves
c) Soft Seat Butterfly Valves
d) Conduit Gate Valves
e) Wedge, Double Disc and Parallel Slide Gate Valves
f) Globe, Screw Down and Disc on Seat Valves
g) Needle Valves
h) Diaphragm Valves
i) Rotary Valves
j) Rack and Pinion Valves
k) Limiting Orifice Valves.

9.3.1 Plug Valves

The plug valve combines simple design with a small pressure loss. It is a rapid shut-off valve in that it can be moved from its maximum flow position to complete shut off by a 90° rotation of the plug. For these reasons it is normal practice in the Gas Industry to use plug valves for the isolation of appliances and for the master control of burners. Such a valve is illustrated in Figure 9.9.

A quadrant cock is a brass plug in a brass body with a sector showing the valve position. These valves are used in gas and air lines up to a pressure of 350 mbar (5 psi). If the valve leaks it is a simple job to strip down the valve, lap the plug into the valve body with fine grinding paste, clean and repack with graphite grease. Brass valves should comply with BS 1552. They can be manufactured with reasonably linear flow opening characteristics.

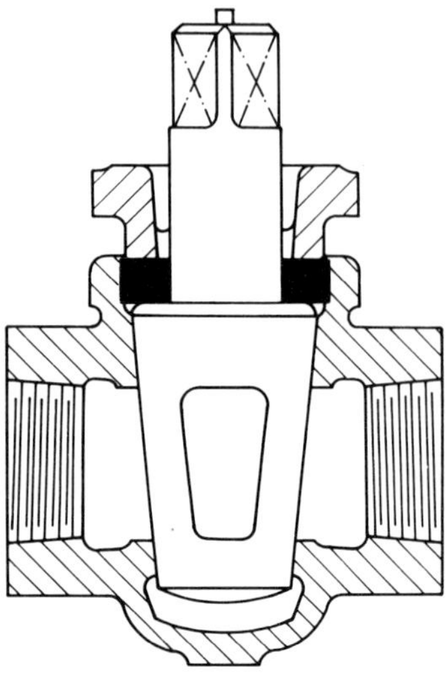

Fig. 9.9 Brass plug valve

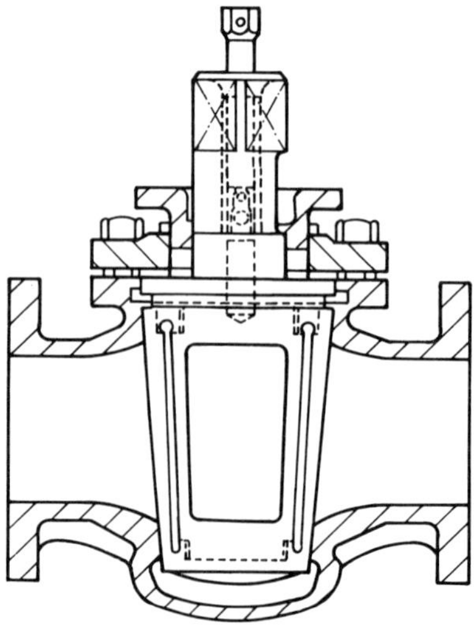

Fig. 9.10 Lubricated taper valve

Lubricated taper plug valves (Figure 9.10)

These valves have a passageway in the centre of the plug through which grease is forced by a lubricant screw while the valve is in service. This grease is forced upwards between the plug and the valve body and effectively forms the valve seal. The more frequently the valve is operated the more frequently the valve should be lubricated. These valves will withstand gas pressures of up to 6 bar (100 psi) when constructed in cast iron and up to 20 bar (300 psi) when constructed of steel. Most plugs and valve spindles in sizes above 75 mm (1½ in) are treated with a LOMU coating. This low friction PTFE reduces the frequency of valve lubrication, and ensures easy operation even after lengthy periods when the valve has been left in the open or closed position, e.g. Serck Audco valve.

With larger valves above 200 mm (8 in) the weight of the plug could be quite high causing the valve to jam in the seat. The valve is therefore inverted at about this size and its weight supported by ball bearings. Large valves of this type are usually operated by a worm gear unit and handwheel.

A further recent development by Serck Audco is the Super H pressure balanced plug valve. Here pressure balance holes in the plug allow gas pressure above and below the plug to prevent the plug from seizing even after long periods of closure, Figure 9.11.

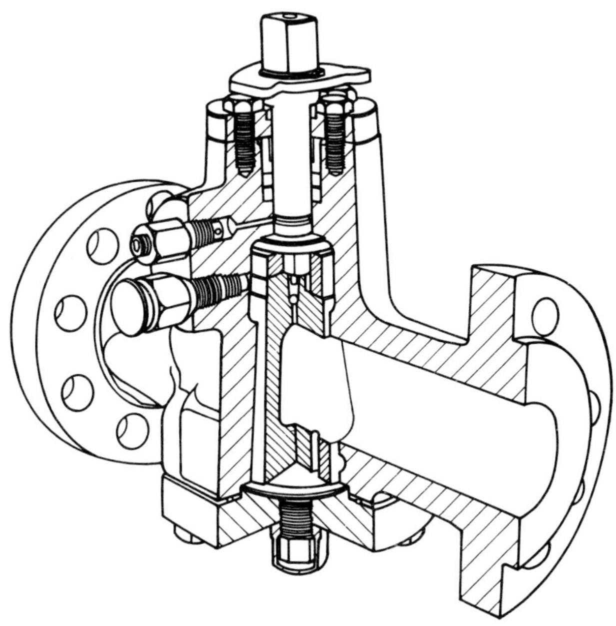

Fig. 9.11 Serck Audco Super H pressure balanced plug valve

All plug valves are capable of mechanical overtravel so that position indicator (proof of closure) switches can be fitted to prove the valve in the closed position. Lubricated and non-lubricated plug valves can be used for plant isolation, burner and zone control, and section isolation. They are also recommended for secondary meter control. The pressure drop through these valves is low and they are fire resistant.

9.3.2 Ball Valves

These are valves in which a ball is turned through 90° between fully open and the fully closed positions. A characteristic of these valves is the smooth, low operating torque. They can be regarded as suitable for all operations in which plug valves are used if fire resistance can be verified. A ball valve is shown in Figure 9.12. Because of the accurately machined and polished ball, larger valves of this type are very expensive, and therefore only used for high pressures.

The ball usually rotates between nylon seats on either side of the ball, but secondary metal seats are available. Little servicing on these valves is normally required. Mechanical overtravel on these valves enables limit switches to be fitted to monitor the closed position of the valve. The pressure drop through these valves is low, because the bore through the ball when open can be the same diameter as the connection to the pipework. The larger valves can have a two piece valve body. Working pressures up to 70 bar (1000 psi) with special designs of this type of valve are not uncommon.

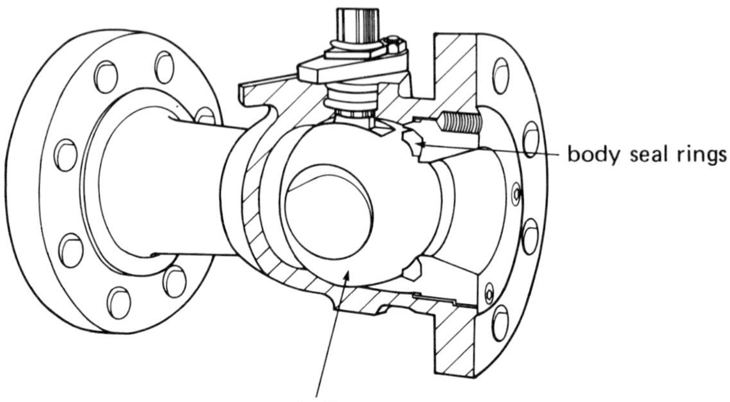

Fig. 9.12 Ball valve

9.3.3 Butterfly Valves

Butterfly valves are frequently used to control air or gas flows when they can be linked together. They can be power operated, their advantages in automatic operations being their simplicity and speed of operation and their low motor requirements due to the ease with which they can be rotated. Butterfly valves offer large port openings and consequently high flow rates.

The flow characteristics of these valves are not linear.

Where complete shut-off is required it is essential that the nitrile rubber lined type of valve is used. This rubber seal can withstand temperatures up to 120 °C (250 °F) and working pressures up to 14 bar (200 psi). The butterfly itself is usually coated with nylon, and there is an interference fit as it forces itself into the nitrile rubber to effect closure of the valve. The Serck Audco slimseal valve is shown in Fig. 9.13. This valve is installed between flanges in the pipework, the flange bolts effectively locating the valve in position. No gasket is necessary as the nitrile rubber is slightly proud of the valve body and this forms the seal. A popular application for these valves is the last isolation valve for example just before the burner on boiler plant. These valves are generally not fire resistant, have a low pressure drop, and require little servicing. If the valve does leak in service it generally cannot be repaired as the rubber seat is bonded to the valve in sizes up to 400 mm (16 in). Larger valves, however, are manufactured with a replaceable nitrile rubber seat. These valves are acceptable as section isolation, plant isolation and burner and zone control valves. When the flange bolts are undone to release the burner train the valve falls out. Some modern slimseal valves have flange bolt legs to obviate this disadvantage.

9.3.4 Conduit Gate Valves

These are valves in which a thin parallel sided gate slides between valve seats. The gate which incorporates a circular port the same diameter as the valve bore, always remains between the valve seats (Figure 9.14). They will give complete shut off and are generally suitable for use at higher pressures than apply to butterfly or plug valves. The gate and seat in which it slides are accurately machined which makes this type of valve relatively expensive. Seats can however be soft, in which case the valve is not fire resistant. The pressure drop is low because the port in the gate is generally the same diameter as the pipework. They are fitted with a handwheel which means that the speed of operation is slow. As with all gate valves, the direction of closing on the valve wheel should be clearly marked especially if the valve is left handed (i.e. clockwise to open). Position indication on these valves can be fitted as an extra. These valves are recommended on secondary meter installations, and acceptable as section and plant isolation valves. As with most gate valves, servicing should check that the valves can open and close freely, and that there is no gas leakage between the valve gland and operating spindle.

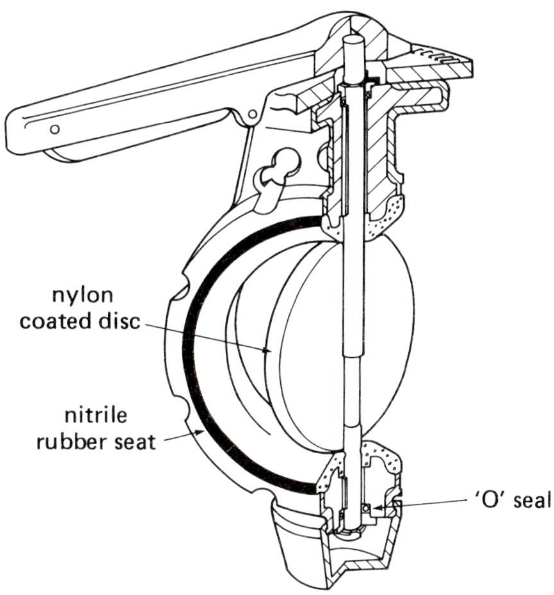

Fig. 9.13 Soft seat butterfly valve

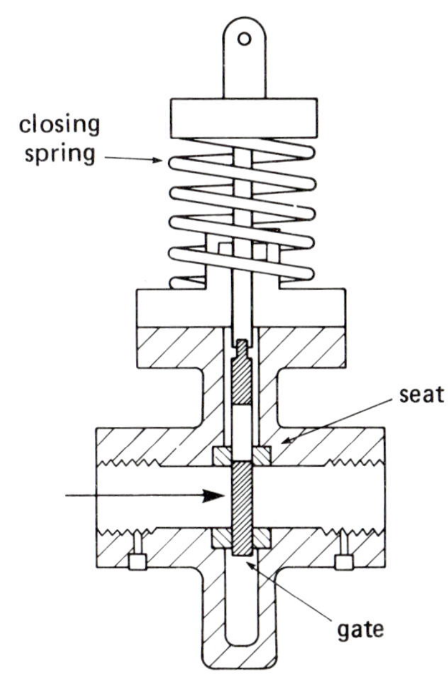

Fig. 9.14 Conduit gate valve closed

These valves can easily be designed with mechanical overtravel and position proving switches fitted, e.g. Maxon Safety Shut off Valves.

9.3.5 Wedge, Double Disc and Parallel Slide Gate Valves

These are valves in which a gate slides between the valve seats to close the valve, being withdrawn from the seats when the valve is open. In the wedge type the gate is wedge shaped and the valve is firmly closed by means of the wedge action between the gate and the seats (Figure 9.15). In the double disc type the gate consists of two discs which are forced apart against the valve seats by means of springs or other mechanisms to effect a tight seal (Figure 9.16). In the parallel slide type the gate slides between parallel seats, closure being assisted by the gas pressure sealing. These valves are expensive due to the accurately machined gates and seats, and are slow in operation as with the conduit gate valves. The pressure drops are low because they have full bore characteristics. The valves are suitable for relatively high working pressures and they are fire resistant as the seats are all metal. Their use is similar to conduit gate valves.

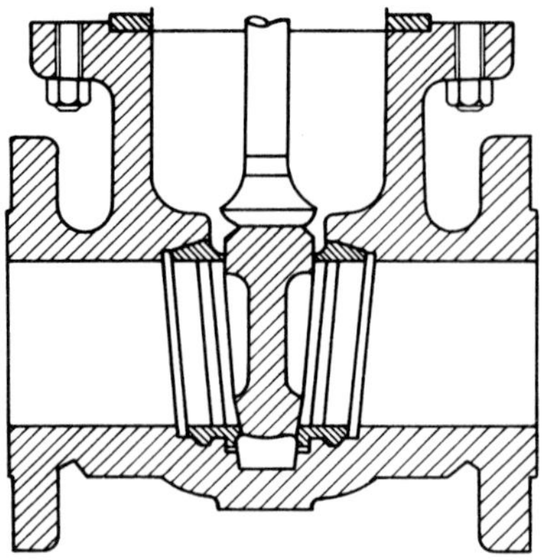

Fig. 9.15 Solid wedge gate valve

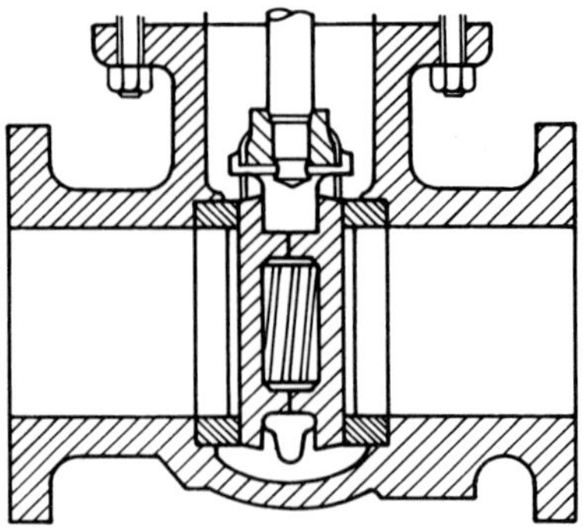

Fig. 9.16 Parallel slide gate valve

The parallel slide valves can be fitted with position proving switches on the valve overtravel.

9.3.6 Globe, Screw Down and Disc on Seat Valves

These are valves in which a disc is screwed down onto the seat to close the valves. The disc may be a mushroom shape (globe valve), Figure 9.17, or a flat disc (disc on seat valve). Figure 9.18. There is an inherent high pressure loss due to the abrupt change in direction of the flow path inside the valve, and the valves are multi-turn in operation resulting in slow speed of operation. The valve seats are generally metal to metal giving good fire resistance, but disc on seat valves usually have soft seats. These valves are not normally used for gas, although they can be made to withstand high pressure. A variation of globe valve is the double beat or equal percentage control valve where the gas flows equally across two valve ports (Figure 9.19).

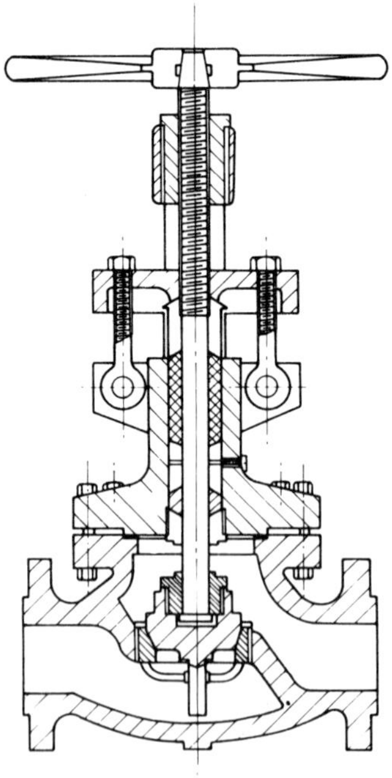

Fig. 9.17 Globe valve

9.3.7 Needle Valves

These are a version of globe valve where the valve is in the form of a tapered plug or needle which locates in a similar tapered seat, (Figure 9.20). The most common application for gas is where very fine and generally low throughput control is required such as on working flame burners.

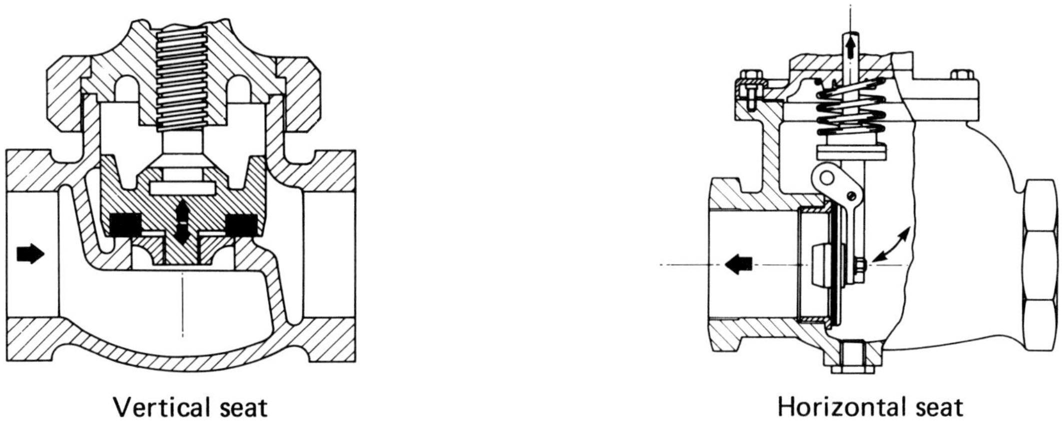

Fig. 9.18 Disc on seat valves

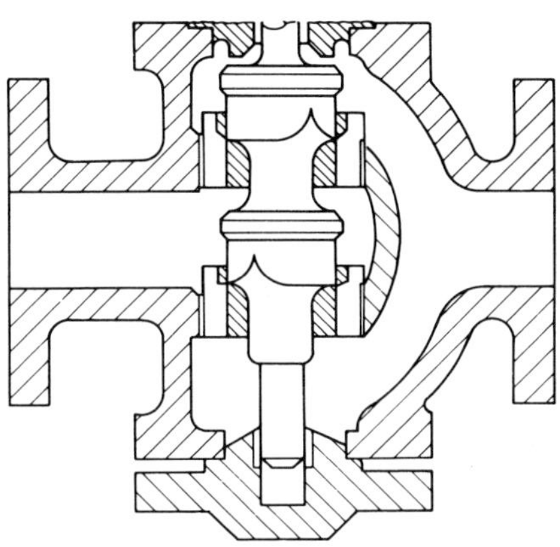

Fig. 9.19 Equal percentage control valve (globe type)

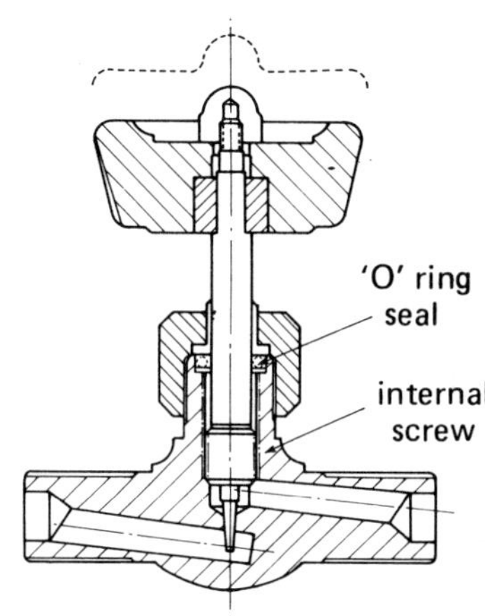

Fig. 9.20 Needle valve

9.3.8 Diaphragm Valves

In these valves a flexible diaphragm is lifted from and lowered onto a streamlined body seat. Valves are also available with straight-through bodies to reduce pressure loss in which the diaphragm seats onto the body bore to give valve closure. If grit or small bodies lodge under the diaphragm, further depression of this resilient material makes closure effective. These valves offer very good control of gas flow with their fine progressive adjustment. For industrial gas use the flexible diaphragm is made of synthetic rubber and reinforced with nylon fibres for strength. Should the diaphragm burst there will be leakage of gas, so this should be replaced if deterioration is apparent on inspection. This can be achieved by removing the cover whilst the valve is still in the pipeline, and should be replaced in any case after 5 years in service.

The most common type incorporates a wheel which gives slow operation. However, lever operated systems are available on sizes up to 18 mm (¾ in) which give rapid operation through 90°. The valves are available with either screwed or flanged connections.

Care should be exercised in the siting of these valves for utilisation applications because the diaphragms are not fire resistant. The latest range of valves manufactured by Saunders Valves Ltd (Figure 9.21) have a plastic sleeve valve position indicator, and a mechanical stop which alleviates the possibility of exerting too much force on the diaphragm thus preventing damage to the diaphragm. They will however withstand pressures up to 16 bar (240 psi) and will be encountered as section isolation and flow trimming valves.

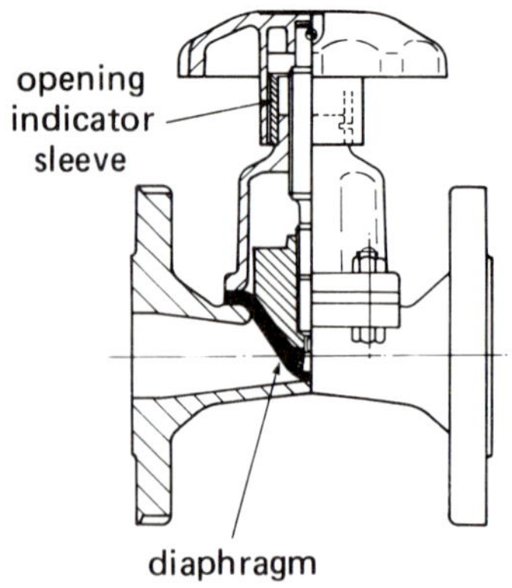

Fig. 9.21 Diaphragm valve

9.3.9 Rotary Gate Valves

These valves have a circular gate pivoted at the centre with a segment equal to approximately half its area forming a port. The gate is rotated through 180° from the fully open to fully closed positions by means of a gear around the gate circumference. They are slow in operation, fire resistant, have a low pressure loss and are available for pressures up to 2 bar (30 psi).

9.3.10 Rack and Pinion Valves

In these a spring loaded single faced gate slides against the valve seat. A pinion on the valve spindle meshes with a rack on the rear of the gate producing rapid operation of the valve. These valves are fire resistant, with a low pressure loss and available for pressures normally up to 70 mbar (1psi).

9.3.11 Limiting Orifice Valves

These are usually small ball or plug valves without an operating lever. They are installed in the gas or air lines to limit the maximum flow available. They are preset by the commissioning engineer to the required flow with an Allen key or special tool to make it difficult to tamper with (e.g. Stordy, Ballofix, Figure 9.22).

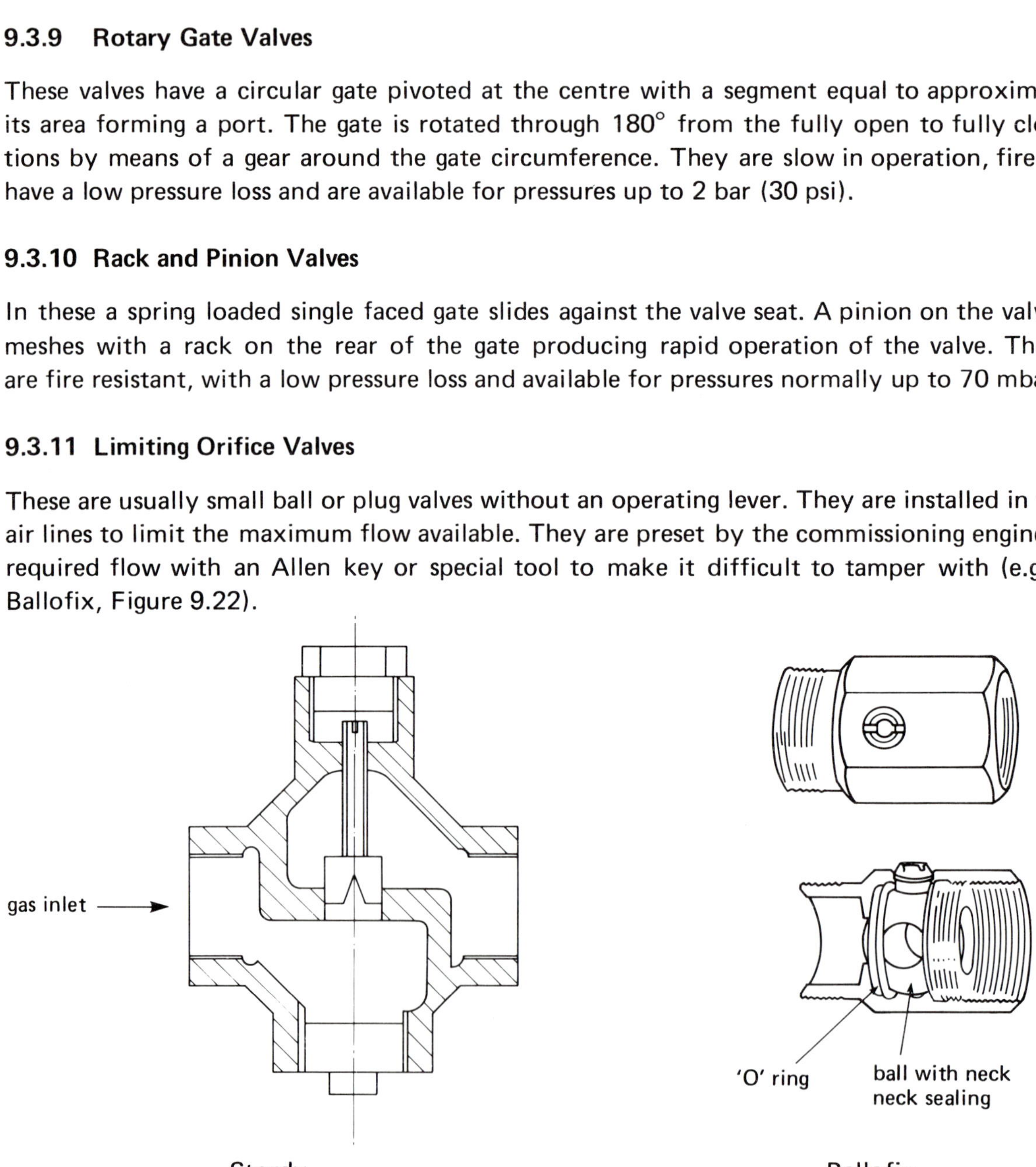

Fig. 9.22 Limiting orifice valves

9.3.12 Valve Applications

Table 1 gives a summary of recommendations for valve applications.

TABLE 1 SUMMARY OF RECOMMENDATIONS FOR VALVE APPLICATIONS

Application	Non-lubricated Plug 3.1	Lubricated Plug 3.1	Ball 3.2	Conduit Gate 3.4	Rotary Gate 3.9	Wedge and Parallel Slide Gate 3.5	Rack and Pinion Gate 3.10	Butterfly 3.3	Globe 3.6	Diaphragm 3.8
Valves Customers own secondary meters	Recommended up to 50 mm (2 in BSP)	Recommended	Recommended	Recommended with position indication otherwise Acceptable	Recommended with position indication otherwise Acceptable	Recommended with position indication otherwise Acceptable	Recommended with position indication otherwise Acceptable	Acceptable	Not recommended	Acceptable
Section Isolation Valves	Recommended up to 50 mm (2 in BSP)	Recommended	Recommended Check for fire resistance where required	Acceptable where fire resistance not required	Acceptable	Acceptable	Acceptable	Acceptable Check for fire resistance where required	Not recommended	Acceptable where fire resistance not required
Buried Valves	Not recommended	Acceptable	Recommended	Recommended	Not recommended	Not recommended	Not recommended	Acceptable Check for dust/debris	Not recommended	Not recommended
Plant Isolation Valves	Recommended up to 50 mm (2 in BSP)	Recommended	Recommended Check for fire resistance where required	Acceptable where fire resistance not required	Acceptable	Acceptable	Acceptable	Acceptable Check for fire resistance where required	Not recommended	Not recommended
Burner and Zone Valves	Recommended up to 50 mm (2 in BSP)	Recommended	Recommended	Not recommended	Not recommended	Not recommended	Not recommended	Acceptable with lever operation	Not recommended	Not recommended
Flow Trimming Valves	Acceptable up to 50 mm (2 in BSP)	Acceptable	Acceptable	Recommended	Recommended	Recommended	Acceptable	Acceptable	Recommended	Acceptable

NOTE: This table should not be used without reference to the main text. It should be noted that the suitability or otherwise of any valve depends on the particular make and construction.

9.4 SAFETY SHUT-OFF VALVES

The safe supply of the gas system to industrial and commercial processes will generally be controlled by electrical safety shut-off valves. The valves are fail safe and should conform to the requirements of BS 5963 Electrically Operated Automatic Gas Shut-Off Valves. The British Gas Certification Testing Scheme at Midlands Research Station and Watson House includes examination of safety shut-off valves from all aspects of safety and the satisfactory performance of valves.

The life tests applied to all Class 1 Fast Opening Valves up to and including 1½ in are 1,000,000 operations. All other valves, i.e. Class 1 Fast Opening above 1½ in, Class 1 Slow Opening (Class 1 High/Low and all Class 2) are tested to 250,000 operations.

Many valves now being manufactured and certificated are Multifunctional Control Valves and Valves with Special Features. These have many advantages for the gas installer. Only one set of end connections are required instead of about six, and they are very compact. Some valves have plug-in electric connections whilst on others the valve shell can be left in the gas line and the working parts replaced. This means that replacement can be carried out by unskilled men.

Multifunctional controls will have at least one Class 1 safety shut-off valve and a separate gas control function, such as a governor, within the same housing.

Valves with special features are those in which the safety shut-off valve actuator performs an additional function such as pressure regulation.

Certification of both the above valves is based upon the examination and testing of their performance as safety shut-off valves to BS 5963. Performance in respect of additional functions, e.g. pressure control is not examined other than to verify that their inclusion does not impair the safety shut off function.

On larger gas installations the safety shut-off valve may be controlled by a pneumatic actuator. These valves are not certificated.

9.4.1 Direct Acting Fast Opening Solenoid Valves

AC mains operated solenoid valves have the advantage of low initial cost and small overall size, but can be noisy in operation.

They are used extensively on gas pilot lines, and manufactured up to about 1½ in size. Above this they are rectified to DC because of the continuous hum and low flux density, Figure 9.23.

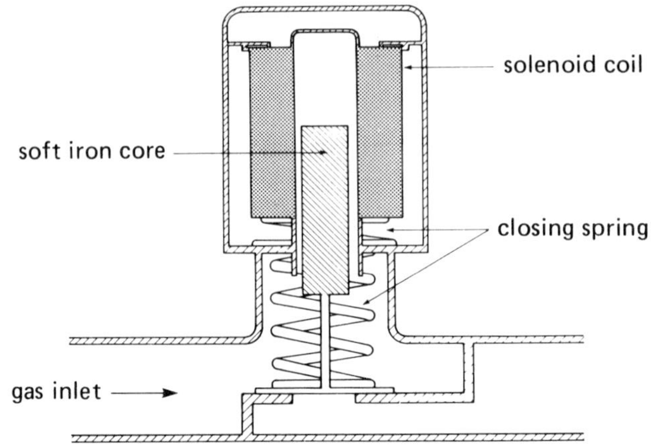

Fig. 9.23 Solenoid valve

9.4.2 Rectified AC Solenoid Valves

Most solenoid valves manufactured today are rectified AC valves. These are much quieter in operation than AC valves and do not suffer from a continuous hum. They have a higher flux density than unrectified AC valves, which means that a stronger valve spring can be employed. There is reduced danger of failure from fatigue because of the reduced liability to chattering.

The most common solenoid valves in use are direct acting. These normally open instantaneously when the solenoid is energised, and when de-energised close instantaneously by the action of spring loading. It can be arranged that these valves operate in the reverse mode i.e. energising of the solenoid closes the valve. These valves then become vent valves. Direct acting solenoid valves create a surge of gas on opening. Consequently they may not be suitable where burner stability is impaired or the surge may adversely affect other adjacent plant. Under these circumstanes slower acting valves should be used.

9.4.3 Direct Acting Slow Opening Solenoid Valves

These valves incorporate a small hydraulic dashpot which forces the oil through a small orifice. This slows the opening of the valve. On some valves this orifice is adjustable to alter the opening times. Sizes range from ⅜ in screwed connections, to 3 in flanged connections, Figure 9.24. Another variation is where the valve can open in two stages. The first stage is fast opening and will give low fire operation, the second stage is slow opening for high fire operation.

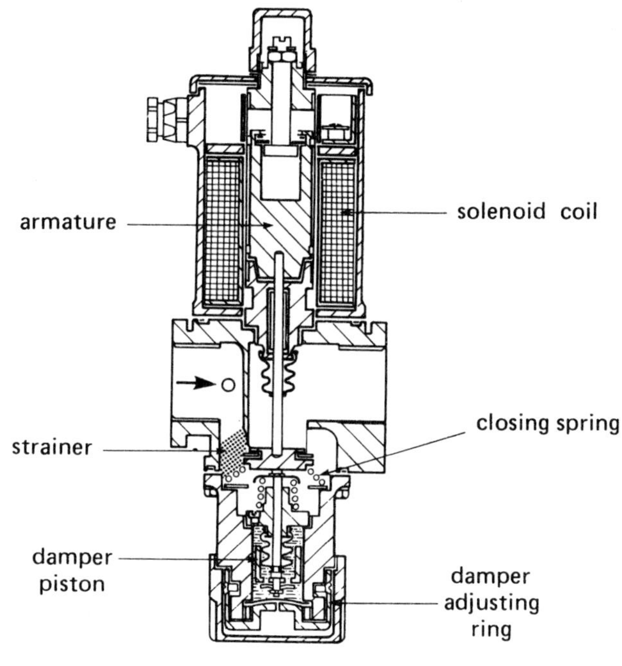

Fig. 9.24 Direct acting slow opening solenoid valve

9.4.4 Slow Opening Motorised Safety Shut-Off Valves

Where strong valve springs are employed to give even greater gas tight closure, rectified AC solenoids are not sufficient to open the valve. These valves can be operated by two distinct modes:

- Electro-hydraulic systems
- Electro-mechanical systems

All these valves close in less than one second, have a minimum specified closing force on the valve seat and in certain cases can be used for high/low/off operation. For safety shut-off purposes valves should comply with BS 5963.

The slow opening feature enables the main flame to be smoothly lit from the pilot flame instead of possibly being blown out.

9.4.5 Electro-Hydraulic Valves

Examples of this type of valve include those manufactured by Honeywell, Kromschroder, Landis & Gyr, Johnson Controls and Black Teknigas. A typical internal layout is illustrated in Figure 9.25A, which shows the Johnson valve.

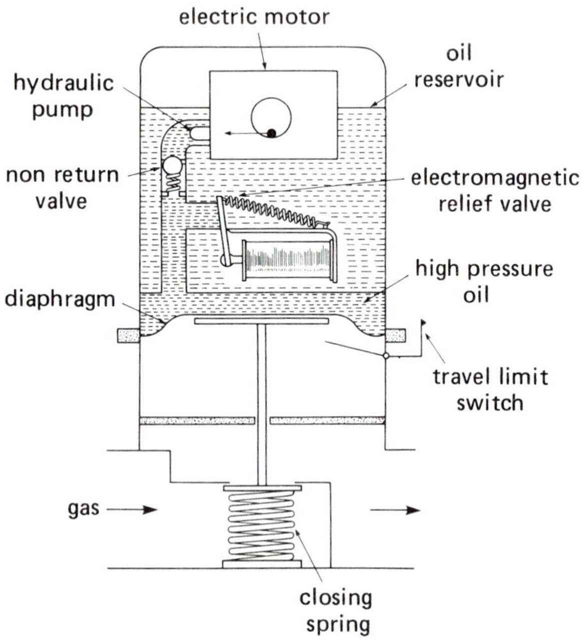

Fig. 9.25A Johnson Hydramotor valve

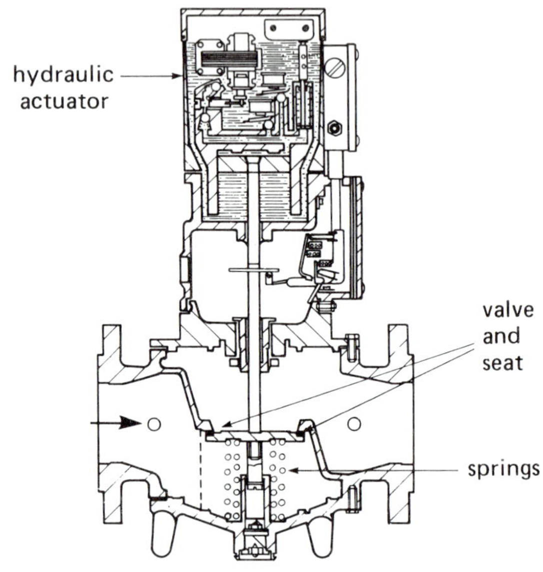

Fig. 9.25B Kromschroder electro hydraulic SSOV

An electric motor driven pump applies hydraulic pressure above the diaphragm which moves the valve stem against the force exerted by a powerful spring below the valve. When electrical power is supplied to the valve the electromagnetic relief valve closes and the motor starts pumping oil through the check valve, The oil forces the diaphragm down and the valve opens until this trips the travel limit switch, which cuts off the motor.

The electromagnetic relief valve remains energised, thus maintaining the pressure above the diaphragm. When the valve is de-energised the electromagnetic valve opens, and oil rapidly escapes from the high pressure region above the diaphragm back to the reservoir, assisted by the action of the spring which thus presses the valve firmly on to its seat.

Positive shut-off for all gases is ensured by employing a soft synthetic valve seat surface. Opening times of the valve are in the order of 10 seconds because of the time required to fill the space above the diaphragm with high pressure fluid.

A further development of this approach is the SKP range of valves manufactured by Landis & Gyr Ltd. These valves can be purely slow opening motorised valves or they can be multi-functional valves with integral governing and air/fuel ratio control. The basic principle of operation is shown in Figure 9.26.

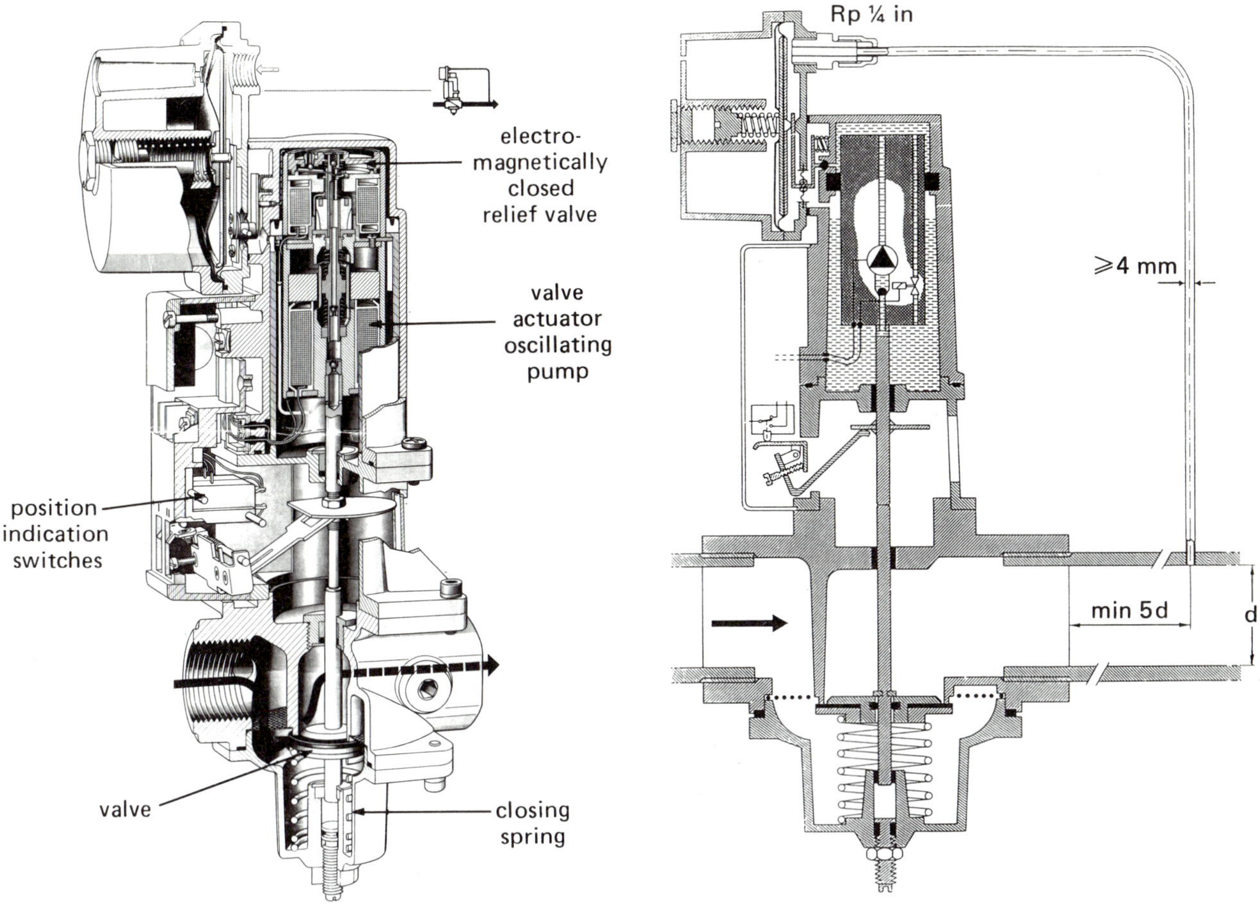

Fig. 9.26 Landis & Gyr SKP valve

When the voltage is supplied to the valve actuator, the oscillating pump pumps oil from the reservoir into the pressure chamber above the piston, causing the piston to move downwards and thus opening the valve. The spring loaded check valve and the electro-magnetically closed relief valve located to the left of the check valve prevent the oil from flowing back into the reservoir. When the piston reaches its equilibrium position the check valve begins to operate as an overload valve since the pressure exerted by the check valve spring is smaller than the oil pressure on the ball. When the power supply to the actuator is disconnected the relief valve opens and the pump stops so that the closing spring of the valve can push the piston upwards.

A disc fitted to the rod of the actuator is used for position indication. The disc with the aid of a lever system also operates the change-over switches of the actuator. The switching positions are infinitely adjustable over the entire stroke e.g the closed position indicator switch is set and sealed and the switches are used for the positioning of the low fire and high fire stroke, or for other stroke dependent control signals. The same actuator is used for all valve sizes. It is mounted into the valve body by means of four screws, and the design of the actuator is such that it can be mounted or replaced while the valve is under pressure.

With high-low operation opening of the valve starts in the same way as on-off. However, when the low fire position is reached, the disc, via the lever system, actuates a microswitch which is adjusted to the low fire rate. The power supply to the pump is now disconnected so that the valve disc remains in its present position. The pump will not be switched on again until the burner control calls for the high fire rate. When the high fire position is reached, the microswitch is activated thus disconnecting the power supply to the pump. When the load controller signals that the low fire position is required, the relief valve is opened until the low fire position is reached. In the event of controlled shut down or lock-out of the burner control, the relief valve is again opened and the actuator returns to the closed position in less than one second.

9.4.6 Electro-Mechanical Valves

One of the simplest of these systems is the Junkers Valve marketed by G Bray & Co Ltd. This is basically an electric motor driving through a gear train and clutch system. This eventually operates an external crank which is directly connected to the valve. On energisation the valve slowly opens through the gear train and on de-energisation the clutch is disengaged and the valve closes under the action of a return spring in less than one second.

Another electromechanical valve on a completely different principle is the Maxon valve shown in Figure 9.27. In addition to the conventional normally closed valve the company manufactures normally open versions as used in double block and vent systems. Both the normally closed and normally open systems are usually supplied with proof of closure switches. These valves are very robust and give extra positive closure on de-energisation, but are relatively expensive.

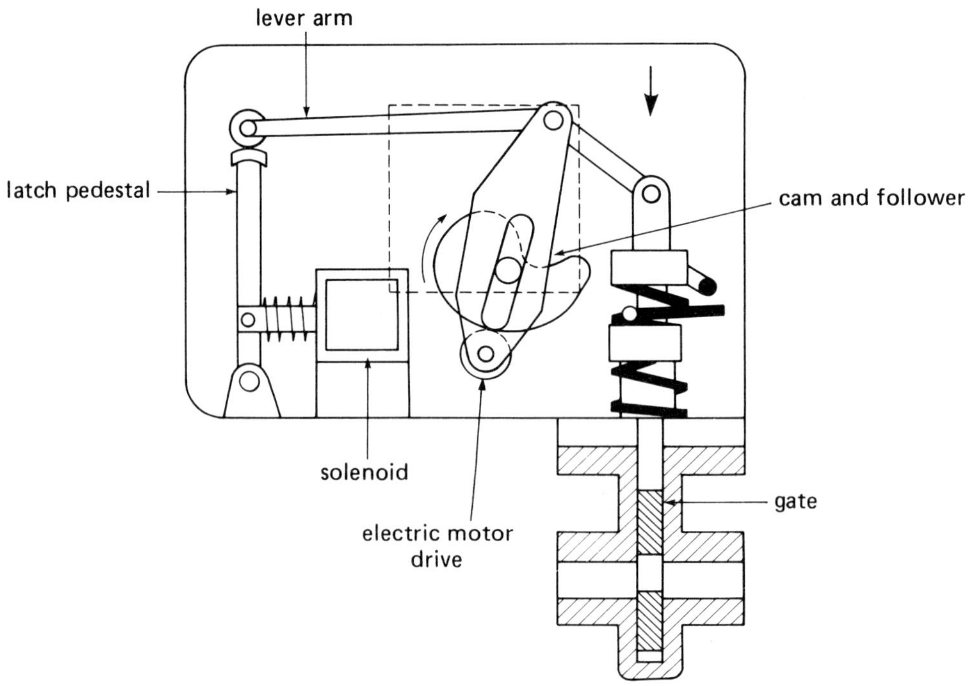

Fig. 9.27 Maxon valve (shown opening)

In this type the electric motor turns a cam. The cam follower is connected by a toggle plate and a lever arm to the valve spindle. A solenoid is incorporated which when energised provides a fulcrum for the lever arm on top of the latch pedestal. When the valve is energised, the latch pedestal is vertical and the lever arm opens the gate valve against the tension of the compression spring. A limit switch controls the amount of valve travel.

When the current is switched off, the solenoid is de-energised, the latch pedestal moves away, allowing the lever arm to drop and the spring to close the gas valve.

9.4.7 Valves fitted with position indicating switches

Several valves are now available with either Proof of Closure switches referred to above or Closed Position Indicator Switches. The proof of closure switch will check the valve to be in the 'mechanical overtravel position', i.e. the valve has mechanically passed the initial closing position to a final closing position such as on the Maxon Gate Valves. The Closed Position Indicator switches are usually fitted to disc-on-seat valves such as solenoids and the vast majority of hydraulically operated valves. The switches check that the valve is nearly closed. These valves comply with the requirements for automatic burner systems above 1 MW and up to 3 MW for system checking. Above 3 MW the double block and vent system using Proof of Closure switches is used or an alternative proving system.

9.4.8 Multi-functional Control Valves and Valves with Special Features

In these valves as the name suggests many of the separate gas controls are housed in one body.

The advantage to the installer is that only one pair of gas connections is necessary, and the overall size of the unit is compact and neat in appearance. Depending on specification, the valve can contain all or some of the following features, i.e. governor, two safety shut off main valves, one of which may be a slow opening and high/low, fully modulating air/gas ratio control and gas pressure switch. Other features can include integral pilot valve(s), add on pressure valve proving, closed position indicator switches and various pressure tapping outlets. A multi-functional control valve will embody at least two of the control devices including one safety shut off valve, Class 1.

As far as maintenance is concerned, removal and replacement can be undertaken by unskilled labour, because many valve bodies can now be left in the gas line, and the working inner parts of the valve removed by undoing set pins or screws after unplugging the electric supply.

At this time in their development the cost of the multi-functional valve is generally more expensive than the individual components of the valve train. Also if one part of the system fails, the whole unit must be replaced. These shortcomings are outweighed to the installer because of the savings in time and space in assembly. The cost, eventually, as they become more widely used should reduce.

These at present are made up to 2 inch BSP although larger valves are being considered. Some examples of these valves will be considered below.

9.4.8.1 Flamtronic Valves

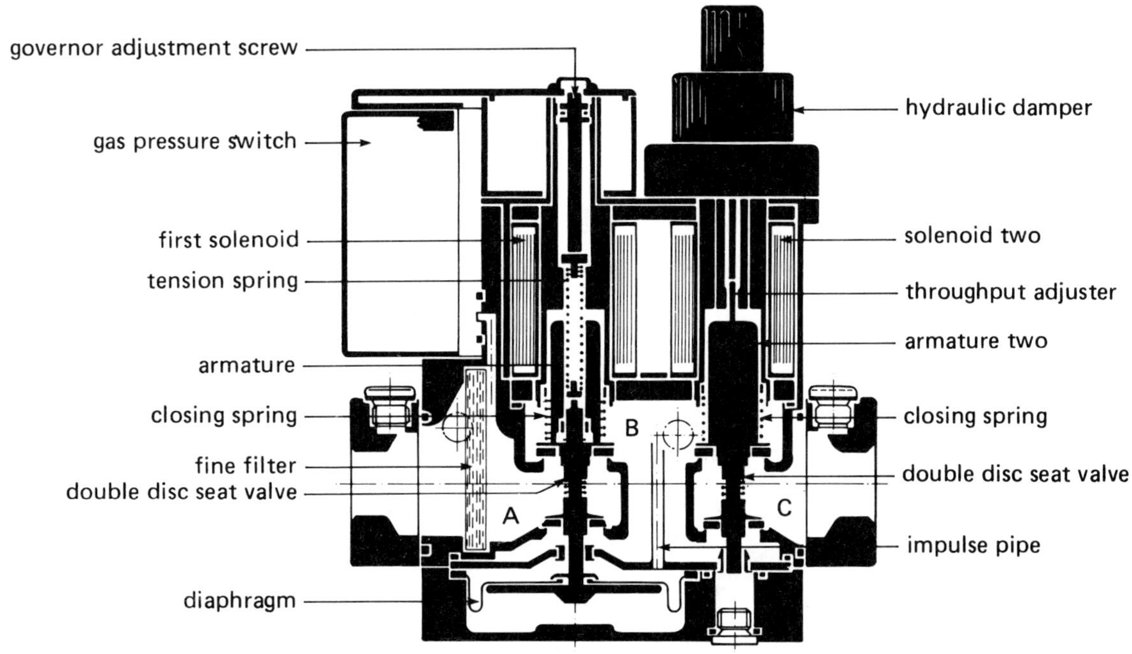

Fig. 9.28 Flamtronic Dungs MBD valve

Dungs valves marketed by Flamtronic Ltd, are shown in Figure 9.28. The gas multi block MBD valve is a complete system, consisting of filter, governor, pressure switch and two safety shut off valves, the second of which can be slow opening. Both valves have a double seat assembly for greater gas flow.

The mode of operation is as follows. When the first solenoid is energised, the armature is lifted against the spring and gas flows through the filter and the double seat valve into chamber B. The double valve now comes under the influence of the tension spring, which has its tension adjusted by its screw. The valve will lift until the gas exerts a downward force on the diaphragm through the impulse pipe. From chamber B the gas flows through the second valve into chamber C. This valve is a conventional solenoid valve, but can be fitted with a hydraulic damper assembly, which gives a slow opening valve. The valve may have a fast initial lift which is adjusted by a screw under the protection cover. The remaining slow opening of the valve is adjusted by a small adjuster screw. This second valve can be a double seat valve. The whole block is equipped with threaded connections suitable for test nipples, and can have a pressure switch as shown.

The complete block can be removed from the gas line by undoing four nuts and withdrawing a DIN connector plug. Screwed flanged adapters are supplied with the valve for assembly.

There is provision under the first and second valves for screwing in a closed position indicator switch.

On the larger valve block there is an optional pilot valve which connects round the second solenoid valve for expanding flame operation. The gas rate around this pilot line is adjusted by a small screw. An optional extra is valve tightness control.

The gas multi block MB ZR valve has similar features to the above but the second valve is high/low. The operation of the first solenoid and governor is identical to the MBD range. The second valve though, which again can be double disc, is operated by two separate solenoids, Figure 9.29, which

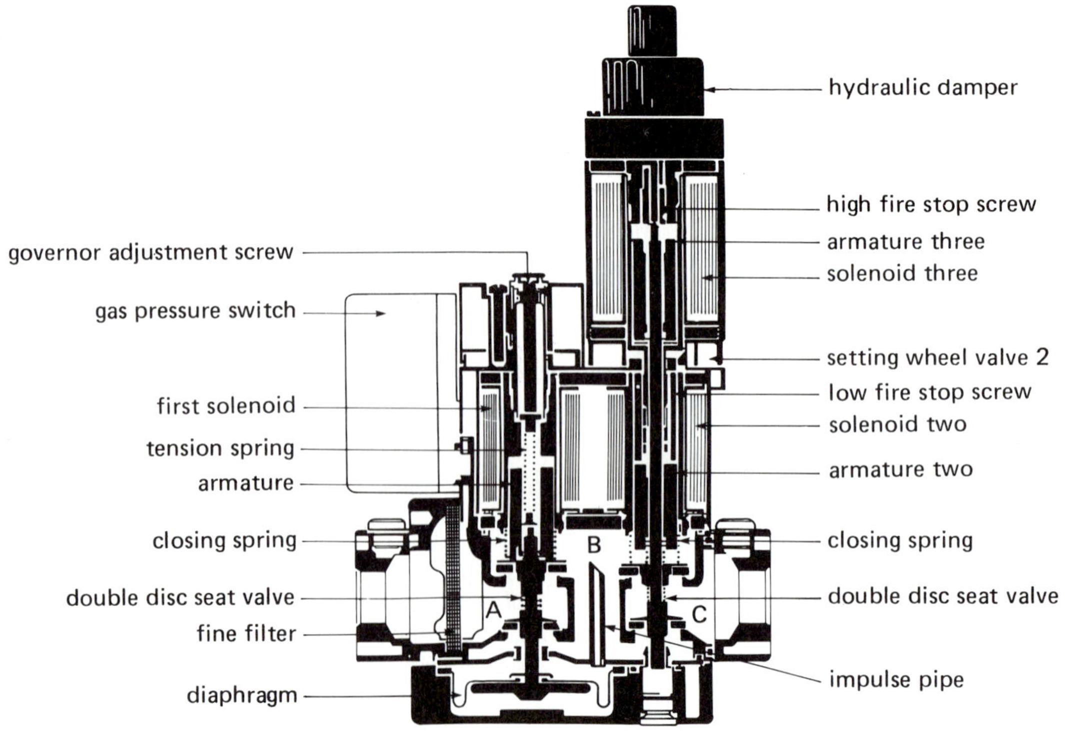

Fig. 9.29 Flamtronic Dungs MB ZR valve

are independent of each other. For low fire operation, solenoid two is energised which partially opens the valve to the adjustable low fire stop screw. When solenoid three is energised, the valve opens to the adjustable high fire stop screw. The slow opening characteristic is adjusted by the fast travel screw under the top cover cap.

9.4.8.2 Inter Albion Valves

Inter Albion Ltd have designed their multi-functional valve systems in a different way on their moduline concept valves. Individual components are bolted together which simplifies replacements within the system. There are over 1000 permutations of valves, governors, ratio controls systems etc.

Figures 9.30 and 9.31 show a general layout of all the possible variations which can be rearranged if necessary.

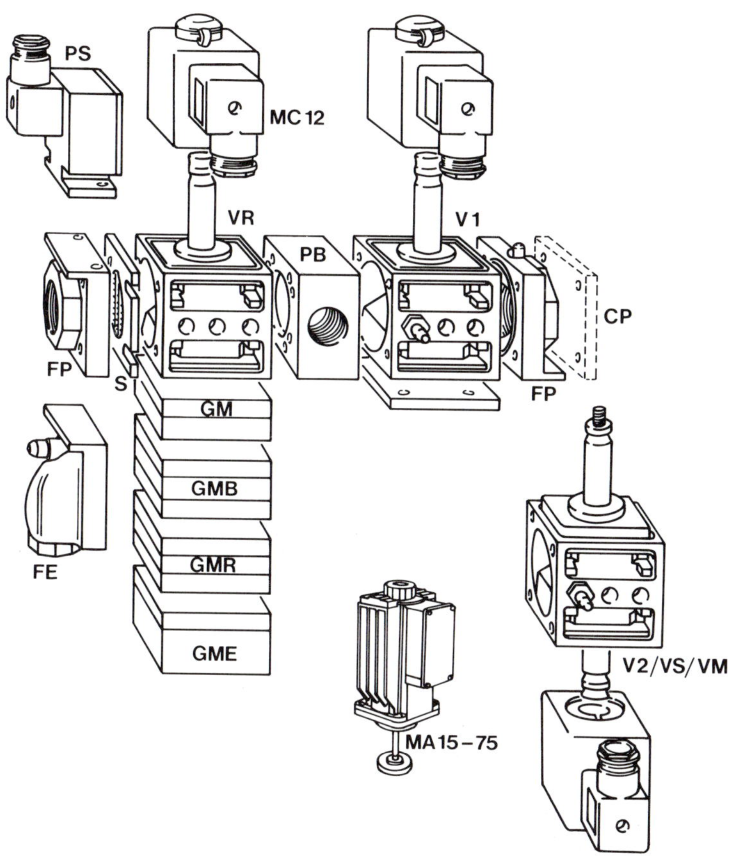

Key

FP or FE	connection flange
S	wire strainer
MC12	solenoid coil
VR	safety-shut off valve with governor
V1	single seat safety-shut off valve
V2	single valve body with two seats in parallel each with separate actuator to give high/low or open/closed control.
VS	similar to V2 but with one small seat for pilot operation
VM	single valve body with 2 seats in series. The first seat closes with a solenoid valve, the second provided for modulating control
PB	pilot body (tapping points for pilots)
GM	simple governor
GMB	air back load governor to give air/fuel ratio control
GMR/GME	pneumatic and electronic ratio controlled governors
MA15-75	solenoid actuator fitted with synthetic oil for heat dissipation and adjustable damping functions
PS	pressure switch
CP	cover plate

Fig. 9.30 Albion valve combinations

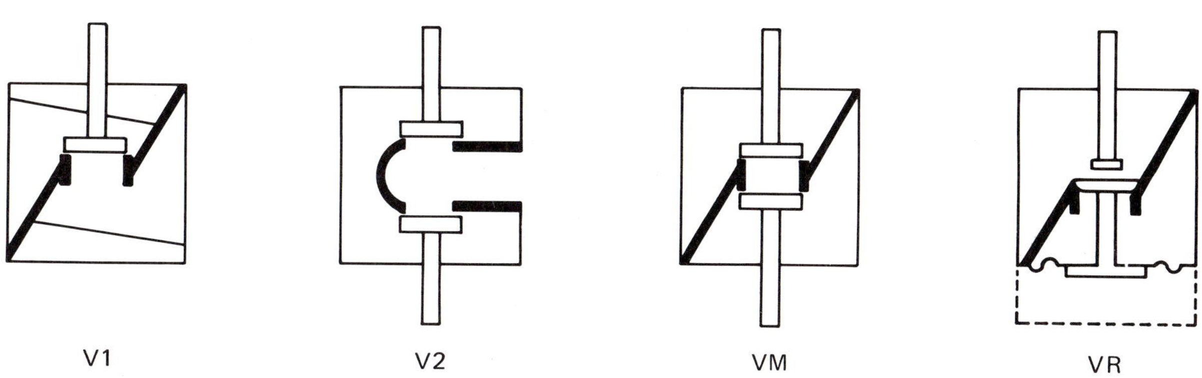

Fig. 9.31 Albion seat variations

Many actuators in the MA15-75 range can be fitted with closed position indicator switches by mechanical contact with the valve stem.

Motorised butterfly throughput adjusters integral with the valve body are available with fast and slow opening Class 1 solenoid valves (Figure 9.32). The butterfly valve is driven by a geared motor which is bolted to the valve body.

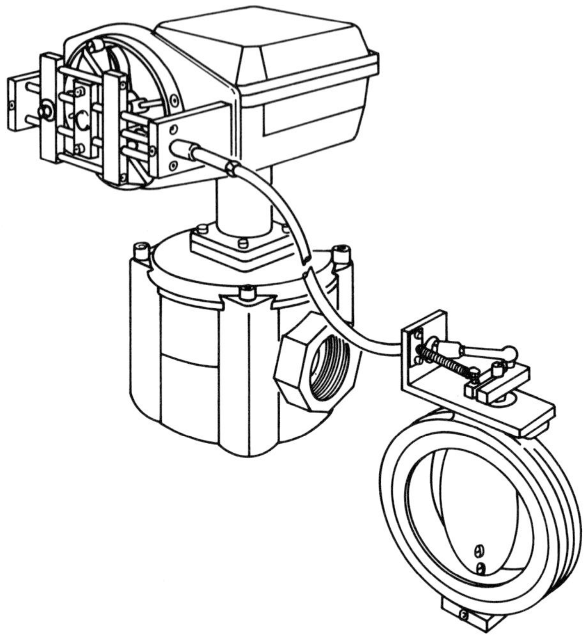

Fig. 9.32 Albion motorised butterfly valve

The operation of the safety shut-off valve/governor (see Figure 9.33) is described below. The plunger disc from the solenoid valve is not connected to the valve disc. When the solenoid is energised, the plunger rises against the spring exposing the inlet to the capillary channel, but the valve remains seated: Gas at inlet pressure flows into the chamber exerting an upthrust on the diaphragm, lifting the valve off the seat and gas flows through the valve. The increasing outlet

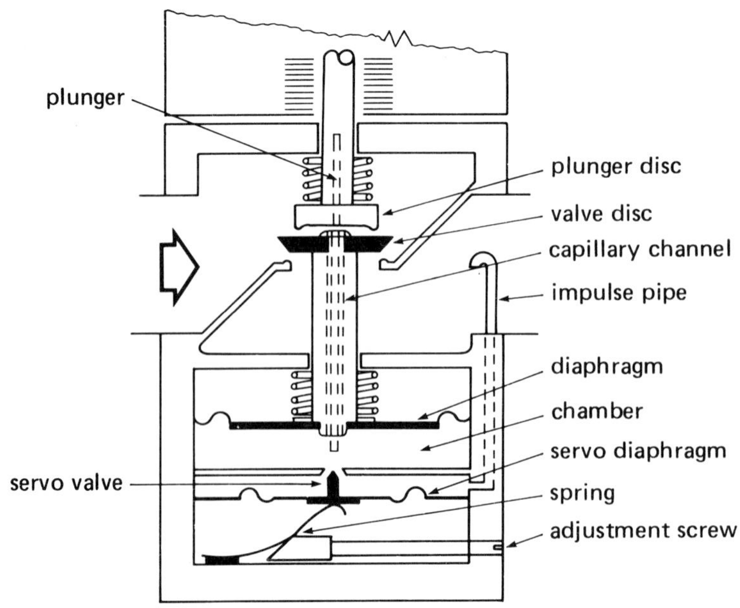

Fig. 9.33 Albion combined governor and SSOV

pressure is transmitted via the impulse pipe to the space above the servo-diaphragm. Downward movement of this diaphragm allows the servo valve to open and gas escapes from the chamber lowering the pressure under the main diaphragm so that the valve disc moves nearer to the seat to regulate the gas flow. The spring supplies the closing force to the servo valve, the tension of which is adjusted by the adjustment screw. This screw sets the governor outlet pressure.

When the solenoid is de-energised the plunger disc descends, forcing the disc on the seat and firmly closing the valve immediately. At the same time the plunger is pushed down to prevent any pressure build up in the chamber.

This governor can be backloaded with air pressure to give simple ratio control (Figure 9.34). The upthrust on the servo-diaphragm is provided by pressure impulse from the combustion air supply. Outlet gas pressure will vary directly in proportion to combustion air pressure. Other systems for air/gas ratio control with pneumatic pressure regulation and electronic pressure regulation are also available.

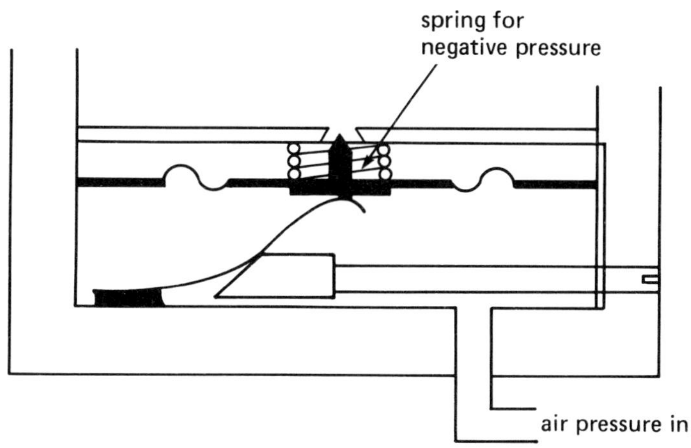

Fig. 9.34 Albion ratio control

9.4.8.3 Kromschroder Valves

The CNL and CN2NL multi-block valves include fast opening Class 2 solenoid valve and governor and slow opening Class 1 solenoid. In addition the second valve shown in Fig. 9.35 includes two valves, high fire and low fire operation, and the system can be fitted with an external pressure switch.

The mode of operation of the CNL valve is that gas passes through filter 1 and when the first valve is energised the spring force is taken off the valve seat and gas passes towards the inlet of the second valve. the first valve now comes under the control of the gas outlet pressure which acts downwards on the diaphragm X which tends to close the valve. This is balanced against the compression spring 7, the adjustment screw Y determining the desired outlet pressure. The gas now passes through the second safety valve which has an initial fast opening, the remaining slow opening being adjusted by a throttle screw.

The CN2NL has two solenoid actuators on the second valve. By adjustment of various screws and rings the first stage maximum opening is adjustable. The second stage opening is also damped to give a slow opening feature.

The CG range of multi-block valves have many features previously described built into them and they can be removed from the gas line by withdrawing four Allen screws from an adapter at each end of the block, and unplugging two electrical adapters. The multi-block valve comprises two Class 1 safety shut-off valves, precision gas governor, accurate air/gas ratio control and a pressure switch. It can also be supplied with bolt on valve tightness proving and start gas solenoid valve which can be fitted across the upstream or downstream safety shut-off valve. Closed position indicator switches may be fitted.

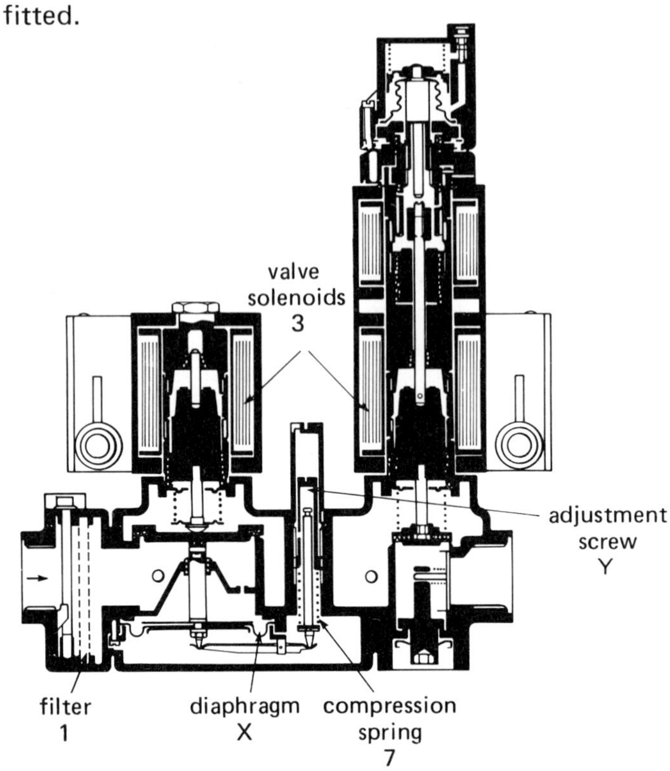

Fig. 9.35 Kromschroder CN2NL valve

The CG-D block valve has a precision governor with start gas rate pressure 2.5 to 10 mbar which will operate for approximately 5 seconds to give a start gas flow rate before the main governor becomes operable. The mode of operation is as follows.

The combination control CG-D is shown in the energised condition in Figures 9.36 and 9.37.

Both valves, being electrically in parallel, are in the closed position when de-energised and are held closed by the armature spring and the weight of the armature. Gas, therefore, cannot pass beyond the first valve.

When the control is energised both armatures lift. The first valve is opened since the valve disc is attached to the armature but the valve disc on the second valve is not mechanically attached to the armature and therefore remains closed. However the valve seal between the armature and the valve disc is opened which permits inlet gas to pass down the tube 1, between the valve disc and lower diaphragm, into the chamber A below the diaphragm. Gas also passes from this area down the vertical impulse tube 2 to chamber C where the pressure causes the diaphragm 3 to lift against the pressure setting arm 3. The initial lift for the start gas rate is set by the adjuster Ps which causes 4 to apply a pressure on the spring at the centre of the diaphragm which controls the opening of the ball valve, 6.

As the pressure in chamber A increases it will lift, thus raising the disc on the second valve. Pressure at the outlet now begins to build up and is applied to chamber B via impulse pipe 7. This presents a downward force on diaphragm 2, balancing the force applied by the spring at 4, which allows the ball valve 6 to open and release pressure from A to the outlet, causing the disc on the second valve to take up a position where it will maintain the start rate pressure previously set by Ps. Because of the restrictor in 2 and the capacity of chamber C there is a delay in the full upward movement of the diaphragm 3, resulting in the start gas rate being held for a period of about 3 to 5 seconds.

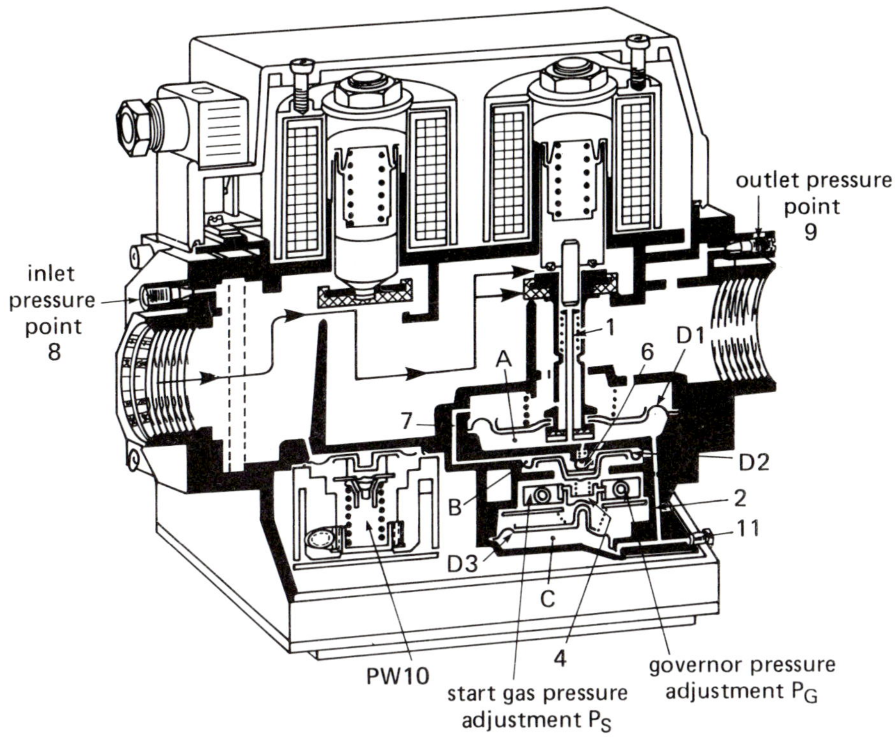

Fig. 9.36 Kromschroder CG-D valve with start gas pressure adjustment

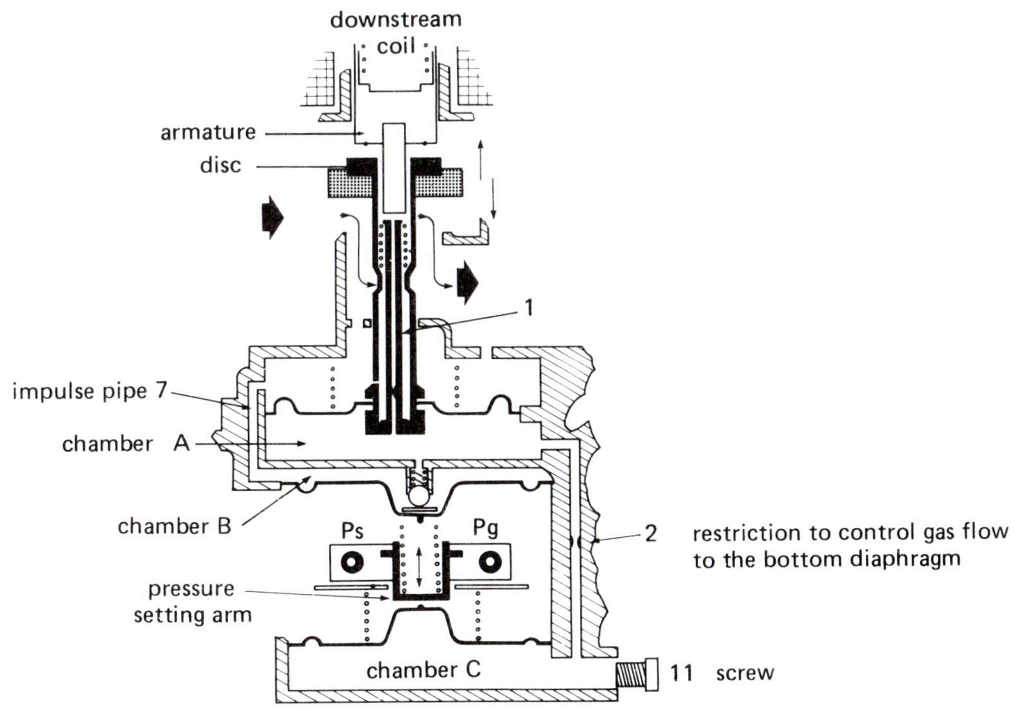

Fig. 9.37 Kromschroder CG-D valve with start gas pressure adjustment – second solenoid

After this delay diaphragm 3 continues to lift until 4 reaches the full rate pressure setting stop set by Pg which increases the spring force on the ball valve. When the pressure in A increases it causes the valve disc to lift higher giving a higher outlet pressure. This pressure is applied to chamber B when diaphragm 2 will take up a position where the downward force balances the upward force of the spring. Sufficient force is now applied to the ball valve to allow it to seal or weep gas to maintain the balance of pressure between A and B to keep the outlet pressure of the control at the preset value.

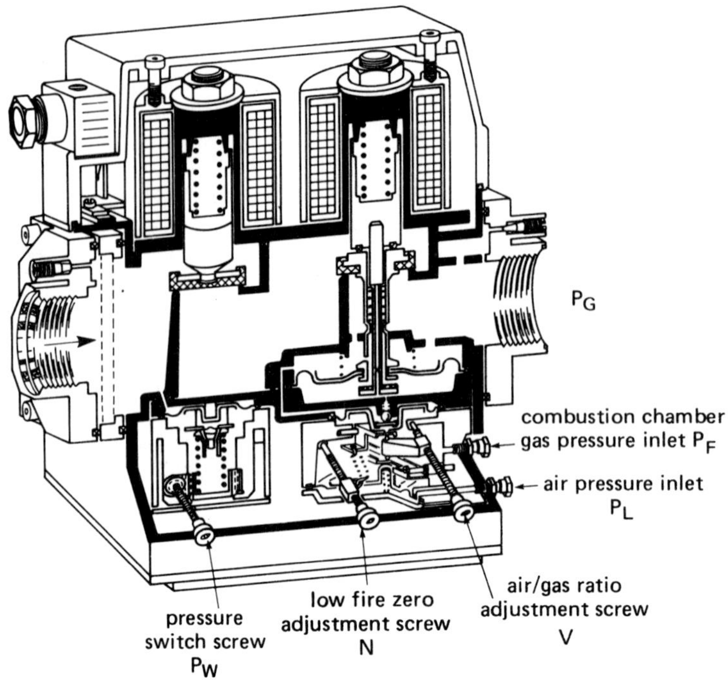

Fig. 9.38 Kromschroder CG-V valve with air/gas ratio control

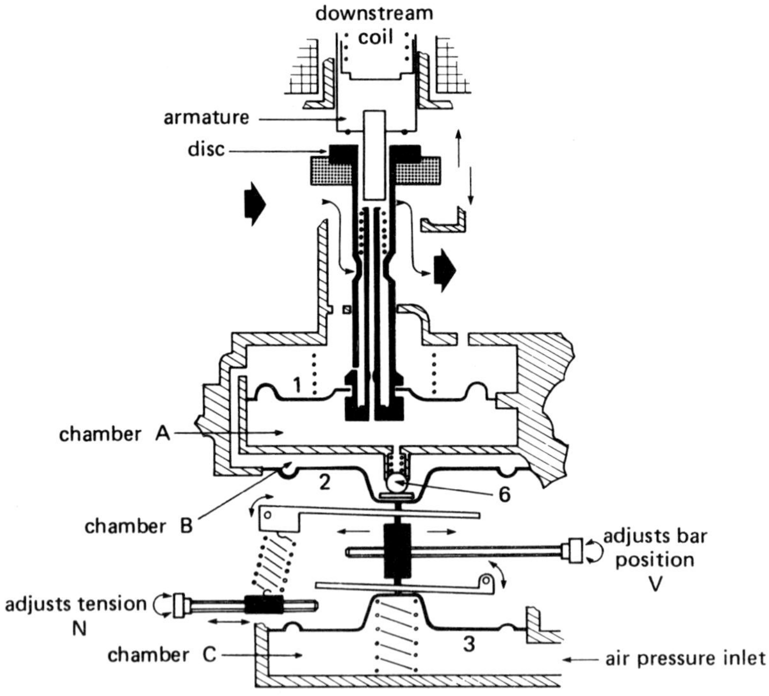

Fig. 9.39 Kromschroder CG-V valve with air/gas ratio control – second solenoid

To enable the start gas rate to be set, screw 11 is removed. This will keep the control in the start rate. Since the small weep of gas passing through the restrictor orifice in 2 will then be released to atmosphere and not applied to diaphragm 3, the diaphragm will not therefore remain in the down position. Screw 11 is then replaced.

A pressure switch 10 is fitted to the inlet side of the first valve and can be wired into the control circuit to prevent operation at below the pressure at which the setting Pg has been adjusted. Pressure points 8 and 9 have been provided to enable the inlet and outlet pressure to be measured.

The CG-V block valve (Figures 9.38 and 9.39) is similar to the above but the start gas governor facility has been replaced with air/gas ratio control. The internal impulse pipe 2 had been replaced by air pressure inlet PL which is taken from a tapping on the combustion line. To compensate for changes in combustion chamber back-pressure, a tapping from here is taken to PF on the valve. The air/gas ratio adjustment is set by screw V, whilst the low fire zero adjustment is set by screw N.

The mode of operation is essentially the same as the CG-D version but the air inlet impulse pipe lifts diaphragm 3 against the tension spring thus closing the ball valve 6. This causes the valve to open giving a higher outlet pressure to chamber B where the downward force balances the upward force of the spring. As the air increases the gas increases in proportion.

The CG-Z block valve has a high/low feature applied to the second valve seat. This is achieved by a second solenoid fitted under the valve block assembly. This replaces the air/gas ratio system on the CG-V valve, but the governing feature remains.

9.4.8.4 Landis & Gyr Valves

In the SKP series of valves the safety-shut off valve function can be combined with a governor and/or air/gas ratio control.

The hydraulic system here is identical to that discussed under Landis & Gyr earlier. In addition there can be a governor diaphragm system bolted on to the unit (see Figure 9.26) to give the governing function.

The outlet pressure represents the actual pressure value which acts on the diaphragm supported by the set-point spring. The movements of the diaphragm are transferred to a lever system which opens and closes a ball valve situated in the by-pass between the pressure side and the reservoir. If the actual value is smaller than the set point, the by-pass is closed so that the actuator can open the valve.

However, if the actual value is near the set point, the by-pass has only a small opening which slows down the movement of the piston because oil flows from the pressure chamber back into the reservoir. If the actual value and the set point are identical, the by-pass is opened to such an extent that the return flow corresponds to the output of the pump, i.e. the movement of the piston comes to a standstill. Since small movements of the diaphragm are sufficient to initiate the movements of the piston, the control of the outlet pressure is very fast and accurate. It follows with this system acting as a governor, the oscillating pump is continually activated, pumping oil through the by-pass ball valve.

The air/gas ratio control systems are shown in Figures 9.40 and 9.41.

With one system the standard actuator is used, and the governor attachment is replaced with a multi-diaphragm device. Air line connections to the governor are from either side of an orifice plate in the air supply line to the burner. An increase in the air flow causes a pressure differential across the diaphragm of the governor, which in turn closes the small ball valve in the by-pass between the pressure side and the reservoir of the valve. More gas then flows through the valve across another orifice plate in the gas line. This pressure differential is applied to another diaphragm of the governor system. This balances the pressure differential exerted by the air, effectively increasing the gas in proportion to the air. This is the Landis & Gyr SKP50 system.

This multi-diaphragm device is similar in operation to the Jeavons 121 air/gas ratio control system. It is suitable for use on systems with varying air back pressures, for example on recuperative burners.

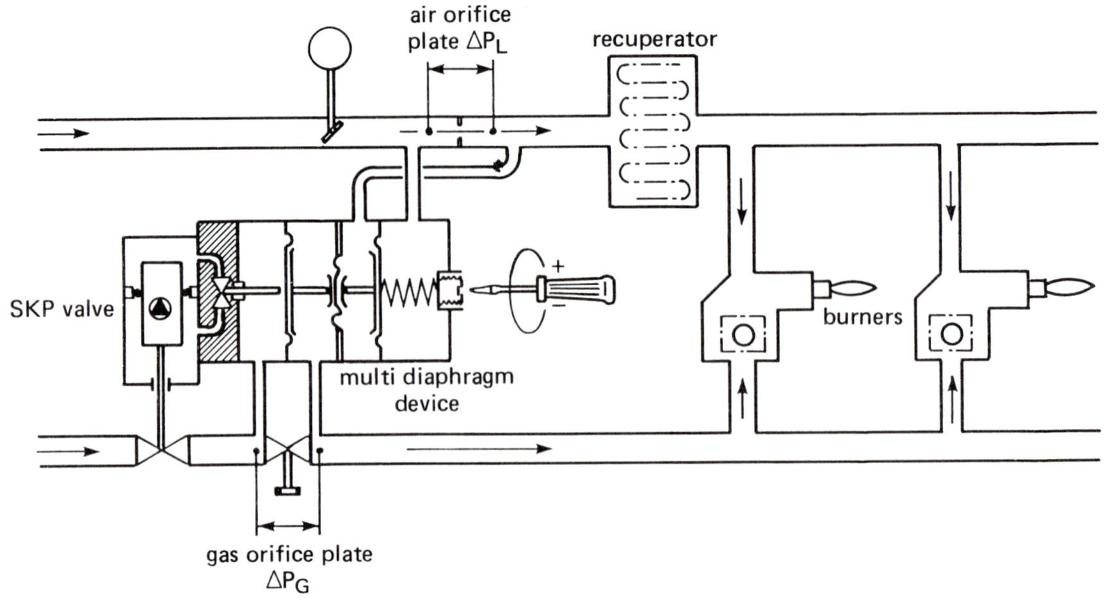

Fig. 9.40 Landis & Gyr SKP50 multi-diaphragm air/gas ratio control

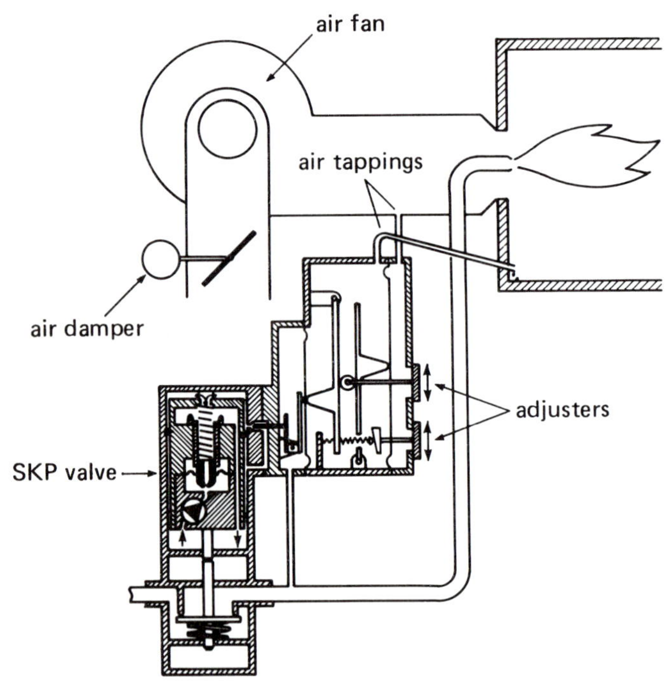

Fig. 9.41 Landis & Gyr SKP70 air loaded air/gas ratio control

The other device is a pressure loading system and can be used for example with packaged burners. Here, a tapping from the air fan moves the diaphragm inside the governor housing increasing the gas in proportion to the air. A combustion chamber pressure loading line is available. This system, known as the SKP70, is described in more detail in section 9.7 – Air/Gas Ratio and Throughput Control Systems.

9.4.9 Pneumatically operated safety shut off valves

On very large burner installations incorporating large ball or plug valves to act as safety shut off valves, electrically operated actuators are not adequate to open/close them. In these cases compressed air at or around 7 bar (100 psi) is used to operate either a piston, diaphragm or valve which is connected to the valve spindle. In some cases the compressed air opens and closes the valve via electrically operated air relay valves, whilst in others a heavy duty spring will close the valve. This is termed 'Fail Safe' because the valves close in the event of air failure. Typical types are the Norbro actuator (Figure 9.42), the Kinetrol vane actuator and the Crane diaphragm actuator.

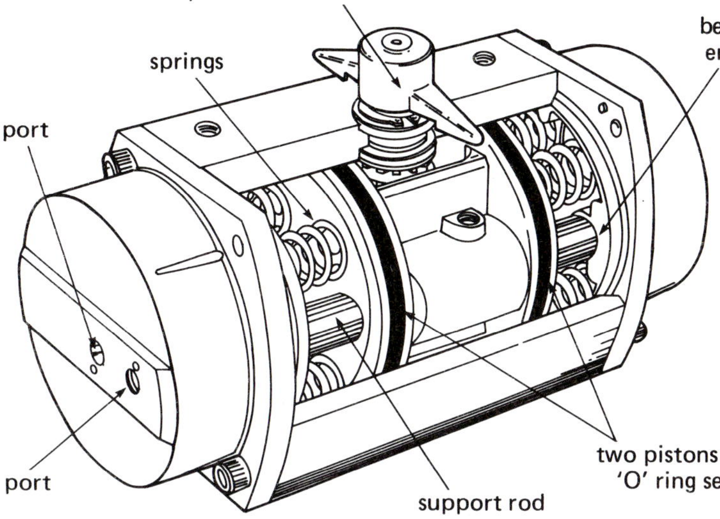

Fig. 9.42 Norbro spring return actuator

9.5 SAFETY SHUT-OFF VALVE SYSTEMS

The British Standard Specification for Industrial Gas Burners of Input Rating 60 kW and above (BS 5885 : Part 1, 1988) requires proving of the main burner safety shut-off valves for burners above 3 MW (10 x 10^6 Btu/h) as do the Codes of Practice for the Use of Gas in High Temperature Plant (British Gas Publication IM/12) and the Code of Practice for the Use of Gas in Low Temperature Plant (British Gas Publication IM/18).

The proving system is installed to automatically verify that gas is not leaking through the safety shut-off valves (SSOVs) and hence into the combustion chamber during the start-up sequence and, if required, on shut-down of the burner(s). It will therefore prove the effective closure of the SSOVs by a method which is more sophisticated than the system check required for burners rated between 1 MW and 3 MW (3.4 to 10 x 10^6 Btu/h). The latter indicates that the valves are in the closed position, by means of closed position indicator switches, but this does not prove that there is no leakage for reasons of, say, swarf on the valve seat.

9.5.1 Valve Proving Systems

Two techniques of valve proving are specified in BS 5885 and IM/18, either

a) detection of small leakage rates, for example, by a pressure proving system, or

b) by safely venting any leakage to atmosphere, for example, by means of a double block and vent system, i.e. two safety shut-off valves in series each fitted with a proof of closure switch and a vent valve fitted with a proof of open switch.

The valve proving check will be incorporated in the logic sequence of the burner enabling the check to be carried out automatically at each start-up and before initiation of the pre-purge, and if appropriate on shut-down. Any failure to prove during a start-up sequence will indicate the safety shut-off system is not leak tight and cause lockout preventing further progression through the sequence. Where a valve proving system additionally carries out a check of the SSOVs on shut-down, failure to prove must give an alarm signal and lockout the burner. It is imperative that the valve proving check is not by-passed in any way during the burner operating sequence. For all means of shut-down or lockout of the burner, the sequence must be designed so that the valve proving check has to be performed at every subsequent start-up. Reference to BS 5885 and IM/18 will identify the requirements to prove the start gas SSOV system in addition to the main SSOV system.

9.5.1.1 Pressure Proving Systems

There are three methods of pressure proving system commonly installed on burner systems requiring valve proving. They are:

- Use of a partial vacuum
- Use of a separate supply of checking gas, usually inert
- Use of the natural gas supply at line pressure.

Each method is arranged to test the pneumatic closure of the SSOVs directly by monitoring pressure(s) or vacuum established within the SSOV system. The equipment performing the tests is connected directly to the pipe cavity between the SSOVs as described below. A vent to atmosphere incorporating a normally open vent valve (open when de-energised) of minimum port diameter 6 mm, should be installed in conjunction with the proving system to meet requirements in BS 5885 and IM/18 (refer to relevant sections in the documents) or to release trapped pressures as part of the operational procedure of the test.

Whilst commercially available proving systems may differ somewhat in their design, the overall principles of operation are similar to those described below.

i) Use of a partial vacuum

Figure 9.43 indicates the layout of controls for this proving technique. The safety shut-off system incorporates an upstream and a downstream SSOV (block valve), a vent line to atmosphere under the control of a normally open vent valve and connections to the vacuum proving unit.

A connection from the pipe cavity is made to a solenoid valve in series with a vacuum pump, the outlet of which is connected into the vent line downstream of the vent valve. A second connection is to vacuum switch(es) to monitor levels of vacuum.

In the shut-down condition the block valves are closed and the vent valve open. Assuming atmospheric pressure is present in the pipe cavity, the test will commence. The vent valve closes and the solenoid valve in the proving unit opens initiating the test sequence as indicated in Figure 9.44. The vacuum pump will then commence evacuation of the cavity to achieve a predetermined level of vacuum, VS1, in a preset time period, monitored by a vacuum switch. Establishment of the required level of vacuum will prove there are no 'large' leaks through the SSOV system. If the vacuum switch is not satisfied the system will go to lockout.

The proving unit continues to monitor the vacuum using either the same or a second switch for a further preset time period to ensure there is no slow (small) leakage through the SSOV system. Providing that the vacuum does not decay by more than a predetermined amount above VS1

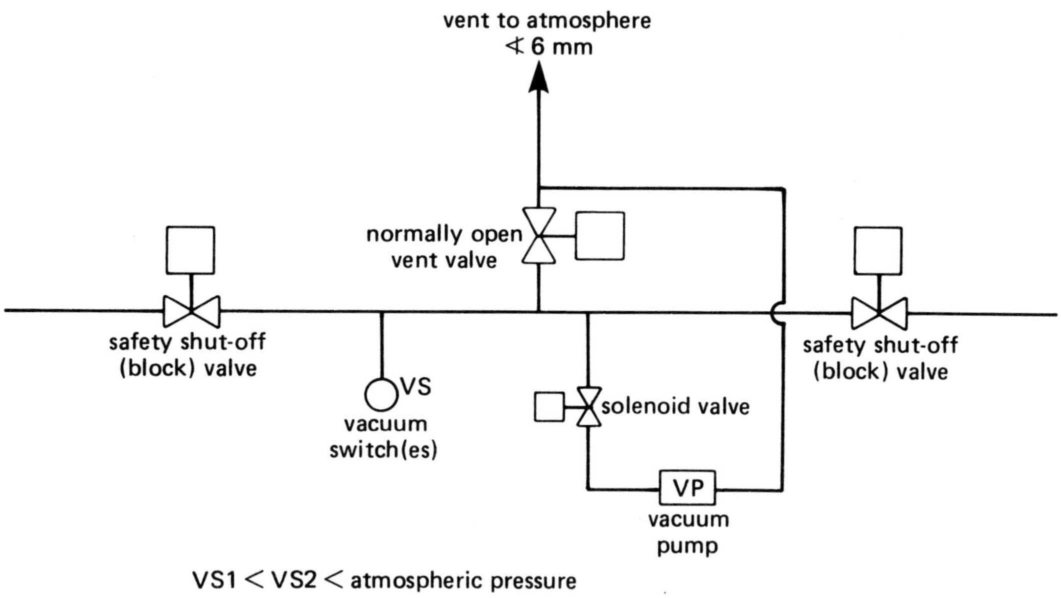

Fig. 9.43 Vacuum proving system

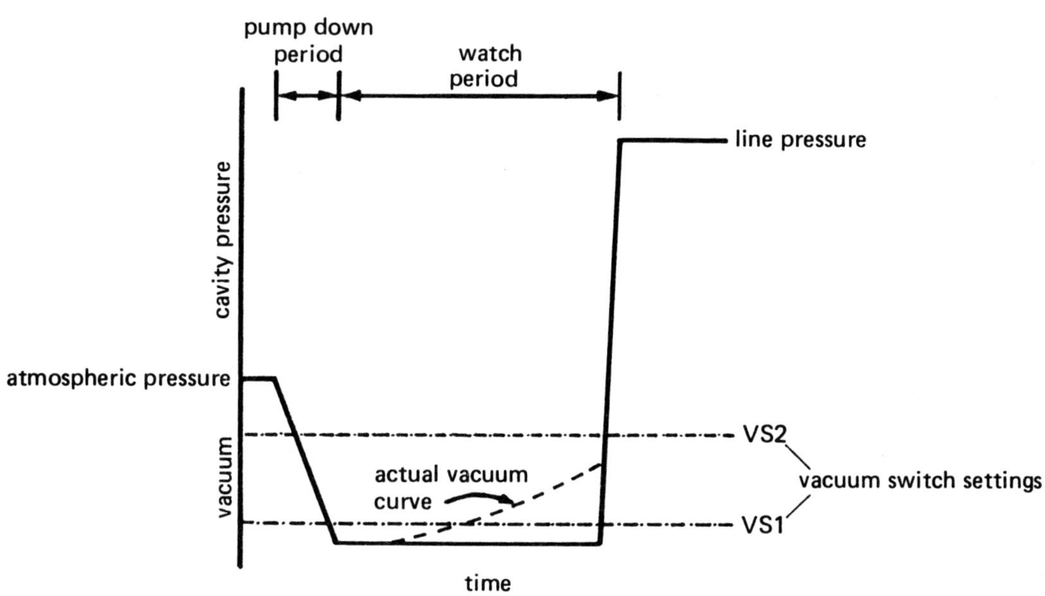

Fig. 9.44 Test sequence – vacuum proving system

and not beyond VS2 acceptable to the proving unit, the SSOV system will be taken as sound and the burner programming unit will be released to carry out its ignition sequence. Again a reduction in vacuum passed the setting of the vacuum monitor will cause lockout and cessation of the pre-purge.

Examples of vacuum proving systems which may be met in the field are:

Kromschroder	–	Proving System
Black Teknigas	–	Provenseal
Johnson Controls	–	Valvac
Landis & Gyr Building Control Ltd	–	Valvegyr LED 2
Maxon Combustion Systems Ltd	–	Provac

ii) Use of separate supply of checking gas, usually inert

The common method of proving a safety shut-off valve system with inert gas uses nitrogen as the checking gas. The layout of the system is shown in Fig. 9.45.

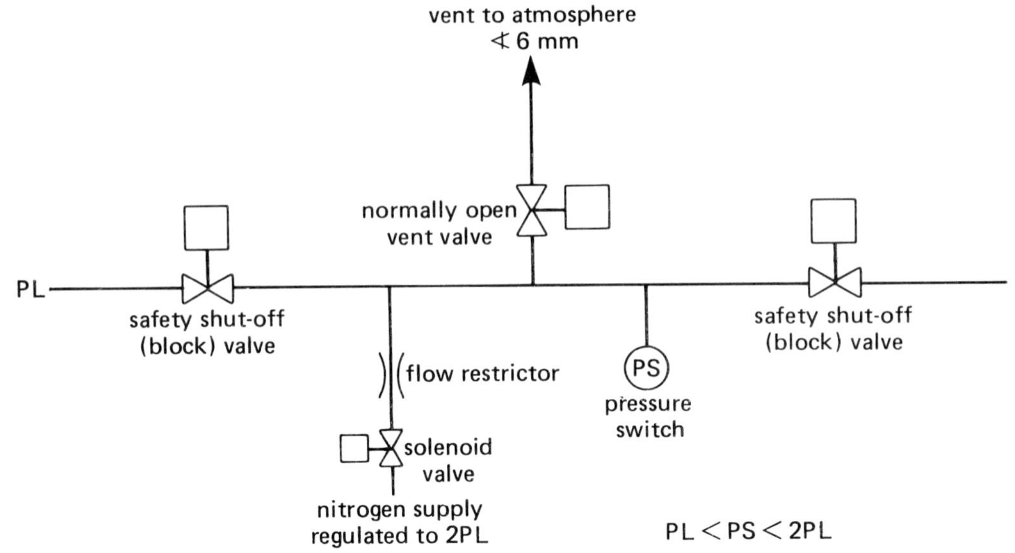

Fig. 9.45 Nitrogen proving system

The basic safety shut-off system of upstream and downstream block valves and normally open vent valve is supplemented by connection of a nitrogen supply into the pipe cavity between the block valves. A pressure switch is also connected into the cavity.

The nitrogen supply is piped from a nitrogen cylinder regulator and second stage regulators providing a two-stage reduction in nitrogen pressure. A solenoid valve and a limiting orifice are also incorporated in the nitrogen supply line before its connection into the pipe cavity. Proving is automatically carried out by initially closing the vent valve and admitting nitrogen into the cavity between the block valves by energising the solenoid valve in the nitrogen supply line. The pressurisation sequence of the test is shown in Fig. 9.46. If the predetermined pressure (normally twice the line pressure, PL, of the natural gas supply) is achieved in a preset period, the SSOV system is taken to be leak tight and the burner start-up sequence is allowed to continue. A failure to generate the set pressure in the time allowed causes lockout.

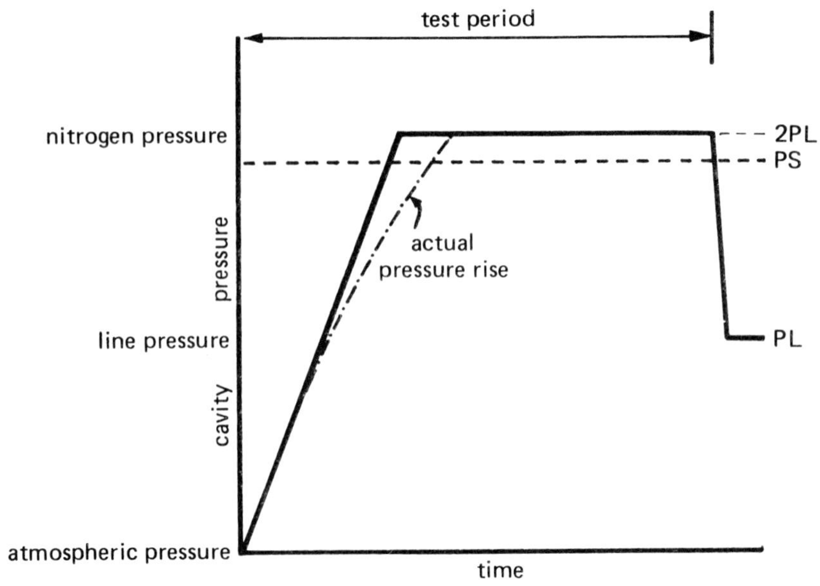

Fig. 9.46 Pressurisation sequence – nitrogen proving system

To prevent the check overcoming any small leak(s) which may be present on the SSOV system, the limiting orifice in the nitrogen supply line, specifically sized to limit the pressure build up, will permit an unacceptable leak to be detected.

Pressure proving by pressurisation is used by Hamworthy Engineering Ltd on gas and dual fuel burner systems which they supply.

A variant on pressure proving by pressurisation utilises a pump in a by-pass around the first safety shut-off valve. This provides a higher pressure supply of natural gas between the two safety shut-off valves for testing of the valves.

An example of this system is the Dungs VDK 200

iii) Use of the natural gas supply at line pressure

A valve proving system using line pressure for pressurisation operates on a sequential principle of opening and closing valves in the SSOV system. The principle is based on the Hegwein system.

An arrangement of the safety shut-off system for sequential valve proving is shown in Figure 9.47. The system consists of an upstream and downstream block valve and a normally open vent valve. Pressures are monitored by pressure switches connected into the pipe cavity between the block valves.

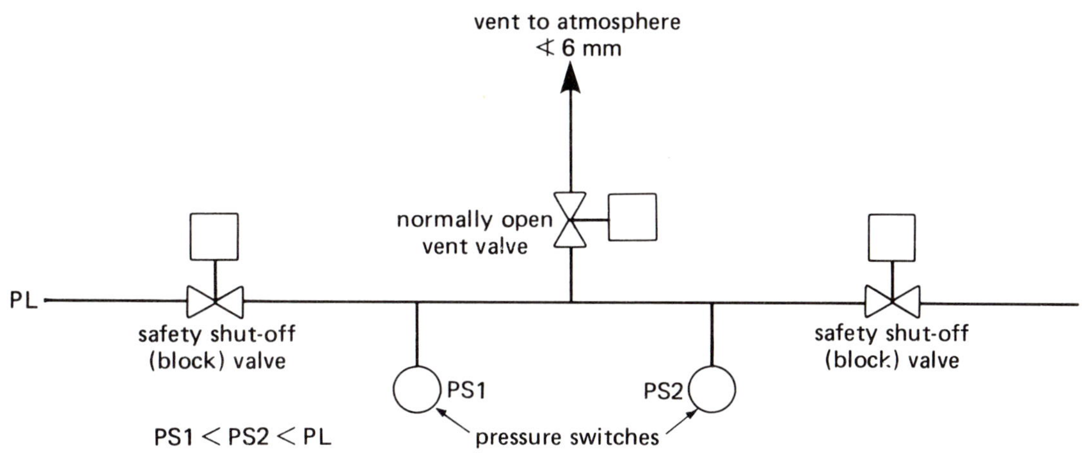

Fig. 9.47 Line pressure proving system

Proving the system on start-up is carried out by initially closing the vent valve, the block valves being closed already, so creating a closed pipe cavity in which atmospheric pressure should be trapped. The sequence of operation is shown in Figure 9.48. The first pressure switch monitors the cavity for a preset time period to determine if there is a build up of pressure. The switch is set to operate at a pressure just above atmospheric. Any pressure build up which is detected will indicate leakage through the upstream block valve and cause lockout of the system preventing progression through the start-up sequence.

The second stage of the test is to check for leakage through the downstream block valve. This is accomplished by momentarily opening a Class 1 solenoid valve in a small bore by-pass around the upstream main block valve and then closing it to admit natural gas at line pressure to the cavity between the block valves. An alternative method is to briefly open and then close the upstream main block valve to pressurise the cavity.

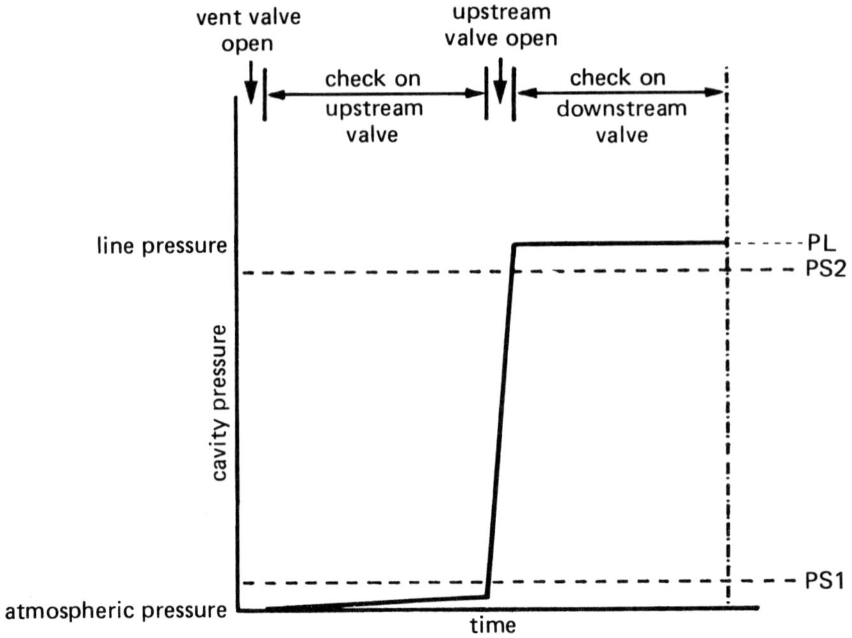

Fig. 9.48 Line pressure proving system - sequence of operation

The second pressure switch will now monitor the cavity for any loss of pressure. It will be set to operate at a pressure just below line pressure. Loss of pressure will indicate the downstream block valve is leaking or possibly that a leak is occurring through the vent valve or pipework cavity connections. Detection of loss of pressure by the pressure switch will cause the system to go to lockout preventing start-up of the burner until the fault is rectified.

An advantage of sequential testing is that in the event of leakage being detected, it may be easier to determine which of the safety shut-off valves is at fault as these are checked for soundness sequentially. Other systems test both upstream and downstream valves simultaneously.

Separate valves can be used solely for pressurisation and venting purposes. These should be to BS 5963. However, an arrangement that can be used to advantage in certain circumstances is to connect the cavity between the main SSOVs into the cavity between the pilot gas or start gas SSOVs. This will simultaneously monitor the soundness of the start gas valves and enable the upstream start gas SSOV to be used for pressurisation purposes. The downstream start gas SSOV can be used for venting purposes, it being permissible to vent the small quantity of gas through the burner into the combustion chamber. Note that such an arrangement cannot be used where dampers are used to restrict the natural ventilation through the flue. A separate vent must be used in such installations.

The Valvegyr LEB1 sequential testing control unit, though no longer made by Landis & Gyr - Billman Ltd, may still be encountered. Gas and dual fuel burners manufactured by H Saacke Ltd, incorporate sequential proving of the safety shut-off valves in the valve train. Other systems include the Dungs DK2F, the Inter Albion LC2 and the Landis & Gyr LDU11.

9.5.1.2 Double Block and Vent Position Proving

This is valve proving by safety venting any leakage to atmosphere by means of a double block and vent system with valves with position proving.

A safety shut-off system for this means of valve proving is shown in Figure 9.49. The system uses the double block and vent principle, having upstream and downstream safety shut-off (block) valves and a vent to atmosphere.

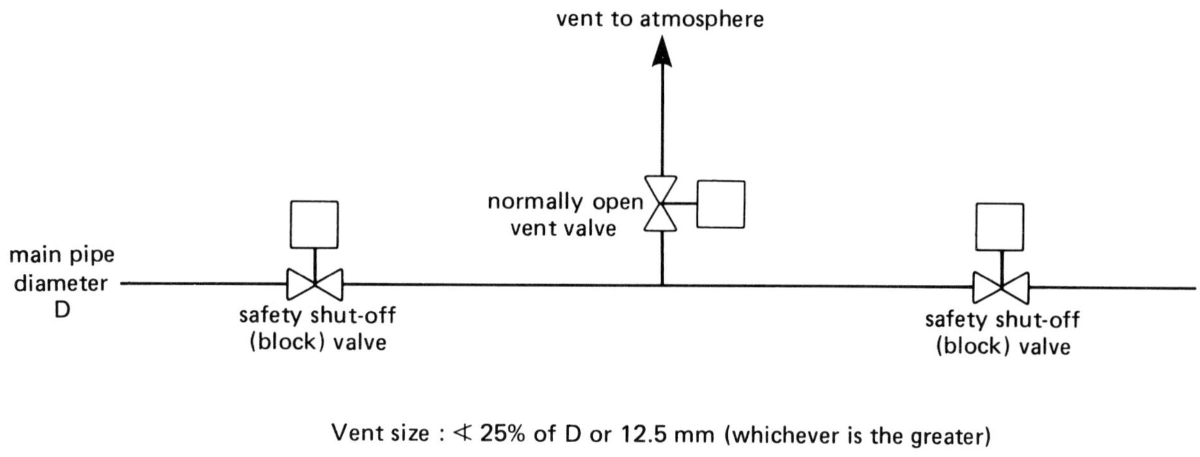

Fig. 9.49 Double block and vent with position proving

The block valves are of a type incorporating proof of closure switches which operate when the valve has closed to a position of mechanical overtravel. Mechanical overtravel is a property of certain types of valve, the stroke of which extends beyond the fully closed position. The operation of a proof of closure switch can thus be used to give an unambiguous signal of complete closure. If the valve is mechanically closed it is reasonable to infer that the valve is pneumatically closed and the tightness of the seal will be ensured by the quality of the valve. An example of this type of valve which has BG certification is the Maxon valve, although ball and plug valves used in conjunction with pneumatic actuators are commonly used for larger burners.

The vent to atmosphere is taken from the pipe cavity between the block valves and incorporates a normally open vent valve. The bore of the vent pipe and the vent valve orifice diameter must be 25% of the effective diameter of the upstream main block valve orifice or 12.5 mm, whichever is the greater.

The technique has the following principle of operation:

When the burner is shut down and the main block valves de-energised any leakage through the upstream block valve will be safely vented to atmosphere so preventing it passing into the combustion chamber. The effective closure of the main block valves is proved electrically by the integral proof of closure switches.

The system automatically proves the safety shut-off at start-up by ensuring the proof of closure switches are made and hence the main block valves closed and the vent valve open by a proof of open switch. If this condition is satisfied it is usual to initially energise and close the vent valve. When the proof of closure switch signifies the vent valve has closed, a signal will be given to allow the main block valves to be opened at the appropriate stage of the light-up sequence.

Where this valve proving technique is used, a check is required not only on start-up but also on shut-down of the burner. This is mandatory because of the relatively larger bore of the vent (c.f. pressure proving where proving on shut-down is optional) which could continuously vent volumes of gas to atmosphere should the upstream block valve be leaking during the shut-down period. On shut-down or lockout of the burner the block valves are closed and the vent valves opened immediately at which time the proof of closure switches in the valves must prove in the required condition. Failure to prove in the correct condition will give an alarm and cause lockout of the burner.

9.5.2. Valve System Check

This system is used for burners between 1 MW and 3 MW (3.4-10x10^6 Btu/h). It uses two Class 1 SSOVs fitted with closed position indicator switches, which indicate when the valve is in the normally closed position. These switches are connected together in series to a master check relay to prove that the SSOVs are mechanically closed before the pre-purge of the burner sequence can commence. No normally open vent valve is used on this system.

9.5.3 Insertion of Valve Proving Systems into Control Systems

Proprietary valve proving systems must be inserted as the last interlock in the burner programming control system. If any interlock opens including the lockout contact of the burner control system, this will cause the valve proving system to be de-energised and have to be reset, and thus prove the valves tight before light up on the burner can be attempted.

9.6 PRESSURE AND FLOW SAFEGUARDS

These systems are to ensure that the gas is at the correct pressure and at the required cleanliness. Too high or low gas pressure is guarded against, and if the air supply fails the system will shut down or lock out. Any extraneous gas is prevented from entering British Gas pipelines by acceptable non-return valves.

9.6.1 Low Pressure Cut-off Devices

The object of these devices is to ensure that when the gas pressure falls below a predetermined value above atmospheric, a valve closes which cannot re-open until all downstream valves have been closed and an adequate gas pressure has been restored. The system can then be restarted manually for example by depressing a button. Low pressure cut off valves once commonly used, have to a large extent been replaced by a low pressure gas switch in conjunction with a certificated safety shut-off valve or valves which is a more positive system.

9.6.1.1 Low pressure cut-off valves

A typical valve is shown in Figure 9.50. This has an auxiliary diaphragm so that inlet pressure does not exert any downward force on the valve. It is reset by a weep of gas through the by-pass orifice when the pressure reset ring is pulled. The size of the orifice is a compromise. It must be small enough to detect a downstream burner valve left open and large enough so that the operator does not have to keep the ring pulled for too long. These valves are fairly slow in closing and should therefore only be used on small installations. They are now seldom fitted and are being superseded by low pressure switches in conjunction with certificated safety shut-off valves, but will be found on older plant.

Low pressure cut-off valves are still manufactured, however, and can be useful in protecting a small area, such as a kitchen or laboratory, against inadvertent admission of gas when a burner valve has been left open. Advice on installation is given in British Gas publication IM/20, Weep By-Pass Pressure Proving Systems.

It should be noted that old weight loaded low pressure cut-off valves will still be found as the sole safety device on older plant. Some of these may be of the automatic reset type which do not require manual intervention to restore a supply when the inlet pressure is restored. Such valves do not meet the necessary requirements for a low pressure cut-off valve and should be replaced.

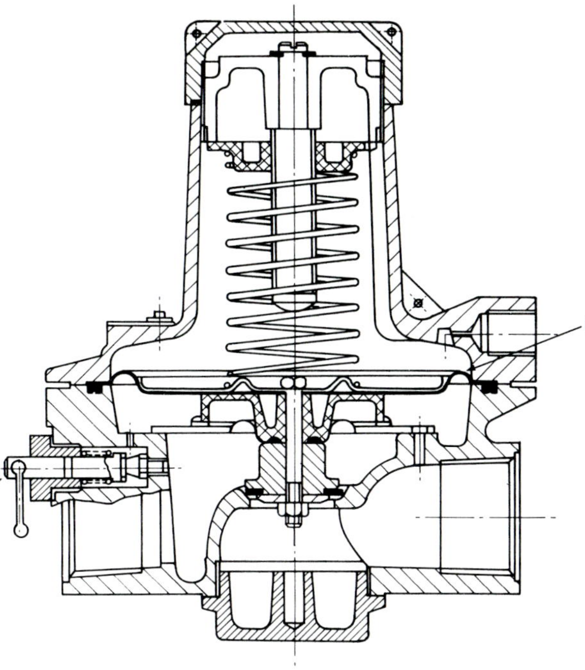

Fig. 9.50 Jeavons J120 Low pressure cut-off valve

Low pressure cut-off valves are unable to fully satisfy the requirements that they will not re-open until all the downstream outlets are completely closed. This is because the outlet pressure, when the weep is opened, may build up to a value sufficient to cause re-opening despite the presence of a small leak downstream of the valve. Full inlet pressure would then be applied to the leak.

For any inlet pressure above that which applied at the outlet will open the valve, there is in fact a size of leak orifice below which an undesired opening of the valve may occur. A smaller weep orifice gives greater protection in such circumstances, since this reduces the leak size above which pressure cannot build up sufficiently to re-open the valve.

It is also desirable that the time required for the outlet pressure to build up sufficiently to open the valve be of short duration; this can be achieved by providing an ample weep orifice. In fact, this conflicts with the requirement for a small weep orifice as discussed above, and a compromise must therefore be sought. A reliable method of sizing the weep orifice is given in British Gas publication IM/20. The re-opening time increases as the enclosed volume downstream of the valve increases.

A low pressure cut-off device offers protection against some of the circumstances which may allow unburnt gas to pass into a combustion chamber. It affords no protection against pilot extinction, other than by gas pressure failure and, therefore, a low pressure cut-off valve must not be regarded as a substitute for a flame failure device.

9.6.1.2 Low pressure cut-off switches

These are used to guard against low pressure in either the gas or air supply. In the event of low pressure the safety shut-off valves will close immediately, usually through a programming control box. The usual type encountered in industrial and commercial gas installations is the diaphragm type with a microswitch illustrated in Figure 9.51 – Kromschroder. The microswitch is held closed by gas or air pressure acting under the diaphragm against the set compression of a spring. If the pressure is lost or falls below a preset value, the spring will force the diaphragm down and the lever on the microswitch will drop causing the contacts to open. Most diaphragm pressure switches can be adjusted for a range of pressures depending on the spring supplied. Adjustment is usually by screwing in or out a setting screw. Some switches are however preset to a set value with no adjustment possible.

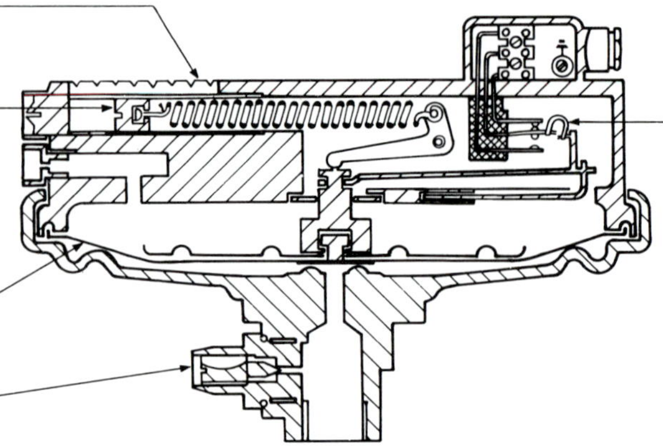

Fig. 9.51 Diaphragm type pressure switch

Advice on selection of diaphragm type switches can be obtained from British Gas publication IM/13 Specification for Pressure Switches for Industrial and Commercial Gas Fired Plant.

Switches are available with wide or narrow operating differentials. A narrow operating differential type for example means that the microswitch will make at a set pressure and the contacts will break again if the pressure drops a very small amount (e.g. less than 1 mbar). Other switches with wide operating differentials will have microswitches which make at a set pressure and will only open again at a much reduced pressure (e.g. make at 75 mbar and break at 35 mbar). Use of the latter type of switch has been made in vacuum proving control boxes.

These switches are often a source of problem on gas control systems. Some of the causes of concern are listed below.

1 Switches can often come 'out of adjustment' caused by plant vibration

2 Diaphragms can rupture causing nil operation of the switch contacts, and nuisance shut-down or lockout.

3 Microswitch contacts can weld together. This could result in valves remaining open for example even when air or gas pressure had been 'lost'.

Regular servicing of these switches is therefore necessary. Normally open contacts should be checked for not welding together, and the pressure switch settings should be checked by reducing operating pressure and observing on a manometer when the switch changes over.

High gas pressure protection is provided by a similar diaphragm switch. In this case, however, normally closed contacts are chosen. If the gas pressure exceeds a preset value, the diaphragm lifts against the compression spring causing the microswitch contacts to open. Most switches of this type are supplied with microswitches which have common, normally closed and normally open connections.

Earlier types of pressure switches were of the mercury tilt type as shown in Figure 9.52. These fell from favour generally because of their high cost and because they could only be installed with the diaphragm in the horizontal position. They also tended not to be adjustable. They are, however, a much more robust type of switch and are frequently encountered when trouble free operation is required and in more arduous locations such as the chemical industry.

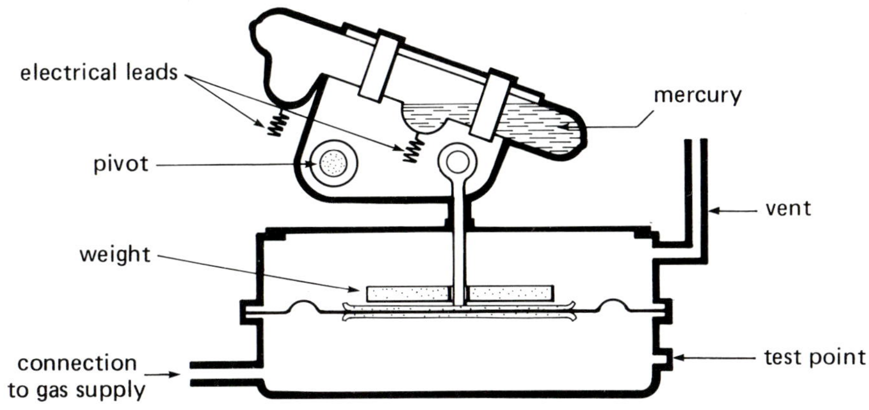

Fig. 9.52 Mercury tilt type low pressure cut-off switch

9.6.1.3 Weep by-pass system

Where flame protection is not employed on high temperature plant, the minimum requirement to comply with IM/12, the Code of Practice for the Use of Gas in High Temperature Plant, is for a low gas pressure protection system employing a weep by-pass. This is shown in Figure 9.53 for high and low pressure gas supplies. If gas pressure is lost or falls below a preset value, the low gas pressure switch contacts will open and the safety shut-off valve(s) will close immediately. To reactivate the system all individual manual main and pilot valves must first be closed. The manual weep push button is then depressed to admit a small volume of gas around the closed safety shut-off valve to pressurise the low gas pressure switch. This in turn will open up the safety shut-off valve enabling the operator to release the weep push button. The individual burner valves can then be opened to light the various burners.

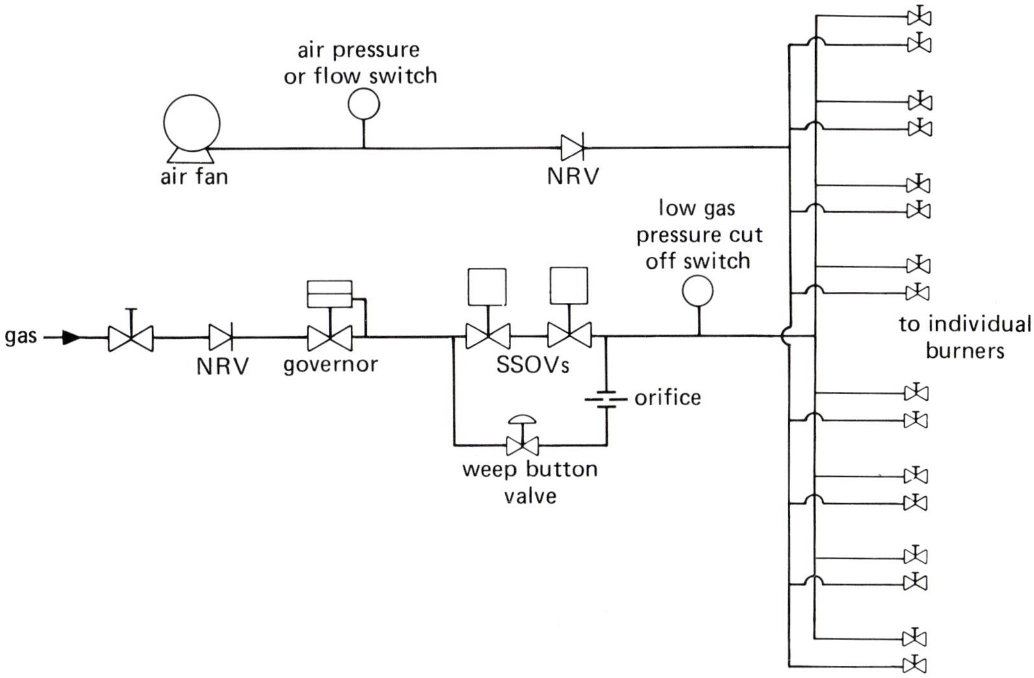

Fig. 9.53 Weep by-pass system

If, however, any valve or pilot valve has been left open or partially open, pressure will escape through it. On depression of the weep reset push button, the pressure switch contacts will not close and the safety shut-off valves will not open. It is important that the limiting orifice is correctly sized and the system designed in conjunction with British Gas Publication IM/20. If the orifice is too large the weep gas flow rate may be sufficient to overcome a small leak passed a partially open burner valve. On the other hand it should be large enough so that the operator does not have to depress the weep push button for more than about 30 seconds. The sizing of this orifice is therefore a compromise, and this system checks the tightness of all burner valves every time it is operated. An alternative to having a weep push-button in the by-pass line is to install a small solenoid valve with limited valve opening. A remote electrical push-button switch when depressed will open the valve. This means that the whole system can be installed remotely instead of possibly dropping small gas lines down to a weep reset button.

Proprietary gas shut-off systems based on the low pressure cut off principle are available for safeguarding for example gas supplies in school laboratories.One such system is the Energy Facilities Management Gasguard Gas Safety System. A reduction in gas pressure including inadvertent closing of upstream manual gas valves will be detected and the gas supply instantly cut off by a safety shut off valve closing. When start up of the system is initiated all burner valves downstream of the system must be closed. A small controlled amount of gas is fed into the system, locked in and isolated. The system is then monitored over a predetermined time period for loss of pressure. Neons in the control panel indicate the condition of the installation and the system only allows a continuous supply of gas if the system is proved gas tight (Figure 9.54). The system is controlled by a master key. Withdrawal of the key when the room is not in use prevents unauthorised use of gas.

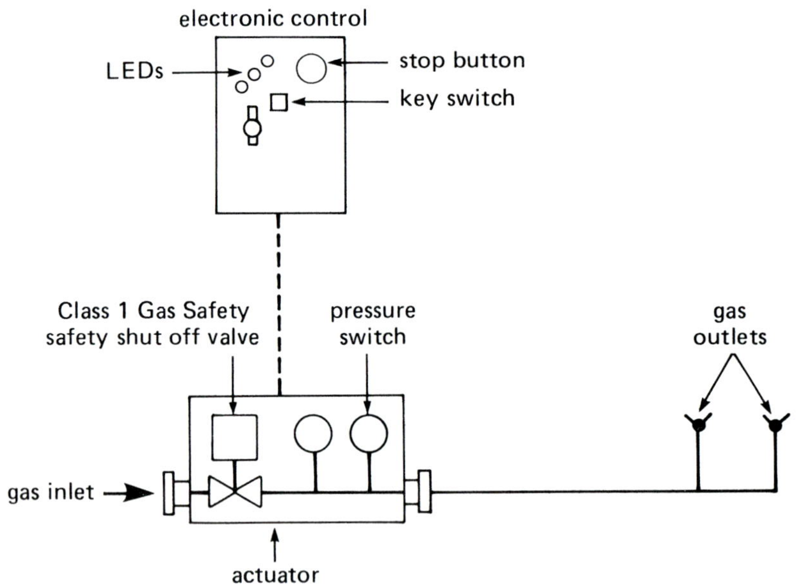

Fig. 9.54 Energy Facilities Management Gasguard System

9.6.1.4 Switched burner valve system

An alternative to the weep by-pass system for low pressure cut off for high temperature plant not equipped with flame protection is the switched burner valve system shown in Fig.ure 9.55. Here, individual burner valves are fitted with microswitches whose contacts are only closed when the burner valves are closed. These switches are connected together in series to the safety shut-off valve(s) or master check relay. The latter will only be energised if all burner valves are completely closed, when the gas pressure passing through the valve will activate the low gas pressure switch.

As this switch is connected in parallel with the individual burner switches, these can in turn be opened to admit gas to the burners. As with the weep by-pass system, if the gas pressure is lost or falls below a preset minimum, the safety shut off valve closes immediately. However, in this case, the initial check is that the valves are mechanically closed which by inference means that they are pneumatically sound also.

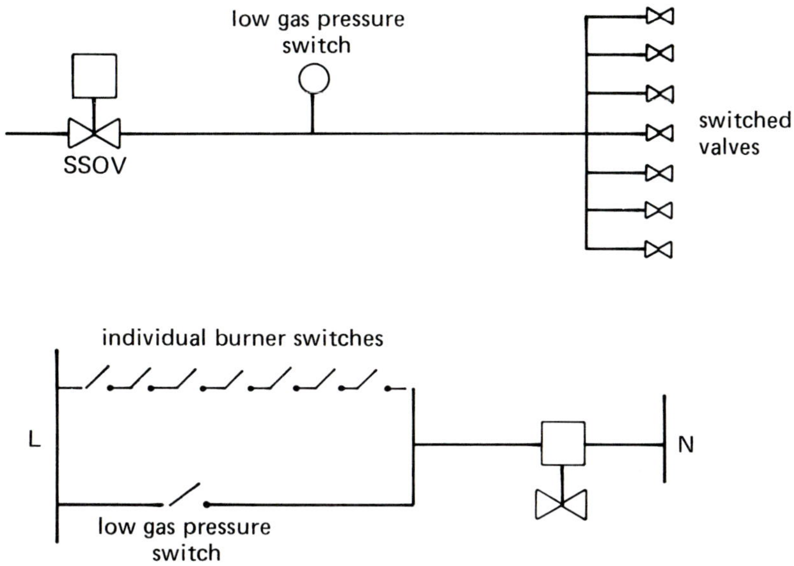

Fig. 9.55 Switched burner valve system

9.6.1.5 Low pressure protection for gas boosters and compressors

Another use for the low gas pressure switches is to prevent electrically driven gas boosters or compressors from causing a reduced or negative pressure on the inlet side of the gas supply, which could cause a number of undesirable or dangerous conditions, such as:-

1 Drawing in the sides of the meter
2 Interfering with other appliances
3 Interfering with other consumers' appliances
4 Drawing in air at other appliances.

Protection against this is a requirement of the Gas Act 1986 (Schedule 5, para 8).

One method of preventing this is with the arrangement shown in Figure 9.56.

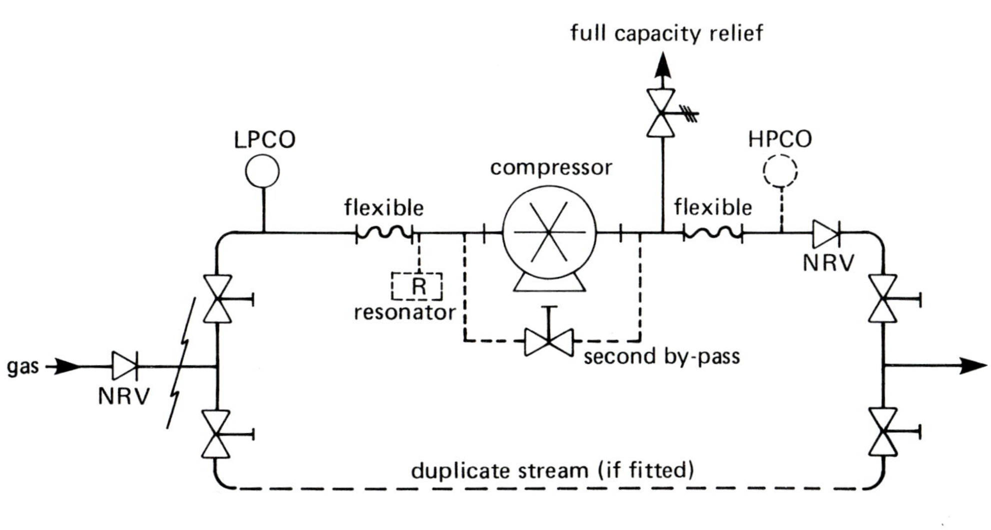

Fig. 9.56 Compressor installation with full capacity relief

A switch is connected in the circuit to electrically-driven gas boosters or compressors. Should the gas pressure at any predetermined point fall below a preset minimum, the switch will break the circuit and stop the compressors.

The pressure switch should be used in conjunction with a push-button switch incorporating a no-volt release, so that current cannot be restored to the motor without manually operating a push-button starter. The switch must be connected directly to the gas supply pipe without any isolation valve. In certain cases it may be necessary to damp the operation of the pressure switch to prevent nuisance shutdown. Advice is given in British Gas Publication IM/16.

9.6.1.6 Factory Mutual Cock Safety Control System

In the USA a group of insurance companies, i.e. the Association of Factory Mutual Fire Insurance Companies, combined to create an Engineering Division responsible for the application of engineering to industrial loss prevention. Of the many techniques developed by the FM Engineering Division, one of particular interest is the FM Cock Safety Control System which may be found on a variety of imported gas fired plants.

The object of the FM Cock System is to ensure that all burner cocks are closed before the main safety shut-off valve can be energised. Like the low pressure cut-off, the system can be used as an aid to correct lighting-up procedure. The system is based upon the use of the approved FM cock which replaces the standard gas cock usually installed at each burner. This cock has an additional passageway through the valve body at right angles to the normal flow path as shown in Figure 9.57. This passage is only opened when the main flow is shut off.

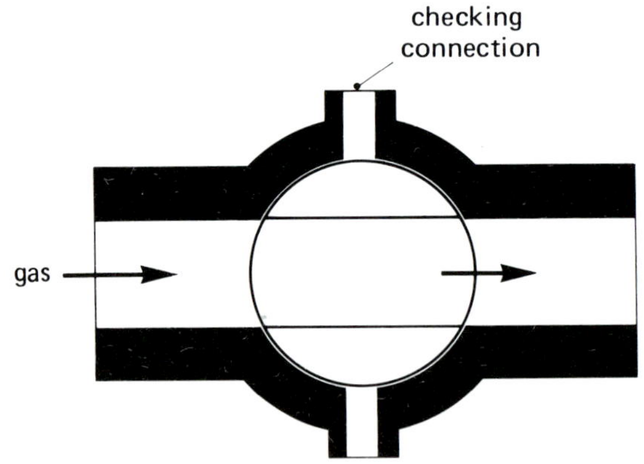

Fig. 9.57 FM Cock

A typical system using the cocks is shown in Figure 9.58. Here a checking fluid (gas or air) is used so that when all the cocks are closed (that is when all the checking passages are open), the pressure of the checking fluid is applied to the pressure switch. The action of this switch is such that when operated it allows a manual reset safety shut-off valve (SSOV) to be opened.

The main gas valve can thus only be opened when all the downstream cocks are shut. At this stage pressure is applied to the low pressure gas switch (LPGS) which allows the FM cocks to be opened. In the event of a low pressure, the LPGS will operate to shut off the SSOV and a complete restart of the above sequence is then required.

The FM system is not very common, being encountered only on American equipment or that installed in premises in the ownership of American companies.

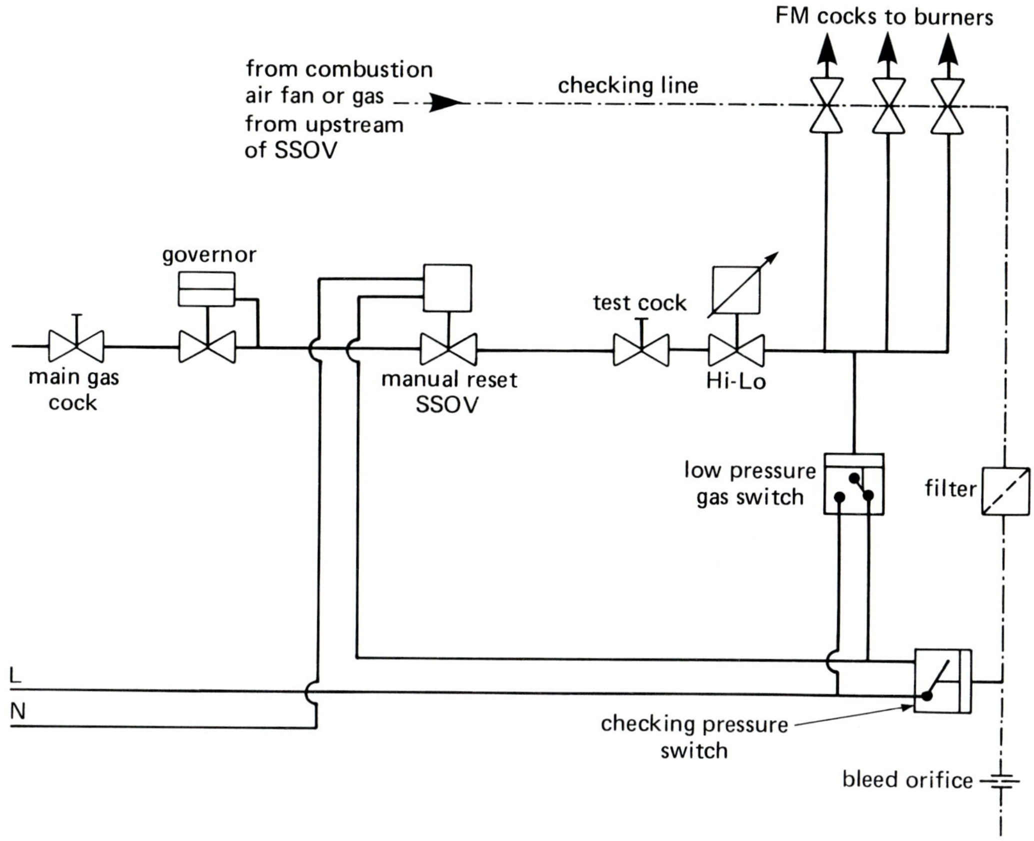

Fig. 9.58 FM Cock system of shut-off

9.6.2 Air Flow Failure Devices

The purpose of air flow failure devices when used in conjunction with gas appliances is to cut off the gas supply should the combustion air flow, or a recirculation air flow fail. There are four main applications of these devices:-

1. Where failure of the air supply would lead to flame extinction and unburned gases passing into the combustion chamber and flues
2. Where failure of the fan or air supply would lead to excessive temperatures in the combustion chamber or heat exchanger of the appliance.
3. Where failure of the fan or air supply would lead to an accumulation of inflammable vapours of explosive concentrations.
4. To ensure that air fans are running before gas can pass to burners.

9.6.2.1 Electrical air pressure switches

Protection against air failure is invariably provided by pressure switches, as discussed in section 9.6.1.2 earlier, and safety shut-off valves complying with BS 5963. For safety reasons it is usually desirable for the pressure switch to be proved in the 'no air' position prior to start-up. This prevents a welded or jammed switch giving an erroneous signal.

The switch is identical to that shown in Fig. 9.51.

9.6.2.2 Electrical air flow switch

Where it is necessary to prove air flow rather than air pressure which is often very low, a vane operated mechanism connected to a mercury tilt switch would be used (see Figure 9.59). It should be noted that in such applications as paint stoving ovens, it is possible for the vane to become jammed with particles of paint and regular maintenance is essential, e.g. Delta.

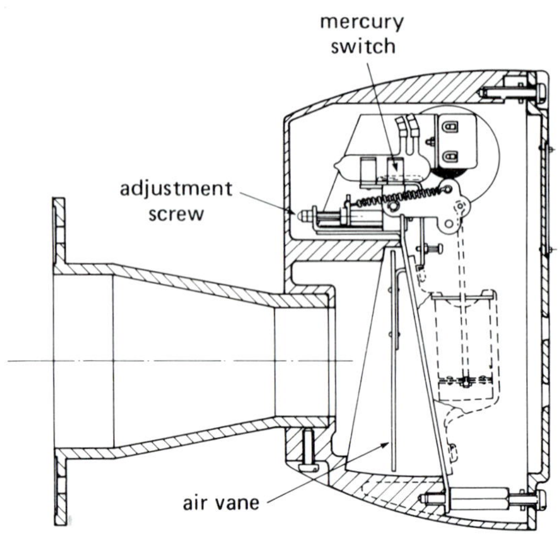

Fig. 9.59 Spring loaded type metal diaphragm-operated mercury switch

9.6.2.3 Centrifugal switches

BS 5885 permits the use of centrifugal switches for certain burner applications. These can be more positive in operation than pressure switches set to change over at the very low air pressures sometimes found with packaged burners. The centrifugal switch must be direction sensitive when used with three-phase fan motors. Centrifugal switches are only permitted with on/off burners of heat input up to 3 MW (10×10^6 Btu/h) with a fixed preset air damper (when fitted): the switch must be connected directly to the fan impeller (see Figure 9.60) and not the fan motor.

The switch works on the principle that at a preset rotational speed of the fan impeller which corresponds to a given volume air flow, small weights on the switch are thrown out by centrifugal force which then press on the microswitch contacts to close them.

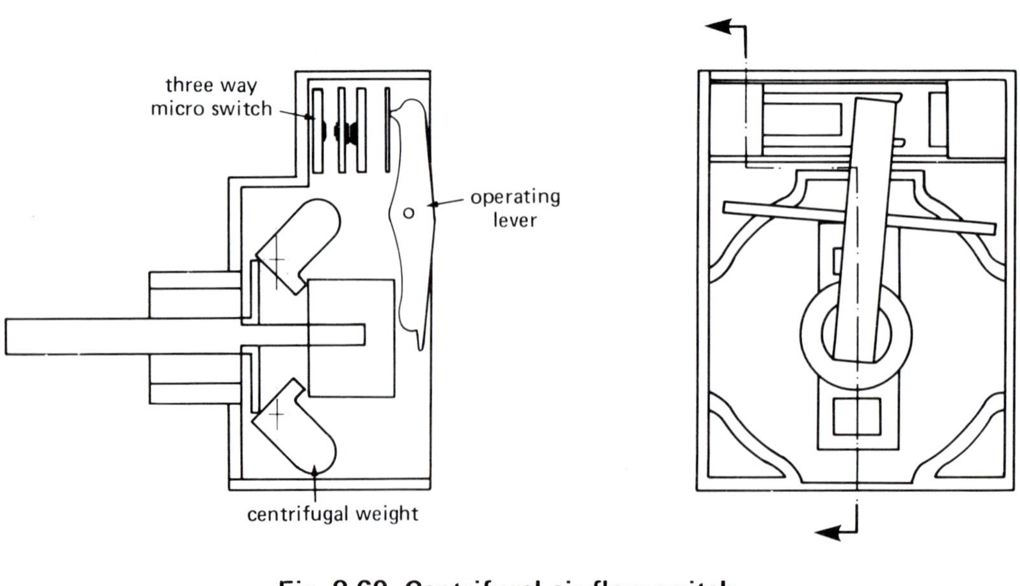

Fig. 9.60 Centrifugal air flow switch

9.6.2.4 Mechanical air flow devices

Various types of mechanical air flow failure devices have been developed in the past. The fact that these did not need any electrical supply, other than to the combustion air fan, could have been an advantage in certain circumstances.

It must be emphasised that such valves, which have only limited closing forces, must be considered obsolete. It is possible that they may be met in the field so that a brief working knowledge of them is useful. Replacement with electrical pressure switches and safety shut off valves as part of an updating exercise is recommended.

One such system is the diaphragm operated gas valve shown in Figure 9.61. This valve is held open by positive or negative air pressure on a diaphragm.

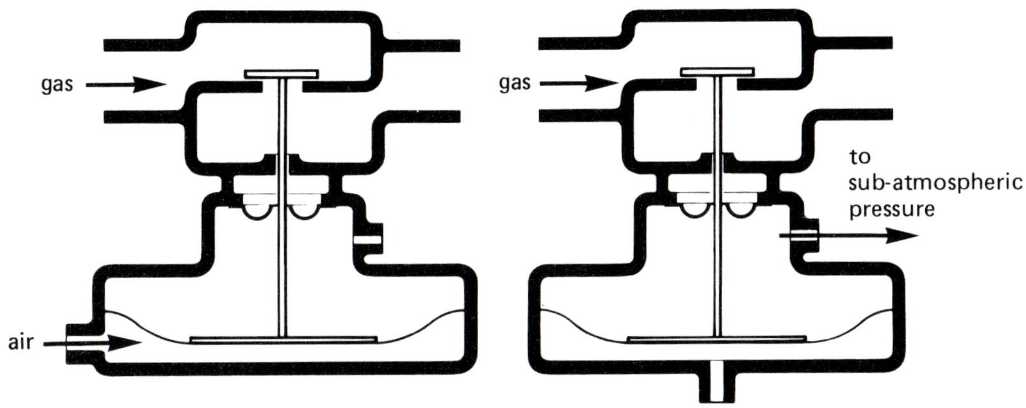

Fig. 9.61 Pressure operated valves

9.6.2.5 Combined air pressure cut-off valves

This system provided protection against both air and gas pressure failure as shown in Figure 9.62. It consists of two diaphragm operated type valves.

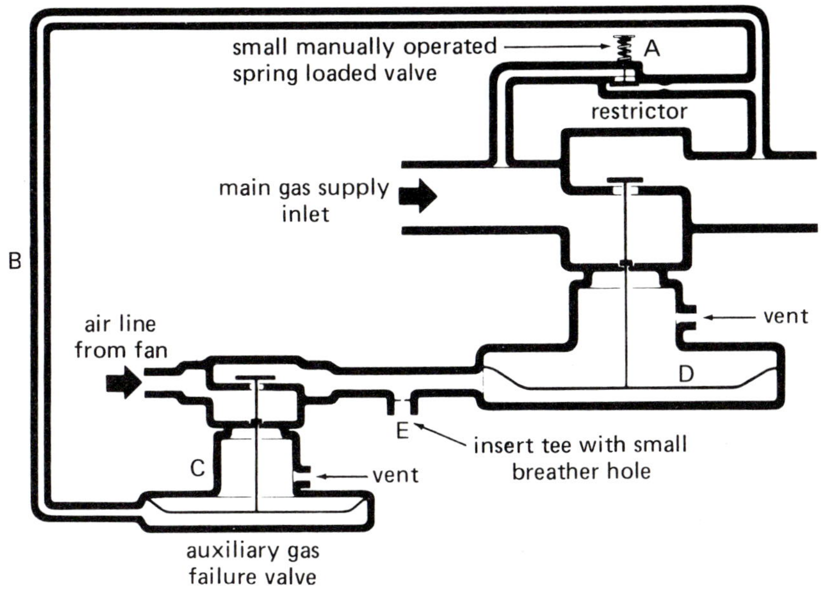

Figure 9.62 Combined fan failure and pressure cut-off valve

The mode of operation is as follows:

To start up the installation, the burner valve must be closed and the fan switch on. The small manually-operated valve (A) is then opened; this causes pressure to be built up in the weep line (B) which is transmitted to the underside of the main diaphragm (C) of the ancillary valve which will then open, thus permitting air pressure to pass to the underside of the main diaphragm (D) of the large fan failure valve, which in turn will open and admit the main gas stream. Correct operation may be confirmed by detection of air leaving the tee (E).

Care should be taken to ensure that the breather hole at (E) does not become blocked by dirt, otherwise the main valve will be held in the open position.

The system thus provides protection against air failure and provides a method of checking that all burner valves are closed prior to start-up of the burners. Protection is also given against failure of gas pressure during operation.

Devices operating on gas or air flow are preferable to those operated by pressure. If a pressure operated device is incorrectly placed in a control line to a burner, closure of any downstream valves will result in flame extinction, but not in operation of the safety device, due to retention of pressure in the gas (or air) line.

Flow operated devices are, however, subject to dust and corrosion problems which may result in the moving vane sticking in the ON position under no-flow conditions.

The combined air/gas failure system was widely used in industrial applications where burners requiring pressure air were used. However, such systems are now obsolete although it is possible that they will still be met in the field. Replacement as part of an updating service is recommended.

Apart from their inherent deficiencies, such as slow speed of response, limited closing force and vulnerability to blockage of important vent holes, the very large size of such systems, in comparison with safety shut-off valves and pressure switches, is a considerable disadvantage.

9.6.3 Non-Return Valves

A non-return valve, sometimes called a back-pressure valve, is designed to allow fluids to pass in one direction only. Any reversal of flow closes the valve instantly.

The function of a non-return valve is:

1) To prevent the admission into the gas service pipe or into any main through which gas is supplied by the supplier, of air from a fan or compressor or of gas not supplied by the supplier.

   It is a requirement of the Gas Act 1986 that, where air from a fan or compressor, or any other type of gas, is used in conjunction with a gas burner, an effective non-return valve must be fitted in the main gas supply. This is to prevent air or any other extraneous gases being admitted into the gas service pipe.

   A non-return valve will not normally be required if a flame protection system with certificated safety shut-off valves is used.

The valves must be capable of withstanding a reverse pressure as follows:

- 7 bar (100 lbf/in^2) on 25 mm (½–1 in) valves
- 2 bar (30 lbf/in^2) on valves 25 to 150 mm (1–6 in)
- 1 bar (15 lbf/in^2) on valves above 150 mm (6 in)

These requirements are from British Gas publication IM/14 – Standard for Non-return Valves

Typical acceptable non-return valves are shown in Figure 9.63. In the valve at (a), gas entering lifts the two leather diaphragms off their seats on the spring loaded valve head so allowing gas to flow to the outlet. In the event of reverse pressure the diaphragms return to their seating, preventing any return flow passed the valve. If the return pressure is increased, the complete metal valve is forced down on to its seating against the spring pressure. This provides an additional seal to withstand the higher back pressure. This valve is only made up to 100 mm (4 in) and must be mounted horizontally.

The valve at (b) is a simple, flap type, disc-on-seat valve with a cast iron body and a knife edge valve support pivot. The valve seat is self-aligning and must be mounted horizontally with the dome uppermost.

The valve at (c) has a pair of nitrile rubber diaphragms which touch around the periphery. In the forward flow, gas forces the diaphragms apart but under reverse pressure the lips of the diaphragms are forced together so preventing gas flow through the valve. This valve can be mounted in any position, but is only manufactured in 12.5 and 25 mm (½ in and 1 in) sizes.

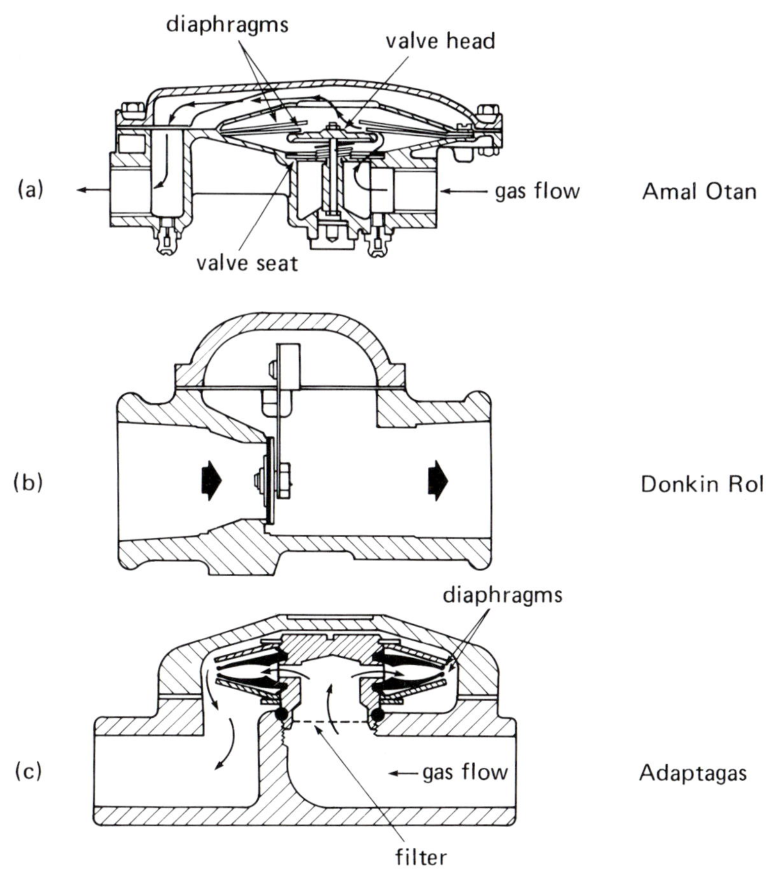

Fig. 9.63 Non-return valves (a) and (c) double diaphragm, (b) roll check valve

No other non-return valves are at present declared acceptable to British Gas for mains protection (as required in paragraph 8 of Schedule 5 of the Gas Act 1986). These three types of valve are also suitable for gas supply protection against oxygen at pressures up to 2 bar (30 lbf/in^2).

2) Fitted in an air line to prevent admission of gas into the air supply and air blowing fan especially when the fan is above the gas valve system.

Where a non-return valve would be unable to withstand the reverse flow application of the pressures in the system of which it is a part (e.g. compressed air or oxygen), a pressure relief should be fitted downstream of the valve so as to limit pressure to that which can safely be withheld by the non-return valve.

On oxygen gas burners with an oxygen or oxygen enriched air supply pressure greater than 2 bar, an appropriate oxygen type non-return valve should be installed in the gas line of the system, see Figure 9.64. This would further be protected by an acceptable non-return valve discussed above.

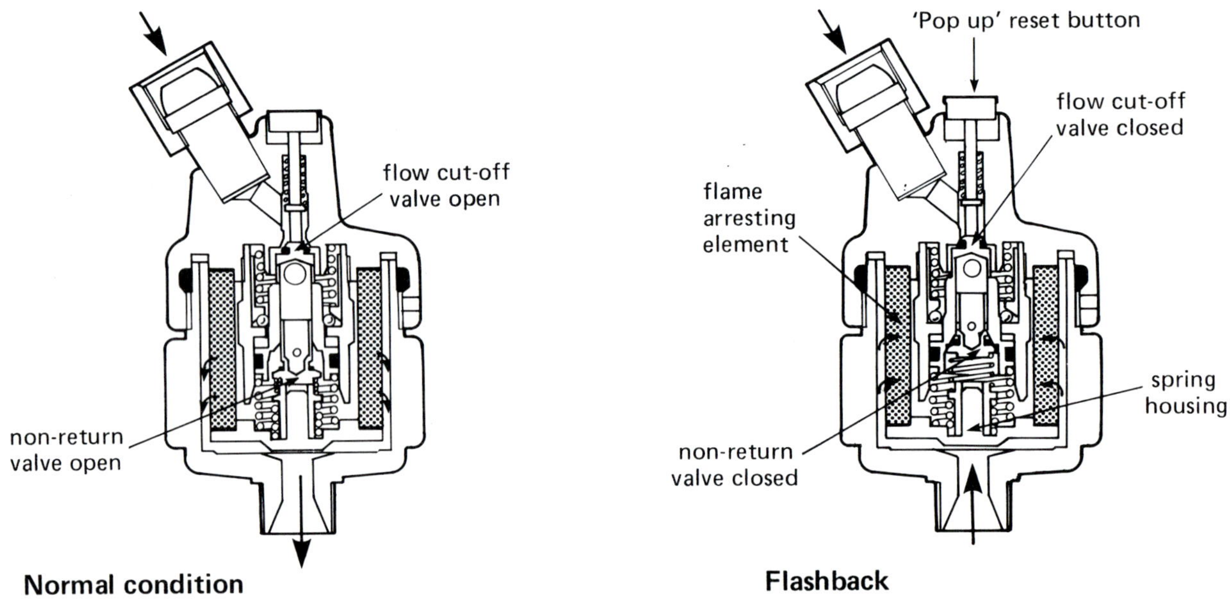

Fig. 9.64 B.O.C. Ltd oxygen type non-return valve

This system, with an oxygen non-return valve is shown in Figure 9.65. Non-return valves for oxygen-gas working are explained in British Gas Publication IM/1.

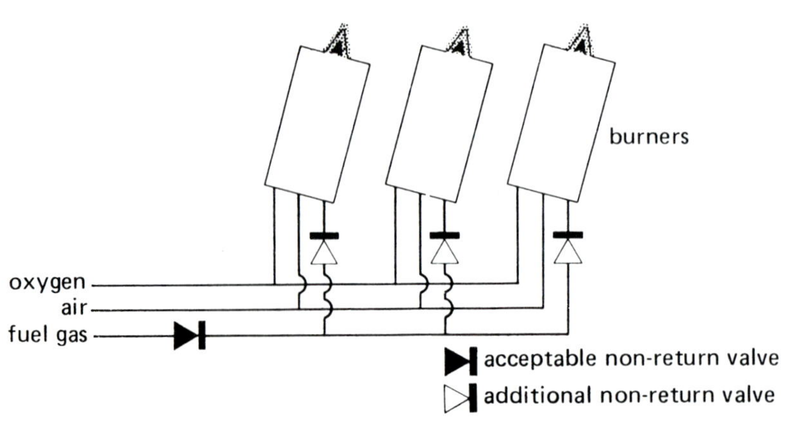

Fig. 9.65 Oxygen and acceptable non-return valves for working flame burners

9.6.4 High Pressure Gas Relief

Where gas is available in premises at pressures higher than that normally available, 21 mbar (8.4 in wg), it may become necessary for suitable safeguards to be incorporated in gas supply systems to prevent pressures increasing beyond the safe working limits of gas appliances, the satisfactory limits depending upon the pressures for which the gas-consuming appliances have been designed and adjusted.

Normally, suitable pressure controlling, venting, or cut-off equipment would be fitted to a consumer's incoming supply, the layout and type of controls being in accordance with gas distribution requirements. However, should it be required to distribute gas within a consumer's premises at a higher pressure than that at which equipment might be designed to operate, and where further reductions in operating pressures are necessary at the point of consumption, then in these circumstances similar control equipment to that used by the distribution section should be incorporated.

Numerous devices are available, each of which can perform one or two basic functions, either by relieving the excess pressure to atmosphere, or by completely isolating the gas supply. In the case of the latter, it is not possible to re-establish flow without manual resetting. In some circumstances soundness testing and purging may be necessary.

9.6.4.1 Slam shut valves

These are used to prevent excessive pressure build up in a pipe line system. A typical device shown in Figure 9.66 is the Donkin 305 valve, which is normally fitted to the inlet side of a governor/regulator. An impulse pipe connects the diaphragm chamber to the outlet side of the regulator and the disc-on-seat valve is kept open against a strong closing spring by a mechanical latch mechanism. When the pressure increases above a predetermined value the diaphragm lifts against the diaphragm spring and releases the mechanical latch via a lever mechanism. The valve slams shut further pressed against the seat by the gas pressure. Resetting can only be achieved manually. A push button valve is provided for pressure equalisation. The valve can be mounted in any orientation.

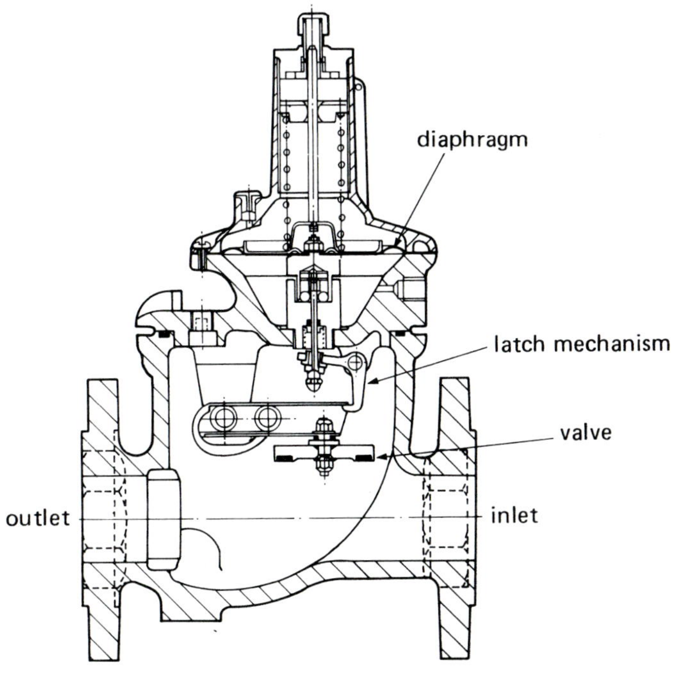

Fig. 9.66 Donkin 305 Slam shut valve

9.6.4.2 Pressure relief devices

Various high and high/low governors have an internal relief valve which will come into operation at approximately 22 mbar (9 in wg) above the outlet pressure setting of the valve. This system was described in detail in section 9.2.2, high pressure and high-low pressure governors.

It should be noted that a gas pressure regulator vent is not normally of adequate size to cope with the full flow of a pressure relieving device. In such circumstances an adequately sized independent vent should be provided and in all cases taken to a point outside the building. Means must be provided to prevent insects, water, or foreign materials from entering the vent pipe. The independent pressure relief valve used for this will have adequate capacity.

Pressure relief valves are installed to protect the burner system from excessive over pressure which could for example be caused by governor failure, and because of this they are generally installed immediately after the governor on a vent line to atmosphere.

The construction is exactly the same as a standard governor, except that the valve is reversed and sits on top of the valve orifice under the influence of a compression spring (Figure 9.67). If the pressure in the vent pipe rises above a preset figure, it communicates with the underside of the diaphragm opening the valve against the spring and venting the gas to outside the building.

In the case of factories located in built-up areas, venting of large volumes of gas via a pressure relieving device would require to be considered very carefully from the standpoint of explosion hazard and/or gassing of personnel.

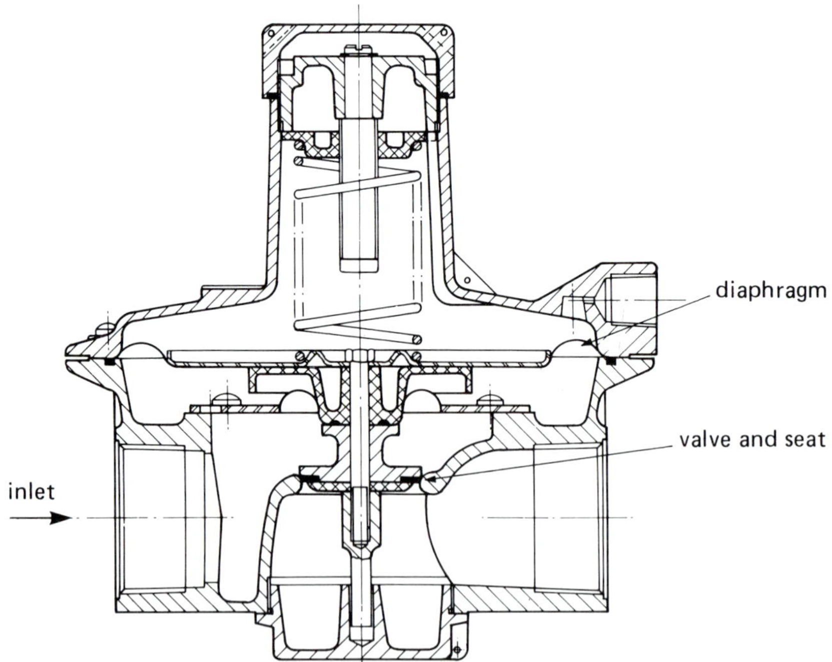

Fig. 9.67 Pressure relief valve

9.6.5 Flame Arrestors

Where pre-mixed gas and air is used in industrial gas fired equipment or wherever there may exist an appreciable volume of explosive mixture, a flame trap should be fitted to prevent the passage of flame back through the connecting pipe system. The exceptions to this rule would be when a restriction on the downstream side of a mixing machine is undesirable, in that, due to the design of the machine, any restriction such as that created by a flame arrestor in the mixture line would cause a change in the air-gas ratio.

A satisfactory flame trap should perform two functions:

- Arrest of the passage of a flame through a pipe system.
- Cut off the fuel supply if a flame reaches the arrestor.

A typical in-line flame arrestor element is shown in Figure 9.68. It consists of several concentric rings of ribbon stainless steel. If the flame flashes back to the trap, the large effective area of ribbon chills thus extinguishing it. It is important to install the arrestors as near as possible to the flame, so that the flame does not have a chance to accelerate sufficiently to go straight through the trap. If this is impracticable, two or more arrestors in series may be necessary.

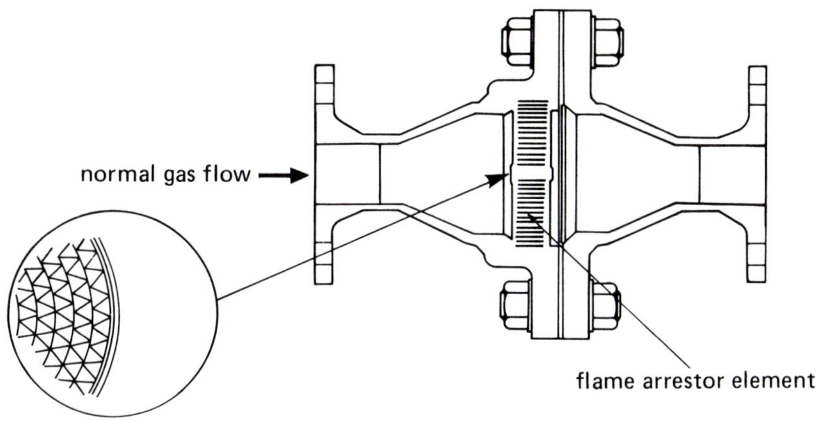

Fig. 9.68 Amal flame arrestor

One system which is available is shown diagrammatically in Figure 9.69. It consists of a sensitive thermocouple, a temperature detector and a relay. The thermocouple is fitted in the downstream side of the flame arrestor in such a position that in the event of ignition at the face of the arrestor, the flame will impinge on the thermocouple.

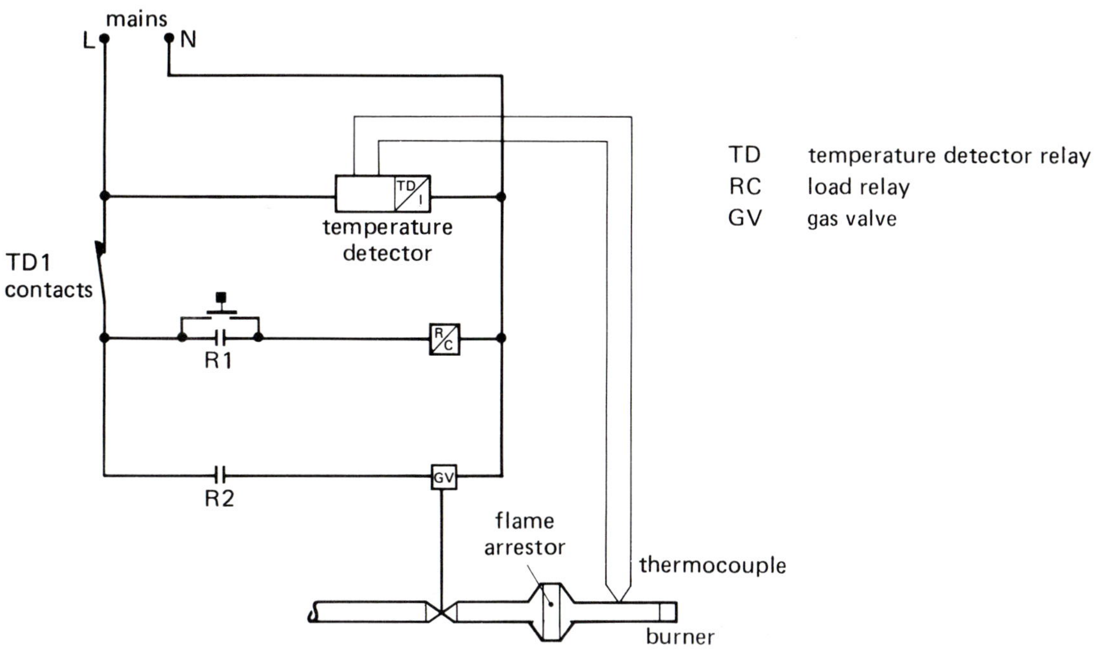

Fig. 9.69 Flame trap

During normal operation, when the thermocouple is cold, the contacts TDI of the temperature detector are closed so that, upon depressing the reset push-button, relay coil R/C is energised closing contacts R1 and R2. R1 continues to hold the relay in after the push-button is released; R2 energises the gas valve GV.

The system remains in this position until a flame is established at the flame arrestor. The thermocouple is heated, temperature detector contact TDI is broken, de-energising relay R/C coil. This opens contacts R1 and R2, closing the gas valve, which cannot be opened until the thermocouple has cooled sufficiently to close contact TDI.

Several thermocouples can be used in series with one relay and switch unit.

Other uses for the arrestors include the end of line vent stand pipe for use in purging gas pipelines. Any source of ignition such as a cigarette end is prevented from entering the purge standpipe by the arrestor element.

9.6.6 Filters

These are fitted on the gas supply to prevent dust or foreign matter in the gas stream from being deposited on control equipment and rendering it inoperative. Normally the filtration provided at the meter in compliance with British Gas publication IM/112 is adequate. Further guidance is given in British Gas publication IM/16. This states that it may be desirable to install a filter to limit the particle size to protect any controls fitted in the installation pipework.

BS 5885 states that a strainer shall be fitted at the inlet to safety shut-off valve systems to prevent the ingress of foreign matter. The largest strainer shall not be greater than 1.5 mm.

The fitting of a filter is especially necessary on high pressure gas lines. They are simple in construction consisting of a casing containing resin bonded wool, glass fibre or other porous material, through which the gas must pass, any foreign matter being retained.

Filters should have the following characteristics:

i) An acceptable pressure loss.
ii) Ability to retain small particles.
iii) High dust holding capacity.

The useful life of the filter under given conditions of dust content in the gas stream is proportional to its dust-holding capacity. As the filter retains the dust the pressure loss will increase until it becomes unacceptable, at which point the filter element must be exchanged.

A typical example is that of a 1½ in BSP filter which has been developed, employing a resin bonded wool. This filter has a retention efficiency of 80% of particles of from 0.2 to 2.0 microns diameter, and a dust holding capacity of 20 grammes of dust of 5 microns diameter. The initial pressure loss of the filter is 1.25 mbar (0.5 in wg) when passing 15.4 m^3/h (550 ft/h) of gas. The acceptable pressure loss would be 2.5 mbar (1 in wg) above the initial pressure loss.

Easy access to the unit is essential to allow for periodic maintenance, the frequency of which will be dependent upon local conditions. This should be determined during the initial period of installation of the filter.

Typical examples include the Jeavons basket type fitted with a 200 micron stainless steel mesh element and the Kromschroder with a filter pad of polyester fleece of 50 micron inserted in an aluminium body (Figure 9.70).

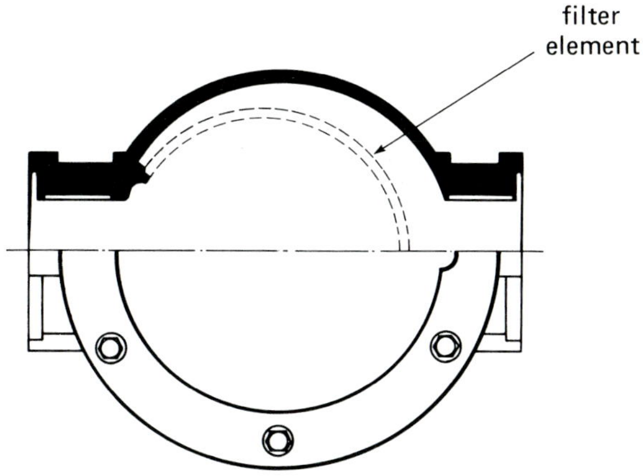

Fig. 9.70 Kromschroder filter

9.7 AIR/GAS RATIO AND THROUGHPUT CONTROL

The function of an air/gas ratio control system is to control combustion quality. The tolerances in combustion quality acceptable for each process application will influence the type of burner system and air/gas ratio control system employed. A moderate degree of control of combustion quality is necessary in order to ensure flame stability and completeness of combustion reactions. In the case of processes where a high excess of air is required, e.g. drying, induced draught burner systems are often employed and in these cases moderate control of air/gas ratio is acceptable. Close control of combustion quality is necessary to achieve maximum thermal efficiency by operating close to stoichiometric conditions, thus minimising flue losses. Limited control may be achieved by natural draught burner systems, which are simple and inexpensive, but are not capable of close air/gas ratio control over a wide throughput range, unless an expensive appliance pressure control system is added. When greater accuracy of combustion quality control is required, forced draught burner systems are used. The descriptions which follow have therefore concentrated on pressure techniques. The precise choice of the air/gas ratio control system will depend on the variation in the firing conditions (e.g. air preheat and back-pressure variations) for which the system has to compensate.

Throughput control is usually necessary where automatic temperature control is required, and also for ignition sequence control. The type of throughput control will depend on the process control mode - on/off; high/low; high/low off; modulating (proportional) being the most commonly employed modes. Other modes are integral; differential; derivative.

Table 9.2 gives guidance on the choice of air/gas ratio control systems suitable for use with various firing conditions and throughput control.

9.7.1 Factors Affecting Air/Gas Ratio Control

Accurate control of air/gas ratio is vital to the safe and efficient operation of gas fired plant. The mass ratio of air and gas is a prime factor affecting combustion products and stack losses. When gas is burned with too much air, fuel is wasted heating the excess air to the flue gas temperatures. When gas is burned with too little air, considerable wastage can occur due to incomplete combustion of the fuel.

TABLE 2 CONTROL QUALITY OF AIR/GAS RATIO CONTROL SYSTEMS

Air/gas ratio control system	Conditions of use					
	Atmospheric		Varying back pressure		Varying air preheat	
	High/low	Modulating	High/low	Modulating	High/low	Modulating
Natural draught techniques						
Basic control layout	C	D	E	E	E	E
Chamber pressure control	B	C	B	C	E	E
Mechanically induced draught technique	C	D	E	E	E	E
Pressure air injection techniques						
Basic control layout	A	B	E	E	E	E
Appliance pressure back-loading	F	F	A, B	B, C	E	E
Mixture pressure back-loading	F	F	A	B	E	E
High pressure gas injection techniques	C	D	E	E	E	E
Pressure air and pressure gas techniques						
Linked valves						
a) Controlled differential pressures	A	B	A	B	A	B
b) Simple linked valves	A	B, C	E	E	E	E
Balanced differential pressure	F	A	A	A	A	A
Pressure divider technique	A	B	A	B	E	E
Adjustable port valves (Selas)	A	A	E	E	E	E
Adjustable flow valves (Maxon)	A	A	E	E	E	E
Linked characterised valves	A	A	E	E	E	E
Balanced differential pressure J121	A	A	B	B	A	A
Thermistor ratio controller	A	A	B	B	A	A
Landis & Gyr SKP ratio controller	A	A	B	B	A	A
Mechanical premixing techniques	A	B	A	B	E	E

A Very close control
B Close control
C Moderate control
D Limited control
E Unsatisfactory control
F Not applicable

a) **Fuel wastage as a function of air/gas ratio**

The loss of sensible heat contained in flue gases depends upon the flue gas temperature and air/gas ratio. Figure 9.71 shows calculated fuel wastage due to increasing stack losses as air/gas ratio deviates from stoichiometric for a range of flue gas temperatures. This is based on the assumptions that the fuel and air are perfectly mixed and combustion processes proceed to completion. It can be seen that deviations in air/gas ratio can have a major effect upon the efficiency of high temperature plant.

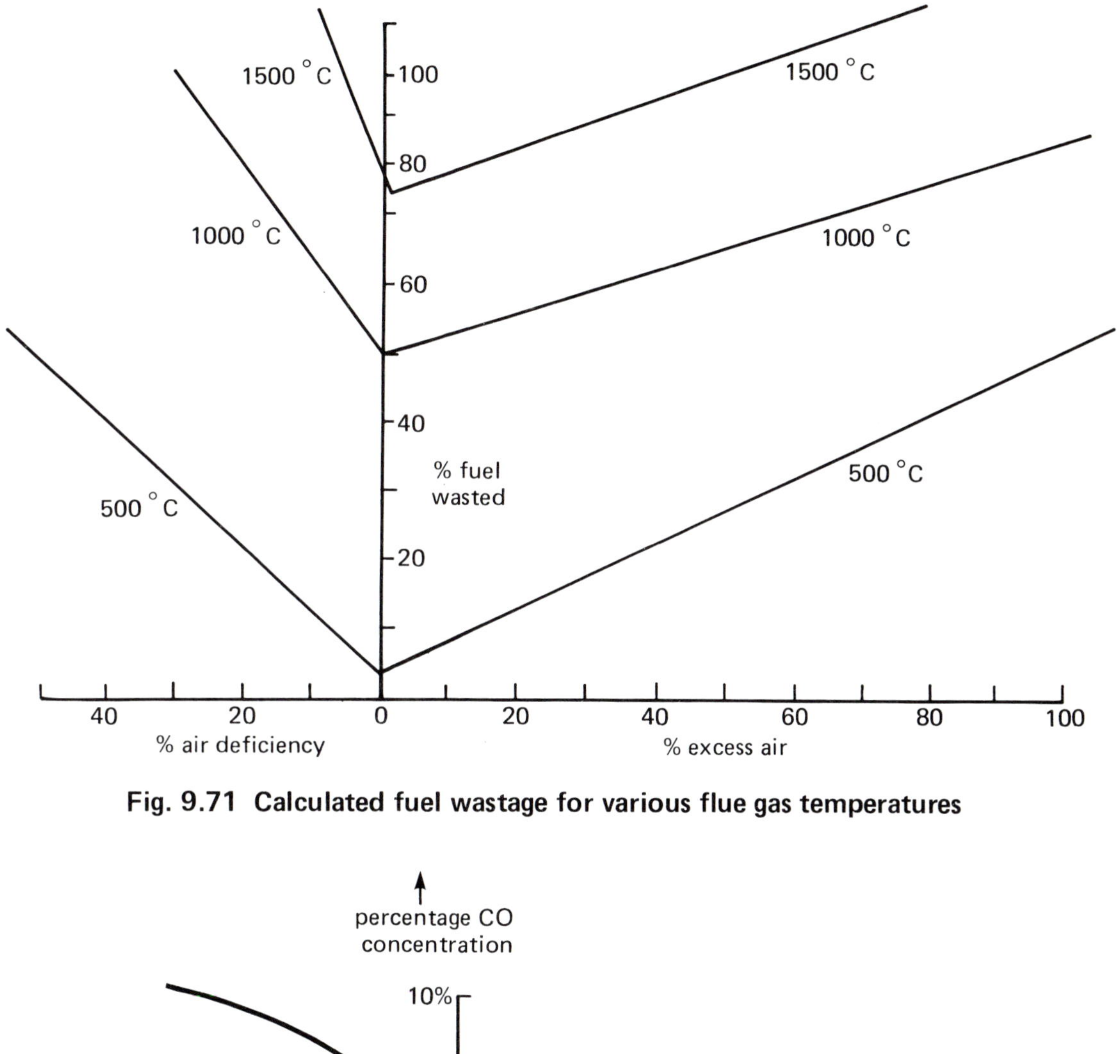

Fig. 9.71 Calculated fuel wastage for various flue gas temperatures

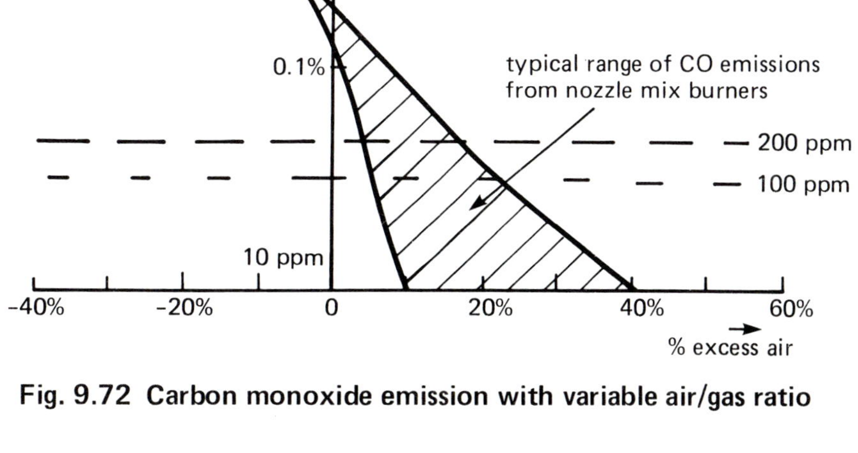

Fig. 9.72 Carbon monoxide emission with variable air/gas ratio

b) Effect of carbon monoxide emission upon ratio setting

In practice, imperfect mixing of fuel and air at the burner means that the air/gas ratio for maximum thermal efficiency has to be set on the excess air side of stoichiometric. Figure 9.72 shows the typical variation of carbon monoxide emission from excess air. This graph is a summary of combustion gas measurements made on various industrial burners. Assuming that the acceptable level of carbon monoxide emission lies somewhere below 100 ppm, it can be seen from the graph that the optimum air/gas ratio setting typically lies in the region of 5 to 23% excess air. Combustion gas analysis has to be performed on the individual plant or appliance in order to determine precisely the minimum amount of excess air that produces an acceptable carbon monoxide emission.

c) Additional factors relating to ratio control

There are many factors, additional to the basic combustion requirements described above, which influence ratio control and which should be taken into consideration when selecting or installing control systems.

i) Burner turndown

As the air and gas flow rates through a burner fall with decreasing firing rate, the mixing of fuel becomes poorer leading to higher carbon monoxide emissions. Thus on modulating or high/low burners with a large turndown, the optimum ratio setting at low fire corresponds to a larger amount of excess air than the high fire setting

Ratio control systems for these applications should be chosen and adjusted accordingly. It is particularly important on modulating systems to ensure that the air/gas ratio varies smoothly and consistently between the high and low firing points.

ii) Convective heat transfer

Excess air has the effect of increasing mass flow rates. This will increase convective heat transfer and may be utilised to advantage for example when good temperature distribution is required.

iii) Combustion air preheat

On installation where a recuperator or recuperative burners are fitted, the maximum combustion air flow rate falls as the air temperature rises. Figure 9.73 shows the variation of air throughput with increasing air temperature assuming constant upstream air pressure and that the air flow restricting orifice is located downstream of the recuperator. As the gas supply to the burner is not preheated, the effect of preheating combustion air is to cause a variable shift in air/gas ratio, dependent upon air temperature. There are several ratio control systems that are capable of compensating for this effect (See section 9.7.6.1).

iv) Other factors

Examples of other factors that directly affect ratio control are:

a. Variation of fuel calorific value
b. Variation of combustion air pressure/combustion air fan speed
c. Variations in atmosphere conditions
d. Wear in mechanical linkages
e Hysteresis in the operation of control systems.

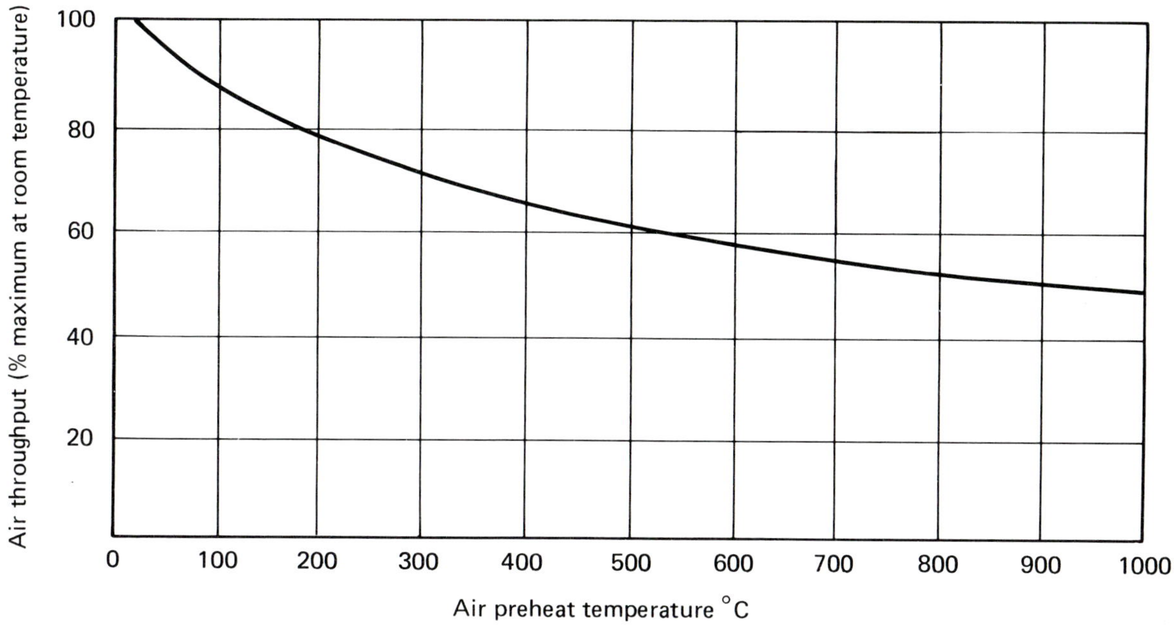

Fig. 9.73 Relative variation of combustion air throughput with air preheat temperature

The excess air level should always be set with a sufficient margin to prevent any of the above factors causing gas rich firing. The more sophisticated ratio control systems are designed to automatically compensate for one or more of the above factors, thus permitting a lower excess air level or increased safety allowance in comparison with simple ratio control systems.

d) Codes and Standards

British Standard BS 5885 and British Gas Codes of Practice IM/12 and IM/18 define general requirements for ratio control systems. It is a principal requirement that all ratio control systems should be designed and constructed so as to minimise the risk of off-ratio firing, particularly where air and gas throughput valves are not directly connected and air and gas flows are not simultaneously altered.

9.7.2 Natural Draught Techniques

9.7.2.1 Basic control layout

This technique employs low pressure gas burners of the neat gas or atmospheric injector type, and secondary air for complete combustion is induced by natural draught. Throughput is controlled by means of a single valve in the gas supply line. Secondary air may be automatically controlled by a mechanical linkage between the gas control valve and a flue damper or secondary air dampers. Neither method is commonly used since both complicate a simple system and the improvement of thermal efficiency from control of air/gas ratio on turndown is marginal.

It must be noted that the damper should be prevented from closing to a point where smothering occurs, preferably by cutting away a portion of the damper plate. In the case of ovens, one-third of the damper should be cut away in accordance with recommendations given in 'Evaporating and other Ovens', Health and Safety booklet HS(G) 16 by Health and Safety Executive.

9.7.2.2 Appliance chamber pressure control technique

This technique, shown schematically in Figure 9.74, is based on maintaining a constant differential pressure between the appliance chamber and atmosphere by applying these pressures across the diaphragm of a regulator. A pressure tapping links the chamber and the regulator, while atmospheric pressure is measured at a point close to the chamber pressure tapping point. Any deviation from the balanced condition results in the appropriate hydraulic, pneumatic, or electric power signal being given to the damper control in the appliance flue, thus restoring the appliance pressure to the balanced condition. This system may be adopted on an appliance in which any variations in the chamber pressure are anticipated, but it is generally used in conjunction with the control scheme already described, for which a constant backpressure is essential in order to achieve a more accurate control of air/gas ratio over the throughput range.

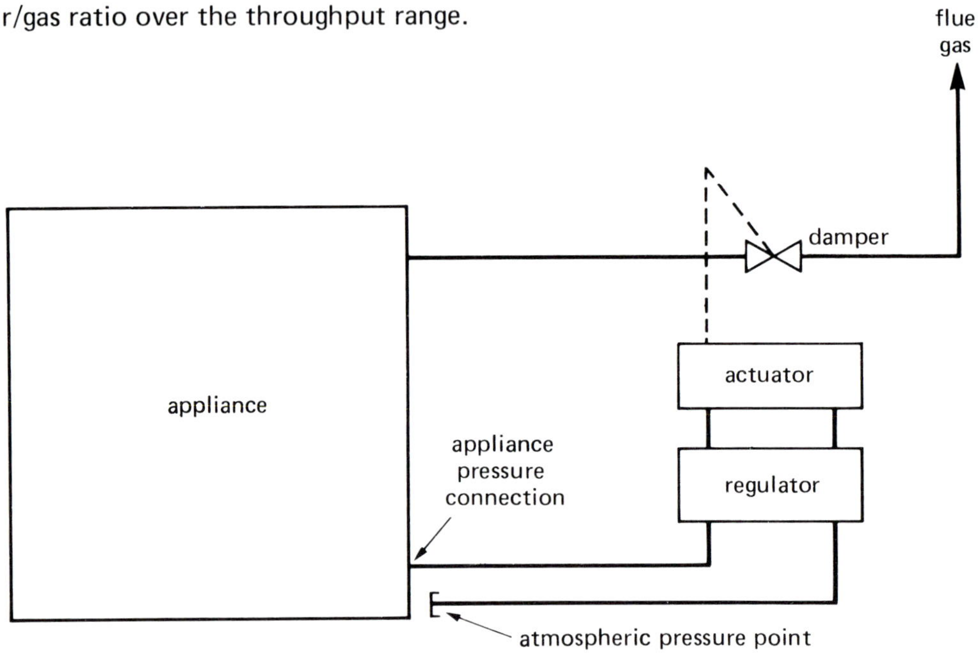

Fig. 9.74 Appliance pressure control

9.7.2.3 Mechanically induced draught technique

This technique, often used where large quantities of flammable vapours may be evolved, employs burners of the neat gas or atmospheric injector type. Secondary air for complete combustion is inducted by natural draught supplemented by an exhaust fan situated in the base of the chimney. The amount of secondary air required for complete combustion is automatically controlled by linking the exhaust fan motor or the flue damper to the throughput control valve in the gas line. Such applications often utilise a large amount of excess air and accurate control of air/gas ratio is not necessary.

It must be noted that any dampers used in the system should be prevented from closing completely as discussed previously.

This technique may be used in conjunction with an appliance chamber pressure control system.

All the systems described rely on the pressure drop across orifices for their control; thus none of them will cope with variations in specific gravity. Since these systems are designed to give constant air/gas ratio, none of them will cope with changes in the air requirement caused by gas quality changes. Gas quality variations only affect the thermal performance to a small extent, but may be important.

In specialised automatic flame heating processes, such as surface hardening, brazing and glass working, the variation in flame structure caused by changes in gas quality and specific gravity may be unacceptable. Satisfactory performance in these cases can often be obtained by employing Wobbe Index and/or Cone Height Control Systems.

9.7.2.4 High pressure gas injection techniques

It is sometimes possible to use gas pressures of 210-350 mbar (3-5 psi) to give a fully aerated mixture pressure for the burner to be stable over a satisfactory turndown range.

Throughput is controlled by means of a single valve in the gas supply to the injector. Suitable valves for throughput control are those which will operate at the line pressure without leakage.

The air/gas ratio is set by means of an air-shutter on the injector. It is maintained reasonably constant over the turndown range, which is usually of the order of 2:1.

9.7.2.5 Relay control valves

A relay control valve, illustrated in Figures 9.75 to 9.77, is used to control air or gas flow and normally takes the form of a horizontal, weighted, flexible diaphragm exposed to the inlet pressure on its underside. From its underside, an inlet weep orifice leads to a chamber above the diaphragm. An outlet weep pipe leads from the top chamber, through an on-off control valve, to release the pressure above the diaphragm. This weep may be bled through the weep shut-off valve to the vicinity of the burner where it is flared off, or in the case of air control valves to the atmosphere. The weep is returned to the manifold downstream for modulation control.

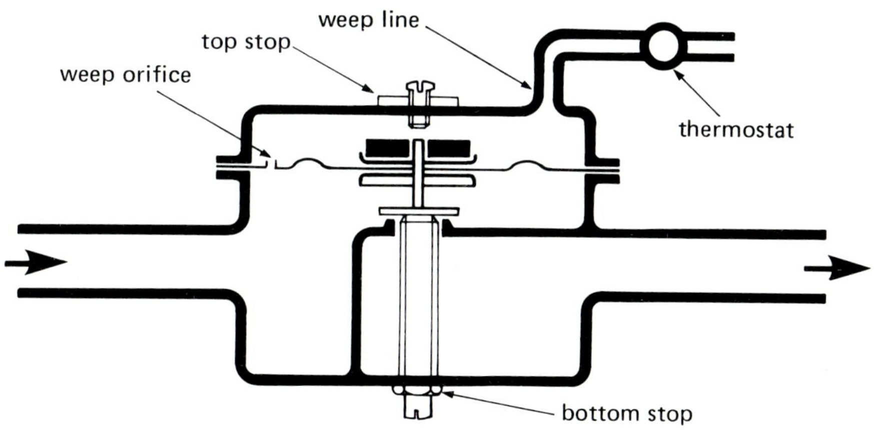

Fig. 9.75 Relay control valve – high-low operation with adjustable low rate and high rate stops

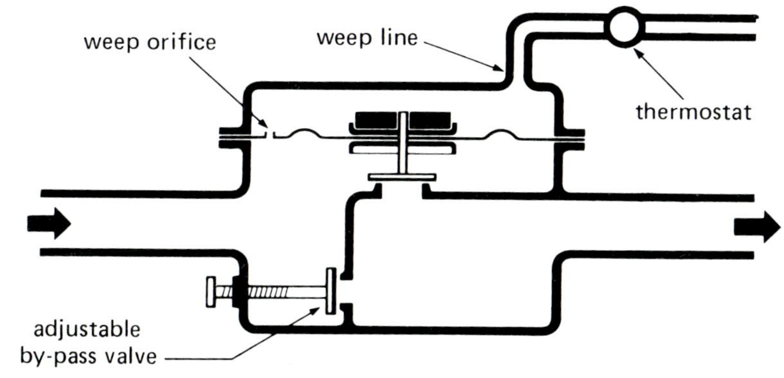

Fig. 9.76 Relay control valve with adjustable by-pass valve

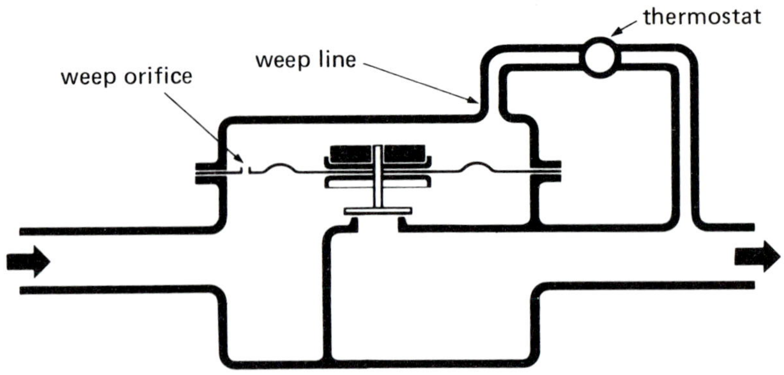

Fig. 9.77 Relay control valve - modulating operation

The underside of the diaphragm is subjected to the full inlet gas pressure which is transmitted to the top chamber via the inlet weep orifice. If the weep shut-off is opened, the pressure in the top chamber is dissipated and so allows the main valve to open fully permitting the maximum flow of gas to the burner(s). When the weep shut-off is closed, the pressure above the diaphragm increases until it equals that below the diaphragm. The weighted diaphragm then gradually falls and closes the valve, thus restricting the flow to the burner(s). The pressure differential across the valve helps to maintain it in the closed position.

The time taken for the valve to open or close is largely dependent on the time taken to pressurise or exhaust the top chamber rather than the speed of movement of the valve. Adjustment of the inlet and outlet weep orifices varies the time required to pressurise or exhaust the top chamber.

For on-off operation the top and bottom stops shown in Fig. 9.75 are fully retracted so that the valve opens and closes fully. The valve bob is covered in neoprene to ensure a complete gas seal in the closed position. The system used in Fig. 9.76 can also be used with the adjustable by-pass valve closed.

For high-low operation the relay control valve, shown in Fig. 9.76, is supplied with an adjustable by-pass. When the shut-off valve is fully closed, the main diaphragm valve closes and the by-pass may be adjusted to regulate the flow to the burner(s). Another fairly common method of producing high-low operation incorporates top and bottom stops for the main diaphragm valve in place of the adjustable by-pass shown.

A modulation effect may be obtained by returning the outlet weep pipe to the manifold downstream of the relay control valve, as shown in Figure 9.77. This arrangement may limit the amount of gas that the relay valve will pass in the open condition.

In the case of on-off or high-low operation the weep pipe is preferably terminated in the vicinity of a protected pilot flame, so that the weep gas may be safely burnt. If possible the flame protection valve (e.g. thermo-electric valve) should be upstream of the relay control valve. Relay valves, however, used with thermo-electric valves with independently fed pilots on several burners of necessity would be situated upstream of the thermo-electric valves.

The most common valves in weep lines of relay control valves are thermostats. However, they may be pressurestats, solenoid valves (which are themselves actuated by thermostats) or clock controlled valves.

A typical rod type thermostat is the Sperryn, shown in Figure 9.78. The sensing element comprises a rod and concentric tube of materials having dissimilar coefficients of expansion. A rise in tempera-

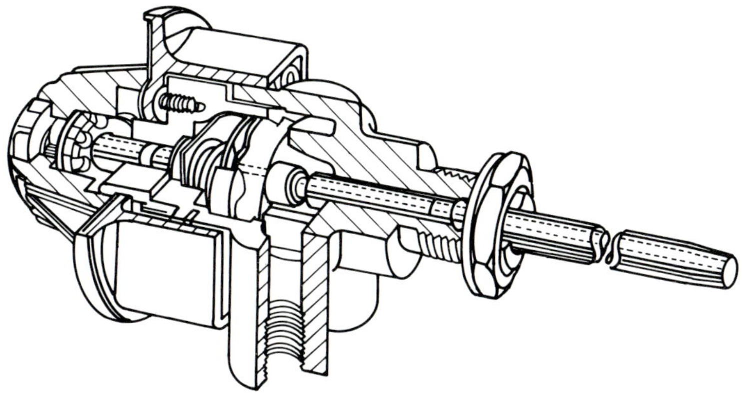

Fig. 9.78 Sperryn rod type thermostat

ture around the sensing element causes the tube to expand to a greater extent than the rod, resulting in the rod moving away from the piston which operates on the valve. The valve then closes to its seat, preventing the flow of gas through the thermostat and operating the main control unit to which it is connected. When the sensing element cools, the rod moves towards the piston, lifting the valve off its seat and reversing the operation of the main control unit. The valve is connected to the spindle by a screw thread. Movement of the spindle to which is attached a temperature setting dial, controls the setting of the thermostat by varying the distance between the valve and its seat.

Relay control valves may be obtained for low pressures up to 30 mbar (12 in wg) with connections between ½ in and 6 in BSP, although 2 in BSP is normally the maximum size readily available.

Relay valves provide a very cheap method of proportional control on small burner systems. Even on larger burners of the natural draught or induced draught type, e.g. drying ovens, relay valves provide an adequate means of high/low control by means of an electrically controlled or mechanically controlled valve in the weep line.

Relay valves may still be found in use as shut-off valves when used in conjunction with a small solenoid valve. Clearly this is undesirable because of the reliance on a diaphragm for positive closure, and the small closing force that is available; these valves cannot meet the requirements of BS 5963, and should be replaced by properly designed safety shut-off valves of Class 1 or 2 as appropriate.

Composite Control Valves

A composite valve consists of a thermo-electric valve, a governor and a relay valve all in the same die-casting, e.g. Sperryn G880 series.

This has the advantage of compactness and only one pair of gas connections is necessary. The disadvantage is that if one component of the unit fails, the whole unit generally has to be replaced. Composite control valves can also include an electrically operated solenoid valve to replace either the governor or the relay valve of the unit. The units are discussed more fully in Chapter 10 on Flame Protection Equipment.

9.7.2.6 Double Duty Valve

The double duty valve is similar in construction to a weep relay valve and operates in a similar manner, albeit in a reversed mode. It is still found on natural draught commercial boilers.

The valve protects the boiler against accident in the event of:

a) failure of the pilot flame
b) low or drastic reduction of gas pressure.

In addition the burner cannot be relit until all gas valves have been shut and the full lighting sequence carried out.

A typical installation incorporating thermocouple flame protection is shown in Figure 9.79. It should be noted that the pilot is unprotected which is an undesirable feature of the system. The double duty valve itself is shown in more detail in Figure 9.80.

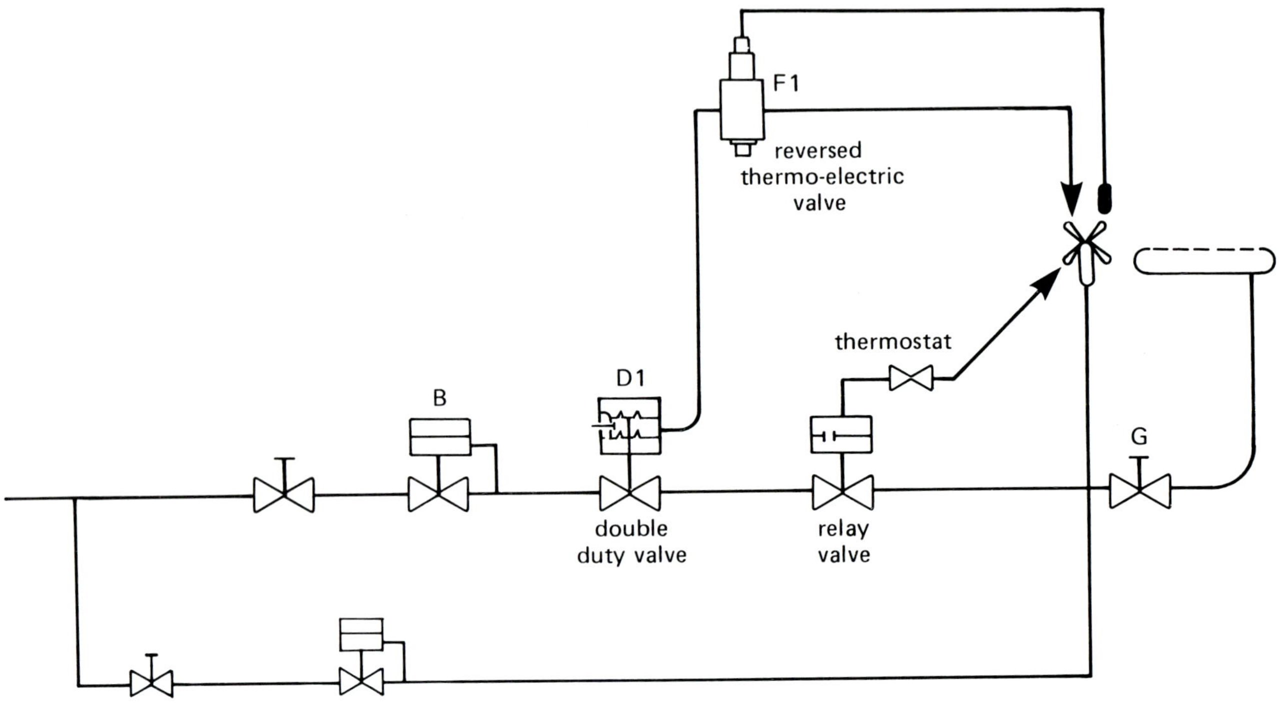

Fig. 9.79 Double duty valve with thermoelectric flame failure

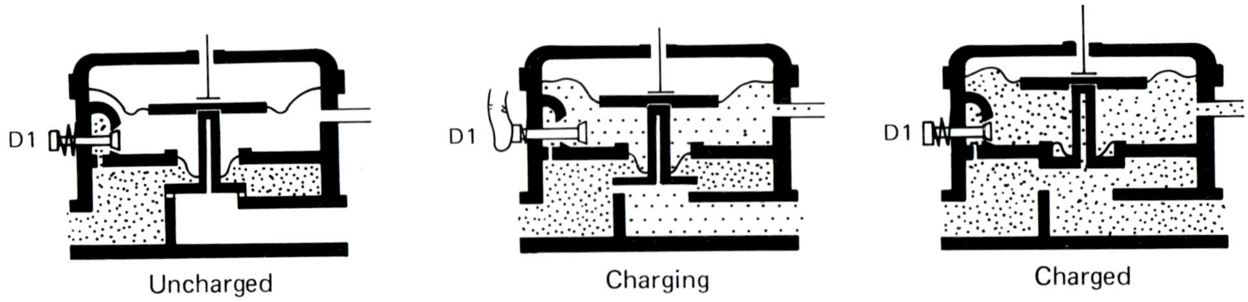

Fig. 9.80 Operation of double duty valve

Operation of the system is as follows:

1 All gas valves are closed with thermostats, clock etc, open.

2 The pilot is ignited and will heat the thermocouple attached to thermo-electric valve, F1.

3 The main gas valve, B, is opened.

4 The thermo-electric valve F1 is operated by its push button. Note that this valve *closes* when energised and hence seals off the outlet from the centre compartment of the double duty valve.

5 The push button, D1, is operated for about 15 seconds. This allows gas to pass through a small restrictor from the inlet of the double duty valve to its centre compartment.

The outlet weep line has been closed by the thermo-electric valve whilst the outlet via the spindle is closed by the burner cock, G. The consequent build up of pressure in the middle compartment raises the main diaphragm together with the spindle and valve so allowing gas to pass to the next control.

6 Opening the burner cock allows smooth ignition of the main burner from the correctly positioned pilot burner.

It is seen that protection is given by means of the double duty valve closing if gas pressure is not maintained in the centre compartment of the valve. Closure will occur in the event of:-

a) Failure of the main gas supply, e.g. by turning off the main gas cock even momentarily; or reduction in main gas supply pressure.

b) Loss of pilot flame. This causes the thermo-electric valve to open, venting the D.D. valve with loss of pressure.

Whenever such a loss of pressure and shut down occur, all cocks must be shut. The D.D. valve must be reprimed as part of the full lighting sequence in order to put the boiler back into operation.

9.7.3 Techniques Employed with Air Blast Burners

9.7.3.1 Basic control system

With the air blast system, the energy in the combustion air supply is used to induce gas into the system at a controlled rate. The system employs only one throughput control valve; variation in the air throughput resulting in variation in the gas throughput. The air blast system is simple, flexible and capable of maintaining a 4:1 turndown.

The basic control system is shown in Figure 9.81. Air from a fan at approximatly 25-75 mbar (10-30 in wg) is allowed to expand through the jet of an air blast injector into a venturi mixing tube. The expanding jet of air entrains gas at atmospheric pressure which is admitted to the mixture tube immediately downstream of the air nozzle. The mixture must be supplied directly to the burner with no restrictions. Providing that the gas is supplied at atmospheric pressure, usually by means of a zero governor, and that the burner is firing into approximately atmospheric pressure, then there is an inherent self-proportioning action which maintains the air/gas ratio substantially constant whilst enabling the throughput to be controlled by the operation of one valve in the air supply line. The gas ratio setting valve is often incorporated into the injector, when it is known as the 'obturator'. In the tunnel burner, injection, mixing and combustion occur together in the tunnel.

Zero governors operate best with an inlet pressure of 5 mbar (2 in wg) and an appliance governor set to this should be incorporated. The proportioning accuracy of the system obviously depends on the performance of the zero governor and on the flow characteristics of the ratio-setting valve. Where the tension of the zero governor is adjustable the ratio may be set independently at low and high flow rates. Changes in friction factor are compensated in this way and a reasonably constant air-gas ratio obtained over the working range.

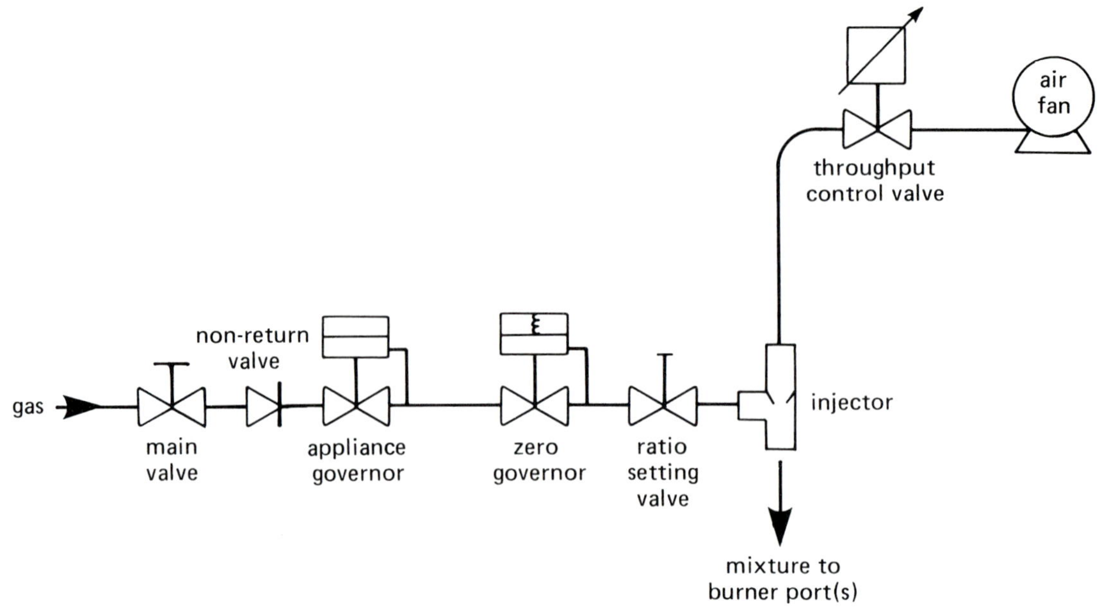

Fig. 9.81 Basic control layout

9.7.3.2 Control scheme with appliance pressure back-loading

Air blast burners are frequently required to fire into combustion chambers in which the pressure differs significantly from atmospheric. A small change in combustion chamber pressure around zero will affect the flow of gas, (which is also at zero) but not the flow of air at, say, 70 mbar (28 in wg). To maintain the self-proportioning action the reference pressure for the appliance and zero governors should be the appliance chamber pressure and not atmospheric pressure. This can be achieved by back-loading the top diaphragms of both the appliance and zero governors with the combustion chamber pressure.

Such a scheme is shown in Figure 9.82. With any back-loading system it is necessary to have the inlet pressure to the zero governor at least 0.75 mbar (0.3 in wg) higher than the backloading pressure.

9.7.3.3 Control scheme with mixture pressure back-loading

Complete compensation for any changes in back-pressure downstream of the mixture manifold is obtained by mixture pressure back-loading. The zero governor and appliance governor are back-loaded with the mixture pressure from a point immediately downstream of the injector. Again it is necessary to have the inlet pressure to the zero governor at least 0.75 mbar (0.3 in wg) higher than the back-loading pressure. As the mixture pressure will be of the order of 18 mbar (7 in wg), it will usually be necessary to use gas boosting, as shown in Figure 9.83.

9.7.3.4 Practical considerations

i) A separate manual on/off valve is always required in the gas line, the ratio-setting valve should not be used for this purpose.

ii) To avoid excessive time lags in automatic systems, the gas control valves and governors should be as close to the burner as practicable, i.e. the first control should be no more than 10 pipe diameters from the burner and there should be no more than 5 pipe diameters between any two consecutive components.

iii) On back-loaded systems both governors and loading lines should be gas-tight.

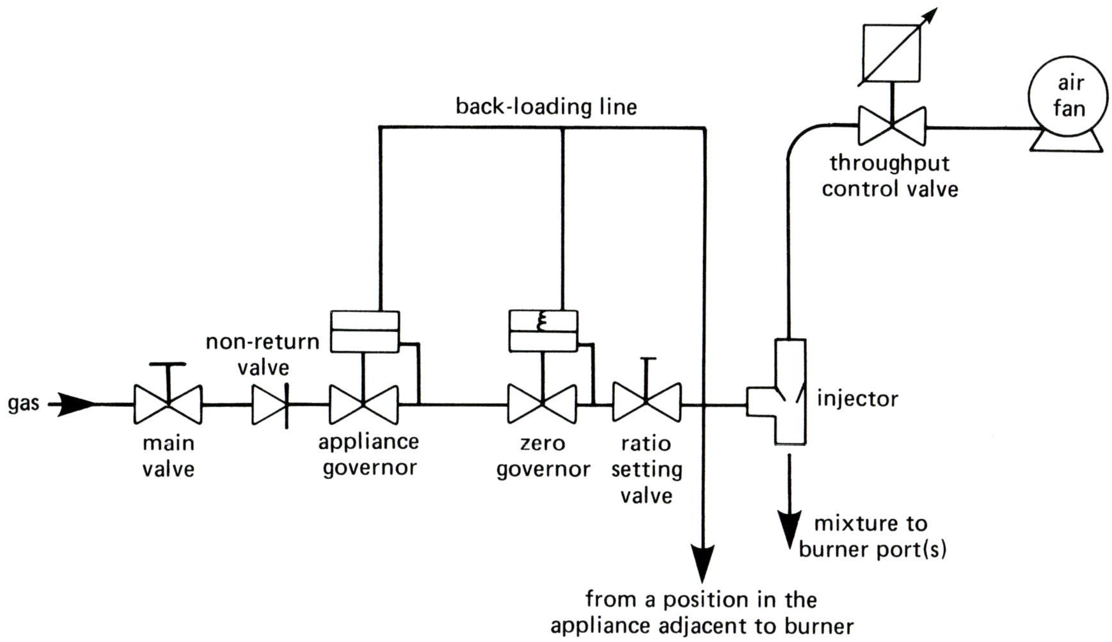

Fig. 9.82 Air blast burner control system - with appliance pressure back-loading

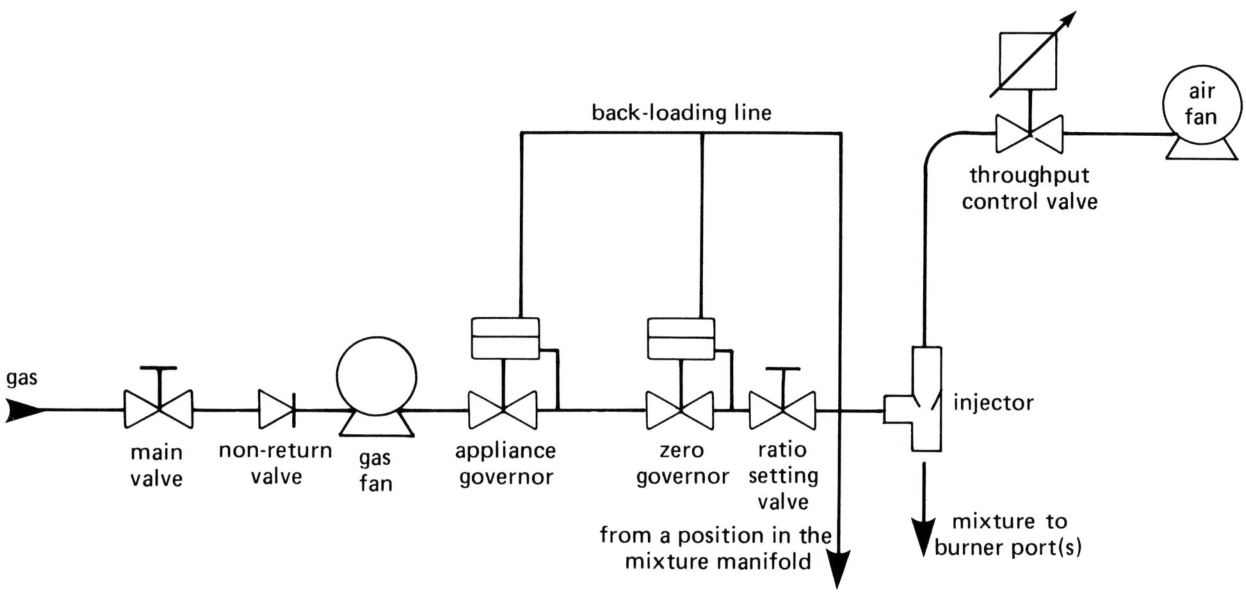

Fig. 9.83 Control layout with mixture pressure back-loading and gas booster

iv) Where one injector and zero governor is used to supply several burners and mixture back-loading is practised, then one or more of the burners may be turned off without affecting the air/gas ratio.

v) A single zero governor can be used to feed several injectors. In this case it may be difficult to set all the burners to the required air/gas ratio over the whole of the turndown ratio since, at the low flow rate, each burner may require a different zero governor setting for the correct air/gas ratio. It is essential that all the burners are firing into conditions of identical back pressure, otherwise considerable variation of air/gas ratio between burners may occur.

vi) With premix burners the air and gas will be intimately mixed before combustion. A sample of this mixture can be burnt in a test burner of known characteristics to ascertain the air/gas ratio. This technique, which cannot be used with nozzle-mix burners, is useful when flue sampling is difficult and there is multi-burner firing etc. Another possibility is to measure the oxygen in the air/gas mixture using an accurate oxygen analyser such as a Servomex, provided that the sampling of an air/gas mixture within the flammable limits does not produce a hazardous situation.

vii) An acceptable non-return valve should be installed after the main isolating valve to the system and preferably before the governor if flame protection and a certificated safety shut off valve is not used.

9.7.4 Techniques Employed with Nozzle Mix Burners

The following techniques are suitable for use with nozzle mix burners where the air and gas supplies can be considered as mixing downstream of any metering orifices and the downstream pressure is common to both supplies.

1. Mechanically Linked Valves

a) Simple linked butterfly valves

The system is shown in Figure 9.84. Variable area butterfly valves in the air and gas supply lines are mechanically linked so that air and gas flows can be altered simultaneously using a single actuator. This system is inexpensive, robust and widely used but only operates satisfactorily within certain limits. It is essential that substantially constant pressure differences are maintained across each valve under all operating conditions. Where changes in back pressure at the burner are small in comparison to the supply pressures, adequate ratio control may be achieved by ensuring that upstream pressures are nearly constant and the pressure drops across the valves are relatively large. The combustion air fan should be chosen so that the constant pressure part of its characteristic corresponds to the desired air flow range.

Linked butterfly valves are available with a slipping clutch mechanism on the gas and air valves with adjustable top and bottom stops, e.g. Industrial Control Valves, so that high/low control is easily achieved. These valves though are not suitable for modulating control as the system will go off-ratio in the middle of the turn down range.

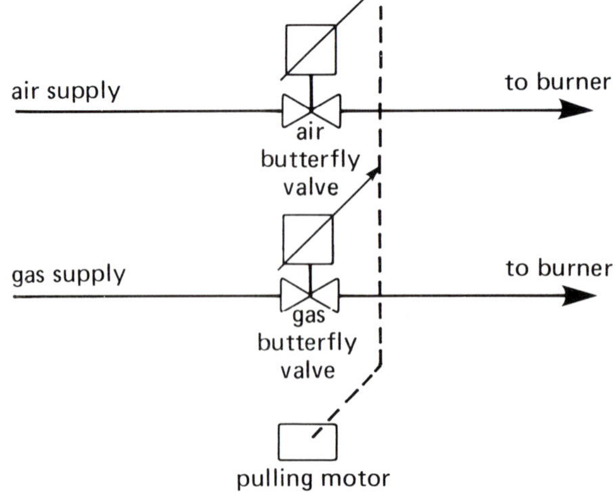

Fig. 9.84 Simple linked butterfly valves

b) Linked valves with individual governors

The incorporation of governors upstream of both the air and gas throughput control valves as shown in Figure 9.85, greatly improves ratio control. Each governor is loaded from a point downstream of the valve, thus maintaining a constant pressure difference across the valve. This system will give satisfactory ratio control when downstream back pressure variations are larger or unequal as encountered on pre-heated air installations. The main disadvantage of this system lies with the additional cost and physical size of the governor that is required for the air supply line.

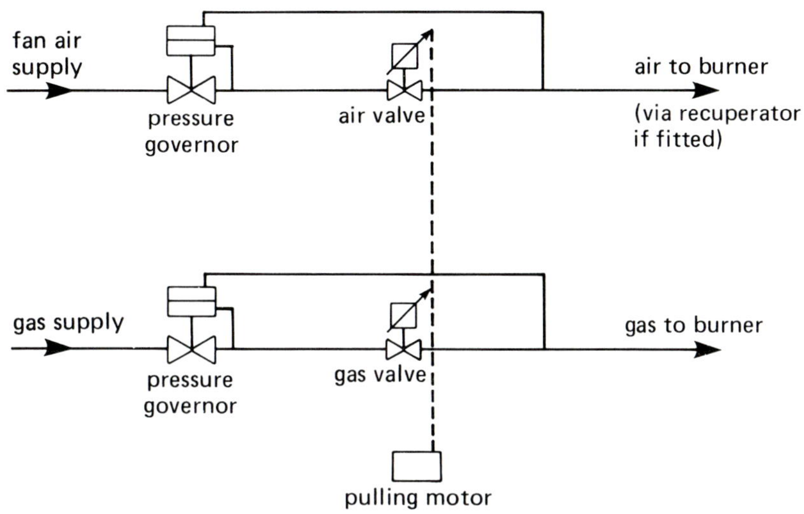

Fig. 9.85 Linked valves with individual governors

2. Proportioning valves

The regulation of air and gas flows to any predetermined ratio over the throughput range by linked air and gas valves may be achieved in either of two ways. Adjustable port valves of matched flow characteristics (usually linear) may be employed for both air and gas in which the relative port area ratio remains essentially constant at a pre-set value over the whole flow range. Alternatively a valve of any characteristic may be used for one stream and its flow characteristic matched by linking with an adjustable flow valve, with multiple settings throughout the flow range, on the other stream.

a) Adjustable port valves

These consist of mechanically linked plug valves on a common spindle, as shown in Fig. 9.86, or separate valves with connected valve levers. Rotation of the valve spindle(s) will alter throughput while adjustment of the valve slides alters the relative port area ratio by varying the height of the rectangular ports. Provision is made for adjustment of the relative positions of the spindle(s), at which each valve opens or closes, to give the lag or lead that may be necessary to compensate for burner instability at all rates. By mounting both square port valves on a common spindle, backlash in the connecting mechanism is eliminated.

The square port valve manufactured by Selas is one example of this type of valve in common use.

Difficulties may occur with this type of valve where one of the ports is adjusted to rather less than square (i.e. rectangular), resulting in poor proportioning over the range of valve travel. Automatic actuation can easily be fitted to these valves.

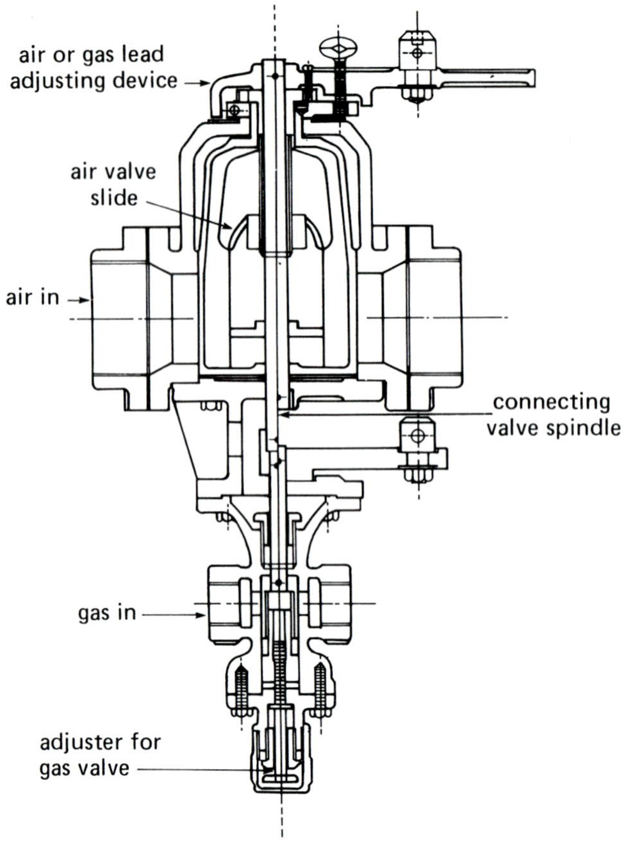

Fig. 9.86 Selas adjustable port valve

b) Adjustable flow valves

One type of adjustable flow valve consists of a cylindrical piston with a rectangular port rotating and reciprocating within the valve body. When the valve lever is moved from low to high position the rectangular opening in the piston is uncovered for flow.. The height of the rectangular port is adjustable at several different points of its travel by adjusting screws which press onto a cam spring which forces the piston down via a rocker arm. Depending on whether more or less flow is desired the adjusting screw is turned to increase or decrease the height of the port opening and thus change the area at that point.

Although these valves are used to advantage for accurate control of gas and air flow (they can be adjusted to produce straight line or other curves as predetermined by the burner requirements), they are not intended to give a tight shut-off. They may be used for operating pressures up to 1.72 bar (25 psi) and temperatures up to 140 °C (300 °F). Automatic actuation can easily be fitted to these valves which are often used for the control of gas and air on steam boiler packaged burners.

Another commonly encountered type is the Maxon Micro-ratio valve. On this a series of adjusting screws form the profile on a cam spring which in turn presses against the valve stem and valve disc. For every position of the air butterfly valve, the appropriate screw is adjusted to give the required gas flow. This facility enables the air/gas ratio to be 'characterised' to any desired profile over the whole burner firing range. See Figure 9.87.

c) Proprietary linked characterised valves

On large boilers, proprietary burner systems such as manufactured by Hamworthy and Saacke will be encountered. These consist of cam profiles adjusted by screws which actuate the butterfly valves in the air and gas lines (Figure 9.88). Sometimes extra adjustment

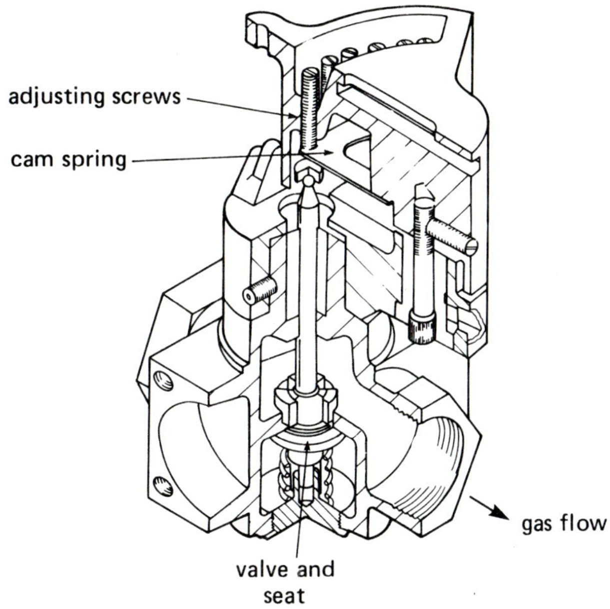

Fig. 9.87 Maxon Micro-ratio valve

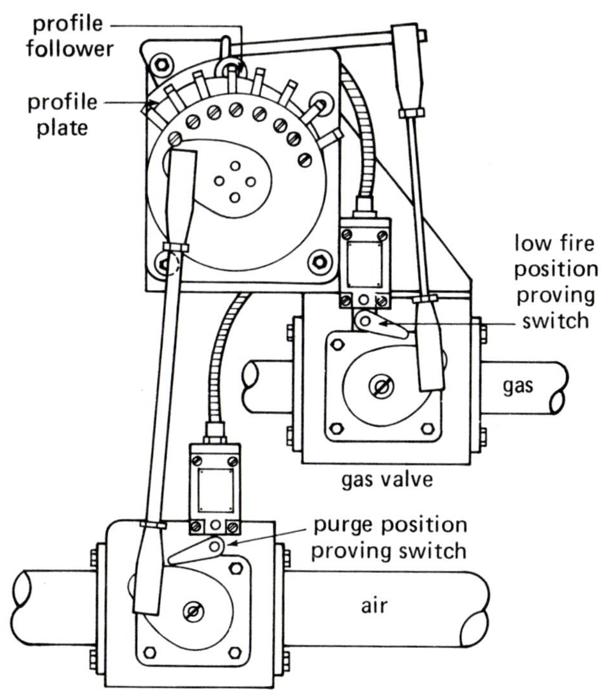

Fig. 9.88 Linked adjustable profile throughput control valves

is afforded by having a cam profile actuating the air and another cam profile actuating the gas valve. In most cases the butterfly valves are connected together by rod linkage or Bowden cable, but sometimes each valve has its own modulating motor and they are electrically interlinked. With the cable system this tends to stretch in use, so regular adjustment is necessary to maintain the correct air/gas ratio. Even with the rod system regular maintenance is required to compensate for wear on linkages, joints, etc.

3. Weep Relay Valves with Solenoid Valves

For a simple high/low control system, relay valves in the gas and air lines, and solenoid valves in the weep lines are used. Top and bottom stops are used for high and low fire respectively, and the solenoid valves on the weep lines are connected to the throughput controller (Fig. 9.89).

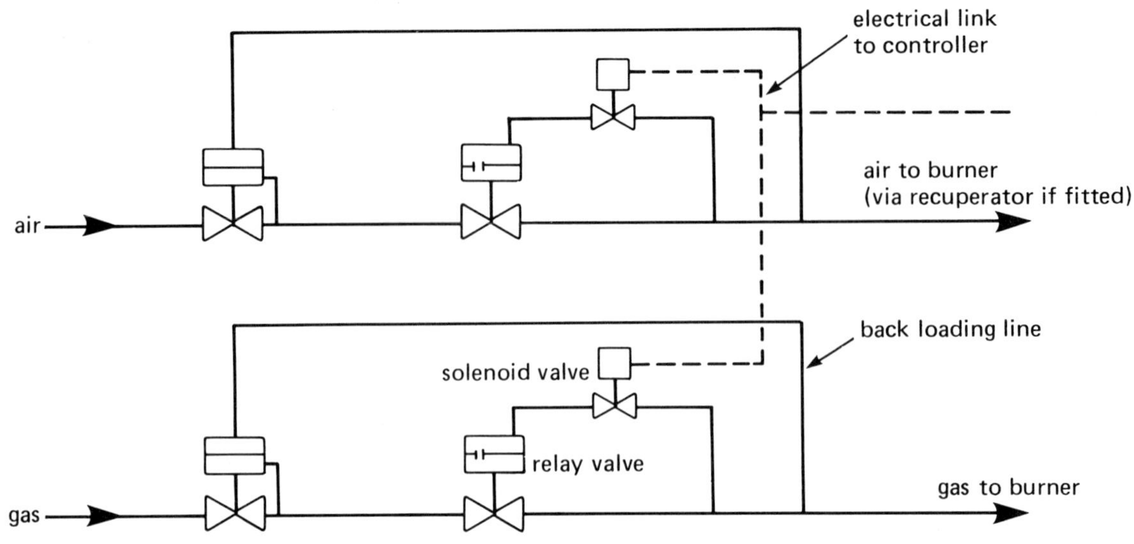

Fig. 9.89 Dual-valve ratio controller using weep relay valves

4. Balanced Pressure Control Systems

The air and gas flows through nozzle mix burners are determined by the pressure differences applied across metering orifices which may be situated at the burner nozzle or in the upstream supply lines. For fixed metering orifices, maintaining the air and gas pressures upstream of the orifices in constant ratio will result in the air/gas ratio at the burner being held constant.

a) Cross loaded governor with pressure divider

This system, shown in Fig. 9.90, is suitable for installations where the air supply pressure is higher than the gas supply pressure. A pressure divider, consisting of fixed and adjustable bleed orifices, is used to load a fraction of the air supply pressure on to a zero governor in the gas line. The adjustable bleed orifice alters the fraction of the air pressure applied to the gas governor and thus acts as the main ratio setting adjustment. Compensation for changes in combustion chamber pressure can be obtained by connecting the outlet of the bleed orifice directly to the combustion chamber rather than to atmosphere. The tension spring adjuster on the zero governor provides a means of introducing an offset or bias into

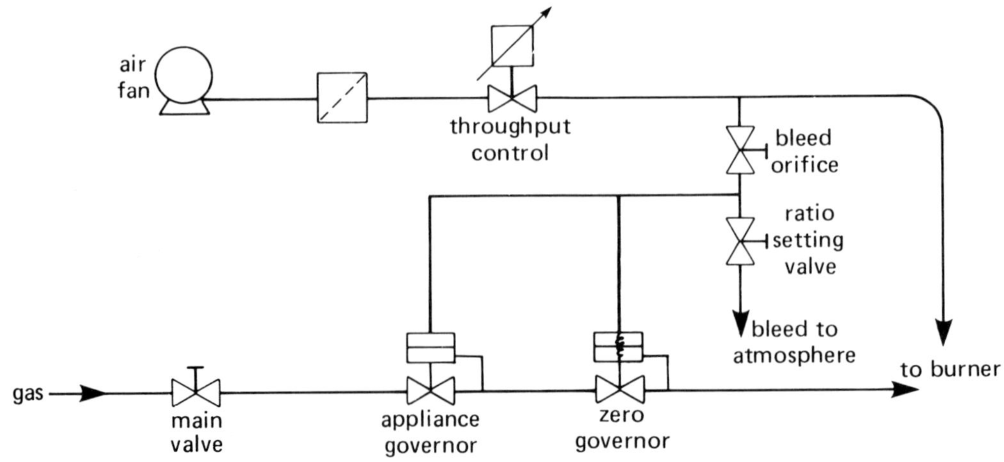

Fig. 9.90 Cross-loaded governor with air pressure divider

the relationship between air and gas pressures as shown graphically in Figure 9.91. The effect of the off set upon the ratio of air/gas pressures is much more pronounced at low fire (low pressures) than at high fire (high pressures). Thus the bleed orifice is used to set the overall air/gas flow ratio at high fire and the tension spring on the zero governor is used to tune air/gas ratio at low fire.

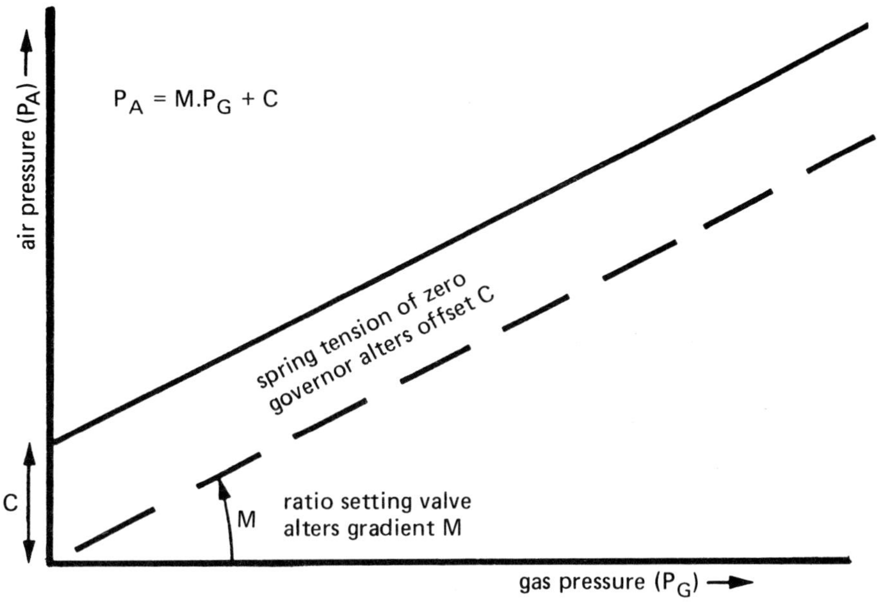

Fig. 9.91 Relationship between air and gas pressures for cross-loaded governor system

b) Cross-loaded governor with equal air and gas pressures

When air and gas supply pressures are equal, an alternative arrangement of the cross-loaded governor technique as shown in Figure 9.92 can be used. This arrangement eliminates the air bleed of the previous arrangement which is susceptible to blockage in dusty environments. As the upstream pressures are equal and the downstream pressures are common, it follows that the system shown in the figure automatically compensates for variations in combustion chamber pressure.

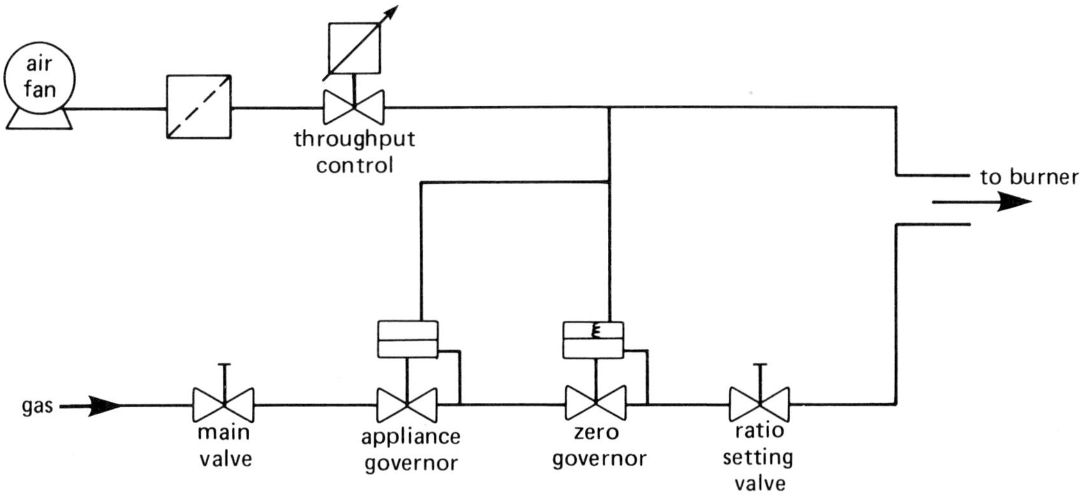

Fig. 9.92 Cross-loaded governor system for equal air and gas supply pressures

c) Proprietary pressure loading systems

A typical commercially available system is the Kromschroder G1 ratio controller shown in Figure 9.93. Here an impulse line from the air supply is taken from a point at least five pipe diameters from downstream of the throughput control valve. The Kromschroder G1 ratio control valve is inverted with a compression spring supporting the weight of the moving parts. This system does not normally require a primary governor, but the pressure drop across the valve is limited to around 100 mbar. The bleed orifice is used to set the overall air/gas flow ratio at high fire and the adjustment screw on the Kromschroder valve is used to tune the air/gas ratio at low fire.

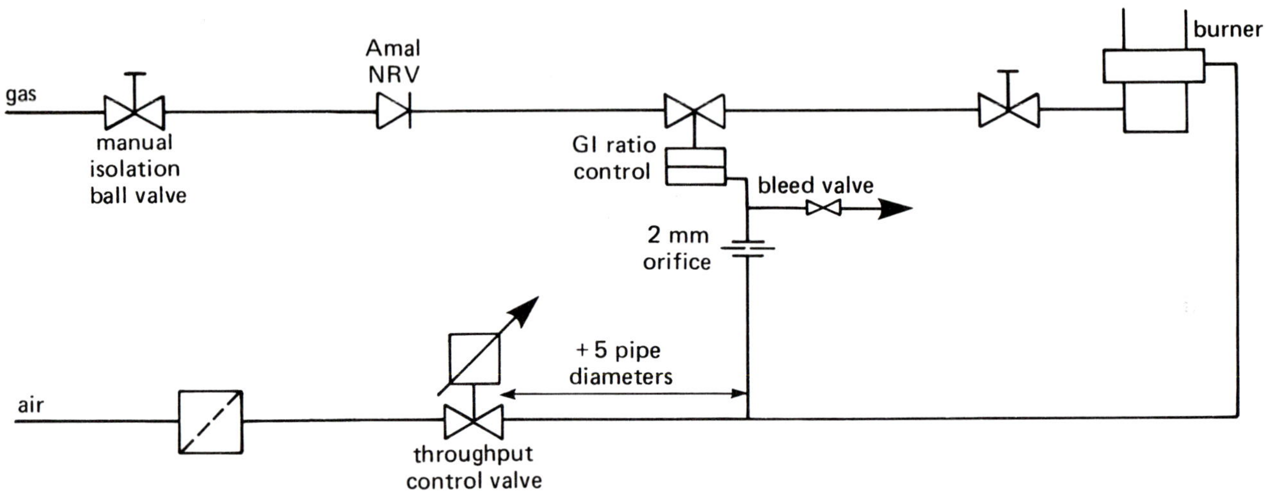

Fig. 9.93 Kromschroder G1 ratio control valve

5. Differential Pressure Systems

The increasing use of recuperators and recuperative burners has led to the development of ratio control systems which employ flow metering techniques such as sensing the differential pressure developed across a metering orifice or valve.

a) Multiple diaphragm differential pressure controller

Examples of this type of controller include the Jeavons J121, Eclipse DPV, Pyronics FCR and North American 7288. Figure 9.94 is a section view of the J121 unit illustrating the construction of multiple diaphragm controllers. The method of installation is shown in Figure 9.95. It should be noted that the throughput control valve should be positioned downstream of the air orifice in order to maintain as constant pressure conditions at the air orifice as possible. Also a constant pressure governor in the gas supply is essential.

The air and gas supply lines incorporate orifice valves where the differential pressure generated across each orifice is proportional to the square of the flow rate. The air differential pressure is applied across the lower main diaphragm of the controller in a direction that produces a downward force opening the valve. The gas differential pressure is applied across the upper main diaphragm of the controller and in a direction that produces an upward force closing the valve. The controller thus reaches an equilibrium position where air and gas differential pressures are equalised and air and gas flows are maintained in proportion.

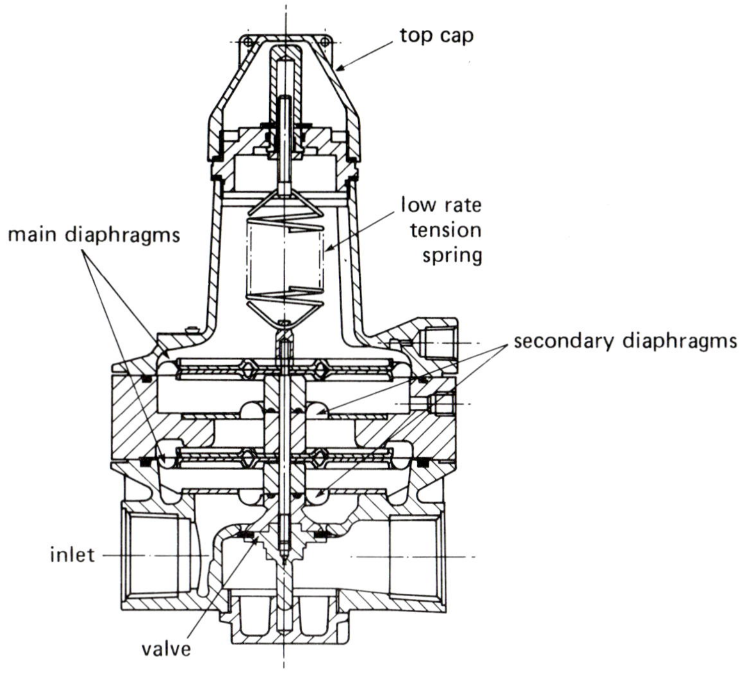

Fig. 9.94 Sectional view of Jeavons J121 multiple diaphragm controller

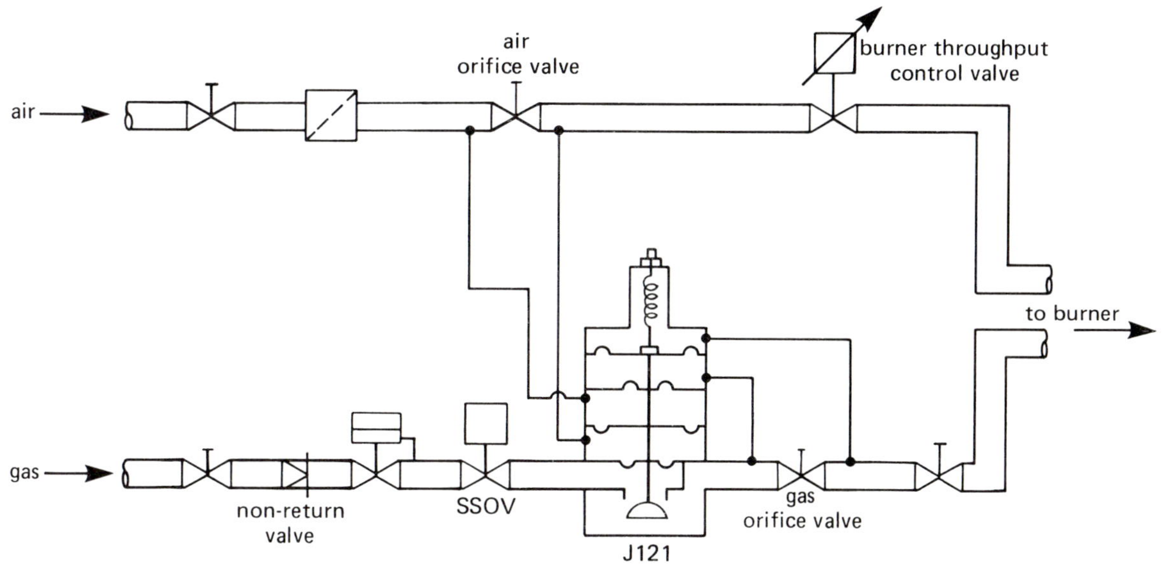

Fig. 9.95 Differential pressure ratio control system using J121 multiple diaphragm controller

Provided the orifice valves are installed upstream of a recuperator, where air temperature is reasonably constant, multiple diaphragm controllers will maintain ratio independent of the fluctuations in back pressure caused by varying air pre-heat. Lockable valves may be used as 'orifices' in the gas and air supplies.

As the relationship between differential pressure and flow through each orifice valve is also a function of valve opening, adjustment of either orifice valve alters the overall air/gas ratio setting. The tension spring in the controller enables an offset to be introduced between air and gas differential pressures and serves as the low fire ratio adjuster.

In setting up a J121 system the air orifice is adjusted to give a suitable differential pressure, say 5 mbar (2 in wg) at high fire, then the air flow is returned to the low fire rate and the low fire gas rate is set by adjusting the tension of the spring. After increasing the air flow to the high fire rate again the gas orifice is adjusted to give the appropriate high fire gas rates. The system is then alternately cycled between high and low fire, making adjustments of the spring tension at low fire and the gas orifice at high fire until a constant air/gas ratio of the desired proportion is maintained over the whole of the turndown range.

Figure 9.96 shows a typical performance curve for the J121 controller. The hysteresis inherent in the controller limits burner turndown to around 4:1 for a 2.5 mbar (1 in wg) pressure difference across orifice valves at high fire.

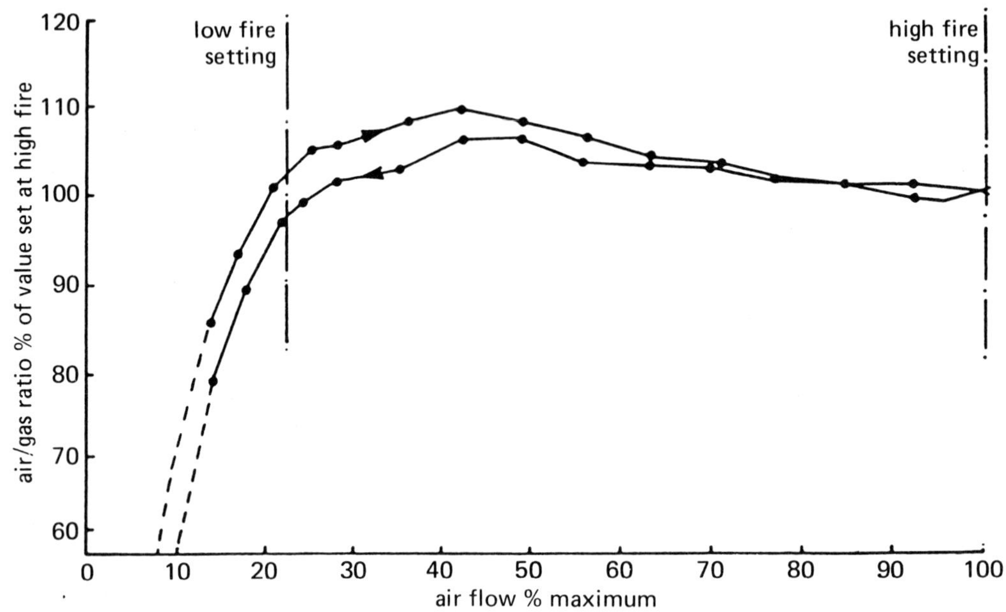

Fig. 9.96 Typical air/gas ratio characteristic for J121 control system

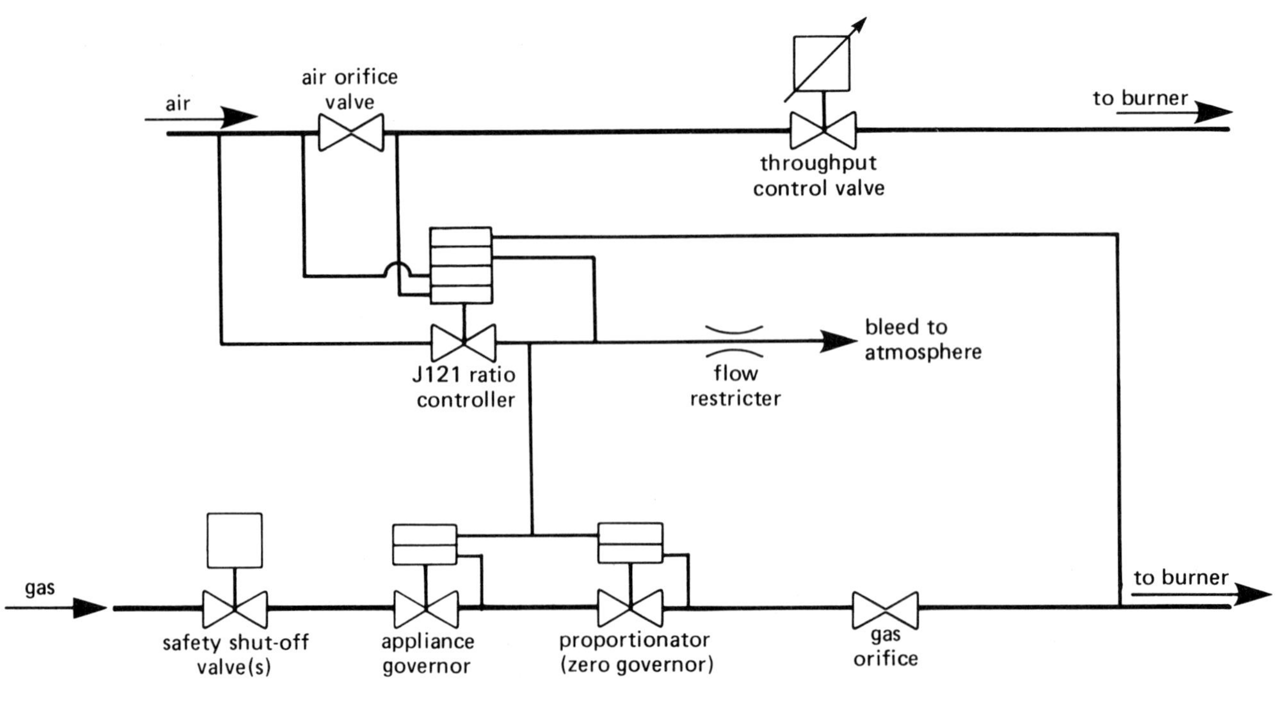

Fig. 9.97 Use of J121 controller as a pilot for a larger zero governor

Double diaphragm controllers are available for 25 mm (1 in) and 40 mm (1½ in) pipe sizes. On large installations, the controller can be used to pilot operate a larger zero governor as shown in Figure 9.97. In this installation, the pressure drop across one of the safety shut-off valves contributes to the differential pressure applied across the gas diaphragm. This arrangement also ensures that the zero governor is in the closed position just prior to start-up.

b) Air differential pressure/gas gauge pressure controllers

With this type of controller, the differential pressure across an orifice in the air line is balanced against the gauge pressure of the gas supply to the burner. The installation is shown in Figure 9.98. The system provides full compensation for back pressure fluctuations in the air supply line and is suitable for air recuperation installations. There is no compensation for back pressure fluctuations in the gas supply line. This method of control requires that the air supply pressure is considerably larger than the gas supply pressure. The method is particularly suitable for low gas pressure installations, since the pressures applied across each of the diaphragms are much larger than the double differential pressure controller and the inherent hysteresis of the controller has a much reduced effect upon the accuracy of the ratio control.

Any multiple diaphragm differential pressure controller can be installed to operate in the above manner.

Kromschroder manufacture a range of controllers (types GIH and GIL) which only operate in the air differential pressure/gas gauge pressure regulating mode.

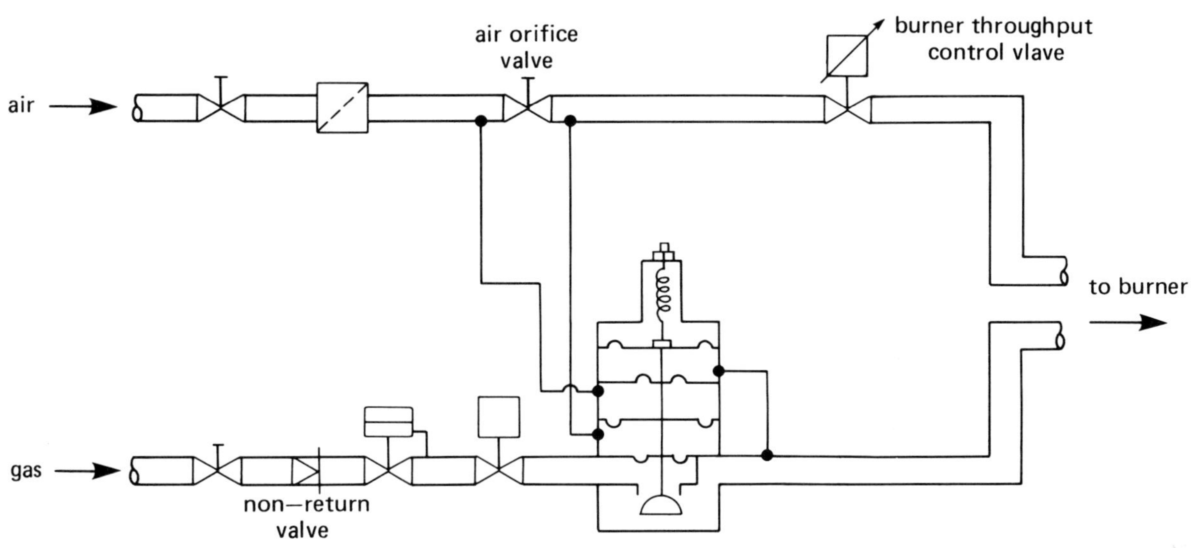

Fig. 9.98 Air differential pressure/gas guage pressure ratio control system

6. Electronic (Thermistor) Ratio Controller

The closeness of control obtainable by an air gas ratio control system, i.e. the level of excess air at which a piece of plant may be set to operate, will be determined by that required for satisfactory combustion, plus some additional allowance. The allowance will be for factors which may cause drift to rich or otherwise unsatisfactory operation. Such factors can include, for example, variations of gas quality, changes of climatic conditions, lost motion in valve linkages and even the effect of electricity supply voltage variations on fan speeds. Where the rates of flow of fuel and air can be measured the effects of many of the variables can be compensated for and the system may be adjusted to operate at lower excess air levels. A system to achieve this is the Thermistor Ratio Controller.

It has been developed at the Midlands Research Station of British Gas and is manufactured under licence by Inter Albion Limited.

i) Operating principle

The controller depends upon shunt flow metering of the air and gas supplies to the burner, using thermistor anemometers as the flow sensors. This arrangement provides a reliable and sensitive method for obtaining an electrical signal dependent upon the air and gas flows to the burner. The layout of the metering system is shown in Figure 9.99. The thermistors are mounted in a holder which is connected to the shunt by-pass lines around suitable orifices in the main air and gas supplies. Filters are provided in the by-passes to protect the thermistors from possible damage from dust in the air and gas flows and the various orifices (or other flow restrictors) are so arranged that, at the required air/gas ratio, the air and gas by-pass flows are the same. This is, in practice, most easily achieved by using adjustable (and lockable) valves as the main flow orifices as shown. The thermistors are connected to anemometer circuits which give voltage outputs dependent upon the air and gas flows past the thermistors and thus upon the main flows.

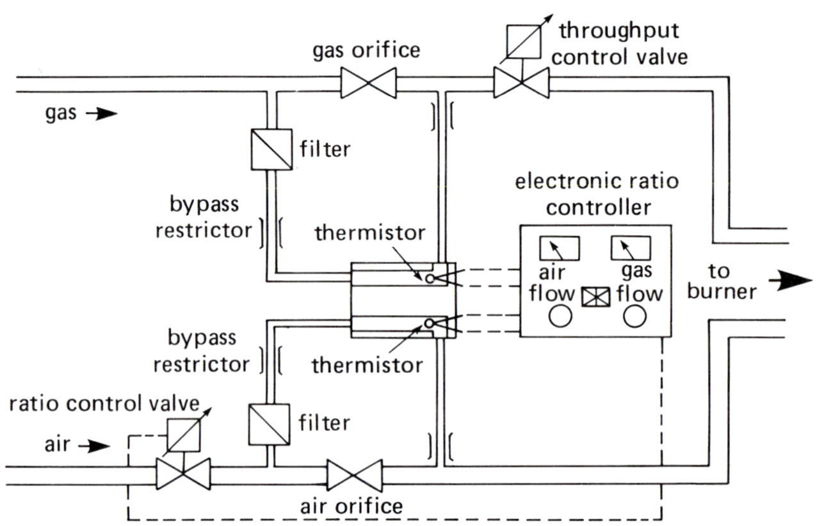

Fig. 9.99 Electronic (thermistor) ratio controller

Within the ratio controller the two anemometer output voltages are compared, and any deviation from equality (that is, from the pre-set air/gas ratio) causes the motorised ratio control valve to be moved to equalise the voltage outputs and thus restore the set ratio. The firing rate of the burner is set by the throughput control valve which is operated from a temperature (or other appropriate) controller for the process. The controller features an out-of-ratio detection circuit with rich and lean alarms which can, if required, be interlocked to the burner management system to cause lockout. As the principle of operation is based upon flow measurement, the electronic ratio controller will automatically compensate for back pressure variations encountered in recuperative installations.

The thermistor flow sensors have properties that enable the electronic system to control ratio accurately over burner turndown of 10:1. Also, as the thermistor sensors respond primarily to mass flow, the operation of the ratio controller is independent of the temperatures of air or gas flowing in the main supply line downstream of a recuperator. Controlling ratio directly on pre-heated air is necessary for individual zone control on multi-zone furnaces fitted with a single recuperator. It also enables the air/gas ratio to be accurately maintained independent of variable air leakage at the recuperator.

Figure 9.99 shows a gas-lead system with the throughput control valve in the gas supply and the ratio control valve in the air supply. The system can, however, equally well be used in an air-lead configuration.

An important feature of the system is the facility, provided electronically, to set the air/gas ratio at low fire independently from that at high fire (which depends on the relative size of the orifices as described above). This facility is required for two reasons, firstly because for correct combustion a burner often requires greater excess air at low fire and secondly because it renders the system relatively insensitive to the need to use orifice plates or carefully matched valves giving an exact square law flow-pressure relationship.

ii) Thermistor flow sensors

The thermistor is a small semi-cconductor bead whose resistance varies greatly with temperature. For use as an anemometer a simple electronic feedback bridge circuit is arranged to maintain the thermistor at constant resistance and thus at a constant temperature (180 °C in this case). The electrical power needed to hold the thermistor at temperature will increase as the fluid velocity passed it increases and therefore the voltage across the thermistor is a function of velocity and can be used as an output signal.

iii) General details of the system

The basic electronic system comprises the thermistors, housed in a standard holder, together with associated by-pass line components and a small control cabinet housing the electronics. The ratio control valve is generally a butterfly valve operated by a standard 24 V a.c. reversing motor. Lockable gate or disc-on-seat valves are usually used as the main line orifices.

The standard thermistor holder houses both gas and air thermistors within a machined perspex block. Appropriate connections for the by-pass lines and for the electrical wiring to the cabinet are provided. The thermistor holder has been progressively developed to provide high sensitivity (to minimise the by-pass line flows) together with low pressure drop, and to minimise turbulence at the thermistors.

The ratio controller cabinet houses the ratio control circuit board together with associated power supplies, potentiometers, indicators (light emitting diodes) and three meters. Two of the meters provide an indication of the air and gas flows. Because these are neither linear nor temperature compensated, they only give a very approximate indication of the flows, but despite this have been found to be valuable during commissioning of the system and for monitoring its operation. The third meter is a centre-zero 'in ratio' meter which displays the output of the first differential amplifier and gives a convenient indication of the correct functioning of the controller and any deviation from the set ratio. Lockable potentiometer controls are provided for the 'low fire ratio' setting and for the gain of the differential amplifier.

It has been found essential, in order to avoid off-ratio firing during transient conditions, to fit the throughput control valve with a slower acting motor than the ratio control valve. For example, referring to Figure 9.99, on a gas lead system the gas throughput control valve might have an '80 second' motor and the air ratio control valve a '40 second' motor. Provided that this precaution is taken the transient deviations from ratio during modulation should be small (less than ± 4% of the set ratio).

7. Landis and Gyr SKP Ratio Controller

Landis and Gyr have developed a novel ratio controller based upon an electrohydraulic safety shut-off valve. The controller, known as the SKP70, is shown schematically in Figure 9.100. The air and gas diaphragms are connected via a complex lever mechanism to act on a by-pass valve within the electrohydraulic valve actuator. The forces acting on the diaphragm servo regulate the positioning of the gas valve so as to maintain a constant ratio of air/gas pressures.

The principle of operation of the SKP70 is functionally equivalent to a cross-loaded governor ratio control system. The lever mechanism connecting the air and gas diaphragms has an adjustable pivot point which alters the balance of the forces on the diaphragms. This enables the ratio of air/gas pressure to be adjusted over the range 0.4:1 to 9:1 without involving ancillary pressure dividing techniques. A spring adjustment on the lever mechanism provides an offset adjustment to the pressure ratio setting. Compensation for combustion chamber pressure fluctuations is provided by the line connecting the underside of both diaphragms to the combustion chamber.

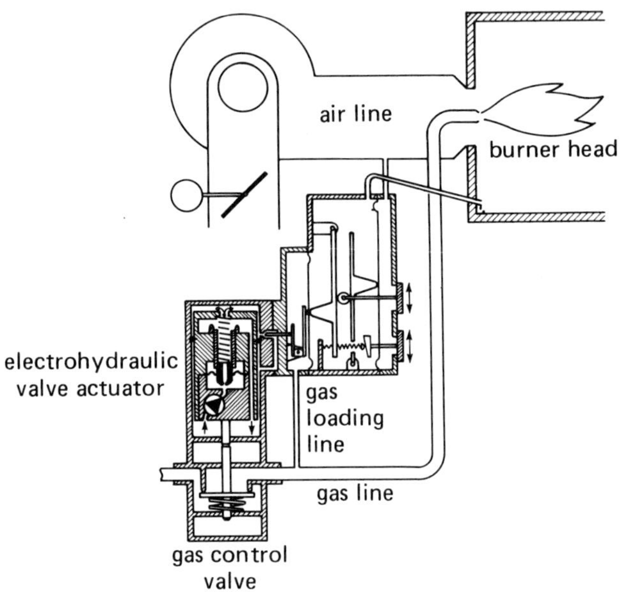

Fig. 9.100 Schematic of Landis and Gyr SKP70 ratio controller

The hysteresis inherent in the design of the SKP70 controller limits the burner turndown flow range to around 4:1 for low pressure (21 mbar) gas supplies. It is advisable to connect the combustion chamber line to the controller on all installations and essential where sub-atmospheric furnace chamber pressures are encountered. The response time of the SKP70 for changes from low to high fire is around 5 seconds which is considerably slower than conventional cross-loaded governor control systems.

For systems with varying back pressure, for example on recuperative burners, the Landis & Gyr SKP50 system can be used. This incorporates an orifice plate in the air supply line to the burner and is described in more detail in section 9.4 on Electrical Safety Shut-off Valves.

9.7.5 Variable Speed Forced Draught Fan System

A.C. Motors as normally fitted to forced draught burner fans provide the lowest cost and simplest method. However, the constant speed A.C. motor requires all the excess fan performance at lower output to be throttled by means of a modulating damper with consequent power loss.

Saacke and other manufacturers of large gas and dual fuel burners use variable speed D.C. motors with thyristor control as shown schematically in Figure 9.101. By varying the speed of the forced draught fan in unison with fuel input it is possible to make considerable savings in electrical energy consumption. Only a small amount of throttling by the modulating damper is required at low speed to provide sufficient air flow at this setting. A further advantage of the system is the exact matching of the fan pressure and system resistance. The full load speed is reduced slightly to exactly match the fan pressure resistance thus power is saved at full power output to the burner as well as at lower outputs.

Other advantages of the system include reduced average noise levels since the fan is much quieter at lower speeds and the system eliminates high current demand at start up. It can be combined with oxygen trim control and can be retrofitted to existing burner systems.

Air flow turndown of 10:1 can be achieved. It is important that the fan speed control system contains appropriate interlocks integrated with the burner management system to ensure that the air flow requirements of BS 5885 and British Gas Code IM/18 are fully satisfied during burner start-up and continuous running.

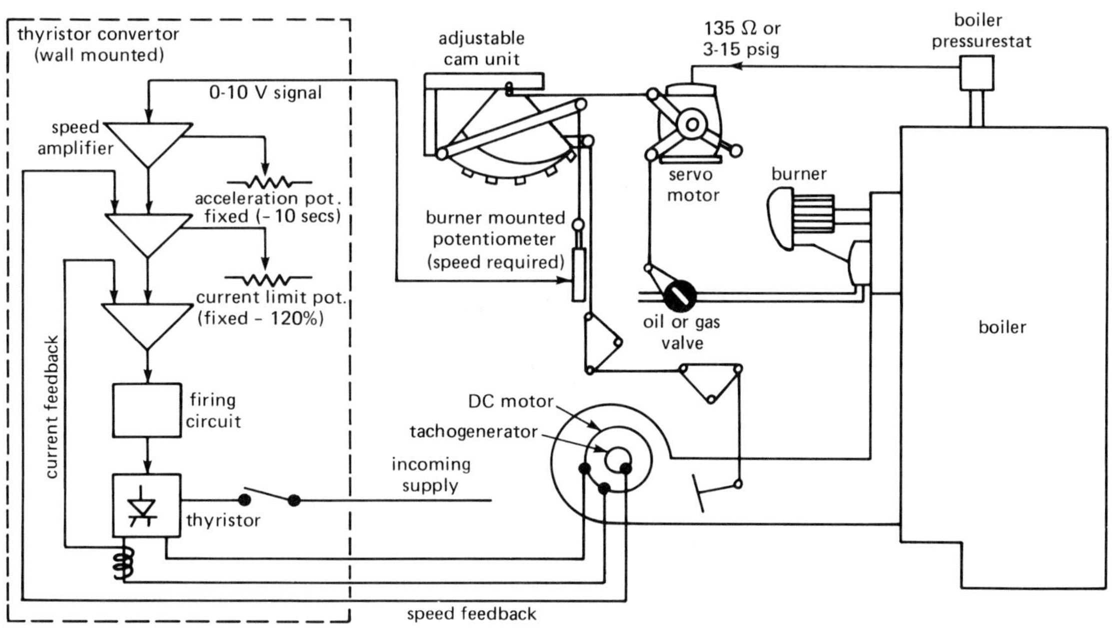

Fig 9.101 Saacke variable fan speed ratio control system for boilers

9.7.6 Ratio Control with Preheated Combustion Air

9.7.6.1 Effects of using preheated air

As the degree of preheat is variable and not predictable, the specific gravity of the combustion air will also be subject to considerable variations. As the types of air/gas ratio control described earlier rely on a pressure differential across an orifice for their control, there will be difficulties with their application to recuperative systems.

With flue mounted recuperators there is the alternative to control either on the flow of cold air into the recuperator, or hot air out, both with single and multi burners. The former option is usually adopted where possible, because most control systems are not suitable for use with hot air. There are also considerable problems with throughput control valves using preheated air. If multi burners are used it is not possible to individually control burners by controlling cold air into the recuperator.

With recuperative burners control is always on the cold inlet air where the possibility of leakage in the recuperator is considerably reduced.

9.7.6.2 Air blast

Air blast premix burners are not normally used with preheated air. Deterioration of the injector jet and venturi would occur if the preheat was considerable.

Design of the injector throat and jet would be difficult if it was necessary to cater for a considerable variation in flow caused by changing air temperatures.

9.7.6.3 Linked valves

With this system the air/gas ratio will vary due to the density change and changes in back-pressure. If, however, the differential pressure across the valve is maintained substantially constant, then a suitable choice of valves enables the air/gas ratio to be maintained within narrow limits.

This may be achieved by means of a suitable pressure governor situated upstream of the valve and backloaded from a pressure tapping downstream as discussed in Section 9.7.4.

9.7.6.4 Multiple diaphragm governor

The simple backload system, i.e. the 'pressure divider' system, is obviously not suitable for use with preheated air, because the air/gas ratio will tend to decrease as air temperature increases. If, however, the control pressure is not the differential across the burner orifice, but the differential across an orifice in the air supply, then a variation on the backloaded governor can be used.

The multiple diaphragm system discussed in 9.7.4 will normally be used upstream of the recuperator, i.e. controlling on cold air. It could be used controlling on preheated air because the static column of air in the two pressure tappings would prevent the governor being overheated. Ratio control may be affected by the difference in density between the flowing air and the static air, and this position will not normally be used.

For larger size of controls it is possible to use the multi-diaphragm governor as a 'pilot' to control larger proportioners as shown in Figure 9.97. This can be regarded as a combination of the multi-diaphragm governor and pressure divider system. It is seen that the controls must be with cold air as there is a flow of combustion air through the governor which is vented to atmosphere.

9.7.6.5 Electronic ratio controller

The electronic ratio controller, which is discussed in section 9.7.4 and shown in Figure 9.99, is suitable for accurate air/gas ratio control with air preheat. This is because the system automatically compensates for changes in air and gas temperature and pressure. The system can be used downstream of a recuperator, although limitations of the air throughput control valve will normally dictate that it is used upstream controlling cold air.

9.7.7. Premix Machines

9.7.7.1 Use of premix machines

Machine premix systems find their widest applications in supplying large numbers of small burners, e.g. working flame burners for the glass industry. In this application the air/gas mixture distributed is well outside the flammability limits. Another important use is for supplying controlled-atmosphere generation plant. In this case the air/gas ratio used is about 2.4/1 for endothermic gases and in the range 6/1 to 10/1 for exothermic gases. Where the mixture is within the flammability limits, great care is needed in system design to eliminate the hazards involved in transporting an explosive mixture (see IM/16). Premix machines are available for air/gas mixtures of up to 12 m^3 s^{-1} (25 000 ft^3/h), and made by manufacturers such as Selas and Keith Blackman.

In all mechanical premixing machines, gas and air are drawn through suitable metering elements into a mixing device and then compressed. The rotary compressor type of premixing machine is the most widely used in industrial gas practice. An essential feature of the system is the ability of the mixing machine to deliver a constant ratio mixture when the back pressure, or the number of burner heads in use, is varied.

There are two basic methods by which the air and gas flows can be metered. The first maintains constant differential pressure across linked valves with suitably matched flow characteristics, usually an adjustable port valve (variable area, constant pressure). The second maintains a constant relation between the differential pressures across fixed metering orifices in the air and gas lines (constant area, variable pressure).

9.7.7.2 Metering by linked valves with constant differential pressure

Air/gas control is achieved by maintaining constant and equal differential pressure across linked adjustable port valves with a constant area ratio between the gas and air ports at all throughputs. The mixer consists of two adjustable-port valves linked mechanically and actuated by a horizontal diaphragm exposed to atmospheric pressure on its underside. This maintains a constant pressure in the mixing chamber downstream of the valve ports by varying the vertical position of the adjustable port valve. In this way it is possible to compensate for change in pressure conditions at the compressor inlet. Gas is drawn into the valve from a zero-governed supply and air is drawn into it from the atmosphere. Normally the air is filtered, in which case a back-loading technique is used, the suction existing at the inlet to the air valve being used to load the diaphragm of the zero governor. This ensures equal inlet pressure to both valves under all conditions. (See Figure 9.102).

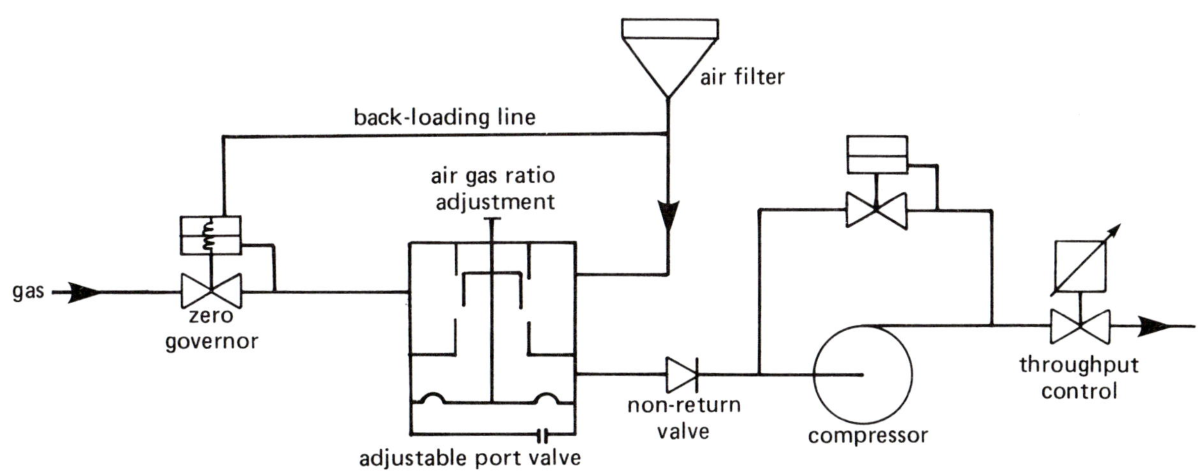

Fig. 9.102 Mechanical premixing machine - variable area/constant pressure

The air/gas ratio is set by rotating the valve plug to adjust the valve port area ratio. The accuracy of the ratio control over the throughput range of the machine is limited by changes that occur in orifice shapes and, hence, discharge coefficients as the valve opens and closes.

Since a rotary compressor is a positive displacement machine, high back pressure may be produced by downstream restrictions and it is necessary to incorporate a delivery pressure governor placed in the by-pass which joins the inlet and outlet of the compressor. The governor spring is adjusted to alter the delivery pressure. If the delivery pressure exceeds that set, the valve opens and allows the mixture to flow back into the compressor inlet.

9.7.7.3 Equalised differential pressure across fixed metering orifices

With this technique a constant air/gas ratio is achieved by maintaining equal differential pressures across the metering orifices. The pressure downstream of the orifices is the compressor inlet pressure, and the upstream pressures are equalised by means of a zero governor in the gas supply. Throughput is controlled by a single valve in the mixture delivery pipe (see Figure 9.103).

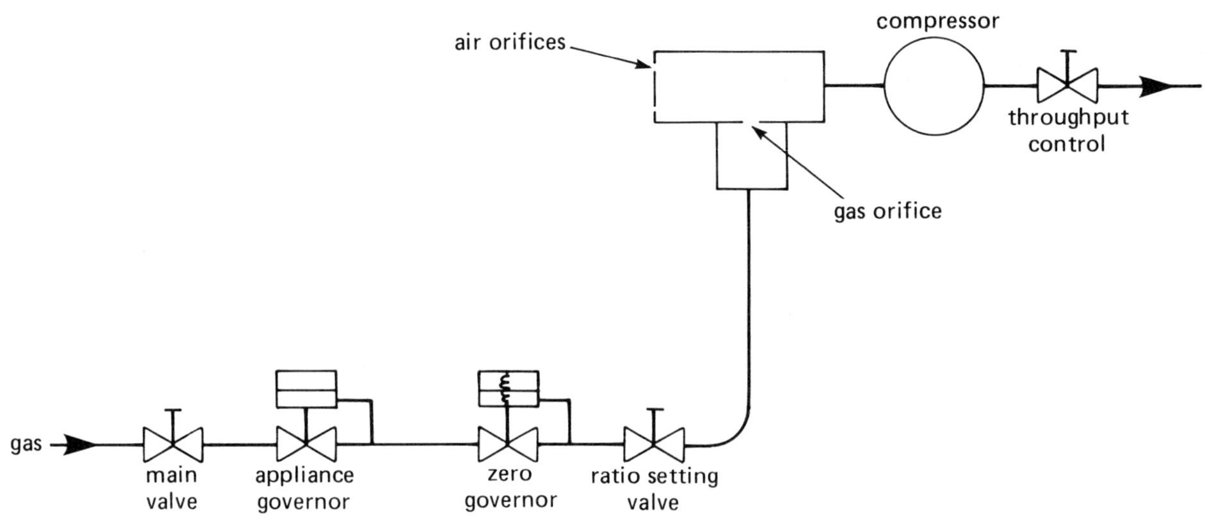

Fig. 9.103 Mechanical premixing machine - constant area/variable pressure

If simple square law orifices are used as the metering elements then the air/gas ratio is dependent upon the relative flow areas of the gas and air orifices. Providing the discharge coefficients of the two orifices remain in a constant ratio over the flow range, then a constant air/gas ratio can be maintained regardless of throughput. Two types of orifice mixer used are the adjustable slot and the multi-orifice, as shown in Figure 9.104. Certain types of crimped-ribbon flame arrestor give linear flow characteristics with operational turndown ranges of 15:1. Such a linear flow mixer is shown in Figure 9.104. As this element is subject to blockage by dust particles, it is advisable to fit an air filter on the inlet side of the element and back-load the zero governor with the suction immediately upstream of the element.

9.7.7.4 Fan premix system

A centrifugal fan can be used in place of a compressor to produce a premix. Variable orifice, constant pressure proportioning or constant area, variable pressure proportioning may be used.

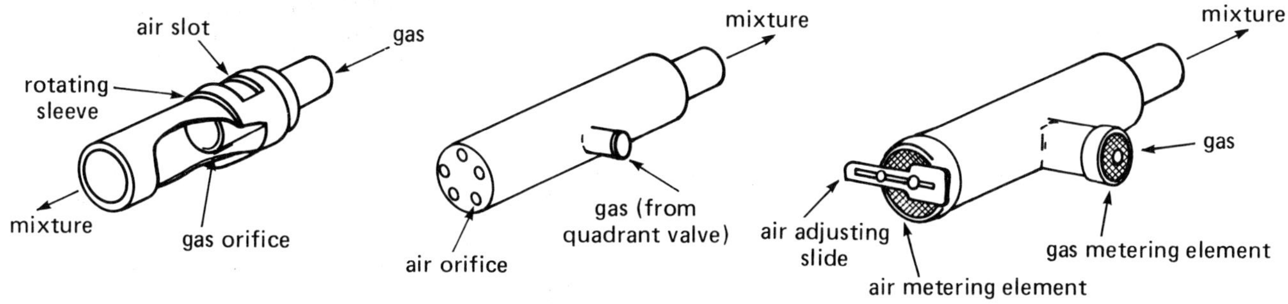

(a) an adjustable slot mixer (b) a multi-orifice mixer (c) a linear-flow mixer

Fig. 9.104 Orifice elements

9.7.7.5 Low pressure protection

The use of a mixing machine of either the compressor or fan type requires the use of low pressure cut-off devices to comply with the Gas Act 1972 (Schedule 4, para 18). Further details are in British Gas Publication IM/16).

9.7.8 Oxygen Trim Feedback Systems

a) In these systems the oxygen content in the flue gases is sensed and a controller regulates the burner combustion air supply to maintain a pre-set air/gas ratio. Oxygen trim is restricted almost entirely to large boiler plant where the degree of close control and relatively small increases in efficiency can be justified as the cost is offset by large total fuel savings. It may, however, be successfully used on other large fuel using equipment.

A look at the composition of combustion products from natural gas (Figure 9.105) reveals that at low levels of excess air carbon monoxide may be present in significant amounts. For boiler plant in particular, this may represent a serious problem if it continues unchecked, and generally the carbon monoxide formation, due largely to imperfect air/gas mixing at the burner, must be minimised and regulated within fine limits. A typical value for maximum CO may be 100 ppm in order to ensure the minimum of damage to the heat exchanger surfaces. The problem for the

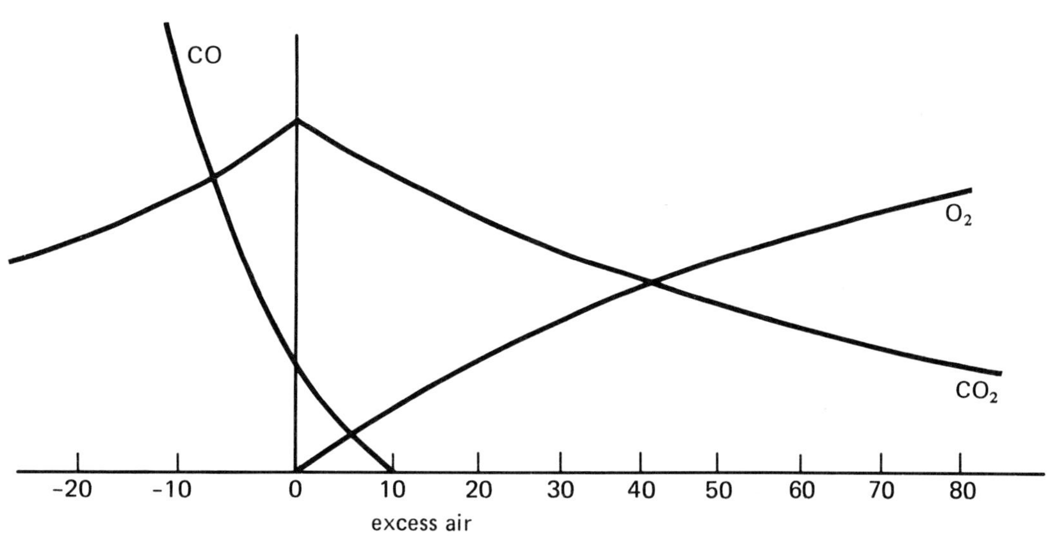

Fig. 9.105 Combustion product composition

combustion engineer is to determine the correct air/gas ratio at the control valve in order to ensure that maximum CO levels are not exceeded between service intervals. The problem is made particularly difficult where modulating control is employed and the air/gas ratio varies along the load line anyway. To make matters worse the level of excess air at which a given CO value is generated will also vary according to the loading. The commissioning engineer, therefore, has to establish not only the correct excess air value to avoid CO formation, but the extent to which this value will vary between low and high fire settings. In addition, a number of other factors may have to be taken into account which further increases the amount of excess air required in order to ensure CO free combustion. Consideration must be taken of control system hysteresis, variation in ambient air temperature and humidity and burner deterioration.

The result of making an allowance for all these factors is that a level of excess air is often set far in excess of the minimum value. This of course results in higher stack losses and a fall off in overall plant efficiency.

If, however, the level of oxygen (excess air) can be accurately monitored during operating periods, then this information may be used to continually re-set the air/gas ratio to a predetermined excess air profile. This allows the plant to operate closer to the theoretical excess air/load line without incurring abnormal levels of CO formation and reducing the heat lost to the stack. This mechanism is known as oxygen trim control.

Although a number of oxygen trim control systems are available, the principles employed are similar in each case and a general description is given below.

The effective trimming of oxygen level requires three fundamental components:

i) knowledge of the ideal oxygen profile

ii) a reliable and accurate measurement of oxygen level at the end of the combustion zone.

iii) an actuator/control mechanism which can interphase with the existing air/gas ratio control system.

b) Burner excess air characteristics

If we assume that the maximum tolerable carbon monoxide level at the reversal chamber of a shell boiler is 100 ppm, then the amount of excess air required to ensure combustion to this degree of completeness will vary as the boiler load varies. This is due to the mixing characteristic of the burner where at the upper end of the scale (approximately 40 to 100% full load) a linear relationship exists but below 40% or so the turbulence at the burner nozzle is reduced and generally speaking greater amounts of air are required to ensure complete combustion. This is shown in Figure 9.106.

Unfortunately this relationship is not constant for all boilers, nor for each of any model of boiler and it is therefore necessary to determine the boiler's individual excess air/load characteristic. This can only be done by direct measurement of carbon monoxide and oxygen levels at the reversal chamber, adjusting the oxygen at a range of firing rates until carbon monoxide reaches the critical value. By this method it is possible to plot the theoretical excess air requirement across the boiler range.

Since these results will ultimately be used to fine tune the effective air/gas ratio, the degree of accuracy employed should be as great as possible in order to ensure that excess air levels are minimised *WITHOUT* exceeding the optimum CO level.

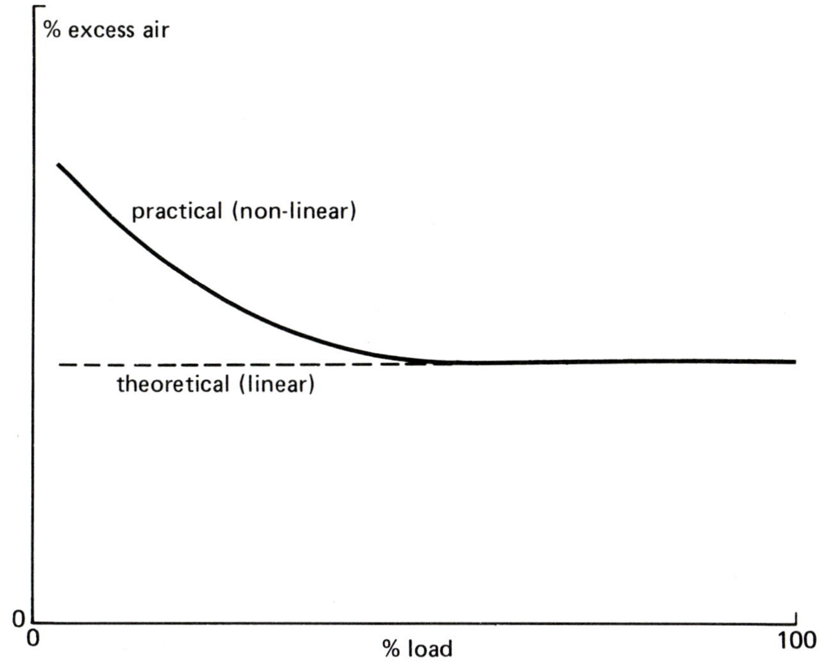

Fig. 9.106 Variation of excess air with load

The correct oxygen level for a burner, as defined by CO breakthrough, is established under various loads. A graph can be plotted as in Figure 9.107, and the oxygen trim controller ideally adjusts the air/gas ratio to follow this curve at all points.

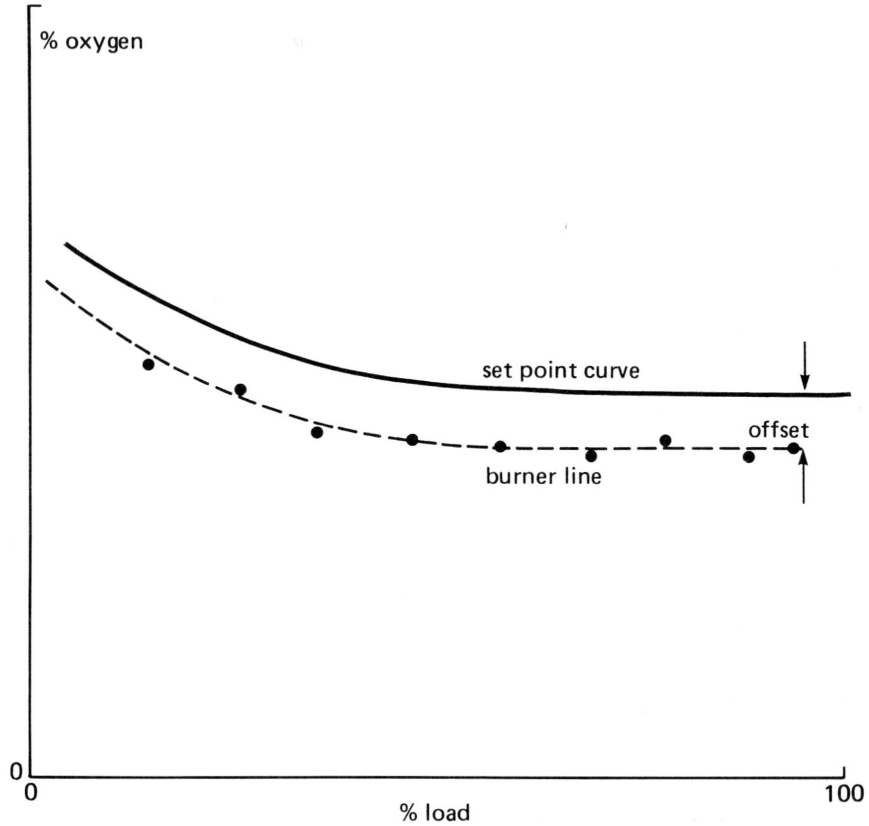

Fig. 9.107 Ideal boiler line

c) Oxygen measurement

Once the burner characteristic is known the oxygen level may be trimmed to reduce excess air in the flue. However, for accuracy, a number of important aspects must be considered:-

- sensor position
- sensor reliability/accuracy
- response time of system.

The position of any measuring sensor is vital if damage to the construction of the boiler is to be avoided. The sensing must be done from a point where the greatest danger exists and must of course coincide with the measurements taken during the characteristic determination exercise. Unless this is so, a false impression may be gained, resulting in abnormally low excess O_2 levels going undetected; resulting in high metal temperatures and damage. In practice oxygen measurement is usually made using a zirconia cell, although paramagnetic cells and solid electrolyte cells are also used. Only the zirconia probe is situated in the flue with consequent minimisation of sampling error.

d) Principles of operation

The zirconia cell is installed in the flue, or in some cases just outside it, so that hot wet flue gases are sampled. No sampling lines, scrubbing, cooling, or condensate removal are therefore necessary. The oxygen reading obtained is a wet analysis, unlike virtually every other method which give dry analyses.

The principle is based on the Nerst equation. The difference in oxygen concentration on the two sides of a zirconium oxide cell develops a voltage.

$$\text{e.m.f.} = \frac{RT}{4F} \ln \frac{P_1\,(O_2)}{P_2\,(O_2)} + C$$

where R = gas constant
F = Faraday constant
T = absolute temperature
P_1 = reference gas partial pressure
P_2 = sample gas partial pressure
C = a constant

In practice this means that if the temperature is held constant and air (20.95% O_2) is used as a reference gas, then the output is proportional to the inverse of the logarithm of the partial pressure of oxygen in the measured gas. As the oxygen concentration decreases, the voltage increases, giving increased sensitivity at low oxygen concentrations. The cell is maintained at a constant temperature, normally 850 °C by means of an electric element, with thermocouple feedback to maintain correct temperature within a narrow band.

e) Safeguards

It is seen that failure of certain parameters will adversely affect the safety of the system. These will include:

i) Analyser power failure
ii) Sample flow (either air or flue gas depending on type) failure

iii) Measured O_2 outside pre-set limits
iv) Zirconia sensor failure
v) Cell temperature out of band
vi) Sensor enclosure temperature out of band
vii) Actuator reaches pre-set limit switch
viii) Broken circuit to cell

A well designed system will contain alarms that, if actuated, will cause the trimming device to drive to neutral, i.e. remove any additional trimming and let the air/gas ratio control revert to the normal lead control only.

The oxygen trim system has certain inherent problems which are overcome to varying extents.

- An assumed CO level is relied on, as determined during the early burner characterisation phase. CO is not actually monitored for most systems.
- Air ingress into the stack below the sampling point will give misleading results.
- There is an inevitable lag between sensing a deviation from set point and correcting it.
- The degree of trim is normally limited. This can usually be preset and for safety reasons limited trim is desirable.

f) Control/actuator mechanism

The control system generally comprises a unit which compares the measured variable (stack oxygen) with the reference point (excess air characteristic). The error value is used to change the position of the actuator (usually the air damper) according to the load on the boiler.

Whatever method of control is employed, the actuator will usually take the form of an additional adjustment to the conventional air flow valve linkage. One system uses a piston arrangement installed within the existing linkage, which is then capable of fine tuning the air flow to the burner. It is of course important that the trim device is not actuated until the burner is lit and running otherwise the purge and ignition flow rates may be adversely affected. The full control outlay is shown in Figure 9.108.

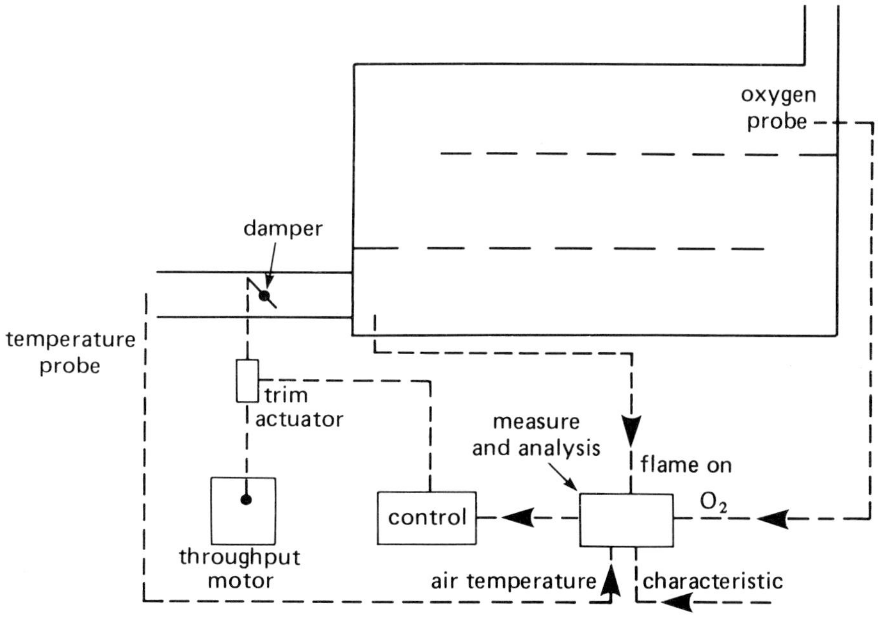

Fig. 9.108 Control actuator mechanism

g) Systems available

Several makes of oxygen trim systems are available. These tend to be generally similar but have different variations and options. These include:-

i) Type of sensor
ii) Method of interlinking the trim damper; a separate damper is sometimes used.
iii) Number of points used to define characteristic curve.
iv) Multi-burner operation.
v) Incorporation of other flue gas parameters including CO and temperature.

Manufacturers include Telegan, Westinghouse, Taylor Servomex, Weishaupt, etc.

h) Cost effectiveness

The capital cost of installing an oxygen trim system will, to a large extent, be independent of the size of the boiler to which it is fitred. This will obviously mean that the payback will be shorter for a larger boiler than a small one. Nevertheless real costs of oxygen trim systems are falling and the minimum size of boiler to which they may be economically applied is falling.

The major factors affecting cost effectiveness are the size of the boiler and its intensity of use as these will determine the amount of fuel used, a percentage of which can be saved by oxygen trim control. Other factors are the accuracy of the existing controls, and the amount of wear and hysteresis existing.

A Dutch study has indicated that paybacks of less than 1 year would be expected with boilers of 1-10 MW (1600-16 000 kg steam/h). Oxygen trim control is therefore particularly attractive, and additional maintenance and operation costs are minimal. American experience suggests that savings of 2-4% are realistic, but also indicates that maintenance costs can be high.

9.7.9 Carbon Monoxide Feedback Systems

a) Advantages

Carbon monoxide can be accurately measured and used as a parameter for controlling air/gas ratio in much the same way as oxygen. The CO content of flue gas is the most precise and least ambiguous indication of efficient combustion. It is a direct measure of the completeness of combustion. CO measurement is an objective measure of combustion efficiency that does not depend on the boiler type, fuel, etc. Oxygen monitoring can be made almost meaningless by ingress of 'tramp' air before the sampling point; and it is here that CO monitoring has an advantage. Carbon monoxide monitors provide an accuracy of ± 10 ppm compared with oxygen monitors ± 0.3%.

b) Principles of operation

The carbon monoxide monitor works on the infra-red absorption principle. An IR beam passes across the full width of the stack and the absorption in the flue gas is compared with a reference sample. The full width of the stack is used to eliminate problems caused by uneven CO concentrations. Obviously fouling of the windows at both transmitting and receiving ends is a problem; this has largely been overcome by the use of continuous air purges. Nevertheless maintenance costs are high, being of the order of twice that for oxygen trim devices. Reliability is fair, but in the harsh conditions of the boiler stack constant calibration is essential.

c) Disadvantages

Although CO increases sharply at low excess air it changes only very gradually at high excess air. This makes it a poor parameter for control. It is normally considered better to adjust the excess air with oxygen trim and then to fine tune with CO measurement. There can be freak problems with CO control, such as sudden brief increases in CO level during increasing load which would tend to lead to an unnecessary increase in air/gas ratio. These can be overcome by characterising the controller to a particular boiler/burner combination, and the use of a microprocessor based controller is essential.

d) Cost effectiveness

Trim systems based on carbon monoxide have been developed but their high installed cost means that they can only be considered for large boiler installations. This high cost must be offset against an increase in efficiency over oxygen trim of only 1% and clearly massive fuel consumptions are necessary to justify the installation.

9.7.10 Motorised Valves and Modulating Control Motors

This section refers to motorised valves for process control, not safety shut off.

9.7.10.1 Electrical systems

The accurate control of a modulating motor which determines the gas rate to a burner generally requires three separate elements, namely potentiometric control, a balancing relay and the drive motor itself.

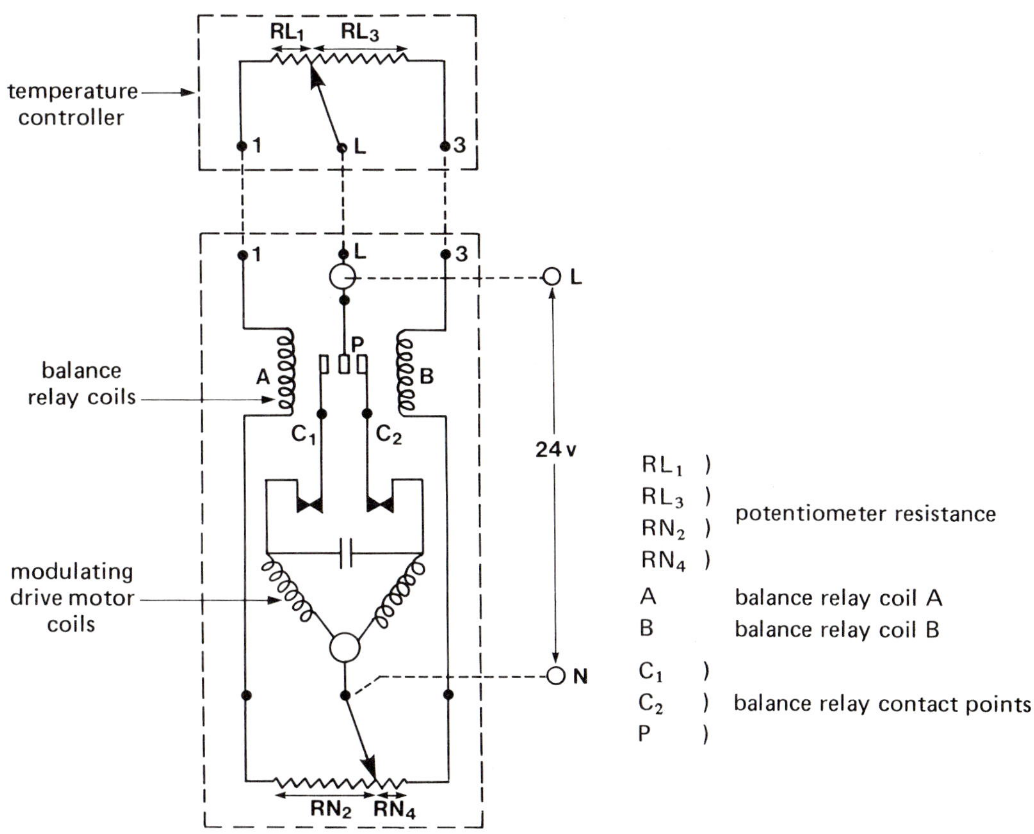

Fig. 9.109 Temperature control potentiometer

The potentiometer or variable resistance converts the measured signal of temperature, pressure or mechanical movement to an electrical signal. The balancing relay consists of two coils and a pendulum contact which directs current to one of the two coils of the modulating motor depending on which way this requires to be driven. One such system is shown in Figure 9.109. In this system the balance relay is incorporated in the drive motor housing and requires five external connections to be made. In other systems the balance relay is part of the temperature controller and then six external connections must be made.

The mode of operation of the system is as follows. Power to the unit is supplied at L and N and can be 240 V or fed via a transformer to give a 24 V AC supply for safety purposes. If the process calls for heat, the potentiometer pointer moves over to point 1 on the slide wire. Since the combined resistance RL_1 and RN_2 on the left side of the system is now less than the combined resistance RL_3 and RN_4 on the right side of the system, more current will flow through coil A in the balance relay. The effect of this will be to create an inducement for point P to move and make contact with point C_1, effectively switching from L to the left hand coil of the modulating motor. This is caused by the influence of the electromagnetic force in coil A similar to the mechanism of a solenoid valve.

The motor will drive to the high fire position moving at the same time a slide wire pointer along a variable resistance towards the right side. When the resistances RL_1 and RN_2 equal the resistances RL_3 and RN_4 the current flowing up coils A and B will be equal and the point P will break from the C_1 contact and come to rest in between C_1 and C_2. The modulating motor will then stop. It will only restart again in either direction if the potentiometer pointer moves in response to changes in temperature, and operates as described above.

In general AC mains motors are very reliable in operation. The rotational speed of the motor is of course geared down to give slow valve movement characteristics. The gearbox oil should be periodically checked, as should all contacts and switches which are integral in the motor housing. The motors can also be used in conjunction with an impulse generator. These are arranged to give pulses over periods of the order 1–10 seconds in each minute and according to which stator in the motor is energised it will run either clockwise or anti-clockwise.

Typical examples of modulating motors are those manufactured by Landis and Gyr/Billman (ME Control Motors) and Honeywell Modulating Motors. On these the primary drive unit is of the split phase capacitor type with a squirrel cage motor which drives through steel gear wheels partially immersed in oil. There is a spring loaded friction brake which prevents overrun by locking the output shaft when the motor current is interrupted. The balance relay is housed inside the motor casing together with the limit switch mechanism and the screw for adjusting the motor travel. The feedback potentiometer and auxillary switches are also inside the removable cover carrying the position indicator. Modulating motors are used in conjunction with a variety of valve mechanisms such as butterfly, plug and ball valves. They are also used to control dampers, etc.

9.7.10.2 Pneumatic or hydraulic systems

A simple pneumatic type is shown in Fig. 9.110, in which the motive force is provided by the action of air pressure on a flexible diaphragm. A return spring is fixed between the underside of the diaphragm and the valve which it operates. The action of applying pressure to the diaphragm causes compression of the spring in proportion to the pressure applied, which is normally between 0.2 and 1.0 bar (3-15 psi) controlled via a pneumatic relay which in turn is operated from the 0.2 to 1.0 bar control air. Alternatively, these valves can be arranged to be normally open, when in the event of control air pressure failure, the valves are opened fully by the spring. Hydraulic or pneumatic cylinders may also be used to provide the motive power for valve actuation.

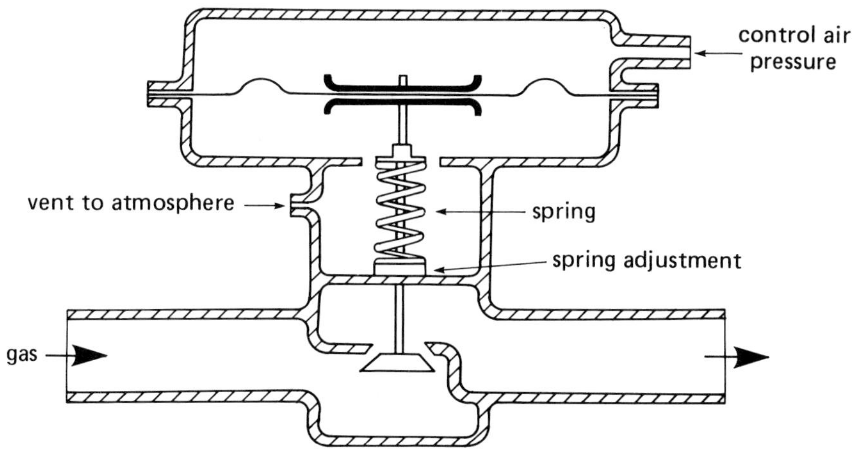

Fig. 9.110 Pneumatically operated gas throughput control valve

9.8 TYPICAL GAS CONTROL SYSTEMS

Having considered all the individual gas controls which are encountered on burner equipment, let us now consider how these are arranged and installed to comply with the relevant Standards and Codes of Practice. Although there will be many slight variations, all systems should comply with the intent of the code which applies to it. For convenience we will consider plant by temperature of operation of the process and by the rating of the installed burners.

9.8.1 Low Temperature Systems

Plant operating below 750 °C at the working chamber walls should comply with the Code of Practice for the Use of Gas in Low Temperature Plant, IM/18.

9.8.1.1 Natural draught burners below 150 KW

These systems are frequently controlled by thermoelectric flame protection and a gas relay valve for temperature control. Figure 9.111 shows direct thermoelectric flame protection for a single gas bar burner without temperature control, whilst Figure 9.112 shows a similar system with separately fed pilots for the thermoelectric valves on several burners with a relay valve for temperature control. In most cases the preferred position for the governor is directly after the main isolating valve. If the pilot burner requires pressure higher than the main burner, the thermoelectric valve in this case may be situated upstream of the governor. Such systems are applied for example to vat and tank heating installations and small batch baking ovens.

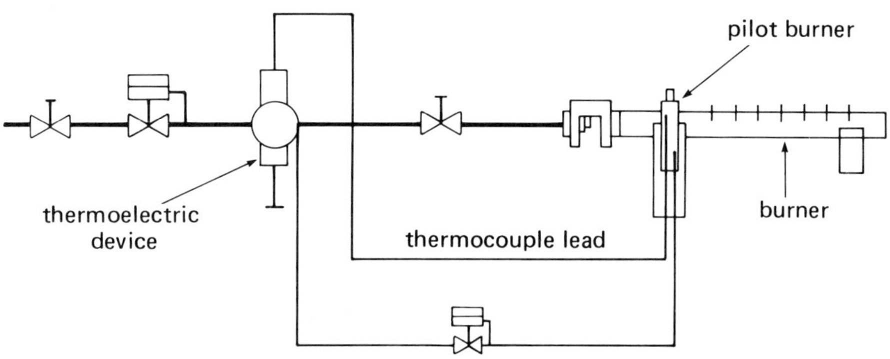

Fig. 9.111 Natural draught system

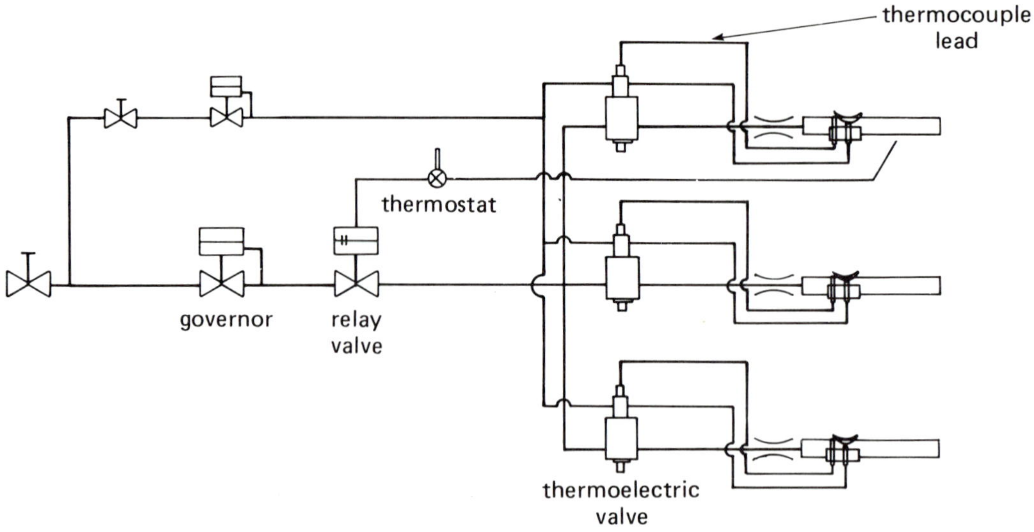

Fig. 9.112 Protection of multiple burners using thermoelectric valves and incorporating separately fed pilot valves

9.8.1.2 Natural draught burners above 150 KW

These systems should be fitted with a flame safeguard to respond to loss of flame in less than one second. This will usually be a flame rectification probe system in conjunction with the appropriate certificated safety shut off valve(s) depending on the rating of the burner (Figure 9.113). In this case gas throughput could be controlled by a relay valve with a small solenoid in the weep line, but could also be performed with a modulating motor in the same position in the gas line. This type of system is commonly found on natural draught boiler installations, small air heaters and ovens.

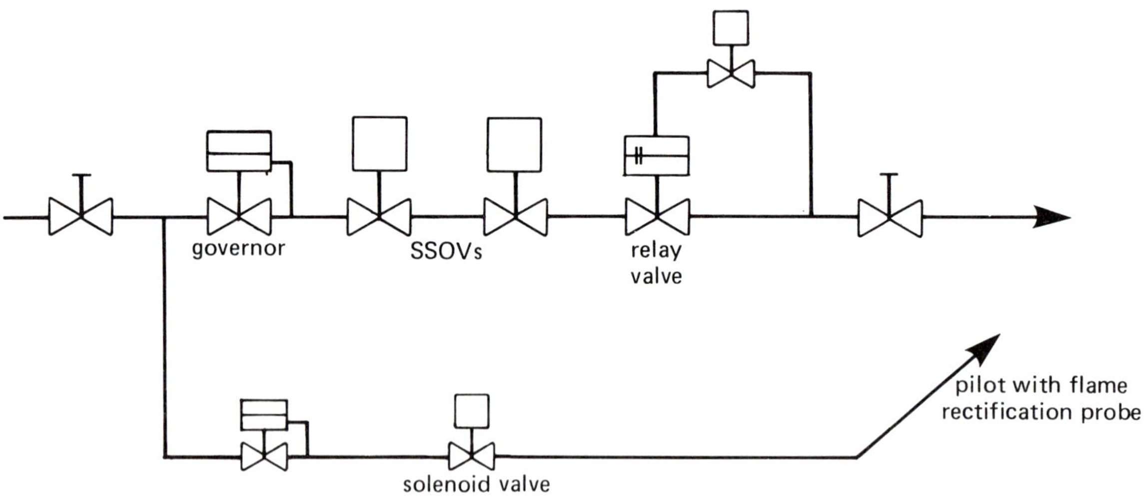

Fig. 9.113 Natural draught burner system in excess of 150 kW

9.8.1.3 Steam tube baking ovens

To guard against a steam tube exploding in these ovens, an electrically operated thermometer is located in each chamber. These are connected in series to a safety shut off valve. In the event of over temperature in any chamber, the safety shut off valve closes immediately (Figure 9.114). A flame safeguard should be fitted instead of low pressure cut off valves as discussed before.

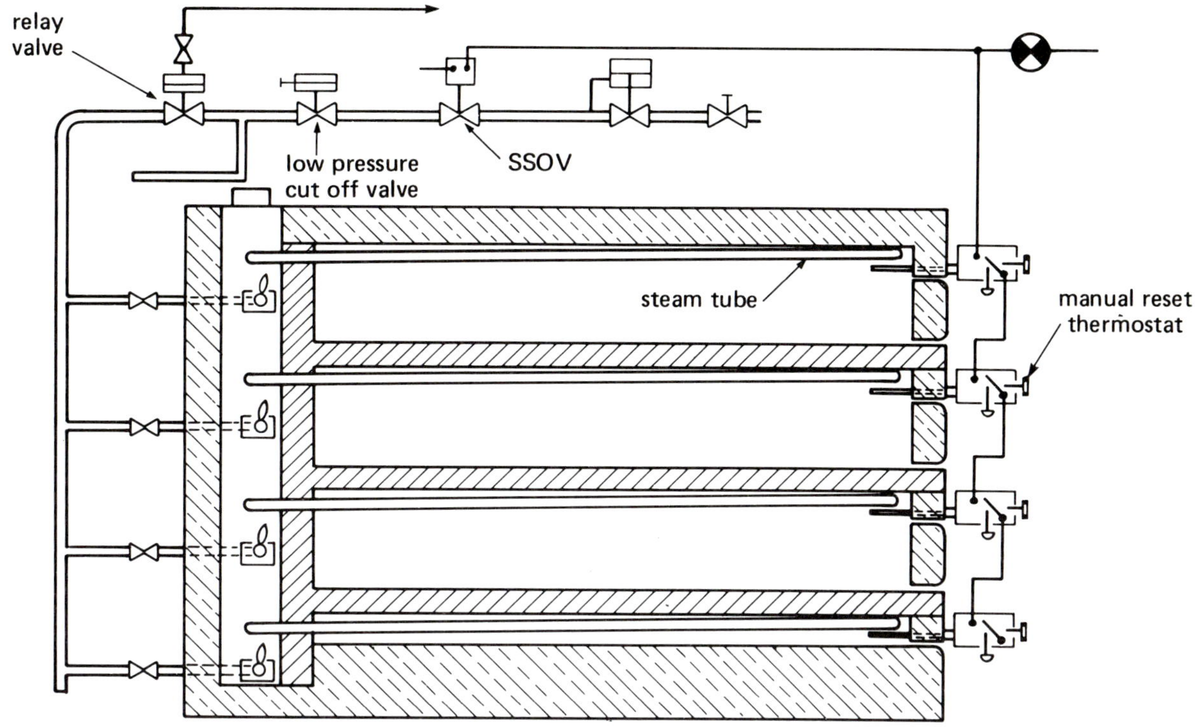

Fig. 9.114 Three deck steam tube oven with overheat thermostatic protection

9.8.1.4 Laboratory protection systems

To guard against gas valves being left open in school laboratories for example, a low pressure cut off system could be installed (Figure 9.115). With this type of arrangement, before gas can be admitted into the laboratory all gas valves must be closed and the appropriate start sequence initiated to pressurise a low gas pressure switch. The safety shut of valve will not open if any gas valve has been left open.

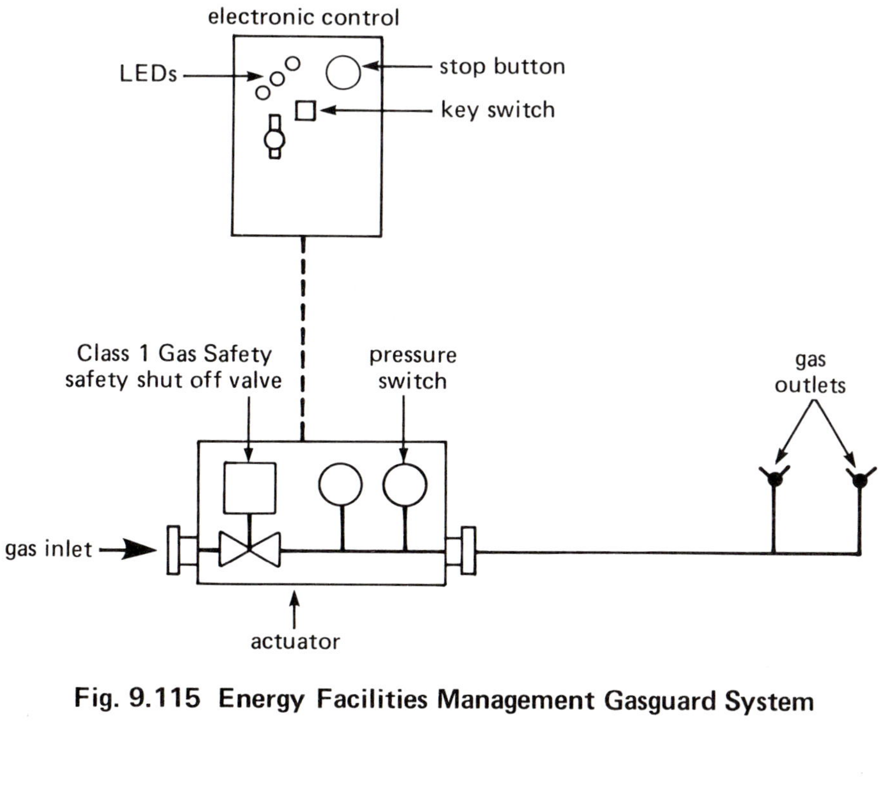

Fig. 9.115 Energy Facilities Management Gasguard System

9.8.1.5 Forced draught/induced draught burners

As with natural draught burners, these systems should comply with the Use of Gas in Low Temperature Plant, IM/18, and be fitted with a flame safeguard system to respond to loss of flame in less than 1 second. Ultra-violet or flame rectification protection will therefore be necessary in conjunction with the appropriate certificated safety shut-off valve(s) as before. Where these burners are automatic, BS 5885 will apply, which covers burners below 60 kW, Part 2, and above 60 kW, Part 1.

A typical valve train layout for automatic burners above 60 kW not fitted with modulating control is shown in Figure 9.116. It should be noted that a low air pressure switch is always necessary but a low gas pressure switch will only be required in special circumstances, The air throughput control damper is usually a flap on the inlet to the fan.

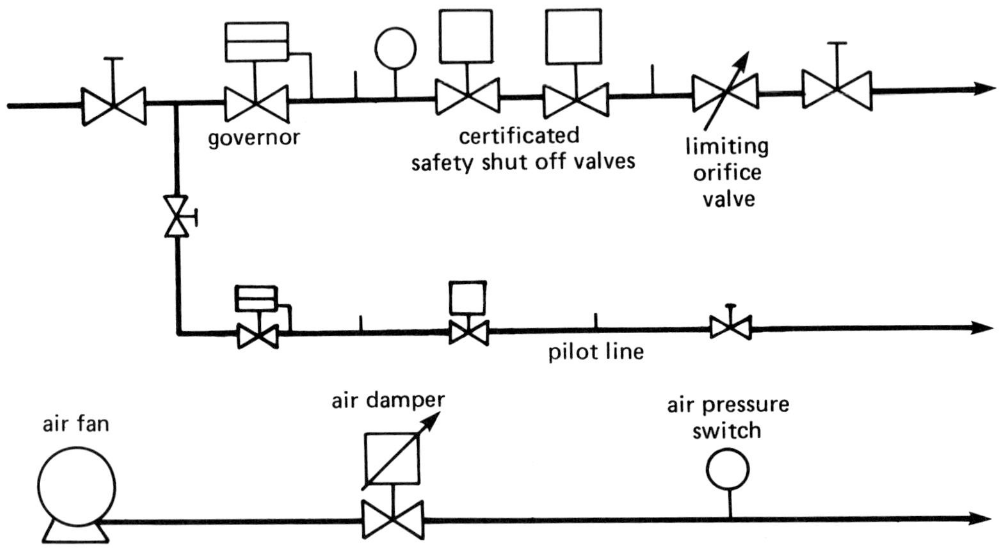

Fig. 9.116 Pipework layout for nozzle mix burner installation over 60 kW

For burners operating above 3 MW which generally implies full modulating control, a valve proving system will be required. A typical system which could be fitted to a large boiler installation is shown in Figure 9.117. High gas pressure protection may be required in certain circumstances.

9.8.2 High Temperature Systems

High temperature plant operating in excess of 750 °C at the combustion and working chamber wall should comply with the Code of Practice for the Use of Gas in High Temperature Plant (Publication IM/12). Flame protection should be fitted whenever reasonably practical.

9.8.2.1 Natural draught systems

Natural draught systems will be similar to those described in low temperature systems. If a flame safeguard is not fitted, a low gas pressure protection system will be required in conjunction with the appropriate safety shut off valves depending on the size of the burner.

9.8.2.2 Forced draught burners.

Where the burners are supplied with air, oxygen, or any other extraneous gas under pressure, a non-return valve that meets the British Gas specification must be fitted. If the only extraneous gas is

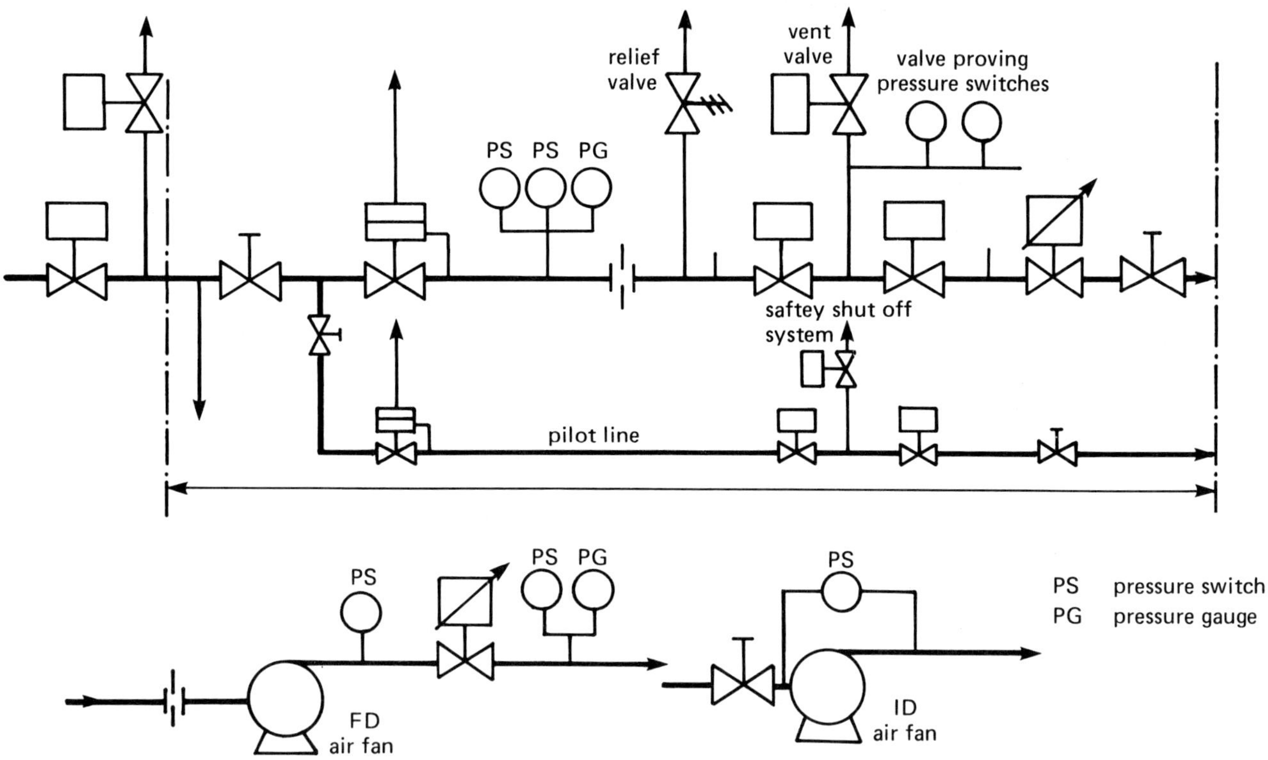

Fig. 9.117 Forced draught burners above 3 MW

air for combustion, a flame safeguard with a response time of less than 1 second and the appropriate safety shut-off valve(s) will fulfil the same requirement as the non-return valve.

A typical layout for small pre-mix (air blast) systems is shown in Figure 9.118. This is a completely manual system without safety shut-off valves and flame protection but typical of the systems found on smaller forge furnaces for example.

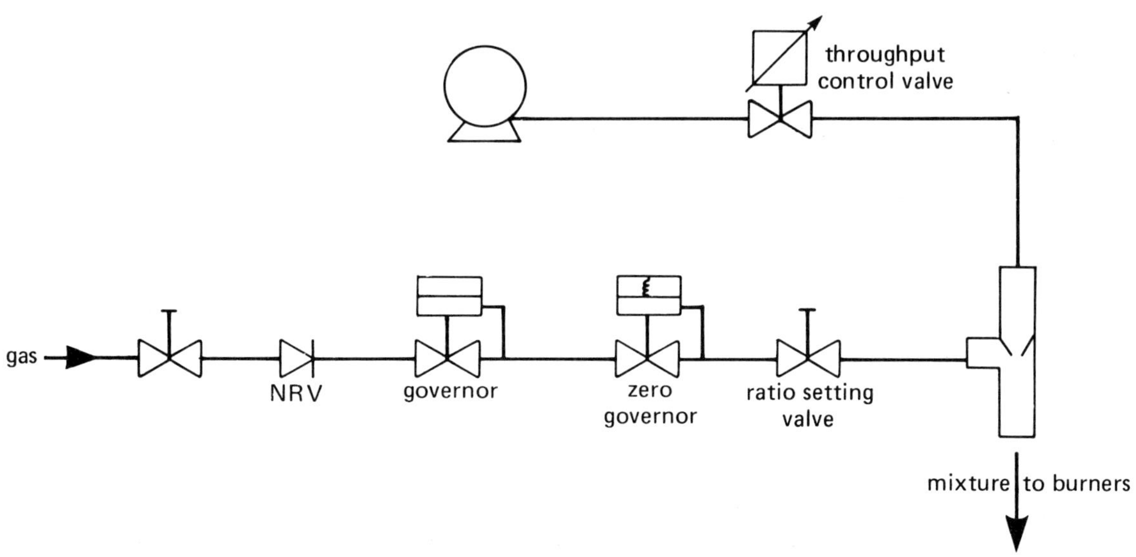

Fig. 9.118 Small air blast burner system

Where several burners are under the control of a pair of safety shut-off valves and no flame protection is fitted, the low pressure cut off system shown in Figure 9.119 is frequently encouraged. An example of this would be a continuous tunnel kiln used for firing ceramic ware with burners on either side of the kiln. In this case, if the fan is situated above the gas manifold, a non-return valve should also be installed in the air manifold to prevent gas from entering the fan casing. The safety shut-off valves in this case could be of the manual type which can be opened by a hand lever when the low gas pressure switch has proved all the burner isolating valves to be closed.

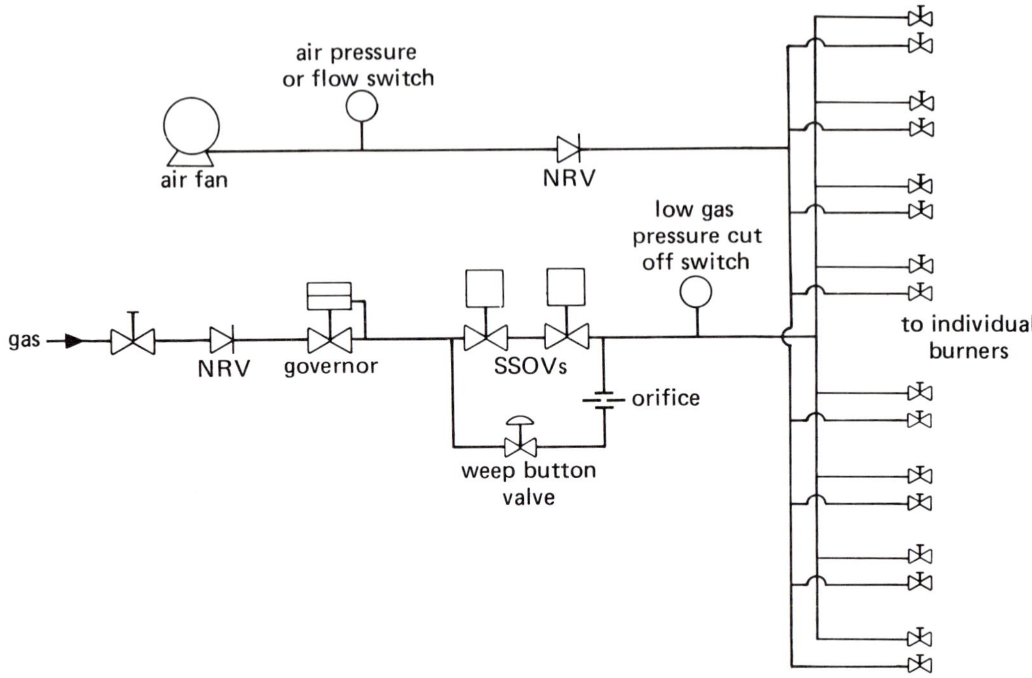

Fig. 9.119 Nozzle mix burner system with low gas pressure cut off

9.8.3 Low Gas Pressure Protection for Gas Compressors, Boosters and Pre-mix Machines.

To prevent the booster or compressor causing a reduced or negative pressure at the meter and in the gas supply system, a low pressure cut off switch shall cause the booster or compressor to shut down on loss of gas pressure. (Gas Act 1972, Schedule 4, Paragraph 18).

A non-return valve shall also be fitted to the inlet and outlet of compressors which would be capable of withstanding the maximum operating pressure. A typical installation incorporating these requirements is shown in Figure 9.56

9.9 EXPLOSION RELIEFS

The probability of serious injury to personnel or damage to plant can be reduced substantially if suitable explosion relief is provided, coupled with adequate securing of doors of the appliance to prevent them becoming missiles in the event of an explosion. The current thinking is that reliefs should be fitted to all solvent evaporating ovens, and consideration be given for the need to fit to low and high temperature ovens and furnaces.

For evaporating ovens the Health and Safety Executive's booklet HS(G)16 'The Prevention of Explosions in Solvent Evaporating Ovens' should be referred to. Guidance is also given in the case of ducts in booklet HS(G)11 'Flame Arrestors and Explosion Reliefs'. An explosion relief should be incorporated in the main evaporating chamber and separate reliefs considered in the indirect fired gas combustion chamber or flueway. The reliefs must be constructed in such a way that they do not form missiles if explosions occur and their strength must be small in order that they will be fully open before the pressure inside can build to a dangerous level. The positioning of reliefs is very important and generally they should be at the back of the oven, unless large sheets are suspended parallel to the back when top relief is preferable. The construction of the relief itself is adequately covered in this booklet, with a sandwich of lightweight insulating material between aluminium foil, held in place by a lightweight metal mesh, as shown in Figure 9.120. Further information is given on the design of explosion reliefs in 'The Investigation of Control of Gas Explosions in Buildings and Heating Plant' by R. J. Harris, British Gas, Published by E & F N Spon, 1983.

Explosion reliefs should be fitted to other low temperature plant such as direct fired ovens of steel box construction whether batch or continuous operation. The design of all new high temperature plant, especially of the low mass module construction type of similar construction to ovens should also incorporate explosion reliefs. External chambers of indirect fired ovens or furnaces should be equipped with these devices. The amalgamation of IM/12 and IM/18 'Codes of Practice for the Use of Gas in High and Low Temperature Plants' into one document includes a statement on explosion reliefs in the appendix.

Explosion reliefs would not normally be required in special cases. These include ovens or furnaces with a steel frame construction and a brick arch roof where the overall strength of the structure including the doors is such that the arch will vent any explosion over pressure. Also plant of very low mechanical strength where the generation of missiles will not occur, and ducts designed to open up to vent explosion over pressure will not require explosion reliefs.

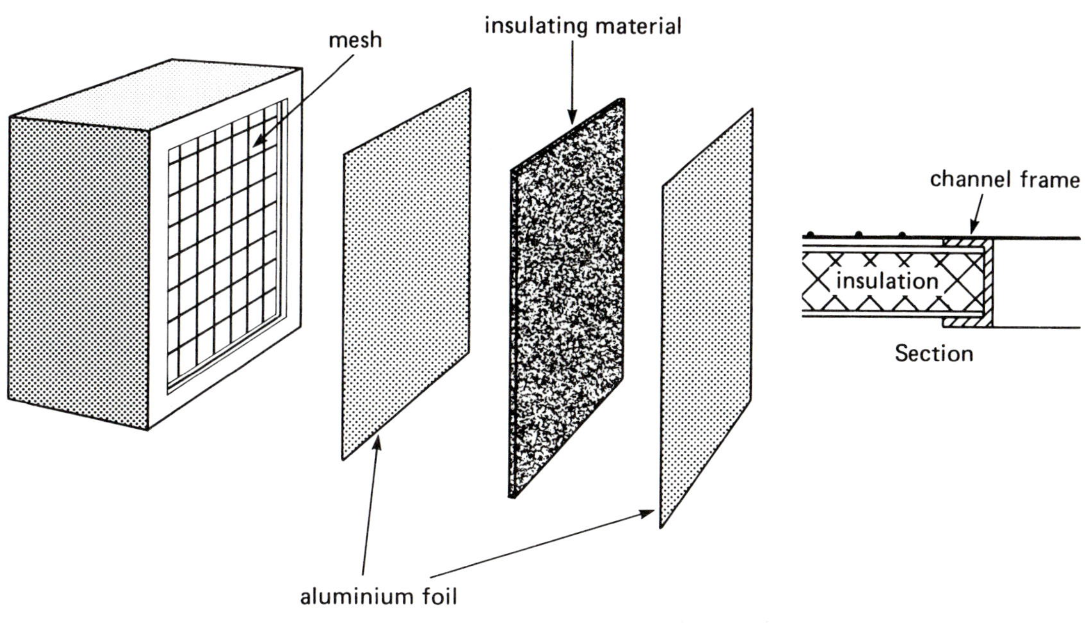

Figure 9.120 Exploded view of an explosion relief

REFERENCES

BRITISH GAS PLC PUBLICATIONS

IM/1 3rd Edition 1989 : Non-return Valves for Oxy-gas Glass Working Burners

IM/12 2nd Edition 1989 : Code of Practice for the Use of Gas in High Temperature Plant

IM/13 1st Edition 1980 : Specification for Pressure Switches for Industrial and Commercial Gas Fired Plant.

IM/14 1st Edition 1980 : Standard for Non-return Valves.

IM/15 1st Edition 1981 : Manual valves – A Guide to Selection for Industrial and Commercial Gas Installations.

IM/16 3rd Edition 1985 : Guidance Notes on the Installation of Gas Pipework, Boosters and Compressors in Customers' Premises.

IM/18 1st Edition 1982 with Amendments 1986 : Code of Practice for the Use of Gas in Low Temperature Plant.

IM/20 1st Edition 1983 : Weep By-pass Pressure Proving Systems.

IM/112 1st Edition 1982 : Industrial and Commercial Metering Installations.

BRITISH STANDARDS

BS 5885 : 1988 Part 1 Specification for Automatic Gas Burners of Input Rating 60 kW and above.

BS 5885 : 1987 Part 2 Specification for Automatic Gas Burners of Input Rating 7.5 kW up to but excluding 60 kW.

BS 5963 : 1981 Specification for Electrically Operated Automatic Gas Shut Off Valves.

BS 1552 : Specification for Manual Valves.

OTHER PUBLICATIONS

Association of Factory Mutual Fire Insurance Companies.

British Gas Certification Testing Scheme.

Gas Act 1972

Gas Act 1986

Flame Arrestors and Explosion Reliefs, Health & Safety Executive Booklet, No. 11.

Evaporating and Other Ovens, Health & Safety Executive Booklet, No. 16.

The Investigation and Control of Gas Explosions in Buildings and Heating Plant by R J Harris, British Gas. E & F N Spon, 1983.

Ignition and Combustion Safeguards 10

CHAPTER 10

IGNITION AND COMBUSTION SAFEGUARDS

10. 1 INTRODUCTION

The ignition of a combustible mixture of air and fuel gas may be accomplished by raising the temperature of a small volume of the mixture to a point at which combustion reactions begin and henceforth become self propagatory. For an air/natural gas mixture the ignition temperature is about 650 °C.

The ignition process may be effected manually or automatically or by a combination of auto/manual control but once initiated combustion must continue safely for as long as the burner is required to be in operation.

Manual operation of burners implies that the ignition sequence and subsequent continued operation of the burner will be conscientiously monitored and controlled by an operator. The condition to be guarded against is the release of an inordinately large volume of unignited fuel gas — from either a failure to achieve immediate ignition or subsequent failure of an established flame — into the combustion space. Such a release of unignited gas may ultimately result in the formation of a flammable mixture of gas and air which may then come accidentally into contact with an ignition source, or may remain in the combustion chamber until such an ignition source is eventually introduced. The resulting detonation may be of a magnitude as will result in damage to the plant and expose persons nearby to hazardous conditions.

It is the function of ignition and combustion safeguards to render a degree of control over the ignition process and to monitor the continued safe operation of the burner in a manner which prevents the accumulation of large volumes of unignited fuel gas; i.e. after the failure of an established flame, burner gas valves must be caused to close as quickly as possible.

The time taken for a hazardous volume of combustible mixture to arise in any given circumstance clearly depends upon:-

i) the rating of the burner

ii) the combustion chamber volume

iii) the degree of natural ventilation through the combustion chamber

iv) the strength of the combustion chamber construction.

However, in order to be able to formalise some common criteria by which ignition and combustion control devices should operate an acceptable minimum strength of combustion chamber must be arrived at and some relationship between volume of flammable mixture and pressure rise upon ignition established.

The majority of combustion chambers are judged to be capable of withstanding a pressure rise of 140 mbar (2 lbf/in^2) and work carried out in Midlands Research Station has shown that there is approximately a one to one relationship between the energy contained in the fuel release prior to ignition and pressure rise occurring at the instant of ignition. In fact, if the pressure rise is not to exceed 140 mbar then the energy release prior to ignition must not exceed 75 kJ/m^3 (2 Btu/ft^3). For most industrial sized burners, therefore, direct ignition of the main flame is not practicable since the maximum permissible energy release would occur in too short a time for reliable ignition to take place, so that the ignition process must be through the prior ignition of a start flame at a low rate. The start gas may be admitted to the main burner or it may be admitted to a small, separate pilot burner.

In industrial burners even the ignition of the start gas may have to be accomplished in a time too short for safe manual ignition if the limitations on energy release during the start of ignition period are not to be exceeded. Hence the development of fully automatic burner controllers by which ignition times are strictly limited. Also, to further limit the pressure rise which may be caused by delayed ignition and to ensure that ignition is always smooth and quiet, it is desirable that start air/gas mixtures are always lean, i.e. less gas in the mixture than the stoichiometric proportion. The Specification for Fully Automatic Gas Burners, BS 5885, for example, requires that the start gas rate is less than 25% of the stoichiometric gas rate for the proved air flow rate at the time of ignition.

Furthermore, to effectively monitor the presence (or absence) of flame, whether from start gas or main gas, implies the use of some unambiguous means of detecting the existence of a flame. Some property of combustion (preferably unique to a flame) must be utilised as an indication of the presence of a flame on a burner. A flame may be used to raise the temperature of a probe, or it may be used as a radiator of energy, or some electrical property of a flame may be used. Flame detection methods are discussed later.

10.2 START FLAMES AND PILOT BURNERS

Except in the case of burners of very small rating in which it may be acceptable to ignite the main flame direct, the first flow of gas to occur during the ignition sequence will be the start gas. The rate of flow of start gas will be low compared to the main flame rate – generally less than 25% of main gas rate. The start gas may be injected into the main burner or start gas may be delivered to a separate 'pilot' burner. Where start gas is injected into the main burner what begins as a small flame, at a low rate upon ignition, subsequently expands to the full fire rate when the main burner valves are opened. When a separate pilot burner is used then cross-lighting must occur from the pilot flame to the main burner when main flame is required. The use of true start flames has obvious advantages in the avoidance of the need for cross-lighting to occur.

10.2.1 Start Flame Classification

a) **Continuous start flame**

A small flame burning constantly throughout the time that a burner is in service, whether the main burner is on or off. Continuous start gas is usually admitted to a separate pilot burner.

b) **Intermittent start flame**

A small flame which is lit prior to ignition of the main flame but is extinguished with the main flame.

c) **Interrupted start gas**

A small flame which is ignited prior to the main flame but which is extinguished after the main flame has been established.

Pilot burner application

In order that a pilot burner may effectively fulfil its purpose any pilot burner must be positioned where its flame will provide immediate and smooth ignition of its associated main flame at any operative thermal input of both pilot and main.

Multiple burners, where there is any possibility that cross ignition between flames will not occur shall be similarly ignited from individual pilot burners, or from ladder pilots.

Pilot burners shall not be positioned where products of combustion can cause interference with start flame stability, such as to affect operation.

A pilot burner must be so sited, screened and protected that draughts will not deflect or alter the start flame such that it will be prevented from igniting the main flame correctly. The flame must remain stable on the burner at any gas pressure to which it may be reasonably subjected.

Start flame stability must not be materially affected by such changes in gas characteristics as may occur.

Mechanical fixing of a pilot burner shall prevent movement which could prevent ignition of the main flame.

Where manual ignition is employed adequate access shall be provided for ignition and provision made for viewing the start flame.

Start gas connections shall be positioned to reduce the possible entry of dust from the supply by avoiding connections from the underside of a gas line or from the base of a vertical pipe connected downward from the main gas pipe.

Start gas connections shall be taken from the main supply fitted downstream of the main burner inlet valve when a flame safeguard is used, but prior to it for unprotected systems.

Start flame turn-down test

Where a flame safeguard is fitted, a test should be made to determine whether the main flame ignites reliably and smoothly when the start flame is turned down to the point where it will just continue to actuate the flame safeguard.

For burners without flame safeguard equipment a start flame turn-down test may use a reduced supply pressure.

As a guide, a start gas pressure of 33% of normal should still provide satisfactory ignition. These tests must obviously be carried out with great care.

Pilot burner design and construction

A pilot burner unit must be mechanically robust and be resistant to the conditions of any environment in which it might be fixed. In particular the flame ports must be resistant to the effects of the flame, to prevent either blockage or enlargement of the ports. This latter is most important when considering small pilot burners. Orifice materials should be made from refractory or corrosion and heat resisting metal.

Start flame ratings

Automatic burners

No start gas rating shall be greater than 25 per cent of the stoichiometric gas rate (at the time of ignition) of the main burner. (The 25 per cent figure is applicable up to a pre-heated air temperature of 400 °C). However for certain appliances, including natural draught boilers and air heaters, more specific requirements exist.

The energy release during the start flame ignition period shall not be more than 53 kJ/m^3 of heating chamber volume (1.4 Btu/ft^3). This assumes that the plant and flueways will withstand a pressure rise of 100 mbar (40 in wg). Where it is known that the plant and flueways will withstand a higher pressure rise, the energy release may be increased proportionally.

Non-automatic burners

No start gas rating shall be greater than 25 per cent of the stoichiometric gas rate (at the time of ignition) of the main burner. The 25 per cent figure is applicable up to a pre-heated air temperature of 400 °C. However for certain appliances, including natural draught boilers and air heaters, more specific requirements exist, primarily for reasons of energy conservation. In non-automatic burner systems, the thermal release during the ignition period is limited by the pilot burner rating and the manually supervised time for which the start valve is maintained open. It is important, therefore, to limit the pilot burner rating to the minimum necessary to ignite the main burner safely and also to provide the operator with instructions on the correct lighting procedure.

In mixed systems where a time controlled period of start flame ignition may be available, but all the other requirements for automatic burners are not met, the start gas thermal release should be as for automatic burners.

10.2.2 Manual Ignition

Where hand ignition by match or taper is used it is most desirable that the point of ignition is within easy reach of and visible from the position of the gas control valve so that the application of the flame to the burner and admission of gas may be carried out simultaneously. If the above locations are not possible it is desirable that ignition be effected by more automatic means.

Hand ignition of start flames is normally permissible, but may be dangerous in restricted combustion chambers. Main flame burners may also be ignited by hand in some circumstances, but ignition by lighting torch is preferable and is strongly recommended with all industrial plant.

Ignition by lighting torch

A lighting torch is a portable burner supplied through a flexible tube, arranged for holding by hand in positions suitable for igniting fixed burners.

Construction of a lighting torch

A lighting torch for industrial uses may typically consist of a length of 3 mm (1/8 in British Standard Pipe) tube with a threaded end for a flexible tube. For natural gas some form of flame retention head is necessary both for aerated or non-aerated burners. The length of torch should enable the operator to provide ignition within easy reach of the supply valve of the burner being lighted.

It may be useful to provide a hook on the torch so that it may be hung out of the way when not being used, but it should be noted that anything which enables the torch to be left in proximity to the main burner, or inside an appliance, should be avoided to prevent uncontrolled and indiscriminate entry of gas.

Ignition torches should have a valve at the fixed supply end of the flexible tube so that the latter will not be so liable to be left continuously under gas pressure.

Application of lighting torches

To minimise any dangers from admission of uncontrolled gas flow into an appliance it is recommended that the gas rate of the torch should be limited, by nozzle or other restrictor, to the reasonable minimum that is necessary for its function. In any event the gas rate of the torch must not be more than 7.3 kW (25 000 Btu/h) or 3% of the main burner gas rate, whichever is the greater.

A suitable notice should be prominently displayed, giving information on lighting procedure.

10.2.3 Electrical Ignition

For the ignition of gas by high voltage spark, a minimum potential in the region of 4 000 to 5 000 volts is necessary, but up to 10 000 volts may be used according to the design of the ignition device. These voltages may be currently obtained in three ways: by mains transformer, electronic pulse system, or piezo-electric generator. Hot wire ignition requiring low voltage may be operated from dry cells or mains transformers, but is rarely met on industrial plant.

Mains transformer spark ignition

A mains ignition transformer consists of a conventional step-up transformer containing independent input and output windings on a laminated iron core enclosed in a pitch-filled steel container or encapsulated in resin.

The high voltage output may be from a single terminal from one end of the secondary winding, the other end being earthed to provide a return path or twin output connections may be provided for use with a spark gap between two insulated electrodes. The mid-point of the secondary winding is then earthed so that the maximum potential above earth is only 50% of the total output voltage. These types of winding are shown in Fig. 10.1 a) and 10.1 b) respectively.

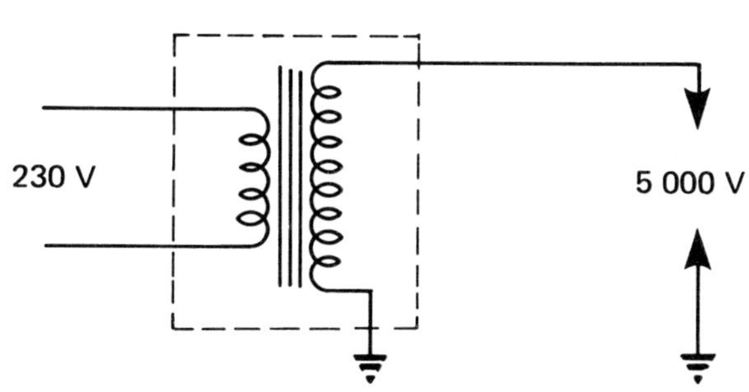

Fig 10.1(a) Mains ignition transformer

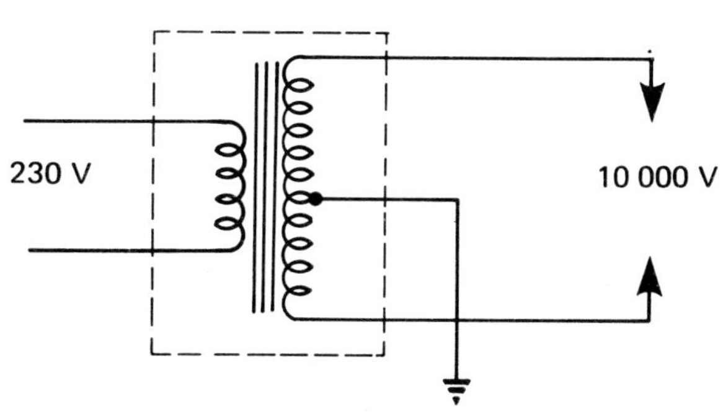

Fig. 10.1(b) Mains ignition transformer — centre earth tapped

The majority of gas ignition systems utilise the former method. It should be noted that it is not possible to obtain sparks simultaneously at two gaps by using the centre earth-tapped type with each electrode sparking to earth. Sparks will be obtained at the shorter gap or at random at both gaps.

The electrical power input required for conventional transformers is usually 100-200 Watts. Intermittent and continuous duty models are available, the former being generally used for gas ignition applications, being adequate for use when the ignition is limited to 5 seconds.

Application of mains voltage transformers

High voltage ignition transformers should be mounted as near to the burner equipment as possible, consistent with avoiding mechanical or thermal damage.

Where high ambient temperatures are unavoidable it is advisable to mount pitch-filled transformers with base horizontally upwards so that the filling will not leak.

The HT leads from the transformers to the ignition electrode may be of metal-braided screened lead or of automobile-type ignition cable with appropriate snap-on end fittings. It is essential that when using the latter an efficient earth return connection be fitted to the transformer case.

It is important that HT wiring should not be run near to any flame detector leads. Even if the latter are screened, sufficient electrical pick-up can be experienced to interfere with the operation of some flame detectors, including a false 'flame present' signal in certain circumstances.

Spark gap length can be from 2.4 mm to 9.5 mm according to the electrode type and burner configuration, 2.5 mm being typical.

All exposed HT leads and connections should be suitably insulated to prevent contact with the conductors by persons or adjacent components.

All electrodes should be mechanically supported so that adjustment requires the use of tools, thus the possibility of unauthorised manual adjustment is minimised.

All insulators should be non-porous. Any spark electrode section subjected to possible flame contact and to spark erosion should be of suitable heat resisting material, and so located that a minimum life of several thousand burner operating hours is obtained.

Electronic pulse spark ignition

By injecting a very high speed pulse of electrical energy into a transformer, which may be an auto-transformer of the automobile ignition coil type, a high voltage spark can be obtained. The input pulse can be conveniently obtained from a solid-state switching circuit containing a silicon-controlled rectifier, triggered repeatedly so that the stored energy in a capacitor is released into the pulse transformer. A schematic arrangement of the system is shown in Figure 10.2.

An isolating transformer is shown at the input to the pulse unit. This is not obligatory for the operation of the pulse system, but is for isolation of the earth return.

A single pulse transformer is shown but, providing the appropriate circuit is used, up to 100 transformers can be operated from one pulse unit. These are now commercially available.

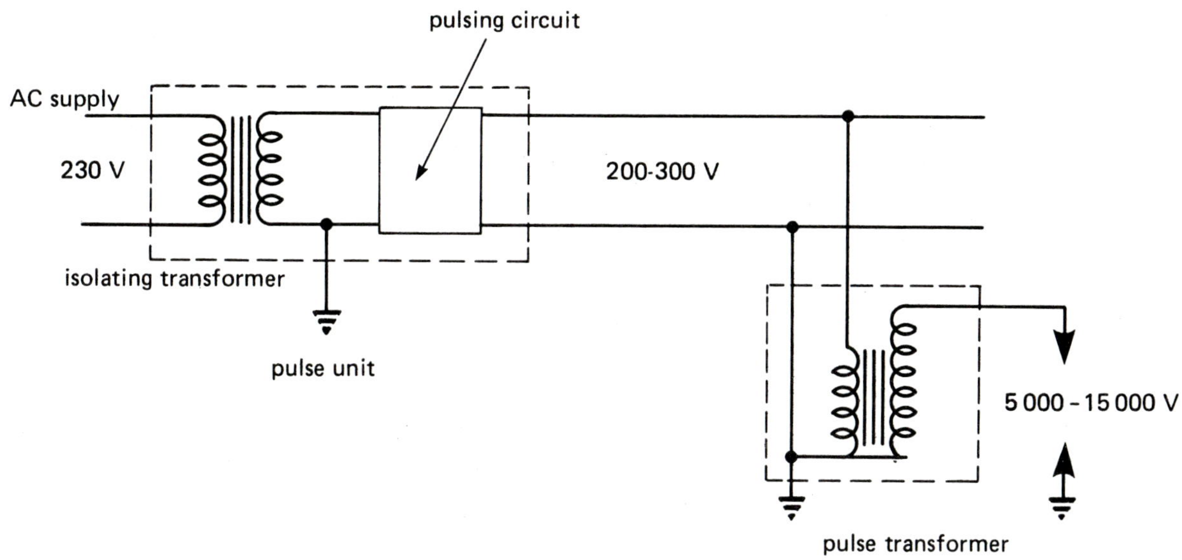

Fig. 10.2 Electronic pulse spark ignition circuit

The electrical power input for pulse ignition units is much lower than for mains transformers because the duration of each spark is in the order of microseconds only. For single spark units not more than 40 Watts and, for multiple units from 2-5 Watts per spark, is representative of actual inputs required.

In practical terms there are limitations in the application of such units. The lower energy provided by the spark means that the electrode gap setting is more critical and where distortion is possible the conventional transformer is to be preferred.

Application of pulse spark system

The minimum energy required to ignite a gas flame is dependent upon the design of the electrode and burner assembly, whether neat gas or air-gas mixture is being ignited, and upon the environment.

The power provided by a conventional mains transformer is adequate for any reasonable application, but care is required when using electronic pulse circuits. A minimum condenser capacity of 1 to 2 microfarads for each connected pulse transformer is recommended. Carbon deposition may occur when igniting neat natural gas and an aerated system is always recommended.

Piezo-electric spark ignition — natural gas

Direct conversion of mechanical energy into electrical energy is utilised in the compression of a synthetic crystal, which produces a potential between the planes of applied pressure during the pressure application interval.

In practice a cylindrical shape is used and compressed between an insulated plate and an earthed surface, as shown in Figure 10.3. The voltage produced will give a spark up to 2.5 mm long. Both compression and subsequent release produce a spark. The actuating force required is quite high and the energy is relatively low. They are currently used for the ignition of some domestic appliances, but are not used industrially.

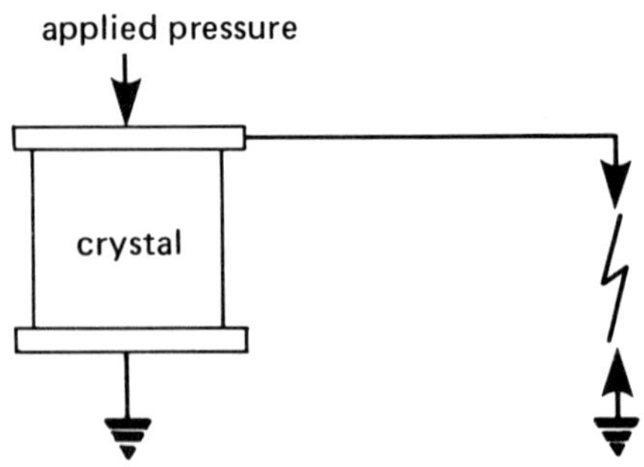

Fig 10.3 Piezo-electric spark generator

Hot-wire ignition

For domestic uses, platinum igniters employ very fine wire heated to a low temperature by a dry cell. Catalytic effect increases the wire temperature sufficiently for ignition. Such igniters are unsuitable for industrial applications.

Nickel chromium wire of heavier gauge is used in conjunction with mains transformers for medium duty applications, mainly heating appliances, and heavy duty units are available as appropriate for other applications.

In general, the application of hot-wire ignition is so critical in respect of location and detail design that its use is limited to standard appliances which can be extensively tested before being marketed. The short life resulting from incorrect application generally excludes their use in other combustion equipment

The ignition of natural gas is difficult by hot-wire because of the higher temperature required with consequent reduction of filament life.

10.3 BASIC IGNITION PROCEDURE FOR INDUSTRIAL GAS EQUIPMENT

The following procedure is intended only to provide a guide to the sequence of valve opening and closing and confirmation of correct burner operation. Modification to suit particular plant or processes may well be required, and therefore this procedure cannot be followed literally in all cases. Where a conveyor, or special electrical or ventilating equipment is involved, the instructions should be modified accordingly.

The general start sequence for burners is as follows:

i) Check that all manually operated burner gas valves and pilot gas valves are closed. If any burner or pilot valve is found to be in the open position, it must be closed and the appliances then ventilated for a time necessary to remove any unburned gas admitted.

ii) Check that doors and flue dampers are open.

iii) Start any exhaust and combustion air fans.

iv) Open the appliance main inlet gas valve. This valve must not be mistaken for the main burner valve.

v) Operate any manually reset air or gas controls.

NB Where there is NO flame safeguard equipment proceed as indicated in Sections 10.3.1 and 10.3.2

10.3.1 Natural Draught Burners Without Flame Safeguards

Apply lighted torch to the burner or ignite permanent pilot and check visually that the igniting flame is stable and adequate to ignite the main flame. Slowly open the main burner gas valve. Check that all burner flames are satisfactory. Repeat for each burner.

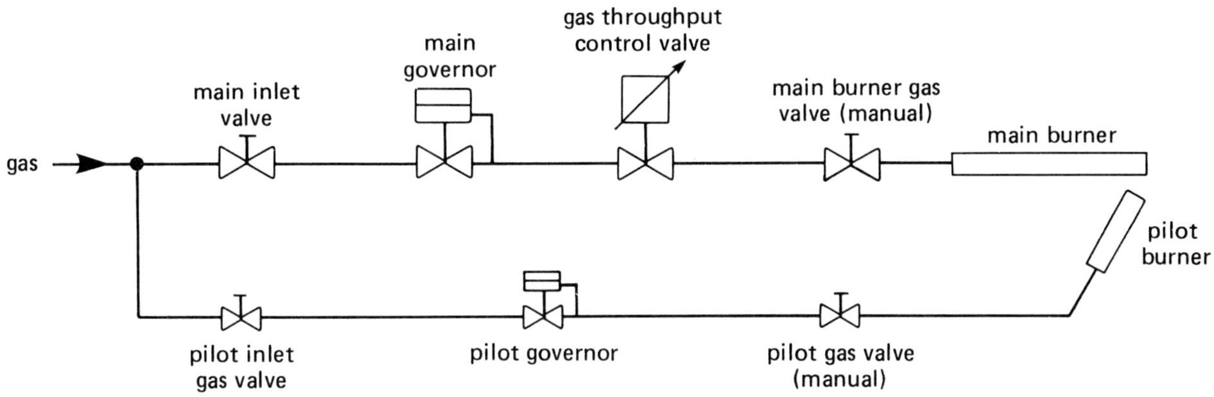

Fig. 10.4 Natural draught burner

10.3.2 Forced Draught Burners Without Flame Safeguards

Fully open the burner air valve and allow sufficient time for purging the combustion chamber. Partially close the burner air valve to the correct ignition rate. Apply lighted torch to burner or ignite permanent pilot and check visually that the igniting flame is stable and adequate to ignite the main flame. Slowly open the main burner valve and check flames during adjustment of gas and air flows to the required rating. For simple control systems it is important that the air/gas ratio is maintained, in particular that gas rich firing is avoided, whilst adjusting air and gas flows. Repeat all steps for multi-burners.

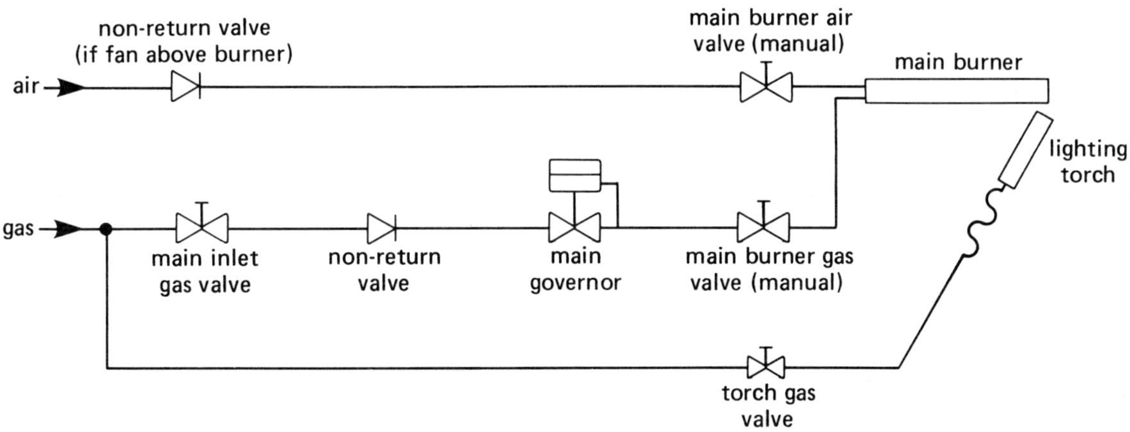

Fig. 10.5 Forced draught burner without flame safeguard

10.3.3 Burners With Flame Safeguards

i) Where there IS flame safeguard equipment, see Figures 10.6 and 10.7, operate in accordance with the appropriate instructions. Check visually pilot flames as in 6) before opening the main burner valve (except for automatic burners having special instructions).

ii) Appliance doors must be closed slowly after ignition of main burners.

iii) Check that all flames continue to burn correctly while adjusting for correct working condition.

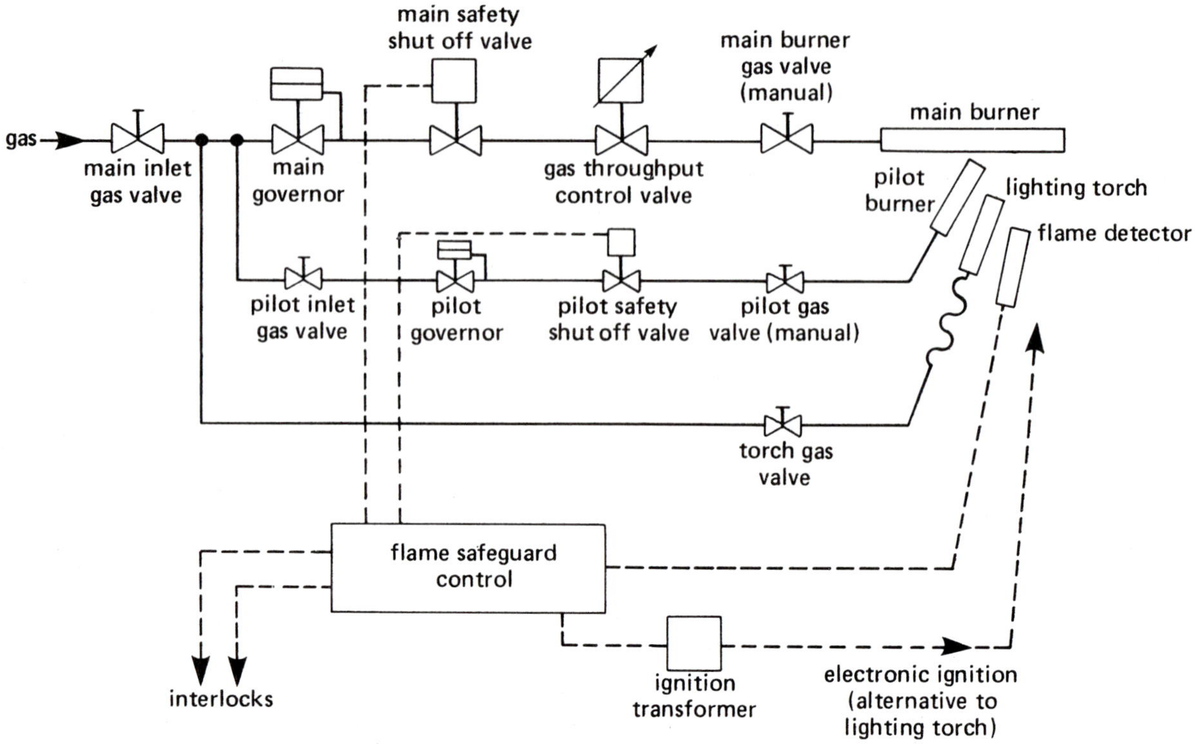

Fig. 10.6 Natural draught burner with flame safeguard

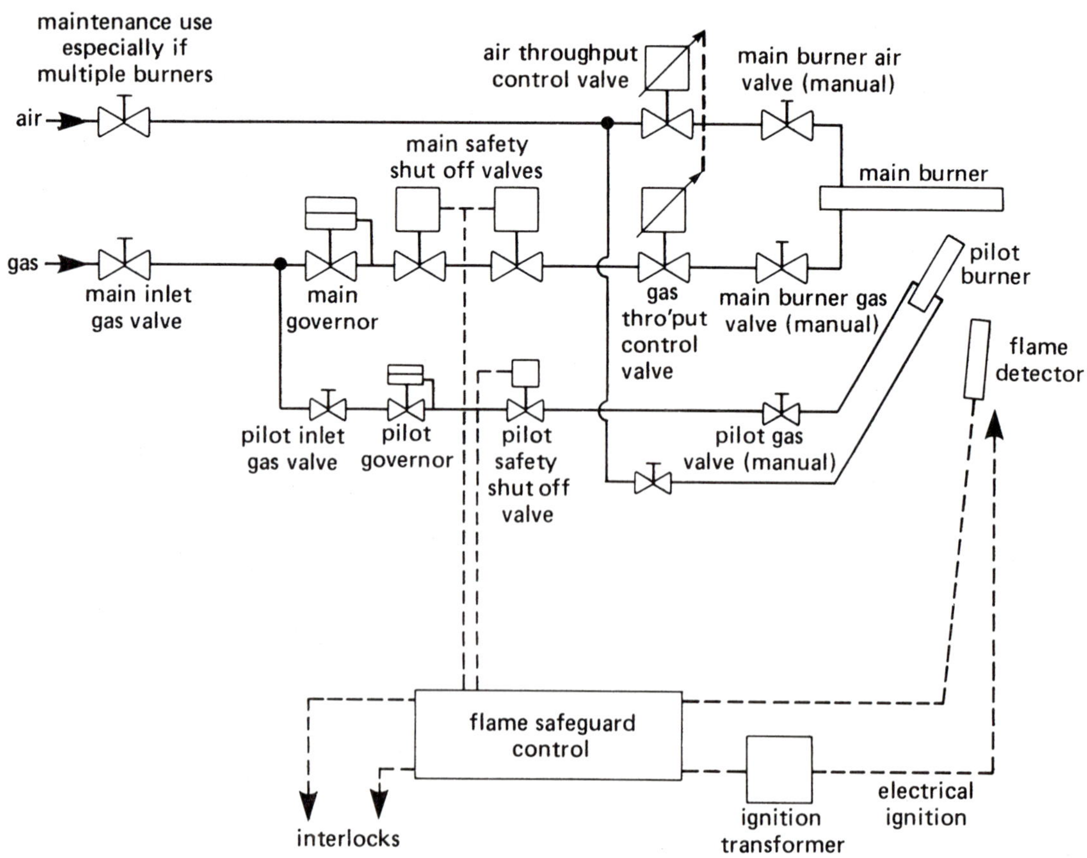

Fig. 10.7 Forced draught burner with flame safeguard

10.4 SHUT DOWN

i) Close each burner gas valve consecutively, followed by the associated air valve, if any, unless post purging is needed for any reason.

ii) Close each pilot valve, followed by the associated air valve, if any.

iii) Close appliance main inlet gas valve, if the stop period is other than temporary.

iv) Stop combustion air and exhaust fans, unless post purging is needed for any reason.

v) Where the process permits, it is advisable to leave the appliance well ventilated.

10.5 COMBUSTION SAFEGUARDS – GENERAL

The majority of dangerous situations in gas-heated industrial equipment are associated with the lighting up operation. For this reason a satisfactory flame protection system shall protect against both:-

i) incorrect lighting procedure and

ii) flame failure from any cause

Some of the devices described in this section, the direct acting thermal expansion and thermo-electric types, are capable in themselves of satisfying these two requirements. The remainder are, in effect, flame detectors and an ancillary device must be used in conjunction with them to control the safety shut-off valves. Complete combustion safeguard systems of this more complex type and multi-burner installations are described later in this chapter.

10.5.1 Indispensable Characteristics

Any flame protection system that does not comply with each of the following requirements is unsatisfactory for industrial use.

Every flame protection system shall:-

i) ensure that the correct lighting procedure, appropriate to the type of burner appliance, is followed.

ii) entirely prevent gas from being supplied to the main burner until a start flame is established

or

prevent the full gas rate from being passed to the main burner until a flame at a lower rate, subject to an appropriate trials for ignition period, is established.

iii) be free from any inherent weakness in design which could give rise to failure to danger, provided each component is fitted correctly.

iv) stop all gas flow to the burners after flame failure and then require manual resetting.

Thus the use of an unprotected pilot in any installation fitted with a flame protection device is unsatisfactory for industrial use.

v) be actuated only by that part of the flame which will always ignite the main flame, i.e. the flame detector shall not be actuated either by a flame which cannot directly light the main flame or by a flame simulating condition. This is dependent upon correct application as well as on correct design.

vi) with the exception of certain thermal expansion devices, be provided with a 'Safe start check' to prevent energising of the gas valve(s) and, where applicable, any electrical ignition device, should a 'flame on' condition be present prior to ignition.

vii) be mechanically and electrically satisfactory and readily serviceable.

10.5.2 Desirable Features

In addition to the indispensable characteristics given above, there are a number of features which are desirable.

Every flame protection system should:-

i) be protected against adjustment by unauthorised persons as far as is practicable

ii) work under all throughput and draught conditions together with all gas characteristic and mixture ratio changes that might normally be expected

iii) not be affected in performance by foreign matter

iv) operate satisfactorily within an ambient temperature range likely to be encountered on a gas fired industrial appliance

v) tolerate reasonable vibration and shock

vi) where mains powered, operate satisfactorily with supply voltage variations of between -15% and +10% of the nominal rating.

vii) where it incorporates a gas carrying component, have an approved capacity within a permitted pressure loss.

10.5.3 Thermal Expansion Devices

The majority of thermal expansion devices are not capable of protecting the pilot and thus are unsuitable for industrial use. Only one type of unimetallic and two types of vapour pressure device at the time of writing are capable of protecting the pilot. However, brief descriptive information is given below on all types of thermal expansion devices, as those which do not protect the pilot may still be in use on some existing installations and hence should be understood, although they should preferably be replaced and the plant updated.

Apart from the unimetallic types, thermal expansion devices are generally slow in action. Their operation is materially affected by the ambient temperature surrounding the flame sensor and consequently they shall be used only on low temperature appliances. The design shall be such that any push button requiring manual depression shall not require depression for more than 40 seconds to achieve opening operation. After flame failure the device shall ensure that, within 90 seconds, all gas flow to the burners is stopped.

Thermal expansion devices are not suitable for industrial or commercial plant and it is unlikely that they will be encountered in the field. They may be, however, met on domestic appliances installed in industrial premises and some knowledge of their working is therefore desirable.

Unimetallic types

These rely on the movement of a metallic tube captive at one end, the mode of operation being similar to the rod-type thermostat. This movement is then used to operate a small valve.

Unimetallic flame devices are robust and, unlike the majority of other thermal expansion devices, relatively rapid in action. Because of the limited movements produced, only small valves in, say, weep lines may be operated by them.

Bimetallic types

Under this heading are included two categories, the one depending on differential longitudinal expansion between a rod and a tube surrounding it, the two being of different materials, and the other in which differential expansion of the two components of a bimetallic strip or disc causes a change in curvature.

The inherent disadvantages of bimetallic elements are:-

i) slowness of reaction

ii) failure due to fatigue

iii) limited working life.

Several devices of this type do not protect the pilot and therefore do not comply with the requirements under Indispensable Characteristics.

Devices of these types are in general cheap and mechanically simple, and are therefore used in those applications where speed of operation is not a premium, e.g. in domestic appliances.

Bimetallic strips, which operate by flexure, are the least satisfactory members of this group, since the direct heating of the bimetallic material is always prone to engender corrosion, and this in turn to result in the device failing to danger. Nevertheless such devices may be encountered on instantaneous water heaters.

Liquid expansion and vapour pressure types

These devices use a phial connected to a bellows, either filled with a liquid such as mercury, or a high vapour pressure liquid such as ether. They fall into two categories:-

i) Those in which the bellows simply operate a gas valve. As they do not protect the pilot these are unsuitable for industrial use.

ii) Those in which the bellows operate a switch or a switch and a gas valve. These can be arranged to protect the pilot. They are used in the same manner as the switching and relay types of thermoelectric valve.

These devices may be encountered on the ovens of domestic cookers. In general, they are mechanically robust in service; but the capillary tube must be handled with care and not sharply deformed on the installation.

Installation and commissioning of thermal expansion devices

Requirements for installation and commissioning of thermal expansion devices are essentially as for thermoelectric devices, see pages 646 and 647.

10.5.4 Thermoelectric Flame Safeguards

These devices are slower in reaction than is really desirable, but they afford a compromise between cost and speed of reaction which makes them suitable for many installations of small size, where the hazards involved in flame failure are not too great. They are considerably more reliable than thermal expansion devices.

Direct acting type

A typical direct acting thermoelectric safeguard is shown in Figure 10.8. When the push button A at the base of the body is pushed upwards, the flow interrupter valve D is moved on to its seating, thus sealing the exit port to the main burner. Further movement of the push button A raises the main valve F, which action admits gas to the pilot burner through the connection E in the body of the device. The pilot burner may then be lit. Upon completion of the movement of the push button, the armature H will be pressed against the electromagnet and, after about 30 seconds, the temperature of the thermocouple is such that sufficient current is generated to energise the electromagnet J which will thus hold the armature H in position and so retain the main gas valve F open. At this stage the push button A may be released, and the reset spring B will force it and the flow interrupter valve D downwards, thus opening the main gasway, allowing gas to pass to the main burner. The output from the thermocouple is typically between 10 and 20 millivolts.

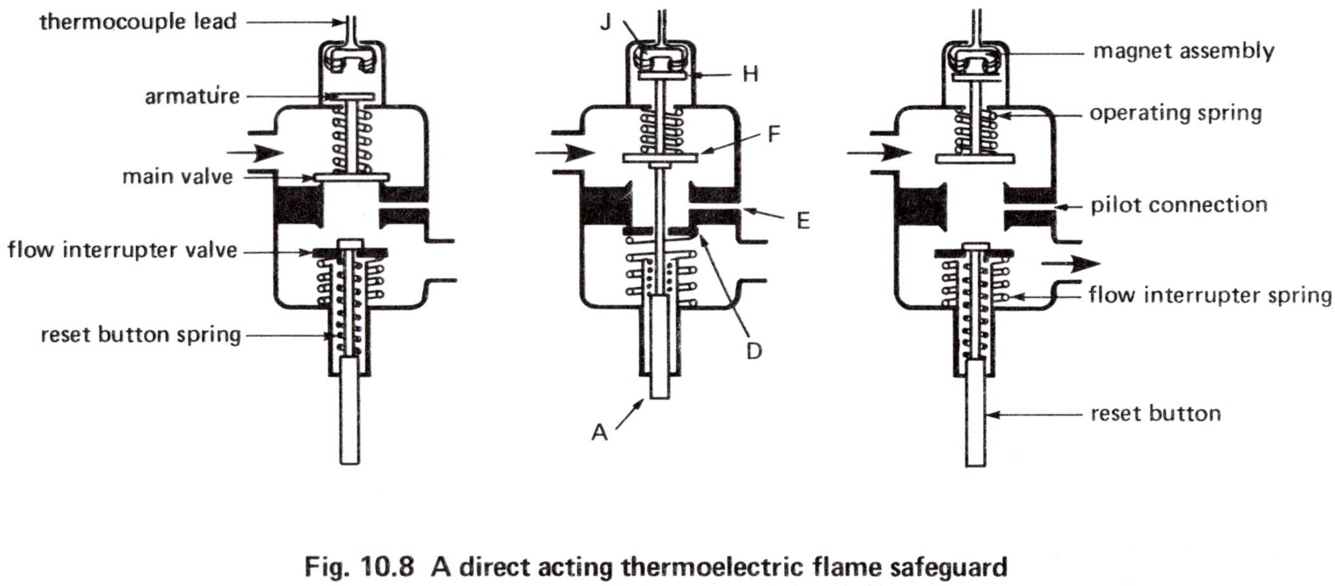

Fig. 10.8 A direct acting thermoelectric flame safeguard

If the gas supply fails, the pilot flame is extinguished, the thermocouple cools and the current falls. The electromagnet J can no longer hold the armature H in position, so that the mains gas valve F is returned to its seating. No gas will flow again until the whole cycle of operation is repeated. Figure 10.9 shows a suitable arrangement for direct protection of the gas supply using this type of thermoelectric device.

Indirect acting types

A further type of device similar in principle to the one in the previous section is shown in Figure 10.10, but there the gas supply to the pilot burner passes, via separate connections, through an independently housed valve mounted on the same spindle as the main valve and armature.

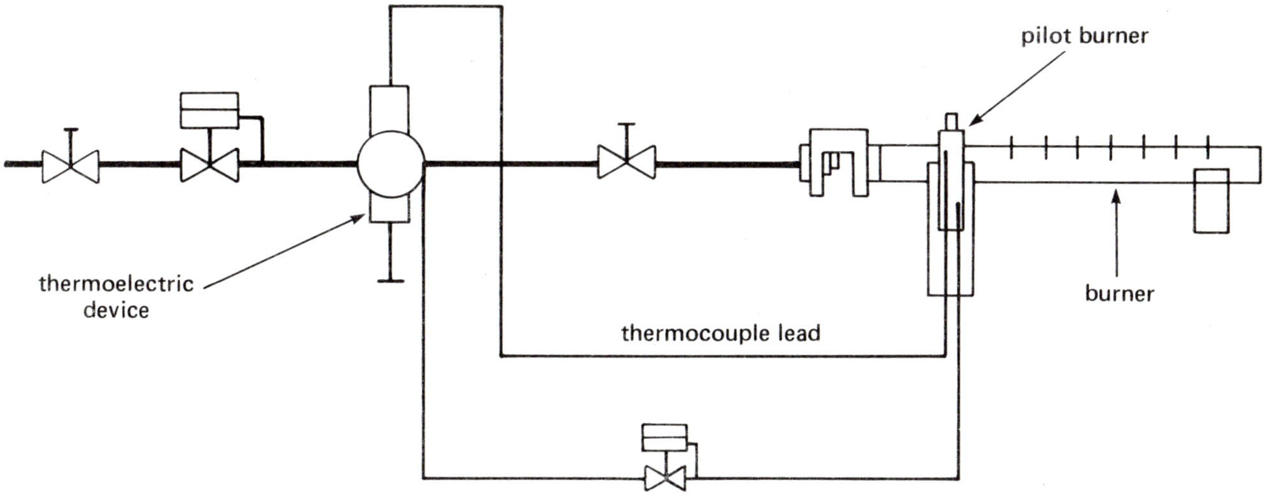

Fig. 10.9 Direct protection of a burner

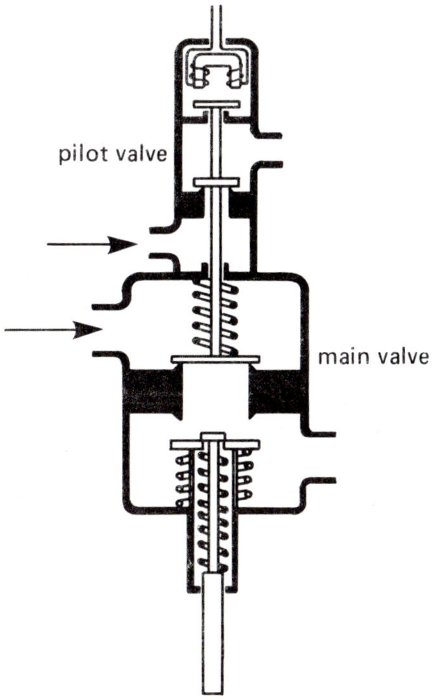

Fig. 10.10 A thermoelectric flame safeguard with a separately fed pilot

A number of systems have evolved in which this type of thermoelectric device is used. Not all of these systems are fully satisfactory, in that they require the use either of an uncontrolled pilot or of a lighting torch to heat the thermocouple. A typical application is the protection of multiple burners with a single relay valve for temperature control purposes as shown in Figure 10.11. An unsatisfactory aspect of this configuration is that the weep from the relay valve is not protected as it is upstream of any thermoelectric valve.

Switching types

These are similar in principle to direct acting thermoelectric devices in that they rely upon the electrical output of a thermocouple to energise an electromagnet. However, instead of directly operating valves in the gas supplies to a burner these units operate electrical contacts capable of

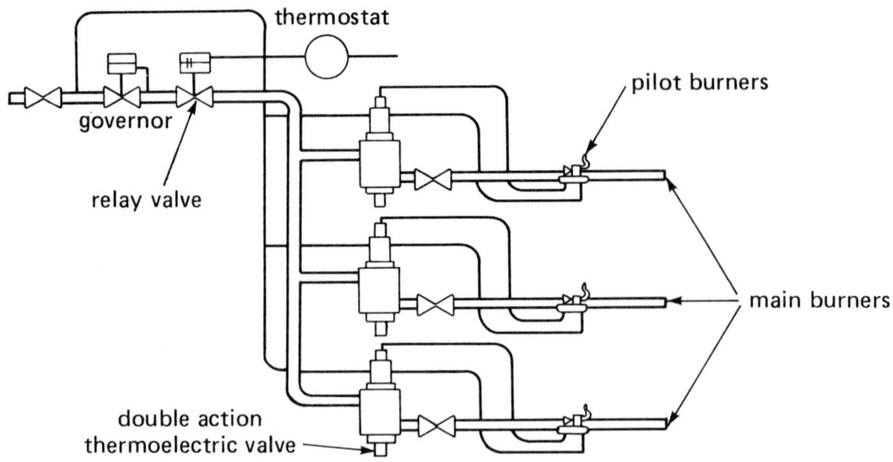

Fig. 10.11 Protection of multiple burners by separately fed pilot valve on each burner

switching mains powered solenoid and safety shut-off valves. The sequence of operations when lighting burners controlled by these switching types is virtually identical to the sequence observed when using direct acting thermoelectric safeguards. The differences being:-

i) the burner valves are electrically powered

ii) some switching types have provision for also controlling spark ignition of the start flame.

Switching types of thermoelectric flame safeguard are now obsolete — their functions having been largely overtaken by electronic flame safeguards — but some examples may be encountered on old plant.

Compact gas controls

Compact gas controls are monobloc devices which incorporate equipment to perform several of the control functions required for the operation of gas burners. Their principal applications are with burners of modest rating such as may be used in domestic central heating boilers, storage water heaters, gas fired air heaters and other appliances of this nature. They provide thermoelectric flame protection, outlet gas pressure governing and switching of the main flame for process temperature control purposes. Control of the main flame may be on/off or it may be high/low/off depending upon the model of the control unit.

Their relevance to this section lies in their incorporation of thermoelectric flame protection.

From the cross-section of a typical compact gas control, Figure 10.12, it will be apparent that the flame protection aspect can be discussed separately from its other functions. Figure 10.13a shows the flame protective part of the unit in the off condition. Depressing the operating button as in Figure 10.13b opens the start gas valve and pushes the latch down which in turn brings the armature of the power unit into contact with the pole faces of the electromagnet. When the start flame is established and the thermocouple is at a sufficiently high temperature, the output from the thermocouple energises the power unit which continues to hold the armature in contact with the pole piece against the compression of the armature spring.

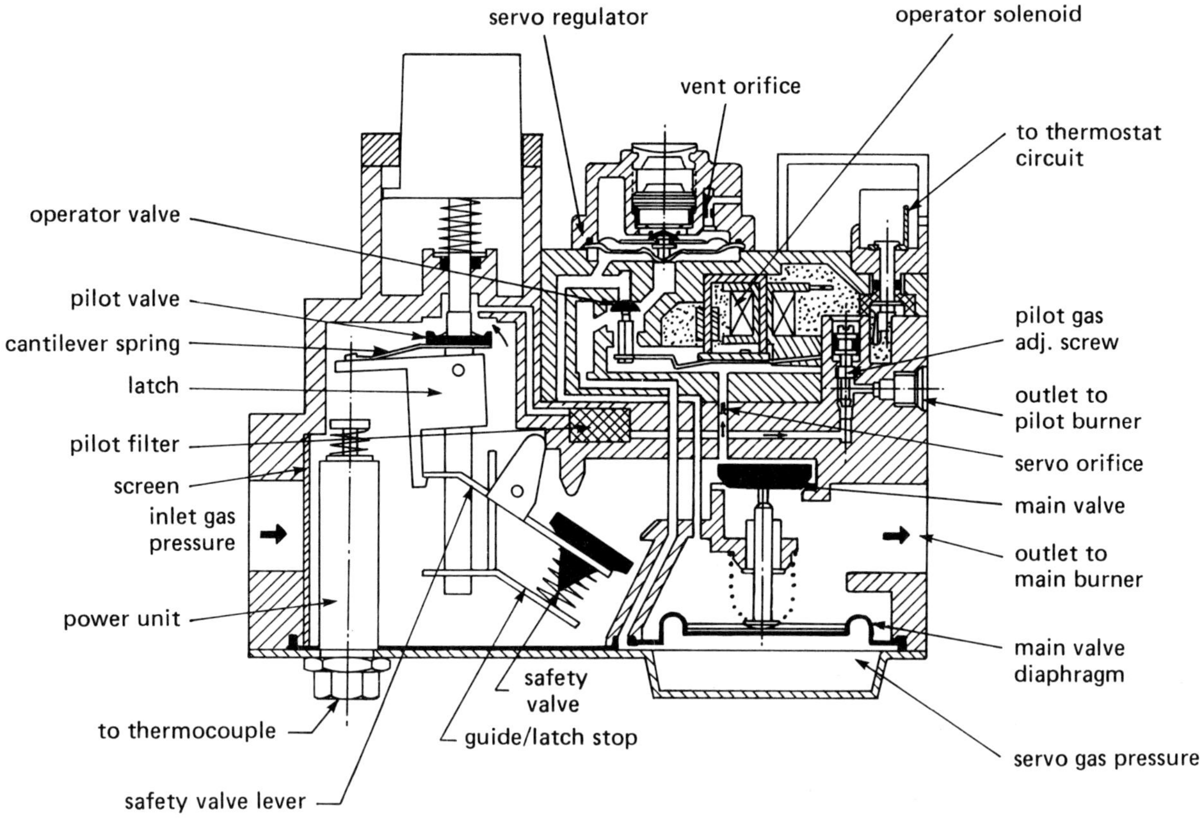

Fig 10.12 A compact gas control valve

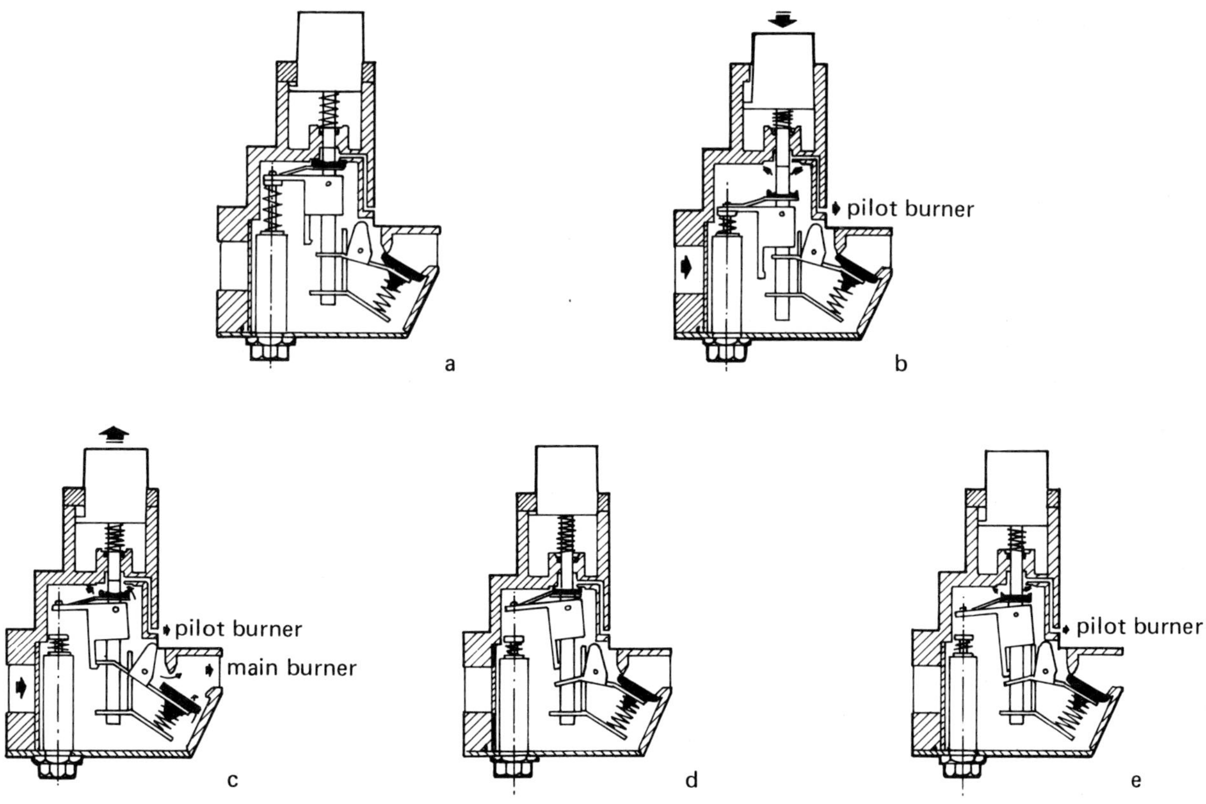

Fig. 10.13 The operation of a compact gas control

Releasing the operating button allows the latch to move upwards and to tilt under the action of the cantilever spring. A hook beneath the latch then engages the safety valve lever and opens the safety valve, allowing gas access to the main valve. Engagement of the hook with the safety valve lever also restrains the latch from further upward movement, thus keeping the start gas valve open, Figure 10.13c. If the plant thermostat is calling for heat the operator solenoid will be energised and in the position shown in Figure 10.12, allowing the main valve to open and pass gas to the main burner. Control of outlet gas pressure is achieved by adjustment of the pressure exerted by of the spring in the servo regulator which in turn is responsible for controlling the position of the main valve. When the plant is at the required temperature and the thermostat is satisfied the operator solenoid will be de-energised causing the operator valve to close off the servo weep gas supply. Pressure beneath the main valve diaphragm is thus lost and the main valve spring closes the main valve and extinguishes the main flame. Start gas continues to flow to maintain a continuous start flame.

A small clockwise turn of the operating button disengages the hook from the latch which permits the safety valve to close under the compression of the safety valve spring – to extinguish the main flame – and the latch to rise until the start gas valve is closed, to extinguish the start flame, Figure 10.13d.

In this condition the power unit will remain energised until the thermocouple cools and to prevent the main valve being opened again, if the button is depressed, a safety stop arrests downward movement of the latch, preventing engagement of the hook with the safety valve lever. Under the condition described, the latch will be tilted from the vertical by the cantilever spring, Figure 10.13e.

When the thermocouple has cooled and its output fallen to an extent whereby it can no longer energise the power unit, the armature will be released and will lift the arm of the latch turning the latch into a vertical position and restoring the unit into a condition ready for restart; i.e. back to the condition shown in Figure 10.13a.

Installation of thermoelectric devices

Points to be observed in installation of these devices are:-

i) The thermocouple must be correctly located with respect to the pilot flame, according to the makers' instruction

ii) The thermocouple should be shielded from radiation from sources other than the pilot flame

iii) Care must be taken to avoid damage to the thermocouple leads, especially at the point where they are connected to the body of the valve. On assembly or reassembly attention should be given to the cleanliness of the electrical connections.

iv) The pilot and thermocouple must be located where they are not subject to excessive draught.

v) The pilot flame must not be in danger of being smothered by products of combustion.

vi) The pilot burner assembly should be rigidly attached to a metal support of substantial cross section which is, if possible, an extension of a colder portion of the appliance. This will enable heat to be conducted away from the thermocouple.

vii) The pipe carrying gas to the pilot burner should be made as rigid as possible, since vibration of this connection has been known to extinguish the pilot.

viii) The gas valve should be mounted sufficiently rigidly to prevent its being affected by undue shock or vibration.

ix) The device must, wherever possible, be so mounted that gravity assists the closing of the gas valves.

x) The portion of the device which carries the press button used in lighting up must be fixed where the operator can, without difficulty, hold the button in and at the same time observe and light the burner.

xi) On all but the smallest installations, a burner cock or cocks should be fitted and kept closed until all other lighting up procedures such as the checking of damper positions and ventilation have been performed.

Where the mains energised units are used, installations should comply with the Electricity (Factories Act) Special Regulations 1908 and 1944 and the Regulations for Electrical Installations (IEE Wiring Regulations) 15th Edition, 1981, published by the Institution of Electrical Engineers.

Also, the satisfactory operation of a thermoelectric flame protection device depends upon the generation of a voltage output by the thermocouple sufficient to reliably power the electromagnet. The output generated by a thermocouple is a function of the difference of temperature between the 'hot' and 'cold' junctions, so that not only must the 'hot' junction be heated to an appropriately high temperature the 'cold' junction must be maintained at an acceptably low temperature.

Difficulty in complying with this requirement may be experienced if, for example, the complete protection device is installed within an insulated cabinet. In some circumstances an extra long thermocouple may be an advantage in permitting the electromagnet unit (and therefore the cold junction) to be sited in a cooler environment than would be so with a thermocouple of standard length.

The rate at which the hot junction of a thermocouple cools on flame failure is also of importance in determining the lapse of time between flame failure and closure of the valve. To keep this time acceptably short the overall temperature of the combustion chamber enclosing the pilot burner and thermocouple assembly should not exceed 250° –300 °C – see also Shut-down test, below.

Commissioning of thermoelectric flame safeguards

On commissioning, the flame protection system shall be examined to ensure that it complies with all the installation requirements listed above. A start flame turn-down test shall be carried out in order to confirm that the installed flame protection device complies with Indispensable Characteristic (e) in the Combustion Safeguards – Section 10.5.1.

Start flame turn-down test

i) Turn off the main burner valve(s).

ii) Reduce the start flame by means of the start gas control valve to a point at which it will only just hold in the flame failure device. Each trial setting of the start flame should be held for three minutes or longer in order to reach thermal equilibrium. (If a valve is not available in the line, disconnect the start gas line at a convenient point and insert a short length of rubber tube having some form of clamp to restrict the flow).

iii) Visually check that the ignition flame is long enough to light the main burner.

iv) Turn on the burner valve(s) and check that ignition is smooth.

NOTE: This test should be carried out at the extremes of draught, throughput and mixture ratio which might occur on the appliance.

A shut-down test shall be made periodically to ensure that the device continues to function correctly.

Shut-down test

i) Bring the appliance up to normal running temperature.

ii) Close the appliance isolation valve.

iii) Note the time taken for the safety shut-off valve to close by listening or by feeling. This time shall not normally exceed 45 seconds but to some extent it depends upon the application and for any given application the appropriate standard should be consulted, e.g. BS 5991, 1989, for Indirect Fired Forced Convection Heaters' specifies a maximum closure time of 60 seconds for main burners up to 150 kW rating.

or

i) Bring the appliance up to normal running temperature.

ii) Carefully close the appliance isolation valve until only a bead of flame remains on each main burner jet. (The pilot flame will then have shortened to an extent where it will not materially heat the thermal element.)

iii) Note the time taken for the beads of flame to start to go out. This time shall not exceed 45 seconds – again with due regard to the application as stated above.

After carrying out the shut-down test the normal shut-down procedure shall be completed by closure of burner valves and isolating valve.

Servicing of thermoelectric flame safeguards

The continued correct operation of all types of device is best ensured by devising and adhering to a routine for closing down the appliance which will test the operation of the flame protection device. Such a routine, which could with advantage be adopted as a normal closing-down routine, would be to turn the gas off at the appliance isolation valve to produce conditions of flame failure and observe that the device has shut down substantially within the time recorded in the shut-down test carried out upon commissioning.

Occasional renewal of the thermocouples is necessary, and it is recommended that these be examined at least once per year.

10.5.5 Electronic Flame Safeguards

The ability to use some physical property of a flame directly as an indication of its continued existence rather than rely upon the ability of a flame to raise the temperature of a probe has some obvious advantages – notably a much reduced time of response. A major objection to the use of thermoelectric devices, for example, is the long time taken to raise the temperature of a thermocouple to a level at which sufficient output is generated to energise the power unit.

Then, again, the time taken for the temperature of the probe to fall to a level at which the unit de-energises after flame failure may also be excessive. This time lapse is a function of the mass of material forming the tip of the thermocouple and although reducing the cross-sectional diameter of the thermocouple will reduce the mass to be heated and therefore reduce the thermal inertia, there is a limit to how much the problem can be alleviated by this means. The use of a too fine gauge thermocouple may, for example, create nuisance problems because of a lack of mechanical robustness in the system.

Systems which make use of some electrical property of a flame, or which use the flame as a radiator of energy are more likely to find widespread applications as flame detectors, if only because of their responses to the onset of combustion or to flame failure are likely to be considerably more rapid than the methods so far considered.

A flame consists of ionized gases and is therefore capable of conducting an electric current. If two electrodes are immersed within a flame and a potential difference applied to the electrodes, then an electric current will pass through the flame between the electrodes. Such a system has possible applications as a method of flame detection but has the disadvantage that a short circuit between the electrodes will result in the continued flow of current irrespective of the existence or non-existence of a flame. It is difficult to render a simple conductive method reliably self-discriminatory against fault conditions and it is therefore not acceptable for the supervision of gas burners. However, under appropriate circumstances a flame may be caused to exhibit a crude rectifying effect upon an alternating current. If an alternating voltage is applied across the flame, a current which approximates to a half-wave rectified signal will flow and may be used as the indicator of the existence of a flame. Systems using this phenomenon are acceptable for supervising the operation of gas burners.

Flames, quite obviously, radiate energy. They radiate visible light, infra-red energy and in the ultra-violet band.

Devices which rely upon the flame radiating visible light are so readily susceptible to interference from ambient light that they are not used. Infra-red and ultra violet detectors do, however, find considerable application as flame detectors.

Rectification flame detection

As previously described, if an alternating potential is applied across two electrodes immersed in a flame, a uni-directional (half-wave rectified) current will flow between the electrodes. This 'DC' component may then be amplified in an electronic amplifier and the amplified signal used to energise a relay in the output circuit of the amplifier. Thus, when a flame is present between the electrodes the relay will be energised and the position of its contacts changed when compared with the relay's non-energised condition. Upon flame failure, or upon the short circuiting of the electrodes, the DC component disappears and the output relay becomes de-energised. A flame rectification system of flame detection is therefore inherently safe in that a fault condition leads to shut-down. The input circuit of the amplifier must have a high impedance to protect the amplifier from damage caused by excessively high fault currents in the event of complete short circuit of the electrodes. Fault currents are typically limited to 0.5 mA, or less. Modern equipment almost invariably uses solid state amplifiers although some valve amplifiers may be still in use. The magnitude of the rectification current in any given application depends upon:-

i) the potential difference across the electrodes; this is fixed by the manufacturer of the equipment,

ii) the difference in area between one electrode and the other,

iii) the calorific value of the fuel gas,

iv) the air to fuel ratio at the burner,

v) the relative position of one of the electrodes in the flame.

Some of the above points require clarification.

The signal strength will be enhanced if one electrode has an area in contact with the flame which is considerably greater than the area of the other electrode. A ratio of four to one is suggested by the manufacturers as being the minimum difference for successful operation. In practice, the burner is generally used as one electrode and a thin rod of heat resisting metal alloy as the other electrode. There is usually little difficulty in obtaining a sufficient difference in area with this arrangement. Conversely, if the ratio is too large the rectification effect appears to diminish; the reasons for which are as yet not satisfactorily explained.

The calorific value of the fuel gas is, of course, a fixed characteristic of the particular fuel in use.

The rod electrode is the one which is movable in relation to the flame. It should be positioned so that it is in contact with a part of the flame where the gases are highly ionised, i.e. where rapid combustion is taking place, usually towards the edge of the flame.

In any given application, the maximum rectification current will be obtained at stoichiometric air/gas ratio. The variation of rectification signal with air/gas ratio is as shown in Figure 10.14. The signal normally lies within the range 2 μA and 50 μA, depending upon the make of equipment, but is more commonly about 7 to 10 μA.

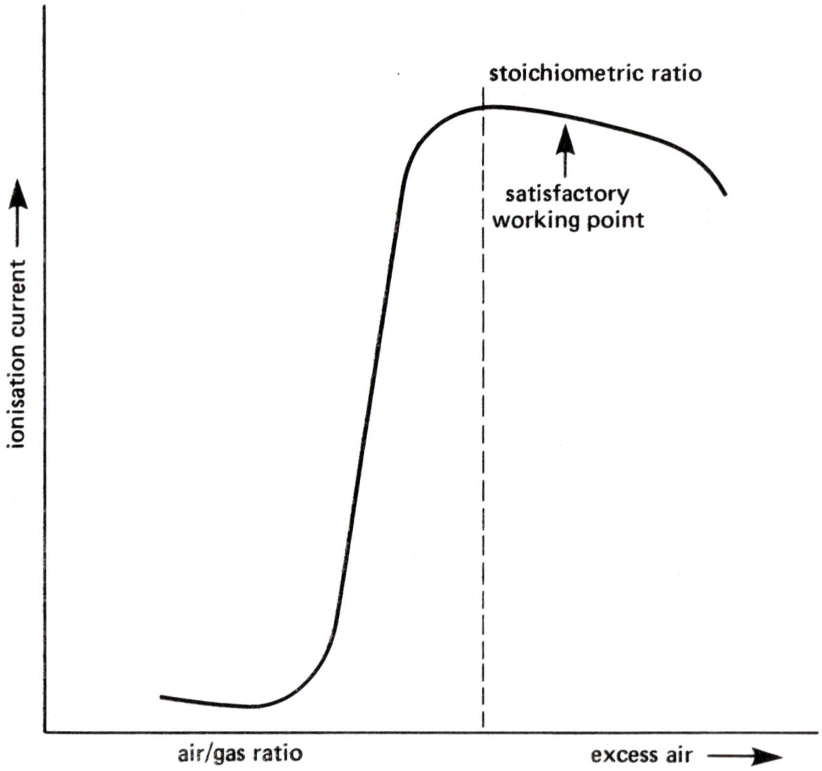

Fig. 10.14 The relationship of rectification current to air/gas ratio

Installation of rectification flame detectors

i) The rectification probe must be well insulated from its supports. The insulation resistance to earth must be at least 100 M ohms. The use of ceramic holders (e.g. Steatite) with long creepage paths is recommended.

ii) Good quality 1.0 mm^2 cables having heat resistant insulation should be used for connecting the flame probe and the burner body (pipework) each to the appropriate terminals of the control box.

iii) The length of connecting leads from the flame probe and burner body to the control should be as short as possible – maximum length 20 metres and kept separate from other cables.

iv) If any part of the burner pipework is used as the return path from the flame to the control, particular attention should be paid to obtaining proper continuity through the pipework.

v) The length of the probe should be as short as is practicable in the circumstances.

vi) The probe must be positioned properly with respect to the flame – towards the edge of the flame and preferably to one side of the burner so that in the event of the probe dropping in use it will fall clear of the burner. Probe positioning may be critical with burners from which the flame front moves or flame size changes with, for example, changing throughputs.

Where the start flame is from a separate pilot burner the probe must be positioned where it will properly detect both the start flame and the main flame.

vii) The probe must be so positioned or protected to avoid 'spark splash' from high voltage ignition sparks. Ignition voltages on the flame detection probe may seriously damage the flame detection amplifier.

viii) The use of a micro-ammeter of suitable range to provide a read-out of ionisation current is recommended. The negative terminal of the meter should be connected to the flame probe and the positive terminal to the probe terminal of the control unit i.e. the meter will be in series with the probe and control unit. Such an arrangement is useful, not only during commissioning of burners using ionisation flame detection, but may well be installed as a permanent feature as an aid to fault finding. Single probe ignition and detection (SPID) systems are used in certain applications such as baking ovens and multiple burner systems. Here one probe is used for spark and flame detection, there is no interval between spark and detection and there is no flame proving period.

Radiation flame detectors

Flames quite clearly radiate energy and it ought, therefore, to be feasible to use the existence of radiation as an indication of the presence of a flame. The radiation occurs in three distinct bands, as heat (infra-red radiation), as visible light and in the ultra-violet part of the spectrum. The use of flames as emitters of visible light is much too susceptible to interference from ambient light to be regarded as a serious method of flame detection and can be summarily dismissed.

Emission in the infra-red and ultra-violet bands can, however, be used successfully and relatively unambiguously to provide a reliable indication of the existence of a flame.

Ultra-violet flame detection

Combustion processes radiate energy quite strongly in the ultra-violet region of the spectrum so that devices which respond to this particular radiation may be used to provide proof of the presence of a flame. There are two types of detector in common use, the Geiger-Muller tube and the symmetrical diode. They are both responsive to radiation in the spectral band 190-270 nm (nm = nanometres) and are thus insensitive to the long wavelength radiation from the hot surfaces of combustion chambers or furnace walls, so that both types will discriminate between radiation from a flame and that from background sources.

Both types of detector consist of a sealed tube of U.V. transmitting glass containing a gas at low pressure and having electrodes within the tube. It is in the type and arrangement of the electrodes in which the two detectors differ.

The Geiger-Muller tube has a central rod anode with a concentric cathode formed from a graphite material deposited onto the inside surface of the envelope, whereas the symmetrical diode has two wire electrodes so shaped and positioned as to form a short parallel section between the wires with a gap of about a millimetre between them. The connections to the electrodes in each type are brought out through the base of the envelope and are sealed through it to maintain its gas tightness. The two types are shown in Figure 10.15.

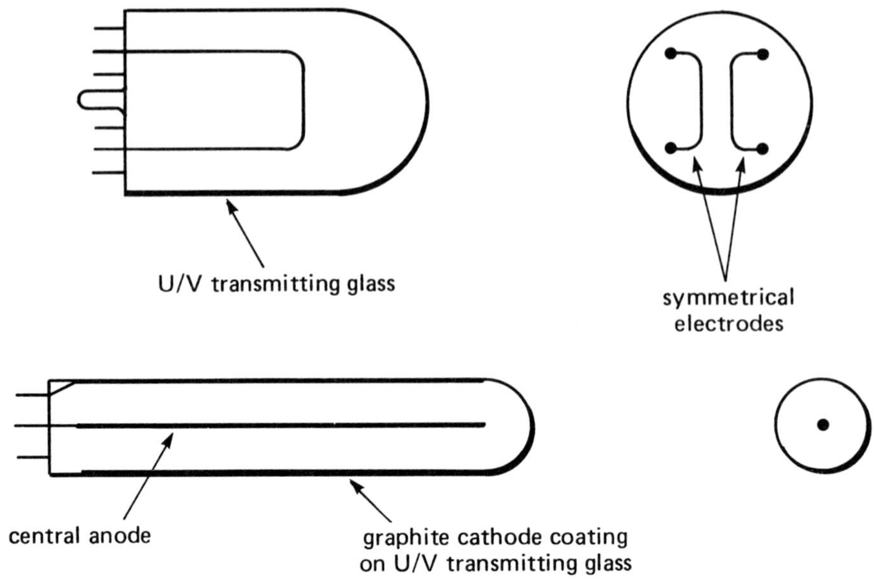

Fig. 10.15 Symmetrical diode and Geiger Muller ultra-violet flame detectors

The symmetrical diode tube is filled at low pressure with a gas which ionises under the influence of ultra-violet radiation so that, with a voltage (usually of the order of normal mains voltage) between the electrodes, the cell will conduct when exposed to radiation from a flame. The parameters of manufacture and operation of the device and of the operation of the associated electronic circuitry are such that ionisation and conduction occur as an avalanche effect only when sufficient ionising radiation falls upon the tube. Also the circuitry is such that conduction when once initiated requires the continued receipt of ionising radiation and cannot become self-sustaining.

The operation and application of the Geiger-Muller tube is similar in many respects to that of the symmetrical diode tube. The principal difference is in the choice of U.V. sensitive material which in the case of the Geiger-Muller tube is the carbon coating of the inside surface of the tube.

The great majority of manufacturers of this kind of flame detection equipment use symmetrical diode devices. It is believed that only one make of flame detection equipment (Honeywell) uses Geiger-Muller tubes.

The output from a U.V. detector will lie in the region of 100 to 250 μamps. In the application of ultra-violet flame detectors certain basic precautions should be observed.

i) The tube should be kept cool (50 °C maximum).

ii) The tube should be kept clean — free from dust deposition or vapour condensation.

iii) Tubes should be limited to a service life of 10 000 hours provided the temperature limitation is met. Higher ambient temperature will shorten tube life.

iv) Care must be exercised in the sighting of tubes to ensure that each tube may only receive radiation from the flame which it is intended to monitor. A U.V. detector cannot differentiate between radiation from one flame or from any other and due attention must be paid to the geometry of the system — especially in multi-burner applications.

v) If a separate pilot burner is used for ignition of the main flame care must be taken to ensure that the line of sight of the U.V. tube interrupts the axis of the main flame at a point where the pilot burner will always produce reliable ignition of the main flame — see Figure 10.16.

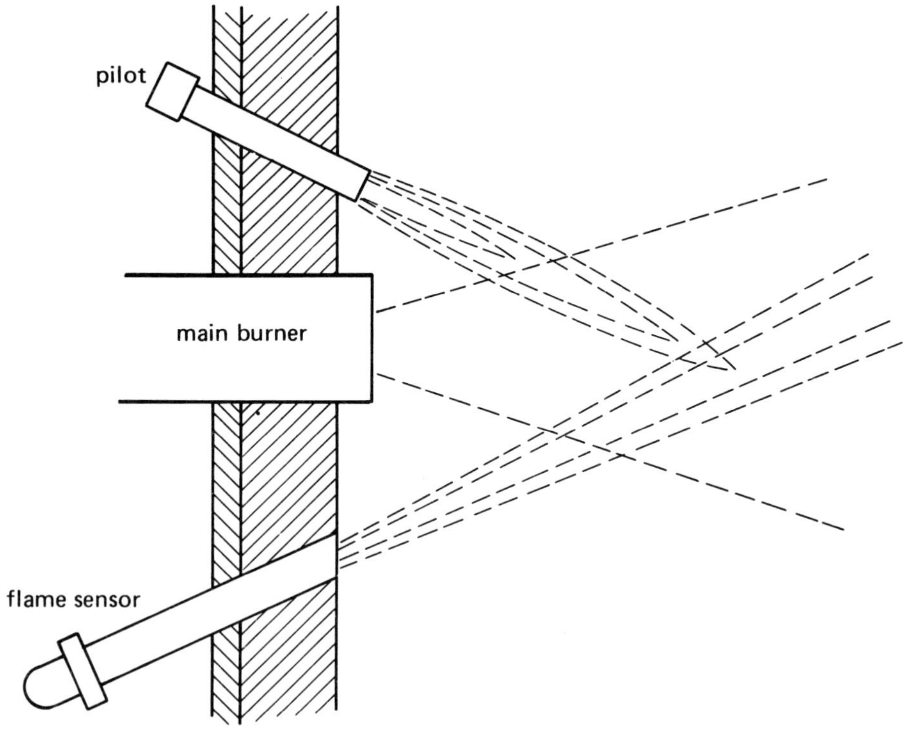

Fig. 10.16 Sighting of ultra-violet flame detectors

vi) Sighting at an angle as near to 90° to the axis of the main flame as circumstances permit is always desirable and sighting should be satisfactory for all the conditions of draught, throughput and air/gas ratio changes that can be reasonably expected to occur.

vii) Ignition sparks are a rich source of ultra-violet radiation and depending upon the associated circuitry and method of operation of the burner control U.V. detectors may require shielding from ignition sparks.

viii) U.V. detector tubes may fail into a condition in which they conduct immediately a voltage is placed across the electrodes, irrespective of the presence of a flame. This condition — the cell is said to be 'soft' — is clearly hazardous and is checked for on each occasion that the burner system is taken through a start-up procedure. To rely on this safe-start check on the cell integrity is satisfactory if the burner operates under on/off control. It is not satisfactory if the burner operates for long periods without experiencing a start-up cycle. Burners operating under the latter conditions should preferably be equipped with self-checking U.V. flame detectors but must at least be shut down manually, periodically, and restarted. Self-checking detectors have a shutter which regularly shields the cell from incident U.V. radiation. The electronics circuitry is designed to recognise that the signal from the detector disappears when the shutter is in position and re-appears when the shutter is withdrawn.

ix) Ultra-violet flame detectors, like other kinds of electronic flame detectors, must be used along with the appropriate burner controllers. The detectors intended for use with controllers of one particular model within a manufacturer's range, are not interchangeable with controllers of different manufacture nor sometimes within the range of models of the same manufacturer. There may, for example, be significant differences in the required energising voltages.

x) Connecting leads should be as short as possible, maximum 20 m, and should preferably be kept separate from other cables.

Infra-red flame detectors

These detectors rely upon materials such as lead sulphide which change their electrical resistance when subjected to infra-red radiation. The magnitude of the change is proportional to the intensity of the radiation to which they are subjected. Thus, if an electrical voltage is applied across a small plate coated with one of these photo-resistive materials a current will flow and the magnitude of the current will depend upon whether the plate is subjected to infra-red radiation of sufficient intensity (as would result from a flame) or not, as the case may be.

The radiation to which these detectors are sensitive lies in the band 1 to 3 μm.

The main objection to the use of infra-red detectors is that other hot surfaces, such as the walls of a combustion chamber, radiate energy in exactly the same waveband as do flames. To render these systems safe, therefore, i.e. to enable them to discriminate between radiation from a flame and from background sources, it is usual to 'tune' the associated amplifier circuit to respond to the flame flicker. The radiation emanating from a flame changes in intensity constantly with the flame flicker, whereas background radiation tends to be of much more constant intensity. The flicker frequency to which these detectors are made to respond usually falls in the band 15 Hz to 40 Hz.

As in the other forms of electronic flame detector, the signal generated by the detector is amplified and used to energise an electromagnetic relay which in turn energises other relays which switch power to the burner isolating valves.

Mainly because of the difficulties associated with obtaining infallible discrimination between signals from a flame and from background sources, the great majority of electronic flame detection used with gas burners are of either the flame rectification or the ultra-violet types. There are, however, two models of infra-red detector from one manufacturer — Electronics Corporation of America — which are certificated for use with gas burners.

The installation requirements for infra-red detectors are broadly similar to those for ultra-violet sensors

10.6 MANUAL, SEMI-AUTOMATIC AND FULLY AUTOMATIC BURNER LIGHT UP, MONITORING AND SHUT DOWN

10.6.1 Manual Control

If the operator could be relied upon to fully perform and monitor the burner light up, run and shut down sequence each and every time and shut down the plant immediately on any failure, this might be the best method. However, operators are not infallible and could never be relied on to perform any of these operations with any certainty of safety. Hence the development of semi-automatic and fully automatic, operator independent control systems. The comparison of all three systems is shown diagrammatically in Figure 10.17.

10.6.2 Semi-Automatic Control

This incorporates some of the features of manual control and fully automatic control described below. Usually a semi-automatic control unit will provide spark ignition and flame detection, but

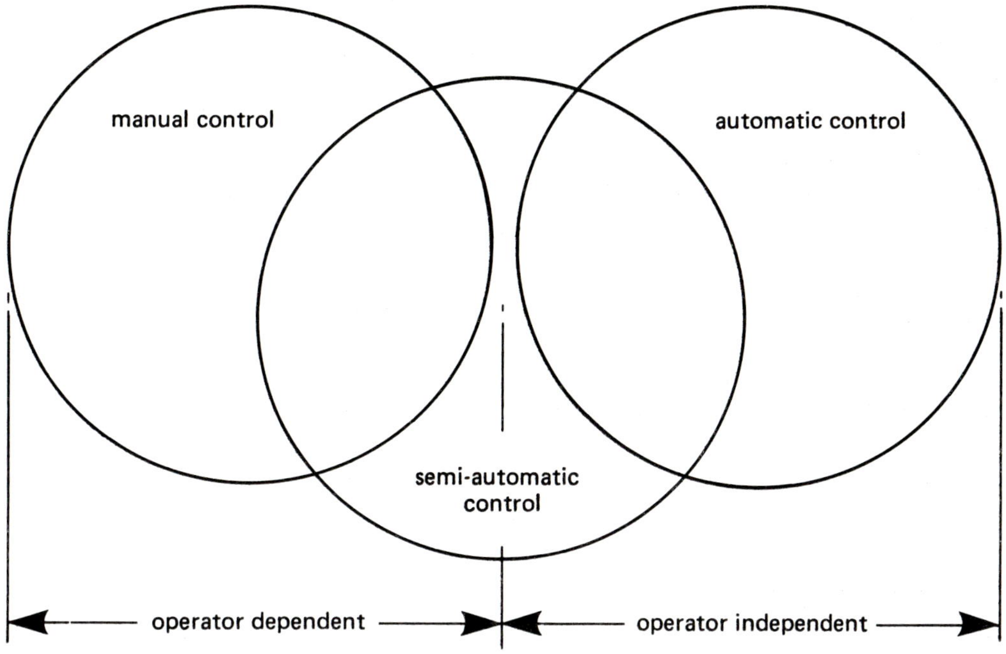

	Manual	Semi Automatic	Automatic
Pilot Ignition	–	Unlimited (or preset)	Maximum 5 s
Pilot Proving	–	None	Minimum of 5 s
Main Flame Establishment	Unlimited	Unlimited (or preset)	Maximum 5 s
Flame Supervision	Varied	Continuous on pilot and main flame	Continuous on pilot and main flame

Fig. 10.17 Comparison between manual, semi-automatic and automatic control

will always require some manual action to bring on the main burner from the shut down condition, even though this may involve releasing the spring loaded start button after the pilot has ignited.

A typical semi-automatic control sequence is:

Pilot or start gas ignition	Operator presses start button. Pilot gas and spark on provided no flame sensed by flame detector.
Pilot proving/main flame ignition	Operator releases start button and provided pilot flame has been detected the main flame will be lit from the pilot. Both flames will remain alight.

This is shown diagrammatically in Figure 10.18.

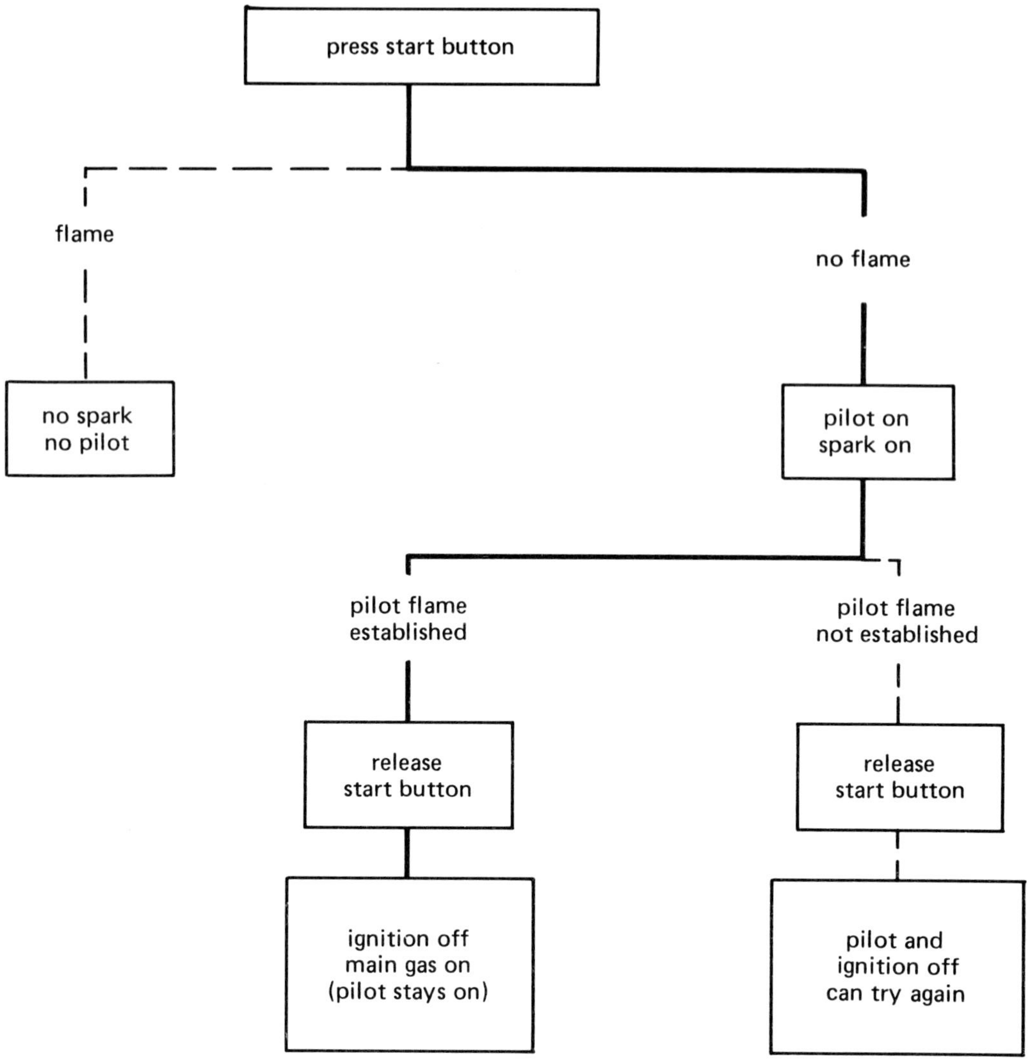

Fig 10.18 Typical light-up sequence for semi-automatic control unit

It must be noted with this system that if the flame is not sensed in a few seconds, the operator should release the start button to prevent the combustion chamber filling up with gas.

A typical wiring diagram utilising a semi-automatic control box is shown in Figure 10.19, where it can be seen there is no light up sequence timer.

10.6.3 Fully Automatic Control

A burner with automatic control is one which will establish a main flame from the shut down condition (i.e. all gas to the burner shut-off) without any manual intervention. On shut down, for example due to a thermostat opening, the programmer returns to the start of the sequence and awaits the contacts closing.

There are two categories of control unit for this, namely certificated programming control boxes for automatic burners above 60 kW complying with BS 5885, Part 1, 1988, and non-certificated control boxes generally used for lower rated automatic burners between 7.5 kW and 60 kW complying with the BS 5885, Part 2, 1987.

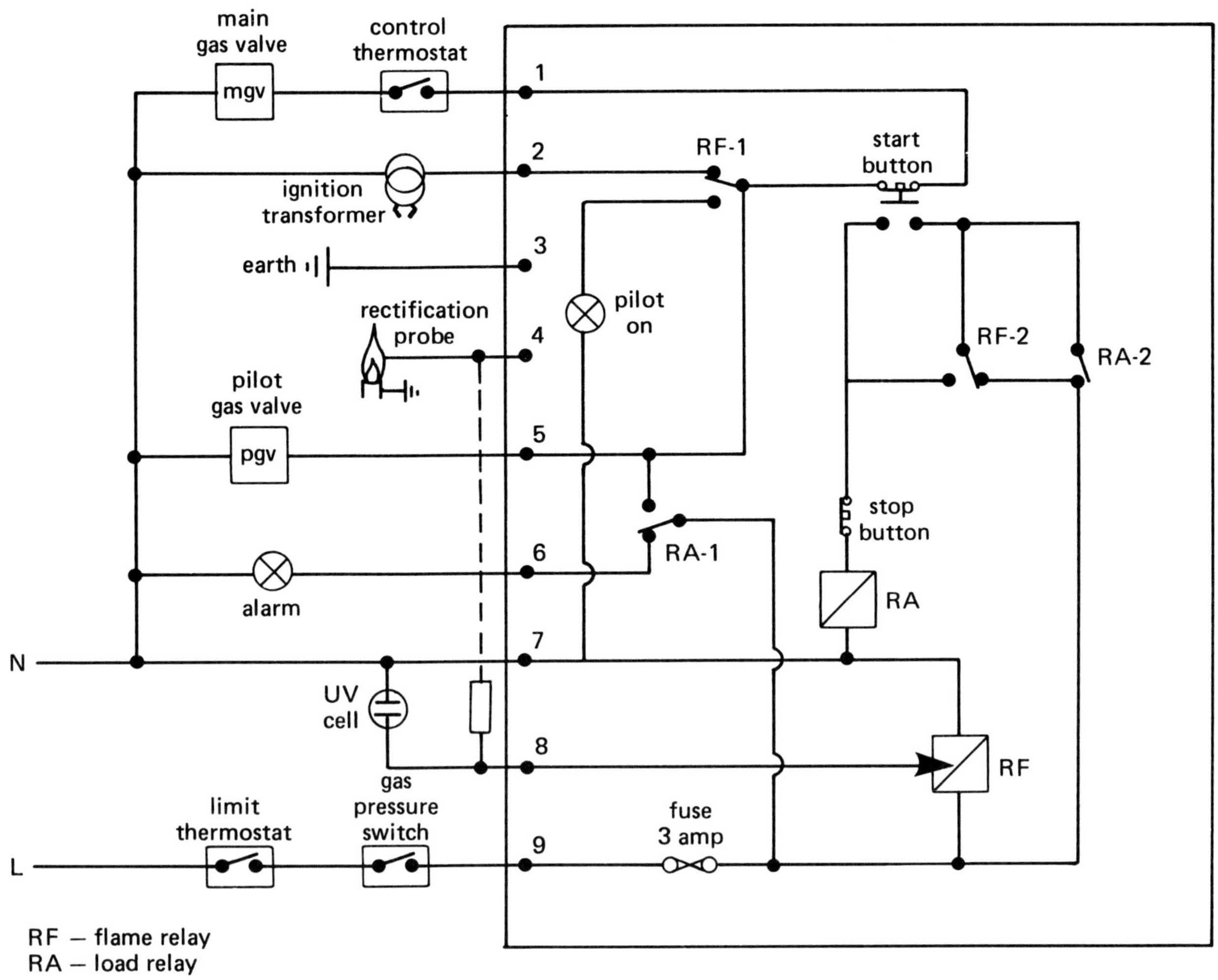

Fig. 10.19 Teknitronic BA1 semi-auto control

We will consider the second category first. These control boxes will be generally cheaper than the certificated units, but nevertheless will cover the important requirements for automatic control and light up. This includes:-

Pre-purge of proved air (by pressure switch) of 10 seconds or to give 5 volume changes of the combustion chamber. Safe start check on flame detector during pre-purge period. Flame detectors to de-energise valves on flame failure in no longer than 1.5 sec. Pilot (start) gas ignition period no longer than 10 secs (pilot gas and spark). Pilot (start) gas proving period of at least 5 seconds (pilot gas no spark). Main flame establishment period no longer than 5 secs (main and pilot burners both alight). Post purge is optional.

A typical control box of this type is the Pactrol CSS Control as fitted to the Ideal Concorde modular boiler each module having an input of 57.7 kW. The control circuit for this boiler is shown in Fig. 10.20.

Fully certificated Programming Control Units to comply with BS 5885, Part 1, 1988, for burners over 60 kW input have had a life test of 100 000 operations carried out by British Gas Midlands Research Station. The system uses either an interrupted pilot system, in which a pilot is used to ignite the main flame and following establishment of the main flame the pilot is extinguished; or an expanding flame system, in which a start gas flame is established as an integral part of the main

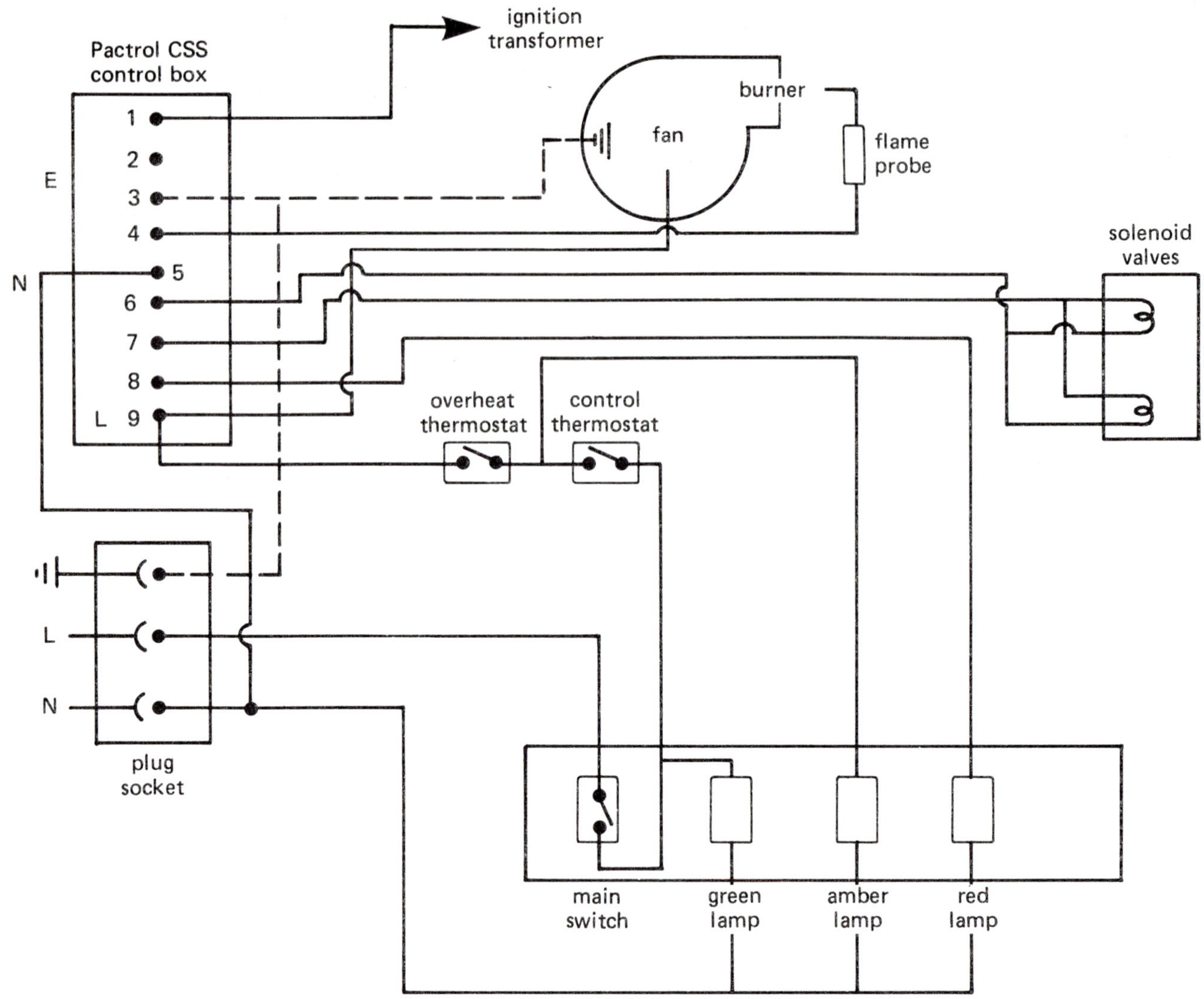

Fig. 10.20 Ideal Concorde wiring diagram

flame at a reduced rate. These systems are shown diagrammatically in Figure 10.21. A control unit designed solely for expanding flame burners should never be used with a separate pilot system, and most certificated units are for use with the interrupted pilot system unless otherwise stated.

The control modes of these units can be either on/off or modulating. The on/off system requires that after the burner shuts down under the control of the thermostat, the burner has to go through the full light up sequence on start up including a full pre-purge. The air flow rate will be set by the commissioning engineer and will remain at that setting for the whole of the sequence. A burner operating in the high/low mode will utilise one of the modulating control boxes. Here, when the thermostat indicates that the process temperature has been satisfied, it will send the second safety shut off valve to the low fire position rather than shutting the system down. The light-up sequence for an on/off burner system to comply with BS 5885, Part 1, 1988, is shown diagrammatically on a bar chart in Figure 10.22

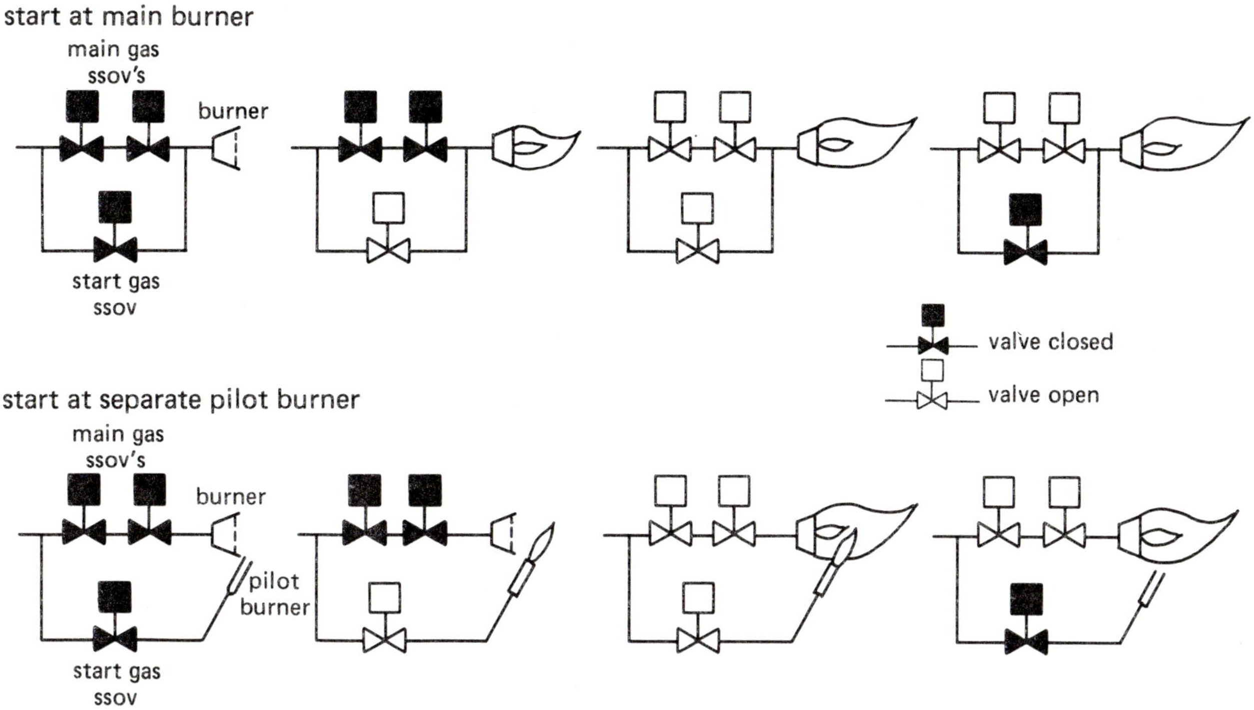

Fig. 10.21 Start flame gas establishment

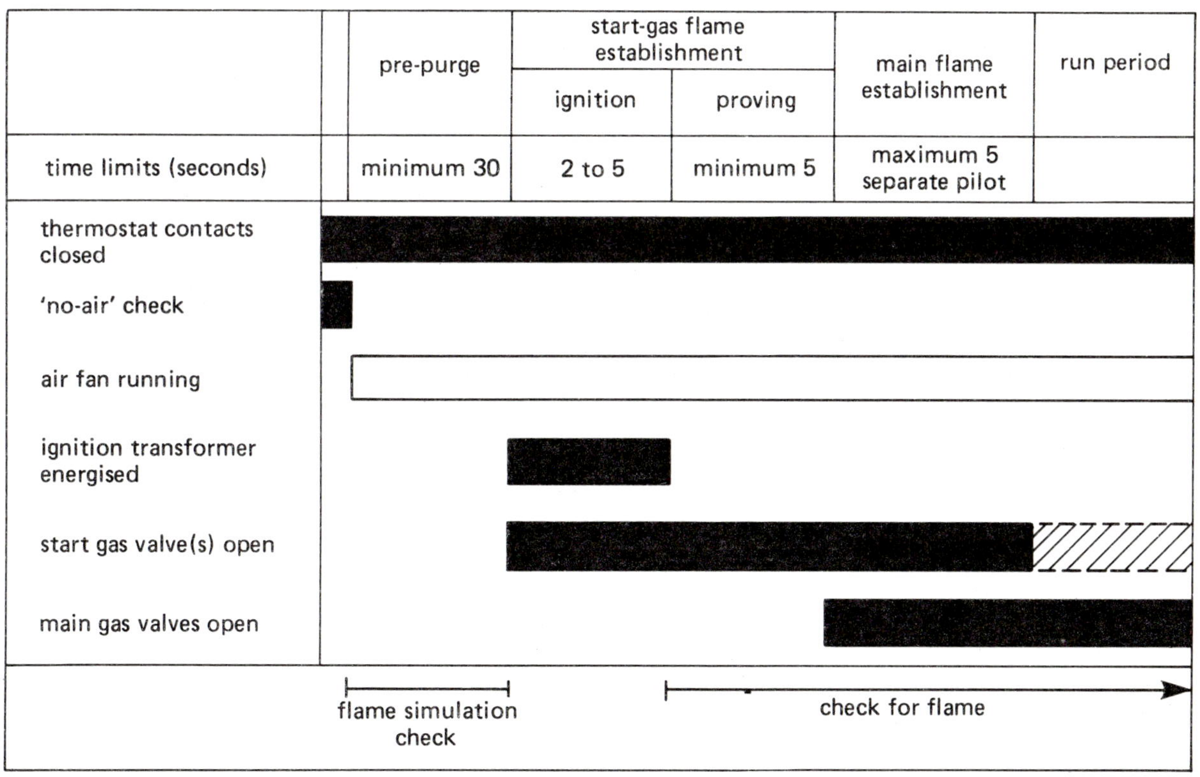

Fig. 10.22 The start up sequence of a fully automatic on/off burner

Light-Up Sequence — On Off Control

'No air' pressure check	To ensure air pressure switch is in the 'no air' position and has not welded together in the 'run' position.
Pre-purge period (air checked by air pressure switch, flow or centrifugal switch)	Fan runs alone for a 'proved air' period for at least 30 seconds (or 5 volume air changes).
Safe start check	In this period there is a 'safe start' check to ensure that there is no flame present or flame simulating condition such as the UV cell has not failed to danger or gone 'soft'.
Start gas or pilot flame ignition period	Pilot and spark are energised together for a period of between 2 and 5 seconds. (Some control boxes might initiate spark before the pilot).
Pilot or start gas flame proving period (2-5 seconds)	Here the spark has been interrupted and the pilot flame must be sensed for the whole of this period to prove that it is stable. No flame or loss of flame at any time will cause lockout.
Main flame ignition period (2-5 seconds)	Here the pilot and the main flame are on together. (Flame must be sensed by flame detector).
Main flame 'run' period	Main flame alone continuously checked by flame detector (interrupted pilot). The pilot only stays on with an intermittent pilot usually fitted to an expanding flame burner.
Post purge period	Fan runs alone to purge out combustion chamber after lockout or shut down of burner. This period is optional.

The Modulating Control System allows the gas and air flows to match the heat demand on the process by modulating the air and gas throughput valves between high and low fire. This type of programming control box will contain all the interlocks required for on-off control with the addition of the following:-

i) The air damper must be proved in the high fire position for the whole of the proved pre-purge period.

ii) The air and gas dampers must be proved in the low fire position before the ignition sequence can commence.

iii The system will be released to full modulation control after the burner has reached the run position.

The light up sequence for this system is shown diagrammatically on a bar chart in Figure 10.23. This is used for the larger type of automatic burner, with an input of over 3 MW to comply with BS 5885, Part 1, 1988.

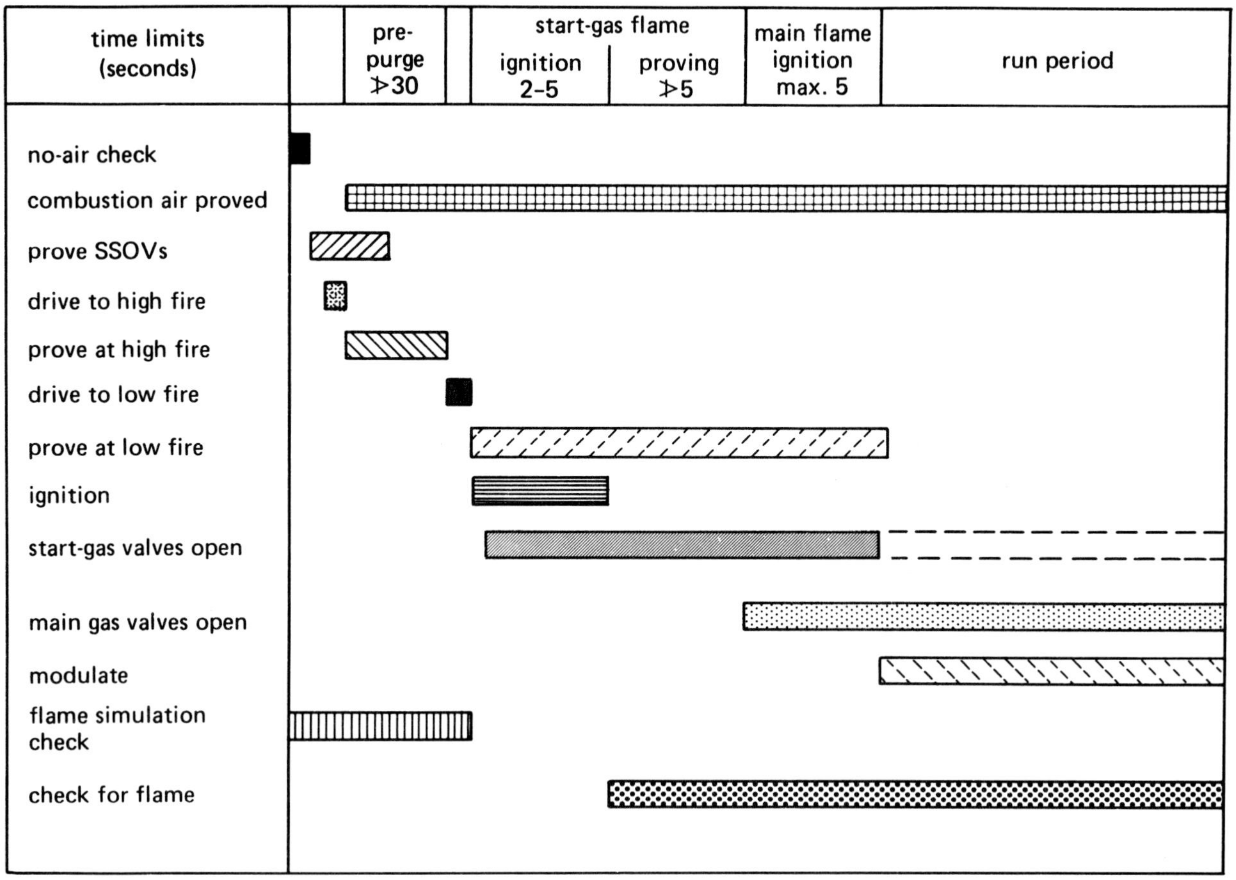

Fig. 10.23 Ignition sequence – modulating burner > 3 MW

It will be noted that this is the same as for the on-off control with the following additions:-

i) Generally, as modulating systems are fitted to larger burners there will be a valve proving check before the sequence can commence.

ii) The air damper must drive to, and be proved in the high fire position for the whole of the pre- purge period.

iii) The air and gas dampers must drive to and be proved in the low fire position before the ignition sequence can commence. The timing sequence motor on the control box will physically stop until this condition is met.

iv) The burner system will be released to full modulation control after the run position is reached.

Figure 10.24 shows the general electrical circuit of the Landis & Gyr LFL modulating programming control unit.

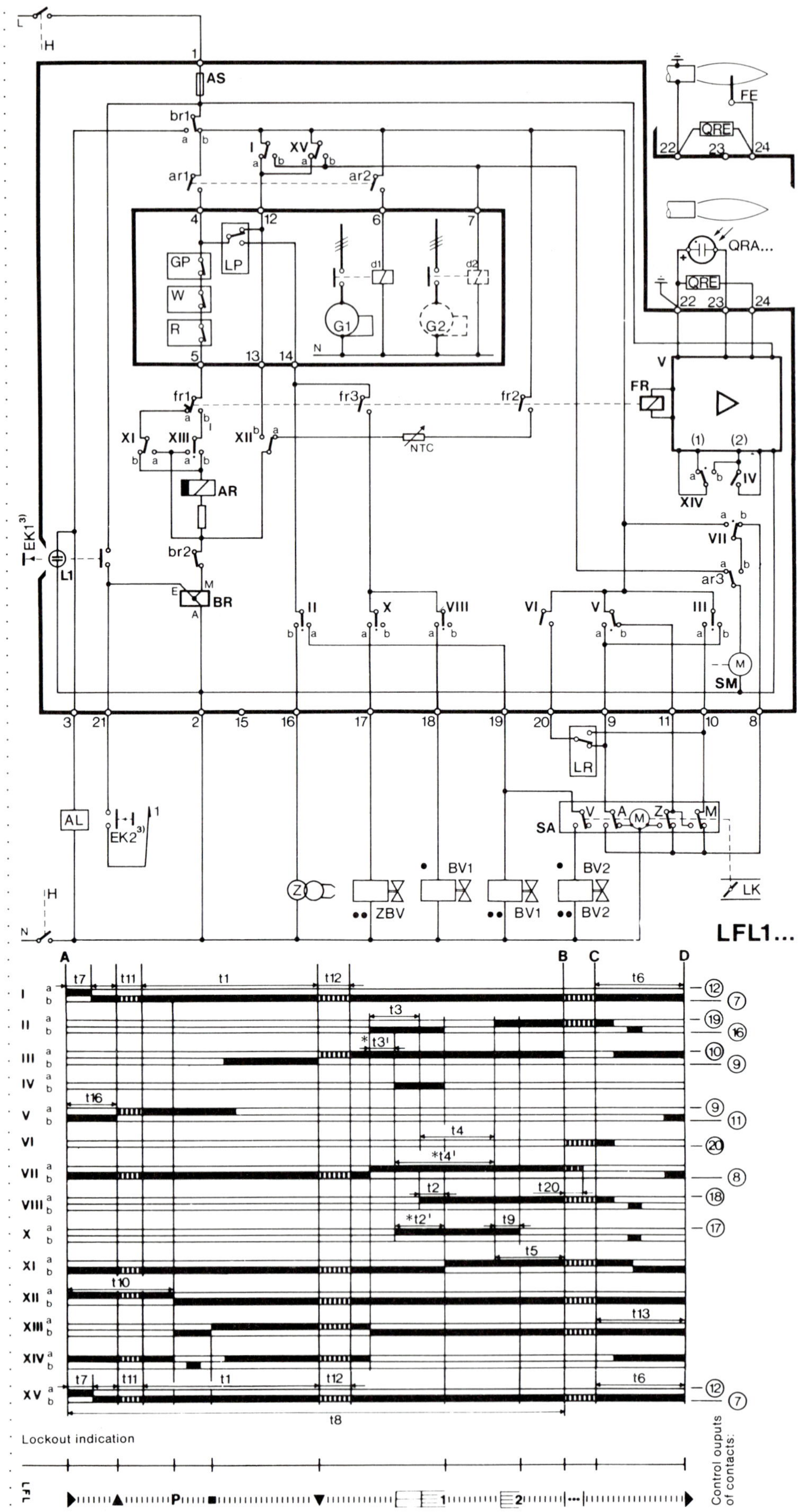

LEGEND

- A Limit switch for OPEN position of air damper
- AL Remote lockout warning device (alarm)
- AR Main relay (load relay) with contacts 'ar'
- AS Unit fuse
- BR Lockout relay with contacts 'bs'
- BV Fuel valve
- d Contactor or relay
- F Flame supervision circuit
- EK Lockout reset button
- FE Ionization electrode
- FR Flame relay with contacts 'fr'
- G... Fan or burner motor
- bv... Limit contacts of gas valves for the CLOSED position test
- GP Gas pressure monitor
- H Mains isolator
- L Lockout warning lamp
- L3 'Burner ready' indication
- LK Air damper
- LP Air pressure monitor
- LR Load controller
- M Auxiliary switch for MIN-position of air damper
- P Control section of control unit
- QRA UV-detector
- QRE Ignition spark detector
- R Control thermostat or pressurestat
- RV Continous fuel valve
- S Fuse
- SA Air damper motor
- SM Sequence switch motor
- V Flame signal amplifier
- V In the air damper motor: auxiliary switch for the release of fuel according to air damper position
- W Limit thermostat or pressurestat
- Z Ignition transformer
- Z In the air damper motor: limit switch for the CLOSED position
- ZBV Ignition fuel valve (Interrupted pilot)

- • Expanding flame burners
- •• Interrupted pilot burners

(1) Input for the increase of the operating voltage necessary for the UV-detector test

(2) Input for the forced energizing or the flame relay during the function test of the flame supervision circuit (contact XIV) and during the safety time t2 (contact IV)

(3) **Press EK for no more than 10 seconds**

- •BV1 (Intermittent pilot)
- ••BV1 (Main valves)
- BV2 (Modulation control valve)

Control programme of the sequence switch

- t1 Pre-purge time
- t2 Safety time
- *t2' 1st safety time
- t3 Pre-ignition time
- *t3' Pre-ignition time
- t4 Interval between tension on terminals 18 and 19
- *t4' Interval between tension on terminals 17 and 19
- t5 Interval between tension on terminals 19 and 20
- t6 Post-purge time
- t7 Interval from start up to tension on terminal 7]
- t8 Duration of the start-up sequence
- *t9 2nd safety time
- t10 Interval up to start of air pressure control
- t11 Running time for the air damper motor to OPEN position
- t12 Running time for the air damper motor to MIN. position
- t13 Permissible after-burn time
- t16 Interval up to open command for air damper
- t20 Interval up to the self-shutdown of the sequence switch

- A Start command
- B Operating position
- C Controlled shut-down
- D End of control programme

* These times are valid for the control and supervision of interrupted pilot burner controls of series 01.

Fig. 10.24 Landis & Gyr LFL modulating programming control unit

10.6.4 Developments in Programming Control Units

With the advance in technology and recent constraints upon manufacturers (cost, alteration of standards, need for flexibility) this has resulted in changes in manufacturing style which subsequently requires a change in technique for fault finding, etc.

Timing sequences for example are now often achieved by using electronic devices rather than electromechanical synchronous motors. These are claimed to be not only more accurate, but more reliable and can be found in some recently certificated boxes.

One recent trend in manufacture is that of designing units to meet the specifications of different countries with only a minimum of modification. Examples of this can be seen in the Landis & Gyr LFM and Honeywell R4140K which both provide ignition spark detection capabilities through a removable or breakable internal link.

10.6.5 Micro processor-based Burner Controls

These boxes designed around one or two dedicated microchips not only perform the automatic start up, supervision and shut down of burners but also offer the user of fired plant many advanced features such as fault diagnostics and flexibility and communication with supervising computers. Because these control units are relatively new and give indication of burner faults by annunciator display windows, flashing lights or other methods, these boxes are at present much more expensive than their electromechanical counterparts. As more of these are utilised, as with personal microcomputers the cost is likely to reduce.

Control boxes currently available

A Honeywell R7241

The functions provided by this unit include automatic burner sequencing, flame supervision, status indication, first out annunciation and self diagnostics.

The first-out annunciation reports the cause of a safety shutdown by a system of flashing lights (with reference to fault code). All field input circuits are monitored including the flame signal amplifier and firing rate position switches. The system distinguishes seven modes of flame failure and detects and annunciates difficult to find intermittent faults.

The self diagnostics distinguishes between field and internal problems. Faults associated with either the flame detection system or the chassis are isolated and coded accordingly.

B Fireye E200

The unit provides the usual burner sequencing and flame monitoring on automatic burners and has a message scroll centre to provide the operator with the burner status and failure mode information. This message centre has a vocabulary of twenty nine different messages which give a readout of burner history, number of cycles and hours of burner operation, and a non-volatile memory allows the control to store this information when the power is off. On safety shutdown or lockout the message centre will advise the operator of the cause of the fault. Self checking for ultra-violet flame detection is built into the unit. An additional expansion module adds the capability of individually displaying any malfunction for an additional sixteen interlock switches.

C Landis & Gyr LGR99

This unit can be adapted to suit local codes, standards and regulations and other plant specific requirements. It can be used to change over a dual fuel burner operation directly, the burner programme can be extended to automatic gas valve proving by plugging in a single programme plug. Air damper positions need no longer be determined by auxiliary switches fitted to the damper actuator, they are reached very accurately with the help of electronic slave control.

During start up and normal burner operation, all important operational states and parameters are indicated with the help of LEDS and digital displays. In the event of faults or lock-out, a 3 digit code provides information not only on the probable cause of the fault but also gives the programme phase during which the fault occurred.

REFERENCES

BRITISH STANDARDS

BS 5885 1988 Part 1 :	Specification for Automatic Gas Burners of Input Rating 60 kW and Above
BS 5885 1989 Part 2 :	Specification for Automatic Packaged Burners with Input Rating 7.5 kW up to but excluding 60 kW.
BS 5991 1989	Specification for Indirect Fired Forced Convection Heaters

OTHER PUBLICATIONS

Regulations for Electrical Installations (IEE Wiring Regulations) 15th Edition, 1981, published by the Institution of Electrical Engineers.

Process Control 11

CHAPTER 11

PROCESS CONTROL

11.1 INTRODUCTION

Process control is the method by which a particular process variable, for example temperature, is maintained at a level determined by such factors as product quality and production rate. It is an area which has seen enormous change in the last forty years with mechanical and pneumatic controllers being challenged first by electronic and then by microprocessor control. The capital cost of solid state controllers relative to other items of plant has fallen so much that they can be cost effective in areas which would have been unthinkable a few years ago. Manual control, with its attendant problems of inconsistency and poor product quality becomes increasingly uncommon and results improve as control is further removed from the operator.

In attempting a review of a very broad field we have necessarily had to be both selective and superficial. The range of process variables with which the control engineer may become involved is extensive. Temperature, pressure, flow, humidity, fluid level, acidity (pH) are some of these variables, and each area has sub-divisions. Flow control involves, amongst others, orifice plates, venturis, turbines and Vortex meters. We have chosen to overlook these and to concentrate on a reasonably comprehensive treatment of temperature measurement on the basis that the gas engineer will be concerned with this most of the time. The other areas are covered in textbooks and the instrument manufacturers are usual extremely helpful, many of whom have introductory guides available.

11.2 PROCESS CONTROL THEORY

11.2.1 Closed Loop Control

The desire to control important process variables is as old as civilised man. In the days of the Pharoahs the water level distribution in a multiple ditch irrigation system would be crudely controlled by a man operating paddles at various positions in the system. A paddle is a wooden board raised or lowered in the ditch to provide control of water flow.

Fig. 11.1 Manual control

If we examine the form of control for an individual paddle we can construct an operation loop (Figure 11.2)

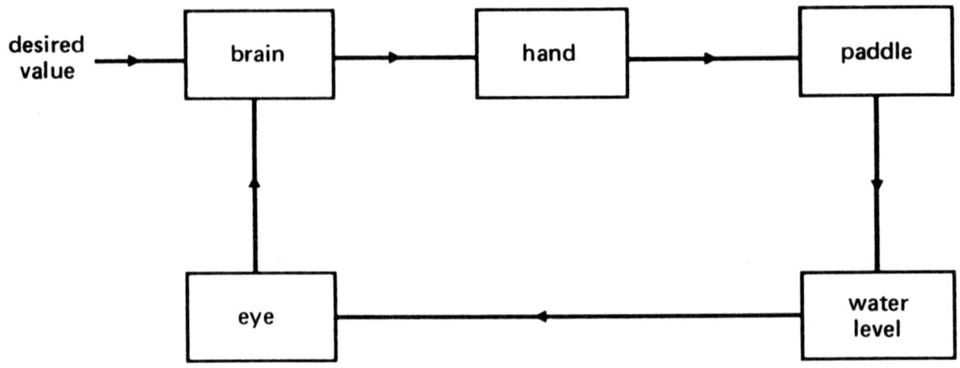

Fig 11.2 Manual control

The operator (man) has been told what water level is required in each ditch (the desired value of the variable). He can manually adjust the appropriate paddle thus changing the water level (the controlled variable) until he sees that he has achieved the required level. This sequence of operations completes a 'loop' and the process variable is under continuous surveillance as long as the operator concentrates and does his job. The system is called CLOSED LOOP CONTROL.

Here the control loop is closed manually by the operator. Alternatively the operator can be replaced by an automatic controller as in Figure 11.3. The controller will compare the controlled variable with the desired value and generate an error signal of the form:

$$e = T_1 - T_2$$

where e = error signal
T_1 = control variable
T_2 = desired value

The controller will then calculate a new value position based on the size of this error signal.

where V = flow or signal magnitude,
$f(e)$ = a function of e

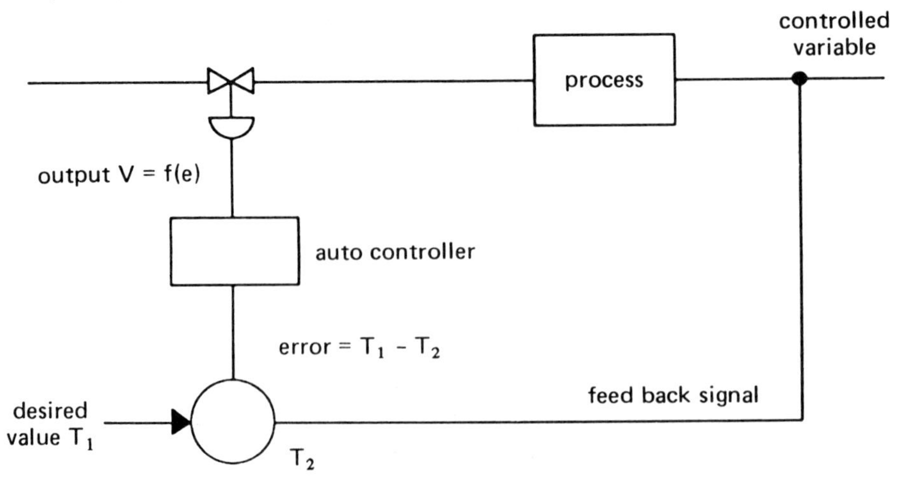

Fig. 11.3 Automatic closed loop control

11.2.2 Control Modes

The control mode describes the way in which the controller output is altered in response to an error signal.

A number of different control modes can be employed in practice. Selection of the most appropriate control mode will depend on various process characteristics and, of course, cost factors.

a) On/off control

In figure 11.1 the operator will initially be able to control the water level simply by fully opening and closing the paddle as the water level varies. This crude form of control is termed 'ON/OFF'. The water level will cycle as in Figure 11.4.

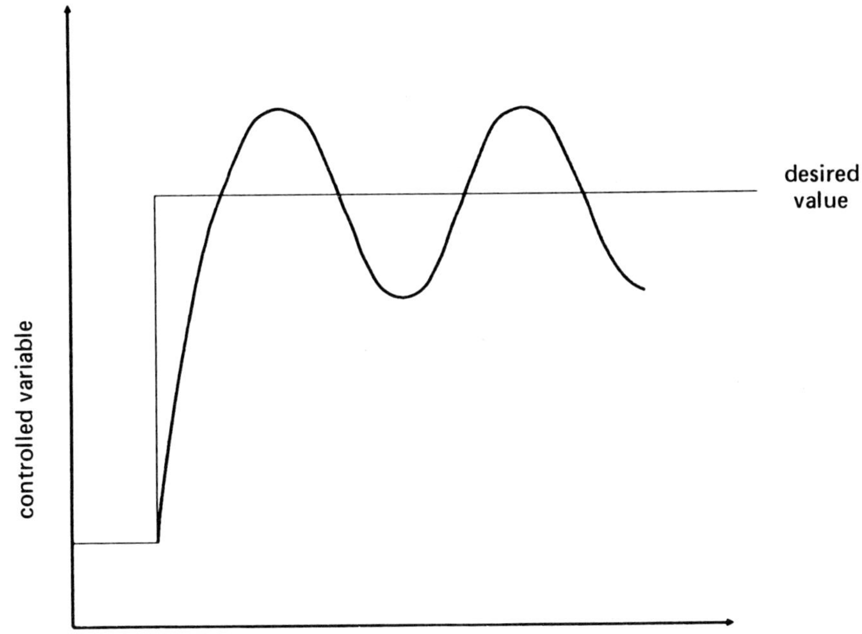

Fig. 11.4 Water level cycle

A domestic thermostat is an example of this system. To prevent excessive cycling such a controller could well have an adjustable dead zone within which the controller will not alter its output.

High/low control is a variant of this method of control. The controlled variable would now exhibit a 'sawtooth' form – Figure 11.5

b) Proportional control

This is a more sophisticated form of control where the control valve is modulated in accordance with the size of the error. Returning to Egypt, the operator soon gains sufficient experience to open or close the paddle to suit water level conditions. He will then adjust the paddle to a position where the flow of water is proportional to the variation in water level. If the level is low the paddle will be partially open, if the level is high the paddle will be partially closed.

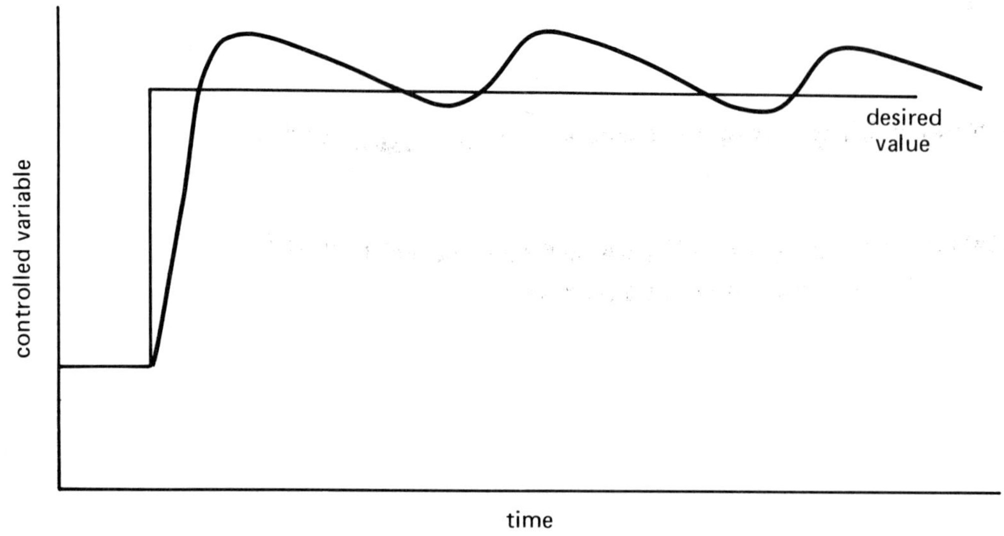

Fig. 11.5 High/low control

This form of control is termed 'PROPORTIONAL' because the controller output is proportional to the size of the error signal. An automatic controller would calculate the valve position according to the equation:-

$$V = K_p.e \qquad 11.1$$

where V = controller output
K_p = proportional gain
e = error signal

The sensitivity of the controller can be altered by adjusting the proportional gain K_p.

Alternatively the controller might have an adjustable proportional band. This produces the same effect as the proportional gain, except that it operates the other way round, i.e. increasing the proportional band will DECREASE the sensitivity.

$$\text{i.e. } K_p = \frac{1}{\text{prop. band}}$$

It is obviously important to establish which system the instrument is using before making any adjustments.

As the operator becomes more experienced in the response of the water level to paddle position, his quality of control will improve. He may, for example, notice that the water level is consistently low, even though he is placing the paddle in a position which would normally be correct. This offset (Figure 11.6) could be brought about by a greater water draw-off than normal. Offset is a characteristic of all proportional action controllers, and another action is required to remove it.

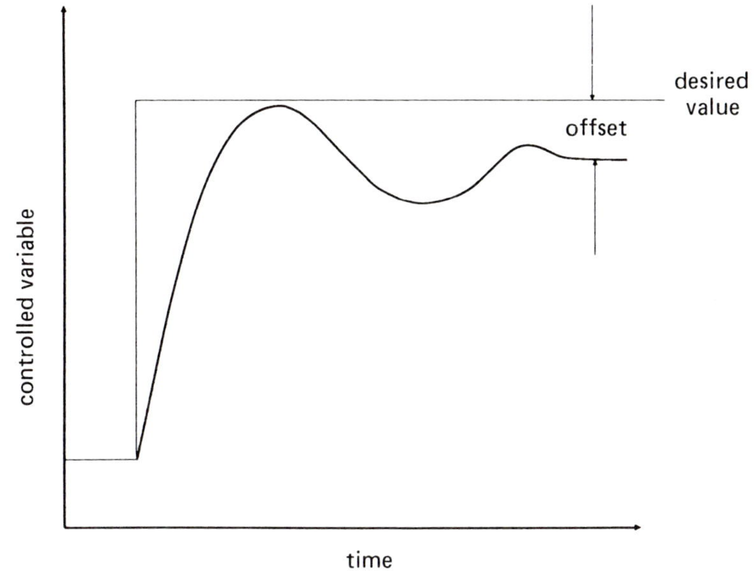

Fig. 11.6 Proportional control

c) Integral control

After a while the operator will notice the persistent error and make a small paddle correction to remove it.

The INTEGRAL control on an automatic controller performs a similar function. The control action is described by the equation:-

$$V = K_I \, e.dt \qquad 11.2$$

where V = controller output
K_I = integral constant
e = error
t = time

As long as there is an error present then the controller will continue to move the valve in a direction which will remove this error. The rate of movement is proportional to K_I, and a high value of K_I is necessary in order that the offset is removed as quickly as possible. Unfortunately increasing K_I also makes the system more unstable.

The integral adjustment on a controller can be in one of two forms:

- Repeats/min – this is equivalent to K_I as described above.
- Integral time – this is equivalent to $1/K_I$ and functions in the opposite direction to K_I. It must be DECREASED to remove the offset more rapidly.

Figure 11.7 shows the response of a P + I system to a load change.

d) Derivative control

Finally, the operator may become so well versed in the reactions of his system that he will learn to 'anticipate' the change in water level and apply a greater paddle movement than would otherwise be the case. For example, if the level is seen to rise rapidly, he will close the paddle further than

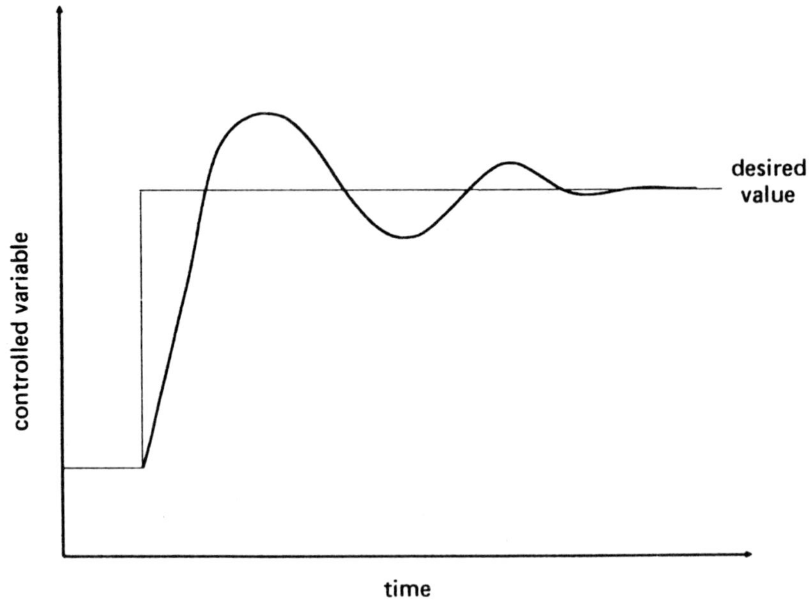

Fig. 11.7 Proportional + integral control

normal because he can see that the level is increasing rapidly and will 'overshoot' the normal paddle position. The operator has thus learned to recognise the 'rate of change' of and not just the size of, the final deviation. This recognition and reaction to the rate of change is termed 'DERIVATIVE' control and is expressed as:-

$$V = -K_d \frac{de}{dt} \qquad 11.3$$

where V = controller output
K_d = derivative constant
e = error
t = time

Any change in the error signal will produce a valve output which will OPPOSE this change i.e. if a furnace is heating up the derivative action will tend to close the valve. Its effect will be to reduce overshoot and to increase system stability

Too much derivative action can be counter productive in that it can produce a 'noisy' signal and increase recovery time.

11.3 AUTOMATIC CONTROLLERS

Industrial controllers occur in a variety of forms. Although pneumatic control is still used, particularly where safety considerations are paramount, the vast majority of controllers will be solid state, usually employing microprocessor technology. Most controllers will have a variety of features largely dependant on the cost. It is suggested that the controller's suitability for an application can be assessed in terms of the following features.

1. Input functions
2. Mode and accuracy of control
3. Output functions
4. Additional features

11.3.1 Controller Features

a) Input functions

This will depend on the application. A temperature controller could be interfaced with a thermocouple, resistance bulb or a radiation pyrometer. Sometimes the controller will be able to accept a variety of inputs. More sophisticated controllers will tend to accept a standard 4-20 mA input. This will require a device which will convert the measuring element output to a form that the controller can accept. This converter will linearise signals from for example, orifice plates or radiation pyrometers.

b) Mode of control

Most controllers nowadays offer Proportional, Integral, Derivative (PID) control as a minimum. If necessary the controller can have a programmed set point so that the process can be made to follow a pre-determined heating cycle.

If the controller is part of a larger system it may need to be able to communicate with a central computer. Again for larger systems controllers are now available which will supervise up to twelve separate control loops.

c) Output functions

These depend upon the nature of the final control element, but generally there are four main types of output.

- *Relay,* to drive contactors or solenoid valves.
- *Solid State Relay (SSR) Drive,* a logic signal used to drive thyristors and solid state relays where fast switching of large currents is necessary.
- *Analogue/Linear,* D.C. volts or mA to drive thyristors or actuators.
- *Valve Motor Drive,* raise and lower signals to switch the A.C. supply to valve actuators. The actuators may incorporate a slide-wire to feed positional information back to the controller.

d) Additional features

Amongst the more common additional features are:-

i) Alarms

Additional relays can be installed to signal abnormal conditions. This is often essential to safeguard against damage arising from power failure or equipment failure.

ii) Self-tuning control

Self-tuning controllers incorporate the latest microprocessor technology and utilise their computing power to continually assess the best tuning constants for the instrument at all times.

Their chief advantages are:

a) No knowledge of process control theory is required in order to apply them. They are simple to use.

b) They ensure the best control under varying load conditions.

c) The instruments automatically compensate for plant wear and ageing.

d) Faster plant start-up and commissioning.

e) They need never be manually tuned.

However, this feature is not quite the 'fit and forget' option it would appear at first sight. Problems can be experienced firstly with non-linear gain and secondly with interaction with other control loops. These problems are currently being investigated by the major instrument manufacturers

11.3.2 Programmable Logic Controllers (PLCs)

The traditional method of engineering control systems for fired plant is based upon the use of relays or solid state modules for implementing sequential control and the use of stand alone single loop controllers for implementing continuous control. This approach enables control systems of widely varying complexity to be assembled at reasonable cost and the repair and maintenance of these types of systems have become well understood by plant engineers and maintenance personnel. However such hardwired systems are inherently inflexible with regard to changes in control action and the large number of individual components and wiring connections involved in a complete control system can make fault finding difficult and time consuming.

Programmable Logic Controls (PLCs) were introduced in the early 1970s as replacements for these hardwired relay control panels. They were immediately accepted in the automotive industry and since then have found countless applications in virtually all other industries.

a) Basic PLC definition

The basic elements of a PLC are shown in Figure 11.8. The programmer is connected temporarily for the purpose of entering the user's programme. Although programming services are available the programme is usually designed and entered by the end user. The programmer can also be used to alter the existing programme or possibly to override part of the control actions.

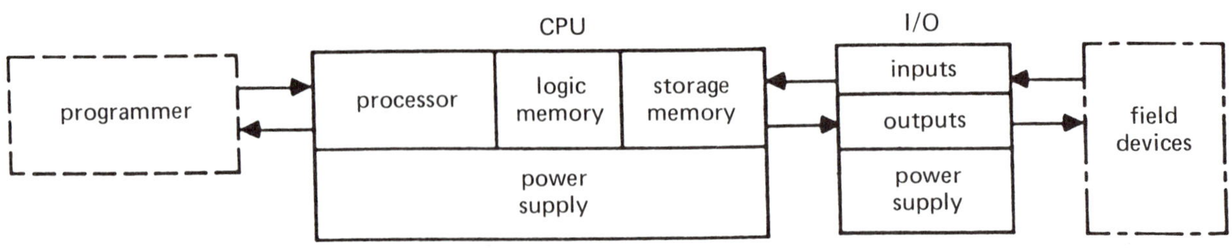

Fig. 11.8 Elements of a PLC

The Central Processing Unit (CPU) is the 'brain' behind the PLCs. Here, all the decisions are made relative to controlling a machine or process. The CPU receives input data, performs logical decisions based upon a stored programme and drives outputs.

The Input/Output structure is a major feature of a PLC. Its purpose is to filter the various signals received from or provided to the exterior components of the control system. These components or field devices can be pushbuttons, limit switches, relay contacts, analogue sensors, selector switches, thumbwheel, (inputs) or motor starters, solenoid valves, indicator lights, light emitting diode (LED) displays, position valves, relay coils, (outputs) etc. In any industrial environment, these I/O signals can contain electrical noise that would cause unreliable operation of the CPU if the noise reaches its circuits. The I/O will protect the CPU from this noise as well as providing it with valuable information.

The field devices are usually selected, supplied and installed by the end user. The type of I/O will normally be determined by the voltage level of the field devices. A wide variety of voltages, current capacity and I/O types are usually available.

b) Benefits of PLCs

Programmable logic controllers offer a number of valuable benefits to industrial users. These benefits can result in savings that exceed the price of the PLC itself, and should be considered whenever selecting an industrial control device. As a result of these benefits the yearly value of PLCs sold worldwide has increased significantly. The advantages of PLCs as compared with other control devices are:

- Less floor space
- Less electrical power required
- Reusable
- Programmable if requirements change
- More reliable
- Easier to maintain
- More flexible, satisfies more applications
- Interfaceable to data processing systems
- Faster system design, components ordered in parallel with logic design.

Recently control equipment manufacturers have enhanced the capabilities of PLCs by adding extra features such as analogue signal inputs and outputs, numerical computation, ports for data communication between controllers or with other computers, feedback control loop functions, fault diagnostics etc. At the same time, many of the practical advantages of PLCs relating to their ease of installation in shop floor environments close to the plant have been retained.

11.3.3 Computer Control

Direct computer control of process variables was introduced to large process plant in the early 1960s. In recent years the increasing availability of cheap microprocessors has resulted in the technique becoming much more widespread. Various configurations are employed. Sometimes the computer will replace conventional PID controllers whilst at other times the computer will supervise the setpoints of the conventional controllers. On larger plants a number of computers could be arranged in a hierarchical fashion.

There is no single reason why computers are to be preferred to conventional controllers, rather a number of benefits could be expected:-

i) All plant data is available at a central point.

ii) Control panel and hence the control room size is reduced.

iii) Records of alarms and operator actions expedite analyses of failures.

iv) Operator interference is less common.

v) All batches of material are identically treated.

vi) Automatic start-up and shutdown of complex plant is facilitated.

vii) The system can be reconfigured more easily.

a) Major Control Functions

All of the functions below can be provided by the control system, although not all will be necessary or provided on every system.

i) Plant data input/output signals (including alarms)

This is provided by electronic hardware which connects external sensors and actuators to the computer. Programmes will examine the state of the inputs, performing alarm checks as necessary. Outputs are calculated at regular intervals and directed to the plant.

ii) Data presentation

The plant inputs can be presented to the operator in a wide variety of forms. Usually they will be displayed on a Visual Display Unit (VDU), and the operator will have the facility to examine a plant overview and to call up particular parts of the process in increasing levels of detail.

A feature of most systems is the ability to store selected parts of this data. This can either be examined on a short-term (24 hours) basis as a trend, or a longer term historical analysis could be performed to examine, for example, quality problems.

iii) Control functions

In its simplest form the computer will undertake continuous control of individual loops as a replacement for conventional PID controllers. The computer can also handle batch processing where a batch of material moves around the plant with different operations being carried out in different vessels. In ideal conditions each batch will receive identical treatment. Sequence control is also offered. However, this requires a detailed description of the process as a series of very small actions. Quite often production of the software for sequence control takes considerable effort.

b) Hardware Considerations

The process control function can be implemented by a range of computers from small microcomputers through to 32 bit superminis. Usually the computer will be only a small part of the system. The rest will be racks, power supplies, I/O devices, displays, cabling etc. There are two broad methods whereby the hardware is arranged.

i) Centralised systems

Basically the computer performs all of the control functions, as shown in Figure 11.9.

ii) Distributed systems

Here the control functions are distributed along a 'data highway' (Figure 11.10). Each control computer (outstation) is physically separated from the main computer and functions as a stand-alone controller. This configuration is common in building energy management systems, where different geographical sites can be linked by radio, telephone or other long distance communications media.

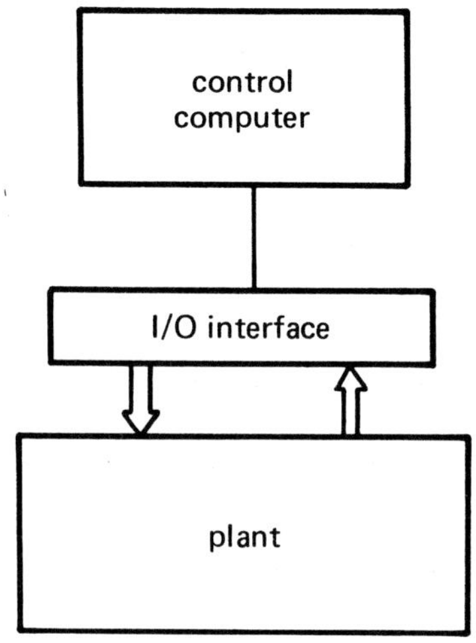

Fig. 11.9 Centralised system

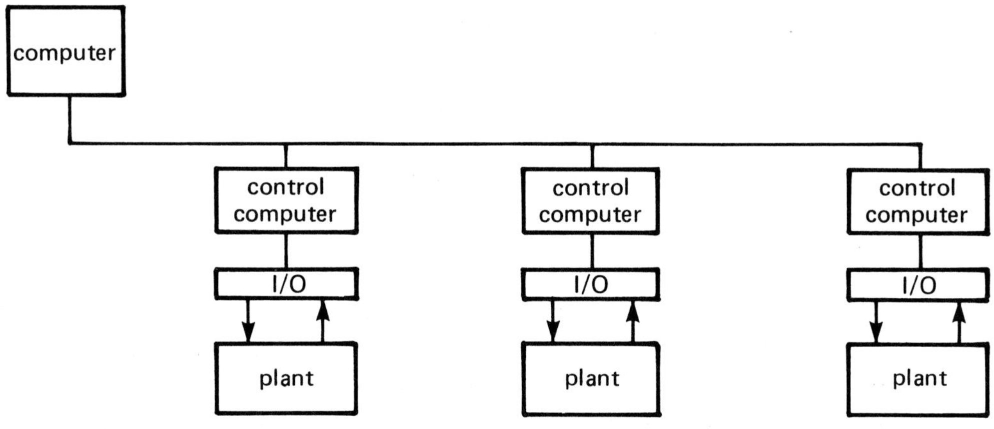

Fig. 11.10 Distributed system

Large plants will have a number of computers performing different functions but in communication with each other. Figure 11.11 is taken from a modern steel plant. The central computer receives information as to setpoints, heating/cooling cycles etc from a supervisory computer. This machine is linked to the materials flow computer, door switches etc and will instruct the control computer when to begin the appropriate control cycle. Operator interference is minimised, consistency, and hence quality is maximised.

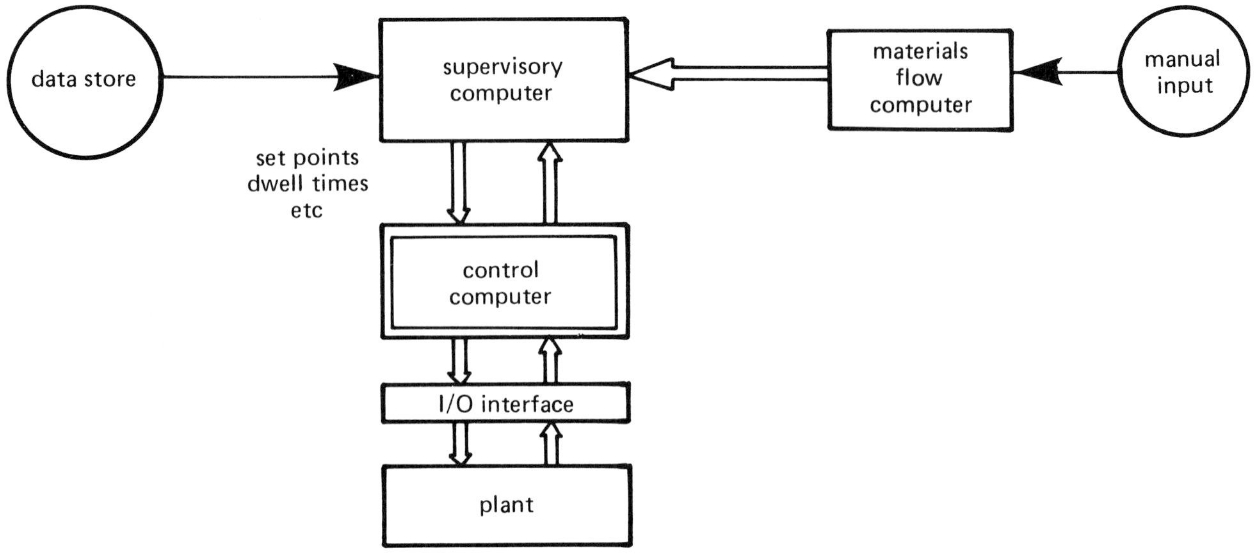

Fig. 11.11 Hierarchical system

11.4 TEMPERATURE MEASUREMENT

Temperature control is the operation most likely to be encountered in the industrial heating sector. The method of control can range from a cheap PID controller to a sophisticated computer control system. Regardless of the method of control the system accuracy can be no greater than that of the measuring element. Consequently great care must be taken in the specification of this element and the advice of a reputable manufacturer should be sought if any doubt exists.

Most applications can be covered by one of the three common measuring devices, which are:-

- Thermocouples
- Resistance thermometer
- Radiation thermometers (pyrometers).

11.4.1 Thermocouples

a) Basic principles

If an electrical conductor is subjected to a temperature gradient the heat flow will cause a flow of electrons which in turn will result in an emf. The size and polarity of the emf will depend upon the material forming the conductor.

In a thermocouple two different materials are exposed to the same temperature difference. Their different temperature/emf characteristics cause a potential difference between the cold ends of the couple.

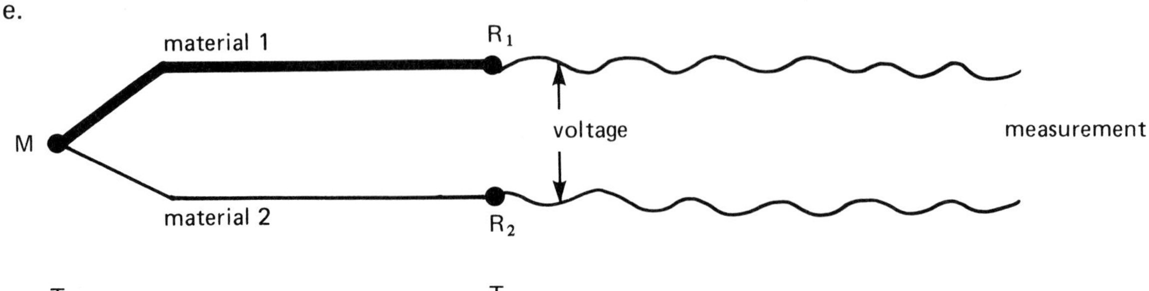

Fig. 11.12 Thermocouple basic principle

The junction M is usually known as the measuring (or hot) junction. R_1 and R_2 are known as the reference junctions. If the reference junction is maintained at a fixed temperature then the measurement junction temperature can be deduced from the value of the voltage between R_1 and R_2.

b) Thermocouple materials

The majority of conducting materials will produce a thermoelectric output but when considerations such as wide temperature range, useful output, and an unambiguous relationship of output with temperature are taken into account, the practical choice is generally reduced. Fortunately, the selection process has been carried out by the thermocouple suppliers and a useful range of metals and alloys is generally available in wire form or as completed sensors to cover a temperature range from - 250 °C to higher than 2000 °C.

This temperature range cannot be covered with a single thermocouple combination. The temperature ranges of some of the more commonly used thermocouples are given in Tables 11.2 and 11.3. These have internationally recognised type designations. In general, the platinum based thermocouple materials are the most stable. They have a useful temperature range from ambient to 2000 °C, although their outputs are low compared with base metal types. Some common types of platinum thermocouples are listed in Table 11.2. The upper temperature limits shown in the Tables are nominal. They may be raised or lowered depending on the conditions, duration of exposure the life and accuracy required, etc.

Table 11.2 : Commonly used Platinum Metal Thermocouples

International type designation	Conductor Material	Temperature range (°C)
K	Pt–13%Rh (+) Pt (-)	0 to +1600
S	Pt–10%Rh (+) Pt (-)	0 to +1550
B	Pt–30%Rh (+) Pt–6%Rh (-)	+100 to +1600 +1750*

* Intermittent use

Among the considerations involved when selecting base metal thermocouple materials (see Table 11.3), will be the temperature range, the sensitivity and compatibility with existing measuring equipment. A thermocouple supplier will assist with the selection for any particular application.

c) Practical thermocouples

In some circumstances it may be expedient for the user to manufacture his own thermocouples but for most applications the thermocouple will be purchased from a proprietary supplier. Frequently

Table 11.3 : Commonly used Base Metal thermocouples

International type designation	Conductor Material		Temperature range (°C)
K	Ni—Cr	(+)	0 to +1100
	Ni—Al	(-)	
T	Cu	(+)	-185 to +300
	Cu—Ni	(-)	
J	Fe	(+)	+20 to +700
	Cu—Ni	(-)	
E	Ni—Cr	(+)	0 to +800
	Cu—Ni	(-)	

the conductors will be mounted in a closed refractory sheath as shown in Figure 11.13. Great care is necessary in the manufacture and installation of the hot junction in order to minimise changes in this area in service.

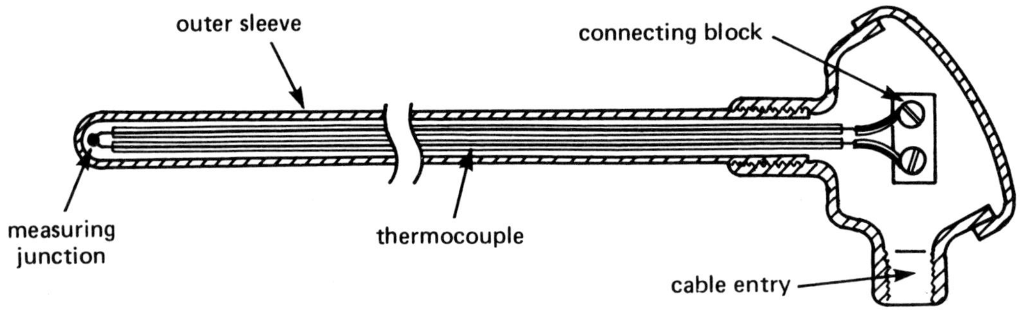

Fig. 11.13 Refractory sheathed thermocouple

An alternative form of construction is using mineral insulated (M.I.) cable where the thermocouple conductors are embedded in a closely compacted, inert mineral powder and surrounded by a metal (e.g. stainless steel or nickel alloy) sheath to form a hermetically sealed assembly. The sheath can form a useful protective cover in many situations. These types of assembly can be obtained with outer diameters ranging from as low as 0.25 mm up to 10.0 mm and lengths can be from a few millimetres to hundreds of metres (Fig. 11.14).

For special applications where a very rapid response is required, it is occasionally advantageous for an M.I. thermocouple to be manufactured with the junction exposed. As this may raise some strength or compatibility considerations, the advice of the supplier should be sought.

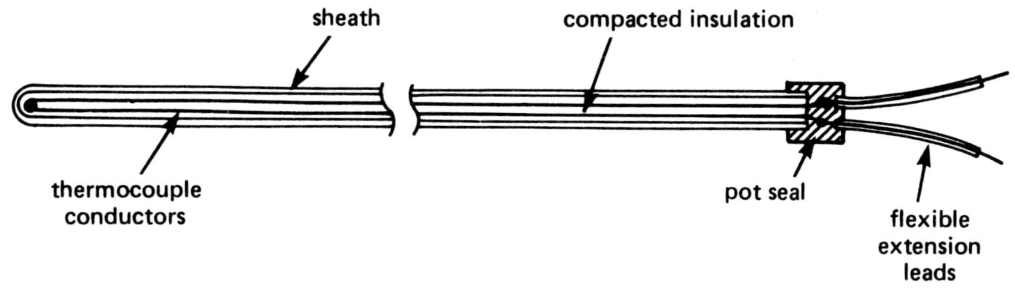

Fig. 11.14 Mineral insulation thermocouple

d) Compensating cables

It is not usual to take the thermocouple junction right back to the reference unit. In this case an extension cable could be employed in the manner shown in Figure 11.15.

The extension cables could be made of the same material as the thermocouple. Alternatively they could be made of quite different materials which would develop the same electrical output as the thermocouple of the limited temperature difference pertaining between the extension plug connection and the reference unit.

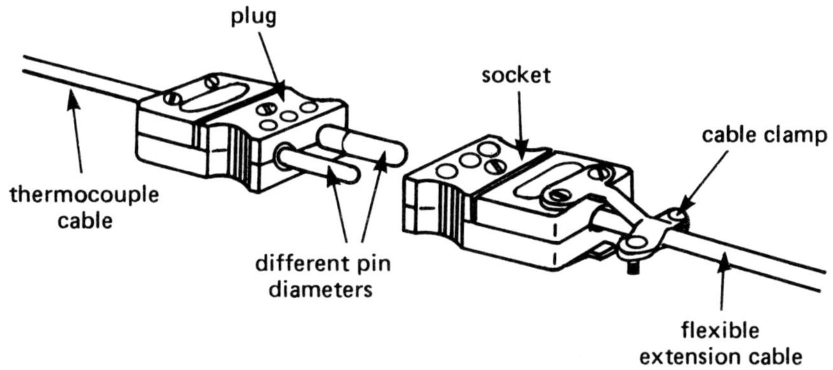

Fig. 11.15 Extension/compensating cables

An obvious use for compensating cable is the prohibitive cost of running a platinum thermocouple all the way back to the reference unit.

e) Measuring methods

i) Direct methods

As the thermocouple is an electrical generator it can be used to energise a galvanometer. The reference junction is formed where the thermocouple joins the galvanometer and variation corrections have to be made to simulate a 0°C reference temperature. Corrections could be electrical compensation, movement compensation or change of scale position.

An important limitation of this technique is that the emf provided by the thermocouple is very small and that as a current is drawn from the measurement circuit accurate work is not possible. With the advent of cheap solid state amplifiers this technique has been largely superseded.

ii) Potentiometric methods

The measuring potentiometer is the classical method for determining voltage. It is shown in simplified diagrammatic form in Figure 11.16. The voltage from the thermocouple is balanced precisely by adjusting the variable resistance, R, until the indicator, I, shows zero current. The variable resistance, or perhaps decades of resistance switches, are calibrated in suitable voltage values. Each measured voltage is converted to temperature using the standard tables for the thermocouple being measured. The method is accurate, reliable, simple to set up and use.

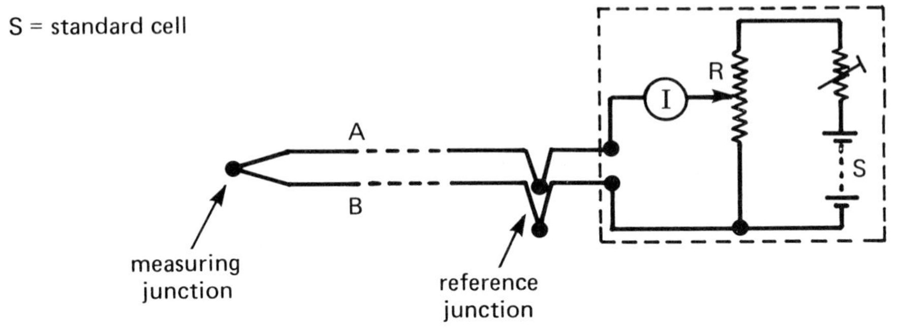

Fig. 11.16 Potentiometric measuring circuit

In portable form the potentiometer can be of considerable value as a known and 'problem free' measuring device and voltage source when checking thermocouple circuits in industrial surroundings where high electrical noise may exist.

The potentiometer principle is also used widely in self-balancing form as the mechanism for recording instruments.

Here the resistance slider is moved mechanically until zero current is detected and a pen attached to the slider is arranged to make a mark on a suitably scaled chart.

iii) Amplifiers

The development of integrated circuits has made miniature, stable amplifiers widely available. Amplifier circuits can be designed with very high input impedance so that when used with thermocouples, virtually zero current flows in the thermocouple circuit and they thus operate with high voltage values. As far as the user is concerned such circuits are often 'hidden' in purpose made instruments either as thermocouple amplifiers or as some of the devices described below.

iv) Digital voltmeters

Advantage has been taken of advances in electronics to produce voltmeters which display the applied voltage in digital form. They are well suited to thermocouple measurement and can indicate voltage changes in steps as low as 1 Volt. Any interference present on the electrical signal can be filtered out. Mains related voltages are suppressed by arranging for the signal integration periods to be precise multiples of the mains frequency.

v) Temperature indicators

The digital principle has been extended to more sophisticated instruments that indicate temperature directly. Such instruments are essentially digital voltmeters, which include an electrical compensation network to correct for the thermocouple reference temperature, and a circuit to correct for non-linear thermocouple characteristics. These instruments range from bench or rack mounted mains operated types that may also be switched to suit different thermocouple combinations, to compact, hand-held battery driven units which are frequently matched to Type K thermocouples.

11.4.2 Resistance Thermometers

a) Basic principles

The electrical resistance of a conductor varies with the temperature. If this relationship is predictable, stable and reasonably linear it can be used as a basis for temperature measurement. A further requirement is that the inherent resistance of the material must not be too low.

Of the major conductors platinum is the most suited to resistance thermometry. It can be drawn into fine wires or strips and large quantities of the material are not necessary. It does, however, need to be kept in the fully annealed condition and various manufacturing techniques are employed to ensure that this is the case.

The relationship of the platinum resistance thermometer with temperature can be represented very closely by a quadratic equation:

$$R_t/R_o = 1 + At + Bt^2$$

Where R_t = the thermometer resistance at temperature t

R_o = the thermometer resistance at 0 °C

t = temperature (°C)

A and B are coefficients determined by calibration.

For commercially produced platinum resistance thermometers, standard tables of resistance versus temperature have been produced based on an R value of 100 ohms and a fundamental interval (R_1 - R_o) of 38.5 ohms.

b) Types of Sensors

Platinum resistance thermometers are available in a wide variety of types and styles. The principle problem is to maintain the platinum in the fully annealed condition, and this necessitates a special mounting method.

The sensors are available fitted into a range of protection tubes. These can be equipped with terminal block for connection to the copper compensating cables leading to the measurement instrument.

As with the thermocouple sensor the protection tube can be fitted into a thermowell for additional protection (Figure 11.17).

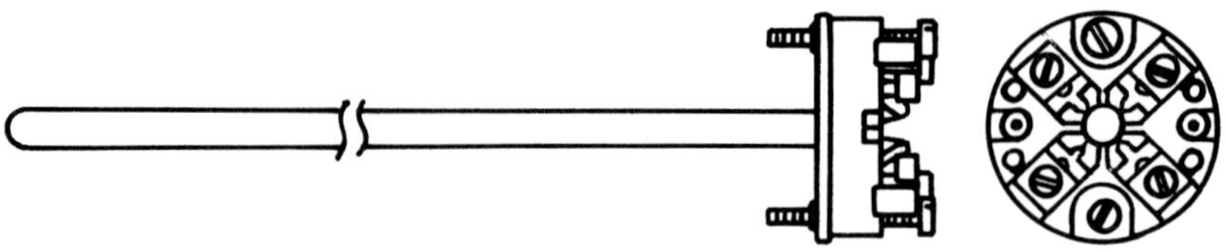

Resistance sensor protection tube

Thermowell

Fig. 11.17 Resistance thermometers

c) Measurement methods

There are two main instruments for determining the resistance of sensors measuring bridges and potentiometers. Measuring bridges are used extensively in the laboratory where the bridge elements may be resistance decades, or tapped inductances in A.C.versions. Highly accurate measurements are also possible using a precision potentiometer. An important requirement with this method is the provision of a stable energising current.

Some commercially produced industrial systems use a type of Wheatstone Bridge. The bridge is not normally balanced by altering resistance values but instead the magnitude of the out-of-balance voltage in a fixed element bridge is used to measure the applied resistance, (Figure 11.18). Alternatively the resistance thermometer might be energised from a constant current source and the voltage developed across it measured in a potentiometer type method.

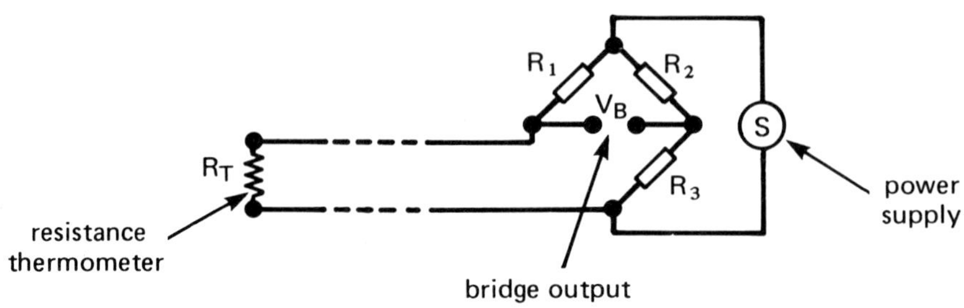

Fig. 11.18 Wheatstone Bridge

The simple two wire connection shown in Figure 11.18 above is used only where high accuracy is not required as the resistance of the connecting wires is always included with that of the sensor.

An alternative arrangement allows for another pair of wires to be carried alongside the thermometer pair. This additional pair is connected together close to the thermometer and the loop formed is introduced into the other side of the bridge circuit. Thus the effects of the two sets of leads tends to cancel, Figure 11.19.

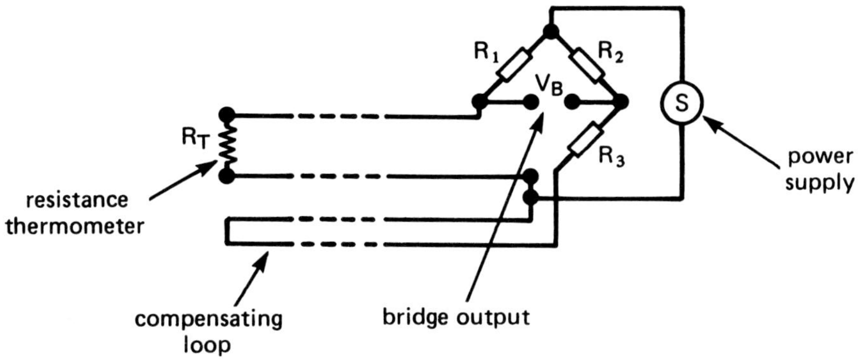

Fig. 11.19 Wheatstone Bridge with compensation

A better scheme is shown in Figure 11.20. Here the two leads to the sensor are on either side of the bridge and thus effectively cancel, while the third lead functions as the extended supply lead to configure the bridge in the form shown below.

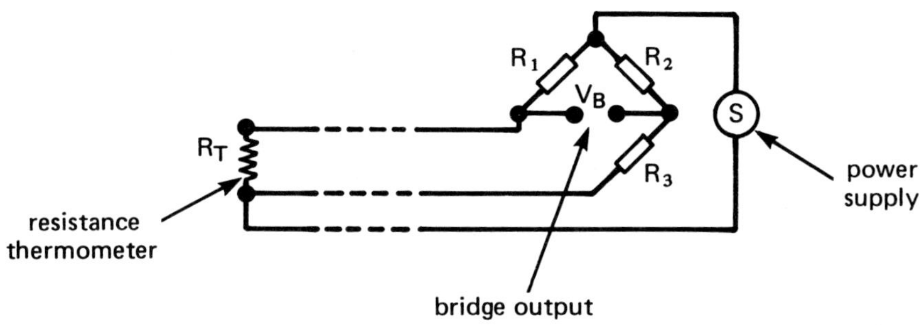

Fig. 11.20 Alternative compensation method

The best method of resistance determination however is by using a full four wire connection scheme, Figure 11.21.

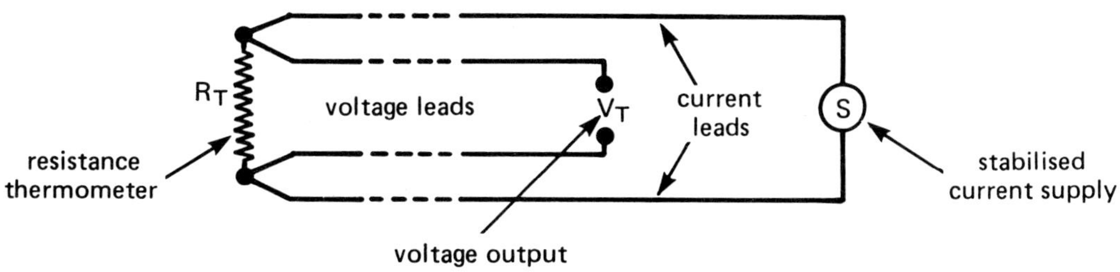

Fig. 11.21 Four wire compensation

The essentials of a voltage based method are shown in Fig. 11.21 where S provides an accurately known current through the sensor R and the voltage developed across the sensor is measured with a high impedance voltmeter or potentiometer. In this way the resistance of the leads has a negligible effect on the measurement.

The four wire connection arrangement can be used to provide cancellation effects with a bridge type measuring technique, Figure 11.22.

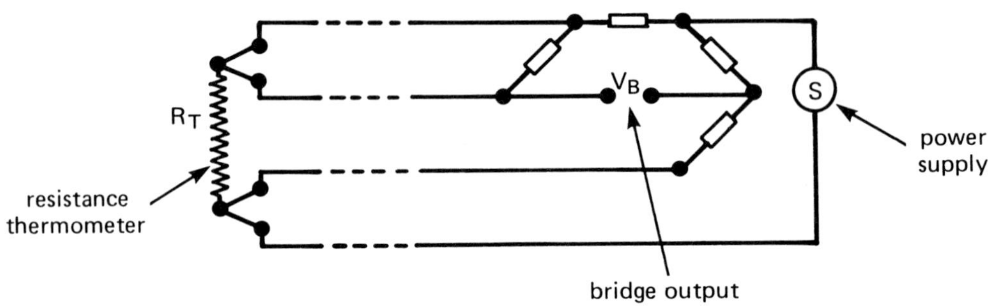

Fig. 11.22 Four wire system plus Wheatstone Bridge

11.4.3 Problems with Direct Temperature Sensors

a) Thermal linkage and heat flow

It is crucial that the sensor takes up the temperature of its immediate environment. It is also important that it takes up this temperature quickly in order to minimise delays in any temperature control circuit. It is therefore necessary to maximise heat transfer to the sensor and to minimise heat transfer along the support and connecting wires.

i) Good thermal linkage of the sensor to its surroundings.

When measuring in solids this includes installing the sensor in a closely fitting hole, considering the use of cements, fillers, high conductivity greases or heat transfer fluids. On surfaces, the use of pads and greases, cements or solders. In fluids, installing in the fastest flowing region, and arranging for the sensor to be in cross-flow if possible. The depth of immersion is important and if the fluids are slow moving with respect to the sensor, external finning may be advantageous.

ii) Heat flow to or from the sensor along the support and connecting wires should be minimised.

The first consideration is to reduce the temperature gradients close to the sensor. This usually implies endeavouring to provide sufficient immersion depth of the sensor into the medium. Further improvements might be made by using pockets and supports with a high axial thermal resistance, e.g. made from thin stainless steel. Additionally, small diameter and low thermal conductivity connecting wires might be used.

b) Gases in ducts

Two particular problems apply here.

i) Radiation to cold surroundings

The sensor will receive heat from the gas stream principally by convection and conduction. The sensor will also be a nett heat radiator to the cooler duct walls and this will result in the sensor taking up a temperature somewhat below that of the gas stream. To reduce effects of this type the sensor casing might have its emissivity reduced or shields might be fitted to intercept the radiation. In extreme cases special measurement techniques (a suction pyrometer) might need to be employed.

ii) Stagnation temperature

As the velocity of a gas flowing over a body increases, the temperature of the layer of gas in contact with the body begins to rise. A temperature measuring sensor is such a body and the measurement of the temperature of fast moving gas flows is complicated by the addition of this dynamic heating effect. The temperature most frequently required is the free stream temperature (i.e. the gas temperature without the dynamic component). This is usually called the static temperature. The static temperature plus the dynamic heating component is called the total temperature. The total temperature is the one most usually measured, using special probes designed virtually to stagnate the gas at the sensor and thus recover the dynamic component. From the measured total temperature the static temperature, T_S, can be derived from calculation using a relationship such as,

$$T_s = T_T[0.5(-1)M + 1]$$

where T_T = the measured total temperature

M = Mach number

Some examples of the temperature rise due to dynamic heating in air at atmospheric pressure are,

1 °C at 45 m/s, 10 °C at 145 m/s and 30 °C at 245 m/s.

11.4.4 Comparison of Resistance and Thermocouple Thermometers

Table 11.4 highlights the major features of the two measurement techniques.

11.4.5 Radiation Thermometers

a) Basic principles

A radiation thermometer assesses the temperature of a body by measuring the electromagnetic radiation it emits over the band of wavelength to which it is sensitive.

This band extends from shorter (lower) wavelength energy in the visible spectrum, to longer infra-red energy up to about 20 micro metres (μm). By focusing attention on this band of wavelengths we will cover the vast majority of surfaces and temperatures encountered in industrial

Table 11.4 : Comparison of Resistance and Thermocouple Techniques

Method	Thermocouple thermometry
Advantages	Wide temperature range. Versatile, e.g sensor can be a robust industrial unit or mineral insulated cable or as ultra fine wires, etc. Simple application, just the junction at the tip needs to take up the required temperature.
Disadvantages	Needs temperature reference. Needs extension cables for long runs. Needs attention to detail for high accuracy

Method	Resistance thermometry
Advantages	Potentially the most accurate method. Simple installation. Needs only copper cables for long runs.
Disadvantages	Needs energising current, Sensor types limited. High accuracy types need careful handling. Sensors larger than thermocouple junctions.

applications. Energy is emitted outside this waveband of 0.2 to 2.0 μm, but exists at such a low level as to be negligible.

Basically the heat radiated by a hot body is given by the Stefan-Boltzmann Law

$$Q = \epsilon AT^4$$

where ϵ = emissivity

Q = heat radiated

A = surface area.

Therefore in order to infer the temperature of a source it is necessary to know the surface emissivity. A perfect emitter (black body) has an emissivity of 1. Other surfaces are less than unity.

Two more concepts are important. These are Planck's Law and Wien's Displacement Law, which are illustrated graphically in Figure 11.23. The family of curves shown in Figure 11.23 are created using Planck's formula, and the peak energy value of each curve can be calculated from Wien's Displacement Law. It can be seen that the peak value is *displaced* towards longer wavelengths as lower target temperatures are considered.

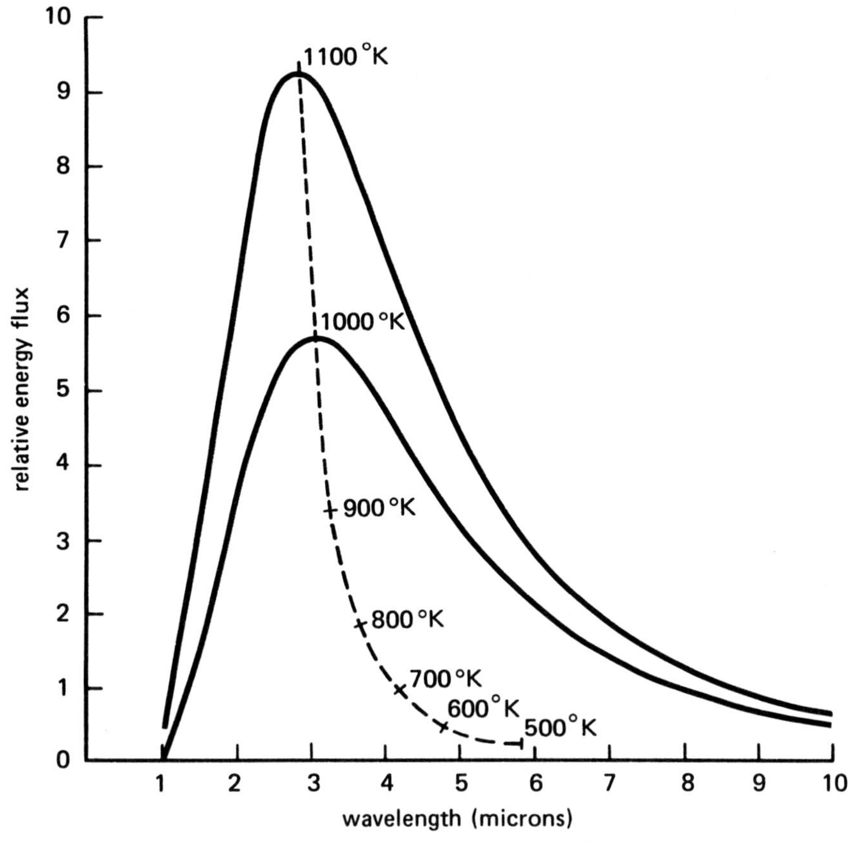

Fig. 11.23 Wien's Displacement Law

Mathematical treatment of the Wien function will show that:-

$$\Delta T \text{ (measurement error)} \propto \text{(wavelength)}$$

Therefore, when making measurements of surfaces in the open (e.g. not in a furnace) ALWAYS USE A THERMOMETER OPERATING AT AS SHORT A WAVELENGTH AS POSSIBLE.

b) Emissivity

For polished metals the emissivity will be as low as 0.05 (particularly at longer wavelengths), whereas for most non-metals the emissivity will be high (typically 0.8 or 0.9 at wavelengths around 1.0 μm). Hence any radiation thermometer pointed at a furnace will read low because it is calibrated to read correctly when viewing a black body source.

Methods of correcting for emissivity

The mathematical method

If the emissivity of the surface is known we can eliminate its effect by multiplying the thermometer output by $1/\epsilon$.

Painting the surface

It is occasionally permissible to paint a surface of low or variable emissivity with a coating of high and constant emissivity material.

Using a ratio (two colour) thermometer

If we view a surface simultaneously with two thermometers operating in two different wavelength bands, we shall obtain two outputs. Assuming that the emissivity is the same for both outputs we should be able to calculate the temperature from the ratio of these two outputs.

In spite of the obvious attractions of this technique the ratio thermometer has found few applications for minimising the errors due to uncertain emissivity, because emissivity tends to vary with wavelength and finally because of cost.

The ratio thermometer does have a place in the measurement of targets partially obscured by dust or fume. However, if the obscuring particles are of a similar size to the wavelength of one of the channels, the technique is unsuccessful.

Selecting a short wavelength thermometer

In section 11.4.5 a) it was shown that the temperature error for a given output whether due to error in assessing emissivity value or any other cause was directly proportional to the wavelength of measurement. Consequently a thermometer operating at as short a wavelength as possible should be used.

c) Reflectivity

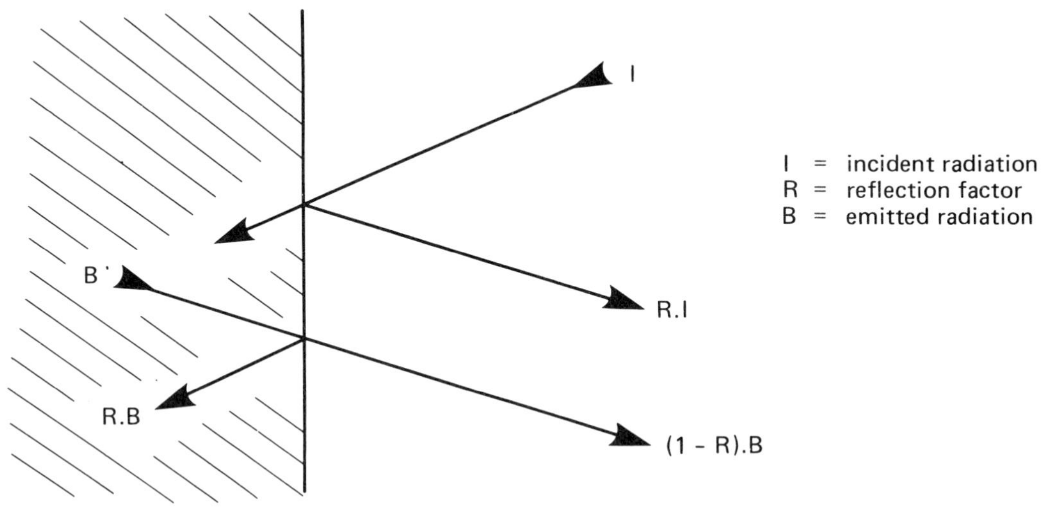

Fig. 11.24 Absorption, emission and reflection

Any radiation incident on a surface will be partially absorbed and partially reflected (Figure 11.24).

Thus the radiation leaving the surface is the sum of the emitted radiation and that reflected, and the thermometer cannot distinguish between them.

The problem of reflectivity is often more difficult than that of emissivity and the best solution requires assessment of the circumstances of each application. Some of the possible combinations are discussed below.

i) A hot body in cool surroundings

The ambient radiation is at a low level and for most thermometers can be neglected. This is certainly true for thermometers operating at short wavelengths but may not be for those operating at long wavelengths. If, for example, a temperature of 1000 °C measured in surroundings of 400 °C, the emissivity being 0.8, typical errors found would be shown as below:-

Table 11.5 : Typical Errors of Thermometers

Thermometer type	Spectral Sensitivity μm	Error °C
SILICON CELL	0.5 – 1.1	0
THERMOPILE	0.7 – 12	+5
PYRO-ELECTRIC	8 – 14	+52

ii) A hot body in surroundings at the same temperature

In this case, $I = B$ and the radiation received is:

$$(1 - R)B + R.B = B$$ (See Figure 11.24)

i.e. is at the black body level. No correction for emissivity is required.

iii) A hot body in hot surroundings which may be somewhat cooler or hotter than the body

Very large errors may now arise if no effort is made to correct for the ambient radiation. This is the most difficult case and no universal solution exists. There are four possible approaches.

In some cases it may be possible to screen off the ambient radiation either by independent screens or one incorporated in the sighting system of the thermometer. In both methods the screen must be kept cool. Usually water cooling is required.

Measure the incident radiation by means of a second thermometer (of, ideally, the same type). The outputs of the two thermometers can then be manipulated to arrive at an accurate surface temperature.

Sometimes the ambient radiation can be ignored if it has a limited spectral range. For example some infra-red heaters have quartz envelopes so that they radiate only at wavelengths less than about 5 m. Hence if we use a thermometer with sensitivity only to wavelengths greater than 5 m the ambient radiation will have no effect on it.

When the surroundings are cooler than the viewed surface it is generally better to use a thermometer working at short wavelengths. If they are hotter, a thermometer working at long wavelengths is preferred.

d) Parts of a radiation thermometer

The essential parts of a radiation thermometer are:

i) a detector which converts the radiation incident on it into a signal usually (though not essentially) electrical in nature.

ii) an optical system which defines the angular field of view of the thermometer and hence its sensitivity. It also determines the size of hot object (target size) that is required. It may also contain a filter to select the desired band of wavelengths to which the thermometer is sensitive. The spectral sensitivity is the product of the spectral transmission of the optical system and the spectral sensitivity of the detector. Either of these may be the limiting factor.

iii) a body to hold these parts.

iv) an electrical connection to convey the output signal to the indicating or recording device. This is usually a removable plug so that the thermometer can be removed easily for cleaning or checking.

There may also be:

v) a pre-amplifier which lifts the detector output from its low level (micro-amps or milli-volts) to the volts level.

vi) in portable models, direct read-out of temperature, which may be analogue or digital. Such models require a battery and often incorporate an emissivity adjustment.

e) Choice of radiation pyrometer

Manufacturers' catalogues show a very wide choice of thermometer available. These have been developed:-

i) As a result of the availability of new detectors, filters, infra-red materials, micro-sized electronics.

ii) As a result of users' demands for greater accuracy, faster response, greater reliability and on more 'difficult' applications.

Detectors currently available commercially include:-

Thermopile (strip type)	—	response time 2 seconds — responsive to all wavelengths
Thermopile (deposited type)	—	response time 60 milliseconds — responsive to all wavelengths
Silicon cell	—	response time 1 millisecond — responsive to 1.2 micrometre
Germanium cell	—	response time 5 milliseconds — responsive to 1.8 micrometre
Lead sulphide cell	—	response time ¼ second — responsive to 2.8 micrometre
Pyro-electric	—	response time ¼ second — responsive to all wavelengths

Filters currently available include:

Cut-on

Cut-off

Spike (narrow band)

Broad band

Optical materials available include:-

Glass — transmission 0.4 to 2.8 micrometre wavelength

Silica — 0.3 to 4 micrometre wavelength

Arsenic trisulphide — 1 to 12 micrometre wavelength

Fluorite — 0.4 to 12 micrometre wavelength

Germanium — 2 to 20 micrometre wavelength

as well as many other 'exotic' (and usually very expensive) materials.

It is clear that a very wide variety of thermometers can be built by the various combinations of the above elements. The most suitable type to be chosen by a consideration of all the factors in the application, which include:-

Target temperature

Target size

Target distance

Target material (especially emissivity)

Intervening atmosphere

Response time required

Accuracy needed.

With experience these factors usually indicate the required type unambiguously: sometimes more than one type could be used and experience will indicate which will be most satisfactory: sometimes the application is just impossible and one or more of the requirements must be relaxed: sometimes the application is very complex and one or more of the requirements must be replaced. On other occasions the application cannot be met with normal considerations.

In general the following guidelines can be applied:

Low temperatures (say below 300 °C) use a 'total' thermometer, i.e. one responding to all wavelengths.

High temperatures (say above 300 °C) use as short a wavelength as possible. This is because for a given error in output (due for instance to uncertainty in emissivity) the temperature error is directly proportional to the wavelength used.

Lightly oxidised metals use as short a wavelength as possible both for the reason given above and because emissivity is much higher at short wavelengths.

f) Applications

A survey of the various industrial requirements highlights the wide ranging capabilities of radiation thermometers. These capabilities now include the ability to measure temperatures as low as minus 50 degrees centigrade, and as high as 3000 degrees centigrade.

i) The Steel Industry

Historically the steel industry has provided the biggest outlet for radiation thermometers, and their applications cover the sinter plants, coke oven and blast furnace right through to the production of steel and finished components.

One example is steel reheating furnaces. Stoppages can occur downstream of the reheat furnace, and despite these delays, the stock must still be delivered to the mill at the right temperature. Overheating in this period means wasted energy (fuel), and underheating means lost time. Hence an accurate knowledge of the temperature of the stock can produce increased efficiency.

A thermocouple would give an accurate reading of the furnace temperature, but would not indicate stock temperature. A *single* radiation thermometer positioned to view the stock in the heating zone would give similarly erroneous results, since it could not distinguish between energy *emitted* from the surface and energy *reflected* from the surface.

This situation can be resolved by employing two sensors; one looking at the billet, and one in a position where it gives an accurate measurement of the source of the reflected component.

When employed as part of a process control system, savings of up to 17% on fuel costs have been produced.

ii) The Glass Industry

Here radiation thermometers are widely used to monitor and control tank roof temperatures, regenerator packing temperature, and indeed the molten glass itself at various points. Thermocouples have been used on many of these applications, but due to the very high temperature of some of the measuring points, the useful life of a thermocouple is reduced due to the contamination of the element. This is particularly true for tank roof measurement where a leading flat glass manufacturer reported drifts in thermocouple output equivalent to 10-70 °C over a period of 1 month.

When incorporated in a control scheme, the tank roof temperature can be used to initiate changes in burner activity and hence fuel efficiency.

At various stages in the glass making process the temperature of the glass itself will be measured. This presents problems as glass is transparent in the visible and near infra-red spectra.

Figure 11.25 shows the transmission and reflection of glass approximately 3 mm thick, together with the atmospheric absorption for a typical installation.

The curve shows that:-

glass is *opaque* in the band 0 to 0.4 μm
glass is *transparent* in the band 0.4 to 2.9 μm
glass is *semi transparent* in the band 2.8 to 4.5 μm
glass is *opaque* in the band 4.5 μm onwards.

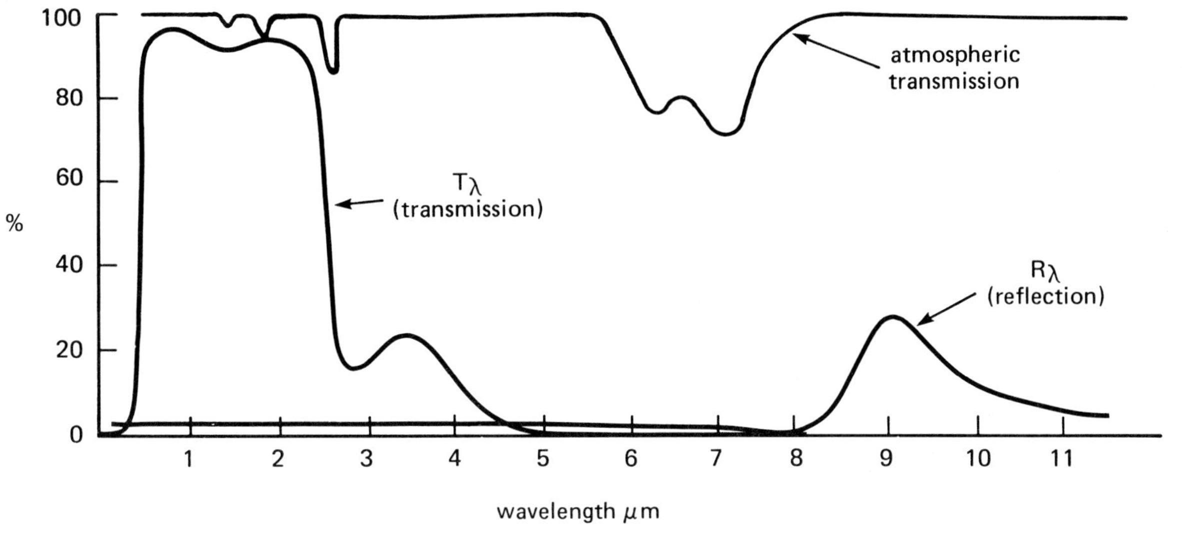

Fig. 11.25 Transmission and reflected radiation for glass

Therefore when measuring the temperature of sheet glass it is vital to choose a thermometer which will not 'see through' it, and one which will not be affected by reflections or atmospheric absorption. From the above figure it can be seen that the only feasible waveband where all these criteria are met is between 4.8 and 5.2 μm. Using the same logic it is possible to deduce that, to measure below the surface of the glass (and hence have an indication of gradients within the glass), a thermometer operating in the waveband 3.2 to 4.0 μm should be selected.

iii) Low temperature process industries

These include textiles, plastics, rubber, paper, human foodstuffs, animal feed, and many more.

One industry which highlights the benefit of the radiation thermometer over the thermocouple is the roadstone industry. Here graded quarry stone (or sometimes blast furnace slag) is coated with bitumen. The process requires the stone to be dried in an inclined rotating drum. The hot stone leaves the drum down a chute, where the abrasive action of the stone rapidly erodes the sheath material and will eventually destroy the thermocouple element. The introduction of radiation thermometers give a rapid response.

The environment of a roadstone plant in particular, and industrial applications in general, is often harsh. High ambient temperatures, excessive quantities of dust or fume, all point to the selection of a radiation thermometer with strong, completely enclosed mountings which incorporate air and water cooling and purging.

11.4.6 Comparison of Thermocouples and Radiation Thermometers

There are many factors which affect the choice between a thermocouple and a radiation thermometer. The relative advantages and disadvantages are listed below and each measuring requirement must be assessed.

RADIATION THERMOMETER

+++	Non-contacting so there is no interference with the process. We measure emitted radiation which is emitted anyway. Very suitable for solids, less so for liquids and gases.
++	Quick response 1 second for thermocouple and 1 millisecond for photon detector.
++	Accurate if properly calibrated and stable if properly maintained.
++	High sensitivity (in terms of %/°C)
+	Have virtually infinite life.
+	No upper temperature limit since the thermometer does not reach the temperature of the hot body.
-	Require line of sight free of obstruction (but flexible fibre optics can be used in some applications).
-	More expensive initially.
-	Somewhat limited range due to considerable non-linearity of output with temperature (but signal can be linearised electronically).
--	Services usually require cooling air/water and purging air.

THERMOCOUPLE

++	Greater range because output nearly linear with temperature.
++	No services required.
+	Can measure temperatures in inaccessible places.
+	Cheaper initially
+	Accurate when new, but deteriorates in use.
-	Low sensitivity (in terms of %/°C)
-	Slower response depending on sheath etc.
-	Have limited life due to contamination etc.
-	Upper temperature limit since the hot junction must reach the temperature of the hot body; limit depends on melting point and contamination.
--	Not suitable for solids; suitable for liquids and gases (with precautions).
---	Contacting so there is some interference with the process.

Key

+++ Great advantage	--- Great disadvantage
++ Moderate advantage	-- Moderate disadvantage
+ Slight advantage	- Slight disadvantage

11.5 CONTROL VALVES AND ACTUATORS

The final part of a closed control loop is likely to be a control valve. Its function is to alter the flow of the process fluid in order to maintain constant pressure, temperature etc, as dictated by the automatic controller. It can be the most costly single item in the control circuit, yet it is often the most neglected. Insufficient attention to the specification of the type and size of valve and neglect of maintenance will threaten the integrity of the whole control circuit, possibly resulting in unstable control and mechanical failure.

11.5.1 Types of Control Valve

Various types of control valve are available to the process engineer. The major types are discussed briefly below. (More detail is given in Chapter 9).

a) Globe

This is probably the valve most commonly employed for the control of liquid flow. Sizes range from 12.5 mm to 700 mm with a wide variety of trims and seating arrangements being available. When used in modulating control the plugs should not operate less than 2% open because of the tendency of the flow to pull the plug onto the seat. Likewise best control is obtained if the plug is not allowed to exceed 85% open.

Three-way versions are available for diverting or mixing applications.

b) Ball

A quarter turn rotary valve consisting of a sphere held captive within a body. A ball has a flow passage through it and can be rotated so as to present either the blank portion of the ball or the passage to the flowstream.

It offers the advantage of high flow capacity and low operating force for high pressure operation and tight shut-off.

The flow capacity is limited to 80% of the line area with a flow characteristic which is essentially equal percentage.

c) Butterfly

A valve having an internal disc or vane which rotates 60° or 90° within the flowstream to modify the process flow.

This valve is very commonly used for the control of gases and, if properly selected, offers the advantages of simplicity, low relative costs, light weight and space saving.

The inherent flow characteristic is equal percentage between 10° and 60° open, although cams can be used with a positioner to generate other characteristics. This design can be used for throttling service or on/off control, and can work adequately with solids or particles within the process fluids. Available in sizes from 50 mm up to 2 metres, but a 3 metre diameter valve is not unusual.

d) Eccentric disc

This type of valve contains an offset plug or disc which describes an eccentric path on opening. The normal rotary slide is between 50° and 60° and an essentially linear characteristic results.

The eccentric motion of the spherical face of the plug, reduces the operating torque requirements. The seating action is positive and the sliding seal-friction, associated with ball valves, is eliminated. Tight shut-off is obtained with relatively low applied force, but may require a long stroke actuator.

The standard flow direction is to the concave side of the disc, reversing the flow results in reduced capacity.

These valves are not suitable on severe cavitation or flashing services, but have a better resistance to cavitation than either the butterfly or ball valves at higher stroke.

e) Self acting regulator

A self acting valve is an automatic process control valve which is operated by means of the line fluid pressure, either upstream or downstream, acting on a spring opposed diaphragm or pilot valve.

A self acting valve is mainly used as a pressure control device and is similar in construction to the single or double seated control valve. Characterisation may be provided by a profile on the valve plug.

11.5.2 Control Valve Selection

There are many types of control valve available and an enormous variation within types. If in doubt the user is advised to seek the advice of a reputable manufacturer. The following points are some of the major factors which need to be considered.

Process fluid

Acids, alkalies and slurries would all demand special materials for the valve internals, as would extreme temperature conditions. Hazardous materials may require a special barrel to provide adequate sealing. Conditions of high temperature and pressure may limit the choice of body connections.

Phase changes in the process fluid (cavitation) can damage the valve and downstream pipework. These effects can be minimised by careful valve selection.

Capacity

The flow capacity of the process, maximum, normal and minimum, is required for the sizing of the valve and to determine the rangeability.

Duty

The type of duty required, whether modulating or simply on/off, and if a three-way valve, whether diverting and mixing characteristics are necessary.

Leakrate

The maximum allowable leakrate or degree of tightness required over the expected life of the valve prior to trim replacement.

Noise level

If noise is considered to be a potential problem then the effect can be minimised by careful valve selection.

Recovery rate

This is a measure of how much flow energy is dissipated through the flow valve. A butterfly or ball valve would lose little pressure overall and would be classed as a high recovery valve. Little cavitation would take place.

A globe valve is a low recovery valve and would be best suited in high pressure drop applications, or where noise is a problem.

11.5.3 Valve Characteristics

The valve characteristic is the relationship between the flow through the valve as a percentage of the valve opening position. It is important to distinguish between the *inherent* characteristic of the valve and *installed* characteristic, as the characteristic will be modified by the pressure drop in the supply pipework.

Three types of inherent characteristic are commonly employed.

i) **Linear** – The capacity is directly proportional to the valve stem travel at constant pressure drop.

ii) **Equal percentage** – Equal increments of rated travel will initially give equal percentage changes of the existing flow, providing a very small opening for plug travel near the seat and very large increases towards the more open position.

iii) **Quick open** – Maximum flow through the valve with minimum travel. Provides a large opening as the plug is lifted from the seat giving maximum flowrate at approximately 30% lift from the valve seat.

11.5.4 Valve Sizing

Accurate valve sizing is crucial to good stability of an automatic control system. A phenomenon often seen is the valve in continual oscillation about a fixed position (hunting). Although usually attributed to inaccurate PID settings, incorrect valve sizing, too large capacity, will make control virtually impossible. The correct procedure would be to determine valve pressure drop.

Once all the process conditions are decided, it is possible to calculate the size of valves required. Although valve sizing is not a particularly difficult task, care should always be taken when selecting the formula to use. Most valve manufacturers base their trim sizes around the 'CV' value. This term is the valve flow coefficient and is a capacity unit derived from the number of U.S. gallons per minute of water at 60 °F which will flow through a valve having a one p.s.i. pressure drop across it. Some manufacturers differentiate between capacity units for liquids, steam and gas service, using a value of CV, CS and Cg accordingly.

e.g. $$CV = \text{Flow}\sqrt{\frac{S.G.}{P}}$$

SG = Specific gravity

P = Valve pressure drop

All the control valve sizing formulae published by the manufacturers are basically the same, but if a high degree of accuracy is required *always* make sure that a particular manufacturer's formula is applied to his valves and his valves alone. This is because many manufacturers include correction factors in their formulae which are only applicable to that type of valve, or valve trim.

11.5.5. Actuators and Accessories

The actuator is the means whereby the controller alters the flow of process fluid. It can be pneumatic, electric or electro-hydraulic. The controller can have various output forms (Section 11.3.1) so that sometimes additional devices will be necessary to convert the controller output into a linear or rotary valve motion.

Various considerations will dictate the selection of the actuator, for example operating force, length of stroke, speed and accuracy of operation etc. The major features of the different actuators are considered briefly below.

a) Pneumatic actuators

i) Diaphragm actuators

These are the most commonly used actuators due to their adaptability to a wide variety of valve types and sizes. They are particularly popular on Petro-chemical plants where fire or explosion hazards exist.

They are generally inexpensive, and will operate from a low pressure air supply, give a fail safe action, and have an adjustable spring to enable high initial or final closing forces to be exerted, and give a linear performance with a fast response.

ii) Piston actuators

These use a high pressure air supply, thus eliminating the need for a supply pressure regulator. They have high thrust output with a fast response.

iii) Power cylinders

These units consist of a cylinder with an internal piston connected to a rod. A position control unit mounted on the cylinder converts the control signal to an output signal for precise positioning of the rod.

Power cylinders are used where very high forces are required and are normally used for dampers and louvres on ventilation systems. They may be also connected to butterfly valves and control valves where long stroke/ H.P. is required.

b) Electro-mechanical actuators

Are capable of being used over long distances, with minimal signal transmission delays. They are immune from the problems of freeze-up in cold ambient conditions. These actuators are very popular with small, low pressure drop applications, e.g. gas burners, H & V applications. However, for more demanding applications they are more complex and expensive, and are used if usually no air supply is available. They have a smooth starting control.

c) Electro-hydraulic actuators

Offer the advantage that they can be placed where there may be no other auxiliary services present. To maintain the stroke, the actuator requires a constant source of pressure, which means the constant use of electric power to pump the hydraulic power fluid.

Generally, they are more expensive, but they provide large stem forces available with no lubrication problems. They are normally used where high stem loadings cause instability.

d) Accessories

Various accessories will be found associated with the valve actuator. Probably the two most common are sometimes employed in conjunction with pneumatic valves.

i) Valve positioners

A positioner must always be used for control accuracy and situations where high imbalance forces caused by the line fluid have to be overcome.

A pneumatic valve positioner is a device which precisely positions the moving parts of a pneumatically operated control valve relative to a pneumatic input.

Positioners can either be purely pneumatic or in some cases, electro-pneumatic. The latter works in exactly the same way as the above description except that the control signal received is electrical.

The controller output is normally calibrated to stroke the control valve with a 3-15 psig signal, (or 4-20 mA in cases of electrical control systems). The positioner with its feed back mechanism, corrects any variation in valve position due to line forces, friction, etc.

Positioners can be used for a variety of applications, such as to increase speed of response of a control valve, split range operation, i.e. where output from one controller is split between two control valves, normally in 3-9 psig and 9-15 psig increments, and reverse control action, i.e. where an increase in controller output signal results in a decrease in positioner output.

ii) Volume boosters

It is possible, by use of a volume booster, to increase the speed of a pneumatically operated control valve. When a volume booster is employed, the positioner output is piped first to the booster. As only about one cubic inch of air is required to position the pilot in the booster, the volume required to operate the booster is very small. The air supply which actuates the control valve comes through the pilot in the booster, and since this pilot has a large capacity, the stroking time for the control valve is substantially reduced. When using an air filter regulator to supply air to the booster, a high capacity type should be used so as not to limit the output capacity of the booster.

11.5.6 Some Installation and Maintenance Considerations

a) Installation

i) Turbulence will be minimised by ensuring a good swage angle both into and out of the valve body,

ii) For the same reason ensure that the valve is at least three pipe diameters before a bend and not less than ten pipe diameters after a bend.

iii) As always ensure good access for maintenance. Remember that heavy valves and actuators will require lifting gear or a gantry.

iv) Heavy actuators will need support if the valve is mounted horizontally.

b) Maintenance

Valve manufacturers provide detailed instructions for the overhaul and maintenance of each type of valve, however, the following general points may be applied to most forms of industrial valve:-

i) Shortly after start-up, periodic field inspection of the valve installation should take place to check for leaks, etc. and satisfactory operation.

ii) Gland packings will usually require adjustment.

iii) Lubricators need checking and re-adjusting.

iv) During operational service, deterioration in performance should be noted so corrective action may be taken during plant shutdown.

v) At time of overhaul, all internals should be inspected and re-worked or replaced as necessary. This especially applies to the seating faces, sliding and guide surfaces. Gaskets, seals, packings, should be replaced as a matter of course.

vi) Actuator parts, i.e. diaphragms, piston seals, etc., should be checked for wear or deterioration. 'Local' instrumentation should be cleaned, inspected and checked for satisfactory operation.

vii) A Repair or Service Record of work undertaken should be completed to serve as a historical record for future planning.

Site Supply and Internal Installation 12

CHAPTER TWELVE

SITE SUPPLY AND INTERNAL INSTALLATIONS

12.1 INTRODUCTION

The provision of gas supplies in the United Kingdom falls into two main sections. The gas service pipe and the governor/primary meter installation is the responsibility of British Gas plc and will be carried out by the Region concerned or contractors working on its behalf. The installation of internal pipework from the outlet side of the primary meter and of all appliances is, on the other hand, the responsibility of the customer.

In both cases the installations must comply with the current Regulations, British Standards and Codes of Practice, and British Gas Specifications, the principal ones of which are listed at the end of this chapter. The documentation is extensive, and this chapter is largely based on British Gas publication IM/16, Guidance Notes on the Installation of Gas Pipework, Boosters and Compressors in Customers' Premises and the Gas Safety Regulations. The latter excludes installations in factories but it is recommended practice to follow their intent for all sites. Binding regulations for factories will be promulgated shortly.

This chapter provides general guidance for the installation and operation of gas distribution pipework and equipment based on current United Kingdom practice; for countries outside the United Kingdom, installations should conform to the current Local Regulations and Standards.

The approved S.I. unit for pressure is the Newton per square meter, N/m^2 or Pascal (Pa). Neither it nor its derivative kN/m^2 (kPa) has proved numerically convenient for practical work and this chapter is standardised on the now generally used bar (100 kN/m^2, 100 kPa, 14.5 lbf/in^2) and mbar (100 N/m^2, 100 Pa, 0.4015 in H_2O).

12.2 SITE SUPPLY

12.2.1 General and Definitions

General

Figure 12.1 shows a schematic layout to identify and indicate the various components involved in a governor/meter installation. In the United Kingdom, British Gas plc is responsible for the installation of all the equipment involved. It is the responsibility of the customer to provide and maintain any housing or security compound that may be required subject to meeting the requirements of British Gas plc.

Service Pipes

The service pipe is the pipe between the British Gas plc main and the meter installation.

Generally the chosen route for the service pipe will follow the shortest distance between the main and the meter installation, but the route may be influenced by such factors as the formation of the ground, the proximity of other buildings and plans for future development. The service pipe shall not be installed under the foundations of a building or under the base of load bearing walls or footings, nor shall it be installed in an unventilated void space.

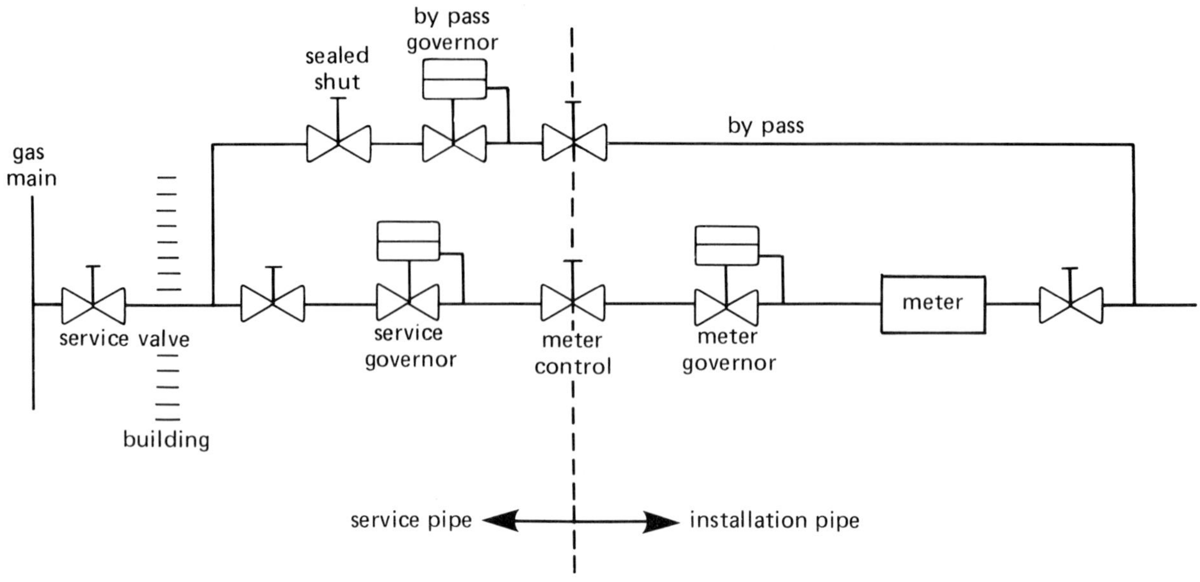

Fig. 12.1 Schematic layout of governor/meter installation

Where the service pipe entering a building passes through any load bearing or cavity wall, or through a floor of solid construction, then it shall be enclosed in a sleeve and sealed at one end with non-setting mastic material.

Service Valve

A valve termed the service valve will be fitted in the service pipe in the following circumstances:-

- Where the service pipe has an internal diameter of 63 mm or more.
- Where the maximum operating pressure in the service pipe exceeds 75 mbar.
- Where a hazardous trade or activity is, or is likely to be, carried out in the premises supplied.
- Where a common service pipe supplies more than one primary meter in a building.

The service valve will be located in the service pipe generally in a position outside but as near as practicable to the property boundary, but in certain circumstances the valve may be fitted within the property boundary in a readily accessible position outside the building. In all cases the valve will be provided with a surface box and cover, and it is desirable that an identification marker giving the location of the valve be affixed either to the nearest building or to an adjacent marker post. This valve is usually the 'property' of the Distribution Department of British Gas and should not be interfered with by the customer.

Meter Location

The meter installation should be sited within the curtilage of the premises as near as practicable to the boundary line. Depending upon the gas supply pressure and the maximum flow rate required, it may be necessary to provide a separate housing or compound near the site boundary, and it is therefore essential to consult with the appropriate Region of British Gas plc before finalising the space reserved for the gas metering installation.

In general, meters should be fitted at ground level in a well ventilated area and should be easily accessible for closing off the supply in emergency, for maintenance of the installation and for meter reading. Normally meters having a capacity in excess of 25 m^3/h are not fitted at high level except in exceptional and agreed circumstances, but where larger meters are fitted at high level it is essential that a permanent means of access and a working platform are provided to ensure safe conditions at all times.

The meter location should be chosen so that the meter is subjected to neither extreme temperatures nor changes in temperature, nor affected by dampness, corrosive atmospheres, chemicals or dirt in general.

In general, a rotary displacement meter will be used on installations where the maximum gas flow rate is in excess of 170 m^3/h.

12.2.2 Installation Arrangements

Meter Control

Every meter installation will include a valve on the meter inlet known as the 'meter control' as required by the Gas Safety Regulations.

Normally the first above-ground valve on the service pipe, on the inlet side of the meter/governor installation, will be the meter control but on larger and high-pressure installations the Region of British Gas plc may nominate one of the other valves on the installation as the meter control. If the meter control is housed in a compound or meter house that is locked the key must always be immediately available.

Available Gas Pressure

Generally the pressure available at the meter outlet in the UK will not exceed 22 mbar but the effect of the pressure loss along the pipework between the meter and the individual appliance isolating valves will reduce the available pressure at the appliance.

If the burner equipment to be installed requires a higher pressure, it will be necessary for the customer to fit a gas booster or compressor to provide the necessary pressure. Any consumer intending to install and use a gas booster, gas compressor or gas engine, or to use either air at pressure or any other gas in conjunction with the gas supplied by British Gas plc, is required to give the Region of British Gas plc fourteen days' written notice of the intended installation. An approved non-return valve should be fitted by the customer if air or any other extraneous gas is used at pressure in conjunction with the natural gas.

Notices

Where a service pipe is installed which supplies more than one primary meter installed in the same premises or in different premises, a notice in permanent form must be prominently mounted on or near each primary meter indicating this fact.

Complete systems

Where the distribution system operates at pressures not exceeding 50 mbar the arrangement shown in Figure 12.2 should be used. Over-pressure protection is not normally necessary but can be provided by a relief valve or slam-shut valve. A locked meter by-pass should be fitted to meters with a capacity greater than 42.5 m^3/h.

A filter is normally positioned upstream of the governor and instrumentation system to provide protection from gas borne dust. In addition, turbine and rotary displacement meters should be fitted with top-hat or skirt-type strainers between the governor and the meter as shown in Figure 12.2.

Refer to British Gas plc publication IM/112 for details of multi-stream governor systems and installation arrangements for medium pressure (75 mbar to 2 bar) and intermediate pressure (2 bar to 7 bar) systems.

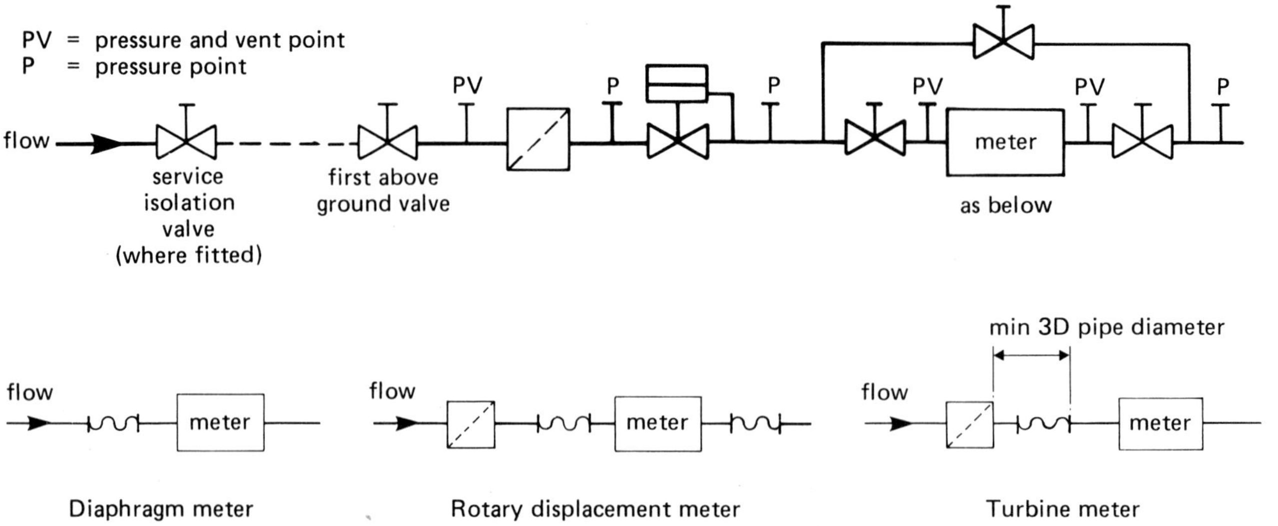

Fig. 12.2 Single stream governor and meter with by-pass, with meter flexible connection and strainer arrangements

12.2.3 Primary Gas Metering and Meter Correction

Gas Meters

Three types of meter can be used for measuring gas throughput for charging purposes, namely:

- Diaphragm meters which have a turn-down range in excess of 50:1 and may be used for all normal loads up to flow rates of 169 m^3/h.
- Rotary displacement meters (RPD) with a turn-down range of at least 30:1.
- Turbine meters having a turn-down range of approximately 15:1.

Because of the limited turn-down range oversizing of RPD and turbine meters should be avoided although pressure losses should be assessed and this may result in some oversizing being necessary.

RPD meter inertia can give rise to temporary under- or over-pressure conditions where large step load changes occur such as with compressors or permanent pilot appliances. This effect can be minimised by the inclusion of a large reservoir of gas between the meter and the appliances.

Turbine meters should not be used to measure flows which are rapidly pulsating, nor should they be used where the total metered gas flow is on/off unless the on time is greater than 30 minutes. This is because the turbine wheel can continue to rotate for some time after the flow through the meter has ceased.

The option of these meter types is discussed fully in Chapter 14.

Meter accuracy

The required accuracy of primary meters, which are the basis on which consumers are charged, is stipulated by the Gas (Meters) Regulations 1983 as follows:

- For diaphragm meters, accuracy to within ±2% over the flow range 2% - 100% Q_{max}*
- For other types, accuracy to within ±2% from Q_{min}* to 20% Q_{max} and to within ±1% from 20% - 100% Q_{max}.

* Q_{max} and Q_{min} = maximum and minimum rated flow capacity respectively.

For billing purposes, primary meter readings are taken at quarterly or monthly intervals depending on the size of the consumption and are the responsibility of the British Gas plc Region. It is important that consumers should also regularly read both primary and secondary meters (either manually or electronically) to maintain effective records. Such records are essential for energy management, internal cost allocation, budgeting etc. and also the early detection of (rare) meter faults.

Shorter term consumptions are often required, for purposes such as commissioning, performance testing and problem solving. If the consumption measured is a fixed load the gas flow rate will be constant and successive meter readings taken over relatively small equal time intervals will be very similar. If, however, the load is fluctuating and the test period prolonged, then only the initial and final meter readings over an accurately timed period are significant, although periodic intermediate readings can be beneficial to show trends or errors.

Meter correction

Gas consumption is measured by volume but it is heat energy for which the customer is charged and the volume of a given mass of gas is markedly affected by temperature and pressure. For example, a therm equivalent of gas in the national distribution grid at 70 bar and 5 °C will occupy about 1.3 ft^3 whereas a therm equivalent of gas through a consumer meter at 20 mbar and 10 °C will record about 92.8 ft^3. Quite small changes can have an appreciable effect: a rise in temperature from 7 °C to 20 °C at constant pressure will increase volume by nearly 5%, and a rise in pressure from 20 mbar to 80 mbar at constant temperature will reduce volumes by over 5%. It is therefore necessary first to fix standard temperature and pressure conditions and secondly to correct metered volumes where actual conditions deviate, as they almost invariably do.

Standard Temperature and Pressure (STP) as defined in BS 350 are 0 °C (or 273.15 K) and 1013.25 mbar in absolute units, and are in general use for technical and scientific purposes. British Gas uses the Imperial Standard Conditions (ISC) which are defined as the pressure at the base of a mercury column 30 inches high at a temperature of 60 ° Fahrenheit at a latitude of 53° North. A temperature of 60 °F is equivalent to 519.67 ° Rankine. The equivalent metric conditions are 15.55 °C (288.7 K) and 1013.7405 mbar. Metric Standard Conditions (MSC) are 15 °C (288.15 K) and 1013.25 mbar, and when defined as 'dry' constitute the Metric Standard Reference Conditions laid down by the International Gas Union; measurements corrected to this standard carry the prefix 'st' e.g. 13.7 stm^3.

The factor for correcting metered volumes to any standard is obtained as follows:-

$$\text{Correction Factor } CF = Pm/Ps \times Ts/Tm \qquad (12.1)$$

Where Pm = pressure at the meter in absolute units
Ps = standard pressure in absolute units
Tm = temperature at the meter in absolute units
Ts = standard temperature in absolute units

Any units may be used to suit the particular circumstances, providing they are consistent for both Pm and Ps, and for both Tm and Ts, because the units cancel out leaving simple numerical ratios, but they must always be absolute.

Absolute temperature is obtained by adding 273.15 to °C or 459.67 to °F. Absolute pressure is obtained by adding the barometer reading to the gauge pressure, which may require unit conversion, e.g. in Hg to mbar. The practice of using a 'standard' atmosphere instead of the true barometer pressure is to be avoided; atmospheric pressure is rarely equal to standard, and hardly ever so at sites well above sea level.

For any given standard, Ps and Ts are constants and formula (12.1) can be simplified to speed up repetitive calculations. For example, with P in mbar abs, and T in K (i.e. °C + 273.15) the factor for correction to British Gas plc standard conditions becomes:-

$$CF = Pm/1013.7405 \times 288.7/Tm = 0.285\ Pm/Tm \qquad (12.2)$$

It is important that the gas pressure and temperature immediately upstream of the meter are measured accurately to enable a realistic correction factor to be derived. Relatively small errors in temperature and pressure can give rise to significant errors in corrected flow rates and measurements should therefore be made to good accuracy by the use of the correct equipment and techniques. For example, thermometers should be inserted into the gas stream rather than strapped to the pipe surface and manometers (and pressure gauges particularly) must be of the correct scale range to allow accurate pressure readings.

Having divided the metered throughput by the time period to establish the flow rate, Qm, then the thermal input can be calculated as follows:

$$\text{Thermal Input } H_I = Qm \times CF \times CV \qquad 12.3$$

where CV = calorific value of gas.

These considerations obviously have a bearing on the sale of gas to consumers but their application is necessarily a compromise between accuracy and economic practicality, bearing in mind the huge number of meters and readings nationwide. Supplies to domestic and the smaller commercial and industrial customers, i.e. those taking not more than 25 000 therms a year, are not corrected. Higher consumptions, which are sold on individually-negotiated contracts (Special Agreements) are corrected as follows:-

- For 25 000 to 100 000 therms annual nominal consumption, correction factors are generalised and based on nominal temperature and pressure at the meter (barometric pressure is adjusted for height above sea level), using British Gas Standard Factors for Temperature and Pressure Correction.
- For consumptions exceeding 100 000 therms annually, current policy is to have automatic on-site correction (applicable to all new installations and progressively being retro-fitted to existing supplies currently on standard correction). This is achieved by fitting permanent sensors at the meter, which transmit temperature, pressure and meter readings to an adjacent electronic device which derives the correction factor, applies it and displays the corrected reading. It is current practice to use true absolute pressure i.e. line pressure plus local barometric pressure.

The foregoing assumes that natural gas behaves as an ideal gas i.e. it follows the classic gas laws and for practical purposes this is true for the temperatures and pressures encountered in commercial and industrial installations. As pressure increases, however, there is a progressive deviation because of the change in gas compressibility and a further correction factor must be applied for line pressures in excess of 2 bar. For supplies metered at over 2 bar, therefore, formula (1) is extended as follows:

$$CF = Pm/Ps \times Ts/Tm \times Fz \qquad 12.4$$

where Fz = Deviation Correction Factor.

The deviation of Fz for correction to ISC is covered in the British Gas booklet of Standard Factors quoted above and is applied in the appropriate circumstances. It may also be incorporated in automatic meter correctors.

It will be appreciated that what is being achieved is correction for variation in density, which incorporates all three factors: temperature, pressure and compressibility. It is feasible to measure density directly and thus obtain an immediate overall factor which can be applied automatically. The equipment is however expensive and complex and only justifiable for the largest of loads. Furthermore.

the 'standard' density to which the flow is to be corrected can itself change because of variations in gas composition and must itself be monitored to ensure the correct reference base.

12.2.4 Electrical Safety Requirements

At present there is no common standard in the United Kingdom covering the zoning of governor/meter and meter installations. The degree of protection required will depend upon the zone classification and reference should be made to BS 5345, Code of Practice for the Selection, Installation and Maintenance of Electrical Apparatus for Use in Potentially Explosive Atmospheres, for guidance on the hazardous area classification. In general, however, for both types of installation where the gas inlet pressure is at or above 75 mbar they are classed as Zone 2, one of the areas defined in the above Standard. For gas inlet pressures below 75 mbar the appropriate British Gas Region should be consulted regarding local regulations and requirements. Special precautions are not normally necessary for electrical equipment used in or at low pressure meter installations. However, equipment mounted in or on meters to provide an electrical output shall be intrinsically safe and suitable for use in a Zone 2 environment.

Currently, all new Rotary Positive Displacement (RPD) and turbine meters incorporate a volume output telemetry signal and this feature is also available on diaphragm meters. This enables the use of data loggers and remote meter reading facilities, providing the equipment is intrinsically safe as already noted. Customers may make use of the electrical outputs from primary gas meters provided the circuitry is installed under the authority of British Gas plc.

12.3 INTERNAL INSTALLATION

12.3.1 General Provisions

In the United Kingdom, British Gas plc is responsible for the installation of the service pipe, but the customer is responsible for all pipework installations from the outlet side of the primary meter installation, known as the installation pipes. The pipe installation must be carried out in accordance with the Gas Safety (Installation and Use) Regulations 1984, the relevant provisions of which are discussed below.

When designing the intended installation pipework, reference should be made to British Gas plc publication IM/16 relating to installation of pipework boosters and compressors. The installation of natural gas and dual fuel engines should follow the Code of Practice IM/17. Publication IM/21 provides guidance notes on the Gas Safety Regulations while the recommendations of BSCP331 Part 3 relating to installation pipes should be followed. It is essential that any electrical installation conforms to the IEE Regulations for Electrical Installations and BS5345 and BS5501 relating to hazardous environments.

12.3.2 Boosters and Compressors

This section includes a summary of the protection required by British Gas plc in order to comply with the requirements of Schedule 5, Paragraph 8 of the Gas Act 1986 relating to equipment installation and mains protection. It is necessary to give British Gas plc fourteen days' written notice of the intended installation of a gas booster, compressor or engine.

Boosters are centrifugal fan type machines which, in single stage form, are capable of producing a maximum pressure lift of approximately 70 mbar depending upon individual design characteristics. Boosters producing pressure lifts in excess of 70 mbar are considered as compressors.

Compressors are either positive displacement type machines which include rotary sliding vane, rotary screw, Rootes, crescent chamber and reciprocating piston types, or single and multi-stage boosters producing pressure lifts in excess of 70 mbar.

Installation

Boosters and compressors must be installed on a firm bed or platform in well ventilated locations preferably near to the equipment being served thus minimising the length of pipework operating at the higher pressures.

Generally boosters and compressors should not be located in rooms intended specifically to house air compressors unless the air inlet of each air compressor is ducted from outside the room to prevent any gas leakage (for example from glands) being drawn into the air compressor inlet and distributed around the compressed air system. Where gas boosters or compressors are installed in special rooms ventilation apertures should be provided at high and low levels opening to the atmosphere. The fitting of gas detectors should not be regarded as a substitute for good ventilation.

To eliminate strain on the booster or compressor and to reduce noise, the inlet and outlet connections should incorporate a short length of flexible metallic tube. Connecting pipework shall be adequately supported independently of the booster or compressor, and where flexible connections are not used accurate alignment is essential. Where the pipework diameter differs from the connections to the booster or compressor, properly designed taper pieces or concentric reducers shall be inserted as close to the unit as is practicable, to prevent turbulence. In certain circumstances, boosters may show some degree of pressure instability or surging at low gas flows. In these instances a small controlled by-pass, for example one incorporating a 25 mm lockshield valve, should be fitted around the booster and adjusted under no-flow conditions to eliminate the instability. In such cases the manufacturers should be consulted. Where either a booster supplies a large volume of outlet pipework, or a multi-stage booster in installed, a non-return valve should be fitted in the booster outlet to prevent a pressure surge back through the booster when turned off. When units are connected in parallel a non-return valve should be fitted at each outlet to prevent operating problems due to recirculation of gas from one unit back through another unit.

Boosters and compressors should be connected to run continuously as long as the equipment served is on demand and modulating burner controls should be used where possible.

Generally, electrical connections and equipment associated with boosters and compressors need not be flameproof; however, the use of electrically safe equipment will depend upon the classification of any hazardous areas: BS 5345 Part 2 should be consulted.

Protection equipment

In accordance with the Gas Act 1986, each booster and compressor must be provided with a low pressure cut-off switch impulsed from the machine gas supply inlet and suitably wired to prevent the booster or compressor causing a reduced or negative pressure at the meter and in the gas supply system. Thus on loss of gas pressure the switch will cause the booster or compressor to shut down with automatic restart on restoration of gas pressure being prevented with a manual interlock.

Where a compressor takes most or all of the metered gas supply it may be necessary to take additional precautions to prevent nuisance shutdowns due to transient depression of the gas supply pressure on start-up. This can be provided by an additional by-pass system or by fully opening the outlet pressure control relief valve on start-up.

A non-return valve, of a type acceptable to British Gas plc, should be fitted to all boosters having a pressure lift in excess of 70 mbar to prevent admission of extraneous gas or compressed air into the service pipe or gas main. This may not be necessary where a single booster is close coupled to a burner system and all meters, governors and components fitted upstream of the booster are capable of withstanding the maximum boosted pressure without risk to British Gas plc mains. In such cases reference must be made to British Gas plc for written permission.

To maintain a constant outlet pressure or to protect the downstream pipework against an excessive outlet pressure it may be necessary to fit a bypass with a relief valve directly between the inlet and outlet connections of positive displacement compressors. Alternatively, where a bypass is not fitted, a full flow capacity relief should be fitted to the compressor outlet to protect the downstream system from the effects of compressor operation against a closed valve. In addition the pipework on the outlet of the compressor should be suitably sized, or a separate pressure vessel fitted, in order to provide a reservoir to smooth out possible pressure fluctuations. A pressure switch should also be fitted to prevent rise of the discharge pressure above a predetermined figure.

Cycling of the flow intake and cylinder off-loading can give rise to continuous inlet pressure fluctuations. Careful consideration should therefore be given to the design of the system as a whole since for example the use of slam-shut valves on meter/governor installations can lead to nuisance shutdowns with rapid load changes, particularly with respect to cylinder off-loading. Fluctuations can be reduced by the use of modulating speed electric motors or a cooled or temperature protected spill return system rather than cylinder off-loading. Thus, turbine meters should not be used to meter gas supplied to installations incorporating a reciprocating compressor that takes more than 20% of the maximum flow rate.

Advice on meter, compressor and non-return valve interactions in order to ensure the correct selection of components and to prevent inaccurate metering should be sought from British Gas plc.

Schematic installation diagrams

Boosters

Typical installation layouts showing the location of components relative to the booster and one another are depicted in Figures 12.3 and 12.4.

Compressors

Typical installation layouts showing the location of components relative to the compressor and one another are depicted in Figures 12.5 and 12.6. The positions for fitting the additional protective devices described in this Section are also indicated.

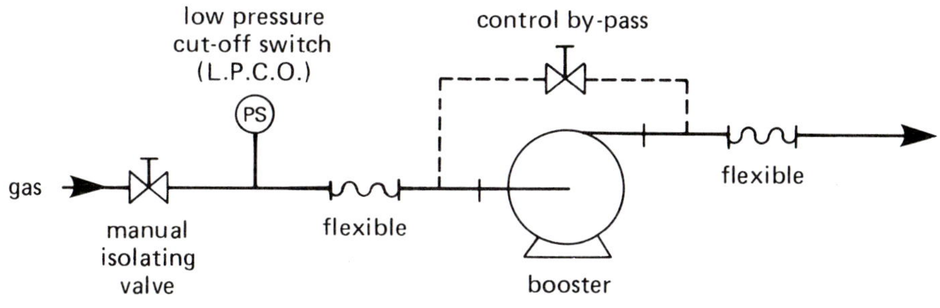

Fig. 12.3 Single Booster Installation

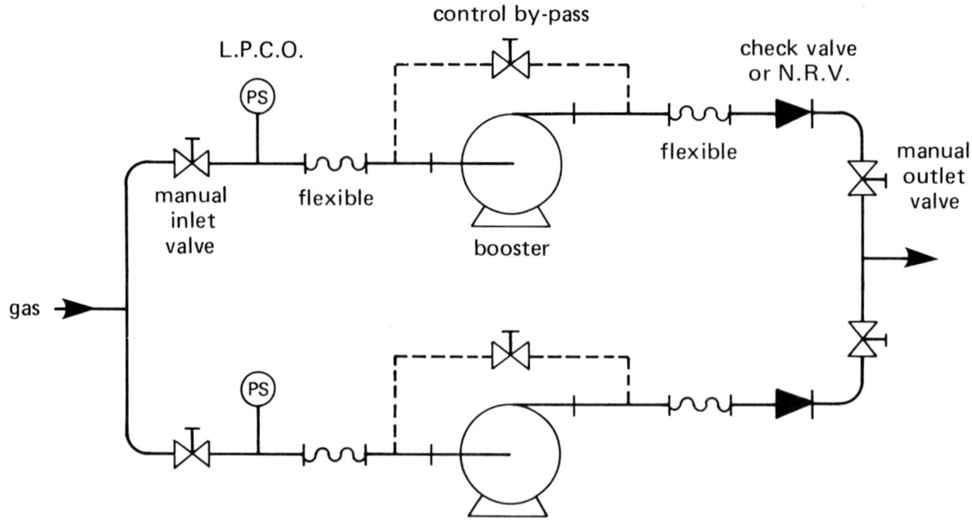

Fig. 12.4 Parallel Booster Installation

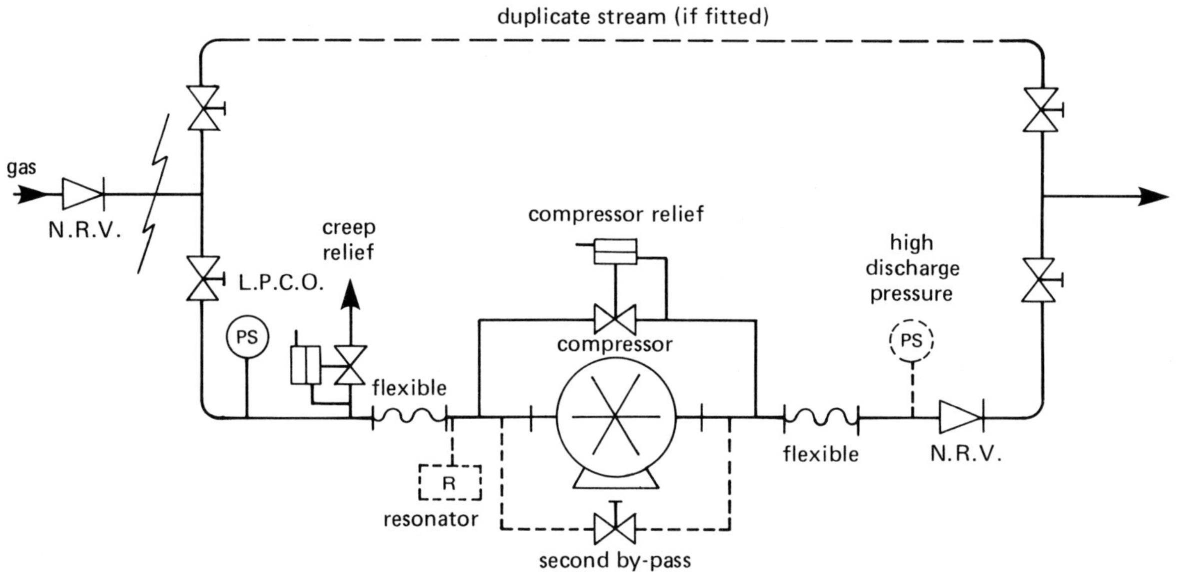

Fig. 12.5 Compressor Installation with Compressor Relief

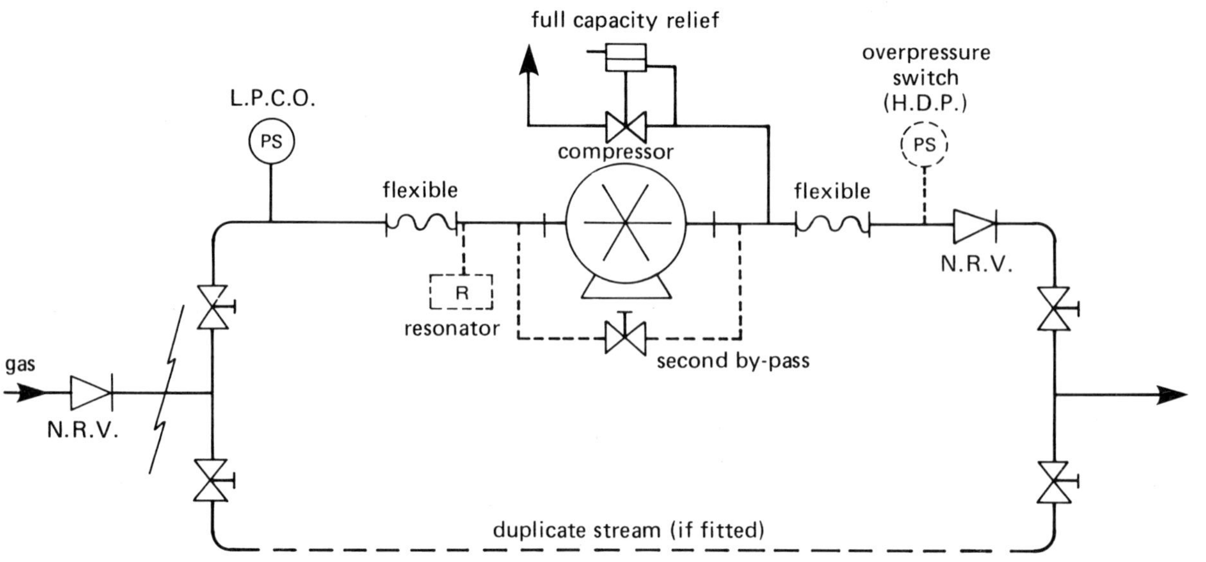

Fig. 12.6 Compressor Installation with Full Capacity Relief

Operating instructions and emergency procedures

British Gas plc publication IM/16 details the requirements for displaying operating instructions and data, together with emergency procedures, adjacent to compressor or compressor type mixer plant.

Machine commissioning

Machines must be commissioned by competent engineers with written commissioning instructions provided by the manufacturer.

Wherever practicable, compressors should be pre-commissioned on air to carry out all the dry run tests on the interlocks, to ensure that the lubricating and cooling system is working correctly and to check or set all relief valve settings. Care should be taken not to overload the machine.

Operation and maintenance

The manufacturer's maintenance instructions and the system designers' instructions should be followed at all times.

The regular servicing of the machines and all ancillaries must be carried out in accordance with the recommendations provided by the manufacturers and system designers.

12.3.3 Pre-Mix Machines – Compressor and Fan Type Mixers

Compressor-type mixers have compressors which incorporate a zero governor on the gas supply and automatically proportion the pre-set air/gas ratio over the operating range.

Fan-type mixers are boosters which incorporate a linked gas valve and air shutter to mix air and gas to a pre-set ratio.

Typical installation layouts showing the location of components relative to the machine and one another are depicted in Figures 12.7 and 12.8.

Installation

The requirements for location, ventilation, mounting, pipe connections and electrical connections are as described for boosters and compressors in Section 12.3.2. In addition, the following apply:-

- Ventilation. The total ventilation area shall be such that the inlet air requirement of the pre-mix machine is supplied with an air velocity through the ventilators not exceeding 2 m/s.
- Mounting. With the smaller units, particularly the fan mixer type, the platforms may be attached to walls by means of brackets, but this may not be suitable for larger models. In all cases the manufacturers should be consulted.
- Pipe Connections. The length of mixture pipework shall be kept as short as possible. With fan-type mixers operating with or close to the limits of flammability at any time during their normal operation the length of the pipework to the burner should not exceed thirty times the pipe diameter (30D) of the machine outlet connection and the pipe volume should not exceed 0.06 m^3 whichever is the smaller.
- Electrical Connections. Where a pre-mix machine produces a mixture within the limits of flammability, all electrical equipment including that on the mixture supply to the burners (for example valves and pressure switches) shall be suitable for use in a Zone 1 area with Group II A gases e.g. Type EE x 'd' or equivalent (see BS 5501 Part 5, Electrical Apparatus for Potentially Explosive Atmospheres Flame Proof Enclosures).

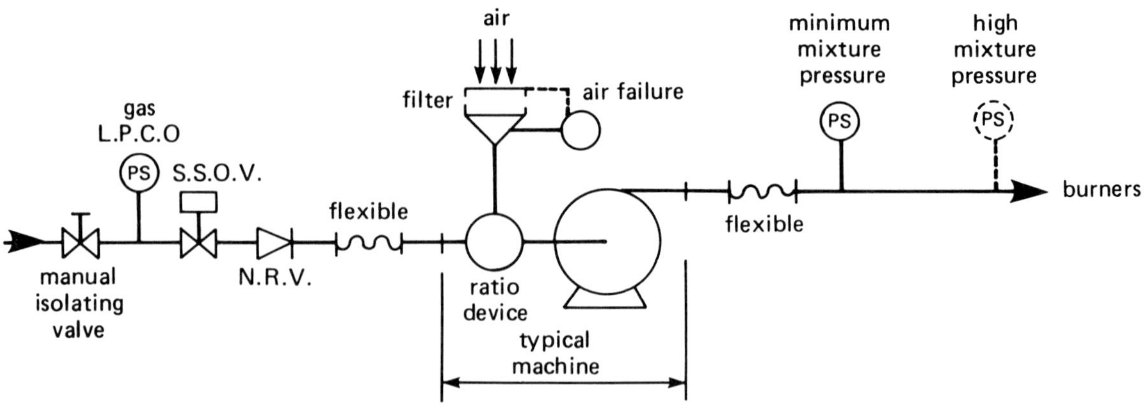

Fig. 12.7 Fan Type Mixer

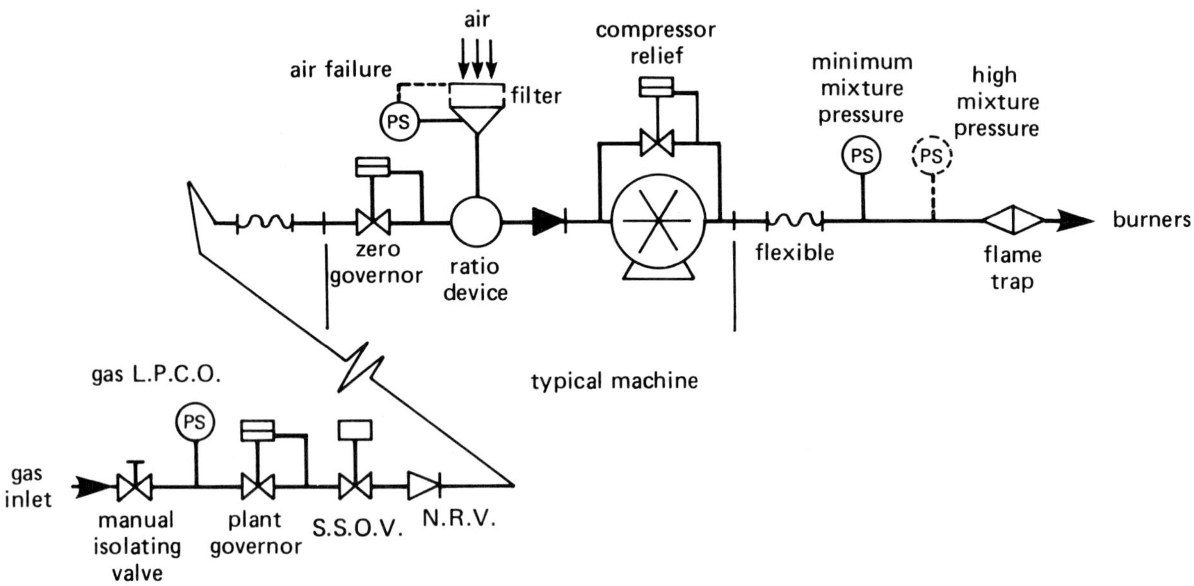

Fig. 12.8 Compressor Type Mixer

Protection equipment

As detailed in Section 12.3.2 low gas pressure cut-off switches and non-return valves are necessary to prevent pressure fluctuations in the supply mains as well as the admission of extraneous gas. Additional protection specific to premix equipment is necessary to ensure safe operation and distribution of flammable mixtures. These requirements are outlined below.

In the event of low gas pressure causing the machine to shut down, the gas supply must also be shut off and require manual resetting. The operation of the safety shut-off valve should be such as to minimise the distribution of mixtures within the limits of flammability during the machine shut-down sequence, for example by delaying valve closure until the machine is almost at rest.

The pre-mix machine must be protected from flashback, for example by the installation of flame traps in the mixture pipework both at the inlet to the burner zones and on the machine outlet. Flame traps should not be inserted in the mixture pipework from fan-type mixers as they may adversely affect the operation of these machines.

A differential pressure switch or similar device should be fitted to stop the mixer and automatically shut off the gas supply in the event of a reduction in the air supply or a blockage of the air inlet. In addition, a pressure switch should be fitted on the mixture outlet from the machine to automatically shut down both the machine and the gas supply in the event of the mixture pressure falling below the pre-set minimum.

When the air/gas mixture required is outside the limits of flammability, means such as mechanical stops should be used to prevent the machine from producing a flammable mixture.

In some cases it may be desirable to install a pressure switch in the mixture line to shut down the machine in the event of excess discharge pressure.

Ancillary equipment, commissioning and operation

The requirements for these are as described in Section 12.3.2 for boosters and compressors. Full details are given in British Gas plc publication IM/16.

12.3.4 Secondary Gas Metering

A secondary gas meter is one inserted in the installation pipework downstream of the main primary meter but upstream of any plant/appliance main isolating valve. A secondary gas meter is used for billing purposes, unlike a check meter which is installed merely to monitor gas usage in a part of industrial or commercial premises. Meter types, accuracy and other requirements are the same as for primary meters detailed in Section 12.2.3.

Installation Requirements

Secondary meters for billing in industrial situations are the responsibility of British Gas plc. If used for the monitoring of gas usage only, this will be the responsibility of the factory occupier. Every meter, together with a meter control, meter bypass, meter governor or other gas fittings to be used in conjunction with a secondary meter must be installed in accordance with Gas Safety Regulations 1984, i.e. as applicable to Primary meters. The relevant provisions are detailed in the British Gas plc publications IM/21 and IM/112.

On completion of the meter installation it should be tested to verify that it is gas tight and installed in accordance with the Gas Safety Regulations. The complete installation, including gas fittings connected to the installation and not previously used, must then be purged: refer to British Gas plc publications IM/2 and IM/5.

Where secondary meters are installed, a notice in permanent form must be prominently mounted on or near the primary meter indicating the number of secondary meters installed in addition to the requirement for displaying the procedure in the event of an escape of gas.

12.4 VENTS

Vents are fitted, for example, to pressure relief governors, safety shut-off valve systems, governor breathers, plant room emergency vents, atmosphere generator vents and commissioning vent valves. Vent pipe terminals are designed to minimise the risk of blockage by foreign matter and the ingress of water and typical examples are illustrated in Figure 12.9.

Vents must be laid and terminated in a safe place, preferably in the open air above roof level. If it is impracticable to terminate above roof level, great care shall be taken to vent into a safe place remote from potential ignition sources and where there is no risk of vented gases accumulating to cause a hazard on entering buildings.

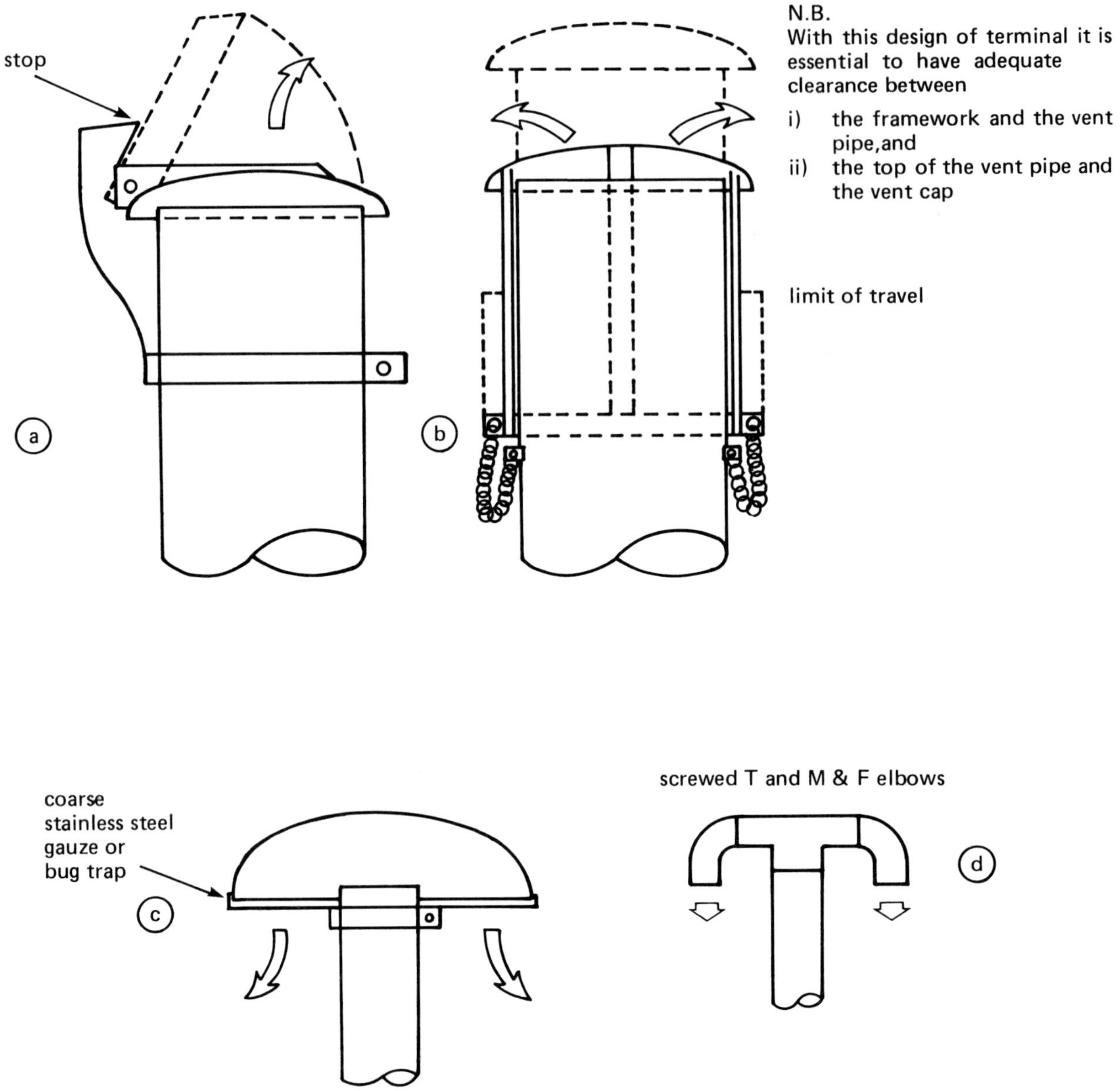

a and b applicable to pressure relief duties

c and d applicable to safety shut-off valve systems

Fig. 12.9 Typical Vent Terminals

Governors with integral relief valves and relief governors must be fitted with vent pipes. In addition, it may be necessary to fit vents to the breather holes of governors without integral relief valves if they are installed in confined spaces. Vent pipes should then be fitted so as not to impair governor performance.

Vents from governor breathers and large relief valves must not be manifolded together or to any other vent. However those from safety shut-off valve systems may be manifolded together and the vents from auxiliary and similar small relief valves may be connected to a common vent stack provided that it is designed to avoid any interaction between the relief valves.

Vent points used to purge and commission pipework and burner controls should be valved, and where not permanently connected to a vent pipe they should be plugged or blanked off prior to the normal operation of the plant: refer to British Gas plc publications IM/2 and IM/16.

Sizing of Emergency Vent Valves

It is suggested that the pipework downstream of the emergency isolating valve should be vented in an emergency to below 1 bar in less than 10 s where practicable. The size of the vent can be calculated from the following formula and table of pressure correction factors, (K):

$$d = 0.04\, KD\sqrt{L(P-1)} \qquad 12.5$$

where d = nominal diameter* of the vent valve (same units as D).
D = nominal diameter of the main gas line (same units as d).
L = length of the main gas line (m). Where the diameter of the main gas line changes, L is the length of pipe of diameter D equivalent to the actual volume of the pipework
P = line pressure (bar)
K = pressure correction factor, as follows:

Line pressure (bar)	1	2	3	4	5
K	1.0	0.9	0.82	0.76	0.71

Irrespective of the above formula the nominal vent valve diameter should never be less than 25 mm or greater than the nominal diameter of the main gas line.

*The port size of the vent valve must not be less than the nominal diameter given by the formula

12.5 INSTALLATION REQUIREMENTS

Installation requirements are governed by the Gas Safety (Installation and Use) Regulations 1984 and guidance notes are provided by British Gas plc publications IM/16 and IM/21. This section summarises the main recommendations.

12.5.1 Gas Flow

Pipework should be sized on the basis of the maximum gas flow rate with an allowance for any possible increase in the load. The maximum gas flow rate in any installation is not necessarily equivalent to the total connected load.

Appropriate calculators are available from various suppliers and are convenient aids for the calculation of gas flows, velocities and pressure drops in pipes. Refer to Chapter 4 for the underlying theory and basic principles and British Gas plc publication IM/16 for tabulated values.

12.5.2 Gas Pressure Losses

The pressure loss in a pipework installation should be assessed with due allowance for the effects of elbows, tees, valves and other fittings.

For low pressure gas supplies metered at 21 mbar the pressure drop between the primary meter and the plant manual isolating valve at maximum flow should not normally exceed 1 mbar.

For gas supplies where the pressure at the outlet of the primary meter is by design greater than 21 mbar, the pressure drop between the primary meter and the plant manual isolating valve at maximum flow should not normally exceed 10% of that pressure.

The buoyancy property of fluids lighter than air is discussed in Chapter 4 and is significant for natural gas which is only about 60% as dense as air. The corresponding altitude effect on gas in vertical pipes is a pressure gain of 0.5 mbar per 10 m of increase in height. This will be particularly relevant in high rise buildings where the pressure gained could be used for the reduction of pipe or control valve sizes and when determining the test pressure required. It is important to remember the converse: there is an equivalent pressure drop for decrease in height.

12.5.3 Gas Velocity

A gas velocity of 45 m/s should not be exceeded on supplies filtered down to 250 μm. In the exceptional case of unfiltered supplies the maximum velocity should not exceed 20 m/s.

These velocities may be exceeded up to a maximum of 75 m/s on the basis of engineering judgement and experience with due regard to the possibility of noise and erosion with high gas velocities and unfiltered supplies.

12.5.4 Gas Filters

It may be desirable to install a filter to limit the particle size to protect any controls fitted in the installation pipework. When fitted, pressure test points should be used to enable checks to be made on the filter performance.

When welding, precautions should be taken to prevent the ingress of debris into pipework during installation. Debris travelling at high velocities within pipes may severely damage valves, filters and other fittings.

12.5.5 Pipe Layout

Installations must be designed to include the valves necessary to provide section isolation as required for soundness testing, for purging and for use in emergency. When designing a new installation consideration should be given to the inclusion of connections for possible extensions.

It is necessary to display a line diagram of the installation pipework for buildings where the gas service to that building is greater than 50 mm or the metering pressure is in excess of 50 mbar. This line diagram (Figure 12.10) should be fitted as near as practicable to each primary meter in the building and should indicate the position of isolating valves, meters, meter controls, pressure test points, condensate receivers and electrical bonding. It is also advisable to display the diagram at the entry to the factory site such as the commissionaire's office or gate house.

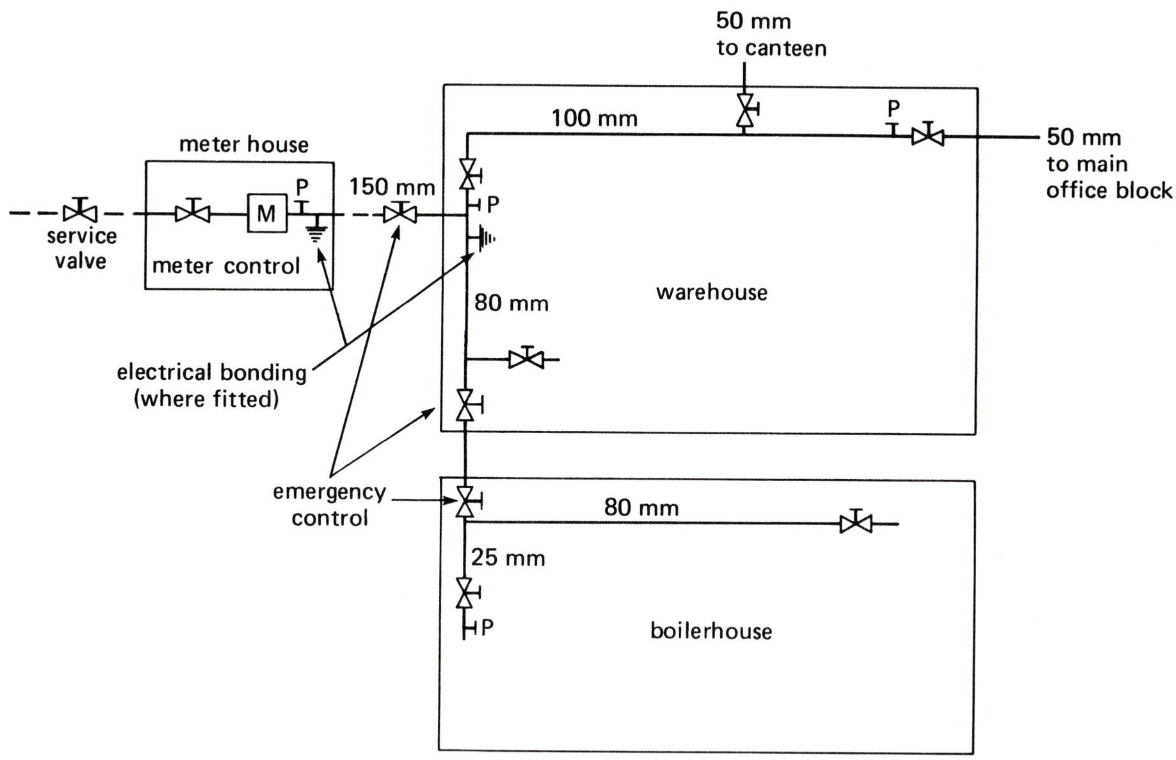

N.B. Copies of this diagram are displayed as follows:

- Meter House
- Gate House
- Engineers

Fig. 12.10 Gas Supply Line Diagram

12.6 INSTALLATION PRACTICE

In the United Kingdom, British Gas plc is responsible for the installation of the service pipe and governor/meter installation, but the customer is responsible for all pipework installations from the outlet side of the primary meter installation, known as the installation pipes. The pipe installation must be carried out in accordance with the Gas Safety Regulations 1984 together with BSCP331 and British Gas plc publications IM/16, IM/5 and IM/2, the relevant provisions of which are discussed below.

12.6.1 Pipework

All installation pipes and pipe fittings must be of good construction and sound material and of adequate strength and size to ensure safe operation. The installation must be carried out by competent persons.

Installation pipes must not be installed under the load bearing foundations of buildings nor under load bearing walls or footings. They must pass through cavity walls by the shortest route, and any pipe passing through solid floors, cavity or fire resistant walls or load bearing walls must be sleeved. The sleeve should be fire resistant with a non-setting mastic material applied at one end between the pipe and the sleeve.

All pipes, and particularly those in a corrosive atmosphere, must be protected against erosion, either by the use of a material inherently resistant to corrosion or by protective coatings or wrappings.

Pipes must not be installed in hazardous areas if they cannot be used with safety, and all pipes must be electrically bonded to any other nearby services to ensure that the gas pipes are the same electrical potential as the electrical earth.

Condensate receivers should be installed where there is a risk of condensate collecting in the pipes due to changes in level in the pipework.

For industrial and commercial premises in a building in which gas is supplied by a service greater than 50 mm to two or more floors, a valve must be fitted in the installation pipe to each floor to which gas is supplied enabling that floor to be isolated. This also applies where a single floor is divided into self-contained areas.

After completion of the installation the pipework and fittings should be pressure tested and purged in accordance with the procedures outlined in British Gas plc publications IM/2 and IM/5/

Pipe supports

All pipes must be properly supported and installed or protected so that there is no undue risk of accidental damage or prejudice to their safety. For instance, installation pipes should not be fitted in electrical intake chambers, transformer rooms and lifts. No pipes should be installed in such a manner as to impair the structure of the building or the fire resistance of any part of its structure.

Flanges and pipe support methods are fully described in British Gas plc publication IM/16.

Materials

The choice of materials for pipework and fittings depends on the application, location and pressure limitations. Generally selection is made from steel, ductile, malleable or cast iron, copper, brass or bronze. Polyethylene pipe can be used for buried pipe installations only out of contact with daylight.

Jointing

The type o,f jointing may be restricted by the grate of pipe used. For example some grades of steel pipe are not suitable for the application of screwed threads and flanges; welding or other fitting methods must be used.

With polyethylene pipe fusion welding or compression couplings, where appropriate, may be used.

Flexible connections

Flexible connections should be considered where it is expected that pipework will be subjected to vibration, expansion or strain. They should not be used where there is a practical alternative and when used their length should be kept to a minimum.

A range of types of flexible couplings are available depending on the type of movement anticipated and the suitability for each installation should be assessed based on engineering judgement and in conjunction with British Gas plc publication IM/16.

Pipework identification

Installation pipes should be identified by painting or banding yellow and the pressure stated where this exceeds 75 mbar. A line diagram of the installation must be provided as shown in Figure 12.10 and detailed in Section 12.5.

Pipework in Ducts

All ducts should have through ventilation to prevent the unsafe accumulation of gases. There should be ventilation openings at each end of ducts to have a free open area equal to 1/150th of the cross sectional area of the duct or 0.05 m^2 whichever is the greater.

12.6.2 Electrical Installation Aspects

Electric supply cables and other services should be spaced at least 25 mm from any gas installation pipe or fitting. Pipework should not be fitted at electrical intake chambers, transformer rooms or lift shafts.

Where electrical bonding of gas pipework is necessary work should be carried out in accordance with IEE Regulations for Electrical Installations.

The protection requirements for electrical installations in hazardous areas and for the provision of output telemetry signals have been outlined in Section 12.2.3 and Section 12.3.2.

12.6.3 Commissioning of Installation on Completion

Commissioning is covered by British Gas plc publications IM/16, IM/5 and IM/2. Summarising the main recommendations, the following items should be checked before the installation may be considered to have been completed:

- The necessary installation drawings have been prepared and are fitted adjacent to the meter.
- The pipework is adequately supported.
- Handles or handwheels, as appropriate, are fitted to all valves and clear indication is provided for the open and closed positions.
- Any necessary cross bonding has been carried out by a qualified electrician.
- The necessary test and purge points are provided, especially at the ends of pipe runs.
- Before soundness testing, all joints on underground pipework which remain exposed are securely anchored if appropriate, and that due care is taken with regard to the safety of personnel in the vicinity of the excavations.
- All pipe ends are spaded, plugged or capped and the meter and other sensitive equipment are similarly protected prior to commencement of the soundness test.
- A satisfactory soundness test is carried out and subsequent connections between new and existing pipework are tested, for example with leak detection fluid (n.b. not soaps or detergents with stainless steel bellows or tubes), prior to painting, wrapping or back filling.
- The appropriate Region of British Gas plc is advised after completion of a satisfactory soundness test to enable any necessary arrangements to be made to turn on the gas supply to the installation.
- Pipework is correctly wrapped or painted and colour coded.
- The installation is satisfactorily purged immediately after a pressure test and any outlets which are not immediately put to use are suitably sealed, for example by plugging or capping-off.
- Any remaining excavations are back filled and reinstated after wrapping or painting any test connections.
- Any surplus backfill and waste materials are removed from site.
- Ancillary equipment such as valves, gas detectors, boosters, compressors etc. have been installed and commissioned in accordance with manufacturer's instructions. Check with leak detection fluid that the high pressure outlet pipework on and from machines is gas tight.

12.6.4 Gas Usage

All plant and equipment connected to the gas installation must be operated in a correct and safe manner.

If, at any time, there is a gas escape then that part of the installation affected must be immediately isolated from the gas supply. The gas supply should not be opened until all necessary steps have been taken to prevent the gas from again escaping, although brief and controlled restoration of supply is permissible to assist in leak detection.

If the meter control valve is turned off in order to isolate an escape of gas and the escape continues, the British Gas plc Region must be informed immediately.

12.6.5 Maintenance and Repair of Installations

An electrical connection must be maintained by means of temporary continuity bonding while a gas pipe, pipe fitting or meter is being removed or replaced, until such work has been completed. The only exception is in the case of a meter when the inlet and outlet form a continuous single pipe. The bond should be not less than 1.2 m of single core tough rubber sheath flexible cord or equivalent having a cross-sectional area of not less than 4 mm^2 (56/.30) 250 volt grade to BS 6500 with robust crocodile clips

On removal of any gas fitting, all the pipes to which it was connected must be capped or plugged with the appropriate pipe fitting.

If any part of the gas installation is altered or replaced, the altered or replaced part must be brought up to the standard of the Gas Safety Regulations.

The installation of replacement meters must comply with the Gas Safety Regulations (see Sections 12.2 and 12.3).

On completion of any work carried out on the installation, it is the responsibility of the person carrying out the work to ensure that the Gas Safety Regulations have been complied with in all respects and that the section on which the work has been carried out is tested to verify that it is gas tight.

REFERENCES

BRITISH GAS PLC PUBLICATIONS

IM/2 2nd Edition 1975 : Purging Procedures for Non-Domestic Gas Installations. (Recommended procedures for introducing and removing gas from gas supply pipework in premises)

IM/5 3rd Edition 1986 : Soundness Testing Procedures for Industrial and Commercial Gas Installations.
Test Record Forms for Soundness Testing. Packs of 25 for New or Existing Installations (specify which). Metric Units only.

IM/6 1st Edition 1974 : Recommendations for the Installation of Low Pressure Cut-off Switches to Conform with the Gas Act. (Describes the application of pressure switches to safeguard gas booster and compressor installations.)

IM/15 1st Edition 1981 : Manual Valves – A Guide to Selection for Industrial and Commercial Gas Installations.

IM/16 4th Edition 1989 : Guidance Notes on the Installation of Gas Pipework, Boosters and Compressors in Customers' Premises. (Incorporates IM/6 as an appendix.)

IM/17 1st Edition 1981 : Code of Practice for natural Gas Fuelled Spark Ignition and Dual-fuel Engines.

IM/21 1st Edition 1985 : A Guide to the Gas Safety Regulations. (A British Gas plc explanation and interpretation of these statutory regulations concerning the use and installation of gas pipes and gas appliances, superseding IM/4.)

IM/112 1st Edition 1982 : Industrial and Commercial Metering Installations (Inlet Pressures not Exceeding 7 bar).

3rd Edition 1984 Standard Factors for Temperature and Pressure Correction

BGC/PS/C6		Ductile iron pipe and pipe fittings.
BGC/PS/PL2	Part 1	Polyethylene pipe.
	Part 2	Polyethylene fittings.
BGC/PS/PL3		Polyethylene pipe fittings.
BGC/PS/V7	Part 1	Pipe markings and valves.
BGC/PS/M7		Industrial and Commercial metering installations.

BRITISH STANDARDS

BS CP331 Code of Practice for the Installation of Pipes and Meters for Town Gas (Applicable to natural gas):
  Part 3 Low Pressure Installations Pipes.
BS CP326 The Protection of Structures Against Lightning.
BS CP413 Ducts for Building Services
BS 10 Flanges and Bolting for Pipes, Valves and Fittings. (Obsolete, see BS 4504).
BS 21 Pipe Threads for Tubes and Fittings Where Pressure-Tight Joints are Made on Threads
BS 143 & 1256 Malleable Cast Iron and Cast Copper Alloy Screwed Pipe Fittings for Steam, Air, Water, Gas, Oil.
BS 476 Fire tests on Building materials and Structures:
  Part 8 Test Methods and Criteria for the Fire Resistance of Elements of Building Construction
BS 746 Specification for Gas Meter Unions and Adaptors

BS 864 Capillary and compression Tube Fittings of Copper and Copper Alloy:
Part 2 Specification for Capillary and Compression Fittings for Copper Tubes.
BS 1387 Steel Tubes and Tubulars Suitable for Screwing to BS 21 Pipe Threads.
BS 1426 & 3461 Surface boxes for Gas and Waterworks Purposes.
BS 1552 Specification for Control Plug Cocks for Low Pressure Gases.
BS 1560 Steel Pipe Flanges and Flanged Fittings for the Petroleum Industry.
BS 1710 Identification of Pipelines.
BS 1740 Wrought Steel Pipe Fittings (screwed BSP thread).
Part 1 Metric Units.
BS 1832 Oil Resistant Compressed Asbestos Fibre Jointing.
BS 2051 Tube and Pipe Fittings for Engineering purposes:
Part 1 Copper and Copper Alloy Capillary and Compression Tube Fittings for Engineering purposes.
BS 2494 Materials for Elastomeric Joint Rings for Pipework and Pipelines.
BS 2640 Class II Oxy-acetylene Welding of Steel Pipelines and Pipe Assemblies For Carrying Fluids.
BS 2815 Compressed Asbestos Fibre Jointing.
BS 2871 Specification for Copper and Copper Alloys, Tubes.
BS 2971 Specification for Class II Arc Welding of Carbon Steel Pipework for Carrying Fluids.
BS 3601 Steel Pipes and Tubes for Pressure Purposes : Carbon Steel with Specified Room Temperature Properties.
BS 3974 Pipe Supports:
Part 1 Pipe Hangers, Slider and Roller Type Supports.
BS 4161 Specification for Gas meters.
BS 4368 Carbon and Stainless Steel Compression Couplings for Tubes.
BS 4375 Unsintered PTFE Tape for Thread Sealing Applications.
BS 4504 Flanges and Bolting for Pipes, Valves and Fittings. Metric Series.
BS 4683 Electrical Apparatus for Explosive Atmospheres:
Part 3 Type of Protection N.
BS 4772 Specification for Ductile iron Pipes and Fittings.
BS 4800 Specification for Paint Colours for Building Purposes.
BS 5292 Specification for Jointing Materials and Compounds for Installation Using Water, Low Pressure Steam or 1st, 2nd and 3rd Family Gases.
BS 5345 Code of Practice for the Selection, Installation and Maintenance of Electrical Apparatus for Use in Potentially Explosive Atmospheres (Other Than Mining Applications or Explosive Processing and Manufacture):
Part 2 Classification of Hazardous Areas.
BS 5501 Electrical Apparatus for Potentially Explosive Atmospheres
Part 5 (Flame Proof Enclosures).
BS 5834 Surface Boxes and Guards for Underground Stopvalves for Gas and Waterworks Purposes:
Part 1 Specification for Guards
BS 5885 Specification for Industrial Gas Burners of Input Rating 60 kW and Above.
BS 6129 Code of Practice for the Selection and Application of Bellows Expansion Joints for use in Pressure Systems:
Part 1 Metallic Bellows Expansion Joints.
BS 6400 Specification for Installation of Domestic Gas Meters (2nd Family Gases).
BS 6500 Insulated Flexible Cords

OTHER PUBLICATIONS

The Gas Act 1986.
The Gas (Meters) Regulations 1983.
The Safety Regulations 1972, with later deletions and amendments, now covering service pipes only.
The Gas Safety (Installation and Use) Regulations 1984, which revoke all (except for Parts I, II, VII and VIII) of the 1972 Regulations.
Health and Safety at Work Act 1974.
Guide to the Use of Flame Arrestors and Explosion Reliefs.
The Building Regulations – latest edition.
Regulations for Electrical Equipment. Inst. of Electrical Engineers, latest edition.

Institution of Gas Engineers

IGE/GM/1 1987. Gas meter installations for Pressures not Exceeding 100 bar.
IGE/GM/3 1977. Gas Measurement : Factors Affecting the Overall Accuracy of Metering Installations.
IGE/TD/3 Main Laying.
IGE/TD/4 Edition 2. Gas Services.

Power Production from Gas 13

CHAPTER 13

POWER PRODUCTION FROM GAS

13.1 INTRODUCTION

There are many situations in industry where electricity is the only feasible energy form at the point of use. Obvious examples are lighting, conveying and industrial processes such as induction heat treatment and arc melting. Given efficient and cost effective methods for the generation on-site of electricity and shaft power and appropriate heat and power demands the market for gas as the primary energy source is potentially large.

Earlier site generation schemes were based on steam and/or gas turbines or in some instances large reciprocating engines. With the development of smaller packages incorporating alternators driven by reciprocating spark ignition gas engines an important market has opened to the Gas Industry. The system economics rely on recovering engine waste heat and on extended operation by satisfying site base load requirements. These small scale gas engines have been applied to direct drive systems such as pumps and air compressors and this application will help to broaden the market share for gas.

The recent development of gas engine based heat pumps has further extended the role of gas-produced shaft power though such systems are only economically attractive where extended operation and moderate temperature differences are assured. In addition the development of steam drying and mechanical vapour recompression based on gas engine and heat pump technology has the potential to significantly improve energy efficiency of a wide number of industrial processes such as drying and evaporation.

While the efficiency of conventional generation methods is often relatively low several new methods of power generation still under development such as fuel cells are giving promising results.

The following sections examine these developments in gas technology together with future longer term developments.

13.2 TOTAL ENERGY

13.2.1 Background

Total energy refers to schemes to supply the energy requirements of a site from a single source. Conventional power stations use the steam cycle to turn heat into electrical energy with an efficiency of up to 35-40%, representing waste heat losses of up to 60%-65% or more: a conventional power generation system is shown in Figure 13.1.

It is possible to design generating systems so that some of the reject heat is recovered for productive use. Electricity and/or mechanical power can be produced from the shaft of a suitable prime mover and the heat in the exhaust gases or in any cooling water may be used for space heating in buildings or for process uses in industry. This will usually mean the provision of hot water or low pressure

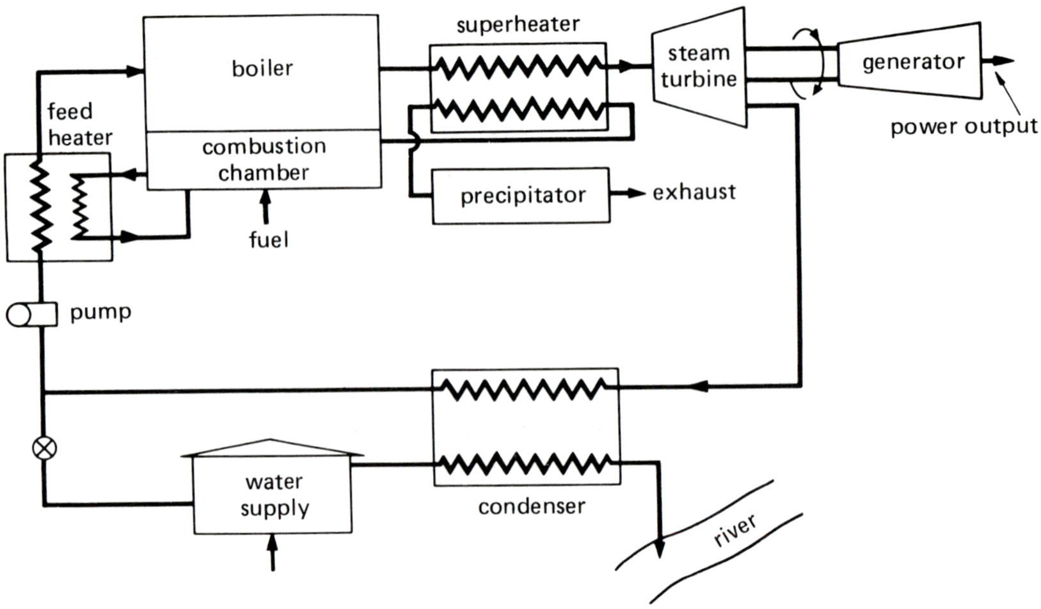

Fig. 13.1 Conventional Power Generation

steam in domestic or commercial applications, and high pressure steam or hot gases on industrial sites.

The use of a single energy complex to produce heat and power is not new. Between the 1950s and the mid-1970s in the UK a number of steam turbine systems with a heat-to-power ratio of between 6:1 and 10:1 were installed in industries such as iron and steel, paper and bulk chemicals. These units were generally well matched to the requirements for heat and electricity therefore making the sites concerned largely independent both of the National Grid Company and of primary heating systems.

Whilst these schemes showed a good economic performance at the time a number of factors have reduced the attractiveness of this approach. The most fundamental change which has occurred in recent years is the reducing price ratio between electricity and fossil fuels, which in turn reduces savings and increases the payback period on plant. In addition during recent years typical site ratios of heat-to-power demand have roughly halved. This is due to both a move away from energy intensive production and to improvements in energy efficiency, particularly in relation to process heating loads. This reduction in the load factor of the plant has led not only to the closure of such plant due to its generally poor turndown performance, but has also discouraged investment in new systems. There has thus been an overall decrease in installed industrial site total energy system capacity and by 1982 only 6% (16 TW h) of the total UK generation capacity of 267 TW h was provided by industrial private generation.

Steam turbines are therefore the main prime movers used in very large plant such as that used for the production of electricity by the Power Generating Companies and in very large factories.

Gas turbine systems which better match present-day site heat-to-power ratios can prove cost effective in industries where total energy schemes have not traditionally been used.

Reciprocating engine systems have also been used as the basis for total energy schemes.

The systems described above are relatively large scale as even the smaller industrial schemes have electrical outputs over 1 MW. In recent years, however, smaller 'micro' Combined Heat and Power (CHP) systems based on industrial reciprocating engines and capable of heating individual buildings have been produced.

13.2.2 General Economics

Total energy is viable provided certain conditions are satisfied. Firstly, the energy should be supplied at a cost saving when compared with conventional supplies such that a significant reduction in overall production or operating costs is achieved. Operation and equipment reliability must be guaranteed, with maintenance services that are both readily available and cost effective.

Since substantial capital investment will be required for any on-site generation scheme, site energy efficiency should first be improved through more conventional cost effective methods. The resulting site heat load and electrical demand should then be assessed in detail to size the system for base load operation.

To establish whether an on-site generation scheme is likely to be profitable one must then compare the cost of producing the various forms of energy on-site with alternative purchased supplies.

The cost of generating power on-site can be expressed as the sum of operating costs and capital charges. Operating costs will include cost of fuel, maintenance and supervision. Capital charges will represent net installed capital cost over that which may be involved in a conventional energy supply system. The value of recoverable heat from a prime mover depends on the process for which it is to be used. In the form of steam or hot water it is equivalent to the cost of the cheapest alternative boiler fuel which may be used. With combustion products for direct use the value depends on whether a premium or crude fuel would be suitable.

These operating costs will depend on the prime mover best matched to meet the site base load and the resultant differences in shaft efficiency and part load characteristics, together with the fuel type and its effect on maintenance costs.

13.2.3 Power Generation by Steam

Since most industrial sites now have heat-to-power ratios of 4:1 or less, steam turbine based systems are therefore unlikely to present a viable total energy option. Nonetheless steam turbine systems are important for large scale electrical generation, up to 1 000 MW was found in some large European district heating schemes and in industries which use large quantities of steam such as agricultural products and sugar manufacturers. The three main types of turbo alternator installations are:

- The straight condensing turbine with an independent supply of process steam.
- The straight back pressure turbine.
- The pass-out condensing turbine.

Condensing Turbines

A large-scale condensing turbine takes steam at about 530 °C and 110 bar boiler pressure and expands it through the nozzles with bleeding to a sub-atmospheric pressure or vacuum where the condenser condenses it to water which returns to the boiler. This is the basis of Power Generating Companies' power station operation, and a well arranged set of this type can produce up to 36% of the net calorific value of the fuel supplied to the plant in the form of electricity. The condensed water is, however, at too low a temperature to be of any value other than as boiler feedwater. Electrical outputs of 1 000 MW or more can be achieved.

The process steam is supplied either from the same boiler, that is at a point in the system before the turbine, or from a separate boiler. Although such a system provides maximum flexibility for meeting varying process steam and electrical modes, it is the least efficient overall.

Back Pressure Turbines

The straight back pressure turbine takes steam from the boiler and produces power by expanding the steam through the turbine stages before finally exhausting the whole of the steam flow to the heating process; the temperature of the exhaust steam typically ranges up to 150 °C. The supply of process steam is thus from a point in the system after the final turbine expansion. This system provides the maximum economy and the simplest installation.

A back pressure turbo-alternator system is most efficient when the power demand corresponds to the output obtainable from the desired steam flow. This is seldom realised in practice and electrical load and process steam requirements often fluctuate widely and are out of step with each other, maximum electrical power often being required at times of low process steam demand. Figure 13.2 shows a back pressure turbine generation system.

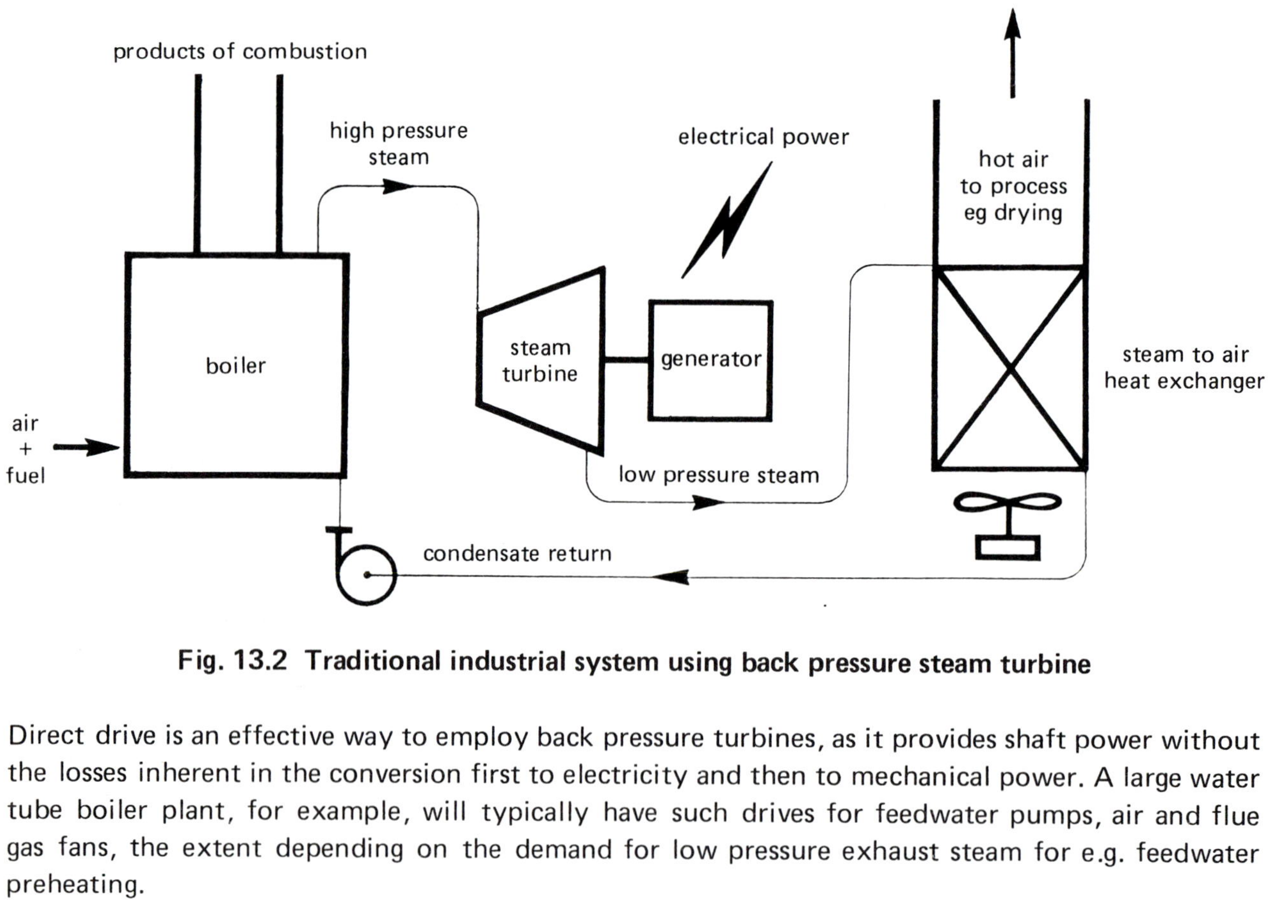

Fig. 13.2 Traditional industrial system using back pressure steam turbine

Direct drive is an effective way to employ back pressure turbines, as it provides shaft power without the losses inherent in the conversion first to electricity and then to mechanical power. A large water tube boiler plant, for example, will typically have such drives for feedwater pumps, air and flue gas fans, the extent depending on the demand for low pressure exhaust steam for e.g. feedwater preheating.

Pass-Out Condensing Turbines

When the process steam demand is small in relation to the electrical demand or if it varies widely throughout the day, a pass-out condensing turbine may provide the best solution. This consists of the combination into a single machine of a back pressure turbine taking steam at the boiler pressure and exhausting to the process pressure followed by a condensing turbine taking additional steam at the process pressure to make up the total electrical load. The turbine is speed governed and a pressure regulator maintains a constant steam pressure at the pass out branch irrespective of fluctuations in the demand for process steam by varying the opening of a pass-through valve controlling the steam flow to the turbine exhaust.

The pass out condensing turbine therefore has the advantage that it can simultaneously control the electrical power output and the pass out pressure with both demands fluctuating. Such turbine plants may be large, often several hundred MW in size and in fact are widely used to operate large-scale European city heating schemes. Efficiency of electrical production is slightly less than that offered by conventional condensing turbine plant.

General

These basic turbine types may be further combined for industries requiring steam at several different pressures. A single machine may thus be pass-out/back pressure, or multi pass-out/condensing.

The steam power concept in this context is only of indirect interest to the gas industry, i.e. only if the boiler plant itself is gas fired. A basic understanding is however desirable because these considerations are directly relevant to, for example, gas turbine combined cycles discussed in Section 13.5.

13.3 PRIME MOVERS

Turbines and engines have now reached a high level of technical development and reliability. The selection will depend on the power range and heat-to-power requirement and on particular characteristics which may be most beneficial. The characteristics to be taken into account for each prime mover are considered below. Steam turbine characteristics have already been discussed in relation to total energy schemes, since the achievable heat-to-power ratio is not generally reflected in present industrial requirements.

13.3.1 Gas Turbines

Gas turbines are particularly convenient for on-site power generation, when compared with steam turbines, because, as internal combustion engines, they need no boilers and condenser. Also, the pressures are much lower than those in steam systems.

The gas turbine is a very reliable machine; it consists mainly of rotating components with an external combustion chamber in which continuous combustion of the fuel takes place. The fuel may be one of a variety of gases or light oils. Filtered air is induced from the atmosphere, compressed to a pressure of 3 to 6 bar, heated in the combustion chamber to a controlled temperature, and then passed with the combustion products out through the turbine blades, causing the shaft to rotate. The shaft speed is high and a gear box is included in the installation to reduce the speed to that suitable for driving an alternator, compressor or pump.

Most gas turbines, particularly those used for industrial duties, are of the single-shaft type, with the air compressor and turbine mounted on a common shaft. However, it is possible to obtain split-shaft units which use two turbine stages, one driving the compressor and the other providing shaft output for the generator drive (or other rotating equipment).

Like all heat engines the efficiency of a gas turbine is governed mainly by the temperature of the hottest part of the cycle. For maximum work output it is therefore desirable to have the combustion chamber gas temperature as high as permitted by the gas turbine materials. At present turbine inlet temperatures are limited to 850 °C to 950 °C, giving overall thermal efficiencies approaching

30%, with some industrial gas turbines operating with efficiencies below 25%. The raising of the turbine inlet temperature could lead to thermal efficiencies of over 40%, eliminating the major disadvantage of the gas turbine when compared with steam turbine plant. The efficiencies of the turbine and compressor are typically in the range 80-90%, and this can also have a significant effect on plant efficiency.

Factors other than size or load affect gas turbine performance, for example the temperature and pressure of the ambient air and the back pressure caused by a waste heat recovery system.

Apart from increasing turbine inlet temperatures, which can only be carried out using blade cooling and/or new materials, the use of intercoolers, reheaters and regenerators can assist.

The application of an intercooler necessitates a two-stage compressor and the turbine work output will be greater than that without intercooling.

Reheating involves at least two turbine stages, with a combustion chamber added between each stage and the turbine work output is greater with reheat.

A third and most important technique for improving the efficiency of the gas turbine, and one which will depend on users' requirements for shaft power and heat, is to use exhaust heat to raise the combustion chamber inlet temperature. This reduces the fuel requirements of the unit, hence increasing the shaft efficiency, but this is at the expense of increasing the overall capital cost per kW of electricity generated.

Incorporation of all the above modification results in the cycle shown in Figure 13.3

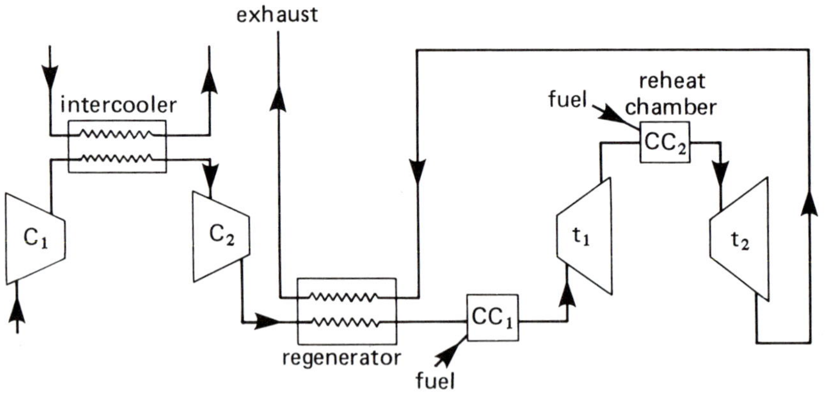

Fig. 13.3 A gas turbine incorporating intercoolers, reheat and regeneration

For base load power generation installations, industrial turbines are readily available in the UK in the size range 0.5 to 6 MW power output, with up to four times as much exhaust heat being available, and in fact most turbines are supplied not as separate prime movers but as complete packages with the equipment that they are to drive.

From the total energy point of view open cycle gas turbines have the advantage that virtually all the thermal energy in the fuel which is not converted into shaft power is available for recovery.

Larger units of not less than 10 MW output employing Rolls-Royce turbines are used by the gas industry to drive gas compressors on the distribution network and units up to 100 MW electrical output are used by the Power Generating Companies for peak load lopping.

The capital cost of gas turbines depends very much on turbine type and power output but is typically £300–400/kW.

In the area of industrial energy applications, gas turbines are used to power compressors, pumps, fans and electricity generators. Their exhaust heat is used for steam generation, process energy, drying, space heating and air conditioning.

Some advantages and disadvantages of a gas turbine are set out below.

Advantages:

- Fast rate of adjustment to load changes.
- Low starting torque and a low degree of mechanical complexity allowing a fast start-up after long periods of idleness.
- Heat is readily available from the exhaust gases at a high temperature in the range 550 to 850 °C.
- Throughout the complete load range a turbine operates at high and constant speed thereby ensuring that electrical generation frequencies can be held to better than 0.25%.
- Low maintenance costs.
- High power to weight ratio.
- No cooling water required.
- Limited exhaust pollution and the exhaust gases may be used directly.

Disadvantages:

- High initial capital cost.
- Low mechanical efficiency, although 30% is achievable from larger machines.
- High pressure gas required (typically 15 bar).
- High frequency noise.

13.3.2 Reciprocating Engines

These machines are particularly suitable for producing shaft power because generally they have higher shaft efficiencies than turbines and offer better part load performance. The idea of gas fuelled reciprocating engines, both spark ignition and dual fuel, for driving various equipment is certainly not new and much operating experience is available. In the UK most of this experience has been with dual-fuel engines which have been installed on sewage works and British Gas plc sites and, most recently, with the introduction of small scale combined heat and power systems. Both diesel and dual-fuel engines operate by compression ignition, the dual-fuel engine requiring 5 to 10% of full load diesel fuel input to initiate combustion when firing with gas as the main fuel, but being capable of operating at 100% diesel fuel whenever necessary. This changeover from one fuel to another can be made easily whilst the engine is on load. Diesel engines are available with outputs up to 9 500 kW.

Spark ignition engines were not readily available in the UK much before the 1980s as the market was minimal. This was due not only to the price of manufactured town gas but also to its high hydrogen content, which severely limited the rating of the engine. Natural gas, however, is an ideal fuel for spark ignition engines, offering the prospects of a high compression ratio although not as high as diesel or dual fuel engines and proven gas engines are now readily available with outputs up to 1 800 kW, and outputs up to 4 000 kW are being introduced.

In general the use of natural gas as a fuel is preferred in industrial applications. It burns cleanly, hence reducing maintenance of the engine as minimal carbon is deposited. Also, no oil dilution occurs and hence lubricant life is improved.

Spark ignition gas engines may be either purpose-built or converted from petrol or diesel engines. Conversion from a diesel engine requires modification of the pistons to correct the compression ratio, introduction of spark plugs into the cylinder head, and changing of the carburation system such that the new unit can homogeneously mix the gas and air in the necessary ratios. Petrol engine conversions also require new carburation systems, and may require some adjustment to the compression ratio. The ignition timing on gas engines is advanced further than on petrol engines because the flame-propagation rates are lower for natural gas.

Typical mechanical efficiencies under full load conditions are up to 39% for diesel engines, 34% for dual fuel units operating on gaseous fuel and 30% for spark ignition engines. These figures are continually being improved by engine manufacturers and particularly so with the use of turbo-charging. In addition, lean burn combustion spark ignition engines are a recent development introduced by a number of manufacturers resulting in improved fuel economy and considerably reduced NO_X emission levels. The heat energy is shared approximately equally between the jacket cooling water and the exhaust gases at up to 600–700 °C and can be recovered if required. It should be remembered that even if the low grade heat from the cooling water is not recovered for any specific heat requirement it still has to be removed – usually either by a water-water heat exchanger with a cooling tower or by using a fan cooled radiator. These engines, as with turbines, are usually obtained in the form of a packaged unit complete with the equipment to be driven.

For industrial applications a water cooled manifold should be fitted which transfer some 10% of the exhaust heat to the jacket. In this case 35% of the fuel consumed by the engine can be recovered from the jacket water as hot water at 85 °C. A further 30% recovery is available from the exhaust gases whilst the remaining 10% is rejected as radiation from the engine block, pipework and ancillary equipment.

High grade heat in the exhaust gases can be recovered by a waste heat boiler and integrated with engine low grade heat recovery providing feedwater preheating.

When heat from the engine is to be used for space or process heating applications, where water at 85 °C or below is required, a circuit similar to that shown in Figure 13.4 can be adopted. The exhaust gas heat exchanger is connected to the engine primary cooling circuit, together with the water cooled manifold and oil cooler if they are fitted. Additional heat may be recovered from the exhaust using a condensing economiser. The engine water pump circulates water through the water cooled manifold and the exhaust gas heat exchanger whenever the engine is running. This provides rapid engine heat-up after a cold start, and reduces exhaust condensate. When the working temperature is reached, heat is released to the secondary circuit by means of a separate shell and tube heat exchanger. If heat is required in the form of hot air, a secondary heat exchanger of finned tube construction (i.e. similar to that used for automotive radiators), can be incorporated. An additional fan for air distribution would then be necessary.

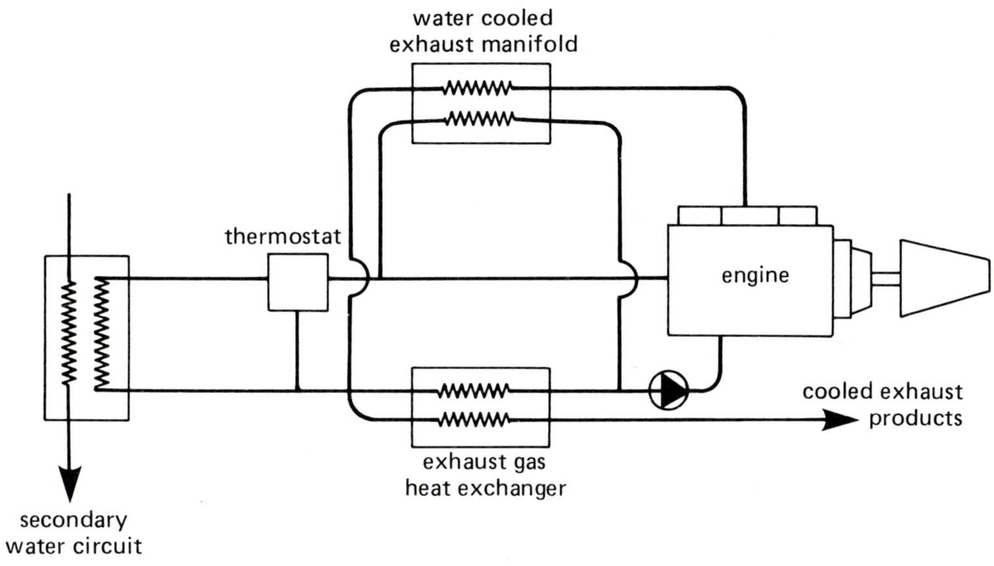

Fig. 13.4 Engine primary cooling circuit

The control system for an engine driven package which is to be fail-safe fulfils two functions:-

- To enable the driven machine to operate unattended yet shut down safely and automatically in the event of specific system faults and to prevent further mechanical damage occurring.
- To automatically control the output of the driven machine to meet maximum demand within the design limits of the equipment.

The first requirement involves the use of a control unit usually employing relay logic, though increasingly micro-electronics based, to supervise signals of engine running, engine overspeed, low engine oil pressure, high gas pressure, low gas pressure, and high engine cooling water temperature in addition to the requirements of the National Grid Company relating to over-voltage, over-frequency and reverse power. This unit forms part of the operating panel which in the event of system failure provides an indication of the first fault conditions to occur.

There are several manufacturers in the UK producing dual-fuel engines over the complete range of sizes from 40 to 4 000 kW, although very few make spark ignition engines of the industrial type; a wide range is however readily available from Europe and the USA.

The cost of reciprocating engine generator sets ranges from about £75-125/kW output dependent on engine type. Waste heat recovery equipment increases these costs considerably. Low grade exhaust heat recovery together with jacket and oil heat recovery units for a reciprocating engine typically at least doubles these costs. Waste heat steam boilers for a reciprocating engine could be about £40/kg/h steam.

Some advantages and disadvantages of a reciprocating engine are listed below:

Advantages:

- High mechanical efficiency which is maintained over a wide load range.
- Moderate initial capital cost.
- The range of power outputs is large from 8 kW up to 1 800 kW for spark ignition engines and up to 9 500 kW for diesel engines.

Disadvantages:

- Requires cooling, even when the recovered heat is not to be used.
- Exhaust gases may contain products of incomplete combustion and therefore cannot be so readily used directly.
- High starting torque.
- High maintenance due to degree of mechanical complexity.
- High lubricating oil consumption.
- Low power to weight ratio, requiring substantial foundations.

13.3.3 Fuel Considerations

The fuels which one would normally associate with on-site generation schemes are LPG, natural gas and all grades of oil, and the relevant characteristics of these are given below. It must be remembered, however, that the price, availability and suitability of the fuel will govern its choice in any on-site generation scheme.

LPG

Has burning characteristics similar to natural gas in gas turbines, but in reciprocating engines down-rating is necessary.

Must be stored at pressure, which is costly, and heat is needed for vaporisation.

Is heavier than air, and adequate precautions must be taken against leakage.

Can be blended with air to produce a mixture compatible with natural gas.

Combusts cleanly, and causes a minimum of gas turbine blade, engine valve and seat or waste-heat boiler corrosion or erosion.

Natural Gas

Is piped supply, no vaporisation is required and requires no storage.

Gives clean combustion products in both engines and gas turbines, with no corrosion or erosion of gas-turbine blade and combustion chamber, engine valves and seats, or waste-heat recovery plant.

Imposes no restrictions on direct use of exhaust gases, and stack heights and temperatures may be reduced.

Is required at pressure to be acceptable for the combustion systems of gas turbines and turbo-charged reciprocating engines; therefore, if a sufficiently high supply pressure is not available pressure boosting on site will be necessary.

Oil Fuels

Have to be stored.

Have to be pumped, and the heavier fuel oils have to be heated, de-watered and cleaned before use in reciprocating engines and some process plant.

Require atomisation in combustion systems.

Suffer from the presence of sulphur and metallic compounds, particularly the heavier grades, and these cause undesirable deposits on, and the erosion and corrosion of, gas-turbine blades, combustion chambers and engine valves, seats and cylinders.

May restrict the direct use of exhaust gases, and fairly extensive cleaning facilities may become necessary on the hot side of waste heat boiler equipment.

Require greater stack heights than other fuels, and control of atmospheric pollution is more of a problem.

Give the highest shaft power efficiency in reciprocating engines, but heavy fractions are not normally considered for gas turbine operation.

13.4 SMALL SCALE COMBINED HEAT AND POWER (CHP) SYSTEMS

'Small scale' or 'micro' CHP may be used to describe engine driven machines which are integrated into host systems linking with the National Grid Company network and heat producing plant, usually boilers. For many years the use of smaller packages incorporating alternators driven by reciprocating spark ignition gas engines have been available. Electricity outputs in the range up to 500 kW make these units economically attractive for base loads in many industrial and commercial market applications provided that the waste heat, recovered as hot water, can be used. The market potential was given a considerable boost by the Energy Act of 1983 which permitted full interfacing between site generators and the main grid supply.

13.4.1 Equipment Selection

Whilst there are novel engine options available and under development, for example the rotary engine, the reciprocating internal combustion engine is the only current shaft efficient prime mover available for small scale CHP. This type of gas engine is available from a number of suppliers in the USA, central Europe and the UK. The majority are spark ignited and developed from automotive or industrial designs.

Typical heat outputs for micro-CHP units are up to 2 to 2½ times the nominal electricity output. The maximum operating water temperatures and pressures at which CHP systems can be used are generally similar to those for low pressure hot water services at approximately 70–85 °C and 3–7 bar.

Engine characteristics and aspects of waste heat recovery have been discussed in detail in 13.3.

The Generator

The generators used in packaged micro-CHP systems have alternating outputs at 415V being either synchronous or asynchronous.

There are three principal ways of operating a micro-CHP system:

- Stand-alone operation, where the particular site electrical load is permanently isolated from the grid and connected only to the CHP unit.
- Parallel operation in which the CHP unit generates electricity to meet the electrical load but is supplemented by the grid should demand be greater than the system output. When the load is smaller than the output, power can be exported to the grid.
- Standby operation, where the unit supplies part of the load if there is a failure of the grid.

The asynchronous generator is mains excited and therefore cannot operate as a stand alone unit nor provide standby power in the event of a grid failure.

The Control System

Micro-CHP units may be set to respond to heat demand or alternatively may respond to electrical requirements.

In the first mode the control requirements of the CHP system are similar to those of a conventional boiler installation. The CHP set should either be configured to preheat return water to the boiler or be operated as a lead boiler in parallel with the site boiler plant.

For control in relation to variations in the electrical load an electronic speed governor is linked to the line current to provide automatic proportional control typically from half to full electrical output. Modulation of electrical output should typically only be necessary for a few hours each day and for the majority of the period of operation the unit will be at full load.

13.4.2 Installation Considerations

Gas engines converted to natural gas can be obtained from several manufacturers. There are also companies who will design, build, install and commission complete small scale CHP projects. The engine must be installed safely, with suitable controls, and match the speed and torque requirements of the driven plant. Similarly, the heat recovery equipment must ensure adequate jacket water cooling at all times, and ideally have a matching use for the available heat. If a gas meter is to be installed this must not be of the turbine type as these are inaccurate when subjected to a pulsing gas flow.

The optimum size of a system depends on several factors:

- The load profile for both electricity and heat.
- Engine sizes available.
- Payback period required.
- Installation and maintenance costs.

Heat demand profiles can generally be estimated from total heating fuel usage, while monitoring for approximately a one week period may be necessary to establish electricity demand profiles.

Fuel tariffs are available from the fuel utility industries. While gas tariffs are uniform throughout the UK regional variations occur in tariffs for electricity and careful analysis of electricity cost is generally required.

Self-excited generator packages can be used to provide standby power in the event of a mains failure and this can increase the cost effectiveness of the investment in terms of standby generating capacity which would otherwise be purely an overhead. However, to justify the investment cost most units installed for this purpose will be sized for near base load duty thereby limiting the usefulness of the plant to the provision of essential services only.

Sequential running of a multi-unit installation is an option for buildings with a high heat and electrical demand such as might occur in a hospital or swimming pool/leisure centre complex. In these situations one or two CHP units may be operated to satisfy a continuous daytime base load whilst a third unit may only operate during the winter period or at times when an excess heat demand exists and there is still a need for some additional electricity. Multiple unit installations have the advantage that some savings can be made against maximum demand surcharges as well as unit electricity costs. It should be possible to ensure that at least one unit is operating in order to reduce the maximum demand which is monitored continuously in ½ hour periods throughout the year, although effectively only incurring a cost penalty during the winter months.

It is important that CHP is correctly integrated into the existing boiler system to gain access to the highest possible heat load. This is particularly relevant if the hot water and space heating circuits are separated. The best option is for the CHP to be the leading 'heat' machine with the boiler as a top up facility. A suitable arrangement would be for the CHP set to preheat return water prior to the boiler. Alternatively the CHP could be operated in parallel with the boiler plant. For maximum running time the CHP set would be sized to produce most of the building's summer hot water requirement. During the winter the CHP continues to provide the domestic hot water while the boiler satisfies the space heating load. If the prevailing electricity tariff is sufficiently high to justify its purchase then a second CHP unit might be used to meet some of this winter space heating.

13.4.3 Engineering Recommendations

Any customer installing or using a gas engine operating on gas supplied by British Gas plc is required under the Gas Act 1986 to give the Region of British Gas fourteen days written notice of the intended installation.

Gas installations are covered by the Building Regulations (Building Standards in Scotland) and by the Gas Safety (Installation and Use) Regulations, 1984.

A British Gas Code of Practice IM/17 has been published to assist customers and contractors on the requirements for the installation of gas engines. Whilst the Code should be used for reference, some main points are worthy of note:

- A gas control train comprising valves, pressure switches, a pressure regulator, and a flexible pipe element is required.
- An air supply providing 4.2 Sm^3/kW h (shaft power) for combustion is required whilst a further 76 Sm^3/kW h may be needed for ventilation.
- An exhaust pipe discharging into open air where there is neither risk of exhaust gases coming into contact with persons nor of entering buildings or other plant through windows, air intakes etc, is required.
- General considerations include isolation of mechanical vibration and acoustic attenuation.

With the advent of the UK Energy Act in 1983 the then Electricity Council issued Engineering Recommendation G 59 which gives the conditions to be met when connection to the grid is required.

The regulations stipulate that the CHP unit must be isolated from the grid under the following conditions:

- If the difference in the declared supply voltage and the generator exceeds ± 10%.
- If the difference in frequency between generator and 50 Hz falls outside the range of +1% to -4%.
- In the event of failure of any one phase in the distribution grid.
- In the event of a loss of the Electricity Company's supply.

If the starter motor of the CHP unit draws its power from the grid other regulations cover the voltage variations (P 13/1) and harmonics (G 5/3) introduced on the grid. The actual interference caused on start-up depends on the grid as well as the CHP unit, but general experience indicates that compliance with G 5/3 and P 13/1 is not a problem with small scale CHP.

13.4.4 Maintenance and Reliability of Gas Engines

Maintenance

Maintenance involves a change of oil and spark plugs, and adjustment of valve clearances at 500 hour intervals in the worst case, although 1 000 hours is considered to be generally achievable. At about 10 000 operating hours the cylinder head will require removal for a top-end overhaul. The ultimate life of the engine before a major overhaul is necessary is probably in excess of 20 000 operating hours.

A maintenance contract or schedule that includes regular combustion tests is recommended; as well as maintaining high operating efficiencies such tests should reduce CO emission levels.

Reliability

Manufacturers of CHP sets offer maintenance contracts for servicing which in some cases include the replacement of components after a predetermined number of running hours. Problems have been encountered on several sites, for instance, the heat recovered has been less than the design value in some cases and mild steel exhaust gas exchangers used in chlorinated swimming pool water circuits are susceptible to metal corrosion. Unscheduled shutdowns have occurred because of problems with engine ignition systems, battery discharging (where fitted), failure of engine protection devices, and failure of micro-processor control components. Most of these problems have been quickly remedied resulting in improved system specification such as more efficient heat exchangers, modified control circuits and electronic engine ignition systems. The occurrence of these problems does, however, emphasise the need for good technical support to customers in the early phases of exploitation.

13.4.5 Economics

Annual cost savings available to a customer as a consequence of installing CHP depend on the capital costs, maintenance charges, prevailing gas and electricity tariffs and the total hours of operation. Capital costs are approximately £400-600 per electrical kW output. Costs for a particular site may deviate significantly from these average values if extensive modifications to the site services are needed to utilise the recovered heat. Maintenance costs vary with type and size of package and the extent of customer participation in servicing schedules and are in the range 0.3p to 1.0p per kW h of electrical output.

Electrical import charges can be computed from the Electricity Company tariffs, and heating costs determined assuming an operating thermal efficiency of the existing site boilers and an average 55% efficiency of heat recovery from the engine. A daytime unit cost of electricity imported from the grid is of the order of 3.5p/kW h but a calculation based on this value will not include possible savings due to the CHP reducing monthly maximum demand and supply availability charges imposed because of demand during the winter months. Over 90% of electricity sold to the industrial and larger commercial markets is sold under a 'Maximum Demand' tariff although Electricity Companies are gradually introducing 'Unit Based' time of day tariffs. The latter have a supply availability charge but the monthly maximum demand surcharge is replaced by a higher unit cost during the winter daytime periods. To determine the effect of either tariff (within the same Electricity Company) on the savings produced by a CHP unit, it is important to establish the existing heat and electrical load profiles of the site which, in the first instance, can be assessed from previous monthly fuel bills.

In order to provide a good economic performance it is essential that the recovered heat is fully utilised, and therefore the daily, weekly and annual patterns of heat load should be established. The savings are only generated during run time and hence continuous operation, i.e. more than 6 000 hours per annum will be most attractive whilst 4 500 hours or less is unlikely to prove economic. It follows that because winter space heating loads typically do not exceed 4 500 hours they are not an appropriate exclusive use for the recovered heat. In addition, space heating loads are often too variable to support base load electricity generation. If an appreciable hot water load can be established for (say) 5 000 hours per annum then CHP would provide attractive annual running cost savings when compared to conventional supply systems.

As the basis for an example, the histograms in Figure 13.5 give the hourly heating and electrical demands, averaged during any one month, for a particular swimming pool complex having a 500 m^2 main pool and smaller training pool. The demand profiles indicate that a CHP set with a 40 kW electrical output and 90 kW of heat recovery could operate as a base load unit for 24 hours per day throughout the year; this would achieve annual savings of £7 000-£8 000 depending on the original electricity purchase tariff. A second unit could also operate during the daytime for 9 months (insufficient heat demand during June to August) if the electrical output were controlled to meet demand up to a total CHP output of 80 kW in any one hour; this would increase the annual saving to £10 000-£12 000.

The installation of a CHP set often coincides with the introduction of other energy saving measures on a particular site. The effect of all these measures must be properly considered, otherwise insufficient load may remain for the CHP unit to operate for the minimum 4 500 daytime hours necessary to give savings compatible with a three year payback on original capital investment. It is worth noting that some revenue can always be achieved by exporting excess electricity generation but dumping excess heat is not recommended practice. Special meters are necessary if the export of electricity is being considered and their installation incurs additional expense. It is therefore generally more cost effective to design a system to use all the generated heat and electricity on site rather than to export it.

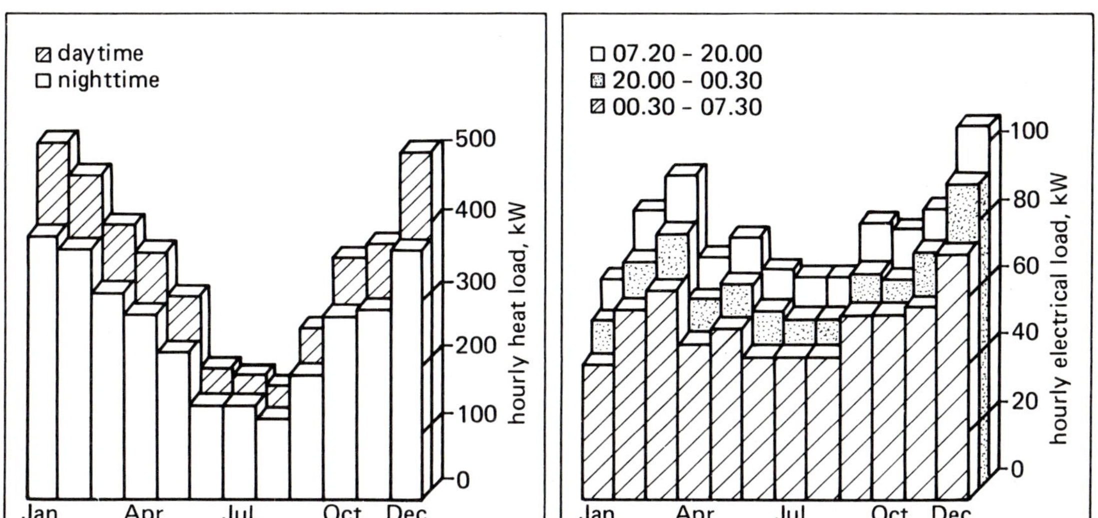

Heat demand profile

Electricity demand profile

Fig. 13.5 Swimming pool energy demand patterns

13.4.6 Applications and Market Potential

The most favourable circumstances for demonstrating the technology and economic viability of CHP within the commercial sector occur in buildings which have a simultaneous need for both heating and electrical power for a large part of the day and for extended periods during the year. Major consumers of both forms of energy include offices and similar premises. Part of this load is met by electricity which also provides additional power for lighting and miscellaneous appliances. Unfortunately the heating requirement for offices is often limited to about 10 hours per day and only for a six to seven month winter period. In this situation CHP may not be economically viable in terms of providing adequate annual savings on fuel costs to offset the capital investment for a CHP package. However, opportunities may exist for a modified CHP scheme incorporating an absorption chiller, which could provide heat during the winter and cooling during the summer. Office complexes with computer suites are particularly suited to this latter technology because of the continuous demand for air conditioning and electrical power.

Medium to large hotels are attractive applications for CHP since a space heating load exists for up to 18 hours/day and there is a high hot water demand for residential and catering needs. The heat to power ratio for a hotel in the winter is typically 3:1 which is well suited to the output of a CHP set which could therefore be sized to satisfy most of the building electrical demand. Hotels with swimming pools are even better applications since the pool water may be heated with low grade heat recovered from a condensing heat exchanger on the gas engine exhaust. This additional heat recovery can improve the CHP overall thermal efficiency to approximately 90%.

The public sector offers the greatest opportunity for exploiting CHP since this area includes hospitals, grouped residential accommodation, university and college campuses, prison and detention centres, swimming pools and leisure centres. These premises require both heat and power for extended periods and at a ratio suited to CHP technology. The potential for CHP must therefore be considerable. A hospital requires space heating for most of the day and substantial quantities of water for up to 20 hours per day. On average the winter heat to power ratio is about 3.5:1 indicating that CHP could be usefully employed in this market sector. A large teaching hospital with student accommodation can have an electrical demand of up to 1 MW which makes it suitable for large scale CHP equipment using bigger engines or gas turbines. However the high capital outlay for these systems is not readily available and investment has at present been restricted to the smaller packages with electrical outputs in the range 40 kW to 85 kW. Several hospitals have been supplied with this size of unit.

Grouped residential accommodation suitable for elderly people operates at relatively high space heating temperatures for up to 18 hours per day with hot water and electricity demands for 14/16 hours per day. The space heating load diminishes during the summer time so that 4 500 hours may be the maximum annual running time for a CHP set. Whilst this reduced operating time may lead to an extended payback period against the installed capital cost, this cost is partly offset by the CHP eliminating the need for a standby generator, the latter being an essential requirement for this type of accommodation. Small packages in the range around the 35 kW would be satisfactory in this market area.

The education sector has a high space heating demand. Universities and colleges with residential accommodation have suitable heat and electrical loads which may extend for a sufficient period during the year to justify the installation of several small CHP sets. Schools are less viable except when a swimming pool is available for using the CHP heat recovery during the summer months. Several schools operate 15 kW CHP sets on this basis and scope for further installations must exist.

Swimming pools and leisure centres probably provide the most suitable conditions for micro CHP with a continuous demand for space heating and hot water and with a fairly constant electrical load for lighting and for driving numerous pumps and fans, resulting in a day time heat to power ratio of between 4 to 5:1. A typical pool of some 400 m^2 area with conventional heat recovery devices such as variable air ventilation and a run around coil already fitted, would still have a sufficient heat and electrical load to justify at least one 40 kW CHP set. A typical installation has four such units generating 160 kW of electricity with 448 kW of recovered heat; three gas boilers provide additional heat when required.

Whilst there will be other applications and circumstances in which small scale CHP sets can be cost effective the most immediate market within the commercial sector therefore includes:

- Hotels
- Hospitals
- Grouped residential accommodation
- University and college campus and some schools
- Prison and detention centres
- Swimming pools and leisure centres

13.4.7 Future Developments in Micro CHP

For most customers the viability of a CHP set is judged on the savings in total fuel bills and the ability to demonstrate a short payback period on the total capital investment. Assuming that the ratio between electricity and gas prices remains at a constant value, then the most effective method of increasing hourly cost savings is to reduce the CHP maintenance costs which at the present time can be as much as 30-50% of the fuel savings. Improved systems reliability and the development of servicing capability throughout the UK are important objectives in the future development of micro-CHP sets.

A further priority is to reduce the capital cost of the package such that payback periods of less than 3 years are the norm rather than the exception. These shorter periods are particularly difficult to achieve if the CHP heat recovery is compared against the higher efficiencies of heat generation now attained with modern gas fired boilers. Reduced capital costs may be realised as numbers being installed increase and manufacturers are able to reduce component costs. Increasing the shaft efficiency of the prime mover may also assist in expanding the market for CHP technology. Assuming that the overall efficiency of the package is maintained, then a higher ratio of electricity generation to heat recovery makes it easier to match the CHP set to the total site electrical demand.

13.5 LARGE SCALE COMBINED HEAT AND POWER (CHP)

Large scale CHP is used here to refer to installations with an electrical generation capacity of 500 kW or more. In general the choice of prime mover for large scale CHP will depend on the heat/power ratio of the site and the base load requirement. Except where there is constantly more heat required than power (i.e. heat/power ratios above 3 or 4:1), reciprocating engines, with their high efficiency of shaft power generation and therefore high ratio value of energy produced to energy consumed, would normally be preferred as prime movers to gas turbines or steam turbines. Gas turbines come into their own at heat/power ratios about 4:1, and increasingly so as the ability to after-fire their exhaust gases at very high efficiency to meet heat/power ratios of up to about 10 or 12:1 is utilised. Pass-out steam sets must also, of course, be considered where high heat/power ratios are required, but, except where this ratio is constantly high, usually above 15:1, steam turbines have not the flexibility of gas turbines. Gas turbines can follow fluctuating load profiles with far greater economy of operation.

13.5.1 Gas Turbine Based CHP Systems

Gas turbines operate at shaft power efficiencies which vary between about 25% for very simple designs to 30% for more complex plant involving intercooling between several stages together with regeneration. From the total energy point of view open cycle gas turbines have the advantage that virtually all the thermal energy in the fuel which is not converted into shaft power is available for recovery. Typical exhaust gas temperatures are in the range 550 °C to 850 °C but it must be remembered that if regenerators or recuperators are used the final exhaust gas temperature will be considerably reduced, possibly down to 270 °C. Although this leads to higher turbine efficiencies the quantity of heat recoverable for use in processes will be significantly lower.

There are several methods of using the waste heat from gas turbines. British Gas document IM/24 gives guidance on the installation of industrial gas turbines, associated gas compressors, etc.

Direct Drying

When natural gas is used as the primary fuel the exhaust gases are clean and can be used directly for process drying requirements. The most common of these are:

- Brick, tile, ceramic and glass manufacture.
- Leather and allied trades.
- Agricultural based establishments e.g. vegetable drying, fish, meal and meatmeal factories.

Waste Heat Boilers and the Provision of Process Steam

The waste heat boiler can be used as a waste heat recovery unit on gas turbines. Hot water waste heat boilers can recover up to 75% of the available heat in the turbine exhaust and can thereby produce an overall efficiency of 80%.

With a high-pressure steam waste heat boiler up to 65% of the available heat can be recovered giving an overall efficiency of 72%.

The use of supplementary firing increases the steam output with very little additional expenditure and releases the end user process from its dependence on the performance of the gas turbine.

The Combined Steam/Gas Cycle

The boiler serving a conventional steam turbine can use large quantities of preheated air from the gas turbine exhaust through making use of the oxygen present for supplementary firing. The operation of the gas turbine will be marginally affected by the back pressure that is imposed upon it. In addition steam turbines cannot accept quick starts and stops like a gas turbine. As the steam turbine machinery takes longer to put into operation than the gas turbine plant it is necessary to have some kind of disconnection device to permit the gas turbine to run on its own until the steam turbine is ready to run. With such a system it is possible to obtain an overall useful power efficiency of more than 40%.

The type of steam turbine used in conjunction with the gas turbine depends on the duty, and the various steam turbine modes have been discussed in 13.2.3. Where electricity generation is of prime importance a condensing steam turbine would be used. With a need for further process heat in the form of steam a back pressure turbine would be required in which the steam exhausts at a relatively high back pressure. Alternatively a passout steam turbine can meet this second requirement: steam is extracted between turbine stages at the required conditions.

13.5.2 Reciprocating Engine Systems

Reciprocating engines have a number of significant advantages over the gas turbine when a substantial mechanical output is required. In general their shaft efficiency is greater than that of the turbine and their cost is less. However the ancillary systems required for reciprocating engines are more complex and the units have a low power to weight ratio. In addition a major difference is that engine heat recovery includes low grade heat as hot water as well as exhaust gases which can provide high grade heat as steam. This hot water may not be required by the site thus reducing the energy savings for the system.

In the UK most of the large reciprocating engine based total energy systems have been installed in sewage works, hospitals and factories, although there are a number used in private power stations serving chemical works.

Many reciprocating engines are available as a package direct from the manufacturer with heat recovery units utilising the engine cooling water and lubricating oil heat. The engine exhaust may then be coupled to waste heat boilers or hot water heaters which also act as exhaust silencers. Units for exhaust waste heat recovery are available for engines ranging in output from 100 kW to 15 MW.

As an alternative to conventional water jacket cooling systems on engines an ebullient system can be used. However the engines capable of utilising ebullient cooling are more expensive than those employing sensible heat removal methods. Additional steam can be provided by utilising a waste heat boiler.

In assessing the economics of heat recovery from reciprocating prime movers maintenance and reliability costs must be carefully estimated. Complete maintenance costs are composed of three basic items:

- Routine maintenance and service cost including make-up oil but excluding the necessary labour cost.
- The cost for performing necessary top end and engine overhauls including the required labour cost.
- The labour costs necessary for routine maintenance and oil additives.

The first two items will vary with the conditions under which the engine is required to operate. The third item is the most variable and will depend on plant location and unit labour costs. Maintenance costs should be based on past experience in the area being considered.

13.6 CHP SYSTEMS FOR DIRECT-DRIVE APPLICATIONS

There is an extensive demand throughout industry for shaft horsepower for which the prime mover is normally an electric motor. An alternative to the packaged CHP set which may help to broaden the market share is a gas engine which replaces an existing electric motor for driving a fan, pump or compressor. Provided that the recovered engine heat can be readily integrated into the site services, then there are distinct advantages in direct drive apart from the obvious avoidance of alternator and electric motor shaft power inefficiency factors inherent in the CHP concept. Direct drive has a lower capital cost per kW of shaft power than a CHP set since the initial investment is limited to the gas engine with its control and heat recovery circuits, part of which may be offset by the purchase price of the alternative electric motor. The engine has a good part-load performance, achieved via variable speed control, which assists in optimising energy usage when full-load output is not required.

13.6.1 Direct-Drive Air Compressors

Whilst gas engines may be used to drive any type of compressor, experience to date is specific to the more common reciprocating and screw machines. In many industrial and commercial applications these compressors will produce between 25–700 litres/second (50–1 500 cubic feet per minute) of air at 7 bar gauge pressure and require shaft power in the range 10–275 kW. The heat produced by air compression can be recovered from the compressor in addition to that from the engine, albeit at a generally lower temperature, typically 50 °C due to oil performance considerations, although with more efficient heat exchangers 80 °C is quite feasible. The provision of waste heat recovery equipment often enables 90% of the energy consumed by the engine to be usefully employed. The potentially high thermal efficiency and the associated fuel economies are the main attractions although there are other benefits over the conventional electric motor drive, namely:

- The speed control devices fitted to gas engines produce good part-load fuel efficiency.
- An option to operate gas engines at a higher speed is often available to permit uprating of the compressor output by up to 10–20%.

Through a number of field trials British Gas plc has gained considerable experience and information on the important factors in the selection of air compressor packages. The essential problem in retrofitting a gas engine to an existing compressor installation is that design details, the drive system, controls and heat exchange systems for example must be individually specified for each unit converted. As a result of collaboration with a large UK manufacturer of packaged air compressors a standard packaged unit has been produced leading the way for marketing of a comprehensive range of systems.

One of the first products of this collaborative venture was a package based on the CompAir Broom-Wade 6 000 series screw compressor and a Perkins G4.236 spark ignition engine which develops 50 kW at 2 000 rev/minute. The compressor is driven through a simple gearbox which increases the shaft speed by a ratio of 2:1. Heat is initially recovered from the engine exhaust gases, lubricating oil and the jacket cooling water by means of a primary water circuit and is then transferred to a secondary circuit through a shell and tube heat exchanger within the package. The secondary water for site use is therefore separated from the engine circuits and provides a source of heat at up to 80 °C. This can be used in a number of ways; to preheat boiler feedwater or to provide space and process heating requirements. More heat is recovered, at 50 °C, by water cooling the compressor. This lower grade heat may be used for washing, process heating or space heating.

The flexibility of the package is increased by electronic engine speed control which varies the operating speed of the engine in proportion to the load. This facility is provided in addition to the normal compressor suction valve which ultimately throttles the compressor output as the load falls. The combination of these two control mechanisms provides superior part load performance of the package in comparison to electric motor driven counterparts, yet retains the inherent safety of conventional equipment.

13.6.2 Direct-Drive Pumps and Air Conditioning Systems

In the UK, many Water Companies use large diesel, dual fuel, and more recently spark ignition engines for pumping water. For smaller sizes there are already diesel and petrol engine driven sets principally for mobile applications. The expertise in operating natural gas engine driven units for stationary duties is therefore only an extension of existing practice. A number of sites have been used to demonstrate this concept of CHP including a swimming pool water circulating pump installed in the West Midlands. In this application the water must be continually filtered, chemically treated, and heated to maintain suitable swimming conditions. A Waukesha gas engine is used for the pump drive, whilst heat exchangers directly transfer the engine jacket and exhaust heat to the pumped pool water. Operating on a continuous basis throughout the year this package would require approximately ten routine services per year. The savings accumulated by displacing an electric load and the value of the recovered heat more than offset the gas engine fuel cost and the maintenance costs such that a simple payback of 2½ years is obtained.

Gas engine driven chiller systems have also been demonstrated but application would be limited to sites such as computer suites where air conditioning would be required continuously and not only during the summer period.

13.6.3 Economics of Direct-Drive Combined Heat and Power Systems

Clearly gas engine driven CHP systems are relatively complex and have higher maintenance costs than their electrical equivalents. However it has been demonstrated that their use can produce significant savings with paybacks of 2½ years. As with all CHP applications, for optimum savings the heat produced simultaneously with the shaft power operation must be utilised for long operating periods, typically 5 000 hours or more per year.

Savings are calculated by comparing the cost of a conventional system based on providing electricity for the compressor drive and gas to provide heat, set against the natural gas requirements of the engine together with the maintenance costs.

These savings are used to offset the additional capital cost of the engine driven unit. Apart from special site variables, the payback period is influenced by two major considerations:

- If an existing electrically driven unit has reached the end of its working life and must be replaced then, since the engine driven unit costs more than its electrical counterpart, reduced running costs should enable the additional outlay to be recovered in less than one year.
- If a serviceable system is removed and replaced with a new engine driven unit it is expected that the investment cost would be recovered in about 2½ years.

13.7 GAS ENGINE DRIVEN HEAT PUMPS

A heat pump is a device which extracts heat from a source that is normally at too low a temperature to be used directly, making it available as useful heat at a higher temperature. Typical sources are heat rejected as process effluent, cooling water, humid air or warm fluid from a chilling process. In addition heat sources at or below ambient temperature can be used, examples being the air, the ground or river water. Heat loads include air and water for space heating or various process fluids such as hot water for washing and hot air for drying. The heat pump can be used to greatest advantage when the temperature lift and the required load temperature are lowest.

Compared to conventional 'passive' heat recovery, the heat pump has three important characteristics:

- It is used for temperature upgrading rather than simple heat recovery; the heat pump can therefore extend the temperature range over which waste heat is recoverable.
- High grade energy is necessary to drive the heat pump so the extra heat is produced at some cost; it is therefore better to think of the heat pump as an efficient heating system rather than a heat recovery device.
- Heat pumps are essentially limited to applications involving temperatures below about 120 °C.

13.7.1 Types of Heat Pump

The most common heat pump operating cycle employs vapour compression which uses the mechanical energy for temperature elevation. There are alternative cycles such as absorption involving a physical chemistry process for temperature elevation, together with modified thermodynamic cycles and thermo-electric processes which are only at the development stage. Only the vapour compression heat pump is considered here because units of this type are more readily available and are likely to be no more costly than the equivalent absorption units.

Vapour compression heat pumps have been designed to operate according to a number of thermodynamic cycles such as those due to Rankine, Otto, Brayton or Stirling. The ideal Carnot efficiency is most closely approached by the Rankine cycle and hence this offers the best performance prospects. Consequently, currently available electric motor and gas engine driven heat pumps employ this cycle. The components used are readily available from air conditioning and refrigeration practice.

The Vapour Compression Cycle

The vapour compression cycle has the advantage that it is proven technology, offering a relatively high efficiency over a reasonable range of temperatures. The basic system is shown in Figure 13.6 and consists of a compressor, expansion device and two heat exchangers. The cycle is closed and the working fluid is commonly a refrigerant such as Freon operating in two phases. A description of the operation, starting with the evaporator, is as follows:

- The temperature of the working fluid in the evaporator is lower than the temperature of the source. Heat is therefore transferred to the working fluid causing it to evaporate.
- The evaporated vapour is then drawn into the compressor and, by the input of shaft power, is compressed to a higher pressure and temperature.
- The vapour leaving the compressor and entering the condenser is at a temperature higher than that of the load to be heated. During its passage through the condenser, therefore, heat is rejected to the load causing the working fluid to condense.
- The high pressure liquid then passes through an expansion device where its pressure and hence temperature are lowered. Low temperature fluid then enters the evaporator to complete the cycle.

A range of refrigerant working fluids are available commercially such as R22, R12 and R114 for low, medium and high temperature applications respectively. Although some heat pumps have operated at temperatures in excess of 120 °C, most are at present limited to less than about 70 °C at the condenser.

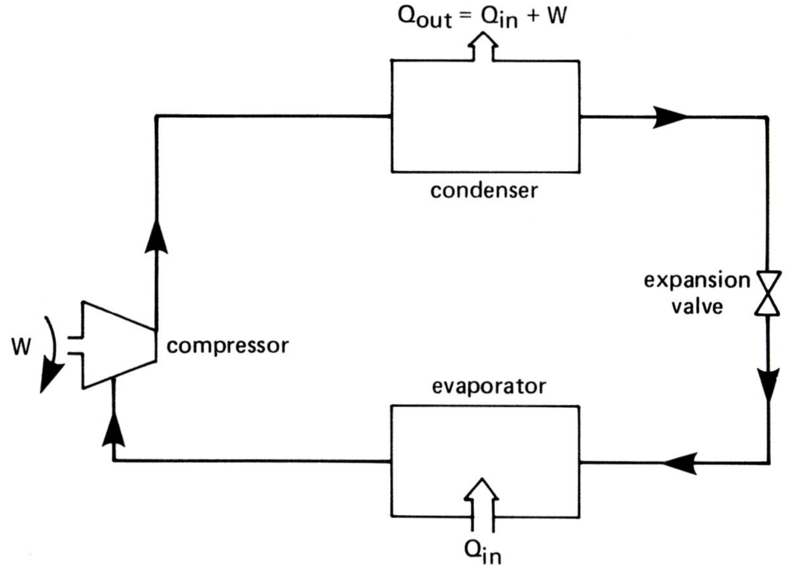

Fig. 13.6 Components used in the vapour compression heat pump

In practice, a heat pump contains more components than are shown in the diagrams although these do not greatly affect the basic heat pump cycle. A number of compressor types are available such as reciprocating, sliding vane, screw and centrifugal. Reciprocating compressors are currently the most common but other types are becoming more popular for heat pump applications.

Gas Engine Driven Heat Pumps

While heat pumps are commonly driven by electric motors, an alternative is to use a reciprocating, spark ignition, internal combustion engine operating on natural gas – see Figure 13.7. The principal advantage of the gas engine is that while shaft power is produced at about the same cost as with the electric motor, a quantity of high grade heat corresponding to about 55% of the energy input can be recovered from the engine jacket and exhaust system. This heat can be used to supplement the heat pump condenser output, resulting in several advantages. First, the size of the heat pump components (compressor, condenser, etc.) will be smaller for a given total heat output for a gas engine heat pump (GEHP) than for an electric motor heat pump (EMHP). Second, the maximum load outlet temperature will be higher for a given refrigerant system design. Alternatively, for a fixed outlet temperature, a higher efficiency will be achieved due to the engine heat providing part of the temperature lift. In cases where heat is required at two distinct levels, an example might be warm air at 45 °C, the GEHP could provide both efficiently. A further advantage is the speed modulation which provides for improved heat pump control and heat output modulation with minimal loss of efficiency.

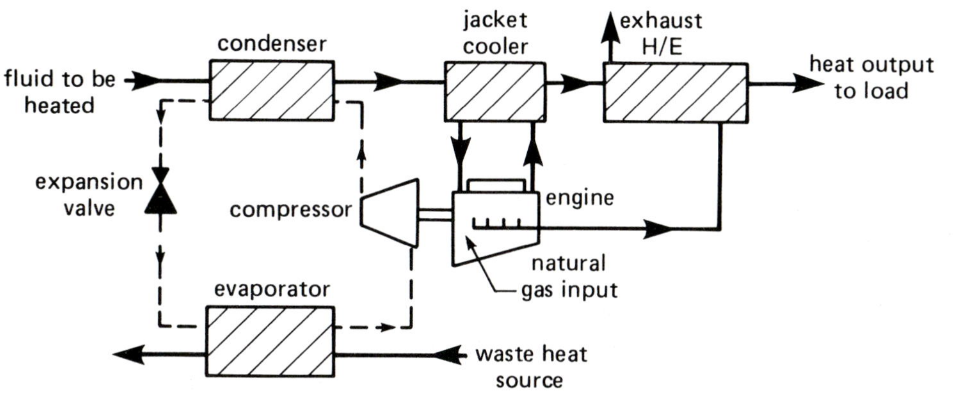

Fig. 13.7 Schematic of the Gas Engine Heat Pump

13.7.2 Heat Pump Performance

Heat pump performance has historically been expressed as the Coefficient of Performance (COP) which is defined as the heat output from the condenser divided by the power required to drive the compressor. COP is of limited value when considering the performance of a heat pump installation since it does not take into account the efficiency of the compressor drive, power for ancillaries or heat supplied by the engine. For this reason, another term, the Performance Effectiveness Ratio (PER), is used and this is defined as:

$$\text{PER} = \frac{\text{useful heat output from the complete installation}}{\text{energy input to the installation}}$$

The PER allows the customer to determine directly the cost of energy delivered by the heat pump based on the unit costs for the input energy (electricity or gas), and thus allows the cost savings compared to other plant to be calculated.

The values of COP and PER for both electric motor heat pumps and gas engine heat pumps vary with the temperature difference between the load and source. PER has the higher value for the gas engine type since part of the temperature lift is being provided by the engine heat recovery. COP is the greater of the two ratios for the electric motor type but it would need to be three or four times as great, as is the typical cost of electricity in relation to that of gas, to achieve the same running costs.

13.7.3 Economics

The economics of heat pumps are affected by the annual energy cost savings achieved, maintenance costs and the capital cost. The annual energy cost savings are a function of a number of factors such as the PER actually achieved, the efficiency of the base heating plant, operating hours and energy prices. Maintenance for gas engine heat pumps has been found to be typically 10% of the engine fuel consumption cost. Capital costs are dependent on the application, rating, ease of incorporation into existing plant, whether or not a developed packaged unit can be used and whether the heat pump is part of a large scheme.

To achieve the best economic case, as many as possible of the following criteria should be satisfied:

- Long operating hours at steady demand.
- Low temperature lifts.
- Moderate upper temperature.
- Matched heat demand and supply of waste heat.
- Existing plant in need of replacement due to age or poor efficiency.
- Conventional heat recovery equipment not suitable.
- Heat demand in excess of minimum economic size (e.g. at least 80 kW for gas engine heat pump).

If most of these criteria can be substantially met, then Performance Effectiveness Ratios of 1.5 to 3.0 can be practically achieved on systems incorporating some passive heat recovery. This equates to energy savings of about 50% to 75%, depending on the efficiency of the previous system. The construction and installation of gas engine heat pumps, although leaning heavily on refrigeration practice, has been well developed now for several years. It is probable that higher performances can be achieved and that there is scope to reduce installed costs. Both of these effects should come about as the market develops and the number sold increases. Dependent upon the trend in future fuel prices, further development should reduce the current payback periods of 3 to 5 years significantly.

13.7.4 Gas Engine Driven Heat Pump Applications

Due to the relatively high capital cost of heat pump installations, they are only economically attractive in applications where long hours of operation and moderate temperature differences can be assured.

Heat pumps have been applied as a means of heat recovery where a low grade source waste heat is of temperature below that required for re-use in part of a process. As discussed, the heat pump is used to elevate temperatures, for example, in a dairy where milk cooling yields low temperature water which can be upgraded for bottle washing. The maximum load temperature from the condenser for a heat pump is around 70 °C when using ambient air at -1 °C as the source. However, where warm effluent is available at say 70 °C then condensing temperatures of 120 °C are achievable.

Heat pump recovery systems have also been installed on heating and dehumidification applications, for example in drying processes and swimming pools.

A specific example is a malt drying kiln (Figure 13.8) where the source and sink may be some distance apart and where the heat extracted by the evaporator is principally latent heat. The heat collected at the evaporator is elevated by the compressor to a temperature of 60 °C at the condenser. The resultant warmed supply air temperature to the kiln is subsequently raised to 72 °C by exhaust and water jacket heat recovery.

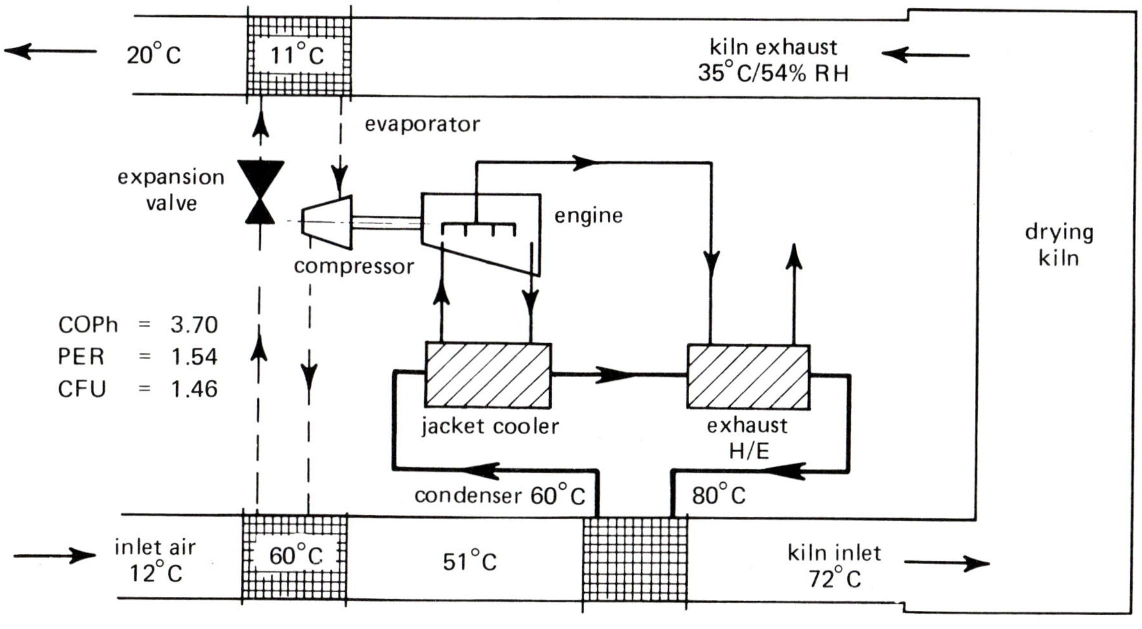

Fig. 13.8 Heat recovery on drying kiln using a gas engine driven heat pump

13.8 MECHANICAL VAPOUR RECOMPRESSION

Drying, evaporation and many other industrial processes reject waste vapour either to atmosphere or a condenser and for maximum energy efficiency it is important to re-use this heat wherever possible. Steam compression is a method of taking waste vapour and upgrading it to a useful level for process heating. Plant should be fully assessed and other methods of heat recovery such as heat exchange should also be considered in order to determine the most appropriate heat recovery system.

There are two common methods of steam compression, thermocompression and mechanical compression.

Thermocompression is achieved by the use of high pressure steam in a venturi to compress waste vapour. This is a viable option under certain circumstances but mechanical compression represents a technique to which gas engine systems can be applied. Mechanical steam or vapour compression makes use of a compressor to raise the pressure and hence the saturation temperature of the vapour.

Significant energy savings are possible as the mechanical energy supplied only delivers the difference in latent heat between the low grade steam and high grade input steam. This is small in comparison with the latent heat of the steam at the lower temperature which would otherwise be wasted.

For example with water vapour exhausted from a drying process at 80 °C the enthalpy is 2 643 kJ/kg. The enthalpy of steam at 140 °C is 2 733 kJ//kg and therefore expenditure of (2 733 - 2 643) kJ = 90 kJ of compression energy can theoretically upgrade 1 kg of waste water vapour at 80 °C to potentially useful steam at 140 °C. In practice, compressor and system losses reduce the overall efficiency of the process.

13.8.1 Mechanical Vapour Recompression System (MVR)

Steam compression represents a type of heat pump system and as such the parameter COP is a useful measure of the effectiveness of the system. As with any heat pump the main constraint on the COP is the temperature lift required across the system. The COP is inversely proportional to temperature difference (ΔT) i.e. a higher COP for a lower ΔT. For a typical compressor efficiency of 65% the COP of a steam compression system varies from 30 at low ΔT (5–10 K) but even with large temperature rises (40–60 K) a COP of 3 to 7 is attainable.

The two basic forms of mechanical steam compressors system are the open or direct cycle and the closed cycle or indirect system.

These are shown schematically in Figures 13.9 and 13.10.

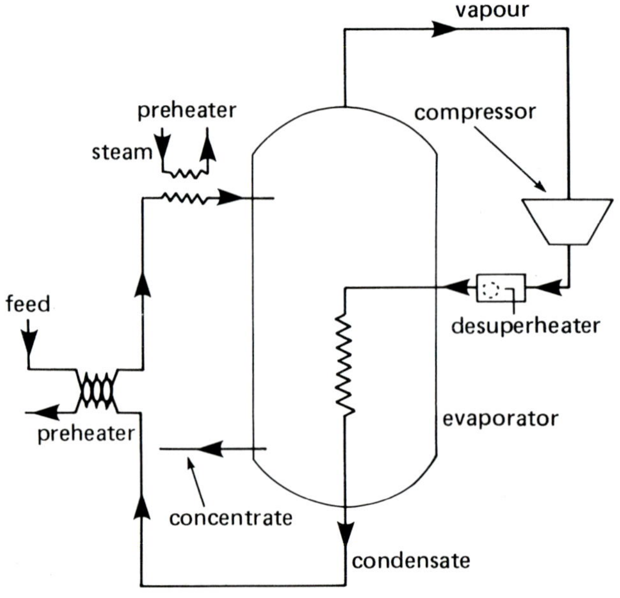

Fig. 13.9 MVR Direct Cycle

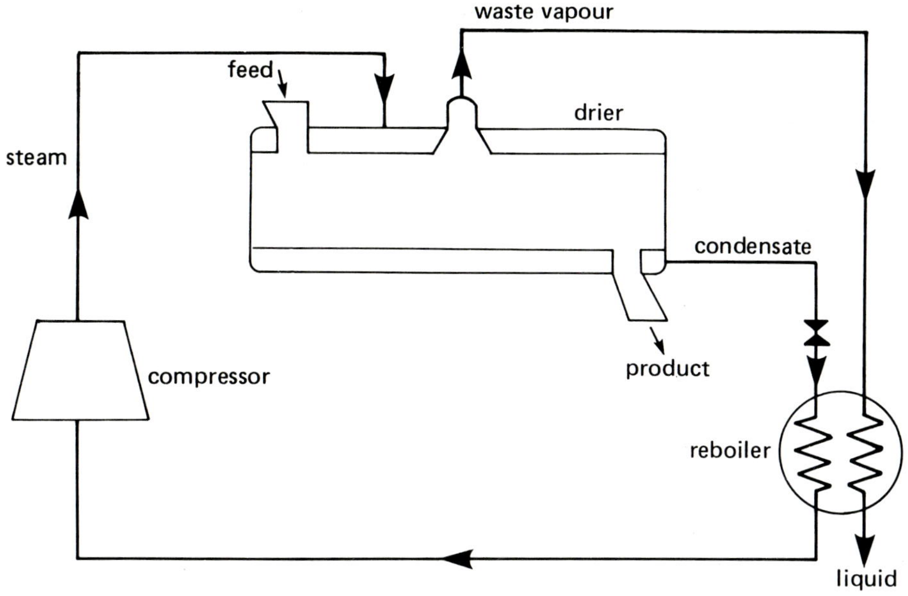

Fig. 13.10 MVR Indirect Cycle

In the open cycle system the waste vapour is taken to a compressor which raises its pressure and thereby saturation temperature. This higher pressure vapour is then used in the heating tubes.

The closed cycle system makes use of a heat exchanger (re-boiler) designed to accept fouling or non-condensible streams such as air. Waste vapour from the process is condensed in the heat exchanger to produce clean steam at slightly sub-atmospheric pressure from the condensate from the process.

To date most work on MVR systems has been carried out by the Electricity Association but there is clearly potential for gas engine driven compressor systems particularly since costs could be reduced further by utilising the heat generated by the engine. Engine waste heat could be used to preheat condensate or to provide supplementary low pressure steam.

13.8.2 Applications and Future Developments

Evaporators

In most plants evaporators are cascaded to enhance the steam economy of the system. MVR can be used as an alternative to multi-effect evaporation (see Figure 13.11). Since temperature differences are usually low in MVR evaporator plant (3–8 K) high COPs approaching 30 can be obtained. The application of open cycle MVR evaporators has expanded rapidly since the mid-1970s and a number of industrial sectors in the UK including the whisky, dairy and foodstuff industries have installed MVR evaporation plant.

Drying

In drying plant the vapour is usually exhausted and its latent heat wasted. Heat can be recovered from the exhaust by a number of methods:

- Exhaust recirculation and/or conventional heat recovery methods: see Chapter 7.
- Heat pump methods, both electrical and gas engine drive, to upgrade the heat available from drier exhausts at low temperatures (40-80 °C).

– Steam recompression drying to provide heat recovery by MVR. If materials were dried in an atmosphere of superheated steam and it were possible to use the compressed exhaust steam as the heating medium then the thermal economy of a multiple-effect evaporator could be achieved. This is an area of extensive research and development and work is in progress by British Gas to assess the benefits and design suitable systems for gas engine driven MVR systems.

13.9 FUEL CELLS

This Chapter has so far described how natural gas can be used to power thermodynamic devices, such as engines or turbines, to produce shaft power which can then be used to drive alternators producing electricity by electro-magnetic means. A more elegant method is to react the oxygen and fuel across an electrolyte, to produce electricity directly. Such a device is called a fuel cell.

Unlike the generation of electrical energy by combustion very little or no heat is produced. There is no conversion of chemical energy into heat and the process is therefore not subject to the effect of Carnot's equation which has a limiting effect upon the actual electricity production efficiency of normal heat engines.

Before dealing with the fuel cell it is appropriate to review some of the present methods of electricity production which make use of the battery in one form or another. These fall into two categories:

– Non re-usable cells, for example, the so called dry batteries.

– The various forms of rechargeable cell using such combinations as lead-acid or nickel-iron. It is the wet cell which is of particular interest to us in the consideration of fuel cells.

The conventional wet cell comprises plates of lead and lead compounds immersed in an acidic electrolyte, or alternatively combinations of nickel and iron plates in an alkaline electrolyte. With the wet cell, when charging current is passed through the plates and electrolyte, chemical changes occur in the lead and lead compounds corresponding to the quantity of electricity fed into the cell. In addition, gases, i.e. hydrogen and oxygen, are given off by the plates. Investigations were therefore directed to the possibility that if oxygen and hydrogen were somehow fed back into the plates of the cell the result should be the generation of an electric current within the cell, and the production of water as a by-product; essentially the reverse of charging a wet cell.

In fact the first investigator to operate a successful hydrogen-oxygen cell was Sir William Grove, who in 1839 reported experiments concerning a gaseous 'voltaic' battery. Until relatively recently, however, the fuel cell tended to be more of a laboratory curiosity rather than a practical possibility for the generation of an electric current. Requirements of the fuel cell are thus supplies of suitable reactant gases and the choice of a suitable electrolyte.

13.9.1 Classification and Types of Fuel Cells

There are three main considerations governing the classification of fuel cells.

Firstly, by temperature of operation, that is High, Medium or Low:

– High temperature cells include those having molten salt electrolyte at temperatures in the region 590 to 650 °C.

– Medium temperature hydrogen-oxygen cells which operate around 200 °C.

- Low temperature cells, which are those operating up to the boiling point of an aqueous electrolyte.

Secondly, by electrolyte type; this can be either an Acidic or an Alkaline system, in either solid or liquid form.

Thirdly, classification by fuel:

- Gaseous, for example hydrogen.
- Liquid, for example alcohol.
- Solid.

13.9.2 Hydrogen/Oxygen Fuel Cells

By far the most common fuel cell is the hydrogen/oxygen system which is available in low-, intermediate- and high-temperature forms with a working temperature from 90 °C to 1 000 °C depending on the electrolyte used.

Reaction Mechanism

The overall chemical reaction in the fuel cell utilising hydrogen and oxygen is the reaction $2H_2 + O_2 = 2H_2O$. The electrode reactions are:

$H_2 = 2H^+ + 2e$	(Hydrogen electrode : anode)
$\frac{1}{2}O_2 + H_2O = 2OH^- - 2e$	(Oxygen electrode : cathode)
$2H^+ + 2OH^- = 2H_2O$	(Water as waste product)

The water formed must not be allowed to pollute the electrolyte and it is necessary therefore that it be removed by being absorbed by the fuel gases passing through the cell and eventually being condensed out. The sensible heat of this water can be used in local space heating applications. It is therefore possible to use this type of cell in moving vehicles, there being no storage capacity required for waste products. It is, however, necessary to provide for storage space for the reactive gases, preferably in a liquid form. The combination of the hydrogen/oxygen cell with an electrolyser would give a system for storing electricity, the gas from the electrolyser being stored under pressure or liquified, then re-combined when required to provide the necessary electrical energy.

A typical hydrogen/oxygen cell consists of two porous metal electrodes with electrolyte being constantly recirculated between their inner faces. The porous electrodes have a matrix of coarse pores with the side facing the electrolyte being coated with a layer of material having finer pores. The reactant gases are under sufficient pressure to displace the liquid electrolyte from the coarse porous structure of each electrode but not from the fine pores, the generation of electrical energy in the fuel cell taking place by means of the electrical ionisation reaction. The voltage available per cell is in the region of 1 volt with the available current being from 250 to several thousand amps per square metre of electrode area.

Hydrogen/oxygen fuel cells are available in four main types.

- Aqueous Acid, where the electrolytes are either diluted sulphuric acid or phosphoric acid. The overall electricity production efficiency presently achievable is 40%.

- Alkaline, where potassium hydroxide or saturated carbonate/bicarbonate solutions are used as the electrolyte. Operating temperatures are low and electrical conversion efficiencies of up to 30% can be attained.

- Molten Carbonate, where a mixture of alkali carbonates and aluminates is kept at a temperature of 650 °C. Higher temperature operation, up to 750 °C, improves the power density of the cell but increases the liability of corrosion. Fuel conversion efficiencies can be as high as 60%.

- Stabilised Zirconium, where solid zirconium oxide at 1 000 °C is used as the electrolyte with conversion efficiencies approaching 60%. These cells have very low capacities, approximately 100 W, and a very short useful life. However this cell is in the development stage and may be attractive in future since waste heat is produced at a high temperature.

Use of Air Instead of Pure Oxygen

In space or submarine applications pure oxygen is the normal choice. Additional difficulties to be expected when using air are not too severe. These include slightly lower theoretical voltage, a higher degree of polarisatiuon on load, pumping losses and some carbonation of the alkaline electrolyte. It can be said that air may possibly replace oxygen whould a rather lower performance be tolerated.

Use of Hydrocarbon Fuels

Some new difficulties arise when attempts are made to use hydrocarbon fuels such as methane, propane and butane in fuel cells in place of hydrogen. For example, hydrocarbons are not so reactive electrochemically as hydrogen in this application and it is therefore usually necessary to raise the operating temperature well above that used by the simple hydrogen/oxygen cells. The reaction products contain CO_2 which will rapidly carbonate an alkaline electrolyte, thus necessitating a molten electrolyte at a high temperature. Acid electrolytes may be used but with obvious additional corrosion problems.

In addition carbon dioxide produced during the reaction cannot easily be separated from the fuel gases whereas by using hydrogen alone the reaction product, steam, is easily removed by condensation. Finally, impurities which may be present such as sulphur in some of the commercially available hydrocarbon fuels may have to be removed to avoid the poisoning of reactive catalyst surfaces of the electrodes.

13.9.3 Indirect Fuel Cells

Practical fuel cells have to rely upon coal, oil or natural gas.

Coal can be converted into fuel gas by means such as the Lurgi gasifier, while fuel gases are produced both at the well-head of oil boreholes and as refinery by-products. Waste gas from sewage works and other sources can also be used as the primary fuel feedstock. The hydrocarbons would be reformed to a mixture of hydrogen and carbon monoxide with further reaction to carbon dioxide.

For steam reforming of natural gas the heat required for the endothermic reaction of methane could be provided by the waste heat produced by the fuel cell. In most fuel cells it is important to reduce to a minimum the carbon monoxide levels as this will adversely affect the fuel cell lifetime. An efficient water gas converter is therefore an essential feature.

A fuel cell and reformer are shown schematically in Figure 13.11.

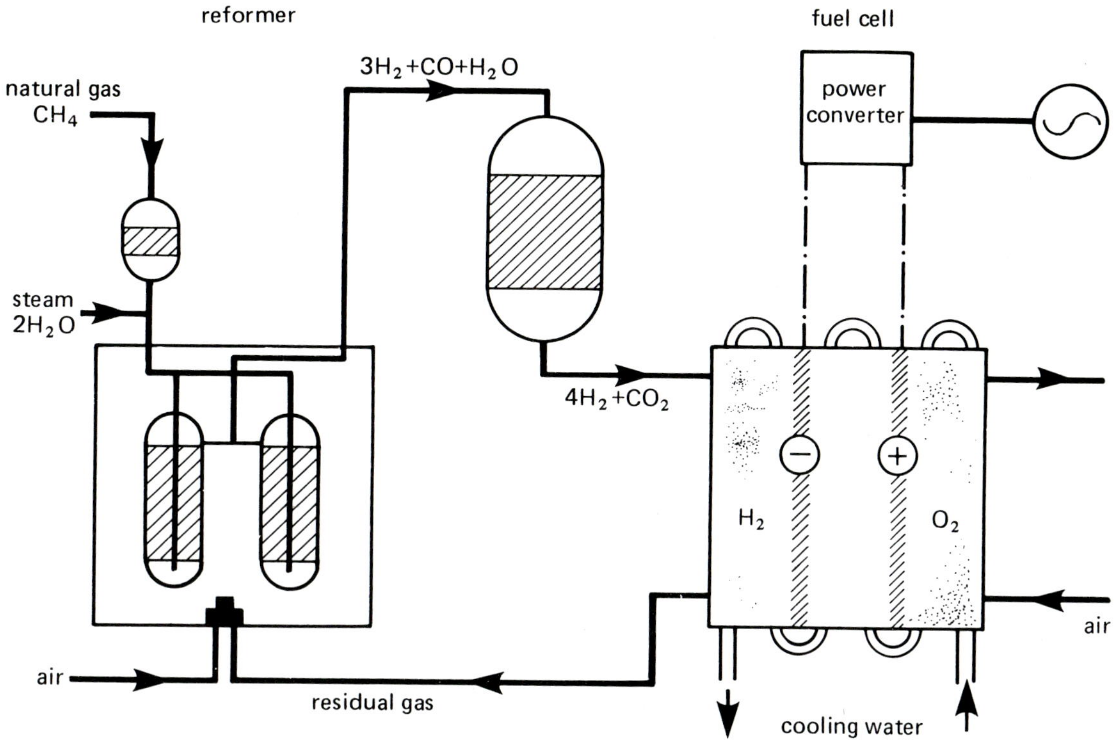

Fig. 13.11 Fuel Cell and Reformer

13.9.4 Fuel Cell Research and Development

Three approaches towards large stationary fuel cells are in progress: these use phosphoric acid, molten carbonate or solid oxide as the electrolyte.

Two distinct markets can be identified. The first of these is CHP of several kilowatts electrical and heat output using natural gas as fuel. The second is multi-megawatt units for power stations using natural gas or naptha as fuel.

12 kW to 40 kW units have been produced in the USA, based on the phosphoric acid electrolyte capable of producing electrical power at 40% efficiency with another 40% of the heat input being recovered to produce l.t.h.w. at 85 °C. Over 100 000 hours operation had been accumulated by 1985.

On the larger scale, two 4.8 MW units have been designed, of which one was installed in New York and one in Tokyo. The Japanese unit has operated only for a few hundred hours. The New York unit cost about $30 million but has never operated, and the unit has now been 'moth-balled'. In spite of these setbacks, an American company, UTC, has formed a company International Fuel Cells (IFC) with Toshiba and they have announced a programme to install twenty-three 11 kW units over a period of a few years.

The Molten Carbonate Fuel Cell was investigated by the former CEGB during the 1960s. Main problems concerned the deterioration of the electrodes and the high vapour pressure of the carbonate melt leading to rapid loss of electrolyte. The programme was abandoned when the intolerance of the system to sulphur was discovered. Interest revived in the USA and work is being carried out by the Institute of Gas Technology. Electrode deterioration is much reduced but electrolyte loss and sulphur poisoning still need further R & D work. The system has the attraction of a potentially high electrical efficiency.

Some of the difficulties of corrosion, fuel gas purity and electrolyte loss can be overcome by the use of a solid electrolyte.

Fuel cell technology has yet to be proven in long term operation. However there are now several types in operation, ranging from a few kilowatts electrical output to several megawatts. Projected costs at the moment are up to £1 500/kWe, with an anticipated reduction to about £500/kWe if technical targets are achieved and market expansion occurs.

Economic analysis of fuel cell based CHP shows that key R & D targets are to reduce capital and maintenance costs and increase electrical efficiency. This latter should aid in lowering capital costs and reduce the dependency of the economic case on heat recovery. Consideration of the present and anticipated electrical performance of a practical range of CHP systems shows that, in the longer term, fuel cells could well out-perform heat engine alternator systems. In addition, they may offer unobtrusive and effluent free operation. Electrical efficiencies upwards of 60% are achievable in principle. Apart from electro-chemistry, other important aspects will be reforming heat transfer, fluid flow, heat recovery, high temperature construction and process controls.

13.10 THERMO-ELECTRIC AND THERMIONIC ELECTRICITY GENERATION

13.10.1 Thermo-Electric Generation

The basic principle of this method of electricity generation is that when a circuit is formed containing a junction of two dissimilar conductors and the junction is heated a current will flow in the circuit (Figure 13.12). This effect was first observed by Seebeck who found that a magnetic needle was deflected when it was held close to a circuit of two dissimilar conductors heated at one end. An associated effect was observed in 1834 by Peltier who found that small quantities of heat were emitted by passing a weak current through junctions of dissimilar metals (Figure 13.13). Seebeck had also, unknown to himself, tested several semi-conductor materials which could have generated electricity from heat at 3% efficiency, a figure comparable to the efficiency of a 19th century steam engine.

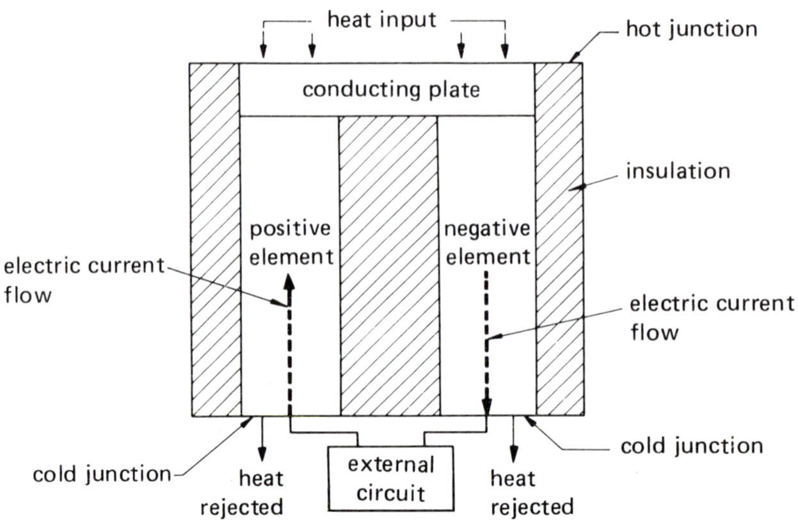

Fig. 13.12 Thermo-electric generator

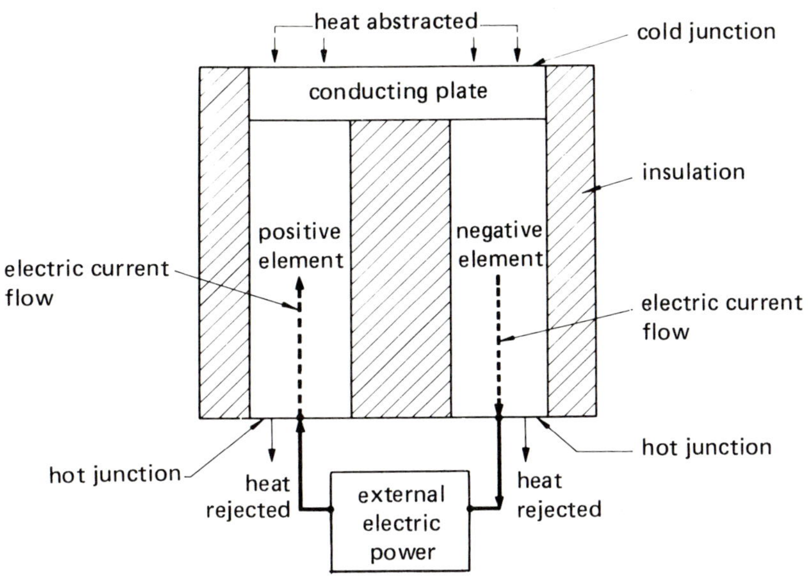

Fig. 13.13 Peltier cooling device

Thermo-Electric Materials

The thermo-electric properties of semi-conductors, for example, lead sulphide (PbS), and other metallic compositions are ten times greater, or more, than those of fully conducting metals, such as copper and iron. By the connection of enough semi-conductors in series, through an external circuit, sufficient power can be made to drive an electrical device. In effect, the action of the applied heat is to 'pump electrons' through the circuit; alternatively, conducting metals such as copper, zinc or lead can be made semi-conductive by doping with, for example, lead sulphide and copper oxide.

The properties of a material necessary for the generation of thermal electricity at high efficiencies are high thermo-electric voltage, low electrical resistance and low thermal conductivity.

Fully conductive metals are not altogether satisfactory because of relatively low generated current and high thermal conductivity. Compounds of metals with non-metals are better since they produce a higher electromotive force (emf) and have lower thermal conductivity but are still not entirely satisfactory due to high electrical resistance. Certain high temperature ceramics and metal halides appear to be capable of giving promising results. Materials have been developed which may be capable of withstanding 1100 °C or more with promising efficiencies.

The addition to thermo-electric materials of small concentrations of positive and negative promoters gives materials of improved electrical conductivity and increased emf generating capacity; these materials are, however, relatively costly and cannot at present be operated at much above 420 °C to 540 °C. Thermo-electrical materials capable of operating in the range 450 °C to 800 °C are under investigation.

Thermo-Electric (Seebeck) Generators

The above term is used for a generator composed of several thermocouples in series. A single couple consists of the two dissimilar materials joined in electrical contact at one end, called the hot junction, and connected to an electric circuit at the other end termed the cold junction. The heat applied to the hot junction is partly converted into electrical energy, the remainder of the heat is conducted to the cold junction. For practical purposes, the thermo-electric generator requires many such couples connected in series to provide a useful voltage output. For example, a single couple may produce 0.022 volts per 50 K temperature difference between the hot and cold junctions, therefore if the maximum temperature of operation is about 600 °C each couple gives 0.24 volts. It would therefore require at least 100 couples to develop 24 volts. However, when power is drawn from a thermo-electric generator only one half of the voltage appears in the external circuit, the other half of the voltage being used in overcoming internal resistances. Thus, in the example cited above about 200 couples would be required to develop 24 volts under actual load conditions.

13.10.2 Thermionic Generation

The principle of a heated surface giving off electrons has been known for some time. The main difficulties preventing the construction of a generator were poor design, lack of suitable electrode materials and the space charge effect.

A practical generator would consist of two metal plates a short distance apart. When the cathode plate is heated from the rear, its surface emits electrons which travel towards the second plate, the anode, which is kept at a relatively low temperature (see Figure 13.14). However, the emitted electrons may meet considerable resistance due to the 'space charge' (an electron cloud near the cathode). The final result is that a current flows from the cathode plate to the anode plate, thence round the external circuit.

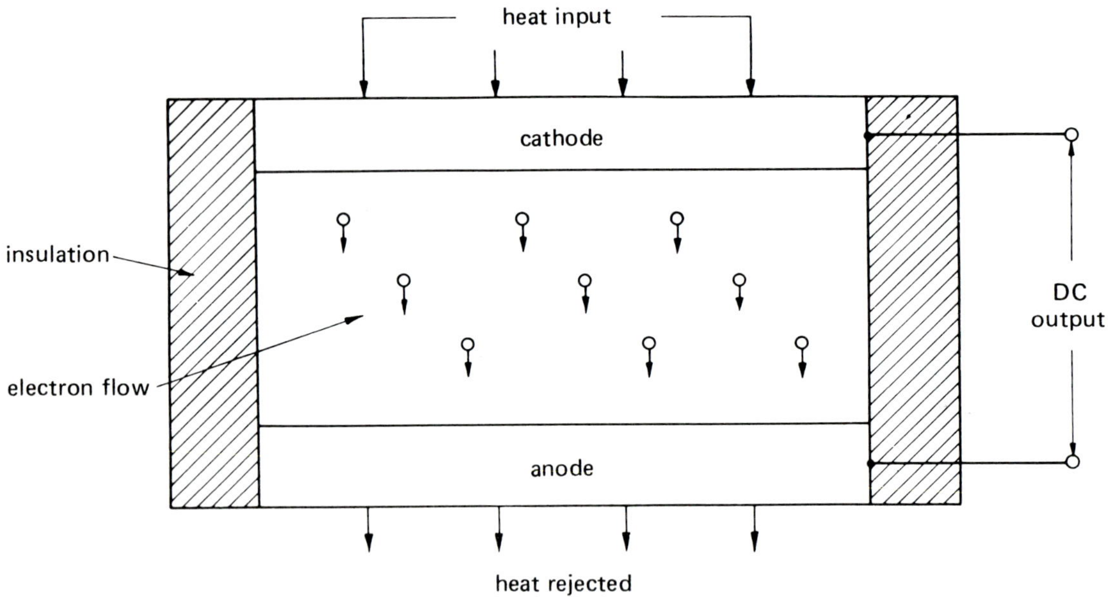

Fig. 13.14 Thermionic generator

A single stage generator may develop 1 to 3 volts at a current density of about 2.3 - 4.7 amps per sq cm (15 to 30 amps per square inch) of electrode surface.

Depending upon electrode composition a thermionic generator operates at cathode temperatures in the range 900 °C to 1300 °C and up to 2000 °C with refractory materials.

Efficiencies of 20 - 25% may ultimately be possible.

13.11 MAGNETO-HYDRODYNAMIC GENERATION

A conventional generator makes use of the rotation of a coil of wire in the magnetic field. Efforts are being made to utilise the inherent thermodynamic efficiency of the very high flame temperatures of burning oil or pulverised coal. In the magneto-hydrodynamic (MHD) process, instead of a solid conductor a jet of ionised gas flows through a magnetic field. High voltage direct current at over 2 000 volts is obtained from electrodes placed in the gas stream. Ionisation requires temperatures in the range of 1 600 to 2 000 °C and in some cases temperatures above this figure. Thermal efficiencies of around 55% have been recorded.

13.11.1 MHD Generator

This comprises a high temperature nozzle which has within it two electrodes to withdraw power (see Figure 13.15). The electrodes are the analogues of brushes in the conventional electricity generator. Electric coils external to the nozzle produce a strong magnetic field across the flowing ionised gas. A typical MHD generation system would contain the following components:

- A motor driven air compressor to develop the pressure needed for flow through the system.
- A pressure burner giving ionised combustion products.
- A specially designed nozzle in which the combustion products reach the high gas velocities needed for power generation in the magnetic field.

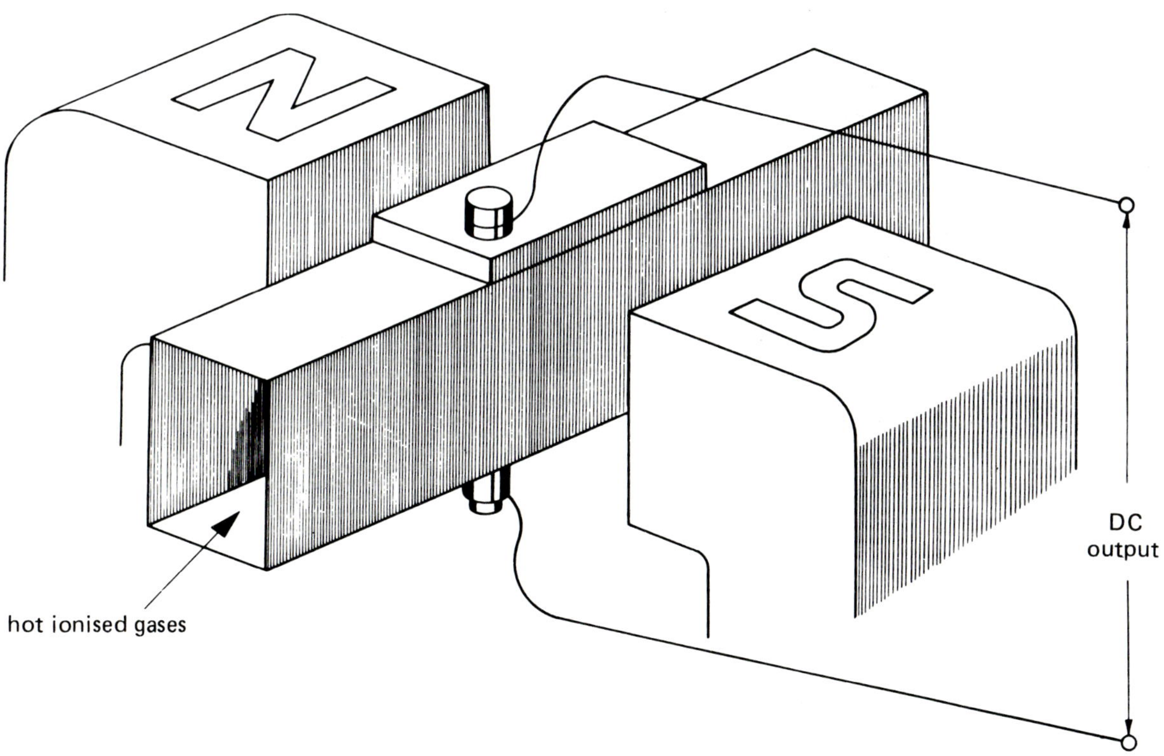

Fig. 13.15 MHD Generator

Since the exhaust gases of the generator are flowing at considerable velocity, they contain large quantities of kinetic energy. In some cases, this can represent as much as 12% of the energy entering the MHD generator. The exhaust gases of the MHD generator may be finally passed through an expansion turbine which can provide up to 74% of the power for the initial air compression (Figure 13.16).

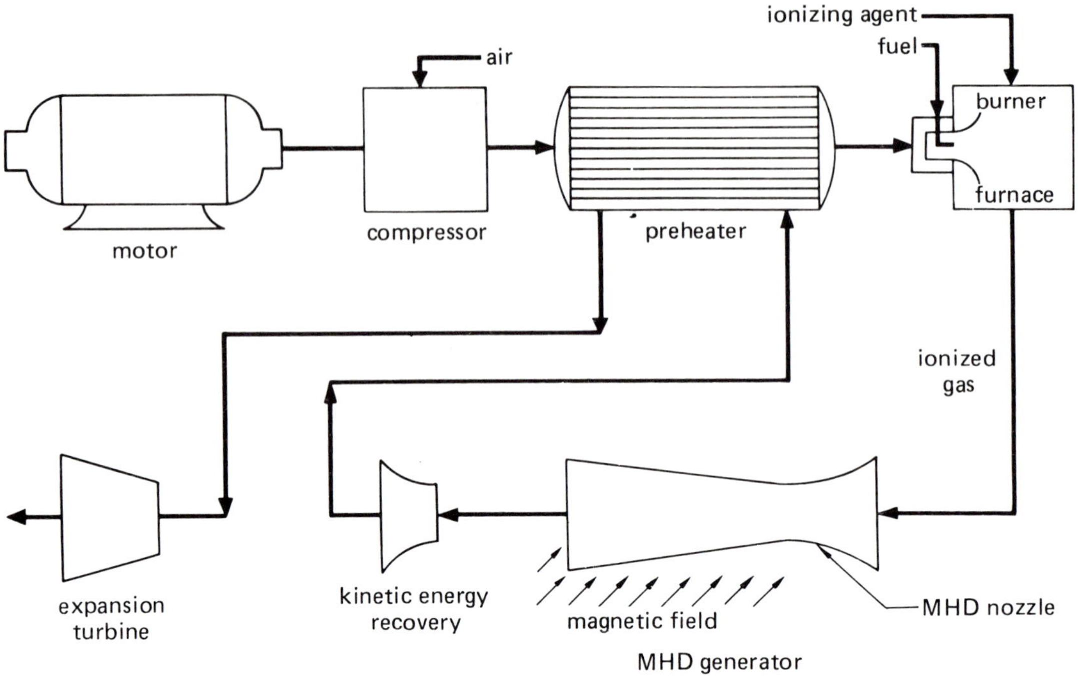

Fig. 13.16 Independent MHD unit

The MHD process may also be used in conjunction with a conventional steam turbine generator system (Fig. 13.17). Steam from a conventional boiler is used to drive a steam turbine which in turn drives a generator and air compressor. Power is available from the generator while the compressed air is used further in the MHD process. The MHD units therefore use the incoming energy in the compressed air while the conventional steam boiler system can utilise a very high proportion of the heat content of the MHD effluent gases.

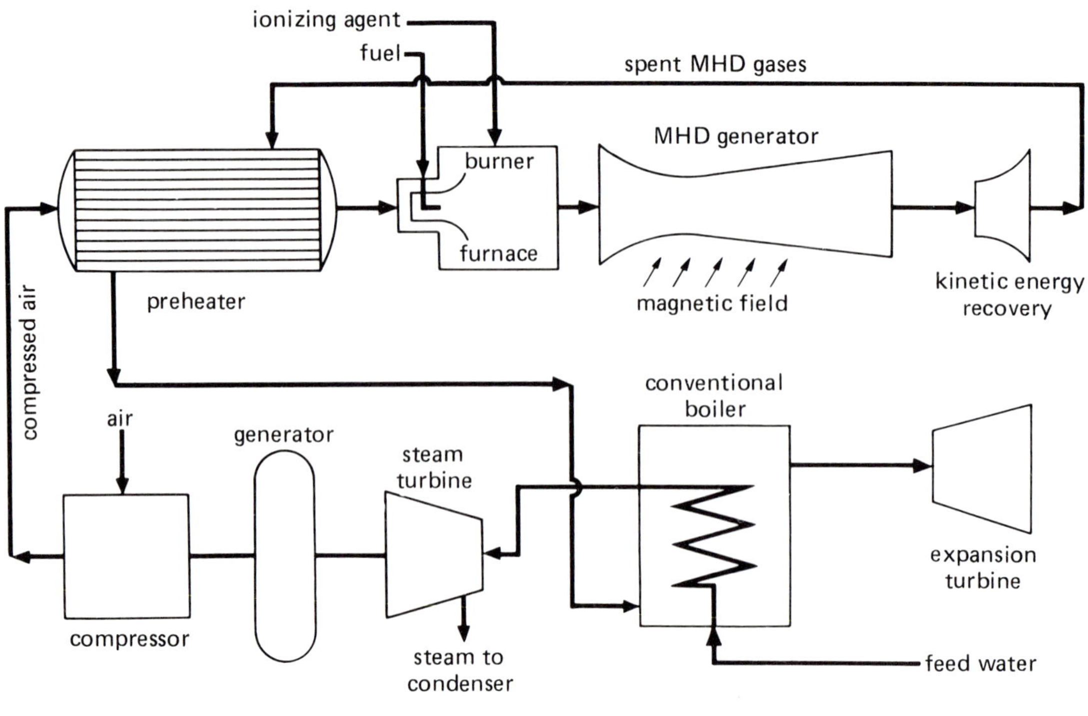

Fig. 13.17 MHD topped conventional plant

13.11.2 Practical problems

Since high temperatures up to 2 750 °C are required to ionize the gas, this gives problems with regard to the choice of materials for the generator and its construction. By adding a seeding material to the high temperature gas stream, operation may proceed at lower temperature but a recovery system for the seed materials is now necessary. Oxygen enrichment of the combustion air may also be required to achieve the necessary temperatures. (Maximum Oxy-NG theoretical flame temperature is around 2 700 °C).

MHD generation is most suitable for large scale stations of sizes about 100 MW. A pilot 70 kW generator using natural gas as a fuel has been operated in the USSR and a 10 MW generator using natural gas seeded with potassium carbonate (K_2CO_3) and expected to operate at 48% efficiency is being developed there; the unit contains a 100 tonne magnet and uses carborundum electrodes. Designs for much larger plants are under way.

13.12 OTHER METHODS OF ELECTRICAL GENERATION

13.12.1 Electro-Gas Dynamics (EGD)

This largely experimental method of electricity generation was originally developed in the USA. The operational principle is similar to that of MHD but uses electrostatic rather than electromagnetic fields to intercept the seeded gas stream. EGD converts thermal pressure energy forced across a corona electrostatic discharge field into very high voltage electricity, in the region of 100 kV (see Figure 13.18). The gas pressures required are 8.5 to 31 bar (120 to 450 lbf/in^2) at temperatures in the region of 1 100 °C. This is the only known method at present of generating high voltage electricity directly. The EGD process will work equally well with any source of energy, i.e. coal, oil, gas or nuclear fuel; however, it is particularly suitable for coal because the ash or dirt in coal combustion gases, normally undesirable in conventional generating systems, carries the electric charge in the EGD energy conversion process and increases the overall efficiency.

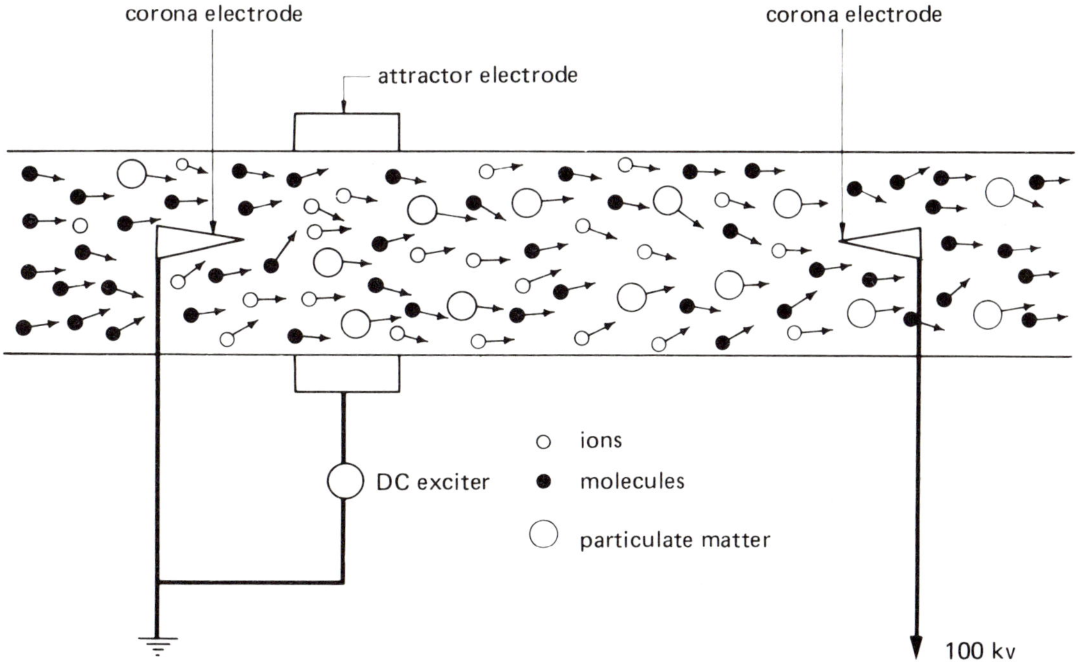

Fig. 13.18 EGD principle

An EGD power station would consist of a compressor which draws air in, compresses it and discharges it under pressure; a combustion chamber in which the fuel is burnt; and the EGD generator which produces high voltage electricity directly from the exhaust gases. The exhaust gases then flow through a precipitator where dirt and other possible pollutant gases are removed before they are exhausted to atmosphere. This system is shown in diagrammatic form in Fig. 13.19.

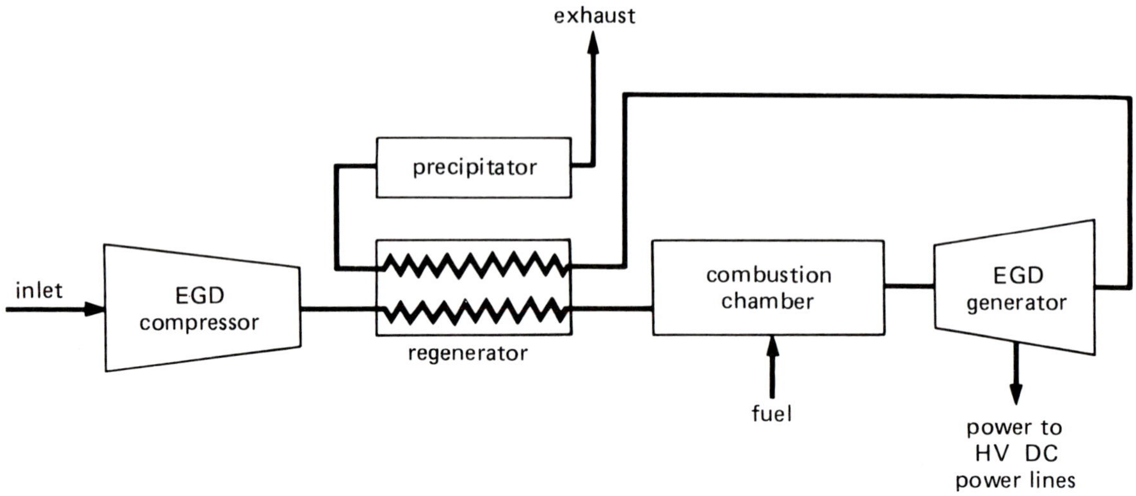

Fig. 13.19 Electro gas dynamics

13.12.2 Electro-Hydrodynamics (EHD)

This system of direct electricity conversion makes use, as its name hydro implies, of liquid droplets (i.e. aerosols) as the energy producing medium. The aerosol is formed as the liquid is forced through a capillary into a high velocity gas stream being discharged across a high intensity electrical field. The charged fluid reaches a collector electrode where the aerosol is discharged and the liquid consolidated, while the electricity flows through the external circuit.

REFERENCES

BRITISH GAS PLC PUBLICATIONS

IM/17: 1st Edition, 1981. Code of Practice for Natural Gas Fuelled Spark Ignition and Dual Fuel Engines

IM/21: 1st Edition, 1985. Guidance Notes for Architects, Builders etc. on the Gas Safety (Installation and Use) Regulations 1984.

IM/24: 1st Edition, 1989. Guidance Notes on the Installation of Industrial Gas Turbines, Associated Gas Compressors etc.

OTHER PUBLICATIONS

Total Energy. Diamant R.M.E., Pergamon Press, Oxford 1970

Total Energy Evaluation and Feasibility Studies. Freeman H. and Horsler A.G., Paper *184* Dec 1975

Engineering Recommendations G5/3 : Limits for Harmonics in U.K. Electricity Supply System. The Electricity Council, 1976.

Fuel Cells for Public Utility and Industrial Power. Noyes R., Noyes Data Corp. Park Ridge, N.J. 1977

Fuel Cells. McDougall A., Macmillan, London 1978.

An Introduction to Electric Power Systems. Harrison J.A., Longman Group Ltd, 1980.

Heat Pump Technology. Cube H.L., Steimie F and Goodall E.G.A., Butterworths, 1981.

Waste Heat Recovery and Internal Combustion Engines in Malting. European Brewing Convention, Parsons N.E. and Marsh J.B., 10th International Congress, Copenhagen, 1981.

Engineering Recommendations P13/1 : Electric motors : Starting Conditions. The Electricity Council, 1982

Heat Pump Systems : A Technology Review. International Energy Agency, Paris, 1982.

The Energy Act 1983. HMSO.

Guidance Notes on the Safety Implications of the Energy Act 1983. ECSB/8403-28, The Electricity Council 1984.

The Gas Safety (Installation and Use) Regulations 1984.

Steam Recompression Drying. Heaton A.V. and Benstead R., 2nd BHRA International Symposium : Large Scale Applications of Heat Pumps. York, 1984.

Energy Conservation Equipment. Diamant R.M.E., Architectural Press Ltd, London, 1984.

Energy Efficiency Technologies for Swimming Pools. Energy Technology Series 3, Energy Efficiency Office, ETSU, Harwell, 1985.

Small Scale Combined Heat and Power. Energy Technology Series, Energy Efficiency Office, ETSU, Harwell, 1985.

Engineering Recommendations G59 : Recommendations for the Connection of Private Generating Plant to the Electricity Board's Distribution System. The Electricity Council, 1985.

Compressed Payback Times with Mechanical Vapour Recompression. Heaton A.V., Chartered Mechanical Engineer, April, 1986.

The Gas Act 1986

The Building Regulations, HMSO 1985 amended 1989.

Instrumentation and Measurement 14

CHAPTER 14

INSTRUMENTATION AND MEASUREMENT

14.1 INTRODUCTION

Instruments and methods for the measurement of performance are essential to ensure safe, economical and reliable plant operation. They range from the simplest manual devices to those used to actuate the complete automatic control of plant.

Test instrumentation (usually of a portable nature) is employed in the performance testing of plant primarily to satisfy the user and the equipment supplier that the specified conditions of design and operation have been achieved. These instruments generally require relatively skilled technical operators, careful handling and regular calibration and are not always suitable for long-term continuous commercial operation.

Permanently installed instrumentation is expected to give satisfactory accuracy and reliability over extended periods of time. The emphasis on dependability and repeatability often demands some compromise in absolute accuracy. However, it should be stated that the accuracy of instrumentation for permanent installation is continually being improved and is approaching, in many areas, that of instruments used for test purposes.

In this Chapter instrumentation available to undertake the measurement of temperature, pressure, fluid flow, the analysis of gases and combustion products, humidity and dewpoint are discussed together with their practical applications.

14.2 TEMPERATURE MEASUREMENT

Heat-affected properties of substances, such as thermal expansion, radiation and electrical effects are used in commercial temperature measuring instruments. These instruments vary in their precision depending on the property utilised and substance used, as well as on the design of the instrument.

14.2.1 Instruments Based on Changes of State

Fusion

For a pure chemical element or compound such as mercury or water, fusion, or change of state from solid to liquid, occurs at a fixed temperature. The melting points of such materials are therefore suitable fixed points for temperature scales.

The fusion of pyrometric cones is widely used in the ceramic industry as a method of measuring high temperatures in refractory heating furnaces. These cones, small pyramids about 5 cm high, are made of selected mixtures of oxides and glass which soften and melt at established temperatures. The cones are similar in nature to ceramic ware, and behave in a manner which indicates what the behaviour of the pottery under similar circumstances is likely to be. The cones are known as Harrison or Seger cones, and by varying their composition a range of temperatures between 600 °C and 2 000 °C may be covered in convenient steps as shown in Table 14.1.

Table 14.1 Softening temperature of Seger cones

The figures given below refer to British Cones with test conditions as BS1902:

Cone No.	End-point °C	Cone No.	End-point °C	Cone No.	End-point °C
022	600	02a	1 060	19	1 520
021	650	01a	1 080	20	1 530
020	670	1a	1 100	26	1 580
019	690	2a	1 120	27	1 610
018	710	3a	1 140	28	1 630
017	730	4a	1 160	29	1 650
016	750	5a	1 180	30	1 670
015a	790	6a	1 200	31	1 690
014a	815	7	1 230	32	1 710
013a	835	8	1 250	33	1 730
012a	855	9	1 280	34	1 750
011a	880	10	1 300	35	1 770
010a	900	11	1 320	36	1 790
09a	920	12	1 350	37	1 825
08a	940	13	1 380	38	1 850
07a	960	14	1 410	39	1 880
06a	980	15	1 435	40	1 920
05a	1 000	16	1 460	41	1 960
04a	1 020	17	1 480	42	2 000
03a	1 040	18	1 500		

A series of cones are placed in the kiln and those of lower temperature will melt and one will just bend over. This cone indicates the temperature of the kiln, confirmed by the fact that the cone of next higher temperature is not affected. In order to obtain maximum accuracy, which is of the order of ±10 °C, the cones must be heated at a controlled rate.

Bars, known as Watkins heat recorders, are used in the same way as cones, their deformation being related to temperature.

Bullers rings are also used to measure kiln temperatures. After firing and cooling, during which the ring contracts, the ring is dropped on to a gauge which registers the temperature corresponding to ring size.

Fusion pyrometers are also made in the form of crayon, paint, and pellets, which indicate a range of established temperatures up to 1 100 °C. The crayon or paint is applied to a cold surface leaving a dull finish mark which melts and changes to a glossy finish when the surface reaches the specified temperature. These marks, therefore, indicate whether or not the surface temperature has reached or exceeded a selected value. The pellets begin to melt at specified temperatures when in contact with a hot surface.

Vaporisation

The vapour pressure of a liquid depends on its temperature. When the liquid is heated to the boiling temperature, the vapour pressure is equal to the total pressure above the liquid surface. Therefore, the boiling points of various pure chemical elements or compounds at standard atmospheric pressure can be used as thermometric fixed points.

The change of vapour pressure with temperature is utilised in the vapour-pressure thermometer, illustrated schematically in Figure 14.1, which consists of a bulb partially filled with liquid and a capillary tube leading from the bulb to a pressure gauge calibrated to read temperature directly, the temperature scale being non-uniform. If the space between the liquid and the pressure gauge is filled only with vapour from the liquid, the pressure will vary directly with the temperature of the liquid in the bulb. The capillary tube may be of considerable length, and its temperature does not affect the reading, but the accuracy of the instrument is affected by variations in atmospheric pressure and by elevation of the bulb above or below the gauge. The accuracy is not affected by changes in ambient temperature as long as the ambient temperature does not oscillate around that of the measuring bulb. The working range of a given instrument is limited and usually lies between -29 °C and +370 °C.

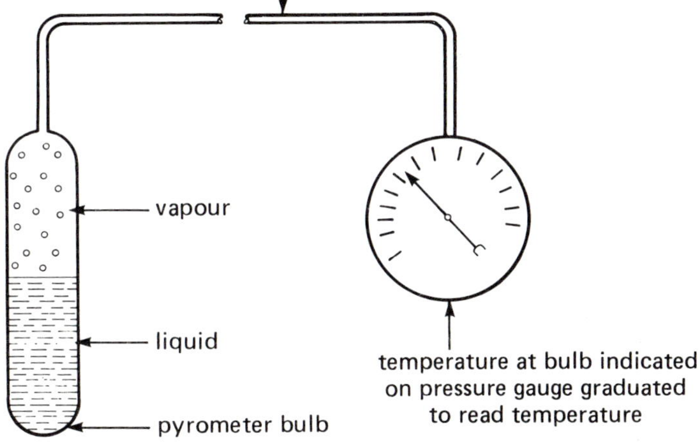

Fig. 14.1 Schematic assembly vapour-pressure thermometer

14.2.2 Instruments Based on Expansion Properties

Most substances expand when heated, and in many cases the amount of expansion is almost directly proportional to the change in temperature. This effect is utilised in various types of thermometers using gases, solids or liquids.

Gases

The volume, pressure and temperature of an 'ideal' gas are related as follows:

$$PV/T = \text{constant}$$

where	P	=	absolute pressure	bar
	V	=	volume	m^3
	T	=	absolute temperature	K

For initial and final conditions 1 and 2 respectively, it therefore follows that:

$$P_1 V_1/T_1 = P_2 V_2/T_2$$

It should be noted that no gas is in fact 'ideal' and there are deviations from the relationship, but it holds good over a reasonable working range for many gases commonly encountered.

Two types of gas thermometer are based on this relationship. In one a constant gas volume is maintained, and changes in pressure are used to measure changes in temperature. In the other a constant pressure is maintained, and changes in volume are used to measure changes in temperature. Very accurate instruments of this type have been developed for laboratory work and the constant-volume type thermometer is widely used commercially.

Nitrogen is commonly used for the gas-filled thermometer in industrial applications. It is suitable for a temperature range of –130 °C to +540 °C. The construction is similar to the vapour-pressure thermometer shown in Figure 14.1 with nitrogen gas replacing the liquid and vapour. Expansion of the heated nitrogen in the bulb increases the pressure in the system and actuates the temperature indicator, the temperature scale being uniform. The capillary tube may be of any length and changes in temperature of the capillary tubing will introduce small errors.

Liquids

The expansion of liquids is used in a thermometer again similar in design to the vapour-pressure instrument shown in Figure 14.1 except that the bulb and capillary tubes are completely filled with liquid. The readings of instruments of this type are subject to error if capillary tubing is subjected to temperature changes although in some instruments a compensation mechanism is incorporated.

The liquid-in-glass thermometer is a simple, direct reading, and conveniently portable instrument, widely used in many activities requiring the determination of temperature. Low-precision thermometers are inexpensive, and instruments of moderate precision are available for laboratory use. This type of thermometer is usually made with a reservoir of liquid in a glass bulb connected directly to a glass capillary tube with graduated markings or with a scale attached. Mercury, the most commonly used liquid, is satisfactory from –40 °C (just above its freezing point) up to about 315 °C if the capillary space above the mercury is evacuated, or up to 480 °C or higher if this space is filled with nitrogen or carbon dioxide under pressure.

Liquid-in-glass thermometers are calibrated either for complete immersion or for partial immersion, and to obtain accurate results should be used accordingly. The use of unprotected glass thermometers is usually restricted to laboratory applications. For more rugged service there are various designs of 'industrial' thermometers, with the bulb and stem protected by a metal casing and usually arranged for use in a thermometer well. With this type of installation, response to rapid changes of temperature is slower than with the unprotected laboratory-type instrument.

Solids

The expansion of solids when heated is applied to temperature measurement in thermometers using a bimetallic strip. Flat ribbons of two different metals with unlike coefficients of thermal expansion, for example brass and Invar, are joined face-to-face by riveting or welding to form a bimetallic strip.

When the strip is heated, the expansion is greater for one side of the double layer than for the other, and the strip bends, if originally flat, or changes its curvature if initially in the spiral form frequently used. Bimetallic strips are widely used in inexpensive household thermometers, many designs of thermostats, and a variety of temperature control and regulating equipment. They are particularly useful for automatic temperature compensation in the mechanisms of other instruments.

14.2.3 Instruments Based on Radiation Properties

All solid bodies emit radiation, the amount being very small at low temperatures and large at high temperatures. The quantity of radiation may be calculated by the Stefan-Boltzmann formula:

$$M = \sigma \epsilon T^4$$

where M = radiant exitance W/m^2
σ = Stefan-Boltzmann constant = 5.67×10^{-8} W/m^2 K^4
ϵ = emissivity of the surface, a dimensionless number between 0 and 1
T = absolute temperature K

At low temperatures the radiation is chiefly in the infra-red range, invisible to the human eye. As the temperature rises, an increasing proportion of the radiation is in shorter wavelengths, becoming visible as a dull red glow at about 540 °C and passing through yellow toward white at higher temperatures. The temperature of hot metals (above 540 °C) can be estimated by their colour and for iron or steel the colour scale is roughly as follows:

Dark red	540 °C
Medium cherry red	680 °C
Orange	900 °C
Yellow	1 010 °C
White	1 200 °C

Two types of temperature-measuring instruments, the optical pyrometer and the radiation pyrometer, are based on the radiating properties of materials.

Optical pyrometers

The most common type in use is the disappearing-filament pyrometer. By sighting the optical pyrometer on a hot object, the brightness of the latter can be compared visually with the brightness of a calibrated source of radiation within the instrument, usually an electrically heated tungsten filament. The current passing through the filament is adjusted until the brightness of the filament matches the brightness of the hot object and is measured with an ammeter whose scale is calibrated in temperature. A red filter may be used to restrict the comparison to a particular wavelength. This instrument is designed for measuring the temperature of surfaces with an emissivity of 1.0 i.e. is a 'black body', which by definition absorbs all radiation incident upon it, reflecting and transmitting none. When accurately calibrated, the pyrometer will give excellent results above 815 °C, provided its use is restricted to the application for which it is designed. Measurement of the temperature of the interior of a uniformly heated enclosure, for example a muffle furnace, is such an application. When used to measure the temperature of a hot object in the open, the optical pyrometer will always read low, the error being small (10 °C) for high-emissivity bodies, such as steel ingots, and considerable (110 to 165 °C) for un-oxidised liquid steel or iron surfaces.

The optical pyrometer has a wide field of application for temperature measurements in heating furnaces and around steel mills and iron foundries. It is of no value for the commercial measurement of gas temperature, since clean gases do not radiate in the visible range.

Radiation pyrometers

Radiation pyrometers also enable surface temperatures to be determined without physical contact. In one type of radiation pyrometer, all radiation from the hot body, regardless of wavelength, is absorbed by the instrument. The heat absorption is measured by the temperature rise of a delicate thermocouple within the instrument onto which the radiation is focussed by a mirror, calibrated to indicate the temperature of the hot surface at which the pyrometer is sighted, on the assumption that the surface emissivity equals 1.0. The hot surface must fill the entire field of view of this instrument.

The radiation pyrometer has been developed into a laboratory research instrument of extreme sensitivity and high precision over a wide range of temperature. The usual industrial-type instrument gives good results above 540 °C when used to measure temperatures of high-emissivity bodies, such as the interiors of uniformly heated enclosures. Since the operation is independent of human judgement, radiation pyrometers may be used as remotely operated indicators or recorders, or in automatic control systems. However, errors in measuring the temperatures of hot bodies with emissivities of less than 1.0, especially if they are in the open, are extremely large.

Radiation pyrometers sensitive to selective wavelengths have been developed and give good results measuring temperatures of bodies or flames utilising, for example, the infra-red band. Infra-red radiation is produced by all matter at temperatures above absolute zero, and a detector sensitive to infra-red, such as lead sulphide, may be used to sense the radiation. Using a system of lenses, a lead sulphide cell, an amplifier and an indicator, temperature measurements may be made of radiating bodies.

A further variation on pyrometers is the so-called 'two-colour' pyrometer, which measures the intensities of two selected wave bands of the visible spectrum emitted by a heated object, computes the ratio of these emitted energies and converts it into a temperature indication. As with the optical pyrometer, indications depend on sighting visible rays, and hence its lower limit is 540 °C; similarly, radiation pyrometers are not capable of determining gas temperatures.

14.2.4 Instruments Based on Electrical Properties

Two classes of widely used temperature measuring instruments, the electrical-resistance thermometer and the thermocouple, are based on the relation of temperature to the electrical properties of metals. Instruments based on these principles form the basis of the many hand held electronic reading units with digital displays that are now widely available.

Resistance thermometer

The electrical-resistance thermometer, used over a range of temperature from −240 °C to 1 000 °C, depends on the increase in electrical resistance of metals with increase in temperature, which is almost in direct proportion. Therefore, if the electrical resistance of a wire of known and calibrated material is measured (by a Wheatstone bridge or other device), the temperature of the wire can be determined.

In the simplest form of this instrument, shown in Figure 14.2A, the reading would be the sum of the resistances of the calibrated wire and the leads connecting this wire to the Wheatstone bridge. This value would thus be subject to error from temperature changes in the leads. Where the greatest accuracy is required, or where the power supply and bridge may be shared as with some data loggers a three or four wire compensated system is used as shown in Figure 14.2B and C respectively. In

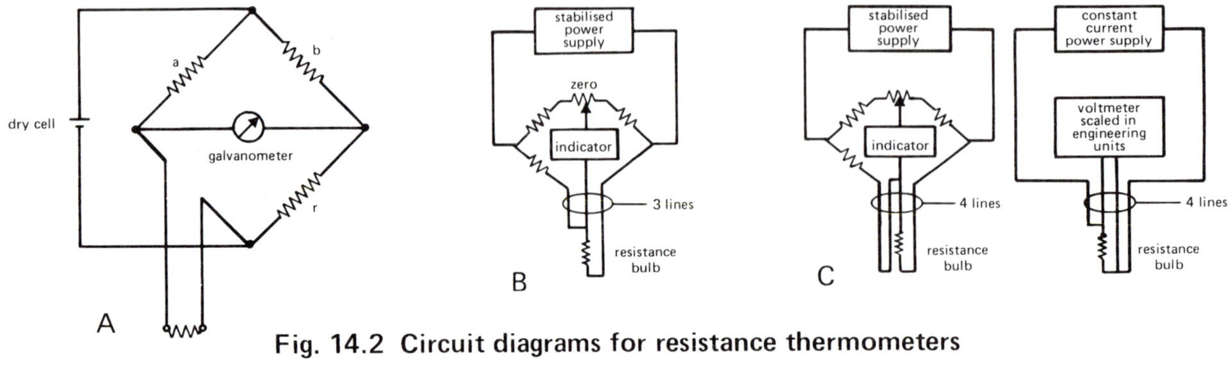

Fig. 14.2 Circuit diagrams for resistance thermometers

order to localize the point of temperature measurement, the resistance wire may be made in the form of a small coil. From room temperature to 120 °C, commercial instruments usually have resistance coils of nickel or copper with platinum being used for higher temperatures and in many high-precision laboratory instruments over a wide range of temperatures.

The electrical-resistance thermometer does not require human judgement and can therefore be used for the remote operation of indicating, recording, or automatic-control instruments. If proper precautions are taken in its use, it is stable and accurate. However, it is less rugged and less versatile than a thermocouple.

A development of the resistance thermometer incorporates thermistors. These consist of an element made from a semiconducting material which has a negative temperature coefficient about ten times greater than that of copper or platinum. They are, therefore, considerably more sensitive than a normal resistance element and as the resistivity of the thermistor material is also much higher than that of any metal, so the size can be made very small, giving a very rapid response. They can be used for temperatures up to about 1 000 °C, but the usual range is lower (-100 °C to 300 °C).

Thermocouple

A thermocouple consists of two electrical conductors of dissimilar materials joined at their extremities to form a circuit. If one of the junctions is maintained at a temperature higher than that of the other, an electromotive force (emf) is set up, producing a flow of current through the circuit, as illustrated in Figure 14.3. This is known as the Seebeck effect.

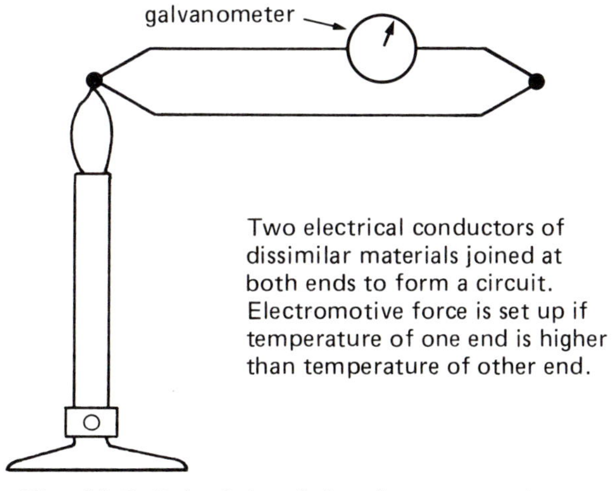

Fig. 14.3 Principle of the thermocouple

The magnitude of the net emf depends on the difference between the temperatures of the two junctions and the materials used for the conductors. No imbalance or net emf will be set up if the two junctions of dissimilar materials are at the same temperature, or if the conductors are of the same material even though the two junctions are at different temperatures. If one junction of a thermocouple, which has been calibrated, is maintained at a known temperature, the temperature of the other junction can be determined by measuring the net emf produced, since this is proportional (almost directly) to the difference in temperature of the two junctions. The relationship between the electrical emf and the corresponding temperature difference between the two junctions has been established by laboratory test throughout the temperature ranges for thermocouple materials in common use. These values are plotted in Figure 14.4 and Table 14.2 lists the figures for 'K' type (chromel/alumel) thermocouples which are very widely used.

The principal advantages of thermocouples in measuring temperatures are their versatility of application, rapidity of response through wide ranges of temperature, high degree of accuracy, durability, accurate reproducibility at relatively low cost, convenience of centralized reading or recording from one or many remote points, and simplicity of application to equipment for control and regulation of temperature.

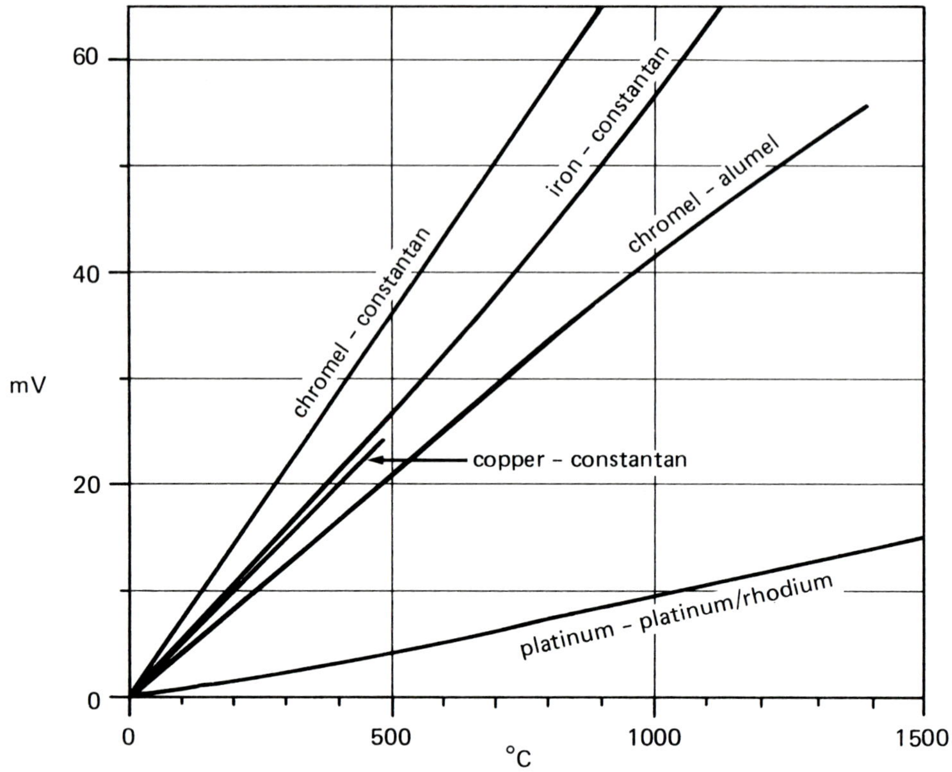

Fig. 14.4 Temperature/emf relationship for some commonly used thermocouples

Table 14.2 Temperature/emf relationship for Chromel/Alumel thermocouples (BS 4937 Part 4) Reference junction at 0 °C

°C	0	-10	-20	-30	-40	-50	-60	-70	-80	-90
					emf in microvolts					
-200	-5 891	-6 035	-6 158	-6 262	-6 344	-6 404	-6 441	-6 458		
-100	-3 553	-3 852	-4 138	-4 410	-4 669	-4 912	-5 141	-5 354	-5 550	-5 730
0	0	-392	-777	-1 156	-1 527	-1 889	-2 243	-2 586	-2 920	-3 242
°C	**0**	**10**	**20**	**30**	**40**	**50**	**60**	**70**	**80**	**90**
					emf in microvolts					
0	0	397	798	1 203	1 611	2 022	2 436	2 850	3 266	3 681
100	4 095	4 508	4 919	5 327	5 733	6 137	6 539	6 939	7 338	7 737
200	8 137	8 537	8 938	9 341	9 745	10 151	10 560	10 969	11 381	11 793
300	12 207	12 623	13 039	13 456	13 874	14 292	14 712	15 132	15 552	15 974
400	16 395	16 818	17 241	17 664	18 088	18 513	18 938	19 363	19 788	20 214
500	20 640	21 066	21 493	21 919	22 346	22 772	23 198	23 624	24 050	24 476
600	24 902	25 327	25 751	26 176	26 599	27 022	27 445	27 867	28 288	28 709
700	29 128	29 547	29 965	30 383	30 799	31 214	31 629	32 042	32 455	32 866
800	33 277	33 686	34 095	34 502	34 909	35 314	35 718	36 121	36 524	36 925
900	37 325	37 724	38 122	38 519	38 915	39 310	39 703	40 096	40 488	40 879
1 000	41 269	41 657	42 045	42 432	42 817	43 202	43 585	43 968	44 349	44 729
1 100	45 108	45 486	45 863	46 238	46 612	46 985	47 356	47 726	48 095	48 462
1 200	48 828	49 192	49 555	49 916	50 276	50 633	50 990	51 344	51 697	52 049
1 300	52 398	52 747	53 093	53 439	53 782	54 125	54 466	54 807		

In the thermocouple circuit illustrated in Figure 14.5, the thermocouple wires of dissimilar materials extend from the hot junction to the cold or 'reference' junction which is usually held at 0 °C. In the potentiometer circuit, copper leads may be run from the reference junction terminals to the measuring instrument without affecting the net emf of the thermocouple. The pair of similar copper leads acts merely as electrical connectors to transfer the emf from the reference junction to the copper terminals of the potentiometer. If the instrument is at a uniform temperature throughout, no emf is set up between the copper conductors and slidewire materials within the potentiometer itself. However, if temperature differences do exist within the instrument, there will be disturbing emfs in the circuit, and the readings will be affected.

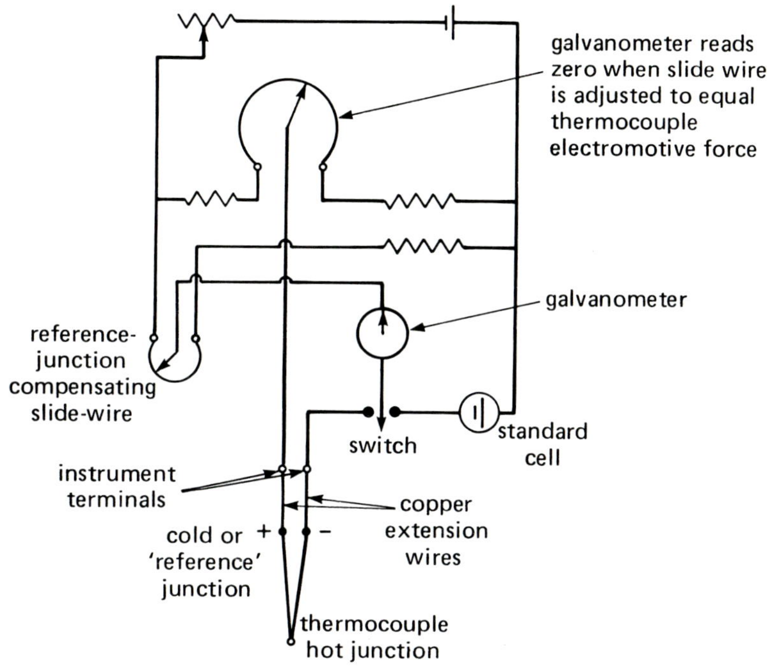

Fig. 14.5 Thermocouple circuit with manually-adjusted reference junction compensation

In the laboratory a reference junction temperature of 0 °C can be achieved using a Dewar flask containing crushed ice. Electrically powered continuously operating chambers are available that can accommodate up to 100 reference junction thermocouples and maintain their temperature at 0 °C ±0.05 °C. In most control and measuring instruments the cold junction compensation technique, as shown in Figure 14.6, is employed to provide accurate temperature measurement.

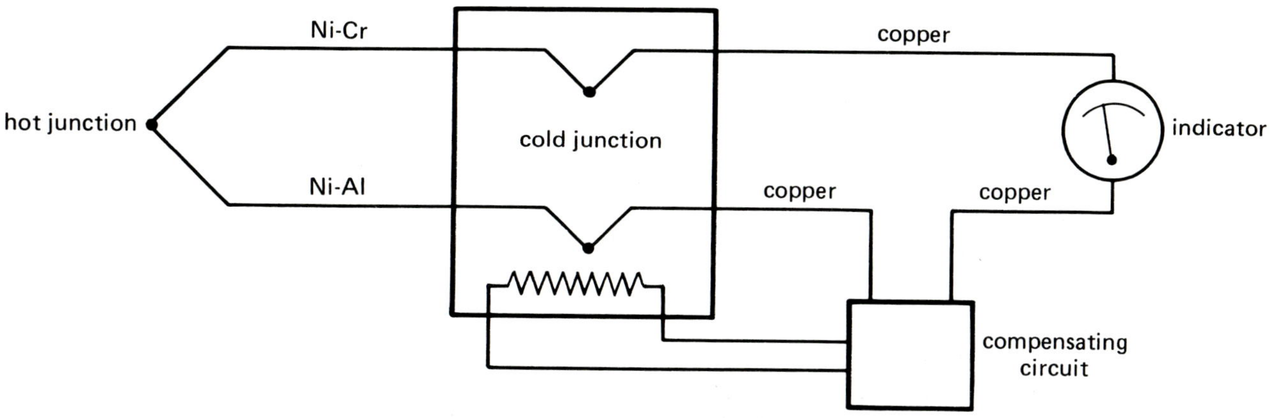

Fig. 14.6 Schematic of cold junction compensation technique used with a Type K thermocouple

Correction for variation of the temperature of the cold junction is provided by an absolute temperature sensor (such as a resistor or thermistor) which is mounted in good thermal contact with the cold junction. The output from the absolute temperature sensor is electronically combined with the thermocouple voltage to compensate for instrument temperature variations and simulate a constant reference cold junction temperature of 0 °C.

Multiple thermocouple circuits

If two or more thermocouples are connected in series, the total net emf at the outside terminals is equal to the sum of the emf developed by the individual couples. Where all of the individual hot junctions and all of the individual cold junctions are maintained at the same respective temperatures, as in the device known as the 'thermopile', this mutiplied value of emf makes it possible to detect and measure extremely small variations in temperature.

Two or more thermocouples may be connected in parallel for the purpose of obtaining a single reading of average temperature. In this case, the resistance of each thermocouple must be the same. The emf across the terminals of such a circuit is the average of all the individual emfs and may be read on a potentiometer normally used for single thermocouples.

To prevent short circuit or current flow between points of differing potential when two or more thermocouples are connected in series or parallel, it is important that both the hot-junction and cold junction terminals of the individual couples be electrically insulated from one another. Also, for multiple-type circuits the simple conversion of average emf to temperature equivalent is strictly true only if the emf-per-degree for the thermocouple materials is constant through the range of temperature of all thermocouple positions.

Selection of thermocouple materials

Combinations of metals and alloys most frequently used for thermocouples are listed in Table 14.3 with their general characteristics and useful temperature range. Selection depends largely on ability to withstand oxidation attack at the maximum service temperature expected, durability depending on the size of wire, use or omission of protection tubes and nature of the surrounding atmosphere. All thermocouple materials tend to deteriorate when exposed at the upper portion of their temperature range to air or gases and when in contact with other materials. Platinum in particular is affected by metallic oxides and by carbon and hydrocarbon gases when used at temperatures above 540 °C and, in the course of time, is subject to calibration drift.

For high-temperature duty in a permanent installation or where destructive contact is likely, service life may be extended, at some sacrifice in rapidity of response, by using closed-end protection tubes of alloy or ceramic material. The arrangement should permit removal of the thermocouple element from the protection tube for calibration and renewal when necessary.

For use during short periods of time, as in some test work, protection tubes may be omitted if calibration is frequent, thus permitting correction of the data. For normal duty within the useful range of the thermocouple selected, the correction for change of calibration is usually negligible.

Sheathed thermocouples

Sheathed magnesium-oxide insulated thermocouples are now in common use. The thermocouple wires are sheathed with inert magnesium-oxide insulation that protects them from the deteriorating effects of the environment. Sheaths can be made of stainless steel or nickel alloy, ensuring

Table 14.3 Common thermocouple materials

Code	Conductor combination	Continuous temperature range °C	Short term temperature range °C	Insulation and sheath colours
T	Copper/copper-nickel	-185 to +300	-250 to +400	+ white - blue sheath blue
J	Iron/copper-nickel	+20 to +70	-180 to +850	+ yellow - blue sheath black
E	Nickel-chromium/ copper-nickel	0 to +800	–	+ brown - blue sheath brown
K	Nickel chromium/ copper aluminium	0 to +1100	-180 to +1350	+ brown - blue sheath red
R	Platinum-10% rhodium/ platinum	0 to +1550	-50 to +1700	–
S	Platinum-13% rhodium/ platinum	0 to +1600	-50 to +1700	–
B	Platinum-30% rhodium/ platinum	450 to +1650	+50 to +1700	–

relatively long life, and resistance to oxidizing, reducing or otherwise corrosive atmospheres. Sheathed thermocouples are available as grounded or non-grounded types illustrated in Figure 14.7. The grounded-type thermocouple has more rapid response to temperature change but cannot be used for connection in series or parallel, because it is grounded to the sheath, and for this purpose the non-grounded type should be used. The grounded-type can be susceptible to separation or parting of the thermocouple wire where long leads at high temperatures are used, whereas the non-grounded type is not affected in this manner.

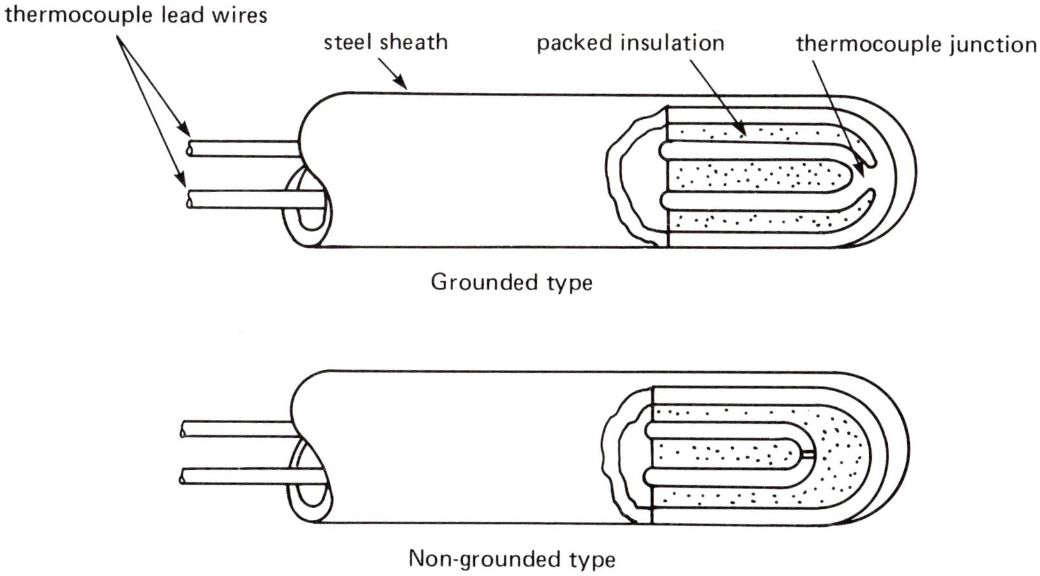

Fig. 14.7 Sheathed thermocouples

Thermocouple and lead wire

There are two classes of wire for thermocouples, the closely standardized and matched 'thermocouple wire', and the less accurate 'compensating lead wire'. For thermocouples of noble metal, extension leads of copper and copper-nickel alloy, which have an emf characteristic close to that of the noble-metal pair, are used to save cost. For thermocouples of base metal, the extension leads in general use, while of the same nominal composition as the thermocouple wires, are less expensive since the control in manufacture and in subsequent calibration need not be as rigorous.

For accuracy, the matched 'thermocouple wire' should be used at the hot junction and continued through the zone of greatest temperature gradient to a point substantially at room temperature, where 'compensating lead wire' may be used for extension to the reference junction. Care should be taken to maintain correct polarity by joining together wires of the same composition; polarity is usually identified by colour code or tracers in the wire covering.

14.2.5 Gas Temperature Measurement

Gas temperature is one of the most important operating parameters required to be measured in the testing or recording of the performance of combustion or process plant, being of equal importance for safe and efficient operation. In the measurement of gas temperature, therefore, care must be taken to make certain that the instrument used indicates the temperature correctly and to interpret the temperature readings to give a true average temperature of the gas stream, which is usually not uniform.

In all cases of gas temperature measurement, the temperature-sensitive element approaches a temperature in equilibrium with the conditions of its environment. While it receives heat primarily by convection transfer from the hot gases in which it is immersed, it is also subject to heat exchange by radiation to and from the surrounding surfaces and by conduction through the instrument itself. If the temperature of the surrounding surfaces is higher or lower than that of the gas, the temperature indicated by the instrument will be correspondingly higher or lower than the gas temperature.

The magnitude of variation from the true temperature of the gas depends on the temperature and velocity of the gas, the temperature of the surroundings, the size of the temperature-measuring element, and the physical construction of the element and its supports. The error typically encountered is illustrated in Figure 14.8.

When gas temperatures below about 500 °C are to be measured, a bare thermocouple, resistance thermometer, mercury-in-glass thermometer or one of the various bulb type thermometers may be used. However, above this temperature suction pyrometers (also known as high velocity thermocouples) must be used. The optical pyrometer and radiation pyrometer are not designed to measure gas temperature and should not be used as results may be extremely misleading.

Suction pyrometer

As previously discussed, a thermocouple placed in a hot gas will indicate a temperature somewhere between that of the gas and its surroundings. In furnace work this can result in large errors, as indicated in Figure 14.8, and the suction pyrometer is designed to minimise these errors.

In a pyrometer of this type, a stream of gas is drawn at high velocity along the length of the thermocouple sheath through an aperture at the far end of the sheath and traverses the hot junction of the

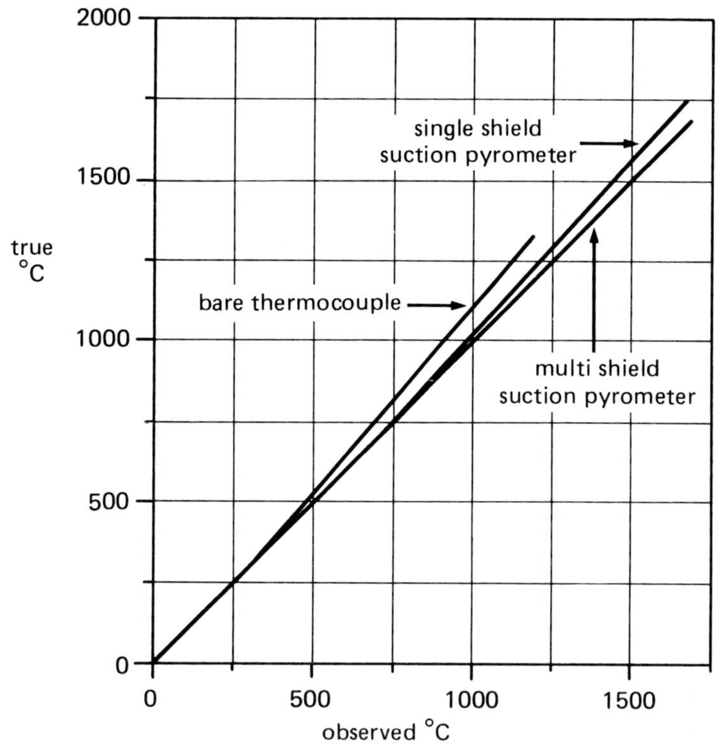

Fig. 14.8 General magnitude of error in observed readings when measuring gas temperatures in boiler cavities with thermocouples

couple, which is provided with an efficient radiation shield. As the velocity of the gas is increased the reading comes progressively closer to the true temperature of the gas, and once a limiting velocity has been reached the residual error becomes small. To eliminate the residual error, when the gas is hotter than the conduit in which it is flowing, electrical heating may be applied to the radiation shield to balance the cooling of the hot junction i.e. to bring the temperature of the shield to that of the gas. When the correct heating rate is used, an increase in the suction velocity has no effect on the reading.

A number of instruments are available to cover temperature ranges up to 1800 °C. The main disadvantages are liability to blockage of the shield system by dirty gases, deterioration of the thermocouple and a slow response speed in some cases.

Suction pyrometers have been developed specifically for industrial gas engineering use to operate either for temperatures up to 1 100 °C, or for temperatures above 1 100 °C.

The metal suction pyrometer is shown in Figure 14.9. Gas is drawn through the stainless steel tube and then passes out of the instrument through the T-piece to the suction line. The emf generated by the thermocouple is fed to a socket housed in the assembly at the rear of the T-piece and compensating cable is used to connect the pyrometer to the measuring instrument. The instrument is calibrated for use at a flow rate of about 30 m/s, a relatively low rate due to the small size of the instrument, which means that it can be widely used on equipment of all sizes. The high response speed of the instrument under suction and non-suction conditions means that it can be used during heating-up periods or for investigating gas streams where there are fluctuations of the order of 100 °C or less per minute. The equipment is portable, robust, is calibrated for use up to 1 100 °C and requires only an AC supply for its operation.

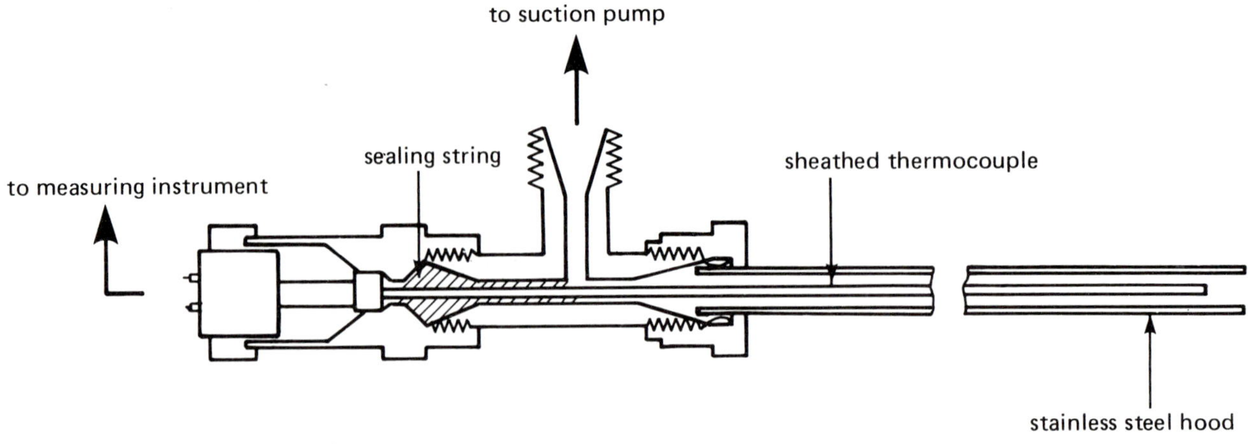

Fig. 14.9 Metal suction pyrometer

The water-cooled metal suction pyrometer is shown in Figure 14.10. This pyrometer was developed as a high-temperature accessory for use with the portable suction pyrometer described above. The instrument is calibrated for use up to 1 600 °C, but by extrapolation this can be extended to 1 800°C. To use the instrument it is only necessary to provide a 14.5 mm clearance hole leading to the position at which a temperature reading is required, insert the pyrometer and take a single reading with suction applied. The true gas-stream temperature corresponding to this reading may then be found from the calibration curve.

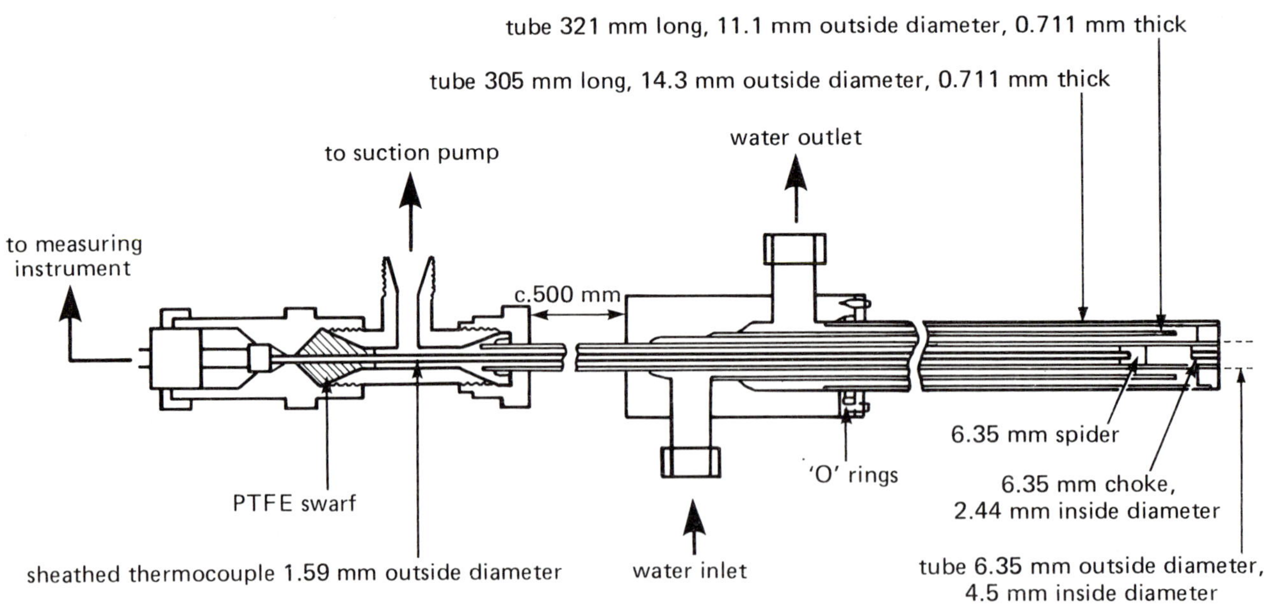

Fig. 14.10 Water cooled metal suction pyrometer

The instrument has a very rapid speed of response which enables any cyclic temperature variations likely to be encountered in normal industrial practice to be measured. The only additional equipment required is that concerned with the supply of water, which may be supplied from the mains, from a tank using a pump, or a pump and reservoir for closed-circuit cooling. The water flow must be sufficient to prevent boiling to avoid instrument damage, but must be high enough to limit any error in indicated temperature caused by the temperature of the water jacket becoming too high.

A number of other methods are available for measuring gas temperature in situations unsuitable for suction pyrometers.

Venturi-pneumatic pyrometer

The Venturi-pneumatic pyrometer is used for applications where dust and temperature conditions are too severe for the suction type or where a rapid response is required. No part of the system attains the hot-gas temperature, the upper temperature limit is fixed by the probe cooling requirements and the level at which the 'perfect-gas' assumptions become invalid. This limit is normally 2 500 °C.

The gas temperature is measured by comparing the gas density at the unknown temperature with its density at a known lower temperature, the density measurements being achieved by venturi restrictions. If ΔP_h and ΔP_c are the pressure drops across the 'hot' and 'cold' venturis respectively, and if gas temperatures at these points are T_h(K) and T_c(K), then:

$$T_h \propto (\Delta P_h/\Delta P_c)\ T_c$$

Schmidt radiation method.

The Schmidt radiation method overcomes the difficulties caused by the partially transparent nature of the gas and provides a measure of the effective temperature and emissivity of the gas. Two readings are taken through the gas with a total-radiation pyrometer, one with a cold background and the other with a hot background at a known radiation temperature. Alternatively, a twin-beam pyrometer can be used, one beam sighted on a hole in the furnace and the other on a hot region of the lining which contains a thermocouple. The temperature obtained is a mean value along the optical path through the gas and this can be unacceptable if large temperature gradients are involved. The accuracy also depends on the emissivity of the gas. The method is usually only preferred if other techniques cannot be used.

14.3 PRESSURE MEASUREMENT

Before considering the actual methods of measuring pressure it is perhaps important to clarify the terms absolute pressure, differential pressure and gauge pressure.

The absolute pressure of a fluid is the difference between the fluid pressure and the pressure in a complete vacuum. Differential pressure refers to the measurement of the difference between any two pressures and gauge pressure is a differential pressure measurement in which one pressure is that of the atmosphere. Absolute pressure is always positive; differential pressure may be positive or negative.

Pressure measuring instruments take various forms, depending on the magnitude of the pressure, the accuracy desired and other conditions. However, they fall into three main categories, namely manometers, gauges and transducers. All three types can be configured to measure negative as well as positive pressures. Digital electronic manometers based on capacitance or piezo-resistance transducers are being increasingly used both for test measurements and in permanent installations for differential pressures and pressures up to 0.25 MPa (2.5 bar).

14.3.1 Manometers

Manometers work on the fundamental principle of balancing applied pressure against a liquid column of known height and density.

Manometers, which may contain a wide variety of fluids depending on the pressure, are capable of high accuracy with careful use. The fluids used vary from those lighter than water for low pressures to mercury for relatively high pressures. For any given pressure, therefore, the height of the liquid column will vary according to the fluid density and the scale must be graduated to suit. The scale will typically give readings in mbar or Pascals, but there are many manometers in current use calibrated in terms of a standard liquid column e.g. inches or millimetres of water or mercury. There are various types of manometers available, as shown in Figure 14.11.

U-Tube manometer

The U-tube is the simplest form of manometer and it consists of transparent U-tube filled with liquid and connected to the point at which the differential pressure is to be measured. The difference in the levels of the liquid will vary according to the pressure acting on one let of the U-tube compared to atmospheric pressure on the other. These should be mounted truly vertically.

Single tube or well type manometer

To avoid the necessity of having to read two liquid levels as in the case of the U-tube, a single tube with its lower end below the surface of the liquid in a closed reservoir may be used. The pressure is applied over the surface of the liquid in the reservoir and the liquid rises in the tube to balance the applied pressure. The cross section of the reservoir is required to be about 100 times that of the tube so that the change in datum level of the reservoir is negligible compared to the height to which the liquid in the tube rises. If the change in reservoir level has to be taken into account then a correction factor is included in the measurement scale. Single tube manometers require careful installation and they should be mounted truly vertical.

Inclined manometer

Inclined or sloping manometers offer greatly increased sensitivity in comparison to vertical manometers. They require very straight bore tubes and need to be accurately aligned at the correct angle. In normal industrial environments the maximum inclination of manometers for reliable operation is a slope of 1 in 20 relative to the horizontal.

A common industrial use for this type of manometer is as a boiler draught gauge.

Micro-manometer

For greater precision in measuring small pressure differentials micro-manometers or Hooke gauges may be used. These instruments employ precision equipment for accurate setting-up and reading, but have generally now been displaced by electronic pressure measuring instruments.

14.3.2 Pressure Gauges

The force produced by a pressure may be measured by balance against a known weight or by the strain or deformation produced in an elastic medium.

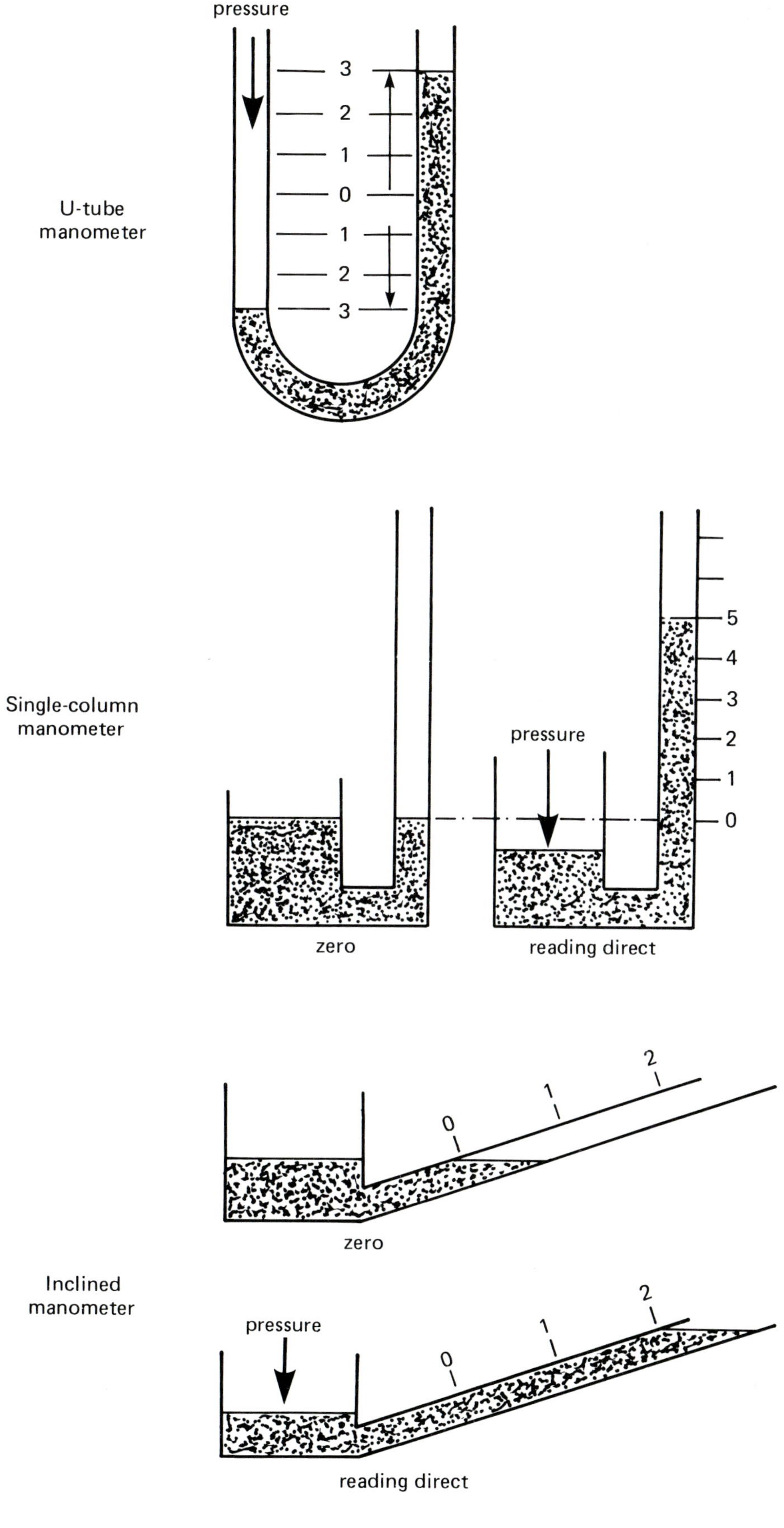

Fig. 14.11 Liquid filled manometers

Measurement by balance against a known weight

A brief description only of these instruments is included since they are not now commonly used and have largely been superseded by transducers.

The pressure gauges operating on this principle are:

- Piston type. In this type of instrument, the force produced on a piston of known area is measured directly by the weight it will support. They can be used up to 550 bar.
- Ring balance type. This type of instrument can be used for the measurement of low differential pressures of the order of 1 mbar.
- Bell type. The force produced by the difference of pressures on the inside and outside of the bell is balanced against a weight or the force produced by the compression of a spring and differential ranges of 0-2.5 mbar and 0-30 mbar are typical at static pressures up to about 10 bar.

Detailed descriptions of these earlier methods can be obtained from older textbooks.

Measurement of strain or deformation

The pressure gauges employing this principle are:

- Bourdon-tube type
- Diaphragm type

These pressure elements are mechanical devices which are deformed by the applied pressure. They possess elasticity and when deformed the stresses establish equilibrium with the applied pressure. The choice and design of the type of element used depends on the magnitude of the pressure to be measured.

Bourdon-tube type

This type of instrument is very widely used. It consists of a narrow-bore tube of elliptical cross-section sealed at one end, as shown diagrammatically in Figure 14.12. The pressure is applied at the other end which is open and fixed. The tube is formed into an arc of a curve, a flat spiral or a helix. When the pressure is applied the effect of the forces is to straighten the tube so that the closed end is displaced. This displacement is magnified and indicated on a circular dial by means of mechanical linkages.

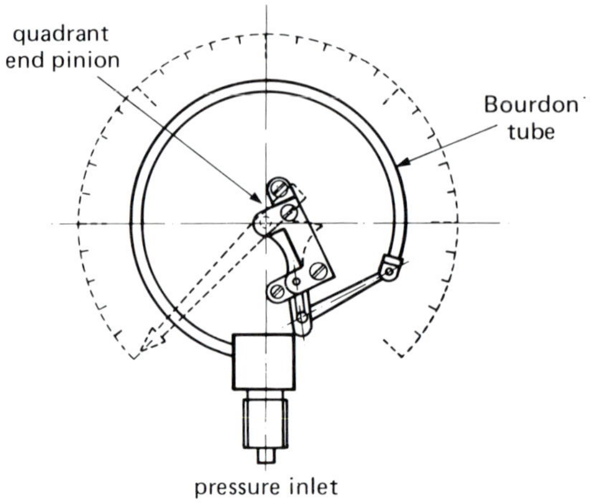

Fig. 14.12 Bourdon-tube pressure gauge

The tubes are made from a variety of materials in a variety of thicknesses. The material chosen depends upon the nature of the fluid whose pressure is being measured and the thickness of the material and on the range of measurement required. The actual dimensions of the tube used will determine the force available to drive the pointer mechanisms, which should be large enough to make any frictional force negligible. Ranges from 1 bar up to 300 bar are available. High-precision types can be produced with a sensitivity of 0.0125% of full scale, an accuracy of ±0.1% of full scale and a temperature effect of 0.15% of the range per 20 °C. Normal working pressure should not be greater than 60% of the maximum pressure indicated. Table 14.4 gives the standard ranges for industrial gauges.

Table 14.4 Standard ranges for industrial gauges (BS1780: Part 2)

Service	Range (bar)	Maximum working pressure for which the gauge is suitable: Steady pressure up to approximately 75% full-scale range (bar)	Maximum working pressure for which the gauge is suitable: Fluctuating pressure up to approximately 65% full-scale range (bar)
Vacuum	-1 to 0		
Combined vacuum and pressure	-1 to +1.5	1.1	1
	-1 to +3	2.2	2
	-1 to +5	4	3.2
	-1 to +9	7	6
Pressure	1	0.8	0.6
	1.6	1.2	1
	2.5	1.8	1.6
	4	3	2.6
	6	4.5	4
	10	7.5	6.6
	16	12	10
	25	18	16
	40	30	26
	60	45	40
	100	75	65
	160	120	100
	250	180	160
	400	300	260
	600	450	400
	1000	750	650

Diaphragm type

The movement of a diaphragm is a convenient method of sensing a pressure differential. The unknown pressure is applied to one side of the diaphragm whose edge is rigidly fixed and the displacement of the centre of the diaphragm is transmitted via a suitable joint and a magnification linkage to the pointer of the instrument. The principle is shown in Figure 14.13. Corrugated discs give a larger displacement and can be conveniently combined to form a capsule, a stack of capsules giving an even greater deflection.

The materials of construction are chosen to suit the application and the instruments can be used above or below atmospheric pressure. They are better than Bourdon gauges for work below 1 bar and can measure fluctuating pressure.

Examples of the lower-range instruments are the altimeter and aneroid barometer. Similar instruments built up from diaphragms, usually about 100 mm diameter and from 0.005 mm thickness upwards, are made to cover ranges from 2.5 mbar up to 4 bar.

An alternative to the diaphragm stack is the bellows element also shown in Figure 14.13 which can be put into instruments to measure differential, absolute or gauge pressure. The 'flexibility' of a bellows is proportional to the number of convolutions and inversely proportional to the wall thickness and modulus of elasticity of the bellows material. These instruments can be used to measure differential pressure in a line having a high static pressure by enclosing the bellows in a high-pressure casing.

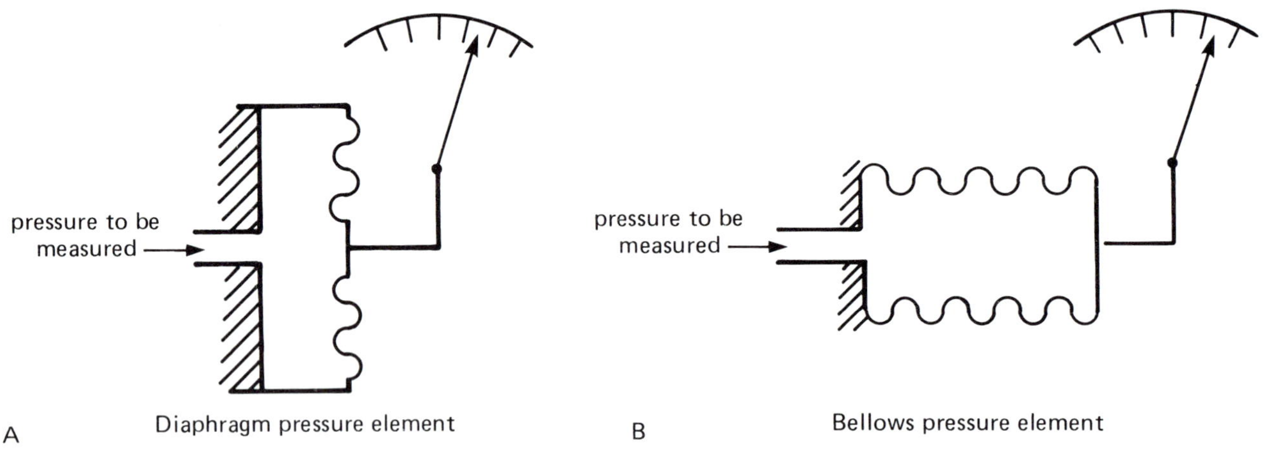

Fig. 14.13 Diaphragm-type pressure gauge

14.3.3 Pressure Transducers

Pressure transducers are devices for changing a pressure signal to an electrical or standard pneumatic signal. They are based on a variety of principles, both mechanical and electrical.

Demands from process industries for more precise control of plant have generated a large market for pressure transducers with electrical output signals. The range of pressure transducers that have been developed to satisfy market requirements is enormous in terms of transducer operating principles, pressure ranges, transducer performance and cost. New designs of pressure transducer based upon mass production semi-conductor technology, where the pressure sensor and associated circuit are etched on to a single silicon chip, are now being manufactured and these offer the prospects for considerable cost reduction.

The main elements in a pressure transducer are a flexible diaphragm or tube across which the pressure difference is applied and a displacement sensor for converting the movement of the diaphragm or tube into a pneumatic or electrical signal.

For process industry applications, where pressure transducers are remote from the control instrumentation, pressure transmitters which feature a 4 to 20 mA current output signal have become an industry standard.

Mechanical transducers

A mechanical pressure transducer consists of a pressure sensing element composed of a beryllium-copper or similar metal capsule, or other device capable of translating applied pressure into mechanical displacement. This displacement is generally applied to the slider of a precision potentiometer producing an electrical signal, or to a device producing a modulated pneumatic signal as shown in Figure 14.14. Typical pressure ranges are -5 mbar to +5 mbar and 0 to 3.4 bar but they will operate at pressures up to 345 bar, the range depending on the design of the element used. Accuracy is normally ± 1%. Other types are available for high temperature operation, up to 300 °C without external cooling or up to 1 100 °C when fitted with special adaptors.

These tranducers are most commonly used as pressurestats, further details of which are given in Section 14.3.4.

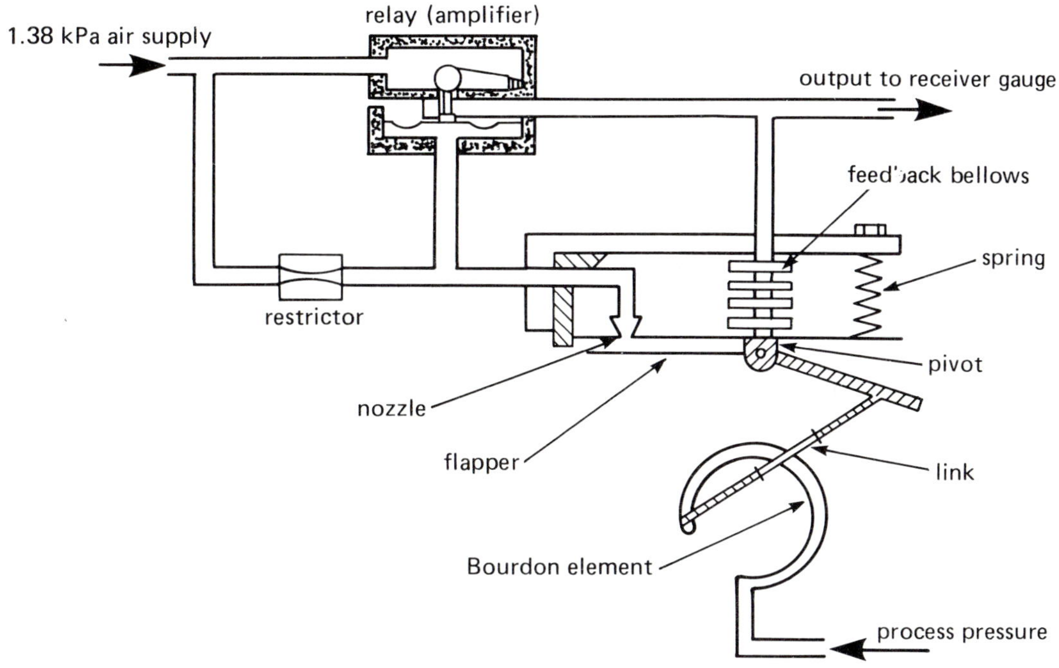

Fig. 14.14 Pneumatic transducers

Electrical transducers

Electrical pressure transducers use a pressure sensing element to convert a pressure signal to an electrical output signal which is proportional to the pressure. The output signal, in millivolts or milliamperes, can be transmitted to a remote recorder or controller. These transducers have a fast speed of response to changes in the measured pressure.

For steady-pressure measurement various types of transducer are available, as shown in Figure 14.15, in which movement of a pressure-actuated diaphragm or bell effects a corresponding change in resistance, reluctance, inductance or capacitance in the pressure sensing element.

In the resistance type a strain gauge is bonded to the diaphragm or stretched between it and a fixed point. Changes of reluctance or inductance are produced by attaching to the diaphragm a piece of magnetic material or an electrical coil respectively, so that diaphragm movements vary the air gap in a magnetic or electric circuit. In every case the effects produced by the diaphragm movements are measured by standard electrical methods.

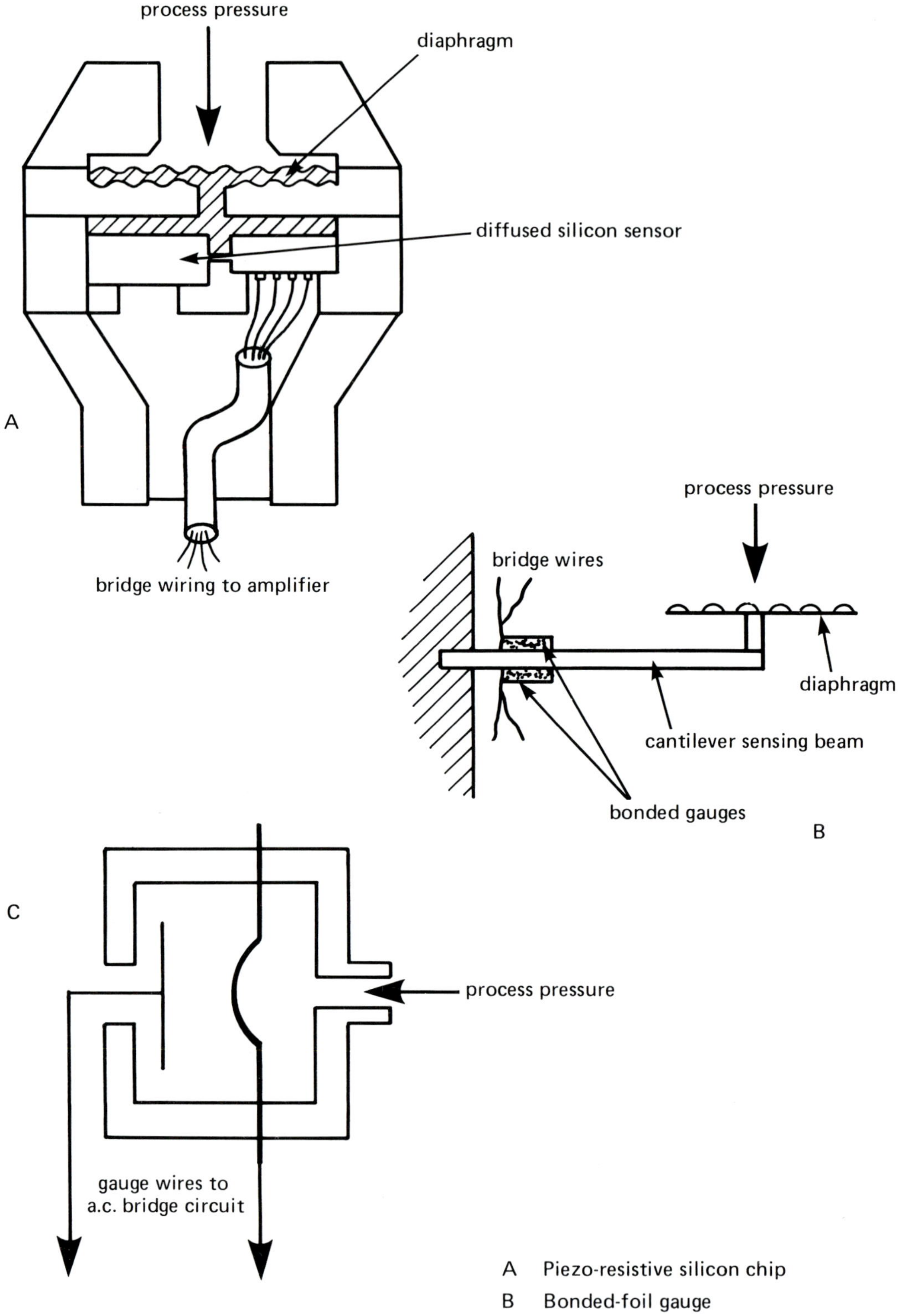

A Piezo-resistive silicon chip
B Bonded-foil gauge
C Capacitive transducer

Fig. 14.15 Electrical transducers

For fluctuating pressures, a small transducer is essential to ensure adequate frequency response and also enable siting close to the pressure sensing element, so avoiding long leads.

Depending on the type and supplier, pressure transducers may have to be recalibrated regularly, usually on an annual basis. In any case such a procedure should be considered as good practice. Most electrical pressure transducers however have built-in compensation for temperature effects. Some types, particularly those used for very low pressures, may be affected by vibration.

They are more convenient for measuring large pressure differences than liquid manometers. Both hand held digital display test instruments and bench units based on semi-conductor electrical transducers are being increasingly used by engineers. Their usefulness on low pressure differences is limited by calibration drift and mechanical friction in some types. For steady applied pressure the lower limit is 0.1 mbar.

Performance characteristics of pressure transducers

Specification of the performance characteristics of pressure transducers is a complex topic. A full appreciation of the terms used to describe performance is needed in order to select the most appropriate transducer for a specific application and in order to make relative comparisons between the performances of commercial transducers. Figure 14.16 illustrates pressure transducer characteristics obtained by recording the output signal as a function of increasing and decreasing applied pressure and is followed by definitions of the more common terms used to describe these characteristics.

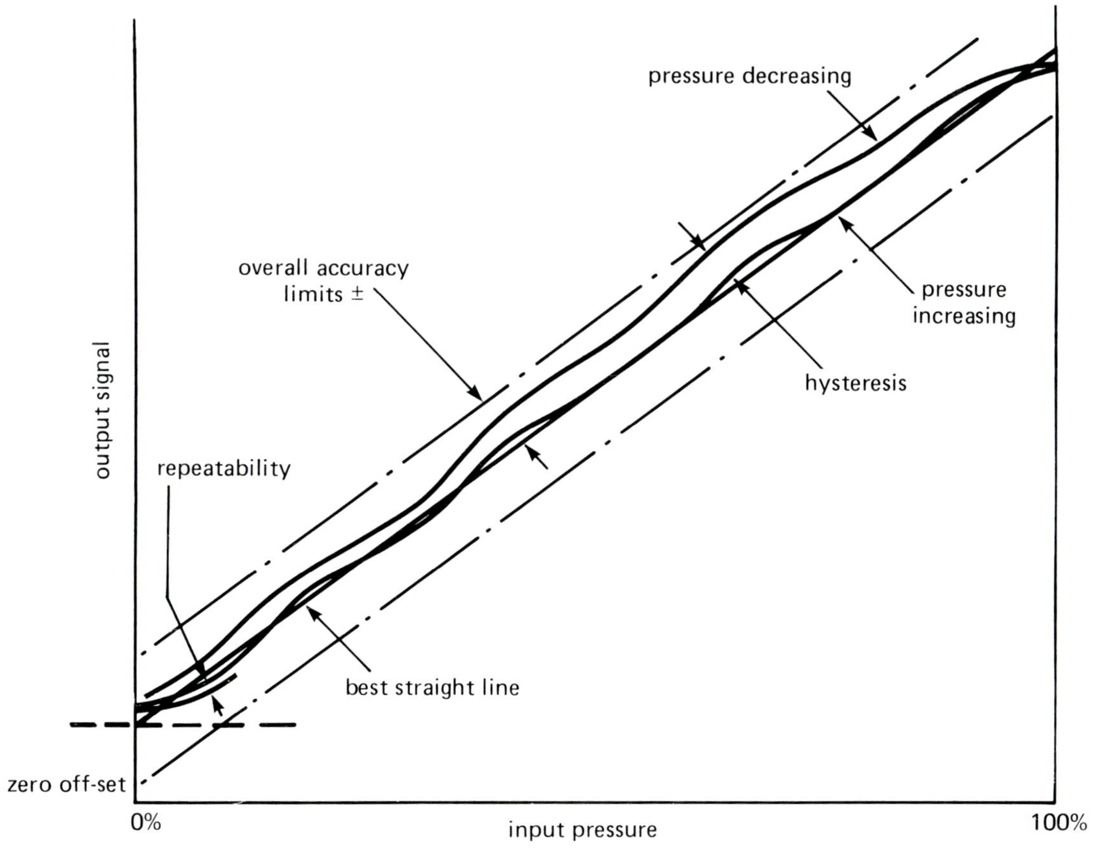

Fig. 14.16 Pressure transducer characteristics

Hysteresis — the maximum difference in output signal between the pressure increasing and pressure decreasing characteristics for the same value of applied pressure.

Best Straight Line — the closest linear relationship between output signal and applied pressure over the full range of the transducer (i.e. the best straight line that can be fitted to the transducer characteristics using regression analysis).

Repeatability — the maximum change in output signal under repeated applications of pressure from the same direction.

Accuracy — the maximum deviation of the pressure increasing and pressure decreasing characteristics from the best straight line. Accuracy includes the combined effects of non-linearity, repeatability and hysteresis.

Hysteresis, repeatability and accuracy are normally expressed in units of ±% full scale output signal (±% FSD). The effect of varying ambient temperature and ageing upon transducer performance can be very important and these are described using the terms following.

Thermal zero shift — change in output signal with varying temperature for zero applied pressure expressed as % FSD per °C.

Thermal span shift — change in output signal with varying temperature for the maximum rated pressure expressed as % FSD per °C.

Long term stability — change in output signal with time during normal operation of the transducer expressed as % FSD per six months.

Non-linearity and temperature drift inherent in the displacement sensors of the pressure transducers are frequently compensated for within the electronic signal processing circuits.

Summary of common types of pressure transducer

Table 14.5 lists the performance characteristics of common types of pressure transducers with pressure ranges from 0-70 mbar to 0-70 bar. The performance characteristics are described using the terms defined above The values quoted are representative of standard industrial grade transducers. Pressure transducers with higher accuracies than those listed in the table are produced in relatively small numbers for special applications (e.g. for use as secondary pressure standards).

Table 14.5 Performance characteristics of common types of pressure transducers

Transducer displacement mechanism	Accuracy ± % FSD	Repeatability ± % FSD	Combined thermal zero and span shifts ± % FSD/°C	Long term Stability ± % FSD/ 6 months
Strain gauge	0.20	0.05	0.02	0.25
Capacitance	0.25	0.05	0.02	0.25
Potentiometer	0.5	0.15	0.02	0.1
LVDT*	0.5	0.05	0.03	0.1
Optical	0.1	0.02	0.01	-
Reluctance	0.5	0.1	0.04	0.1

*Linear variable displacement transducer

14.3.4 Pressurestats

Although pressurestats are not measuring devices as such, they do operate on the same principles as the pressure transducers discussed above and can be used to control valves. They are most commonly used to maintain steam pressure in a boiler system by regulating the quantity of fuel passing to the burner in relation to the steam pressure, and may be mechanical or electrical. Mechanical pressurestats are illustrated in Figure 14.17 and are of the following types:

- Direct acting. This type provides a main gas valve directly controlled by steam pressure.
- Indirect acting. In this type the steam pressure either closes a valve in the weep from the top of a relay valve, or opens a valve in the weep to the top of a relay valve. This device may be combined with a relay control valve.

Electrical pressurestats are classed similarly to the mechanical type:

- Direct acting. This type is an electrically-operated valve controlling the main gas flow.
- Indirect acting. In this type there is a solenoid valve in the weep line from the top of the relay valve.

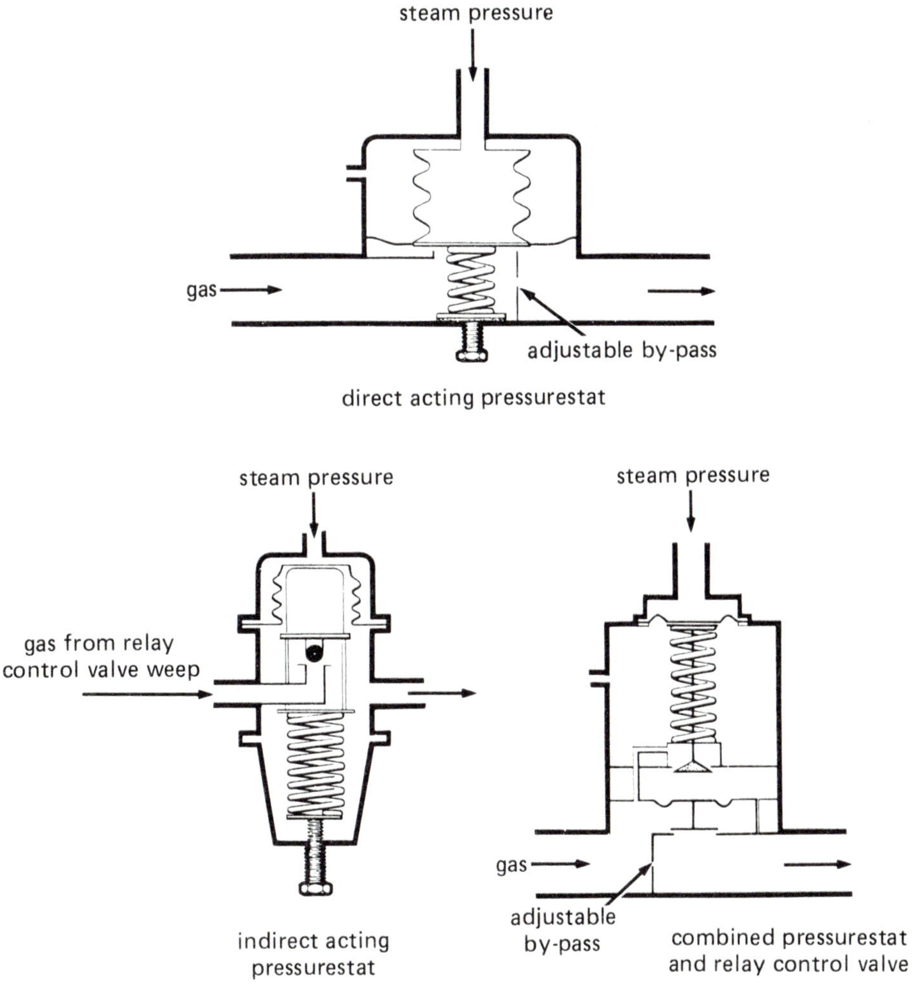

Fig. 14.17 Three types of mechanical pressurestat

14.4 FLOW METERING

The measurement of liquid and gaseous flow rates is so commonplace that its fundamental importance is self-evident. A very wide range of equipment has been developed, embracing a variety of techniques, and the process is a continuing one.

There are two fundamental techniques: direct and inferential. The first is the measurement of the actual volume passing in unit time. The second is the measurement of a flow-related characteristic of the fluid, which is subsequently converted by mechanical and/or electrical means to flow units; analogous to temperature measurement by thermocouple where millivolts are measured but °C are displayed.

Attempts at further classification on a scientific basis are often made but invariably result in confusion, particularly as meters often combine two or more measurement principles. In this Section, the practical approach of classification by method has been adopted.

It is often insufficiently appreciated that all the basic flow metering devices commonly encountered give results in terms of fluid density at the measuring point and further correction is almost invariably required to convert the figures to standard volumetric or mass units. This is often a fixed correction factor incorporated in the mechanical linkage or electrical circuitry, but further manual correction is required if the metered fluid density varies significantly from that at which the meter was calibrated. Continuous automatic correction is increasingly being used as the very real financial and operational benefits of greater accuracy are perceived, assisted by the advent of low-cost reliable electronic equipment

Flow measurement may be required either as flow rate or as a quantity, the latter being the time integral of the former. For control purposes the rate meter is most important, and in control applications the performance of any subsequent control system is critically dependent on the accuracy and repeatability of the basic flow measurement. For stock transfer measurement and costing, direct quantity measurement by means of a bulk meter is more applicable.

14.4.1 Volumetric Displacement

Volumetric displacement meters are essentially flow totalisers and measure the actual volume of fluid passed through each instrument at line conditions in a given time. They are also known as positive-displacement meters and commercially referred to as bulk meters.

These mechanical devices are best suited for the measurement of clean fluids since grit or other suspended matter can cause wear of their moving parts, affecting accuracy. In most cases more maintenance is needed than for other volumetric devices and these meters tend to cause pulsations in the flow. They can, however, be used to meter very low flow rates, operate with a wide range of fluids, achieve good accuracy and direct readout can readily be obtained.

The operating principles of volumetric displacement meters are briefly outlined below.

Semi-rotary piston meter (liquids)

The semi-rotary piston meter, as shown in Figure 14.18 consists essentially of a hollow piston mounted eccentrically within a cylinder. The piston is split and its movement is such that when oscillating it can slide along the division plate which separates the incoming from the outgoing liquid. Liquid enters through the inlet 'A' and, depending on the position of the piston, flows into and around the piston. As the pressure of the liquid increases, it forces the piston to move round the cylinder displacing all of the liquid in front of it through outlet 'B' until the piston itself is also emptied. The pressure of the flowing liquid now moves the piston back to the inlet where it is refilled and another cycle repeated.

This type of meter, which has a rangeability usually within 10:1 to 20:1, can be used to measure flows between 6×10^{-7} and 7.5×10^{-2} m^3/s. Different materials of manufacture can be used and therefore a variety of clean liquids such as water, oil and acids can be metered. The liquids are usually metered at pressures between atmospheric and 20 bar and the meter can be used at temperatures up to 150 °C. When correctly maintained the meters are accurate within ±1% of output reading.

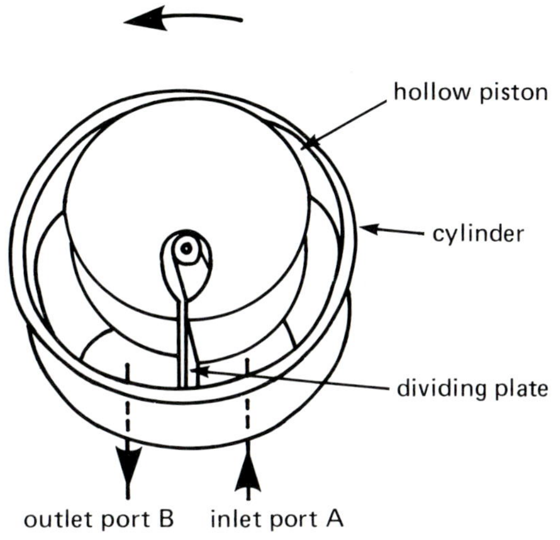

Fig. 14.18 Semi-rotary piston meter

Reciprocating piston meter (liquids)

Several types of reciprocating piston meter are available. Figure 14.19 illustrates a meter with two cylinders and a double-ended piston, but meters with as many as five pistons are often used. In the position shown, cylinder 'A' is exhausting while cylinder 'B' is being charged with liquid. As liquid fills cylinder 'B' it moves the piston and slide valve towards the right, forcing out the liquid in cylinder 'A'. When the slide valve has closed the inlet to cylinder 'B', cylinder 'A' then charges and the liquid in cylinder 'B' is in turn exhausted.

These meters find most frequent use at flowrates within the range of 6×10^{-7} to 7.5×10^{-2} m^3/s. They can be used at pressures up to 10 bar and temperatures as high as 180 °C. They usually have a rangeability between 10:1 and 20:1 and can attain accuracies of ±1% of output reading.

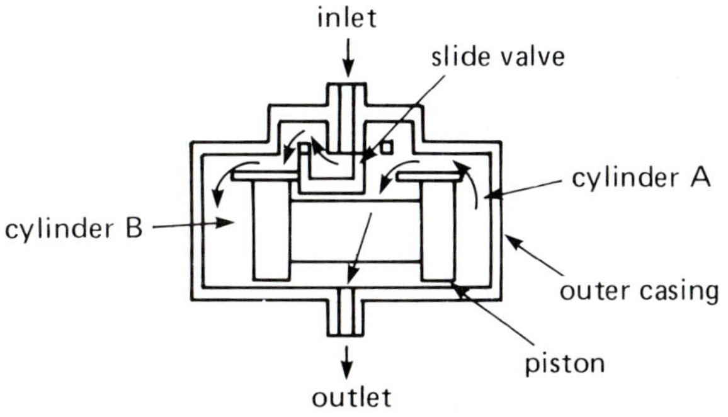

Fig. 14.19 Reciprocating piston meter

Nutating disc meter (liquids)

In the nutating disc meter a disc is installed in an enclosed volume as shown in Figure 14.20. The disc nutates as liquid enters the device so that for each nutation cycle a discrete volume of liquid flows through the outlet. Each cycle is recorded by a shaft connected to a readout system.

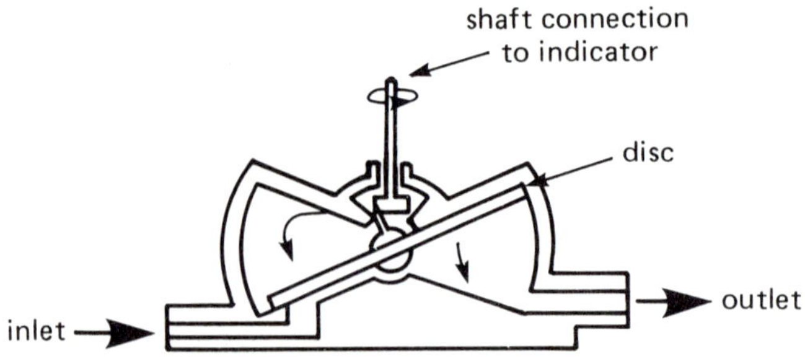

Fig. 14.20 Nutating Disc Meter

This is one of the most commonly used types of mechanical displacement meter as it is simple to manufacture and generally requires little maintenance. The maximum temperatures and pressures at which nutating disc meters are used are approximately 120 °C and 30 bar and they can be used to measure flows from a 2×10^{-6} to 7.5×10^{-2} m^3/s. The meters have a rangeability of 10:1 and accuracies within ±1% of reading can be achieved.

Rotating vane meter (liquids and gases)

Basically, the rotating vane meter consists of a set of vanes arranged inside a casing, as shown in Figure 14.21. As fluid enters the meter it impinges on one of the vanes causing the drum to rotate. As the drum rotates each compartment, which is enclosed by two vanes and the casing walls, is filled and, in turn, emptied through the outlet. A second rotor, driven by the vanes through timing gears acts as a gate to allow the vanes to return from the meter outlet to the meter inlet without letting fluid bypass the metering annulus.

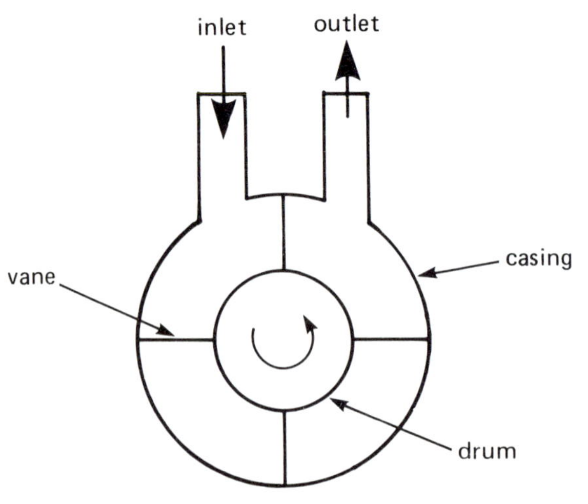

Fig. 14.21 Basic rotating vane meter

Various configurations of rotating vane meter are available and a version used for gas measurement is the vane-and-gate or rotary-piston type, illustrated in Fig. 14.22. Depending upon the make of the meter the vanes and gate rotate in the same or opposite directions.

Fig. 14.22 Rotary-vane meter

Liquids usually metered are petroleum products at flows between 1.5×10^{-3} and 0.5 m^3/s and pressures and temperatures are generally limited to below 80 bar and 250 °C respectively. For gases, the meters have rangeabilities up to 25:1, and flows between 10^{-3} and 10^{-1} m^3/s can be measured at line conditions. With gases, the meters can operate with pressures up to 90 bar and temperatures up to 60 °C, with an accuracy within ±1% of reading.

In the installed condition, the slippage through the meter is extremely small which is a reflection of the accuracy of the manufacturing methods. It is, therefore, essential to use a 200 micron strainer or filter on the meter inlet, and also on the outlet if flow is upwards.

When installing these meters it is important that they are accurately levelled, supported as recommended, filled with the appropriate oil in the gear housings, and the pipework is clear of debris. The use of a flange adaptor on the meter inlet pipework greatly eases the cleaning and replacement of the strainer.

Wet gas meter

The wet gas meter consists of a horizontally disposed drum divided into compartments, as shown in Figure 14.23. The drum is free to rotate in a water bath which is filled to a level just above the axis of rotation. The gas to be measured enters the drum at the centre and, as a compartment is filled, the drum rotates. When the compartment is filled the inlet to the compartment is sealed by water and the inlet to the next compartment then opens and the drum continues to rotate, water entering the first compartment with the contained gas being expelled through the outlet. When calibrated, one revolution of the drum therefore displaces a known volume of gas.

The gas passing through the meter should be saturated with water vapour and flowrates and pressures should be such that no water displacement errors are caused. In this context if the drum rotates too quickly the water level will bank up on one side of the meter.

One of the main disadvantages of this meter is that to measure large flowrates, meters of considerable dimensions must be used. For instance, for a meter rated for a maximum delivery of 9.5×10^{-3} m^3/s an instrument 1.2 m long and 1.05 m high is likely to be required.

Meters of this type are used to meter gases at flowrates between 2.5×10^{-6} and 4×10^{-3} m^3/s. The pressure and temperature of the gas are usually close to ambient, with the pressure at the meter being generally less than 25 mbar. Under carefully controlled conditions accuracies of some ±0.25% of reading can be achieved and meters with rangeabilities up to 10:1 can be obtained.

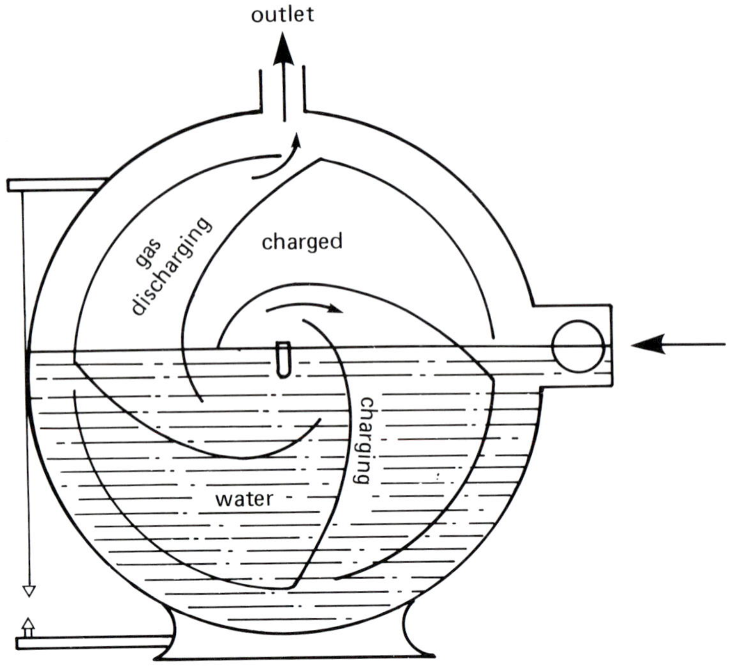

Fig. 14.23 Wet gas meter

Diaphragm meter (gases)

The diaphragm meter is the most commonly used for the sale of natural gas to domestic and commercial consumers in the United Kingdom. Figure 14.24 illustrates a meter at four stages in its operating cycle. The device consists of a housing containing four chambers, two of which, 'B' and 'C', are enclosed by bellows which expand and contract as they are charged and exhausted respectively. Flow into and out of the chambers is by means of slide valves. The volume of gas passed through the meter is obtained through a linkage arrangement which connects the diaphragms to a mechanical readout system which counts the number of displacements.

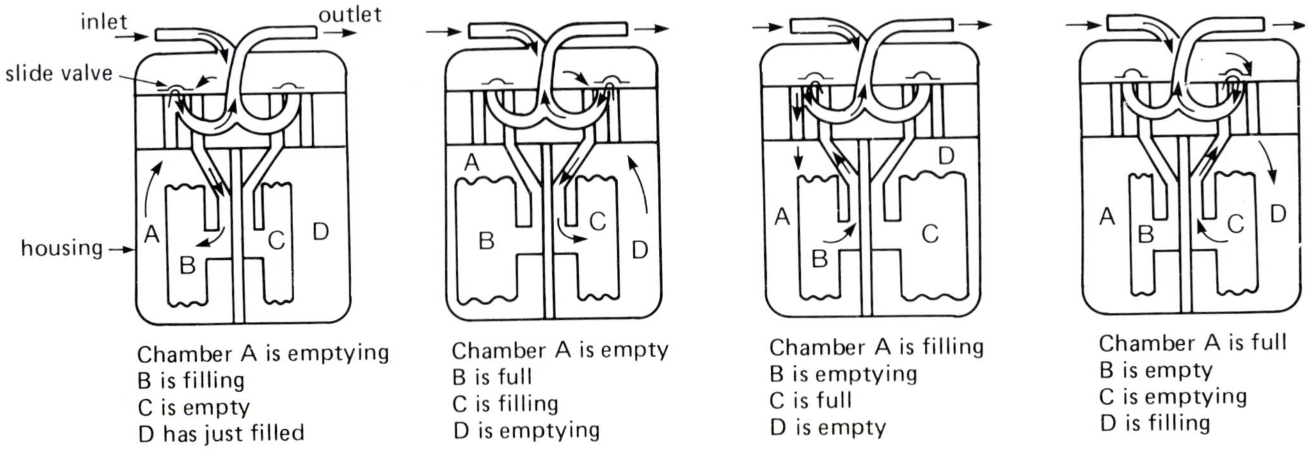

Fig. 14.24 Diaphragm meter – stages of operation

Equipment that will automatically register correction for variation in pressure and temperature is available and provision to take this equipment is built into aluminium-case meters although it should be noted that this is not common in the United Kingdom. Tinplate and steel-case meters are not easily adapted to take correcting equipment.

Remote-reading systems have recently been developed and most meter designs are capable of being modified to take the necessary index unit if at any time in the future a remote-reading system is operated. Modern steel-case meters with pulse outputs can readily be used with electronic correctors.

With diaphragm meters accuracies of ±1% can be attained. The meters measure gas at rates ranging from 5 to 10^{-4} to 10^{-1} m^3/s. The older type tin-case meters are being replaced by steel-case or die-cast aluminium-case meters and are usually only suitable for pressures up to 50 mbar. Specialised steel-case meters are manufactured for use at pressures up to 10 bar. Gas temperatures must be below 60 °C and the meters have a rangeability of at least 50:1 and typically over 100:1.

These meters are popular since they can be cheaply manufactured and measure volume directly. They are, however, only suitable for non-corrosive gases and cannot be used for large flowrates.

Lobed rotary displacement meter (gases)

The lobed rotary displacement meter, known also as a Rootes-type meter, is outlined in Figure 14.25. Two figure-of-eight lobed rotors are geared together and intermesh closely as they are driven, in the direction shown, by the gas flowing through the system. On each rotation a calibrated volume is swept out and flow is totalised from the number of rotor cycles. The clearances between the various parts are as little as 0.05 mm for all positions of the impellers on smaller meters. On very large low-pressure meters the clearances may exceed 0.2 mm.

Like most other positive displacement meters, pulsations to the flow are introduced and their operation can be seriously affected by oil or dirt particles in the gas stream. Hence, as with rotary

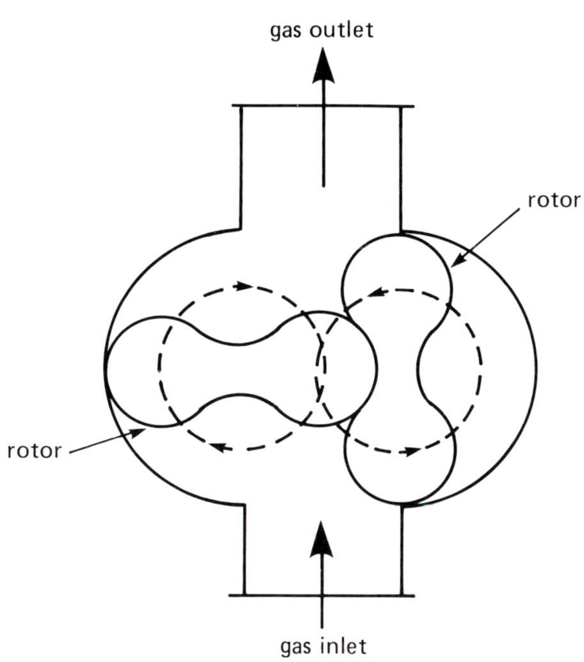

Fig. 14.25 Rotary displacement meter

vane meters and others of this type, filtration of the gases is essential, as is correct mounting and maintenance. Since, however, each system is designed so that leakage and slip are extremely small, accuracies of better than ± 1% are attainable with clean gases.

Meters are used at pressures up to 80 bar but temperatures do not generally exceed 60 °C. Rangeabilities of up to 25:1 can be achieved with flowrates in the range from 7 x 10^{-3} to 2 m^3/s at line conditions.

14.4.2 Variable Area

Meters using variable area methods give a direct indication of the rate of flow at the instant of observation. They are, therefore, well suited for process control applications although, when the fluid density is different from that at calibration, the meter reading must be corrected. While not used for such high flows as pressure differential systems, the special upstream and downstream piping requirements for standard orifice places, nozzles and venturi tubes are not necessary.

The most common forms of variable area meters are tapered tube and float meters, cylinder and piston meters and orifice and plug meters. In these systems flow cross-sectional area is varied by positioning an area changing element in the flow line. The element can move such that forces exerted on it by the flow are exactly balanced by gravity or mechanical means. The position of the element gives an indication of the rate of flow through the meter.

Variable area meters are manufactured to cover flowrates which range from near zero to 1.5 m^3/s for gases and 0.1 m^3/s for liquids. Special instruments to handle fluids at pressures up to 35 bar and temperatures up to 350 °C can be obtained.

Tapered tube and float (liquids and gases)

This is the most widely used form of variable area meter and is shown in Figure 14.26. The fluid flows upwards in a vertical tube which is tapering downwards. The flow supports a float which is often provided with slots cut slantwise, the slots enabling the flow to rotate the float to give central stability.

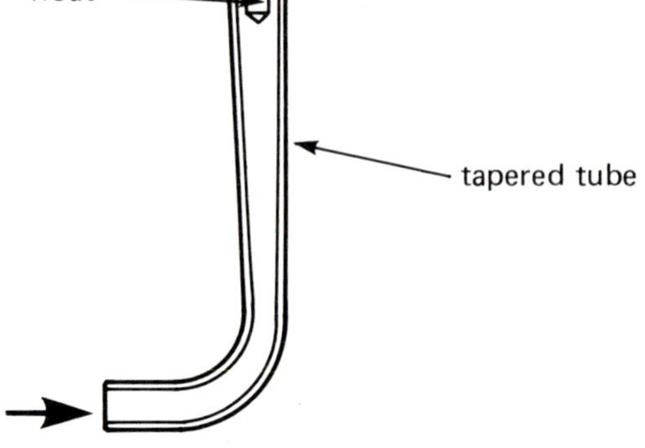

Fig. 14.26 Tapered tube and float meter (Rotameter)

At a given flowrate, the velocity of the fluid varies along the length of the tube since the tube is tapered. The float will therefore assume an equilibrium position, at a height within the tube dependent on the rate of flow provided the flowrate is within the capacity of the meter. If the tube is made of a transparent material, such as glass, the equilibrium position of the float can be observed directly against a scale, usually inscribed on the tube wall, graduated in rate-of-flow units. In larger meters and in meters designed for high working pressures, the metering tube is made of metal and such means as magnetic couplings are used to give an indication of the float position. Accuracies can be of the order of ± 1% of indicated reading and rangeabilities of 10:1 can be obtained.

Cylinder and piston (liquids and gases)

Figure 14.27 shows the essential feature of a typical cylinder and piston meter. A cylinder with a large number of equally sized orifices spaced in a uniform helical pattern is located as shown. The moving element is a loose fitting piston, the movement of which may be regulated by a spring or by a weight and dashpot. The number of holes exposed is in direct proportion to the travel of the piston and therefore to the rate of flow. A rod attached to the top of the piston moves within a sight glass containing a graduated scale.

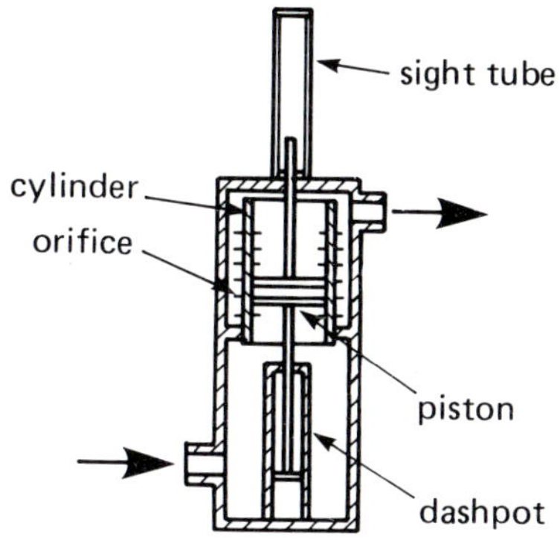

Fig. 14.27 Cylinder and piston meter

Orifice and plug (liquids and gases)

An orifice and plug meter is shown in Figure 14.28. With flow a guided tapered plug rises vertically within the base of the orifice plate, increasing the annular area between the plug and the orifice. As with other variable area meters, the position of the tapered plug indicates the rate of flow.

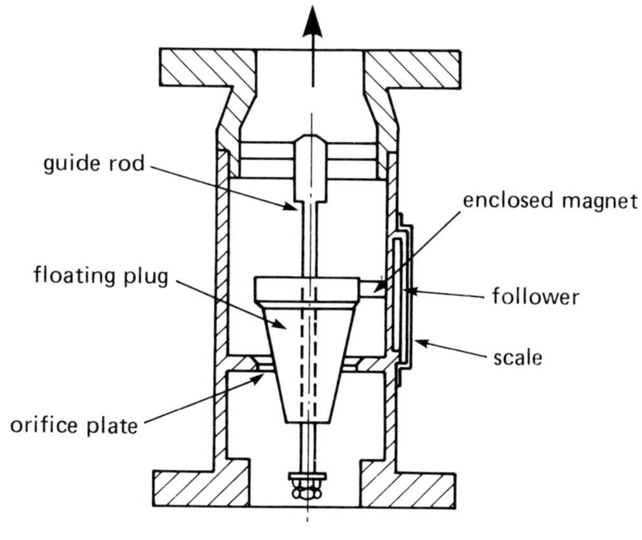

Fig. 14.28 Orifice and plug meter

14.4.3 Pressure Differential Methods

Pressure differential methods utilise a variety of devices, with orifice plates, nozzles and venturi tubes being the most commonly used basic elements; and are illustrated in Figure 14.29. These elements introduce a constriction in the flow line which causes the velocity of the fluid to increase; applying Bernoulli's equation, this causes its kinetic energy to increase and its pressure therefore to fall. The differential pressure so created across the element is measured and converted into the corresponding volumetric or mass flow units. As the flowrate is proportional to the square root of the differential pressure, accuracy falls off sharply below about 30% rated capacity and rangeability is thus relatively poor. This is overcome in meters such as those manufactured by Gervase Instruments Ltd., where the orifice plate aperture may be varied according to the flowrate and a rangeability of 10:1 and better is achieved. Alternatively with multiple transducers a turndown of 10:1 and better can be achieved by extending the pressure measurement range.

Because of their wide usage, a considerable amount of data is available on the performance of orifice plates, nozzles and venturi tubes and standards have been produced in several countries. BS1042 sets out a very thorough treatment of the theory and operation of orifice plates, together with a guide to installation. If the recommendations in these standards are followed, flow measurement accuracies within ± 2% can be attained for each device, without calibration. While improved accuracies can be obtained through calibration such accuracies cannot be maintained in practice without taking rigorous precautions, particularly with regard to the siting of the element. The following points should be considered:

- Location in pipework with regard to bends or changes in cross-section. Wherever possible the specified lengths of straight pipework both before and after the element should be observed.
- Dimensions and conditions of surface piping before and after element.
- If necessary, approach straightening vanes may be required.
- Location and type of pressure tappings.
- Position of element relative to direction of fluid flow.
- Type and arrangement of piping from primary element to differential pressure measuring instrument.

Orifice plate (liquids and gases)

An orifice plate is a thin flat plate having a central hole. Various shapes of hole inlet edge are used but the sharp-edged plate of the type illustrated in Figure 14.29A is the most common. Although this illustration shows corner pressure tappings, various other tapping arrangements may be incorporated. Since the geometry of the orifice plate is simple and is relatively cheap to manufacture, this element is the most widely used pressure differential device. Pressure recovery is, however, low, long pipe runs are usually required and slight damage or wear to the inlet edge can cause marked changes in performance, causing errors of up to 5%.

The elements have a flow rangeability of approximately 3:1 using single transducers and are most commonly used at pipe Reynolds numbers between 10^4 and 10^7. Measurements with orifice plates are reported for liquids at pressures up to 100 bar and temperatures up to 550 °C. For gases, temperatures are usually within the range −50 °C to 250 °C and measurements have been carried out at pressures at high as 550 bar.

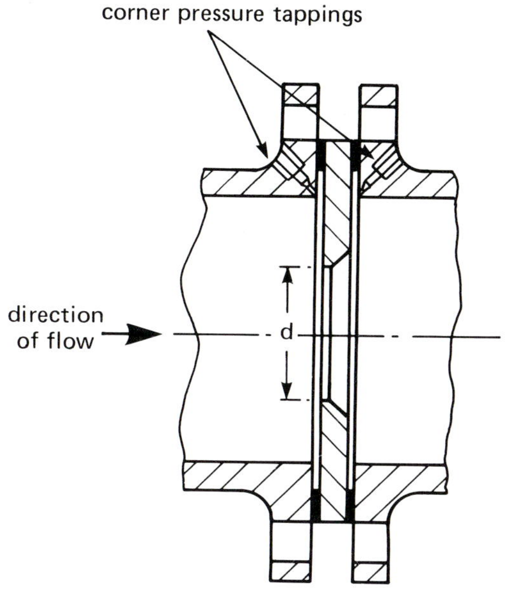

A Orifice plate

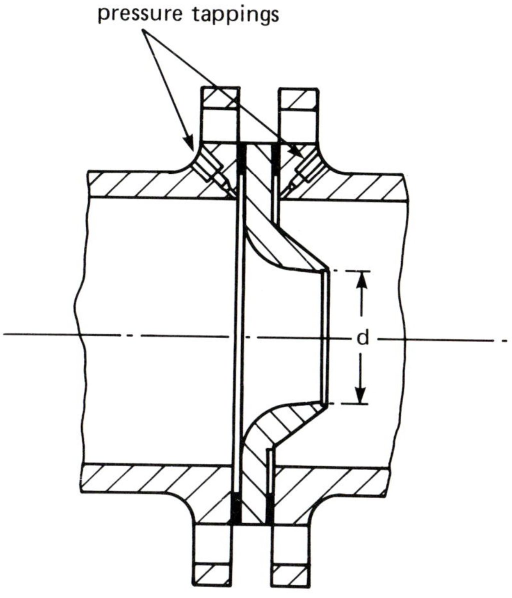

B Nozzle

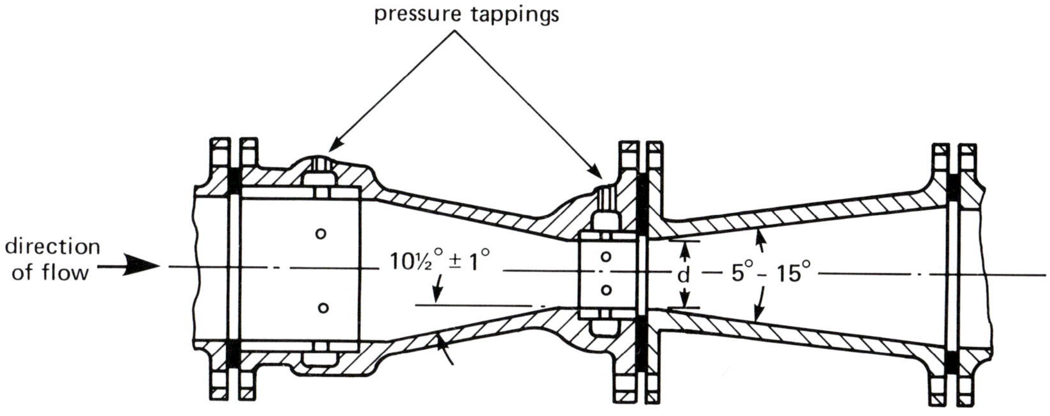

C Venturi tube

Fig. 14.29 Pressure differential devices

Orifice meters are extremely important to the British gas industry. Natural gas is currently bought and the largest gas sales made using these type of meters. Ultrasonic flow metering methods however, are being developed to measure gas flows in the national transmission system.

Nozzle (liquids and gases)

Nozzles have a shaped convergent entry usually followed by a short cylindrical throat. Like orifices, nozzles of varying geometries are used and Figure 14.29B shows the standard nozzle. Since the performances of nozzles are less affected by wear than orifice plates they are more suitable for handling dirty fluids and their discharge coefficients are higher. Their manufacture is however, more difficult and expensive and they cannot be as readily installed or removed from the line.

Measurements on both gases and liquids are reported at pressures up to 100 bar but while liquids have been metered at temperatures of 250 °C, gas temperatures have generally been within the range -40 °C to 100 °C. Nozzle accuracies and rangeabilities are similar to those of orifice plates, with widest use at pipe Reynolds numbers between 10^4 and 10^6.

Venturi tube (liquids and gases)

The standard venturi tube, shown in Figure 14.29C, since it has a diffuser exit section, gives excellent pressure recovery particularly when compared with nozzles or orifice plates which cause substantial pressure losses. Like the nozzle it is more suitable than orifice plates for metering dirty liquids but of the three basic pressure differential measuring elements, it is the most expensive to manufacture and the most difficult to install.

The ranges of pressure, temperature and rangeability covered by venturi tubes are similar to those of nozzles but standard meters experience most use at pipe Reynolds numbers between 10^5 and 10^6.

14.4.4 Full Flow Velocity

While the methods of measurement covered in this Section are basically dependent on fluid velocity, they are usually used to determine volumetric flowrate. To obtain volumetric flowrate the mean fluid velocity is simply multiplied by the cross-sectional area of the fluid flow and the meters can readily be calibrated in volumetric units.

Electromagnetic methods (liquids)

The operating principle of the electromagnetic flowmeter is based on Faraday's law of electromagnetic induction, which states that if a conductive fluid flows through a magnetic field which is perpendicular to the direction of motion, an electromotive force is induced in a direction perpendicular to both field and motion. The magnitude of the induced voltage is proportional to the flow velocity. The voltage is generally measured using two insulated electrodes set into the wall of the pipe, as shown in Figure 14.30.

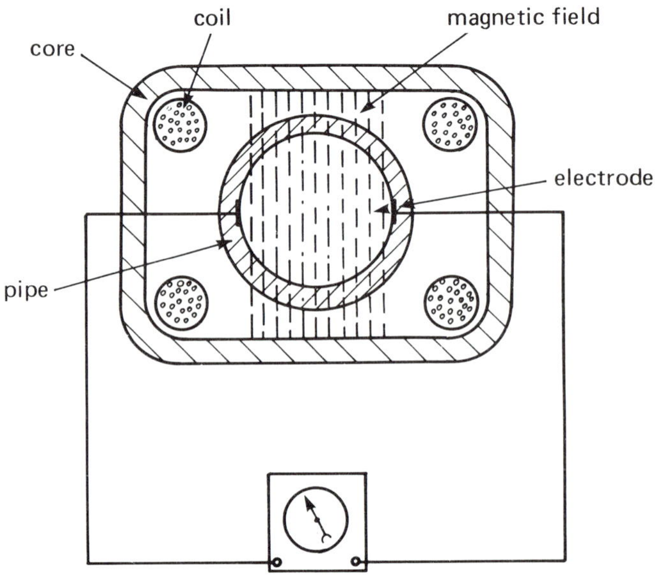

Fig. 14.30 Section through an electromagnetic flowmeter (schematic)

Since the meter consists of a length of parallel pipe, around which a powerful electro-magnet is wound, the device presents no obstruction to the flow and therefore no additional head losses are encountered. Its main disadvantage is that it can only be used with conducting fluids and consequently is not suitable for gas measurement.

With calibration, accuracies of ±1% of indicated flow over a 10:1 flow range of operation are attainable and meters suitable for temperatures up to 200 °C and pressures up to 10 bar are readily available. The method can be used for all flow velocities below 40 m/s and meters with diameters from 0.0025 to 3 m are manufactured.

Turbine meter (liquids and gases)

The turbine meter incorporates a free-running bladed rotor mounted coaxially on the pipe centre line within an appropriate housing, as shown in Figure 14.31.

The rotor, which usually has helical blades, is driven by the flowing fluid and sweeps almost the entire annular area with an angular velocity which, over its working range, is proportional to the mean fluid velocity in the line. The speed of rotation is therefore also proportional to the volumetric flowrate.

To determine the speed of rotation a pick-up coil is utilised. In certain meters the tips of each of the rotor blades contain magnetic material and as a blade tip passes the coil a pulse in a magnet housed within the coil is initiated. In other meters, magnets are embedded in the blades and each time they pass the coil a pulse is produced. Flowrate is determined by noting the frequency of the pulses. Turbine meters used for low pressure gas normally have a mechanical index driven via a magnetic coupling from the rotor.

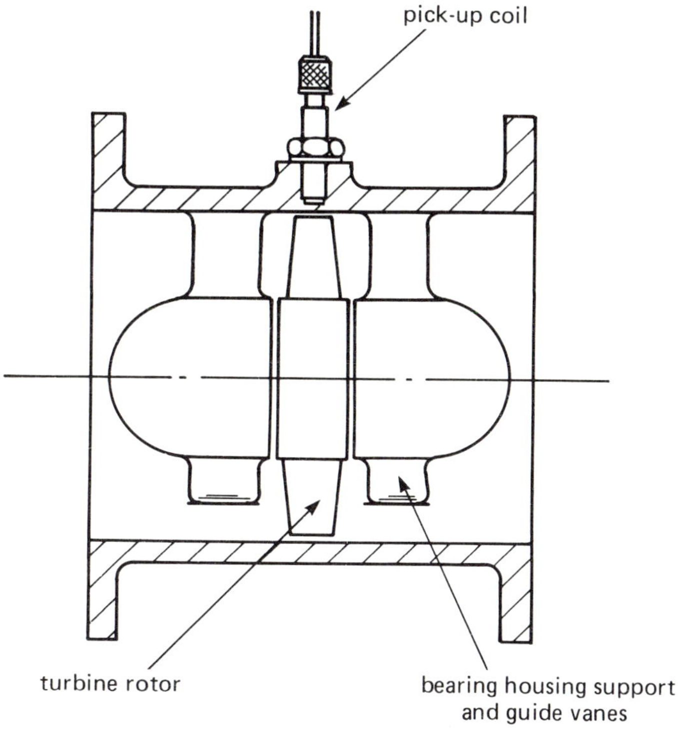

Fig. 14.31 Turbine flowmeter

Turbine meters show excellent repeatability and they have considerably wider rangeabilities than differential pressure meters since readout is directly proportional to velocity and not velocity squared. Unlike rotary displacement meters, if turbine meters fail and cease to rotate, flow will still continue.

The turndown increases proportionally with the square root of the density ratio. Thus for gas measurement at 2 MPa (20 bar), the turndown ratio may be as high as 100:1 compared with 15:1 at millibars gauge working pressure.

Meters are available for use with most fluids although they are not recommended for handling slurries or liquids where the content of suspended matter is high, and a suitable strainer is recommended. They also cause a resistance in the flow line and calibration can be affected by changes in inlet velocity profile. Important factors which inhibit the use of turbine meters in certain gas measuring applications are associated with the relatively slow response to gas flow causing over run following flow shut-off. The effect of pulsations could also give rise to errors. Pulsations will tend to cause the meter to read high while swirls will produce an error dependent on the swirl angle. These difficulties can be dealt with by regular inspection and spin tests and by the use of flow straighteners and profiles.

Liquid meters are manufactured in sizes from 0.025 to 0.6 m in diameter and will operate, depending on size, at pressures up to 30 MPa (300 bar). Flowrates from 1×10^{-7} to 1 m^3/s can be measured and the meters usually have rangeabilities between 10:1 and 20:1. Materials of construction can be used to enable operation at temperatures within the range -200 to 600 °C and, with calibration, accuracies of better than ±0.25% of the flowrate can be attained. The meter can measure given flowrates with repeatabilities of better than ±0.1%.

Meters for gas measurement at low pressure are available in nominal pipe sizes of around 0.025 m diameter having a range of 5×10^{-4} to 7×10^{-3} m/s, up to 1.0 m diameter with a maximum capacity of 12.5 m^3/s. Various pressure ratings are available from 0.25 MPa (2.5 bar) to 30 MPa (300 bar). The meter accuracy is ±1%, dependent on an established velocity profile being achieved at the meter inlet. The relatively small size, large volume capacity and rangeability of gas turbine meters makes them very suitable for measuring large volume flow. They are compact and light in weight, which makes them easily moveable. In addition to the usual instrumentation, turbine meters can be supplied with correcting devices identical to those used on other meters.

Rotary inferential meters (gases)

The rotary inferential gas meter is designed specifically as a secondary meter for use in industrial and commercial applications. The meter consists of a cast-iron housing within which an aluminium anemometer fan rotates on a vertical shaft which in turn transmits the drive to a counter mechanism that integrates the flow, as shown in Figure 14.32. These meters are of rugged construction and are suitable for most non-corrosive gases.

The meters, which must be installed in horizontal pipework, are available to measure flows from 1×10^{-3} to 5.5×10^{-2} m^3/s at pressures up to 2.7 bar. Accuracies of ±2% are claimed with a rangeability of 10:1.

14.4.5 Point Velocity Methods

These methods are designed to measure velocity as the basis for temporary measurements or testing purposes. Their application is generally limited to larger pipelines or ductwork or for the assessment

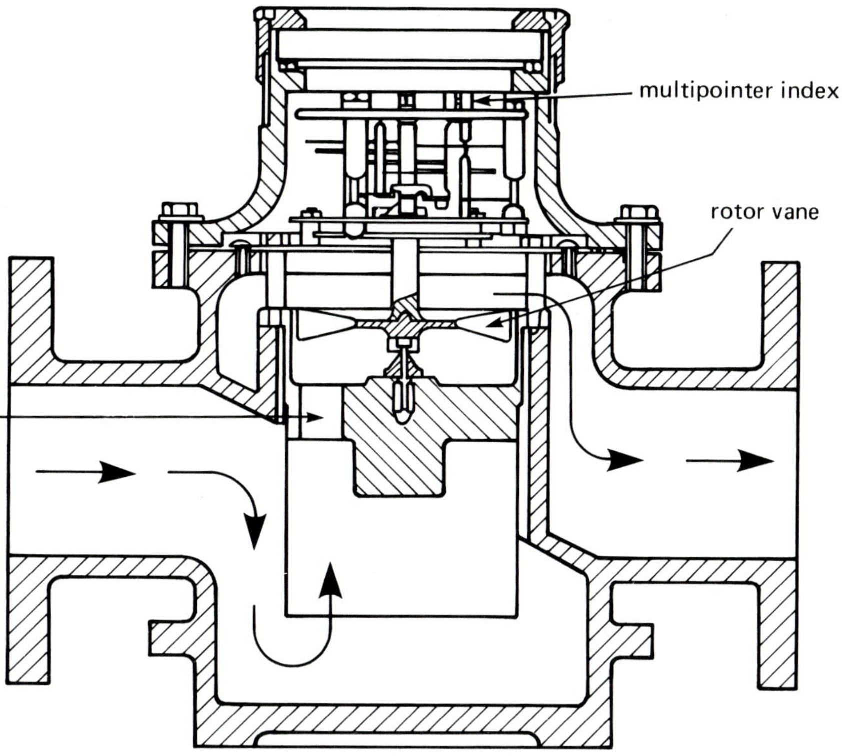

Fig. 14.32 Rotary inferential gas meter

of air movement in buildings. The flowrate in pipelines may be estimated by measuring a number of point velocities at discrete positions and integrating these over the flow cross section using such numerical techniques as the log-linear and tangential rules.

When converting localised flow velocities into overall volumetric flowrates, the following factors must be taken into account

- The effect of reduced cross-sectional area at the measurement plane or 'blockage factor' due to the introduction of the probe into the pipeline.
- Asymmetry of the velocity profile at the measurement plane.
- Estimation of the internal cross-sectional area of the pipe.

On industrial installations, the measurement of overall volumetric flowrates using insertion flowmeters is likely to have a tolerance of around ±5% due to the above factors, in addition to any inaccuracy in velocity measurement by the flowmeter.

Hot-wire and hot-film anemometers (liquids and gases)

In a hot-wire anemometer the sensing element is a fine metal wire, usually nickel or platinum, suspended between two metal supports or prongs, as shown in Figure 14.33. In the hot-film device the wire is replaced by a thin metal film deposited on an insulating substrate. For both instruments the sensor is positioned at the end of the probe and connected by insulated lead wires to the anemometer readout and control system.

In these devices a heating current is supplied to the wire or film which forms part of an arm of a Wheatstone bridge circuit. The sensor is held at a constant temperature higher than that of the

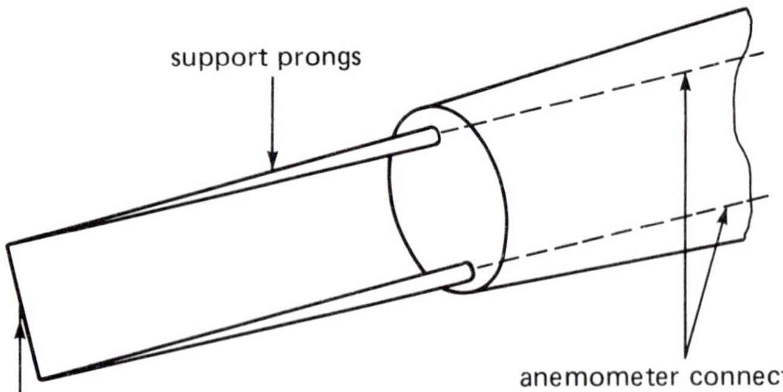

Fig. 14.33 Hot-wire anemometer probe

fluid being measured and since heat transfer from the wire is dependent on fluid velocity, flow velocity may be determined from the current requirements of the sensor provided the resistance of the sensing element does not vary. Alternatively, the wire or film current may be held constant and the fluid velocity determined by measuring the variations in resistance of the sensing element.

These anemometers are used extensively to analyse the velocity microstructures of turbulent gas or liquid flows. They have several distinct advantages over other local velocity meters, particularly their small size, rapid response and high sensitivity at low flowrates.

However, hot-wire probes are fragile and their characteristics are affected by dust deposition and other surface contamination. While shielded probes in which these disadvantages are largely overcome have been developed, the response time of a shielded element is inevitably slower than that of an unshielded wire. These instruments are used for measurements in air at velocities between 0.1 to 500 m/s and temperatures up to 150 °C. High temperature probes can be obtained for measurements in air at temperatures up to approximately 750 °C. Wire sensors may also be used in most non-conducting liquids at low velocities over a range of about 0.01 to 5 m/s.

Hot-film probes are considerably more robust that hot-wire probes but their response is slower. They are used for measurements in liquids over a velocity range of approximately 0.01 to 25 m/s and in gases at high velocities up to 500 m/s. For both hot-wire and hot-film anemometers accuracies of better than ±1% can be attained under favourable conditions.

Current meters and vane anemometers (liquids and gases)

Current meters and vane anemometers operate on the same basic principles as turbine meters. Propeller type current meters are widely used for large-scale water flow measurements while vane anemometers are mainly used for measuring air speeds in large ducts or ventilating shafts.

A current meter, as shown in Figure 14.34, comprises a propeller, its axle and bearings, and is held in an independent frame which contains the contact mechanism and necessary electrical connections for determining the rotational velocity of the propeller. Meters with diameters from 0.02 to 0.125 m can readily be obtained. Standard meters are used for velocities from approximately 0.2 to 5 m/s and when determining the flowrate in circular conduits measurements are normally carried out at discrete positions along at least top diameters. Gauging is performed either with meters sliding along a fixed support or with several meters at fixed positions within the conduit. Typical accuracy of velocity measurement with these devices is ±2%.

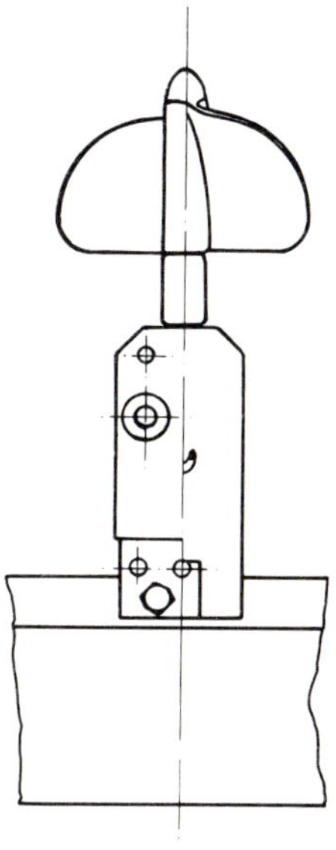

Fig. 14.34 Current meter

Although the operating principles of vane anemometers are the same as those of current meters their construction is different, as shown in Figure 14.35. The vane anemometer is composed of a number of vanes fixed on radial arms which are attached to a common spindle. The spindle is mounted in low friction bearings and its speed of rotation may be determined using either photocells, gearing or capacitance transducers coupled to an appropriate read-out system. Vane anemometers vary in diameter from approximately 0.005 to 0.4 m and can measure air speeds from 0.15 to 80 m/s. They have rangeabilities between 10:1 and 20:1 and accuracies of ±2% can be achieved in point velocity measurement. These devices are mainly used to measure air speeds at conditions close to ambient.

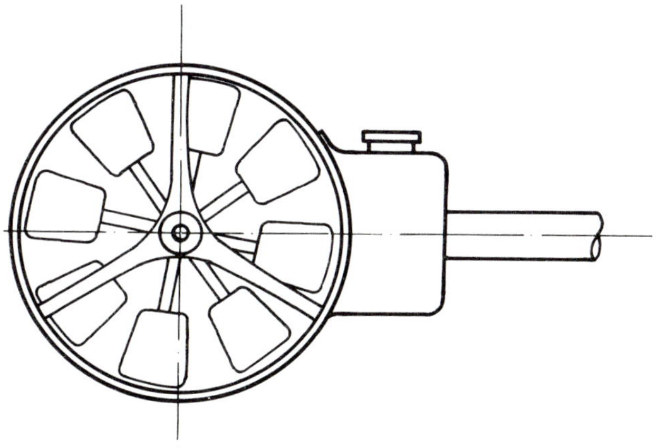

Fig. 14.35 Vane anemometer

While wear affects the initial calibration of current meters and vane anemometers, and they are usually less accurate than other point velocity devices, they are much more robust than hot-wire or hot-film anemometers and are more readily coupled to electric read-out systems than pitot tubes.

Insertion flowmeters (liquids and gases)

Insertion flowmeters are used to measure air and gas flows through pipelines at high and low pressures. The flowmeter is usually inserted at right angles to the axis of the main pipeline through a stub pipe or T piece fitted with an isolation valve and compression seal. The insertion flowmeter measures flow velocity at localised positions across the diameter of the pipeline.

The insertion turbine flowmeter, shown in Figure 14.36, is suitable for low pressure applications. The turbine is supported by low friction ball race bearings and all wetted parts are manufactured from stainless steel. The rotation of the turbine head is detected by an electronic proximity sensor in the head of the flowmeter and the frequency of the output signal is proportional to localised flow velocity. Insertion turbine meters are available for entry through nominal 25 mm bore openings into pipes 100 mm diameter or larger. Accuracies of ±2% are obtainable for typical flow ranges of 0.75 to 20 m/s, and sensing head operating temperatures range from -40 °C to 150 °C.

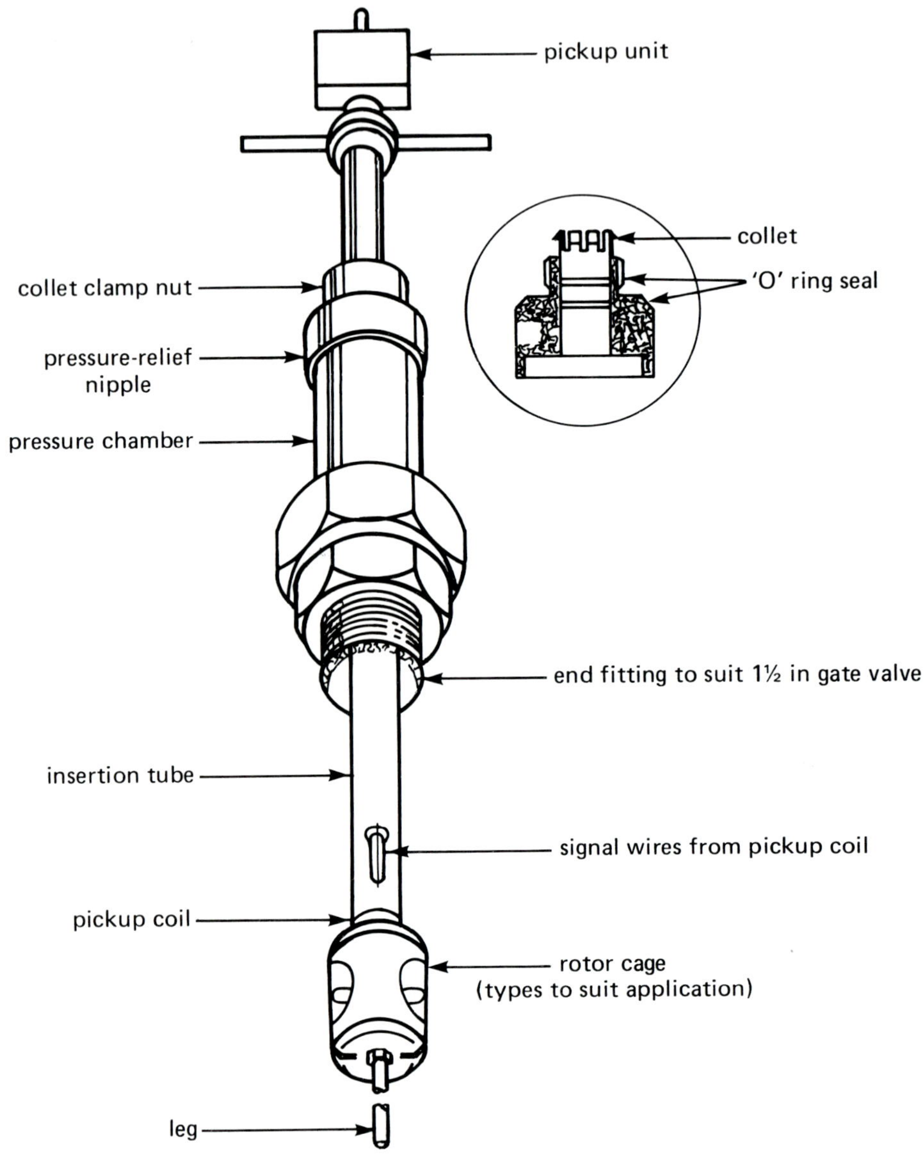

Fig. 14.36 Insertion turbine flowmeter

The insertion vortex flowmeter, shown in Figure 14.37, employs the vortex shedding principle of flow measurement. Fluid flow around the bluff body in the head aperture generates vortices which are detected by sensors mounted in the head downstream of the bluff body. The frequency of vortex generation is directly proportional to the local fluid velocity under turbulent flow conditions. The vortex meter shown can be inserted through a nominal 32 mm opening into a pipe of 150 mm or larger. A typical flow range is 0.25 to 25 m/s with accuracies of ±2%, and the operating temperature range for the head sensor is typically -30 °C to 200 °C.

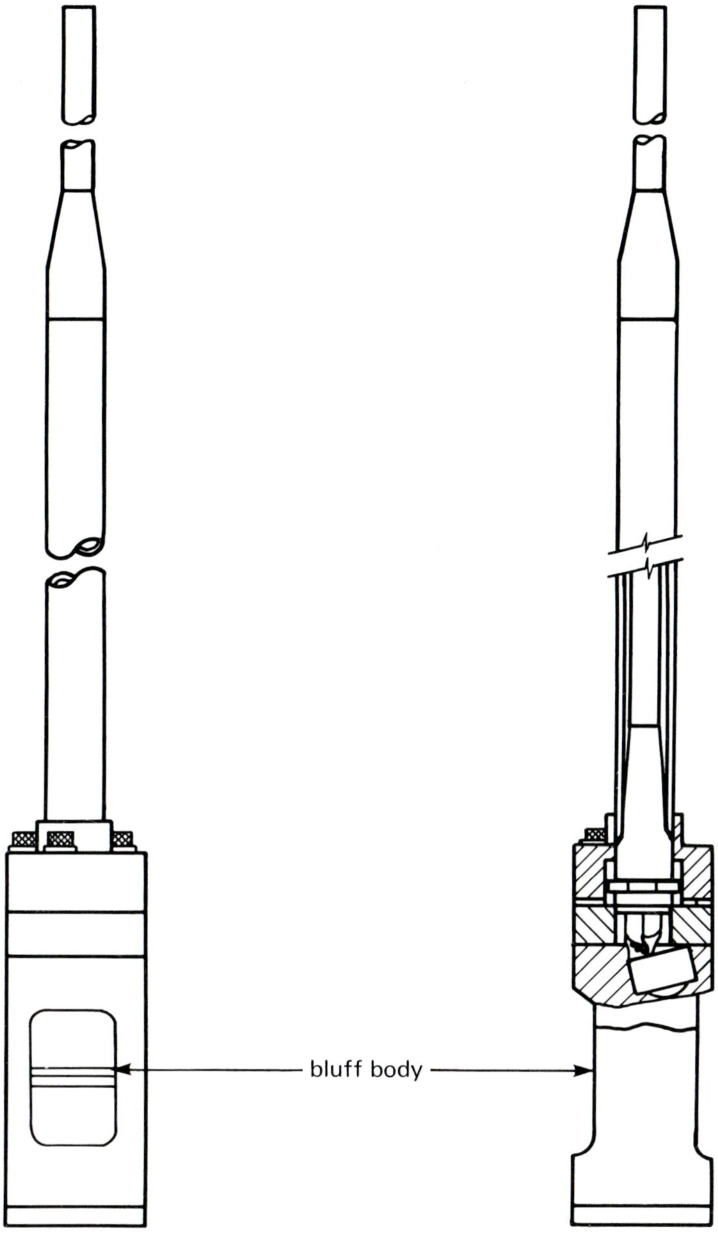

Fig. 14.37 Insertion vortex shedding flowmeter

Bulk vortex flowmeters are also available for pipelines from 0.05 m to 0.2 m diameter for liquid flow from 1×10^{-2} to 0.2 m^3/s. Gas flows ranging from 5×10^{-2} to 30 m^3/s can be measured with a turndown ratio of typically 10:1.

Pitot tubes (liquids and gases)

The pitot tube is a pressure differential meter which measures the difference between local total or stagnation pressure and the local static pressure, to give the dynamic pressure from which the fluid velocity can be derived. As with other local velocity devices, flowrate may be determined by the integration of a number of point velocities across the pipe cross section.

The pitot tube is a useful device for checking large meters in situ and has the advantage that only extremely small head losses are incurred. The main disadvantage, like that of other pressure differential devices, is that since the velocity is proportional to the square root of the differential, pressure reading at low velocities is very small.

There are many types of pitot tubes, each designed for a specific purpose. The probe may comprise a pitot-static combination in one head, as shown in Figure 14.38, or a separate forward-facing orifice in conjunction with a wall static tapping, as shown in Figure 14.39. The probe shown in Figure 14.38 was designed at the National Physical Laboratory and is one of the most widely used designs since its characteristics are reliably predictable. For gases, pitot tubes are most used within the velocity range 10 to 60 m/s and have been used to measure flows at pressures up to 100 bar. However, although water-cooled probes have been developed for high temperature flow measurement, gas temperatures are generally within the range -40 °C to 70 °C. Liquid velocities are usually within 1 to 10 m/s with pressures and temperatures close to ambient. If measurements are carefully performed accuracies of better than ±1% can be achieved under controlled conditions, but normally would be nearer ±5%.

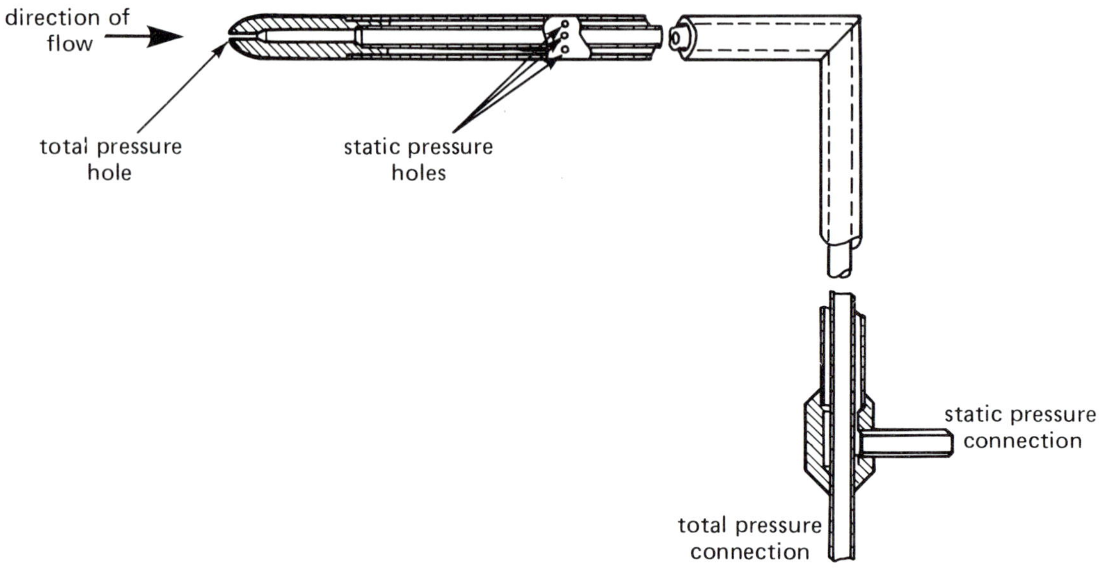

Fig. 14.38 NPL Modified ellipsoidal nosed standard pitot-static tube

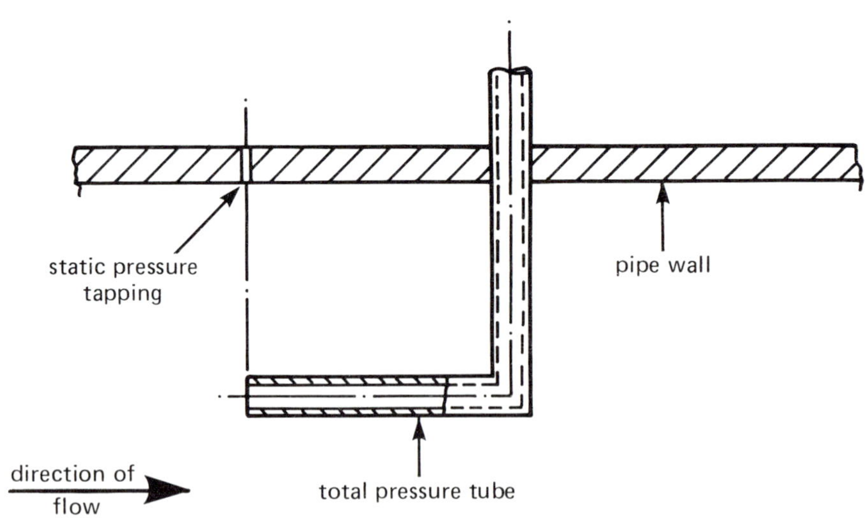

Fig. 14.39 Total pressure tube and wall static tapping pitot tube

Averaging pitot flowmeters (gases)

A schematic of an averaging pitot flow sensor is shown in Figure 14.40. The impact ports facing upstream are located in equal areas of cross-section of the pipeline diameter. The rear port facing downstream senses a pressure less than the static pressure due to the suction effect of fluid flow, and the pressure difference generated between the front and rear ports is around twice that produced by the general purpose pitot tube under similar flow conditions. Each averaging pitot tube requires individual calibration to determine the flow coefficient accurately. Averaging pitot sensors are available for insertion into pipelines of 50 mm diameter and larger. As with the general purpose pitot tubes, the pressure differentials in low pressure gases at low flow velocities are extremely small and the range and accuracy of flow measurement are very dependent upon the type of manometer or pressure transducer employed. The accuracy of the pitot sensor is ±1%.

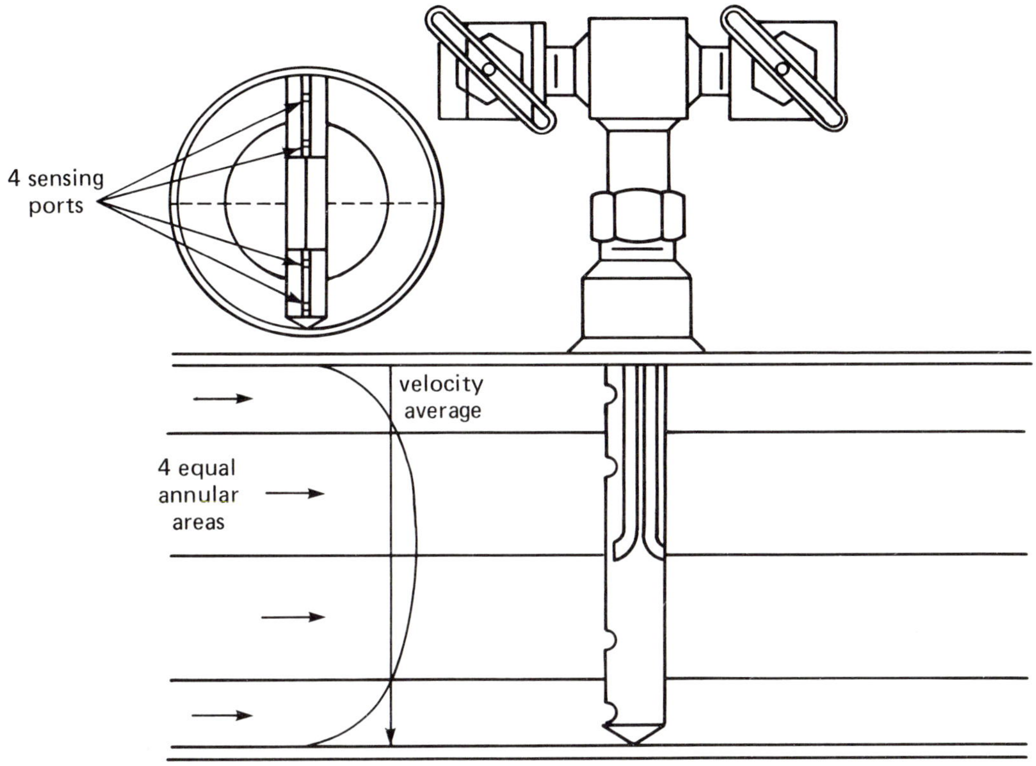

Fig. 14.40 Insertion averaging pitot flowmeter

14.4.6 Meter Correction

Continuous automatic correction is increasingly used for metering of commercial and industrial gas supplies.

Automatic correctors can achieve pressure and/or temperature correction and can correct for compressibility either by the inclusion of a fixed factor or by actual correction. Correctors all require an output drive from the meter and this may be by means of direct mechanical linkage or by indirect electromechanical (or electrical) means.

Automatic pressure/temperature correctors are available in two basic types: mechanical and electronic. Their claimed accuracies are ±1 per cent absolute and ±0.25 per cent full scale deflection respectively. Mechanical types may be step or continuous integrators, whereas electronic correctors are continuous integrators.

Automatic pressure/temperature/compressibility correctors are also available in two basic types: the trapped sample and the density cell. The trapped sample type operates by trapping a specific mass of gas in a flexible container and measuring its deflection when subjected to gas line pressure and temperature. A correction accuracy of ±1 per cent absolute is possible provided the line gas characteristics do not differ significantly from those of the trapped sample. The density-cell type utilizes a density transducer mounted in a small gas-sampling line and has a claimed accuracy of within 0.2 per cent full scale deflection and is discussed in more detail in Section 14.4.7.

14.4.7 Mass Flow Measurement

Like other flow measuring procedures a considerable number of mass flow measurement methods have been developed and are now available and it is not possible to cover them all.

Inferential methods of mass flow determination are the most widely used. Until recently such mass flow measurements tended to be obtained from a knowledge of pressure and temperature, with an assumed density relation. Advances in electronic circuitry have resulted in the development of true mass flowmeters based on density-corrected volumetric displacement metering.

Direct mass flow devices are based on the measurement of fluid momentum.

Mass flow of gases can also be measured by critical flow nozzles.

This section describes the above categories of mass flowmeter in current use. For information on other methods, such as thermal flowmeters, refer to more specialised textbooks.

Density compensated systems

In density compensated systems volumetric fluid flowrate and density at line conditions are measured and, by multiplying flowrate by density, mass flowrate is obtained. Any of the previously described volumetric flowmeters may be used to determine flow and Figure 14.41 outlines a typical installation using an orifice plate.

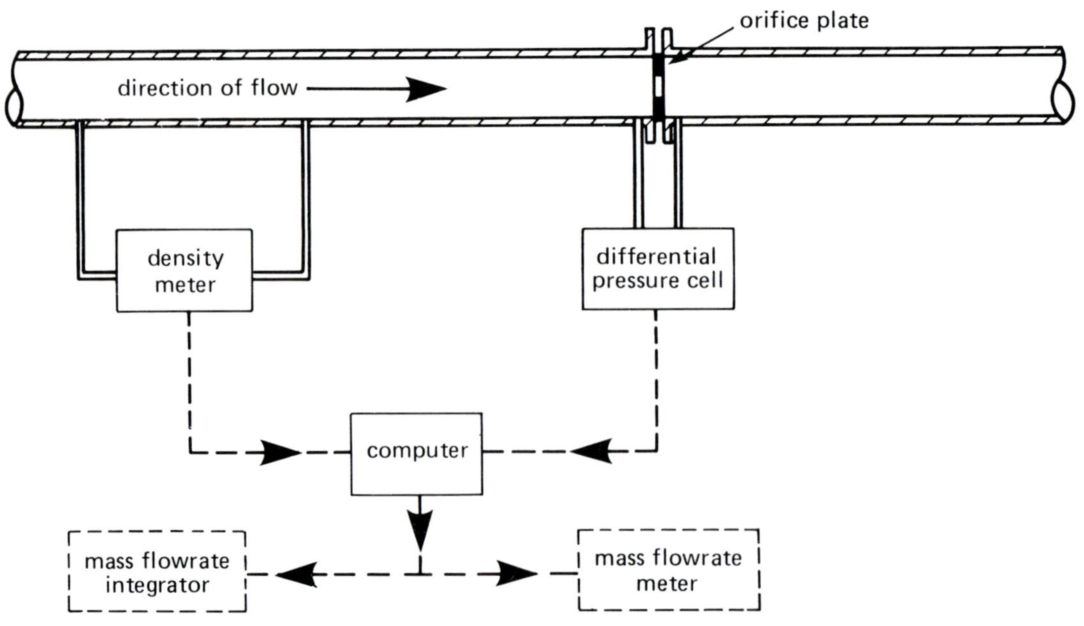

Fig. 14.41 Density compensated system

Various types of density meters are manufactured and these include vibrating element, buoyant beam and spinner type devices. Standard meters are available to measure liquids at pressures up to

70 bar and temperatures in the range -30 to 100 °C with an accuracy estimated to be within ±0.01%. For gases, the temperature range is -10 to 85 °C for standard meters and measurements can be made at pressures up to 140 bar; accuracies of some 0.2% of reading are claimed.

Density measuring elements can be installed inside or outside the pipeline but while the former is usually preferable it can prove inconvenient for flowmeters such as orifice plates or turbine meters where calibration is affected by changes in the upstream flow pattern, and this must be carefully considered before installing an element in the flow line.

Density compensated flow measurement systems are now widely used and have the advantage that they can utilise well-tried and tested flowmeters. In many cases, however, corrections still need to be made to account for changes in fluid viscosity, pressure and velocity as well as density. Theoretically, for true mass flowmeters corrections for these parameters need not be applied and true mass devices usually give faster response to changes in flowrate.

Direct mass flowmeters (liquids and gases)

A true mass flowmeter is one in which the reaction of the basic sensing element is dependent on the mass flowrate of fluid through the device and instruments which operate on the axial flow transverse-momentum principle are the most widely used type of mass flowmeter in the United Kingdom. The operation of an instrument which was first introduced by Orlando and Jennings may be explained using Figure 14.42, where the main elements of the meter are shown.

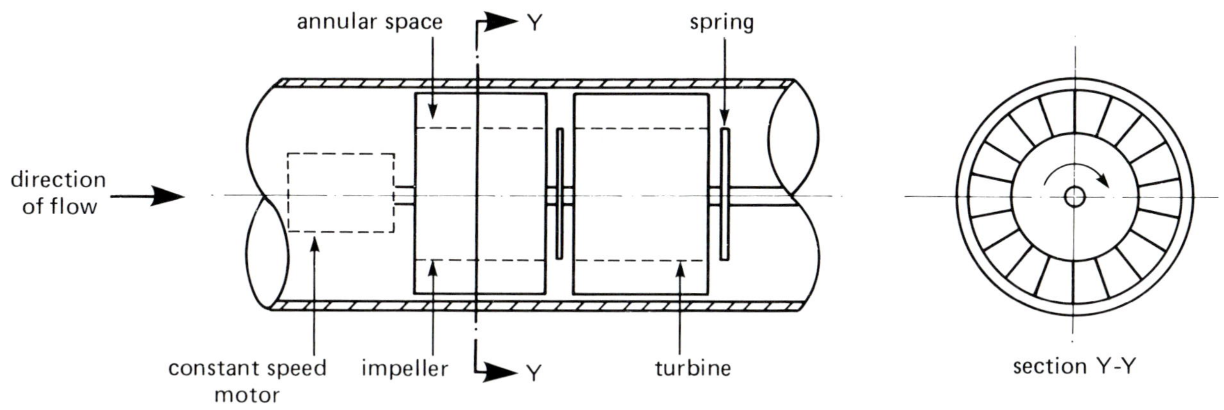

Fig. 14.42 Direct mass flowmeter

The annular space in the impeller is fitted with axial vanes and the impeller is rotated at constant speed by a driving motor. The turbine is similar to the impeller and free to rotate about the same axis but is constrained against rotation by a calibrated spring. The fluid enters the annular space of the impeller, is caused to swirl and is passed to the turbine with an angular velocity equal to that of the impeller. Since, on leaving the turbine, all angular velocity has been removed from the fluid the torque produced on the turbine will be proportional to the mass flow. The angle through which the turbine is turned against the control spring may be measured and transmitted to a remote indicator and, since this angle is proportional to the torque, this measurement gives a measure of mass flow.

These devices give a direct indication of mass flowrate and flow totalisation can also be readily achieved. Provided they are operated within the density ranges specified by the manufacturers, changes in density do not affect performance. The meters are, however, expensive and require rotating seals as well as an accurate constant speed driving motor. Axial-flow transverse-momentum meters can be used for both gases and liquids and accuracies of better than ±1% of reading can be achieved over a 10:1 flow range. The instruments meter fluids at pressures up to 100 bar and temperatures within the range -30 °C to 50 °C. They can measure liquid flows of 0.25 to 40 kg/s and gas flowrates range from 3 x 10^{-2} to 7 kg/s.

Critical flow nozzle for gases

A critical nozzle may be used to determine accurately the mass flow of a gas in a pipeline. The device is designed so that the pressure drop between inlet and throat is such that mass flow of a given upstream pressure is a maximum and sonic velocity exists at the throat. At pressure drops above this limit, mass flow through the meter is only a function of upstream pressure, the physical properties of the fluid and the area ratio, being independent of downstream pressure. Two flowrate equations may be used depending on whether inlet static or inlet stagnation properties are used. The converging entry of the nozzle may have several shapes, e.g. conical, quadrant, and bell-mouth.

The accuracy of the device depends largely on the accuracy with which pressure and temperature, as well as discharge coefficient, can be determined and it appears probable that circular-arc venturis of the type evaluated by Smith and Matz, shown in Figure 14.43, will eventually be adopted as one of a range of ISO standard devices. Such instruments are likely to experience widest use at throat Reynolds numbers above 10^5 and in pipes with diameters between 0.025 and 0.3 m. Upstream pressures will be greater than 100 mbar and nozzle inlet temperatures will be close to ambient. It is estimated that for general use flowrates may be measured to an accuracy of better than ±1% if specified conditions are maintained.

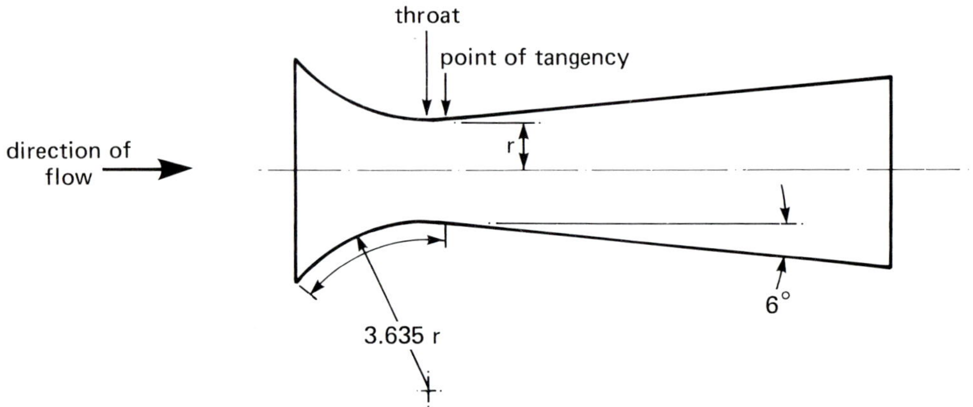

Fig. 14.43 Smith-Matz circular-arc venturi

The ranges covered by critical nozzles are mainly limited by the minimum pressure ratio required to establish sonic flow, and also the high pressure losses incurred in many installations. Turndown is also limited unless a series of nozzles is used. However, critical flow venturis have now been developed in which head losses across the device are less than 5% of the upstream pressure. It should also be noted that since critical flowmeters are unaffected by downstream pressure variations they make excellent flow controllers as in most applications flowrate varies almost linearly with nozzle upstream pressure.

14.4.8 New Developments in Metering Techniques

Advances in transducer and electronics technology have led to the development of a range of advanced metering techniques. New methods are being continually developed and thus only a limited number of techniques are discussed in order to illustrate the scope and direction of meter development in the 1980s.

Multiple ultrasonic metering

This metering technique is based on a series of pairs of transducers aligned on opposite sides of a pipe and angled to the direction of flow. The difference in transit times of the resultant two pulses generated and detected by each pair of transducers enables the velocity of the flowing gas to be calculated in that plane. The bulk speed of the gas is measured by the use of up to four pairs of transducers. Meters are currently only applicable to high pressure large pipe diameters from 0.1 to 1.0 m. Accuracies of ±1% can be achieved over a very wide flow range (over 80:1). The technique is in use on the gas distribution network where the performance and reliability are being assessed with a view to replacing other types of mechanically operated meter.

Doppler velocity meters

This method is based on the principle of measuring the Doppler Shift of either laser radiation or ultrasound from scattering from entrained particles or air bubbles within the gas stream. The frequency of the scattered waves is increased or decreased in direct proportion to the local rate of motion. The basis of the laser Doppler velocity meter is shown in Figure 14.44. Optical anemometer systems are now becoming widely used in fluid mechanics and combustion research.

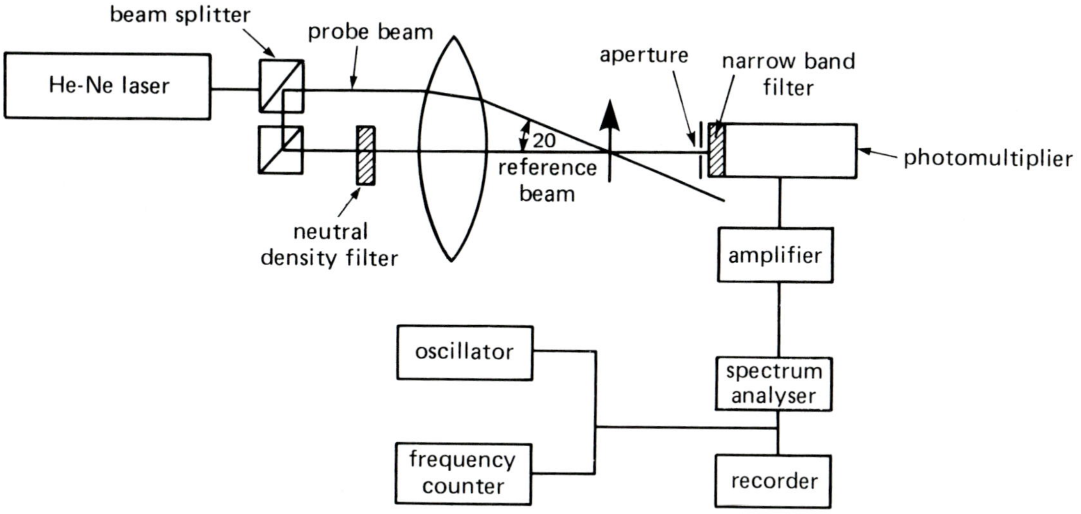

Fig. 14.44 Laser Doppler velocity meter

Meters based on ultrasound operation are non-intrusive and can be readily installed without interrupting the process flow. They are designed for flow velocities of up to 10 m/s at temperatures of -20 to 125 °C.

Solid state and optical fibre sensors

Solid state pressure sensors are well established for use with orifice plates and other differential pressure devices. The development and use of solid state sensors is becoming increasingly important and for example meter systems are now available based on the calorimetric principle, through the use of solid state sensors.

Optical fibre sensors can be used for the measurement of flow by monitoring the deflection of a variable area orifice plate. The development of intensity modulated and wavelength modulated optical sensors is likely to provide more sophisticated and intrinsically safe measurement methods.

14.4.9 Selection and Commissioning of Meters

Selection

The function of a flow system, together with the operating conditions, defines the meter's design specification, which will include the following:

- Performance targets including continuity of measurement, flow rate accuracy and integrated flow accuracy.
- Maximum flow capacity.
- Operating load profile.
- Upstream and downstream control equipment characteristics.
- Maximum allowable pressure drop
- Operating fluid pressure and temperature ranges
- Fluid composition.
- Read-out requirements.
- Available power supplies.
- Space available.

The main characteristics of flow meters are detailed in Table 14.6 to enable comparison and assist in the selection of metering equipment for gas and liquid flow measurement.

Commissioning

In the commissioning of meters it is very important always to follow the instructions provided with the meter, but it should also be appreciated that major damage can be caused by overspeeding during initial purging or pressurisation of the pipework. Where a valve is placed on each side of the meter it is recommended that the upstream valve should first be opened and then the downstream valve should be cracked open very slowly to pressurise the pipework. Particularly in the case of orifice plates, rotary inferential, turbine, rotary or displacement and diaphragm meters, and where a downstream valve is not fitted adjacent to the meter, extra care must be taken in the operation of the upstream valve. Where the gas pressure is above 2 bar consideration should be given to pressurising the downstream pipework through a small bore valve which bypasses the inlet valve in order that the rate of pressurisation does not exceed 350 mbar/s.

14.5 COMBUSTION PRODUCTS ANALYSIS

The techniques available for analysing gases are numerous and include chemical absorption, spectrochemistry, mass spectrometry, chromatography, and the determination of thermal conductivity, heat of reaction and paramagnetism. Much of the available equipment is complicated, of a highly sophisticated nature and often rather bulky and is used in laboratories and research stations.

However, for the on-site performance evaluation of combustion plant, the requirement for combustion product analysis is most commonly restricted to the measurement of the concentrations of

Table 14.6 Flowmeter Characteristics

Meter type	Flow range m^3/s	Maximum operating conditions bar	°C	Rangeability	Accuracy ±%
GAS METERS					
Volumetric Displacement					
Rotary vane/piston	10^{-3} to 10^{-1}	90	60	25:1	1
Wet gas	2.5×10^{-6} to 4×10^{-3}	0.025	50	10:1	0.25
Diaphragm					
Tin-case	5×10^{-4} to 10^{-1}	0.05	60	100:1	1
Steel-case		10			
Rotary displacement	7×10^{-3} to 2	0.08	60	25:1	1
Variable Area					
Rotameter	Up to 0.25	5	100	10:1	1 to 5
Differential Pressure					
Orifice plate	10^4 to 10^7 R_e	550	250	3:1*	1 to 2
Full Flow Velocity					
Turbine	5×10^{-4} to 10	2.5 to 300	60	15:1 (low pressure) 100:1 (20 bar)	1 to 2
Rotary inferential	10^{-3} to 5×10^{-2}	2	60	10:1	2
Vortex shedding	5×10^{-2} to 30	20	200	10:1	2
LIQUID METERS					
Volumetric Displacement					
Semi-rotary piston	6×10^{-7} to 7.5×10^{-2}	20	150	20:1	1
Reciprocating piston	6×10^{-7} to 7.5×10^{-2}	10	180	20:1	1
Nutating disc	2×10^{-6} to 7.5×10^{-2}	30	120	10:1	1
Rotating vane	1.5×10^{-3} to 0.5	80	250	20:1	1
Variable Area					
Rotameter	Up to 0.1	25	250	10:1	1 to 5
Differential Pressure					
Orifice plate	10^4 to 10^7 R_e	100	550	3:1*	1 to 2
Full Flow Velocity					
Electromagnetic	Up to 3 m dia pipework	10	200	10:1	1
Turbine	10^{-7} to 1	300	600	20:1	0.25
Vortex shedding	10^{-2} to 0.2	20	200	10:1	2

* 10:1 or greater with multiple pressure transducers.

carbon dioxide, carbon monoxide and oxygen, with the addition on occasions of sulphur dioxide and oxides of nitrogen. There may also be occasions when it is necessary to measure the methane concentrations in gases, often for safety considerations.

The equipment to be used for this type of on-site duty must be readily portable, robust, reliable and of acceptable accuracy, with a reasonable response time. A range of instruments that generally satisfy these requirements, to a lesser or greater degree, is described in this Section; in many cases, the equipment may also be applied to the analysis of fuel gases or other gaseous mixtures.

14.5.1 Flue Gas Chemical Analysis

Fyrite gas analyser

Gas analysed (separate analysers):	O_2 and CO_2
Accuracy:	analysed % concentration ±0.5%

The Fyrite gas analyser, shown in Figure 14.45, is a small, easily portable instrument for the measurement of either carbon dioxide or oxygen contents of combustion gases and which relies upon the absorption of the required component from the gas sample by liquid solutions.

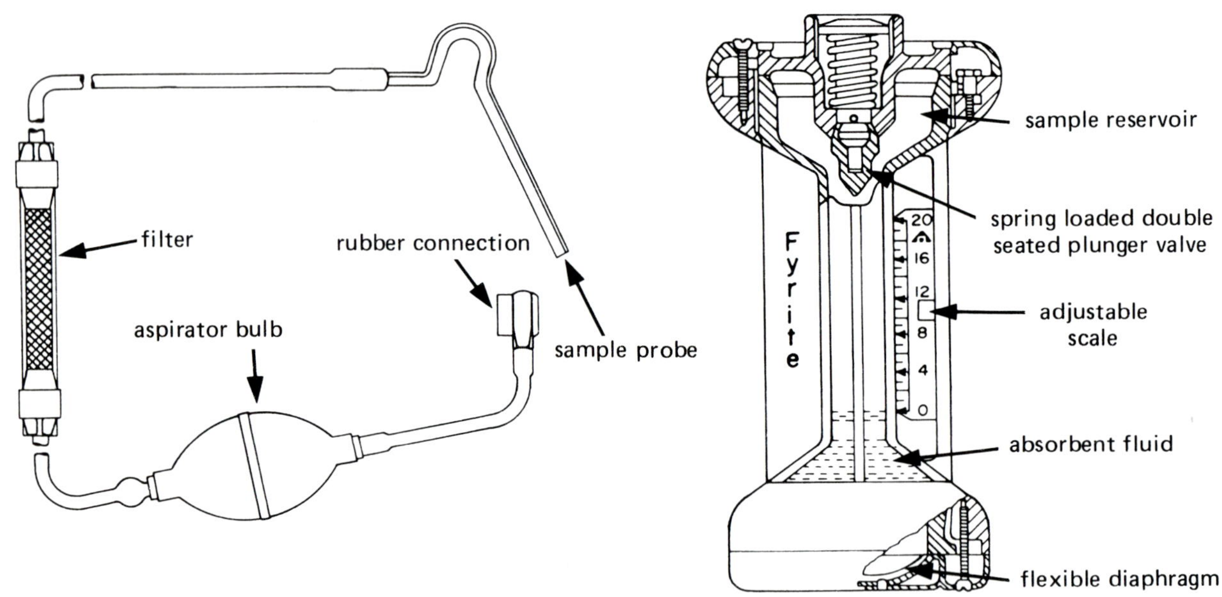

Fig. 14.45 Fyrite analyser.

The decrease in sample volume due to the selective absorption of one constituent of the gas mixture causes the liquid level to rise up the calibrated tube when the unit is returned to its original position. The percentage concentration is read directly off the scale. Two versions of the analyser are available, one for carbon dioxide measurement and one for oxygen measurement. Its use has declined due to the inconvenience associated with wet manual chemical techniques and the increased availability of portable electronic instruments based on electrochemical sensors.

Drager tube analysers

Gases analysed with separate tubes:	over 100 types
Range (various tubes):	O_2 5%-23% CO_2 0.01%-60% CO 5 ppm-7% SO_2 0.1 ppm-2000 ppm NO_x 0.5 ppm-5000 ppm
Accuracy:	various, but generally ±10-15% of reading

The Drager gas analyser, shown in Figure 14.46, is a relatively simple device consisting of a hand held bellows suction pump which draws the gas to be analysed through a small glass tube filled with a granular absorbent solid which contains an indicating agent. The reaction between the gas constituent being measured and the absorbent produces a colour change in the indicator. The length of the stain indicates the concentration of the particular component, being read-off directly against a scale carried on each tube. There is a different type of tube for each individual gas to be measured and the device is capable of determining concentrations of a very wide range of gases and vapours. A new tube is required for each indiviual analysis.

Drager tubes are generally used for spot checks to give an indication of concentration, not as an accurate measurement method.

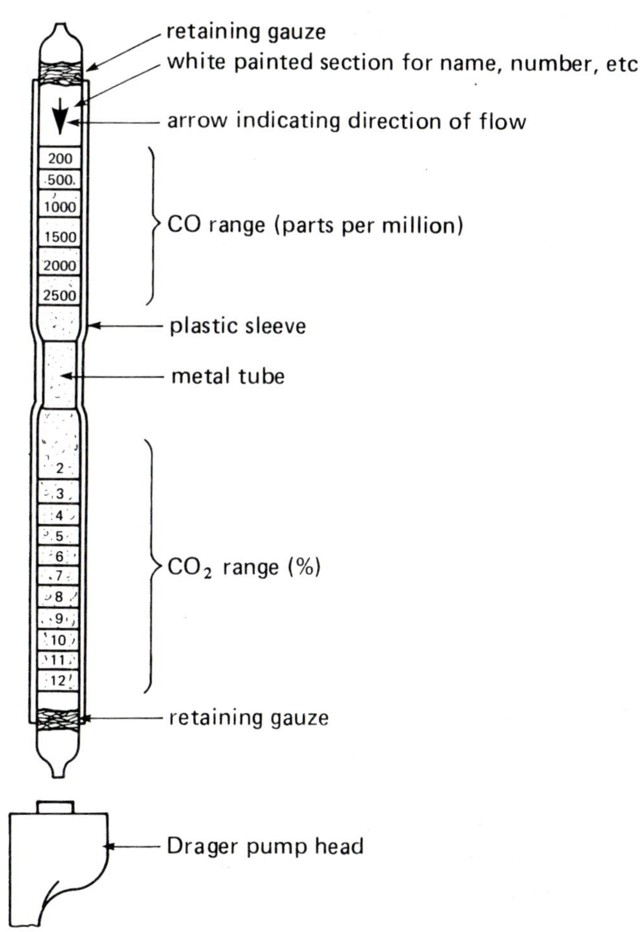

Fig. 14.46 Drager tube analyser

14.5.2 Paramagnetic Oxygen Analyser

The portable instruments so far described for gas analysis all have their disadvantages which become most pronounced in the case of oxygen measurements. The reaction between oxygen and the absorbing solution in the Fyrite is slow, taking some appreciable time to reach completion. The Fyrite has the additional disadvantage of restricted area of contact between the liquid and the gas. The Drager detector tubes for oxygen do not give reliable readings in the presence of carbon monoxide.

Both the Drager and the Fyrite instruments have a restricted scale length which does not make for high accuracy. This limitation applies to all measurements, not just to oxygen determination. Similary, assessment of the end of the stain in the Drager tubes is somewhat indeterminate. This too applies equally well to all measurements.

Paramagnetic oxygen analysers overcome the above shortcomings to a great extent. These analysers measure the magnetic susceptibility of the sample gas. Only very few gases, including oxygen and the oxides of nitrogen, exhibit a positive magnetic susceptibility or paramagnetism. The majority of gases exhibit a negative magnetic susceptibility or diamagentism. Thus paramagnetic analysers enable oxygen concentrations to be measured accurately with little interference from other gases in flue gases and process gas streams.

The widely used paramagnetic analysers are the Taylor Servomex range based upon the Munday cell shown schematically in Figure 14.47. The cell contains a dumb-bell consisting of two nitrogen filled spheres suspended on a platinum ribbon in a strong non-uniform magnetic field. The presence of oxygen in the sample gas passing through the cell causes the dumb-bell to deflect to a new position. The deflection is detected by an optical system employing twin photocells. The output signal from the photocells is amplified and fed to a coil around the dumb-bell so as to restore the dumb-bell to its original position. The current in this feedback loop is a measure of the paramagnetism of the gas sample in the cell.

The Servomex analyser may be used to analyse a continuously flowing sample or a static sample, the latter being the preferred method of operation. The instrument has an inlet and an outlet connection and the gas sample is transferred from a sampling probe to the inlet of the instrument by means of an aspirator bulb or mechanical pump system.

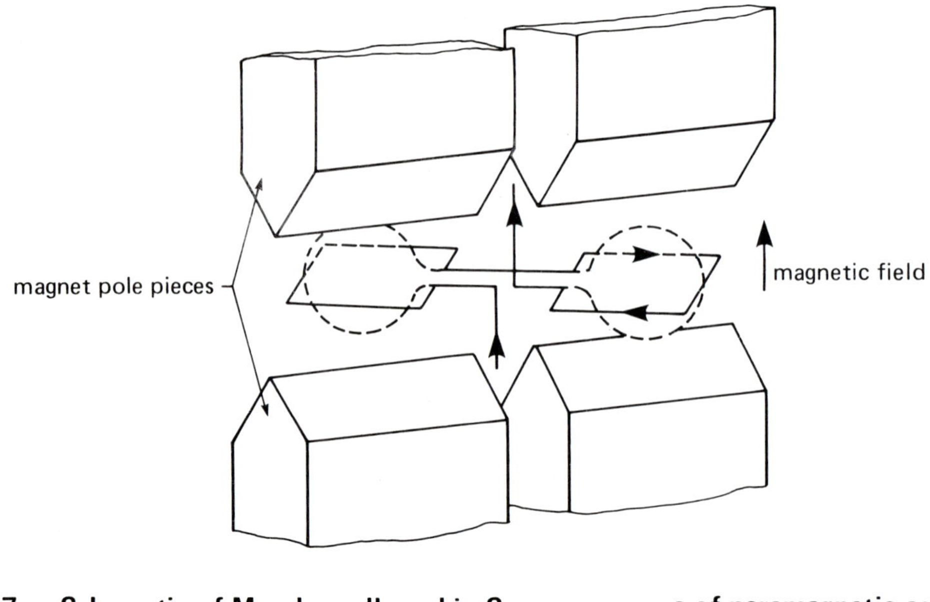

Fig. 14.47 Schematic of Munday cell used in Servomex range of paramagnetic oxygen analysers

The advantages of the paramagnetic analysers are linear response (hence easy calibration using nitrogen and ambient air for zero and span settings respectively) and a response time of a few seconds.

The accuracy of registration of magnetic oxygen analysers is, however, seriously affected by water vapour and by sample gases at high temperature. The sample for analysis should therefore be thoroughly dried by passing through a liquid trap and over a dessicant (dry silica gel), through a filter to prevent ingress of solid particles and cooled to below 100 °C, but lower temperatures are preferable. The sample temperature should not change during a measurement by more than 5 °C otherwise the accuracy of measurement deteriorates. Internally heated Servomex analysers are available for continuous analysis applications.

Typical performance (portable units)

Ranges of measurement	0-5%, 0-10%, 0-25%, 0-100% O_2
Sensitivity to NO relative to O_2	0.42:1.0
Sensitivity to NO_2 relative to O_2	0.28:1.0
Accuracy	±1% range

14.5.3 Zirconia Oxygen Analyser

The zirconia oxygen sensor was originally developed for the monitoring of heat treatment furnace atmospheres. Subsequently the sensor was used for in-situ oxygen measurements on flue gases and is now being used in portable oxygen analysers.

A zirconia sensor is manufactured using yttrium stabilised zirconia. When heated to a temperature in excess of 600 °C this material conducts oxygen ions, the conductivity increasing exponentially with temperature. An oxygen concentration cell is formed by mounting an yttrium stabilised zirconia disc in the centre of a tube of the same material, coating each side of the disc with platinum and mounting this assembly in a small temperature controlled tubular furnace as shown in Figure 14.48.

When the sample and reference gases in the sensor unit have the same oxygen partial pressure there is no potential difference across the zirconia disc and therefore no output from the sensor unit.

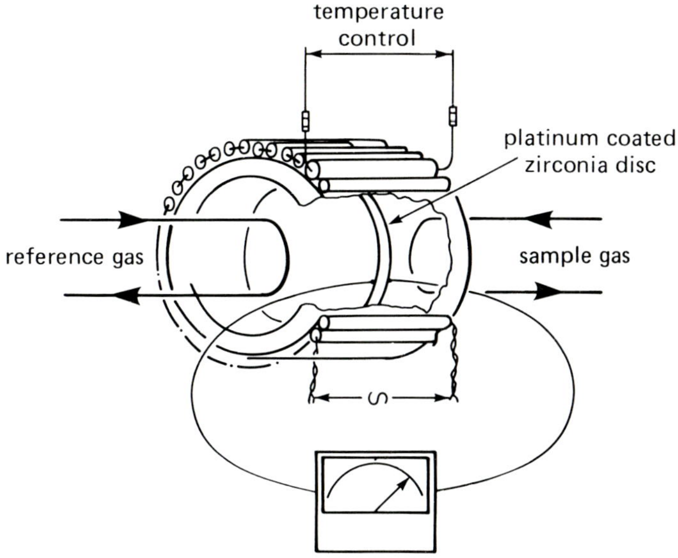

Fig 14.48 Section of Zirconia oxygen sensor

As the oxygen partial pressure is reduced on the sample side of the disc, a potential difference occurs and an output is obtained from the sensor.

Manufacturers of zirconia oxygen probes mount their sensor in various types of probe assemblies. However, they all have to ensure that both a source reference air, normally the atmosphere, and the hot flue gases are passed to the zirconia oven.

The major advantages of the zirconia sensor are its exclusive response to oxygen, a very wide measurement range, fast response and the fundamental relationship between sensor voltage and oxygen concentration independent of sensor geometry or construction details. Because the sensor is mounted in the flue, the O_2 reading is under actual flue gas conditions and is thus the % O_2 in the *wet* flue gases.

Typical performance

Measurement range	1 part in 10^{-25} to 20.9% O_2
Accuracy for complete instrument over range 0.1% -20.9% O_2	±3% of reading, ±0.1% O_2

In recent years the advantages of the zirconia sensor in relation to oxygen monitoring have been used to develop automatic burner trimming systems for boilers to ensure accurate control of air/fuel ratio settings.

14.5.4 Electrochemical Cell Analyser

Electrochemical sensors of the fuel cell type have been developed at the City University, London, for measuring oxygen and carbon monoxide concentrations. Electrochemical cells are used in many portable flue gas analysers including among others, those manufactured by Neotronics, Kane-May, Bacharach, Crowcon and Anglo-Nordic.

Electrochemical cells employ capillary diffusion barrier technology. The supply of sample gas to the sensing electrodes of the cell is controlled by a diffusion barrier so that the supply rate is relatively insensitive to pressure and temperature variations. Selective chemical reaction of the appropriate constituent (O_2 or CO) in the sample gas at the electrodes of the cell produces an electrical current proportional to the concentration of the constituent gas. The rate of chemical reaction within the cell is fast in comparison to the rate of gaseous diffusion through the capillary barrier. The current generated by the cell is converted into a voltage signal via an external load resistor connected between the anode and cathode of the cell.

Electrochemical cells have a finite lifetime and need to be continually replaced on instruments that are only occasionally used. The oxygen cells exhibit a low cross-sensitivity to the presence of other common gases but the carbon monoxide cells are, however, sensitive to several other common gases. Cells are also available for the detection of methane.

Typical performance, O_2 cell instruments

Range of measurement	0-25% O_2
Accuracy	measured % concentration ±0.5%
Cell shelf lifetime	6-9 months

Typical performance, CO cell instruments

Range of measurement	0–4000 ppm
Sensitivity to H_2S relative to CO	3.5:1
Sensitivity to H_2 relative to CO	0.4:1
Accuracy	±8% reading ±20 ppm
Cell shelf lifetime	18–24 months

14.5.5 Thermal Conductivity Gas Analyser

The thermal conductivity technique is employed in flue gas analysers such as the Kent Kathet, Anagas and Siemens WLF for the measurement of carbon dioxide concentrations. The basis of the technique is that the thermal conductivities of air, oxygen, nitrogen and carbon monoxide agree to within 4% of each other at ambient temperatures whereas the thermal conductivity of carbon dioxide is considerably (around 40%) lower.

Instrumentation for measuring thermal conductivity usually consists of a bridge circuit of identical heated filaments exposed to the sample gas and reference air. A typical circuit is shown in Figure 14.49. The presence of carbon dioxide in the sample gas decreases the heat loss from the filament causing an increase in filament temperature and electrical resistance. The filaments are enclosed in a high thermal conductivity enclosure to minimise errors due to radiation and convection from the filaments.

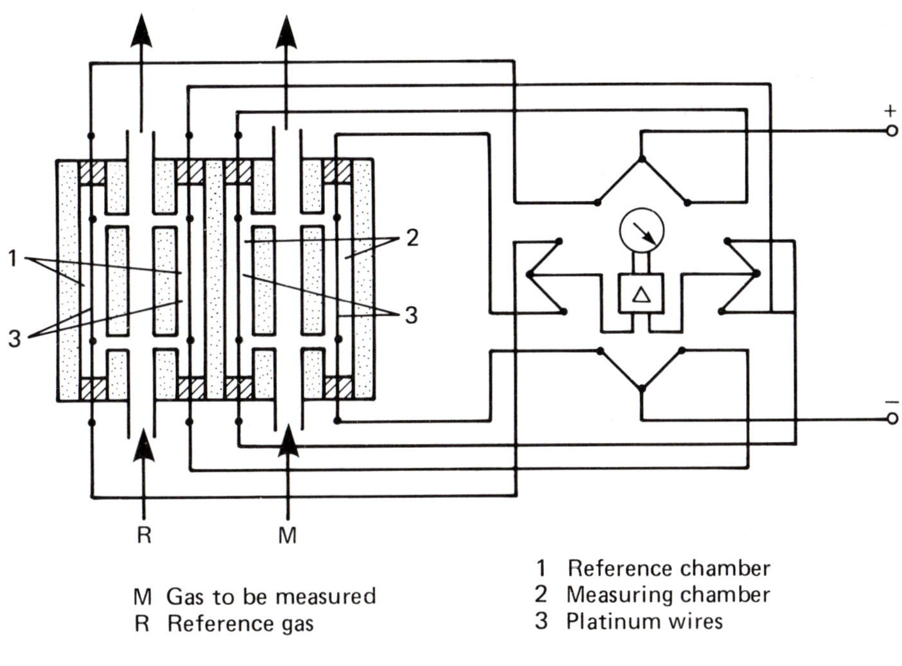

Fig. 14.49 Schematic of thermal conductivity analyser cell and measurement circuit

The accuracy of the technique is susceptible to interference from hydrogen and water vapour. The thermal conductivity of hydrogen is around seven times that of air and the thermal conductivity of water vapour is comparable to that of carbon dioxide. Care must be taken to ensure that the dew-point of the sample gas remains close to that of the reference air. Some instruments incorporate facilities for equalising the humidities of the sample gas and reference air prior to their passage through the thermal conductivity chamber. Correction for the influence of hydrogen is required when sampling gas rich atmospheres.

Typical performance

Range of measurement	0–15% CO_2
Resolution	±0.1% CO_2

14.5.6 Non-Dispersive Infra-Red Analyser

Non-dispersive infra-red (NDIR) analysis has become established as one of the more accurate techniques for flue gas analysis. The technique offers high sensitivity combined with high selectivity and can be used for short term or continuous gas monitoring.

The principles of infra-red analysis are based upon the fact that molecules of gases and vapours containing at least two different atoms (for example CO, SO_2, CH_4 but not H_2, O_2, N_2,) absorb infra-red radiation in the wavelength region 2.5 to 16 microns in a manner that is specific to the atomic composition of the molecule. Each type of molecule therefore exhibits a unique absorption spectrum (and a complementary unique transmission spectrum) to the passage of broad-band infra-red radiation through the gas. NDIR instruments use techniques for selectively enhancing response to the spectrum of the constituent gas whose concentration is to be measured and decreasing sensitivity to the spectra of other gases that may be present. Flue gases are cooled and dried prior to measurement and thus analysis on the dry basis is obtained.

The degree of absorption of infra-red radiation depends on the number of molecules present. Thus absorption depends not only upon the volume concentration of the constituent gas but also upon gas pressure and the path length of radiation passing through the gas sample.

NDIR analysers suitable for flue gas analysis can be classified into the following types:

- The Luft Analyser. Originally developed in the 1940s, this analyser is widely used and has excellent sensitivity but possesses practical disadvantages in terms of physical size, warm-up time and susceptibility to mechanical vibrations.
- Improved analysers incorporating some of the basic principles of the Luft instrument but with new methods of radiation detection to overcome some of the practical disadvantages of the Luft analyser.
- Compact lightweight analysers employing interference filters for wavelength selection and solid state detectors. These are generally less sensitive than Luft analysers and are more influenced by cross-interference effects.
- Cross stack analysers mounted directly in the flue duct without the need for sampling systems.

Luft analyser

A schematic of the Luft analyser is shown in Figure 14.50. Infra-red radiation emitted from the hot filament sources is 'chopped' by a rotating shutter so that it alternatively passes through the reference gas cell and sample gas cell. The reference cell contains a non-absorbing background gas. Radiation emerging from the cells enters the Luft detector which contains the constituent gas to be analysed in two chambers separated by a thin diaphragm. Within the detector, the radiation is selectively absorbed by the constituent gas causing expansion and movement of the diaphragm. Thus the detector is specifically tuned to the spectrum of the gas to be analysed.

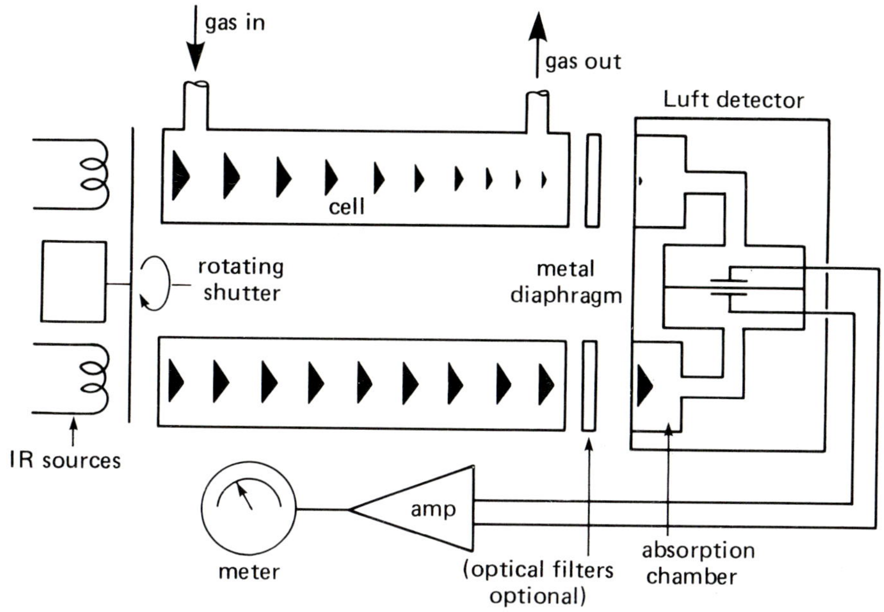

Fig. 14.50 Schematic of Luft infra-red gas analyser

When the sample gas contains no constituent gas, a similar level of radiation is measured by each side of the Luft detector. When constituent gas is present in the sample cell, infra-red radiation is absorbed and the levels of radiation detected from the sample cell and reference cell are no longer equal. The movement of the diaphragm in the detector is normally measured by a capacitance transducer.

Luft analysers can achieve a very high sensitivity with low cross-sensitivities in the presence of other gases. Most commercial Luft analysers are designed for laboratory work rather than field work. Consequently the analysers tend to be large and slow to reach thermal equilibrium after switch on. The Luft detector is susceptible to the influence of external mechanical vibrations.

Typical performance

Gas analysed	heteroatomic molecules e.g. CO, SO_2, CH_4, NO_2
Minimum concentration range	0–200 ppm
Maximum concentration range	0–100%
Accuracy	±2% of range
Warm-up time	1–2 hours

BINOS analyser

A schematic of the BINOS infra-red analyser is shown in Figure 14.51. The general arrangement is similar to the Luft analyser with beams of infra-red radiation passing through the sample cell and reference cell.

The radiation detector is filled with the constituent gas to be analysed but pressure pulsations within the detector are converted into an electrical signal by a micro-flow sensing device rather than a capacitance transducer. The BINOS detector has a much faster response time than the Luft detector and this permits high frequency modulation of the radiation with the rotating chopper

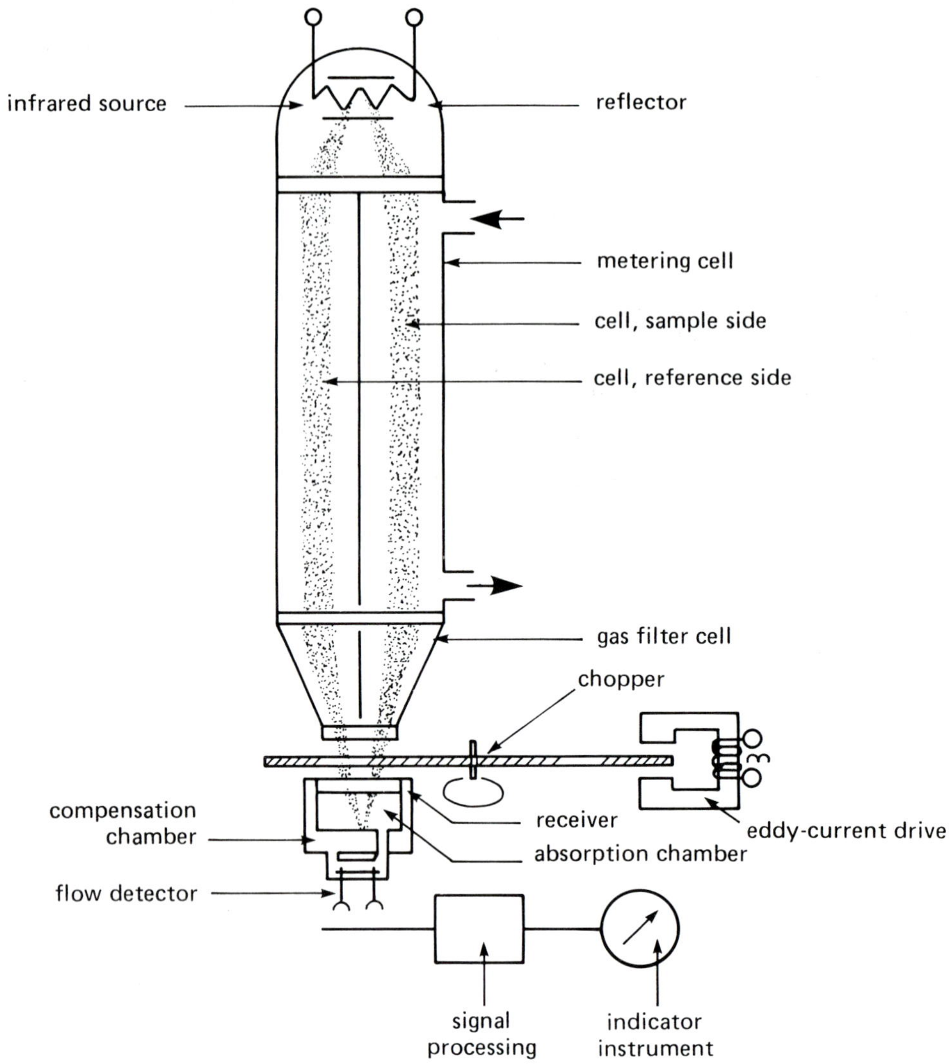

Fig. 14.51 Schematic of BINOS infra-red gas analyser

incorporating techniques that automatically compensate for thermal instability of the radiation source.

In comparison to the Luft analyser, the BINOS analyser is lighter, approximately half the weight, has a rapid warm-up time and is far less sensitive to external vibrations.

Typical performance

Gases analysed	heteroatomic molecules
Minimum concentration range	0–200 ppm
Maximum concentration range	0–100%
Accuracy	± 2% of range
Warm-up time	2 minutes

ADC RF series analyser

The principles of the Analytical Development Company RF series analyser are shown in Figure 14.52. Infra-red radiation from the hot source passes through a rotating modulator wheel incorporating two sealed gas cells. One cell contains the constituent gas and the other cell contains a reference inert gas. After modulation, the radiation passes through a filter, then through the single gas sample tube and is finally focussed on a solid-state infra-red detector. The rotating gas cells in the modulator wheel alternatively filter the radiation incident on the sample tube. Thus the radiation passing through the sample gas alternates between being sensitive to absorption by the constituent gas and being insensitive to absorption. The concentration of constituent gas in the sample tube can be found from the ratio of modulated output signals from the infra-red detector.

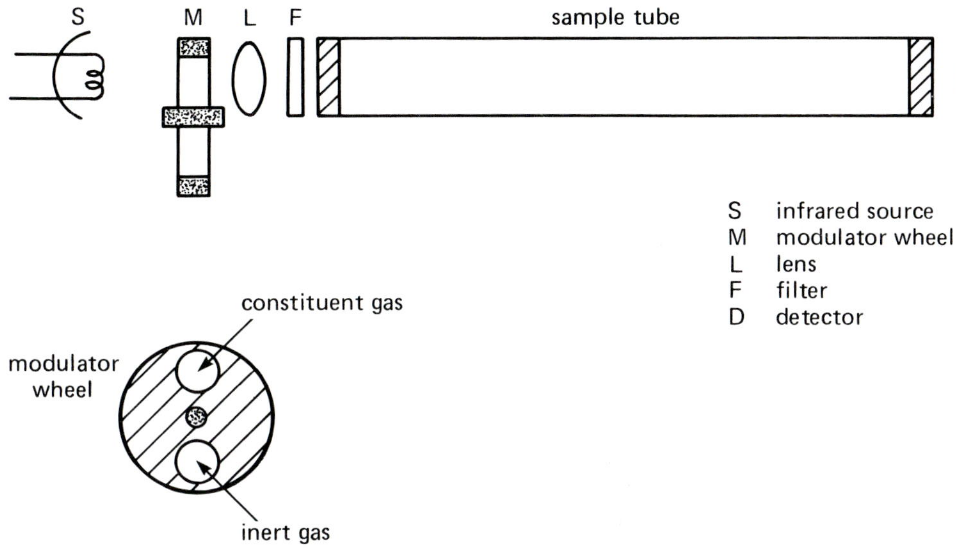

Fig. 14.52 Schematic of ADC RF series infra-red gas analyser

The RF series analysers exhibit operating characteristics that are more suitable for field work than traditional Luft type analysers.

Typical performance

Gases analysed	heteroatomic molecules e.g. CO, SO_2, CH_4, NO_2
Minimum concentration range	0–200 ppm
Maximum concentration range	0-100%
Accuracy	± 2% of range
Warm-up time	usable after five minutes

Interference filter NDIR analysers

Portable NDIR analysers employing narrow band interference filters for wavelength selection and solid state detectors are particularly suitable for field work. These analysers are usually single channel instruments where the filter is tuned to one of the main absorption bands of the constituent gas in a part of the spectrum relatively free from interference due to other gases. The analyser is zeroed by temporarily diverting the sample gas through a chemical absorbent that removes the constituent gas. The absence of a reference channel tenders the instruments susceptible to long term thermal drifts. Both CO and CO_2 possess strong absorption bands in the 4 to 5 micron wavelength region that are reasonably free from cross-interference effects. In general these instruments are considerably less sensitive and less selective than the Luft analyser.

Typical performance

Gases analysed	CO, CO_2, hydrocarbons
Minimum concentration range	0–1%
Maximum concentration range	0–100%
Short term accuracy	±2% of range
Warm-up time	30 minutes

CO cross stack analysis

The cross stack analyser consists of four units; an Infra-red Source which projects a beam of radiation across the duct, a Receiver Unit which detects any absorption of this radiation due to the presence of CO, a Signal Processor which computes CO concentration and a Power Supply which supplies power to the system. Figure 14.53 shows details of the Infra-red Source and Receiver Unit. The cross-duct system eliminates the need for sampling systems and enables continuous, on-line CO measurements to be made in-situ on the flue ducting or stack with high reliability. An accuracy of ± 30 ppm CO is possible with this type of system.

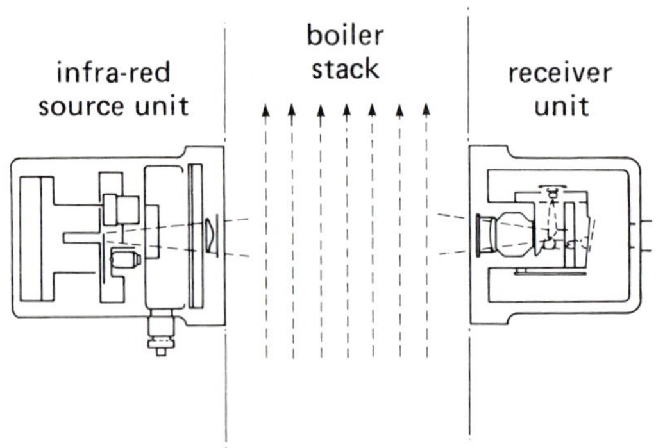

Fig. 14.53 Schematic of CO cross stack analyser

14.5.7 Analysis Techniques for NO_X and SO_2

Measurement of nitrogen oxides (NO_X) and sulphur oxides is an increasingly important aspect of flue gas analysis with the necessity to reduce these atmospheric pollutants.

Methods can be based on a variety of techniques such as chemical reaction/electrochemical cells, gas chromatography, infra-red or UV absorption or fluorescence/luminescence.

Luminescence methods are significantly more sensitive and there is a general perference for their use in flue gas analysis. The response time is in the order of 1-2 seconds and full scale deflection can range from 0.01 to 10 000 ppm. However the instruments are relatively high cost and are generally used only in the laboratory for detailed atmosphere analysis.

Luminescence occurs when molecules are excited by interaction with electromagnetic radiation. Fluorescence describes the phenomenon if the subsequent release of energy is immediate, and forms the basis for SO_2 analysis. If the excitation energy is obtained from the chemical energy of reaction the process is luminescence which is used for NO analysis. Analysis of NO_2 is by conversion by heat to the oxide. The fundamental principles of fluorescence measurement are shown in Figure 14.54

Any conventional fluorimeter can be converted to do chemiluminescent work by screening the exciting light source so that luminescence can be produced by chemical reaction

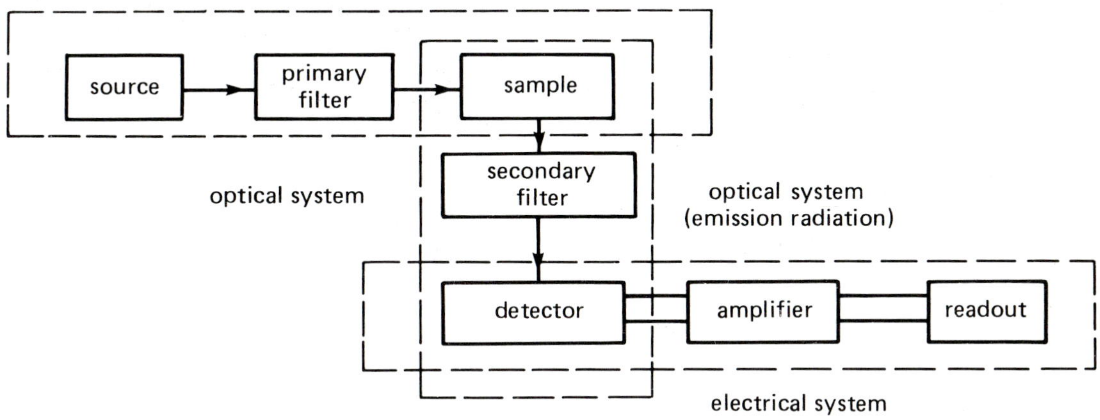

Fig. 14.54 Schematic of the optical components of a typical filter fluorimeter

Sampling of gases at elevated temperatures through a metal probe containing an excess quantity of reducing species can lead to loss of NO and NO_2. In addition metal probes at high temperatures will also tend to act as converters thus changing the relative concentrations of NO and NO_2.

14.5.8 Calibration of Gas Analysis Instrumentation

Research grade gases and gas mixtures are normally used as reference standards in the calibration of gas analysis instrumentation. These gases are supplied in standard industrial gas cylinders.

Virtually any mixture of research grade gases, excluding explosive mixtures, can be supplied. For example 4000 ppm CO, 15% CO_2 with balance N_2 can be obtained in one cylinder for the calibration of flue gas analysers. A certificate of the chemical analysis of gas mixtures should be obtained from the supplier for each cylinder that is to be used for calibration purposes.

The calibration of oxygen analysers for flue gas measurements is normally performed on ambient air and on oxygen free gas for the span and zero adjustments respectively.

Tolerances on reference grade gas mixtures

Concentrations above 10%	±1% of quoted analytical composition.
Concentrations below 10%	±2% of quoted analytical composition.

14.5.9 Methane Determination in Air

The analysers discussed above have been primarily concerned with the analysis of flue gas products. The Gasco Seeker, however, is a portable instrument for measuring methane concentrations in air/natural gas mixtures at atmospheric pressure. LIDAR is a state-of-the-art system for remotely measuring the concentration of natural gas released into the atmosphere.

Gascoseeker

This instrument was developed by British Gas in conjunction with the manufacturers Gas Measurement Industries Ltd. The unit has three concentration ranges: 0-10% lower explosive limit (LEL), 0-100% LEL and 0-100% Volume Gas.

On the % LEL ranges, the unit operates by measuring the heat produced by combustion of the flammable part of the test atmosphere. On the % Volume Gas range, the instrument operates by measuring the thermal conductivity of the test atmosphere.

A hand operated aspirator is used to draw the test atmosphere through a hydrophobic (water removing) filter prior to passing through the analysis chamber. Depending upon the scale range selected, the sample is analysed by one of two pairs of Pellistor cells forming the arms of a Wheatstone bridge circuit. The imbalance of the bridge circuit is displayed on a meter calibrated directly in the scale range units. Sintered bronze flash-back arrestors isolate the analysis chamber from the surrounding atmosphere.

Gascoseekers are also available with oxygen measurement combined with gas analysis already discussed and with digital readings.

Typical performance

Measurement ranges	0-10% LEL, 0-100% LEL, 0-100% Gas
Accuracy	±3% reading ±1% range (LEL scales)
	±1% reading ±1% range (Gas scales)

LIDAR

The British Gas light detection and ranging system (LIDAR) operates by transmitting through the atmosphere short-duration high energy pulses of UV laser light. As the beam propagates it interacts with the constituents of the atmosphere in a number of ways, some of the light is scattered directly off solid particles of dust, some light may be absorbed, and a small proportion of the light is scattered with a frequency shift which is characteristic of the molecules in its path. It is the relatively high concentrations of gas in air (2-20%) during the release or venting of gas which allows the use of Raman scattering to detect and measure methane concentrations.

To allow the equipment to be used at any British Gas site, the whole LIDAR System, complete with 500 mm diameter telescope and laser mounting, together with the associated microcomputer, is mounted on a 10 tonne truck. The previously used method of gas cloud analysis utilised sensor probes mounted between masts.

Typical performance

Measurement range	2-20%
Distance range	100 – 1 000 m
Accuracy of measurement	±10%
Spatial resolution along a line of sight	1.4 m

14.5.10 Gas Sampling

Gas sampling is fraught with many pitfalls. Ideally the sample withdrawn from a duct or enclosure should be representative of the whole gas from which it has been taken. It should be identical in composition and rate of flow from time to time and from point to point in the gas to be sampled.

The problems associated with gas sampling are:

– Stratification, such that the gas does not have constant composition across the entire section of the duct. The larger the duct and the lower the gas velocity, the more likely it is that stratification will be a problem. Large differences between the densities of the components of a gas stream will help to maintain stratification.

- Air in-leakage, e.g. at defective portions of a flue at negative pressure will cause non-constant composition of the flue gas mixture. A layer of cold air close to a flue wall may persist for considerable distances along flue ways. Air entering a flue downstream of the point of combustion registers as air in excess of combustion requirements but, of course, will not have taken part in the combustion process. Care must also be taken to ensure that there are no air in-leakages along the sample line, especially at the joints.
- Variations in total gas flow will lead to discrepancies in the sample where sampling is being carried out over a period of time unless the rate of sample withdrawal is made to vary proportionally with the flue gas flow rate (isokinetic sampling). This is of some importance in the continuous analysis of gases.
- Changes in composition may occur in the time which elapses between a sample being withdrawn and the analysis being performed. These changes may be caused by combustion occurring in the sampling system, either naturally or catalytically in contact with materials of the sampling probe, or chemical reactions may occur between a component of the gas and the materials of the sampling system, e.g. oxygen and carbon monoxide will react with copper or iron at high temperature.
- Condensation, caused by lowering the temperature of the gas sample to below its dewpoint, may result in the loss of a water soluble component, or in the blockage of the sampling system, because of the collection of the dust burden in the condensed moisture. The need for cooling the sample must be emphasised firstly to quench reactions that may be taking place between constituents of the sample gas and secondly to avoid damage to analysing equipment. The gas samples for paramagnetic oxygen and infra-red analysers must be dried by passing the sample through silica drying columns and the use of water separators or knockout pots.
- Time lags in the sampling system may occur if the exhausting device is slow or if the volume of the sampling system is large. If the composition of the flue gas is varying in a cycle time which is short compared with the time-lag in the sampling and analysis system a truly representative sample may not reach the analyser.

Other problems may arise from tarry deposits in sampling lines, corrosion, leakage at joints in the sampling system and a heavy dust burden in the gases.

Where continuous sampling and analysis are to be performed it will generally be feasible to sample from a single point only and that point must be selected with some care if the problems enumerated are to be avoided. It is likewise necessary to avoid dead spaces to ensure that the point of sampling is being continuously swept by the gas to be analysed. Points in a flue or duct system where reasonable turbulence exists are to be preferred for sampling because the likelihood of stratification is thereby reduced. Reduction in flue cross-sectional area, operation of fans, bends, all assist in breaking up stratification patterns. Sampling on bends themselves, however, should be avoided because of the probably existence of dead spaces.

It is worthwhile pointing out that in ducts of square or rectangular cross-section the flow of a fluid takes up a roughly circular profile so that the corners of ducts of these cross-sections will not be subject to an appreciable flow and errors of dead space will again arise if sampling is carelessly performed.

The foregoing remarks have been made with gas composition analysis in mind but they are just as valid where many other measurements are concerned. In the measurement of gas temperatures, for example, the avoidance of dead spaces, stratification and air in-leakage are just as important as in composition analysis.

14.5.11 Gas Sampling Probes

Uncooled metal probes

Maximum operating temperature for mild steel or copper tube is 300 °C, above this temperature oxidation effects may cause errors.

Some alloy steels are satisfactory up to 1 100 °C where there is no likelihood of combustion or other chemical reaction taking place between constituents of the flue gas.

The stability of alloy steel heat-resisting probes is greatly influenced by the nature of the flue gases, whether oxidizing or reducing, and by their sulphur content. The makers' recommendations should be sought.

Water-cooled metal probes

Water-cooled probes are required to quench chemical reaction between constituents of the flue gases, particularly combustion which may be catalysed by the material of the probe. Such probes are also necessary, of course, to preserve the integrity of the probe at elevated temperatures.

Refractory probes

Refractory probes generally use materials of fused silica, porcelain, aluminous porcelain, mullite or recrystallized alumina. All have low mechanical strength and all are sensitive to thermal shock, with the exception of fused silica.

If used continuously above temperatures of 900 °C silica proves are subject to devitrification but they may be used intermittently up to temperatures of 1 500 °C. They are subject to fluxing by ash particles.

Glazed porcelain probes are satisfactory up to 1 400 °C but like silica probes are subject to fluxing.

Aluminous porcelain probes are physically and chemically stable up to 1 500 °C. Mullite probes are even more stable than the porcelain and are therefore more resistant to fluxing. They are also more refractory and are in consequence suitable for continuous use up to 1 700 °C.

Re-crystallized alumina probes, because of their high purity, are inert to the fluxes normally occurring in combustion gases and are suitable for continuous use up to 1 900 °C.

Glass probes

These should be made of boro-silicate glass (e.g. Pyrex) but should not be used at temperatures above 450 °C.

14.6 HUMIDITY AND DEWPOINT MEASUREMENT

14.6.1 Properties of Vapour/Gas Mixtures

The characteristics and behaviour of a mixture of a vapour and one or more gases are of considerable industrial and commercial importance. For example, combustion products containing water vapour and sulphur oxides arising from the fuel and/or added within the plant being fired, can cause corrosion if the temperature falls below the point at which sulphurous or sulphuric acid starts condensing out. By far the most important case, however, because of its widespread application, is air containing water vapour and this Section is basically confined to that case and its associated measurement techniques.

Humidity is usually expressed as 'relative humidity' (RH). Relative humidity is the humidity present relative to the maximum possible humidity at the same temperature. It is defined as the ratio of the actual water vapour pressure to the saturation vapour pressure at the temperature and is normally expressed as a percentage. The higher the temperature, the greater the amount of water that can exist as vapour in the air, which is why a drying process, for example, is accelerated when heated air is used. This also means that if the temperature rises without increasing moisture content, the RH falls, and vice versa.

Dewpoint is defined as the temperature whose saturation vapour pressure equals the actual vapour pressure of the gas. In practice it is the temperature at which dew begins to condense out if the gas is cooled. A dew-point lower than the freezing point of water is usually called a frostpoint.

RH and dewpoint, often with additional properties such as density and enthalpy, are plotted against temperature to produce a 'psychrometric chart'. These charts are widely available and cover all the temperature and pressure ranges generally encountered. Figure 14.55 is an example of such a chart, with density and enthalpy omitted for clarity. Taking the point marked, arrow (a) points to dry bulb temperature, (b) to wet bulb temperature, (c) to RH, (d) to moisture content and (e) to dewpoint temperature (i.e. wet bulb temperature at 100% RH). It will be seen that if any two of the first four quantities are determined, a point can be fixed on the chart and the other three then found. Psychrometric charts may also be used for typical combustion products, because like air they are predominantly nitrogen, providing due caution is exercised and the limits of accuracy appreciated.

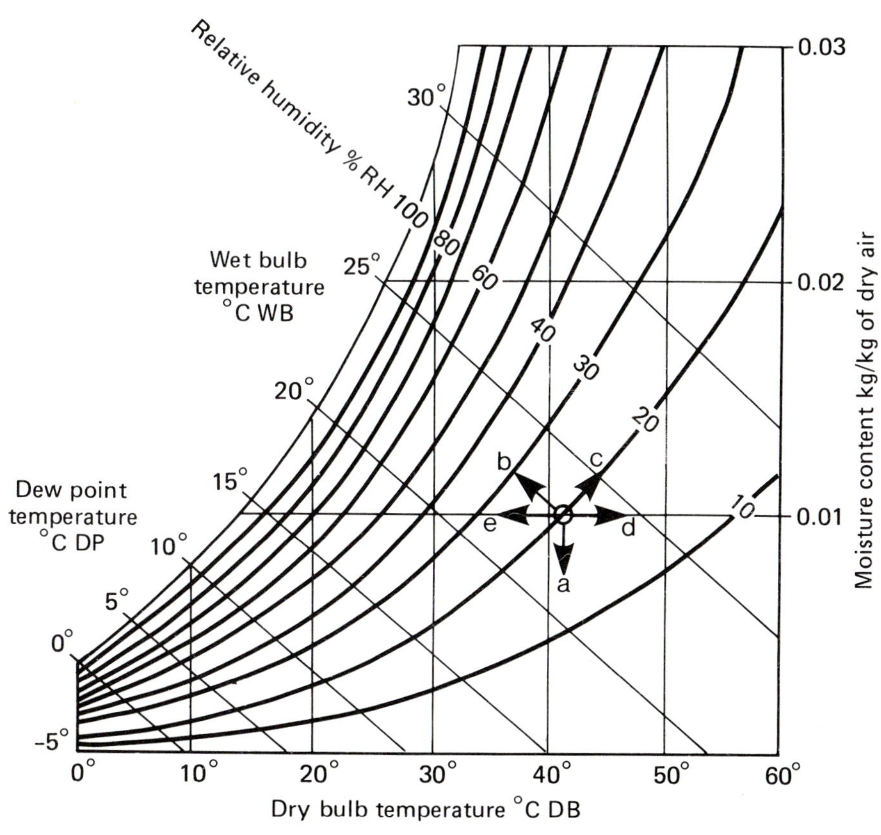

Fig. 14.55 Psychrometric chart

14.6.2 Application

The ability to measure and control the amount of water vapour in gaseous atmospheres is essential in many fields, both commercial and industrial.

In air conditioned buildings control of humidity, as well as temperature, is necessary if acceptable space conditions are to be achieved and maintained, both for the comfort of personnel and for the satisfactory operation of the modern electronic equipment and computers now found in all areas of commerce and industry. In the manufacture and storage of a variety of goods, relative humidity is equally important.

Drying is probably the commonest industrial application for humidity and dewpoint equipment and both operational performance and energy efficiency are dependent on good measurement and control.

In the field of heat treatment, the dewpoint of the furnace atmosphere is now well established as a practical measure of its carbon potential. For this reason dewpoint measurement is widely used for controlling the furnace atmosphere for such heat treatments as carburising and hardening.

14.6.3 Instrumentation – Relative Humidity

The terms hygrometer and psychrometer are nornmally applied to relative humidity measuring instruments.

The most commonly available RH instruments on the market at present are:

- Wet and dry bulb hygrometers.
- Hair hygrometers.
- Conductive and capacitive hygrometers.
- Infra-red hygrometers.

These types of hygrometers will be discussed in more detail in the following sections. Other RH sensors available include hygrometers based upon the zirconia cell, nuclear magnetic resonance and neutron moderation. These more specialised hygrometers will not be discussed because of their limited application.

Wet and dry bulb hygrometer

In its basic form this type of instrument consists of two thermometers, one of which measures the ambient temperature, the other being covered with a damp wick from which water evaporates, so lowering the bulb temperature.

Such instruments depend upon a certain air flow across the wet and dry thermometer bulbs in order to obtain accurate results. This flow must not be less than 5 m/s to ensure that the results obtained are not affected by the geometry of the wet bulb. As a result of this, these instruments are in several forms, each designed to maintain an air flow across the two thermometer bulbs.

The Assman Hygrometer, shown in Figure 14.56, consists of two mercury thermometers in a frame. The bulbs are situated in a duct through which air is drawn by means of a small fan mounted at the top. The instrument is used in conjunction with tables to derive RH from the two temperatures.

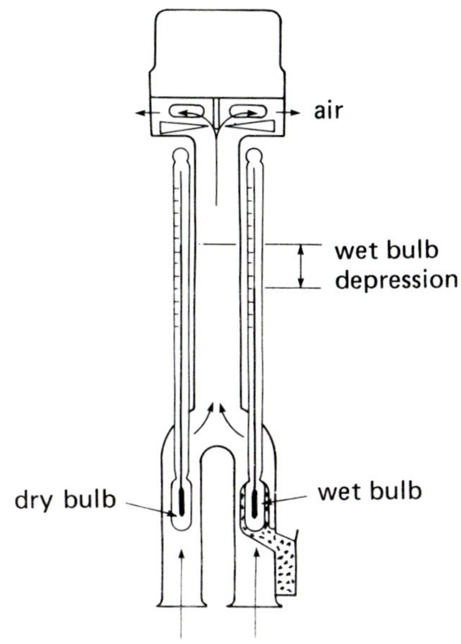

Fig. 14.56 Assman Hygrometer

By using resistance thermometers or thermistors together with modern electronic circuitry direct readings of % RH can be obtained in addition to wet and dry bulb temperatures. As this system operates from the fundamental principle of psychrometry, it does not need the frequent calibration checks that solid state sensors need and at the same time provides a compact hand held instrument. Accuracy is generally limited by the electronic circuitry and is better than ±2% RH.

The hygrometers described above are not generally suitable for industrial use, being designed for laboratory or meteorological purposes. Industrial types generally consist of more robust equipment and usually have facilities for a continuous record of wet and dry bulb temperatures, utilising a two-pen recorder.

Hair hygrometer

The hair hygrometer, illustrated schematically in Figure 14.57, uses a primary sensing element composed of a number of strands of human hair fastened at one end and connected at the other end through a mechanical linkage to an instrument pen or pointer. Because of the hygroscopic quality of hair, an increase in relative humidity causes an expansion of the primary element, allowing a spring to move the instrument pen or pointer up-scale. A decrease in relative humidity causes the hair element to contract, thus forcing the pen or pointer downscale.

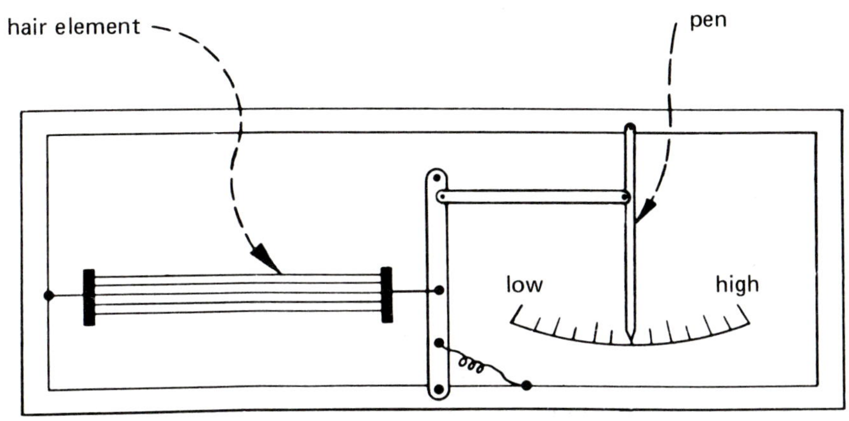

Fig. 14.57 Schematic of hair hygrometer

Once properly calibrated, the hair hygrometer provides a direct reading of per cent relative humidity, with a uniformly graduated scale or chart with an accuracy of ±5% RH. It is commonly used in combination with a temperature pen in a single portable instrument – useful, for example, in checking the operation of air-conditioning equipment in textile mills, food processing plants, and other places where relative humidity is a factor. It can also be made to switch on power to operate fans, spray nozzles, dampers or dehumidifiers.

The hair element is also used in non-indicating Humid-U-Stat controllers which are equipped with a temperature compensator and a one-pipe pneumatic control system. To enable this controller to respond more rapidly to changes in humidity, an air sampling box is employed to draw room air through a filter and over the hair element by means of a small motor-driven blower.

Other sensing elements used in this type of hygrometer include cotton threads and nylon bands.

Electrical hygrometer

Fast, accurate and direct readings of relative humidity can be obtained by monitoring the electrical properties (e.g. capacitance, resistance) of certain moisture-sensitive compounds.

Resistance hygrometer

Variation in ambient relative humidity can produce a change of resistance in certain materials e.g. carbon powder, ceramic materials or lithium chloride. If one of these materials is applied as a film over an insulating substrate and is terminated by metal electrodes it may be used as a humidity sensor. The sensor normally forms one arm of a bridge circuit which is ac powered and the output of the sensor is linearised to give a direct reading of relative humidity. The usual operating range is 10 to 95% RH.

Systematic calibration is necessary since the sensor calibration can vary with time, and can be affected by contamination and, in particular, by exposure to humidity and temperature extremes. Calibration checks against saturated salt solutions are usually adopted. Accuracy of the order of ±2% RH is possible in this way for operating temperatures up to 125 °C.

Capacitive hygrometer

The capacitive humidity sensor can be imagined as a condenser which has a dielectric medium, a polymer and moisture causes a change in the condenser's capacitance. The capacity changes are processed electronically and are converted to a voltage signal proportional to the relative humidity as shown schematically in Figure 14.58.

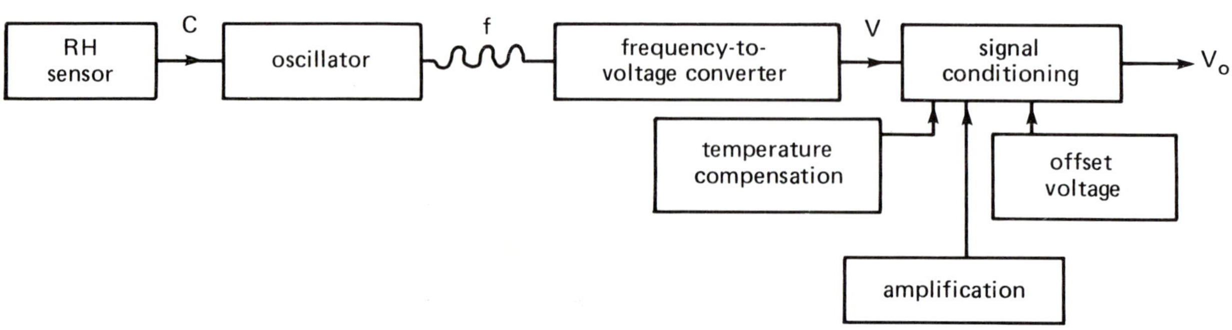

Fig. 14.58 Block diagram illustrating operation of capacitance hygrometer

Polymer engineering has resulted in the development of polymers which give good linearity over the range of 0–100% RH as well as a negligible thermal coefficient and which can be used at temperatures of up to 150 °C.

Sensor manufacturers have attempted to:

- Reduce the size of the sensor to speed up the response time.
- Increase the sensor's capacitance to enable the operating frequency to be kept low which reduces the effect of stray capacitance and makes possible the use of longer lead wires.

An accuracy of the order of ±2% RH is possible with this type of sensor. A typical sensor is shown in Figure 14.59.

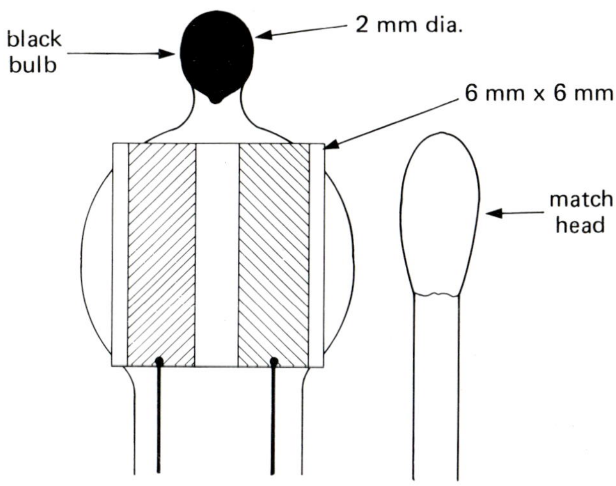

Fig. 14.59 Typical capacitance hygrometer sensor

Infra-red hygrometer

The principle of the infra-red hygrometer is shown in Figure 14.60. The reference and sample gases do not absorb radiation to the same extent and this results in an unequal heating effect in the detectors causing a pressure differential which acts on a diaphragm. The resultant variation in electrostatic capacity can be used to indicate or record the absolute humidity. Accuracy is typically ±1% RH but instrumentation is expensive and is generally not suited to industrial use.

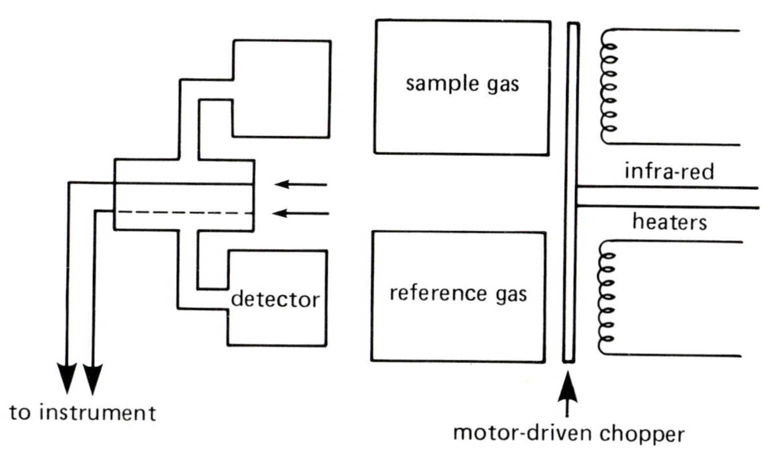

Fig. 14.60 Infra-red hygrometer

Primary gravimetric hygrometer

The primary gravimetric hygrometer is an instrument in which the moisture content of the air is determined by separating the moisture and air and measuring the mass of each component.

Recently such an instrument, capable of measuring the moisture content of air at temperatures from −50 to 50 °C, was installed at the National Physical Laboratory to provide a national standard for the calibration of hygrometers. A two pressure humidity generator or a dewpoint hygrometer can be used as a transfer standard. Wet and dry bulb psychrometry provides a convenient calibration system. In the two pressure system the gas sample is saturated at a constant temperature and then expanded to a lower pressure. The principle of operation is based on Dalton's Law, in that the ratio of partial pressures varies directly with the total pressure. The precision of the system is governed by the accuracy of the pressure and temperature measurement, and by the maintenance of isothermal conditions. The basis of the dewpoint hygrometer is discussed in Section 14.6.4.

14.6.4 Instrumentation–Dewpoint

Dewpoint Techniques

Chilled mirror optical dewpoint calibration techniques have been in use for approximately 200 years. The principle of operation, shown in Figure 14.61 indicates that when the saturation temperature, or dewpoint, occurs, and the gas mixture is saturated with respect to water or ice, the rate of water molecules leaving the atmosphere and condensing on the chilled surface is the same as the rate of water molecules leaving the chilled surface and re-entering the atmosphere. At equilibrium saturation, the water vapour partial pressure of the condensate is equal to the water vapour partial pressure of the gas atmosphere. To establish this dynamic equilibrium at the mirror surface, it is necessary to precisely cool and control the mirror at the saturation temperature. A temperature element is then placed in thermal contact with the mirror and the mirror temperature utilised directly at the dewpoint, or saturation, temperature.

Historically, the cooling of the mirror surface has been accomplished with acetone and dry ice, liquid CO_2, mechanical refrigeration and, more recently, by thermoelectric heat pumps. Detection of the condensation has been observed visually, and the equilibrium cooling manually controlled. Newer versions utilise optical photo-resistor detection designs to automatically control the surface at the dewpoint or frost point. The temperature instrumentation has included the entire spectrum from glass bulb thermometers to all types of electrical temperature elements.

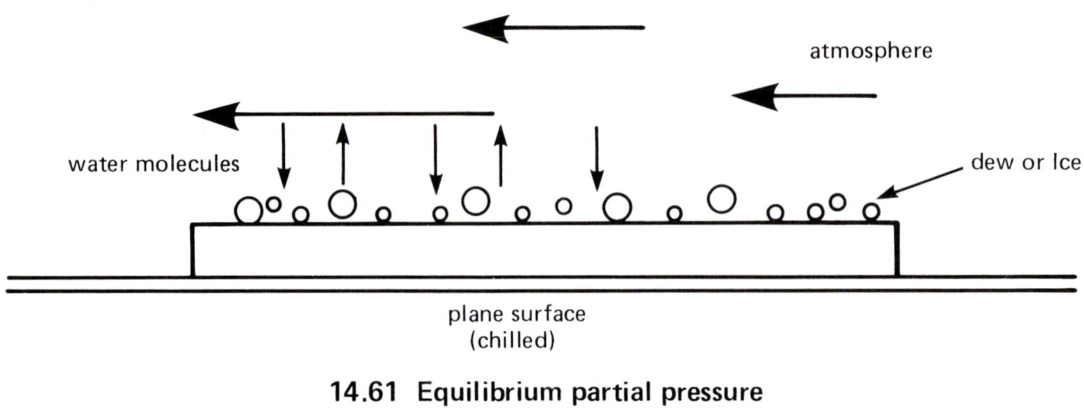

14.61 Equilibrium partial pressure

The manually cooled, visually observed optical hygrometer is commonly known as the dew cup. It is a relatively inexpensive technique, and when operated by an experienced and skilled technician, it is quite accurate. However, it does suffer from some limitations:

- It is not a continuous measurement.
- It is operator dependent; readings may vary from operator to operator.
- Versions using expendable coolants require replacement supplies.

All of these difficulties are overcome by the thermoelectrically cooled, optically observed dewpoint hygrometer. An instrument of this type is shown in Figure 14.62. The mirror surface is chilled to the dewpoint by a thermoelectric cooler while a continuous sample of the atmosphere gas is passed over the mirror. The mirror is illuminated by a light source, and observed by a photo-detector bridge network. As condensate forms on the mirror, the change in reflectance is detected by a reduction in the direct reflected light level received by the photo-detector due to the light-scattering effect of the individual dew molecules. This light reduction forces the optical bridge toward a balance point, reduces the input error signal to the amplifier, and proportionally controls the drive from the power supply to the thermoelectric cooler. This continuously maintains the mirror at a temperature at which a constant thickness dew layer is retained. Independently, embedded within the mirror, a temperature measuring element measures the dewpoint temperature directly.

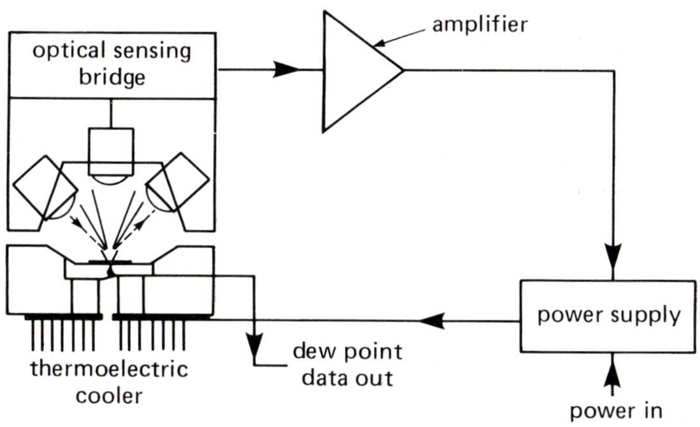

Fig. 14.62 Principle of operation of dewpoint hygrometer

This type of hygrometer is continuous measuring, direct reading in dewpoint (absolute humidity), and is ideally suited for both process monitoring and control. It is however, a relatively expensive instrument because of its sophistication and highly accurate primary measuring technique. An accuracy of ±0.5 °C can be expected with this instrument over the range -70 °C to +50 °C dewpoint.

Conductivity-type dewpoint meter

A conductivity-type dewpoint meter is illustrated in Figure 14.63 and consists of a sample chamber containing an electrically insulating material, one side of which is cooled or refrigerated. Two probes are situated on the cooled surface, and as moisture condenses on the surface they measure the resultant change in electrical conductivity. A suitable control unit is used with these conductivity probes in order to control the amount of coolant fed to the underside of the surface. The surface is therefore maintained at the temperature at which condensation is just taking place. A thermocouple or resistance bulb is attached to the surface in order to measure its temperature and coupled to a suitable indicator or recorder. The temperature of the surface is the dewpoint temperature of the gas sample.

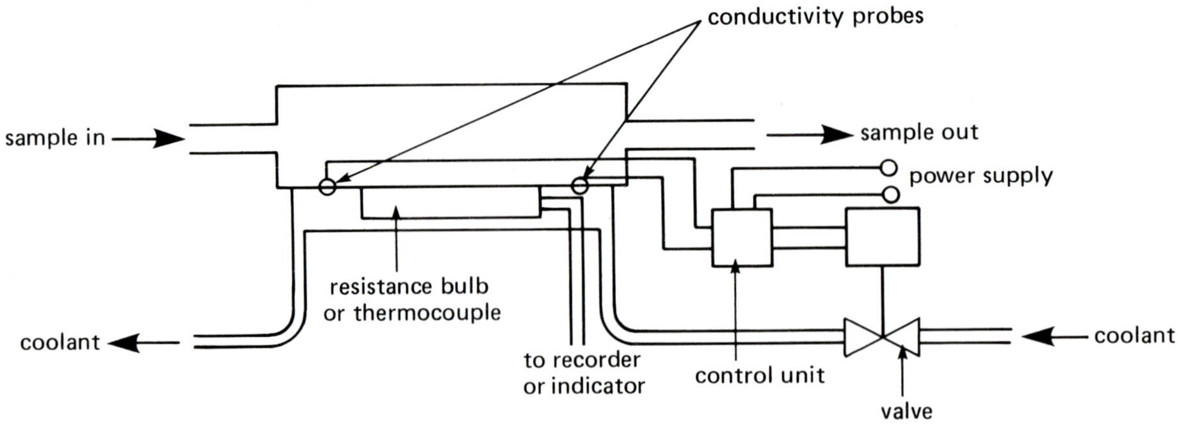

Fig. 14.63 Conductivity-type dewpoint meter

Solid state dewpoint sensor

A technique has been devised that operates on the same principle as the cooled mirror design, but replaces the mirror with a solid state sensor. This sensor consists of a series of conducting bands placed on a substrate and the system detects when condensation occurs by monitoring the sensor for any tracking that occurs between these conducting bands. The system, shown schematically in Figure 14.64, can measure the dewpoint to an accuracy of ±1 °C.

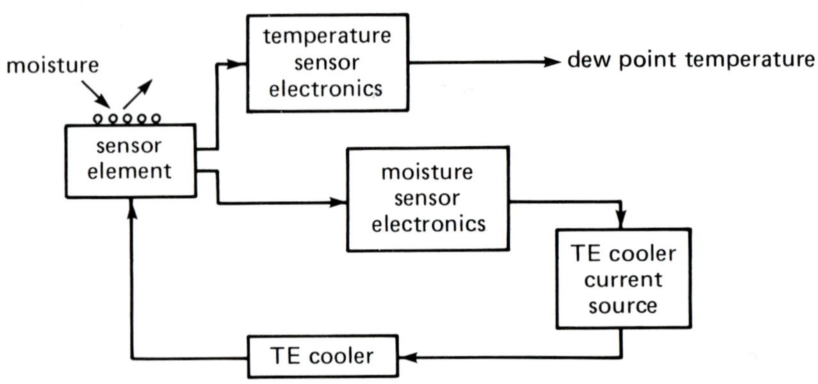

Fig. 14.64 Schematic operation of solid state dewpoint system

14.6.5 Humidity/Dewpoint Controller

Humidity controllers based upon wet and dry bulb hygrometers

Since instruments of this type cannot measure relative humitity directly, they are not easy to use for relative humidity control purposes. Although such instruments can be used to control the amount of water vapour in an air stream by means of the wet bulb they cannot give a constant relative humidity, since the air stream temperature as measured by the dry bulb may vary. This type of controller is, therefore, limited in its application for relative humidity work. However, it can be used for dewpoint temperature control for large air volumes, e.g. in air conditioning plant for large office blocks.

In order to use this type of instrument for relative humidity control purposes, both the wet and dry bulb temperatures must be taken into account. One method of doing this is by means of a ratio controller, such a controller being used to maintain the ratio of wet bulb to dry bulb temperature constant. By this means the value of the relative humidity may be controlled over a limited range of dry bulb temperatures. However, if wide fluctuations of dry bulb temperature are experienced, the system will not be able to maintain a constant relative humidity due to the fact that the relation of relative humidity to the ratio of wet and dry bulb temperatures is not constant. A block diagram of this system is shown in Figure 14.65.

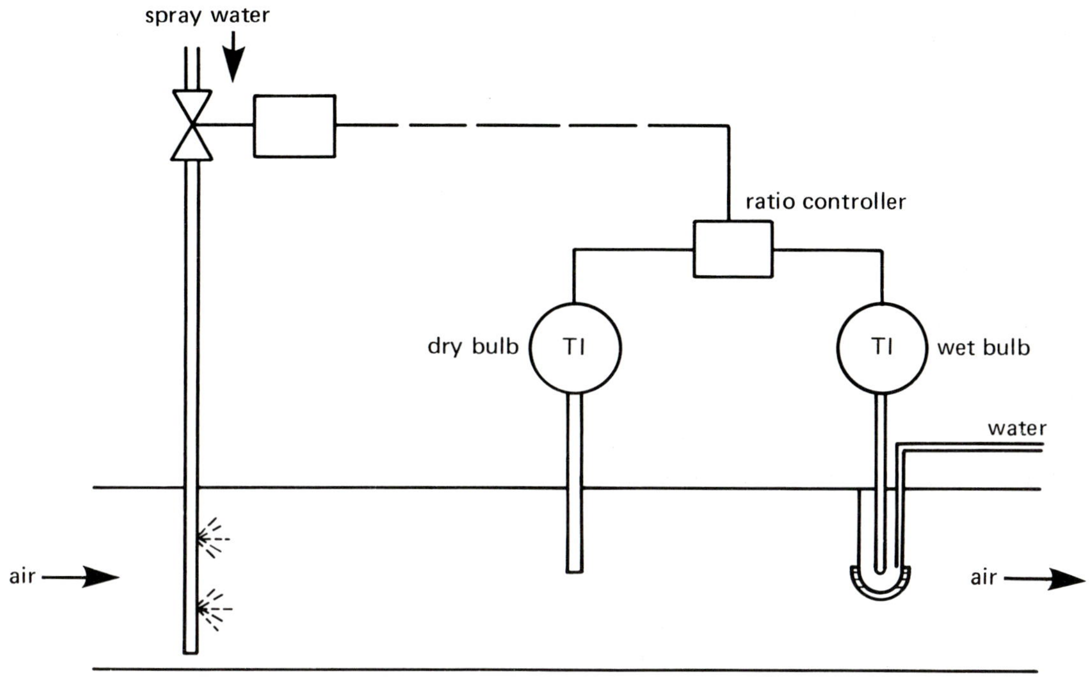

Fig. 14.65 Wet and dry bulb humidity controller

A disadvantage of these instruments is that they require a great deal of maintenance.

Electrical resistance thermometer type, wet and dry bulb controller

A controller based upon electrical resistance wet and dry bulb thermometers has been developed. Due to the increased sensitivity of the small resistance bulb type of instrument compared with mercury-in-steel thermometers, and the fact that the relative humidity is indicated directly, it follows that the humidity of a process can be controlled accurately.

The aspirated wet and dry bulb unit is housed in a steel case, fitted with a dust filter to prevent contamination of the wet bulb wick and a water supply system which maintains the correct water level. It is recommended that distilled water be used and under normal conditions the water container must be refilled every 2 to 3 weeks.

Relative humidity is measured and recorded, and on/off control contacts can be fitted to the indicator or recorder or a more complex system can be used such as three-term control. The accuracy of control for a typical installation depends upon the type of control fitted, the controller being used to operate a motor and valve or damper assembly in order to control the process.

Capacitance humidity control system for a drying process

By placing an adjustable damper within the flue exhaust of the drying system, it is possible to control the amount of recirculation and hence humidity within the drying system. The required stages to control the damper are shown in the block diagram in Figure 14.66.

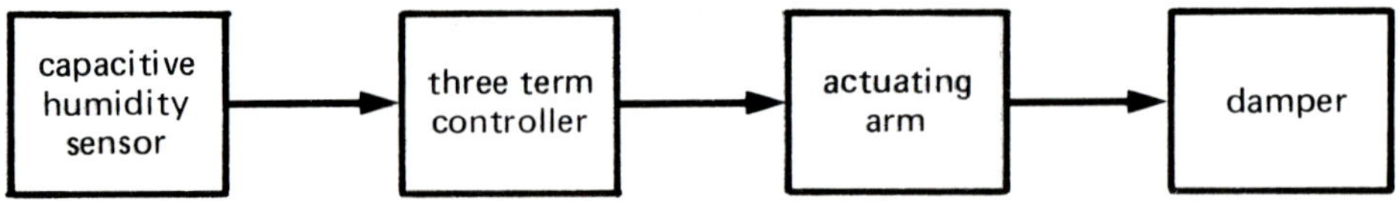

Fig. 14.66 Control strategy for humidity control system

Once the humidity has been measured by the sensor, the controller will maintain the humidity to a set point by varying the actuating arm which is connected to the damper. In certain systems, on/off control may be more applicable than modulating control.

14.7 COMPUTERISED ENERGY MONITORING SYSTEMS

A wide range of monitoring equipment employing microprocessors is available to enable the continuous monitoring and logging of all types of measured data. Such monitors normally directly interface with sensors providing an electrical output, for example thermocouples, transducers and pulse transmitters.

Data loggers range from small battery powered single channel instruments to sophisticated building and energy management systems which in addition to their control functions, allow the monitoring and logging of large volumes of data, often from remote sites, at a central point.

As with the larger management systems many of the smaller data loggers are equipped with printers which enable hard-copy of the monitored readings to be obtained.

To permit interfacing the computerised monitors, some metering equipment (for example flow meters) is manufactured with electrical pick-ups. A range of sensors is, however, available to enable monitoring to be carried out without having to install special metering equipment. In addition to the familiar temperature sensors, clip-on transducers for the measurement of electrical load and optical sensors which can read rotating dials, needles, fingers or counters have been developed. The latter are particularly useful in reading electrical and gas meters.

There is now a growing number of low cost data loggers available. These instruments have been designed with a varied selection of tasks in mind. Their common characteristic is that they are not as powerful in terms of speed, memory and flexibility as more expensive computer based systems.

In purchasing a low cost data logger there is an inevitable compromise. The user sacrifices computing power but gains a cost reduction and some less obvious advantages. Low cost instruments, being physically small and in some cases entirely battery powered, are generally much quicker to install and easier to operate. They can often be purchased with pre-wired sensors which simply need to be plugged in to be ready to run. Limited memory capacity and a small number of sensors can help to avoid the temptation to collect far more data than is needed or than is possible to use.

The basic principles of these systems are discussed in this Section to illustrate their capabilities. Emphasis is placed on low cost data logger systems rather than more expensive sophisticated equipment since use of the latter may be justified where detailed analysis of a complex system is required but can be quite inappropriate for monitoring fuel consumption and other simple tasks.

14.7.1 Retrieval of Information

Normally, a data logger is required to record temperatures, pressures, humidity, sensor output, voltages, or gas and electricity consumptions. The logger is expected to present this information on demand in an understandable format either as a list of 'raw' readings or with some analysis of the data. This might be averaging or totalising, or perhaps a graph to reveal trends. Two examples of the tasks that might be performed by a data logger are given below.

For example, an engineer may wish to establish a relationship between outside air temperature and gas consumption in order to identify the proportion of gas load being used for heating. For this task the logger would need to record one or more ambient air temperatures once an hour, at most, possibly with daily averaging and totalising.

Another equally suitable task would be monitoring the efficiency of a billet heating furnace in terms of gas used per billet. This would require totalising of gas consumption by the hour, or perhaps more frequently, and counting billets entering the furnace using a suitably mounted switch. The engineer might also require one or two thermocouple inputs to record furnace temperature.

14.7.2 Communication

It is necessary to communicate with the data logger for three reasons – to give it information on the number and type of sensors fitted or the rate at which data is to be recorded, to give commands such as 'start recording', 'stop recording' or 'display data', and to extract the recorded data.

Two basic types of low cost data logger are commonly used and are illustrated schematically in Figures 14.67 and 14.68. Firstly, there are stand alone instruments which have their own communication devices built in (e.g. a keyboard and printer or LCD screen). Secondly, there are instruments which are partly or entirely dependent on a line to a computer, so that configuration of the data logger or handling of data are controlled by computer software. This arrangement can be extremely

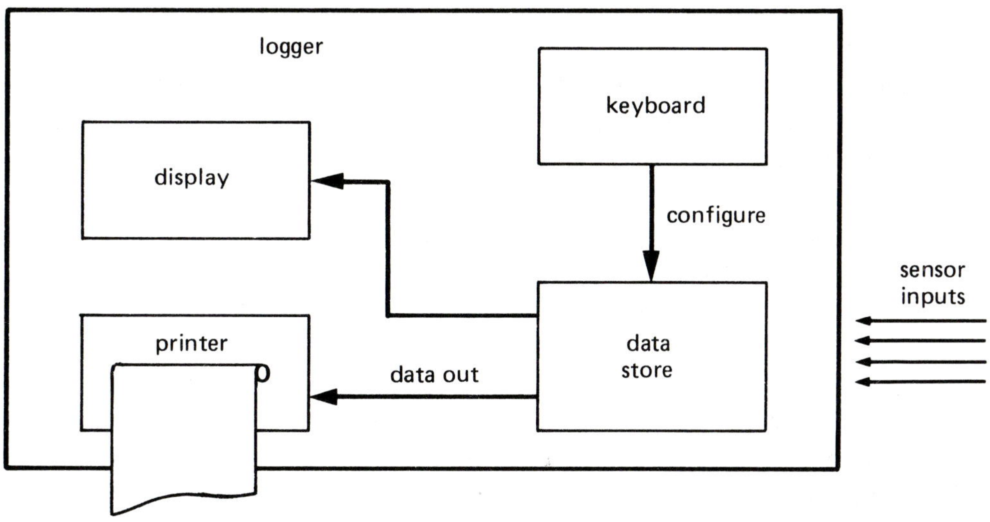

Fig. 14.67 Typical arrangement of a stand-alone data logger with built-in keyboard and either display or printer

versatile if the manufacturer provides software for portable computers currently in production. Typically such a computer is battery powered and has a built-in LCD screen, printer and micro-cassette recorder. This type of computer makes it possible to collect and examine date from data logging instruments in situ.

Sometimes this type of data logger is configured using just two or three keys, which are used to select options from a built-in menu. For this there obviously has to be some way, usually a small LCD screen, to display the options available.

Output of information, either details of configuration or recorded data, can be by a built-in display or printer or by a computer link as shown in Figure 14.68.

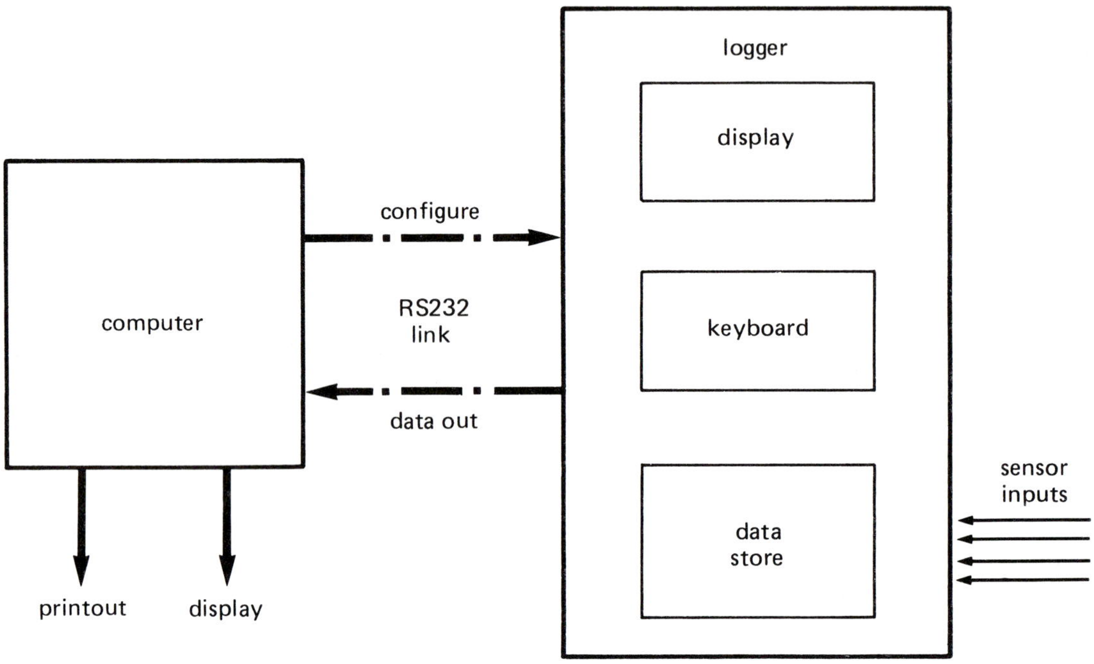

Fig. 14.68 Typical arrangement of a computer dependent data logger which might also have a built-in keyboard or display

14.7.3 Future Developments

In the future we can expect to see many new low cost data loggers emerge. Continuing advances in microchip technology are already producing faster speeds and huge memories on a single chip, while prices plummet. It seems inevitable that data loggers will become more powerful. It is to be hoped that they do not also become too unwieldy for the ordinary engineer.

REFERENCES

BRITISH STANDARDS

BS 1041 : 1985	Code for Temperature Measurement.
BS 1042 : 1984	Measurement of Fluid Flow in Closed Conduits.
BS 1339 : 1981	Definitions, Formulae and Constants Relating to the Humidity of the Air.
BS 1756 : 1977	Methods for Sampling and Analysis of Flue Gases.
BS 1780 : 1985	Specification for Bourdon Tube Pressure and Vacuum Gauges.
BS 1794 : 1984	Specification for Chart Ranges for Temperature Recording Instruments.
BS 1900 : 1985	Specification for Secondary Reference Thermometers.
BS 1904 : 1984	Specification for Industrial Platinum Resistance Thermometer Sensors.
BS 2082 : 1954	Code for Disappearing-filament Optical Pyrometers.
BS 3048 : 1958	Code for the Continuous Sampling and Automatic Analysis of Flue Gases : Indicators and Recorders.
BS 3292 : 1988	Specification for Direct Reacting Hygrometers.
BS 4161 :	Specification for Gas Meters:
Part 5 : 1990	Positive Displacement Meters for Pressures up to 7 bar.
Part 6 : 1987	Rotary Displacement and Turbine Meters for Gas Pressures up to 100 bar.
Part 7 : 1983	Mechanical Volume Correctors.
BS 4833 : 1986	Schedule for Hygrometric Tables for Use in the Testing and Operation of Environmental Enclosures.
BS 4937 : 1986	International Thermocouple Reference Tables.
BS 5074 : 1985	Specification for Short and Long Solid Stem Thermometers for Precision Use.
BS 5248 : 1990	Specification for Aspirated Hygrometers.
BS 5309 :	Methods for Sampling Chemical Products:
Part 1 : 1976	Introduction and General Principles
Part 2 : 1976	Sampling of Gases.
BS 5728 : 1986	Measurement of Flow of Cold Potable Water in Closed Conduits.
BS 5857 : 1986	Methods for Measurement of Fluid Flow in Closed Conduits, using Tracers.
BS 6175 : 1982	Specification for Temperature Transmitters with Electrical Outputs.
BS 6199 : 1986	Methods for Measurement of Liquid Flow in Closed Conduits Using Weighing and Volumetric Methods.

OTHER PUBLICATIONS

Practical Fluorescence. Guilbrault G.G., Marcel Dekker, New York 1973.

Engineering Measurements and Instrumentations. Adams L. English Universities Press, 1975.

Efficient use of Energy. Dryden I.G.C. (Ed.) IPC Science & Technology Press, 1975.

Methods of Air Sampling and Analysis. Katz M. (Ed.) American Public Health Association, 2nd Edition, 1977.

Industrial Gas Utilisation : Engineering Principles and Practice. Pritchard R, Guy J.J., Conner N.E. Bowker, 1977

Engineering Instrumentation and Control. Bolton w. Butterworth, 1980

Measurements. Collett C.V. and Hope A.D. Pitman, 2nd Edition, 1983.

Appendix: SI Units and Conversions

APPENDIX

SI UNITS AND CONVERSIONS

A 1 INTRODUCTION

This book employs SI units and symbols throughout, with usage generally conforming to the appropriate British Standards (which are also International Standards).

These Standards are designed to meet virtually every requirement of science and technology and are therefore extremely comprehensive. Most organisations, not requiring such detail, develop their own internal standards, exercising such options as are offered by the official Standards and sometimes modifying them. This approach has been adopted for this book to ensure consistency and clarity, the few variants from the Standards are explained in A 4.

The SI system was adopted by international conference as far back as 1960, to meet the pressing need for a coherent standard resulting from the rapidly increasing internationalisation of science and technology. Its implementation, however, has been far from universal and SI co-exists, especially in the UK, with the fps (foot pound second) and cgs (centimetre gramme second) systems. Conversion factors and tables will be needed for a long time yet.

This appendix briefly describes SI both in general and as applied in this book, with a selection of the most relevant conversion factors.

A 2 SI BASICS

A 2.1 Units

There are seven base units, from which all other units are derived. The base units are shown in Table A1.

Table A1 Base Units

Quantity	Name of base SI unit	Symbol
length	metre	m
mass	kilogram	kg
time	second	s
electric current	ampere	A
thermodynamic temperature	kelvin	K
amount of substance	mole	mol
luminous intensity	candela	cd

Table A2 gives examples of derived units and how they are derived from the base units or from other derived units.

Table A2 Examples of Derived Units

Quantity	Special name of derived SI unit	Symbol	Expressed in terms of base SI units or in terms of other derived SI units
force	newton	N	1 N = 1 kg m/s^2
pressure, stress	pascal	Pa	1 Pa = 1 N/m^2
energy, work quantity of heat	joule	J	1 J = 1 N m
power	watt	W	1 W = 1 J/s
Celsius temperature	degree Celsius	°C	1 °C = 1 K

A 2.2 Multiples

Decimal multiples and sub-multiples are expressed as prefixes attached to the unit symbol, e.g. 10 km = 10 000 m, 1 μm = 0.000001 m. Table A3 lists the Standard prefixes and their symbols.

Table A3 Multiples and Sub-multiples

Factor	Prefixes	Symbol
10^{18}	exa	E
10^{15}	peta	P
10^{12}	tera	T
10^{9}	giga	G
10^{6}	mega	M
10^{3}	kilo	k
10^{2}	hecto	h
10	deca	da
10^{-1}	deci	d
10^{-2}	centi	c
10^{-3}	milli	m
10^{-6}	micro	μ
10^{-9}	nano	n
10^{-12}	pico	p
10^{-15}	femto	f
10^{-18}	atto	a

A 2.3 RECOMMENDED USAGE

	Examples: Right	Examples: Wrong
Unit symbols are *not* abbreviations and thus never require full stops except for grammatical reasons, e.g. when occurring at the end of a sentence. They remain the same in the singular and in the plural.	0.5 kg 25 kg	0.5 kg. 25 kgs
Numbers are separated by one space from the symbol. There is no space between the symbol and multiple.	5 Pa 4 km	5Pa 4k m
* Compound symbols formed by multiplying unit symbols have the latter separated by a space	kW h	kWh
* Compound symbols formed by division use the oblique, not 'per' or the negative exponent.	kg/s N/m	kg per s $N\ m^{-1}$
Areas and volumes formed from linear symbols always use the exponent, not the abbreviation.	m^3	cu.m.

* Selected on grounds of clarity from a number of options offered by the Standards.

A 3 SYMBOLS USED IN THE BOOK

Roman Alphabet

Symbol	Quantity	Units
A	Area	m^2
a	Acceleration	m/s^2
b	Breadth	m
c_d	Discharge coefficient	—
c_r	Coefficient of friction	—
c_p	Specific heat capacity at constant pressure	J/(kg K)
c_v	Specific heat capacity at constant volume	J/(kg K)
D, d	Diameter	m
d	Thickness	m
d	Relative density	—
E	Energy	J
F	Force	N
G	Thermal conductance	W/K
Gr	Grashof number	—
g	Acceleration due to gravity	m/s^2
H	Enthalpy	J
h	Calorific value	J/kg
h	Surface heat transfer coefficient	$W/(m^2\ K)$
h	Height	m
K	Overall heat transfer coefficient	$W/(m^2\ K)$
k	Thermal conductivity	W/(m K)
l	Length	m
M	Radiant exitance	W/m^2
M	Thermal insulance	$(m^2\ K)/W$
m	Mass	kg
Nu	Nusselt number	—
Pr	Prandtl number	—
p	Pressure	Pa
Q	Quantity of heat	J
q	Density of heat flow rate	W/m^2

Symbol	Quantity	Units
q_m	Mass flow rate	kg/s
q_v	Volume flow rate	m^3/s
R	Universal Gas Constant	J/(mol K)
R_G	Gas constant	J/(kg K)
R	Thermal resistance	K/W
Ra	Rayleigh number	—
Re	Reynolds number	—
r	Radius	m
s	length of path	m
T	Thermodynamic (absolute) temperature	K
t	Celsius temperature	°C
u	Velocity	m/s
V	Volume	m^3
v	Specific volume	m^3/kg
W	Work	J
z	Compressibility factor	—

Greek Alphabet

α	Absorptivity	—
α	Coefficient of expansion	K^{-1}
Δ	Difference (between two values of quantity following)	—
γ	Specific heat capacity ratio, c_p/c_v	—
ϵ	Emissivity	—
ζ	Velocity pressure loss factor	—
Λ	Perimeter	m
λ	Wavelength	m
μ	Dynamic viscosity	Pa s
ν	Kinematic viscosity	m^2/s
ρ	Density	kg/m^3
Σ	Summation (of quantities following)	—
σ	Stefan-Boltzmann constant	$W/(m^2 K^4)$
τ	Shear stress	N/m^2
Φ	Energy flow rate	W

A 4 VARIANTS FROM STANDARDS

A 4.1 Specific Energy

The Standards reserve 'specific' for properties per unit of mass only, and 'specific energy' or 'specific enthalpy' can signify calorific value, latent heat or sensible heat. 'Specific' is therefore omitted for volumetric quantities, giving rise to expressions such as 'heat content, volume basis'.

This seems both ambiguous and cumbersome, and this book follows almost universal practice in using, for both unit mass and unit volume, the following:

- Calorific value
- Latent heat
- Specific heat capacity.

A 4.2 Pressure

Bar and mbar are used throughout, except for formulae and calculations where Pascals must be used to maintain consistency of units.

A 4.3 Typeface for Symbols

The Standards specify when upright or italic type should be used for symbols and their subscripts. This is unnecessarily complex for the limited range of symbols in the book and has been disregarded.

A 5 CONVERSION FACTORS

I LENGTH

	mm	cm	in	ft	yd	m	km	mile
mm	* 1	* 0.1	0.0393701	3.2808×10^{-3}	1.093×10^{-3}	* 10^{-3}		
cm	* 10	* 1	0.393701	0.032808	0.010936	* 0.01		
in	* 25.4	* 2.54	* 1	0.83333	0.027778	*0.0254		
ft	* 304.8	* 30.48	* 12	* 1	0.333333	*0.3048	*3.048×10^{-4}	1.894×10^{-4}
yd	* 914.4	* 91.44	* 36	* 3	* 1	*0.9144	*9.144×10^{-4}	5.682×10^{-4}
m	* 1000	* 100	39.3701	3.28084	1.09361	* 1	* 10^{-3}	6.214×10^{-4}
km	* 10^6	*100000	30370.1	3280.84	1093.61	* 1000	* 1	0.621371
mile	1.60934×10^6	160934	* 63360	* 5280	* 1760	1609.34	1.60934	* 1

Note: Starred numbers are exact conversions

II AREA

	mm^2	cm^2	in^2	ft^2	yd^2	m^2	acre	ha	km^2	$mile^2$
mm^2	1	0.01	1.550×10^{-3}	1.076×10^{-5}	1.196×10^{-6}	10^{-6}				
cm^2	100	1	0.1550	$1.076.10^{-3}$	1.196×10^{-4}	10^{-4}				
in^2	645.16	6.4516	1	6.944×10^{-3}	7.716×10^{-4}	6.452×10^{-4}				
ft^2	92903	929	144	1	0.1111	0.09290	2.30×10^{-5}	9.29×10^{-6}	9.29×10^{-8}	3.587×10^{-8}
yd^2	836127	8361	1296	9	1	0.8361	2.066×10^{-4}	8.361×10^{-5}	8.361×10^{-7}	3.228×10^{-7}
m^2	10^6	10000	1550	10.764	1.196	1	2.471×10^{-4}	10^{-4}	10^{-6}	3.861×10^{-7}
acre				43560	4840	4047	1	0.4047	4.047×10^{-3}	1.562×10^{-3}
ha				107639	11960	10000	2.471	1	0.01	3.861×10^{-3}
km^2				1.0764×10^7	1.196×10^6	10^6	247.1	100	1	0.3861
$mile^2$				2.7878×10^7	3.0976×10^6	2.590×10^6	640	259.0	2.590	1

100 m^2 = 1 acre = 0.01 ha

III VOLUME

	mm^3	ml	in^3	l*	US gal	UK gal	ft^3	yd^3	m^3
mm^3	1	10^{-3}	6.1024×10^{-5}	10^{-6}	2.642×10^{-7}	2.200×10^{-7}	3.531×10^{-8}	1.308×10^{-9}	10^{-9}
ml	1000	1	0.061026	10^{-3}	2.642×10^{-4}	2.200×10^{-4}	3.532×10^{-5}	1.308×10^{-6}	10^{-6}
in^3	16387	16.39	1	1.01639	4.329×10^{-3}	3.605×10^{-3}	5.787×10^{-4}	2.143×10^{-5}	1.639×10^{-5}
l*	10^6	1000	61.026	1	0.2642	0.2200	0.03532	1.308×10^{-3}	10^{-3}
US gal	3.785×10^6	3785	231.0	3.785	1	0.8327	0.1337	4.951×10^{-3}	3.785×10^{-3}
UK gal	4.546×10^6	4546	277.4	4.546	1.201	1	0.1605	5.946×10^{-3}	4.546×10^{-3}
ft^3	2.832×10^7	2.832×10^4	1728	28.32	7.4805	6.229	1	0.03704	0.02832
yd^3	7.6456×10^8	7.6453×10^5	46656	764.53	202.0	168.2	27	1	0.76456
m^3	10^9	10^6	61024	1000	264.2	220.0	35.31	1.308	1

* Since 1964, the litre has been defined as exactly 1 dm^3 instead of 1.000028 dm^3 as defined in 1901.

1 US barrel = 42 US gal = 34.97 UK gal
1 fluid oz = 28.41 ml
1 UK pint = 568.2 ml
1 litre = 1.760 UK pints

IV MASS

	g	oz	lb	kg	cwt	US ton (short ton)	t (tonne)	UK ton
g	1	0.035274	2.2046×10^{-3}	10^{-3}				
oz	28.3495	1	0.0625	0.028350				
lb	453.592	16	1	0.453592	8.9286×10^{-3}	5.00×10^{-4}	4.5359×10^{-4}	4.4643×10^{-4}
kg	10^{3}	35.2740	2.20462	1	0.019684	1.1023×10^{-3}	10^{-3}	9.8421×10^{-4}
cwt	50802.3	1792	112	50.8023	1	0.056	0.05080	0.05
US ton (short ton)	907185	32000	2000	907.185	17.8571	1	0.907185	0.892857
t(tonne)	10^{6}	35273.9	2204.62	1000	19.6841	1.10231	1	0.984207
UK ton	1.01605×10^{6}	35840	2240	1016.05	20	1.12	1.01605	1

V DENSITY

	kg/m^3	lb/ft^3	lb/UK gal	g/cm^3	UK ton/yd^3	lb/in^3
kg/m^3	1	0.062428	0.010022	10^{-3}	7.5248×10^{-4}	3.6046×10^{-5}
lb/ft^3	16.0185	1	0.160544	0.0160185	0.0120536	5.7870×10^{-4}
lb/UK gal^3	99.776	6.22884	1	0.099776	0.075080	3.6046×10^{-3}
g/cm^3	1000	62.4280	10.0224	1	0.75248	0.036127
UK ton/yd^3	1328.94	82.9630	13.3192	1.32894	1	0.048011
lb/in^3	27679.9	1728	277.419	27.6799	20.8286	1

VI VELOCITY

	mm/s	ft/min	km/h	ft/s	mile/h	m/s
mm/s	* 1	0.19685	* 3.6×10^{-3}	3.281×10^{-3}	2.237×10^{-3}	* 10^{-3}
ft/min	* 5.08	1	0.018288	0.016667	0.01136	* 5.08×10^{-3}
km/h	277.778	54.6086	* 1	0.911344	0.621371	0.277778
ft/s	* 304.8	* 60	* 1.09728	* 1	0.681818	* 0.3048
mile/h	* 447.04	* 88	* 1.609344	1.46667	* 1	* 0.44704
m/s	* 1000	196.850	* 3.6	3.28084	2.23694	* 1

VII MASS RATE OF FLOW

	lb/h	kg/h	g/s	lb/min	t/h	UK ton/h	lb/s	kg/s
lb/h	1	0.4536	0.1260	0.01667	$4.536x10^{-4}$	$4.464x10^{-4}$	$2.778x10^{-4}$	$1.260x10^{-4}$
kg/h	2.205	1	0.2778	0.03674	$1x10^{-3}$	$9.842x10^{-4}$	$6.124x10^{-4}$	$2.778x10^{-4}$
g/s	7.937	3.6	1	0.1323	$3.6x10^{-3}$	$3.543x10^{-3}$	$2.205x10^{3}$	10^{-3}
lb/min	60	27.216	7.560	1	$2.722x10^{-2}$	$2.678x10^{-2}$	$1.667.10^{-2}$	$7.56x10^{-3}$
t/h	2205	1000	277.8	36.74	1	0.9842	0.6124	0.2778
UK ton/h	2240	1016	282.2	37.33	1.016	1	0.6222	0.2822
lb/s	3600	1633	453.6	60	1.633	1.607	1	0.4536
kg/s	7937	3600	1000	132.3	3.6	3.543	2.205	1

VIII VOLUME RATE OF FLOW

	litres/h	ml/s	gal/h	l/min	gal/min	m^3/h	ft^3/min	l/s	ft^3/s	m^3/s
litres/h	1	0.2778	0.2200	0.01667	3.666×10^{-3}	10^{-3}	5.886×10^{-4}	2.778×10^{-4}	0.810×10^{-6}	2.778×10^{-7}
ml/s	3.6	1	0.7919	0.0600	0.01320	3.6×10^{-3}	2.119×10^{-3}	10^{-3}	3.532×10^{-5}	10^{-6}
gal/h	4.546	1.263	1	0.07577	0.01667	$4.546 . 10^{-3}$	2.676×10^{-3}	1.263×10^{-3}	4.460×10^{-5}	1.263×10^{-6}
l/min	60	16.67	13.20	1	0.2200	0.0600	0.03531	0.01667	5.886×10^{-4}	1.667×10^{-5}
gal/min	272.8	75.77	60	4.546	1	0.2728	0.1605	0.07577	2.676×10^{-3}	7.577×10^{-5}
m^3/h	1000	277.8	220.0	16.67	3.666	1	0.5886	0.2778	0.810×10^{-3}	2.778×10^{-4}
ft^3/min	1699	471.9	373.7	28.31	6.229	1.699	1	0.4719	0.01667	4.719×10^{-4}
l/s	3600	1000	792	60	13.20	3.6	2.119	1	0.03531	10^{-3}
ft^3/s	1.019×10^5	2.832×10^4	2.242×10^4	1699	373.7	101.9	60	28.32	1	0.02832
m^3/s	3.6×10^6	10^6	7.919×10^5	6×10^4	1.320×10^4	3600	2119	1000	35.31	1

gal = UK or Imperial gallon
1 US gal = 0.833 UK gal

IX PRESSURE

	Pa	mbar	mm Hg	in H_2O	kPa	in Hg	lbf/in^2	kgf/cm^2	bar	atm
Pa	1	0.0100	7.501×10^{-3}	4.015×10^{-3}	10^{-3}	2.953×10^{-4}	1.450×10^{-4}	1.020×10^{-5}	10^{-5}	9.869×10^{-6}
mbar	100	1	0.7501	0.4015	0.1000	0.02953	0.01450	1.020×10^{-3}	10^{-3}	9.869×10^{-4}
mm Hg	133.3	1.333	1	0.5352	0.1333	0.03937	0.01934	1.360×10^{-3}	1.333×10^{-3}	1.316×10^{-3}
in H_2O	249.1	2.491	1.868	1	0.2491	0.07356	0.03613	2.540×10^{-3}	2.491×10^{-3}	2.458×10^{-3}
kPa	1000	10	7.501	4.015	1	0.2953	0.1450	0.01020	0.0100	9.869×10^{-3}
in Hg	3386	33.86	25.40	13.60	3.386	1	0.4912	0.03453	0.03386	0.03342
lbf/in^2	6895	68.95	51.71	27.68	6.895	2.036	1	0.07031	0.06895	0.06805
kgf/cm^2	9.807×10^4	980.7	735.6	393.7	98.07	28.96	14.22	1	0.9807	0.9678
bar	10^5	1000	750.1	401.5	100	29.53	14.50	1.020	1	0.9869
atm	1.013×10^5	1013	760.0	406.8	101.3	29.92	14.70	1.033	1.013	1

1 kgf/cm^2 = 1 kp/cm^2 = 1 technical atmosphere = 14.22 lbf/in^2
1 torr = 1 mm Hg (to within 1 part in 7 million)

X ENERGY, WORK, HEAT

	J	kJ	Btu	kcal	MJ	hp h	kw h	therm	GJ
J	1	10^{-3}	9.478×10^{-4}	2.388×10^{-4}	10^{-6}	3.725×10^{-7}	2.778×10^{-7}	$0.478.10^{-9}$	10^{-9}
kJ	1000	1	0.9478	0.2388	10^{-3}	3.725×10^{-4}	2.778×10^{-4}	0.478×10^{-6}	10^{-6}
Btu	1055.1	1.0551	1	0.2520	1.055×10^{-3}	3.930×10^{-4}	2.931×10^{-4}	10^{-5}	1.055×10^{-6}
kcal	4186.8	4.1868	3.9683	1	4.187×10^{-3}	1.560×10^{-3}	1.163×10^{-3}	3.968×10^{-5}	4.187×10^{-6}
MJ	10^{6}	1000	947.82	238.85	1	0.3725	0.2778	9.478×10^{-3}	10^{-3}
hph	2.6845×10^{6}	2684.5	2544.4	641.19	2.6845	1	0.7457	0.02544	2.6845×10^{-3}
kwh	3.6000×10^{6}	3600	3412.1	859.84	3.600	1.3410	1	0.03412	3.6×10^{-3}
therm	1.0551×10^{8}	1.0551×10^{-5}	100 000	25200	105.51	39.301	29.307	1	0.10551
GJ	10^{9}	10^{6}	9.4782×10^{5}	2.3885×10^{5}	10^{3}	372.5	277.8	9.478	1

1 thermie = 1.163 kW h = 4.186 MJ = 999.7 kcal = 3967.1 Btu

XI POWER, HEAT FLOW RATE

	Btu/h	W	kcal/h	hp	kW	MW
Btu/h	1	0.2931	0.2520	3.930×10^{-4}	2.931×10^{-4}	2.93×10^{-7}
W	3.4121	1	0.8598	1.341×10^{-3}	10^{-3}	10^{-6}
kcal/h	3.9683	1.163	1	1.560×10^{-3}	1.163×10^{-3}	1.16×10^{-6}
hp	2544	745.70	641.19	1	0.7457	7.457×10^{-4}
kW	3412.1	1000	859.8	1.3410	1	10^{-3}
MW	3.4121×10^{6}	10^{6}	8.598×10^{6}	1341	1000	1

1 W = 1 J/s
1 cal/s = 3.6 kcal/h
1 ton of refrigeration = 3517 W = 12000 Btu/h

XII SPECIFIC ENERGY (CALORIFIC VALUE, ENTHALPY)

A. MASS BASIS

	kJ/kg	Btu/lb	kcal/kg	MJ/kg
kJ/kg	1	0.4299	0.2388	10^{-3}
Btu/lb	2.326	1	0.5556	2.326×10^{-3}
kcal/kg	4.187	1.8	1	4.187×10^{-3}
MJ/kg	1000	429.9	238.8	1

B. VOLUME BASIS – CONSTANT REFERENCE CONDITIONS

	kJ/m^3	$kcal/m^3$	Btu/ft^3	MJ/m^3
kJ/m^3	1	0.2388	0.02684	1×10^{-3}
$kcal/m^3$	4.187	1	0.1124	4.187×10^{-3}
Btu/ft^3	37.26	8.899	1	0.03726
MJ/m^3	1000	238.8	26.84	1

1 therm (10^5 Btu)/UK gal = 23208 MJ/m^3

1 thermie/litre = 4185 MJ/m^3

C. VOLUME BASIS – VARIOUS REFERENCE CONDITIONS

		MJ/m^3 15 °C 1013.25 mbar		Btu/ft^3 60 °F, 30 in Hg (IT calorie)		Btu/ft^3 60 °F, 30 in Hg (15 ° calorie)	
		Dry	Sat	Dry	Sat	Dry	Sat(1)
MJ/m^3 15 °C 1013.25 mbar	Dry(2)	1	0.9832	26.80	26.33	26.81	26.34
	Sat	1.017	1	27.26	26.78	27.27	26.79
Btu/ft^3 60 °F in Hg (International calorie basis)	Dry	0.03731	0.03669	1	0.9826	1.000	0.9829
	Sat	0.03797	0.03734	1.018	1	1.018	1.000
Btu/ft^3 60 °F 30 in Hg (15° calorie)	Dry	0.03730	0.03667	0.9997	0.9823	1	0.9826
	Sat	0.03796	0.03732	1.017	0.9997	1.018	1

(1) Imperial Standard Condition (ISC)
(2) Metric Standard Condition (MSC)

XIII THERMAL CONDUCTANCE, U VALUE

	$W/(m^2\ K)$	$kcal/(m^2\ h\ K)$	$Btu/(ft^2\ h\ ^\circ F)$
$W/(m^2\ K)$	1	0.8598	0.1761
$kcal/(m^2\ h\ K)$	1.163	1	0.2048
$Btu/(ft^2\ h\ ^\circ F)$	5.678	4.882	1

XIV THERMAL CONDUCTIVITY

	$Btu\ in/(ft^2\ h\ ^\circ F)$	$W/(m\ K)$	$kcal/(m\ h\ K)$
$Btu\ in/(ft^2\ h\ ^\circ F)$	1	0.1442	0.1240
$W/(m\ K)$	6.933	1	0.8598
$kcal/(m\ h\ K)$	8.064	1.163	1

XV DENSITY OF HEAT FLOW RATE

	W/m^2	$kcal/(m^2 h)$	$Btu/(ft^2 h)$
W/m^2	1	0.8598	0.3170
$kcal/(m^2 h)$	1.163	1	0.3687
$Btu/(ft^2 h)$	3.155	2.712	1

XVI SPECIFIC HEAT CAPACITY

	kJ (kg K)	Btu/(lb °F)	kcal/(kg K)
kJ (kg K)	1	0.2388	0.2388
Btu/(lb °F)	4.1868	1	1
kcal/(kg K)	4.1868	1	1

REFERENCES

BS 350: Conversion Factors and Tables

- Part 1: 1974 (1983) Basis of tables. Conversion factors
- Part 2: Supplement No. 1: 1967 (PD6203) Additional tables for SI conversions.

BS 5555: 1981 (1983): Specification for SI Units and Recommendations for the Use of their Multiples and of Certain Other Units

(Identical with International Standard ISO 1000 – 1981).

BS 5775: Specification for Quantities, Units and Symbols:
(Identical with International Standard ISO 31)

- Part 0: 1982 General principles
- Part 1: 1979 Space and time
- Part 2: 1979 Periodic and related phenomena
- Part 3: 1979 Mechanics
- Part 4: 1979 Heat
- Part 5: 1980 Electricity and magnetism
- Part 6: 1982 Light and related electromagnetic radiation
- Part 7: 1979 Acoustics
- Part 8: 1982 Physical chemistry and molecular physics
- Part 9: 1982 Atomic and nuclear physics
- Part 10: 1982 Nuclear reactions and ionizing radiations
- Part 11: 1979 Mathematical signs and symbols for use in the physical sciences and technology
- Part 12: 1982 Dimensionless parameters
- Part 13: 1982 Solid state physics

SI Units and Conversion Factors for Use in the British Gas Industry. British Gas plc and the Society of British Gas Industries. 1971, revised 1972.

Index